College Algebra

Instructor's Edition
College Algebra

Robert Blitzer
Miami-Dade Community College

PRENTICE HALL
Upper Saddle River, NJ 07458

Library of Congress Cataloging-in-Publication Data

Blitzer, Robert.
    College algebra / Robert Blitzer.
       p. cm.
    Includes index.
    ISBN 0-13-399940-8   (SE : hardcover). -- ISBN 0-13-660259-2 (IE)
    1. Algebra. I. Title.
    QA152.2.B584   1998
    512.9--dc21   97-22702
                                            CIP

Acquisitions Editor: Sally Denlow
Editorial Assistant/Supplements Editor: April Thrower
Assistant Vice President of Production and Manufacturing: David W. Riccardi
Executive Managing Editor: Kathleen Schiaparelli
Senior Managing Editor: Linda Mihatov Behrens
Manufacturing Manager: Trudy Pisciotti
Manufacturing Buyer: Alan Fischer
Director of Marketing: John Tweeddale
Marketing Manager: Patrice Jones
Marketing Assistant: Patrick Murphy
Creative Director: Paula Maylahn
Art Manager: Gus Vibal
Text/Cover Design and Project Management: Elm Street Publishing Services, Inc.
Art Studio: Academy Artworks/Laurel Technical Services
Photo Researcher: Clare Maxwell
Cover Image: Giacomo Balla "Abstract Speed, The Car Has Passed" 1913. Tate Gallery, London.
               Photo: John Webb. Art Resource, NY. © VAGA
Compositor: Preparé/Emilcomp

 © 1998 by Prentice-Hall, Inc.
Simon & Schuster / A Viacom Company
Upper Saddle River, New Jersey 07458

All rights reserved. No part of this book may be reproduced, in any form or by any means,
without permission in writing from the publisher.

Printed in the United States of America
10 9 8 7 6 5 4 3 2 1

ISBN 0-13-660259-2 (Instructor's Edition)

ISBN 0-13-399940-8 (Student Edition)

Prentice-Hall International (UK) Limited, *London*
Prentice-Hall of Australia Pty. Limited, *Sydney*
Prentice-Hall Canada Inc., *Toronto*
Prentice-Hall Hispanoamericana, S.A., *Mexico*
Prentice-Hall of India Private Limited, *New Delhi*
Prentice-Hall of Japan, Inc., *Tokyo*
Simon & Schuster Asia Pte. Ltd., *Singapore*
Editora Prentice-Hall do Brasil, Ltda., *Rio de Janeiro*

Contents

Matrices and Linear Systems 519

Conic Sections and Nonlinear Systems 667

Sequences, Series, and Probability 757

APPENDIX   Review Problems Covering the Entire Book   A1

Preface

College Algebra is written for students who need either a liberal arts college algebra course or a prerequisite college algebra course for a calculus sequence. Although the book's prerequisites chapter (Chapter P) reviews the basic concepts of algebra, students using this book should have the equivalent of two years of high school algebra.

The book has three fundamental goals: First, to help students acquire a solid foundation in college algebra, preparing them for other college courses such as calculus, business calculus, finite mathematics, and computer science; second, to show students how algebra can model and solve authentic real-world problems; and third, to enable students to develop problem-solving skills, fostering critical thinking, within a varied and interesting setting.

Key Pedagogical Features

The book's features are based on the recommendations and standards for college mathematics set forth by the National Council of Teachers of Mathematics and the American Mathematical Association of Two-Year Colleges. Consequently, the emphasis of the book is on problem solving, graphing, functions, mathematical modeling, technology, discovery approaches, critical thinking, collaborative learning, and contemporary applications that use real data. These applications are listed in the index of applications that appears inside the front covers. The features of this book are designed to support the three fundamental goals mentioned above.

Functions and Modeling. The book's emphasis is on the use of formulas and functions that describe interesting and relevant quantitative relations in the real world. Functions are introduced in Chapter 2, with an integrated functional approach emphasized throughout the book.

Graphing and Technology. Graphing and graphing utilities are discussed in the book's opening prerequisites chapter. Most examples and exercises use graphs to explore relationships between data and to provide ways of visualizing a problem's solution. Although the use of graphing utilities is optional, they are utilized in Using technology boxes to enable students to visualize, discover, and explore algebraic concepts. The use of graphing utilities is also reinforced in the technology problems appearing in the problem sets for those who want this option. With the book's early introduction to graphing, students can look at the calculator screens in the Using technology boxes and gain an increased understanding of an example's solution even if they are not using a graphing utility in the course.

Problem Solving. The emphasis of the book is on learning to use the language of algebra as a tool for solving problems related to everyday life. Problem-solving steps introduced in the third section in Chapter 1 are explicitly described, and they are used regularly. Since students have such difficulty translating word problems, a great deal of attention has been placed on translating the words and phrases of verbal models into algebraic equations. The extensive collection of applications, including environmental and financial issues, promotes the problem-solving theme and demonstrates the usefulness of the mathematics to students.

Extensive Application to Geometric Problem Solving. Problem solving using geometric formulas and concepts is emphasized throughout the book. Included whenever possible are geometric models that allow students to visualize algebraic formulas.

Detailed Step-by-Step Explanations. Illustrative examples are presented one step at a time. No steps are omitted, and each step is clearly explained. Color is used in the mathematics to show students precisely where the explanation applies.

Collaborative Learning and Chapter Projects. Most problem sets contain group activity problems. These collaborative activities give students the opportunity to work cooperatively as they think and talk about mathematics. The chapter projects at the end of each chapter use challenging and interesting applications of mathematics that not only stand alone as ways to stimulate class discussions on a variety of topics, but also cultivate an interest in independent explorations of mathematics on the Worldwide Web. Using the Worldwide Web, with links to many countries, as well as links to art, music, and history, students are encouraged to develop a multicultural, multidisciplinary approach to the study of algebra.

Interactive Learning. Discover for yourself exercises encourage students to actively participate in the learning process as they read the book. This feature encourages students to read with a pen in hand and interact with the text. Through the discovery exercises, they can explore problems in order to better understand them and their solutions.

Study Tips. Study tip boxes offer suggestions for problem solving, point out common student errors, and provide informal tips and suggestions. These invaluable hints appear in abundance throughout the book.

Enrichment Essays. Interspersed throughout the book are enrichment essays—many utilizing contemporary fine art—that germinate from ideas appearing in expository sections. Providing historical information and interdisciplinary connections, topics of some of the enrichment essays include relativity and the clock paradox, recursive processes, viruses of extraordinary virulence, carbon dating and artistic development, cubist art and inequalities, Halley's comet, and the Hubble Space Telescope.

Contemporary Fine Art. Algebra and fine art enable us to view the world in new and exciting ways. An extensive collection of contemporary, thought-provoking images selected by the author provides visual commentary to the book's unique collection of contemporary applications. The art adds an aesthetic sense to the book's pages, while visually reminding students of how algebra is connected to the whole spectrum of learning.

Extensive and Well-Organized Problem Sets. An extensive collection of problems is included in a problem set at the end of each section. Problem sets are organized into seven categories, making it easier to select homework problems consistent with the role of graphing utilities, critical thinking, collaborative learning, and writing in mathematics in your vision of the course. The categories include:

- *Practice Problems:* These problems give students an opportunity to practice the concepts that have been developed in the section.
- *Application Problems:* These problems include many relevant, up-to-date applications that will provoke student interest. Many of these problems offer students the opportunity to construct mathematical models from data.
- *True-False Critical Thinking Problems:* Several true-false problems that take students beyond the routine application of basic algebraic concepts are included in nearly every problem set. The true-false format is less intimidating than a more open-ended format, helping students gain confidence in divergent thinking skills.
- *Technology Problems:* These problems enable students to use graphing utilities to explore algebraic concepts and relevant mathematical models.
- *Writing in Mathematics:* These exercises are intended to help students communicate their mathematical knowledge by thinking and writing about algebraic topics.
- *Critical Thinking Problems:* This category contains the most challenging exercises in the problem sets. These open-ended problems were written to explore concepts while stimulating student thinking.
- *Group Activity Problems:* There are enough of these problems in each chapter to allow instructors to use collaborative learning quite extensively as an instructional format. It is hoped that many of these problems will result in interesting group discussions.

Chapter Introductions. Chapter introductions present fine art that is related either to the general idea of the chapter or to an application of algebra contained within the chapter.

Learning Objectives. Learning objectives open every section. The objectives are reinstated in the margin at their point of use.

Example Titles. All examples have titles so that students immediately see the purpose of each example.

Chapter Summaries. Inclusive summaries appear at the conclusion of each chapter, helping students to bring together what they have learned after reading the chapter.

Review Problems. A comprehensive collection of review problems follows the summary at the end of each chapter. A chapter test that focuses the review problems so that students can see if they are prepared for an actual class test follows the more extensive set of review problems. In addition, Chapters 2–6 conclude with cumulative review problems. Cumulative review problems covering the entire book appear in the appendix.

Supplements for the Instructor

Printed Supplements

Instructor's Edition　(0-13-660259-2)　Consists of the complete student text, with a special Instructor's answer section at the back of the text containing answers to all exercises.

Instructor's Solutions Manual　(0-13-746629-3)
- Step-by-step solutions for every even-numbered exercise.
- Step-by-step solutions (even and odd) of the Chapter Review Problems, Chapter Tests, and Cumulative Reviews.

Test Item File　(0-13-761925-1)

Media Supplements

TestPro3 Computerized Testing
IBM Single-User　(0-13-746918-7)
MAC　(0-13-746877-6)
- Allows instructors to generate tests or drill worksheets from algorithms keyed to the text by chapter, section, and learning objective.
- Instructors select from thousands of test questions and hundreds of algorithms which generate different but equivalent equations.
- A user-friendly expression-building toolbar, editing and graphing capabilities are included.
- Customization toolbars allow for customized headers and layout options which provide instructors with the ability to add or delete workspace or add columns to conserve paper.

Supplements for the Student

Printed Supplements

Student's Solution Manual　(0-13-746869-5)
- Contains complete step-by-step solutions for every odd-numbered exercise
- Contains complete step-by-step solutions for all (even and odd) Chapter Review Problems, Chapter Tests, and Cumulative Reviews.

Life on the Internet: Mathematics
- Free guide which provides a brief history of the Internet, discusses the use of the Worldwide Web, and describes how to find your way within the Internet and how to find others on it. Contact your local Prentice Hall representative for *Life on the Internet: Mathematics*.

NY Times Themes of the Times
- A free newspaper, created new each year, from Prentice Hall and *The New York Times*
- Interesting and current articles on mathematics
- Invites discussion and writing about mathematics

Media Supplements

MathPro Tutorial Software
- Fully networkable tutorial package for campus labs or individual use
- Designed to generate practice exercises based on the exercise sets in the text
- Algorithmically driven, providing the student with unlimited practice
- Generates graded and recorded practice problems with optional step-by-step tutorial
- Includes a complete glossary including graphics and cross-references to related words

Videotapes (0-13-746934-9)
- Instructional tapes in a lecture format featuring worked-out examples and exercises taken from each section of the text.
- Presentation by Professors Michael C. Mayne and (Biff) John D. Pietro of Riverside Community College in Riverside, California.

Acknowledgments

I wish to express my appreciation to all the reviewers for their helpful criticisms and suggestions. In particular I would like to thank:

Howard Anderson	*Skagit Valley College*
John Anderson	*Illinois Valley Community College*
Michael H. Andreoli	*Miami Dade Community College— North Campus*
Warren J. Burch	*Brevard Community College*
Alice Burstein	*Middlesex Community College*
Sandra Pryor Clarkson	*Hunter College*
Sally Copeland	*Johnson County Community College*
Robert A. Davies	*Cuyahoga Community College*
Ben Divers, Jr.	*Ferrum College*
Irene Doo	*Austin Community College*
Charles C. Edgar	*Onondaga Community College*
Susan Forman	*Bronx Community College*
Gary Glaze	*Eastern Washington University*
Jay Graening	*University of Arkansas*
Robert B. Hafer	*Brevard Community College*
Mary Lou Hammond	*Spokane Community College*
Donald Herrick	*Northern Illinois University*
Beth Hooper	*Golden West College*
Tracy Hoy	*College of Lake County*
Gary Knippenberg	*Lansing Community College*
Mary Koehler	*Cuyahoga Community College*
Hank Martel	*Broward Community College*
John Robert Martin	*Tarrant County Junior College*
Irwin Metviner	*State University of New York at Old Westbury*
Allen R. Newhart	*Parkersburg Community College*
Peg Pankowski	*Community College of Allegheny County— South Campus*
Nancy Ressler	*Oakton Community College*
Gayle Smith	*Lane Community College*

Dick Spangler	*Tacoma Community College*
Janette Summers	*University of Arkansas*
Robert Thornton	*Loyola University*
Lucy C. Thrower	*Francis Marion College*
Andrew Walker	*North Seattle Community College*

Additional acknowledgments are extended to Professor John (Biff) Pietro and Professor Michael C. Mayne of Riverside Community College, for creating the videotapes for each section of the book; Donna Gerken of Miami-Dade Community College, for writing the chapter projects; George Seki, Terry Lovelace, Cindy Trimble, and the mathematicians at Laurel Technical Services, for the Herculean task of solving all the book's problems, preparing the answer section and the solutions manuals, as well as serving as accuracy checker; Amy Mayfield, whose meticulous work as copy editor put me at my syntactical best; Clare Maxwell, photo researcher, for pursuing permissions for the book's extensive contemporary art collection and helping me to obtain wonderful replacement art when the permissions were not forthcoming; Paula Maylahn and Gus Vibal, for contributing to the book's wonderful look; the team of graphic artists at Academy Artworks, whose superb illustrations provide visual support to the verbal portions of the text; Frank Weihenig of Preparé, Inc. the book's compositor, for inputting hundreds of pages with hardly an error; and especially, Ingrid Mount of Elm Street Publishing Services, whose talents as supervisor of production kept every aspect of this very complex project moving through its many stages.

I am also grateful to Tony Palermino my development editor, who contributed invaluable edits and suggestions that resulted in a finished product that is both accessible and up to date. His influence on the book is extraordinary, with the pace in the text and problem sets a result of his remarkable talents.

Finally, I wish to thank the terrific team at Prentice Hall. Special acknowledgements go to Editor-in-Chief Jerome Grant and Acquisitions Editor Sally Denlow who collectively shaped this book into the culmination of my algebra series, set forth the initial vision for the series' design, and urged me onward in my quest to create the first algebra textbooks with an extensive collection of art. Credit is also due to Linda Behrens, managing editor, and Alan Fischer, manufacturing buyer, for keeping an ever-watchful eye on the production process, and Patrice Jones, marketing manager, and Patrick Murphy, marketing assistant, for their outstanding sales force and very impressive marketing efforts.

As I have done in all the books of this series, I must conclude by extending my heartfelt thanks to the gifted artists who gave me permission to share their exciting work, and, ultimately, their humanity within the pages of this book.

Robert Blitzer

The pages that follow highlight the features of the book discussed in the Preface.

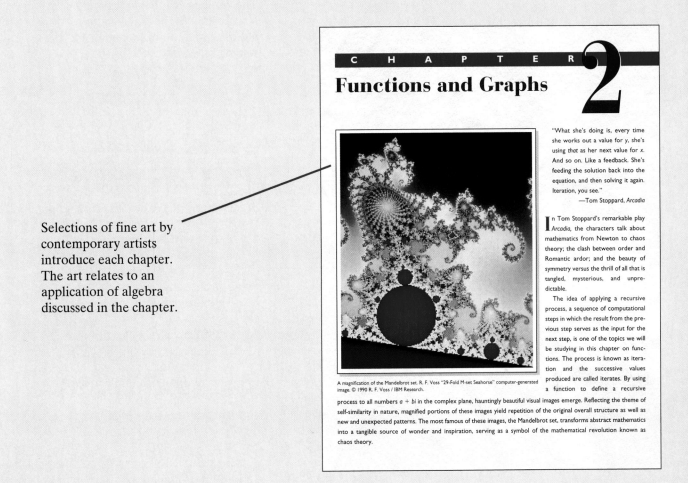

Selections of fine art by contemporary artists introduce each chapter. The art relates to an application of algebra discussed in the chapter.

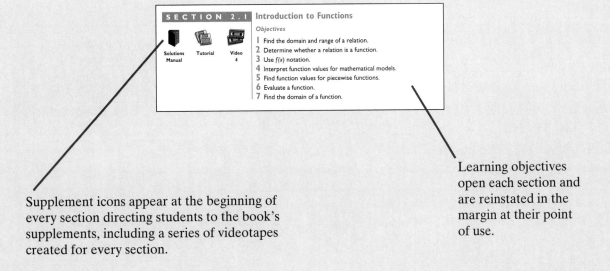

Supplement icons appear at the beginning of every section directing students to the book's supplements, including a series of videotapes created for every section.

Learning objectives open each section and are reinstated in the margin at their point of use.

EXAMPLE 7 Einstein's Special Relativity Models

Einstein's equations relate the mass, time, and length at rest (M_0, T_0, L_0) to the mass, time, and length (M, T, L) for a velocity v. The speed of light, c, is approximately 186,000 miles per second.

$$M = \frac{M_0}{\sqrt{1 - \frac{v^2}{c^2}}} \qquad T = \frac{T_0}{\sqrt{1 - \frac{v^2}{c^2}}} \qquad L = L_0 \sqrt{1 - \frac{v^2}{c^2}}$$

How fast would a futuristic starship that is 600 meters tall have to move so that it would measure only 84 meters tall from the perspective of an observer who looks up and sees the ship pass by?

Solution

600 meters

Starship at Rest

84 meters

Change in starship's dimensions when moving at 99% of light's speed when viewed by an observer

Figure 1.13

$L = L_0 \sqrt{1 - \frac{v^2}{c^2}}$	This is Einstein's model for length.
$84 = 600 \sqrt{1 - \frac{v^2}{c^2}}$	Substitute L_0 (length at rest) = 600 and L (length at the velocity we must determine) = 84.
$0.14 = \sqrt{1 - \frac{v^2}{c^2}}$	Divide both sides by 600.
$0.0196 = 1 - \frac{v^2}{c^2}$	Square both sides.
$0.0196c^2 = c^2 - v^2$	Multiply both sides by c^2. Remember that c, the speed of light, is approximately 186,000 miles per second.
$v^2 = 0.9804c^2$	Isolate v^2 on the left.
$v = \sqrt{0.9804c^2}$	Assume that the starship is moving upward, so take only the positive root.
$v \approx 0.99c$	

If the 600-meter tall futuristic starship moves at 99% the speed of light (approximately 184,140 miles per second), it will measure only 84 meters tall from the perspective of our stationary viewer. A before and after picture of the starship is shown in Figure 1.13. Observe that lengths perpendicular to the direction of motion are unchanged by the effect of speed, so the width of the starship remains unchanged. ∎

The emphasis of the book is on learning to use the language of algebra as a tool for modeling and solving relevant problems.

Average and Minimum Salaries for Major League Baseball Players

Figure 2.19

Source: Based on statistics from the National Baseball League Players Association

EXAMPLE 8 Modeling Average Baseball Salaries

The graph in Figure 2.19 shows average and minimum salaries for major league baseball players. Write the equation that models the data for the line representing average salaries from 1990 through 1994. Then extrapolate from the data and predict the average salary for a major league baseball player in the year 2000.

Solution

The graph indicates that the average salary in 1990 was $0.6 million, so we will use the data points (1990, 0.6) and (1994, 1.3). To write the point-slope equation of the line, we need (x_1, y_1), a fixed point, and m, the slope. *Either* ordered pair can be taken as (x_1, y_1). Using the smaller numbers, we have

$$x_1 = 1990 \quad \text{and} \quad y_1 = 0.6.$$

EXAMPLE 2 Carbon-14 Dating: The Dead Sea Scrolls

Carbon-14 (C-14) decays exponentially with a half-life of approximately 5715 years, meaning that after 5715 years a given amount of C-14 will have decayed to half the original amount.

a. Find the exponential decay model for C-14.
b. In 1947 earthenware jars containing what are known as the Dead Sea Scrolls were found by an Arab Bedouin herdsman. Analysis indicated that the scroll wrappings contained 76% of their original C-14. Estimate the age of the Dead Sea Scrolls.

Solution

a. We use the exponential decay model and the information about the half-life of C-14 to find k.

$A = A_0 e^{kt}$	This is the exponential decay model.
$\frac{A_0}{2} = A_0 e^{k5715}$	After 5715 years ($t = 5715$), $A = \frac{A_0}{2}$ (since the amount present (A) is half the original amount (A_0).)
$\frac{1}{2} = e^{5715k}$	Divide both sides of the equation by A_0.
$\ln \frac{1}{2} = \ln e^{5715k}$	Take the natural logarithm of both sides.
$\ln \frac{1}{2} = 5715k$	$\ln e^x = x$
$k = \frac{\ln \frac{1}{2}}{5715} \approx -0.000121$	Solve for k.

Substituting for k in the decay model, the model for C-14 is $A = A_0 e^{-0.000121t}$.

Illustrative examples are presented one step at a time, with each step clearly explained. Titles enable students to immediately see each illustrative example's purpose.

Discover for yourself

The quadratic inequalities in Examples 1 and 2 had solution sets consisting of a single interval and the union of two intervals. However, other possibilities can occur.

The graph of $y = 4x^2 - 8x + 7$ is shown in Figure 1.24. Use the graph to solve $4x^2 - 8x + 7 > 0$ and $4x^2 - 8x + 7 < 0$.

The graph of $y = x^2 - 4x + 4$ is shown in Figure 1.25. Use the graph to solve $x^2 - 4x + 4 > 0$ and $x^2 - 4x + 4 \leq 0$.

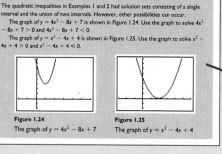

Figure 1.24
The graph of $y = 4x^2 - 8x + 7$

Figure 1.25
The graph of $y = x^2 - 4x + 4$

Discover for yourself exercises encourage students to actively participate in the learning process as they read the book.

Study tip

The major problem generally encountered in the study of permutations and combinations is the ability to distinguish between each in a given real world situation. Remember that *order is important* when considering a *permutation*. Different orderings or arrangements produce different permutations. On the other hand, *order is not important* when considering a *combination*. A change in order does not produce a new combination.

Permutations	Combinations
$_nP_r = \frac{n!}{(n-r)!}$	$\binom{n}{r} = \frac{n!}{(n-r)!r!}$
Different orderings produce different permutations.	Different orderings do not produce different combinations.
Helpful words: Arrangement, Order, Schedule	Helpful words: Selection, Group, Committee, Subcommittee, Subset

Study tip boxes offer suggestions for problem solving and point out common student errors.

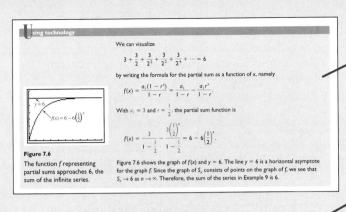

Using technology

We can visualize

$$3 + \frac{3}{2} + \frac{3}{2^2} + \frac{3}{2^3} + \frac{3}{2^4} + \cdots = 6$$

by writing the formula for the partial sum as a function of x, namely

$$f(x) = \frac{a_1(1 - r^x)}{1 - r} = \frac{a_1}{1 - r} - \frac{a_1 r^x}{1 - r}.$$

With $a_1 = 3$ and $r = \frac{1}{2}$, the partial sum function is

$$f(x) = \frac{3}{1 - \frac{1}{2}} - \frac{3\left(\frac{1}{2}\right)^x}{1 - \frac{1}{2}} = 6 - 6\left(\frac{1}{2}\right)^x.$$

Figure 7.6 shows the graph of $f(x)$ and $y = 6$. The line $y = 6$ is a horizontal asymptote for the graph f. Since the graph of S_n consists of points on the graph of f, we see that $S_n \to 6$ as $n \to \infty$. Therefore, the sum of the series in Example 9 is 6.

Figure 7.6
The function f representing partial sums approaches 6, the sum of the infinite series.

Graphing utility screens enable students to visualize concepts. Visualization is enhanced through optional technology exercises in each problem set.

Enrichment essays connect algebra to the whole spectrum of learning.

E N R I C H M E N T   E S S A Y

Carbon Dating and Artistic Development

Carbon dating is a method for estimating the age of organic material. Until recently, the oldest prehistoric cave paintings known were from the Lascaux cave in France. Charcoal from the site was analyzed, and the paintings were estimated to be 15,505 years old.

The artistic community was electrified by the discovery in 1995 of spectacular cave paintings in a limestone cavern near Lascaux. Carbon dating showed that the images, created by artists of remarkable talent, were 30,000 years old, making them the oldest cave paintings ever found. The artists seemed to have used the cavern's natural contours to heighten a sense of perspective. The quality of the painting suggests that the art of early humans did not mature steadily from primitive to sophisticated in any simple linear fashion.

Jean Clottes, France's foremost expert on prehistoric cave art, was moved by the sophisticated imagery of the 30,000-year-old work. "I remember standing in front of the paintings and being profoundly moved by the artistry," he exclaimed. "Tears were running down my cheeks. I was witnessing one of the world's great masterpieces."

Jean-Marie Chauvet / Sygma

TABLE 4.2	Transformations Involving Exponential Functions	
Transformation	**Equation**	**Description**
Horizontal translation	$y = b^{x+c}$	Shifts the graph of $y = b^x$ to the left c units if $c > 0$ and to the right c units if $c < 0$.
Stretching or shrinking	$y = cb^x$	Stretches the graph of $y = b^x$ for $c > 1$ and shrinks the graph of $y = b^x$ for $0 < c < 1$, multiplying y-coordinates of $y = b^x$ by c.
Reflecting	$y = -b^x$	Reflects the graph of $y = b^x$ in the x-axis.
	$y = b^{-x}$	Reflects the graph of $y = b^x$ in the y-axis.
Vertical translation	$y = b^x + c$	Shifts the graph of $y = b^x$ upward c units if $c > 0$ and downward c units if $c < 0$.

Summary boxes organize and highlight key ideas.

Geometry is integrated throughout the book.

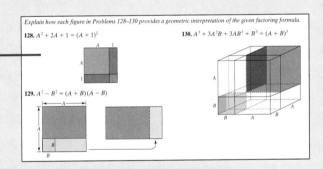

Explain how each figure in Problems 128–130 provides a geometric interpretation of the given factoring formula.

128. $A^2 + 2A + 1 = (A + 1)^2$

129. $A^2 - B^2 = (A + B)(A - B)$

130. $A^3 + 3A^2B + 3AB^2 + B^3 = (A + B)^3$

Problem sets are organized into seven categories, making it easier to select homework problems consistent with the role of graphing utilities, critical thinking, collaborative learning, and writing exercises in your vision of the course.

Practice Problems

1. The graph of a function f is shown in the figure.

a. State the values of $f(-6)$, $f(0)$, and $f(3)$.

b. State the intervals on which f is increasing and on which f is decreasing.

c. State the minimum value of f. At what value of x does this occur?

d. If the domain of f is $(-\infty, \infty)$, what is the function's range?

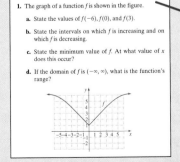

Practice Problems give students the chance to practice and master the concepts developed in each section.

Application Problems

67. The graph represents the probability of two people in the same room sharing a birthday as a function of the number of people in the room. Call the function f.

a. Explain why f has an inverse that is a function.

b. Describe in practical terms the meaning of $f^{-1}(0.25)$, $f^{-1}(0.5)$, and $f^{-1}(0.7)$.

Application Problems include relevant, up-to-date applications, offering students the opportunity to construct mathematical models from data.

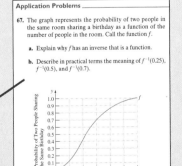

True–False Critical Thinking Problems

58. Which one of the following is true?
    a. If the coordinates of a point satisfy the inequality $xy > 0$, then (x, y) must be in quadrant I.
    b. The ordered pair $(2, 5)$ satisfies $3y - 2x = -4$.
    c. If a point is on the x-axis, it is neither up nor down, so $x = 0$.
    d. None of the above is true.

59. Morphine is widely used in the United States to treat severe pain. Its physiological effects, however, differ for nonaddicts (graph a) and addicts (graph b). Both graphs show the concentration of morphine in the blood following an intravenous injection of 10 milligrams to relieve severe pain. The concentration is measured in micrograms per milliliter (μg/ml). Which one of the following is true based on the graphs?
    a. Nonaddicts receive narcotic effects only when the bloodstream concentration is above 1 μg/ml.
    b. The addict receives narcotic effects for up to 2 hours.
    c. The concentration at which the nonaddict experiences no effects from the drug is the same at which the addict starts experiencing symptoms of withdrawal.
    d. The morphine concentration in the blood over 24 hours differs for nonaddicts and addicts.

True-False Critical Thinking Problems encourage thinking skills in a format that is not intimidating to most students.

Technology Problems

53. Use a graphing utility to graph $y = \frac{1}{x}$, $y = \frac{1}{x^3}$, and $\frac{1}{x^5}$ in the same viewing rectangle. For odd values of n, how does changing n affect the graph?

54. Use a graphing utility to graph $y = \frac{1}{x^2}$, $y = \frac{1}{x^4}$, and $y = \frac{1}{x^6}$ in the same viewing rectangle. For even values of n, how does changing n affect the graph?

55. The model $f(x) = \frac{72,900}{100x^2 + 729}$ describes the 1992 unemployment rate (as a percent) in the United States as a function of years of education (x).
    a. Use a graphing utility to graph the model for $x \geq 0$.
    b. What does the shape of the graph indicate about increasing levels of education?
    c. Explain why f is one-to-one. What is $f^{-1}(3.6)$ and what practical information does this provide?
    d. Is there an education level that leads to guaranteed employment? How is your answer indicated on the graph?

Technology Problems enable students to use graphing utilities to explore algebraic concepts and mathematical models.

Writing in Mathematics

98. Consider a position function $s(t)$ for *Freedom 7*, the spacecraft that carried the first American into space in 1961. Total flight time was 15 minutes, and the spacecraft reached a maximum altitude of 116 miles. Is the function $s(t)$, expressing height in miles as a function of time (t, in minutes) one-to-one? Explain your answer in a sentence or two.

99. Discuss similarities between f and f^{-1} and real numbers and their multiplicative inverses.

100. Describe how to draw a picture (similar to the one shown here) in which the function is not one-to-one.

A one-to-one function   has an inverse.

Writing exercises enable students to communicate their mathematical knowledge by thinking and writing about algebraic topics.

Critical Thinking Problems

72. If $f(x) = b^x + 1$, show that $\frac{1}{f(x)} + \frac{1}{f(-x)} = 1$.

73. If $f(x) = 3^x$, show that $\frac{f(x + h) - f(x)}{h} = 3^x\left(\frac{3^h - 1}{h}\right)$.

74. The half-life of a substance that is decaying exponentially is the time required for half of the substance to disintegrate. Use the figure to estimate the half-life of each substance whose exponential decay curve is shown.

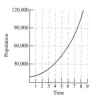

76. The hyperbolic cosine and hyperbolic sine functions are defined by
$$\cosh x = \frac{e^x + e^{-x}}{2} \quad \text{and} \quad \sinh x = \frac{e^x - e^{-x}}{2}.$$
    a. Prove that $(\cosh x)^2 - (\sinh x)^2 = 1$.
    b. Prove that the hyperbolic cosine is an even function.
    c. Prove that the hyperbolic sine is an odd function.

Critical Thinking Problems encourage students to explore concepts while stimulating rigorous, divergent thinking.

Group Activity Problem

94. As we know, the planets in our solar system revolve around the sun in elliptical orbits with the sun at one focus. The eccentricities and the lengths of the major axes for these elliptical orbits are given in the following table.

Planet	Eccentricity	Length of Major Axis (in millions of kilometers)
Mercury	0.206	116
Venus	0.007	216
Earth	0.017	299
Mars	0.093	456
Jupiter	0.048	1557
Saturn	0.056	2854
Uranus	0.047	5738
Neptune	0.008	8996
Pluto	0.249	11,800

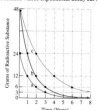

Ptolemaic Universe, 16th Century. Italy 16th c., Portulan of Admiral Coligny: Zodiaque, Ms. 700/1602 fol. 4v and 5r. Chantilly, Musée Conde. Giraudon / Art Resource, NY.

Find an equation for the orbit of each of the nine planets around the sun. Then consider the equations in pairs and solve the resulting systems to determine whether or not any two of the planets in our solar system could collide.

Group Activity Problems allow you to use collaborative learning as a teaching strategy.

CHAPTER PROJECT
Jurassic Math: A Matter of Scale

Looking at numbers describing the physical attributes of a dinosaur is not quite the same as watching a movie filled with fast-moving predators, or seeing a full-scale reproduction in a museum. We may understand an equation in an intellectual sense, yet still feel we are missing some connection to make it real. There is always a human side to presenting information, making it accessible at a level other than an abstract, intellectual understanding. The mathematics in this chapter represents a large amount of information in a compact, symbolic form, but has something been lost in the translation?

For example, judging the size or scale of something is not an exercise in fantasy; it is an important skill for many people involved in the design, construction, or alteration of the landscape around us. To give more information to a client, architects create models of their plans complete with familiar sizes of people and objects. Engineers create models of differing scales as they design new cars, planes, or other objects.

Chapter projects at the end of each chapter use challenging and interesting applications of mathematics as a collaborative learning tool, and cultivate mathematical exploration on the Worldwide Web.

Chapter Review

SUMMARY

1. Basic Ideas about Functions

a. A function is a correspondence between a first set, the domain, and a second set, the range, such that each element in the domain corresponds to exactly one element in the range.

b. Using function notation $y = f(x)$, the letter f names the function, and $f(x)$, read "f of x" or "f at x" rep-

resents an expression for the value of the function at x. The value y, the dependent variable, is calculated after selecting a value for x, the independent variable.

c. The domain of a function is the set of all values of the independent variable for which the function is defined. If a function is defined by an algebraic

Inclusive summaries appear at the conclusion of each chapter so that students can bring together what they have learned.

Chapters 2–6 conclude with cumulative review problems.

CUMULATIVE REVIEW PROBLEMS (CHAPTERS P–4)

In Problems 1–2, simplify.

1. $\dfrac{3}{4x^2 + 4x + 1} + \dfrac{x + 3}{2x^2 - x - 1} - \dfrac{2}{x - 1}$

2. $\dfrac{1 - \dfrac{2}{x}}{1 - \dfrac{3}{x} + \dfrac{2}{x^2}}$

REVIEW PROBLEMS

In Problems 1–13, identify each conic and sketch its graph by hand. For ellipses and hyperbolas, find the foci. For parabolas, identify the vertex, focus, and directrix. Use a graphing utility to verify all graphs.

1. $y^2 + 8x = 0$

2. $4x^2 + y^2 = 16$

3. $9x^2 - 16y^2 - 144 = 0$

4. $x^2 + 16y = 0$

5. $(y - 2)^2 = -16x$

6. $\dfrac{(x - 1)^2}{16} + \dfrac{(y + 2)^2}{9} = 1$

7. $\dfrac{(x - 2)^2}{25} - \dfrac{(y + 3)^2}{16} = 1$

8. $(x - 4)^2 = 4(y + 1)$

9. $4x^2 - y^2 - 8x - 4y - 16 = 0$

10. $4x^2 - 40x - y + 102 = 0$

11. $4x^2 + 9y^2 + 24x - 36y + 36 = 0$

12. $y^2 - 4x - 10y + 21 = 0$

13. $4x^2 - y^2 + 8x + 4y + 4 = 0$

Review problems appear throughout the book. Each chapter contains review problems.

CHAPTER 3 TEST

In Problems 1–2, use the vertex and intercepts to graph each quadratic function. Give the equation of the axis of symmetry.

1. $f(x) = (x - 1)^2 - 4$

2. $f(x) = 2x^2 - 4x - 6$

3. The function $f(x) = -x^2 + 46x - 360$ models the daily profit ($f(x)$, in hundreds of dollars) for a company that manufactures x VCRs daily. How many VCRs should be manufactured each day to maximize profit? What is the maximum daily profit?

4. A rectangular plot of land along a river is to be fenced along 3 sides using 600 feet of fencing. No fencing is to be placed along the river's edge. Find the dimensions that maximize the enclosed area. What is the maximum area?

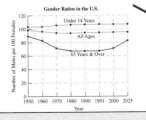

7. The graph of $f(x) = 6x^3 - 19x^2 + 16x - 4$ is shown in the figure.

a. Based on the graph of f, find the integral root of the equation $6x^3 - 19x^2 + 16x - 4 = 0$.

b. Use synthetic division to find the other two roots of $6x^3 - 19x^2 + 16x - 4 = 0$.

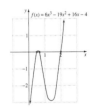

5. Consider the function $f(x) = x^3 - 5x^2 - 4x + 20$.

a. Use factoring to find all zeros of f.

b. Use the Leading Coefficient Test and the zeros of f to graph the function.

6. Use end behavior to explain why the graph at the top of the next column cannot be the graph of $f(x) = x^5 - x$. Then use intercepts to explain why the graph cannot represent $f(x) = x^5 - x$.

Each chapter contains a chapter test so that students can prepare for an actual in-class exam.

Solving Equations and Inequalities

Systematic procedures for solving certain equations and inequalities are an important component of algebra. Problems 1–21 give you the opportunity to review these procedures. Solve each problem, expressing irrational solutions in simplified form and imaginary solutions in the form $a + bi$.

1. $\dfrac{2}{x + 1} + \dfrac{3}{2x - 3} = \dfrac{6x + 1}{2x^2 - x - 3}$

2. $x(4x - 11) = 3$

3. $(x^2 - 5)^2 - 3(x^2 - 5) - 4 = 0$

4. $\sqrt{2x + 3} - \sqrt{4x - 1} = 1$

5. $x^{1/2} - 6x^{1/4} + 8 = 0$

6. $\dfrac{x - 3}{4} + \dfrac{x + 2}{3} \leq 2$

7. $-2 < 8 - 5x < 7$

8. $|2x + 1| \leq 1$

9. $6x^2 - 6 < 5x$

10. $\dfrac{1 - x}{3 + x} < 4$

11. $x^3 - x^2 - 4x + 4 = 0$

12. $3x^3 + 4x^2 - 7x + 2 = 0$

13. $e^{14 - 7x} - 53 = 24$

14. $e^{2x} - 10e^x + 9 = 0$

15. $\log_2(x + 1) + \log_2(x - 1) = 3$

16. $\ln(3x) + \ln(x + 2) = \ln 9$

17. (Solve using matrices.)
$\begin{aligned} x - y + z &= 17 \\ -4x + y + 5z &= -2 \\ 2x + 3y + z &= 8 \end{aligned}$

18. (Solve for x only using Cramer's rule.)
$\begin{aligned} x + 2y - z &= 1 \\ x + 3y - 2z &= -1 \\ 2x - y + z &= 6 \end{aligned}$

19. $\dfrac{x^2}{4} + \dfrac{y^2}{16} = 1$

$x - y = 2$

20. $4x^2 + 3y^2 = 48$

$3x^2 + 2y^2 = 35$

21. $3x^2 - 2xy + 3y^2 = 34$

$x^2 + y^2 = 17$

Graphs and Graphing

Throughout the book, we have used graphs to help visualize a problem's solution. We have also studied graphing equations and inequalities in the rectangular coordinate system. Problems 22–36 focus on problem solving with graphs.

22. Use the graphs shown in the figure to describe how the graph of $h(x) = (x + 2)^3 + 1$ is obtained from the graph of $f(x) = x^3$. Include the graph of g in your description.

Gender Ratios in the U.S.

Cumulative review problems covering the entire book appear in the Appendix.

To the Student

The process of learning mathematics requires that you do at least three things—read the book, work the problems, and get your questions answered if you are stuck. This book has been written so that you can learn directly from its pages. All concepts are carefully explained, important definitions and procedures are set off in boxes, and worked-out examples that present solutions in a step-by-step manner appear throughout. Study tip boxes offer hints and suggestions, and often point out common errors to avoid. Discovery boxes encourage you to actively participate in the learning process as you read the book. A great deal of attention has been given to show you the vast and unusual applications of algebra in order to make your learning experience both interesting and relevant. As you begin your studies, I would like to offer some specific suggestions for using this book and for being successful in algebra.

- *Read the book.*
  a. Begin with the chapter introduction. Enjoy the art while you obtain a general idea of what the chapter is about.
  b. Move on to the objectives and the introduction to a particular section. The objectives will tell you exactly what you should be able to do once you have completed the section. Each objective is restated in the margin at the point in the section where the objective is taught.
  c. At a slow and deliberate pace, read the section with pen (or pencil) in hand. Move through the illustrative examples with great care. These worked-out examples provide a model for doing the problems in the problem sets. Be sure to read all the hints and suggestions in the Study tip boxes. Your pen is in hand for the Discover for yourself exercises that are intended to encourage you to actively participate in the learning process as you read the book. The Discover for yourself exercises let you explore problems in order to understand them and their solutions better, so be sure not to jump over these valuable discovery experiences in your reading.
  d. Enjoy the Enrichment essays and the contemporary art that is intended to make your reading more interesting and show you how algebra is connected to the whole spectrum of learning.

As you proceed through the reading, do not give up if you do not understand every single word. Things will become clearer as you read on and see how various procedures are applied to specific worked-out examples.

- *Work problems every day and check your answers.* The way to learn mathematics is by *doing* mathematics, which means by *solving problems.* The more problems you work, the better you will become at solving problems which, in turn, will make you a better algebra student.

    a. Work the assigned problems in each problem set. Problem sets are organized into seven categories. Minimally, you should work all odd-numbered problems in the first two categories (Practice Problems and Application Problems). Answers to most odd-numbered problems are given in the back of the book. Once you have completed a problem, be sure to check your answer. If you made an error, find out what it was. Ask questions in class about homework problems you don't understand.

    b. Problem sets also include critical thinking problems, technology problems, writing exercises, and group activity learning experiences. Don't panic! You are not expected to work every problem, or even all the odd-numbered problems, in each problem set. This vast collection of problems provides options for your learning style and your instructor's teaching methods. You may be assigned some problems from one or more of these categories. Problems in the critical thinking categories are the most difficult, intended to stimulate your ability to think and reason. Thinking about a particular question, even if you are confused and somewhat frustrated, can eventually lead to new insights.

- *Prepare for chapter exams.* After completing a chapter, study the summary, work assigned problems from the chapter review problems, and work all the problems in the chapter test.

- *Review continuously.* Working review problems lets you remember the algebra you learned for a much longer period of time. Cumulative review problems appear at the end of each chapter, beginning with Chapter 2. The book's appendix contains review problems covering the entire course. By working the appendix problems assigned by your professor, you will be able to bring together the procedures and problem-solving strategies learned throughout the course.

- *Attend all lectures.* No book is intended to be a substitute for the valuable insights and interactions that occur in the classroom. In addition to arriving for a lecture on time and prepared, you might find it helpful to read the section that will be covered in class beforehand so that you have a clear idea of the new material that will be discussed.

- *Use the supplements that come with this book.* A solutions manual that contains worked-out solutions to the book's odd-numbered problems and all review problems, as well as a series of videotapes created for every section of the book, are among the supplements created to help you learn algebra. Ask your instructor what supplements are available and where you can find them.

Algebra is often viewed as the foundation for more advanced mathematics. It is my hope that this book will make algebra accessible, relevant, and an interesting body of knowledge in and of itself.

Discover for yourself

What conclusion can you draw from these circle graphs about student success and attending lectures?

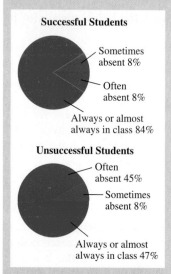

Successful Students

Sometimes absent 8%

Often absent 8%

Always or almost always in class 84%

Unsuccessful Students

Often absent 45%

Sometimes absent 8%

Always or almost always in class 47%

Source: *The Psychology of College Success: A Dynamic Approach,* by permission of H. C. Lindgren, 1969

P

Prerequisites: Fundamental Concepts of Algebra

This chapter reviews topics from intermediate algebra that are prerequisites for the study of college algebra. Since the ideas of algebra build in a logical and orderly manner from the fundamentals, it is important to have your feet planted firmly on the ground. By reviewing and mastering topics about real numbers, graphing, exponents, polynomials, factoring, fractional expressions, and complex numbers, you will be ready to advance in your ability to use algebra to solve problems and describe our world in a significant way.

Using algebra to meaningfully describe our world is the primary focus of this book.

René Magritte, "Le Modele Rouge" © 1998 C. Herscovici, Brussels / Artists Rights Society (ARS), New York. Private Collection. Art Resource, NY.

Study tip

You can use the review exercises at the end of this chapter as a pretest to determine any weaknesses that you might have with the prerequisite topics that are reviewed in the chapter.

SECTION P.1 The Real Number System

Solutions Manual   Tutorial   Video 1

Objectives

1 Recognize subsets of the real numbers.
2 Use set-builder and interval notations.
3 Find unions and intersections of sets.
4 Evaluate absolute value.
5 Use absolute value to express distance.

Numbers and their properties have intrigued humankind since the beginning of civilization. In this section, we describe the subsets of the real numbers, using the real number line to order these numbers.

1 Recognize subsets of the real numbers.

The Set of Real Numbers

The sets that make up the real numbers are summarized in Table P.1. We refer to these sets as *subsets* of the real numbers, meaning that all elements in each subset are also elements in the set of real numbers.

When we combine the set of all irrational numbers with the set of all rational numbers, we obtain the set of all *real numbers*. Figure P.1 shows a diagram of the subsets of the real numbers.

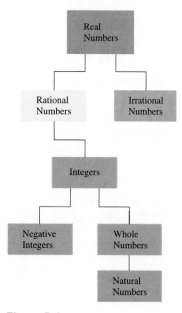

Figure P.1

The real number system

TABLE P.1	**Important Subsets of the Real Numbers**	
Name	**Description**	**Examples**
Natural numbers	$\{1, 2, 3, 4, 5, \ldots\}$ These numbers are used for counting and are also called the *counting numbers*.	87 2,369 $\sqrt{36}$ (or 6)
Whole numbers	$\{0, 1, 2, 3, 4, 5, \ldots\}$ The whole numbers add 0 to the set of natural numbers.	0, 87 $\frac{18}{3}$ (or 6)
Integers	$\{\ldots, -5, -4, -3, -2, -1, 0, 1, 2, 3, 4, 5, \ldots\}$ The integers add the opposites of the natural numbers to the set of whole numbers.	$-87, -14,$ $-\sqrt{25}$ (or -5), 0, 14, 87

TABLE P.I *(cont.)* Important Subsets of the Real Numbers		
Name	**Description**	**Examples**
Rational numbers	These numbers can be expressed as an integer divided by a nonzero integer: $\frac{a}{b}$; a and b are integers; $b \neq 0$. In decimal form, rational numbers either terminate or repeat.	$\frac{3}{4}$ (or 0.75) ($a = 3, b = 4$) $-\frac{3}{11} = -0.2727... = -0.\overline{27}$ ($a = -3, b = 11$) $87 = \frac{87}{1}$ ($a = 87, b = 1$)
Irrational numbers	This is the set of numbers whose decimal representations do not repeat and do not terminate. Irrational numbers cannot be expressed as an integer divided by an integer.	$\sqrt{2} \approx 1.414214$ $-\sqrt{3} \approx -1.73205$ $\pi \approx 3.142$ $-\frac{\pi}{2} \approx -1.571$

Study tip

The symbol $\approx$ is read "is approximately equal to." The reason that

$$\sqrt{2} \approx 1.414214$$

is that $(1.414214)^2$ is not exactly 2.

EXAMPLE I Classifying Real Numbers

List the numbers in the set

$$\{-7, -\tfrac{3}{4}, 0, 0.\overline{6}, \sqrt{5}, \pi, 6\tfrac{1}{4}, 7.3, \sqrt{81}\}$$

that belong to each subset of the real numbers:

a. Natural numbers
b. Whole numbers
c. Integers
d. Rational numbers
e. Irrational numbers
f. Real numbers

Solution

a. Natural numbers: The only natural number in the set is $\sqrt{81}$ since $\sqrt{81} = 9$. (9 multiplied by itself is 81.)
b. Whole numbers: The whole numbers consist of the natural numbers and 0. The elements of the set that are whole numbers are 0 and $\sqrt{81}$.
c. Integers: The integers consist of the natural numbers, 0, and the negatives of the natural numbers. The elements of the set that are integers are $\sqrt{81}$, 0, and -7.
d. Rational numbers: All numbers in the set that can be expressed as the quotient of two integers are rational numbers. The rational numbers are -7 ($-7 = \frac{-7}{1}$), $-\frac{3}{4}$, 0 ($0 = \frac{0}{1}$), $0.\overline{6}$ ($0.\overline{6} = 0.6666 ... = \frac{2}{3}$), $6\frac{1}{4}$ ($6\frac{1}{4} = \frac{25}{4}$), 7.3 ($7.3 = 7\frac{3}{10} = \frac{73}{10}$), and $\sqrt{81}$ ($\sqrt{81} = \frac{9}{1}$).
e. Irrational numbers: The irrational numbers in the set are $\sqrt{5}$ ($\sqrt{5} \approx 2.236$) and π ($\pi \approx 3.14$). Both $\sqrt{5}$ and π are only approximately equal to 2.236 and 3.14, respectively. In decimal form, $\sqrt{5}$ and π neither terminate nor have repeating patterns.
f. Real numbers: All the numbers in the set are real numbers. ∎

The Real Number Line

The *real number line* is a graph used to represent the set of real numbers (Figure P.2). An arbitrary point, called the *origin*, is labeled 0; units to the right of the origin are *positive* and units to the left of the origin are *negative*.

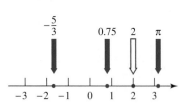

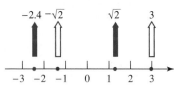

Every real number corresponds to exactly one point on the real number line.

Figure P.2

The real number line

Every point on the real number line corresponds to exactly one real number.

Figure P.3

A one-to-one correspondence between real numbers and points on a number line

Figure P.3 illustrates that every real number corresponds to a point on the number line and every point on the number line corresponds to a real number. For this reason, we say that there is a *one-to-one correspondence* between all the real numbers and all points on a real number line. The real number corresponding to a particular point on the line is called the *coordinate* of the point. If you draw a point on the real number line corresponding to a real number, you are *plotting* the real number.

Ordering the Real Numbers

The real number line is useful in demonstrating the *order* of real numbers. We say that the real number *a is less than* real number *b*, written $a < b$, if *a* is to the left of *b* on the number line. Equivalently, *b is greater than a*, written $b > a$, if *b* is to the right of *a* on the number line (Figure P.4).

The symbols $<$ and $>$ are sometimes combined with an equal sign.

Figure P.4

a is less than *b*: *a* is to the left of *b*

The symbols $\leq$ and $\geq$		
Symbols	**Meaning**	**Examples**
$a \leq b$	*a* is less than or equal to *b*.	$3 \leq 7$ (because $3 < 7$)
		$7 \leq 7$ (because $7 = 7$)
$b \geq a$	*b* is greater than or equal to *a*.	$7 \geq 3$ (because $7 > 3$)
		$-5 \geq -5$ (because $-5 = -5$)

Study tip

A right triangle with two sides of length 1 and a third side of length $\sqrt{2}$ can be used to plot $\sqrt{2}$ on a number line.

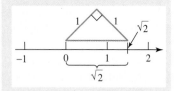

2 Use set-builder and interval notations.

Intervals

Inequalities are used to describe subsets of real numbers, called *intervals*. For instance, the inequality $x \leq 3$ describes all real numbers less than or equal to 3. Using *set-builder* notation, we write

$$\{x \,|\, x \leq 3\}$$

reading this as "the set of all real numbers *x* such that *x* is less than or equal to 3." Using *interval* notation, we write

$$(-\infty, 3].$$

Figure P.5

The interval $(-\infty, 3]$

As shown in Figure P.5, the square bracket at 3 shows that 3 *is* to be included. The negative infinity symbol $-\infty$ does not represent a real number. It indicates that the interval includes all real numbers less than or equal to 3, extending indefinitely to the left.

Let's consider another example of a subset of real numbers that can be expressed in both set-builder and interval notations. The inequality $x > -2$ describes all real numbers greater than -2. Using *set-builder* notation, we write

$$\{x \,|\, x > -2\}$$

reading this as "the set of all real numbers x such that x is greater than -2." Using *interval* notation, we write

$$(-2, \infty).$$

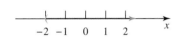

Figure P.6

The interval $(-2, \infty)$

As shown in Figure P.6, the parenthesis at -2 indicates that -2 is *not* included in the interval. The infinity symbol ∞ does not represent a real number. It indicates that the interval extends indefinitely to the right.

Some inequalities do not extend indefinitely in positive or negative directions. For example, the inequality $-2 < x \le 4$ indicates that $x > -2$ and $x \le 4$. This *double* or *compound* inequality denotes all real numbers between -2 and 4, excluding -2 but including 4. Using set-builder notation, we write

$$\{x \,|\, -2 < x \le 4\}$$

reading this as "the set of all real numbers x such that x is greater than -2 and less than or equal to 4." Using interval notation, we write

$$(-2, 4].$$

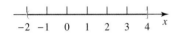

Figure P.7

The interval $(-2, 4]$

Figure P.7 illustrates that the parenthesis at -2 excludes -2 from the set and the square bracket at 4 includes 4 as a member of the set.

Table P.2 lists the nine possible types of intervals.

TABLE P.2	**Intervals on the Real Number Line**	

Let a and b be real numbers such that $a < b$.

Interval Notation	Set-Builder Notation	Graph
(a, b)	$\{x \,|\, a < x < b\}$	
$[a, b]$	$\{x \,|\, a \le x \le b\}$	
$[a, b)$	$\{x \,|\, a \le x < b\}$	
$(a, b]$	$\{x \,|\, a < x \le b\}$	
(a, ∞)	$\{x \,|\, x > a\}$	
$[a, \infty)$	$\{x \,|\, x \ge a\}$	
$(-\infty, b)$	$\{x \,|\, x < b\}$	
$(-\infty, b]$	$\{x \,|\, x \le b\}$	
$(-\infty, \infty)$	$\mathbb{R}$ (set of all real numbers)	

EXAMPLE 2 **Intervals and Inequalities**

Express the intervals in terms of inequalities and graph:

a. $(-1, 4]$ **b.** $[2.5, 4]$ **c.** $(-4, \infty)$

Solution

a. $(-1, 4] = \{x \mid -1 < x \leq 4\}$

b. $[2.5, 4] = \{x \mid 2.5 \leq x \leq 4\}$

c. $(-4, \infty) = \{x \mid x > -4\}$ ∎

Phrases such as "at most" or "at least" indicate that inequalities are used in everyday language. Example 3 shows how certain English expressions can be written in both set-builder and interval notations.

EXAMPLE 3 **The English Language and Inequalities**

Rewrite each English sentence in the left column in both set-builder and interval notations.

Solution

	Set-Builder Notation	Interval Notation
x is less than 5.	$\{x \mid x < 5\}$	$(-\infty, 5)$
x is greater than or equal to 3.	$\{x \mid x \geq 3\}$	$[3, \infty)$
x lies between -2 and 5, excluding -2 and 5.	$\{x \mid -2 < x < 5\}$	$(-2, 5)$
x is greater than or equal to 2 and less than 7.	$\{x \mid 2 \leq x < 7\}$	$[2, 7)$
x is nonnegative. "Nonnegative" indicates a number is greater than or equal to 0.	$\{x \mid x \geq 0\}$	$[0, \infty)$
x is at most 4. "At most" means $\leq$.	$\{x \mid x \leq 4\}$	$(-\infty, 4]$
x is at least 2. "At least" means $\geq$.	$\{x \mid x \geq 2\}$	$[2, \infty)$
x is positive but not more than 5. "Not more than" means $\leq$.	$\{x \mid 0 < x \leq 5\}$	$(0, 5]$

∎

3 Find unions and intersections of sets.

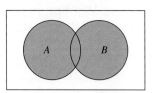

$A \cup B$ is the set of all elements in A or B, or in both A and B.

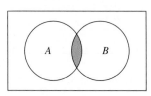

$A \cap B$ is the set of elements common in both A and B.

Union and Intersection of Sets

If A and B are sets, then their *union*, $A \cup B$, is the set containing all elements that are in A or in B (or in both A and B). Their *intersection*, $A \cap B$, is the set containing all elements that are in both A and B.

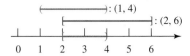

Figure P.8

$(1, 4) \cap [2, 6] = [2, 4)$

Figure P.9

$(-2, \infty) \cap (3, \infty) = (3, \infty)$

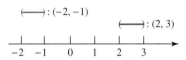

Figure P.10

$(-2, -1) \cap (2, 3) = \emptyset$

tudy tip

The intersection of two intervals is the region where their graphs overlap.

Figure P.11

$(1, 4) \cup [2, 6] = (1, 6]$

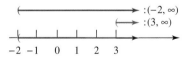

Figure P.12

$(-2, \infty) \cup (3, \infty) = (-2, \infty)$

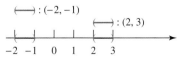

Figure P.13

$(-2, -1) \cup (2, 3)$ is not a single interval

Union and intersection of sets

Union

$A \cup B$ is the set of all elements that are in set A or in set B, or in both A and B.

Intersection

$A \cap B$ is the set of all elements that are in set A *and* in set B.

EXAMPLE 4   Finding the Union and Intersection of Sets

Express each set in interval notation:

a. $(1, 4) \cap [2, 6]$ **b.** $(-2, \infty) \cap (3, \infty)$
c. $(-2, -1) \cap (2, 3)$ **d.** $(1, 4) \cup [2, 6]$
e. $(-2, \infty) \cup (3, \infty)$ **f.** $(-2, -1) \cup (2, 3)$

Solution

a. To find $(1, 4) \cap [2, 6]$, we graph each set above the number line and then locate the interval on the number line common to both. Figure P.8 illustrates that numbers common to both intervals begin at 2 and extend up to but not including 4. Thus,

$(1, 4) \cap [2, 6] = [2, 4)$.

b. Figure P.9 illustrates that numbers common to both intervals are all real numbers greater than 3. Thus,

$(-2, \infty) \cap (3, \infty) = (3, \infty)$.

c. Figure P.10 shows that there are no numbers common to the two intervals. Their intersection is a set with no elements, called the *empty set* or the *null set*, denoted by $\emptyset$. Thus,

$(-2, -1) \cap (2, 3) = \emptyset$.

In parts (d)–(f), the union of the two intervals consists of the real numbers that are in either or both of the intervals.

d. To find $(1, 4) \cup [2, 6]$, we graph each set above the number line and then locate the interval on the number line that unites or joins these intervals. As shown in Figure P.11, numbers in either or both intervals are all real numbers greater than 1, extending up to and including 6. Thus,

$(1, 4) \cup [2, 6] = (1, 6]$.

e. Figure P.12 illustrates that uniting both intervals shown above the number line results in all real numbers greater than -2. Thus,

$(-2, \infty) \cup (3, \infty) = (-2, \infty)$.

f. Figure P.13 indicates that the union of the intervals consists of real numbers that are either in $(-2, -1)$ or $(2, 3)$. We cannot express the answer as a single interval. Using interval notation, the best we can do is to write

$(-2, -1) \cup (2, 3)$.

Using set-builder notation, we could express this union as

$$\{x \mid -2 < x < -1 \quad \text{or} \quad 2 < x < 3\}. \qquad \blacksquare$$

4 Evaluate absolute value.

Absolute Value and Distance

The symbol $|x|$ is read "the absolute value of x." Geometrically, the absolute value of x is the distance between both the real number x and the origin on the real number line (Figure P.14). Algebraically, we can define the absolute value of the real number x as follows.

Figure P.14

Absolute value as the distance from 0

Definition of absolute value

$$|x| = \begin{cases} x & \text{if } x \geq 0 \\ -x & \text{if } x < 0 \end{cases}$$

The case $x \geq 0$ tells us that the absolute value of a nonnegative number is the number itself. Thus,

$$|5| = 5 \qquad |\pi| = \pi \qquad \left|\frac{1}{3}\right| = \frac{1}{3} \qquad |0| = 0$$

The case $x < 0$ tells us that the absolute value of a real number is never negative. For example,

$$|-3| = -(-3) = 3 \qquad |-\pi| = -(-\pi) = \pi \qquad \left|-\frac{1}{3}\right| = -\left(-\frac{1}{3}\right) = \frac{1}{3}$$

Study tip

If x is a negative number, such as -3, then $-x$ is a positive number. Thus, the absolute value of a real number is either positive or zero. Zero is the only number whose absolute value is zero.

EXAMPLE 5   Evaluating Absolute Value

Evaluate by rewriting as an equivalent expression without absolute value bars:

a. $|13|$ **b.** $|-19|$ **c.** $|\sqrt{3} - 1|$ **d.** $|2 - \pi|$

e. $\dfrac{|x|}{x}$ if $x > 0$ **f.** $\dfrac{|x|}{x}$ if $x < 0$

Using technology

Most graphing calculators have an $\boxed{\text{ABS}}$ (absolute value) key for evaluating absolute value expressions. (Consult your manual for the location of this key.) To evaluate $|-19|$, use these keystrokes:

$\boxed{\text{ABS}} \boxed{(-)} \, 19 \, \boxed{\text{ENTER}}$

When evaluating an expression such as $|2 - \pi|$, use parentheses:

$\boxed{\text{ABS}} \boxed{(} \, 2 \boxed{-} \, \pi \boxed{)} \, \boxed{\text{ENTER}}$

Solution

a. $|13| = 13$

b. $|-19| = -(-19) = 19$

c. Since $\sqrt{3} \approx 1.7$, the expression inside the absolute value bars is positive. The absolute value of a positive number is the number itself. Thus,

$$|\sqrt{3} - 1| = \sqrt{3} - 1.$$

d. Since $\pi \approx 3.14$, the number inside the absolute value bars is negative. The absolute value of x when $x < 0$ is $-x$. Thus,

$$|2 - \pi| = -(2 - \pi) = \pi - 2.$$

e. If $x > 0$, then $|x| = x$. Thus,

$$\frac{|x|}{x} = \frac{x}{x} = 1.$$

f. If $x < 0$, then $\left|x\right| = -x$. Thus,

$$\frac{|x|}{x} = \frac{-x}{x} = -1.$$ ∎

Next, we list several basic properties of absolute value. Each of these properties can be derived from the definition of absolute value.

Discover for yourself

Verify the first three properties in the box on the right for $a = 4$ and $a = -5$. Verify properties 4–6 for each of the following pairs of values:

$a = 4, b = 5; a = -4, b = -5;$

$a = 4, b = -5; a = -4, b = 5$

When does equality occur in the triangle inequality and when does inequality occur? Verify your observation with additional number pairs.

Properties of absolute value

For all real numbers a and b,

1. $\left|a\right| \geqslant 0$ **2.** $\left|-a\right| = \left|a\right|$

3. $a \leqslant \left|a\right|$

4. $\left|ab\right| = \left|a\right|\left|b\right|$ **5.** $\left|\dfrac{a}{b}\right| = \dfrac{\left|a\right|}{\left|b\right|}, \quad b \neq 0$

6. $\left|a + b\right| \leqslant \left|a\right| + \left|b\right|$ (called the triangle inequality)

Distance on a Number Line

If a and b are any real numbers, the distance between a and b is the absolute value of their difference.

Distance between two points on the real number line

The distance between real numbers a and b is

$$\left|a - b\right| = \left|b - a\right|.$$

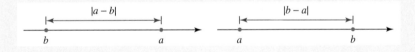

5 Use absolute value to express distance.

EXAMPLE 6   **Absolute Value and Distance**

Express the distance between the numbers -5 and 3 using absolute value. Then evaluate the absolute value.

Solution

The distance between -5 and 3 can be expressed in two equivalent ways, namely,

$$\left|-5 - 3\right| = \left|-8\right|$$
$$\left|3 - (-5)\right| = \left|8\right|$$

We evaluate these absolute values as

$$\left|-8\right| = 8$$
$$\left|8\right| = 8$$

Figure P.15 verifies that the calculation is correct. ∎

Figure P.15

The distance between -5 and 3 is 8.

EXAMPLE 7   **Absolute Value and Distance**

Rewrite using absolute value: The distance between x and the origin is 2. What are the possible values for x?

Solution

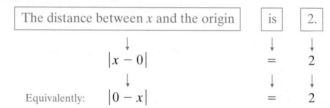

The distance between x and the origin		is	2.
↓		↓	↓
$\lvert x - 0 \rvert$		$=$	2
↓		↓	↓
Equivalently: $\lvert 0 - x \rvert$		$=$	2

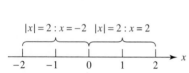

$\lvert x \rvert = 2 : x = -2$ $\lvert x \rvert = 2 : x = 2$

Figure P.16

$\lvert x \rvert = 2$ means $x = 2$ or $x = -2$

Using $\lvert x \rvert = 2$, there are two numbers whose distance from the origin on the number line is 2, namely, -2 and 2, as shown in Figure P.16. Generalizing from this example, if c is any positive number and $\lvert x \rvert = c$, then $x = c$ or $x = -c$. ∎

PROBLEM SET P.1

Practice Problems

*In Problems 1–4, indicate which numbers in the given set are **a.** Natural numbers **b.** Whole numbers **c.** Integers **d.** Rational numbers **e.** Irrational numbers*

1. $\{-10, -\sqrt{2}, -\frac{3}{4}, 0, \frac{4}{5}, \sqrt{4}, \pi, 7, \frac{18}{2}, 100\}$

2. $\{-5, -\sqrt{9}, -\frac{4}{3}, 0, \frac{2}{3}, \sqrt{7}, 2\pi, \frac{30}{2}, 90\}$

3. $\{-\sqrt[3]{8}, \frac{0}{3}, \sqrt[3]{7}, \sqrt{\frac{4}{9}}, 1.\overline{126}\}$

4. $\{-\sqrt[3]{125}, \frac{0}{6}, \sqrt[4]{7}, \sqrt{\frac{9}{25}}, 3.\overline{47}\}$

In Problems 5–18, express each interval in terms of an inequality and graph the interval on a number line.

5. $(1, 6]$ **6.** $(-2, 4]$ **7.** $[-5, 2)$ **8.** $[-4, 3)$

9. $[-3, 1]$ **10.** $[-2, 5]$ **11.** $(2, \infty)$ **12.** $(3, \infty)$

13. $[-3, \infty)$ **14.** $[-5, \infty)$ **15.** $(-\infty, 3)$ **16.** $(-\infty, 2)$

17. $(-\infty, 5.5)$ **18.** $(-\infty, 3.5]$

In Problems 19–34, express each English phrase in both set-builder and interval notations.

19. x is less than 6.

20. x is greater than 2.

21. x is greater than or equal to -1.

22. x is less than or equal to -3.

23. x lies between 5 and 12, excluding 5 and 12.

24. x lies between -4 and 7, excluding -4 and 7.

25. x lies between 2 and 13, excluding 2 and including 13.

26. x lies between -7 and 2, including -7 and excluding 2.

27. x is at most 6.

28. x is at least 3.

29. x is at least 2 and at most 5.

30. x is at least -3 and at most 2.

31. x is not more than 60.

32. x is not more than 32.

33. x is negative and at least -2.

34. x is not positive and at least -5.

Express each set in Problems 35–54 in interval notation.

35. $(1, 5) \cap [2, 7]$ **36.** $(1, 3) \cap [2, 6]$ **37.** $(-1, \infty) \cap (5, \infty)$

38. $(-3, \infty) \cap (7, \infty)$ **39.** $(-\infty, 2) \cap (-\infty, 4)$ **40.** $(-\infty, 1) \cap (-\infty, 5)$

41. $(-\infty, 3] \cap (-\infty, 6)$ **42.** $(-\infty, 7] \cap (-\infty, 10)$ **43.** $(-3, -1) \cap [2, 4]$

44. $(-2, 0) \cap [3, 5]$ **45.** $(1, 5) \cup (2, 7)$ **46.** $(1, 3) \cup [2, 6]$

47. $(-1, \infty) \cup (5, \infty)$ **48.** $(-3, \infty) \cup (7, \infty)$ **49.** $(-\infty, 2) \cup (-\infty, 4)$

50. $(-\infty, 1) \cup (-\infty, 5)$ **51.** $(-\infty, 3] \cup (-\infty, 6)$ **52.** $(-\infty, 7) \cup (-\infty, 10)$

53. $(-3, -1) \cup [2, 4]$ **54.** $(-2, 0) \cup [3, 5]$

In Problems 55–68, evaluate each expression.

55. $|300|$

56. $|705|$

57. $|-203|$

58. $|-109|$

59. $|12 - \pi|$

60. $|7 - \pi|$

61. $|\pi - 12|$

62. $|\pi - 7|$

63. $|\sqrt{2} - 5|$

64. $|\sqrt{5} - 13|$

65. $\frac{-3}{|-3|}$

66. $\frac{-7}{|-7|}$

67. $||-3| - |-7||$

68. $||-5| - |-13||$

In Problems 69–80, express the distance between the given numbers using absolute value. Then find the distance by evaluating the absolute value expression.

69. 2 and 17

70. 4 and 15

71. -2 and 5

72. -6 and 8

73. -19 and -4

74. -26 and -3

75. -3.6 and -1.4

76. -5.4 and -1.2

77. $-\frac{3}{10}$ and $\frac{2}{5}$

78. $-\frac{7}{10}$ and $\frac{4}{5}$

79. $-9\frac{1}{4}$ and $-6\frac{1}{2}$

80. $-12\frac{1}{5}$ and $-2\frac{3}{4}$

In Problems 81–88, use absolute value notation to describe each situation.

81. The distance between x and the origin is 7.

82. The distance between x and the origin is 13.

83. x is at least 6 units from the origin.

84. x is at most 6 units from the origin.

85. The distance between mile marker 72 and mile marker 99.

86. The distance between mile marker 54 and mile marker 76.

87. The distance between y and 7 is no more than 3.

88. The distance between y and -4 is at least 10.

Application Problems

The bar graph shows the average number of times U.S. adults report to have sex per year. In Problems 89–94, you are given the number of times a particular person in an age group claims to have sex per year. Find the amount by which this number differs from the average number shown in the graph.

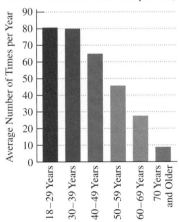

Sex through the Years, by Age
(As Self-reported, the Average Number of Times U.S. Adults Have Sex per Year)

Source: National Opinion Research Council. Adapted from *The Macmillan Visual Almanac.*

	x: An estimate of the average number of times U.S. adults in the designated age group have sex per year	*y:* The number of times a particular person in the age group has sex per year	$|x - y|$
89. Ages 18–29		180	
90. Ages 30–39		60	
91. Ages 40–49		30	
92. Ages 50–59		100	
93. Ages 60–69		2	
94. Ages 70 and older		85	

The bar graph shows the top ten reasons for doctors' office visits each year in the United States. If x represents the number of visits in millions, write the reason or reasons described in Problems 95–102.

95. $x > 25$

96. $x > 30$

97. x is in the interval $[10, 20)$.

98. x is in the interval $[15, 25]$.

99. x is at least 20.

100. x is at most 20.

101. $15 \leqslant x < 30$

102. $20 \leqslant x \leqslant 25$

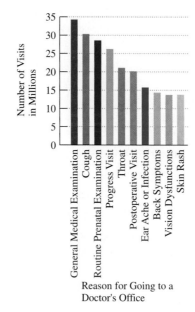

Source: National Center for Health Statistics, U.S. Department of Health and Human Services

True–False Critical Thinking Problems

103. Which one of the following is true?

  a. There are at least two whole numbers that are not natural numbers.

  b. Using interval notation, the set $\{x \mid x < 6\}$ can be expressed as $[-\infty, 6)$.

  c. If a is a positive number and b is a negative number, then $|a + b| = |a| + |b|$.

  d. The least natural number can be expressed as $\left| \dfrac{27 - 5}{5 - 27} \right|$.

104. Which one of the following is true?

  a. Some integers are not rational numbers.

  b. Irrational numbers cannot be located on a real number line because in decimal form they neither terminate nor repeat.

  c. The graph of the union of two intervals can never be the same as the graph of the intersection of the same two intervals.

  d. If $x < -1$, then $\dfrac{|x + 1|}{x + 1}$ simplifies to -1.

Technology Problems

In Problems 105–106, use a calculator to order the real numbers from smallest to largest.

105. $\dfrac{4243}{3000}, \dfrac{873}{618}, \dfrac{55}{37}, \sqrt{2}, \dfrac{112}{80}$

106. $2.23\overline{20}, \sqrt{5}, \dfrac{37,961}{17,000}, \dfrac{111,525}{50,000}, \sqrt{17} - \sqrt{13}$

Writing in Mathematics

107. Consider the rational number $\dfrac{a}{b}$. Suppose that a is a natural number that does not vary, but b gets large. Describe what happens to $\dfrac{a}{b}$.

108. Consider the rational number $\dfrac{a}{b}$. Suppose that a is a natural number that does not vary, but b gets close to zero. Describe what happens to $\dfrac{a}{b}$.

Critical Thinking Problems _____

109. Give an example of two different irrational numbers a and b whose product is a rational number.

110. Is there a positive number that is closest to zero on the number line? Explain your answer.

S E C T I O N   P . 2

Solutions   Tutorial   Video
Manual   1

Properties of Algebraic Expressions; Exponents

Objectives

1 Evaluate algebraic expressions.
2 State the property illustrated by an equation.
3 Use properties of exponents.
4 Use scientific notation.

The power of algebra lies in its ability to describe the world in a compact, symbolic manner. When we say "let b be any real number," we condense the symbols for all real numbers into one all-embracing symbol. By writing b^n, where n is an exponent, a symbol is attached to a symbol, condensing all repeated multiplications for all numbers into one compact space. By writing that all the solutions to $ax^2 + bx + c = 0$ are

$$\frac{-b \pm \sqrt{b^2 - 4ac}}{2a}$$

we once again use a compact, symbolic style to present a formula for solving infinitely many different equations. Algebra lets us say infinitely many things at once!

Algebra's power is based on an understanding of algebraic expressions.

Algebraic Expressions

In algebra we use letters, called *variables*, to represent numbers. An expression that combines constants and variables using the operations of addition, subtraction, multiplication, or division, as well as one that contains powers or roots, is called an *algebraic expression*. The following are examples of algebraic expressions.

$$4x^3 - 7x^2 + 5 \qquad\qquad 5(x - y) - (3x - 2y)$$

$$2L + 2W \qquad\qquad \frac{-b + \sqrt{b^2 - 4ac}}{2a}$$

The *terms* of an algebraic expression are separated by addition. For example,

$$4x^3 - 7x^2 + 5 = 4x^3 + (-7x^2) + 5$$

contains three terms: $4x^3$ and $-7x^2$ are the *variable terms* and 5 is the *constant term*. The variable factor of a variable term is the *literal factor*, and the numerical factor is the *numerical coefficient*, or simply *coefficient*. For example, the numerical coefficient of $4x^3$ is 4 and the numerical coefficient of $-7x^2$ is -7.

Evaluating an algebraic expression involves substituting given numerical values for each variable in the expression.

1 Evaluate algebraic expressions.

EXAMPLE 1   Evaluating Algebraic Expressions

Evaluate:

a. $(25t^2 + 125t) \div (t^2 + 1)$ for $t = 3$

b. $\dfrac{-b + \sqrt{b^2 - 4ac}}{2a}$ for $a = 2$, $b = 9$, and $c = -5$

Solution

Expression	Value of the Variable (s)	Substitute	Evaluating the Expression
a. $(25t^2 + 125t) \div (t^2 + 1)$	$t = 3$	$(25 \cdot 3^2 + 125 \cdot 3) \div (3^2 + 1)$	$\begin{aligned} &= (25 \cdot 9 + 125 \cdot 3) \div (9 + 1) \\ &= (225 + 375) \div (9 + 1) \\ &= 600 \div 10 \\ &= 60 \end{aligned}$
b. $\dfrac{-b + \sqrt{b^2 - 4ac}}{2a}$	$a = 2$	$\dfrac{-9 + \sqrt{9^2 - 4(2)(-5)}}{2(2)}$	$= \dfrac{-9 + \sqrt{81 - (-40)}}{4}$
	$b = 9$		$= \dfrac{-9 + \sqrt{121}}{4}$
	$c = -5$		$= \dfrac{-9 + 11}{4} = \dfrac{2}{4} = \dfrac{1}{2}$ ∎

2 State the property illustrated by an equation.

Algebra's Basic Rules

The basic properties of algebra, listed below, apply to real numbers, variables, and algebraic expressions.

Algebra's basic properties

For each property listed below, a, b, and c represent real numbers, variables, or algebraic expressions.

Property	Example	Description
Commutative Property of Addition $a + b = b + a$	$2x + x^2 = x^2 + 2x$	Order does not matter in addition.
Commutative Property of Multiplication $ab = ba$	$x^2(4 + x) = (4 + x)x^2$	Order does not matter in multiplication.
Associative Property of Addition $(a + b) + c = a + (b + c)$	$(x^2 + x) + 3 = x^2 + (x + 3)$	When we add three numbers (or variables), it does not matter which two we add first.

Associative Property of Multiplication $(ab)c = a(bc)$	$(3 \cdot 7) \cdot 4 = 3 \cdot (7 \cdot 4)$	When we multiply three numbers (or variables), it does not matter which two we multiply first.
Distributive Property $a(b + c) = ab + ac$ $(b + c)a = ba + ca$	$2 \cdot (7 + 4) = 2 \cdot 7 + 2 \cdot 4$ $(4x + y) \cdot 3 = 4x \cdot 3 + y \cdot 3$	When we multiply a number and a sum, we can equivalently multiply the number by each of the terms in the sum and then add the results.
Additive Identity Property $a + 0 = a$ $0 + a = a$	$5x^2 + 0 = 5x^2$ $0 + 7xy = 7xy$	Zero can be deleted from a sum.
Multiplicative Identity Property $a \cdot 1 = a$ $1 \cdot a = a$	$1 \cdot x = x$ $13y^2 \cdot 1 = 13y^2$	One can be deleted from a product.
Additive Inverse Property $a + (-a) = 0$ $(-a) + a = 0$	$17 + (-17) = 0$ $(-4y^3) + 4y^3 = 0$	The sum of a number and its additive inverse gives 0, the additive identity.
Multiplicative Inverse Property If $a \neq 0$, $a \cdot \frac{1}{a} = 1$ $\frac{1}{a} \cdot a = 1$	$7 \cdot \frac{1}{7} = 1$ $(x^2 + 13)\left(\dfrac{1}{x^2 + 13}\right) = 1$	The product of a nonzero number and its multiplicative inverse gives 1, the multiplicative identity.

Algebra's nine basic properties apply to the operations of addition and multiplication. Subtraction and division are defined in terms of addition and multiplication.

Subtraction and division as inverse operations

Let a and b represent real numbers.

Subtraction: $a - b = a + (-b)$

We call $-b$ the *additive inverse* or *opposite* of b.

Division: $a \div b = a \cdot \frac{1}{b}$, where $b \neq 0$

We call $\frac{1}{b}$ the *multiplicative inverse* or *reciprocal* of b. The quotient of a and b, $a \div b$, can be written in the form $\frac{a}{b}$, where a is the *numerator* and b the *denominator* of the fraction.

Study tip

The distributive property can be extended to cover more than two terms.

$a(b_1 + b_2 + b_3 + \cdots + b_n)$

$= ab_1 + ab_2 + ab_3 + \cdots + ab_n$

and

$a(b_1 - b_2 - b_3 - \cdots - b_n)$

$= ab_1 - ab_2 - ab_3 - \cdots - ab_n$

Because subtraction is defined in terms of adding an inverse, the distributive property can be applied to subtraction:

$$a(b - c) = ab - ac$$

$$(b - c)a = ba - ca$$

ENRICHMENT ESSAY

Variables and Pronouns

The English language contains all the abstraction and complexity of the compact, symbolic language used in algebra. Consider, for example, the following sentence: "A man once gave his wife something so offensive that she tossed it into her shopping bag and refused to talk about it again, although he occasionally asks her where it is." If we use letters to represent the unknowns, the sentence becomes: "x gave x's wife y a z, and y found z so offensive that y tossed z into a shopping bag and refused to talk about z to x, although x occasionally asks y where z is." This second form of the sentence replaces the variable pronouns with letters, but also contains English words.

Until the 13th century, all algebraic formulas and equations were written out in verbal sentences. Later, abbreviations replaced some of the words. Algebra's current notation, originating near the beginning of the 17th century with the work of the French mathematician René Descartes, utilizes compact symbols without the use of words. In this notation, variables such as x represent any number, with every number condensed into one all-embracing symbol. Using modern notation, mathematicians write x^n, where n is an exponent, a symbol attached to a symbol that condenses all repeated multiplications for all numbers into one compact space.

The symbolic notation used in algebra is the result of a slow and tedious evolution of ideas and symbols. It wasn't until the late 16th century that the French mathematician François Viète came upon the idea of using letters such as x, y, and z to represent numbers in much the same way that English pronouns represent nouns.

Duane Hanson, "Couple with Shopping Bags" © 1976.

Additional Properties of Algebra

The definition of subtraction

$$a - b = a + (-b)$$

states that to subtract b from a, we add the inverse or *negative of b* to a. The following list summarizes the basic properties of negatives.

Study tip

Properties 5 and 6 state that expressions within parentheses that are preceded by a negative can be simplified by dropping the parentheses and changing the sign of every term inside the parentheses.

$-(3x^4 - 2x^2 + 5x - 6)$

$= -3x^4 + 2x^2 - 5x + 6$

Properties of negatives

Let a and b represent real numbers, variables, or algebraic expressions.

Property	Examples
1. $(-1)a = -a$	$(-1)3 = -3$
2. $-(-a) = a$	$-(-8) = 8$
	$-(-5x^2y) = 5x^2y$
3. $(-a)b = -(ab)$	$(-5)8 = -(5 \cdot 8)$ or -40
$\quad a(-b) = -(ab)$	$7(-3) = -(7 \cdot 3)$ or -21
4. $(-a)(-b) = ab$	$(-6)(-9) = 6 \cdot 9$ or 54
5. $-(a + b) = -a - b$	$-(3 + 11x) = -3 - 11x$
6. $-(a - b) = -a + b$	$-(3x^2 - 7x) = -3x^2 + 7x$
$\qquad = b - a$	$= 7x - 3x^2$

The additive identity property for zero,

$$a + 0 = 0 + a = a$$

was included in our list of algebra's nine basic properties. The following list summarizes other properties of zero.

Properties of zero

Let a and b represent real numbers, variables, or algebraic expressions.

Property	**Examples**
1. $a + 0 = a$	$17x + 0 = 17x$
2. $a - 0 = a$	$14x^2 - 0 = 14x^2$
3. $a \cdot 0 = 0$	$9x^3 \cdot 0 = 0$
$\ 0 \cdot a = 0$	
4. $\dfrac{0}{a} = 0, a \neq 0$	$\dfrac{0}{7x} = 0,$ if $x \neq 0$
5. $\dfrac{a}{0}$ is undefined.	$\dfrac{7x}{0}$ is undefined.
	$\dfrac{3}{x-4}$ is undefined if $x = 4$.
6. If $ab = 0$, then $a = 0$ or $b = 0$. (The "or" includes the possibility that both factors may be 0.)	$(x - 3)(x - 7) = 0$ $x - 3 = 0$ or $x - 7 = 0$

Equations

Throughout this book, we will be working with equations. An *equation* is a statement of equality between two expressions. The basic statement

$$a = b$$

("a is equal to b" or "a equals b") means that a and b represent the same real number.

We will frequently use the following properties when working with equations.

Properties used when working with equations

Let $a, b,$ and c represent real numbers, variables, or algebraic expressions.

Property	**Examples**
Addition Property If $a = b$, then $a + c = b + c$.	If $x - 7 = 5$, then $x - 7 + 7 = 5 + 7$ (so $x = 12$).
Subtraction Property If $a = b$, then $a - c = b - c$.	If $x + 7 = 5$, then $x + 7 - 7 = 5 - 7$ (so $x = -2$).
Multiplication Property If $a = b$, then $ac = bc$.	If $\dfrac{x}{4} = 5$, then $\dfrac{x}{4} \cdot 4 = 5 \cdot 4$ (so $x = 20$).

Division Property

If $a = b$, then $\dfrac{a}{c} = \dfrac{b}{c}$, $c \neq 0$.

If $-4x = 12$, then

$\dfrac{-4x}{-4} = \dfrac{12}{-4}$ (so $x = -3$).

Cancellation Properties

If $a + c = b + c$, then $a = b$.

If $a - c = b - c$, then $a = b$.

If $ac = bc$ and $c \neq 0$, then $a = b$.

If $\dfrac{a}{c} = \dfrac{b}{c}$ and $c \neq 0$, then $a = b$.

If $x + 7 = 3 + 7$, then $x = 3$.

If $2x + 6 - x^2 = 10 - x^2$, then
$2x + 6 = 10$.

If $x \cdot 4 = 8 \cdot 4$, then $x = 8$.

If $\dfrac{3y}{17} = \dfrac{15}{17}$, then $3y = 15$.

3 Use properties of exponents.

Integer Exponents

Repeated multiplication of the same factor can be expressed using an exponent.

Definition of a natural number exponent

If b is a real number and n is a natural number,

Exponent
$$b^n = b \cdot b \cdot b \cdot \,\cdots\, \cdot b$$

Base b appears as a factor n times.

b^n is read "the nth power of b" or "b to the nth power." Thus, the nth power of b is defined as the product of n factors of b. Furthermore, $b^1 = b$.

Using technology

Here are the graphing calculator keystrokes for Example 2.

a. 7^2: 7 ⌃ 2 ENTER
  or 7 x^2 ENTER

b. $(-3)^4$: ((−) 3)
  ⌃ 4 ENTER

c. -3^4: (−) 3 ⌃ 4 ENTER

d. $(-2)^3 \cdot 3^2$: ((−) 2)
  ⌃ 3 × 3 ⌃ 2 ENTER

EXAMPLE 2   Evaluating Exponential Expressions

Evaluate:

a. 7^2 **b.** $(-3)^4$ **c.** -3^4 **d.** $(-2)^3 \cdot 3^2$

Solution

a. $7^2 = 7 \cdot 7 = 49$ 7^2 is read "7 squared."

b. $(-3)^4 = (-3)(-3)(-3)(-3) = 81$

c. $-3^4 = -(3 \cdot 3 \cdot 3 \cdot 3) = -81$ The negative of a number is taken to a power only when the negative is inside the parentheses.

d. $(-2)^3 \cdot 3^2 = (-2)(-2)(-2)(3)(3)$ $(-2)^3$ is read "negative 2 cubed."
  $= -8 \cdot 9 = -72$ ∎

The major properties of exponents are summarized below.

Properties of exponents

For each property listed on the next page, a and b represent real numbers, variables, or algebraic expressions. (All denominators and bases are nonzero.) The exponents, m and n, represent integers.

Property	**Example**	**Description**
1. $b^n \cdot b^m = b^{n+m}$	$2^5 \cdot 2^3 = 2^{5+3} = 2^8$	When multiplying expressions with the same base, add the exponents.
2. $\dfrac{b^n}{b^m} = b^{n-m}$	$\dfrac{y^{13}}{y^6} = y^{13-6} = y^7$	When dividing expressions with the same base, subtract the exponents.
3. $b^{-n} = \dfrac{1}{b^n} = \left(\dfrac{1}{b}\right)^n$	$x^{-9} = \dfrac{1}{x^9} = \left(\dfrac{1}{x}\right)^9$	When a negative integer appears as an exponent, switch the position of the base (from numerator to denominator or from denominator to numerator) and make the exponent positive.
4. $\dfrac{1}{b^{-n}} = b^n$	$\dfrac{1}{y^{-2}} = y^2$	
5. $b^0 = 1, b \neq 0$	$(3x^2 + 7)^0 = 1$	Any nonzero number or expression to the 0 power is 1.
6. $(b^n)^m = b^{nm}$	$(x^5)^{-6} = x^{5(-6)} = x^{-30} = \dfrac{1}{x^{30}}$	To raise a power to a new power, multiply the exponents.
7. $(ab)^n = a^n b^n$	$(-2y)^4 = (-2)^4 y^4 = 16y^4$	When a product is raised to a power, raise each factor to the power.
8. $(b_1 b_2 b_3 \cdots b_n)^n = b_1^n b_2^n b_3^n \cdots b_n^n$	$(-3xy)^3 = (-3)^3 x^3 y^3 = -27x^3 y^3$	
9. $\left(\dfrac{a}{b}\right)^n = \dfrac{a^n}{b^n}$	$\left(-\dfrac{3}{x}\right)^3 = \dfrac{(-3)^3}{x^3} = \dfrac{-27}{x^3}$	When a quotient is raised to a power, raise the numerator and the denominator to the power.

EXAMPLE 3   **Using the Properties of Exponents**

Simplify:

a. $(-3x^4 y^5)^3$ **b.** $(-7xy^4)(-2x^5 y^6)$ **c.** $\dfrac{30a^7 b^4}{-6a^3 b}$

Solution

a. $(-3x^4 y^5)^3 = (-3)^3 (x^4)^3 (y^5)^3$ Raise each factor in the product to the power.
$\qquad\qquad\quad = (-3)^3 x^{4\cdot3} y^{5\cdot3}$ Multiply powers to powers.
$\qquad\qquad\quad = -27x^{12} y^{15}$ $(-3)^3 = (-3)(-3)(-3) = -27$

b. $(-7xy^4)(-2x^5 y^6) = (-7)(-2)xx^5 y^4 y^6$ Group factors with the same base.
$\qquad\qquad\qquad\quad = 14x^{1+5} y^{4+6}$ When multiplying expressions with the same base, add the exponents.
$\qquad\qquad\qquad\quad = 14x^6 y^{10}$ Simplify.

c. $\dfrac{30a^7 b^4}{-6a^3 b} = \left(\dfrac{30}{-6}\right)\left(\dfrac{a^7}{a^3}\right)\left(\dfrac{b^4}{b}\right)$ Group factors with the same base.

$\qquad\qquad = -5a^{7-3} b^{4-1}$ When dividing expressions with the same base, subtract the exponents.

$\qquad\qquad = -5a^4 b^3$ Simplify.   ∎

Example 3 illustrates the meaning of an exponential expression in simplest form. An exponential expression is simplified when no parentheses appear, no powers are raised to powers, and each base occurs only once. An exponential expression must also contain no negative exponents if it is in simplified form. This involves moving factors with negative exponents from the numerator to the denominator (or vice versa) by changing the sign of the exponent in the last step of the simplification.

EXAMPLE 4   **Simplifying Expressions with Negative Exponents**

Simplify:

a. $\left(\dfrac{4x^2}{y^3}\right)^{-3}$ **b.** $\left(\dfrac{25x^2y^4}{-5x^6y^{-8}}\right)^2$ **c.** $(16x^5y^{10})(-2x^{-3}y^4)^{-3}$

Solution

Discover for yourself

There is often more than one way to simplify an exponential expression. Example 4a can be solved by starting with

$\left(\dfrac{4x^2}{y^3}\right)^{-3} = \left(\dfrac{y^3}{4x^2}\right)^3.$

In Example 4b, you can begin by rewriting y^{-8} as $\dfrac{1}{y^8}$ inside the parentheses. Try solving all parts of Example 4 by a different method from the one shown.

a. $\left(\dfrac{4x^2}{y^3}\right)^{-3} = \dfrac{4^{-3}(x^2)^{-3}}{(y^3)^{-3}}$ Raise each factor inside the parentheses to the -3 power.

$= \dfrac{4^{-3}x^{-6}}{y^{-9}}$ Multiply powers to powers.

$= \dfrac{y^9}{4^3x^6}$ Move factors with negative exponents from the numerator to the denominator (or vice versa) by changing the sign of the exponent.

$= \dfrac{y^9}{64x^6}$ $4^3 = 4 \cdot 4 \cdot 4 = 64$

b. $\left(\dfrac{25x^2y^4}{-5x^6y^{-8}}\right)^2 = (-5x^{2-6}y^{4-(-8)})^2$ Simplify inside parentheses by subtracting exponents with the same base.

$= (-5x^{-4}y^{12})^2$

$= (-5)^2(x^{-4})^2(y^{12})^2$ Square each factor inside the parentheses.

$= 25x^{-8}y^{24}$ Multiply powers to powers.

$= \dfrac{25y^{24}}{x^8}$ $b^{-n} = \dfrac{1}{b^n},$ so $x^{-8} = \dfrac{1}{x^8}.$

Study tip

Be careful when simplifying $(-2)^{-3}$. Change only the sign of the exponent and not the base when moving -2 to the denominator.

Correct: $(-2)^{-3} = \dfrac{1}{(-2)^3}$

Incorrect: $(-2)^{-3} = \dfrac{1}{2^3}$

c. $(16x^5y^{10})(-2x^{-3}y^4)^{-3}$

$= (16x^5y^{10})(-2)^{-3}(x^{-3})^{-3}(y^4)^{-3}$ Raise each factor in the second parentheses to the -3 power.

$= (16x^5y^{10})(-2)^{-3}x^9y^{-12}$ Multiply powers to powers.

$= 16(-2)^{-3}(x^5x^9)(y^{10}y^{-12})$ Group factors with the same base.

$= 16(-2)^{-3}x^{14}y^{-2}$ When multiplying expressions with the same base, add the exponents.

$= 16 \cdot \dfrac{1}{(-2)^3} \cdot x^{14} \cdot \dfrac{1}{y^2}$ Move factors with negative exponents from the numerator to the denominator by changing the sign of the exponent.

$= 16 \cdot \dfrac{1}{-8} \cdot x^{14} \cdot \dfrac{1}{y^2}$ $(-2)^3 = (-2)(-2)(-2) = -8$

$= \dfrac{-2x^{14}}{y^2}$ Simplify.   ∎

EXAMPLE 5   **Simplifying an Exponential Expression**

Simplify: $\dfrac{(-2x^3y^2z)^4}{-8xy^{-3}z^4}$

Solution

$$\frac{(-2x^3y^2z)^4}{-8xy^{-3}z^4} = \frac{(-2)^4(x^3)^4(y^2)^4z^4}{-8xy^{-3}z^4}$$

Raise each factor inside the parentheses to the fourth power.

$$= \frac{16x^{12}y^8z^4}{-8xy^{-3}z^4}$$

Multiply powers to powers.

$$= \frac{16}{-8} \cdot \frac{x^{12}}{x} \cdot \frac{y^8}{y^{-3}} \cdot \frac{z^4}{z^4}$$

Group factors with the same base.

$$= -2x^{12-1}y^{8-(-3)}z^{4-4}$$

When dividing expressions with the same base, subtract the exponents.

$$= -2x^{11}y^{11}z^0$$

Simplify.

$$= -2x^{11}y^{11}$$

$z^0 = 1$, and 1 can be omitted in a product. ■

Discover for yourself

Cover the second and third columns in the Study tip with a sheet of paper. Describe the error in the left column and then write the correction on your own. Then compare your description and correction with the one in the chart.

Study tip

Try to avoid the following common errors that can occur when simplifying exponential expressions.

Incorrect	Description of Error	Correct
$b^3b^4 = b^{12}$	Exponents should be added, not multiplied.	$b^3b^4 = b^7$
$3^n \cdot 3^m = 9^{n+m}$	The common base should be retained, not multiplied.	$3^n \cdot 3^m = 3^{n+m}$
$\dfrac{5^{16}}{5^4} = 5^4$	Exponents should be subtracted, not divided.	$\dfrac{5^{16}}{5^4} = 5^{12}$
$(4a)^3 = 4a^3$	Both factors should be cubed.	$(4a)^3 = 64a^3$
$b^{-n} = -\dfrac{1}{b^n}$	Only the exponent should change sign.	$b^{-n} = \dfrac{1}{b^n}$
$(a+b)^{-1} = \dfrac{1}{a} + \dfrac{1}{b}$	The exponent applies to the entire expression $a + b$.	$(a+b)^{-1} = \dfrac{1}{a+b}$

4 Use scientific notation.

Scientific Notation

We frequently encounter very large and very small numbers. For example, a light-year is the number of miles that light travels in one year, which is approximately

5,880,000,000,000 miles.

The large number of zeros in this number makes it difficult to read, write, and say. *Scientific notation* gives us a compact way to display and say this number. In scientific notation, a light-year is approximately 5.88×10^{12} miles.

In general, a scientific notation numeral has the form $c \times 10^n$, where $1 \le c < 10$ and n is an integer. It is customary to use $\times$ rather than a dot to indicate multiplication in scientific notation.

> ### Definition of scientific notation
>
> A scientific notation numeral appears as the product of two factors. The first factor is a number greater than or equal to 1 but less than 10. The second factor is base 10 raised to a power that is an integer.

EXAMPLE 6   Converting from Scientific to Decimal Notation

Rewrite in decimal notation:

a. 1.7×10^3 **b.** 2.31×10^0 **c.** 7.43×10^{-2} **d.** 6.153×10^{-4}

Solution

a. $1.7 \times 10^3 = 1.7 \times 1000 = 1700$
b. $2.31 \times 10^0 = 2.31 \times 1 = 2.31$

c. $7.43 \times 10^{-2} = 7.43 \times \dfrac{1}{10^2} = 7.43 \times \dfrac{1}{100} = 0.0743$

d. $6.153 \times 10^{-4} = 6.153 \times \dfrac{1}{10^4} = 6.153 \times \dfrac{1}{10,000} = 0.000\ 615\ 3$ ∎

Observe that when multiplying by 10 to a power, we move the decimal point the same number of places as the exponent of 10. If the exponent is *positive*, we move the decimal point in the first factor to the *right*. If the exponent is *negative*, we move the decimal point in the first factor to the *left*.

You can change the mode setting of a graphing calculator so that numbers are displayed in scientific notation. (Consult your manual.) Once you're in the scientific notation mode, simply enter the number and press ENTER.

Discover for yourself

Study the scientific notation numerals for the numbers in Example 7. Describe a process for converting a positive number to scientific notation without using a calculator.

EXAMPLE 7   Converting to Scientific Notation on a Graphing Calculator

Use a graphing calculator to express in scientific notation:

a. 82,000,000,000   **b.** 0.000 000 000 000 000 000 160 2   **c.** 3.7284

Solution

Number		Display	Number in Scientific Notation
a. 82,000,000,000	ENTER	8.2 E 10	8.2×10^{10}
b. 0.000 000 000 000 000 000 160 2	ENTER	1.602 E −19	1.602×10^{-19}
c. 3.7284	ENTER	3.7284 E 0	3.7284×10^0

∎

Using technology

Even if you do not set your graphing calculator to a scientific notation mode, it will automatically switch to scientific notation when displaying large or small numbers that exceed the display range. For example use your graphing calculator to multiply

(79,000) (3,400,000,000).

The display shows

2.686E14

indicating that the product is 2.686×10^{14}. If you set your calculator to the scientific notation mode, answers to all computations will be displayed in scientific notation even if they do not exceed the display range.

In the Discover for yourself, were you able to state the following procedure for converting a number into scientific notation?

Writing a number in scientific notation

1. Count the number of places n that the decimal point must be moved to obtain the number c, where $1 \leq c < 10$.

2. If the given number is greater than or equal to 1, then its scientific notation is $c \times 10^n$. If the given number is between 0 and 1, then its scientific notation is $c \times 10^{-n}$.

Applying this procedure to the first two parts of Example 7 enables us to write the numbers in scientific notation as follows.

$$8\,2,\,0\,0\,0,\,0\,0\,0,\,0\,0\,0. = 8.2 \times 10^{10}$$

$$0.\,0\,0\,0\,0\,0\,0\,0\,0\,0\,0\,0\,0\,0\,0\,0\,0\,1\,6\,0\,2 = 1.602 \times 10^{-19}$$

An Application: Black Holes in Space

The concept of a black hole, a region in space where matter appears to vanish, intrigues scientists and nonscientists alike. Scientists theorize that when massive stars run out of nuclear fuel, they begin to collapse under the force of their own gravity. As the star collapses, its density increases. In turn, the force of gravity increases so tremendously that even light cannot escape from the star. Consequently, it appears black.

A mathematical formula, called the Schwarzschild formula, describes the critical value to which the radius of a massive body must be reduced for it to become a black hole. This formula forms the basis of our next example.

Jasper Johns, "Mirror's Edge" 1992, oil on canvas, 66×44 in. Photo courtesy Leo Castelli Gallery. © Jasper Johns / Licensed by VAGA, New York 1998.

EXAMPLE 8   An Application of Scientific Notation

Use the Schwarzschild formula

$$R_s = \frac{2GM}{c^2}$$

where

R_s = Radius of the star (in meters) that would cause it to become a black hole

M = Mass of the star (in kilograms)

G = A constant, called the gravitational constant

$$= 6.7 \times 10^{-11}\, \frac{m^3}{kg \cdot s^2}$$

c = Speed of light

$$= 3 \times 10^8\, \text{meters per second}$$

to determine to what length the radius of the sun must be reduced for it to become a black hole. The sun's mass is approximately 2×10^{30} kilograms.

Solution

$$R_s = \frac{2GM}{c^2}$$

$$= \frac{2 \times 6.7 \times 10^{-11} \times 2 \times 10^{30}}{(3 \times 10^8)^2} \qquad \text{Substitute the given values.}$$

$$= \frac{(2 \times 6.7 \times 2) \times (10^{-11} \times 10^{30})}{(3 \times 10^8)^2} \qquad \text{Apply the commutative and associative properties.}$$

$$= \frac{26.8 \times 10^{-11+30}}{3^2 \times (10^8)^2} \qquad \text{Apply the properties of exponents. Use the rule for multipying exponential expressions in the numerator. Use the rule for raising a product to a power in the denominator.}$$

$$= \frac{26.8 \times 10^{19}}{9 \times 10^{16}} \qquad \text{Use } (b^n)^m = b^{nm} \text{ on the second factor in the denominator.}$$

$$= \frac{26.8}{9} \times 10^{19-16} \qquad \frac{b^n}{b^m} = b^{n-m}$$

$$\approx 2.978 \times 10^3$$

$$= 2978$$

Although the sun is not massive enough to become a black hole (its radius is approximately 700,000 kilometers), the Schwarzchild model theoretically indicates that if the sun's radius were reduced to approximately 2978 meters, that is, about $\frac{1}{235,000}$ its present size, it would become a black hole.   ■

PROBLEM SET P.2

Practice Problems

In Problems 1–12, evaluate each expression for the given value(s) of the variable(s).

1. $x^2 + 5x + 9$; $x = 4$

2. $x^2 + 9x + 6$; $x = 10$

3. $x^2 - 7x + 5$; $x = -3$

4. $2x^2 - 9x - 3$; $x = -2$

5. $3x^3 - 5x^2 - x + 6$; $x = -2$

6. $2x^3 - 4x^2 - x - 7$; $x = -1$

7. $\dfrac{x+1}{x-1}$; $x = -1$

8. $\dfrac{1 - (x-2)^2}{1 + (x-2)^2}$; $x = -1$

9. $\dfrac{2x^2 - 1}{2x^3 + 1}$; $x = -\dfrac{1}{2}$

10. $\dfrac{3^{x+1} - \dfrac{1}{3^{x+1}}}{3^{x+1} + \dfrac{1}{3^{x+1}}}$; $x = -1$

11. $\dfrac{-b + \sqrt{b^2 - 4ac}}{2a}$; $a = 4, b = -20, c = 25$

12. $\dfrac{-b + \sqrt{b^2 - 4ac}}{2a}$; $a = 3, b = 5, c = -2$

In Problems 13–24, state the property that is illustrated by each equation.

13. $7 \cdot 6 = 6 \cdot 7$

14. $3 + (7 + 9) = (3 + 7) + 9$

15. $a(b + c) = (b + c)a$

16. $a(b + c) = a(c + b)$

17. $4(x + 5) = 4x + 20$

18. $\dfrac{1}{(x + 3)}(x + 3) = 1$; $x \neq -3$

19. $(a + b)(c + d) = a(c + d) + b(c + d)$ **20.** $(x + 4) + [-(x + 4)] = 0$

21. $(x + 3) + 0 = x + 3$

22. $1(y + 15) = y + 15$

23. $\dfrac{3}{\sqrt{2}} = \dfrac{3}{\sqrt{2}} \cdot \dfrac{\sqrt{2}}{\sqrt{2}}$

24. $\dfrac{\sqrt{3} + 7}{\sqrt{5} + 2} = \dfrac{\sqrt{3} + 7}{\sqrt{5} + 2} \cdot \dfrac{\sqrt{5} - 2}{\sqrt{5} - 2}$

In Problems 25–38, use properties of algebra to write each expression without parentheses.

25. $7(x + 3y)$

26. $6(5a - 4b)$

27. $\frac{1}{3}(-12x)$

28. $-\frac{1}{3}(3y)$

29. $4x^2(3x^3 - 2x)$

30. $(5y^4 - 2y)(-6y^3)$

31. $-(-13x)$

32. $(-13x)(2y)$

33. $(-5a)(-4b)$

34. $(-20a)\left(-\frac{1}{20}\,b\right)$

35. $-(2x^2 - 5x - 6)$

36. $-(-6x^4 - 9x + 17)$

37. $\frac{1}{3}(3x) + [(4y) + (-4y)]$

38. $-\frac{3}{4}\left(-\frac{4}{3}\,y\right) + [(6x) + (-6x)]$

Evaluate each exponential expression in Problems 39–56.

39. $5^2 \cdot 2$

40. $6^2 \cdot 2$

41. $(-2)^5$

42. $(-2)^4$

43. -2^4

44. -5^3

45. $2^2 \cdot 2^3$

46. $\dfrac{3^5}{3^2}$

47. $\dfrac{4^8}{4^5}$

48. $3^3 \cdot 3^2$

49. $(2^3)^2$

50. $(3^2)^3$

51. $\dfrac{2}{2^{-4}}$

52. $64(-2)^{-5}$

53. $\dfrac{3^{-2}}{3^{-1}}$

54. $\dfrac{2^{-5}}{2^{-3}}$

55. $(3^{19} \cdot 3^{25})^0$

56. $(7^{23} \cdot 7^{96})^0$

Simplify each exponential expression in Problems 57–94.

57. $(4x^2)^3$

58. $(6x^3)^2$

59. $(-3x^4)(-2x^7)$

60. $(-11x^5)(-9x^{12})$

61. $\dfrac{25a^{12}}{5a^6}$

62. $\dfrac{30a^{20}}{10a^{10}}$

63. $\dfrac{14b^7}{7b^{14}}$

64. $\dfrac{20b^{10}}{10b^{20}}$

65. $(2x^2)^3(4x^3)^{-1}$

66. $(5x^3)^2(10x^2)^{-2}$

67. $(-9x^{-2})^{-3}$

68. $(-6x^{-4})^{-3}$

69. $(-2x^3y^4)^5$

70. $(-3x^2y^3)^3$

71. $(-6x^2y^3)(-4xy^5)$

72. $\left(\dfrac{3}{4}\,xy^5\right)\left(-\dfrac{4}{3}\,x^2y^3\right)$

73. $\dfrac{25a^{13}b^4}{-5a^2b^3}$

74. $\dfrac{35a^{14}b^6}{-7a^7b^3}$

75. $(-4x^2)^3(2x^3)^{-1}$

76. $(4x^{-4})(3x^2)$

77. $\left(\dfrac{5x^3}{y^2}\right)^{-2}$

78. $\left(\dfrac{3x^4}{y^7}\right)^{-4}$

79. $(-3x^2y^{-5})^{-2}$

80. $(-4x^{-3}y^2)^{-3}$

81. $(5x^3y^4)^2(-3x^7y^{11})$

82. $(2r^3s^4)^3(-5rs^{-14})$

83. $(4ab^3)^3(-3a^{-5}b^8)$

84. $(12x^4y^7)(-2x^{-2}y^5)^{-2}$

85. $(54r^3s^9)(-3r^2s^{-4})^{-3}$

86. $(-x^2y^3z^{-6})(x^4y^{-5}z^3)^{-5}$

87. $(-a^{-2}b^3c)(ab^{-1}c^{-4})^{-3}$

88. $(2x^2y^3)^2(4x^{-3}y^6)^{-2}$

89. $(3x^{-3}y^{-4}z)^3(3xy^{-5}z)^2(-3x^{-4}z^{12})$

90. $\dfrac{24a^{-5}bc^2}{-10a^3b^{-7}c^2}$

91. $\dfrac{-27x^{-8}y^4z}{15x^4y^4z^{-4}}$

92. $\dfrac{(-2xy^3z^2)^5}{16x^{-2}yz^{10}}$

93. $\dfrac{(-4x^4yz^3)^3}{-4x^2y^{-3}z^9}$

94. $\dfrac{(2ab^{-2}c^3)^{-2}}{(4ab^3c^2)^{-3}}$

In Problems 95–98, express each number in decimal notation.

95. 7.13×10^5

96. 5.024×10^9

97. 3.07×10^{-8}

98. 6.573×10^{-7}

In Problems 99–108, express each number in scientific notation.

99. 96,500,000

100. 167,300,000

101. 7,361,000,000,000

102. 5,024,000,000,000,000

103. 7.53

104. 9.04

105. 0.00016

106. 0.0000037

107. 0.007253

108. 0.009621

In Problems 109–112, use scientific notation to perform each indicated computation. Write the answer in scientific notation.

109. $(0.00037)(8,300,000)$

110. $(0.025)(9,400,000,000,000)$

111. $(4,200,000,000,000) \div (14,000)$

112. $(0.00000000124) \div (3,100,000)$

Application Problems

113. The algebraic expression $0.0014x^2 - 0.1529x + 5.855$ describes the percent of income that one contributes to charities, where x represents annual income in thousands of dollars. Evaluate the expression for $x = 10$ and describe what this means in practical terms.

114. The algebraic expression $0.78x^2 + 76.7x + 4449$ describes world population (in millions) x years after 1980. Evaluate the expression for $x = 10$ and describe what this means in practical terms.

115. The algebraic expression $2\pi r^2 + 2\pi rh$ describes the surface area of a cylinder, where r is the radius and h is the height. Evaluate the expression for $r = 6$ inches and $h = 10$ inches. Express the answer in terms of π, and state the cylinder's surface area with the appropriate unit of measure.

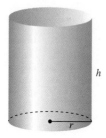

116. The figure on the right shows the path of a person who rows across a river (A to B) and then runs along the river bank on the opposite side (B to C). With the distances shown in the figure, and with a rowing rate of 6

kilometers per hour and a running rate of 8 kilometers per hour, the algebraic expression

$$\frac{\sqrt{x^2 + 9}}{6} + \frac{8 - x}{8}$$

describes the total time (in hours) rowing and running. The figure indicates that the running distance is represented by $8 - x$. Evaluate the total time expression for $x = 4$. What is the total time rowing and running and where does the person come ashore?

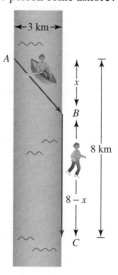

In Problems 117–120, write each number in scientific notation.

117. A light-year: 5,880,000,000,000 miles

118. Number of pounds of dust from the atmosphere that inundate Earth on a daily basis: 26,000,000.

119. Diameter of a hydrogen atom: 0.0000001016 centimeter

120. One micron: 0.00003937 inch

Use scientific notation to answer Problems 121–126.

121. The mass of Earth is approximately

$$5,976,000,000,000,000,000,000,000,000 \text{ grams}$$

and the mass of the hydrogen atom is

$$0.000\,000\,000\,000\,000\,000\,000\,001\,66 \text{ gram.}$$

If Earth were composed only of hydrogen atoms, how many hydrogen atoms would it contain?

122. If the length of a hydrogen atom is 0.000 000 03 millimeter and the average human foot measures 200 millimeters, how many times as large as the hydrogen atom is the human foot?

123. The tallest tree known is the Howard Libbey redwood in Redwood Grove, California. The tree measures 100,000 millimeters. The distance from Earth to the sun is approximately 150,000,000,000,000 millimeters. How many of the tallest redwood trees would it take to span this distance?

124. Among the planets of the solar system, Pluto is the most distant from the sun, approximately 4.6×10^9 miles. How many seconds does it take the light of the sun to reach Pluto if light travels 1.86×10^5 miles per second?

125. Our galaxy measures approximately 1.2×10^{17} kilometers across. If a space vehicle were capable of moving at half the speed of light (approximately 1.5×10^5 kilometers per second), how many years would it take for the vehicle to cross the galaxy?

126. Pouiseville's law states that the speed of blood (S, in centimeters per second) located r centimeters from the central axis of an artery is

$$S = (1.76 \times 10^5)[(1.44 \times 10^{-2}) - r^2].$$

Find the speed of blood at the central axis of this artery.

True–False Critical Thinking Problems

127. Which one of the following is not true for all real numbers?
 a. $3(x^4y^3) = (3x^4)y^3$
 b. $2x(y^4 + z^3) = 2x(z^3 + y^4)$
 c. $2x(y^4 + z^3) = 2x(y^3 + z^4)$
 d. $-(-6x^2y^3) = 6x^2y^3$

128. Which one of the following is not true for all real numbers?
 a. $(x^2 + y^3)(2z + 7) = (7 + 2z)(y^3 + x^2)$
 b. $5(x - 2y + 3z^2) = 15z^2 - 10y + 5x$
 c. $x(4z) = (x \cdot 4)z = (4x)z$
 d. $7x + (yz) = (7x + y)(7x + z)$

129. Which one of the following is not true?
 a. $2^{-1} + 2^{-1} = 1$
 b. $3^{100} = 9^{50}$
 c. $9^7 9^7 = 81^{14}$
 d. $(0.25)^{-1} = 4$

130. Which one of the following is true?
 a. $\dfrac{3^{10}}{3^{-14}} = 3^{-4}$
 b. $-5^{-2} = \dfrac{1}{25}$
 c. $\left(\dfrac{2}{3}\right)^{-2} = \left(\dfrac{3}{2}\right)^2$
 d. $534.7 = 5.347 \times 10^3$

Writing in Mathematics

131. If $b \neq 0$, we know that $b^{-n} = \dfrac{1}{b^n}$ and $\dfrac{1}{b^{-n}} = b^n$. Express this idea strictly in words rather than using the compact, symbolic notation of algebra.

132. Describe the differences, if any, among, b^2, b^{-2}, $-b^2$, and $(-b)^2$.

133. Discuss one advantage of expressing a number in scientific notation over decimal notation.

Critical Thinking Problems

134. If $b^A = MN$, $b^C = M$, and $b^D = N$, what is the relationship among A, C, and D?

135. Which pair of numbers is closer together: 10^7 and 10^{43} or 10^{200} and 10^{201}? Explain your answer.

Solutions Manual	Tutorial	Video

SECTION P.3 Graphs and Graphing Utilities

Objectives

1 Plot ordered pairs in rectangular coordinates.
2 Interpret information given by graphs.
3 Find the distance between two points.
4 Find the midpoint of a line segment.
5 Graph equations in rectangular coordinates.
6 Use a graph to determine intercepts.
7 Graph equations with a graphing utility.

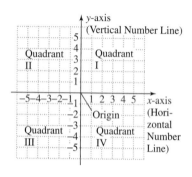

Figure P.17

The rectangular coordinate system

In Section P.1, we saw that any real number could be represented by a point on the number line. We are now ready to graph ordered pairs of real numbers in a plane called the *rectangular coordinate system*. The ideas discussed in this section are those of the French mathematician and philosopher René Descartes. For this reason, the rectangular coordinate system is also called the *Cartesian coordinate plane*.

Points and Ordered Pairs

Descartes used two number lines that intersect at right angles at their zero points (Figure P.17). Each line is called a *coordinate axis*. The horizontal axis is the *x-axis* and the vertical axis is the *y-axis*. The point of intersection of these axes is the *origin*. The horizontal and vertical number lines divide the plane into four quarters, called *quadrants*, numbered in a counterclockwise direction beginning with the upper right. Points on the axes are not in any quadrant.

Each point in the plane corresponds to a unique *ordered pair* (x, y) of real numbers. Examples of such pairs are $(4, 2)$ and $(-5, -3)$. The first number in each pair, called the *x-coordinate*, denotes the distance and direction from the origin along the *x*-axis. The second number, called the *y-coordinate*, denotes vertical distance and direction along a line parallel to the *y*-axis or along the *y*-axis itself. Figure P.18 shows how we locate, or plot, the points corresponding to the ordered pairs $(4, 2)$ and $(-5, -3)$. The phrase "the point corresponding to the ordered pair $(-5, -3)$" is often abbreviated "the point $(-5, -3)$."

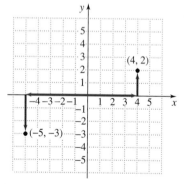

Figure P.18

Plotting points

 Plot ordered pairs in rectangular coordinates.

EXAMPLE I **Plotting Points in a Rectangular Coordinate System**

Plot the points: $(-2, 3)$, $(5, 4)$, $(-3, -4)$, $(4, -1)$, $(0, 0)$, $(-6, 0)$, and $(0, -5)$

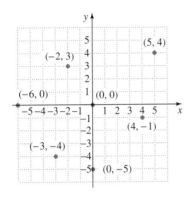

Figure P.19

Plotting points

Solution

See Figure P.19. We plot the points in the following way:

$(-2, 3)$: 2 units left, 3 units up (in quadrant II)
$(5, 4)$: 5 units right, 4 units up (in quadrant I)
$(-3, -4)$: 3 units left, 4 units down (in quadrant III)
$(4, -1)$: 4 units right, 1 unit down (in quadrant IV)
$(0, 0)$: 0 units right or left, 0 units up or down (at the origin)
$(-6, 0)$: 6 units left, 0 units up or down (on the x-axis)
$(0, -5)$: 0 units right or left, 5 units down (on the y-axis). ■

Study tip

Any point on the x-axis has a y-coordinate of 0, and any point on the y-axis has an x-coordinate of 0.

2 Interpret information given by graphs.

Study tip

The *median* of a set of numbers that is arranged in order is the number in the middle or the average of the two numbers in the middle.

Interpreting Information Given by Graphs

Magazines and newspapers often display information using graphs in the first quadrant of rectangular coordinate systems.

EXAMPLE 2 **The Age at which Americans Marry**

Figure P.20 shows the median age at which people in the United States marry for the first time. The data for the years 1997 through 2000 are projections.

a. Estimate the coordinates for point A and interpret the coordinates in practical terms.
b. For the period shown in the graph, what is the minimum age at which men marry for the first time? Approximately when did this occur?

The Median Age at which Americans Marry for the First Time

Figure P.20

Source: U.S. Bureau of the Census, *Statistical Abstract*, 1997.

Solution

(Be sure to turn back a page so that you can see both parts of this solution.)

a. A reasonable estimate for the coordinates of point A is $(1930, 21)$. This means that in 1930 women in the United States married for the first time at age 21.
b. For the period shown in the graph for the men, point B is the lowest point on the graph. With approximate coordinates of $(1972, 22)$, this means that the minimum age at which men married for the first time was 22, and this occurred in 1970. ∎

The Distance Formula

We can use the Pythagorean Theorem to find the distance between two points in the Cartesian coordinate plane. Recall that the Pythagorean Theorem involves a triangle with a right angle, called a *right triangle*. The side opposite the right angle is called the *hypotenuse*. The other two sides are called *legs*. Using Figure P.21, the Pythagorean Theorem can be expressed as

$$a^2 + b^2 = c^2.$$

In any right triangle, the sum of the squares of the legs is equal to the square of the hypotenuse.

Now we are ready to find the distance between two points $P_1(x_1, y_1)$ and $P_2(x_2, y_2)$ in the plane. Using absolute value to express the distance between two points on a number line, the distance between P_1 and P_3 shown in Figure P.22 is $|x_2 - x_1|$. Similarly, the distance between P_3 and P_2 on a vertical line is $|y_2 - y_1|$. By the Pythagorean Theorem,

$$d^2 = |x_2 - x_1|^2 + |y_2 - y_1|^2$$

$$d = \sqrt{|x_2 - x_1|^2 + |y_2 - y_1|^2}$$

$$d = \sqrt{(x_2 - x_1)^2 + (y_2 - y_1)^2}$$

This result is the *distance formula*.

---

3 Find the distance between two points.

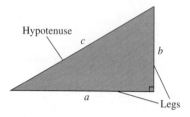

Figure P.21

The Pythagorean Theorem:
$a^2 + b^2 = c^2$

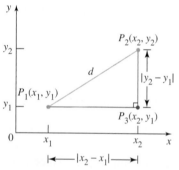

Figure P.22

Finding the distance between P_1 and P_2

Figure P.23

Superimposing rectangular coordinates over a map

The distance formula

The distance d between the points (x_1, y_1) and (x_2, y_2) in the Cartesian coordinate plane is

$$d = \sqrt{(x_2 - x_1)^2 + (y_2 - y_1)^2}.$$

EXAMPLE 3 **Using the Distance Formula**

A rectangular coordinate system with coordinates in miles is superimposed over a map (Figure P.23). Bangkok has coordinates $(-115, 170)$ and Phnom Penh has coordinates $(65, 70)$. How long will it take an airplane averaging 400 miles per hour to fly directly from one city to the other?

Discover for yourself

Work Example 3 with

$(x_1, y_1) = (65, 70)$

and

$(x_2, y_2) = (-115, 170)$.

Does the distance formula still give the same result? What can you conclude?

Solution

The distance between the cities is

$$d = \sqrt{(x_2 - x_1)^2 + (y_2 - y_1)^2}$$ This is the distance formula.

$$= \sqrt{[65 - (-115)]^2 + (70 - 170)^2}$$ Let $(x_1, y_1) = (-115, 170)$ and $(x_2, y_2) = (65, 70)$.

$$= \sqrt{(180)^2 + (-100)^2}$$ Simplify.

$$= \sqrt{32{,}400 + 10{,}000}$$

$$= \sqrt{42{,}400}$$

$$\approx 206$$

Because the cities are approximately 206 miles apart, the flight will take $\frac{206}{400}$, or 0.52 of an hour (approximately 31 minutes). ■

4 Find the midpoint of a line segment.

The Midpoint Formula

The distance formula can be used to prove a formula for finding the midpoint of a line segment between two given points. The formula is stated below and its proof is left for the problem set.

The midpoint formula

Consider a line segment whose endpoints are (x_1, y_1) and (x_2, y_2). The coordinates of the segment's midpoint are

$$\left(\frac{x_1 + x_2}{2}, \frac{y_1 + y_2}{2} \right).$$

To find the midpoint, take the average of the two x-values and of the two y-values.

EXAMPLE 4   **Using the Midpoint and Distance Formulas**

The *median* of a triangle is the line segment from any vertex to the midpoint of the opposite side. If the vertices of a triangle are $(2, 3)$, $(-1, -1)$, and $(3, -4)$, what is the length of the median from $(2, 3)$?

Solution

The situation is illustrated in Figure P.24. We can find the length of median $P_1 M$ by first finding the coordinates of M.

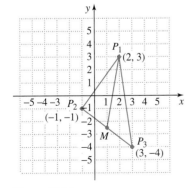

Figure P.24

Median $P_1 M$

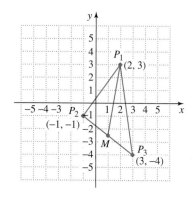

Figure P.24 is repeated here so that you don't have to turn back a page.

$$M = \left(\frac{x_1 + x_2}{2}, \frac{y_1 + y_2}{2}\right)$$

This is the midpoint formula.

$$= \left(\frac{-1 + 3}{2}, \frac{-1 + (-4)}{2}\right)$$

Apply the formula with $(x_1, y_1) = (-1, -1)$ and $(x_2, y_2) = (3, -4)$.

$$= (1, -2.5)$$

Simplify.

Now that we have the coordinates of midpoint M, we are ready to find the length of the designated median.

$$P_1M = \sqrt{(x_2 - x_1)^2 + (y_2 - y_1)^2}$$

This is the distance formula.

$$= \sqrt{(2 - 1)^2 + [3 - (-2.5)]^2}$$

Apply the formula with $(x_1, y_1) = (1, -2.5)$ and $(x_2, y_2) = (2, 3)$.

$$= \sqrt{1 + 30.25}$$

Simplify.

$$\approx 5.59$$

The length of the median from $(2, 3)$ is approximately 5.59 linear units. ■

5 Graph equations in rectangular coordinates.

Graphs of Equations

A relationship between two quantities can be expressed as an *equation in two variables*, such as

$$y = 2x - 4.$$

A point (a, b) *satisfies* the equation if the equation is true when a is substituted for x and b is substituted for y. For example, the point $(3, 2)$ satisfies $y = 2x - 4$ because $2 = 2(3) - 4$ or $2 = 2$ is a true statement. We call $(3, 2)$ a *solution* or *solution point* of the equation. The *graph of the equation* is the set of all solution points.

> ### The graph of an equation
>
> The *graph* of an equation in x and y is the set of all points (x, y) in the rectangular coordinate system that satisfy the equation.

The rectangular coordinate system allows us to visualize relationships between two variables by connecting any equation in two variables with a geometric figure, its graph. One way to graph an equation is by plotting points.

> ### Graphing an equation by plotting points
>
> **1.** Make a table of some coordinates that satisfy the equation.
> **2.** Plot these ordered pairs and connect them with a smooth curve or line.

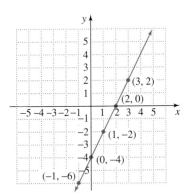

Figure P.25

The graph of $y = 2x - 4$

Study tip

Arrows are used on a graph to show that the graph continues indefinitely in the indicated direction.

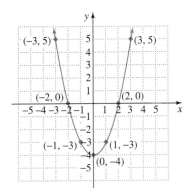

Figure P.26

The graph of $y = x^2 - 4$

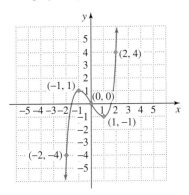

Figure P.27

The graph of $y = x^3 - 2x$

EXAMPLE 5 **Graphing Equations Using the Point-Plotting Method**

Use point-plotting to sketch the graph of each equation:

a. $y = 2x - 4$ **b.** $y = x^2 - 4$ **c.** $y = x^3 - 2x$ **d.** $y = \sqrt[3]{x}$ **e.** $y = |x|$

Solution

a. $y = 2x - 4$

We first construct a table of coordinates by selecting negative, zero, and positive values of x, and calculating the corresponding values of y.

x	$y = 2x - 4$	(x, y)
-1	$y = 2(-1) - 4 = -6$	$(-1, -6)$
0	$y = 2 \cdot 0 - 4 = -4$	$(0, -4)$
1	$y = 2 \cdot 1 - 4 = -2$	$(1, -2)$
2	$y = 2 \cdot 2 - 4 = 0$	$(2, 0)$
3	$y = 2 \cdot 3 - 4 = 2$	$(3, 2)$

Now we plot these points and connect them as shown in Figure P.25. It appears that the graph of $y = 2x - 4$ is a straight line. We'll be discussing the graphs of such equations in detail in Section 2.2.

b. $y = x^2 - 4$

Once again, we begin with a table of coordinates.

x	$y = x^2 - 4$	(x, y)
-3	$y = (-3)^2 - 4 = 5$	$(-3, 5)$
-2	$y = (-2)^2 - 4 = 0$	$(-2, 0)$
-1	$y = (-1)^2 - 4 = -3$	$(-1, -3)$
0	$y = 0^2 - 4 = -4$	$(0, -4)$
1	$y = 1^2 - 4 = -3$	$(1, -3)$
2	$y = 2^2 - 4 = 0$	$(2, 0)$
3	$y = 3^2 - 4 = 5$	$(3, 5)$

As shown in Figure P.26, we plot the seven points and connect them with a smooth curve. In general, the graph of any equation in the form $y = ax^2 + bx + c$ is a *parabola*. We will discuss such equations extensively in Section 3.1.

c. $y = x^3 - 2x$

The following table shows some of the points that satisfy this equation.

x	$y = x^3 - 2x$	(x, y)
-2	$y = (-2)^3 - 2(-2) = -8 + 4 = -4$	$(-2, -4)$
-1	$y = (-1)^3 - 2(-1) = -1 + 2 = 1$	$(-1, 1)$
0	$y = 0^3 - 2 \cdot 0 = 0 - 0 = 0$	$(0, 0)$
1	$y = 1^3 - 2 \cdot 1 = 1 - 2 = -1$	$(1, -1)$
2	$y = 2^3 - 2 \cdot 2 = 8 - 4 = 4$	$(2, 4)$

We plot the five points and connect them with a smooth curve, as shown in Figure P.27.

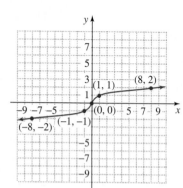

Figure P.28

The graph of $y = \sqrt[3]{x}$.

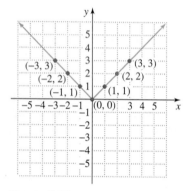

Figure P.29

The graph of $y = |x|$.

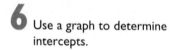

Use a graph to determine intercepts.

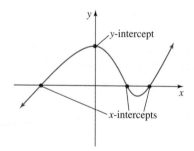

Figure P.30

At all x-intercepts, $y = 0$, and at all y-intercepts, $x = 0$.

d. $y = \sqrt[3]{x}$

In this table of coordinates, we've chosen values of x that are perfect cubes. The graph of $y = \sqrt[3]{x}$ is shown in Figure P.28.

x	$y = \sqrt[3]{x}$	(x, y)
-8	$y = \sqrt[3]{-8} = -2$	$(-8, -2)$
-1	$y = \sqrt[3]{-1} = -1$	$(-1, -1)$
0	$y = \sqrt[3]{0} = 0$	$(0, 0)$
1	$y = \sqrt[3]{1} = 1$	$(1, 1)$
8	$y = \sqrt[3]{8} = 2$	$(8, 2)$

e. $y = |x|$

As before, we first make a table of coordinates.

x	$y = |x|$	(x, y)
-3	$y = |-3| = 3$	$(-3, 3)$
-2	$y = |-2| = 2$	$(-2, 2)$
-1	$y = |-1| = 1$	$(-1, 1)$
0	$y = |0| = 0$	$(0, 0)$
1	$y = |1| = 1$	$(1, 1)$
2	$y = |2| = 2$	$(2, 2)$
3	$y = |3| = 3$	$(3, 3)$

In Figure P.29 we plot these points and use them to sketch the graph of the equation. It appears that the graph of $y = |x|$ is two straight lines that meet at a $\vee$. ∎

Intercepts

The points, if any, where a graph intersects the x- and y-axes are called *intercepts* and are particularly important. Figure P.30 illustrates four such points. The x-coordinates of the points where a graph intersects the x-axis are called the x-*intercepts* of the graph. The x-intercepts are found by setting $y = 0$ in the graph's equation and solving for x. The y-coordinates of the points where a graph intersects the y-axis are called the y-*intercepts* of the graph. The y-intercepts are found by setting $x = 0$ in the graph's equation and solving for y.

Intercepts

Intercepts	How They Are Determined	What They Look Like
x-intercepts: The x-coordinates of the points where an equation's graph intersects the x-axis.	Set y equal to 0 in the equation, and solve for x.	

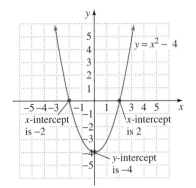

Figure P.31

Intercepts of $y = x^2 - 4$

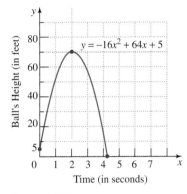

Figure P.32

The height of a ball at different times

Discover for yourself

Use the graph in Figure P.32 and the equation to determine how high the ball goes.

7 Graph equations with a graphing utility.

y-intercepts:
The *y*-coordinates of the points where an equation's graph intersects the *y*-axis.

Set *x* equal to 0 in the equation, and solve for *y*.

The graph of $y = x^2 - 4$, obtained in Example 5b, is shown again in Figure P.31. The *y*-intercept is -4, which can be found by setting $x = 0$ in the graph's equation.

$$y = 0^2 - 4 = -4$$

Saying that the *y*-intercept is -4 means that the graph passes through $(0, -4)$.

The graph indicates that there are two *x*-intercepts, -2 and 2. Equivalently, the graph passes through $(-2, 0)$ and $(2, 0)$.

EXAMPLE 6 An Application of Intercepts

A ball is tossed directly upward. The height of the ball (y, in feet) at any time (x, in seconds) is given by the equation

$$y = -16x^2 + 64x + 5.$$

The graph of the equation is shown in Figure P.32.

a. What is the *y*-intercept and what does this mean in practical terms?

b. What is a reasonable approximation of the *x*-intercept and what does this mean in practical terms?

Solution

a. The graph crosses the *y*-axis at 5, so the *y*-intercept is 5. This means that the graph passes through $(0, 5)$. At time 0, height is 5 feet. This tells us that the ball was tossed from a height of 5 feet.

b. The graph crosses the *x*-axis at approximately 4.1, so the *x*-intercept is about 4.1. The graph passes through $(4.1, 0)$, so at time 4.1 seconds, the height is 0 feet. In practical terms, this means that the ball hits the ground after about 4.1 seconds. ∎

Graphing Utilities

Graphing calculators or graphing software packages for computers are referred to as graphing utilities or graphers. The point-plotting method is used by all graphing utilities. A graphing utility displays only a portion of the Cartesian coordinate plane, called a *viewing rectangle*, a *viewing window*, or a *viewing screen*. The viewing rectangle is determined by six values: the minimum *x*-value (Xmin), the maximum *x*-value (Xmax), the *x*-scale (Xscl), the minimum *y*-value (Ymin), the maximum *y*-value (Ymax), and the *y*-scale (Yscl). By entering these six values into the graphing utility, you set the *range* of the viewing rectangle, which is the boundary of the screen.

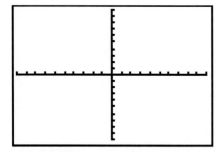

Figure P.33

A standard viewing rectangle for many graphing utilities is $[-10, 10] \times [-10, 10]$.

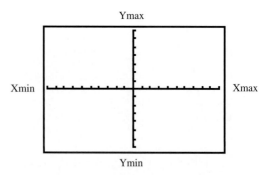

Figure P.34

The viewing rectangle [Xmin, Xmax] by [Ymin, Ymax]

The standard viewing rectangle for many graphing utilities is shown in Figure P.33. This viewing rectangle can be described as $[-10, 10]$ by $[-10, 10]$. On most graphing utilities, the display screen is two-thirds as high as it is wide. By using a *square setting*, you can make the x and y tick marks be equally spaced, which does not occur in the standard viewing rectangle.

In general, the viewing rectangle [Xmin, Xmax] by [Ymin, Ymax] is the set of all points in the Cartesian coordinate plane satisfying Xmin $\leq x \leq$ Xmax and Ymin $\leq y \leq$ Ymax, as shown in Figure P.34. Our work throughout the book will focus on an understanding of equations so that you can determine a viewing window that shows the most important parts of an equation's graph.

Graphing an Equation in x and y Using a Graphing Utility

> **Graphing an equation in x and y using a graphing utility**
>
> 1. If necessary, solve the equation for y in terms of x.
> 2. Enter the equation into the graphing utility.
> 3. Use the standard viewing rectangle or set the range to determine a viewing rectangle that will show a complete picture of the equation's graph.
> 4. Start the graphing utility.

EXAMPLE 7   **Graphing Equations with a Graphing Utility**

Use a graphing utility to graph **a.** $y = 2x - 4$, **b.** $y = x^2 - 4$, **c.** $y = \sqrt[3]{x}$, and **d.** $y = |x|$ in a standard (Xmin $= -10$, Xmax $= 10$, Xscl $= 1$, Ymin $= -10$, Ymax $= 10$, Yscl $= 1$) viewing rectangle.

Solution (Figures appear on page 37.)

a. Figure P.35 shows the graph of $y = 2x - 4$, entered as Y$_1$ $\boxed{=}$ 2X $\boxed{-}$ 4 on many graphing utilities. The line we drew by hand for this equation in Example 5a was straight. Why does the line here seem somewhat bent?

b. Figure P.36 shows the graph of $y = x^2 - 4$, entered as Y$_1$ $\boxed{=}$ X $\boxed{\wedge}$ 2 $\boxed{-}$ 4. Notice that powers are entered using $\boxed{\wedge}$, the power or exponentiation key.

c. Figure P.37 shows the graph of $y = \sqrt[3]{x}$. The equation is entered as Y$_1$ $\boxed{=}$ 3 $\boxed{\sqrt[x]{}}$ X. (Consult your manual for the location of the root key, $\boxed{\sqrt[x]{}}$.) You may remember from your work in intermediate algebra that $x^{1/3}$ means $\sqrt[3]{x}$. Thus, the equation $y = \sqrt[3]{x}$ can also be entered as Y$_1$ $\boxed{=}$ X $\boxed{\wedge}$ $\boxed{(}$ 1 $\boxed{\div}$ 3 $\boxed{)}$.

ENRICHMENT ESSAY

Cartesian Coordinates: An Historical Perspective

The beginning of the 17th century was a time of innovative ideas and enormous intellectual progress in Europe. English theatergoers enjoyed a succession of exciting new plays by Shakespeare. William Harvey proposed the radical notion that the heart was a pump for blood rather than the center of emotion. Galileo, with his new-fangled invention called the telescope, supported the theory of Polish astronomer Copernicus that the sun, not the Earth, was the center of the solar system. Monteverdi was writing the world's first grand operas. French mathematicians Pascal and Fermat invented a new structure of mathematics known as the theory of probability.

Into this arena of intellectual electricity stepped French aristocrat René Descartes (1596–1650). Descartes, propelled by the creativity surrounding him, felt that it was his destiny to discover a method that would bring together all thought and knowledge using the deductive system of mathematics. Beginning with a simple foundation of rules (axioms), Descartes believed that all truth regarding nature could be proved in much the same way that the ancient Greek mathematicians proved geometric theorems.

The idea that all knowledge should be presented in the framework of mathematical reasoning appeared in Descartes' book, *A Discourse on the Method of Rightly Conducting the Reason and Seeking Truth in the Sciences*. Descartes concluded his book with three specific examples of how the method could be applied. The third example, a 106-page footnote called *La Géométrie (The Geometry)*, involved the development of a new branch of mathematics that brought together arithmetic, algebra, and geometry in a unified way—a way that visualized numbers as points on a graph, equations as geometric figures, and geometric figures as equations. This new branch of mathematics, called *analytic geometry*, established Descartes as one of the founders of modern thought and among the most original mathematicians and philosophers of any age.

d. Figure P.38 shows the graph of $y = |x|$, entered as Y_1 = ABS X. (Consult your manual for the location of the ABS [absolute value] key.).   ■

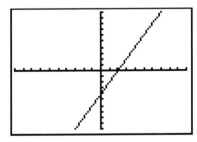

Figure P.35

The graph of $y = 2x - 4$

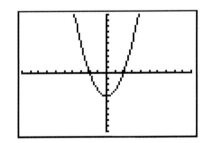

Figure P.36

The graph of $y = x^2 - 4$

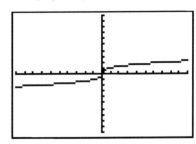

Figure P.37

The graph of $y = \sqrt[3]{x}$

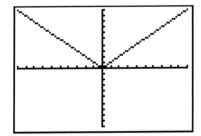

Figure P.38

The graph of $y = |x|$

PROBLEM SET P.3

Practice Problems

In Problems 1–10, **a.** *Plot the points.* **b.** *Find the distance between the points.* **c.** *Find the midpoint of the line segment joining the points.*

1. $(4, -3)$, $(-6, 2)$
2. $(6, -3)$, $(-4, -5)$
3. $(3, 2)$, $(6, 7)$
4. $(-3, 7)$, $(5, 7)$

5. $(0, 0)$, $(5, -12)$
6. $(\frac{7}{4}, -\frac{3}{2})$, $(\frac{5}{2}, -\frac{3}{4})$
7. $(0, -3)$, $(-3, 3)$
8. $(-3, 6)$, $(3, 4)$

9. $(1, -2)$, $(-3, 6)$
10. $(5, 7)$, $(2, 3)$

11. Plot the following ordered pairs: $(5, 7)$, $(1, 10)$, and $(-3, -8)$. Then find the perimeter of the triangle whose vertices are the three points.

12. Plot the points $(2, 3)$, $(-1, -1)$, and $(3, -4)$. Then prove that the triangle connecting these points is isosceles (has two sides that are equal in measure).

13. If the vertices of a triangle are $(5, 7)$, $(1, 10)$, and $(-3, -8)$, what is the length of the median from $(5, 7)$?

14. If the vertices of a triangle are $(2, 3)$, $(-1, -1)$, and $(3, -4)$, what is the length of the median from $(2, 3)$?

The converse of the Pythagorean Theorem states that for a triangle with sides of lengths a, b, and c, if $a^2 + b^2 = c^2$, then the triangle is a right triangle. Plot the points in Problems 15–16 and show that they form the vertices of a right triangle. Then find each triangle's area.

15. $(2, 1)$, $(-1, -5)$, $(4, 0)$
16. $(2, 2)$, $(3, -1)$, $(-3, -3)$

17. Show that the points $A(-4, -6)$, $B(1, 0)$, and $C(11, 12)$ are collinear (lie along a straight line) by showing that the distance from A to B plus the distance from B to C equals the distance from A to C.

18. The vertices of a quadrilateral are $(-1, -5)$, $(-8, 2)$, $(7, 10)$, and $(4, -3)$. Find the length of the diagonals.

In Problems 19–48, use the point-plotting method to graph the given equation. For each graph, identify all intercepts. Use a graphing utility to verify your graph. (Use the standard viewing rectangle.)

19. $y = 3x - 6$
20. $y = 4x - 8$
21. $y = -2x + 4$
22. $y = -2x + 6$

23. $y = x^2 - 1$
24. $y = x^2 - 9$
25. $y = 9 - x^2$
26. $y = 1 - x^2$

27. $y = x^2 + x - 6$
28. $y = x^2 - 4x + 3$
29. $y = x^2 - 2x$
30. $y = x^2 + x$

31. $y = -x^2 + x + 1$
32. $y = -x^2 + 2x - 2$
33. $y = x^3 + 2x$
34. $y = x^3 - 3x$

35. $y = \sqrt{x}$
36. $y = \sqrt{x} - 1$
37. $y = \sqrt{x} - 2$
38. $y = \sqrt{x} + 4$

39. $y = |2x|$
40. $y = |3x|$
41. $y = |x + 1|$
42. $y = |x + 2|$

43. $y = \sqrt[3]{x} + 1$
44. $y = \sqrt[3]{x} + 2$
45. $y = |x - 1|$
46. $y = |x - 2|$

47. $y = \sqrt[3]{x} - 1$
48. $y = \sqrt[3]{x} - 2$

Application Problems

49. Women are taking part in sports as never before. The graph shows participation in high school athletics by women (in thousands). Estimate the coordinates of point A and interpret the coordinates in practical terms.

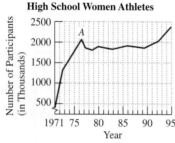

High School Women Athletes

Source: Women's Sports Foundation

50. The graph on the right shows the height (y, in feet) of a ball dropped from the top of the Empire State Building at different times.

a. What is a reasonable estimate of the height of the Empire State Building?

b. What is a reasonable estimate, to the nearest tenth of a second, of how long it takes for the ball to hit the ground?

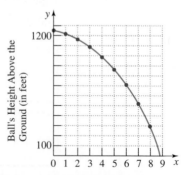

Use the figure, which shows the number of accident-related deaths in the United States at home and at work, to answer Problems 51–53.

51. For the period shown, what is a reasonable estimate of the greatest number of accidental deaths that occurred at home? When did this occur?

52. For the period shown, what is a reasonable estimate of the least number of accidental deaths that occurred at home? When did this occur?

53. Describe the trend in the number of accidental deaths that occurred at work.

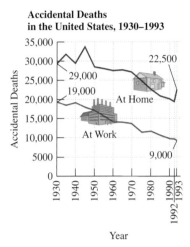

Source: National Safety Council. Adapted from *The Macmillan Visual Almanac.*

A baseball is tossed straight up into the air by a person who is six feet tall. The height of the ball above the ground (y, in feet) depends upon how long the ball is in the air (x, in seconds) and is described by the equation

$$y = -16x^2 + 64x + 6.$$

In this equation, x = 0 corresponds to the instant that the ball is thrown. The graph of this equation is shown in the figure below on the left. Use the figure to answer Problems 54–56.

54. What is the maximum height of the ball? When does the ball reach this height?

55. What is a reasonable estimate of the *x*-intercept for the graph? What happens to the ball for this value of *x*?

56. What is the *y*-intercept for this graph? Explain what this has to do with the height from which the ball was thrown.

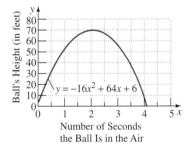

57. A photographic light meter is used to measure the brightness of a shining flashlight on a wall. The equation

$$y = \frac{1}{x^2}$$

describes the intensity of brightness (*y*) in terms of the flashlight's distance from the wall (*x*). The graph of this equation is shown in the figure below. As the distance from the light to the wall doubles, by what fraction does the light intensity decrease?

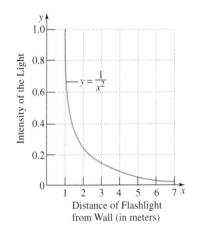

True–False Critical Thinking Problems

58. Which one of the following is true?
    a. If the coordinates of a point satisfy the inequality $xy > 0$, then (x, y) must be in quadrant I.
    b. The ordered pair $(2, 5)$ satisfies $3y - 2x = -4$.
    c. If a point is on the x-axis, it is neither up nor down, so $x = 0$.
    d. None of the above is true.
59. Morphine is widely used in the United States to treat severe pain. Its physiological effects, however, differ for nonaddicts (graph a) and addicts (graph b). Both graphs show the concentration of morphine in the blood following an intravenous injection of 10 milligrams to relieve severe pain. The concentration is measured in micrograms per milliliter (μg/ml). Which one of the following is true based on the graphs?
    a. Nonaddicts receive narcotic effects only when the bloodstream concentration is above 1 μg/ml.
    b. The addict receives narcotic effects for up to 2 hours.
    c. The concentration at which the nonaddict experiences no effects from the drug is the same at which the addict starts experiencing symptoms of withdrawal.
    d. The morphine concentration in the blood over 24 hours differs for nonaddicts and addicts.

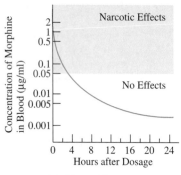

(a) Nonaddicts

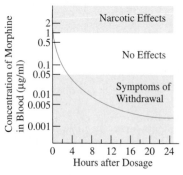

(b) Addicts

Injecting 10 milligrams of morphine to relieve pain

Technology Problems

60. Use a graphing utility and a standard viewing rectangle to graph each of the following equations. What do you observe?

    $y = 2x + 3$, $\quad y = 2x + 1$, $\quad y = 2x$,

    $y = 2x - 1$, $\quad y = 2x - 3$

61. Use a graphing utility and a standard viewing rectangle to graph each of the following equations. What do you observe?

    $y = 4x + 1$, $\quad y = 2x + 1$, $\quad y = -2x + 1$,

    $y = -4x + 1$

62. Use a graphing utility to graph each pair of equations in the same viewing rectangle.
    a. $y = x \cdot x^2$ $\quad$ and $\quad y = x^3$
    b. $y = x^2 \cdot x^3$ $\quad$ and $\quad y = x^5$
    c. $y = \dfrac{x^5}{x^3}$ $\quad$ and $\quad y = x^2$

    d. What do you observe? What generalizations can you make regarding properties of exponents?

A graph of an equation is a complete graph if it shows all of the important features of the graph. Use a graphing utility to graph the equations in Problems 63–66 in each of the given viewing rectangles. Then choose which viewing rectangle gives a complete graph.

63. $y = x^2 + 10$
    a. $[-5, 5]$ by $[-5, 5]$ $\qquad$ b. $[-10, 10]$ by $[-10, 10]$
    c. $[-10, 10]$ by $[-50, 50]$
64. $y = 0.1x^4 - x^3 + 2x^2$
    a. $[-5, 5]$ by $[-8, 2]$ $\qquad$ b. $[-10, 10]$ by $[-10, 10]$
    c. $[-8, 16]$ by $[-16, 8]$

65. $y = \sqrt{x + 18}$
    a. $[-10, 10]$ by $[-10, 10]$ $\qquad$ b. $[-50, 50]$ by $[-10, 10]$
    c. $[-10, 10]$ by $[-50, 50]$
66. $y = x^3 - 30x + 20$
    a. $[-10, 10]$ by $[-10, 10]$ $\qquad$ b. $[-10, 10]$ by $[-50, 50]$
    c. $[-10, 10]$ by $[-50, 100]$

In Problems 67–69, find a complete graph of each equation in the same viewing rectangle.

67. $y = |x|$, $y = |x - 1|$, $y = |x - 2|$, $y = |x + 3|$

68. $y = x^2$, $y = x^2 - 1$, $y = x^2 - 2$, $y = x^2 + 3$

69. $y = x^2$, $y = (x - 1)^2$, $y = (x - 2)^2$, $y = (x + 3)^2$

In Problems 70–73, use a graphing utility to graph each equation. Then use the $\boxed{\text{TRACE}}$ *feature to move the cursor along the curve to approximate the variable coordinate(s) of the solution point to two decimal places. In some cases, you may need to use the* $\boxed{\text{ZOOM}}$ *feature to obtain the desired accuracy.*

70. $y = \sqrt{6 - x}$ **a.** $(2, y)$ **b.** $(x, 2)$

71. $y = x^2 - 4x + 3$ **a.** $(-0.5, y)$ **b.** $(x, 3)$

72. $y = x^3 - 4x$ **a.** $(2, y)$ **b.** $(x, -2)$

73. $y = x^3(x - 2)$ **a.** $(2.5, y)$ **b.** $(x, 4.5)$

74. The formula $y = -3.1x^2 + 51.4x + 4024.5$ approximately describes the average number of cigarettes (y) per year consumed by Americans 18 and older x years after 1950. The formula describes reality from the years 1950 through 1990, or from $x = 0$ through $x = 40$.

   a. Construct a table of coordinates using $x = 0, 5, 10, 15, 20, 30,$ and 40. Find the corresponding values for y using a calculator and rounding to the nearest integer. Then graph the equation by hand.

   b. Use a graphing utility with the following range setting to check your hand-drawn graph.

     Xmin = 0, Xmax = 40, Xscl = 1, Ymin = 0,

     Ymax = 5000, Yscl = 1000

   c. Change the range setting so that Xmax = 45. Use your graphing utility to graph the equation for cigarette consumption with this new range setting. Based on the shape of the graph, does the given equation accurately describe reality from $x = 40$ through $x = 45$? What does this tell you about equations that attempt to describe variables over long periods of time?

75. The formulas

$$y = 67.0166(1.00308)^x$$

and

$$y = 74.9742(1.00201)^x$$

describe the life expectancy (y) for white males and females, respectively, in the United States whose present age is x.

   a. Use a graphing utility to graph both equations in the same viewing rectangle. Enter the formulas as

     Y_1 $\boxed{=}$ 67.0166 $\boxed{\times}$ 1.00308 $\boxed{\wedge}$ X

     Y_2 $\boxed{=}$ 74.9742 $\boxed{\times}$ 1.00201 $\boxed{\wedge}$ X

     with the following range setting:

     Xmin = 0, Xmax = 80, Xscl = 1, Ymin = 0,

     Ymax = 100, Yscl = 1

   b. Use the $\boxed{\text{TRACE}}$ feature of your utility to trace along each curve until you reach $x = 40$. What is the corresponding value for y on y_1 and y_2? How long can a 40-year-old white male expect to live? How long can a 40-year-old white female expect to live?

   c. Explain how each graph indicates that life expectancy increases as one gets older. Use the $\boxed{\text{TRACE}}$ feature to reinforce your observation.

Writing in Mathematics _____

76. The figure shown here depicts actual data points for cigarette consumption in the United States. How does your graph in Problem 74 compare to the one shown here? What does this tell you about mathematical equations that describe variables in the real world?

Cigarette Consumption per U.S. Adult

Source: U.S. Department of Health and Human Services.

77. Using a random sample of white Americans (from which we can generalize), sociologists Lawrence Bobo and James Kluegel (1991) studied the extent of prejudice in the United States. Details of some of their research are shown in the graph. Describe the trend shown by the graph in terms of education and prejudice. What is misleading about the scale on the horizontal axis?

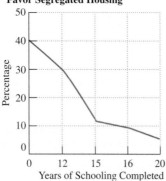

Percentage of White Americans Who Favor Segregated Housing

Source: Bobo and Kluegel, 1991.

78. The parentheses used to represent an ordered pair are also used to represent an interval. If you are reading a section in a math book and (3, 4) is mentioned, how will you know if they are discussing the point (3, 4) or the interval for which $3 < x < 4$?

Critical Thinking Problems

79. Show that the points $A\ (1, 1 + d)$, $B\ (3, 3 + d)$, and C $(6, 6 + d)$ are collinear (lie along a straight line) by showing that the distance from A to B plus the distance from B to C equals the distance from A to C.

80. Prove the midpoint formula by using the following procedure.

   a. Show that the distance between $(x_1,\ y_1)$ and $\left(\dfrac{x_1 + x_2}{2}, \dfrac{y_1 + y_2}{2}\right)$ is equal to the distance between (x_2, y_2) and $\left(\dfrac{x_1 + x_2}{2}, \dfrac{y_1 + y_2}{2}\right)$.

   b. Use the procedure from Problem 79 to show that the points $(x_1,\ y_1)$, $\left(\dfrac{x_1 + x_2}{2}, \dfrac{y_1 + y_2}{2}\right)$, and $(x_2,\ y_2)$ are collinear.

S E C T I O N   P . 4   # Radicals and Rational Exponents

Solutions Manual   **Tutorial**   **Video I**

Objectives

1 Find *n*th roots.

2 Simplify radical expressions.

3 Rationalize denominators.

4 Combine radicals.

5 Evaluate expressions with rational exponents.

6 Simplify expressions with rational exponents.

7 Reduce the index of a radical.

The number of plant species on the various islands of the Galápagos chain is given by

$$S = 28.6 \sqrt[3]{A}$$

in which A is the area of the island in square miles and S is the number of plant species on the island. Radical expressions frequently appear in descriptions of the physical world. In this section, we review such expressions and the use of rational exponents to indicate radicals.

▌ Find nth roots.

Even and Odd nth Roots

Roots in general are related to repeated factors. For example,

$$2^3 = 2 \cdot 2 \cdot 2 = 8$$

so 2 is the *cube root* of 8, written

$$\sqrt[3]{8} = 2.$$

As another example,

$$3^4 = 3 \cdot 3 \cdot 3 \cdot 3 = 81$$

so 3 is the *fourth root* of 81, written

$$\sqrt[4]{81} = 3.$$

Let's generalize from these examples, beginning with the definition of even roots.

> **Definition of even roots**
>
> If n is an even natural number and $b > 0$, $\sqrt[n]{b}$ (the nth root of b) is the positive real number whose nth power is b. Thus $\sqrt[n]{b} = a$ if $a^n = b$.
>
> $$(\sqrt[n]{b})^n = b \quad \text{when} \quad b > 0$$

EXAMPLE 1 **Finding nth Roots When n Is an Even Number**

Find each root, or state that the root is not a real number.

a. $\sqrt[4]{81}$ **b.** $-\sqrt[4]{81}$ **c.** $\sqrt[4]{-81}$ **d.** $\sqrt[6]{64}$ **e.** $(\sqrt[4]{6})^4$ **f.** $\sqrt[4]{\frac{81}{16}}$

Solution

a. $\sqrt[4]{81} = 3$ because $3^4 = 81$.

b. $-\sqrt[4]{81} = -(\sqrt[4]{81}) = -3$

c. $\sqrt[4]{-81}$ is not a real number. There is no real number we can raise to the fourth power and obtain -81.

d. $\sqrt[6]{64} = 2$ because $2^6 = 64$.

e. $(\sqrt[4]{6})^4 = 6$ because $(\sqrt[n]{b})^n = b$.

f. $\sqrt[4]{\frac{81}{16}} = \frac{3}{2}$ because $(\frac{3}{2})^4 = \frac{81}{16}$.   ∎

If n is an odd natural number, it is not necessary to restrict b to positive real numbers.

Using technology

Verify the results of Examples 1 and 2 using a graphing calculator, if possible. First consult your manual to locate the root ($\boxed{\sqrt[x]{}}$) key.

Verifying Example 1a: $\sqrt[4]{81}$

4 $\boxed{\sqrt[x]{}}$ 81 $\boxed{\text{ENTER}}$

Verifying Example 2c: $\sqrt[5]{-32}$

5 $\boxed{\sqrt[x]{}}$ $\boxed{(-)}$ 32 $\boxed{\text{ENTER}}$

Definition of odd roots

If n is an odd natural number, $\sqrt[n]{b}$ is the positive or negative real number whose nth power is b. Thus, $\sqrt[n]{b} = a$ if $a^n = b$.

$$(\sqrt[n]{b})^n = b$$

EXAMPLE 2   Finding nth Roots when n Is an Odd Number

Find each root:

a. $\sqrt[3]{27}$ **b.** $\sqrt[3]{-64}$ **c.** $\sqrt[5]{-32}$ **d.** $(\sqrt[7]{-6})^7$ **e.** $\sqrt[7]{-\frac{1}{128}}$

Solution

a. $\sqrt[3]{27} = 3$ because $3^3 = 27$.
b. $\sqrt[3]{-64} = -4$ because $(-4)^3 = -64$.
c. $\sqrt[5]{-32} = -2$ because $(-2)^5 = -32$.
d. $(\sqrt[7]{-6})^7 = -6$ because $(\sqrt[n]{b})^n = b$.
e. $\sqrt[7]{-\frac{1}{128}} = -\frac{1}{2}$ because $(-\frac{1}{2})^7 = -\frac{1}{128}$.   ∎

Discover for yourself

If n is even or odd, the nth root of 1 is 1.

$$\sqrt{1} = 1, \quad \sqrt[3]{1} = 1, \quad \sqrt[4]{1} = 1, \quad \sqrt[5]{1} = 1, \quad \sqrt[6]{1} = 1, \quad \text{etc.}$$

In general,

$$\sqrt[n]{1} = 1.$$

Can you think of another number all of whose nth roots equal the number itself?

In the Discover for yourself, did you discover that the number is 0, a number with many special properties?

The nth root of zero

If n is even or odd,

$$\sqrt[n]{0} = 0.$$

Mathematicians have a special vocabulary related to even and odd roots.

The vocabulary of radicals

Given $\sqrt{b}, \sqrt[3]{b}, \sqrt[4]{b}, \sqrt[5]{b},$ and so on, the numbers 2, 3, 4, 5, ... are called the *index* of the root. Observe that an index of 2, indicating the square root, is omitted. $\sqrt{}$ is called the *radical sign*. The expression under the radical sign, b, is the *radicand*. The expressions $\sqrt{b}, \sqrt[3]{b}, \sqrt[4]{b},$ and $\sqrt[5]{b}$ are *radicals*.

Before turning to properties of radicals, let's take a moment to summarize our discussion of nth roots.

---

nth roots

1. If n is even: $\sqrt[n]{b} = a$ $(b > 0)$ if $a^n = b$.
   a. Every positive number has two nth roots that are opposites of each other. For example, the fourth roots of 16 are 2 and -2.
   b. If $b < 0$, then $\sqrt[n]{b}$ is not a real number.
2. If n is odd: $\sqrt[n]{b} = a$ if $a^n = b$.
   a. Every real number has one nth root. For example, $\sqrt[3]{8} = 2$ and $\sqrt[3]{-27} = -3$.
   b. The nth root of a positive number is positive, and the nth root of a negative number is negative.
3. If n is even or odd: $\sqrt[n]{0} = 0$.

The following properties will be used to simplify radical expressions.

Properties of nth roots

Let b and c be real numbers, variables, or algebraic expressions, and let m and n be natural numbers. In each property, the indicated roots are real numbers.

Property	**Example**
1. $\sqrt[n]{b} \cdot \sqrt[n]{c} = \sqrt[n]{bc}$	$\sqrt[3]{4} \cdot \sqrt[3]{2} = \sqrt[3]{4 \cdot 2} = \sqrt[3]{8} = 2$
2. $\dfrac{\sqrt[n]{b}}{\sqrt[n]{c}} = \sqrt[n]{\dfrac{b}{c}}, \quad c \neq 0$	$\dfrac{\sqrt[5]{36}}{\sqrt[5]{4}} = \sqrt[5]{\dfrac{36}{4}} = \sqrt[5]{9}$
3. $(\sqrt[n]{b})^n = b$	$(\sqrt[3]{5})^5 = 5$
4. If n is even, $\sqrt[n]{b^n} = \lvert b \rvert$	$\sqrt{(-6)^2} = \lvert -6 \rvert = 6$
$\quad$ If n is odd, $\sqrt[n]{b^n} = b$	$\sqrt[3]{(-6)^3} = -6$
5. $\sqrt[m]{\sqrt[n]{b}} = \sqrt[mn]{b}$	$\sqrt[3]{\sqrt{5}} = \sqrt[3 \cdot 2]{5} = \sqrt[6]{5}$

Study tip

The nth root of a sum is *not* equal to the sum of the nth roots:
$$\sqrt[n]{b + c} \neq \sqrt[n]{b} + \sqrt[n]{c}$$
Show that this is the case by letting $n = 2$, $b = 9$, and $c = 16$.

2 Simplify radical expressions.

Simplified Radical Form

Properties of nth roots are important because they allow us to simplify radicals. A simplified radical expression satisfies four conditions.

Conditions for simplified radical form

A radical expression with index n is in simplest form, or is simplified completely, whenever:

1. The radicand contains as few factors as possible. The radicand should contain no factors that are nth powers.
2. There are no fractions in the radicand.
3. There are no radicals in the denominator.
4. The index of the radical is reduced.

The first condition for simplified form requires that we remove as many factors from the radicand as possible, using

$$\sqrt[n]{b^n} = |b| \qquad \text{when } n \text{ is even}$$

and

$$\sqrt[n]{b^n} = b \qquad \text{when } n \text{ is odd}$$

EXAMPLE 3   Simplifying Radicals

Write each radical in simplified form by factoring the radicand and removing as many factors as possible.

a. $\sqrt[3]{54}$ **b.** $\sqrt{48x^3}$ **c.** $\sqrt[3]{-16y^4}$ **d.** $\sqrt[4]{32x^4y^6}$

Solution

a.

$$\sqrt[3]{54} = \sqrt[3]{27 \cdot 2}$$

Observe that 27 is a *perfect cube* because its cube root is an integer: $\sqrt[3]{27} = 3$.

$$= \sqrt[3]{3^3 \cdot 2}$$

The index is 3, so we want radicand factors that are cubes.

$$= \sqrt[3]{3^3} \cdot \sqrt[3]{2}$$

$$= 3\sqrt[3]{2} \qquad \sqrt[n]{b^n} = b \text{ if } n \text{ is odd.}$$

b.

$$\sqrt{48x^3} = \sqrt{16x^2 \cdot 3x} \qquad \text{Notice that } 16x^2 \text{ is a perfect square factor.}$$

$$= \sqrt{(4x)^2 \cdot 3x}$$

$$= \sqrt{(4x)^2} \cdot \sqrt{3x}$$

$$= |4x| \cdot \sqrt{3x} \qquad \sqrt[n]{b^n} = |b|, \text{ if } n \text{ is even.}$$

$$= 4|x|\sqrt{3x} \qquad |4x| = |4||x| = 4|x|$$

c.

$$\sqrt[3]{-16y^4} = \sqrt[3]{-8y^3 \cdot 2y} \qquad \text{The perfect cube factor is } -8y^3.$$

$$= \sqrt[3]{(-2y)^3 \cdot 2y}$$

$$= \sqrt[3]{(-2y)^3} \cdot \sqrt[3]{2y}$$

$$= -2y\sqrt[3]{2y} \qquad \sqrt[n]{b^n} = b \text{ if } n \text{ is odd.}$$

d.

$$\sqrt[4]{32x^4y^6} = \sqrt[4]{16x^4y^4 \cdot 2y^2}$$

The index is 4, so we want radicand factors that are perfect fourth powers.

$$= \sqrt[4]{(2xy)^4 \cdot 2y^2}$$

$$= \sqrt[4]{(2xy)^4} \cdot \sqrt[4]{2y^2}$$

$$= |2xy|\sqrt[4]{2y^2} \qquad \sqrt[n]{b^n} = |b| \text{ if } n \text{ is even.}$$

$$= 2|xy|\sqrt[4]{2y^2}$$

∎

Donald Judd "Untitled" 1967, green lacquer / galvanized iron, 12 units at $9 \times 40 \times 31$ in. Courtesy Joseph Helman Gallery, New York. © Estate of Donald Judd / Licensed by VAGA, New York, NY.

3 Rationalize denominators.

The second and third conditions for simplified radical form do not permit radicals in a denominator. Such radical expressions are simplified by multiplying both the numerator and the denominator by a factor chosen to eliminate any radicals in the denominator. If the denominator contains a square root, you can multiply the numerator and the denominator by the square root expression in the denominator:

$$\frac{1}{\sqrt{a}} = \frac{1}{\sqrt{a}} \cdot 1 = \frac{1}{\sqrt{a}} \cdot \frac{\sqrt{a}}{\sqrt{a}} = \frac{\sqrt{a}}{a}$$

Because we are multiplying by 1, we are not changing the value of the radical expression. Observe that the denominator in the last fraction contains no radical. Removing radicals from a denominator without changing the value of a radical expression is called *rationalizing the denominator*.

EXAMPLE 4   **Rationalizing a Single-Term Denominator**

Simplify by rationalizing the denominator: $\dfrac{7}{2\sqrt{6}}$

Solution

We multiply the numerator and the denominator by $\sqrt{6}$ because $\sqrt{6}\cdot\sqrt{6}=6$.

$$\frac{7}{2\sqrt{6}}=\frac{7}{2\sqrt{6}}\cdot\frac{\sqrt{6}}{\sqrt{6}}=\frac{7\sqrt{6}}{2(6)}=\frac{7\sqrt{6}}{12}$$ ■

Things get a bit more complicated when higher roots appear in the denominator. If the denominator is of the form $\sqrt[n]{a^m}$ with $m<n$, we can rationalize the denominator by multiplying the numerator and the denominator by $\sqrt[n]{a^{n-m}}$.

$$\frac{1}{\sqrt[n]{a^m}}=\frac{1}{\sqrt[n]{a^m}}\cdot 1=\frac{1}{\sqrt[n]{a^m}}\cdot\frac{\sqrt[n]{a^{n-m}}}{\sqrt[n]{a^{n-m}}}=\frac{\sqrt[n]{a^{n-m}}}{\sqrt[n]{a^{m+n-m}}}=\frac{\sqrt[n]{a^{n-m}}}{\sqrt[n]{a^n}}=\frac{\sqrt[n]{a^{n-m}}}{a}$$

EXAMPLE 5   **Rationalizing Single-Term Denominators with Higher Roots**

Simplify by rationalizing the denominator:

a. $\sqrt[3]{\dfrac{2}{9}}$ **b.** $\dfrac{7}{\sqrt[5]{8}}$

Solution

a. We begin by writing the fraction as $\dfrac{\sqrt[3]{2}}{\sqrt[3]{9}}$. We note that the denominator can be expressed as $\sqrt[3]{3^2}$. We now multiply the numerator and the denominator by $\sqrt[3]{3}$ because $\sqrt[3]{3^2}\cdot\sqrt[3]{3}=\sqrt[3]{3^3}=3$.

$$\sqrt[3]{\frac{2}{9}}=\frac{\sqrt[3]{2}}{\sqrt[3]{9}}=\frac{\sqrt[3]{2}}{\sqrt[3]{3^2}}\cdot\frac{\sqrt[3]{3}}{\sqrt[3]{3}}=\frac{\sqrt[3]{6}}{\sqrt[3]{3^3}}=\frac{\sqrt[3]{6}}{3}$$

b. We begin by expressing the fraction as $\dfrac{7}{\sqrt[5]{2^3}}$. We will multiply the numerator and the denominator by $\sqrt[5]{2^2}$ because $\sqrt[5]{2^3}\cdot\sqrt[5]{2^2}=\sqrt[5]{2^5}=2$.

$$\frac{7}{\sqrt[5]{8}}=\frac{7}{\sqrt[5]{2^3}}\cdot\frac{\sqrt[5]{2^2}}{\sqrt[5]{2^2}}=\frac{7\sqrt[5]{2^2}}{\sqrt[5]{2^5}}=\frac{7\sqrt[5]{4}}{2}$$ ■

If the denominator is of the form $a+b\sqrt{c}$, we multiply the numerator and the denominator by the *conjugate radical* $a-b\sqrt{c}$. The product of $a+b\sqrt{c}$ and $a-b\sqrt{c}$ does not contain a radical.

Study tip

As we have seen, to rationalize the denominator with square roots, you can multiply the numerator and the denominator by the number in the denominator:

$$\frac{1}{\sqrt{2}}=\frac{1}{\sqrt{2}}\cdot\frac{\sqrt{2}}{\sqrt{2}}=\frac{\sqrt{2}}{2}$$

However, this does not work with higher roots.

Incorrect:

$$\frac{1}{\sqrt[3]{2}}=\frac{1}{\sqrt[3]{2}}\cdot\frac{\sqrt[3]{2}}{\sqrt[3]{2}}=\frac{\sqrt[3]{2}}{2}$$

Correct:

$$\frac{1}{\sqrt[3]{2}}=\frac{1}{\sqrt[3]{2}}\cdot\frac{\sqrt[3]{2^2}}{\sqrt[3]{2^2}}=\frac{\sqrt[3]{2^2}}{\sqrt[3]{2^3}}$$
$$=\frac{\sqrt[3]{4}}{2}$$

$$(a + b\sqrt{c})(a - b\sqrt{c})$$

$$= a(a - b\sqrt{c}) + b\sqrt{c}(a - b\sqrt{c}) \quad \text{Distribute each term in the first parentheses over the second parentheses.}$$

$$= a^2 - ab\sqrt{c} + ab\sqrt{c} - (b\sqrt{c})^2 \quad \text{Distribute.}$$

$$= a^2 - (b\sqrt{c})^2 \quad \text{Simplify.}$$

$$= a^2 - b^2 c$$

The significance of this last step is that it does not contain a radical.

Multiplying by a conjugate radical

$$(a + b\sqrt{c})(a - b\sqrt{c}) = a^2 - (b\sqrt{c})^2 = a^2 - b^2 c$$

EXAMPLE 6 **Rationalizing Denominators Containing Two Terms**

Rationalize the denominator:

a. $\dfrac{21}{4 - \sqrt{2}}$ **b.** $\dfrac{20}{\sqrt{7} + \sqrt{3}}$

Discover for yourself

In Example 6a, show that multiplying the given fraction by

$$\dfrac{4 - \sqrt{2}}{4 - \sqrt{2}} \quad \text{or} \quad \dfrac{\sqrt{2}}{\sqrt{2}}$$

does not eliminate the radical in the denominator. In Example 6b, show that multiplying by

$$\dfrac{\sqrt{7} + \sqrt{3}}{\sqrt{7} + \sqrt{3}}$$

does not eliminate the denominator's radicals.

Solution

a. $\dfrac{21}{4 - \sqrt{2}} = \dfrac{21}{4 - \sqrt{2}} \cdot \dfrac{4 + \sqrt{2}}{4 + \sqrt{2}}$ Multiply the numerator and the denominator by the conjugate radical of the denominator.

$$= \dfrac{21(4 + \sqrt{2})}{4^2 - (\sqrt{2})^2} \quad (a + b\sqrt{c})(a - b\sqrt{c}) = a^2 - (b\sqrt{c})^2$$

$$= \dfrac{21(4 + \sqrt{2})}{16 - 2} \quad (\sqrt{2})^2 = 2. \text{ In general, } (\sqrt[n]{b})^n = b.$$

$$= \dfrac{21(4 + \sqrt{2})}{14}$$

$$= \dfrac{3(4 + \sqrt{2})}{2} \quad \text{Simplify, dividing the numerator and the denominator by 7.}$$

The simplified radical can be expressed as $\dfrac{12 + 3\sqrt{2}}{2}$.

b. $\dfrac{20}{\sqrt{7} + \sqrt{3}} = \dfrac{20}{\sqrt{7} + \sqrt{3}} \cdot \dfrac{\sqrt{7} - \sqrt{3}}{\sqrt{7} - \sqrt{3}}$ Multiply the numerator and the denominator by the denominator's conjugate radical.

$$= \dfrac{20(\sqrt{7} - \sqrt{3})}{(\sqrt{7})^2 - (\sqrt{3})^2} \quad (\sqrt{a} + \sqrt{b})(\sqrt{a} - \sqrt{b}) = (\sqrt{a})^2 - (\sqrt{b})^2$$

$$= \dfrac{20(\sqrt{7} - \sqrt{3})}{7 - 3} \quad (\sqrt[n]{b})^n = b$$

$$= \dfrac{20(\sqrt{7} - \sqrt{3})}{4}$$

$$= 5(\sqrt{7} - \sqrt{3}) \quad \text{Simplify. The answer can be equivalently expressed as } 5\sqrt{7} - 5\sqrt{3}.$$

4 Combine radicals.

Combining Radicals

Radical expressions can be added and subtracted if they are *like radicals*, meaning they have the same index and the same radicand. We can use the distributive property to add or subtract like radicals. For example,

$$7\sqrt{2} + 5\sqrt{2} = (7 + 5)\sqrt{2} = 12\sqrt{2}$$

$$19\sqrt[3]{5} - 12\sqrt[3]{5} = (19 - 12)\sqrt[3]{5} = 7\sqrt[3]{5}$$

Although at first glance many radicals might not appear to be like radicals, simplification may enable us to add or subtract. Let's see how this works.

EXAMPLE 7 Combining Radicals

Combine:

a. $7\sqrt{18} + 5\sqrt{8} - 6\sqrt{2}$

b. $12\sqrt{\dfrac{1}{3}} - 4\sqrt{\dfrac{1}{12}}$

Solution

a. $7\sqrt{18} + 5\sqrt{8} - 6\sqrt{2}$

$\quad = 7\sqrt{9\cdot2} + 5\sqrt{4\cdot2} - 6\sqrt{2}$ Begin by simplifying $\sqrt{18}$ and $\sqrt{8}$.

$\quad = 7\cdot3\sqrt{2} + 5\cdot2\sqrt{2} - 6\sqrt{2}$ $\sqrt{9\cdot2} = \sqrt{3^2\cdot2} = 3\sqrt{2}$ and
$\quad\quad\quad\quad\quad\quad\quad\quad\quad\quad\quad\quad\quad\quad\quad \sqrt{4\cdot2} = \sqrt{2^2\cdot2} = 2\sqrt{2}.$

$\quad = 21\sqrt{2} + 10\sqrt{2} - 6\sqrt{2}$ Multiply.

$\quad = (21 + 10 - 6)\sqrt{2}$ Use the distributive property.

$\quad = 25\sqrt{2}$ Combine numerical terms.

b. $12\sqrt{\dfrac{1}{3}} - 4\sqrt{\dfrac{1}{12}}$

$\quad = 12\cdot\dfrac{1}{\sqrt{3}} - 4\cdot\dfrac{1}{\sqrt{12}}$

$\quad = \dfrac{12}{\sqrt{3}}\cdot\dfrac{\sqrt{3}}{\sqrt{3}} - \dfrac{4}{\sqrt{12}}\cdot\dfrac{\sqrt{3}}{\sqrt{3}}$ Rationalize denominators. The smallest number that produces a perfect square in the second denominator is $\sqrt{3}$, but you can also multiply by $\dfrac{\sqrt{12}}{\sqrt{12}}$.

$\quad = \dfrac{12\sqrt{3}}{3} - \dfrac{4\sqrt{3}}{6}$ $\sqrt{3}\cdot\sqrt{3} = (\sqrt{3})^2 = 3$ and $\sqrt{12}\cdot\sqrt{3} = \sqrt{36} = 6.$

$\quad = 4\sqrt{3} - \dfrac{2\sqrt{3}}{3}$ Simplify.

$\quad = \left(4 - \dfrac{2}{3}\right)\sqrt{3}$ Use the distributive property.

$\quad = \dfrac{10\sqrt{3}}{3}$ $4 - \dfrac{2}{3} = \dfrac{12}{3} - \dfrac{2}{3} = \dfrac{10}{3}$ ∎

Discover for yourself

In Examples 4–7, use a calculator to find a decimal approximation for the given number. Then find a decimal approximation for the simplified form of the number. The decimal approximations should be the same, which is a way of checking the simplification.

5 Evaluate expressions with rational exponents.

Rational Numbers as Exponents

We are now ready to give meaning to expressions that contain rational numbers as exponents, such as $27^{1/3}$. You can discover the significance of rational exponents by first working the Discover for yourself with your graphing utility.

Discover for yourself

Use your graphing utility to graph each of the following pairs of equations in the same $[-100, 100]$ by $[-10, 10]$ viewing rectangle.

$$y = x^{1/2} \text{ and } y = \sqrt{x}; \qquad y = x^{1/3} \text{ and } y = \sqrt[3]{x}; \qquad y = x^{1/4} \text{ and } y = \sqrt[4]{x}$$

Based on the graphs of the equations in each pair, generalize and give a definition for $b^{1/n}$.

In the Discover for yourself, did you obtain identical graphs for $y = b^{1/n}$ and $y = \sqrt[n]{b}$? Thus $b^{1/n} = \sqrt[n]{b}$, and rational exponents indicate roots.

Definition of rational exponents

If n is a positive integer and $\sqrt[n]{b}$ is a real number, then $b^{1/n}$ is defined to be the principal nth root of b.

$$b^{1/n} = \sqrt[n]{b}$$

If m is a positive integer that has no common factors with n, then

$$b^{m/n} = (b^m)^{1/n} = \sqrt[n]{b^m} \quad \text{and} \quad b^{m/n} = (b^{1/n})^m = (\sqrt[n]{b})^m.$$

We refer to $b^{m/n}$ as the *exponential form* and $\sqrt[n]{b^m}$ or $(\sqrt[n]{b})^m$ as the *radical form*. Observe that the denominator n in $b^{m/n}$ denotes the index of the root and m represents a power.

$$b^{\overset{\text{Power}}{m}/\underset{\text{Index}}{n}} = \sqrt[n]{b^m} = (\sqrt[n]{b})^m$$

Study tip

The expression

$$b^{m/n}$$

is not defined if $\sqrt[n]{b}$ is not a real number. This can lead to some strange results. For example,

$$(-125)^{2/6}$$

is not defined because $\sqrt[6]{-125}$ is not a real number. However,

$$(-125)^{1/3}$$

is defined because $\sqrt[3]{-125} = -5$.

EXAMPLE 8   Evaluating Expressions Containing Rational Exponents

Evaluate:

a. $64^{2/3}$ **b.** $16^{5/2}$ **c.** $(-125)^{4/3}$ **d.** $-125^{4/3}$

Solution

a. $64^{2/3} = (\sqrt[3]{64})^2 = 4^2 = 16$

b. $16^{5/2} = (\sqrt{16})^5 = 4^5 = 1024$

c. $(-125)^{4/3} = (\sqrt[3]{-125})^4 = (-5)^4 = 625$

d. $-125^{4/3} = -(\sqrt[3]{125})^4 = -(5)^4 = -625$

■

Using technology

Here are the graphing utility keystroke sequences for Example 8.

a. $64^{2/3}$: 64 $\boxed{\wedge}$ $\boxed{(\!(}$ 2 $\boxed{\div}$ 3 $\boxed{)}$ $\boxed{\text{ENTER}}$

b. $16^{5/2}$: 16 $\boxed{\wedge}$ $\boxed{(\!(}$ 5 $\boxed{\div}$ 2 $\boxed{)}$ $\boxed{\text{ENTER}}$

c. $(-125)^{4/3}$: Some graphing utilities do not evaluate $b^{m/n}$ when b is negative. Enter part (d) as $[(-125)^{1/3}]^4$:

$\boxed{(\!(}$ $\boxed{(\!(}$ $\boxed{(-)}$ 125 $\boxed{)}$ $\boxed{\wedge}$ $\boxed{(\!(}$ 1 $\boxed{\div}$ 3 $\boxed{)}$ $\boxed{)}$ $\boxed{\wedge}$ 4 $\boxed{\text{ENTER}}$

d. $-125^{4/3}$: $\boxed{(-)}$ 125 $\boxed{\wedge}$ $\boxed{(\!(}$ 4 $\boxed{\div}$ 3 $\boxed{)}$ $\boxed{\text{ENTER}}$

6 Simplify expressions with rational exponents.

The properties for integer exponents also apply to rational exponents. For example, by extending the property

$$b^{-n} = \frac{1}{b^n}$$

to rational exponents, we can evaluate an expression such as $81^{-3/4}$ as follows:

$$81^{-3/4} = \frac{1}{81^{3/4}} = \frac{1}{(\sqrt[4]{81})^3} = \frac{1}{3^3} = \frac{1}{27}$$

Example 9 illustrates how we can use properties of exponents to simplify expressions containing rational exponents. Once again, only positive exponents will appear in the final result.

EXAMPLE 9 **Simplifying Expressions Involving Rational Exponents**

Simplify:

a. $(4x - 3)^{4/3}(4x - 3)^{-1/3}$

b. $\dfrac{x^2 + 1}{(x^2 + 1)^{-1/2}}$

c. $\dfrac{32x^{1/2}}{16x^{3/4}}$

d. $\left(\dfrac{7x^{3/5}}{4y^{1/3}}\right)^3$

Study tip

Remember that division by zero is undefined. This means that

$$(4x - 3)^{-1/3} \quad \text{or} \quad \frac{1}{(4x - 3)^{1/3}}$$

must exclude the value of x for which

$$(4x - 3)^{1/3} = 0$$

Thus, x cannot equal $\frac{3}{4}$.

Solution

a. $(4x - 3)^{4/3}(4x - 3)^{-1/3}$

$= (4x - 3)^{4/3 + (-1/3)}$ $b^n \cdot b^m = b^{n+m}$; Add exponents.

$= (4x - 3)^1$

$= 4x - 3$

b. $\dfrac{x^2 + 1}{(x^2 + 1)^{-1/2}}$

$= (x^2 + 1)^{1-(-1/2)}$ $\dfrac{b^n}{b^m} = b^{n-m}$; Subtract exponents.

$= (x^2 + 1)^{3/2}$

c. $\dfrac{32x^{1/2}}{16x^{3/4}}$

$= \dfrac{32}{16}x^{1/2-3/4}$ $\dfrac{b^n}{b^m} = b^{n-m}$; Subtract exponents.

$= 2x^{-1/4}$

$= \dfrac{2}{x^{1/4}}$ $b^{-n} = \dfrac{1}{b^n}$

d. $\left(\dfrac{7x^{3/5}}{4y^{1/3}}\right)^3$

$= \dfrac{(7x^{3/5})^3}{(4y^{1/3})^3}$ $\left(\dfrac{a}{b}\right)^n = \dfrac{a^n}{b^n}$; Cube the numerator and the denominator.

$= \dfrac{7^3(x^{3/5})^3}{4^3(y^{1/3})^3}$ $(ab)^n = a^nb^n$; Cube each factor in the numerator and the denominator.

$= \dfrac{7^3x^{3/5\cdot3}}{4^3y^{1/3\cdot3}}$ $(b^n)^m = b^{nm}$; Multiply exponents.

$= \dfrac{343x^{9/5}}{64y}$ ■

Study tip

As with integer exponents, an expression with rational exponents is in simplified form if:
1. No parentheses appear.
2. Each base appears once.
3. Only positive exponents appear.

7 Reduce the index of a radical.

The final condition for simplified radical form is that the index of the radical is reduced. Rational exponents can be used to reduce the index, as shown in Example 10.

EXAMPLE 10   **Reducing the Index of a Radical**

Simplify:

a. $\sqrt[9]{b^3}$ **b.** $\sqrt[6]{16x^2y^4}$ **c.** $7\sqrt[3]{25} + 2\sqrt[9]{5^6}$ **d.** $\sqrt[3]{\sqrt{8}}$

Solution

a. $\sqrt[9]{b^3} = b^{3/9}$ Rewrite in exponential form; $\sqrt[n]{b^m} = b^{m/n}$.

$= b^{1/3}$ Reduce $\frac{3}{9}$.

$= \sqrt[3]{b}$ Rewrite in radical form; $b^{1/n} = \sqrt[n]{b}$.

b. $\sqrt[6]{16x^2y^4} = \sqrt[6]{2^4x^2y^4}$

$= (2^4x^2y^4)^{1/6}$ Rewrite in exponential form.

$= (2^4)^{1/6}(x^2)^{1/6}(y^4)^{1/6}$ Raise each factor to the $\frac{1}{6}$th power; $(abc)^n = a^nb^nc^n$.

$= 2^{2/3}x^{1/3}y^{2/3}$ $(b^n)^m = b^{nm}$; Multiply exponents and reduce.

$= (2^2xy^2)^{1/3}$ $a^nb^nc^n = (abc)^n$

$= \sqrt[3]{4xy^2}$ Rewrite in radical form.

c. $7\sqrt[3]{25} + 2\sqrt[9]{5^6}$

$= 7\sqrt[3]{25} + 2\cdot5^{6/9}$ Reduce the index of $\sqrt[9]{5^6}$ by first rewriting in exponential form; $\sqrt[n]{b^m} = b^{m/n}$.

$= 7\sqrt[3]{25} + 2\cdot5^{2/3}$ Reduce $\frac{6}{9}$.

$= 7\sqrt[3]{25} + 2\sqrt[3]{25}$ $5^{2/3} = \sqrt[3]{5^2} = \sqrt[3]{25}$

$= (7+2)\sqrt[3]{25}$ Apply the distributive property.

$= 9\sqrt[3]{25}$

Study tip

Reducing an index is possible when every factor in the radicand is written with an exponent, and the index and these exponents share a common factor.

Discover for yourself

In Example 10b, describe a method for going directly from

to

$\sqrt[3]{2^2xy^2}$

mentally.

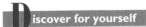

iscover for yourself

Work Example 10d by using exponential form:

$$\sqrt[3]{\sqrt{8}} = ((2^3)^{1/2})^{1/3}$$

d. $\sqrt[3]{\sqrt{8}} = \sqrt[3]{\sqrt[2]{8}}$

$\qquad = \sqrt[6]{8}$

$\qquad = \sqrt[6]{2^3}$

$\qquad = 2^{3/6}$

$\qquad = 2^{1/2}$

$\qquad = \sqrt{2}$

$\sqrt[m]{\sqrt[n]{b}} = \sqrt[mn]{b}$

Notice that the index (6) and the exponent (3) have a common factor.

Write in exponential form; $\sqrt[n]{b^m} = b^{m/n}$

Reduce the fraction.

Rewrite in radical form.   ∎

PROBLEM SET P.4

Practice Problems

In Problems 1–18, evaluate each expression or state that the expression is not a real number. Do not use a calculator.

1. $\sqrt{49}$

2. $\sqrt{100}$

3. $\sqrt[3]{27}$

4. $\sqrt[3]{125}$

5. $\sqrt[3]{-8}$

6. $\sqrt[3]{-64}$

7. $\sqrt{-4}$

8. $\sqrt{-64}$

9. $\sqrt[3]{\dfrac{8}{125}}$

10. $\sqrt[3]{\dfrac{8}{27}}$

11. $\sqrt[5]{-1}$

12. $\sqrt[7]{-1}$

13. $-\sqrt{\dfrac{4}{9}}$

14. $\sqrt{\dfrac{9}{25}}$

15. $\sqrt[5]{-\dfrac{1}{32}}$

16. $(\sqrt[5]{-3})^5$

17. $(\sqrt[3]{-8})^3$

18. $\sqrt[6]{\dfrac{1}{64}}$

In Problems 19–34, write each radical in simplified form by factoring the radicand and removing as many factors as possible.

19. $\sqrt[3]{16}$

20. $\sqrt[3]{250}$

21. $\sqrt[5]{64}$

22. $\sqrt[4]{32}$

23. $\sqrt{200y^3}$

24. $\sqrt{40x^3}$

25. $\sqrt{20xy^3}$

26. $\sqrt{45x^4y}$

27. $\sqrt{75xy^2z^5}$

28. $\sqrt{18x^2y^3z}$

29. $\sqrt[3]{32x^3}$

30. $\sqrt[3]{16xy^3}$

31. $\sqrt[3]{-32xy^5z^6}$

32. $\sqrt[3]{-16x^4y^5z^2}$

33. $\sqrt[4]{48y^7}$

34. $\sqrt[4]{32x^{11}}$

In Problems 35–52, simplify by rationalizing the denominator.

35. $\dfrac{1}{\sqrt{7}}$

36. $\dfrac{2}{\sqrt{10}}$

37. $\sqrt{\dfrac{3}{5}}$

38. $\sqrt{\dfrac{3}{7}}$

39. $\sqrt[3]{\dfrac{7}{2}}$

40. $\sqrt[5]{\dfrac{7}{16}}$

41. $\dfrac{6}{\sqrt[3]{2}}$

42. $-\dfrac{8}{\sqrt[3]{2}}$

43. $-\dfrac{6}{\sqrt[5]{8}}$

44. $-\dfrac{16}{\sqrt[5]{4}}$

45. $\dfrac{20}{5 - \sqrt{3}}$

46. $\dfrac{17}{6 - \sqrt{2}}$

47. $\dfrac{13}{\sqrt{11} + 3}$

48. $\dfrac{3}{\sqrt{7} + 3}$

49. $\dfrac{6}{\sqrt{5} + \sqrt{3}}$

50. $\dfrac{12}{\sqrt{7} + \sqrt{3}}$

51. $\dfrac{11}{\sqrt{7} - \sqrt{3}}$

52. $\dfrac{13}{\sqrt{5} - \sqrt{3}}$

In Problems 53–62, simplify where possible and then combine radicals.

53. $\sqrt{50} + \sqrt{18}$

54. $\sqrt{28} + \sqrt{63}$

55. $3\sqrt{18} - 5\sqrt{50}$

56. $4\sqrt{12} - 2\sqrt{75}$

57. $3\sqrt{8} - \sqrt{32} + 3\sqrt{72} - \sqrt{75}$

58. $3\sqrt{54} - 2\sqrt{24} - \sqrt{96} + 4\sqrt{63}$

59. $8\sqrt{\dfrac{1}{2}} - \dfrac{1}{2}\sqrt{8}$

60. $\sqrt{\dfrac{5}{9}} + \sqrt{\dfrac{9}{5}}$

61. $16\sqrt{\dfrac{5}{8}} + 6\sqrt{\dfrac{5}{2}}$

62. $4\sqrt{\dfrac{1}{12}} + 18\sqrt{\dfrac{1}{27}}$

In Problems 63–72, evaluate each expression without using a calculator.

63. $36^{1/2}$

64. $121^{1/2}$

65. $8^{1/3}$

66. $(-27)^{1/3}$

67. $125^{2/3}$

68. $8^{2/3}$

69. $(-27)^{4/3}$

70. $\left(\frac{27}{125}\right)^{4/3}$

71. $32^{-4/5}$

72. $16^{-5/2}$

In Problems 73–92, simplify each expression.

73. $(7y^{1/3})(2y^{1/4})$

74. $(3x^{2/3})(4x^{3/4})$

75. $(3x^{3/4})(-5x^{-1/2})$

76. $(-6x^{2/3})(-8x^{-1/4})$

77. $\dfrac{20x^{1/2}}{5x^{1/4}}$

78. $\dfrac{72y^{3/4}}{9y^{1/3}}$

79. $\dfrac{80y^{1/6}}{10y^{1/4}}$

80. $\dfrac{-60y^{1/3}}{20y^{3/4}}$

81. $(2x^{1/5}y^2z^{2/5})^5$

82. $(3x^{1/4}y^3z^{3/4})^4$

83. $(25x^4y^6)^{1/2}$

84. $(125x^9y^6)^{1/3}$

85. $(16xy^{1/4}z^{2/3})^{1/4}$

86. $(27x^{1/4}y^{2/3}z^3)^{1/3}$

87. $\left(\dfrac{2x^{1/4}}{5y^{1/3}}\right)^3$

88. $\left(\dfrac{7x^{2/3}}{3y^{1/2}}\right)^2$

89. $\left(\dfrac{x^3}{y^5}\right)^{-1/2}$

90. $\left(\dfrac{a^5}{b^{-4}}\right)^{-1/5}$

91. $\left(\dfrac{27a^{-3}}{64b^{-3}}\right)^{-1/3}$

92. $\left(\dfrac{81x^{-8}y^6}{25x^4y^{-10}}\right)^{-1/2}$

In Problems 93–106, simplify by reducing the index of the radical.

93. $\sqrt[4]{b^2}$

94. $\sqrt[6]{b^4}$

95. $\sqrt[9]{(x-1)^6}$

96. $\sqrt[8]{(x-1)^2}$

97. $\sqrt[4]{x^2y^2}$

98. $\sqrt[10]{x^5y^5}$

99. $\sqrt[9]{2^3x^3y^6}$

100. $\sqrt[12]{2^4x^4y^8}$

101. $\sqrt[9]{27x^3y^6}$

102. $\sqrt[4]{81x^2y^6}$

103. $5\sqrt[9]{16} + \sqrt[9]{8}$

104. $2\sqrt{45} + \sqrt[4]{25}$

105. $\sqrt[3]{\sqrt{125}}$

106. $\sqrt{\sqrt{32}}$

Application Problems

107. The formula for the frequency (f) of a note is $f = (2L)^{-1}P^{1/2}m^{-1/2}$, where L is the length of the string, P is the stretching force on the string, and m is the mass of one centimeter of the string. Write the formula in simplified radical form.

108. On a piano, the frequency of the A note above middle C is 440 vibrations per second. The frequency f_n of a note n notes above A should be $f_n = 440 \cdot 2^{n/12}$.

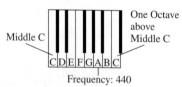

Middle C

One Octave above Middle C

CDEFGABC

Frequency: 440

a. At what frequency should a piano tuner set the A that is one octave (12 notes) above the A above middle C?

b. What should be the frequency of middle C, nine notes below A?

109. Archaeologists use the amount of carbon-14 left in a fossil to estimate the fossil's age. The approximate number of milligrams (A) of carbon-14 left in a fossil after 5000 years can be found using the formula $A = A_0(2.7)^{-3/5}$, where A_0 is the initial amount of carbon-14 in the organism. Find the amount of carbon-14 left in a 5000-year-old organism that originally contained 500 milligrams of carbon-14.

110. Meteorologists use the formula $D = 0.07d^{3/2}$ to determine the duration (D, in hours) of a storm whose diameter is d miles. Determine the duration of a storm that is 9 miles in diameter.

111. The formula

$$r = \left(\frac{A}{P}\right)^{1/t} - 1$$

describes the annual rate of return (r) of an investment of P dollars that is worth A dollars after t years. What is the annual rate of return, to the nearest tenth of a per-

cent, on a condominium that is purchased for $60,000 and sold 5 years later for $80,000?

True–False Critical Thinking Problems

112. Which one of the following is true?

  a. Neither $(-8)^{1/2}$ nor $(-8)^{1/3}$ represent real numbers.

  b. $\sqrt{x^2 + y^2} = x + y$

  c. $8^{-1/3} = -2$

  d. $16^{1/4} = 4^{1/2}$

113. Which one of the following is true?

  a. $\sqrt[3]{x^3 + y^3} = x + y$

  b. $\sqrt[4]{16} = \pm 2$

  c. $\left(\dfrac{9}{16}\right)^{-5/2} - \left(\dfrac{1000}{27}\right)^{4/3} = -\dfrac{28,976}{243}$

  d. $b^{-1/n} = \dfrac{1}{b^n}$

Technology Problems

To graph $y = x^{m/n}$ on a graphing utility, enter the equation as $y = (x^m)^{1/n}$ or $y = (x^{1/n})^m$. Many graphing utilities do not store the definition of $x^{m/n}$, so in effect you are telling the utility the definition. Use this idea to graph each equation in Problems 114–116.

114. Graph $y = x^{2/3}$ on a graphing utility by graphing both $y_1 = (x^2)^{1/3}$ and $y_2 = (x^{1/3})^2$.

115. Graph $y = x^{3/4}$ on a graphing utility by graphing both $y_1 = (x^3)^{1/4}$ and $y_2 = (x^{1/4})^3$.

116. Graph $y_1 = x^{2/3}$ and $y_2 = (x^{1/3})^2$ in the same viewing rectangle. Are the graphs that are displayed the same? If not, what difference do you observe?

117. Rationalize the denominator: $\dfrac{2}{\sqrt{x} + 1}$. Then use a graphing utility to graph $y_1 = \dfrac{2}{\sqrt{x} + 1}$ and $y_2 = \dfrac{2(\sqrt{x} - 1)}{x - 1}$, showing that the two graphs are identical.

Critical Thinking Problems

118. Find the exact value of $\sqrt{13 + \sqrt{2} + \dfrac{7}{3 + \sqrt{2}}}$ without the use of a calculator.

119. Press the square root key repeatedly on your calculator, and then enter the number 2. Describe the two emerging patterns. (One pattern is far less obvious than the other.)

120. Place the correct symbol, $>$ or $<$, in the box between each of the given numbers. *Do not use a calculator.* Then check your result with a calculator.

  a. $5^{1/2} \ \Box \ 5^{-2}$

  b. $3^{1/2} \ \Box \ 3^{1/3}$

  c. $\left(\dfrac{1}{2}\right)^{1/3} \ \Box \ \left(\dfrac{1}{2}\right)^{1/4}$

  d. $\sqrt{7} + \sqrt{18} \ \Box \ \sqrt{7 + 18}$

121. Rationalize the denominator: $\dfrac{1}{1 + \sqrt{3} + \sqrt{5}}$.

122. If $x + y = 2z$, simplify:

$$\left[\dfrac{(3^{x-y})^y(3^{y-z})^{z-x}}{(3^{z+y})^{z-y}}\right]^{1/z}$$

| SECTION P.5 | Polynomials |

Solutions Manual   **Tutorial**   **Video**

Objectives

1 Write a polynomial in standard form.

2 Determine the degree of a polynomial.

3 Add and subtract polynomials.

4 Multiply polynomials.

5 Use FOIL in polynomial multiplication.

6 Use special products in polynomial multiplication.

In this section, we review basic ideas about polynomials and their operations.

The Vocabulary of Polynomials

A *polynomial* is an algebraic expression consisting of a finite number of terms with whole number exponents on the variables. Some examples of polynomials are

$$3x + 7, \ 5x^4 - 9x^2 + 3x + 14, \text{ and } 3x^2y - 4xy - 17.$$

The first two examples are called *polynomials in x* and the third is a *polynomial in x and y*. A polynomial containing exactly one term is called a *monomial*, one with exactly two terms is a *binomial*, and one with exactly three terms is a *trinomial*. Thus, $7x^2$ is a monomial, $7x^2 + 11x$ a binomial, and $7x^2 + 11x + 13$ a trinomial.

The formal definition of a polynomial in one variable illustrates its general form.

> ### Definition of a polynomial in x
>
> A *polynomial in x* is an algebraic expression of the form
> $$a_n x^n + a_{n-1} x^{n-1} + a_{n-2} x^{n-2} + \cdots + a_1 x + a_0,$$
> where $a_n, a_{n-1}, a_{n-2}, \ldots, a_1$ and a_0 are real numbers, $a_n \neq 0$, and n is a nonnegative integer. The polynomial is of *degree n*, a_n (read "a sub n") is the *leading coefficient*, and a_0 is the *constant term*.

A polynomial in one variable is written in *descending powers* of the variable when the exponents on the terms decrease from left to right. Written with descending powers of x, the polynomial is said to be in *standard form*. Three polynomials in standard form and their degrees are shown below.

Polynomial	Standard Form	Degree
$5x - 3 + 7x^2$	$7x^2 + 5x - 3$	2
$7x^2 - 6x^5 - 4 + 2x^3$	$-6x^5 + 2x^3 + 7x^2 - 4$	5
17	$17 \ (17 = 17x^0)$	0

If a polynomial contains more than one variable, the degree of a term is the sum of the exponents on all the variables. For example, the degree of $5x^2y^7$ is $2 + 7$ or 9. The degree of the polynomial is the greatest of the degrees of any of its terms.

EXAMPLE 1 The Degree of a Polynomial in Two Variables

Find the degree of the polynomial:
$$7x^6y^2 - 5x^4y^5 - 7xy + 3$$

Solution

We begin by finding the degree of each of the polynomial's four terms.

Term	$7x^6y^2$	$-5x^4y^5$	$-7xy$	3 or $3x^0y^0$
Degree	$6 + 2 = 8$	$4 + 5 = 9$	$1 + 1 = 2$	$0 + 0 = 0$

S tudy tip

Algebraic expressions whose variables do not contain whole number exponents, such as

$3x^{-2} + 7$

and

$5x^{3/2} + 9x^{1/2} + 2$

are not polynomials.

S tudy tip

A monomial is either a constant or the product of a constant and variables to whole number exponents.

1 Write a polynomial in standard form.

2 Determine the degree of a polynomial.

S tudy tip

Since

$0 = 0x = 0x^2 = 0x^3 = 0x^4,$

and so on,

a polynomial consisting only of the term zero has no degree.

ENRICHMENT ESSAY

Names, Words, and Meanings

In language, words such as *polynomial* and *degree* (*of a polynomial*) have general agreed-upon definitions. In our work, we are sharing these definitions so that communication within the polynomial realm is possible. However, other words—proper names such as Bob, Alice, Samantha—do not have a meaning other than denoting a particular person.

With the flair of a mathematician defining words such as *polynomial, terms,* and *degree,* Lewis Carroll (*Alice's Adventures in Wonderland* and *Through the Looking Glass*) has his character, Humpty Dumpty, inverting and reversing ordinary linguistic conventions.

> *"Don't stand chattering to yourself like that,"* Humpty Dumpty said, looking at her for the first time, *"but tell me your name and your business."*
>
> *"My* name *is Alice, but—"*
>
> *"It's a stupid name enough!"* Humpty Dumpty interrupted impatiently. *"What does it mean?"*
>
> *"Must a name mean something?"* Alice asked doubtfully.
>
> *"Of course it must,"* Humpty Dumpty said with a short laugh; *"my name means the shape I am—and a good handsome shape it is, too. With a name like yours, you might be any shape, almost."*
>
> *Alice was too much puzzled to say anything; so after a minute Humpty Dumpty began again. "They've a temper, some of them—particularly verbs; they're the proudest—adjectives you can do anything with, but not verbs—however, I can manage the whole lot of them! Impenetrability! That's what I say!"*
>
> *"Would you tell me please,"* said Alice, *"what that means?"*
>
> *"Now you talk like a reasonable child,"* said Humpty Dumpty, looking very much pleased. *"I meant by 'impenetrability' that we've had enough of that subject, and it would be just as well if you'd mention what you mean to do next, as I suppose you don't mean to stop here all the rest of your life."*
>
> *"That's a great deal to make one word mean,"* Alice said in a thoughtful tone.
>
> *"When I make a word do a lot of work like that,"* said Humpty Dumpty, *"I always pay it extra."*
>
> *"Oh!"* said Alice. She was too much puzzled to make any other remark.

Robert Arneson "Klown" 1978, glazed ceramic, $37 \times 19 \times 19$ in. Collection of Des Moines Art Center, Des Moines, Iowa. Purchase with funds from the Gardner and Florence Call Cowles Foundation, Des Moines Art Center permanent collection. Photo courtesy of George Adams Gallery, New York. © Estate of Robert Arneson / Licensed by VAGA, New York, NY.

The degree of this polynomial in x and y is 9, the greatest of the degrees of any of its terms.

3 Add and subtract polynomials.

Operations with Polynomials

If we want to add or subtract polynomials, we apply the commutative and associative properties to group terms having the same variables to the same powers, called *like terms* or *similar terms*. We then use the distributive property to combine the similar terms. For example, $-7x^4y^5$ and $2x^4y^5$ are similar terms and their sum is

$$-7x^4y^5 + 2x^4y^5 = (-7 + 2)x^4y^5 = -5x^4y^5.$$

EXAMPLE 2 **Adding and Subtracting Polynomials**

Perform the indicated operations and simplify:

a. $(-9x^3 + 7x^2 - 5x + 3) + (13x^3 + 2x^2 - 8x - 6)$
b. $(7x^3 - 8x^2 + 9x - 6) - (2x^3 - 6x^2 - 3x + 9)$

Solution

a. $(-9x^3 + 7x^2 - 5x + 3) + (13x^3 + 2x^2 - 8x - 6)$

$\quad = (-9x^3 + 13x^3) + (7x^2 + 2x^2)$ Group similar terms.

$\qquad + (-5x - 8x) + (3 - 6)$

$\quad = 4x^3 + 9x^2 + (-13x) + (-3)$ Combine similar terms.

$\quad = 4x^3 + 9x^2 - 13x - 3$ Do you find that you're working all these steps mentally?

> ### Study tip
>
> You can also arrange similar terms in columns and combine vertically:
>
> $$\begin{array}{r} 7x^3 - 8x^2 + 9x - 6 \\ -2x^3 + 6x^2 + 3x - 9 \\ \hline 5x^3 - 2x^2 + 12x - 15 \end{array}$$
>
> The similar terms can be combined mentally by adding their coefficients.

b. $(7x^3 - 8x^2 + 9x - 6) - (2x^3 - 6x^2 - 3x + 9)$

$\quad = (7x^3 - 8x^2 + 9x - 6) + (-2x^3 + 6x^2 + 3x - 9)$ Rewrite subtraction as addition of the additive inverse. Be sure to change the sign of each term inside parentheses preceded by the negative sign.

$\quad = (7x^3 - 2x^3) + (-8x^2 + 6x^2)$ Group similar terms.

$\qquad + (9x + 3x) + (-6 - 9)$

$\quad = 5x^3 + (-2x^2) + 12x + (-15)$ Combine similar terms.

$\quad = 5x^3 - 2x^2 + 12x - 15$ ∎

4 Multiply polynomials.

The product of two polynomials is the polynomial obtained by multiplying each term of one polynomial by each term of the other polynomial and then combining similar terms.

EXAMPLE 3 **Multiplying a Binomial and a Trinomial**

Multiply: $(2x + 3)(x^2 + 4x + 5)$

Solution

$\quad (2x + 3)(x^2 + 4x + 5)$

$\quad = 2x(x^2 + 4x + 5) + 3(x^2 + 4x + 5)$ Use the distributive property to multiply the trinomial by each term of the binomial.

$\quad = 2x^3 + 8x^2 + 10x + 3x^2 + 12x + 15$ Use the distributive property.

$\quad = 2x^3 + 11x^2 + 22x + 15$ Combine similar terms. ∎

Another method for solving Example 3 is to use a vertical format similar to that used for multiplying whole numbers:

$$
\begin{array}{r}
x^2 + 4x + 5 \\
2x + 3 \\
\hline
\end{array}
$$

Similar terms are written in the same column.

$$
\begin{array}{r}
3x^2 + 12x + 15 \quad \longleftarrow \ 3(x^2 + 4x + 5) \\
2x^3 + 8x^2 + 10x \quad \longleftarrow \ 2x(x^2 + 4x + 5) \\
\hline
2x^3 + 11x^2 + 22x + 15 \quad \text{Combine similar terms.}
\end{array}
$$

The vertical pattern used in the Study tip becomes convenient if the polynomials we wish to multiply each contain three terms or more.

EXAMPLE 4   **Multiplying Polynomials Using a Vertical Format**

Multiply: $(x^2 - 5x + 3)(3x - 4 + 2x^2)$

Solution

$$
\begin{array}{r}
x^2 - 5x + 3 \\
2x^2 + 3x - 4 \qquad \text{Write the polynomial in standard form.} \\
\hline
-4x^2 + 20x - 12 \quad \longleftarrow \ -4(x^2 - 5x + 3) \\
3x^3 - 15x^2 + 9x \quad \longleftarrow \ 3x(x^2 - 5x + 3) \\
2x^4 - 10x^3 + 6x^2 \qquad \longleftarrow \ 2x^2(x^2 - 5x + 3) \\
\hline
2x^4 - 7x^3 - 13x^2 + 29x - 12 \quad \text{Add similar terms.} \qquad \blacksquare
\end{array}
$$

5 Use FOIL in polynomial multiplication.

The Product of Two Binomials: FOIL

The product of two binomials occurs quite frequently in algebra. The product can be found using a method called FOIL, which is based on the distributive property. For example, we can find the product of the binomials $3x + 2$ and $4x + 5$ as follows:

$$
\begin{aligned}
(3x + 2)(4x + 5) &= 3x(4x + 5) + 2(4x + 5) \\
&= 3x(4x) + 3x(5) + 2(4x) + 2(5) \\
&= 12x^2 + 15x + 8x + 10
\end{aligned}
$$

Before combining like terms, let us consider the origin of each of the four terms in the sum.

Origin of	Terms of $(3x + 2)(4x + 5)$	Result of Multiplying Terms
$12x^2$	$(3x + 2)(4x + 5)$	$(3x)(4x) = 12x^2$ First terms
$15x$	$(3x + 2)(4x + 5)$	$(3x)(5) = 15x$ Outer terms
$8x$	$(3x + 2)(4x + 5)$	$(2)(4x) = 8x$ Inner terms
10	$(3x + 2)(4x + 5)$	$(2)(5) = 10$ Last terms

The product is obtained by combining these four results.

$$(3x + 2)(4x + 5) = 12x^2 + 15x + 8x + 10$$

$$= 12x^2 + 23x + 10$$

We see, then, that two binomials can be quickly multiplied by using the FOIL method, in which F represents the product of the *first* terms in each binomial, O represents the product of the *outside* or *outer* terms, I represents the product of the two *inside* or *inner* terms, and L represents the product of the *last* or second terms in each binomial.

$$
\underset{O}{\underbrace{\overset{F \quad\quad L}{(3x + 2)}\underset{I}{(4x + 5)}}} = 12x^2 + 15x + 8x + 10
$$

$$= 12x^2 + 23x + 10$$

EXAMPLE 5   **Using the FOIL Method**

Multiply using the FOIL method: $(5x^2 - 6)(4x^2 + 2)$

Study tip

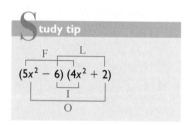

$$
\underset{O}{\underbrace{\overset{F \quad L}{(5x^2 - 6)}\underset{I}{(4x^2 + 2)}}}
$$

Solution

$$(5x^2 - 6)(4x^2 + 2) = \overset{F}{(5x^2)(4x^2)} + \overset{O}{(5x^2)(2)} + \overset{I}{(-6)(4x^2)} + \overset{L}{(-6)(2)}$$

$$= 20x^4 + 10x^2 - 24x^2 - 12$$

$$= 20x^4 - 14x^2 - 12 \qquad \blacksquare$$

EXAMPLE 6   **Using the FOIL Method**

Multiply using the FOIL method: $(7x^3 - 2x)(3x^3 - 5x)$

Solution

$$(7x^3 - 2x)(3x^3 - 5x) = \overset{F}{(7x^3)(3x^3)} + \overset{O}{(7x^3)(-5x)} + \overset{I}{(-2x)(3x^3)} + \overset{L}{(-2x)(-5x)}$$

$$= 21x^6 - 35x^4 - 6x^4 + 10x^2$$

$$= 21x^6 - 41x^4 + 10x^2 \qquad \blacksquare$$

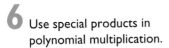 Use special products in polynomial multiplication.

Although it's convenient to memorize these forms, the FOIL method can be used on all five examples in the box. To cube $x + 4$, you can first square $x + 4$ using FOIL and then multiply this result by $x + 4$. In short, you do not necessarily have to utilize these special formulas. What is the advantage of knowing and using these forms?

Special Products

There are several products that occur so frequently that it's convenient to memorize the form or pattern of these formulas.

Special products

Let A and B represent real numbers, variables, or algebraic expressions.

Special Product	**Example**
Sum and Difference of Two Terms	
$(A + B)(A - B) = A^2 - B^2$	$(2x + 3)(2x - 3) = (2x)^2 - 3^2$
	$= 4x^2 - 9$
Squaring a Binomial	
$(A + B)^2 = A^2 + 2AB + B^2$	$(y + 5)^2 = y^2 + 2 \cdot y \cdot 5 + 5^2$
	$= y^2 + 10y + 25$
$(A - B)^2 = A^2 - 2AB + B^2$	$(3x - 4y)^2 = (3x)^2 - 2 \cdot 3x \cdot 4y + (4y)^2$
	$= 9x^2 - 24xy + 16y^2$
Cubing a Binomial	
$(A + B)^3 = A^3 + 3A^2B + 3AB^2 + B^3$	$(x + 4)^3 = x^3 + 3x^2(4) + 3x(4)^2 + 4^3$
	$= x^3 + 12x^2 + 48x + 64$
$(A - B)^3 = A^3 - 3A^2B + 3AB^2 - B^3$	$(x - 2)^3 = x^3 - 3x^2(2) + 3x(2)^2 - 2^3$
	$= x^3 - 6x^2 + 12x - 8$

The special form for

$(A + B)^2$

can be interpreted geometrically.

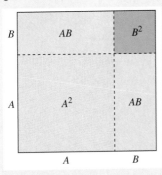

The area of the large rectangle is

$(B + A)(A + B) = (A + B)^2$.

The sum of the areas of the four smaller rectangles is

$AB + B^2 + A^2 + AB$

$= A^2 + 2AB + B^2$.

The area of the large rectangle equals the sum of the areas of the four smaller rectangles:

$(A + B)^2 = A^2 + 2AB + B^2$

EXAMPLE 7 **Multiplying the Sum and Difference of Two Terms**

Find the product of $(6x^2y + 5y)$ and $(6x^2y - 5y)$.

Solution

$(6x^2y + 5y)(6x^2y - 5y)$

$= (6x^2y)^2 - (5y)^2$ The product of the sum and difference of the same two terms is the difference of their squares.

$= 36x^4y^2 - 25y^2$ Square all factors.

EXAMPLE 8 **Squaring and Cubing Binomials**

Find the product:

a. $(4x^2y + xy)^2$ **b.** $(3x - 4y)^3$

Solution

$(A \ + \ B)^2 \ = \ A^2 \ + \ 2 \ A \ \cdot \ B \ + \ B^2$

a. $(4x^2y + xy)^2 = (4x^2y)^2 + 2(4x^2y)(xy) + (xy)^2$

$= 16x^4y^2 + 8x^3y^2 + x^2y^2$

$$(A - B)^3 = A^3 - 3 A^2 \cdot B + 3 A \cdot B^2 - B^3$$

b. $(3x - 4y)^3 = (3x)^3 - 3(3x)^2(4y) + 3(3x)(4y)^2 - (4y)^3$
$$= 27x^3 - 3(9x^2)(4y) + 3(3x)(16y^2) - (64y^3)$$
$$= 27x^3 - 108x^2y + 144xy^2 - 64y^3 \qquad \blacksquare$$

Special products can sometimes be used to find the products of certain trinomials, as illustrated in Example 9.

EXAMPLE 9   Using the Special Products

Find the products:

a. $(7x + 5 + 4y)(7x + 5 - 4y)$ **b.** $(3x + y + 1)^2$

Discover for yourself

Describe how to work Examples 7–9 without using the special products.

Solution

a. By grouping the first two terms within each of the parentheses, we can find the product using the form for the sum and difference of two terms:

$$(A + B) \cdot (A - B) = A^2 - B^2$$

$$[(7x + 5) + 4y] \cdot [(7x + 5) - 4y] = (7x + 5)^2 - (4y)^2$$

$$= (7x)^2 + 2 \cdot 7x \cdot 5 + 5^2 - (4y)^2$$

$$= 49x^2 + 70x + 25 - 16y^2$$

Discover for yourself

Work Example 9b by grouping $y + 1$:

$[3x + (y + 1)]$

Do you still get the same answer?

b. We can group the terms so that the pattern for the square of a binomial can be applied:

$$(A + B)^2 = A^2 + 2 \cdot A \cdot B + B^2$$

$$[(3x + y) + 1]^2 = (3x + y)^2 + 2 \cdot (3x + y) \cdot 1 + 1^2$$

$$= 9x^2 + 6xy + y^2 + 6x + 2y + 1 \qquad \blacksquare$$

PROBLEM SET P.5

Practice Problems

In Problems 1–8, perform the indicated operations. Write the resulting polynomial in standard form and indicate its degree.

1. $(-6x^3 + 5x^2 - 8x + 9) + (17x^3 + 2x^2 - 4x - 13)$

2. $(-7x^3 + 6x^2 - 11x + 13) + (19x^3 - 11x^2 + 7x - 17)$

3. $(17x^3 - 5x^2 + 4x - 3) - (5x^3 - 9x^2 - 8x + 11)$

4. $(18x^4 - 2x^3 - 7x + 8) - (9x^4 - 6x^3 - 5x + 7)$

5. $(5x^2 - 7x - 8) + (2x^2 - 3x + 7) - (x^2 - 4x - 3)$

6. $(8x^2 + 7x - 5) - (3x^2 - 4x) - (-6x^3 - 5x^2 + 3)$

7. $(4x^2 - 5xy + 6y^2) + (7x^2 + 2xy - 4y^2) - (8x^2 - xy - 3y^2)$

8. $(2x^3 - 2x^2y + 3xy^2 + 4y^3) - (4x^2y - 3xy^2 - 5y^3) + (3x^3 - x^2y + 5xy^2 - y^3)$

In Problems 9–26, find each product.

9. $(x + 1)(x^2 - x + 1)$
10. $(x + 5)(x^2 - 5x + 25)$
11. $(y^2 + 7y - 3)(3y^2 - y + 2)$
12. $(2y^2 + 5y - 4)(y^2 - 3y + 2)$
13. $(z^2 + 2)(2z^3 + z^2 - 4z)$
14. $(z^2 + 3)(4z^3 - z^2 + 3z)$
15. $(xy + 2)(x^2y^2 - 2xy + 4)$
16. $(x^2y^4 + 3)(x^4y^8 - 3x^2y^4 + 9)$
17. $(5x + 3)(7x + 1)$
18. $(9x - 4)(3x + 5)$
19. $(9y^3 - 5)(2y^3 + 7)$
20. $(6y^3 - 7)(3y^3 + 4)$
21. $(3x^2y + 4xy)(2x^2y - 5xy)$
22. $(7x^2y + 2xy)(5x^2y - xy)$
23. $(9x^2 + 4)(9x^3 - 4)$
24. $(9y^2 + 5)(9y^3 - 5)$
25. $(2x + 1)(x - 2)(1 - x)$
26. $(2x - 1)(x + 3)(2 - x)$

In Problems 27–56, use the special products to find each product.

27. $(5y + 3)(5y - 3)$
28. $(3y + 7)(3y - 7)$
29. $(4x^2y + 5x)(4x^2y - 5x)$
30. $(5x^3y + 3x)(5x^3y - 3x)$
31. $(-2x + y)(y + 2x)$
32. $(-3a^4b^2 + 5c)(3a^4b^2 + 5c)$
33. $(x + 6)^2$
34. $(x - 5)^2$
35. $(x - 6)^3$
36. $(x + 5)^3$
37. $(3x^2y + 2xy)^2$
38. $(4xy^2 + xy)^2$
39. $(3x + 2y)^3$
40. $(2x - 5y)^3$
41. $(2x^2y - xy)^3$
42. $(4x^2y + xy)^3$
43. $(x^4 + 2x^5)^3$
44. $(3x^6 - x^7)^3$
45. $(3x + 7 + 5y)(3x + 7 - 5y)$
46. $(5x + 7y + 2)(5x - 7y - 2)$
47. $[5y - (2x + 3)][5y + (2x + 3)]$
48. $[8y + (7 - 3x)][8y - (7 - 3x)]$
49. $(2x + y + 1)^2$
50. $(5x + 1 + 6y)^2$
51. $[(3x - 1) + y]^2$
52. $[(2x - y) + 8]^2$
53. $[(3x - 1) + y][(3x - 1) - y]$
54. $[(2x - y) + 8][(2x - y) - 8]$
55. $(x + y - 3)(x - y + 3)$
56. $(x^2 + 3y - 1)(x^2 - 3y + 1)$

Application Problems

In Problems 57–59, express the area of each shaded region as a polynomial in standard form.

57.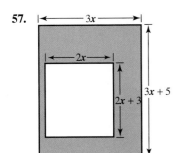

58.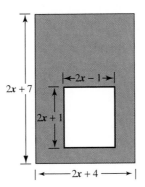

59.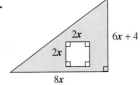

In Problems 60–63, express the area of each plane figure as a polynomial in standard form.

60.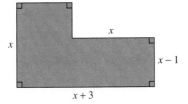

61.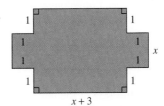

62.

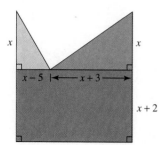

63.

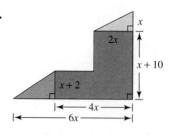

64. Write a polynomial in standard form that expresses:
  a. the difference between the volumes of the two solids shown in the figure.
  b. the difference between the surface areas of the two solids.

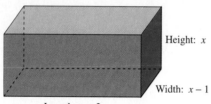   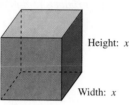

65. Write a polynomial in standard form expressing the difference between the volume and the surface area of the open box in the figure shown.

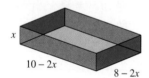

66. The number of people who catch a cold t weeks after January 1 is $C = 5t - 3t^2 + t^3$ and the number of people who recover t weeks after January 1 is $R = t - t^2 + \frac{1}{3}t^3$. Write a polynomial in standard form for the number of people who are still ill with a cold t weeks after January 1.

67. A rock on Earth and a rock on the moon are thrown into the air with a velocity of 48 feet per second by a 6-foot person. The height reached by the rock (in feet) after t seconds is given by

$$h_{\text{Earth}} = -16t^2 + 48t + 6$$

$$h_{\text{Moon}} = -2.7t^2 + 48t + 6$$

Write a polynomial in standard form expressing the difference between moon height and Earth height after t seconds.

True–False Critical Thinking Problems

68. Which one of the following is true?
  a. $4x^3 + 7x^2 - 5x + 2x^{-1}$ is a polynomial in standard form of degree 3 and contains four terms.
  b. If two polynomials of degree 3 are added, their sum must be a polynomial of degree 3.
  c. There is no such thing as a polynomial in x having four terms, written in descending powers, and lacking a third-degree term.
  d. Suppose a square garden has area represented by $9x^2$ square feet. If one side is made 7 feet longer and the other side is made 2 feet shorter, then the trinomial that represents the area of the larger garden is $9x^2 + 15x - 14$ square feet.

69. Which one of the following is true?
  a. Every algebraic expression is a polynomial.
  b. When simplified, the following expression does not result in a natural number:

$$(y - z)(y + z - x) + (z - x)(z + x - y)$$
$$+ (x - y)(x + y - z)$$

  c. $(x^2 + 8 - 4x)(x^2 + 8 + 4x) = x^4 + 16x^2 + 64$
  d. The special product for $(A + B)(A - B)$ cannot be used to find the product of $\sqrt{a + b} + \sqrt{a - b}$ and $\sqrt{a + b} - \sqrt{a - b}$ because the two factors are not polynomials.

Technology Problems

70. Use a graphing utility to graph $y = (x - 2)^2$ and $y = x^2 - 4$ in the same viewing rectangle. Is $(x - 2)^2 = x^2 - 4$?

71. Use a graphing utility to graph $y = (x + 2)^2$ and $y = x^2 + 4x + 4$ in the same viewing rectangle. What do the graphs indicate about $(x + 2)^2$ and $x^2 + 4x + 4$?

72. Use a graphing utility to show that $(x - 1)(x + 3) = x^2 + 2x - 3$.

73. Use a graphing utility to show that $(x - 1)^3 \neq x^3 - 1$.

74. Use a graphing utility to determine which of the following statements is/are false.

   a. $x^3 - 1 = (x - 1)(x^2 + x + 1)$

   b. $x^4 - 1 = (x^2 + 1)(x + 1)(x - 1)$

   c. $x^2 + 1 = (x + 1)^2$

   d. $x^4 - 2x^2 + 1 = (x^2 - 1)^2$

Critical Thinking Problems

For Problems 75–76, represent the volume of each figure as a polynomial in standard form.

75.

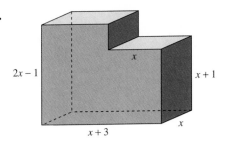

76.

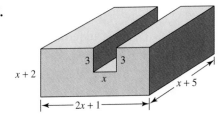

77. Fill in the numbers in fifth, sixth, and seventh positions in the number sequences on the right by using the emerging pattern. Then write a polynomial in x for the number that appears in the xth position.

   a. 0, 3, 8, 15, ____, ____, ____,

   b. 2, 10, 30, 68, ____, ____, ____,

78. Simplify: $(y^n + 2)(y^n - 2) - (y^n - 3)^2$.

SECTION P.6 # Factoring

Solutions   Tutorial   Video
Manual   2

Objectives

1 Factor out the greatest common factor of a polynomial.

2 Factor by grouping.

3 Factor special forms.

4 Factor trinomials.

5 Use a factoring strategy.

6 Factor expressions with fractional and negative exponents.

In general, factoring involves the reverse of the multiplication process. For example, because we know that $(2x + 1)(3x - 2) = 6x^2 - x - 2$, we can also say that $6x^2 - x - 2$ can be expressed as

$$6x^2 - x - 2 = (2x + 1)(3x - 2).$$

The process of writing the polynomial $6x^2 - x - 2$ as a product is called *factoring*. The *factors* of $6x^2 - x - 2$ are $2x + 1$ and $3x - 2$.

In this section, we review basic factoring techniques.

Introduction

Factoring is the process of writing a polynomial as the product of two or more polynomials. We generally *factor over the set of integers*, meaning that the numerical coefficients in the factors are integers. For example, we will not consider the following type of factoring as factoring over the integers:

$$x^2 - 5 = (x + \sqrt{5})(x - \sqrt{5}).$$

We say that $x^2 - 5$ cannot be factored using integer coefficients. Such polynomials are *irreducible over the integers*, or *prime*.

If the coefficients of a polynomial are rational numbers, we will include factors with rational coefficients. Thus, the factorization of $\frac{1}{9}x^2 - 25$ over the rational numbers is given by

$$\frac{1}{9}x^2 - 25 = (\tfrac{1}{3}x + 5)(\tfrac{1}{3}x - 5).$$

The goal in factoring a polynomial is to use one or more factoring techniques until each of the polynomial's factors is prime or irreducible. In this situation, the polynomial is said to be *factored completely*.

We will now discuss basic factoring techniques for factoring polynomials.

◗ **Factor out the greatest common factor of a polynomial.**

Common Factors

In any factoring problem, the first step is to look for the *greatest common factor* (GCF)—that is, an expression of the highest degree that divides each term of the polynomial. The distributive property in the reverse direction

$$ab + ac = a(b + c)$$

can be used to factor out the greatest common factor.

Study tip

You will often find the GCF of a polynomial by inspection. You can also use these steps:

1. Find the coefficient of the GCF, the largest integer that divides all numerical coefficients.
2. List, to the lowest power to which it occurs, each variable factor that occurs in every term of the polynomial.
3. The GCF is the product of the GCF of the coefficients and the variable factors from step 2.

EXAMPLE 1   **Factoring Out the GCF**

Factor:

a. $3x^4 + 6x^3 - 15x^2$ **b.** $20x^4y^2 + 10x^3y^3 - 5xy^4$

c. $x^2(x + 3) + 5(x + 3)$

Solution

a. We begin by determining the GCF.
   $3, 6,$ and -15 have 3 as the GCF.
   $x^4, x^3,$ and x^2 have x^2 as the GCF.
   Thus, the GCF of the three terms in the polynomial is $3x^2$.
   $3x^4 + 6x^3 - 15x^2$
   $= 3x^2(x^2) + 3x^2(2x) + 3x^2(-5)$ Express each term with the GCF as a factor.
   $= 3x^2(x^2 + 2x - 5)$ Factor out the GCF.

b. Again, we begin by determining the GCF.
   $20, 10,$ and -5 have 5 as the GCF.
   $x^4, x^3,$ and x have x as the GCF.
   $y^2, y^3,$ and y^4 have y^2 as the GCF.
   Thus, the GCF of the three terms in the polynomial is $5xy^2$.

$$20x^4y^2 + 10x^3y^3 - 5xy^4$$
$$= 5xy^2(4x^3) + 5xy^2(2x^2y) + 5xy^2(-y^2) \qquad \text{Express each term with the GCF as a factor.}$$
$$= 5xy^2(4x^3 + 2x^2y - y^2) \qquad \text{Factor out the GCF.}$$

c. In this situation, the GCF is the common binomial factor $(x + 3)$. We factor out this common factor as follows.

$$x^2(x + 3) + 5(x + 3) = (x + 3)(x^2 + 5) \qquad \text{Factor out the common binomial factor.} \quad \blacksquare$$

2 Factor by grouping.

Factoring by Grouping

Some polynomials have only a GCF of 1. However, by a suitable rearrangement of the terms, it still may be possible to factor. This process, called *factoring by grouping*, is illustrated in Examples 2 and 3.

EXAMPLE 2 **Factoring by Grouping**

Factor: $x^3 + 4x^2 + 3x + 12$

Solution

Group terms that have a common factor:

$$\boxed{x^3 + 4x^2} + \boxed{3x + 12}$$

$$\uparrow \qquad\qquad \uparrow$$
$$\text{Common} \qquad \text{Common}$$
$$\text{factor is } x^2. \qquad \text{factor is 3.}$$

Discover for yourself

In Example 2 group the terms as follows:

$$(x^3 + 3x) + (4x^2 + 12)$$

Factor out the GCF from each group and complete the factoring process. Describe what happens. What can you conclude?

We now factor the given polynomial as follows.

$$x^3 + 4x^2 + 3x + 12$$
$$= (x^3 + 4x^2) + (3x + 12) \qquad \text{Group terms with common factors.}$$
$$= x^2(x + 4) + 3(x + 4) \qquad \text{Factor out the GCF from the grouped terms. The remaining two terms have } x + 4 \text{ as a common binomial factor.}$$
$$= (x + 4)(x^2 + 3) \qquad \text{Factor } (x + 4) \text{ out of both terms.}$$

Thus, $x^3 + 4x^2 + 3x + 12 = (x + 4)(x^2 + 3)$. Check the factorization by multiplying the right side using the FOIL method, obtaining the polynomial on the left. $\quad \blacksquare$

Using technology

If a polynomial contains one variable, you can check your factorization by graphing the given polynomial and the factored form of the polynomial in the same viewing rectangle. The graphs should be identical.

We did this for Example 2, graphing

$$y_1 = x^3 + 4x^2 + 3x + 12$$

and

$$y_2 = (x + 4)(x^2 + 3)$$

$y_1 = x^3 + 4x^2 + 3x + 12$

$y_2 = (x + 4) \cdot (x^2 + 3)$

$[-5, 5]$ by $[-50, 50]$ Yscl = 5

as shown on the right. The graphs are the same, so $x^3 + 4x^2 + 3x + 12 = (x + 4)(x^2 + 3)$.

iscover for yourself

In Example 3, group the terms as follows:

$$(x^2y^2 + 5x^2) + (-8y^2 - 40).$$

Complete the factoring process by factoring x^2 from the first two terms and -8 from the last two terms. Do you get the same answer as in Example 3? Describe what happens if you factor 8, rather than -8, from the second grouping.

3 Factor special forms.

EXAMPLE 3   Factoring by Grouping

Factor: $x^2y^2 - 40 - 8y^2 + 5x^2$

Solution

Other than 1, the first two terms and the last two terms have no common factor. Let's rearrange the terms and try a different grouping.

$$x^2y^2 - 40 - 8y^2 + 5x^2$$

$= (x^2y^2 - 8y^2) + (5x^2 - 40)$ Group the terms that have a GCF. The first two terms have a GCF of y^2 and the last two terms have a GCF of 5.

$= y^2(x^2 - 8) + 5(x^2 - 8)$ Factor out the GCF from the grouped terms. The remaining two terms have a GCF of $x^2 - 8$.

$= (x^2 - 8)(y^2 + 5)$ Factor $(x^2 - 8)$ out of both terms.

Thus, $x^2y^2 - 40 - 8y^2 + 5x^2 = (x^2 - 8)(y^2 + 5)$. Check by multiplying the right side using the FOIL method, obtaining the polynomial on the left.   ∎

Factoring Special Forms

Certain forms of polynomials occur so frequently that the factors of these forms should be memorized.

Factoring special polynomials forms

Factoring Special Binomials

Factored Form	Example
Difference of Two Squares	
$A^2 - B^2 = (A + B)(A - B)$	$x^2 - 25 = x^2 - 5^2$
	$= (x + 5)(x - 5)$
Sum of Two Cubes	
$A^3 + B^3 = (A + B)(A^2 - AB + B^2)$	$x^3 + 125 = x^3 + 5^3$
	$= (x + 5)(x^2 - x \cdot 5 + 5^2)$
	$= (x + 5)(x^2 - 5x + 25)$
Difference of Two Cubes	
$A^3 - B^3 = (A - B)(A^2 + AB + B^2)$	$x^3 - 8 = x^3 - 2^3$
	$= (x - 2)(x^2 + x \cdot 2 + 2^2)$
	$= (x - 2)(x^2 + 2x + 4)$

Factoring Perfect Square Trinomials

Factored Form	Example
$A^2 + 2AB + B^2 = (A + B)^2$	$x^2 + 10x + 25$
	$= x^2 + 2 \cdot x \cdot 5 + 5^2$
	$= (x + 5)^2$
$A^2 - 2AB + B^2 = (A - B)^2$	$x^2 - 10x + 25$
	$= x^2 - 2 \cdot x \cdot 5 + 5^2$
	$= (x - 5)^2$

EXAMPLE 4   **Factoring the Difference of Two Squares**

Factor completely:

a. $9x^2 - 16$ **b.** $25x^2 - 36y^2$ **c.** $100y^2 - (3x + 1)^2$

Solution

We begin by rewriting each polynomial as the difference of two squares.

$$A^2 \quad - \quad B^2 \quad = (A \ + B) \ (A \ - B)$$

a.
$$9x^2 - 16 = (3x)^2 \ - \quad 4^2 \quad = (3x + 4)(3x - 4)$$

b.
$$25x^2 - 36y^2 = (5x)^2 \ - \ (6y)^2 \ = (5x + 6y)(5x - 6y)$$

c.
$$100y^2 - (3x + 1)^2 = (10y)^2 - (3x + 1)^2 = [10y + (3x + 1)][10y - (3x + 1)]$$
$$= (10y + 3x + 1)(10y - 3x - 1) \quad \blacksquare$$

In certain situations, we can apply the technique for factoring the difference of two squares more than once.

EXAMPLE 5   **Factoring the Difference of Two Squares More Than Once**

Factor completely: $81x^4 - 16$

Solution

$$A^2 \ - B^2 \ = \ (A \ + B) \ (A \ - B)$$
$$81x^4 - 16 = (9x^2)^2 - 4^2 = (9x^2 + 4)(9x^2 - 4)$$

The factorization is not complete because the second factor is the difference of two squares.

$$A^2 \ - B^2 \ = \qquad (A \ + B)(A \ - B)$$
$$(9x^2 + 4)(9x^2 - 4) = (9x^2 + 4)[(3x)^2 - 2^2] = (9x^2 + 4)(3x + 2)(3x - 2)$$

The complete factorization of the polynomial is given by

$$81x^4 - 16 = (9x^2 + 4)(3x + 2)(3x - 2). \qquad \blacksquare$$

EXAMPLE 6   **Factoring the Sum and Difference of Two Cubes**

Factor:

a. $27x^3 + 1$ **b.** $x^6 - 8$

Solution

We begin by expressing the polynomials as the sum and the difference of two cubes, respectively. Then we use the appropriate factoring formula.

$$A^3 + B^3 = (A + B)(A^2 - A B + B^2)$$

a. $27x^3 + 1 = (3x)^3 + 1^3 = (3x + 1)[(3x)^2 - 3x \cdot 1 + 1^2]$ Use the factoring formula for $A^3 + B^3$.

$$= (3x + 1)(9x^2 - 3x + 1)$$ Simplify.

$$A^3 - B^3 = (A - B)(A^2 + A B + B^2)$$

b. $x^6 - 8 = (x^2)^3 - 2^3 = (x^2 - 2)[(x^2)^2 + x^2 \cdot 2 + 2^2]$ Use the factoring formula for $A^3 - B^3$.

$$= (x^2 - 2)(x^4 + 2x^2 + 4)$$ Simplify.   ∎

tudy tip

Here's how to recognize a perfect square trinomial:

1. First and last terms are positive perfect squares.
2. The middle term is twice the product of the square roots of the first and last terms.

EXAMPLE 7   **Factoring Perfect Square Trinomials**

Factor completely:

a. $9y^4 + 12y^2 + 4$ **b.** $16x^2 - 8x + 1$

Solution

We first express each polynomial in the form of a perfect square trinomial, and then we use the factoring formula.

$$A^2 + 2 A B + B^2 = (A + B)^2$$

a. $9y^4 + 12y^2 + 4 = (3y^2)^2 + 2(3y^2)(2) + 2^2 = (3y^2 + 2)^2$

$$A^2 - 2 A B + B^2 = (A - B)^2$$

b. $16x^2 - 8x + 1 = (4x)^2 - 2(4x)(1) + 1^2 = (4x - 1)^2$ ∎

The terms of a polynomial can be grouped in several ways. A successful grouping that will lead to a factorization is frequently based on one or more of the special forms.

EXAMPLE 8   **Using Grouping to Obtain Special Forms**

Factor completely:

a. $x^2 + 8x + 16 - 25a^2$ **b.** $x^3 + y + y^3 + x$

Solution

a. $x^2 + 8x + 16 - 25a^2$

$= (x^2 + 8x + 16) - 25a^2$ Group the perfect square trinomial.

$= (x + 4)^2 - 25a^2$ Factor the perfect square trinomial.

$= (x + 4)^2 - (5a)^2$ Express the polynomial as the difference of two squares.

$= (x + 4 + 5a)(x + 4 - 5a)$ Factor the difference of two squares.

b. $x^3 + y + y^3 + x$

$= (x^3 + y^3) + (x + y)$ Group the two terms involving the sum of two cubes.

$= (x + y)(x^2 - xy + y^2) + (x + y)$ Factor $x^3 + y^3$, the sum of two cubes.

$= (x + y)(x^2 - xy + y^2 + 1)$ Factor out $x + y$, the common factor. ∎

4 Factor trinomials.

Factoring Trinomials

To factor a trinomial of the form $ax^2 + bx + c$, a little trial and error may be necessary.

A strategy for factoring $ax^2 + bx + c$

(Assume, for the moment, that there is no GCF.)

1. Find two First terms whose product is ax^2:

$(\Box\, x + \quad)(\Box\, x + \quad) = ax^2 + bx + c$

2. Find two Last terms whose product is c:

$(x + \Box)(x + \Box) = ax^2 + bx + c$

3. By trial and error, perform steps 1 and 2 until the sum of the Outer product and Inner product is bx:

$(\Box\, x + \Box)(\Box\, x + \Box) = ax^2 + bx + c$

       ⌐ I ⌐

     ⌐ O ⌐

     (sum of O + I)

If no such combinations exist, the polynomial is prime.

EXAMPLE 9 **Factoring Trinomials Whose Leading Coefficients Are 1**

Factor:

a. $x^2 + 6x + 8$ **b.** $x^2 + 3x - 18$

Solution

a. The factors of the first term are x and x:

$(x \qquad)(x \qquad)$

To find the second term of each factor, we must find two numbers whose product is 8 and whose sum is 6. From the table in the margin, we see that 4 and 2 are the required integers. Thus,

$x^2 + 6x + 8 = (x + 4)(x + 2)$ or $(x + 2)(x + 4)$.

b. We begin with

$x^2 + 3x - 18 = (x \qquad)(x \qquad)$.

Factors of 8	8, 1	4, 2	−8, −1	−4, −2
Sum of Factors	9	6	−9	−6

This is the desired sum.

Factors of -18	$18, -1$	$-18, 1$	$9, -2$	$-9, 2$	$6, -3$	$-6, 3$
Sum of Factors	17	-17	7	-7	3	-3

This is the ⎯⎯⎯ desired sum.

To find the second term of each factor, we must find two numbers whose product is -18 and whose sum is 3. From the table in the margin, we see that 6 and -3 are the required integers. Thus,

$$x^2 + 3x - 18 = (x + 6)(x - 3) \text{ or } (x - 3)(x + 6). \quad ■$$

EXAMPLE 10 **Factoring Trinomials Whose Leading Coefficients Are Not 1**

Factor:

a. $2x^2 + 7x + 3$ **b.** $6x^2 + 19x - 7$

Solution

a. First we factor the first term, $2x^2$. Let's agree that when the first term in the trinomial is positive, we will use only positive factors of that first term. Thus, the only possibility for factors is $(2x)(x)$ (we exclude $(-2x)(-x)$).

$$2x^2 + 7x + 3 = (2x \quad)(x \quad)$$

We must still complete the factorization by filling in the blanks with the correct numbers.

Next, we must find two last terms whose product is 3. The possibilities are $(3)(1)$ or $(-3)(-1)$. Because the signs in the given trinomial are all positive, we will use 3 and 1. Thus, the trinomial has the possible factors

$$(2x + 3)(x + 1)$$

$$(2x + 1)(x + 3).$$

Since we are factoring $2x^2 + 7x + 3$, the sum of the outer and inner products must be $7x$.

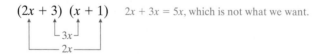

$(2x + 3)\ (x + 1)$ $2x + 3x = 5x$, which is not what we want.

$(2x + 1)\ (x + 3)$ $6x + x = 7x$, which is the trinomial's middle term.

Thus, the desired factorization is

$$2x^2 + 7x + 3 = (2x + 1)(x + 3) \quad \text{or} \quad (x + 3)(2x + 1).$$

b. $6x^2 + 19x - 7$ has factors for $6x^2$ of $6x$ and x or $3x$ and $2x$, so we begin with

$$(6x \quad)(x \quad) \quad \text{or} \quad (3x \quad)(2x \quad).$$

The pairs of factors of -7 are 7 and -1 or -7 and 1. This leads to eight possible factorizations.

sing technology

The graphs of

$$y_1 = 2x^2 + 7x + 3$$

and

$$y_2 = (2x + 1)(x + 3)$$

are shown in the same viewing rectangle. With identical graphs, we can conclude that $2x^2 + 7x + 3 = (2x + 1)(x + 3)$.

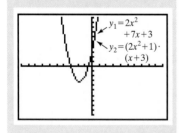

Using technology

The graphs of

$$y_1 = 6x^2 + 19x - 7$$

and

$$y_2 = (3x - 1)(2x + 7)$$

shown in a $[-5, 2]$ by $[-25, 10]$ viewing rectangle, are the same. Thus, $6x^2 + 19x - 7 = (3x - 1)(2x + 7)$.

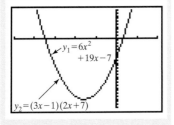

Possible Factors of $6x^2 + 19x - 7$	Sum of Outer and Inner Products (should equal $19x$)
$(6x + 7)(x - 1)$	$-6x + 7x = x$
$(6x - 7)(x + 1)$	$6x - 7x = -x$
$(6x + 1)(x - 7)$	$-42x + x = -41x$
$(6x - 1)(x + 7)$	$42x - x = 41x$
$(3x + 7)(2x - 1)$	$-3x + 14x = 11x$
$(3x - 7)(2x + 1)$	$3x - 14x = -11x$
$(3x + 1)(2x - 7)$	$-21x + 2x = -19x$
$(3x - 1)(2x + 7)$	$21x - 2x = 19x$ ← This is the desired sum.

The last factorization gives the required middle term, $19x$. Thus,

$$6x^2 + 19x - 7 = (3x - 1)(2x + 7). \qquad ■$$

Study tip

You can minimize trial and error by using factoring by grouping in the factoring of a trinomial. For example, to factor

$$6x^2 + 19x - 7$$

multiply the leading coefficient a and the constant c, giving $6(-7)$ or -42. Now find the factors of -42 whose sum is 19. The numbers are 21 and -2, and the middle term can be expressed in terms of these numbers.

Factoring $ax^2 + bx + c$ by Grouping

1. Multiply a and c.
2. Find the factors of ac whose sum is b.
3. Rewrite the middle term, bx, as a sum or difference using the factors from step 2.
4. Factor by grouping.

An Example: $6x^2 + 19x - 7$

$6(-7) = -42$

21 and -2 are factors of -42 whose sum is 19.

$6x^2 + 19x - 7$

$= 6x^2 + 21x - 2x - 7$

$= 3x(2x + 7) - 1(2x + 7)$

$= (2x + 7)(3x - 1)$ or $(3x - 1)(2x + 7)$

5 Use a factoring strategy.

A Strategy for Factoring

The following strategy provides an order for factoring polynomials completely.

A factoring strategy: factoring polynomials over the integers

1. Is there a common factor? If so, factor out the GCF.
2. Is the polynomial a binomial? If so, can it be factored by one of the following special forms?

$$A^2 - B^2 = (A + B)(A - B) \qquad \text{Difference of two squares}$$

$$A^3 + B^3 = (A + B)(A^2 - AB + B^2) \qquad \text{Sum of two cubes}$$

$$A^3 - B^3 = (A - B)(A^2 + AB + B^2) \qquad \text{Difference of two cubes}$$

3. Is the polynomial a trinomial? If it is not a perfect square trinomial, use the trial and error method of Examples 9 and 10. If it is a perfect square trinomial, use one of the special forms:

$$A^2 + 2AB + B^2 = (A + B)^2 \quad \text{or} \quad A^2 - 2AB + B^2 = (A - B)^2$$

4. Does the polynomial contain four or more terms? If so, try factoring by grouping.
5. Is the polynomial factored completely? If a factor with more than one term can be factored further, do so.

EXAMPLE 11   **Factoring Using the General Factoring Strategy**

Factor completely:

a. $2a^3 + 8a^2 + 8a$ **b.** $4x^2 - 16x - 20$

c. $x^2(x - 5) - 4(x - 5)$ **d.** $9b^2x - 16y - 16x + 9b^2y$

Solution

a. $2a^3 + 8a^2 + 8a$
$= 2a(a^2 + 4a + 4)$ Factor out the GCF.
$= 2a(a + 2)^2$ Factor the perfect square trinomial.
b. $4x^2 - 16x - 20$
$= 4(x^2 - 4x - 5)$ Factor out the GCF.
$= 4(x + 1)(x - 5)$ Factor the remaining trinomial.
c. $x^2(x - 5) - 4(x - 5)$
$= (x - 5)(x^2 - 4)$ Factor out the GCF.
$= (x - 5)(x + 2)(x - 2)$ Factor the difference of two squares.
d. $9b^2x - 16y - 16x + 9b^2y$
$= (9b^2x + 9b^2y) + (-16x - 16y)$ Rearrange the terms for factoring by grouping.
$= 9b^2(x + y) + (-16)(x + y)$ Factor within each group.
$= (x + y)(9b^2 - 16)$ Factor out the common binomial.
$= (x + y)(3b + 4)(3b - 4)$ Factor the difference of two squares.   ∎

6 Factor expressions with fractional and negative exponents.

Factoring Algebraic Expressions Containing Fractional and Negative Exponents

Factoring techniques can be applied to expressions that are not polynomials.

EXAMPLE 12   **Factoring Variables with Negative Exponents**

Factor:

a. $2x^{3/2} - 16x^{1/2} + 30x^{-1/2}$ **b.** $x(x + 1)^{-3/4} + (x + 1)^{1/4}$

Solution

a. The GCF is the product of 2 and the x with the *smallest* of the exponents in all three terms, namely, $x^{-1/2}$.

$$2x^{3/2} - 16x^{1/2} + 30x^{-1/2}$$ 　　　　The GCF is $2x^{-1/2}$.

$$= 2x^{-1/2}(x^2) - 2x^{-1/2}(8x) + 2x^{-1/2}(15)$$ 　Each term is expressed with the common factor.

$$= 2x^{-1/2}(x^2 - 8x + 15)$$ 　　　Factor out the GCF.

$$= 2x^{-1/2}(x - 3)(x - 5)$$ 　　Factor completely by factoring the remaining trinomial.

$$= \frac{2(x - 3)(x - 5)}{x^{1/2}}$$ 　　　$b^{-n} = \dfrac{1}{b^n}$

b. We factor out the $x + 1$ with the smallest exponent, that is, $(x + 1)^{-3/4}$.

$$x(x + 1)^{-3/4} + (x + 1)^{1/4}$$ 　　The GCF is $(x + 1)^{-3/4}$.

$$= (x + 1)^{-3/4}x + (x + 1)^{-3/4}(x + 1)$$ 　Each term is expressed with the common factor.

$$= (x + 1)^{-3/4}[x + (x + 1)]$$ 　　Factor out the GCF.

$$= \frac{2x + 1}{(x + 1)^{3/4}}$$ 　　　Simplify. ∎

EXAMPLE 13 **Factoring an Expression Containing Negative Exponents**

Factor: $(x^2 - 9)(2x - 1)^{2/3} - (x^2 - 9)^2(2x - 1)^{-1/3}$

Solution

The two terms are separated by a subtraction sign. The GCF involves common factors in each term with an exponent equal to the smaller of the exponents of these factors. Thus, the GCF is $(x^2 - 9)(2x - 1)^{-1/3}$.

$$(x^2 - 9)(2x - 1)^{2/3} - (x^2 - 9)^2(2x - 1)^{-1/3}$$

$$= (x^2 - 9)(2x - 1)^{-1/3}[(2x - 1)^{3/3}]$$
$$\quad - (x^2 - 9)(2x - 1)^{-1/3}[(x^2 - 9)]$$ 　　Express each term with the common factor. (This step is often done mentally.)

$$= (x^2 - 9)(2x - 1)^{-1/3}[(2x - 1) - (x^2 - 9)]$$ 　Factor out the GCF.

$$= (x^2 - 9)(2x - 1)^{-1/3}(2x - 1 - x^2 + 9)$$ 　Simplify inside the brackets.

$$= \frac{(x^2 - 9)(-x^2 + 2x + 8)}{(2x - 1)^{1/3}}$$ 　　Continue simplifying.

$$= \frac{(x + 3)(x - 3)(-1)(x^2 - 2x - 8)}{(2x - 1)^{1/3}}$$ 　Factor $x^2 - 9$. Factor -1 from the trinomial to make subsequent factoring easier.

$$= \frac{(x + 3)(x - 3)(-1)(x - 4)(x + 2)}{(2x - 1)^{1/3}}$$ 　Factor the trinomial.

$$= -\frac{(x + 3)(x - 3)(x - 4)(x + 2)}{(2x - 1)^{1/3}}$$ 　Simplify. ∎

PROBLEM SET P.6

Practice Problems

In Problems 1–14, factor out the GCF.

1. $18x + 27$

2. $16x - 24$

3. $3x^2 + 6x$

4. $4x^2 - 8x$

5. $9x^4 - 18x^3 + 27x^2$

6. $6x^4 - 18x^3 + 12x^2$

7. $12x^2y - 8xy^2$

8. $33x^2 - 22xy^4$

9. $7x^3y^2 + 14x^2y - 42x^5y^3 + 21xy^4$

10. $36pq^3 - 12p^3q^2 + 60p^2q^4 - 24p^4q^3$ **11.** $x^2(x - 3) + 12(x - 3)$

12. $x^2(2x + 5) + 17(2x + 5)$

13. $2x(x - 5)^2 - 3y(x - 5)$

14. $4x(x - 3)^2 - 2y(x - 3)$

In Problems 15–24, factor by grouping.

15. $x^3 - 2x^2 + 5x - 10$

16. $x^3 - 3x^2 + 4x - 12$

17. $x^3 - x^2 + 2x - 2$

18. $x^3 + 6x^2 - 2x - 12$

19. $3x^3 - 2x^2 - 6x + 4$

20. $x^3 - x^2 - 5x + 5$

21. $a^2c + 5ac + 2a + 10$

22. $3y^2 + 4yz + 24y + 32z$

23. $4x^2y + 16xy - x - 4$

24. $5x^2 + 15 - x^2y - 3y$

In Problems 25–36, factor the difference of two squares.

25. $x^2 - 100$

26. $x^2 - 144$

27. $36x^2 - 49$

28. $64x^2 - 81$

29. $9x^2 - 25y^2$

30. $36x^2 - 49y^2$

31. $9y^2 - (2x + 1)^2$

32. $16y^2 - (3x - 1)^2$

33. $x^4 - 81$

34. $x^4 - 16$

35. $9x^4 - y^6$

36. $81x^4 - 49y^6$

In Problems 37–44, factor the sum or difference of two cubes.

37. $x^3 - 8$

38. $x^3 - 64$

39. $y^3 + 27$

40. $y^3 + 125$

41. $64d^3 + 125$

42. $8d^3 + 27$

43. $125 - 8p^3d^3$

44. $64 - 27y^3z^3$

In Problems 45–72, factor each trinomial or state that the trinomial is prime.

45. $x^2 + 8x + 15$

46. $x^2 + 10x + 16$

47. $x^2 - 8x + 15$

48. $x^2 - 5x + 6$

49. $x^2 + x - 30$

50. $x^2 + 14x - 32$

51. $x^2 - 3x - 28$

52. $x^2 - 4x - 21$

53. $x^2 + 4x + 4$

54. $x^2 + 6x + 9$

55. $x^2 - 8x + 16$

56. $x^2 - 10x + 25$

57. $25x^2 - 20x + 4$

58. $9x^2 - 12x + 4$

59. $2x^2 + 9x + 7$

60. $5x^2 + 56x + 11$

61. $4y^2 + 9y + 2$

62. $8y^2 + 10y + 3$

63. $10x^2 + 19x + 6$

64. $5x^2 + 17x + 6$

65. $8y^2 - 18y + 9$

66. $9y^2 - 30y + 25$

67. $4y^2 - 27y + 18$

68. $6y^2 - 23y + 15$

69. $12y^2 - 19y - 21$

70. $16y^2 - 6y - 27$

71. $x^2 + 13x + 9$

72. $x^2 + 17x + 12$

In Problems 73–102, factor each polynomial completely or state that the polynomial is prime.

73. $2y^3 + 8y^2 + 8y$

74. $3y^3 - 6y^2 + 3y$

75. $4x^2 - 16x - 20$

76. $3x^2 - 12x - 36$

77. $3x^4 + 24x$

78. $2x^4 - 250x$

79. $x^2(x - 5) - 4(x - 5)$

80. $x^2(x + 3) - 81(x + 3)$

81. $x^2 + 10x + 25 - 36a^2$

82. $x^2 - 8x + 16 - 49a^2$

83. $c^3 - 16c$

84. $y^3 - y$

85. $9b^2x - 16y - 16x + 9b^2y$

86. $cx^2 + cx + dxy + dy$

87. $x^2 + 25$

88. $x^2 + 1$

89. $y^3 - 2y^2 - 4y + 8$

90. $y^3 - 2y^2 - y + 2$

91. $16x - 2x^4$

92. $7x^5y - 7xy^5$

93. $8x^2 + 40x + 50$

94. $9b^3 - 9b^6$

95. $x^2(x^3 + y^3) - z^2(x^3 + y^3)$

96. $12a^3b - 12ab^3$

97. $3ax + 4bx - 15a - 20b$

98. $63y^2 + 30y - 72$

99. $r^3 - s^3 + r - s$

100. $7by^4 - 7b$

101. $a^6b^6 - a^3b^3$

102. $x^3y^6 - x^3$

In Problems 103–116, factor each algebraic expression.

103. $x^{3/2} - x^{1/2}$

104. $x^{3/4} - x^{1/4}$

105. $4x^{-2/3} + 8x^{1/3}$

106. $12x^{-3/4} + 6x^{1/4}$

107. $3x^{3/2} - 9x^{1/2} + 6x^{-1/2}$

108. $x^{1/2} + 2x^{-1/2} + x^{-3/2}$

109. $(x + 3)^{1/2} - (x + 3)^{3/2}$

110. $(x^2 + 4)^{3/2} + (x^2 + 4)^{7/2}$

111. $(x + 5)^{-1/2} - (x + 5)^{-3/2}$

112. $(x^2 + 3)^{-2/3} + (x^2 + 3)^{-5/3}$

113. $(4x - 1)^{1/2} - \frac{1}{3}(4x - 1)^{3/2}$

114. $-8(4x + 3)^{-2} + 10(5x + 1)(4x + 3)^{-1}$

115. $2x(x^2 + 3)^{-1/2} - x^3(x^2 + 3)^{-3/2}$

116. $\frac{1}{2}(x - 5)^{-1/2}(x + 5)^{-1/2} - \frac{1}{2}(x + 5)^{1/2}(x - 5)^{-3/2}$

True–False Critical Thinking Problems

117. Which one of the following is true?

    a. Because $x^2 + 1$ is irreducible over the integers, it follows that $x^3 + 1$ is also irreducible.

    b. One correct factored form for $x^2 - 4x + 3$ is $x(x - 4) + 3$.

    c. $x^3 - 64 = (x - 4)^3$

    d. None of the above is true.

118. Which one of the following is true?

    a. $x^2 + 49 = (x + 7)(x + 7)$

    b. $x^3 + 3x^2a + 3xa^2 + a^3$ can be factored as $(x + a)^3$.

    c. If a rectangular room has a floor space of $2x^2 + 17x + 35$ square feet, and if the room's width is $x + 5$ feet, then there are at least two different polynomials that represent the room's length in feet.

    d. None of the above is true.

Technology Problems

Factorizations for algebraic expressions in one variable can be verified using a graphing utility. For example, $6x^2 + 19x - 7 = (3x - 1)(2x + 7)$ can be checked by graphing $y_1 = 6x^2 + 19x - 7$ and $y_2 = (3x - 1)(2x + 7)$ in the same viewing rectangle, obtaining identical graphs. In Problems 119–121, use a graphing utility to check whether the given factorization is correct.

119. $x^3 - 5x^2 + 3x - 15 = (x^2 - 3)(x + 5)$

120. $x^3 - 1 = (x - 1)(x^2 - x + 1)$

121. $8 + (x - 2)^3 = x(x^2 - 6x + 12)$

In Problems 122–125, factor completely and verify your result with a graphing utility.

122. $x^2 + 3x - 4$

123. $x^2 - 6x + 9$

124. $x^4 - 1$

125. $x^3 + 1$

126. Computer algebra systems such as *Mathematica, Maple,* and *Derive* will factor polynomials. Graphing calculators, such as the TI-92 with *Derive,* will also allow you to enter a polynomial and use the [F2] *Algebra* toolbar menu to display its factored form. Use a computer system or a calculator that can perform symbolic manipulations to verify some of your factorizations in Problems 1–116.

127. Suppose that you are using a graphing utility without the capability described in Problem 126. You have factored $x^2 - 9a^2 + 12x + 36$ as $(x + 6 + 3a)(x + 6 - 3a)$ and would like to verify your factorization by graphing each side of the equation

$$x^2 - 9a^2 + 12x + 36 = (x + 6 + 3a)(x + 6 - 3a),$$

obtaining identical graphs. However, your graphing utility can only graph in one variable. What procedure might you use to convince yourself that your factorization is correct?

Writing in Mathematics

Explain how each figure in Problems 128–130 provides a geometric interpretation of the given factoring formula.

128. $A^2 + 2A + 1 = (A + 1)^2$

130. $A^3 + 3A^2B + 3AB^2 + B^3 = (A + B)^3$

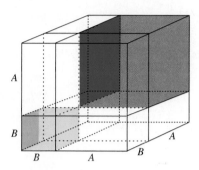

129. $A^2 - B^2 = (A + B)(A - B)$

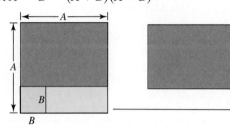

Critical Thinking Problems

131. a. Factor $x^4 + 2x^2y^2 + y^4$.
   b. Factor $x^4 + x^2y^2 + y^4$ by adding and subtracting a term, using part (a) as a guide.

132. Factor $x^6 - y^6$ by two different methods. Use part (b) of Problem 131 to show that the factorizations are the same.

133. Factor completely: $x^4 - y^4 - 2x^3y + 2xy^3$.

134. Factor completely: $(x + y + z)^3 - x^3 - y^3 - z^3$.

135. Express $x^3 + x + 2x^4 + 4x^2 + 2$ as the product of two polynomials of degree 2.

SECTION P.7 **Rational Expressions**

Solutions Manual   **Tutorial**   **Video 2**

Objectives

1 Reduce rational expressions.
2 Multiply and divide rational expressions.
3 Add and subtract rational expressions.
4 Simplify complex rational expressions.

Just as rational numbers play a significant role in the study of arithmetic, fractional expressions and rational expressions frequently appear in algebra. In this section, we review operations with these expressions.

Basic Definitions

A *rational expression* is any expression that can be written as the quotient of two polynomials. The following are examples of rational expressions:

$$\frac{5x^2}{x^2 + 1} \qquad \frac{x^3 - 7x^2 - 18x}{x^3 + 3x^2 + 2x} \qquad \frac{3x^2 + 12xy - 15y^2}{6x^3 - 6xy^2}$$

Study tip

Fractional expressions that do not contain polynomials in the numerator or denominator are not rational expressions. The fractional expression

$$\frac{\sqrt{x^2 + 3}}{4x^3 + 5x^2 - 7}$$

is not a rational expression because $\sqrt{x^2 + 3}$ is not a polynomial.

▌ Reduce rational expressions.

tudy tip

The procedure for simplifying rational expressions lets us *cancel* identical *nonzero* factors in the numerator and denominator.

If $b \neq 0$ and $c \neq 0$;

$$\frac{ac}{bc} = \frac{a \cdot \cancel{c}}{b \cdot \cancel{c}} = \frac{a}{b}$$

Simplifying Rational Expressions

A rational expression is *simplified* or in *reduced form* if its numerator and denominator have no common factors other than 1 or −1. The following procedure can be used to simplify rational expressions.

Simplifying rational expressions

1. Factor the numerator and denominator completely.
2. Divide both the numerator and denominator by the common factors.

EXAMPLE 1 **Simplifying Rational Expressions**

Simplify:

a. $\dfrac{x^3 - 7x^2 - 18x}{x^3 + 3x^2 + 2x}$ **b.** $\dfrac{x^3 - 125}{5x - x^2}$

Solution

a. $\dfrac{x^3 - 7x^2 - 18x}{x^3 + 3x^2 + 2x}$ This is the given rational expression.

$= \dfrac{x(x^2 - 7x - 18)}{x(x^2 + 3x + 2)}$ Factor out the GCF.

$= \dfrac{x(x - 9)(x + 2)}{x(x + 1)(x + 2)}$ Factor completely. Observe that $x \neq 0$, $x \neq -1$, and $x \neq -2$. A rational expression is undefined if the denominator equals 0.

$= \dfrac{\cancel{x}(x - 9)\cancel{(x + 2)}}{\cancel{x}(x + 1)\cancel{(x + 2)}}$ Divide by the common factors.

$= \dfrac{x - 9}{x + 1}$ Reduce. Although it's obvious that $x \neq -1$, we should restate from above that $x \neq 0$ and $x \neq -2$.

Notice that the restricted values of x in the original expression are also restricted in the reduced fraction, even if these values do not cause division by 0 in the rational expression's simplified form.

b. Before simplifying, we should observe that the quotient of two polynomials that have exactly opposite signs and are additive inverses is −1.

$$\frac{a - b}{b - a} = \frac{-(-a + b)}{b - a} = \frac{-\cancel{(b - a)}}{1\cancel{(b - a)}} = -\frac{1}{1} = -1$$

Discover for yourself

In Example 1a and 1b, evaluate the given rational expression and its simplified form for a value of x of your choice. What do you observe? How can this be used to convince yourself that the simplification is correct?

$$\frac{x^3 - 125}{5x - x^2}$$

This is the given rational expression.

$$= \frac{(x - 5)(x^2 + 5x + 25)}{x(5 - x)}$$

Factor completely. The numerator is the difference of two cubes, $x^3 - 5^3$. Observe that $x \neq 0$, and $x \neq 5$.

$$= \frac{\overset{(-1)}{\cancel{(x - 5)}}(x^2 + 5x + 25)}{x\cancel{(5 - x)}}$$

Two polynomials that have opposite signs and are additive inverses have a quotient of -1.

$$= -\frac{x^2 + 5x + 25}{x}$$

Although it's obvious that $x \neq 0$, we should restate from above that $x \neq 5$.   ∎

Study tip

Avoid these common errors in simplifying rational expressions:

| **Incorrect!** | **Incorrect!** | **Incorrect!** | **Incorrect!** |

$$\frac{\cancel{x} + 4}{\cancel{x}} = 4 \qquad \frac{\overset{x}{\cancel{x^2}} + 8}{\cancel{x}} = x + 8 \qquad \frac{\cancel{3}}{\cancel{3} + x} = \frac{1}{x} \qquad \frac{x^2\cancel{y} + 4x}{\cancel{y}} = x^2 + 4x$$

You can only divide out, or cancel, common factors that are parts of products. Only when expressions are multiplied can they be factors, so if you can't factor, then don't try to cancel.

Correct:

$$\frac{x^3 + 1}{x + 1} = \frac{(x + 1)(x^2 - x + 1)}{1(x + 1)} = x^2 - x + 1$$

Incorrect:

$$\frac{\overset{x^2}{\cancel{x^3}} + \cancel{x}}{\cancel{x} + \cancel{1}} = x^2 \text{ or } x^2 + 1$$

In Example 1b we used the property that $\dfrac{-a}{b} = -\dfrac{a}{b}$. Properties of fractions, useful in performing operations with rational expressions, are summarized below.

Properties of rational expressions

Let $a, b, c,$ and d be real numbers, variables, or algebraic expressions such that $b \neq 0$ and $d \neq 0$.

Property

1. *Rules of Signs*

$$\frac{-a}{b} = \frac{a}{-b} = -\frac{a}{b}$$

$$\frac{-a}{-b} = \frac{a}{b}$$

2. *Reducing Fractions*

$$\frac{ac}{bc} = \frac{a}{b} \quad (c \neq 0)$$

3. *Writing Equivalent Fractions*

$$\frac{a}{b} = \frac{ac}{bc} \quad (c \neq 0)$$

Example

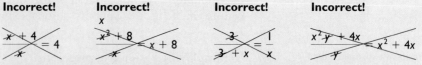

$$\frac{-x}{5} = -\frac{x}{5}$$

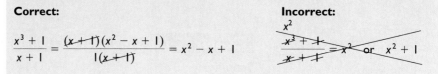

$$\frac{2\cancel{(x + 2)}}{\cancel{(x + 2)}(x + 3)} \quad (x \neq -2, \, x \neq -3)$$

$$= \frac{2}{x + 3} \quad (x \neq -2)$$

$$\frac{2}{x + 3} = \frac{2(x + 2)}{(x + 3)(x + 2)} \quad (x \neq -2)$$

SECTION P.7 RATIONAL EXPRESSIONS **81**

4. *Multiplying Fractions*

$$\frac{a}{b} \cdot \frac{c}{d} = \frac{ac}{bd}$$

$$\frac{3}{x-1} \cdot \frac{x-2}{x^2+5} = \frac{3(x-2)}{(x-1)(x^2+5)}$$

5. *Dividing Fractions*

$$\frac{a}{b} \div \frac{c}{d} = \frac{a}{b} \cdot \frac{d}{c} = \frac{ad}{bc} \quad (c \neq 0)$$

$$\frac{9}{x} \div \frac{5}{7} = \frac{9}{x} \cdot \frac{7}{5} = \frac{63}{5x}$$

6. *Adding or Subtracting Fractions with Like Denominators*

$$\frac{a}{b} \pm \frac{c}{b} = \frac{a \pm c}{b}$$

$$\frac{5x-1}{x+2} - \frac{2x-1}{x+2}$$

$$= \frac{5x-1-(2x-1)}{x+2} = \frac{5x-1-2x+1}{x+2}$$

$$= \frac{3x}{x+2}$$

7. *Adding or Subtracting Fractions with Unlike Denominators*

$$\frac{a}{b} \pm \frac{c}{d} = \frac{ad \pm bc}{bd}$$

$$\frac{x}{x-1} + \frac{3}{x-2}$$

$$= \frac{x(x-2)+(x-1)\cdot 3}{(x-1)(x-2)}$$

$$= \frac{x^2-2x+3x-3}{(x-1)(x-2)} = \frac{x^2+x-3}{(x-1)(x-2)}$$

8. *Proportion Principle*

If $\frac{a}{b} = \frac{c}{d}$, then $ad = bc$.

If $ad = bc$, then $\frac{a}{b} = \frac{c}{d}$.

If $\frac{x}{3} = \frac{4}{7}$, then $x \cdot 7 = 3 \cdot 4$. Equivalently,

$7x = 12$.

---

2 Multiply and divide rational expressions.

Operations with Rational Expressions

Multiplication and division can be accomplished using properties 4 and 5 of rational expressions.

EXAMPLE 2 **Multiplying and Dividing Rational Expressions**

Perform the indicated operations:

a. $\dfrac{2x^2-13x-7}{x^2-6x-7} \cdot \dfrac{x^2-x-2}{2x^2-5x-3}$ **b.** $\dfrac{x^3+8}{x^3-5x} \div \dfrac{x+2}{5-x^2}$

Study tip

To multiply rational expressions:
1. Factor all numerators and denominators completely.
2. Cancel all common factors in any numerator and denominator.
3. Multiply remaining factors in the numerator. Do the same in the denominator.

Solution

a. $\dfrac{2x^2-13x-7}{x^2-6x-7} \cdot \dfrac{x^2-x-2}{2x^2-5x-3}$

This is the given multiplication problem.

$= \dfrac{(2x+1)(x-7)}{(x+1)(x-7)} \cdot \dfrac{(x-2)(x+1)}{(2x+1)(x-3)}$

Factor and reduce.

$= \dfrac{x-2}{x-3}$

From the previous step, notice that $x \neq -1, x \neq 7, x \neq -\frac{1}{2}$, and $x \neq 3$.

Study tip

To divide rational expressions:
1. Change the division to multiplication and invert the rational expression that follows the division symbol ÷.
2. Use the procedure in the previous Study tip.

b. $\dfrac{x^3 + 8}{x^3 - 5x} \div \dfrac{x + 2}{5 - x^2}$ This is the given division problem.

$= \dfrac{x^3 + 8}{x^3 - 5x} \cdot \dfrac{5 - x^2}{x + 2}$ Multiply by the reciprocal of the divisor.

$= \dfrac{\cancel{(x + 2)}(x^2 - 2x + 4)}{x\cancel{(x^2 - 5)}} \cdot \dfrac{\overset{-1}{\cancel{(5 - x^2)}}}{\cancel{x + 2}}$ Factor and reduce. Observe that $x \neq 0, x \neq \pm\sqrt{5}$, and $x \neq -2$.

$= -\dfrac{x^2 - 2x + 4}{x}$ Multiply the remaining factors. ∎

Rational expressions with like denominators can be added or subtracted using property 6 of rational expressions, namely, adding or subtracting numerators and placing this expression over the common denominator. For example,

$$\dfrac{2x - 1}{x^2 - 2x + 2} - \dfrac{3x - 7}{x^2 - 2x + 2} = \dfrac{2x - 1 - (3x - 7)}{x^2 - 2x + 2} = \dfrac{2x - 1 - 3x + 7}{x^2 - 2x + 2} = \dfrac{-x + 6}{x^2 - 2x + 2}.$$

3 Add and subtract rational expressions.

Rational expressions that have no common factors in their denominators can be added or subtracted using property 7, namely,

$$\dfrac{a}{b} \pm \dfrac{c}{d} = \dfrac{ad \pm bc}{bd}.$$

The least common denominator (LCD) is the product of the distinct factors in the two denominators.

EXAMPLE 3   **Subtracting Rational Expressions Having No Common Factors in Their Denominators**

Subtract: $\dfrac{x + 2}{2x - 3} - \dfrac{4}{x + 3}$

Solution

The LCD is the product of the distinct factors in each denominator, namely

$(2x - 3)(x + 3).$

We can therefore use property 7 as follows:

$$\underset{\downarrow}{\dfrac{a}{b}} - \underset{\downarrow}{\dfrac{c}{d}} = \underset{\downarrow}{\dfrac{ad - bc}{bd}}$$

$$\dfrac{x + 2}{2x - 3} - \dfrac{4}{x + 3} = \dfrac{(x + 2)(x + 3) - (2x - 3)4}{(2x - 3)(x + 3)}$$ Observe that $a = x + 2$, $b = 2x - 3, c = 4$, and $d = x + 3$.

$$= \dfrac{x^2 + 5x + 6 - (8x - 12)}{(2x - 3)(x + 3)}$$ Multiply.

$$= \dfrac{x^2 + 5x + 6 - 8x + 12}{(2x - 3)(x + 3)}$$

$$= \dfrac{x^2 - 3x + 18}{(2x - 3)(x + 3)}$$ Combine like terms in the numerator. ∎

When adding and subtracting rational expressions that have different denominators with one or more common factors in the denominators, it is more efficient to first find the LCD, which is the least common multiple of the denominators. This can be done by factoring each denominator and taking the product of distinct factors, using the highest power that appears in any of the factors.

EXAMPLE 4 **Adding Rational Expressions with Different Denominators**

Add: $\dfrac{x + 3}{x^2 + x - 2} + \dfrac{2}{x^2 - 1}$

Solution

We find the LCD by first factoring the denominators.

$$x^2 + x - 2 = (x + 2)(x - 1)$$

$$x^2 - 1 = (x + 1)(x - 1) \quad \text{Notice that the denominators share the factor } x - 1.$$

The LCD is the product of the distinct factors, namely, $(x + 2)(x - 1)(x + 1)$. We now write each rational expression in terms of this LCD by inserting into the numerator and denominator the extra factor required to form the LCD.

$$\frac{x + 3}{x^2 + x - 2} + \frac{2}{x^2 - 1}$$

$$= \frac{x + 3}{(x + 2)(x - 1)} + \frac{2}{(x + 1)(x - 1)} \quad \begin{array}{l}\text{Factor denominators. The LCD, the prod-}\\ \text{uct of each factor the greatest number of}\\ \text{times it occurs in any one factorization, is}\\ (x + 2)(x - 1)(x + 1).\end{array}$$

$$= \frac{(x + 3)(x + 1)}{(x + 2)(x - 1)(x + 1)} + \frac{2(x + 2)}{(x + 2)(x - 1)(x + 1)} \quad \begin{array}{l}\text{Rewrite each rational}\\ \text{expression with the LCD.}\\ \text{Multiply the numerator}\\ \text{and the denominator by}\\ \text{whatever extra factors are}\\ \text{required to form the LCD.}\end{array}$$

$$= \frac{(x + 3)(x + 1) + 2(x + 2)}{(x + 2)(x - 1)(x + 1)} \quad \text{Add numerators, putting this sum over the LCD.}$$

$$= \frac{x^2 + 4x + 3 + 2x + 4}{(x + 2)(x - 1)(x + 1)} \quad \text{Multiply in the numerator.}$$

$$= \frac{x^2 + 6x + 7}{(x + 2)(x - 1)(x + 1)} \quad \text{Combine like terms in the numerator.} \quad \blacksquare$$

EXAMPLE 5 **Adding and Subtracting Rational Expressions with Different Denominators**

Combine: $\dfrac{3y + 2}{y - 5} + \dfrac{4}{3y + 4} - \dfrac{7y^2 + 24y + 28}{3y^2 - 11y - 20}$

Study tip

To add and subtract rational expressions that have different denominators with shared factors:

1. Find the LCD by factoring the denominators and taking the product of distinct factors. Use the highest power that appears in any of the factors.

2. Write all rational expressions in terms of the LCD by inserting into the numerator and denominator of each whatever extra factors are needed to form the LCD.

3. Add or subtract the numerators, placing the resulting expression over the LCD.

4. If necessary, simplify the resulting rational expression.

Solution

Because $3y^2 - 11y - 20 = (3y + 4)(y - 5)$, the LCD is $(3y + 4)(y - 5)$.

$$\frac{3y + 2}{y - 5} + \frac{4}{3y + 4} - \frac{7y^2 + 24y + 28}{3y^2 - 11y - 20}$$

$$= \frac{(3y + 2)(3y + 4)}{(3y + 4)(y - 5)} + \frac{4(y - 5)}{(3y + 4)(y - 5)} - \frac{7y^2 + 24y + 28}{(3y + 4)(y - 5)} \quad \begin{array}{l}\text{Express all ratio-}\\\text{nal expressions}\\\text{with the LCD.}\end{array}$$

$$= \frac{(3y + 2)(3y + 4) + 4(y - 5) - (7y^2 + 24y + 28)}{(3y + 4)(y - 5)} \quad \begin{array}{l}\text{Add and subtract}\\\text{as indicated.}\end{array}$$

$$= \frac{9y^2 + 18y + 8 + 4y - 20 - 7y^2 - 24y - 28}{(3y + 4)(y - 5)}$$

$$= \frac{2y^2 - 2y - 40}{(3y + 4)(y - 5)} \quad \begin{array}{l}\text{Combine like terms in}\\\text{the numerator.}\end{array}$$

$$= \frac{2\cancel{(y - 5)}(y + 4)}{(3y + 4)\cancel{(y - 5)}} \quad \begin{array}{l}\text{Factor and simplify.}\end{array}$$

$$= \frac{2(y + 4)}{3y + 4} \qquad \blacksquare$$

4 Simplify complex rational expressions.

Complex Rational Expressions

Complex rational expressions have numerators or denominators containing one or more rational expressions. Complex rational expressions can be simplified by multiplying the numerator and the denominator by the LCD of all the fractions in the expression.

| **EXAMPLE 6** | **Simplifying Complex Rational Expressions** |

Simplify:

a. $\dfrac{\dfrac{1}{x + h} - \dfrac{1}{x}}{h}$ **b.** $\dfrac{\dfrac{1}{x - 2} - \dfrac{1}{x - 3}}{1 + \dfrac{1}{x^2 - 5x + 6}}$ **c.** $\dfrac{x^{-2} - y^{-2}}{x^{-1} + y^{-1}}$

Solution

a. $\dfrac{\dfrac{1}{x + h} - \dfrac{1}{x}}{h}$

The LCD of the denominators is $x(x + h)$. (The h in the denominator has a denominator understood to be 1.)

$$= \frac{\left(\dfrac{1}{x + h} - \dfrac{1}{x}\right)x(x + h)}{h\,x(x + h)}$$

Multiply the numerator and denominator by $x(x + h)$. Because $\dfrac{x(x + h)}{x(x + h)} = 1$ ($x \neq 0$, $x \neq -h$), we are not changing the complex fraction.

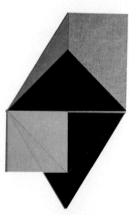

Dorothea Rockburne "Noli Me Tangere" 1976, gesso, oil paint, blue pencil and varnish on linen, 55 in. H × 34 in. W. © 1998 Dorothea Rockburne / Artists Rights Society (ARS), New York.

$$= \frac{\dfrac{1}{x+h} \cdot x(x+h) - \dfrac{1}{x} \cdot x(x+h)}{h \cdot x(x+h)}$$

Use the distributive property in the numerator.

$$= \frac{x - (x+h)}{hx(x+h)}$$

Simplify: $\dfrac{1}{\cancel{x+h}} \cdot x\cancel{(x+h)} = x$ and $\dfrac{1}{\cancel{x}} \cdot \cancel{x}(x+h) = x+h$.

$$= \frac{x - x - h}{hx(x+h)}$$

Perform the indicated operation in the numerator.

$$= \frac{-h}{hx(x+h)}$$

This fraction can be reduced.

$$= -\frac{1}{x(x+h)}$$

Reduce to lowest terms. ($h \neq 0$)

b. $\dfrac{\dfrac{1}{x-2} - \dfrac{1}{x-3}}{1 + \dfrac{1}{x^2 - 5x + 6}}$

$$= \frac{\dfrac{1}{x-2} - \dfrac{1}{x-3}}{1 + \dfrac{1}{(x-2)(x-3)}}$$

Factor. The LCD of the denominators is $(x-2)(x-3)$.

$$= \frac{\left(\dfrac{1}{x-2} - \dfrac{1}{x-3}\right)}{\left(1 + \dfrac{1}{(x-2)(x-3)}\right)} \cdot \frac{(x-2)(x-3)}{(x-2)(x-3)}$$

Multiply the numerator and the denominator by the LCD of the denominators.

$$= \frac{\dfrac{1}{x-2}(x-2)(x-3) - \dfrac{1}{x-3}(x-2)(x-3)}{1(x-2)(x-3) + \dfrac{1}{(x-2)(x-3)}(x-2)(x-3)}$$

Use the distributive property.

$$= \frac{(x-3) - (x-2)}{(x-2)(x-3) + 1}$$

Cancel identical factors in numerators and denominators.

$$= \frac{x - 3 - x + 2}{x^2 - 3x - 2x + 6 + 1}$$

Perform the indicated operations.

$$= \frac{-1}{x^2 - 5x + 7} \quad \text{or} \quad -\frac{1}{x^2 - 5x + 7}$$

Combine like terms.

c. $\dfrac{x^{-2} - y^{-2}}{x^{-1} + y^{-1}}$

$$= \frac{\dfrac{1}{x^2} - \dfrac{1}{y^2}}{\dfrac{1}{x} + \dfrac{1}{y}}$$

Since $b^{-n} = \dfrac{1}{b^n}$, then $x^{-2} = \dfrac{1}{x^2}$ and $x^{-1} = \dfrac{1}{x}$. Similarly, the terms involving y are rewritten without negative exponents.

Discover for yourself

A second method for simplifying a complex rational expression is to combine its numerator and denominator into single fractions. Then find the quotient by inverting the denominator and multiplying. Use this method to simplify the complex fractions in Example 6.

$$= \frac{\left(\dfrac{1}{x^2} - \dfrac{1}{y^2}\right)x^2y^2}{\left(\dfrac{1}{x} + \dfrac{1}{y}\right)x^2y^2}$$

The LCD of the denominators is x^2y^2. Multiply the numerator and denominator by x^2y^2. ($x \ne 0$ and $y \ne 0$)

$$= \frac{\dfrac{1}{x^2} \cdot x^2y^2 - \dfrac{1}{y^2} \cdot x^2y^2}{\dfrac{1}{x} \cdot x^2y^2 + \dfrac{1}{y} \cdot x^2y^2}$$

Use the distributive property.

$$= \frac{y^2 - x^2}{xy^2 + x^2y}$$

Simplify.

$$= \frac{\cancel{(y + x)}(y - x)}{xy\cancel{(y + x)}}$$

Factor.

$$= \frac{y - x}{xy}$$

Reduce. ($x \ne -y$) ∎

Example 7 involves a fractional expression containing radicals. Although this is not a rational expression, we can use the procedure for simplifying complex rational expressions to rewrite this expression as a single fraction.

EXAMPLE 7 **Simplifying a Fractional Expression Containing Radicals**

Simplify: $\dfrac{\sqrt{9 - x^2} + \dfrac{x^2}{\sqrt{9 - x^2}}}{9 - x^2}$

Solution

$$\frac{\sqrt{9 - x^2} + \dfrac{x^2}{\sqrt{9 - x^2}}}{9 - x^2}$$

The LCD of the denominators is $\sqrt{9 - x^2}$.

$$= \frac{\sqrt{9 - x^2} + \dfrac{x^2}{\sqrt{9 - x^2}}}{9 - x^2} \cdot \frac{\sqrt{9 - x^2}}{\sqrt{9 - x^2}}$$

Multiply the numerator and the denominator by $\sqrt{9 - x^2}$.

$$= \frac{\sqrt{9 - x^2}\sqrt{9 - x^2} + \dfrac{x^2}{\sqrt{9 - x^2}}\sqrt{9 - x^2}}{(9 - x^2)\sqrt{9 - x^2}}$$

Use the distributive property in the numerator.

$$= \frac{(9 - x^2) + x^2}{(9 - x^2)^{3/2}}$$

In the denominator:

$$(9 - x^2)^1(9 - x^2)^{1/2} = (9 - x^2)^{1 + 1/2}$$
$$= (9 - x^2)^{3/2}$$

$$= \frac{9}{\sqrt{(9 - x^2)^3}}$$

Since the original expression was in radical form, write the denominator in radical form. ∎

Discover for yourself

A second method for solving Example 7 is to use exponential form.

$$\frac{(9 - x^2)^{1/2} + x^2(9 - x^2)^{-1/2}}{9 - x^2}$$

Simplify this expression by factoring $(9 - x^2)^{-1/2}$ from the numerator. Use properties of exponents and make the final answer look like the one in Example 7.

PROBLEM SET P.7

Practice Problems

In Problems 1–10, reduce each rational expression to lowest terms.

1. $\dfrac{3x - 9}{x^2 - 6x + 9}$

2. $\dfrac{4x - 8}{x^2 - 4x + 4}$

3. $\dfrac{x^2 - 12x + 36}{4x - 24}$

4. $\dfrac{x^2 - 8x + 16}{3x - 12}$

5. $\dfrac{y^2 + 7y - 18}{y^2 - 3y + 2}$

6. $\dfrac{y^2 - 4y - 5}{y^2 + 5y + 4}$

7. $\dfrac{y^2 - 9y + 18}{y^3 - 27}$

8. $\dfrac{y^3 - 8}{y^2 + 2y - 8}$

9. $\dfrac{x^3 + x^2 - 20x}{x^3 + 2x^2 - 15x}$

10. $\dfrac{x^3 - 9x}{x^3 + 5x^2 + 6x}$

In Problems 11–24, multiply or divide as indicated.

11. $\dfrac{x^2 - 5x + 6}{x^2 - 2x - 3} \cdot \dfrac{x^2 - 1}{x^2 - 4}$

12. $\dfrac{x^2 + 5x + 6}{x^2 + x - 6} \cdot \dfrac{x^2 - 9}{x^2 - x - 6}$

13. $\dfrac{x^3 - 8}{x^2 - 4} \cdot \dfrac{x + 2}{3x}$

14. $\dfrac{x^2 + 6x + 9}{x^3 + 27} \cdot \dfrac{1}{x + 3}$

15. $\dfrac{x^2 - 1}{(x - 1)^2} \cdot \dfrac{x^3 - 1}{x + 1}$

16. $\dfrac{x^2 - 1}{x^2 - x + 1} \cdot \dfrac{x^3 + 1}{x^2 + 2x + 1}$

17. $\dfrac{2x^2 + 9x - 35}{6x^2 - 13x - 5} \cdot \dfrac{3x^2 + 10x + 3}{x^2 + 10x + 21}$

18. $\dfrac{x^2 + x - 12}{x^2 + x - 30} \cdot \dfrac{x^2 + 5x + 6}{x^2 - 2x - 3} \cdot \dfrac{x^2 + 7x + 6}{x + 3}$

19. $\dfrac{x^2 - 25}{2x - 2} \div \dfrac{x^2 + 10x + 25}{x^2 + 4x - 5}$

20. $\dfrac{x^2 - 4}{x^2 + 3x - 10} \div \dfrac{x^2 + 5x + 6}{x^2 + 8x + 15}$

21. $\dfrac{x^2 + 5x}{x + 5} \div \dfrac{1}{x^3 - 25x}$

22. $\dfrac{y - 1}{y^2 - 3y} \div \dfrac{1}{y^3 - 9y}$

23. $\dfrac{x^4 - 1}{2x} \div \dfrac{x^3 + x}{x^2}$

24. $\dfrac{1 - x^2}{6x + 6} \div \dfrac{x^4 - 1}{6x^2 + 6}$

In Problems 25–44, add or subtract as indicated.

25. $\dfrac{4x - 10}{x - 2} - \dfrac{x - 4}{x - 2}$

26. $\dfrac{x^2 - 4x}{x^2 - x - 6} - \dfrac{x - 6}{x^2 - x - 6}$

27. $\dfrac{3}{x + 4} + \dfrac{6}{x + 5}$

28. $\dfrac{8}{x - 2} + \dfrac{2}{x - 3}$

29. $\dfrac{3}{x + 1} - \dfrac{3}{x}$

30. $\dfrac{4}{x} - \dfrac{3}{x + 3}$

31. $\dfrac{2x}{x + 2} + \dfrac{x + 2}{x - 2}$

32. $\dfrac{3x}{x - 3} - \dfrac{x + 4}{x + 2}$

33. $\dfrac{x + 5}{x - 5} + \dfrac{x - 5}{x + 5}$

34. $\dfrac{x + 3}{x - 3} + \dfrac{x - 3}{x + 3}$

35. $\dfrac{4}{x^2 + 6x + 9} + \dfrac{4}{x + 3}$

36. $\dfrac{3}{5x + 2} + \dfrac{5x}{25x^2 - 4}$

37. $\dfrac{3x}{x^2 + 3x - 10} - \dfrac{2x}{x^2 + x - 6}$

38. $\dfrac{x}{x^2 - 2x - 24} - \dfrac{x}{x^2 - 7x + 6}$

39. $\dfrac{y + 3}{y^2 - y - 2} - \dfrac{y - 1}{y^2 + 2y + 1}$

40. $\dfrac{y - 1}{y^2 + 3y + 2} - \dfrac{y + 7}{y^2 + 5y + 6}$

41. $\dfrac{2}{y^2 - 4} - \dfrac{3}{y^2 - 4y + 4} + \dfrac{4}{y^2 + y - 2}$

42. $\dfrac{y - 3}{y - 2} + \dfrac{7 - 4y}{2y^2 - 9y + 10} - \dfrac{y + 1}{2y - 5}$

43. $\dfrac{4x^2 + x - 6}{x^2 + 3x + 2} - \dfrac{3x}{x + 1} + \dfrac{5}{x + 2}$

44. $\dfrac{6x^2 + 17x - 40}{x^2 + x - 20} + \dfrac{3}{x - 4} - \dfrac{5x}{x + 5}$

Simplify each complex rational expression in Problems 45–62.

45. $\dfrac{\dfrac{x}{3} - 1}{x - 3}$

46. $\dfrac{\dfrac{x}{4} - 1}{x - 4}$

47. $\dfrac{1 + \dfrac{1}{x}}{3 - \dfrac{1}{x}}$

48. $\dfrac{8 + \dfrac{1}{x}}{4 - \dfrac{1}{x}}$

49. $\dfrac{\dfrac{1}{x} + \dfrac{1}{y}}{x + y}$

50. $\dfrac{1 - \dfrac{1}{x}}{xy}$

51. $\dfrac{c - \dfrac{c}{c + 3}}{c + 2}$

52. $\dfrac{b - 3}{b - \dfrac{3}{b - 2}}$

53. $\dfrac{\dfrac{3}{y - 2} - \dfrac{4}{y + 2}}{\dfrac{7}{y^2 - 4}}$

54. $\dfrac{\dfrac{y}{y - 2} + 1}{\dfrac{3}{y^2 - 4} + 1}$

55. $\dfrac{\dfrac{1}{x + 1}}{\dfrac{1}{x^2 - 2x - 3} + \dfrac{1}{x - 3}}$

56. $\dfrac{\dfrac{6}{x^2 + 2x - 15} - \dfrac{1}{x - 3}}{\dfrac{1}{x + 5} + 1}$

57. $\dfrac{\dfrac{1}{x^3 - y^3}}{\dfrac{1}{x - y} - \dfrac{1}{x^2 + xy + y^2}}$

58. $\dfrac{\dfrac{1}{x^3 - 125}}{\dfrac{1}{x^2 - 25} - \dfrac{1}{x^2 + 5x + 25}}$

59. $\dfrac{\dfrac{1}{(x + h)^2} - \dfrac{1}{x^2}}{h}$

60. $\dfrac{\dfrac{x + h}{x + h + 1} - \dfrac{x}{x + 1}}{h}$

61. $\dfrac{2x^{-1} + y^{-1}}{xy}$

62. $\dfrac{x^{-1} - 2(x + 2)^{-1}}{x^{-2}}$

Simplify each fractional expression in Problems 63–68.

63. $\dfrac{\dfrac{1}{\sqrt{x + h}} - \dfrac{1}{\sqrt{x}}}{h}$

64. $\dfrac{\sqrt{y} - \dfrac{1}{4\sqrt{y}}}{\sqrt{y}}$

65. $\dfrac{\dfrac{y^2}{\sqrt{y^2 + 2}} - \sqrt{y^2 + 2}}{y^2}$

66. $\dfrac{\sqrt{5 - y^2} + \dfrac{y^2}{\sqrt{5 - y^2}}}{5 - y^2}$

67. $\dfrac{4y(y^2 - 9)y^{2/3} - 2y^{-1/3}(y^2 - 9)^2}{(y^{2/3})^2}$

68. $\dfrac{y^3(1 - y^2)^{-1/2} + 2y(1 - y^2)^{1/2}}{y^4}$

Application Problems

69. The polynomial $-0.14t^2 + 0.51t + 31.6$ describes the U.S. population (in millions) age 65 and older t years after 1990. The polynomial $0.54t^2 + 12.64t + 107.1$ describes the total yearly cost of Medicare (in billions of dollars) t years after 1990. Write a rational expression that describes the average cost of Medicare per person age 65 or older t years after 1990. Then find this average cost for the years 1995 through 1998.

70. The average speed on a round-trip commute having a one-way distance d is given by the complex rational expression

$$\dfrac{2d}{\dfrac{d}{r_1} + \dfrac{d}{r_2}}$$

in which r_1 and r_2 are the speeds on the outgoing and return trips, respectively. Simplify the expression. Then find the average speed for a person who drives from home to work at 30 miles per hour and returns on the same route averaging 20 miles per hour. Explain why the answer is not 25 miles per hour.

71. If three resistors with resistances R_1, R_2, and R_3 are connected in parallel, their combined resistance R is given by the equation

$$R = \frac{1}{\dfrac{1}{R_1} + \dfrac{1}{R_2} + \dfrac{1}{R_3}}.$$

Simplify the complex rational expression on the right. Then find R when R_1 is 4 ohms, R_2 is 8 ohms, and R_3 is 12 ohms.

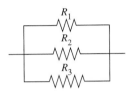

72. Write a rational expression in simplified form that describes how many times greater the area of the rectangle is than the area of the triangle in the figure at the top of the next column.

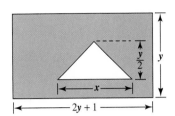

73. Write a rational expression in simplified form (and in terms of π) that describes how many times greater the area of the rectangle is than the area of the circle in the figure.

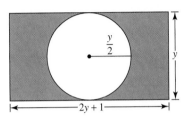

True–False Critical Thinking Problems

74. Which one of the following is true?

a. $\dfrac{x^2 - 25}{x - 5} = x - 5$

b. $\dfrac{x}{y} \div \dfrac{y}{x} = 1$, if $x \neq 0$ and $y \neq 0$.

c. The LCD needed to find $\dfrac{1}{x} + \dfrac{1}{x + 3}$ is $x + 3$.

d. $\dfrac{x}{7} - 1 = \dfrac{x - 7}{7}$ for all real numbers x.

75. Which one of the following is true?

a. $\dfrac{1}{a} + \dfrac{1}{b} = \dfrac{1}{a + b}$

b. $\dfrac{a + b}{a} = b$

c. $\dfrac{a}{b + c} = \dfrac{a}{b} + \dfrac{a}{c}$

d. The rational expression

$$\frac{x^2 - 16}{x - 4}$$

is not defined for $x = 4$. However, as x gets closer and closer to 4, the value of the expression approaches 8.

Technology Problems

76. a. Show that $\dfrac{x^2 - x - 6}{x + 2} = x - 3$.

b. Use a graphing utility to graph $y_1 = \dfrac{x^2 - x - 6}{x + 2}$ and $y_2 = x - 3$ in the same viewing rectangle. Do the graphs appear to be identical? Don't be fooled by

this; if $x = -2$ then $\dfrac{x^2 - x - 6}{x + 2}$ is undefined, which is not shown on the graph. For all values of x other than -2, $\dfrac{x^2 - x - 6}{x + 2} = x - 3$. (Verify this fact using the $\boxed{\text{TRACE}}$ feature.)

In Problems 77–78, use a graphing utility to answer each question.

77. Is $\dfrac{x}{3} + \dfrac{1}{6}$ equivalent to $\dfrac{2x + 1}{6}$ or $\dfrac{x + 2}{6}$?

78. Is $\dfrac{1}{x} + \dfrac{1}{2}$ equivalent to $\dfrac{1}{x + 2}$ or $\dfrac{x + 2}{2x}$ (assuming $x \neq 0$)?

In Problems 79–83, perform the indicated operations and support your work with a graphing utility.

79. $\dfrac{x^2 - 4}{2 - x}$

80. $\dfrac{2}{x - 1} \div \dfrac{4}{x^2 - 1}$

81. $\dfrac{x}{2} + \dfrac{x}{5}$

82. $6 - \dfrac{5}{x + 3}$

83. $\dfrac{x}{x - 5} - \dfrac{5}{5 - x}$

Critical Thinking Problems

84. Perform the indicated operations:

$$\frac{1}{x^n - 1} - \frac{1}{x^n + 1} - \frac{1}{x^{2n} - 1}.$$

85. Simplify:

$$\left(\frac{a^{-2} + b^{-2}}{a^{-2} - b^{-2}} - \frac{a^{-2} - b^{-2}}{a^{-2} + b^{-2}} \right) \div \left[\left(\frac{a + b}{a - b} + \frac{a - b}{a + b} \right) \left(\frac{a^2}{b^2} + \frac{b^2}{a^2} - 2 \right) \right]^{-1}.$$

86. Find the product of the following 199 factors:

$$\left(1 - \frac{1}{2} \right) \left(1 - \frac{1}{3} \right) \left(1 - \frac{1}{4} \right) \cdots \left(1 - \frac{1}{x + 1} \right) \cdots \left(1 - \frac{1}{200} \right).$$

87. Simplify $(1 + b^{x-y})^{-1} + (1 + b^{y-x})^{-1}$ to obtain the smallest of the natural numbers.

Group Activity Problem

88. As a group, find out as much as you can about computer algebra systems such as *Mathematica*, *Maple*, and *Derive*, or graphing calculators with such systems built in (such as the TI-92). Use such a system to solve five of the practice problems that have appeared in Problem Sets P.5 through P.7. Within your group, debate the issue of whether or not graphing utilities capable of symbolic manipulation should eliminate the necessity for learning algebra's many manipulative skills, such as operations with radicals and rational expressions.

Solutions Manual   **Tutorial**   **Video 2**

SECTION P.8   Complex Numbers

Objectives

1 Find powers of *i*.

2 Add and subtract complex numbers.

3 Multiply complex numbers.

4 Divide complex numbers.

5 Perform operations with roots of negative numbers.

6 Plot complex numbers.

Some of the numbers that appear in algebra, particularly as the solutions of equations, are not real numbers. In this section, we expand the set of real numbers to include the square roots of negative numbers.

The Imaginary Unit *i*

In Chapter 1, we'll be studying equations whose solutions involve the square roots of negative numbers. Since the square of a real number is never negative, there is no real number x such that $x^2 = -1$. To provide a setting in which such equations have solutions, mathematicians invented an expanded system of numbers, the complex numbers. The imaginary number i, defined to be a solution to the equation $x^2 = -1$, is the basis of this new set.

The imaginary unit *i*

The imaginary unit i is defined as

$$i = \sqrt{-1}, \quad \text{where} \quad i^2 = -1.$$

Using the imaginary unit i, we can express the square root of any negative number as a real multiple of i. For example,

$$\sqrt{-25} = i\sqrt{25} = 5i.$$

We can check this result by squaring $5i$ and obtaining -25.

$$(5i)^2 = 5^2 i^2 = 25(-1) = -25$$

A new system of numbers, called *complex numbers*, is based on adding multiples of i, such as $5i$, to the real numbers.

Complex numbers

The set of all numbers in the form

$$a + bi$$

with real numbers a and b, and i, the imaginary unit, is called the set of *complex numbers*. The real number a is called the *real part*, and the real number b is called the *imaginary part* of the complex number $a + bi$. If $a = 0$ and $b \neq 0$, then the complex number bi is called a *pure imaginary number* (Figure P.39).

René Magritte "Portrait" (Le Portrait) 1935, oil on canvas, 28 7/8 × 19 7/8 in. The Museum of Modern Art, New York. Gift of Kay Sage Tanguy. Photograph © 1997 The Museum of Modern Art, New York. © 1998 C. Herscovici, Brussels / Artists Rights Society (ARS), New York.

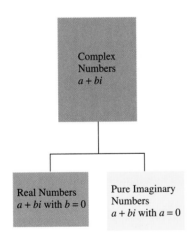

Figure P.39

The complex number system

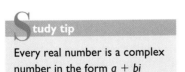

Study tip

Every real number is a complex number in the form $a + bi$ with $b = 0$.

A complex number is said to be *simplified* if it is expressed in the *standard form* $a + bi$. If b is a radical, we usually write i before b. For example, we write $7 + i\sqrt{5}$ rather than $7 + \sqrt{5}i$, which could easily be confused with $7 + \sqrt{5i}$.

Expressed in standard form, two complex numbers are equal if and only if their real parts are equal and their imaginary parts are equal.

tudy tip

The connective "if and only if" between two statements p and q means

  If p then q.
  AND
  If q then p.

The definition for equality of complex numbers is really two statements in one:

1. If $a + bi = c + di$, then $a = c$ and $b = d$.
2. If $a = c$ and $b = d$, then $a + bi = c + di$.

Find powers of i.

Equality of complex numbers

$a + bi = c + di$ if and only if $a = c$ and $b = d$.

Powers of i

The fact that $i^2 = -1$ can be used to find higher powers of i.

Discover for yourself

Replace i^2 with -1 and express each of the following in terms of $1, -1, i,$ or $-i$.

$$i^1 = i, \quad i^2 = -1, \quad i^3 = i^2 \cdot i = \underline{\quad}, \quad i^4 = (i^2)^2 = \underline{\quad},$$

$$i^5 = (i^2)^2 i = \underline{\quad}, \quad i^6 = (i^2)^3 = \underline{\quad}, \quad i^7 = (i^2)^3 i = \underline{\quad}, \quad i^8 = (i^2)^4 = \underline{\quad}$$

In the Discover for yourself, did you observe that the powers of i rotate through the four numbers $i, -1, -i,$ and 1?

Powers of i

$i^1 = i$	$i^5 = i$	$i^9 = i$	$i^{13} = i$
$i^2 = -1$	$i^6 = -1$	$i^{10} = -1$	$i^{14} = -1$
$i^3 = -i$	$i^7 = -i$	$i^{11} = -i$	$i^{15} = -i$
$i^4 = 1$	$i^8 = 1$	$i^{12} = 1$	$i^{16} = 1$

Although we can use the rotating pattern to simplify higher powers of i, an easier method is to write these powers in terms of i^2 and then replace i^2 by -1.

EXAMPLE 1 **Finding Powers of i**

Simplify:

a. i^{12} **b.** i^{39} **c.** i^{50} **d.** i^{1883}

Discover for yourself

Since $i^4 = 1$, another method for simplifying higher powers of i is to write them in terms of i^4. Use this method to simplify the numbers in Example 1.

Solution

a. $i^{12} = (i^2)^6 = (-1)^6 = 1$
b. $i^{39} = i^{38}i = (i^2)^{19}i = (-1)^{19}i = (-1)i = -i$
c. $i^{50} = (i^2)^{25} = (-1)^{25} = -1$
d. $i^{1883} = i^{1882}i = (i^2)^{941}i = (-1)^{941}i = (-1)i = -i$ ∎

Operations with Complex Numbers

The form of a complex number $a + bi$ is like the binomial $a + bx$. Consequently, we can add, subtract, and multiply complex numbers using the same methods we used for binomials, remembering that $i^2 = -1$.

2 Add and subtract complex numbers.

Adding and subtracting complex numbers

1. $(a + bi) + (c + di) = (a + c) + (b + d)i$
In words, this says that you add complex numbers by adding their real parts, adding their imaginary parts, and expressing the sum as a complex number.

2. $(a + bi) - (c + di) = (a - c) + (b - d)i$
In words, this says that you subtract complex numbers by subtracting their real parts, subtracting their imaginary parts, and expressing the difference as a complex number.

EXAMPLE 2   **Adding and Subtracting Complex Numbers**

Perform the indicated operations, writing the result in standard form.

a. $(5 - 11i) + (7 + 4i)$ **b.** $(-5 + 7i) - (-11 - 6i)$
c. $4 - (-7 + 3i) + (-11 + 6i)$

Study tip

The following examples, using the same integers as in Example 2, show how operations with complex numbers are just like operations with polynomials.

a. $(5 - 11x) + (7 + 4x)$
   $= 12 - 7x$
b. $(-5 + 7x) - (-11 - 6x)$
   $= -5 + 7x + 11 + 6x$
   $= 6 + 13x$
c. $4 - (-7 + 3x) + (-11 + 6x)$
   $= 4 + 7 - 3x - 11 + 6x$
   $= 3x$

Solution

a. $(5 - 11i) + (7 + 4i)$
$= 5 - 11i + 7 + 4i$ Remove the parentheses.
$= 5 + 7 - 11i + 4i$ Group real and imaginary terms.
$= (5 + 7) + (-11 + 4)i$
$= 12 - 7i$ Do you find that you're working all steps mentally?

b. $(-5 + 7i) - (-11 - 6i)$
$= -5 + 7i + 11 + 6i$ Remove the parentheses.
$= -5 + 11 + 7i + 6i$ Group real and imaginary terms.
$= (-5 + 11) + (7 + 6)i$
$= 6 + 13i$

c. $4 - (-7 + 3i) + (-11 + 6i)$
$= 4 + 7 - 3i - 11 + 6i$ Remove the parentheses.
$= 4 + 7 - 11 - 3i + 6i$ Group the real and imaginary terms.
$= (4 + 7 - 11) + (-3 + 6)i$
$= 0 + 3i$ Equivalently, the answer can be expressed as $3i$. ∎

3 Multiply complex numbers.

Multiplication of complex numbers is performed the same way as multiplication of polynomials, using the distributive property and the FOIL method. After completing the multiplication, we replace i^2 with -1. This idea is illustrated in the next example.

EXAMPLE 3   **Multiplying Complex Numbers**

Find the products:

a. $4i(3 - 5i)$ **b.** $(7 - 3i)(-2 - 5i)$ **c.** $(4 + 3i)(4 - 3i)$

Solution

a. $4i(3-5i) = 4i(3) - 4i(5i)$ Distribute $4i$ throughout the parentheses.

$\qquad\qquad = 12i - 20i^2$ Multiply.

$\qquad\qquad = 12i - 20(-1)$ Replace i^2 with -1.

$\qquad\qquad = 12i + 20$ Simplify. In standard form, the product can be expressed as $20 + 12i$.

b. $(7 - 3i)(-2 - 5i)$

$\qquad\quad\ \ \ \overset{\text{F}\quad\ \text{O}\quad\ \ \text{I}\quad\ \ \text{L}}{}$

$\qquad = -14 - 35i + 6i + 15i^2$ Use the FOIL method.

$\qquad = -14 - 35i + 6i + 15(-1)$ $i^2 = -1$

$\qquad = -14 - 15 - 35i + 6i$ Group real and imaginary terms.

$\qquad = -29 - 29i$ Combine real and imaginary terms.

c. $(4 + 3i)(4 - 3i)$

$\qquad\quad\ \ \ \overset{\text{F}\quad\ \ \text{O}\quad\ \ \text{I}\quad\ \ \text{L}}{}$

$\qquad = 16 - 12i + 12i - 9i^2$ Use the FOIL method.

$\qquad = 16 - 9(-1)$ $i^2 = -1$

$\qquad = 25$ We can also write $25 + 0i$. ∎

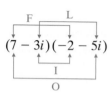

$(7 - 3i)(-2 - 5i)$

with labels F, L, I, O

Discover for yourself

Work Example 3c using

$(A + B)(A - B) = A^2 - B^2$

with

$A = 4$ and $B = 3i$.

Which method do you find easier?

Discover for yourself

It is not necessary to memorize the definitions for addition and subtraction of complex numbers to perform these operations. The same is true for multiplication, which is defined by

$(a + bi)(c + di) = (ac - bd) + (ad + bc)i.$

Show that multiplying $(a + bi)(c + di)$ by the FOIL method gives the preceding result.

---

4 Divide complex numbers.

Complex Conjugates and Division

In Example 3c, we found that

$(4 + 3i)(4 - 3i) = 25.$

Thus, the product of two complex numbers can be a real number. Generalizing from this situation, let's multiply $a + bi$ and $a - bi$.

$\qquad\qquad\qquad\quad \overset{\text{F}\qquad\ \text{O}\qquad\ \text{I}\qquad\ \text{L}}{}$

$(a + bi)(a - bi) = a^2 - abi + abi - b^2i^2$ Use the FOIL method.

$\qquad\qquad\qquad = a^2 - b^2(-1)$ $i^2 = -1$

$\qquad\qquad\qquad = a^2 + b^2$ Notice that this product eliminates i.

For the complex number $a + bi$, we define its *complex conjugate* to be $a - bi$. The multiplication of complex conjugates results in a real number.

Conjugate of a complex number

The complex conjugate of the number $a + bi$ is $a - bi$, and the complex conjugate of $a - bi$ is $a + bi$. The multiplication of complex conjugates gives a real number.

$(a + bi)(a - bi) = a^2 + b^2$

$(a - bi)(a + bi) = a^2 + b^2$

EXAMPLE 4   **Multiplying Complex Conjugates**

Multiply:

a. $(3 + 4i)(3 - 4i)$ **b.** $(7 - 6i)(7 + 6i)$

Solution

$$
\begin{array}{cccccc}
(a & + & bi) & (a & - & bi) & = & a^2 & + & b^2 \\
\downarrow & & \downarrow & \downarrow & & \downarrow & & \downarrow & & \downarrow
\end{array}
$$

a. $(3 + 4i)(3 - 4i) = 3^2 + 4^2$
$$= 9 + 16$$
$$= 25$$

$$
\begin{array}{cccccc}
(a & - & bi) & (a & + & bi) & = & a^2 & + & b^2 \\
\downarrow & & \downarrow & \downarrow & & \downarrow & & \downarrow & & \downarrow
\end{array}
$$

b. $(7 - 6i)(7 + 6i) = 7^2 + 6^2$
$$= 49 + 36$$
$$= 85 \qquad ∎$$

Discover for yourself

Consider the division problem

$$\frac{7 + 4i}{2 - 5i}.$$

What must be done to both the numerator and the denominator so that the denominator contains only a real number?

In the Discover for yourself, did you observe that multiplying the numerator and the denominator by $2 + 5i$ (the conjugate of the denominator) results in a real number in the denominator? We see how conjugates are used when dividing complex numbers in Example 5.

EXAMPLE 5   **Using Conjugates to Divide Complex Numbers**

Divide: $7 + 4i$ by $2 - 5i$

Solution

We first write the problem as $\dfrac{7 + 4i}{2 - 5i}$. The conjugate of the denominator, $2 - 5i$, is $2 + 5i$, so we multiply the numerator and the denominator by $2 + 5i$.

$$\frac{7 + 4i}{2 - 5i} = \frac{(7 + 4i)}{(2 - 5i)} \cdot \frac{(2 + 5i)}{(2 + 5i)}$$

Multiply the numerator and the denominator by the conjugate of the denominator.

$$
= \frac{\overset{\text{F}\quad\text{O}\quad\text{I}\quad\text{L}}{14 + 35i + 8i + 20i^2}}{2^2 + 5^2}
$$

Use the FOIL method in the numerator and $(a - bi)(a + bi) = a^2 + b^2$ in the denominator.

$$= \frac{14 + 43i + 20(-1)}{29}$$

Replace i^2 by -1.

$$= \frac{-6 + 43i}{29}$$

Combine real terms in the numerator.

$$= -\frac{6}{29} + \frac{43}{29}i$$

Observe that the quotient is expressed in the form $a + bi$, with $a = -\frac{6}{29}$ and $b = \frac{43}{29}$. ∎

Using conjugates to divide complex numbers

When dividing two complex numbers, if necessary express the indicated division as a fraction and then multiply the numerator and the denominator by the complex conjugate of the denominator.

EXAMPLE 6 **Dividing Complex Numbers**

Divide and simplify to the form $a + bi$: $\dfrac{10 + 6i}{5i}$

Solution

The conjugate of $0 + 5i$ is $0 - 5i$, or $-5i$, so we will multiply the numerator and the denominator by $-5i$.

$$\frac{10 + 6i}{5i} = \frac{10 + 6i}{5i} \cdot \frac{-5i}{-5i}$$

Multiply by 1, using the conjugate of the denominator.

$$= \frac{-50i - 30i^2}{-25i^2}$$

Multiply, using the distributive property in the numerator.

$$= \frac{-50i - 30(-1)}{-25(-1)}$$

$i^2 = -1$

$$= \frac{-50i + 30}{25}$$

Multiply.

$$= \frac{-50}{25}i + \frac{30}{25}$$

To write the quotient in the form $a + bi$, divide each term in the numerator by 25.

$$= \frac{6}{5} - 2i$$

Reverse the order of the terms and simplify.

The quotient is in the form $a + bi$, with $a = \frac{6}{5}$ and $b = -2$. ∎

---

5 Perform operations with roots of negative numbers.

Roots of Negative Numbers

The square of $4i$ and the square of $-4i$ both result in -16.

$$(4i)^2 = 16i^2 = 16(-1) = -16 \qquad (-4i)^2 = 16i^2 = -16$$

Consequently, in the complex number system -16 has two square roots, namely, $4i$ and $-4i$. We call $4i$ the *principal square root* of -16.

Principal square root of a negative number

For any positive number real number b, the *principal square root* of the negative number $-b$ is defined by

$$\sqrt{-b} = i\sqrt{b}.$$

Study tip

Do not apply the properties

$$\sqrt{b}\,\sqrt{c} = \sqrt{bc} \quad \text{and} \quad \frac{\sqrt{b}}{\sqrt{c}} = \sqrt{\frac{b}{c}}$$

to the pure imaginary numbers because these properties can only be used when b and c are positive.

Correct:	**Incorrect:**
$\sqrt{-25}\,\sqrt{-4} = i\sqrt{25}\,i\sqrt{4}$	$\sqrt{-25}\,\sqrt{-4} = \sqrt{(-25)(-4)}$
$\quad = (5i)(2i)$	$= \sqrt{100}$
$\quad = 10i^2$	$= 10$
$\quad = 10(-1)$	
$\quad = -10$	

One way to avoid confusion is to *represent all imaginary numbers in terms of i before adding, subtracting, multiplying, or dividing.*

EXAMPLE 7   **Operations Involving Square Roots of Negative Numbers**

Perform the indicated operations and write the result in standard form:

a. $\sqrt{-18} - \sqrt{-8}$ **b.** $(-1 + \sqrt{-5})^2$

c. $\dfrac{-25 + \sqrt{-50}}{15}$ **d.** $\sqrt{-12}\,(\sqrt{-3} - \sqrt{2})$

Study tip

The first thing to do in each part of Example 7 is to express $\sqrt{-b}$ as $i\sqrt{b}$.

Solution

a. $\sqrt{-18} - \sqrt{-8} = i\sqrt{18} - i\sqrt{8} = i\sqrt{9 \cdot 2} - i\sqrt{4 \cdot 2} = 3i\sqrt{2} - 2i\sqrt{2} = i\sqrt{2}$

$$\begin{array}{ccccccccc}
(A & + & B)^2 & = & A^2 & + & 2 & A & B & + & B^2 \\
\downarrow & & \downarrow & & \downarrow & & \downarrow & \downarrow & \downarrow & & \downarrow
\end{array}$$

b. $(-1 + \sqrt{-5})^2 = (-1 + i\sqrt{5})^2 = (-1)^2 + 2(-1)(i\sqrt{5}) + (i\sqrt{5})^2$

$$= 1 - 2i\sqrt{5} + 5i^2$$
$$= 1 - 2i\sqrt{5} + 5(-1)$$
$$= -4 - 2i\sqrt{5}$$

c. $\dfrac{-25 + \sqrt{-50}}{15}$

$= \dfrac{-25 + i\sqrt{50}}{15}$ $\sqrt{-b} = i\sqrt{b}$

$= \dfrac{-25 + 5i\sqrt{2}}{15}$ $\sqrt{50} = \sqrt{25 \cdot 2} = 5\sqrt{2}$

$$= \frac{-25}{15} + \frac{5i\sqrt{2}}{15} \qquad \text{Write the complex number in standard form.}$$

$$= -\frac{5}{3} + i\frac{\sqrt{2}}{3} \qquad \text{Simplify.}$$

d. $\sqrt{-12}\,(\sqrt{-3} - \sqrt{2})$

$$= i\sqrt{12}\,(i\sqrt{3} - \sqrt{2}) \qquad \sqrt{-b} = i\sqrt{b}$$

$$= i^2\sqrt{36} - i\sqrt{24} \qquad \text{Apply the distributive property.}$$

$$= (-1)(6) - 2i\sqrt{6} \qquad \sqrt{24} = \sqrt{4\cdot 6} = 2\sqrt{6};\ i^2 = -1$$

$$= -6 - 2i\sqrt{6} \qquad \text{Express the product in standard form.} \qquad ∎$$

Figure P.40

The complex plane

Plotting Complex Numbers

The *complex plane*, shown above in Figure P.40, consists of a horizontal axis, called the *real axis*, and a vertical axis, called the *imaginary axis*. Every complex number corresponds to a point in the complex plane and every point in the complex plane corresponds to a complex number.

6 Plot complex numbers.

EXAMPLE 8   Plotting Complex Numbers

Plot in the complex plane:

a. $3 + 4i$ **b.** $-1 - 2i$ **c.** -3 **d.** $-4i$

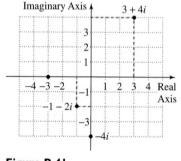

Figure P.41

Plotting complex numbers

Solution (See Figure P.41)

a. The complex number $3 + 4i$ is plotted in the same way we plot $(3, 4)$ in the Cartesian coordinate plane. We move 3 units to the right on the real axis and 4 units up parallel to the imaginary axis.

b. The complex number $-1 - 2i$ corresponds to the point $(-1, -2)$ in the Cartesian coordinate plane.

c. Since $-3 = -3 + 0i$, this number corresponds to the point $(-3, 0)$. We plot -3 by moving 3 units to the left on the real axis.

d. Since $-4i = 0 - 4i$, we plot $-4i$ by moving 4 units down on the imaginary axis. ∎

PROBLEM SET P.8

Practice Problems

In Problems 1–6, simplify each expression.

1. i^{24} **2.** i^{36} **3.** i^{75} **4.** i^{37}

5. i^{49} **6.** i^{53}

In Problems 7–14, add or subtract as indicated and write the result in standard form.

7. $(7 + 2i) + (1 - 4i)$ **8.** $(-2 + 6i) + (4 - i)$ **9.** $(3 + 2i) - (5 - 7i)$

10. $(-7 + 5i) - (-9 - 11i)$ **11.** $6 - (-5 + 4i) - (-13 - 11i)$ **12.** $7 - (-9 + 2i) - (-17 - 6i)$

13. $6 - (5 - 2i\sqrt{32}) - (11 - 5i\sqrt{8})$ **14.** $7 - (3 + 7i\sqrt{8}) - (-10 - 2i\sqrt{50})$

In Problems 15–28, find each product and write the result in standard form.

15. $-3i(7i - 5)$ **16.** $-8i(2i - 7)$ **17.** $(-5 + 4i)(3 + 7i)$ **18.** $(-4 - 8i)(3 + 9i)$

19. $(7 - 5i)(-2 - 3i)$ **20.** $(8 - 4i)(-3 + 9i)$ **21.** $(3 + 5i)(3 - 5i)$ **22.** $(2 + 7i)(2 - 7i)$

23. $(-5 + 3i)(-5 - 3i)$ **24.** $(-7 - 4i)(-7 + 4i)$ **25.** $(2 + 3i)^2$ **26.** $(5 - 2i)^2$

27. $\left(\dfrac{1}{2} + \dfrac{\sqrt{3}}{2}i\right)^2$ **28.** $\left(\dfrac{\sqrt{3}}{2} - \dfrac{1}{2}i\right)^2$

In Problems 29–42, divide and express the result in standard form.

29. $\dfrac{2}{3 - i}$ **30.** $\dfrac{3}{4 + i}$ **31.** $\dfrac{2i}{1 + i}$ **32.** $\dfrac{5i}{2 - i}$

33. $\dfrac{8i}{4 - 3i}$ **34.** $\dfrac{-6i}{3 + 2i}$ **35.** $\dfrac{2 + 3i}{2 + i}$ **36.** $\dfrac{3 - 4i}{4 + 3i}$

37. $\dfrac{-4 + 7i}{-2 - 5i}$ **38.** $\dfrac{-3 - i}{-4 - 3i}$ **39.** $\dfrac{3 - i}{2i}$ **40.** $\dfrac{4 + i}{3i}$

41. $\dfrac{-4 + 7i}{-5i}$ **42.** $\dfrac{-3 - i}{-3i}$

In Problems 43–52, perform the indicated operations and write the result in standard form.

43. $(8 + 9i)(2 - i) - (1 - i)(1 + i)$ **44.** $(2 - 3i)(1 - i) - (3 - i)(3 + i)$

45. $\dfrac{4}{(2 + i)(3 - i)}$ **46.** $\dfrac{6}{(2 + i)(1 - i)}$ **47.** $(1 - 3i)^3$ **48.** $(1 + 2i)^3$

49. $i^4(1 + i^3)$ **50.** $6i^4(2 - i^5)$ **51.** i^{-17} **52.** i^{-13}

In Problems 53–68, perform the indicated operations and write the result in standard form.

53. $\sqrt{-64} - \sqrt{-25}$ **54.** $\sqrt{-81} - \sqrt{-144}$ **55.** $5\sqrt{-16} + 3\sqrt{-81}$ **56.** $5\sqrt{-8} + 3\sqrt{-18}$

57. $(-2 + \sqrt{-4})^2$ **58.** $(-5 - \sqrt{-9})^2$ **59.** $(-3 - \sqrt{-7})^2$ **60.** $(-2 + \sqrt{-11})^2$

61. $\dfrac{-8 + \sqrt{-32}}{24}$ **62.** $\dfrac{-12 + \sqrt{-28}}{32}$ **63.** $\dfrac{-6 - \sqrt{-12}}{48}$ **64.** $\dfrac{-15 - \sqrt{-18}}{33}$

65. $\sqrt{-8}\,(\sqrt{-3} - \sqrt{5})$ **66.** $\sqrt{-12}\,(\sqrt{-4} - \sqrt{2})$ **67.** $(3\sqrt{-5})(-4\sqrt{-12})$ **68.** $(3\sqrt{-7})(2\sqrt{-8})$

In Problems 69–76, plot each complex number in the complex plane.

69. $-3 + i$ **70.** $-4 - 2i$ **71.** $1 - 5i$ **72.** $2 - 3i$

73. i **74.** $-3i$ **75.** 2 **76.** -4

True-False Critical Thinking Problems

77. Which one of the following is true?

    a. Some irrational numbers are not complex numbers.

    b. $i^{-1} = i$

    c. $\dfrac{7 + 3i}{5 + 3i} = \dfrac{7}{5}$

    d. In the complex number system, $x^2 + y^2$ (the sum of two squares) can be factored as $(x + yi)(x - yi)$.

78. Which one of the following is true?

    a. $\dfrac{20}{3 - 2i} = \dfrac{20}{3 - 2i} \cdot \dfrac{3 + 2i}{3 + 2i} = \dfrac{60 + 40i}{9 - 4} = 12 + 8i$

    b. $(\sqrt{-5} + 4)(i - \sqrt{-7})$
        $= i\sqrt{-5} - \sqrt{-5}\sqrt{-7} + 4i - 4\sqrt{-7}$
        $= i \cdot i\sqrt{5} - \sqrt{35} + 4i - 4i\sqrt{7}$
        $= -\sqrt{5} - \sqrt{35} + 4i - 4i\sqrt{7}$

    c. $i^7(1 + i^3) = 0$

    d. $\dfrac{i^4 + i^3 + i^2 + i + 1}{1 + i} = \dfrac{1}{2} - \dfrac{1}{2}i$

Writing in Mathematics

Explain the error in Problems 79–82.

79. $-\sqrt{-100} = -(-\sqrt{100}) = \sqrt{100} = 10$

80. $(\sqrt{-9})^2 = \sqrt{-9}\sqrt{-9} = \sqrt{81} = 9$

81. $\sqrt{-9} + \sqrt{-16} = \sqrt{-25} = i\sqrt{25} = 5i$

82. The only square root of -1 is i.

Critical Thinking Problems

83. Show that $\sqrt[3]{1} = -\dfrac{1}{2} + \dfrac{\sqrt{3}}{2}i$ by showing that $\left(-\dfrac{1}{2} + \dfrac{\sqrt{3}}{2}i\right)^3 = 1$.

84. Perform the indicated operations: $\dfrac{1+i}{1+2i} + \dfrac{1-i}{1-2i}$.

85. Simplify: i^{2n+1}.

86. Simplify: $\left(\dfrac{i}{1+i} + \dfrac{2+i}{2-i}\right)i$.

Group Activity Problem

87. The statement *p if and only if q* is two statements in one. The two statements are

    If *p* then *q*.
    AND
    If *q* then *p*.

Group members should explain whether the connective *if and only if* is used correctly or incorrectly in each of the following sentences.

  a. A number is positive if and only if that number is greater than zero.

  b. A person is in Washington D.C. if and only if that person is in the United States.

  c. Shakespeare is the author if and only if the play is *Macbeth*.

  d. *x* is not greater than 5 if and only if *x* is less than or equal to 5.

  e. The team won if and only if they did not lose.

  f. *a* = *b* if and only if *b* = *a*.

  g. A triangle has two sides equal in measure if and only if it has two angles equal in measure.

Group members should now write eight sentences, four which illustrate the correct use of *if and only if* and four that illustrate its misuse. Consult a general mathematics textbook to find two other ways to express *if and only if*. Express your five correct sentences using these equivalent connectives.

CHAPTER PROJECT

Jurassic Math: A Matter of Scale

Looking at numbers describing the physical attributes of a dinosaur is not quite the same as watching a movie filled with fast-moving predators, or seeing a full-scale reproduction in a museum. We may understand an equation in an intellectual sense, yet still feel we are missing some connection to make it real. There is always a human side to presenting information, making it accessible at a level other than an abstract, intellectual understanding. The mathematics in this chapter represents a large amount of information in a compact, symbolic form, but has something been lost in the translation?

For example, judging the size or scale of something is not an exercise in fantasy; it is an important skill for many people involved in the design, construction, or alteration of the landscape around us. To give more information to a client, architects create models of their plans complete with familiar sizes of people and objects. Engineers create models of differing scales as they design new cars, planes, or other objects. For the ultimate model, technological advances make it possible to put on a headset and use virtual reality to walk through a building and investigate how the design will fit into a human scale.

We can explore how the presentation of information affects the assimilation of information. In other words, how do we make the information *real*. We will use something familiar in an intellectual sense, but unfamiliar to most people in a realistic sense: dinosaurs. Given only a list of numbers describing the height and length of a dinosaur, would you be able to make realistic judgments about how this historic creature would fit into your world? Could a Tyrannosaurus Rex peer in the window of a three-story building? Could a Stegosaurus fit in your classroom? How many parking spaces would a Brontosaurus fill?

1. In a small group, select a dinosaur to study. Write the approximate dimensions of the dinosaur in terms of height and length in an uncommon unit of measurement, such as inches, centimeters, or a fraction of a mile. Ask several people to make a comparative estimate of the size of the dinosaur from these measurements. For example, you might ask, "Is it bigger than a car?" or "Is it taller than this house?" You may tell the person if the dinosaur walked on two or four legs, but do not reveal the name of the dinosaur. Record the answers from several sources and compare the results with the rest of your group. How accurate were the answers?

2. Conduct another set of interviews, this time giving the information in a more typical form, such as feet or yards. How does the accuracy of the new estimates compare to the ones in Problem 1? Share your results with other groups in the class. Does the actual size of the dinosaur seem to affect the estimates? In other words, does the size of larger dinosaurs seem harder to estimate accurately than smaller ones or vice versa?

3. Reverse the process of giving out information. Give someone a description of the size of a dinosaur by explaining it in terms of familiar objects. For example, "It is as long as three cars put together." Ask for an estimate in numerical terms. How do these estimates compare with the others? Would you expect different results if you had given the name of the dinosaur?

4. Use a variety of methods to show how a dinosaur would fit into the reality of your school by placing a dinosaur in familiar settings. Your group may use one or more of the following ideas or create your own:
   - Take a photograph of a familiar part of your campus and place a silhouette of your dinosaur, sized to scale, in the picture.
   - Begin with a picture of your dinosaur and insert silhouettes of familiar objects, drawn to scale.
   - Use a small plastic model of your dinosaur and place it in a cardboard model of some part of the campus, constructed to scale.
   - Create a life-sized measurement of your dinosaur. Use the gym or a field and mark the length of the dinosaur on the ground while using a helium balloon to show the height of the head.

5. Discuss with the class how well the methods in Problem 4 presented the reality of a dinosaur. Which style of presentation would be preferred if you stayed inside a classroom? Which presentation would be preferred if a life-sized reproduction was constructed? The weight of a dinosaur was never mentioned. What kind of effect would the weight have on understanding the scale of a dinosaur?

6. Use the numerical data on dinosaurs in each group to put together some of the different styles of equations found in Example 5 on pages 33–34. Each group should be able to create at least three equations, one linear, one whose graph is a parabola, and one with a radical, which give numerical solutions matching the values used for a particular dinosaur. Would solving an equation to obtain the length of a dinosaur give you a better understanding of its length?

Worldwide Web Resources

 Go to the Prentice Hall website (http://www.prenhall.com/blitzer) to access other locations on the Internet that will allow you to further explore the concepts presented in this project.

Chapter Review

S U M M A R Y

(Starting with Chapter 1, a complete summary will be provided at the end of each chapter. For this chapter on reviewing fundamental concepts, we simply refer you back to some of the significant tables and definitions presented throughout the prerequisites survey.)

1. The Real Numbers
   a. Subsets of the real numbers are defined in Table P.1 on pages 2–3.

   b. Interval notation is summarized in Table P.2 on page 5.

2. Algebraic Expressions and Exponents

  a. Algebra's nine basic properties appear in the box on pages 14–15.

  b. Properties of negatives are summarized in the box on page 16.

  c. Properties of zero are summarized in the box on page 17.

  d. Properties that are useful in solving equations are found in the box on pages 17–18.

  e. Properties of exponents and examples are found in the box on pages 18–19.

  f. The definition of scientific notation is given in the box on page 22.

3. Graphing

  a. The distance formula appears in a box on page 30.

  b. The midpoint formula appears in a box on page 31.

  c. Graphing equations by plotting points is reviewed in the box on page 32.

  d. Intercepts are reviewed in the box on pages 34–35.

  e. An introduction to graphing utilities appears on pages 35–37.

4. Radicals and Rational Exponents

  a. Properties of nth roots with examples appear in the box on page 45.

  b. The conditions for simplified radical form are given on page 45.

  c. Rational exponents are defined in the box on page 50.

5. Polynomials

  a. The definition of a polynomial in x is given in the box on page 56.

  b. Operations with polynomials are discussed beginning on page 57.

  c. Special products that are useful in polynomial multiplication are given in the box on page 61.

6. Factoring

  a. A general factoring strategy is shown in the box on pages 73–74.

  b. A discussion of factoring techniques for expressions that are not polynomials begins on page 74.

7. Rational Expressions

  a. Properties of rational expressions with examples appear in the box on pages 80–81.

  b. A discussion of simplifying complex rational expressions begins on page 84.

8. Complex Numbers

  a. Complex numbers are defined in the box on page 91.

  b. Powers of i are shown in the box on page 92.

  c. Operations with complex numbers are presented beginning on page 92.

  d. The use of conjugates to divide complex numbers appears in the box on page 96.

  e. A method for plotting complex numbers is shown on page 98.

REVIEW PROBLEMS OR FUNDAMENTAL ALGEBRA SKILLS DIAGNOSTIC TEST

You can use these review problems, like the review problems at the end of each chapter, to test your understanding of the chapter's topics. However, you can also use these problems as a prerequisite test to check your mastery of the fundamental algebra skills needed in this book. Here are some suggestions for using these review problems as a diagnostic test:

  1. *Work through all 74 items at your own pace.*

  2. *Use the answer section in the back of the book to check your work.*

  3. *If your answer differs from that in the answer section or if you are not certain how to proceed with a particular item, turn to the section and the illustrative example given in parentheses at the end of each problem. Study the step-by-step solution of the example that parallels the review problem, and then try working the problem again. If you feel that you need more assistance, study the entire section in which the example appears and work on a selected group of problems in the problem set for that section.*

  1. Indicate which numbers in the following set are **a.** Natural numbers **b.** Whole numbers **c.** Integers **d.** Rational numbers **e.** Irrational numbers (Section P.1; Example 1)

$$\left\{ -13, \ -\sqrt{16}, \ -\sqrt{5}, \ -\frac{2}{3}, \ 0, \ \frac{\pi}{3}, \ 1.\overline{27}, \ \sqrt[4]{16} \right\}$$

In Problems 2–4, express each interval in terms of an inequality, and graph the interval on a number line. (Section P.1; Example 2)

2. $(-2, 3]$ **3.** $[-1.5, 2]$ **4.** $(-1, \infty)$

In Problems 5–8, express each set in interval notation. (Section P.1; Example 4)

5. $(1, 3) \cap [2, 7]$ **6.** $(-1, \infty) \cap (5, \infty)$ **7.** $(1, 3) \cup [2, 7]$ **8.** $(-1, \infty) \cup (5, \infty)$

9. Rewrite each expression without absolute value bars. (Section P.1; Example 5)

    a. $\left| \sqrt{2} - 1 \right|$ **b.** $\left| 3 - \sqrt{17} \right|$

10. Express the distance between the numbers -17 and 4 using absolute value. Then evaluate the absolute value. (Section P.1; Example 6)

11. Evaluate: $6x^3 - 5x^2 - 7x + 9$ for $x = -2$. (Section P.2; Example 1)

In Problems 12–13, state the property that justifies each statement. (Section P.2; Box on pages 14–15)

12. $\dfrac{1}{(x - 3)}(x - 3) = 1 \ (x \neq 3)$ **13.** $3 + (9 + 6) = (3 + 9) + 6$

14. Evaluate: $(-2)^4 - 3^2$. (Section P.2; Example 2)

Simplify each exponential expression in Problems 15–18. (Section P.2; Examples 3–5)

15. $(-2x^4 y^3)^3$ **16.** $(-5x^3 y^2)^3 (-2x^{-11} y^{-3})$

17. $\left(\dfrac{2x^{-2}}{3y^3} \right)^{-4}$ **18.** $\left(\dfrac{-10x^2 y}{20xy^{-2}} \right)^{-3}$

In Problems 19–20, write each number in scientific notation. (Section P.2; Box on page 23 and example following the box)

19. $98{,}000{,}000{,}000{,}000$ **20.** 0.000362

21. The graph shows the number of juveniles in the United States arrested for marijuana possession from 1980 through 1995.

    a. What are the coordinates for point A? What does this mean in practical terms?

    b. For the period shown in the graph, in approximately what year were arrests for marijuana possession at a minimum? Approximately how many arrests were there during that year? (Section P.3; Examples 1 and 2)

22. Plot $(-3, 7)$ and $(5, 7)$. Then find the distance between the points and the midpoint of the line segment joining the points. (Section P.3; Examples 3 and 4)

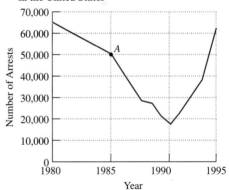

Source: FBI.

In Problems 23–27, use the point-plotting method to graph the given equation. Then use a graphing utility to verify your hand-drawn graph. (Section P.3; Example 5)

23. $y = -3x + 6$ **24.** $y = x^2 - 1$ **25.** $y = x^3 - x$ **26.** $y = \sqrt[3]{x + 1}$

27. $y = -|x|$ **28.** Evaluate: $\sqrt[3]{-27} + \sqrt[5]{32}$. (Section P.4; Examples 1 and 2)

29. Write in simplified form by factoring the radicand and removing as many factors as possible: $\sqrt[3]{16x^3 y^8}$. (Section P.4; Example 3)

In Problems 30–32, simplify by rationalizing the denominator. (Section P.4; Examples 4–6)

30. $\dfrac{15}{2\sqrt{5}}$

31. $\sqrt[3]{\dfrac{1}{3}}$

32. $\dfrac{14}{\sqrt{7}-\sqrt{5}}$

In Problems 33–34, simplify by combining the radicals. (Section P.4; Example 7)

33. $5\sqrt{18}+3\sqrt{8}-\sqrt{2}$

34. $16\sqrt{\dfrac{5}{8}}+6\sqrt{\dfrac{5}{2}}$

35. Evaluate without using a calculator: $8^{2/3}-32^{-4/5}$. (Section P.4; Example 8).

In Problems 36–37, simplify each algebraic expression. (Section P.4; Example 9)

36. $(7x^{1/3})(4x^{1/4})$

37. $\dfrac{80y^{3/4}}{-20y^{1/5}}$

In Problems 38–39, simplify by reducing the index of the radical. (Section P.4; Example 10)

38. $\sqrt[8]{16x^6y^2}$

39. $8\sqrt[4]{2}+2\sqrt[8]{2^{10}}$

40. Perform the indicated operations. Write the resulting polynomial in standard form and indicate its degree:

$$(16x^3-2x^2-5x+3)-(-2x^3-x^2+4x-1)+(3x^3-7x^2-x+1)$$

(Section P.5; Examples 1 and 2)

In Problems 41–42, find each product. (Section P.5; Examples 3–6)

41. $(x^2+3x-5)(2x^2-x+4)$

42. $(9x-7)(4x+3)$

In Problems 43–44, use the special products to find each product. (Section P.5; Examples 8–9)

43. $(4x-3y)^3$

44. $[5y-(2x+1)]\ \ [5y+(2x+1)]$

In Problems 45–57, factor each expression completely.

45. $60x^6y^3-20x^4y^2+10xy^5$ (Section P.6; Example 1)

46. $x^3-7x^2+9x-63$ (Section P.6; Example 2)

47. x^4-81y^4 (Section P.6; Examples 4–5)

48. x^3-1000 (Section P.6; Example 6)

49. $4x^2-20x+25$ (Section P.6; Example 7)

50. $x^2+6x+9-4a^2$ (Section P.6; Example 8)

51. $8x^2+2x-15$ (Section P.6; Examples 9–10)

52. $4a^3+32$ (Section P.6; Example 11)

53. $x^2(x-3)-9(x-3)$ (Section P.6; Example 11)

54. $2xy-2x^{10}y$ (Section P.6; Example 11)

55. x^4-x^3-x+1 (Section P.6; Example 11)

56. $16x^{-3/4}+32x^{1/4}$ (Section P.6; Example 12)

57. $(x^2-4)(x^2+3)^{-1/2}-(x^2-4)^2(x^2+3)^{-3/2}$ (Section P.6; Example 13)

58. Simplify: $\dfrac{6x^2+7x+2}{2x^2-9x-5}$ (Section P.7; Example 1)

In Problems 59–66, perform the indicated operations, and simplify if possible.

59. $\dfrac{y^4-81}{y^2+9}\cdot\dfrac{4y-20}{y^2-8y+15}$ (Section P.7; Example 2)

60. $\dfrac{x^2+9x+20}{x^2-16}\div\dfrac{x^2+5x}{4x-16}$ (Section P.7; Example 2)

61. $\dfrac{3x}{x+2}+\dfrac{x}{x-2}$ (Section P.7; Example 3)

62. $\dfrac{x}{x^2-9}+\dfrac{x-1}{x^2-5x+6}$ (Section P.7; Example 4)

63. $\dfrac{4}{a^2+a-2}-\dfrac{2}{a^2-4}+\dfrac{3}{a^2-4a+4}$ (Section P.7; Example 5)

64. $\dfrac{\dfrac{1}{y+5}+1}{\dfrac{6}{y^2+2y-15}-\dfrac{1}{y-3}}$ (Section P.7; Example 6)

65. $\dfrac{\sqrt{25-x^2}+\dfrac{x^2}{\sqrt{25-x^2}}}{25-x^2}$ (Section P.7; Example 7)

66. Simplify: i^{23}. (Section P.8; Example 1)

In Problems 67–70, perform the indicated operations and write the result in standard form.

67. $3 - (-5 + 11i) + (-17 + 4i)$ (Section P.8; Example 2)

68. $(7 - 5i)(2 + 3i)$ (Section P.8; Example 3)

69. $\dfrac{3 - 4i}{4 + 2i}$ (Section P.8; Example 5)

70. $\dfrac{5 + i}{3i}$ (Section P.8; Example 6)

In Problems 71–73, perform the indicated operations and write the result in standard form. (Section P.8; Example 7)

71. $\sqrt{-32} - \sqrt{-18}$

72. $(-2 + \sqrt{-11})^2$

73. $\dfrac{-18 - \sqrt{-43}}{30}$

74. Plot $-2 - 3i$ in the complex plane. (Section P.8; Example 8)

CHAPTER P TEST

1. Express $(-\infty, 3]$ in terms of an inequality and graph the interval on a number line.

2. Express $(2, 5) \cap [3, 6]$ in interval notation.

3. Find the distance between $(2, 9)$ and $(6, 3)$ in simplified radical form.

4. Graph $y = x^2 - 3$.

5. What property is illustrated by $3 + (7 + 4) = 3 + (4 + 7)$?

Simplify each expression in Problems 6–14.

6. $\left(\dfrac{x^2 y^{-2}}{3}\right)^{-3}$

7. $\dfrac{1}{\sqrt{2} + 1}$

8. $\dfrac{8}{\sqrt[3]{16}}$

9. $4\sqrt{50} - 3\sqrt{18}$

10. $\sqrt{3} + \sqrt{\dfrac{1}{3}}$

11. $\sqrt[6]{16x^4 y^2}$

12. $(x + 5y)^2 - (x + 2y)(x - 7y)$

13. $[7y - (4x + 1)][7y + (4x + 1)]$

14. $\dfrac{4x - x^2}{3x^2 - 7x - 20}$

In Problems 15–18, factor completely.

15. $x^4 - 16$

16. $2x^3 + 128y^3$

17. $x^3 - 4x^2 - x + 4$

18. $(x^2 - 9)(x^2 + 1)^{-1/2} - (x^2 - 9)(x^2 + 1)^{-3/2}$

In Problems 19–22, perform the operations, and simplify if possible.

19. $(y^3 - 8) \div \dfrac{3y^2 + 6y + 12}{3y - 7}$

20. $\dfrac{x + 3}{x^2 - 1} + \dfrac{2}{x} - \dfrac{3}{x - 1}$

21. $\dfrac{1 - \dfrac{x}{x + 2}}{1 + \dfrac{1}{x}}$

22. $\dfrac{2x\sqrt{x^2 + 5} - \dfrac{2x^3}{\sqrt{x^2 + 5}}}{x^2 + 5}$

In Problems 23–25, perform the operations and write the result in standard form.

23. $(6 - 7i)(2 + 5i)$

24. $\dfrac{5}{2 - i}$

25. $14i - (3 + 2\sqrt{-49})$

Equations, Inequalities, and Mathematical Models

Salvador Dalì "The Persistence of Memory" 1931, oil on canvas, 9 1/2 × 13 in. (24.1 × 33 cm). The Museum of Modern Art, New York. Given anonymously. Photograph © 1998 The Museum of Modern Art, New York. © 1998 Demart Pro Arte ®, Geneva / Artists Rights Society (ARS), New York.

"If you could stop every atom in its position and direction, and if your mind could comprehend all the actions thus suspended, then if you were really, *really* good at algebra you could write the formula for all the future; and although nobody can be so clever as to do it, the formula must exist just as if one could."

Tom Stoppard, *Arcadia*

Albert Einstein's (1879–1955) special theory of relativity destroyed fixed notions of time and space. Mass, length, and time are not fixed entities in Einstein's model of reality, but rather are relative to each of us.

The complexities of special relativity can be simplified using algebraic equations. Equations that describe the physical world are called mathematical models and are indispensable as we grasp at the possibilities of nature. Solving algebraic equations is essential in understanding the vastly complicated world just beyond our reach.

S E C T I O N   I . I

Solutions   Tutorial   Video
Manual   3

Linear Equations

Objectives

1 Solve linear equations in one variable.
2 Solve equations involving rational expressions.
3 Solve linear equations involving absolute value.
4 Solve problems modeled by linear equations.
5 Solve linear models for a specified variable.

In this section, we establish the basic techniques for solving equations containing variables raised only to the first power, called *linear equations*. We also consider the solution of linear equations involving both absolute value and rational expressions. These techniques are applied as we answer questions about real world phenomena described in formulas known as mathematical models.

Equations and Solution Sets

An *equation* is a statement indicating that two algebraic expressions are equal. Examples of equations in x include

$$2x - 3 = 5 \quad x^2 - 3x = 2 \quad 2x + 3x = 5x.$$

The third equation in the list is true no matter what value the variable x represents. An equation that is true for all real numbers for which both sides are defined is called an *identity*.

The first two equations in our list are true for only some of the real numbers. Equations that are true for some (or possibly none) of the real numbers are called *conditional equations*.

Solving an equation in x means determining all values of x which when substituted into the equation result in a true statement. Such values are *solutions* or *roots* of the equation. These values are said to *satisfy* the equation. The set of all such solutions is called the equation's *solution set*.

Conditional equations are neither true nor false until we choose a value for x. For example, the equation $2x - 3 = 5$ becomes a true statement if we substitute 4 for x, resulting in $2(4) - 3 = 5$ (or $8 - 3 = 5$), a true statement. Since 4 is the only number that results in a true statement for this equation, the equation's solution set is {4}.

Solving Linear Equations in One Variable

The most common type of conditional equation, and perhaps the simplest, is a *linear* equation or *first-degree* equation in one variable.

Mel Bochner "Vertigo" 1982, charcoal, conte crayon and pastel on sized canvas, fixed with magna varnish, 108×74 in. Albright–Knox Art Gallery, Buffalo, New York, Charles Clifton Fund, 1982.

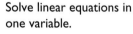

Study tip

The only number that satisfies

$$2x - 3 = 5$$

is 4 because 4 is the only number that when doubled and then decreased by 3 leaves 5.

Solve linear equations in one variable.

Definition of linear equation

A *linear* or *first-degree* equation in one variable x is an equation that can be written in the standard form

$$ax + b = c$$

where a and b are real numbers and $a \neq 0$.

An example of a linear equation in one variable is $4x + 12 = 0$.

The equations $4x + 12 = 0$ and $4x = -12$ and $x = -3$ are *equivalent equations* because they all have the same solution set, namely $\{-3\}$. To solve a linear equation in x, we use a series of transformations into equivalent equations. Our final equivalent equation should be in the form

$$x = d$$

where d is a real number. By inspection, we can see that the solution set for this equation is $\{d\}$.

In generating equivalent equations, we will be using the following principles.

Generating equivalent equations

An equation can be transformed into an *equivalent equation* by one or more of the following operations.

		Given Equation	**Equivalent Equation**
1.	Simplify an expression by removing grouping symbols and combining like terms.	$3(x - 6) = 6x - x$	$3x - 18 = 5x$
2.	Add (or subtract) the same real number or variable expression on *both* sides of the equation.	$3x - 18 = 5x$	$3x - 18 - 3x = 5x - 3x$ $-18 = 2x$
3.	Multiply (or divide) on *both* sides of the equation by the same *nonzero* quantity.	$-18 = 2x$	$\dfrac{-18}{2} = \dfrac{2x}{2}$ $-9 = x$
4.	Interchange the two sides of the equation.	$-9 = x$	$x = -9$

Study tip

As you study the transformations shown on the right, notice that we've actually solved the equation

$3(x - 6) = 6x - x.$

After isolating the variable term on the right, obtaining

$-18 = 2x$

we solved for the variable by dividing both sides by 2.

The following example illustrates the use of these transformation principles to solve linear equations.

EXAMPLE 1   Solving a Linear Equation

Solve for x: $4(2x + 1) - 29 = 3(2x - 5)$

Study tip

3. Isolate the variable and solve.
4. Check.

...tion

$\quad 1) - 29 = 3(2x - 5)$ This is the given equation.

$\quad 4 - 29 = 6x - 15$ Apply the distributive property, thereby removing parentheses.

$\quad 25 = 6x - 15$ Simplify by combining numerical terms.

$8x - 25 - 6x = 6x - 15 - 6x$ Collect variable terms on the left by subtracting $6x$ from both sides. This step is usually done mentally.

Discover for yourself

Solve the equation in Example 1 by collecting terms with the variable on the right and numerical terms on the left. What do you observe?

$$2x - 25 = -15 \qquad \text{Simplify.}$$

$$2x - 25 + 25 = -15 + 25 \qquad \text{Collect numerical terms on the right by adding 25 to both sides. Once again, this step is usually done mentally.}$$

$$2x = 10 \qquad \text{Simplify.}$$

$$\frac{2x}{2} = \frac{10}{2} \qquad \text{Solve for } x \text{ by dividing both sides by 2, usually done mentally.}$$

$$x = 5 \qquad \text{Simplify.}$$

Check

The proposed solution, 5, should be checked in the original equation:

$$4(2x + 1) - 29 = 3(2x - 5) \qquad \text{This is the original equation.}$$

$$4(2 \cdot 5 + 1) - 29 \stackrel{?}{=} 3(2 \cdot 5 - 5) \qquad \text{Substitute 5 for } x. \text{ When checking, be sure to work each side separately.}$$

$$4(11) - 29 \stackrel{?}{=} 3(5) \qquad \text{Perform operations in grouping symbols.}$$

$$44 - 29 \stackrel{?}{=} 15 \qquad \text{Multiply.}$$

$$15 = 15 \checkmark \qquad \text{This true statement indicates that 5 is the solution.}$$

The check verifies that the solution set of the given linear equation is {5}.   ∎

Using technology

A graphing utility can transform the algebraic solution into a visual solution that can be used to check the solution obtained algebraically.

To find the solution of $4(2x + 1) - 29 = 3(2x - 5)$ on a graphing utility, we use the fact that two expressions with the same value have a difference equal to zero. Use the $\boxed{y(x) =}$ key to enter the following equations:

$y_1 \boxed{=} 4(2x + 1) - 29$ This is the left side of the given equation.

$y_2 \boxed{=} 3(2x - 5)$ This is the right side of the given equation.

$y_3 \boxed{=} y_1 - y_2$ Also enter the difference of the two sides. Your graphing utility manual should explain how to do this.

Now turn off the highlighting on the equals sign for y_1 and y_2. On some utilities, you accomplish this by moving the cursor to y_1 and pressing $\boxed{\text{SELCT}}$, and then do the same for y_2. On others, you'll need to put the cursor on the equal sign and press $\boxed{\text{ENTER}}$. As always, consult your manual.

You will plot the graph of $y_1 - y_2$ and determine where it crosses the x-axis. The solution of $4(2x + 1) - 29 = 3(2x - 5)$ is the x-intercept of the graph of

$$y_3 = \boxed{4(2x + 1) - 29} - \boxed{3(2x - 5)}$$

This is y_1.   This is y_2.

tudy tip

The solution to

$$4(2x + 1) - 29 = 3(2x - 5)$$

is 5, an integer, so it's easy to see this solution using a standard zoom setting on a graphing utility. If the solution is not an integer, TRACE and ZOOM features can be used to verify an equation's solution.

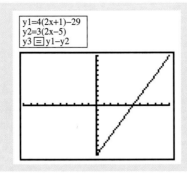

The utility draws only the expression with the highlighting on the equal sign. Use a standard zoom setting to obtain the graph.

The graph of $y_1 - y_2$ crosses the x-axis at 5. The solution set is $\{5\}$.

Equations Involving Rational Expressions

2 Solve equations involving rational expressions.

If an equation involves rational expressions, we can clear the equation of all fractions by multiplying both sides by the LCD of the fractions. We then solve for the variable. Examples 2 through 4 illustrate this procedure. Although the original equations are not linear, they become linear equations upon simplification.

EXAMPLE 2 **Solving an Equation with Rational Expressions**

Solve: $\dfrac{2}{3x} + \dfrac{1}{4} = \dfrac{11}{6x} - \dfrac{1}{3}$

Solution

The LCD for the rational expressions is $12x$. Since we must multiply by a nonzero number to obtain an equivalent equation, we multiply both sides by $12x$ with the restriction that $x \neq 0$.

$$\frac{2}{3x} + \frac{1}{4} = \frac{11}{6x} - \frac{1}{3}$$

This is the given equation; each side is undefined if $x = 0$.

$$12x\left(\frac{2}{3x} + \frac{1}{4}\right) = 12x\left(\frac{11}{6x} - \frac{1}{3}\right)$$

Multiply both sides by the LCD. $x \neq 0$

$$12x \cdot \frac{2}{3x} + 12x \cdot \frac{1}{4} = 12x \cdot \frac{11}{6x} - 12x \cdot \frac{1}{3}$$

$$8 + 7x - 8 = 22 - 8$$

Subtract 8 from both sides.

$$7x = 14$$

Simplify.

$$\frac{7x}{7} = \frac{14}{7}$$ Divide both sides by 7.

$$x = 2$$ Simplify.

Check

We check our solution by substituting 2 into the original equation or by using a graphing utility. Note that the initial restriction that $x \neq 0$ is met. The solution set is $\{2\}$. ∎

Our next example illustrates that multiplying both sides of an equation by an algebraic expression does not necessarily lead to an equivalent equation.

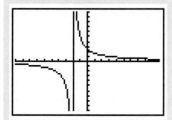

Using technology

To find the solution to Example 3 on your graphing utility, enter y_1, y_2, and y_3, highlighting only y_3.

$$y_1 = \frac{1}{x - 2} + \frac{4}{x + 2}$$

$$y_2 = \frac{4}{x^2 - 4}$$

$$y_3 = y_1 - y_2$$

The graph shown above appears to contain a vertical line drawn through $x = -2$. The line results when the utility connects a point just to the left of -2 with a point just to the right of -2. Now use a ⌈DRAW DOT⌉ rather than a ⌈DRAW LINE⌉ format. As shown below, the resulting graph consists of points that are not connected. The graph does not cross the x-axis, verifying that the original equation has no solution.

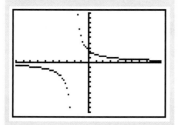

EXAMPLE 3 An Equation Having No Solution

Solve: $\dfrac{1}{x - 2} + \dfrac{4}{x + 2} = \dfrac{4}{x^2 - 4}$

Solution

$$\frac{1}{x - 2} + \frac{4}{x + 2} = \frac{4}{x^2 - 4}$$ This is the given equation.

$$\frac{1}{x - 2} + \frac{4}{x + 2} = \frac{4}{(x + 2)(x - 2)}$$ Factor $x^2 - 4$ using $A^2 - B^2 = (A + B)(A - B)$.

$$(x + 2)(x - 2)\left[\frac{1}{x - 2} + \frac{4}{x + 2}\right] = (x + 2)(x - 2)\left[\frac{4}{(x + 2)(x - 2)}\right]$$

Multiply both sides by $(x + 2)(x - 2)$, the LCD. Since we cannot multiply by 0, $x + 2 \neq 0$ and $x - 2 \neq 0$. Equivalently, $x \neq -2$ and $x \neq 2$.

$$(x + 2)(x - 2) \cdot \frac{1}{x - 2} + (x + 2)(x - 2) \cdot \frac{4}{x + 2}$$

$$= (x + 2)(x - 2) \cdot \frac{4}{(x + 2)(x - 2)}$$ Use the distributive property.

$$x + 2 + 4(x - 2) = 4$$ Simplify.

$$x + 2 + 4x - 8 = 4$$ Apply the distributive property.

$$5x - 6 = 4$$ Combine like terms.

$$5x = 10$$ Add 6 to both sides.

$$x = 2$$ Divide both sides by 5.

Check

The proposed solution, 2, is *not* a solution because of the restriction that $x \neq 2$. Let's see what happens when we substitute 2 for x in the original equation.

$$\frac{1}{x - 2} + \frac{1}{x + 2} = \frac{4}{x^2 - 4}$$ This is the original equation.

Study tip

Check all proposed solutions in rational equations. Eliminate any value of a variable that makes any denominator of the equation equal 0.

$$\frac{1}{2-2} + \frac{1}{2+2} \overset{?}{=} \frac{4}{2^2 - 4}$$ Substitute 2 for x.

$$\frac{1}{0} + \frac{1}{4} \overset{?}{=} \frac{4}{0}$$

The expressions $\frac{1}{0}$ and $\frac{4}{0}$ are undefined, so the proposed solution does not check in the original equation. There is *no solution* to this equation. The solution set contains no elements and is called the *null set* (written $\emptyset$), or the *empty set*. ∎

Just as some equations have no solutions, others have infinitely many real numbers in their solution sets. The next example involves such an equation.

EXAMPLE 4 **An Equation Having Infinitely Many Solutions**

Solve: $\dfrac{1}{x-4} - \dfrac{1}{x+4} = \dfrac{8}{x^2 - 16}$

Solution

$$\frac{1}{x-4} - \frac{1}{x+4} = \frac{8}{x^2 - 16}$$ This is the given equation.

$$\frac{1}{x-4} - \frac{1}{x+4} = \frac{8}{(x+4)(x-4)}$$ Factor $x^2 - 16$.

$$(x+4)(x-4)\left[\frac{1}{x-4} - \frac{1}{x+4}\right] = (x+4)(x-4)\left[\frac{8}{(x+4)(x-4)}\right]$$

Multiply by $(x+4)(x-4)$, the LCD. ($x \neq 4$ and $x \neq -4$)

$$x + 4 - (x-4) = 8$$ Apply the distributive property and simplify.

$$8 = 8$$

Study tip

There are situations where an algebraic analysis makes more sense than using a graphing utility. In Example 4, a graphing utility will show nothing other than the axes, since the solution set contains all real numbers on

We have a trivial statement of equality in the final step, namely, $8 = 8$. The left and right sides of the original equation are equal for all permissible values of the variable. Because of the restriction that $x \neq 4$ and $x \neq -4$ (be

$$2x - 3 = 11 \quad \text{and} \quad 2x - 3 = -11.$$

Rewriting an absolute value equation without absolute value bars

If c is positive,

$$|ax + b| = c \quad \text{is equivalent to} \quad ax + b = c \quad \text{or} \quad ax + b = -c$$

EXAMPLE 5   **Solving an Equation Involving Absolute Value**

Solve: $|2x - 3| = 11$

Solution

$|2x - 3| = 11$		This is the given equation.
$2x - 3 = 11 \quad$ or $\quad 2x - 3 = -11$		Rewrite the equation without absolute value bars.
$2x = 14 \qquad\qquad 2x = -8$		Add 3 to both sides of each equation.
$x = 7 \qquad\qquad\quad x = -4$		Divide both sides of each equation by 2.

Check

$|2x - 3| = 11$		This is the original equation.
$|2(7) - 3| \overset{?}{=} 11 \quad |2(-4) - 3| \overset{?}{=} 11$		Substitute the proposed solutions.
$|11| \overset{?}{=} 11 \qquad\quad |-11| \overset{?}{=} 11$		Perform operations inside the absolute value bars.
$11 = 11 \checkmark \qquad\quad 11 = 11 \checkmark$		These true statements indicate that 7 and -4 are solutions.

The solution set is $\{-4, 7\}$.   ∎

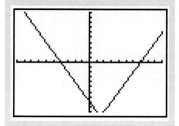

Using technology

(Consult your manual for the location of the ⎡ABS⎤, absolute value, key.) For Example 5, enter y_1, y_2, y_3, highlighting only y_3.

$y_1 = $ ⎡ABS⎤ $(2x - 3)$

$y_2 = 11$

$y_3 \boxed{=} y_1 - y_2$

The graph of y_3 crosses the x-axis at -4 and 7, verifying that $\{-4, 7\}$ is the solution set.

Our next equation involves two absolute value expressions. In this case, $|e| = |f|$ *if and only if* the expressions in absolute value are equal to each other ($e = f$) or the expressions in absolute value are negatives of one another ($e = -f$). In terms of linear equations, we can state this observation as follows.

Rewriting an absolute value equation containing two absolute values

The equation $|ax + b| = |cx + d|$ is equivalent to

$$ax + b = cx + d \quad \text{or} \quad ax + b = -(cx + d).$$

EXAMPLE 6   **An Absolute Value Equation with Two Absolute Values**

Solve: $|3x - 1| = |x + 5|$

Using technology

For Example 6, enter y_1, y_2, and y_3, highlighting only y_3.

$$y_1 = \boxed{\text{ABS}}\, (3x - 1)$$
$$y_2 = \boxed{\text{ABS}}\, (x + 5)$$
$$y_3 \boxed{=}\, y_1 - y_2$$

The graph of y_3 intersects the x-axis at -1 and 3, verifying that $\{-1, 3\}$ is the solution set.

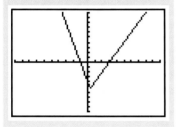

Solution

$\lvert 3x - 1 \rvert = \lvert x + 5 \rvert$		This is the given equation.

$3x - 1 = x + 5 \quad$ or $\quad 3x - 1 = -(x + 5)$ — Rewrite without absolute value bars.

$3x - 1 = x + 5 \qquad\qquad 3x - 1 = -x - 5$ — Use the distributive property in the second equation, thereby removing parentheses.

$2x - 1 = 5 \qquad\qquad 4x - 1 = -5$ — Subtract x from both sides of the first equation and add x to both sides of the second equation.

$2x = 6 \qquad\qquad 4x = -4$ — Add 1 to both sides of each equation.

$x = 3 \qquad\qquad x = -1$ — Divide both sides by 2 and 4, respectively.

Check

$$\lvert 3x - 1 \rvert = \lvert x + 5 \rvert$$
This is the original equation.

$$\lvert 3(3) - 1 \rvert \overset{?}{=} \lvert 3 + 5 \rvert \qquad \lvert 3(-1) - 1 \rvert \overset{?}{=} \lvert -1 + 5 \rvert$$
Substitute the proposed solutions.

$$\lvert 8 \rvert \overset{?}{=} \lvert 8 \rvert \qquad\qquad \lvert -4 \rvert \overset{?}{=} \lvert 4 \rvert$$
Perform operations inside the absolute value bars.

$$8 = 8 \checkmark \qquad\qquad 4 = 4 \checkmark$$
These true statements indicate that 3 and -1 are solutions.

The solution set is $\{-1, 3\}$. ∎

Discover for yourself

Must every absolute value equation have two solutions? Use the technique in Example 6 to solve $\lvert x \rvert = \lvert x - 1 \rvert$. Describe what happens.

4 Solve problems modeled by

Mathematical Models

rising undeniably. The major source of additional CO_2 is the burning of fossil fuels related to energy consumption.

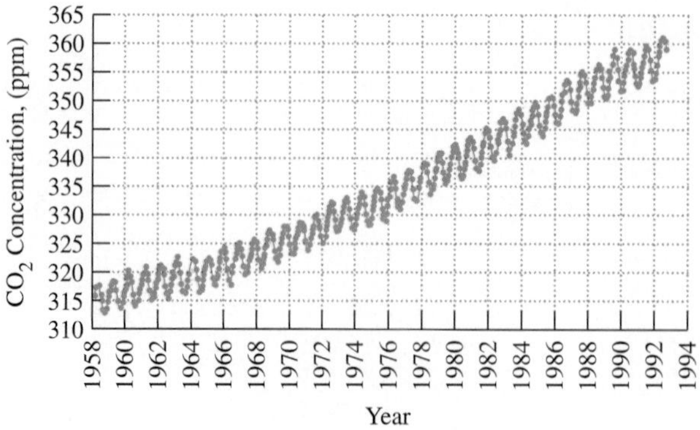

Figure 1.1

Atmospheric carbon dioxide has been rising.

Figure 1.1 shows atmospheric CO_2 concentration in parts per million from 1958 through 1994. Although the CO_2 concentration fluctuates between winter and summer, mathematicians have modeled these data points using

$$C = 1.44t + 280$$

where C represents CO_2 concentration in parts per million (ppm) t years after 1939. The preindustrial CO_2 concentration of 280 ppm remained fairly constant until World War II. In what year was the concentration 25% more than the preindustrial level?

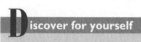

iscover for yourself

How well does the model

$$C = 1.44t + 280$$

describe the data in Figure 1.1? Since the data fluctuates yearly, you might want to use a yearly average based on high and low concentrations. Do this for at least five years, comparing the actual data with the atmospheric CO_2 level predicted by the model.

Solution

25% more than the preindustrial level of 280 ppm is

$$280 + 25\% \text{ of } 280 = 280 + (0.25)(280) = 350.$$

We can find when concentration was 350 ppm by using the given model, substituting 350 for C and solving for t:

$C = 1.44t + 280$	The given model relates CO_2 concentration to time (t years after 1939).
$350 = 1.44t + 280$	Substitute 350 for C.
$70 = 1.44t$	Subtract 280 from both sides.
$\dfrac{70}{1.44} = t$	Divide both sides by 1.44.
$49 \approx t$	t is approximately equal to ($\approx$) 49.

Since t represents the number of years after 1939, we have

$$1939 + 49 = 1988.$$

Thus, according to the model, CO_2 concentration was 350 ppm in 1988. Figure 1.2 on page 117 indicates that this is not a bad approximation of the actual data, although 350 ppm appears to be a low level in the 1988 seasonal fluctuation.

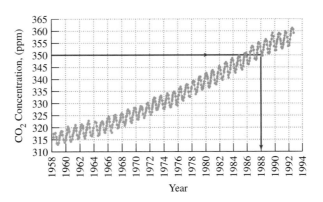

Figure 1.2

CO_2 Concentration: 350 ppm in 1988

5 Solve linear models for a specified variable.

An important skill in mathematics involves solving a formula or a mathematical model for a specified variable. Let's see exactly how this is done.

EXAMPLE 8 **Solving a Mathematical Model for a Specified Variable**

Consider the straight-line depreciation model

$$V = C - \frac{C - S}{L} N$$

where

L = the useful life of an asset (in years)

C = the cost of the asset when it was new

S = the salvage value of the asset at the end of its useful life

V = the value of the asset at the end of year N

a. Solve the formula for C.

$$LV = LC - (CN - SN) \qquad \text{Use the distributive property.}$$

ENRICHMENT ESSAY

The Monsters of Malignancy

In Chapter 2, we will see how mathematicians use data points that appear to lie along a line to write an equation for the line of best fit. Using data from numerous countries that depict daily fat intake (F, in grams) and deaths per 100,000 women (D) from breast cancer, researchers derived the model

$$D = 0.2F - 1.$$

The death rate of U.S. women from breast cancer is 19 per 100,000. Use the model to estimate the daily fat intake in the United States.

Although the linear model provides useful information, it does not convey the way cancer intrudes into the ordinary course of our lives. One woman in ten will be afflicted with breast cancer; 40,000 women in the United States will die of the disease this year.

Nancy Fried "Mourning" 1986, terra cotta, $8 \times 5 \times 3\ 1/2$ in. Photo courtesy DC Moore Gallery.

$LV = LC - CN + SN$	Distribute -1 throughout the parentheses.
$LV - SN = LC - CN$	To solve for C, isolate terms with C on one side (the right) by subtracting SN from both sides.
$LV - SN = C(L - N)$	Since we cannot combine terms on the right, factor out C so that it is isolated.
$\dfrac{LV - SN}{L - N} = C$	Solve for C by dividing both sides by $L - N.\ L \neq N$

When solving a formula or model for a specified variable in terms of other variables appearing in the model, we do not write the solution in set notation. Thus, $C = \dfrac{LV - SN}{L - N}$.

b. $C = \dfrac{LV - SN}{L - N}$ This is the result from part (a).

$= \dfrac{6(15,000) - 3000(3)}{6 - 3}$ The useful life is six years ($L = 6$). The salvage value is $3000 ($S = 3000$). The value at the end of the third year is $15,000 ($V = 15,000$ when $N = 3$). Substitute these values into the equation.

$= 27,000$ Use a calculator.

The cost of the automobile when new was $27,000.   ∎

PROBLEM SET 1.1

Practice Problems

Use solution set notation to express all answers in this problem set. Check results algebraically or with a graphing utility. Solve each linear equation in Problems 1–16.

1. $5x - 8 = 72$

2. $6x - 3 = 63$

3. $11x - (6x - 5) = 40$

4. $5x - (2x - 10) = 35$

5. $2x - 7 = 6 + x$

6. $3x + 5 = 2x + 13$

7. $7x + 4 = x + 16$

8. $13x + 14 = 12x - 5$

9. $3(x - 2) + 7 = 2(x + 5)$

10. $2(x - 1) + 3 = x - 3(x + 1)$

11. $3(x - 4) - 4(x - 3) = x + 3 - (x - 2)$

12. $2 - (7x + 5) = 13 - 3x$

13. $16 = 3(x - 1) - (x - 7)$

14. $5x - (2x + 2) = x + (3x - 5)$

15. $25 - [2 + 5y - 3(y + 2)] = -3(2y - 5) - [5(y - 1) - 3y + 3]$

16. $45 - [4 - 2y - 4(y + 7)] = -4(1 + 3y) - [4 - 3(y + 2) - 2(2y - 5)]$

Solve each fractional equation in Problems 17–40 by multiplying both sides by the LCD, thereby clearing fractions.

17. $\dfrac{3x}{5} = \dfrac{2x}{3} + 1$

18. $\dfrac{x}{2} = \dfrac{3x}{4} + 5$

19. $\dfrac{x}{4} = 2 + \dfrac{x - 3}{3}$

20. $5 + \dfrac{x - 2}{3} = \dfrac{x + 3}{8}$

21. $\dfrac{x + 1}{3} = 5 - \dfrac{x + 2}{7}$

22. $\dfrac{3x}{5} - \dfrac{x - 3}{2} = \dfrac{x + 2}{3}$

23. $\dfrac{4}{y} - 3 = \dfrac{5}{2y}$

24. $\dfrac{5}{y} - 4 = \dfrac{10}{3y}$

25. $\dfrac{2}{x} - \dfrac{3}{5} = -\dfrac{1}{2}$

26. $\dfrac{3}{x} - \dfrac{1}{4} = \dfrac{1}{3}$

27. $\dfrac{x - 2}{2x} + 1 = \dfrac{x + 1}{x}$

28. $\dfrac{4}{x} = \dfrac{9}{5} - \dfrac{7x - 4}{5x}$

29. $\dfrac{x}{x + 1} + \dfrac{2}{x} = 1$

30. $\dfrac{x}{x + 5} + \dfrac{3}{x} = 1$

31. $\dfrac{3}{x - 2} + \dfrac{2}{x + 3} = \dfrac{5}{x^2 + x - 6}$

32. $\dfrac{1}{x + 2} + \dfrac{5}{x - 4} = \dfrac{6}{x^2 - 2x - 8}$

33. $\dfrac{6}{y} - \dfrac{3}{y^2 - y} = \dfrac{7}{y - 1}$

34. $\dfrac{4}{x + 5} + \dfrac{2}{x - 5} = \dfrac{32}{x^2 - 25}$

35. $\dfrac{x - 3}{x - 2} + \dfrac{x + 1}{x + 3} = \dfrac{2x^2 - 15}{x^2 + x - 6}$

36. $\dfrac{2y - 14}{y^2 + 3y - 28} + \dfrac{y - 2}{y - 4} = \dfrac{y + 3}{y + 7}$

37. $\dfrac{7}{y + 5} + 2 = \dfrac{2 - y}{y + 5}$

38. $\dfrac{3y}{y - 3} = 1 + \dfrac{9}{y - 3}$

39. $\dfrac{1}{x + 5} - \dfrac{2}{x - 3} = \dfrac{2x + 2}{x^2 + 2x - 15}$

40. $\dfrac{1}{x - 4} - \dfrac{5}{x + 2} = \dfrac{6}{x^2 - 2x - 8}$

Solve each equation in Problems 41–52 by first rewriting each equation as two equations without absolute value bars.

41. $|2x - 5| = 13$

42. $|2x - 1| = 15$

43. $|2y + 3| + 3 = 20$

44. $|4y - 5| - 8 = 3$

45. $\left|-\frac{5}{2}y + 4\right| + 3 = 15$

46. $\left|-\frac{3}{4}y + 1\right| + 4 = 7$

47. $|2x - 3| = |x + 6|$

48. $|4 - 2x| = |8 - x|$

49. $|5x + 2| = |4x + 7|$

50. $|2x - 5| = |-6x - 5|$

51. $|18x - 16| = |26 - 10x|$

52. $|9x - 12| = |10 - 13x|$

concentration in parts per million (ppm) t years after 1939. When will the concentration be 50% greater than that of the preindustrial concentration of 280 ppm?

(P) of lost hikers found in search and rescue missions when members of the search team walk parallel to one another separated by a distance of d yards. If a search and rescue team finds 70% of lost hikers, what is the parallel distance of separation between members of the search party?

63. A woman's height (h) is related to the length of the femur (f) (the bone from the knee to the hip socket) by the mathematical model $f = 0.432h - 10.44$. Both h and f are measured in inches. A partial skeleton is found of a woman in which the femur is 16 inches long. Police find the skeleton in an area where a woman slightly over 5 feet tall has been missing for over a year. Can the partial skeleton be that of the missing woman? Explain.

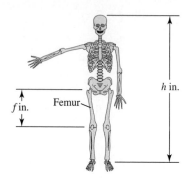

64. The mathematical model $T = 0.143L - 1.18$ relates the total length (L) and the tail length (T) of females of the snake species *Lampropeltis polyzona*, where both variables are measured in millimeters. What is the tail length of such a female snake with a total length of 457 mm?

65. The formula $d = 5000c - 525,000$ models the relationship between the annual number of deaths (d) in the United States from heart disease and the average adult cholesterol level (c, in milligrams per deciliter of blood). In 1990, 500,000 Americans died from heart disease. What was the average cholesterol level at that time? If the United States could reduce its average cholesterol level to 180, how many lives could be saved compared to 1990?

66. The mathematical model $y = -0.358709x + 256.835$ is used to project the world record for the mile run (y, in seconds), where x is the last two digits of any year. In what year was the world record for the mile run 224.91 seconds?

True–False Critical Thinking Problems

67. Which one of the following is true?

   a. The equation $-7x = x$ has no solution.

   b. The equations $\dfrac{x}{x-4} = \dfrac{4}{x-4}$ and $x = 4$ are equivalent.

   c. The equations $3y - 1 = 11$ and $3y - 7 = 5$ are equivalent.

   d. If a and b are any real numbers, then $ax + b = 0$ always has only one number in its solution set.

68. Which one of the following is true?

   a. The equation $\dfrac{1}{x-1} - \dfrac{1}{x+1} = \dfrac{2}{x^2-1}$ is an example of an equation with infinitely many solutions.

   b. Solving an equation with a graphing utility is always preferable to using algebraic methods.

   c. Some linear equations can only be solved with a graphing utility.

   d. If a and b are any negative real numbers, then $ax + b = 0$ has no solution if $|a| = |b|$.

Technology Problems

Estimate the solution of each equation in Problems 69–76 using a graphing utility. If solutions do not appear to be integers, use the [TRACE] *and* [ZOOM] *features to estimate solutions to the nearest tenth.*

69. $18 - 45x - 4 = 13x + 24 - 56x$

70. $8x - 3[2 - (x + 4)] = 4(x - 2)$

71. $0.2x + 4.5 = 1.2x$

72. $\dfrac{2x}{3} - \dfrac{3}{5} = -\dfrac{1}{3} - \dfrac{2x}{5}$

73. $|2x - 3| = 7$

74. $\dfrac{2}{3x + 6} = \dfrac{1}{6} - \dfrac{1}{2x + 4}$

75. $\dfrac{x + 6}{x + 3} - 2 = \dfrac{3}{x + 3}$

76. $|3x + 2| = |5x - 2|$

77. Consider the equation $\dfrac{x+3}{6} - \dfrac{2x-3}{9} = \dfrac{5}{6} - \dfrac{x}{18}$.

  a. Substitute each of the following numbers for x: -3, 0, 3, and 9. In each case, compute (by hand or with your calculator) both sides of the equation. What seems to be true about this linear equation?

  b. Solve the equation algebraically. Describe what happens and draw a conclusion.

  c. Use your graphing utility to solve the equation. Describe what appears on the viewing screen.

  d. Change the $\boxed{\text{FORMAT}}$ on the viewing screen by deleting the axes $\boxed{\text{Axes off}}$. Press $\boxed{\text{GRAPH}}$. What do

you see? Write a description relating what appears on the screen to the solution set of this equation.

78. The method that we discussed for solving linear equations on a graphing utility involves calling the left side of the equation y_1, the right side y_2, and then graphing $y_1 - y_2$. How might you solve the equation using only y_1 and y_2? Try this on Problem 69, graphing both y_1 and y_2. How is the equation's solution determined? (You should already have the solution using the previous method.) Try this method on a few of the exercises in Problems 69–76. Which method do you prefer? Why?

Writing in Mathematics

79. Under what conditions does a rational equation have no solution?

80. Refer to the model described in Problem 62.

  a. The model indicates that the chance of lost hikers being found increases as searchers in any area get closer to one another. If this is the case, why not simply have almost no distance separating members of the search team to obtain near certainty of finding the lost hikers?

  b. The model was determined from actual data collected by search and rescue teams in the American Southwest. How does this limit the usefulness of the model?

81. The optimum heart rate is the rate that a person should achieve during exercise for the exercise to be most beneficial. The formula $r = 0.6(220 - a)$ models optimum heart rate (r, in beats per minute), where a represents age. Write and solve a word problem using this model.

Critical Thinking Problems

In Problems 82–84, solve for x.

82. $\dfrac{x^2 + 2x - 15}{x^2 - 4x + 3} \div \dfrac{x^2 + x - 20}{3x^2 - 3} = -3$

83. $6 + |2 - 4x| = 3$

84. $\dfrac{x-C}{D} + \dfrac{x-D}{C} + \dfrac{x-1}{CD} = \dfrac{2}{C} + \dfrac{2}{D} + 2$

ocean level, use the given models to determine by what year the beachside community will be flooded.

Global Annual Temperature

60.0

seawall. If it is known that a global temperature increase of 1.8° Fahrenheit corresponds to a one-foot rise in

$V = C - \dfrac{C-S}{L} N.$

SECTION 1.2 Quadratic Equations

Solutions
Manual

Tutorial

Video
3

Objectives

1 Solve quadratic equations by factoring.
2 Solve quadratic equations by the square root method.
3 Solve quadratic equations by completing the square.
4 Solve quadratic equations using the quadratic formula.
5 Use the discriminant to determine kinds of solutions.
6 Solve problems modeled by quadratic equations.

In this section, we establish the basic techniques for solving equations containing variables raised to the second power, called *quadratic equations*. (The adjective *quadratic* is derived from the Latin word *quadrus* meaning "square.") We then apply these techniques to answering questions about variables contained in mathematical models that have evolved from real world data.

The Standard Form of a Quadratic Equation

An equation that contains a variable with an exponent of 2, but no higher power, is called a quadratic equation.

> **Definition of a quadratic equation**
>
> A *quadratic equation* in x is an equation that can be written in the standard form
>
> $$ax^2 + bx + c = 0$$
>
> where a, b, and c are real numbers with $a \neq 0$. A quadratic equation in x is also called a *second-degree polynomial equation* in x.

This Babylonian tablet shows a quadratic equation and its solution written sometime between 1900 and 1600 B.C. Yale University Library—Babylonian Collection.

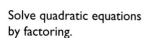

1 Solve quadratic equations by factoring.

Solving Quadratic Equations by Factoring

If the trinomial $ax^2 + bx + c$ can be factored, then $ax^2 + bx + c = 0$ can be solved by using the *zero product principle*.

Study tip

In the definition of a quadratic equation, we must state $a \neq 0$. If we allowed a to equal 0 in

$$ax^2 + bx + c = 0$$

the resulting equation would be

$$bx + c = 0$$

which is linear, and not quadratic.

> **The zero product principle**
>
> Let A and B be real numbers, variables, or algebraic expressions. If $AB = 0$, then $A = 0$ or $B = 0$ or A and B are both 0. In words, this says that if a product is zero, then at least one of the factors is equal to zero.

EXAMPLE I **Solving Quadratic Equations by Factoring**

Solve by factoring, using the zero product principle:

a. $2x^2 + 7x = 4$

b. $4x^2 - 2x = 0$

c. $8x(2x - 1) = -1$

d. $\dfrac{4}{x + 1} - 1 = \dfrac{3}{x + 2}$

U **sing technology**

The solutions to

$$ax^2 + bx + c = 0$$

correspond to the x-intercepts
for the graph of

$$y = ax^2 + bx + c$$

because the x-intercepts are the
points at which $y = 0$.

a. $2x^2 + 7x = 4$

Graph $y = 2x^2 + 7x - 4$ in a
$[-5, 2]$ by $[-11, 2]$ viewing
rectangle. The x-intercepts
are -4 and $\frac{1}{2}$, verifying
$\{-4, \frac{1}{2}\}$ as the solution set.

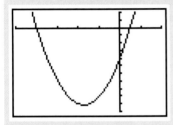

b. $4x^2 - 2x = 0$

Graph $y = 4x^2 - 2x$ in a
$[-2, 2]$ by $[-2, 2]$ viewing
rectangle. The x-intercepts
are 0 and $\frac{1}{2}$, verifying $\{0, \frac{1}{2}\}$ as
the solution set.

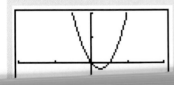

Solution

a.

$2x^2 + 7x = 4$	This is the given equation.
$2x^2 + 7x - 4 = 0$	To use the zero product principle, subtract 4 from each side, writing the quadratic equation in standard form. One side of a quadratic equation must be 0 when solving by factoring.
$(2x - 1)(x + 4) = 0$	Factor on the left.
$2x - 1 = 0 \quad$ or $\quad x + 4 = 0$	Set each factor equal to 0, using the zero product principle.
$x = \dfrac{1}{2} \quad$ or $\quad x = -4$	Solve each of the two resulting equations.

Check $\dfrac{1}{2}$:   **Check -4:**

$$2x^2 + 7x = 4 \qquad\qquad\qquad 2x^2 + 7x = 4$$

$$2(\tfrac{1}{2})^2 + 7(\tfrac{1}{2}) \overset{?}{=} 4 \qquad\qquad 2(-4)^2 + 7(-4) \overset{?}{=} 4$$

$$\tfrac{1}{2} + \tfrac{7}{2} \overset{?}{=} 4 \qquad\qquad\qquad 32 + (-28) \overset{?}{=} 4$$

$$4 = 4 \checkmark \qquad\qquad\qquad\qquad 4 = 4 \checkmark$$

The solution set is $\{-4, \frac{1}{2}\}$.

b.

$4x^2 - 2x = 0$	This is the given equation.
$2x(2x - 1) = 0$	Factor.
$2x = 0 \quad$ or $\quad 2x - 1 = 0$	Set each factor equal to 0.
$x = 0 \quad$ or $\quad x = \dfrac{1}{2}$	Solve the resulting equations.

Checking both values in the original equation, we confirm that the solution
set is $\{0, \frac{1}{2}\}$.

c.

$8x(2x - 1) = -1$	This is the given equation.
$16x^2 - 8x = -1$	Apply the distributive property on the left.
$16x^2 - 8x + 1 = 0$	Add 1 to each side.
$(4x - 1)(4x - 1) = 0$	Factor.
$4x - 1 = 0$	Set the repeated factor equal to 0.
$x = \dfrac{1}{?}$	

(continued) $4(x + 2) - (x + 1)(x + 2) = 3(x + 1)$ Simplify.

Using technology (cont.)

The x-intercept is $\frac{1}{4}$, verifying $\{\frac{1}{4}\}$ as the solution set.

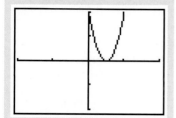

d. $\dfrac{4}{x+1} - 1 = \dfrac{3}{x+2}$

Graph

$$y = \dfrac{4}{x+1} - 1 - \dfrac{3}{x+2}.$$

Use a $\boxed{\text{DRAW DOT}}$ format. The x-intercepts are -3 and 1, verifying $\{-3, 1\}$ as the solution set.

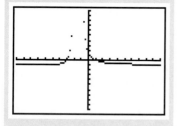

2 Solve quadratic equations by the square root method.

$$4x + 8 - x^2 - 3x - 2 = 3x + 3 \qquad \text{Multiply.}$$
$$-x^2 + x + 6 = 3x + 3 \qquad \text{Combine like terms on the left.}$$
$$0 = x^2 + 2x - 3 \qquad \text{Write the equation in standard form.}$$
$$0 = (x + 3)(x - 1) \qquad \text{Factor on the right.}$$
$$x + 3 = 0 \quad \text{or} \quad x - 1 = 0 \qquad \text{Set each factor equal to 0.}$$
$$x = -3 \qquad\qquad x = 1 \qquad \text{Solve the resulting equations.}$$

Check both solutions in the original equation to verify that both satisfy the equation. Neither solution interferes with the conditions that $x \neq -1$ and $x \neq -2$. The solution set is $\{-3, 1\}$. ■

Solving Quadratic Equations by the Square Root Method

Quadratic equations of the form $u^2 = d$, where $d > 0$ and u is an algebraic expression, can be solved by the *square root method*. To use this method, we isolate the squared expression on one side of the equation and then take the square root of both sides. Let's take a moment to use factoring to verify that $u^2 = d$ has two solutions.

$$u^2 = d \qquad \text{This is the given equation.}$$
$$u^2 - d = 0 \qquad \text{Set the equation equal to 0.}$$
$$(u + \sqrt{d})(u - \sqrt{d}) = 0 \qquad \text{Factor.}$$
$$u + \sqrt{d} = 0 \quad \text{or} \quad u - \sqrt{d} = 0 \qquad \text{Set each factor equal to 0.}$$
$$u = -\sqrt{d} \qquad\qquad u = \sqrt{d} \qquad \text{Solve the resulting equations.}$$

Since the solutions differ only in sign, we can write them in abbreviated notation as $u = \pm\sqrt{d}$.

Now that we have verified these solutions, we can solve $u^2 = d$ directly by taking square roots. This process is called *the square root method* (or *extracting square roots*).

The square root method

If u is an algebraic expression and d is a positive real number, then $u^2 = d$ has exactly two solutions:

If $u^2 = d$, then $u = \sqrt{d}$ or $u = -\sqrt{d}$.

Equivalently:

If $u^2 = d$, then $u = \pm\sqrt{d}$.

$\boxed{\textbf{EXAMPLE 2}}$ **Solving Quadratic Equations by the Square Root Method**

Solve by the square root method:

a. $4x^2 = 20$ **b.** $(4x - 2)^2 = 8$

Solution

a. $4x^2 = 20$ This is the given equation.

$\quad\quad x^2 = 5$ Divide both sides by 4.

$\quad\quad\quad x = \pm\sqrt{5}$ Apply the square root method, with $u = x$ and $d = 5$.

The solution set is $\{-\sqrt{5}, \sqrt{5}\}$. A graphing utility can approximate these solutions and can reinforce what we found algebraically. Using the $\boxed{\text{ZOOM}}$ and $\boxed{\text{TRACE}}$ features, the x-intercepts of $y = 4x^2 - 20$ are approximately -2.2 and 2.2. Try checking $-\sqrt{5}$ in the original equation.

b. $(4x - 2)^2 = 8$ This is the given equation.

$\quad 4x - 2 = \pm\sqrt{8}$ Apply the square root method, with $u = 4x - 2$ and $d = 8$.

$\quad 4x - 2 = \pm 2\sqrt{2}$ $\sqrt{8} = \sqrt{4}\sqrt{2} = 2\sqrt{2}$

$\quad\quad\quad 4x = 2 \pm 2\sqrt{2}$ Add 2 to both sides.

$\quad\quad\quad\quad x = \dfrac{2 \pm 2\sqrt{2}}{4}$ Divide both sides by 4. These values can be simplified.

$\quad\quad\quad\quad\quad = \dfrac{2(1 \pm \sqrt{2})}{4}$ Factor 2 from the numerator.

$\quad\quad\quad\quad\quad = \dfrac{1 \pm \sqrt{2}}{2}$ Divide numerator and denominator by 2.

The solution set is $\left\{ \dfrac{1 + \sqrt{2}}{2}, \dfrac{1 - \sqrt{2}}{2} \right\}$.   ■

Discover for yourself

Try checking

$$\frac{1 + \sqrt{2}}{2}$$

by substituting this value for x:

$$(4x - 2)^2 = 8$$

$$\left[4\left(\frac{1 + \sqrt{2}}{2}\right) - 2 \right]^2 \overset{?}{=} 8$$

$$[2(1 + \sqrt{2}) - 2]^2 \overset{?}{=} 8$$

Now complete the check by simplifying the expression on the left.

3 Solve quadratic equations by completing the square.

Solving Quadratic Equations by Completing the Square

When completing the square, our goal is to express a quadratic equation in the form $(x + d)^2 = e$ so we may then apply the square root method. To change a quadratic equation in standard form

$$ax^2 + bx + c = 0$$

to an equivalent equation in the form

Study tip

You can visualize the process of completing the square. The area of the blue region is

EXAMPLE 3   **Solving Quadratic Equations by Completing the Square**

Solve by completing the square:

a. $x^2 - 6x + 2 = 0$ **b.** $2x^2 + 5x - 4 = 0$

Solution

Before completing the square, we try factoring $x^2 - 6x + 2$ and $2x^2 + 5x - 4$ since factoring is easier and faster than completing the square. Both expressions are prime, so we must use a method other than factoring.

a. $x^2 - 6x + 2 = 0$ — This is the given equation.

$x^2 - 6x = -2$ — Subtract 2 from both sides to isolate the binomial $x^2 - 6x$.

$x^2 - 6x + 9 = -2 + 9$ — Complete the square. Take half the coefficient of x and square this result: $\frac{1}{2}(-6) = -3$ and $(-3)^2 = 9$. Notice that 9 is added to both sides to keep the equation balanced.

$\frac{1}{2}(-6) = -3$
and $(-3)^2 = 9$

$(x - 3)^2 = 7$ — Factor $x^2 - 6x + 9$, which is a perfect square trinomial.

$x - 3 = \pm\sqrt{7}$ — Apply the square root method.

$x = 3 \pm \sqrt{7}$ — Add 3 to both sides.

The solution set is $\{3 + \sqrt{7}, 3 - \sqrt{7}\}$.

b. $2x^2 + 5x - 4 = 0$ — This is the given equation.

$2x^2 + 5x = 4$ — Isolate variable terms, adding 4 to both sides.

$x^2 + \frac{5}{2}x = 2$ — Since it is easier to complete the square when the coefficient of x^2 is 1, divide both sides by 2.

$x^2 + \frac{5}{2}x + \frac{25}{16} = 2 + \frac{25}{16}$ — Complete the square. Since

$\frac{1}{2}(\frac{5}{2}) = \frac{5}{4}$
and $(\frac{5}{4})^2 = \frac{25}{16}$

$\frac{1}{2}\left(\frac{5}{2}\right) = \frac{5}{4}$ and $\left(\frac{5}{4}\right)^2 = \frac{25}{16}$, add $\frac{25}{16}$ to both sides.

$\left(x + \frac{5}{4}\right)^2 = \frac{57}{16}$ — Factor the left-hand side.

$x + \frac{5}{4} = \pm\frac{\sqrt{57}}{4}$ — Apply the square root method.

$x = -\frac{5}{4} \pm \frac{\sqrt{57}}{4}$ — Subtract $\frac{5}{4}$ from both sides.

The solution set is $\left\{\dfrac{-5 + \sqrt{57}}{4}, \dfrac{-5 - \sqrt{57}}{4}\right\}$. ∎

4 Solve quadratic equations using the quadratic formula.

Solving Quadratic Equations Using the Quadratic Formula

As we have seen, the method of completing the square leads to an equivalent equation that has a perfect square trinomial on one side. Once the coefficient of x^2 is 1, we add the square of half the coefficient of x to both sides of the equation. This number completes the square.

Applying this method to any quadratic equation, we can derive a formula that can be used to solve all quadratic equations.

$$ax^2 + bx + c = 0$$

This is the standard form of a quadratic equation. Remember that $a \neq 0$.

$$ax^2 + bx = -c$$

Isolate variable terms, subtracting c from both sides.

$$x^2 + \frac{b}{a}x = -\frac{c}{a}$$

Divide both sides by a so that the leading coefficient is 1.

$$x^2 + \frac{b}{a}x + \left(\frac{b}{2a}\right)^2 = -\frac{c}{a} + \left(\frac{b}{2a}\right)^2$$

Complete the square by adding the square of half the coefficient of x to both sides.

$$\left(x + \frac{b}{2a}\right)^2 = -\frac{c}{a} + \frac{b^2}{4a^2}$$

Factor the left side. Check this result by squaring $x + \dfrac{b}{2a}$.

René Magritte "Portrait de P.G. van Hecke" 1928, oil on canvas. Herscovici /Art Resource, NY.

$$\left(x + \frac{b}{2a}\right)^2 = -\frac{4ac}{4a^2} + \frac{b^2}{4a^2}$$

The LCD on the right side is $4a^2$.

$$\left(x + \frac{b}{2a}\right)^2 = \frac{b^2 - 4ac}{4a^2}$$

Combine algebraic fractions on the right.

$$x + \frac{b}{2a} = \pm\sqrt{\frac{b^2 - 4ac}{4a^2}}$$

Apply the square root method.

$$x + \frac{b}{2a} = \pm\frac{\sqrt{b^2 - 4ac}}{2a}$$

Simplify on the right. $\sqrt{4a^2} = 2|a|$. Because $\pm 2|a|$ represents the same numbers as $\pm 2a$, the absolute value is omitted.

$$x = -\frac{b}{2a} \pm \frac{\sqrt{b^2 - 4ac}}{2a}$$

Subtract $\dfrac{b}{2a}$ from both sides.

$$x = \frac{-b \pm \sqrt{b^2 - 4ac}}{2a}$$

Express the right side with a common denominator.

a. $2x^2 + 2x - 1 = 0$ **b.** $3y^2 = 2y - 4$

Solution

a. $2x^2 + 2x - 1 = 0$ This is the given equation.

$$a = 2 \quad b = 2 \quad c = -1$$

$$x = \frac{-b \pm \sqrt{b^2 - 4ac}}{2a}$$ Use the quadratic formula with $a = 2, b = 2$, and $c = -1$.

$$= \frac{-2 \pm \sqrt{2^2 - 4(2)(-1)}}{2(2)}$$ Substitute these values.

$$= \frac{-2 \pm \sqrt{4 - (-8)}}{4}$$ Perform the indicated operations under the radical.

$$= \frac{-2 \pm \sqrt{12}}{4}$$

$$= \frac{-2 \pm 2\sqrt{3}}{4}$$ $\sqrt{12} = \sqrt{4}\sqrt{3} = 2\sqrt{3}$

$$= \frac{2(-1 \pm \sqrt{3})}{4}$$ Factor out 2 from the numerator.

$$= \frac{-1 \pm \sqrt{3}}{2}$$ Divide numerator and denominator by 2.

The solution set is $\left\{ \dfrac{-1 + \sqrt{3}}{2}, \dfrac{-1 - \sqrt{3}}{2} \right\}.$

b. $3y^2 = 2y - 4$ This is the given equation.

$3y^2 - 2y + 4 = 0$ Write the equation in standard form.

$$a = 3 \quad b = -2 \quad c = 4$$

$$y = \frac{-b \pm \sqrt{b^2 - 4ac}}{2a}$$ Use the quadratic formula with $a = 3, b = -2$, and $c = 4$.

$$= \frac{-(-2) \pm \sqrt{(-2)^2 - 4(3)(4)}}{2(3)}$$ Substitute these values.

$$= \frac{2 \pm \sqrt{-44}}{6}$$ Since the number under the radical sign is negative, the solutions will not be real numbers.

$$= \frac{2 \pm 2i\sqrt{11}}{6}$$ $\sqrt{-44} = \sqrt{4(11)(-1)}$
$\qquad\qquad = 2i\sqrt{11}$

$$= \frac{2(1 \pm i\sqrt{11})}{6}$$ Factor 2 from the numerator.

$$= \frac{1 \pm i\sqrt{11}}{3}$$ Divide numerator and denominator by 2.

You can check that these solutions are correct using operations with complex numbers. The solutions are complex conjugates and the solution set is
$$\left\{ \frac{1 + i\sqrt{11}}{3}, \frac{1 - i\sqrt{11}}{3} \right\}. \qquad \blacksquare$$

Using technology

Use your graphing utility to explore the solutions to Example 4.

a. $2x^2 + 2x - 1 = 0$
Graph $y = 2x^2 + 2x - 1$ in a $[-5, 5]$ by $[-5, 10]$ viewing rectangle. Since

$$\frac{-1 + \sqrt{3}}{2} \approx 0.37$$

and

$$\frac{-1 - \sqrt{3}}{2} \approx -1.37,$$

you should be able to use the ZOOM and TRACE features to verify these values for the x-intercepts.

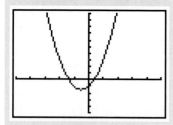

b. $3y^2 = 2y - 4$
Graph $y = 3x^2 - 2x + 4.$

The graph has no x-intercepts, indicating that

$$3x^2 - 2x + 4 = 0$$

has no real solutions. However, our algebraic approach indicated two nonreal complex solutions.

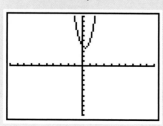

5   Use the discriminant to determine kinds of solutions.

The Discriminant

The quantity $b^2 - 4ac$ that appears under the radical sign in the quadratic formula is called the *discriminant*. In Example 4a the discriminant was 12, a positive number that is not a perfect square, and the equation had two solutions that were irrational numbers. In Example 4b the discriminant was -44, a negative number, and consequently the equation had solutions involving the imaginary number i. In this case our graph had no x-intercepts.

These observations are generalized in Table 1.1.

TABLE 1.1 The Discriminant and the Nature of the Solutions to $ax^2 + bx + c = 0$		
Discriminant $D = b^2 - 4ac$	**Kinds of Solutions** to $ax^2 + bx + c = 0$	**Graph of** $y = ax^2 + bx + c$
$D > 0$	2 real solutions	Two x-intercepts
$D = 0$	1 real solution (a double or repeated solution)	One x-intercept
$D < 0$	0 real solutions 2 complex solutions	No x-intercepts

Using technology

The graph of

$$y = 20x^2 - 200x + 640$$

was obtained with a graphing utility using the following range setting:

Xmin = 0, Xmax = 10,

Xscl = 1,

Ymin = 0, Ymax = 800,

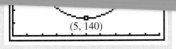

(5, 140)

$$0 = 20t^2 - 200t + 540$$

$$0 = t^2 - 10t + 27 \qquad \text{To simplify things, divide both sides by 20.}$$

Rather than solving $t^2 - 10t + 27 = 0$, we will compute the discriminant. Notice that $a = 1, b = -10$, and $c = 27$.

$$b^2 - 4ac = (-10)^2 - 4(1)(27) = 100 - 108 = -8$$

The discriminant is negative. This indicates that $100 = 20t^2 - 200t + 640$ (or, equivalently, $t^2 - 10t + 27 = 0$) has no real solutions. In terms of the problem's question, this means that the concentration will never get as low as 100 bacteria per cubic centimeter. ∎

---

tudy tip

Which Method to Use When Solving a Quadratic Equation

We have now considered four methods for solving quadratic equations: factoring, the square root method, completing the square, and the quadratic formula. The method of completing the square will be useful later in the book when working with conic sections, but its value here was to derive the quadratic formula. We will no longer need it for solving quadratic equations. Table 1.2 shows which method is most efficient for the various forms of quadratic equations.

TABLE 1.2   Determining the Most Efficient Technique to Use When Solving a Quadratic Equation

Description and Form of the Quadratic Equation	Most Efficient Solution Method	Example
$ax^2 + c = 0$ The quadratic equation has no linear (x) term.	Solving for x^2 and the square root method	$4x^2 - 7 = 0$ $4x^2 = 7$ $x^2 = \dfrac{7}{4}$ $x = \pm\dfrac{\sqrt{7}}{2}$
$u^2 = d; u$ is a linear expression.	The square root method	$(x + 4)^2 = 5$ $x + 4 = \pm\sqrt{5}$ $x = -4 \pm \sqrt{5}$
$ax^2 + bx + c = 0$ and $ax^2 + bx + c$ can be obviously factored.	Factoring and the zero product principle	$3x^2 + 5x - 2 = 0$ $(3x - 1)(x + 2) = 0$ $3x - 1 = 0 \quad$ or $\quad x + 2 = 0$ $x = \frac{1}{3} \qquad\qquad x = -2$
$ax^2 + bx + c = 0$ and $ax^2 + bx + c$ cannot be factored or the factoring is too difficult.	The quadratic formula: $x = \dfrac{-b \pm \sqrt{b^2 - 4ac}}{2a}$	$x^2 - 2x - 6 = 0$ $x = \dfrac{2 \pm \sqrt{4 - 4(1)(-6)}}{2(1)}$ $= \dfrac{2 \pm \sqrt{28}}{2} = \dfrac{2 \pm \sqrt{4}\sqrt{7}}{2}$ $= \dfrac{2 \pm 2\sqrt{7}}{2} = \dfrac{2(1 \pm \sqrt{7})}{2}$ $= 1 \pm \sqrt{7}$

---

6 Solve problems modeled by quadratic equations.

Quadratic Mathematical Models

Creating mathematical models that fit real world data often results in models in the form $y = ax^2 + bx + c$, called *quadratic models*. In developing formulas that model data, mathematicians strive for both accuracy and simplicity.

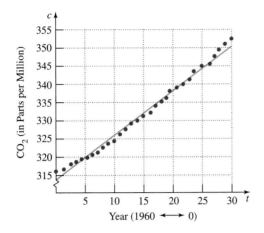

Figure 1.3

Fitting a linear model to data

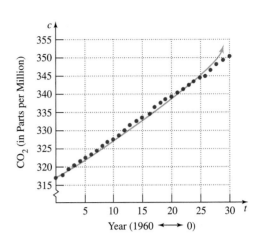

Figure 1.4

Fitting a quadratic model to data

Although linear models are easier to work with than quadratic models, in certain situations they are less accurate. For example, the points shown in Figure 1.3 represent data showing carbon dioxide concentration (C, in ppm) t years after 1960. The linear model $C = 1.24t + 313.6$ has been fitted to the data. Now look at Figure 1.4, in which the quadratic model

$$C = 0.018t^2 + 0.70t + 316.2$$

has been fitted to the data. The quadratic model appears to be a better fit in that its graph passes through more of the data points than does the linear model.

These models form the basis of Example 6.

EXAMPLE 6   Carbon Dioxide Concentration Revisited

Use the linear and quadratic models discussed above to predict in what year

Jordan Massengale "Heaven–Day" 1996, pastel on paper, 16 × 24 in.

$$C = 0.018t^2 + 0.70t + 316.2$$ This is the given quadratic model.

$$470 = 0.018t^2 + 0.70t + 316.2$$ Substitute 470 for C.

$$0 = 0.018t^2 + 0.70t - 153.8 \qquad \text{Subtract 470 from both sides.}$$

$$a = 0.018 \quad b = 0.70 \quad c = -153.8$$

$$t = \frac{-b \pm \sqrt{b^2 - 4ac}}{2a} \qquad \text{Use the quadratic formula.}$$

$$= \frac{-0.70 \pm \sqrt{(0.70)^2 - 4(0.018)(-153.8)}}{2(0.018)} \qquad \text{Substitute the appropriate values.}$$

$$= \frac{-0.70 \pm \sqrt{11.5636}}{0.036} \qquad \text{Perfom operations under the radical.}$$

$$t \approx -114 \quad \text{or} \quad t \approx 75 \qquad \text{Use a calculator.}$$

Since the model begins with a value of $t = 0$ corresponding to the year 1960, we reject -114. The quadratic model predicts a CO_2 concentration of 470 ppm 75 years after 1960, or in the year 2035.

Because the quadratic model seems to fit the data better than the linear model, scientists predict that the CO_2 concentration in the Earth's atmosphere in the year 2035 will be 470 parts per million. ∎

PROBLEM SET 1.2

Practice Problems

Solve each equation in Problems 1–14 by factoring.

1. $x^2 - 3x - 10 = 0$

2. $x^2 - 13x + 36 = 0$

3. $x^2 = 8x - 15$

4. $x^2 = -11x - 10$

5. $6x^2 + 11x - 10 = 0$

6. $9x^2 + 9x + 2 = 0$

7. $3x^2 - 2x = 8$

8. $4x^2 - 13x = -3$

9. $3x^2 + 7x = 0$

10. $5x^2 - 9x = 0$

11. $2x(x - 3) = 5x^2 - 7x$

12. $16x(x - 2) = 8x - 25$

13. $7 - 7x = (3x + 2)(x - 1)$

14. $10x - 1 = (2x + 1)^2$

Solve each equation in Problems 15–28 by the square root method.

15. $3x^2 = 27$

16. $5x^2 = 45$

17. $5x^2 + 1 = 51$

18. $3x^2 - 1 = 47$

19. $(x + 2)^2 = 25$

20. $(x - 3)^2 = 36$

21. $(3x + 2)^2 = 9$

22. $(4x - 1)^2 = 16$

23. $(5x - 1)^2 = 7$

24. $(8x - 3)^2 = 5$

25. $(3x - 4)^2 = 8$

26. $(2x + 8)^2 = 27$

27. $(x + 3)^2 = -16$

28. $(x + 5)^2 = -9$

Solve each equation in Problems 29–36 by completing the square.

29. $x^2 - 4x - 11 = 0$

30. $x^2 - 2x - 5 = 0$

31. $2x^2 - 7x + 4 = 0$

32. $9x^2 - 6x - 4 = 0$

33. $4x^2 - 4x - 3 = 0$

34. $2x^2 - x - 6 = 0$

35. $2x^2 - 6x + 5 = 0$

36. $2x^2 + 2x + 5 = 0$

Solve each equation in Problems 37–48 using the quadratic formula.

37. $3x^2 - 3x - 4 = 0$

38. $5x^2 + x - 2 = 0$

39. $4y^2 = 2y + 7$

40. $3y^2 = 6y - 1$

41. $x^2 - 6x + 10 = 0$

42. $x^2 - 2x + 17 = 0$

43. $(w + 2)^2 = 2(5w - 2)$

44. $(w - 12)(w + 7) = (3 - w)(w + 4)$

45. $y^2 - 5 = 2(y + 3) - 4$

46. $3y^2 + 11 = 2y(2y - 1)$

47. $1.5x^2 - 7.3 = 10.2$

48. $3.2x^2 + 7.6x = 9.1$

Solve each equation in Problems 49–82 by the method of your choice.

49. $(x - 3)(x + 3) = 12$

50. $(2x - 3)(3x + 1) = 17$

51. $8y^2 - 3 = 0$

52. $12y^2 - 5 = 0$

53. $(5z - 1)(2z + 3) = 3z - 3$

54. $(2z - 1)(z - 1) = (z - 1)(z + 1)$

55. $2x^2 - 5x - 12 = -5x$

56. $5x^2 - 2x + 4 = 8 - 2x$

57. $(3x - 4)^2 = 81$

58. $(2x - 3)^2 = 121$

59. $2.4y^2 - 12.72y + 3.6 = 0$

60. $2.3y^2 - 13.15y + 8.07 = 0$

61. $3w^2 - 4w + 2 = 0$

62. $w^2 - 2w + 4 = 0$

63. $(y - 3)(y + 4) + 2 = y - 3$

64. $(y - 2)(y - 5) = (2y - 7)y$

65. $\dfrac{2}{y + 1} + \dfrac{3y}{2} = 4$

66. $\dfrac{3y}{y + 1} + \dfrac{2}{y - 2} = 5$

67. $\dfrac{2}{z + 4} - \dfrac{3}{z + 1} = 4$

68. $\dfrac{3}{z + 2} - \dfrac{5}{z - 2} = 2$

69. $10^{-4}x^2 + 2(10^{-3})x + 10^{-2} = 0$

70. $10^{-4}x^2 + 3(10^{-3})x + 10^{-2} = 0$

71. $\sqrt{2}x^2 + x - 10\sqrt{2} = 0$

72. $2\sqrt{5}x^2 - x = 2\sqrt{5}$

73. $(x + 3)^2 = (x - 5)^2$

74. $(x + 4)^2 = (x - 1)^2$

75. $(x + 1)^2 = 9x^2$

76. $(x - 2)^2 = 16x^2$

77. $(2x + 3)^2 - 9 = 0$

78. $(3x - 1)^2 - 121 = 0$

79. $(3x - 1)^2 = (x + 5)^2$

80. $(5x + 2)^2 = (x - 4)^2$

81. $\left| x^2 + 2x - 36 \right| = 12$

82. $\left| x^2 + 6x + 1 \right| = 8$

Application Problems

83. The number of inmates (I, in thousands) in federal and state prisons in the United States is modeled by $I = 2x^2 + 22x + 320$, where x represents the number of years after 1980. In what year does the model predict that the number of inmates will reach 1560 thousand?

84. The average cost of a new car (C, in dollars) t years after 1970 is approximated by the model $C = 30.5t^2 + 4192$. In what year will the average cost of a car reach $25,000?

85. The model

86. The population (P, in millions) of the United States t decades after 1800 is approximated by the model $P = 0.6942t^2 + 6.183$, where $0 \le t \le 9$. Use this model to determine when the population of the United States was 17,290,200. What population is predicted by this model for the present time? How far off is this prediction? What does this tell you about mathematical models?

87. The model $M = 0.0075t^2 - 0.2676t + 14.8$ provides an estimation of fuel efficiency of passenger cars (M, measured in miles per gallon) t years after 1940. Environmentalists are pressuring automobile manufacturers for

Use the discriminant to answer the questions in Problems 89–91.

89. As shown in the figure, a ball is thrown upward from the top of a 200-foot building with a velocity of 40 feet per second. The height (s, in feet) of the ball above the ground after t seconds is given by the model $s = -16t^2 + 40t + 200$.

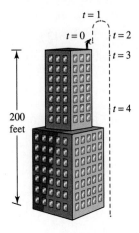

a. Will the ball ever reach a height of 240 feet?

b. Will the ball ever reach a height of 220 feet? How often does this occur? What does this mean in terms of the path that the ball follows?

c. Will the ball ever reach a height of 225 feet? How often does this occur? Explain what this means in terms of the path that the ball follows.

90. The daily profit (y, in dollars) from the sale of x units of a product is given by $y = -x^2 + 180x - 4500$. A newly hired advertising executive has promised the company that a unique advertising campaign will result in daily sales of \$3800. What does the discriminant indicate about this person's future with the company? Explain.

91. The model $P = -5I^2 + 80I$ describes the power (P) of an 80-volt generator subject to a current (I, in amperes) of electricity. Can there ever be enough current to generate 340 volts of power?

True–False Critical Thinking Problems

92. Which one of the following is true?

a. The equation $(2x - 3)^2 = 25$ is equivalent to $2x - 3 = 5$.

b. Every quadratic equation has two distinct numbers in its solution set.

c. A quadratic equation whose coefficents are real numbers can never have a solution set containing one real number and one complex nonreal number.

d. The equation $ax^2 + c = 0$ cannot be solved by the quadratic formula.

93. Which one of the following is true?

a. Some quadratic equations can be solved by factoring but not by the quadratic formula.

b. Some equations in the form $ax^2 + c = 0$ can be solved by factoring.

c. If $(x - 5)(2x + 3) = 0$ then $x = 5$ or $x = \frac{3}{2}$.

d. The quadratic equation whose solution set is $\{-4, -2\}$ has a negative discriminant because both solutions are negative.

Technology Problems

94. Graph the model of Problem 87 ($y = 0.0075x^2 - 0.2676x + 14.8$) using the following range setting:

Xmin = 0, Xmax = 80, Xscl = 20,

Ymin = 0, Ymax = 50, Yscl = 10

Move the cursor along the graph, using the $\boxed{\text{TRACE}}$ and $\boxed{\text{ZOOM}}$ features to estimate the year in which automobile fuel efficiency was the poorest. What was the gas mileage in that year? Find pictures of the most popular cars for that year and write a sentence relating the most popular cars of the time with the fuel efficiency indicated by the model's graph.

95. Graph the model of Problem 89 ($y = -16x^2 + 40x + 200$) using the following range setting:

Xmin = 0, Xmax = 5, Xscl = 1,

Ymin = 0, Ymax = 240, Yscl = 24

Use the graph with values of time in seconds along the x-axis ($0 \leqslant x \leqslant 5$) and values of the height of the ball along the y-axis to answer each of the three parts of Problem 89.

96. From elementary school through the university level, total school expenditures in the United States y, in millions of dollars, x years after 1975 are approximated by the model

$$y = 459.5x^2 + 9170.3x + 107{,}298.$$

Graph the model using the following range setting:

Xmin = 0, Xmax = 20, Xscl = 2,

Ymin = 0, Ymax = 500,000, Yscl = 50,000

A journalist wrote that total school expenditures in the United States have been increasing *steadily* each year since 1975. Does the graph of the expenditure model seem to support this statement? Explain.

97. In Problems 94–96, you were asked to graph three quadratic models in the form $y = ax^2 + bx + c$. Range settings were given. Try graphing each of the models in the standard range setting and see what happens. How might you determine the range settings for the models in Problems 94–96 if they were not given? What information might you want to know about the graph of $y = ax^2 + bx + c$ to help determine appropriate range settings?

98. Graphing utilities are capable of providing the solutions to any polynomial equation. A quadratic equation can be thought of as a polynomial equation of degree (or order) 2. If your utility has a key labeled POLY (polynomial equations), try using it to solve $2.7x^2 - 27x - 6 = 0$. As soon as you enter order = 2, a quadratic equation ($a_2 x \wedge 2 + a_1 x + a_0 = 0$) should appear in the viewing screen. Enter the coefficients ($a_2 = 2.7$, $a_1 = -27$, $a_0 = -6$) and then press SOLVE. The solutions ($x_1 \approx 10.2174919475$, $x_2 \approx -0.2174919475$) should appear on the screen. Use the POLY feature to solve some of the quadratic equations in this problem set, including those whose solutions are not real numbers. How does your graphing utility indicate nonreal solutions?

Writing in Mathematics

99. Problem 98 indicates that graphing utilities can instantly provide the solutions to any quadratic equation. Do you think that this eliminates the need for students to learn the four algebraic methods for solving quadratic equations? If you were teaching youngsters, would you use calculators instead of explaining the steps involved in addition, subtraction, multiplication, and division of whole numbers? Are your answers the same to both questions? If not, explain the difference you see between the two situations.

100. Describe the relationship between the discriminant and the graph of $y = ax^2 + bx + c$.

101. If $(x + 2)(x - 4) = 0$ indicates that $x + 2 = 0$ or $x - 4 = 0$, explain why $(x + 2)(x - 4) = 6$ does not mean $x + 2$

103. What method would you use to solve each of the following quadratic equations? Explain, in each case, why you would select that method.

a. $x^2 - 5x + 6 = 0$

b. $x^2 - 6x + 2 = 0$

c. $x^2 - 5 = 0$

d. $(x + a)^2 - d^2 = 0$

104. In the first two sections of this chapter, we discussed linear and quadratic equations. Browse through the entire chapter and you'll quickly see that it's almost impossible to randomly select a page that does not contain an equation. In spite of this, Stephen Hawking, talking about his book, *A Brief History of Time*, wrote, "Some-

$$\frac{}{2a} \qquad \frac{}{2a} \qquad \frac{}{a}$$

b. Use part (a) to write a statement about the product and sum of the solutions of $ax^2 + bx + c = 0$.

tion. Explain how you can use part (a) with your written statement in part (b) to check the solutions. Is the solution set correct?

In Problems 106–107, find the value(s) of k for which each equation will have exactly one solution.

106. $kx^2 + kx + 2 = 0$

107. $3x^2 + (5\sqrt{k})x + 6 = 0$

108. Write a quadratic equation in standard form whose solution set is $\{-3, 5\}$.

109. Verify that $\dfrac{-b + \sqrt{b^2 - 4ac}}{2a}$ is a solution of $ax^2 + bx + c = 0$ by substituting the expression into the quadratic equation.

110. a. Show that

$$ax^2 + bx + c$$
$$= a\left(x - \frac{-b + \sqrt{b^2 - 4ac}}{2a}\right)\left(x - \frac{-b - \sqrt{b^2 - 4ac}}{2a}\right)$$

by multiplying on the right (using the FOIL method) to obtain the trinomial on the left.

b. Use your result from part (a) to derive the quadratic formula.

c. Use your result from part (a) to factor $x^2 + 8x + 144$ and $3x^2 + 10x + 6$.

111. Solve for x in terms of d and e:

$$x^2 - \left(\sqrt{\frac{d}{e}} + \sqrt{\frac{e}{d}}\right)x + 1 = 0$$

112. For what values of b and c will the solution set of $x^2 + bx + c = 0$ be $\{b, c\}$?

113. Consider the quadratic equation $x^2 + bx + c = 0$, where $b \neq 0$ and $c \neq 0$.

a. If one solution is four times the other, show that $\dfrac{b^2}{c} = \dfrac{25}{4}$.

b. If one solution is k times the other, show that $\dfrac{b^2}{c} = \dfrac{(k + 1)^2}{k}$ $(k \neq -1$ and $k \neq 0)$.

SECTION 1.3   # Linear and Quadratic Modeling

Objectives

Solutions   Tutorial   Video
Manual   3

1 Use linear equations to solve problems.

2 Use quadratic equations to solve problems.

A Problem-Solving Strategy

In the first two sections of this chapter, the models for atmospheric carbon dioxide, depreciation, bacteria concentration, and breast cancer–fat intake correlation were given. Using these models, we were able to solve real world problems. We focus now on how models can be constructed from verbal conditions that are implied in a given situation. Since these problems are a bit more difficult, we offer the following five-step strategy for solving problems.

> **Strategy for problem solving**
>
> *Step 1.* Read the problem carefully. Attempt to state the problem in your own words and state what the problem is looking for. Let x (or any variable) represent one of the quantities in the problem.
>
> *Step 2.* If necessary, write expressions for any other unknown quantities in the problem in terms of x.
>
> *Step 3.* Form a verbal model of the problem's conditions and then write an equation in x that translates the verbal model.
>
> *Step 4.* Solve the equation written in step 3 and answer the question in the problem.
>
> *Step 5.* Check your answer against the verbal conditions of the original problem, possibly using a graphing utility to further explore the geometric significance of the answer.

Use linear equations to solve problems.

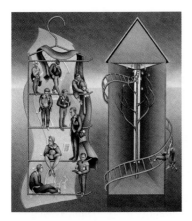

Wendy Seller "Dissension" 1993, oil on canvas, 34 in. H × 30 in. W. Courtesy of the Pepper Gallery, Boston, MA. In the Collection of Robert and Elayne Simandl. Photo by Clements/Howcroft, Boston.

Constructing Linear Models

Let's see how we can use this five-step strategy by turning to a number of specific examples.

EXAMPLE 1 **The Impact of Gender on Education and Income**

Numerous variables affect the relationship between education and income. A simplified form of two such models focuses on the effect of gender in these relationships. Yearly income for men increases by $1600 with each year of education; men with no education earn $6300 yearly. For women, yearly income increases by $1200 for each year of education; women with no education earn $2100 yearly. Using these models, how many years of education must a woman have to earn the same yearly salary as a man with 11 years of education?

Discover for yourself

1. Use the fact that men earn $6300 yearly plus $1600 for each year of education to compute the yearly earnings for a man with 11 years of education.
2. Use the fact that women earn $2100 yearly plus $1200 for each year of education to write an algebraic expression for the yearly earnings for a woman with x years of education. Use the same method as in step 1.
3. Read the example again and explain how you can use your work in steps 1 and 2 to solve the problem.

Solution

Steps 1 and 2. Represent unknowns in terms of x.

Let x = the years of education needed by a woman to earn the same yearly salary as at man with 11 years of education.

Step 3. Write an equation that models

$$23,900 = 2100 + 1200x \qquad \text{Multiply and add } \ldots$$

$$21,800 = 1200x \qquad \text{Subtract 2100 from both sides.}$$

$$\frac{21,800}{1200} = x \qquad \text{Divide both sides by 1200.}$$

$$x \approx 18.2 \qquad \text{Perform the computation. Remember that } \approx \text{ means "is approximately equal to."}$$

The models show that a woman must have approximately 18.2 years of education to earn the same salary as a man with 11 years of education.

Step 5. Check the proposed solution in the original wording of the problem.

Discover for yourself

Take a few minutes to check the solution. Explain how the graphs shown in Figure 1.5 approximately illustrate the solution.

Years of Education	Predicted Average Personal Wages for Men
0	$6300
1	$7900
2	$9500
3	$11,100
4	$12,700
5	$14,300
6	$15,900
7	$17,500
8	$19,100
9	$20,700
10	$22,300
11	$23,900
12	$25,500
13	$27,100
14	$28,700
15	$30,300
16	$31,900
17	$33,500
18	$35,100
19	$36,700
20	$38,300

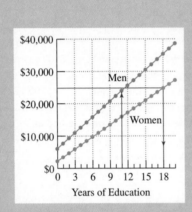

Figure 1.5

Average personal wages for women and men by years of education

Years of Education	Predicted Average Personal Wages for Women
0	$2100
1	$3300
2	$4500
3	$5700
4	$6900
5	$8100
6	$9300
7	$10,500
8	$11,700
9	$12,900
10	$14,100
11	$15,300
12	$16,500
13	$17,700
14	$18,900
15	$20,100
16	$21,300
17	$22,500
18	$23,700
19	$24,900
20	$26,100

EXAMPLE 2   Selecting a Heating System

Costs for two different kinds of heating systems for a three-bedroom home are given below.

System	Cost to Install	Operating Cost per Year
Solar	$29,700	$150
Electric	$5000	$1100

After how many years will total costs for solar heating and electric heating be the same? What will be the cost at that time?

Solution

Steps 1 and 2. Represent unknowns in terms of x.

Step 3. Write an equation that models the verbal conditions.

Let x = the number of years after which total costs for the two systems will be the same.

We begin by constructing a model for the total cost for each system.

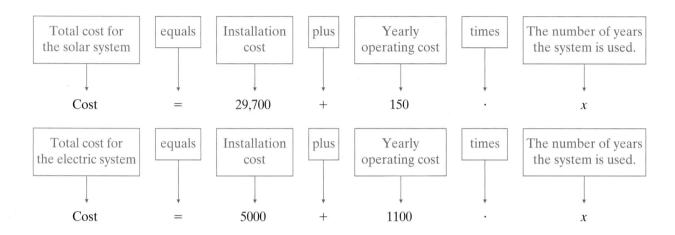

Our total cost models are

$$\text{Solar} = 29{,}700 + 150x$$

$$\text{Electric} = 5000 + 1100x$$

Now that we have modeled the conditions, we model the problem's first question:

After how many years will

Step 5. Check the proposed solution in the original wording of the problem.

After 26 years, the total costs for the two systems will be the same. To answer the problem's second question, we substitute 26 into our total cost models.

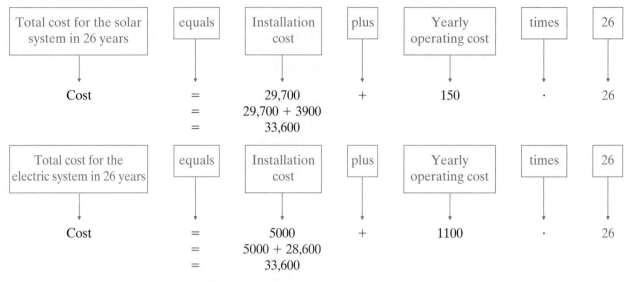

Total cost for the solar system in 26 years	equals	Installation cost	plus	Yearly operating cost	times	26

$$\text{Cost} \quad = \quad 29{,}700 \quad + \quad 150 \quad \cdot \quad 26$$
$$= \quad 29{,}700 + 3900$$
$$= \quad 33{,}600$$

Total cost for the electric system in 26 years	equals	Installation cost	plus	Yearly operating cost	times	26

$$\text{Cost} \quad = \quad 5000 \quad + \quad 1100 \quad \cdot \quad 26$$
$$= \quad 5000 + 28{,}600$$
$$= \quad 33{,}600$$

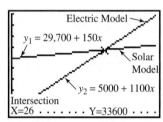

Figure 1.6

After 26 years, solar and electric systems have the same total costs: $33,600.

After 26 years, both systems will cost $33,600.

We can visualize this solution by graphing the two models in the same rectangular coordinate system. Representing total costs by y, the models are $y = 29{,}700 + 150x$ and $y = 5000 + 1100x$.

Figure 1.6 shows the graphs obtained on a graphing utility using

$$\text{Xmin} = 0, \text{Xmax} = 40, \text{Xscl} = 2, \text{Ymin} = 5000,$$

$$\text{Ymax} = 50{,}000, \text{Yscl} = 5000.$$

The graphs intersect at $(26, 33{,}600)$, verifying that after 26 years total costs for both systems will be $33,600. ∎

EXAMPLE 3 **Selecting a Mortgage Plan**

Table 1.3 on page 141 gives the monthly mortgage payment for each $1000 of mortgage at various mortgage rates for different periods of time. For example, for a mortgage of $80,000 for 30 years at 8.5%, the monthly payment for principal and interest would be 80 times $7.69 (shown in pink in Table 1.3) or $615.20. The same mortgage payment for 30 years at 7.25% results in monthly payments for principal and interest of 80 times $6.83 (also shown in pink in Table 1.3) or $546.40.

Suppose you are purchasing a home and are considering the following two options for an $80,000 mortgage.

Option 1:
8.5% interest for a 30-year loan, 1 point, (a point is a one-time charge of 1% of the mortgage amount), and a $300 loan application fee

Option 2:
7.25% interest for a 30-year loan, 5 points, and a $1500 loan application fee

Approximately how many months will it take for the total cost of each mortgage option to be the same?

Solution

Steps 1 and 2. Represent unknowns in terms of x.

Let x = the number of months for the total cost for the two mortgage options to be the same.

TABLE 1.3	Any Bank, USA: Equal Monthly Payment to Amortize a Loan of $1000								
Rate	**Payment for a Mortgage Period (years) of:**				**Rate**	**Payment for a Mortgage Period (years) of:**			
(%)	**15**	**20**	**25**	**30**	**(%)**	**15**	**20**	**25**	**30**
4.500	7.65	6.33	5.56	5.07	8.625	9.93	8.76	8.14	7.78
4.625	7.71	6.39	5.63	5.14	8.750	10.00	8.84	8.23	7.87
4.750	7.78	6.46	5.70	5.22	8.875	10.07	8.92	8.31	7.96
4.875	7.84	6.53	5.77	5.29	9.000	10.15	9.00	8.40	8.05
5.000	7.91	6.60	5.85	5.37	9.125	10.22	9.08	8.48	8.14
5.125	7.97	6.67	5.92	5.44	9.250	10.30	9.16	8.57	8.23
5.250	8.04	6.73	6.00	5.52	9.375	10.37	9.24	8.66	8.32
5.375	8.10	6.81	6.07	5.60	9.500	10.45	9.33	8.74	8.41
5.500	8.17	6.88	6.14	5.68	9.625	10.52	9.41	8.83	8.50
5.625	8.24	6.95	6.22	5.76	9.750	10.60	9.49	8.92	8.60
5.750	8.30	7.02	6.29	5.84	9.875	10.67	9.57	9.00	8.69
5.875	8.37	7.09	6.37	5.92	10.000	10.75	9.66	9.09	8.78
6.000	8.44	7.16	6.44	6.00	10.125	10.83	9.74	9.18	8.87
6.125	8.51	7.24	6.52	6.08	10.250	10.90	9.82	9.27	8.97
6.250	8.57	7.31	6.60	6.16	10.375	10.98	9.90	9.36	9.06
6.375	8.64	7.38	6.67	6.24	10.500	11.06	9.99	9.45	9.15
6.500	8.71	7.46	6.75	6.32	10.625	11.14	10.07	9.54	9.25
6.625	8.78	7.53	6.83	6.40	10.750	11.21	10.16	9.63	9.34
6.750	8.85	7.60	6.91	6.49	10.875	11.29	10.24	9.72	9.43
6.875	8.92	7.68	6.99	6.57	11.000	11.37	10.33	9.81	9.53
7.000	8.99	7.76	7.07	6.66	11.125	11.45	10.41	9.90	9.62
7.125	9.06	7.83	7.15	6.74	11.250	11.53	10.50	9.99	9.72
7.250	9.13	7.91	7.23	6.83	11.375	11.61	10.58	10.08	9.81
7.375	9.20	7.98	7.31	6.91	11.500	11.69	10.67	10.17	9.91
7.500	9.28	8.06	7.39	7.00	11.625	11.77	10.76	10.26	10.00
7.625	9.35	8.14	7.48	7.08	11.750	11.85	10.84	10.35	10.10
7.750	9.42	8.21	7.56	7.17	11.875	11.93	10.93	10.44	10.20
7.875	9.49	8.29	7.64	7.26	12.000	12.01	11.02	10.54	10.29
8.000	9.56	8.37	7.72	7.34	12.125	12.09	11.10	10.63	10.39
8.125	9.63	8.45	7.81	7.43	12.250	12.17	11.19	10.72	10.48
8.250	9.71	8.53	7.89	7.52	12.375	12.25	11.28	10.82	10.58
8.375	9.78	8.60	7.97	7.61	12.500	12.33	11.37	10.91	10.68
8.500	9.85	8.68	8.06	7.69					

Step 3. Write an equation that models the verbal conditions.

As we did in the previous example, we construct models for the total cost for each mortgage option over x months, setting the models equal to each other.

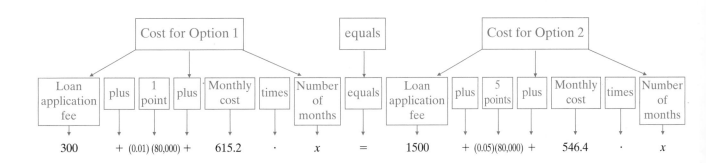

Cost for Option 1   equals   Cost for Option 2

Loan application fee — plus — 1 point — plus — Monthly cost — times — Number of months — equals — Loan application fee — plus — 5 points — plus — Monthly cost — times — Number of months

$300 \quad + (0.01)(80{,}000) + \quad 615.2 \quad \cdot \quad x \quad = \quad 1500 \quad + (0.05)(80{,}000) + \quad 546.4 \quad \cdot \quad x$

Step 4. Solve the equation and answer the problem's question.

$$300 + (0.01)(80{,}000) + 615.2x = 1500 + (0.05)(80{,}000) + 546.4x$$

This is the equation implied by the given conditions.

$$1100 + 615.2x = 5500 + 546.4x$$

Multiply and add as indicated.

$$1100 + 68.8x = 5500$$

Subtract 546.4x from both sides.

$$68.8x = 4400$$

Subtract 1100 from both sides.

$$x = \frac{4400}{68.8}$$

Divide both sides by 68.8.

$$\approx 64$$

After approximately 64 months or $\frac{64}{12} = 5\frac{1}{3}$ years, the total cost of both mortgage options will be the same. Notice that option 1 involves less money up front but greater monthly payments. Prior to 64 months, option 1 will result in lower total costs. After that time, option 2 makes more sense. Knowing how long the house will be kept would help you select the loan with the lowest total costs. ■

Step 5. Check the proposed solution in the original wording of the problem.

Take a few minutes to check the solution.

iscover for yourself

Use a graphing utility to visualize the solution to Example 3. Graph $y_1 = 1100 + 615.2x$ and $y_2 = 5500 + 546.4x$ (the simplified models) in the same viewing rectangle. Use a setting such as

   Xmin = 0, Xmax = 70, Xscl = 1,

   Ymin = 0, Ymax = 45,000, Yscl = 300

Then use the ⎢TRACE⎥ or ⎢ISECT⎥ feature to find the point of intersection of the two lines. The x-coordinate of the intersection point should be approximately 64. What is the y-coordinate and what does this represent?

In our next example, we must once again use common sense to write a sentence that will then, phrase by phrase, be translated into an equation.

EXAMPLE 4   **Modeling Implied Conditions**

A two-year union contract guarantees an immediate 6% raise and a 5% raise after one year. How much must an employee be earning weekly at the start of the contract if that employee's weekly salary increases $60 over the two-year life of the contract?

iscover for yourself

One way to familiarize yourself with the problem is to select a particular number and use that number to help understand the problem's conditions. Suppose the worker is making $600 per week at the start of the contract.

1. After the 6% increase, the worker will earn $600 plus 6% of $600. How much is this new salary?

Fernand Léger "Les constructeurs" / Musée Leger, © 1998 Artists Rights Society (ARS), New York / ADAGP, Paris.

2. After the 5% increase, the worker will earn the salary computed in step 1 plus 5% of this salary. How much is this salary with the second increase?

3. What is the increase over the two-year life of the contract? Subtract the original salary ($600) from the salary that you computed in step 2.

4. Read Example 4 again. Based on your answer in step 3, what is a reasonable estimate for the answer to the problem?

5. Rework steps 1 and 2 if the original salary is *x* dollars rather than $600.

Solution

Steps 1 and 2. Represent unknowns in terms of *x*.

Let x = the employee's weekly salary at the start of the contract. An implied verbal model in this situation is

Step 3. Write an equation that models the verbal conditions.

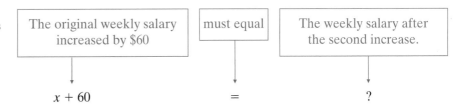

We must find an algebraic expression for the phrase on the right.

Since there is an immediate 6% raise, the salary after the first raise increases by 6% of what it was originally.

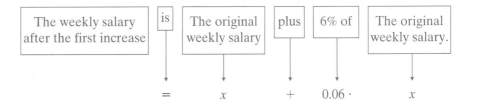

Thus, the weekly salary after the first increase is represented by $x + 0.06x$.

Since there is a 5% raise after a year, the salary after the second raise increases by 5% of the salary after the first raise.

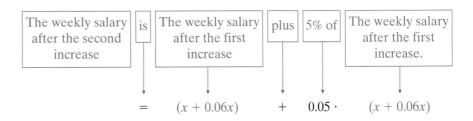

Thus, the weekly salary after the second increase is represented by

$$(x + 0.06x) + 0.05(x + 0.06x).$$

We can now complete the equation implied by our original boxed expression.

The original weekly salary increased by $60	must equal	The weekly salary after the second increase.

$$x + 60 \qquad = \qquad (x + 0.06x) + 0.05\,(x + 0.06x)$$

Step 4. Solve the equation and answer the problem's question.

$x + 60 = (x + 0.06x) + 0.05(x + 0.06x)$ This is the equation implied by the problem's conditions.

$x + 60 = x + 0.06x + 0.05x + 0.003x$ Use the distributive property on the right.

$x + 60 = 1.113x$ Combine like terms on the right.

$60 = 0.113x$ Subtract x from both sides.

$x = \dfrac{60}{0.113} \approx 530.97$ Divide both sides by 0.113.

Step 5. Check the proposed solution in the original wording of the problem.

An employee must be earning approximately $530.97 per week for the weekly salary to increase by $60 over the life of the contract. Let's see if this answer makes sense. With the 6% raise, the new weekly salary is

$$530.97 + (0.06)(530.97)$$

or approximately $562.83. With the 5% raise, the new weekly salary is

$$562.83 + (0.05)(562.83)$$

or approximately $590.97. Notice that $590.97 is a $60 increase over the original weekly salary of $530.97.   ■

2 Use quadratic equations to solve problems.

Constructing Quadratic Models

We continue using our five-step strategy. The only difference is that the equation that models the verbal conditions will now be quadratic.

EXAMPLE 5 Constructing a Quadratic Model

An owner of a large diving boat that can carry as many as 70 people charges $10 per passenger to groups of between 15 and 20 people. If more than 20 divers charter the boat, the fee per passenger is decreased by 10 cents for every person over 20. How many people must there be in a diving group to generate an income of $323.90?

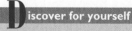

iscover for yourself

Before solving Example 5 using algebra, let's consider a few specific values for the number of people in the group and the resulting income.

Number of People in the Diving Group	Cost per Person	Income = (Number of People) × (Cost per Person)
20	$10.00	(20)($10.00) = $200
21	$10.00 − $0.10 = $9.90	(21)($9.90) = $207.90
22	$10.00 − $0.20 = $9.80	(22)($9.80) = $215.60
23	$10.00 − $0.30 = $9.70	(23)($9.70) = $223.10

Do you see a pattern forming? If x represents the number of people beyond 20 (so there are $20 + x$ people in the diving group), write an algebraic expression for the income.

   Take a moment to compute the income if there are 30 people in the group. Repeat this computation for 50 divers. What is a reasonable estimate for the number of divers to generate $323.90?

Solution

Steps 1 and 2. Represent unknowns in terms of x.

Since the cost per person is computed differently depending on whether the number of divers is less than or equal to 20 or greater than 20,

   let x = the number of people beyond 20 to generate $323.90 of income.

Thus,

   $20 + x$ = the total number of people in the group.

Step 3. Write an equation that models the verbal conditions.

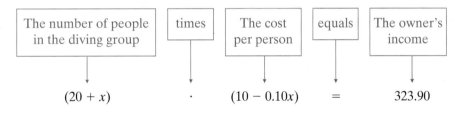

The number of people in the diving group	times	The cost per person	equals	The owner's income
$(20 + x)$	$\cdot$	$(10 - 0.10x)$	$=$	323.90

Step 4. Solve the equation and answer the problem's question.

$$(20 + x)(10 - 0.10x) = 323.90$$

This is the equation implied by the problem's conditions.

$$200 + 8x - 0.10x^2 = 323.90$$

Multiply using FOIL.

$$-0.10x^2 + 8x - 123.9 = 0$$

Write the quadratic equation in standard form.

$$a = -0.10 \quad b = 8 \quad c = -123.9$$

$$x = \frac{-b \pm \sqrt{b^2 - 4ac}}{2a}$$

Write the quadratic formula and then let $a = -0.10$, $b = 8$, and $c = -123.9$.

$$= \frac{-8 \pm \sqrt{8^2 - 4(-0.10)(-123.9)}}{2(-0.10)}$$

Substitute the given numbers.

$$= \frac{-8 \pm \sqrt{14.44}}{-0.20}$$

The positive discriminant (14.44) indicates that the equation will have two real solutions.

$$x = \frac{-8 + 3.8}{-0.20} \quad \text{or} \quad x = \frac{-8 - 3.8}{-0.20} \qquad \sqrt{14.44} = 3.8$$

$$= 21 \qquad\qquad\qquad = 59 \qquad\qquad \text{Simplify.}$$

Since x represents the number of passengers above 20, we see that $41(20 + 21)$ or $79(20 + 59)$ people will generate an income of \$323.90. However, we were told that the boat can carry as many as 70 passengers. Thus, 79 divers are too many, and the required number of people is 41.

Step 5. Check the proposed solution in the original wording of the problem.

With 41 passengers (21 above 20), the cost per person is reduced by $21(0.10)$ or \$2.10. Each person pays \$10.00 − \$2.10 or \$7.90. The income is $41(7.90) =$ \$323.90, which checks with the given conditions.   ∎

Using technology

The graph of $y = -0.10x^2 + 8x + 200$, the model for income with $20 + x$ divers in the group, was obtained with a graphing utility using the following range setting:

Xmin = 0, Xmax = 50, Xscl = 5,

Ymin = 0, Ymax = 400, Xscl = 40.

We can use the graph and the ⎡TRACE⎤ feature to find how many divers should be on the boat to maximize income. The graph indicates that income increases up to 40 divers (meaning that $20 + 40$ or 60 passengers will maximize income). After 40, income starts to decrease, as shown in the table.

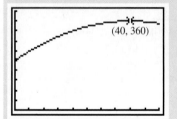
(40, 360)

Number of People in the Diving Group	Cost per Person	Income
59 ($x = 39$)	\$6.10	359.90
60 ($x = 40$)	\$6.00	360.00 (maximum)
61 ($x = 41$)	\$5.90	359.90

Geometric Modeling

Solving geometry problems usually requires a knowledge of basic geometric ideas and formulas. This can range from knowing basic formulas for perimeter, area, and volume to recognizing properties of similar triangles. Often we are asked to use logical reasoning based on a geometric figure.

A knowledge of the formulas in Table 1.4 on page 147 will be useful in solving problems involving perimeter, area, and volume.

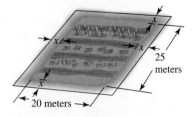

x

x x 25 meters

x

20 meters

Figure 1.7

EXAMPLE 6   Problem Solving Using Geometric Formulas

A path of uniform width is constructed around the outside of a 20 meter by 25 meter rectangular garden (Figure 1.7). Find how wide the path should be if there is enough brick to cover 196 square meters.

TABLE 1.4 Common Formulas for Area, Perimeter, and Volume

Square
$A = s^2$
$P = 4s$

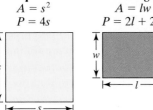

Rectangle
$A = lw$
$P = 2l + 2w$

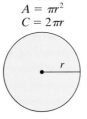

Circle
$A = \pi r^2$
$C = 2\pi r$

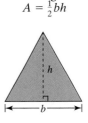

Triangle
$A = \frac{1}{2}bh$

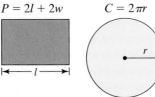

Trapezoid
$A = \frac{1}{2}h(a + b)$

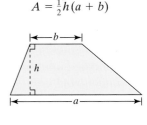

Cube
$V = s^3$

Rectangular Solid
$V = lwh$

Circular Cylinder
$V = \pi r^2 h$

Sphere
$V = \frac{4}{3}\pi r^3$

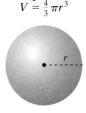

Cone
$V = \frac{1}{3}\pi r^2 h$

Wassily Kandinsky
"Swinging" (Schaukeln),
1925. Tate Gallery, London.
Photo: John Webb. Art
Resource, NY.

Discover for yourself

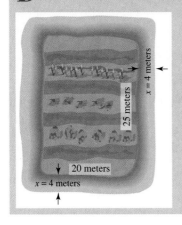

Suppose that the width of the path in Example 6 is 4 meters.

1. What are the dimensions of the rectangle consisting of the garden and the path combined? What is the area of this rectangle?

2. What is the area of the rectangle consisting of just the garden?

3. Find the area of the path by using the areas you found in steps 1 and 2.

4. Repeat steps 1–3 if the width of the path is x meters rather than 4 meters.

5. Based on your answer in step 3, since the area of the path is 196 square meters, what is a reasonable estimate of its width?

Solution

Steps 1 and 2. Represent unknowns in terms of x.

Let x = the width of the path. The situation is illustrated in Figure 1.8 on page 148. In the Discover for yourself, were you able to observe that although the path is not a rectangle, its area can be determined by subtracting the area of the rectangular garden from the area of the rectangle containing the garden and path combined?

Step 3. Write an equation that models the verbal conditions.

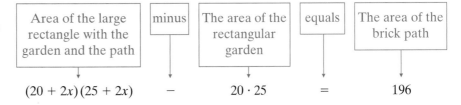

Area of the large rectangle with the garden and the path	minus	The area of the rectangular garden	equals	The area of the brick path
$(20 + 2x)(25 + 2x)$	$-$	$20 \cdot 25$	$=$	196

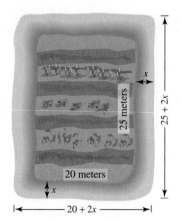

Figure 1.8

Step 4. Solve the equation and answer the problem's question.

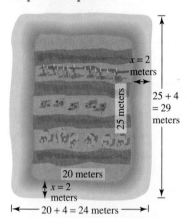

Figure 1.9

Step 5. Check the proposed solution in the original wording of the problem.

$$(20 + 2x)(25 + 2x) - 20 \cdot 25 = 196$$ This is the equation modeling the path's area.

$$500 + 90x + 4x^2 - 500 = 196$$ Multiply using the FOIL method.

$$90x + 4x^2 = 196$$ Simplify.

$$4x^2 + 90x - 196 = 0$$ Write the quadratic equation in standard form.

$$2x^2 + 45x - 98 = 0$$ Divide both sides by 2.

$$(x - 2)(2x + 49) = 0$$ Factor.

$$x - 2 = 0 \quad \text{or} \quad 2x + 49 = 0$$ Set each factor equal to 0.

$$x = 2 \qquad x = -\frac{49}{2}$$ Solve the resulting equations.

Since the path's width must be positive, its width should be 2 meters. The area of the rectangle with the garden and path combined (Figure 1.9) is $24 \cdot 29 = 696$ square meters. The area of the garden is $20 \cdot 25 = 500$ square meters. Thus, the area of the path is $696 - 500 = 196$ square meters, which checks with the problem's conditions. ∎

Using technology

A graphing utility enables us to visualize the solution to Example 6. In graphing the model representing the area of the path, namely,

$$y = (20 + 2x)(25 + 2x) - 20(25),$$

we used the following range setting:

Xmin = 0, Xmax = 3, Xscl = 1,

Ymin = 0, Ymax = 300, Yscl = 4

Using the ⎡TRACE⎤ feature to trace along the graph, we find that when $x = 2$, $y = 196$. This means that when the width of the path is 2 meters, its area is 196 square meters. This serves as another way of checking our solution to the problem.

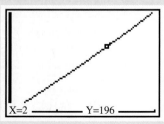

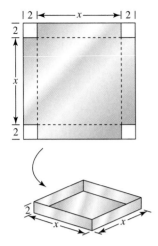

Figure 1.10

Steps 1 and 2. Represent unknowns in terms of x.

Step 3. Write an equation that models the verbal conditions.

Step 4. Solve the equation and answer the problem's question.

Step 5. Check the proposed solution in the original wording of the problem.

EXAMPLE 7   **Problem Solving Using Geometric Formulas**

A machine produces open boxes using square sheets of metal. Figure 1.10 illustrates that the machine cuts equal-sized squares measuring 2 inches on a side from the corners and then shapes the metal into an open box by turning up the sides. If each box must have a volume of 200 cubic inches, find the size of the length and width of the open box.

Solution

Let $x =$ the length and the width of the open box. The volume of the box is to be 200 cubic inches and can be found by taking the product of its length, width, and height.

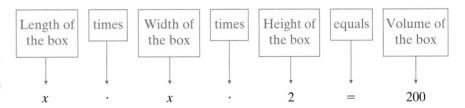

Length of the box	times	Width of the box	times	Height of the box	equals	Volume of the box
x	$\cdot$	x	$\cdot$	2	$=$	200

$$2x^2 = 200 \qquad \text{This is the equation modeling the volume of the open box.}$$

$$x^2 = 100 \qquad \text{Divide both sides by 2.}$$

$$x = \pm\sqrt{100} = \pm 10 \qquad \text{Apply the square root method.}$$

Since the open box must have positive dimensions, it should have a length and width of 10 inches on a side. The volume of the open box is $10(10)(2) = 200$ cubic inches, which checks with the problem's condition. ∎

Our final example has a solution that depends on knowing some basic facts about similar triangles.

Study tip

To model situations involving similar triangles, here's what you should know.

1. $\triangle ABC$ is similar to $\triangle DEF$ means that their respective angles have equal measure ($m \angle A = m \angle D$, $m \angle C = m \angle F$, $m \angle B = m \angle E$) and that sides opposite these angles (called corresponding sides) are proportional:

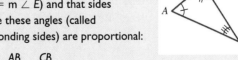

$$\frac{AC}{DF} = \frac{AB}{DE} = \frac{CB}{FE}$$

2. Informally speaking, similar triangles have the same shape but not necessarily the same size.

3. If two angles of one triangle are equal in measure to two angles of a second triangle, then the triangles are similar.

EXAMPLE 8   **Using Similar Triangles to Solve a Problem**

A water tank has the shape of an inverted right circular cone of altitude 12 feet and base radius 6 feet. (Figure 1.11). How much water is in the tank when the water is 8 feet deep?

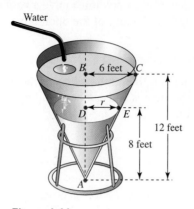

Figure 1.11

A conical water tank

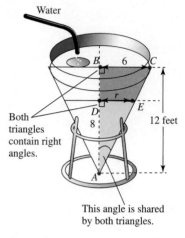

This angle is shared by both triangles.

Figure 1.12

A cross section of the water tank

Solution

The amount of water in the tank can be determined by finding the volume of the cone that is filled with water, shown as the smaller cone on the bottom in Figure 1.11. Since the volume of a cone is $V = \frac{1}{3}\pi r^2 h$ and $h = 8$, we must find r to answer the question.

Let $r =$ the radius of the cone that is filled with water. Figure 1.12 shows a cross section of Figure 1.11. $\triangle ABC$ is similar to $\triangle ADE$ because $\angle B$ has the same measure as $\angle D$ (both measure 90°) and the two triangles share $\angle A$. With two angles of $\triangle ABC$ equal in measure to two angles of $\triangle ADE$, the triangles are similar and the lengths of their corresponding sides are proportional.

Study tip

An equation with a single fraction on each side can be cleared of fractions by *cross-multiplying*. This is equivalent to multiplying by the LCD on both sides:

$$\frac{6}{r} = \frac{3}{2}$$

$$6 \cdot 2 = 3 \cdot r$$

$$12 = 3r$$

In general, if $\dfrac{a}{b} = \dfrac{c}{d}$, then

$$ad = cb.$$

$$\frac{BC}{DE} = \frac{AB}{AD} \qquad \text{Corresponding sides of similar triangles are proportional.}$$

$$\frac{6}{r} = \frac{12}{8} \qquad \text{This is the equation that models the problem's similar triangles.}$$

$$\frac{6}{r} = \frac{3}{2} \qquad \text{Simplify.}$$

$$2r \cdot \frac{6}{r} = \frac{3}{2} \cdot 2r \qquad \text{Multiply both sides by } 2r\text{, the LCD.}$$

$$12 = 3r \qquad \text{Simplify.}$$

$$4 = r \qquad \text{Divide both sides by 3.}$$

We have determined that the radius of the cone that is filled with water is 4 feet. We now can determine its volume.

$$v = \frac{1}{3}\pi r^2 h = \frac{1}{3}\pi(4)^2(8) = \frac{128}{3}\pi$$

When the water is 8 feet deep, there are $\frac{128}{3}\pi$ cubic feet of water in the tank, which is approximately 134 cubic feet. ∎

PROBLEM SET 1.3

Practice and Application Problems

1. In 1990, the average cost of an advertisement during the Super Bowl was $700,000. On the average, this cost has increased by $60,000 each year. Using this model, in what year will the cost of an advertisement during the Super Bowl be $1,480,000? (*Hint:* Let x = the number of years after 1990 when this will occur.)

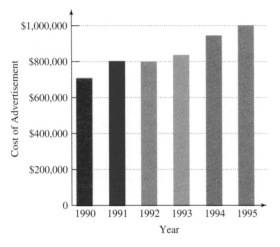

Source: *USA Today,* November 3, 1994

2. The bus fare in a city is $1.25. People who use public transportation have the option of purchasing a monthly coupon book for $21.00. With the coupon book, the fare is reduced to $0.50. How many times in a month must the bus be used so that the total monthly cost without the coupon book is the same as the total monthly cost with the coupon book?

3. A person purchasing a home is considering two options for a $50,000 mortgage.

Option 1:
7.5% interest on a 30-year loan and no points

Option 2:
7.125% interest on a 30-year loan with 3 points

Use the table on page 141 to determine how long it will take for the total cost of each mortgage to be the same. How much will be saved by choosing the 7.125% option over 20 years?

4. A person purchasing a home is considering two options for a $60,000 mortgage.

Option 1:
9.0% interest on a 30-year loan, no points, and no application fee

Option 2:
8.5% interest on a 30-year loan, 2 points, and a $200 application fee

Use the table on page 141 to determine how long it will take for the total cost of each mortgage to be the same. How much will be saved by choosing the 8.5% option over 20 years?

5. A two-year union contract guarantees an immediate 7% raise and a 3% raise after one year. How much must an employee be earning at the start of the contract if that employee's weekly salary increases by $50 over the two-year life of the contract?

6. A two-year union contract guarantees an immediate 4% raise and a 5% raise after a year. How much must an employee be earning at the start of the contract if that employee's weekly salary increases by $70 over the two-year life of the contract?

7. A grapefruit tree produces approximately 200 grapefruits per year if no more than 16 trees are planted per acre. For each additional tree planted per acre, the yield per tree decreases by 10 grapefruits per year.

   a. If x represents the number of grapefruit trees per acre beyond the basic 16, write a mathematical model for the number of grapefruits produced per year in terms of x.

   b. Approximately how many trees per acre should be planted to generate 3230 grapefruits yearly?

8. Acme Industries can sell 500 radios priced at $60 each. With each $2 reduction in price per radio, an additional 50 radios can be sold.

   a. If x represents the number of $2 reductions, write a mathematical model for the manufacturer's income in terms of x.

   b. How many $2 reductions should be applied to result in an income of $38,250?

9. A job pays an annual salary of $33,150, which includes a holiday bonus of $750. If paychecks are issued twice a month, what is the gross amount for each paycheck?

10. A store is reducing each of its items by 35%. Find the original price of a computer selling for $780.00.

The graph shows differences between the average incomes of full-time workers ages 25 and older. Problems 11–12 are based on this graph.

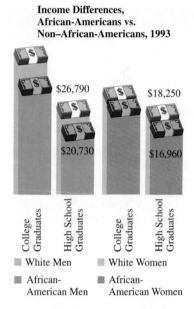

Income Differences, African-Americans vs. Non–African-Americans, 1993

$26,790

$20,730

$18,250

$16,960

■ White Men   ■ White Women

■ African-American Men   ■ African-American Women

Source: U.S. Bureau of the Census

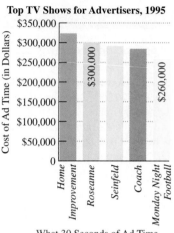

Top TV Shows for Advertisers, 1995

Cost of Ad Time (in Dollars)

$350,000
$300,000
$250,000
$200,000
$150,000
$100,000
$50,000
0

Home Improvement | Roseanne | Seinfeld | Coach | Monday Night Football

$300,000

$260,000

What 30 Seconds of Ad Time Costs on Top-Ranked Programs

Source: Based on statistics from *Advertising Age* magazine

11. Among male college graduates, whites earn $9350 more than blacks. The average annual income for the two groups is $39,015. Find the annual income for white and black male college graduates.

12. Among female college graduates, whites earn $2390 more than blacks. The average annual income for the two groups is $29,325. Find the annual income for white and black female college graduates.

13. An advertiser is considering running 30 seconds of ad time on the top-ranked programs *Home Improvement, Seinfeld,* and *Coach.* As shown in the graph at the top of the next column, a 30-second ad on *Home Improvement* costs $30,000 more than *Seinfeld,* and a 30-second ad on *Seinfeld* costs $5000 more than *Coach.* The total cost to run the 3 ads is three and a half times the cost of running a 30-second ad on *Monday Night Football.* What is the cost of a 30-second ad on each of the three programs?

14. The graph to the right shows the ten major league baseball players with the greatest number of runs as of the end of 1994. Ty Cobb had 71 more runs than Babe Ruth, who tied with Hank Aaron. Pete Rose had 9 fewer runs than both Ruth and Aaron. The four players had a combined total of runs that exceeded four times Frank Robinson's career total by 1442. Find the number of runs for Cobb, Ruth, Aaron, and Rose.

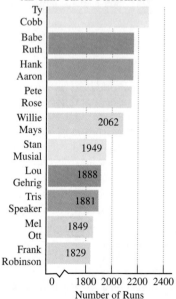

Top 10 Major League Baseball All-Time Career Performers

Ty Cobb

Babe Ruth

Hank Aaron

Pete Rose

Willie Mays — 2062

Stan Musial — 1949

Lou Gehrig — 1888

Tris Speaker — 1881

Mel Ott — 1849

Frank Robinson — 1829

0   1800 2000 2200 2400
Number of Runs

*Figures calculated as of the end of the abortive 1994 baseball season

Source: Based on statistics from Major League Baseball/Adapted from *The Macmillan Visual Almanac,* © 1996, Simon & Schuster

The position model $s = -16t^2 + v_0t + s_0$ describes the height of an object that is falling or is vertically projected into the air. In this formula, s represents the position (in feet) above the ground at time t (in seconds), v_0 represents the original velocity of the object (in feet per second), and s_0 represents the original position above the ground (in feet) of the object. Use this model to answer Problems 15–16.

15. A ball is thrown vertically upward from the top of a 200-foot building with an initial velocity of 40 feet per second. The figure on the right shows that the ball first moves upward, reaches a maximum height above the ground, and then begins to fall.

    a. How long does it take for the ball to pass by the roof of the building?

    b. How long does it take for the ball to hit the ground?

    c. The graph of the position model is shown below. Find the points on the graph corresponding to your solutions in parts (a) and (b).

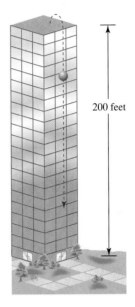

200 feet

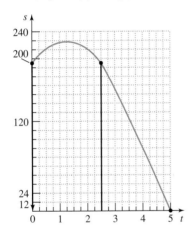

16. A set of car keys is accidentally dropped (not thrown) from the top of a 400-foot skyscraper.

    a. How long does it take for the keys to hit the ground?

    b. Sound travels at approximately 1100 feet per second. The person who dropped the keys shouts "Danger—falling object!" to warn people below. Can a person on the street hear the warning and get out of the way of the falling keys in time to avoid being hit?

Revenue for a company is the money received for the sale of x units of a product, and is found by multiplying the number of units sold by the price per unit. A company's profit is the difference between its revenue and its cost of producing x units of a product. Use these ideas to answer Problems 17–18.

17. Electric Products Incorporated manufactures and sells microwave ovens. They find that they can sell x ovens per hour at a price of $400 - 25x$ per oven, where $0 \leq x \leq 10$. The company expenses amount to $5x^2 + 40x + 600$ dollars per hour. How many ovens should be produced and sold to generate a profit of $450 per hour?

18. Sunshine Energy manufactures and sells solar collector cells. They find that they can sell x cells per day at a price of $100 - 0.05x$ per cell, where $250 \leq x \leq 800$. (Observe how the demand for the product decreases as the price increases. If the price is $100 ($100 = 100 - 0.05x$), then $x = 0$, which means they can't sell any cells

at that price. On the other hand, if the price is $20 ($20 = 100 - 0.05x$), then $x = 1600$, which means they can sell 1600 cells at $20 per unit). Company expenses amount to $4000 per day plus $60 - 0.01x$ for each cell manufactured. How many cells should be produced and sold to generate a profit of $6000 per day?

19. The length of a rectangle is 3 feet longer than the width. If the area is 54 square feet, find the rectangle's dimensions.

20. The length of a rectangle is 9 feet longer than 3 times the width. If the area is 210 square feet, find the rectangle's dimensions.

21. A brick path of uniform width is constructed around the outside of a 6- by 10-yard rectangular garden. Find how wide the path should be if there is enough brick to cover 132 square yards.

22. A rectangular garden measuring 80 by 60 feet has its area doubled by a border of uniform width along both shorter sides and one longer side. Find the width of the border.

In Problems 23–26, find the value of x for the given area of the shaded region.

23.

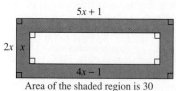

Area of the shaded region is 30 square yards

26.

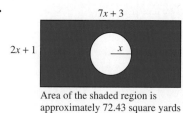

Area of the shaded region is approximately 72.43 square yards

24.

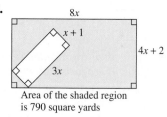

Area of the shaded region is 790 square yards

25.

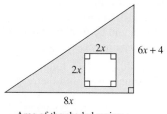

Area of the shaded region is 228 square yards

27. A picture is 20 inches long and 25 inches high. The picture is placed on a mat so that the mat's uniformly wide strip shows as a border for the picture. The perimeter of the mat is 114 inches. Find the width of the mat's strip that surrounds the picture.

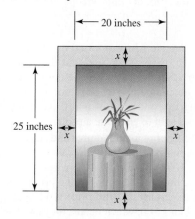

28. A gardener has 96 feet of fencing to enclose a rectangular garden and a 2-foot wide surrounding border. If the length of the garden is 3 times its width, what are its dimensions?

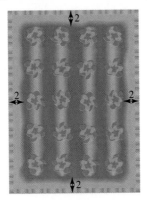

29. A machine produces open boxes using rectangular pieces of metal measuring 15 centimeters by 25 centimeters. The machine cuts equal-sized squares from each corner and then shapes the metal into an open box by turning up the sides. If the area of the bottom of the box is 231 square centimeters, find the length of each side of the square that is cut from the four corners.

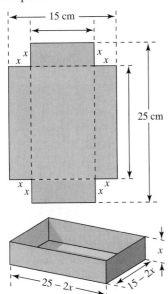

30. An open box is to be made using a rectangular piece of cardboard whose length is 4 centimeters greater than its width. (See the figure at the top of the next column.) Equal squares measuring 3 centimeters on a side are cut from each corner of the cardboard and the sides that remain are folded up to form the box. If the volume of the resulting box is 180 cubic centimeters, find the dimensions of the original piece of cardboard.

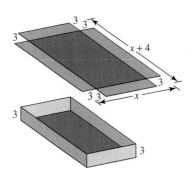

31. A person 5 feet tall is standing 18 feet from the base of a streetlight. If the light is at the top of a 16-foot pole, how long is the person's shadow that is cast from the streetlight?

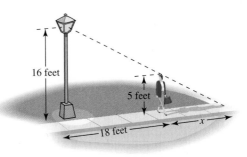

32. A person 6 feet tall is standing 10 feet from the base of a streetlight. If the light is at the top of an 18-foot pole, how long is the person's shadow that is cast from the streetlight?

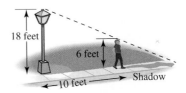

33. A water tank has the shape of an inverted right circular cone of altitude 20 feet and base radius 5 feet. How much water is in the tank when the water is 8 feet deep?

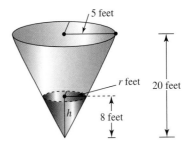

34. A water tank has the shape of an inverted right circular cone of altitude 6 meters and base radius 3 meters. How much water is in the tank when the water is 3 meters deep?

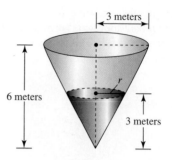

In Problems 35–38, the area of each plane figure is given. Find the value of x.

35.

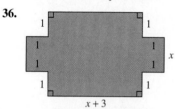

Area: 48 square meters

36.

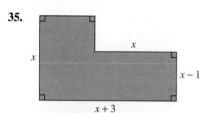

Area: 66 square meters

37.

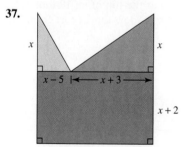

Area: 76 square meters

38.

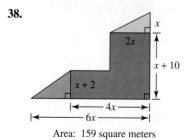

Area: 159 square meters

39. The Norman window in the figure to the right of Problem 35 has a rectangular base and a semicircular top. If 2500 square centimeters of glass are used in making the window, what is the area of each of the four smaller rectangular regions?

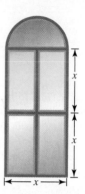

40. The figure shows a rectangle that is 8 meters by 6 meters. If the area of the cross is equal to the sum of the areas of the four smaller rectangles, what is the width of the cross?

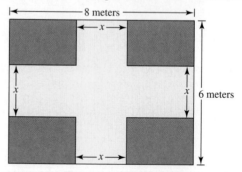

41. A storage bin consists of a cylinder and a roof that is a cone. The height of the roof is half the height of the cylinder, and the total volume of the bin is 11,200 π cubic feet. Find the height of the cylinder.

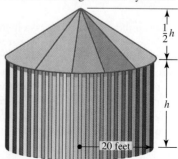

42. By how many meters should the radius of a 10-meter circle be increased so that the area increases by 5π square meters?

43. When drawing a rectangle, many artists and architects use a certain ratio of the length to the width to achieve a satisfying visual effect. This ratio, called the golden ratio, can be determined by answering the following question from Euclid's *Elements*:

Divide a line segment such that the ratio of the larger part to the whole is equal to the ratio of the smaller part to the larger part.

a. Use Euclid's statement and the figure below to set up a proportion. Solve the proportion for x. The exact value of the golden ratio is x to 1.

b. Which part of the figure in the next column do you find most appealing? Explain what this means in terms of the ratio that you determined in part (a).

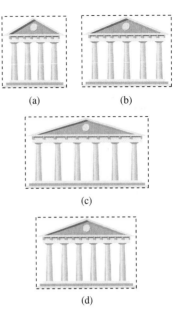

(a)   (b)

(c)

(d)

44. A long rectangular sheet of metal, 12 centimeters wide, is to be made into a rain gutter by turning up two sides at right angles to the sheet.

a. If x represents the length of each side that is turned up to form the rain gutter, write a model in terms of x that describes the cross-sectional area of the gutter.

b. How many centimeters of metal should be turned up to give the gutter a capacity of 16 square centimeters?

Technology Problems

45. A rectangular field is to be fenced on three sides with 1000 meters of fencing. As shown in the figure, the fourth side is a straight river's edge that will not be fenced.

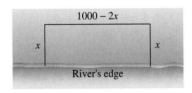

a. Find the dimensions of the field so that the area of the enclosure is 120,000 square meters.

b. Use a graphing utility to graph the area model that you obtained in part (a). Experiment with the range setting until your graph looks like the one shown in the figure on the right. Then trace along the graph and find the point whose coordinates correspond to your algebraic solution in part (a).

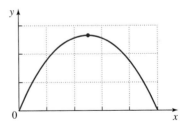

c. What is a reasonable estimate of the maximum area that can be enclosed? Trace along the graph and verify this estimate by finding the coordinates of the highest point on the graph. What are the dimensions of the rectangle of maximum area?

d. Does the rectangle enclosed in part (a) make the best use of the 1000 meters of fencing? Explain.

46. Consider the collection of all rectangles whose perimeter is 150 feet, represented by the green rectangle on page 158.

a. Write a model for the area of all such rectangles.

b. What values of x are meaningful in this situation? Why?

c. Use a graphing utility to determine the smallest viewing rectangle that will give a relatively complete algebraic picture of the area model and graph the model. (The shape of the graph in Problem 45 should be helpful.)

d. Describe the behavior of the graph that appears in the viewing rectangle, relating this behavior to the area of rectangle.

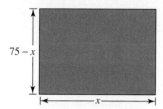

47. A rock on the moon and a rock on Earth are thrown into the air with an initial velocity of 48 feet per second by a 6-foot person. The position of the rock above the ground (s, in feet) after t seconds is described by

$$s_{\text{Moon}} = -2.7t^2 + 48t + 6$$

$$s_{\text{Earth}} = -16t^2 + 48t + 6$$

a. Graph both position models.

b. Describe differences that you observe in the graphs. How do you account for these differences?

c. Use the graphs to determine how much higher the rock is on the moon than on Earth after 3 seconds. Use the models to verify this observation.

48. The demand for a product tends to decrease as the price increases. A company finds that they can sell x units of a product at a price of $50 - 0.0001x$ dollars per unit.

a. Graph the *demand equation* $y = 50 - 0.0001x$ and describe what the graph indicates in terms of various selling prices and the number of units sold at that price.

b. Graph the revenue model. (Revenue = the number of units sold times the price per unit.)

c. Use the TRACE feature to move the cursor along the graph to approximate the number of units that will maximize revenue. At maximum revenue, what is the price per unit?

d. Use the graph to indicate a revenue amount that will never be obtained.

e. Use the graph to determine the number of units and the corresponding price that yield a revenue of $960,000. If you were to determine the price of this product, what price would you charge and why?

49. Use a graphing utility to graph the model for the cross-sectional area obtained in Problem 44. Does 16 square centimeters give the rain gutter its maximum capacity?

Writing in Mathematics

50. Write and solve an algebraic word problem similar to Problems 11 and 12 on page 152 using the figures for high school graduates shown in the graph.

51. Describe the changes that would take place in solving part (a) of Problem 45 on page 157 if the rectangular field is to be enclosed away from the river with fencing used along all sides of the rectangle.

52. Why should an algebraic word problem be checked using the verbal conditions given in the problem rather than the equation that is modeled from these verbal conditions?

53. Reread the description of the position model that precedes Problem 15 on page 153. Explain the difference between the following situations in terms of writing the appropriate position models.

A set of keys is thrown upward from a 300-foot building with an initial velocity of 20 feet per second.

A set of keys is accidently dropped from a 300-foot building.

Critical Thinking Problems

54. The price of an item is increased by 50%. When the item does not sell, it is reduced by 50% of the increased price. If the final price is $280, what was the original price? When the price is increased by 50% and then reduced by 50%, what percent represents the net reduction in price?

55. Four meters of wire is used to form a circle and a square. Let x represent the amount of wire used to form the square and let $4 - x$ represent the amount of wire used to form the circle.

a. Write a model in terms of x representing the total area enclosed by the square and the circle.

b. A graph of the model in part (a) indicates that the maximum total area that can be enclosed is approximately 1.273 square meters. Find the side of the square and the radius of the circle so that the total area model in part (a) results in 1.273 square meters. Explain what this means about how to use the wire so that a maximum area is enclosed.

56. A Norman window has a rectangular base and a semi-circular top. The perimeter of the Norman window shown in the figure is 16 meters.

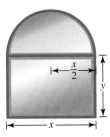

a. Write a model for the area of the Norman window using one variable.

b. Find x and y so that the area of the Norman window is 16 square meters.

57. Use the description of the position model that precedes Problem 15 on page 153 to answer this problem:

A ball is dropped from a 576-foot skyscraper. With what initial velocity should a second ball be thrown straight downward 3 seconds later so that the two balls hit the ground at the same time?

Group Activity Problem _____

58. This problem is intended as a group learning experience and is appropriate for groups of three to five people. Before working on the various parts of the problem, reread the description of the position model that precedes Problem 15 on page 153.

a. Drop a ball from a height of 3 feet, 6 feet, and 12 feet. Record the number of seconds it takes for the ball to hit the ground.

b. For each of the three initial positions, use the position model to determine the time required for the ball to hit the ground.

c. What factors might result in differences between the times that you recorded and the times indicated by the model?

d. What appears to be happening to the time required for a free-falling object to hit the ground as its initial position is doubled? Verify this observation algebraically and with a graphing utility.

e. Repeat part (a) using a sheet of paper rather than a ball. What differences do you observe? What factor seems to be ignored in the mathematical model?

f. What is meant by the acceleration of gravity and how does this number appear in the position model for a free-falling object?

S E C T I O N 1 . 4	**Other Types of Equations**

Objectives

Solutions Manual   **Tutorial**   **Video 3**

1 Solve polynomial equations by factoring.
2 Solve equations that are quadratic in form.
3 Solve radical equations.
4 Solve equations with rational exponents.
5 Solve applied modeling problems.

Our focus in this section is on solving certain kinds of polynomial equations and on equations involving radicals and rational exponents.

	Polynomial Equations

Solve polynomial equations by factoring.

The linear and quadratic equations that we studied in the first two sections of this chapter can be thought of as polynomial equations of degree 1 and 2, respectively. Some polynomial equations of degree greater than 2 can be solved by factoring.

tudy tip

In solving $3x^4 = 27x^2$, be careful not to divide both sides by x^2. If you do, you'll lose 0 as a solution. In general, do not divide both sides of an equation by a variable since that variable might take on the value 0 and you cannot divide by 0.

sing technology

a. $3x^4 = 27x^2$

Use your graphing utility to graph $y = 3x^4 - 27x^2$ using the following range setting:

Xmin = −6, Xmax = 6,

Xscl = 1, Ymin = −100,

Ymax = 100, Yscl = 50.

The x-intercepts are −3, 0, and 3, corresponding to the equation's solutions.

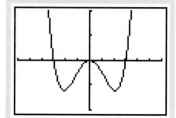

b. $2x^3 + 3x^2 - 8x - 12 = 0$

Graph
$y = 2x^3 + 3x^2 - 8x - 12$
using the range setting

Xmin = −5, Xmax = 5,

Xscl = 1, Ymin = −20,

Ymax = 5, Yscl = 5

Using the $\boxed{\text{TRACE}}$ and $\boxed{\text{ZOOM}}$ features for further clarification, you can see that the x-intercepts are −2, −1.5, and 2, corresponding to the equation's solutions.

(continued)

2 Solve equations that are quadratic in form.

EXAMPLE 1   **Solving Polynomial Equations by Factoring**

Solve by factoring:

a. $3x^4 = 27x^2$ **b.** $2x^3 + 3x^2 - 8x - 12 = 0$ **c.** $x^4 - 8x^2 - 9 = 0$

Solution

a.

$3x^4 = 27x^2$	This is the given equation.
$3x^4 - 27x^2 = 0$	Set this fourth-degree polynomial equation equal to 0.
$3x^2(x^2 - 9) = 0$	Factor $3x^2$ from each term.
$3x^2 = 0$ or $x^2 - 9 = 0$	Set each factor equal to 0.
$x^2 = 0 \qquad x^2 = 9$	Solve the resulting equations.
$x = 0 \qquad x = \pm 3$	Use the square root method to solve $x^2 = 9$.

The three solutions, 0, −3, and 3, can be checked by substitution into the original equation to verify that the solution set is $\{-3, 0, 3\}$.

b.

$2x^3 + 3x^2 - 8x - 12 = 0$	This is the given equation.
$x^2(2x + 3) - 4(2x + 3) = 0$	Factor x^2 from the first two terms and −4 from the last two terms.
$(2x + 3)(x^2 - 4) = 0$	Factor out $2x + 3$.
$2x + 3 = 0$ or $x^2 - 4 = 0$	Set each factor equal to 0.
$2x = -3 \qquad x^2 = 4$	Solve for x.
$x = -\dfrac{3}{2} \qquad x = \pm 2$	

The solution set is $\left\{-2, -\dfrac{3}{2}, 2\right\}$. Observe that the original third-degree (cubic) equation has three solutions. Check these solutions in the original equation.

c.

$x^4 - 8x^2 - 9 = 0$	This is the given equation.
$(x^2 - 9)(x^2 + 1) = 0$	Factor.
$x^2 - 9 = 0$ or $x^2 + 1 = 0$	Set each factor equal to 0.
$x^2 = 9 \qquad x^2 = -1$	Solve for x.
$x = \pm\sqrt{9} \qquad x = \pm\sqrt{-1}$	Use the square root method.
$= \pm 3 \qquad = \pm i$	

The solution set is $\{-3, 3, -i, i\}$.   ■

Polynomial Equations That Are Quadratic in Form

The equation of Example 1c is *quadratic in form*. Its *structure* is quadratic, which can be seen by replacing x^2 by t.

$$x^4 - 8x^2 - 9 = 0 \qquad \text{This is the equation of Example 1c.}$$
$$(x^2)^2 - 8x^2 - 9 = 0 \qquad \text{Replace } x^2 \text{ with } t.$$
$$t^2 - 8t - 9 = 0 \qquad \text{This is a quadratic equation in } t.$$

Before we solve a polynomial equation that is quadratic in form, let's consider the following definition.

Using technology (cont.)

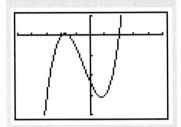

c. $x^4 - 8x^2 - 9 = 0$

Graph $y = x^4 - 8x^2 - 9$
using the range setting

Xmin = -5, Xmax = 5,

Xscl = 1, Ymin = -30,

Ymax = 30, Yscl = 5

The real solutions (-3
and 3) are the only ones that
appear as x-intercepts.

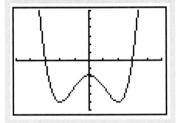

Definition of an equation that is quadratic in form

An equation is *quadratic in form* if it can be written in the form

$$at^2 + bt + c = 0$$

where t is an algebraic expression and $a \neq 0$.

This definition implies that we introduce a substitution so that the quadratic-in-form equation becomes a quadratic equation in t. Let's see how this is done.

EXAMPLE 2 **Solving a Polynomial Equation That Is Quadratic in Form**

Solve: $(x^2 + 3x)^2 - 8(x^2 + 3x) - 20 = 0$

Solution

If we let $t = x^2 + 3x$, the equation can be written as $t^2 - 8t - 20 = 0$ which, in turn, can be solved by factoring or the quadratic formula.

$(x^2 + 3x)^2 - 8(x^2 + 3x) - 20 = 0$ This is the given equation.

$\qquad t^2 \qquad -8 \quad t \qquad -20 = 0$ Let $t = x^2 + 3x$.

We can solve this quadratic equation in t by factoring.

$(t - 10)(t + 2) = 0$ Factor.

$t - 10 = 0$ or $t + 2 = 0$ Set each factor equal to 0.

$\qquad t = 10 \qquad\qquad t = -2$ Solve for t.

The solution process is incomplete because we must solve the original equation for x. However, because we know that $t = x^2 + 3x$, we can replace t with $x^2 + 3x$ in our work up to this point.

$\qquad t = 10 \qquad$ or $\qquad t = -2$ This was our last step.

$x^2 + 3x = 10 \quad$ or $\quad x^2 + 3x = -2$ Because $t = x^2 + 3x$, replace t with $x^2 + 3x$.

$x^2 + 3x - 10 = 0 \qquad\qquad x^2 + 3x + 2 = 0$ Write each quadratic equation in standard form.

$(x + 5)(x - 2) = 0 \qquad\qquad (x + 2)(x + 1) = 0$ Factor.

$x + 5 = 0 \quad$ or $\quad x - 2 = 0 \quad x + 2 = 0 \quad$ or $\quad x + 1 = 0$ Set each factor equal to 0.

$\qquad x = -5 \qquad\qquad x = 2 \qquad\qquad x = -2 \qquad\qquad x = -1$ Solve for x.

The solution set of the original equation is $\{-5, 2, -2, -1\}$, which can be verified by substitution or with a graphing utility. ∎

Solution procedure for equations that are quadratic in form

Solving equations that are quadratic in form

$$ax^n + bx^{n/2} + c = 0$$

1. Introduce the substitution $t = x^{n/2}$.
2. Use this substitution to express $ax^n + bx^{n/2} + c = 0$ as a quadratic equation in t.

$$a(x^{n/2})^2 + bx^{n/2} + c = 0$$
$$at^2 + bt + c = 0$$

3. Solve the quadratic equation for t.
4. Solve for x by substituting the value(s) of t into $x^{n/2} = t$.

3 Solve radical equations.

Equations Involving Radicals

Solving equations involving radicals often involves raising both sides of the equation to the same power. All solutions of the original equation are also solutions of the resulting equation. However, the resulting equation may have some extra solutions that do not satisfy the original equation. Because the resulting equation may not be equivalent to the original equation, the solution process always must be completed by substituting all possibilities in the original equation to determine whether they are really solutions. Let's see exactly how this works.

EXAMPLE 3   Solving an Equation Involving a Radical

Solve: $x + \sqrt{26 - 11x} = 4$

Solution

The graph of

$$y = x + \sqrt{26 - 11x} - 4$$

is shown in a $[-10, 3]$ by $[-4, 3]$ viewing rectangle. The x-intercepts are -5 and 2, verifying $\{-5, 2\}$ as the solution set.

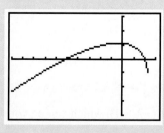

To solve this equation algebraically, we can isolate the radical expression $\sqrt{26 - 11x}$ on one side of the equation. By squaring both sides of the equation, we can then eliminate the square root.

$x + \sqrt{26 - 11x} = 4$	This is the given equation.
$\sqrt{26 - 11x} = 4 - x$	Isolate the radical by subtracting x from both sides.
$(\sqrt{26 - 11x})^2 = (4 - x)^2$	Square both sides.
$26 - 11x = 16 - 8x + x^2$	
$0 = x^2 + 3x - 10$	Write the quadratic equation in standard form.
$0 = (x + 5)(x - 2)$	Factor.
$x + 5 = 0 \quad$ or $\quad x - 2 = 0$	Set each factor equal to 0.
$x = -5 \qquad\qquad x = 2$	Solve for x.

Complete the solution process by checking both proposed solutions.

$$x + \sqrt{26 - 11x} = 4$$

This is the original equation.

Check -5:

$$-5 + \sqrt{26 - 11(-5)} \stackrel{?}{=} 4$$

$$-5 + \sqrt{81} \stackrel{?}{=} 4$$

$$-5 + 9 \stackrel{?}{=} 4$$

$$4 = 4 \checkmark$$

Check 2:

$$2 + \sqrt{26 - 11(2)} \stackrel{?}{=} 4 \quad \text{Substitute } -5 \text{ and } 2, \text{ respectively.}$$

$$2 + \sqrt{4} \stackrel{?}{=} 4 \quad \text{Simplify.}$$

$$2 + 2 \stackrel{?}{=} 4$$

$$4 = 4 \checkmark \quad \text{Both } -5 \text{ and } 2 \text{ are solutions.}$$

The solution set is $\{-5, 2\}$. ∎

The algebraic solution of radical equations with two or more square root expressions involves isolating a radical, squaring both sides, and then repeating this process. Let's consider an example containing two square root expressions.

EXAMPLE 4   Solving an Equation Involving Two Radicals

Solve: $\sqrt{3x + 1} - \sqrt{x + 4} = 1$

Solution

$$\sqrt{3x + 1} - \sqrt{x + 4} = 1 \qquad \text{This is the given equation.}$$

$$\sqrt{3x + 1} = \sqrt{x + 4} + 1 \qquad \begin{array}{l}\text{Isolate the radicals by adding}\\ \sqrt{x + 4} \text{ to both sides.}\end{array}$$

$$(\sqrt{3x + 1})^2 = (\sqrt{x + 4} + 1)^2 \qquad \text{Square both sides.}$$

$$3x + 1 = x + 4 + 2\sqrt{x + 4} + 1 \qquad \begin{array}{l}\text{On the right, use } (A + B)^2\\ = A^2 + 2AB + B^2.\end{array}$$

$$3x + 1 = x + 5 + 2\sqrt{x + 4} \qquad \begin{array}{l}\text{Combine numerical terms on}\\ \text{the right.}\end{array}$$

$$2x - 4 = 2\sqrt{x + 4} \qquad \begin{array}{l}\text{Isolate } 2\sqrt{x + 4}, \text{ the radical}\\ \text{term, by subtracting } x + 5 \text{ from}\\ \text{both sides.}\end{array}$$

$$x - 2 = \sqrt{x + 4} \qquad \text{Divide both sides by 2.}$$

$$(x - 2)^2 = (\sqrt{x + 4})^2 \qquad \text{Square both sides.}$$

$$x^2 - 4x + 4 = x + 4 \qquad \text{Multiply.}$$

$$x^2 - 5x = 0 \qquad \begin{array}{l}\text{Write the quadratic equation in}\\ \text{standard form by subtracting}\\ x + 4 \text{ from both sides.}\end{array}$$

$$x(x - 5) = 0 \qquad \text{Factor.}$$

$$x = 0 \quad \text{or} \quad x - 5 = 0 \qquad \text{Set each factor equal to 0.}$$

$$x = 0 \qquad\qquad x = 5 \qquad \text{Solve for } x.$$

Complete the solution process by checking both proposed solutions.

Check 0:

$$\sqrt{3x + 1} - \sqrt{x + 4} = 1$$

Check 5:

$$\sqrt{3x + 1} - \sqrt{x + 4} = 1$$

Using technology

The graph of

$$y = \sqrt{3x + 1} - \sqrt{x + 4} - 1$$

has only one x-intercept at 5. This verifies that the solution set for the given equation is $\{5\}$.

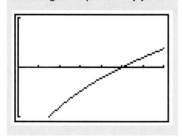

$$\sqrt{3 \cdot 0 + 1} - \sqrt{0 + 4} \stackrel{?}{=} 1 \qquad\qquad \sqrt{3 \cdot 5 + 1} - \sqrt{5 + 4} \stackrel{?}{=} 1$$

$$\sqrt{1} - \sqrt{4} \stackrel{?}{=} 1 \qquad\qquad \sqrt{16} - \sqrt{9} \stackrel{?}{=} 1$$

$$1 - 2 \stackrel{?}{=} 1 \qquad\qquad 4 - 3 \stackrel{?}{=} 1$$

$$-1 = 1 \text{ False} \qquad\qquad 1 = 1 \checkmark$$

This final step in the solution process indicates that 0 is not a solution. The solution set is {5}. ∎

Equations with Rational Exponents

4 Solve equations with rational exponents.

We turn now to equations in the form $x^{m/n} = k$ in which m and n are positive integers, $\frac{m}{n}$ is in lowest terms, and k is a real number. To solve such equations, we raise both sides to the $\frac{n}{m}$th power.

Study tip

Remember the definition of $b^{m/n}$:

$$b^{m/n} = \sqrt[n]{b^m} = (\sqrt[n]{b})^m.$$

As shown on the right, it is not necessary to insert the $\pm$ symbol when the numerator of the exponent is odd. An odd index has only one root.

If m is even: | **If m is odd:**

$$x^{m/n} = k \qquad\qquad x^{m/n} = k$$

$$(x^{m/n})^{n/m} = \pm k^{n/m} \qquad (x^{m/n})^{n/m} = k^{n/m}$$

$$x = \pm k^{n/m} \qquad\qquad x = k^{n/m}$$

EXAMPLE 5 Solving Equations Involving Rational Exponents

Solve:

a. $3x^{3/4} - 6 = 0$ **b.** $x^{2/3} - \dfrac{3}{4} = -\dfrac{1}{2}$ **c.** $(x^2 - 4x + 7)^{3/2} - 2 = 6$

Using technology

The solution, $2^{4/3}$, to the equation

$$3x^{3/4} - 6 = 0$$

can be verified by graphing

$$y = 3x^{3/4} - 6$$

Since

$$2^{4/3} = \sqrt[3]{2^4} \approx 2.52$$

the graph indicates that the x-intercept is approximately 2.52. (Use the $\boxed{\text{TRACE}}$ and $\boxed{\text{ZOOM}}$ features.)

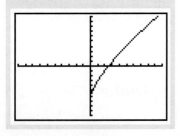

Solution

Our goal in each solution is to isolate the expression with the rational exponent, which is equivalent to isolating the radical expression.

a. $3x^{3/4} - 6 = 0$ This is the given equation; we will isolate $x^{3/4}$.

$\qquad\quad 3x^{3/4} = 6$ Add 6 to both sides.

$\qquad\quad\ x^{3/4} = 2$ Divide both sides by 3.

$\ (x^{3/4})^{4/3} = 2^{4/3}$ Raise both sides to the $\frac{4}{3}$th power. Since $\frac{m}{n} = \frac{3}{4}$ and m is odd, the $\pm$ symbol is not necessary.

$\qquad\qquad x = 2^{4/3}$ $(x^{3/4})^{3/4} = x^{3/4 \cdot 4/3} = x^1 = x$

Check

$\qquad 3x^{3/4} - 6 = 0$ This is the original equation.

$\ 3(2^{4/3})^{3/4} - 6 \stackrel{?}{=} 0$ Substitute the proposed solution.

$\qquad 3 \cdot 2 - 6 \stackrel{?}{=} 0$ $(2^{4/3})^{3/4} = 2^{4/3 \cdot 3/4} = 2^1 = 2$

$\qquad\qquad 0 = 0 \checkmark$ Thus, $2^{4/3}$ is a solution.

The solution set is $\{2^{4/3}\}$.

b. $x^{2/3} - \dfrac{3}{4} = -\dfrac{1}{2}$ This is the given equation.

$$x^{2/3} = \frac{1}{4} \qquad \text{Add } \frac{3}{4} \text{ to both sides.}$$

$$(x^{2/3})^{3/2} = \pm\left(\frac{1}{4}\right)^{3/2} \qquad \text{Raise both sides to the } \frac{3}{2} \text{ power. Since } \frac{m}{n} = \frac{2}{3} \text{ and } m \text{ is even, the } \pm \text{ symbol is necessary.}$$

$$x = \pm\frac{1}{8} \qquad (\tfrac{1}{4})^{3/2} = (\sqrt{\tfrac{1}{4}})^3 = (\tfrac{1}{2})^3 = \tfrac{1}{8}$$

Take a moment to verify that the solution set is $\left\{-\frac{1}{8}, \frac{1}{8}\right\}$.

c.
$$\begin{aligned}
(x^2 - 4x + 7)^{3/2} - 2 &= 6 & \text{This is the given equation.}\\
(x^2 - 4x + 7)^{3/2} &= 8 & \text{Add 2 to both sides.}\\
[(x^2 - 4x + 7)^{3/2}]^{2/3} &= 8^{2/3} & \text{Raise both sides to the } \tfrac{2}{3} \text{ power.}\\
x^2 - 4x + 7 &= 4 & 8^{2/3} = (\sqrt[3]{8})^2 = 2^2 = 4\\
x^2 - 4x + 3 &= 0 & \text{Write the quadratic equation in standard form.}\\
(x - 1)(x - 3) &= 0 & \text{Factor.}
\end{aligned}$$
$$\begin{aligned}
x - 1 = 0 \quad \text{or} \quad x - 3 &= 0 & \text{Set each factor equal to 0.}\\
x = 1 \qquad\qquad x &= 3 & \text{Solve for } x.
\end{aligned}$$

Both proposed solutions can be shown to satisfy the original equation, indicating that the solution set is $\{1, 3\}$. ∎

Some equations involving rational exponents are quadratic in form and can be solved using an appropriate substitution. Example 6 illustrates this.

EXAMPLE 6 **Solving an Equation Containing Rational Exponents That Is Quadratic in Form**

Solve: $5x^{2/3} + 11x^{1/3} + 2 = 0$

Solution

$$5x^{2/3} + 11x^{1/3} + 2 = 0 \qquad \text{This is the given equation.}$$
$$5(x^{1/3})^2 + 11(x^{1/3}) + 2 = 0$$
$$5t^2 \quad + 11t \quad + 2 = 0 \qquad \text{Let } t = x^{1/3}.$$

We can now solve this quadratic equation in t by factoring.

$$\begin{aligned}
(5t + 1)(t + 2) &= 0 & \text{Factor.}\\
5t + 1 = 0 \quad \text{or} \quad t + 2 &= 0 & \text{Set each factor equal to 0.}\\
t = -\frac{1}{5} \qquad\qquad t &= -2 & \text{Solve for } t.\\
x^{1/3} = -\frac{1}{5} \qquad\qquad x^{1/3} &= -2 & \text{Replace } t \text{ with } x^{1/3}.\\
(x^{1/3})^3 = \left(-\frac{1}{5}\right)^3 \qquad (x^{1/3})^3 &= (-2)^3 & \text{Solve for } x \text{ by cubing both sides of each equation.}\\
x = -\frac{1}{125} \qquad\qquad x &= -8
\end{aligned}$$

Check these values to verify that the solution set is $\left\{-\frac{1}{125}, -8\right\}$. ∎

Using technology

We can confirm the solutions to

$$(x^2 - 4x + 7)^{3/2} - 2 = 6$$

by graphing

$$y = (x^2 - 4x + 7)^{3/2} - 8.$$

The x-intercepts are 1 and 3, verifying that $\{1, 3\}$ is the solution set.

Study tip

It is easy to recognize an equation containing rational exponents that is quadratic in form. When the equation is written in descending powers, the exponent on the middle term is half the exponent on the leading term.

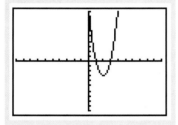

5 Solve applied modeling problems.

Rene Magritte, Belgian (1898–1967) "Time Transfixed" 1938, oil on canvas, 147 × 98.7 cm. Joseph Winterbotham Collection, 1970.426. Photograph © 1997, The Art Institute of Chicago, All Rights Reserved. © 1998 C. Herscovici, Brussels / Artists Rights Society (ARS), New York.

Mathematical Models

We have seen a model that describes the position of free-falling objects, derived by Isaac Newton (1642–1727). In Newton's view of the world, mass, length, and time appear identical to every person. By contrast, in Albert Einstein's special theory of relativity, time slows down, mass increases, and length in the direction of motion decreases from the point of view of an observer watching an object moving at a velocity close to the speed of light. Einstein's theory, verified with experiments in atomic physics, forms the basis of Example 7.

EXAMPLE 7 **Einstein's Special Relativity Models**

Einstein's equations relate the mass, time, and length at rest (M_0, T_0, L_0) to the mass, time, and length (M, T, L) for a velocity v. The speed of light, c, is approximately 186,000 miles per second.

$$M = \frac{M_0}{\sqrt{1 - \dfrac{v^2}{c^2}}} \qquad T = \frac{T_0}{\sqrt{1 - \dfrac{v^2}{c^2}}} \qquad L = L_0 \sqrt{1 - \frac{v^2}{c^2}}$$

How fast would a futuristic starship that is 600 meters tall have to move so that it would measure only 84 meters tall from the perspective of an observer who looks up and sees the ship pass by?

Solution

$L = L_0 \sqrt{1 - \dfrac{v^2}{c^2}}$ This is Einstein's model for length.

$84 = 600 \sqrt{1 - \dfrac{v^2}{c^2}}$ Substitute L_0 (length at rest) = 600 and L (length at the velocity we must determine) = 84.

$0.14 = \sqrt{1 - \dfrac{v^2}{c^2}}$ Divide both sides by 600.

$0.0196 = 1 - \dfrac{v^2}{c^2}$ Square both sides.

$0.0196c^2 = c^2 - v^2$ Multiply both sides by c^2. Remember that c, the speed of light, is approximately 186,000 miles per second.

$v^2 = 0.9804c^2$ Isolate v^2 on the left.

$v = \sqrt{0.9804c^2}$ Assume that the starship is moving upward, so take only the positive root.

$v \approx 0.99c$

600 meters

Starship at Rest

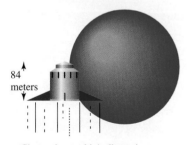

84 meters

Change in starship's dimensions when moving at 99% of light's speed when viewed by an observer

Figure 1.13

If the 600-meter tall futuristic starship moves at 99% the speed of light (approximately 184,140 miles per second), it will measure only 84 meters tall from the perspective of our stationary viewer. A before and after picture of the starship is shown in Figure 1.13. Observe that lengths perpendicular to the direction of motion are unchanged by the effect of speed, so the width of the starship remains unchanged. ∎

tudy tip

The Pythagorean Theorem

$$a^2 + b^2 = c^2$$

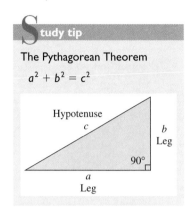

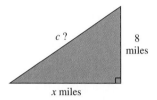

Figure 1.14

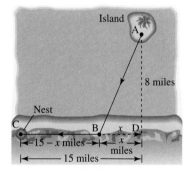

Figure 1.15

Steps 1 and 2. Represent unknowns in terms of x.

Step 3. Write an equation that models the verbal conditions.

tudy tip

The equation is modeled from the implied condition that

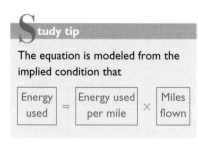

Step 4. Solve the equation and answer the problem's question.

In Example 7, we were given the model needed to solve the problem. Often we must create a model based on a problem's conditions, and sometimes these conditions are not given explicitly. Furthermore, in many situations we must bring a knowledge of some basic idea to the problem to obtain a model. This is the situation in Example 8, where modeling depends on the Pythagorean Theorem.

A right triangle like the one in Figure 1.14 will be used to model the problem. We will need to obtain an expression for the hypotenuse in terms of x. We can apply the Pythagorean Theorem:

$$(\text{Hypotenuse})^2 = (\text{Leg})^2 + (\text{Leg})^2$$
$$c^2 = x^2 + 8^2$$

$$c^2 = x^2 + 64$$

Since the length of the hypotenuse must be positive, we take only the positive square root.

$$c = \sqrt{x^2 + 64}$$

EXAMPLE 8   **Creating a Mathematical Model**

A bird flies from point A on an island that is located 8 miles off shore from point D (see Figure 1.15). Traveling over the water from point A to point B, the bird uses 15 kilocalories (kcal) of energy per mile. The bird then flies from point B to its nest at point C, using 10 kcal of energy per mile over land. Find the location distance from point B to point D so that precisely 240 kcal of energy is used during the entire flight.

Solution

Let

$$x = \text{the distance from point } B \text{ to point } D$$

$$\sqrt{x^2 + 64} = \text{the distance that the bird flies over water}$$
$$\text{(by the Pythagorean Theorem)}$$

$$15 - x = \text{the distance that the bird flies over land}$$

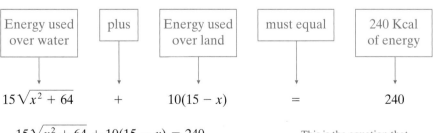

$$15\sqrt{x^2 + 64} + 10(15 - x) = 240 \qquad \text{This is the equation that models the given conditions.}$$

$$15\sqrt{x^2 + 64} = 240 - 10(15 - x) \qquad \text{Isolate the radical.}$$

$$15\sqrt{x^2 + 64} = 90 + 10x \qquad \text{Simplify on the right.}$$

$$3\sqrt{x^2 + 64} = 18 + 2x$$ To keep the numbers smaller, divide both sides by 5.

$$(3\sqrt{x^2 + 64})^2 = (18 + 2x)^2$$ Square both sides.

$$9(x^2 + 64) = 324 + 72x + 4x^2$$ Multiply.

$$9x^2 + 576 = 324 + 72x + 4x^2$$ Use the distributive property on the left.

$$5x^2 - 72x + 252 = 0$$ Write the quadratic equation in standard form.

$$(x - 6)(5x - 42) = 0$$ Factor (or use the quadratic formula.)

$$x - 6 = 0 \quad \text{or} \quad 5x - 42 = 0$$ Set each factor equal to 0.

$$x = 6 \quad \text{or} \quad x = \tfrac{42}{5} = 8.4$$ Solve the resulting equations.

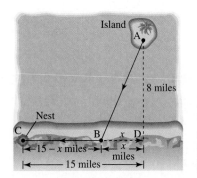

Island
A

8 miles

Nest
C

B   x   D

|←15 − x miles→|← x miles →|

|←——— 15 miles ———→|

Look again at Figure 1.15, repeated at the left. The two values for x mean that the bird should reach land and end its flight over water at point B either 6 miles or 8.4 miles from point D so that precisely 240 kcal of energy is used during the entire flight. In other words, the distance from point B to point D is either 6 or 8.4 miles. ∎

Step 5. Check the proposed solution in the original wording of the problem.

Discover for yourself

Check these solutions using the given condition of 240 kcal of energy. What is the distance from A to C? (Use the Pythagorean Theorem.) How much energy is needed for that direct water route?

Using technology

The graph of

$$y = 15\sqrt{x^2 + 64} + 10(15 - x)$$

shows that the energy the bird uses in flight depends on how far (x) from point D it reaches land. The graph on the left, obtained using the range setting

   Xmin = 0, Xmax = 15, Xscl = 1, Ymin = 0, Ymax = 300, Yscl = 15

and tracing along the curve, shows the energy used if the bird reaches land 15 miles from point D. This means that the entire flight is over water, and 255 kcal of energy must be used. The most efficient path in terms of using the least amount of energy can be determined by finding the lowest point on the curve. Using the minimum value feature of the graphing utility, this point is shown on the right. In practical terms, if the bird comes ashore approximately 7.2 miles from point D, a minimum energy of approximately 239.4 kcal is expended on the entire flight.

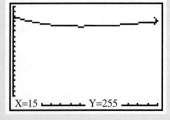

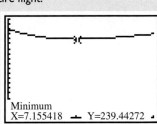

ENRICHMENT ESSAY

Einstein's Model and the Clock Paradox

Einstein's model

$$T = \frac{T_0}{\sqrt{1 - \dfrac{v^2}{c^2}}}$$

relates the passage of time T at a velocity v to the passage of time at rest, T_0. In the model, c is the speed of light, approximately 186,000 miles per second. If a futuristic starship moves at a velocity approaching the speed of light ($v \approx c$), the denominator in the model is close to zero, resulting in very large values for time. This means that time passage for those of us waiting on Earth is much faster than that for the space travelers. The travelers return to Earth in what seems to them to be a few years, though on Earth many centuries have passed. This effect, called the clock paradox, means they return to a futuristic world in which their friends and loved ones would be long dead.

Edward Kienholtz (1927–1994) "The Wait" 1964–65. Tableau: wood, fabric, polyester resin, flock, metal, bones, glass, paper, leather, varnish, b&w photographs, taxidermed cat, live parakeet, wicker and plastic. 13 units, overall: 80 × 160 × 34 in. (203.2 × 406.4 × 213.4 cm). Gift of the Howard and Jean Lipman Foundation, Inc. 66.49a–m. Photograph copyright © 1997: Whitney Museum of American Art, New York. Photography by: Jerry L. Thompson, N.Y. 1994.

PROBLEM SET 1.4

Practice Problems

Find the solution set for each equation in Problems 1–48. Check your proposed solutions by substituting into the original equation or by using a graphing utility.

1. $3x^4 = 48x^2$

2. $5x^4 = 20x^2$

3. $3x^3 + 2x^2 - 12x - 8 = 0$

4. $4x^3 - 12x^2 - 9x + 27 = 0$

5. $2x - 3 = 8x^3 - 12x^2$

6. $x + 1 = 9x^3 + 9x^2$

7. $4y^3 - 2 = y - 8y^2$

8. $9y^2 + 8 = 4y + 18y^2$

9. $x^4 - 10x^2 + 9 = 0$

10. $x^4 - 8x^2 + 15 = 0$

11. $6x^4 - 7x^2 + 2 = 0$

12. $8x^4 - 14x^2 + 3 = 0$

13. $3x^3 - 12x^2 - 15x = 0$

14. $4x^3 - 8x^2 - 12x = 0$

15. $(x^2 - x)^2 - 14(x^2 - x) + 24 = 0$

16. $(x^2 - 2x)^2 - 11(x^2 - 2x) + 24 = 0$

17. $\left(y - \dfrac{8}{y}\right)^2 + 5\left(y - \dfrac{8}{y}\right) - 14 = 0$

18. $\left(y - \dfrac{10}{y}\right)^2 + 6\left(y - \dfrac{10}{y}\right) - 27 = 0$

19. $6\left(\dfrac{2w}{w - 3}\right)^2 = 5\left(\dfrac{2w}{w - 3}\right) + 6$

20. $4\left(\dfrac{w + 1}{w - 1}\right)^2 = 8\left(\dfrac{w + 1}{w - 1}\right) - 3$

21. $\sqrt{3x + 18} = x$

22. $\sqrt{20 - 8x} = x$

23. $x - \sqrt{2x + 5} = 5$

24. $x - \sqrt{x + 11} = 1$

25. $\sqrt{2y - 3} - \sqrt{y - 2} = 1$

26. $\sqrt{2y + 3} + \sqrt{y - 2} = 2$

27. $\sqrt{3\sqrt{x} + 1} = \sqrt{3x - 5}$

28. $\sqrt{1 + 4\sqrt{x}} = 1 + \sqrt{x}$

29. $\sqrt{6x - 2} = \sqrt{2x + 3} - \sqrt{4x - 1}$

30. $\sqrt{x + 4} + \sqrt{x - 1} = \sqrt{x - 4}$

31. $\sqrt{5x + 1} = \sqrt{3x + 4} + \sqrt{x - 6}$

32. $\sqrt{2x - 5} = \sqrt{3x - 5} - \sqrt{x - 6}$

33. $(x - 4)^{3/2} = 8$

34. $(x + 5)^{3/2} = 8$

35. $(x - 4)^{2/3} = 16$

36. $(x + 5)^{2/3} = 4$

37. $(y^2 - y - 4)^{3/4} - 2 = 6$

38. $(y^2 - 3y + 3)^{3/2} - 1 = 0$

39. $(11y^2 - 18)^{1/4} = 14$

40. $(7y^2 - 12)^{1/4} = y$

41. $(-2x + 1)^{-5/3} = -\dfrac{1}{32}$

42. $(3y + 1)^{-3/5} = -\dfrac{1}{8}$

43. $2x^{2/3} + 3x^{1/3} - 2 = 0$

44. $2x^{2/3} + 7x^{1/3} - 15 = 0$

45. $2x - 3x^{1/2} + 1 = 0$

46. $x + 3x^{1/2} - 4 = 0$

47. $(z - 2)^{1/2} = 11(z - 2)^{1/4} - 18$

48. $(z - 1)^{1/2} = 2(z - 1)^{1/4} + 15$

Application Problems

49. The model $v = 2\sqrt{3m}$ gives an approximate relationship between the velocity of a car (v, in miles per hour) that skids to a stop on wet pavement and the length of the resulting skid marks (m, in feet). Approximately how long are the skid marks left by a car traveling 50 miles per hour when the brakes are applied?

50. Shallow-water wave motion can be modeled by $c = \sqrt{gH}$ where c is the wave velocity in feet per second, H is the water depth in feet, and g, the acceleration of gravity, is 32 feet per second squared. If the wave velocity is 24 feet per second, find the water's depth.

Use Einstein's model

$$L = L_0 \sqrt{1 - \frac{v^2}{c^2}}$$

which relates length at rest (L_0) to length (L) at velocity v to answer Problems 51–52. The speed of light, c, is approxmately 186,000 miles per second.

186,000 Miles Per Second
Is Not Just A Good Idea.
It's The Law.

51. How fast would a futuristic starship that is 600 meters tall have to move so that it would measure only half this length from the perspective of an observer who looks up and sees the ship?

52. How fast would a futuristic starship that is 600 meters tall have to move so that its length would be diminished by 75% from the perspective of an observer who looks up and sees the ship?

Use Einstein's models

$$M = \frac{M_0}{\sqrt{1 - \dfrac{v^2}{c^2}}} \quad \text{and} \quad T = \frac{T_0}{\sqrt{1 - \dfrac{v^2}{c^2}}}$$

which relate mass and time at rest (M_0, T_0), to mass and time (M, T) at velocity v, to answer Problems 53–54. The speed of light, c, is approximately 186,000 miles per second.

53. How fast would a 120-pound person have to travel for that person's mass to double as perceived by an observer at rest?

54. A clock on board a futuristic starship traveling at $\frac{9}{10}$ of the speed of light measures a 10-second time period. How long is this time period from the perspective of an observer who looks up and sees the ship pass by?

55. Laser Records marketing research department determines that weekly demand for a boxed set of CDs by the Jumping Artichokes depends on the price per set, approximated by the model $p = 30 - \sqrt{0.01x + 1}$, where p represents the price of the set and x represents the number of sets sold each week at price p.

  a. At what price will there be no demand for the CD sets?

  b. Approximately how many CD sets will sell weekly at a price of $27.76?

The amount of money in an account (A) when interest is compounded n times a year is given by the model

$$A = P\left(1 + \frac{r}{n}\right)^{nt}.$$

In this formula, A is the balance in the account, P is the amount originally deposited, r is the annual interest rate (in decimal form), n is the number of compoundings per year, and t is the time in years. Use this formula to answer Problems 56–57.

56. A total of $12,000 is invested. After five years, the balance in the account is $18,726.11. If the interest is compounded quarterly, what was the annual interest rate for this investment?

57. A total of $2500 is invested. After 10 years, the balance in the account is $8155.09. If the interest is compounded quarterly, what was the annual interest rate for this investment?

58. Two vertical poles of lengths 6 feet and 8 feet stand 10 feet apart (see the figure). A cable reaches from the top of one pole to some point on the ground between the poles and then to the top of the other pole. Where should this point be located to use 18 feet of cable?

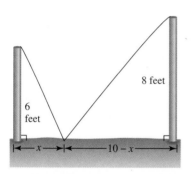

59. Twelve miles separate towns *A* and *B*, located 5 miles and 3 miles, respectively, from a major expressway. Two new roads are to be built from *A* to the expressway and then to *B*. (See the figure.)

  a. Find *x* if the length of the new roads is 16 miles.

  b. Write a verbal description for the road crew telling them where to position the new roads based on your answer to part (a).

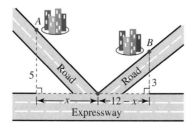

60. A box is in the shape of a cube. If the width is increased by 9 inches and the length is decreased by 4 inches, then the volume of the new box is 180 cubic inches. What are the dimensions of the cubic box?

61. A rectangular box has a volume of 6 cubic feet. Its length is 3 feet greater than its height, and its width is 1 foot less than its height. What are the dimensions of the box?

62. Two spherical paperweights of radius 3 inches and 4 inches are to be melted and reformed into a larger sphere. Find the radius of the larger sphere.

63. A spherical container has a radius of 6 inches. By how much should the radius be increased if the volume of the new container is to be increased by 288π cubic inches?

64. A box with a square base has a height that is 2 inches greater than the length of the base. The box is surrounded by a 1-inch thick layer of packing material. If 130 cubic inches of packing material surrounds the box, what are the dimensions of the box?

65. A power station is located on the shoreline of a 6-mile-wide river. A factory is 12 miles downstream on the other side of the river. It costs $130 thousand per mile to run power lines under water and $80 thousand per mile to run power lines over the land. How far from the factory should the power line come ashore if the cost of the project is $1575 thousand?

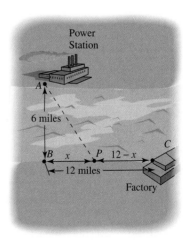

True–False Critical Thinking Problems

66. Which one of the following is true?

a. Squaring both sides of $\sqrt{y+4} + \sqrt{y-1} = 5$ leads to $y + 4 + y - 1 = 25$, an equation with no radicals.

b. The equation $(x^2 - 2x)^9 - 5(x^2 - 2x)^3 + 6 = 0$ is quadratic in form and should be solved by letting $t = x^2 - 2x$.

c. If a radical equation has two proposed solutions and one of these values is not a solution, the other value is also not a solution.

d. None of these statements is true.

67. Which one of the following is true?

a. The polynomial equation $x^{2/3} - 9x^{1/3} + 8 = 0$ can be solved by letting $t = x^{1/3}$, resulting in $t^2 - 9t + 8 = 0$.

b. If $\left| x^3 - 6x^2 \right| = 5x$, then $x^3 + 6x^2 = 5x$ or $x^3 + 6x^2 = -5x$.

c. The solution of $2x^{1/4} - 3x^{1/2} + 3 = 0$ can be found by letting $t = x^{1/2}$ and $t^2 = x^{1/4}$.

d. None of these statements is true.

Technology Problems

In Problems 68–74, use a graphing utility to estimate the real solutions of each equation, accurate to two decimal places.

68. $15x^3 + 14x^2 - 3x - 2 = 0$

69. $x^3 + x^2 - 4x - 4 = 0$

70. $2x^3 - 3x^2 - 11x + 6 = 0$

71. $3x^3 + 7x^2 - 22x - 8 = 0$

72. $x + \sqrt{26 - 11x} = 4$

73. $2x^{\frac{2}{3}} + 4x^{\frac{1}{3}} + 1.8 = 0$

74. $3\sqrt{2} - \dfrac{2}{\sqrt{x}} = 2$

75. a. The model $N = -143x^3 + 1810x^2 - 187x + 2331$ approximates the number of new AIDS cases in the United States x years after 1983, where $0 \leq x \leq 7$. Graph this model using the following range setting:

$$\text{Xmin} = 0,\ \text{Xmax} = 7,\ \text{Xscl} = 1,\ \text{Ymin} = 0,$$

$$\text{Ymax} = 50{,}000,\ \text{Yscl} = 5000$$

b. Describe what the graph of the model visually indicates about the spread of AIDS.

c. In what year were approximately 40,000 cases reported?

d. Construct a bar graph based on the graph of the model.

e. How many new cases does the model indicate there were last year? Look up the actual number of new cases reported for that year. What does this tell you about mathematical models?

76. Use a graphing utility to graph the model that represents the total amount of cable needed for the project in Problem 58. Then use the minimum value feature of the utility to determine approximately where the point should be located to use the least amount of cable.

77. Use a graphing utility to graph the formula that models the total length of the new roads in Problem 59. Then use the minimum value feature of the utility to determine the shortest road that meets the problem's requirements.

78. Use a graphing utility to graph the model that represents the total cost of running the power line under the river and over the land in Problem 65. Then trace along the curve or use the minimum value feature to explain the significance of the cost of the project given in the problem.

79. A person rows in a straight line across a 2-mile wide channel from point A to point D and then walks from point D to a house located at point C. The figure shows that the distance from point B to the house at point C is 6 miles. The rowing rate is 3 miles per hour and the walking rate is 5 miles per hour.

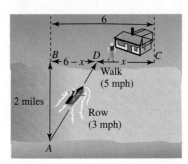

a. Use the fact that time is distance divided by rate to show that the total time spent rowing and walking, represented by y, is modeled by

$$y = \frac{\sqrt{4 + (6 - x)^2}}{3} + \frac{x}{5}$$

b. Suppose that the total time spent walking and rowing is 1 hour and 44 minutes, or about $\frac{26}{15}$ hours. Use your graphing utility to graph the model in part (a), using the following range setting:

$$\text{Xmin} = 0,\ \text{Xmax} = 6,\ \text{Xscl} = 1,\ \text{Ymin} = 1.7,$$

$$\text{Ymax} = 2.1,\ \text{Yscl} = 1$$

Then trace along the curve until $y = \frac{26}{15}$. What is the corresponding value of x? Where, then, should the person come ashore in terms of distance from the house? What is the significance of this rowing–walking path in terms of the time required to reach the house?

c. Verify part (b) algebraically by solving the equation

$$\frac{\sqrt{4 + (6 - x)^2}}{3} + \frac{x}{5} = \frac{26}{15}$$

$$\left(1 \text{ hour and } 44 \text{ minutes} = 1\frac{44}{60} = 1\frac{11}{15} = \frac{26}{15} \text{ hours}\right)$$

Writing in Mathematics

80. We have seen that squaring both sides of an equation does not always produce an equivalent equation. Is this true when cubing both sides of an equation? Explain.

81. Reread Problem 55 on page 170. Suppose you are writing a report on the relationship between the price of the CDs and the units that will sell. Assume that other than yourself, everyone in the company is innumerate and cannot understand the model $p = 30 - \sqrt{0.01x + 1}$. Consequently, you must describe the relationship between p and x strictly using words. Write such a description, minimizing the use of mathematical terminology.

82. Give an example of an equation that is easier to solve algebraically than with a graphing utility. Give an example of an equation that is easier to solve with a graphing utility than algebraically. What conclusion can you draw?

83. Describe what happens to time and length in Einstein's models (see page 166) as the velocity of futuristic starship approaches the speed of light ($v \approx c$). What does this mean in practical terms for a person returning from a 20-year cruise at nearly the speed of light?

84. Reread Problems 49, 50 and 55. Two of these examples contain models that are the "real thing," and one contains a model that was artificially constructed to provide further practice and drill in solving equations. Which contains the latter? How can you tell?

Critical Thinking Problems

85. Solve for x: $\dfrac{(x - 3)^{1/2}(x - 4)^{1/5}}{(x - 7)^{1/3}} = 0$

86. Solve for x: $x^{5/6} + x^{2/3} - 2x^{1/2} = 0$

Solve each equation in Problems 87–89 for real values of y.

87. $y^{1/3} + 4^{1/4} = \dfrac{2}{2 - \sqrt{2}}$

88. $\sqrt{\dfrac{y - 30}{y}} + 4\sqrt{\dfrac{y}{y - 30}} = 5$

89. $\sqrt{\sqrt{y} + \sqrt{c}} + \sqrt{\sqrt{y} - \sqrt{c}} = \sqrt{2\sqrt{y} + 2\sqrt{d}}$

90. A beam is to be cut from a circular log of radius 1 foot. Let x and y represent half the length and half the width of a rectangular cross section of the beam.

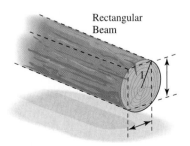

Rectangular
Beam

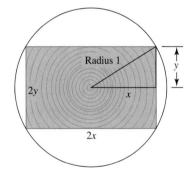

Radius 1

$2y$

x

$2x$

y

a. Write an equation in terms of x and y for the area of the beam's cross section.

b. Apply the Pythagorean Theorem to the small right triangle in the figure shown above and write an equation for y in terms of x.

c. Substitute the equation from part (b) into the area model of part (a), representing the area of the beam's cross section in terms of x alone.

d. Use the model from part (c) to determine the value of x that will result in an area of 2 square feet. What are the dimensions of the cross section?

e. Use a graphing utility to graph the area model from part (c) and explain the significance of your answer to part (d).

Solutions | Tutorial | Video
Manual | | 3

SECTION 1.5   Linear Inequalities

Objectives

1 Solve linear inequalities.
2 Solve applied modeling problems.
3 Solve compound inequalities.
4 Solve inequalities involving absolute value.

In this section, we establish the basic techniques for solving inequalities containing variables raised only to the first power, called *linear inequalities*. We also consider the solution of linear inequalities involving absolute value.

| Solve linear inequalities.

Linear Inequalities and Their Properties

We know that a linear equation in x can be expressed in the form $ax + b = c$. A similar definition describes a linear inequality in one variable.

> **Definition of a linear inequality**
>
> A *linear inequality* in one variable x is any inequality that can be written in one of the forms
>
> $$ax + b < c \qquad ax + b \leq c \qquad ax + b > c \qquad ax + b \geq c$$
>
> where a, b, and c are real numbers and $a \neq 0$.

An example of a linear inequality in one variable is $2x + 3 < 17$, in which $a = 2$, $b = 3$, and $c = 17$. Because the greatest power on the variable is one, a linear inequality is also called a first-degree inequality.

Any value of the variable for which an inequality is true is called a *solution*. The *solution set* of the inequality is the set of all such solutions. The solutions are said to *satisfy* the inequality.

To solve a linear inequality, we use a series of transformations into equivalent inequalities (inequalities that have the same solution set). Our final equivalent inequality should be in the form

$$x > d \qquad x < d \qquad x \geq d \qquad x \leq d.$$

We can use properties of inequalities to isolate the variable. These properties are similar to the properties of equality that we used to solve linear equations. There are, however, two important exceptions. Multiplying or dividing both sides of an inequality by a negative number reverses the sense (or direction) of the inequality. To illustrate, let's take three true inequality statements and multiply both sides by -3.

$$7 < 10 \qquad\qquad -2 < 6 \qquad\qquad -6 < -1$$
$$\downarrow \qquad\qquad\qquad \downarrow \qquad\qquad\qquad \downarrow$$
$$-3(7) > -3(10) \qquad -3(-2) > -3(6) \qquad -3(-6) > -3(-1)$$
$$-21 > -30 \qquad\qquad 6 > -18 \qquad\qquad 18 > 3$$

In each case, the resulting inequality symbol points in the opposite direction from the original one.

When both sides of an inequality are multiplied or divided by a negative number, the direction of the inequality symbol is reversed.

To isolate the variable in a linear inequality, we use the same basic techniques used in solving linear equations. The following properties are used to create equivalent inequalities.

Properties of inequalities

Property	The Property in Words	Example
Addition and Subtraction Properties If $a < b$, then $a + c < b + c$. If $a < b$, then $a - c < b - c$.	If the same quantity is added to or subtracted from both sides of an inequality, the resulting inequality is equivalent to the original one.	$2x + 3 < 7$ Subtract 3: $2x + 3 - 3 < 7 - 3$ Simplify: $2x < 4$
Positive Multiplication and Division Properties If $a < b$ and c is positive, then $ac < bc$ If $a < b$ and c is positive, then $\dfrac{a}{c} < \dfrac{b}{c}$.	If by the same positive quantity we multiply or divide both sides of an inequality, the resulting inequality is equivalent to the original one.	$2x < 4$ Divide by 2: $\dfrac{2x}{2} < \dfrac{4}{2}$ Simplify: $x < 2$
Negative Multiplication and Division Properties If $a < b$ and c is negative, then $ac > bc$. If $a < b$ and c is negative, then $\dfrac{a}{c} > \dfrac{b}{c}$.	If by the same negative quantity we multiply or divide both sides of an inequality, the result is an equivalent inequality in which the inequality symbol is reversed.	$-4x < 20$ Divide by -4 and reverse the sense of the inequality: $\dfrac{-4x}{-4} > \dfrac{20}{-4}$ Simplify: $x > -5$

Study tip

The properties of inequality are true if $<$ and $>$ are replaced by $\leq$ and $\geq$.

EXAMPLE 1   Solving a Linear Inequality

Solve and graph the solution set on a number line: $3 - 2x < 11$

Solution

$$3 - 2x < 11 \qquad \text{This is the given inequality.}$$

$$3 - 2x - 3 < 11 - 3 \qquad \text{Subtract 3 from both sides.}$$

$$-2x < 8 \qquad \text{Simplify.}$$

$$\frac{-2x}{-2} > \frac{8}{-2} \qquad \text{Divide both sides by } -2 \text{ and reverse the sense of the inequality.}$$

$$x > -4 \qquad \text{Simplify.}$$

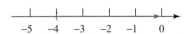

Figure 1.16

The graph of $x > -4$

Using technology

To solve Example 2 on your graphing utility, begin by entering each side of the given inequality:

$$y_1 \boxed{=} 7(x + 4) - 13$$

$$y_2 \boxed{=} 12 + 13(3 + x)$$

Now enter y_3, the difference of y_1 and y_2:

$$y_3 \boxed{=} y_1 - y_2$$

To solve

$$7(x + 4) - 13 < 12 + 13(3 + x)$$

we must find where $y_1 - y_2 < 0$, or where y_3 lies below the x-axis. Turn off the high-lighting on the equal sign for y_1 and y_2, and graph y_3. The graph of y_3 shows that $y_1 - y_2 < 0$ (lies below the x-axis) if $x > -6$, confirming our algebraic solution.

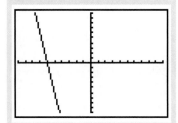

Figure 1.17

The graph of $x > -6$

The solution set consists of all real numbers that are greater than -4, expressed as $\{x \mid x > -4\}$. The interval notation for the solution set is $(-4, \infty)$. The graph is shown on a number line in Figure 1.16.   ■

Discover for yourself

We cannot check all members of an inequality's solution set, but we can take a few values to get an indication of whether it is correct. In Example 1, we found that the solution set to $3 - 2x < 11$ is $(-4, \infty)$. Show that -3 satisfies the given inequality, whereas -5 does not.

EXAMPLE 2   **Solving a Linear Inequality**

Solve and graph the solution set on a number line:

$$7(x + 4) - 13 < 12 + 13(3 + x)$$

Solution

$7(x + 4) - 13 < 12 + 13(3 + x)$	This is the given inequality.
$7x + 28 - 13 < 12 + 39 + 13x$	Simplify each side and apply the distributive property.
$7x + 15 < 51 + 13x$	Combine numerical terms.

We will collect variable terms on the left and constant terms on the right.

$7x + 15 - 13x < 51 + 13x - 13x$	Subtract $13x$ from both sides.
$-6x + 15 < 51$	Simplify.
$-6x + 15 - 15 < 51 - 15$	Subtract 15 from both sides.
$-6x < 36$	Simplify.
$\dfrac{-6x}{-6} > \dfrac{36}{-6}$	Divide both sides by -6 and reverse the sense of the inequality.
$x > -6$	Simplify.

The solution set consists of all real numbers that are greater than -6. The interval notation for this solution set is $(-6, \infty)$.

The graph of the solution set on a number line is shown in Figure 1.17.   ■

Discover for yourself

The addition and subtraction properties of inequality provide us with the option of collecting all terms with the variable on the right and all numbers on the left. Use this option to solve

$$7x + 15 < 51 + 13x$$

and compare your solution to that of Example 2.

2 Solve applied modeling problems.

Modeling with Linear Inequalities

Example 3 illustrates how we can apply our five-step strategy for problem solving to an inequality situation. The example involves simple interest. The annual simple interest that an investment earns is given by the formula

$$I = Pr$$

where I is the simple interest, P is the amount invested (called the *principal*), and r is the simple interest rate.

EXAMPLE 3 **An Application: Simple Interest**

A person inherits $16,000 with the stipulation that for the first year the money must be invested in two stocks paying 6% and 8% annual interest, respectively. What is the greatest amount that can be invested at 6% if the total interest earned for the year is to be at least $1180?

Solution

Steps 1 and 2. Represent unknowns in terms of x.

Let

$$x = \text{Amount invested at } 6\%$$

$$16,000 - x = \text{Amount invested at } 8\%$$
$$\text{(since the total amount invested was \$16,000)}$$

The total yearly interest must be at least $1180.

Step 3. Write an inequality that models the verbal conditions.

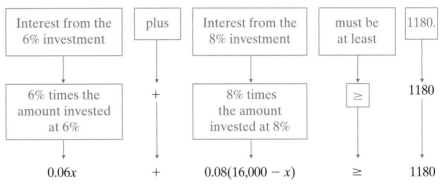

Step 4. Solve the inequality and answer the problem's question.

$0.06x + 0.08(16,000 - x) \geqslant 1180$	This is the inequality that models the given conditions.
$6x + 8(16,000 - x) \geqslant 118,000$	Multiply by 100 to clear decimals (optional).
$6x + 128,000 - 8x \geqslant 118,000$	Use the distributive property.
$-2x + 128,000 \geqslant 118,000$	Combine like terms.
$-2x \geqslant -10,000$	Subtract 128,000 from both sides.
$x \leqslant 5000$	Divide by -2 and reverse the sense of the inequality.

Step 5. Check.

Thus, at most $5000 can be invested at 6% if the total yearly interest is to be at least $1180. Suppose that $5000 is invested at 6%. Thus, $16,000 - $5000 or $11,000 is invested at 8%. The interest is

$$0.06(5000) + 0.08(11,000)$$

or \$1180. Try this again, with \$3000 invested at 6% (at most \$5000 can be invested at this rate) and \$13,000 invested at 8%. As specified by the problem's conditions, the yearly interest should be at least \$1180. ∎

3 Solve compound inequalities.

Solving Compound Inequalities

Two inequalities are often written as a *compound inequality*. For example, consider a hypothetical situation involving the United Cable Television Company. Let x represent the number of customers served by the company. The weekly profit of the company is 50 times the number of customers served less \$100. The company has *two* constraints placed upon it:

Study tip

Notice that

   $50x - 100 > 5000$

and

   $5000 < 50x - 100$

mean the same thing. In both inequalities, the inequality sign branches open in the direction of $50x - 100$, the quantity that is greater than 5000. For this reason, we can rewrite

   $50x - 100 > 5000$

and

   $50x - 100 < 30,000$

as

   $5000 < 50x - 100 < 30,000.$

1. Its weekly profits must be greater than \$5000, or the corporation will be sold by the stockholders.
2. Its contract with the local officials restricts weekly profits to less than \$30,000.

Since profits are $50x - 100$ (50 times the number of customers minus \$100), we have

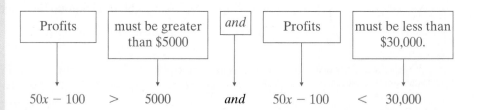

We can express these conditions uslng a compound (or double) inequality as follows:

$$5000 < 50x - 100 < 30,000$$

This form permits us to solve the two inequalities together. Our goal is to isolate x by itself in the middle. Here's how it is done:

$5000 < 50x - 100 < 30,000$	This is the given compound inequality.
$5100 < 50x < 30,100$	Add 100 to all three parts.
$\dfrac{5100}{50} < \dfrac{50x}{50} < \dfrac{30,100}{50}$	Divide all three parts by 50.
$102 < x < 602$	Simplify.

In the context of the original situation, this means that to not be sold or to avoid contract violation, the company must have more than 102 customers *and* fewer than 602 customers.

Our next example involves a compound inequality.

EXAMPLE 4 **Solving a Compound Inequality**

Solve: $\dfrac{2}{3} \le \dfrac{5 - 3x}{2} < \dfrac{3}{4}$

Solution

$$\dfrac{2}{3} \le \dfrac{5 - 3x}{2} < \dfrac{3}{4}$$ This is the given compound inequality; our goal is to isolate x in the middle.

$$12\left(\dfrac{2}{3}\right) \le 12\left(\dfrac{5 - 3x}{2}\right) < 12\left(\dfrac{3}{4}\right)$$ Clear fractions by multiplying each part by 12.

$$8 \le 6(5 - 3x) < 9$$ Multiply.

$$8 \le 30 - 18x < 9$$ Use the distributive property.

$$-22 \le -18x < -21$$ Subtract 30 from each part.

$$\dfrac{-22}{-18} \ge x > \dfrac{-21}{-18}$$ Divide each part by -18 and reverse the sense of the inequality.

$$\dfrac{11}{9} \ge x > \dfrac{7}{6}$$ Simplify the fractions.

This inequality is correct but awkward to interpret. It can be rewritten with the lesser rational number, $\frac{7}{6}$, on the left as

$$\dfrac{7}{6} < x \le \dfrac{11}{9}.$$

The solution set in interval notation is $(\frac{7}{6}, \frac{11}{9}]$. The graph of the solution set is shown on a number line in Figure 1.18. ∎

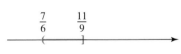

Figure 1.18

The graph of $\dfrac{7}{6} < x \le \dfrac{11}{9}$

4 Solve inequalities involving absolute value.

Solving Inequalities Involving Absolute Value

In our prerequisites chapter, we interpreted $|x|$ as the distance between x and the origin on a number line. Using this interpretation, Table 1.5 shows how we can rewrite absolute value inequalities without absolute value bars.

TABLE 1.5 Rewriting Absolute Value Inequalities Without Absolute Value Bars

Example	Rewritten Without Absolute Value	Generalization
$\|x\| < 2$ The distance between x and the origin is less than 2. (number line: -2, 0, 2)	All real numbers in $(-2, 2)$ have distances from the origin that are less than 2, so $\|x\| < 2$ means $-2 < x < 2$.	If c is a positive real number, $\|x\| < c$ means $-c < x < c$.
$\|x\| > 3$ The distance between x and the origin is greater than 3. (number line: -3, 0, 3)	All real numbers in $(3, \infty)$ or $(-\infty, -3)$ have the distances from the origin that are greater than 3, so $\|x\| > 3$ means $x > 3$ or $x < -3$.	If c is a positive real number, $\|x\| > c$ means $x > c$ or $x < -c$.

We can use these relationships to solve inequalities involving absolute value. Both relationships involve rewriting absolute value inequalities as equivalent inequalities without absolute value bars. We can apply these relationships to linear inequalities as follows.

Rewriting absolute value inequalities without absolute value bars

If c is positive,

1. $\left| ax + b \right| < c$ means $-c < ax + b < c.$
2. $\left| ax + b \right| > c$ means $ax + b > c \text{ or } ax + b < -c.$

These rules are valid if $<$ is replaced by $\leq$ and $>$ is replaced by $\geq$.

Once the absolute value inequality has been rewritten as an equivalent inequality without absolute value bars, the solutions can be found using the methods for solving linear inequalities. This idea is illustrated in our next two examples.

Using technology

The solution to Example 5 can be verified graphically by entering each side of the given inequality on a graphing utility.

$$y_1 = \boxed{\text{ABS}}\,((3X + 3)/5)$$

$$y_2 = 3$$

$$y_3 \boxed{=} y_1 - y_2$$

Highlighting only y_3, we must find where $y_3 < 0$. The graph of y_3 lies below the x-axis for $-6 < x < 4$, visually confirming our algebraic solution.

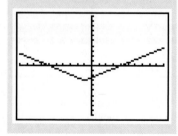

EXAMPLE 5   **Solving an Absolute Value Inequality with $<$**

Solve: $\left| \dfrac{3x + 3}{5} \right| < 3$

Solution

$\left\| \dfrac{3x + 3}{5} \right\| < 3$	This is the given inequality.
$-3 < \dfrac{3x + 3}{5} < 3$	Rewrite without absolute value bars. $\left\|ax + b\right\| < c$ means $-c < ax + b < c.$
$-15 < 3x + 3 < 15$	Multiply each part by 5.
$-18 < 3x < 12$	Subtract 3 from each part.
$-6 < x < 4$	Divide each part by 3.

All real numbers between -6 and 4, excluding -6 and 4, satisfy the original inequality. Using interval notation, the solution set is $(-6, 4)$. The graph of the solution set is shown in Figure 1.19.   ∎

Figure 1.19

The graph of $-6 < x < 4$

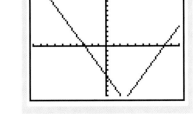

Using technology

You can use a graphing utility to verify the solution to Example 6. Enter

$Y_1 = \boxed{\text{ABS}} (-2x + 5)$

$Y_2 = 11$

$Y_3 \boxed{=} Y_1 - Y_2$

and highlight only y_3. The graph of y_3 shows that $y_3 \geq 0$ (lies on or above the x-axis) if $x \leq -3$ or $x \geq 8$, visually confirming our algebraic solution.

Figure 1.20

The graph of $x \leq -3$ or $x \geq 8$

Meret Oppenheim "Object (Le Dejeuner en fourrure)" 1936, fur-covered cup, saucer and spoon; cup 4 3/8 in. diameter; saucer 9 3/8 in. diameter; spoon 8 in. long; overall height 2 7/8 in. The Museum of Modern Art, New York. Purchase. Photograph © 1997 The Museum of Modern Art, New York. © 1998 Artists Rights Society (ARS), New York / Pro Litteris, Zurich.

EXAMPLE 6   **Solving an Absolute Value Inequality with $\geq$**

Solve: $\left| -2x + 5 \right| \geq 11$

Solution

$\left| -2x + 5 \right| \geq 11$		This is the given inequality.
$-2x + 5 \geq 11 \quad$ or $\quad -2x + 5 \leq -11$		Rewrite without absolute value bars. $\left| ax + b \right| \geq c$ means $ax + b \geq c$ or $ax + b \leq -c$.
$-2x \geq 6 \qquad\qquad -2x \leq -16$		Subtract 5 from both sides of both inequalities.
$x \leq -3 \qquad\qquad x \geq 8$		Divide both sides of both inequalities by -2.

Using interval notation, the solution set is $(-\infty, -3]$ or $[8, \infty)$. This can be expressed as $(-\infty, -3] \cup [8, \infty)$ where the symbol $\cup$ (the union symbol) denotes the set of numbers in $(-\infty, -3]$ *or* in $[8, \infty)$. The graph of the solution set is shown on the number line in Figure 1.20. ■

Modeling with Absolute Value Inequalities

Example 7 shows how we can apply absolute value inequalities.

EXAMPLE 7   **Modeling Defective Products**

The percentage (p) of defective products manufactured by a company is described by

$$\left| p - 0.35\% \right| \leq 0.16\%$$

or, equivalently

$$\left| p - 0.0035 \right| \leq 0.0016 .$$

If 100,000 products are manufactured and the company offers a $6 refund for each defective product, describe the company's cost of refunds.

Solution

We'll start with what seems most obvious, namely:

The company's cost of refunds	equals	$6 refund for each defective product	times	The number of defective products.
R	$=$	6	$\cdot$	$?$

We now need an expression for the number of defective products.

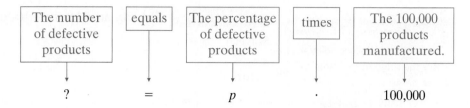

$$? \qquad = \qquad p \qquad \cdot \qquad 100{,}000$$

We can find p by solving the given inequality for this variable.

$\lvert p - 0.0035 \rvert \leq 0.0016$	This is the given inequality.
$-0.0016 \leq p - 0.0035 \leq 0.0016$	Rewrite without absolute value bars.
$0.0019 \leq p \leq 0.0051$	Add 0.0035 to each part.

In percent notation, the percentage of defective products ranges from 0.19% to 0.51%. This will give us a range for the number of defective products.

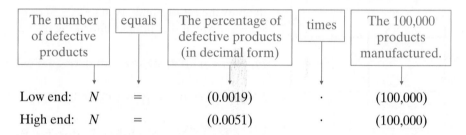

Low end: N = (0.0019) · (100,000)

High end: N = (0.0051) · (100,000)

Equivalently,

$$(0.0019)(100{,}000) \leq N \leq (0.0051)(100{,}000).$$

Performing the multiplications, we see that

$$190 \leq N \leq 510,$$

meaning that the number of defective products is greater than or equal to 190 but less than or equal to 510 for every 100,000 products manufactured. Since the company refunds $6 for each defective product, we can now describe the company's cost of refunds.

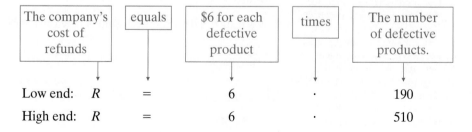

Low end: R = 6 · 190

High end: R = 6 · 510

Performing the multiplications, we see that

$$1140 \leq R \leq 3060.$$

For every 100,000 products manufactured, the cost of refunds ranges from $1140 to $3060. ∎

P R O B L E M   S E T   1 . 5

Practice Problems

Solve each linear inequality in Problems 1–20 and graph the solution set on a number line.

1. $5x + 11 < 26$

2. $2x + 5 < 17$

3. $3x - 7 \geqslant 13$

4. $8x - 2 \geqslant 14$

5. $-9x \geqslant 36$

6. $-5x \leqslant 30$

7. $8x - 11 \leqslant 3x - 13$

8. $18x + 45 \leqslant 12x - 8$

9. $4(x + 1) + 2 \geqslant 3x + 6$

10. $8x + 3 > 3(2x + 1) + x + 5$

11. $2x - 11 < -3(x + 2)$

12. $-4(x + 2) > 3x + 20$

13. $1 - (x + 3) \geqslant 4 - 2x$

14. $5(3 - x) \leqslant 3x - 1$

15. $\dfrac{x}{4} - \dfrac{3}{5} \leqslant \dfrac{x}{2} + 1$

16. $\dfrac{3x}{10} + 1 \geqslant \dfrac{1}{5} - \dfrac{x}{10}$

17. $1 - \dfrac{x}{2} > 4$

18. $7 - \dfrac{4}{5}x < \dfrac{3}{5}$

19. $4(3y - 2) - 3y < 3(1 + 3y) - 7$

20. $3(y - 8) - 2(10 - y) > 5(y - 1)$

Solve each inequality in Problems 21–38 by isolating the variable by itself in the middle. Graph the solution set on a number line.

21. $6 < x + 3 < 8$

22. $7 < x + 5 < 11$

23. $-3 \leqslant x - 2 < 1$

24. $-6 < x - 4 \leqslant 1$

25. $-9 < 3x \leqslant -6$

26. $-24 \leqslant 4x < -20$

27. $-9 \leqslant -3x \leqslant -6$

28. $-24 \leqslant -4x \leqslant -20$

29. $-11 < 2x - 1 \leqslant -5$

30. $3 \leqslant 4x - 3 < 19$

31. $-3 \leqslant \dfrac{2}{3}x - 5 < -1$

32. $-6 \leqslant \dfrac{1}{2}x - 4 < -3$

33. $-8 < 2 - 3x < 10$

34. $-3 \leqslant 1 - 2x \leqslant 5$

35. $-5 \leqslant 3 - 2x < 18$

36. $0 < 3(-x - 5) - 9 \leqslant 6$

37. $-1 < \dfrac{1 - 4y}{3} \leqslant 1$

38. $\dfrac{2}{3} < \dfrac{5 - 3y}{-2} \leqslant \dfrac{3}{4}$

Solve each inequality in Problems 39–68 by first rewriting each one as an equivalent inequality without absolute value bars. Graph the solution set on a number line.

39. $|x| < 3$

40. $|x| < 5$

41. $|x - 1| \leqslant 2$

42. $|x + 3| \leqslant 4$

43. $|2x - 6| < 8$

44. $|3x + 5| < 17$

45. $|2(x - 1) + 4| \leqslant 8$

46. $|3(x - 1) + 2| \leqslant 20$

47. $\left| \dfrac{2y + 6}{3} \right| < 2$

48. $\left| \dfrac{3(x - 1)}{4} \right| < 6$

49. $|x| > 3$

50. $|x| > 5$

51. $|x - 1| \geqslant 2$

52. $|x + 3| \geqslant 4$

53. $|3x - 8| > 7$

54. $|5x - 2| > 13$

55. $\left| \dfrac{2x + 2}{4} \right| \geqslant 2$

56. $\left| \dfrac{3x - 3}{9} \right| \geqslant 1$

57. $\left| 3 - \dfrac{2}{3}y \right| > 5$

58. $\left| 3 - \dfrac{3}{4}y \right| > 9$

59. $3|y - 1| + 2 \geqslant 8$

60. $-2|4 - x| \geqslant -4$

61. $3 < |2y - 1|$

62. $5 \geqslant |4 - y|$

63. $12 < \left| -2x + \dfrac{6}{7} \right| + \dfrac{3}{7}$

64. $1 < \left| y - \dfrac{11}{3} \right| + \dfrac{7}{3}$

65. $4 + \left| 3 - \dfrac{y}{3} \right| \geqslant 9$

66. $\left| 2 - \dfrac{y}{2} \right| - 1 \leqslant 1$

67. $\left| \dfrac{y+1}{2} - \dfrac{y-1}{3} \right| \leq 1$ **68.** $\left| \dfrac{4(y-1)}{3} + \dfrac{3(y-2)}{4} \right| \leq 2$

Application Problems

69. A person inherits $25,000 with the stipulation that the first year the money must be invested in two stocks paying 9% and 12% annual interest, respectively. What is the greatest amount that can be invested at 9% if the total interest for the year is to be at least $2250?

70. A person inherits $18,750 with the stipulation that the first year the money must be invested in two stocks paying 10% and 12% annual interest, respectively. What is the greatest amount that can be invested at 10% if the total interest for the year is to be at least $2117?

71. Inclusive of a 6.5% sales tax, the amount a person can spend on a car cannot exceed $17,466. What is the most expensive car that can be purchased?

72. Inclusive of a 6.5% sales tax, the amount a person can spend on a television cannot exceed $788.10. What is the most expensive television that can be purchased?

73. If the cost of producing x units of a commodity is $C = 600(2x + 4) + 600 + 600x$ and the revenue received for the x units is $R = 1850x$, describe the number of units that must be sold to achieve a profit.

74. If the cost of producing x units of a commodity is $C = 400(3x - 1) + 700$ and the revenue received for the x units is $R = 1250x$, describe the number of units that must be sold to achieve a profit.

75. Membership in a fitness club costs $500 yearly plus $1 per hour spent working out. A competing club charges $440 yearly plus $1.74 per hour. How many hours must a person work out yearly to make membership in the first club a better deal?

76. A local bank charges $8 per month plus 5¢ per check. The credit union charges $2 per month plus 8¢ per check. How many checks should be written each month to make the credit union a better deal?

77. A company that manufactures small clocks has fixed costs of $75,000 per month. It costs the company $3 to manufacture each clock. If the clocks sell for $18 each, how many should be manufactured and sold monthly to make a profit?

78. A company that manufactures graphing calculators has fixed costs of $65,000 per month. It costs the company $20 to manufacture each calculator. If the calculators sell for $85 each, how many should be manufactured and sold monthly to make a profit?

79. The figure at the top of the next column shows a square with four identical equilateral triangles attached. If the

perimeter of the figure is at least 24 centimeters and at most 36 centimeters, write an inequality to describe possible values for k.

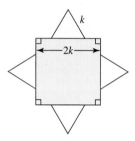

80. The figure below shows a right triangle surmounted by a trapezoid. If the combined area of the triangle and trapezoid is at most 125 square feet and at least 68 square feet, write an inequality to describe possible values for x.

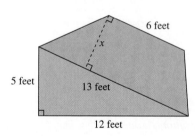

81. The daily production (x) at a refinery is described by $|x - 2,560,00| \leq 135,000$, where x represents the number of barrels of oil produced. Find the high and low production levels.

82. If a coin is tossed 100 times, we would expect approximately 50 of the outcomes to be heads. It can be demonstrated that a coin is unfair if h, the number of outcomes that result in heads, satisfies $\left| \dfrac{h - 50}{5} \right| \geq 1.645$. Describe the number of outcomes that determine an unfair coin that is tossed 100 times.

83. A pool measuring 10 meters by 20 meters is surrounded by a path of uniform width. The perimeter of the rectangle composed of the pool and path combined does not exceed 100 meters and is no less than 76 meters. Write an inequality that describes the width of the path.

84. A rectangle is surmounted by an equilateral triangle in such a way that the rectangle shares its longer side with the triangle. The shorter side of the rectangle is 3 meters less than a side of the triangle. If the perimeter of the figure composed of the rectangle and triangle combined does not exceed 54 meters and is no less than 34 meters, write an inequality that describes the length of a side of the triangle.

85. The bar graph shows the number of people in the United States with various disorders of mental illness. If x represents millions of people, what disorders are described by $11 \leqslant 3x - 4 \leqslant 56$ cases?

Mental Illness in America, by Disorder

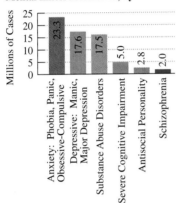

Source: National Institute of Mental Health.

86. The figure in the next column shows a bar graph representing the daily volume of the New York Stock Exchange for five days during a particular week. If x

represents the volume of stock traded, then for how many days is the inequality $|x - 280| > 40$ true, where each side of the inequality is measured in millions?

Volume of Stock Traded on New York Stock Exchange

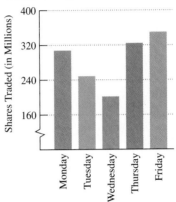

87. The percentage (p) of defective products manufactured by a company is given as $|p - 0.3\%| \leqslant 0.2\%$. If 100,000 products are manufactured and the company offers a $5 refund for each defective product, describe the company's cost of refunds.

88. Brozek's model for calculating the percent of a person's body fat (B) is $B = \dfrac{4.57}{d} - 4.142$, where d is that person's body density. For a particular person, body density is described by $|d - 1.0572| < 0.0061$. Describe the range for the percent of that person's body fat.

True–False Critical Thinking Problems _____

89. Which one of the following is true?

a. The first step in solving $|2x - 3| > -7$ is to rewrite the inequality as $2x - 3 > -7$ or $2x - 3 < -7$.

b. The smallest real number in the solution set of $2x > 6$ is 4.

c. All irrational numbers satisfy $|x - 4| > 0$.

d. None of these statements is true.

90. Which one of the following is true?

a. The first step in solving $|x| + 3 < 7$ is to rewrite the inequality as $-7 < x + 3 < 7$.

b. The greatest irrational number in the solution set of $-2x \geqslant -8$ is π.

c. If your age is 32 and the difference between your age and my age (x) is at least 12 years, then $|x - 32| \geqslant 12$ is an accurate model of the given verbal conditions.

d. None of these statements is true.

Technology Problems _____

In Problems 91–97, solve each inequality using a graphing utility.

91. $-5(x - 1) > 4x - 13$

92. $2(x + 1) > x - 3$

93. $|3 - 4x| \geqslant 13$

94. $|4(x - 1) - 5x| < 4$

95. $3x - [2 - 2(x - 1)] \leqslant 5 - 2[3x - 2(2x - 3)]$

96. $-2.5 < 0.75x + 1 < 6$

97. $(3x + 8)(9 - 2x) > 1 - 2(x - 8)(3x - 4)$

98. In solving the inequality $7(x + 4) - 13 < 12 + 13(3 + x)$, we used the graph of y_3 in which

$$y_1 = 7(x + 4) - 13$$

$$y_2 = 12 + 13(3 + x)$$

$$y_3 \boxed{=} y_1 - y_2$$

The solution $(-6, \infty)$ is found by determining where the graph of y_3 lies below the x-axis.

In Problems 101–102, solve using a graphing utility.

101. The formula for converting Celsius temperature (C) to Fahrenheit (F) is $F = \frac{9}{5}C + 32$. If Fahrenheit temperature exceeds $77°$, what does this mean in terms of Celsius temperature?

Use your graphing utility to devise another method for solving the inequality using only y_1 and y_2. Which method do you prefer?

99. Solve $2(x + 1) \geq 3 + 2x$ using a graphing utility. Then solve the inequality algebraically and explain what's happening.

100. Solve $5(x + 1) \geq 3 + 5x$ using a graphing utility. Then solve the inequality algebraically and explain what's happening.

102. The operating cost for a moving truck is $2300 per year plus 30 cents per mile. What number of miles will result in an annual cost that does not exceed $6800?

Writing in Mathematics

103. Under what conditions does $|ax + b| < c$ have one solution? no solutions? Describe why this is so.

104. Since there are conditions for which $|ax + b| < c$ has no solution, do such conditions exist for $|ax + b| > c$? Explain.

105. Suppose that you are tutoring a student in an introductory algebra course.

    a. In solving $|x + 3| > 2$, you write that $x + 3 > 2$ or $x + 3 < -2$. Your student asks, "Where did the absolute value bars go?" What is your response?

    b. In solving $|x + 3| > 2$, your student writes $-2 > x + 3 > 2$. If the inequality is then solved for x, how would you explain that the resulting inequality cannot possibly be correct?

Critical Thinking Problems

106. Solve $|x - 3| + |x + 2| < 11$ by considering three cases: $x < -2$, $-2 \leq x < 3$, and $x \geq 3$.

107. a. Why is $-|a| \leq a \leq |a|$?

    b. Use the fact that $-|a| \leq a \leq |a|$ and $-|b| \leq b \leq |b|$ adding respective members of both inequalities, to

prove that $|a + b| \leq |a| + |b|$. This result is called the *triangle inequality*.

108. If $|x - 1| < 0.2$ and $|y - 2| < 0.02$, show that $|xy - 2| \leq 0.424$. (*Hint:* First verify that $xy - 2 = x(y - 2) + 2(x - 1)$. Then use the triangle inequality proved in Problem 107.)

SECTION 1.6 # Polynomial and Rational Inequalities

Objectives

1 Solve quadratic inequalities.
2 Solve higher-order inequalities.
3 Solve applied modeling problems.
4 Solve rational inequalities.

Solutions Manual   **Tutorial**   **Video 3**

In this section, we establish the basic techniques for solving inequalities containing polynomials and rational expressions.

▍ Solve quadratic inequalities.

Solving Quadratic Inequalities

We know that a quadratic equation can be expressed in the form $ax^2 + bx + c = 0$. A similar definition describes a quadratic inequality in one variable.

Definition of a quadratic inequality

A *quadratic inequality* is any inequality that can be put in one of the forms

$$ax^2 + bx + c < 0 \qquad ax^2 + bx + c > 0$$
$$ax^2 + bx + c \leqslant 0 \qquad ax^2 + bx + c \geqslant 0$$

where a, b, and c are real numbers and $a \neq 0$.

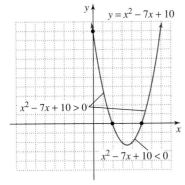

Figure 1.21

The blue portion of

$$y = x^2 - 7x + 10$$

represents

$$x^2 - 7x + 10 > 0$$

Graphing utilities are extremely useful in terms of estimating solutions of quadratic inequalities. Consider, for example,

$$x^2 - 7x + 10 > 0.$$

We can analyze this inequality geometrically by graphing $y = x^2 - 7x + 10$. The graph is drawn by hand in Figure 1.21 because we want to emphasize a particular portion of the curve, but you can obtain this graph on a graphing utility using a $[-2, 8]$ by $[-3, 7]$ viewing rectangle.

Because we are interested in the inequality $x^2 - 7x + 10 > 0$, focus on the portion of the graph of $y = x^2 - 7x + 10$ where $y > 0$ (indicated by the blue part of the curve). This occurs where the graph lies above the x-axis, namely, when $x < 2$ or $x > 5$. Thus, the solution set to $x^2 - 7x + 10 > 0$ is $(-\infty, 2) \cup (5, \infty)$.

Notice that Figure 1.21 also gives the solution set for $x^2 - 7x + 10 < 0$, which occurs when the graph lies below the x-axis, between $x = 2$ and $x = 5$. Thus $(2, 5)$, is the solution set for $x^2 - 7x + 10 < 0$.

To solve either $x^2 - 7x + 10 > 0$ or $x^2 - 7x + 10 < 0$ algebraically, we can begin by finding values of x for which $x^2 - 7x + 10 = 0$. These values, 2 and 5, are called the *critical numbers* of the inequality and divide the x-axis into three *test intervals*,

$$(-\infty, 2), (2, 5), \text{ and } (5, \infty).$$

To solve the inequality $x^2 - 7x + 10 > 0$, we need only to test one value from each of these test intervals.

This gives us a procedure for solving quadratic inequalities.

Procedure for solving quadratic inequalities

1. Express the inequality in the form

$$ax^2 + bx + c > 0 \quad \text{or} \quad ax^2 + bx + c < 0.$$

2. Solve the equation $ax^2 + bx + c = 0$. The real solutions are the *critical numbers*.

3. Locate these critical numbers on a number line, thereby dividing the number line into *test intervals*.
4. Choose one representative number for each test interval. If substituting that value into the original inequality produces a true statement, then all real numbers in the test interval belong to the solution set. The solution set is the union of all such test intervals.
5. The graph of the solution set on a number line usually appears as

This procedure is valid if $<$ is replaced by $\leq$ and $>$ is replaced by $\geq$.

EXAMPLE I   Solving a Quadratic Inequality

Solve and graph the solution set on a real number line: $2x^2 + x < 15$

Solution

$2x^2 + x < 15$	This is the given inequality.
$2x^2 + x - 15 < 0$	Write the inequality in standard form.
$2x^2 + x - 15 = 0$	Solve the related quadratic equation.
$(2x - 5)(x + 3) = 0$	Factor.
$2x - 5 = 0 \quad$ or $\quad x + 3 = 0$	Set each factor equal to 0.
$x = \frac{5}{2} \qquad\qquad x = -3$	Solve for x. The critical numbers -3 and $\frac{5}{2}$. These values are located on a number line. The intervals are $(-\infty, -3)$, $(-3, \frac{5}{2})$, and $(\frac{5}{2}, \infty)$.

Now we will take one representative number from each test interval and substitute that number into the original inequality.

Test Interval	Representative Number	Substitute into $2x^2 + x < 15$	Conclusion
$(-\infty, -3)$	-4	$2(-4)^2 + (-4) \overset{?}{<} 15$ $28 < 15$ False	$(-\infty, -3)$ does not belong to the solution set.
$\left(-3, \frac{5}{2}\right)$	0	$2 \cdot 0^2 + 0 \overset{?}{<} 15$ $0 < 15$ True	$\left(-3, \frac{5}{2}\right)$ belongs to the solution set.
$\left(\frac{5}{2}, \infty\right)$	3	$2 \cdot 3^2 + 3 \overset{?}{<} 15$ $21 < 15$ False	$\left(\frac{5}{2}, \infty\right)$ does not belong to the solution set.

Our analysis shows that the solution set is $(-3, \frac{5}{2})$, represented on a number line in Figure 1.22. ∎

Using technology

The solution set for

$2x^2 + x < 15$

or, equivalently

$2x^2 + x - 15 < 0$

can be verified with a graphing utility. The graph of $y = 2x^2 + x - 15$ was obtained using a $[-10, 10]$ by $[-16, 6]$ viewing rectangle. The graph is negative (lies below the x-axis) for x between -3 and $\frac{5}{2}$, verifying $(-3, \frac{5}{2})$ as the solution set.

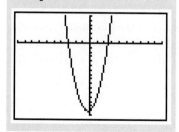

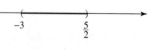

Figure 1.22

Study tip

In solving $2x^2 + x - 15 < 0$, or, equivalently, $(2x - 5)(x + 3) < 0$, an option to the use of test points is to use *sign analysis* on the product $(2x - 5)(x + 3)$.

$(2x - 5) < 0$	$(2x - 5) < 0$	$(2x - 5) > 0$
$(x + 3) < 0$	$(x + 3) > 0$	$(x + 3) > 0$

$$\xrightarrow[\quad -3 \quad\quad\quad \frac{5}{2} \quad\quad]{\bullet \quad\quad\quad \bullet}$$

If $x < -3 : (2x - 5) < 0$ and $(x + 3) < 0$

If $-3 < x < \frac{5}{2} : (2x - 5) < 0$ and $(x + 3) > 0$

If $x > \frac{5}{2} : (2x - 5) > 0$ and $(x + 3) > 0$

This analysis indicates that $(2x - 5)(x + 3)$ is negative in the interval $\left(-3, \frac{5}{2}\right)$, confirming our previous solution.

EXAMPLE 2 Solving A Quadratic Inequality

Solve and graph the solution set on a real number line: $x^2 \geqslant 4x - 2$

Solution

$x^2 \geqslant 4x - 2$	This is the given inequality.
$x^2 - 4x + 2 \geqslant 0$	Write the inequality in standard form.
$\underset{\underset{a=1}{\uparrow}}{x^2} - \underset{\underset{b=-4}{\uparrow}}{4x} + \underset{\underset{c=2}{\uparrow}}{2} = 0$	Solve the related quadratic equation.
$x = \dfrac{-b \pm \sqrt{b^2 - 4ac}}{2a}$	Use the quadratic formula with $a = 1, b = -4,$ and $c = 2$.
$= \dfrac{-(-4) \pm \sqrt{(-4)^2 - 4(1)(2)}}{2(1)}$	Substitute the given numbers.
$= \dfrac{4 \pm \sqrt{8}}{2}$	Simplify under the radical sign.
$= \dfrac{4 \pm 2\sqrt{2}}{2}$	$\sqrt{8} = \sqrt{4}\sqrt{2} = 2\sqrt{2}$
$= 2 \pm \sqrt{2}$	Divide numerator and denominator by 2. The critical numbers are $2 - \sqrt{2} \approx 0.6$ and $2 + \sqrt{2} \approx 3.4$.

Approximately 0.6   Approximately 3.4

$$\xrightarrow[\quad 2-\sqrt{2} \quad\quad 2+\sqrt{2} \quad]{}$$

Locate these values on a number line.

The intervals are $(-\infty, 2 - \sqrt{2})$, $(2 - \sqrt{2}, 2 + \sqrt{2})$, and $(2 + \sqrt{2}, \infty)$. Now we will take one representative number from each test interval and substitute that number into the original inequality.

Test Interval	Representative Number	Substitute into $x^2 \geqslant 4x - 2$	Conclusion
$(-\infty, 2 - \sqrt{2})$	0	$0^2 \overset{?}{\geqslant} 4 \cdot 0 - 2$ $0 \geqslant -2$ True	$(-\infty, 2 - \sqrt{2})$ belongs to the solution set.
$(2 - \sqrt{2}, 2 + \sqrt{2})$	1	$1^2 \overset{?}{\geqslant} 4 \cdot 1 - 2$ $1 \geqslant 2$ False	$(2 - \sqrt{2}, 2 + \sqrt{2})$ does not belong to the solution set.
$(2 + \sqrt{2}, \infty)$	4	$4^2 \overset{?}{\geqslant} 4 \cdot 4 - 2$ $16 \geqslant 14$ True	$(2 + \sqrt{2}, \infty)$ belongs to the solution set.

Our analysis shows that $(-\infty, 2 - \sqrt{2})$ and $(2 + \sqrt{2}, \infty)$ belong to the solution set. Because we were given that $x^2 \geqslant 4x - 2$, the values of x satisfying $x^2 = 4x - 2$ (equivalently, $x^2 - 4x + 2 = 0$) are also included in the solution. Thus, the solution set is $(-\infty, 2 - \sqrt{2}] \cup [2 + \sqrt{2}, \infty)$, shown on a number line in Figure 1.23.

Figure 1.23
The graph of $x \leqslant 2 - \sqrt{2}$ or $x \geqslant 2 + \sqrt{2}$ ∎

iscover for yourself

The quadratic inequalities in Examples 1 and 2 had solution sets consisting of a single interval and the union of two intervals. However, other possibilities can occur.

The graph of $y = 4x^2 - 8x + 7$ is shown in Figure 1.24. Use the graph to solve $4x^2 - 8x + 7 > 0$ and $4x^2 - 8x + 7 < 0$.

The graph of $y = x^2 - 4x + 4$ is shown in Figure 1.25. Use the graph to solve $x^2 - 4x + 4 > 0$ and $x^2 - 4x + 4 \leqslant 0$.

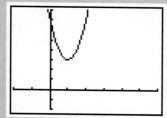 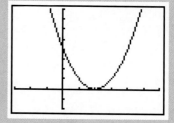

Figure 1.24
The graph of $y = 4x^2 - 8x + 7$

Figure 1.25
The graph of $y = x^2 - 4x + 4$

iscover for yourself

Try solving $4x^2 - 8x + 7 > 0$ and $4x^2 - 8x + 7 < 0$ algebraically. First solve

$$4x^2 - 8x + 7 = 0.$$

There should be no real solutions, so the test interval is the entire number line because there are no critical numbers.

Now see what happens if you solve $x^2 - 4x + 4 > 0$ and $x^2 - 4x + 4 \leqslant 0$ algebraically. First solve

$$x^2 - 4x + 4 = 0.$$

There should be only one solution, so the number line is divided into two test intervals.

How can the discriminant be useful in this exercise?

In the Discover for yourself, did you observe that some quadratic inequalities are satisfied by all real numbers, some are satisfied by no real numbers, and some are satisfied by only a single number?

For example, the graph of $y = 4x^2 - 8x + 7$ (Figure 1.24) lies above the x-axis (is positive) everywhere, so the solution set of $4x^2 - 8x + 7 > 0$ consists of the entire set of real numbers, $(-\infty, \infty)$. The graph of $y = 4x^2 - 8x + 7$ is not less than zero (does not lie below the x-axis) for any value of x, so the solution set of $4x^2 - 8x + 7 < 0$ is the empty set, $\emptyset$.

Now look at the graph of $y = x^2 - 4x + 4$ in Figure 1.25. The graph lies above the x-axis for all values of x except at $x = 2$. Thus, the solution set for $x^2 - 4x + 4 > 0$ is $(-\infty, 2) \cup (2, \infty)$. The graph of $y = x^2 - 4x + 4$ is never less than 0 and is equal to 0 only if $x = 2$. Thus, the solution set of $x^2 - 4x + 4 \leqslant 0$ consists of the single real number $\{2\}$.

Our next example involves anticipating what will appear in the viewing rectangle of a graphing utility for a given equation.

EXAMPLE 3 Anticipating the Graph of an Equation

Suppose that you decide to graph $y = \sqrt{27 - 3x^2}$ using a graphing utility. For what values of x would you anticipate seeing the graph?

Solution

We anticipate seeing the graph for all x-values that cause $\sqrt{27 - 3x^2}$ to be a real number. Because $\sqrt{27 - 3x^2}$ is a real number only if $27 - 3x^2$ is nonnegative, the graph will appear if $27 - 3x^2 \geqslant 0$.

$27 - 3x^2 \geqslant 0$	This is the inequality that answers the given question.
$27 - 3x^2 = 0$	Solve the related quadratic equation.
$27 = 3x^2$	Add $3x^2$ to both sides.
$9 = x^2$	Divide both sides by 3.
$\pm 3 = x$	Apply the square root method. Critical numbers are -3 and 3.

Locate these values on a number line.

The intervals are $(-\infty, -3)$, $(-3, 3)$, and $(3, \infty)$. We will take a representative number from each test interval and substitute that number into the original inequality.

Test Interval	Representative Number	Substitute into $27 - 3x^2 \geqslant 0$	Conclusion
$(-\infty, -3)$	-4	$27 - 3(-4)^2 \overset{?}{\geqslant} 0$ $-21 \geqslant 0$ False	$(-\infty, -3)$ does not belong to the solution set.
$(-3, 3)$	0	$27 - 3 \cdot 0^2 \overset{?}{\geqslant} 0$ $27 \geqslant 0$ True	$(-3, 3)$ belongs to the solution set.
$(3, \infty)$	4	$27 - 3 \cdot 4^2 \overset{?}{\geqslant} 0$ $-21 \geqslant 0$ False	$(3, \infty)$ does not belong to the solution set.

Because we were given that $27 - 3x^2 \geqslant 0$, values of x satisfying $27 - 3x^2 = 0$ are also included in the solution. Thus, the solution set is the *closed interval* $[-3, 3]$. The graph of $y = \sqrt{27 - 3x^2}$ shown in Figure 1.26 confirms our anticipation of seeing a graph appear for $-3 \leqslant x \leqslant 3$. (The graphing utility shows gaps when the graph appears nearly vertical.) ∎

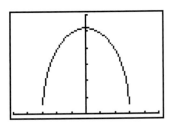

Figure 1.26

The graph of $y = \sqrt{27 - 3x^2}$ appears for $-3 \leqslant x \leqslant 3$.

2 Solve higher-order inequalities.

Solving Polynomial Inequalities of Degree Greater Than 2

Methods for solving quadratic inequalities may be applied to some higher-order polynomial inequalities.

> **EXAMPLE 4** **Solving a Third-Degree Polynomial Inequality**

Solve: $x^3 + x^2 - 4x - 4 > 0$

Solution

$x^3 + x^2 - 4x - 4 > 0$	This is the given inequality.
$x^3 + x^2 - 4x - 4 = 0$	Solve the related polynomial equation.
$x^2(x + 1) - 4(x + 1) = 0$	Factor by grouping.
$(x + 1)(x^2 - 4) = 0$	Complete the factoring.
$x + 1 = 0$ or $x^2 - 4 = 0$	Set each factor equal to 0.
$x = -1$ $x^2 = 4$	Solve for x.
$x = \pm 2$	The critical numbers are -2, -1, and 2.

Locate the critical numbers on a number line.

The intervals are $(-\infty, -2)$, $(-2, -1)$, $(-1, 2)$, and $(2, \infty)$. Once again we will take a representative number from each test interval and substitute that number into the original inequality.

Using technology

The solution of
$$x^3 + x^2 - 4x - 4 > 0$$
can be found using a graphing utility. The graph of
$$y = x^3 + x^2 - 4x - 4$$
is positive (lies above the x-axis) for $-2 < x < -1$ and for $x > 2$.

Test Interval	Representative Number	Substitute into $x^3 + x^2 - 4x - 4 > 0$	Conclusion
$(-\infty, -2)$	-3	$(-3)^3 + (-3)^2 - 4(-3) - 4 \overset{?}{>} 0$ $-10 > 0$ False	$(-\infty, -2)$ does not belong to the solution set.
$(-2, -1)$	-1.5	$(-1.5)^3 + (-1.5)^2 - 4(-1.5) - 4 \overset{?}{>} 0$ $0.875 > 0$ True	$(-2, -1)$ belongs to the solution set.
$(-1, 2)$	0	$0^3 + 0^2 - 4 \cdot 0 - 4 \overset{?}{>} 0$ $-4 > 0$ False	$(-1, 2)$ does not belong to the solution set.
$(2, \infty)$	3	$3^3 + 3^2 - 4 \cdot 3 - 4 \overset{?}{>} 0$ $20 > 0$ True	$(2, \infty)$ belongs to the solution set.

Our analysis shows that the solution set is $(-2, -1) \cup (2, \infty)$, represented on a number line in Figure 1.27.

Figure 1.27

Some polynomial inequalities cannot be easily solved by algebraic techniques but can be handled with a graphing utility.

EXAMPLE 5 **Estimating the Solution of a Polynomial Inequality with a Graphing Utility**

Solve: $x^3 - 3x - 1 > 0$

Solution

We begin by graphing $y = x^3 - 3x - 1$, as shown in Figure 1.28. We need to find values of x where the graph is positive (lies above the x-axis). Using $\boxed{\text{TRACE}}$ and $\boxed{\text{ZOOM}}$ features (shown partly for one of the three critical numbers in Figure 1.29), we see that the solution set is approximately

$$(-1.53, -0.35) \cup (1.88, \infty).$$

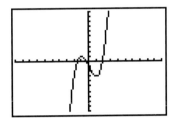

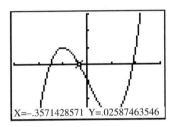

X=−.3571428571 Y=.02587463546

Figure 1.28

The graph of $y = x^3 - 3x - 1$

Figure 1.29

The graph of $y = x^3 - 3x - 1$ using the $\boxed{\text{TRACE}}$ and $\boxed{\text{ZOOM}}$ features ∎

3 Solve applied modeling problems.

Marcel Duchamp (American, born France, 1887–1968) "Nude Descending a Staircase, No. 2" 1912, oil on canvas, 58 × 35 in. Philadelphia Museum of Art: The Louise and Walter Arensberg Collection. Color transparency by Graydon Wood, 1994. © 1998 Artists Rights Society (ARS), New York/ADAGP, Paris/Estate of Marcel Duchamp.

Modeling with Polynomial Inequalities

In Problem Set 1.3 you worked with the following position model for a falling object.

Position model for a free-falling object near earth's surface

An object that is falling or vertically projected into the air has its height in feet above the ground modeled by

$$s = -16t^2 + v_0 t + s_0$$

where s is the height (in feet), v_0 is the original velocity (initial velocity) of the object (in feet per second), t is the time that the object is in motion (in seconds), and s_0 is the original height (initial height) of the object (in feet).

In Example 6 we apply the solution process for a quadratic inequality to solve a problem about the position of a free-falling object.

$t = 0$
$s_0 = 176$
$V_0 = 96$

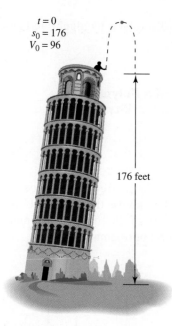

176 feet

Figure 1.30

Throwing a ball from 176 feet with a velocity of 96 feet per second

---

Using technology

The graphs of

$$y_1 = -16x^2 + 96x + 176$$

and

$$y_2 = 176$$

were obtained with a graphing utility using the following range setting:

Xmin = 0, Xmax = 8, Xscl = 1,

Ymin = 0, Ymax = 320, Yscl = 32

The graph shows that $y_1 > y_2$ (height exceeds that of the tower) for $0 < x < 6$.

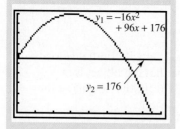

$y_1 = -16x^2 + 96x + 176$

$y_2 = 176$

---

4 Solve rational inequalities.

EXAMPLE 6 Using the Position Model

A ball is thrown vertically upward from the top of the Leaning Tower of Pisa (176 feet high) with an initial velocity of 96 feet per second (Figure 1.30). During what time period will the ball's height exceed that of the tower?

Solution

$$s = -16t^2 + v_0 t + s_0 \qquad \text{This is the position model for a free-falling object.}$$

$$s = -16t^2 + 96t + 176 \qquad \text{Since } v_0 \text{ (initial velocity)} = 96 \text{ and } s_0 \text{ (initial position)} = 176, \text{ substitute these values into the model.}$$

When will the ball's height exceed that of the tower?

$$-16t^2 + 96t + 176 \qquad > \qquad 176$$

$$-16t^2 + 96t + 176 > 176 \qquad \text{This is the inequality implied by the problem's question. We must find } t.$$

$$-16t^2 + 96t > 0 \qquad \text{Subtract 176 from both sides.}$$

$$-16t^2 + 96t = 0 \qquad \text{Solve the related quadratic equation.}$$

$$-16t(t - 6) = 0 \qquad \text{Factor.}$$

$$-16t = 0 \quad \text{or} \quad t - 6 = 0 \qquad \text{Set each factor equal to 0.}$$

$$t = 0 \qquad\qquad t = 6 \qquad \text{Solve for } t. \text{ The critical numbers are 0 and 6.}$$

Locate these values on a number line, with $t \geqslant 0$.

The intervals are $(0, 6)$ and $(6, \infty)$, although the time interval should not extend to infinity but rather to the value of t when the ball hits the ground. (By setting $-16t^2 + 96t + 176$ equal to zero, we find $t \approx 7.47$; the ball hits the ground after approximately 7.47 seconds.)

At this point, with intervals of $(0, 6)$ and $(6, 7.47)$, it seems geometrically obvious that the ball's height exceeds that of the tower during its first 6 seconds. We verify this in the usual manner.

Test Interval	Representative Number	Substitute into $-16t^2 + 96t > 0$	Conclusion
$(0, 6)$	1	$-16 \cdot 1^2 + 96 \cdot 1 \overset{?}{>} 0$ $\quad 80 > 0 \text{ True}$	$(0, 6)$ belongs to the solution set.
$(6, 7.47)$	7	$-16 \cdot 7^2 + 96 \cdot 7 \overset{?}{>} 0$ $\quad -112 > 0 \text{ False}$	$(6, 7.47)$ does not belong to the solution set.

The ball's height exceeds that of the tower between 0 and 6 seconds, excluding $t = 0$ and $t = 6$. ∎

Solving Rational Inequalities

Inequalities that involve quotients can be solved in the same manner as polynomial inequalities. For example, the inequalities

$$(x + 3)(x - 7) > 0 \quad \text{and} \quad \frac{x + 3}{x - 7} > 0$$

are very similar in that both are positive under the same conditions. Each of these inequalities must have two positive factors

$$x + 3 > 0 \quad \text{and} \quad x - 7 > 0$$

or two negative factors

$$x + 3 < 0 \quad \text{and} \quad x - 7 < 0.$$

Consequently, we can once again use critical numbers to divide the number line into test intervals and select one representative number in each interval to determine whether that interval belongs to the solution set.

EXAMPLE 7 Solving a Rational Inequality

Solve: $\dfrac{x + 1}{x + 3} < 2$

Solution

Our first step is to obtain 0 on one side of the inequality.

$\dfrac{x + 1}{x + 3} < 2$	This is the given inequality.
$\dfrac{x + 1}{x + 3} - 2 < 0$	Subtract 2 from both sides, obtaining 0 on the right.
$\dfrac{x + 1}{x + 3} - \dfrac{2(x + 3)}{x + 3} < 0$	Prepare to subtract fractions, writing the 2 as an equivalent fraction with $x + 3$ as the denominator.
$\dfrac{x + 1 - 2(x + 3)}{x + 3} < 0$	Subtract rational expressions.
$\dfrac{-x - 5}{x + 3} < 0$	Simplify the numerator.
$-x - 5 = 0 \qquad x + 3 = 0$	Set the numerator and denominator equal to 0. These are the values that make the previous quotient zero or undefined.
$x = -5 \qquad x = -3$	Solve for x. The critical numbers are -5 and -3.

Locate these values on a number line.

We've used open dots because -5 cannot be a solution (it makes $\dfrac{-x - 5}{x + 3}$ equal zero and we want solutions that make the fraction less than zero) and -3 cannot be a solution (it makes the denominator zero).

The intervals are $(-\infty, -5)$, $(-5, -3)$, and $(-3, \infty)$. Now we will take one representative number from each test interval and substitute that number into the original inequality.

Study tip

It is incorrect to begin the solution of

$\dfrac{x + 1}{x + 3} < 2$

by multiplying both sides by $x + 3$ to clear fractions. The problem with this is that because $x + 3$ contains a variable and can be positive or negative (depending on the value of x), we do not know whether or not to reverse the sense of the inequality. Instead, as in solving quadratic inequalities, begin the solution process by obtaining 0 on one side of the inequality.

Discover for yourself

Since $(x + 3)^2$ is positive, it is possible to solve

$\dfrac{x + 1}{x + 3} < 2$

by first multiplying both sides by $(x + 3)^2$ (where $x \neq -3$). This will not reverse the sense of the inequality and will clear the fraction. Try using this solution method and compare it to the one on the right.

Using technology

To solve

$$\frac{x + 1}{x + 3} < 2 \text{ or } \frac{x + 1}{x + 3} - 2 < 0$$

on your graphing utility, begin by graphing

$$y = \frac{x + 1}{x + 3} - 2.$$

Now find values of x for which $y < 0$. The values of x for which the graph lies below the x-axis are

$$x < -5 \quad \text{or} \quad x > -3.$$

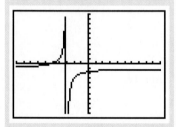

Test Interval	Representative Number	Substitute into $\frac{x + 1}{x + 3} < 2$	Conclusion
$(-\infty, -5)$	-6	$\frac{-6 + 1}{-6 + 3} \overset{?}{<} 2$ $\frac{5}{3} < 2$ True	$(-\infty, -5)$ belongs to the solution set.
$(-5, -3)$	-4	$\frac{-4 + 1}{-4 + 3} \overset{?}{<} 2$ $3 < 2$ False	$(-5, -3)$ does not belong to the solution set.
$(-3, \infty)$	0	$\frac{0 + 1}{0 + 3} \overset{?}{<} 2$ $\frac{1}{3} < 2$ True	$(-3, \infty)$ belongs to the solution set.

Our analysis indicates that the solution set is $(-\infty, -5) \cup (-3, \infty)$, shown on a number line in Figure 1.31.

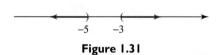

Figure 1.31

Before considering another example, let's summarize the process involved in the solution of rational inequalities.

> **Procedure for solving rational inequalities**
>
> 1. Write the inequality so that one side is 0.
> 2. If necessary, combine terms on the other side into a single quotient.
> 3. Locate points on the number line for which the quotient is zero (numbers making the numerator zero) or undefined (numbers making the denominator zero). These critical numbers will divide the number line into test intervals.
> 4. Choose one representative number for each test interval. If substituting that value into the original inequality produces a true statement, then all real numbers in the test interval belong to the solution set. The solution set is the union of all such test intervals.

EXAMPLE 8 Solving a Rational Inequality

Solve: $\dfrac{3}{x - 1} - \dfrac{2}{x + 1} \geq 1$

Solution

$$\frac{3}{x - 1} - \frac{2}{x + 1} \geq 1 \qquad \text{This is the given inequality.}$$

$$\frac{3}{x-1} - \frac{2}{x+1} - 1 \geq 0$$

Subtract 1 from both sides, obtaining 0 on the right.

$$\frac{3(x+1)}{(x-1)(x+1)} - \frac{2(x-1)}{(x-1)(x+1)} - \frac{(x-1)(x+1)}{(x-1)(x+1)} \geq 0$$

Prepare to combine terms as a single quotient. The LCD is $(x-1)(x+1)$.

$$\frac{3(x+1) - 2(x-1) - (x-1)(x+1)}{(x-1)(x+1)} \geq 0$$

Subtract as indicated in the numerator.

$$\frac{3x + 3 - 2x + 2 - x^2 + 1}{(x-1)(x+1)} \geq 0$$

Multiply.

$$\frac{-x^2 + x + 6}{(x-1)(x+1)} \geq 0$$

Simplify.

$$-x^2 + x + 6 = 0 \qquad (x-1)(x+1) = 0$$

Find critical numbers by setting numerator and denominator equal to 0.

The solution for

$$\frac{3}{x-1} - \frac{2}{x+1} - 1 \geq 0$$

can be verified on a graphing utility by graphing

$$y = \frac{3}{x-1} - \frac{2}{x+1} - 1$$

and finding where $y \geq 0$. The graph lies on or above the x-axis for $-2 \leq x < -1$ or $1 < x \leq 3$. Experiment with various range settings until this becomes relatively clear.

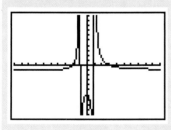

$$x^2 - x - 6 = 0 \quad x - 1 = 0 \quad \text{or} \quad x + 1 = 0$$

$$(x - 3)(x + 2) = 0$$

$$x = 3 \quad \text{or} \quad x = -2 \qquad x = 1 \quad \text{or} \qquad x = -1$$

The critical numbers are $-2, -1, 1,$ and 3.

Locate these values on a number line.

The open circles at -1 and 1 indicate that they cannot be solutions. (Why not?) The intervals are $(-\infty, -2)$, $(-2, -1)$, $(-1, 1)$, $(1, 3)$, and $(3, \infty)$. By testing a representative number from each interval, you can determine that the original inequality is satisfied in $(-2, -1)$ and $(1, 3)$. Furthermore, since the simplified form of the inequality is

$$\frac{-x^2 + x + 6}{(x-1)(x+1)} \geq 0$$

we must include values of x that cause the expression on the left to equal 0. These were the values that result in zero in the numerator ($-x^2 + x + 6 = 0$), which we found to be -2 and 3. We can conclude that the solution set of the inequality is $[-2, -1) \cup (1, 3]$, shown on a number line in Figure 1.32.

Figure 1.32

PROBLEM SET 1.6

Practice Problems

Solve each inequality in Problems 1–48 and graph the solution set on a real number line. Express each solution set in interval notation.

1. $x^2 - 8x > -15$

2. $x^2 - 3x > 10$

3. $2x^2 + 5x \leq 3$

4. $2x^2 - 7x \leq -5$

5. $x^2 + 8x + 13 \geq 0$

6. $x^2 - 2x - 17 \geq 0$

7. $2x^2 - 7x + 4 < x - 1$

8. $2x^2 - 5x + 4 < x + 1$

9. $2x^2 - 7x - 3 \leq 9 - 7x$

10. $(x - 3)(x + 1) \leq 1$

11. $x^2 + 2x + 5 > 0$

12. $x^2 - 8x + 19 > 0$

13. $2x^2 + 3x + 4 \leq 0$

14. $3x^2 + 4x + 3 \leq 0$

15. $2(x - 1) + 8 < 2(x + 3)(x - 5)$

16. $(x - 2)(x + 3) - 2 < 2(x + 2)(x - 2)$

17. $x^2 - 4x + 4 \geq 0$

18. $x^2 - 6x + 9 \geq 0$

19. $x^3 + 2x^2 - x - 2 \geq 0$

20. $x^3 + 2x^2 - 4x - 8 \geq 0$

21. $x^3 - 3x^2 - 9x + 27 < 0$

22. $x^3 + 7x^2 - x - 7 < 0$

23. $9x^3 + 8 > 4x + 18x^2$

24. $4x^3 - 2 > x - 8x^2$

25. $x^3 + x^2 + 4x + 4 > 0$

26. $x^3 - x^2 + 9x - 9 > 0$

27. $(x - 2)^3 > 0$

28. $(x - 3)^3 > 0$

29. $\dfrac{x + 3}{x - 7} > 0$

30. $\dfrac{x - 4}{x + 3} > 0$

31. $\dfrac{x + 3}{x + 4} < 0$

32. $\dfrac{x + 5}{x + 2} < 0$

33. $\dfrac{x + 3}{x - 2} \leq 2$

34. $\dfrac{x - 2}{x + 2} \leq 2$

35. $\dfrac{x + 4}{2x - 1} \geq 3$

36. $\dfrac{4x + 2}{2x - 3} \geq 2$

37. $\dfrac{3}{x + 3} > \dfrac{3}{x - 2}$

38. $\dfrac{1}{x + 1} > \dfrac{2}{x - 1}$

39. $\dfrac{3}{x - 1} - \dfrac{x}{x + 1} > 1$

40. $\dfrac{2}{x} - \dfrac{2}{x + 1} > 1$

41. $\dfrac{2x + 1}{x - 1} < 1 + \dfrac{2}{x - 3}$

42. $\dfrac{2x}{x + 5} < \dfrac{1}{5} - \dfrac{x - 1}{x - 5}$

43. $\dfrac{x}{x - 2} \geq 2 + \dfrac{3}{x + 1}$

44. $x \geq 4 + \dfrac{10}{x - 1}$

45. $\dfrac{x^2 - x - 2}{x^2 - 4x + 3} > 0$

46. $\dfrac{x^2 - 3x + 2}{x^2 - 2x - 3} > 0$

47. $|x^2 + 2x - 36| > 12$

48. $|x^2 + 6x + 1| > 8$

Application Problems

Use the position model

$$s = -16t^2 + v_0 t + s_0 \ (v_0 = \text{initial velocity}, s_0 = \text{initial position}, t = \text{time})$$

to answer Problems 49–52.

49. A projectile is fired straight upward from ground level with an initial velocity of 80 feet per second. During what interval of time will the projectile's height exceed 96 feet?

50. A projectile is fired straight upward from ground level with an initial velocity of 128 feet per second. During what interval of time will the projectile's height exceed 128 feet?

51. A ball is thrown upward with a velocity of 64 feet per second from the top edge of a building 80 feet high. For how long is the ball higher than 96 feet?

52. A diver leaps into the air at 20 feet per second from a diving board that is 10 feet above the water. For how

many seconds is the diver at least 12 feet above the water?

53. The sum of the first n natural numbers is given by

$$1 + 2 + 3 + \cdots + n = \frac{n^2 + n}{2}.$$

What values of n will produce a sum that is less than 66?

54. The total cost of producing x radios per hour is $C = \frac{1}{4}x^2 + 35x + 25$, and the price at which each radio can be sold is given by $50 - \dfrac{x}{2}$. What should be the hourly output to obtain a profit that exceeds $47 per radio?

55. The cost of removing $p\%$ of the bacteria from a river is given by the model

$$C = \frac{4p}{100 - p}$$

where C is measured in hundreds of thousands of dollars. If less than \$600,000 is spent, what percentage of the bacteria can be removed?

56. The average cost per unit $(\overline{C})$ of producing x units of a product is modeled by

$$\overline{C} = \frac{150{,}000 + 0.25x}{x}.$$

How many units must be produced so that the average cost of producing each unit does not exceed \$1.75?

57. A pool measuring 30 feet by 20 feet is to have a brick path of uniform width placed completely around it. If no more than 336 square feet of brick are to be used, describe possibilities for the path's width.

58. Identical squares are removed from the four corners of an 8-by-11-centimeter rectangular sheet of cardboard. The sides are turned up to form a box with no lid. Find the possible lengths of the sides of the removed squares if the area of the bottom does not exceed 50 square centimeters.

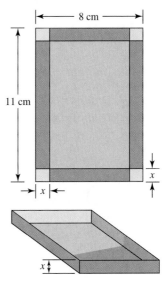

True–False Critical Thinking Problems

59. Which one of the following is true?
a. The solution set to $x^2 > 25$ is $(5, \infty)$.

b. The inequality $\dfrac{x - 2}{x + 3} < 2$ can be solved by multiplying both sides by $x + 3$, resulting in the equivalent inequality $x - 2 < 2(x + 3)$.

c. $(x + 3)(x - 1) > 0$ and $\dfrac{x + 3}{x - 1} > 0$ have the same solution set.

d. None of these statements is true.

60. Which one of the following is true?
a. The solution set to $(2x - 3)(x + 2) < 0$ is $(-\infty, -2) \cup (1.5, \infty)$.

b. Every quadratic inequality contains at least one real number in its solution set.

c. The solution set to $\dfrac{x - 3}{(x + 4)^2} > 0$ is $(3, \infty)$.

d. None of these statements is true.

Technology Problems

Use a graphing utility to estimate the solution of each inequality in Problems 61–66.

61. $12x^3 - 6x^2 - 24x + 18 > 0$

62. $x^4 - 5x^3 + 6x^2 + 4x - 8 \leqslant 0$

63. $\dfrac{x}{x + 3} > 4$

64. $\dfrac{4}{x} \leqslant x$

65. $\dfrac{x^2 - 1}{x^2 + 1} \geqslant 0$

66. $\dfrac{6}{x - 1} - \dfrac{6}{x} \geqslant 1$

67. The cost (y) of removing $x\%$ of pollutants in a lake is given by the mathematical model

$$y = \frac{80{,}000x}{100 - x}.$$

a. If less than \$320,000 is spent, what percent of the lake's pollutants can be removed? (Solve algebraically.)

b. Verify your algebraic solution by graphing the model on your graphing utility using the following range setting:

Xmin = 0, Xmax = 100, Xscl = 10, Ymin = 0,

Ymax = 1,000,000, Yscl = 100,000

c. What does the graph indicate about removing all of the lake's pollutants?

68. In a study of the winter moth in Nova Scotia, the number of eggs (N) in a female moth depended on her abdominal width (W, in millimeters), approximated by $N = 14W^3 - 17W^2 - 6W + 34$, where $1.5 \leqslant W \leqslant 3.5$. Graph the model on your graphing utility and use the TRACE and ZOOM features to describe the abdominal width of a moth with more than 46 eggs.

69. The concentration of a particular drug in the bloodstream is approximated by the model

$$C = \frac{3t}{t^2 + t + 1}$$

where C is the concentration in milligrams per liter t hours after the drug is orally administered.

a. Graph the model on your graphing utility using the following range setting:

$$\text{Xmin} = 0, \text{Xmax} = 10, \text{Xscl} = 1,$$

$$\text{Ymin} = 0, \text{Ymax} = 1, \text{Yscl} = 0.1$$

b. If the concentration exceeds 0.6 milligram per liter, the drug is effective. Use the graph to determine the time interval in which this occurs.

c. Experiment with the range settings for Xmax and Xscl, describing what happens to the concentration with increased time. From the behavior of the graph, when, if ever, is the concentration zero?

Suppose that you decide to graph each equation in Problems 70–81 using a graphing utility. For what values of x would you anticipate seeing the graph? After finding each interval algebraically, verify your result by using a utility to graph the equation.

70. $y = \sqrt{x^2 - x - 2}$

71. $y = \sqrt{x^2 - 7x + 10}$

72. $y = \dfrac{1}{\sqrt{x^2 - x - 2}}$

73. $y = \dfrac{1}{\sqrt{x^2 - 7x + 10}}$

74. $y = \sqrt[3]{x^2 - x - 2}$

75. $y = \sqrt[3]{x^2 - 7x + 10}$

76. $y = \sqrt{\dfrac{x + 1}{x - 3}}$

77. $y = \sqrt{\dfrac{x + 2}{x - 5}}$

78. $y = \dfrac{1}{\sqrt{3x^3 + 3x^2}}$

79. $y = \dfrac{1}{\sqrt{2x^4 + 6x^3}}$

80. $y = \sqrt{x^2 - x + 15}$

81. $y = \dfrac{1}{\sqrt{16x^2 + 24x + 9}}$

Writing in Mathematics

In Problems 82–85, describe how to solve each inequality by inspection. What is the solution set for each?

82. $(x - 2)^2 > 0$?

83. $\dfrac{1}{(x - 2)^2} > 0$

84. $-\dfrac{1}{(x - 2)^2} > 0$

85. $(x - 2)^2 \geqslant 0$

86. Suppose you are a math tutor and your student brings you the following problem. First solve the problem and then describe how you would explain the solution to your student.

A piece of wire 6 decimeters long is cut into two pieces and each piece is then bent to form a square. If the length of one piece is represented by x, describe the values of x that will result in the two squares having a combined area that exceeds 20 square decimeters.

Critical Thinking Problems

87. For what values of c is $x = 1$ a solution of $\dfrac{c + 2x}{c + 5x} \leqslant 1$?

88. For what values of k will $x^2 + kx + k = 0$ have two real numbers in its solution set?

89. Write a quadratic inequality whose solution set is $[-3, 5]$.

90. Write a rational inequality whose solution set is $(-\infty, -4) \cup [3, \infty)$.

91. Solve $(x - 1)^{-2} + 2(x - 1)^{-1} + 3 \geqslant 0$ by letting $t = (x - 1)^{-1}$.

92. Solve for x: $b^2x^{-2} - 2bx^{-1} + 1 > 0$.

CHAPTER PROJECT

Space-time and the Fourth Dimension

In Section 1.4, we introduced Einstein's equations relating mass, time, and length from his theory of special relativity, and looked at some interesting problems in space and time. The theory of special relativity is concerned with the motions of objects in space.

1. In your mind, create a three-dimensional coordinate system by looking at a corner of the room. Let the *xy*-plane be the floor beneath your feet, and another axis, the *z*-axis, extend up from the floor to the ceiling. Think of the positive *x*-axis as running along the bottom of the left-hand wall, the positive *y*-axis as running along the bottom of the right-hand wall and the positive *z*-axis as running up and down where the two walls meet. The origin of our coordinate system is the corner at the bottom where the walls meet the floor. Sit quietly for a few seconds. Relative to this coordinate system, did you move? Picture a coordinate system centered on the sun. Sit quietly for a few seconds. Did you move with respect to the sun-centered coordinate system? In general, when sitting perfectly still, how could you answer the question, "Are you moving?"

Einstein's theory of general relativity is concerned with the structure, or the geometry, of the universe as a whole. In order to describe the universe, Einstein discovered that he needed four variables: three variables to locate an object in space and a fourth variable describing time. This system is known as *space-time*.

2. In a random pattern, wave a pencil in the air. Try this same motion with a small flashlight in a darkened room. Concentrate on the trail in space created by the motion. How could you describe the existence of the trail you follow with your eyes? To help answer that question, move the pencil or light in a repeating figure eight pattern. Why doesn't the pencil or light ever hit itself as it moves through the pattern? To see the world in terms of space-time, walk through a mall and try to picture people walking as trails in space-time.

3. If we attempt to view the universe as points in space-time, we may ask the question, "What causes the illusion of time passing?" Some philosophers have argued that the passage of time is a myth, that each moment of our lives exists already and we are, in fact, existing in each moment simultaneously. They state it is only our memories keeping an "order" to time. There is no such thing as free will because each moment in space-time already exists and has always existed. On the other hand, every possible choice may also exist. We exist as many different "selves," each of them following a different choice. This is called the *branching universe model*.

To view a point in Einstein's space-time, we would need four coordinates to plot, three coordinates for space and one for time. Since we are three-dimensional beings, how can we imagine a four-dimensional graph? One interesting approach to visualizing four dimensions is to consider an analogy of a two-dimensional being struggling to understand three dimensions. This approach first appeared in a book called *Flatland* by Edwin Abbott written around 1884.

Flatland is a world existing on a flat, two-dimensional plane. The beings in Flatland are geometric objects such as squares and triangles. A house in Flatland would look like a blueprint or a line drawing to us. If we were to draw a closed circle around a being in Flatland, they would be imprisoned in a cell with no way to see out or escape because there is no way to move up and over the line. For a two-dimensional being moving only on a plane, the idea of up would be incomprehensible. We could explain that up means moving in a new direction, perpendicular to the two dimensions they know, but it would be similar to someone telling us we can move in the fourth dimension by traveling perpendicular to our three dimensions.

Obtain a copy of *Flatland* from your library and read this early science fiction story.

4. Imagine a Flatland square sitting in its house in a closed room. You begin speaking to the square. How would the square describe where the voice was coming from? You try to show the square the existence

of the third dimension by pushing a ball through the plane where the square sits. What would the square see as the ball is being pushed through its plane? If you pushed your hand through the plane, what would the square see? You look down from above, and tell the square what is happening in the next room. In fact, you can tell the square what is happening inside its body. There is no place in Flatland closed to you. How would you interpret this analogy with a four-dimensional being intruding in our world?

5. Using resources on the Worldwide Web or your library, discover a representation of the four-dimensional version of a cube called a hypercube or tesseract. Does this representation help to visualize the fourth dimension? Would it help to make a three-dimensional model of the hypercube to study? Share your findings with the class.

6. Imagine pushing a three-dimensional cube through a plane. Can you sketch what you would see if you could only look in two dimensions? Change the orientation of the cube slightly and repeat the push through a plane. How do your new sketches differ from the first set? Would a two-dimensional being understand a cube by looking at the different sets of sketches? Contrast this exercise using a cube with one using a sphere.

These two different approaches to understanding the fourth dimension, one where it is time, the other where it is another spatial dimension, are not incompatible. Time may always be viewed as one more dimension added on, no matter how many spatial dimensions exist. In fact, some modern theories now use as many as ten dimensions to explain the structure of the universe.

Worldwide Web Resources

Go to the Prentice Hall website (http://www.prenhall.com/blitzer) to access other locations on the Internet that will allow you to further explore the concepts presented in this project.

Chapter Review

SUMMARY

1. **Solving Linear Equations**

   a. If necessary, clear fractions, multiplying both sides by the LCD.

   b. Simplify each side.

   c. Add (or subtract) the same real number or variable expression on both sides to collect all terms with the variable on one side and all other terms on the other side.

   d. Multiply (or divide) both sides of the equation by the same nonzero quantity and solve for the variable.

   e. Check the proposed solution by substituting or with a graphing utility.

2. **Solving Linear Equations Involving Absolute Value**

   a. If c is positive, rewrite $|ax + b| = c$ as $ax + b = c$ or $ax + b = -c$ and solve each of the resulting equations for x.

   b. If $|ax + b| = |cx + d|$, rewrite as $ax + b = cx + d$ or $ax + b = -(cx + d)$ and solve each of the resulting equations for x.

3. **Solving Formulas and Linear Mathematical Models for a Specified Variable**

   These equations often contain more than one variable. Follow the procedure for solving linear equations. If the terms containing the specified variable cannot be combined as like terms, factor the variable from these terms. Then solve for the specified variable by dividing both sides of the equation by the factor that appears with the specified variable.

4. **Solving Quadratic Equations by Factoring**

   a. The standard form of a quadratic equation is
   $$ax^2 + bx + c = 0 \qquad (a \neq 0).$$

   b. Some quadratic equations can be solved by factoring.

1. Write the equation in standard form.

2. Factor $ax^2 + bx + c$.

3. Apply the zero product principle, setting each factor equal to 0.

4. Solve the equations in step 3.

5. Check proposed solutions by substitution or with a graphing utility.

5. **Solving Quadratic Equations by the Square Root Method**

   a. The square root method states that if $u^2 = d$, then $u = \pm\sqrt{d}$, where u is an algebraic expression.

   b. The square root method works well on quadratic equations in the form $ax^2 + c = 0$. Solve for x^2 and then apply the method.

   c. The method works well on equations of the form $(ax + b)^2 = d$, resulting in $ax + b = \pm\sqrt{d}$.

6. **Solving Quadratic Equations by Completing the Square**

   a. Write the equation in the form $ax^2 + bx = c$.

   b. If $a \neq 1$, divide both sides by a.

   c. Complete the square by adding the square of half the coefficient of the x-term to both sides.

   d. Factor the left side and simplify the right side, if possible.

   e. Apply the square root method and solve.

7. **Solving Quadratic Equations by the Quadratic Formula**
   The formula is derived by completing the square.

   If $ax^2 + bx + c = 0$, then $x = \dfrac{-b \pm \sqrt{b^2 - 4ac}}{2a}$.

8. **The Discriminant**

   The discriminant, $b^2 - 4ac$, determines the nature of the solutions to a quadratic equation.

   a. If $b^2 - 4ac$ positive, the equation has two real solutions.

   b. If $b^2 - 4ac = 0$, the equation has one rational solution.

   c. If $b^2 - 4ac < 0$, the solutions are not real numbers. There are two nonreal complex solutions.

9. **A Problem-Solving Strategy**

   1. Read the problem and let x represent one unknown quantity.

   2. If necessary, represent other unknowns in terms of x.

   3. Form a verbal model of the problem's conditions and then write an equation in x that translates the verbal model.

   4. Solve the equation and answer the problem's question.

   5. Check the answer using the verbal conditions given in the problem.

10. **Polynomial Equations**

    a. Some polynomial equations of degree greater than 2 can be solved by factoring.

    b. Real solutions to an nth-degree polynomial equation can be found (or approximated) by using a graphing utility. The solutions correspond to the x-intercepts.

    c. Some polynomial equations are quadratic in form in that they can be written in the form $at^2 + bt + c = 0$ where t is a polynomial and $a \neq 0$.

11. **Solving Equations That Are Quadratic in Form**
    Introduce the substitution $t = x^{\frac{n}{2}}$ in the equation $ax^n + bx^{\frac{n}{2}} + c = 0$. Solve $at^2 + bt + c = 0$ for t. Find x by substituting the value(s) of t into $x^{\frac{n}{2}} = t$.

12. **Solving Equations Involving Radicals**

    a. Isolate one of the radical terms on one side of the equation.

    b. Raise both sides of the equation to a power that is the same number as the index of the radical.

    c. Solve the resulting equation.

    d. Substitute all proposed solutions in the original radical equation, eliminating solutions that do not check.

13. **Solving Equations in the Form $x^{\frac{m}{n}} = k$**

    a. Raise both sides to the $\dfrac{n}{m}$ power.

    b. If m is even: $x = \pm k^{\frac{n}{m}}$
       If m is odd: $x = k^{\frac{n}{m}}$

14. **Solving Linear Inequalities**
    Use the procedure for solving linear equations. However, when multiplying or dividing by a negative number, reverse the sense of the inequality.

15. **Solving Linear Inequalities Involving Absolute Value**
    If c is positive:

    a. $|ax + b| < c$ means $-c < ax + b < c$.

    b. $|ax + b| > c$ means $ax + b > c$ or $ax + b < -c$.
       These rules are valid if $<$ is replaced by $\leqslant$ and $>$ is replaced by $\geqslant$.

16. **Solving Quadratic Inequalities**

    a. Express the inequality in the form $ax^2 + bx + c > 0$ or $ax^2 + bx + c < 0$

    b. Solve the equation $ax^2 + bx + c = 0$. The real solutions are *critical numbers*.

    c. Locate these critical numbers on a number line, thereby dividing the number line into test intervals.

    d. Choose one representative number for each test interval. If substituting that value into the original inequality produces a true statement, then all real numbers in the test interval belong to the solution set.

    e. The solution set is the union of all such test intervals.

17. Solving Rational Inequalities

a. Write the inequality so that one side is 0.

b. If necessary, combine terms on the other side into a single quotient.

c. Locate points on a number line for which the quotient is zero (numbers making the numerator zero) or undefined (numbers making the denominator

zero). These critical numbers will divide the number line into test intervals.

d. Choose one representative number for each test interval. If substituting that value into the original inequality produces a true statement, then all real numbers in the test interval belong to the solution set. The solution set is the union of all such test intervals.

REVIEW PROBLEMS

Find the solution set for each equation in Problems 1–31. Check your answer by substitution or with a graphing utility.

1. $x - 4(2x - 7) = 3(x + 6)$

2. $\frac{1}{2}(3x + 1) = \frac{1}{3}(3 - x) + \frac{1}{6}(7x + 3)$

3. $3 - [2x - (2x - 1)] = 5x - [6x + 4(5 - x)]$

4. $\frac{9}{4} - \frac{1}{2x} = \frac{4}{x}$

5. $\frac{7}{x - 5} + 2 = \frac{x + 2}{x - 5}$

6. $\frac{1}{x - 1} - \frac{1}{x + 1} = \frac{2}{x^2 - 1}$

7. $\frac{1}{x^2 - x - 2} = \frac{1}{x^2 - 3x + 2} - \frac{2}{x^2 - 1}$

8. $|3x - 2| = 4$

9. $\left|1 - \frac{3}{4}x\right| + 4 = 7$

10. $|3x - 5| = |5x + 3|$

11. $x(x - 3) - 12 = 2x(x + 2)$

12. $\frac{12}{x} + 1 = \frac{12}{x - 1}$

13. $3x^2 - 6x = 0$

14. $6x^2 - 5 = 0$

15. $(3x - 6)^2 = 27$

16. $2x^2 - 2x + 2 = 7$

17. $\frac{2x}{x + 2} = 1 - \frac{3}{x + 4}$

18. $\frac{x}{2} + \frac{x^2}{2} = \frac{-3}{4}$

19. $2x^4 = 50x^2$

20. $2x^3 - x^2 - 18x + 9 = 0$

21. $x^4 - 5x^2 + 4 = 0$

22. $(x^2 + 2x)^2 = 5(x^2 + 2x) - 6$

23. $5 + \sqrt{5x - 1} = x$

24. $\sqrt{5x} - \sqrt{2x - 1} = 2$

25. $\sqrt{x + 9} - \sqrt{x + 16} + \sqrt{x + 1} = 0$

26. $x^{\frac{2}{3}} - 1 = 8$

27. $3x^{\frac{3}{4}} - 24 = 0$

28. $(x^2 - x - 4)^{\frac{3}{4}} - 2 = 6$

29. $x^{\frac{1}{2}} + 3x^{\frac{1}{4}} - 10 = 0$

30. $3x^{\frac{2}{3}} - 5x^{\frac{1}{3}} + 2 = 0$

31. $\left(x + \frac{12}{x}\right)^2 - 15\left(x + \frac{12}{x}\right) + 56 = 0$

Solve each inequality in Problems 32–47, expressing the solution set in interval notation. Graph the solution set on a real number line. Verify your solution sets with a graphing utility.

32. $3 - (4 - x) \geq 3x - (x + 5)$

33. $4(x - 2)(x + 3) < (2x + 3)(2x - 3)$

34. $\frac{x}{3} - \frac{3}{4} - 1 > \frac{x}{2}$

35. $7 < 2x + 3 \leq 9$

36. $1 \leq 1 - (5x - 2) < 11$

37. $|3(x - 1) + 2| < 20$

38. $\left|1 - \frac{4x}{3}\right| \geq 5$

39. $|3 - 2x| - 8 < 4$

40. $2x^2 + 7x \leq 4$

41. $2x^2 > 6x - 3$

42. $x^2 + 3x + 8 \leq 0$

43. $x^2 + 2x > -6$

44. $x^3 + 2x^2 - 9x - 18 > 0$

45. $x^4 + 4x^3 + 3x^2 < 0$

46. $\frac{x + 3}{x - 4} < 5$

47. $\frac{x}{x + 1} - \frac{2}{x + 3} \leq 1$

In Problems 48–49, solve each formula for the specified variable.

48. $A = 2LW + 2WH + 2LH$, for H

49. $R_T = \frac{R_1 R_2}{R_1 + R_2}$, for R_1

50. Chamberlain's model describes the number of years (N) that you should drive your car before purchasing a new one. The formula states

$$N = \frac{GMC}{(G - M)DP}$$

where G = new car's gas mileage, M = your present car's gas mileage, C = the cost of the new car, D = number of miles driven yearly, and P = price of gasoline per gallon.

a. Solve the formula for G.

b. Use part (a) to write and solve a problem involving gas mileage that should be expected from a new car. You'll need to provide values for each variable, except G, in the model.

51. The models $C = 1.24t + 313.6$ and $C = 0.018t^2 + 0.70t + 316.2$ both describe carbon dioxide concentration (C,

in ppm) t years after 1960. Use each model to predict in what year carbon dioxide concentration will be 400 ppm.

52. The model

$$d = x + \frac{x^2}{20}$$

describes the distance (d, in feet) that it takes for a car to come to a complete stop once the brakes are applied, where x is the car's speed in miles per hour. If a car travels 175 feet once the brakes are suddenly applied, what was its speed prior to the brakes being hit?

Use the position model $s = -16t^2 + v_0t + s_0$, where v_0 represents initial velocity, s_0 represents initial position, and t represents time to answer Problems 53–54.

53. A diver jumps directly upward from a diving board that is 32 feet above the water with an initial velocity of 16 feet per second. How long does it take for the diver to hit the water?

54. A person standing close to the edge of an 80-foot cliff throws a rock directly upward with an initial velocity of 64 feet per second.

a. Write the formula that models the position of the rock above the ground at any time t.

b. Use the discriminant to determine if the rock will ever reach a height of 150 feet.

c. Verify part (b) using a graphing utility.

55. Costs for two different kinds of heating systems for a three-bedroom house are given below.

System	Cost to Install	Operating Cost per Year
Electric	$5000	$1100
Gas	$12,000	$700

After how long will total costs for electric heating and gas heating be the same? What will be the cost at that time?

56. A person purchasing a house is considering two options for a $40,000 mortgage:

Option 1:
8.5% interest on
a 30-year loan
with no points

Option 2:
8% interest on
a 30-year loan
with 3 points

Use the table on page 141 to determine how long it will take for the total cost of each mortgage to be the same.

57. A two-year union contract guarantees an immediate 4% raise and a 7% raise after a year. How much must an

employee be earning at the start of the contract if that employee's weekly salary increases by $45.12 over the two-year life of the contract?

58. An owner of a large diving boat that can carry as many as 50 people charges $8 per passenger to groups of between 5 and 10 people. If more than 10 divers charter the boat, the fee per passenger is decreased by 20 cents times the number of people over 10. How many people must there be in a diving group to generate an income of $117.80?

59. a. In 1993, there were 42.3 million Americans who were never married. This is a 98% increase over the number for 1970. How many Americans in 1970 were never married?

b. The number of married Americans in 1993 is 170% greater than the number of never-married Americans for the same year. How many married Americans were there in 1993?

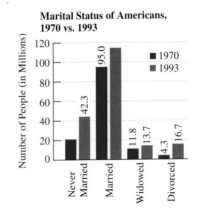

Marital Status of Americans, 1970 vs. 1993

Source: U.S. Bureau of the Census

Revenue for a company is the money received for the sale of x units of a product, and is found by multiplying the number of units sold by the price per unit. A company's profit is the difference between its revenue and its cost of producing x units of a product. Use these ideas to answer Problem 60.

60. A publishing company spends $10,000 to set up the press for publishing a book that costs the company $8 for each book printed. They find that they can sell x books at a price of $20 - \dfrac{x}{1000}$ dollars per book. How many books should be printed and sold to generate a profit of $26,000?

61. A billboard is 15 feet longer than it is high and has space for 324 square feet of advertising. What are the billboard's dimensions?

62. A swimming pool has a length that is $2\frac{1}{2}$ times longer than its width. The pool is surrounded by a brick path that has a uniform width of 5 feet. The path requires 1150 square feet of brick. Determine the pool's dimensions.

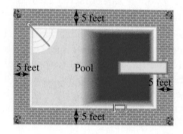

63. A rectangular field is to be fenced on three sides with 2400 feet of fencing. As shown in the figure, the fourth side is a straight river's edge that will not be fenced.

a. If the area of the enclosure is 720,000 square feet, find the field's dimensions.

b. Use a graphing utility to graph the area model obtained in part (a). Experiment with the range setting until your graph resembles the one shown in the figure at the top of the next column. Does the rectangle enclosed in part (a) make the best use of the 2400 feet of fencing? Explain.

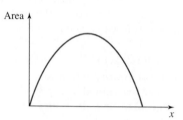

64. An open box is made from a square piece of cardboard by cutting 3-inch squares from each corner and turning up the sides. If the volume of the box is 108 cubic inches, what are the dimensions of the original piece of cardboard?

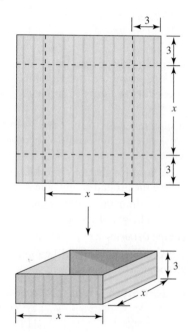

65. The figure shows a triangle drawn inside a square. If the area of the shaded region is 368.5 square centimeters, find the dimensions of the square. Be sure to check your value of x with the given condition about the area of the shaded region.

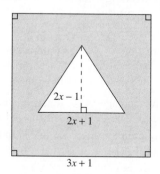

66. If the area of the rectangular base of the solid shown in the figure is 40 square inches, find the volume of the solid.

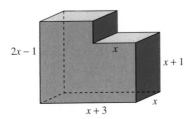

67. A water tank has the shape of an inverted right circular cone of altitude 8 feet and base radius 4 feet. How much water is in the tank when the water is 6 feet deep?

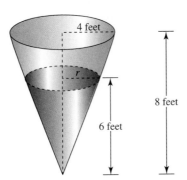

68. Use Einstein's model

$$L = L_0 \sqrt{1 - \frac{v^2}{c^2}}$$

which relates length at rest (L_0) to length (L) at velocity v, to answer this question. (The speed of light, c, is approximately 186,000 miles per second.) How fast would a futuristic starship that is 500 meters tall have to move so that its length would be diminished by 80% from the perspective of a stationary observer who looks up and sees the ship?

69. Use Einstein's model

$$T = \frac{T_0}{\sqrt{1 - \frac{v^2}{c^2}}}$$

which relates time at rest (T_0) to time (T) at velocity v, to answer this question. A clock on board a futuristic starship traveling at 80% of the speed of light measures a 20-second time period. How long is this time period from the perspective of a stationary observer who looks up and sees the ship?

70. The demand equation for a product is $p = 60 - \sqrt{0.01x + 1}$, where x is the number of units demanded per day and p is the price per unit. Find the demand if the price is set at $57.00.

71. A person rows in a straight line across a 5-mile wide channel from point P to point C, and then walks from point C to point B, as shown in the figure. The distance from A to B is 6 miles. The rowing rate is 2 miles per hour, and the walking rate is 6 miles per hour.

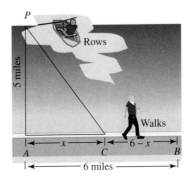

a. Since

$$\text{Time} = \frac{\text{Distance}}{\text{Rate}}$$

we can use the figure and the given information to set up a model for the total time spent rowing and walking. Refer to the figure and fill in the missing portions in the modeling process shown below.

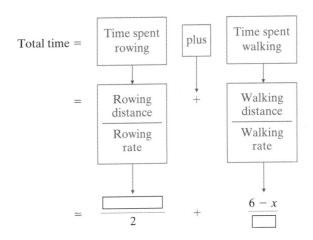

b. If the total time spent rowing and walking is $\frac{11}{3}$ hours (3 hours and 40 minutes), use your total time model from part (a) to find x. How far from point B does this person come ashore?

72. The remaining life expectancies shown in the graph on page 208 indicate that a male child born in 1993 could expect to live to age 72.1; a female to age 78.9. A 70-year old male in 1993 could expect to live another 12.4 years; a female, another 15.5 years. The formula $E = \sqrt{0.66A^2 - 110.55A + 4680.24}$ is an approximate model for some of the data in the graph, where A represents current age and E stands for remaining life expectancy (in years).

a. Is the formula a better model for males or for females?

b. If $E = 10$, find A and describe what your result means in terms of the variables modeled by the formula.

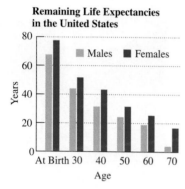

Remaining Life Expectancies in the United States

Source: Department of Health and Human Services, 1993

73. A person inherits $10,000 with the stipulation that for the first year the money must be invested in two stocks paying 8% and 12% annual interest, respectively. What is the greatest amount that can be invested at 8% if the total interest for the year is to be at least $950?

74. In 1984, approximately 1644 thousand turntables were sold in the United States. This number has decreased by around 82 thousand each year since then. On the other hand, annual sales of compact disc players has increased by 496 thousand each year, with approximately 284 thousand units sold in 1984. What was the first year in which the sales of compact disc players exceeded those of turntables?

75. A projectile is fired vertically upward from ground level with an initial velocity of 48 feet per second. During what time period will the projectile's height exceed 32 feet? (Use the position model described prior to Problem 53 on page 205.)

76. Suppose that you decide to graph $y = \sqrt{x^2 + 7x + 10}$ using a graphing utility. For what values of x would you anticipate seeing the graph? Verify this algebraic analysis by graphing the equation.

The graph indicates that the U.S. has the world's highest incarceration rate. If x represents the incarceration rate per 100,000 population, list the country or countries that satisfy each inequality in Problems 77–78.

77. $437 \le 4x - 7 \le 1229$

78. $|x - 320| > 80$

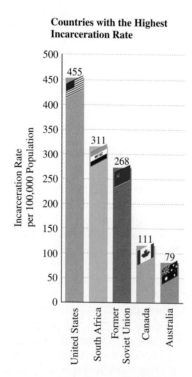

Countries with the Highest Incarceration Rate

Source: FBI Uniform Crime Report

CHAPTER I TEST

In Problems 1–12, find each solution set and check by substitution or with a graphing utility.

1. $7(x - 2) = 4(x + 1) - 21$

2. $|2x + 1| = |x + 3|$

3. $2x^2 - 3x - 2 = 0$

4. $(3x - 1)^2 = 75$

5. $x(x - 2) = 4$

6. $\dfrac{x - 1}{2} + \dfrac{2x}{x + 1} = 0$

7. $4x^2 = 8x - 5$

8. $x^3 - 4x^2 - x + 4 = 0$

9. $(x^2 - 3x)^2 - 2(x^2 - 3x) - 8 = 0$

10. $2\sqrt{x + 2} - \sqrt{2x + 3} = 3$

11. $x^{1/3} - 9x^{1/6} + 8 = 0$

12. $5x^{3/2} - 10 = 0$

Solve each inequality in Problems 13–17. Express the answer in interval notation and graph the solution on a real number line.

13. $x - 4[1 - (2x + 1)] \le 3x - 2$

14. $-3 \le \dfrac{2x + 5}{3} < 6$

15. $|3x + 2| \ge 3$

16. $x^2 < x + 12$

17. $\dfrac{2x + 1}{x - 3} > 3$

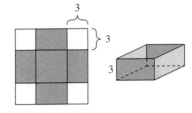

18. The monthly benefit (B) of a retirement plan is given by

$$B = \frac{2}{5}w + \frac{1}{125}n$$

where

$w =$ an employee's average monthly salary

$n =$ the number of years an employee worked for the company

  a. Solve the formula for n.

  b. Find the number of years an employee whose average monthly salary is \$800 must work to receive a monthly benefit of \$512.

19. The life expectancy for a person born in the United States in 1950 is 65 years for a man and 71 years for a woman. This has increased by approximately 0.16 year for each year of birth after 1950 for a man and by 0.2 year for a woman. In what year of birth will a woman's life expectancy exceed a man's by 10 years?

20. A computer online service charges a flat monthly rate of \$25 or a rate of \$10 plus 25 cents for each hour spent online. How many hours online each month will result in the same charge for the two billing options?

21. With a 9% raise, a physical therapist will earn \$34,880 annually. What is the therapist's salary prior to this raise?

22. An open-top box is made from a square piece of cardboard by cutting 3-inch squares from each corner and then folding up the sides. (See the figure at the top of the next column.) If the box is to hold 243 cubic inches, what size cardboard should be used?

23. An L-shaped sidewalk connects two apartment buildings. The path connecting the buildings is 340 meters, although most residents cut diagonally across the lawn, a distance of 260 meters. How long are the lengths of the two legs of the existing path?

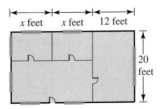

24. A water tank has the shape of an inverted right circular cone of altitude 8 feet and base radius 4 feet. How much water is in the tank when the water is 3 feet deep?

25. The floor plan shown is to have an area of at least 1000 square feet and at most 1160 square feet. Solve for x and then write a sentence that describes possible lengths x of the two rooms on the upper left shown in the figure.

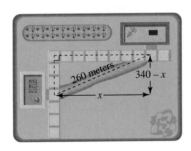

Functions and Graphs

A magnification of the Mandelbrot set. R. F. Voss "29-Fold M-set Seahorse" computer-generated image. © 1990 R. F. Voss / IBM Research.

"What she's doing is, every time she works out a value for y, she's using *that* as her next value for x. And so on. Like a feedback. She's feeding the solution back into the equation, and then solving it again. Iteration, you see."

—Tom Stoppard, *Arcadia*

In Tom Stoppard's remarkable play *Arcadia,* the characters talk about mathematics from Newton to chaos theory; the clash between order and Romantic ardor; and the beauty of symmetry versus the thrill of all that is tangled, mysterious, and unpredictable.

The idea of applying a recursive process, a sequence of computational steps in which the result from the previous step serves as the input for the next step, is one of the topics we will be studying in this chapter on functions. The process is known as iteration and the successive values produced are called iterates. By using a function to define a recursive process to all numbers $a + bi$ in the complex plane, hauntingly beautiful visual images emerge. Reflecting the theme of self-similarity in nature, magnified portions of these images yield repetition of the original overall structure as well as new and unexpected patterns. The most famous of these images, the Mandelbrot set, transforms abstract mathematics into a tangible source of wonder and inspiration, serving as a symbol of the mathematical revolution known as chaos theory.

SECTION 2.1

Introduction to Functions

Solutions Manual

Tutorial

Video 4

Objectives

1 Find the domain and range of a relation.
2 Determine whether a relation is a function.
3 Use $f(x)$ notation.
4 Interpret function values for mathematical models.
5 Find function values for piecewise functions.
6 Evaluate a function.
7 Find the domain of a function.

In this section, we introduce functions and their notation. The mathematical models from Chapter 1 will be equivalently expressed using functional notation. We will be using the function concept throughout the remainder of the book.

1 Find the domain and range of a relation.

Relations

The man who opened the gateway to higher mathematics, René Descartes, did so because his rectangular coordinate system made it possible to construct graphs for sets of ordered pairs. The mathematical term for a set of ordered pairs is a *relation*.

> **Definition of a relation**
>
> A *relation* is any set of ordered pairs. The set of all first components of the ordered pairs is called the *domain* of the relation, and the set of all second components is called the *range* of the relation.

EXAMPLE 1 **Analyzing Atmospheric Carbon Dioxide as a Relation**

The first component of the ordered pairs given below represents the number of years after 1939. The second component represents carbon dioxide concentration in parts per million (ppm). Find the domain and range of this relation.

$$\{(0, 280), (1, 281.44), (2, 282.88), (3, 284.32), (4, 285.76)\}$$

Solution

The domain is the set of all first components, and the range is the set of all second components. Thus, the domain is $\{0, 1, 2, 3, 4\}$ and the range is $\{280, 281.44, 282.88, 284.32, 285.76\}$. ∎

The relation in Example 1, represented by a set of ordered pairs, can also be represented by the following.

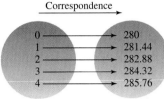

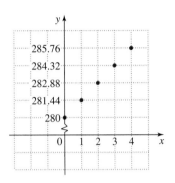

Figure 2.1

The graph of a relation

1. A table of coordinates

x	0	1	2	3	4
y	280	281.44	282.88	284.32	285.76

2. A graph (see Figure 2.1)
3. An equation: $y = 1.44x + 280$, for $0 \leq x \leq 4$, and x is a whole number
4. A rule: For every whole number from 0 to 4 inclusively, multiply by 1.44 and add 280 to obtain its corresponding value.

In Example 1, each member of the domain corresponds to exactly one member of the range. This is not the case in our next example.

EXAMPLE 2 **Analyzing a Relation**

Find the domain and range of:

$$\{(0, 0), (1, 1), (1, -1), (4, 2), (4, -2), (9, 3), (9, -3)\}$$

Solution

The domain is the set of all first components, and the range is the set of all second components.

Domain: $\{\ 0,\quad 1,\qquad\quad 4,\qquad\quad 9\qquad\quad\}$

Relation: $\{(0, 0),\ (1, 1),\ (1, -1),\ (4, 2),\ (4, -2),\ (9, 3),\ (9, -3)\}$

Range: $\{\ 0\ ,\quad 1\ ,\quad -1\ ,\quad 2\ ,\quad -2\ ,\quad 3\ ,\quad -3\ \}$

The range can equivalently be written as $\{-3, -2, -1, 0, 1, 2, 3\}$. As shown in the margin, three members of the domain each correspond to more than one member of the range. ∎

The relation in Example 2, represented by a set of ordered pairs, can also be represented by the following.

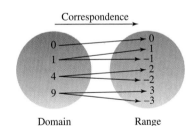

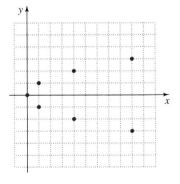

Figure 2.2

The graph of a relation

1. A table of coordinates

x	0	1	1	4	4	9	9
y	0	1	-1	2	-2	3	-3

2. A graph (see Figure 2.2)
3. A rule: For the integers 0, 1, 4, and 9, assign to each its positive and negative square root.

$$0 \to \pm\sqrt{0} = 0 \quad 1 \to \pm\sqrt{1} = \pm 1 \quad 4 \to \pm\sqrt{4} = \pm 2 \quad 9 \to \pm\sqrt{9} = \pm 3$$

4. An equation: $y = \pm\sqrt{x}$, for $x = 0, 1, 4, 9$

2 Determine whether a relation is a function.

same first
components

$(1, 1)$ and $(1, -1)$

different second
components

Not ordered pairs of a function

Functions

A relation in which each member of the domain corresponds to exactly one member of the range is called a *function*. The relation in Example 1 is a function. However, the relation in Example 2 is not a function because three members of the domain correspond to more than one member of the range. Equivalently, a function is a relation in which no two ordered pairs have the same first component and different second components. The ordered pairs of Example 2, $(1, 1)$ and $(1, -1)$, or $(4, 2)$ and $(4, -2)$, are not ordered pairs of a function.

Definition of a function

A *function* is a correspondence between a first set, called the *domain*, and a second set, called the *range*, such that each element in the domain corresponds to *exactly one* element in the range.

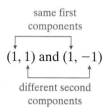

Study tip

It is helpful to think of the *domain* as the set of a function's *inputs* and the *range* as the set of a function's *outputs*.

Alice Neel (1900–1984) "The Soyer Brothers" 1973, oil on canvas, 60 × 46 in. (152.4 × 116.8 cm.) Purchase, with funds from Arthur M. Bullowa, Sydney Duffy, Stewart R. Mott and Edward Rosenthal. 74.77. Photograph © 1997: Whitney Museum of American Art, New York. Photography by Geoffrey Clements.

3 Use $f(x)$ notation.

EXAMPLE 3   **Determining Whether a Correspondence is a Function**

Determine whether each correspondence is a function.

Domain	**Correspondence**	**Range**
a. The set of all states	Each state's population at a given point in time	A set of positive numbers
b. The set of real numbers	Each number's cube	The set of real numbers
c. The set of people in a town with uncles	Each person's uncle	A set of men

Solution

a. The correspondence is a function. At a given point in time, each state has exactly one population.

b. The correspondence is a function. Each real number has only one cube. We can express this function by an equation: $y = x^3$.

c. The correspondence is not a function. Some people have more than one uncle.   ■

Using Function Notation

Functions are usually given in terms of equations, rather than as sets of ordered pairs. These equations are expressed in a special notation. You can understand this notation if you think of a function as a machine into which you put members of the domain. The machine is programmed with a rule or equation for the correspondence between inputs and outputs. Consequently, the machine gives you a member of the range (the output).

Input x

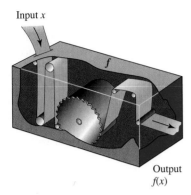

Output
f(x)

Figure 2.3

A function as a machine with inputs and outputs

tudy tip

The notation f(x) does *not* mean "f times x." The notation describes the value of the function at x.

As shown in Figure 2.3, the letter f is frequently used to name functions. The input is represented by x and the output by $f(x)$. The special notation $f(x)$, read "f of x" or "f at x" represents an expression for the value of the function at x.

Functions that are given as equations use the $f(x)$ notation. For example,

$$f(x) = 1.44x + 280$$

describes the function f that takes an input x, multiplies it by 1.44, and then adds 280.

input

$$f(x) = 1.44x + 280 \qquad \text{We read this as } f \text{ of } x \text{ equals } 1.44x + 280.$$

Multiply   Then
the input   add
by 1.44.   280.

For example, to find $f(3)$ (f of 3), we take the input 3, multiply it by 1.44, and add 280 to get 284.32. Here's how it looks in function notation:

$$f(x) = 1.44x + 280 \qquad \text{This is the given function.}$$

$$f(3) = 1.44 \cdot 3 + 280 \qquad \text{To find } f(3), \text{ substitute 3 for } x. \text{ The input is 3.}$$

$$= 284.32$$

Since f of 3 is 284.32, when the machine's input is 3, its output is 284.32. When the domain of the function has the value 3, the corresponding value for the range is 284.32. Since this function models carbon dioxide concentration x years after 1939, this means that 3 years after 1939, in 1942, carbon dioxide concentration was 284.32 ppm.

The function $f(x) = 1.44x + 280$ can also be expressed as $y = 1.44x + 280$, replacing $f(x)$ by y. The value of y, the *dependent variable*, is calculated after selecting a value for x, the *independent variable*.

The notation involving $f(x)$ is more compact than the notation with y. Here's an example:

a. If $f(x) = 1.44x + 280$, then $f(3) = 284.32$.
b. If $y = 1.44x + 280$, then the value of y is 284.32 when x is 3.

4 Interpret function values for mathematical models.

Modeling with Functions

Table 2.1 at the top of the next page presents examples of mathematical models expressed in the special notation for functions. Throughout this text, we will use functions that model the physical world.

iscover for yourself

As you study Table 2.1, cover the last column and describe what the result in the previous column means. Then uncover the column and see if your description is accurate.

TABLE 2.1 Mathematical Models in Function Notation

Mathematical Model and What it Describes	The Model in Function Notation	Finding an Indicated Function Value	What the Function Value Means
$y = 553.37x + 27{,}966$ Teachers' annual salaries x years after 1985	$f(x) = 553.37x + 27{,}966$	$\begin{aligned} f(8) &= 553.37(8) + 27{,}966 \\ &= 32{,}392.96 \end{aligned}$	Eight years after 1985 (in 1993) teachers' average annual salary in the U.S. was approximately $32,393.
$M = 0.0075t^2 - 0.2676t + 14.8$ An estimation of fuel efficiency of passenger cars (M, measured in miles per gallon) t years after 1940	$M(t) = 0.0075t^2 - 0.2676t + 14.8$ or $M(x) = 0.0075x^2 - 0.2676x + 14.8$ or $f(x) = 0.0075x^2 - 0.2676x + 14.8$	$\begin{aligned} M(17) &= 0.0075(17)^2 - 0.2676(17) + 14.8 \\ &\approx 12.4 \end{aligned}$	In 1957 (17 years after 1940), cars averaged about 12.4 miles per gallon.
$I = (20 + x)(10 - 0.10x)$ Income (in dollars) for a diving boat where x represents the number of passengers above 20 on the boat	$I(x) = (20 + x)(10 - 0.10x)$ or $f(x) = (20 + x)(10 - 0.10x)$	$\begin{aligned} I(10) &= (20 + 10)(10 - 0.10 \cdot 10) \\ &= 270 \end{aligned}$	Income is $270 with $20 + 10$, or 30, people on the diving boat.
$L = 600\sqrt{1 - \dfrac{v^2}{c^2}}$ Length of a 600-meter tall starship moving at velocity v from the perspective of an observer at rest ($c \approx 186{,}000$ miles per second)	$L(v) = 600\sqrt{1 - \dfrac{v^2}{c^2}}$ or $L(x) = 600\sqrt{1 - \dfrac{x^2}{c^2}}$ or $f(x) = 600\sqrt{1 - \dfrac{x^2}{c^2}}$	$\begin{aligned} L(167{,}400) &= 600\sqrt{1 - \dfrac{(167{,}400)^2}{(186{,}000)^2}} \\ &= 600\sqrt{1 - (0.9)^2} \\ &\approx 262 \end{aligned}$	Moving at 167,400 miles per second (or 90% the speed of light), the starship will measure approximately 262 meters tall from the perspective of a stationary observer.

5 Find function values for piecewise functions.

Sometimes, data is modeled by two or more equations. When these equations are written in function notation, such functions are called *piecewise functions*.

EXAMPLE 4 **A Piecewise Function Defined by Two Equations**

The function

$$f(x) = \begin{cases} 0.0005x^2 + 0.025x + 8.8 & \text{if } 0 \le x < 30 \\ 0.0202x^2 - 1.58x + 39.2 & \text{if } x \ge 30 \end{cases}$$

Art by Henry L. Small, Ceramic caricature of a classic Chevy.

describes the average number of miles (in thousands) that each automobile in the United States was driven x years after 1940. Find and interpret: **a.** $f(15)$ **b.** $f(50)$

Solution

a. To find $f(15)$, we let $x = 15$. Because $x = 15$ is less than 30, we use the first line of the piecewise function.

$f(x) = 0.0005x^2 + 0.025x + 8.8$ This is the function's equation for $0 \leq x < 30$.

$f(15) = 0.0005(15)^2 + 0.025(15) + 8.8$ Replace x with 15.

$= 9.2875$ Use a calculator and evaluate on the right.

This means that 15 years after 1940 (in 1955) each automobile in the United States was driven 9.2875 thousand miles, or approximately 9288 miles.

b. To find $f(50)$, we let $x = 50$. Because $x = 50$ is greater than 30, we use the second line of the piecewise function.

$f(x) = 0.0202x^2 - 1.58x + 39.2$ This is the function's equation for $x \geq 30$.

$f(50) = 0.0202(50)^2 - 1.58(50) + 39.2$ Replace x with 50.

$= 10.7$ Evaluate on the right.

This means that 50 years after 1940 (in 1990) each automobile in the United States was driven 10.7 thousand miles, or approximately 10,700 miles. ■

6 Evaluate a function.

Evaluating Functions

We have seen that to evaluate a function f at a number, we substitute that number for x in the definition of f. In the same way, we can replace x with a variable or with an algebraic expression. For example, to find $f(a + h)$, we substitute $a + h$ for each occurrence of x in the definition of f.

EXAMPLE 5   Evaluating a Function

If $f(x) = 3x^2 + 4x - 5$, evaluate:

a. $f(a)$ **b.** $f(-a)$ **c.** $f(a + h)$ **d.** $\dfrac{f(a + h) - f(a)}{h}, h \neq 0$

Solution

a. $f(x) = 3x^2 + 4x - 5$ This is the given function.

$f(a) = 3a^2 + 4a - 5$ To find f of a, replace each occurrence of x with a.

b. $f(x) = 3x^2 + 4x - 5$ This is the given function.

$f(-a) = 3(-a)^2 + 4(-a) - 5$ Find f of $-a$ by replacing

$= 3a^2 - 4a - 5$ x with $-a$.

c. $f(x) = 3x^2 + 4x - 5$ This is the given function.

$f(a + h) = 3(a + h)^2 + 4(a + h) - 5$ Replace x with $a + h$.

$= 3(a^2 + 2ah + h^2) + 4(a + h) - 5$ Square $a + h$.

$= 3a^2 + 6ah + 3h^2 + 4a + 4h - 5$ Multiply.

d. $\dfrac{f(a + h) - f(a)}{h}$

$= \dfrac{(3a^2 + 6ah + 3h^2 + 4a + 4h - 5) - (3a^2 + 4a - 5)}{h}$

> Substitute the evaluations from parts (c) and (a).

$= \dfrac{3a^2 + 6ah + 3h^2 + 4a + 4h - 5 - 3a^2 - 4a + 5}{h}$

> Remove parentheses and change the sign of each term in the second parentheses.

$= \dfrac{(3a^2 - 3a^2) + (4a - 4a) + (-5 + 5) + 6ah + 3h^2 + 4h}{h}$

> Group like terms.

$= \dfrac{6ah + 3h^2 + 4h}{h}$

> Simplify.

$= \dfrac{h(6a + 3h + 4)}{h}$

> Factor h from the numerator.

$= 6a + 3h + 4$

> Cancel identical factors of h in the numerator and denominator. ∎

7 Find the domain of a function.

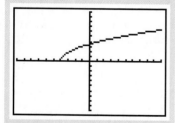

Using technology

For parts (a) through (c), we can graph each function and visually determine the values of x for which the function is defined.

a. $f(x) = \sqrt{3x + 12}$

The graph of

$y = \sqrt{3x + 12}$

appears only for $x \geq -4$, verifying $[-4, \infty]$ as the domain.

b. $g(x) = \dfrac{3}{x^2 + 2x - 15}$

Using a standard viewing rectangle, the graph of g is a bit difficult to see. Instead, we've graphed the denominator $y = x^2 + 2x - 15$, excluding

(continued)

Finding the Domain of a Function

We have seen that the domain of a function is the set of all values taken on by the independent variable (often represented by x). Some mathematical models have their domains explicitly given along with the function. However, the domain may be *implied* by the algebraic expression that defines a function.

The implied domain of a function

The domain of a function is the set of all real numbers for which the expression defining the function is defined and equal to a real number. This is called the *implied domain* of the function. The implied domain must exclude real numbers that cause division by zero and real numbers that result in an even root of a negative number.

EXAMPLE 6 **Finding the Domain of a Function**

Find the domain of each function:

a. $f(x) = \sqrt{3x + 12}$

b. $g(x) = \dfrac{3}{x^2 + 2x - 15}$

c. $h(x) = \dfrac{1}{\sqrt{2x^2 + 7x - 4}}$

d. $A = \pi r^2$ (area of a circle)

Solution

In each case, the domain is the set of all real numbers for which the formula defines a real number and for which it makes sense when interpreted in the physical world.

Using technology (cont.)

values that make the denominator 0. These values occur when the graph crosses the x-axis, at −5 and 3. Although the graph is partly cut off, we can see that the domain of g is the set of all real numbers excluding −5 and 3.

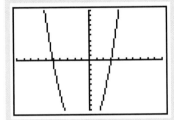

c. $h(x) = \dfrac{1}{\sqrt{2x^2 + 7x - 4}}$

Once again, the graph of h is difficult to see, so we instead graph $y = 2x^2 + 7x - 4$. Since $2x^2 + 7x - 4$ must be greater than 0, the graph lies above the x-axis for $x < -4$ or $x > \frac{1}{2}$, which also describes the domain.

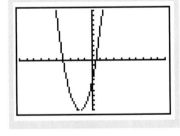

a. $f(x) = \sqrt{3x + 12}$ This is the given function.

$3x + 12 \geqslant 0$ Since we want $\sqrt{3x + 12}$ to equal a real number, the quantity under the radical sign must be nonnegative.

$3x \geqslant -12$ Because the domain is the set of values taken on by x, we now solve for x.

$x \geqslant -4$ Divide both sides by 3.

The domain of $f(x) = \sqrt{3x + 12}$ is the set of all real numbers greater than or equal to −4, represented by $[-4, \infty)$.

b. $g(x) = \dfrac{3}{x^2 + 2x - 15}$ This is the given function.

$g(x) = \dfrac{3}{(x + 5)(x - 3)}$ We must exclude x-values that yield zero in the denominator. Factoring the denominator, we find that we must exclude $x = -5$, and $x = 3$.

The domain of $g(x) = \dfrac{3}{x^2 + 2x - 15}$ is the set of all real numbers excluding −5 and 3. This can be represented as $(-\infty, -5) \cup (-5, 3) \cup (3, \infty)$.

c. $h(x) = \dfrac{1}{\sqrt{2x^2 + 7x - 4}}$ This is the given function.

$2x^2 + 7x - 4 > 0$ We want $\sqrt{2x^2 + 7x - 4}$ to equal a *positive* real number. (Why can't $2x^2 + 7x - 4$ equal 0?)

At this point we must solve the resulting quadratic inequality using methods of Section 1.6. Solve $2x^2 + 7x - 4 = 0$ using factoring:

$(x + 4)(2x - 1) = 0$

$x = -4 \quad \text{or} \quad x = \frac{1}{2}$

The test intervals are $(-\infty, -4)$, $(-4, \frac{1}{2})$, and $(\frac{1}{2}, \infty)$. Using a representative number from each test interval, we find that the solution is $(-\infty, -4) \cup (\frac{1}{2}, \infty)$.

The domain of $h(x) = \dfrac{1}{\sqrt{2x^2 + 7x - 4}}$ is $(-\infty, -4) \cup (\frac{1}{2}, \infty)$.

d. The area of a circle, given by $A = \pi r^2$, is a function of the radius. We can equivalently write $A(r) = \pi r^2$, choosing positive values for the radius r. The domain of the function is the set of all real numbers r such that $r > 0$, represented by $(0, \infty)$.   ■

PROBLEM SET 2.1

Practice Problems

Determine which of the relations in Problems 1–8 are not functions. Give the domain and range for each relation.

1. $\{(1, 2), (3, 4), (5, 5)\}$

2. $\{(4, 5), (6, 7), (8, 8)\}$

3. $\{(3, 4), (3, 5), (4, 4), (4, 5)\}$

4. $\{(5, 6), (5, 7), (6, 6), (6, 7)\}$

5. $\{(-3, -3), (-2, -2), (-1, -1), (0, 0)\}$

6. $\{(-7, -7), (-5, -5), (-3, -3), (0, 0)\}$

7. $\{(1, 4), (1, 5), (1, 6)\}$

8. $\{(4, 1), (5, 1), (6, 1)\}$

Find the function values in Problems 9–14.

9. $f(x) = 4x - 6$

   a. $f(0)$ **b.** $f(-2)$ **c.** $f(\frac{1}{2})$ **d.** $f(a)$ **e.** $f(-a)$ **f.** $f(2a)$

10. $f(x) = 3x - 5$

   a. $f(0)$ **b.** $f(-4)$ **c.** $f(-\frac{2}{3})$ **d.** $f(a)$ **e.** $f(-a)$ **f.** $f(2a)$

11. $f(x) = x^2 - 3x + 7$

   a. $f(0)$ **b.** $f(4)$ **c.** $f(-4)$ **d.** $f(-x)$ **e.** $f(a + h)$

12. $f(x) = 4x^2 - 2x + 6$

   a. $f(0)$ **b.** $f(-3)$ **c.** $f(3)$ **d.** $f(-x)$ **e.** $f(a + h)$

13. $g(x) = x^2 + \dfrac{1}{x^2}$

   a. $g(1)$ **b.** $g(-\frac{1}{2})$ **c.** $g(a)$ **d.** $g(-a)$

14. $g(x) = x^3 + \dfrac{1}{x^3}$

   a. $g(1)$ **b.** $g(-\frac{1}{2})$ **c.** $g(a)$ **d.** $g(-a)$

In Problems 15–22, find: **a.** $f(a)$ **b.** $f(a + h)$ **c.** $\dfrac{f(a + h) - f(a)}{h}$ **d.** $f(a) + f(h)$

15. $f(x) = 4x + 2$ **16.** $f(x) = 3x - 1$ **17.** $f(x) = 2x^2 - 3x + 5$ **18.** $f(x) = 3x^2 - 7x - 2$

19. $f(x) = 6$ **20.** $f(x) = 7$ **21.** $f(x) = \dfrac{1}{x}$ **22.** $f(x) = \dfrac{1}{x + 1}$

In Problems 23–26, find: **a.** $f(-x)$ **b.** $f(x + h)$ **c.** $\dfrac{f(x + h) - f(x)}{h}$

23. $f(x) = 6x - 5$ **24.** $f(x) = -4x + 3$ **25.** $f(x) = 2x^2 - 4x - 1$ **26.** $f(x) = -3x^2 + 2x + 8$

Find the function values in Problems 27–28.

27. $f(x) = \begin{cases} x + 3 & \text{if } x \geq -3 \\ -(x + 3) & \text{if } x < -3 \end{cases}$

   a. $f(0)$ **b.** $f(-6)$ **c.** $f(-3)$ **d.** $f(-\pi)$

28. $f(x) = \begin{cases} x + 5 & \text{if } x \geq -5 \\ -(x + 5) & \text{if } x < -5 \end{cases}$

   a. $f(0)$ **b.** $f(-6)$ **c.** $f(-5)$ **d.** $f(-\sqrt{10})$

In Problems 29–50, find the domain of each function.

29. $f(x) = 4x^2 - 3x + 1$ **30.** $f(x) = 8x^2 - 5x + 2$ **31.** $g(x) = \dfrac{3}{x - 4}$

32. $g(x) = \dfrac{2}{x + 5}$ **33.** $h(x) = \dfrac{4}{x^2 + 11x + 24}$ **34.** $h(x) = \dfrac{5}{6x^2 + x - 2}$

35. $f(t) = \dfrac{3}{t^2 + 4}$ **36.** $f(t) = \dfrac{5}{t^2 + 9}$ **37.** $f(x) = \sqrt{x - 3}$

38. $f(x) = \sqrt{x + 2}$ **39.** $f(x) = \sqrt[3]{x - 3}$ **40.** $f(x) = \sqrt[3]{x + 2}$

41. $g(t) = \dfrac{1}{\sqrt[4]{t - 3}}$ **42.** $g(t) = \dfrac{1}{\sqrt[4]{t + 2}}$ **43.** $g(x) = \sqrt{x^2 + 2x - 8}$

44. $g(x) = \sqrt{6x^2 + 13x - 5}$ **45.** $h(r) = \dfrac{\sqrt{r - 1}}{r - 3}$ **46.** $h(r) = \dfrac{\sqrt{r - 2}}{r - 5}$

47. $f(x) = \dfrac{3}{\sqrt{3x^2 - 7x + 2}}$

48. $f(x) = \dfrac{5}{\sqrt{6x^2 - x - 2}}$

49. $r(v) = \sqrt[4]{1 - v^2}$

50. $r(v) = \sqrt[4]{16 - v^2}$

Application Problems

51. The function $N(x) = -143x^3 + 1810x^2 - 187x + 2331$ describes the number of new AIDS cases in the United States x years after 1983, where $0 \leqslant x \leqslant 7$. Find and interpret $N(5)$.

52. The function $C(t) = 0.018t^2 + 0.70t + 316.2$ describes carbon dioxide concentration (in ppm) t years after 1960. Find and interpret $C(30)$.

53. The function $I(x) = (30 + x)(20 - 0.05x)$ describes income (in dollars) for a diving boat where x represents the number of passengers above 30 on the boat. Find and interpret $I(20) - I(10)$.

54. The function $s(t) = -16t^2 + 96t + 176$ describes the position of a free-falling body above the ground (in feet) t seconds after being thrown upward from the top of a 176-foot tall building. Find and interpret $s(5) - s(4)$.

55. The function $L(v) = 400\sqrt{1 - \dfrac{v^2}{c^2}}$ describes the length of a 400-meter tall starship moving at velocity v from the perspective of an observer at rest. (c, the speed of light, is approximately 186,000 miles per second.) Find and interpret $L(148,800)$.

56. The function $L(v) = 800\sqrt{1 - \dfrac{v^2}{c^2}}$ describes the length of an 800-meter tall starship moving at velocity v from the perspective of an observer at rest. (c, the speed of light, is approximately 186,000 miles per second.) Find and interpret $L(93,000)$.

57. A particle is moving along a horizontal number line. Its position on the number line at any time t (in seconds) is given by the function $s(t) = t^3 + t^2 - t - 1$. If average velocity is defined as change in position divided by change in time, find $\dfrac{s(4) - s(2)}{4 - 2}$ and interpret the result in terms of average velocity.

58. A particle is moving along a horizontal number line. Its position on the number line at any time t (in seconds) is given by the function $s(t) = t^3 - 3t^2 + 5$. If average velocity is defined as change in position divided by change in time, find $\dfrac{s(5) - s(2)}{5 - 2}$ and interpret the result in terms of average velocity.

59. The average price of a mobile home in the United States t years after 1984 is modeled by the piecewise function

$$p(t) = \begin{cases} 1753.6t + 19,503.6 & \text{if } 0 \leqslant t \leqslant 5 \\ 81.11t^2 + 19,838.8 & \text{if } 6 \leqslant t \leqslant 14 \end{cases}$$

Find and interpret $p(12) - p(2)$.

60. During a particular year, the taxes owed by a married person filing separately with an adjusted gross income of x dollars is given by the piecewise function

$$T(x) = \begin{cases} 0.15x & \text{if } 0 \leqslant x < 17,900 \\ 0.28(x - 17,900) + 2685 & \text{if } 17,900 \leqslant x < 43,250 \\ 0.31(x - 43,250) + 9783 & \text{if } x \geqslant 43,250 \end{cases}$$

Find and interpret $T(70,000) - T(40,000)$.

True–False Critical Thinking Problems

61. Which one of the following is true?

a. Any set of only two ordered pairs must be a function.

b. If $f = \{(1, 1), (2, 2), (3, 4)\}$, then $f(6) = 7$.

c. If $f(x) = x^2$, then $f(x + h) = x^2 + h$.

d. For people filing a single return, federal income tax is a function of adjusted gross income.

62. Which one of the following is true?

a. If $f = \{(1, 1), (2, 4), (3, 9), (5, 25)\}$, and $g(x) = x^2$, then f and g represent the same function.

b. If $f(x) = x^2$, then $f(a + h) = a^2 + h$.

c. If $f = \{(1, 1), (2, 2), (3, 4)\}$, then $f(1) = 2$.

d. If a set of ordered pairs that represents a function has all of its components reversed, the new set might not be a function.

63. Which one of the following is true?

a. If $f(t) = \dfrac{t - 3}{t + 3}$, then $f(0) = -3$.

b. If $f(x) = 2x + 1$, then $f(1 + 3) = f(1) + f(3)$.

c. If $f(x) = 4x^2$, then $f(2x) \neq 2f(x) \neq f(x^2) \neq [f(x)]^2$.

d. If $f(x) = \dfrac{x - b}{x + b}$, then $f(5b) \neq 2f(2b)$.

64. Which one of the following is true?

 a. If $f(x) = 2x - 3$, then $f(x^2) = [f(x)]^2$.

 b. If $f(x) = \sqrt{x}$, then $f(1 + 3) = f(1) + f(3)$.

 c. If $f(x) = 3x^2$, then $f\left(\dfrac{x}{2}\right) \neq \dfrac{f(x)}{2} \neq 2f(x) \neq f(2x)$.

 d. If $f(x) = 2x^2 - x$ and $g(x) = 3$, then there are no real numbers x_0 such that $f(x_0) = g(x_0)$.

65. Which one of the following is true?

 a. There are no integers m and n such that $y^m = x^n$ defines y as a function of x.

 b. If $f(1) = 1, f(2) = 1 - 2, f(3) = 1 - 2 + 3, f(4) = 1 - 2 + 3 - 4$, and so on, then $f(20) + f(40) - f(65) = 3$.

 c. If $f(x) = $ the digit in the 10^{-4} place when x is expressed in decimal notation, then $f(\frac{18}{19}) < f(\sqrt{17})$.

 d. None of the above is true.

Technology Problems

Use a graphing utility to find the domain of each function in Problems 66–73.

66. $f(x) = \dfrac{2}{x^2 - 8x + 15}$

67. $f(x) = \dfrac{4}{x^2 - 8x + 12}$

68. $g(x) = \sqrt{6 - 2x}$

69. $g(x) = \sqrt{15 - 3x}$

70. $h(x) = \dfrac{4}{\sqrt{x^2 - 7x + 10}}$

71. $h(x) = \dfrac{3}{\sqrt{x^2 + 5x + 4}}$

72. $\phi(x) = \sqrt{4x^2 - 8x + 7}$

73. $\phi(x) = \dfrac{3}{\sqrt{x^2 - 4x + 4}}$

74. Consider the function $L(x) = $ the exponent to which 10 must be raised to obtain x.

 a. Fill in the table in the next column.

x	$\frac{1}{100}$	$\frac{1}{10}$	1	$\sqrt{10}$	10	100
$L(x)$						

 b. Use a graphing utility to graph $y = \log x$. Use the following range setting:

 $\text{Xmin} = 0, \text{Xmax} = 10, \text{Xscl} = 1,$

 $\text{Ymin} = -3, \text{Ymax} = 3, \text{Yscl} = 1$

 Use the $\boxed{\text{TRACE}}$ feature and experiment with the range setting to show the ordered pairs in the previous table as points on the graph of $y = \log x$. Convince yourself that $y = \log x$ is the same function as the one defined by $L(x)$.

 c. Use your conclusion in part (b) to write an equation in x and y that expresses the verbal rule used to define $L(x)$.

75. The sum of two numbers is 6. One of the numbers is represented by x.

 a. Express the product P as a function of x.

 b. Graph the function by replacing $P(x)$ by y.

 c. Describe one bit of useful information revealed by the graph of $P(x)$ that is not obvious by looking at its equation.

Writing in Mathematics

In Problems 76–79, determine whether the correspondence is a function. Explain your answer.

Domain	Correspondence	Range
76. The set of houses in a town	Each house's area	A set of positive numbers
77. The air temperature each day at noon at an ocean beach over a one-week period	The number of people at the beach each day at noon over the one-week period	A set of positive numbers
78. The set of numbers representing the weight of each person in a town	The number of years of education for each person in the town	A set of nonnegative numbers
79. The set of real numbers	The square of each number	A set of nonnegative numbers

80. For people filing a single return, federal income tax is a function of adjusted gross income because for each value of adjusted gross income there is a unique tax to be paid. On the other hand, the price of a house is not a function of the lot size on which the house sits because houses on same-sized lots can sell for many different prices.

a. Describe two everyday situations between variables that are functions.

b. Describe two everyday situations between variables that are not functions.

Critical Thinking Problems

81. Suppose that a relation assigns to every natural number the prime number that is nearest to it on a number line.

   a. Is the relation a function? Explain your answer.

   b. If the relation is not a function, how might the assignment be changed slightly so that it becomes a function?

82. Give an example of an equation that does not define y as a function of x but that does define x as a function of y.

83. Give an example of a function f defined by an equation in x

   a. Whose domain is $[-3, \infty)$.

   b. Whose domain is $(-\infty, -3) \cup (-3, 5) \cup (5, \infty)$.

c. Describe the process that you used to determine the function in parts (a) and (b).

84. If $f(x) = ax^2 + bx + c$ and $r_1 = \dfrac{-b + \sqrt{b^2 - 4ac}}{2a}$, find $f(r_1)$ without doing any algebra and explain how you arrived at your result.

85. If $f(x) = \dfrac{cx + d}{ax - c}$, find $f\left(\dfrac{cx + d}{ax - c}\right)$ and simplify.

86. If $f(x) =$ the number of prime numbers $\leq x$, where x is a positive integer, find two numbers a and b such that $f(a) = 2f(b)$.

87. Find a function $g(t)$ such that $g(x)\,g(y) - g(xy) = x + y$.

SECTION 2.2 Linear Functions and Slope

Solutions Manual **Tutorial** **Video 4**

Objectives

1 Compute a line's slope.
2 Interpret slope as average rate of change.
3 Write the point-slope form of a line.
4 Write the slope-intercept form of a line.
5 Write the general form of a line.
6 Recognize equations for horizontal and vertical lines.
7 Write equations for lines parallel and perpendicular to a given line.
8 Model data with linear functions.
9 Model using direct variation.

When we attempt to fit a curve to a set of data, our goal is to find a function whose graph fits the data well. In this section, we consider data that fall along a straight line, using linear functions to model the data and make predictions.

An Introduction to Linear Functions

Now that we have introduced functions and their notation, we turn to a special class of functions called *linear functions*.

> **Definition of a linear function**
>
> A function that can be expressed in the form
>
> $$f(x) = mx + b$$
>
> is called a *linear function*.

The graph of a linear function is a line. The value b is related to an important point on the line and the value m is significant in terms of the line's steepness.

Study tip

If you are not using a graphing utility in this course, be sure to graph the functions in the Discover for yourself by hand with a table of coordinates.

Discover for yourself

Use a graphing utility to graph the linear functions in each column in the same viewing rectangle. Use the $\boxed{\text{ZOOM SQUARE}}$ feature to avoid possible distortions. In what way are the graphs the same? How does the value of m (the coefficient of x) affect the graphs? What affect does b (the constant term) have on the graphs?

$$f(x) = x + 2 \qquad\qquad f(x) = -x + 2$$
$$g(x) = 2x + 2 \qquad\qquad g(x) = -2x + 2$$
$$h(x) = 3x + 2 \qquad\qquad h(x) = -3x + 2$$

In the Discover for yourself, did you observe that the graph of $f(x) = mx + b$ passes through the point $(0, b)$ and that b is the line's y-intercept? Did you also observe that the value m is responsible for the steepness or slant of the line? For the lines in which m is positive, there is an upward slant from left to right and for the lines in which m is negative there is a downward slant from left to right. This is illustrated in Figure 2.4.

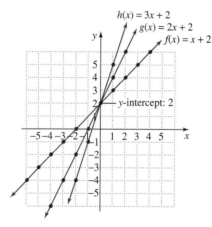

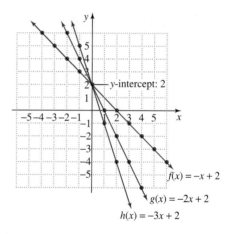

Figure 2.4

When m is positive, the lines slant upward
When m is negative, the lines slant downward

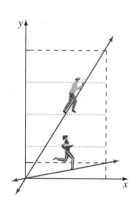

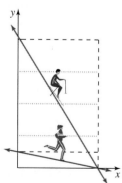

Figure 2.5

Lines with various steepnesses

The Slope of a Line

An important characteristic of a line is the angle that the line makes with the x-axis (see Figure 2.5). Some lines are quite steep, forming an angle close to 90°. Other lines rise quite slowly, forming an angle close to 0°.

Mathematicians have developed a useful measure of the steepness of a line, called the *slope* of the line. Slope compares the vertical change (the *rise*) to the horizontal change (the *run*) encountered when moving from one fixed point to another along the line. To calculate the slope of a line, mathematicians use a ratio comparing the change in y (the rise) to the change in x (the run).

Compute a line's slope.

Definition of slope

The *slope* of the line through the distinct points (x_1, y_1) and (x_2, y_2) is

$$\frac{\text{Change in } y}{\text{Change in } x} = \frac{\text{Rise}}{\text{Run}} = \frac{y_2 - y_1}{x_2 - x_1}$$

where $x_2 - x_1 \neq 0$.

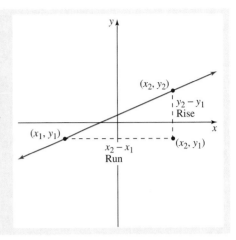

The letter m—the first letter in the French verb *monter*, meaning to go up or to climb—is used to denote slope.

The quantity $x_2 - x_1$ in the definition of slope represents the amount by which x changes in moving from (x_1, y_1) to (x_2, y_2) along the line. This *change in x* is represented by the symbol Δx (read *delta x*). Similarly, the change in y, namely, $y_2 - y_1$, is represented by Δy. Using these ideas (see Figure 2.6), we can represent slope by

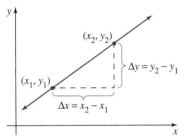

Figure 2.6

Slope as Δy divided by Δx

$$m = \frac{y_2 - y_1}{x_2 - x_1} = \frac{\Delta y}{\Delta x}.$$

Study tip

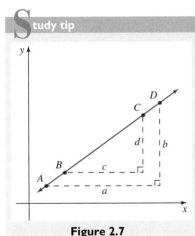

Figure 2.7

The slope of a line does not depend on which two points on the line are used in the computation. This is shown in Figure 2.7, in which the two right triangles are similar. Because corresponding sides of similar triangles are proportional,

$$\frac{b}{a} = \frac{d}{c}.$$

Since $\dfrac{b}{a}$ is slope calculated using points A and D, and $\dfrac{d}{c}$ is slope calculated using points B and C, the values obtained for slope are equal regardless of what two points are used on the line.

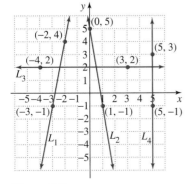

Figure 2.8

Four lines and their slopes

EXAMPLE 1   Using the Definition of Slope

Calculate and compare the slopes of the four lines shown in Figure 2.8.

Solution

L_1: **Line L_1 passes through $(-3, -1)$ and $(-2, 4)$.**
Let $(x_1, y_1) = (-3, -1)$ and $(x_2, y_2) = (-2, 4)$. By definition, the slope is

$$m = \frac{y_2 - y_1}{x_2 - x_1} = \frac{4 - (-1)}{-2 - (-3)} = \frac{5}{1} = 5.$$

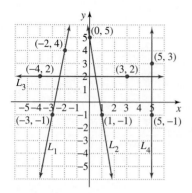

Figure 2.8 (repeated)

Discover for yourself

In Example 1, let (x_1, y_1) represent the point that we called (x_2, y_2) and let (x_2, y_2) represent the point that we called (x_1, y_1). Find the slope for lines L_1, L_2, and L_3. What can you conclude? What happens if you call one point (x_1, y_2) and the other (x_2, y_1) subtract in one order in the numerator and then in the opposite order in the denominator?

2 Interpret slope as average rate of change.

This means that for line L_1, for every 1 unit we move to the right (the change in x) the line rises 5 units (the change in y).

L_2: **Line L_2 passes through (1, −1) and (0, 5).**
Let $(x_1, y_1) = (1, -1)$ and $(x_2, y_2) = (0, 5)$.

$$m = \frac{y_2 - y_1}{x_2 - x_1} = \frac{5 - (-1)}{0 - 1} = \frac{6}{-1} = -6$$

Slope is the difference in y-values divided by the difference in x-values.

L_3: **Line L_3, a horizontal line, passes through (−4, 2) and (3, 2).**

$$m = \frac{2 - 2}{3 - (-4)} = \frac{0}{7} = 0$$

A horizontal line is neither increasing nor decreasing from left to right, and its slope is 0.

L_4: **Line L_4, a vertical line, passes through (5, −1) and (5, 3).**

$$m = \frac{3 - (-1)}{5 - 5} = \frac{4}{0}$$

which is undefined. The slope of a vertical line is not defined. ∎

Example 1 contains the four basic situations that can be encountered about the slope of a line. These situations can be generalized as follows.

The slope of a line

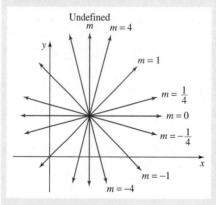

1. Lines with positive slope ($m > 0$) rise (increase) from left to right.
2. Lines with negative slope ($m < 0$) fall (decrease) from left to right.
3. Lines with zero slope ($m = 0$) are horizontal.
4. Lines with undefined slope are vertical.

Modeling with Slope

Slope is defined as $\dfrac{\Delta y}{\Delta x}$, the ratio of a change in y to a corresponding change in x. Consequently, in applied situations slope can be thought of as the *average change in y per unit of change in x*, where y is a function of x. This idea is illustrated in Example 2.

EXAMPLE 2 **Slope as the Average Rate of Change**

World population started a rapid growth phase in the early 1800s and has grown sixfold in the last 200 years (see Figure 2.9 on the next page).

ENRICHMENT ESSAY

Slope and the Streets of San Francisco

San Francisco's Filbert Street has a slope of 0.613, meaning that for every horizontal change of 100 feet, the street ascends 61.3 feet vertically. The street is too steep to pave and is only accessible by wooden stairs. Legend has it that the cottages along this street with its 31.5° angle of inclination were once used to "shanghai" sailors. "To shanghai" was a verb coined in early San Francisco. (Look up "shanghai" in a dictionary if you are not sure of its meaning.)

Carol Simowitz / San Francisco Convention and Visitors Bureau.

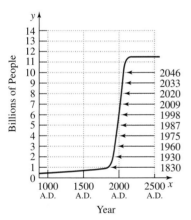

Figure 2.9

World population

Source: Population Reference Bureau

a. Compute the slope of the line drawn through the years 1830 and 1930. Interpret this value in terms of average rate of change.

b. Repeat part (a) for the years 1987 and 1998.

Solution

a. For 1830 and 1930, we use the values from the graph, namely, $(1830, 1)$ and $(1930, 2)$. (The y-coordinate) represents the world population in billions for each of these years.) We then apply the definition of slope.

$$m = \frac{2 - 1}{1930 - 1830} = \frac{1}{100} = 0.01$$

Between 1830 and 1930, the average world population grew at a rate of 0.01 billion people (or 10 million people) each year.

b. For 1987 and 1998, we again use the values from the graph, namely, $(1987, 5)$ and $(1998, 6)$. We then apply the definition of slope.

$$m = \frac{6 - 5}{1998 - 1987} = \frac{1}{11} \approx 0.09$$

Between 1987 and 1930, the average world population grew at a rate of 0.09 billion people (or approximately 90 million people) each year. ■

3 Write the point-slope form of a line.

The Point-Slope Form of a Line

If a particular point on a line and the slope of a line are known, we can write an equation that describes the line. Let (x_1, y_1) represent a fixed point on the line. Let m represent the slope of the line. Let (x, y) stand for all ordered

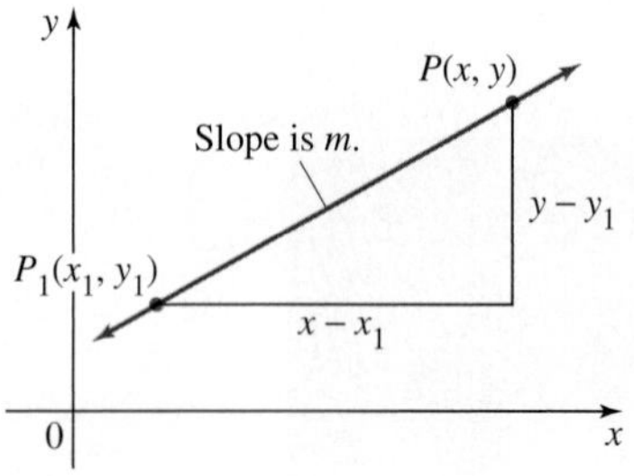

Figure 2.10

A line passing through (x_1, y_1) with slope m

pairs corresponding to points along the line other than (x_1, y_1). The situation is illustrated in Figure 2.10. By definition,

$$\frac{y - y_1}{x - x_1} = m.$$

If we multiply both sides by $x - x_1$, we obtain

$$y - y_1 = m(x - x_1).$$

This discussion leads to the following statement.

Point-slope form of the equation of a line

The point-slope equation of the line that passes through the point (x_1, y_2) and has slope m is

$$y - y_1 = m(x - x_1).$$

EXAMPLE 3 **Writing the Point-Slope Equation of a Line**

Write the point-slope equation of the line passing through $(-1, 3)$ with slope 4. Solve the equation for y and sketch the line.

Solution

We use the point-slope equation of a line with $m = 4$, $x_1 = -1$, and $y_1 = 3$.

$y - y_1 = m(x - x_1)$ This is the point-slope equation.

$y - 3 = 4[x - (-1)]$ Substitute the given values.

$y - 3 = 4(x + 1)$ We now have the point-slope equation for the given line.

$y - 3 = 4x + 4$ Solve for y; apply the distributive property on the right.

$y = 4x + 7$ Add 3 to both sides.

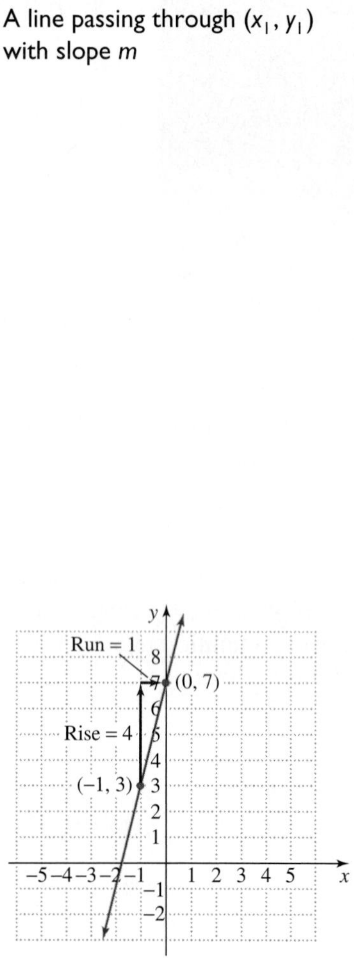

Figure 2.11

The line passing through $(-1, 3)$ with slope 4

Graph the line by first plotting the point $(-1, 3)$. Since the slope is 4, this means that a rise of 4 units corresponds to a run of 1 unit. (Equivalently, when we move 1 unit to the right, the line rises 4 units.) This enables us to sketch the line shown in Figure 2.11. ∎

Using technology

Use a graphing utility to graph $y = 4x + 7$ and obtain a graph similar to the one shown in Figure 2.11. If you use the ⎍ZOOM SQUARE⎍ feature, horizontal and vertical tick marks will have equal spacing, and the slope should visually appear to equal 4.

4   Write the slope-intercept form of a line.

The Slope-Intercept Form of a Line

We have seen that linear functions can be expressed in the form

$$f(x) = mx + b.$$

Now it's time to verify that m, the coefficient of x, is the slope of the line, and the constant term b is the y-intercept. We can do this by writing an equation of the line having slope m and y-intercept b.

EXAMPLE 4   **Deriving the Slope-Intercept Equation of a Line**

Write an equation of the nonvertical line that has slope m and y-intercept b.

Solution

As shown in Figure 2.12, since the y-intercept is b, the line intersects the y-axis at $(0, b)$. We use the point-slope equation of a line with $x_1 = 0$ and $y_1 = b$.

$$y - y_1 = m(x - x_1) \quad \text{This is the point-slope equation.}$$

$$y - b = m(x - 0) \quad \text{Since } (x_1, y_1) = (0, b), \text{ we let } x_1 = 0 \text{ and } y_1 = b. \text{ The slope is } m.$$

$$y - b = mx \quad \text{Multiply on the right.}$$

$$y = \underset{\text{Slope}}{m}x + \underset{y\text{-intercept}}{b} \quad \text{Solve for } y.$$

■

The last equation in Example 4 is called the *slope-intercept form* of the equation of a line.

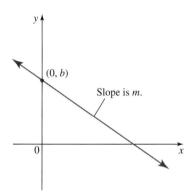

Figure 2.12

A line with slope m and y-intercept b

Slope-intercept form of the equation of a line

The slope-intercept equation of the line that has slope m and y-intercept b is

$$y = mx + b.$$

The slope-intercept equation, $y = mx + b$, can be expressed in function notation, replacing y with $f(x)$.

$$f(x) = mx + b$$

EXAMPLE 5   **Using the Slope-Intercept Equation**

Find the slope and y-intercept of each line. Then graph the line.

a. $4x + y = 2$ **b.** $2x - 3y + 6 = 0$

Solution

We solve each equation for y. In the form $y = mx + b$, the coefficient of x is the slope of the line, and the constant term is its y-intercept.

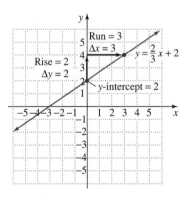

Figure 2.13

The line with y-intercept 2 and slope −4

a. Solving $4x + y = 2$ for y, we obtain $y = -4x + 2$. Thus, the slope is -4 and the y-intercept is 2. Since the line's y-intercept is 2, the line intersects the y-axis at $(0, 2)$. We plot this point. We graph the slope, -4, using

$$m = \frac{\Delta y}{\Delta x} = \frac{-4}{1}.$$

The graph of $4x + y = 2$ or, equivalently, $y = -4x + 2$, is shown in Figure 2.13.

b. We begin by writing the equation in the form $y = mx + b$.

$2x - 3y + 6 = 0$	This is the given equation.
$2x + 6 = 3y$	To isolate the y-term, add $3y$ on both sides.
$3y = 2x + 6$	Reverse the two sides. (This step is optional.)
$y = \dfrac{2}{3}x + 2$	Divide both sides by 3.

From the slope-intercept form of the line's equation, we see that the slope is $m = \frac{2}{3}$ and the y-intercept is $b = 2$. Starting at $(0, 2)$ and using

$$m = \frac{\Delta y}{\Delta x} = \frac{2}{3}$$

we graph the line in Figure 2.14.

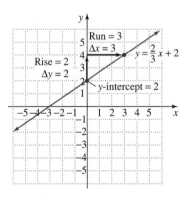

Figure 2.14

The line with y-intercept 2 and slope $\frac{2}{3}$ ∎

⑤ Write the general form of a line.

Other Forms of a Line's Equation

In Example 5b, we wrote the slope-intercept form of a line whose equation was given as

$$2x - 3y + 6 = 0.$$

This form of the line is called its *general equation*.

6 Recognize equations for horizontal and vertical lines.

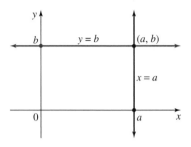

Figure 2.15

Equations of horizontal and vertical lines

Straight-line graphs, with the exception of vertical lines, are graphs of linear functions.

7 Write equations for lines parallel and perpendicular to a given line.

Discover for yourself

Use a graphing utility to graph the following equations in the same viewing rectangle.

$$y_1 = \frac{2}{3}x + 2$$

$$y_2 = \frac{2}{3}x$$

$$y_3 = \frac{2}{3}x - 1$$

What is true about the slopes of the three lines? What do you observe about their graphs? In general, what appears to be true about the graphs of lines with equal slopes?

General form of the equation of a line

Every line has an equation that can be written in the *general form*

$$Ax + By + C = 0$$

where A and B are not both zero.

Finally, let's consider horizontal and vertical lines. If a line is horizontal, its slope is zero ($m = 0$), so the equation $y = mx + b$ becomes $y = b$, where b is the y-intercept. This is illustrated in Figure 2.15.

A vertical line has slope that is undefined. However, Figure 2.15 illustrates that the x-coordinate of every point on the vertical line is a. Thus we can write its equation as $x = a$, where a is the x-intercept.

Vertical and horizontal lines

The equation of a horizontal line through (a, b) is $y = b$. The line's y-intercept is b.

The equation of a vertical line through (a, b) is $x = a$. The line's x-intercept is a.

Parallel and Perpendicular Lines

Two nonintersecting lines that lie in the same plane are parallel. If two lines do not intersect, the ratio of the vertical change to the horizontal change is the same for each line. Because two parallel lines have the same "steepness," they must have the same slope.

Slope and parallel lines

1. If two nonvertical lines are parallel, then they have the same slope.
2. If two distinct nonvertical lines have the same slope, then they are parallel.
3. Two vertical lines, both with undefined slopes, are parallel.

EXAMPLE 6 **Writing Equations of a Line Parallel to a Given Line**

Write an equation of the line passing through $(-3, 2)$ and parallel to the line whose equation is $y = 2x + 1$. Express the equation in point-slope form, slope-intercept form, and general form.

Solution

Figure 2.16 on page 232 should help us to analyze the problem. We begin with the point-slope form. We see that $(x_1, y_1) = (-3, 2)$ since this is a point on the

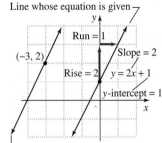

Figure 2.16

Writing equations of a line parallel to a given line

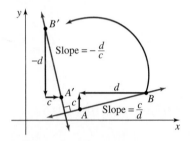

Figure 2.17

Slopes of perpendicular lines

line. Since the slope of the line whose equation is given is 2, then $m = 2$ for the line whose equation we must write. Parallel lines have the same slope.

$$y - y_1 = m(x - x_1)$$ This is the point-slope equation.

$$y - 2 = 2[x - (-3)]$$ $(x_1, y_1) = (-3, 2)$ and $m = 2$.

$$y - 2 = 2(x + 3)$$ This is the point-slope form of our problem's equation.

$$y - 2 = 2x + 6$$ Apply the distributive property. By solving for y we will obtain the slope-intercept form.

$$y = 2x + 8$$ Add 2 to both sides. This is the slope-intercept form $(y = mx + b)$.

$$2x - y + 8 = 0$$ Subtract y from both sides. This is the general form $(Ax + By + C = 0)$.   ∎

The relationship between the slopes of perpendicular lines is not as obvious as that between parallel lines. Figure 2.17 shows line AB, with a slope of $\frac{c}{d}$. Rotate line AB 90° to the left to obtain line $A'B'$ perpendicular to line AB. The figure indicates that the rise and the run of the new line are reversed from the original line, but the rise is now negative. This means that the slope of the new line is $-\frac{d}{c}$. Notice that the product of the slopes of the two perpendicular lines is -1:

$$\left(\frac{c}{d}\right)\left(-\frac{d}{c}\right) = -1.$$

This relationship holds for all perpendicular lines and is summarized below.

Slope and perpendicular lines

1. If two nonvertical lines are perpendicular, then the product of their slopes is -1.
2. If the product of the slopes of two lines is -1, then the lines are perpendicular.
3. A horizontal line having zero slope is perpendicular to a vertical line having undefined slope.

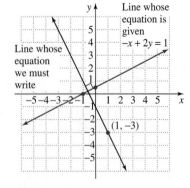

Figure 2.18

Writing equations of a line perpendicular to a given line

An equivalent way of stating this relationship is to say that one line is perpendicular to another line if its slope is the *negative reciprocal* of the slope of the other.

EXAMPLE 7 **Writing Equations of a Line Perpendicular to a Given Line**

Write an equation of the line passing through $(1, -3)$ and perpendicular to the line whose equation is $-x + 2y = 1$. Again, express the equation in point-slope form, slope-intercept form, and general form.

Solution

The situation is illustrated in Figure 2.18. We begin with the point-slope equation. We must find two things: a point on the line [which, again, we have:

Discover for yourself

Use a graphing utility to graph

$$y_1 = \frac{1}{2}x + \frac{1}{2}$$

and

$$y_2 = -2x - 1$$

in the same viewing rectangle. Do the lines appear to be perpendicular? Now use the ZOOM SQUARE setting. What do you observe?

$(1, -3)]$ and its slope. Because the line through $(1, -3)$ is to be perpendicular to the line whose equation is $-x + 2y = 1$, its slope is the negative reciprocal of the slope of the given equation. We can find the slope of $-x + 2y = 1$ by using slope-intercept form. We solve the equation for y.

$$-x + 2y = 1 \qquad \text{This is the given equation.}$$

$$2y = x + 1 \qquad \text{Add } x \text{ to both sides.}$$

$$y = \frac{1}{2}x + \frac{1}{2} \qquad \text{Obtain slope-intercept form by dividing both sides by 2.}$$

$$\qquad\quad\downarrow\qquad\downarrow$$
$$\text{Slope} \quad y\text{-intercept}$$

The slope of the line whose equation we must write is -2, the negative reciprocal of $\frac{1}{2}$. With $(x_1, y_1) = (1, -3)$ and $m = -2$, we obtain

$$y - y_1 = m(x - x_1)$$

$$y - (-3) = -2(x - 1) \qquad (x_1, y_1) = (1, -3) \text{ and } m = -2.$$

$$y + 3 = -2(x - 1) \qquad \text{This is the point-slope form.}$$

Solving for y, we obtain

$$y + 3 = -2x + 2$$

$$y = -2x - 1 \qquad \text{This is the slope-intercept form.}$$

Since general form $(Ax + By + C = 0)$ is usually written with $A > 0$, we add $2x + 1$ to both sides, obtaining

$$2x + y + 1 = 0 \qquad \text{This is the general form.} \qquad \blacksquare$$

8 Model data with linear functions.

Linear Modeling

Our next example illustrates how we can use real world data to develop a mathematical model. Once the model is developed, we can *interpolate* and *extrapolate* from the data. *Interpolation* refers to finding numerical values *between* actual data measurements. *Extrapolation* refers to finding numerical values *outside* the actual data measurements.

Average and Minimum Salaries for Major League Baseball Players

Figure 2.19

Source: Based on statistics from the National Baseball League Players Association

EXAMPLE 8 **Modeling Average Baseball Salaries**

The graph in Figure 2.19 shows average and minimum salaries for major league baseball players. Write the equation that models the data for the line representing average salaries from 1990 through 1994. Then extrapolate from the data and predict the average salary for a major league baseball player in the year 2000.

Solution

The graph indicates that the average salary in 1990 was $0.6 million, so we will use the data points (1990, 0.6) and (1994, 1.3). To write the point-slope equation of the line, we need (x_1, y_1), a fixed point, and m, the slope. *Either* ordered pair can be taken as (x_1, y_1). Using the smaller numbers, we have

$$x_1 = 1990 \quad \text{and} \quad y_1 = 0.6.$$

Now we must calculate the slope.

$$m = \frac{y_2 - y_1}{x_2 - x_1} = \frac{1.3 - 0.6}{1994 - 1990} = \frac{0.7}{4} = 0.175$$

This indicates that average salaries between 1990 and 1994 increased at a rate of $0.175 million, or $175,000 each year.

Now we are ready to use the point-slope form of a line to model the data.

$y - y_1 = m(x - x_1)$	This is the point-slope equation.
$y - 0.6 = 0.175(x - 1990)$	Use the values from above: $x_1 = 1990$, $y_1 = 0.6$, $m = 0.175$
$y - 0.6 = 0.175x - 348.25$	Multiply on the right.
$y = 0.175x - 347.65$	Solve for y. This is the slope-intercept form of the line's equation.

This is the equation that models the data from 1990 through 1994. We now extrapolate from the data. We predict the average salary for the year 2000 by substituting 2000 for x and solving for y.

$$y = 0.175(2000) - 347.65 = 2.35$$

The model predicts that the average salary for a major league baseball player in the year 2000 will be $2.35 million. ∎

Study tip

Baseball salaries are a function of time, expressed in function notation as

$f(x) = 0.175x - 347.65.$

Using this notation, we have

$f(2000) = 0.175(2000) - 347.65$

$\qquad = 2.35$

which indicates salaries of $2.35 million by the year 2000.

Scatter Plots and Regression Lines

Mathematical models can be obtained for a set of data points that appear to fall nearly on a straight line. For example, Figure 2.20 shows a collection of points representing men's and women's winning times (in seconds) in the Olympic 100-meter freestyle swimming race. This collection of points is called a *scatter plot*. The line that best fits the data is called the *regression line*, and a regression line is shown for both the women and the men.

A graphing utility will give you the equation of a line that best fits a set of data. Details for this type of linear modeling are discussed under Technology Problems in Problem Set 2.2.

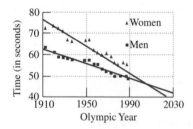

Figure 2.20

9 Model using direct variation.

Modeling Using Direct Variation

The linear model $y = mx$ that has a y-intercept of zero is useful in modeling situations described by the language of *variation*. If $y = mx$, we say that y *varies directly* as x, or that y is *proportional* to x.

> **Direct variation**
>
> y *varies directly* as x if there exists some nonzero constant m such that
>
> $y = mx.$
>
> We also say that y is *proportional* to x. The number m is called the *constant of variation* or the *constant of proportionality*.

EXAMPLE 9   Modeling Using Direct Variation

The volume of gas (V) at constant pressure is directly proportional to the temperature (T). A certain gas has a volume of 250 cubic meters at a temperature of 30 Kelvin.

a. Find a model that gives volume in terms of pressure.
b. Use this model to find the volume of the gas at a temperature of 90 Kelvin.

Solution

a. $V = mT$ Translate "volume is directly proportional to temperature" into a linear model. (Equivalently: Volume *varies directly* as temperature.)

$250 = m \cdot 30$ Find m. We are given that when $V = 250$, $T = 30$.

$\dfrac{25}{3} = m$ Solve for m.

$V = \dfrac{25}{3} T$ Substitute the value of m into the original equation.

The model that gives volume in terms of pressure is $V = \dfrac{25}{3} T$.

b. $V = \dfrac{25}{3} T$ We now want the value for V if $T = 90$.

$V = \dfrac{25}{3}(90)$ Substitute 90 for T into the model.

$= 750$

At a temperature of 90 Kelvin, the volume of the gas is 750 cubic meters. ■

PROBLEM SET 2.2

Practice Problems

In Problems 1–6, find the slope of the line that passes through the given pair of points.

1. $(7, 3)$ and $(6, 9)$

2. $(8, 5)$ and $(5, 2)$

3. $(-3, 2)$ and $(-7, -6)$

4. $(2, -4)$ and $(-3, 2)$

5. $(-12, -2)$ and $(7, -2)$

6. $(-4, -3)$ and $(2, -3)$

7. Compute the slopes of the three lines shown in the figure.

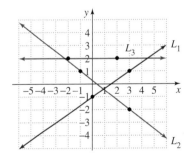

8. Compute the slopes of the three lines shown in the figure.

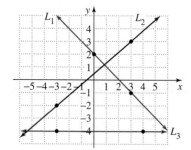

9. Use the figure to make the following lists.

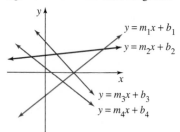

a. List the slopes m_1, m_2, m_3, and m_4 in order of decreasing size.

b. List the number b_1, b_2, b_3, and b_4 in order of decreasing size.

10. Use the figure to make the following lists.

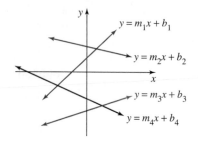

a. List the slopes m_1, m_2, m_3, and m_4 in order of decreasing size.

b. List the number b_1, b_2, b_3, and b_4 in order of decreasing size.

For Problems 11–22, use the given conditions to write an equation for each line in point-slope form, slope-intercept form, and general form.

11. Passing through $(-5, 6)$ with slope 5

12. Passing through $(3, -4)$ with slope -2

13. Passing through $(6, -1)$ and $(3, 2)$

14. Passing through $(3, -8)$ and $(5, 0)$

15. Passing through $(-1, -3)$ and $(7, -2)$

16. Passing through $(1, 4)$ and $(-2, 3)$

17. Passing through $(-8, -10)$ and parallel to the line whose equation is $y = -4x + 3$

18. Passing through $(-2, -7)$ and parallel to the line whose equation is $y = -5x + 4$

19. Passing through $(4, -7)$ and perpendicular to the line whose equation is $x - 2y - 3 = 0$

20. Passing through $(5, -9)$ and perpendicular to the line whose equation is $x + 7y - 12 = 0$

21. Having an x-intercept of -3 and perpendicular to the line passing through $(0, 0)$ and $(6, -2)$

22. Perpendicular to the line whose equation is $3x - 2y = 4$ with the same y-intercept

Find the slope and y-intercept of each line in Problems 23–40. Then graph the line.

23. $y = 2x + 1$

24. $y = 3x + 2$

25. $y = -2x + 1$

26. $y = -3x + 2$

27. $y = \frac{3}{4}x - 2$

28. $y = \frac{3}{4}x - 3$

29. $3x + y = 2$

30. $4x + y = 3$

31. $f(x) = -\frac{3}{2}x$

32. $f(x) = -\frac{4}{3}x$

33. $3x - y + 2 = 0$

34. $4x - y + 3 = 0$

35. $3x - 4 = 5$

36. $5x - 2 = 0$

37. $3y + 6 = 0$

38. $4y + 12 = 0$

39. $y + \sqrt{7} = \sqrt{7}$

40. $y + \pi = \pi$

Application Problems

41. As shown in the graph at the top of the next page, from 1975 through 1990 U.S. health-care costs increased from $132 billion to $671 billion.

a. Compute the slope of the line drawn through 1975 and 1980. Interpret this value in terms of average rate of change.

b. Repeat part (a) for the line drawn through 1985 and 1990.

U.S. Health-Care Costs, 1975–1990

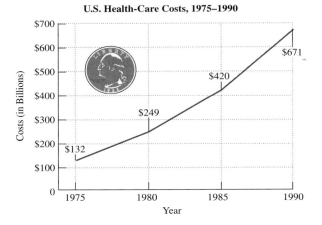

Source: U.S. Health Care Financing Administration

42. The graph below shows violent crime arrests per 100,000 U.S. juveniles ages 10–17 from 1965 through 1990.

a. Find a reasonable estimate for the number of arrests per 100,000 juveniles in 1970. Then use this estimate to compute the slope of the line drawn through 1965 and 1970. Interpret this value in terms of average rate of change.

b. Find a reasonable estimate for the number of arrests per 100,000 juveniles in 1980 and 1985. Then use these estimates to compute the slope of the line drawn through 1980 and 1985. Interpret this value in terms of average rate of change.

Juvenile Arrests, 1965–1990
(Violent Crime Arrests per 100,000 Juveniles, Ages 10–17)

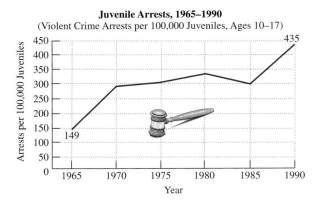

Source: U.S. Department of Justice; Federal Bureau of Investigation

43. The first graph in the next column shows the number of motor-vehicle deaths in the United States from 1975 through 1993.

a. Through which two years specified on the horizontal

axis can a line be drawn with the greatest slope? What is a reasonable estimate for the slope of this line? Interpret this value in terms of average rate of change.

b. Through which two years specified on the horizontal axis can a line be drawn with a slope that is negative and greatest in absolute value? What is a reasonable estimate for the slope of this line? Interpret this value in terms of average rate of change.

c. What is misleading about the scale on the horizontal axis?

Motor Vehicle Deaths in the U.S., 1975–1993

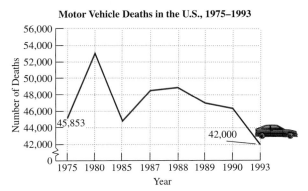

Source: National Safety Council

44. The graph below shows high school dropout rates for three groups. For each group, identify the greatest time interval where a line with slope approximately 0 can be drawn. What does this mean in terms of dropout rate?

**High School Dropout Rates by
Race or Ethnic Group, 1983–1991**

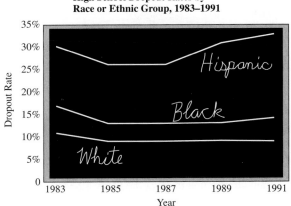

Source: American Federation of Teachers

45. The graph in the top left column on page 238 indicates the number of U.S. companies on the Internet from 1991 through 1994. Write the equation that models the data for a line drawn through 1991 and 1994. Then extrapolate from the data and predict the number of U.S. companies on the Internet for the year 2000.

Number of U.S. Companies on the Internet, 1991–1994

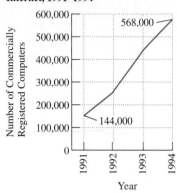

Source: Statistics based on IBM user survey; *USA Today* research

46. The graph below indicates the number of accidental deaths in the U.S. at work and at home from 1930 through 1993. Write the equation that models the data for a line drawn through 1930 and 1993 for the number of accidental deaths at work. Then use this equation to predict the year in which work-related accidental deaths will reach zero.

Accidental Deaths in the United States

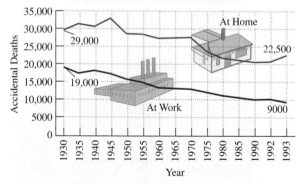

Source: National Safety Council

47. Spending on drug control has skyrocketed in the United States, as shown in the graph at the top of the next column. Draw a line through the years 1981 and 1994 and write the equation of the line through these two data points. Then extrapolate from the data and predict the amount that the government will spend on its drug-control interception budget in the year 2000. Interpolate from the data and use your model to predict drug-control spending for 1988. How close does this come to the number shown in the graph?

U.S. Government Spending on Drug Interceptions and Seizures

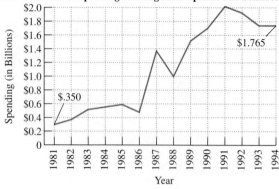

Source: Office of Management and Budget

48. The number of women on active duty in the U.S. military is shown in the graph. Draw a line through the years 1981 and 1994 and write the equation of the line through these two data points. Then extrapolate from the data and predict the number of women on active duty in the year 2000. Interpolate from the data and use the model to predict the number of women on duty for 1992. How close does this come to the number shown in the graph?

Military Women on Active Duty, 1981–1994

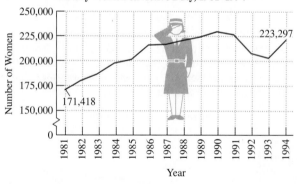

Source: U.S. Department of Defense

49. A person's starting salary in 1995 was $26,500 with yearly increases of $2500. Write a linear model for salary (y) x years after 1995.

50. A company purchases an office machine for $9300 in 1995. The machine is expected to last for 6 years with a yearly loss in value of $1200. Write a linear model for the value of the machine (V) t years after 1995.

51. The cost (C) of an airplane trip is directly proportional to the number of miles in the trip (M). A 3000-mile trip costs $400.

  a. Find a model that gives cost in terms of mileage.

b. Use this model to determine the number of miles that can be covered at a cost of $20.

52. The state of Pennsylvania has a state income tax that is directly proportional to a resident's gross income. A person working in Pennsylvania with a gross yearly income of $21,000 paid $441 in state taxes.

    a. Find a model that gives state income tax as a function of gross income.

    b. Use the model to determine state taxes for a yearly income of $36,000.

53. Running at a constant speed, a runner's distance varies directly as time. If a runner covers 1 mile in 6 minutes, what distance will be covered in 45 minutes?

54. The distance that a spring is stretched varies directly as the force pulling on the spring. For a particular spring, a pull of 4 pounds results in a stretch of 5 inches. How far will a pull of 13 pounds stretch the spring?

True–False Critical Thinking Problems

55. Which one of the following is true?

    a. The slope of the line through $(5, 5)$ and $(7, 7)$ is $\frac{7}{5}$.

    b. The graph of $y = 5$ in the Cartesian coordinate plane is the single point $(0, 5)$.

    c. Every line in the Cartesian coordinate plane has an equation in point-slope form.

    d. The line whose equation is $x - 3y + 10 = 0$ passes through the point $(-1, 3)$ and has slope $\frac{1}{3}$.

56. Which one of the following is true?

    a. A linear function with nonnegative slope has a graph that rises from left to right.

    b. The equations $y = 4x$ and $y = -4x$ have graphs that are perpendicular lines.

    c. The line whose equation is $5x + 6y - 30 = 0$ passes through the point $(6, 0)$ and has slope $-\frac{5}{6}$.

    d. The graph of $y = 7$ in the Cartesian coordinate plane is the single point $(7, 0)$.

Technology Problems

Use a graphing utility to graph the three equations in Problems 57–60 on the same viewing rectangle.

57. $y = 3x, y = -3x, y = \frac{1}{3}x$

58. $y = \frac{3}{4}x, y = -\frac{3}{4}x, y = \frac{3}{4}x + 1$

59. $y = -\frac{1}{3}x, y = -\frac{1}{3}x + 2, y = 3x - 4$

60. $y = x - 5, y = x + 2, y = -x + 4$

61. In parts (a) through (d), solve each equation for y and use a graphing utility to graph the line. For each of the original equations, write down the x- and y-intercepts determined from the graph.

    a. $\dfrac{x}{3} + \dfrac{y}{2} = 1$

    b. $\dfrac{x}{-3} + \dfrac{y}{-2} = 1$

    c. $\dfrac{x}{4} + \dfrac{y}{7} = 1$

    d. $\dfrac{x}{-4} + \dfrac{y}{-7} = 1$

    e. Based on parts (a)–(d), describe the graph of $\dfrac{x}{a} + \dfrac{y}{b} = 1$.

Your graphing utility will give you the equation of a line that best fits a set of data. The line of best fit is the one in which the sum of the squares of the vertical distances from each point to the line are at a minimum. The line is called the regression line. Problems 62–64 involve regression lines.

62. The data below is from an article in the *Journal of Environmental Health,* May–June 1965, Volume 27, Number 6, pages 883–897. Radioactive wastes seeping into the Columbia River have exposed citizens of eight Oregon counties and the city of Portland to radioactive contamination. The table lists the number of cancer deaths per 10,000 people for the eight Oregon counties and the city of Portland.

County	Index x	Cancer Deaths per 100,000 Residents y
Sherman	1.3	114
Wasco	1.6	138
Umatilla	2.5	147
Morrow	2.6	130
Gilliam	3.4	130
Hood River	3.8	162
Columbia	6.4	178
Clatsop	8.3	210
City of Portland	11.6	208

a. Use the statistical menu of your graphing utility to enter the nine data points ((1.3, 114) through (11.6, 208)).

b. Use the draw menu and the scatter plot capability to draw a scatter plot of the data points.

c. Select the linear regression option. Your utility should give you the regression coefficients, a and b, for the regression equation (usually in the form $y = ax + bx$). You may also be given a correlation coefficient, r. Values of r close to 1 indicate that the points can be described by a linear relationship and the regression line has a positive slope. Values of r close to -1 indicate that the points can be described by a linear relationship and the regression line has a negative slope. Values of r close to 0 indicate no linear relationship between the variables.

d. Use the appropriate sequence (consult your manual) to graph the regression equation on top of the points in the scatter plot.

63. The National Center for Health Statistics issued the following data for years of life expected at birth for American women and men.

Year	Years of Life Expected at Birth
1950	68.2
1960	69.7
1965	70.2
1970	70.8
1975	72.6
1980	73.7
1985	74.7
1990	75.4

a. Let $x =$ the number of years after 1950 and $y =$ years of life expected at birth. Use the statistical

menu of your graphing utility to enter the eight (x, y) data points.

b. Use the draw menu and the scatter plot capability to draw a scatter plot of the data points.

c. Select the linear regression option and find the equation of the regression line, as well as the correlation coefficient.

d. Use the appropriate sequence to graph the regression equation on top of the points in the scatter plot.

e. Use the regression line's equation to predict years of life expected at birth for the year 2010.

f. In what year will Americans be expected to live 90 years from the time of their birth?

64. The following table presents the men's and women's winning times (in seconds) in the Olympic 100-meter freestyle swimming race from 1912 to 1988.

Year	Men's	Women's
1912	63.2	72.2
1920	61.4	73.6
1924	59.0	72.4
1928	58.6	71.0
1932	58.2	66.8
1936	57.6	65.9
1948	57.3	66.3
1952	57.4	66.8
1956	55.4	62.0
1960	55.2	61.2
1964	53.4	59.5
1968	52.2	60.0
1972	51.22	58.59
1976	49.99	55.65
1980	50.40	54.79
1984	49.80	55.92
1988	48.63	54.93

Figure 2.20 on page 234 shows the graphs of the best-fitting lines (the regression lines) through the points representing the data values for the men and women.

a. Use the statistical menu of your graphing utility to enter the data values in the table for the men's winning times.

b. Select the linear regression option. Your utility should give you the regression coefficients, a and b, for the regression equation (usually in the form $y = ax + b$) and the correlation coefficient r. What is the value of r? Can you explain what this value means?

c. Repeat parts (a) and (b) for the data values in the table for the women's winning times.

d. Graph the equations of both regression lines using your graphing utility. Graph the equations in the same viewing rectangle. Use the $\boxed{\text{ISECT}}$ or the $\boxed{\text{TRACE}}$ feature to find the coordinates of the intersection point.

e. What does the intersection point mean in terms of the women's time and the men's time? What predic- tions could be made about the swimming race in the Olympic years to the right of the intersection point?

Writing in Mathematics

65. What is a reasonable estimate for the slope of each of the six lines designated by L_1 through L_6 in the figure? Describe how you determined a value for the slope of each line.

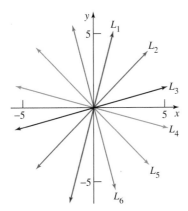

66. Which of the following variables will have a linear rela- tionship over time? Which will not? Explain how you arrived at your answers.

a. U.S. health-care expenditure

b. The number of people with AIDS in the world

c. The price for admission to a movie

d. Population of Canada

e. Population of third-world countries

f. Cost of a computer

g. Position of a free-falling body thrown vertically upward

h. Value of an original painting by Salvador Dali

i. Value of an office copying machine

67. A grasshopper that is $1\frac{1}{4}$-inches long can jump 28 inches. Could this information be used to create a linear model based on the idea that jumping ability is directly pro- portional to body length? Could this linear model be used to predict how high a person who is $5'10''$ can jump? Explain.

Critical Thinking Problems

68. Use the figure to answer the following.

a. Find the equation of the line from C to the midpoint of $\overline{AB}$, called the *median*.

b. Find the equation of the line from C that is perpen- dicular to $\overline{AB}$, called the *altitude*.

c. Find the equation of the line passing through the midpoint of $\overline{AB}$ and perpendicular to $\overline{AB}$, called the *perpendicular bisector* of $\overline{AB}$.

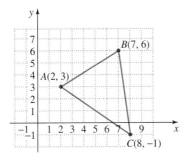

d. Repeat part (a) for the other two medians of the tri- angle.

e. Repeat part (b) for the other two altitudes of the triangle.

f. Repeat part (c) for the other two perpendicular bisectors of the triangle's sides.

69. Repeat all parts of Problem 68 for the triangle shown below.

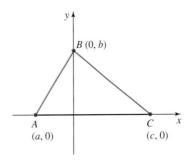

70. Find all points P on the graph of $y = x^2 + 4$ so that the line between P and $(0, 1)$ is perpendicular to the line whose equation is $y = -\frac{1}{4}x + 2$.

71. A line with negative slope has a *y*-intercept of 5. The area of the triangle in the first quadrant formed by the line and the two coordinate axes is 15 square units. Write the point-slope and slope-intercept equations of the line.

72. Prove that the equation of a line passing through $(a, 0)$ and $(0, b)$ $(a \neq 0, b \neq 0)$ can be written in the form $\dfrac{x}{a} + \dfrac{y}{b} = 1$. Why is this called the *intercept form* of a line?

Group Activity Problem _____

73. This problem is a "contest" problem. Work on the problem in groups of 3 to 5 people. Use as many suggestions as possible from group members to solve the problem.

A line with negative slope passes through the point $(2, \frac{6}{5})$. The area of the triangle in the first quadrant

formed by the line and the two coordinates axes is 5 square units. Write the general form of the line's equation.

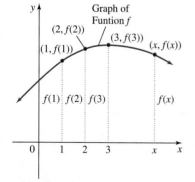

SECTION 2.3 | **Graphs of Relations and Functions**

Solutions Manual   Tutorial   Video 4

Objectives

1 Graph functions and relations.
2 Use the vertical line test to identify functions.
3 Graph circles and write their equations.
4 Graph piecewise functions.

René Descartes' rectangular coordinate system made it possible to construct graphs for relationships between interconnected variables, usually designated by *x* and *y*. In this section, we apply Descartes' ideas to generate relatively accurate pictures of relations and functions.

 Graph functions and relations.

Graphs of Relations and Functions

A useful method for visualizing a function's behavior is from its graph.

> **Graphs of relations and functions**
>
> The graph of a relation is the graph of its ordered pairs. The graph of a function *f* is the set of points $(x, f(x))$ such that *x* is in the domain of *f*.

Figure 2.21

The graph of a function

The graph of function *f* is shown in Figure 2.21. The *y*-coordinate of any point (x, y) on the graph is $y = f(x)$. This means that the value of $f(x)$ is the height of

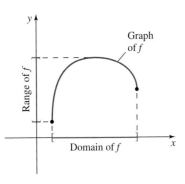

Figure 2.22

A function's domain and range

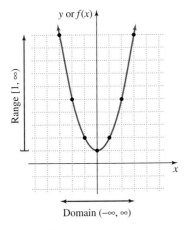

Figure 2.23

The graph of $f(x) = x^2 + 1$

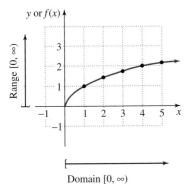

Figure 2.24

The graph of $f(x) = \sqrt{x}$

the graph above the value x. Figure 2.22 illustrates that we can picture a function's domain on the x-axis and its range on the y-axis.

One way to graph a relation or a function is by plotting ordered pairs.

EXAMPLE 1   **Graphing Functions by Plotting Ordered Pairs**

Graph each function and use the graph to specify the function's domain and range.

a. $f(x) = x^2 + 1$ **b.** $f(x) = \sqrt{x}$

Solution

a. The graph of $f(x) = x^2 + 1$ is, by definition, the graph of $y = x^2 + 1$. We begin by setting up a table of coordinates.

x	$f(x) = x^2 + 1$	(x, y) or $(x, f(x))$
-3	$f(-3) = (-3)^2 + 1 = 10$	$(-3, 10)$
-2	$f(-2) = (-2)^2 + 1 = 5$	$(-2, 5)$
-1	$f(-1) = (-1)^2 + 1 = 2$	$(-1, 2)$
0	$f(0) = 0^2 + 1 = 1$	$(0, 1)$
1	$f(1) = 1^2 + 1 = 2$	$(1, 2)$
2	$f(2) = 2^2 + 1 = 5$	$(2, 5)$
3	$f(3) = 3^2 + 1 = 10$	$(3, 10)$

Now we plot the seven points and draw a smooth curve through them, as shown in Figure 2.23. Notice that the vertical axis can be labeled y or $f(x)$. The graph of f has a cuplike shape, with a domain consisting of all real numbers, represented by $(-\infty, \infty)$. The range consists of real numbers greater than or equal to 1, represented by $[1, \infty)$.

b. The graph of $f(x) = \sqrt{x}$ is, by definition, the graph of $y = \sqrt{x}$. We construct a table of coordinates, plot the resulting points, and draw a smooth curve through them to obtain the graph of f shown in Figure 2.24.

x	$f(x) = \sqrt{x}$	(x, y) or $(x, f(x))$
0	$f(0) = \sqrt{0} = 0$	$(0, 0)$
1	$f(1) = \sqrt{1} = 1$	$(1, 1)$
2	$f(2) = \sqrt{2} \approx 1.4$	$(2, \sqrt{2}) \approx (2, 1.4)$
3	$f(3) = \sqrt{3} \approx 1.7$	$(3, \sqrt{3}) \approx (3, 1.7)$
4	$f(4) = \sqrt{4} = 2$	$(4, 2)$
5	$f(5) = \sqrt{5} \approx 2.2$	$(5, \sqrt{5}) \approx (5, 2.2)$

The graph indicates that both the domain and the range of $f(x) = \sqrt{x}$ consist of 0 and all positive real numbers, represented by $[0, \infty)$.   ■

The functions in Example 2 can be graphed by either plotting points or by using our knowledge of graphing linear functions.

EXAMPLE 2   **Graphing Functions Using y-Intercepts and Slopes**

Graph each function and use the graph to specify the function's domain and range.

a. $f(x) = |x|$ **b.** $f(x) = \dfrac{x^2 - 9}{x - 3}$

Solution

a. The graph of $f(x) = |x|$ is, by definition, the graph of the equation $y = |x|$. We can construct a table of coordinates that satisfy $y = |x|$ or we can use the definition of absolute value.

Using coordinates:

x	$f(x) = |x|$	(x, y) or $(x, f(x))$
-3	$f(-3) = |-3| = 3$	$(-3, 3)$
-2	$f(-2) = |-2| = 2$	$(-2, 2)$
-1	$f(-1) = |-1| = 1$	$(-1, 1)$
0	$f(0) = |0| = 0$	$(0, 0)$
1	$f(1) = |1| = 1$	$(1, 1)$
2	$f(2) = |2| = 2$	$(2, 2)$
3	$f(3) = |3| = 3$	$(3, 3)$

Using absolute value:

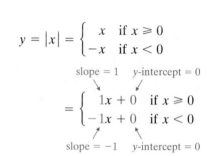

We can either plot the ordered pairs or graph the two lines whose equations are

$$y = x \quad \text{(graphed for } x \geq 0, \text{ on and to the right of the } y\text{-axis)}$$

and

$$y = -x \quad \text{(graphed for } x < 0, \text{ to the left of the } y\text{-axis)}.$$

The graph of $f(x) = |x|$ has a V-shape, shown in Figure 2.25. The domain of f consists of all real numbers, represented by $(-\infty, \infty)$. The range consists only of nonnegative real numbers, represented by $[0, \infty)$.

b. We can obtain the graph of $f(x) = \dfrac{x^2 - 9}{x - 3}$ by first simplifying the expression that defines the function. The exclusion of x-values that yield zero in the denominator means that 3 must be excluded from the domain of f.

$$f(x) = \frac{x^2 - 9}{x - 3} = \frac{(x + 3)(x - 3)}{x - 3} = x + 3 \quad \text{(where } x \neq 3\text{)}$$

Thus we graph f by graphing the equation $y = x + 3$ with $x \neq 3$. We can plot points or we can use the fact that $y = x + 3$ has a y-intercept of 3 and slope 1. The graph of f is shown in Figure 2.26. The open dot above 3 excludes this value from the graph, indicating that $(3, 6)$ does not belong to the graph of f. The domain of f is the set of all real numbers excluding 3, represented by $(-\infty, 3) \cup (3, \infty)$. The range of f is the set of all real numbers excluding 6, represented by $(-\infty, 6) \cup (6, \infty)$. ∎

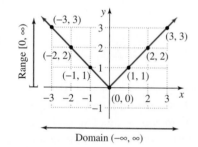

Figure 2.25

The graph of $f(x) = |x|$

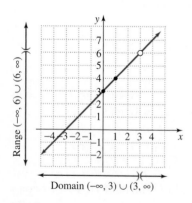

Figure 2.26

The graph of $f(x) = \dfrac{x^2 - 9}{x - 3}$

Graphing Relations That Are Not Functions: The Vertical Line Test

In Section 2.1, we saw that not every equation in x and y defines y as a function of x. Relations that are not functions can still be graphed by plotting points.

> **EXAMPLE 3**   **Graphing a Relation That Is Not a Function**

Graph the relation given by $x = y^2$ and use the graph to specify the relation's domain and range.

Solution

We can obtain the graph of $x = y^2$ by using the square root method and solving the equation for y.

$$y = \pm\sqrt{x}$$

Notice that $\pm$ indicates that corresponding to every nonzero value of x there are two values of y. Therefore, the relation does not define y as a function of x. However, we can still begin by setting up a table of coordinates.

x	$y = \pm\sqrt{x}$	(x, y)
0	$y = \pm\sqrt{0} = 0$	$(0, 0)$
1	$y = \pm\sqrt{1} = \pm 1$	$(1, 1)$ and $(1, -1)$
2	$y = \pm\sqrt{2} \approx \pm 1.4$	$(2, \sqrt{2}) \approx (2, 1.4)$ and $(2, -\sqrt{2}) \approx (2, -1.4)$
3	$y = \pm\sqrt{3} \approx 1.7$	$(3, \sqrt{3}) \approx (3, 1.7)$ and $(3, -\sqrt{3}) \approx (3, -1.7)$
4	$y = \pm\sqrt{4} = \pm 2$	$(4, 2)$ and $(4, -2)$
5	$y = \pm\sqrt{5} \approx \pm 2.2$	$(5, \sqrt{5}) \approx (5, 2.2)$ and $(5, -\sqrt{5}) \approx (5, -2.2)$

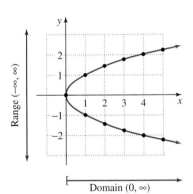

Figure 2.27

The graph of $x = y^2$

By plottings points and drawing a smooth curve through them, we obtain the graph of $x = y^2$ shown in Figure 2.27. The domain is $[0, \infty)$ and the range is the set of all real numbers, $(-\infty, \infty)$.   ∎

2 Use the vertical line test to identify functions.

> **D**iscover for yourself
>
> Look at the graphs in Examples 1–3 (Figure 2.23 through Figure 2.26). How is the graph of $x = y^2$ in Figure 2.27 different from the other graphs? How can this difference be illustrated using vertical lines that intersect each graph?

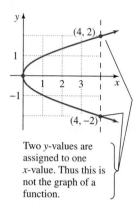

Two y-values are assigned to one x-value. Thus this is not the graph of a function.

Figure 2.28

The graph of a relation that is not a function

In the Discover for yourself, did you observe that you can draw a vertical line that intersects the graph of $x = y^2$ in two points? You could not do this with any of the graphs in Examples 1 and 2. The vertical line drawn through the two points $(4, 2)$ and $(4, -2)$ on the graph of $x = y^2$ (Figure 2.28) indicates that the same x-value, 4, is assigned two different y-values, 2 and -2. Recall that if y is a function of x, then at most one y-value corresponds to a given x-value. For each a in the domain of a function there should be *only one* point $(a, f(a))$ on the graph. Equivalently, each vertical line drawn through the graph of the function must intersect the graph in *at most* one point. We can use this *vertical line test* to determine if a graph defines y as a function of x.

The vertical line test for functions

If any vertical line intersects the graph of a relation in more than one point, the relation does not define y as a function of x.

The vertical line test is illustrated in the following box.

Using the vertical line test

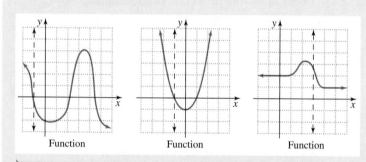

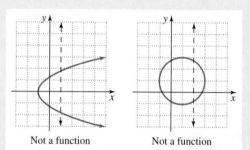

No vertical line intersects the graph more than once, so y is a function of x.

At least one vertical line intersects the graph more than once, so y is not a function of x.

Using technology

The function $N(x) = 0.036x^2 - 2.8x + 58.14$ approximates the number of deaths per year per thousand people (N) for people who are x years old, where $40 \leqslant x \leqslant 60$. Use a graphing utility to graph the function. Points along the graph can be described in practical terms:

$N(40) \approx 4$　　Approximately 4 people per 1000 who are 40 years old die annually.

$N(60) \approx 20$　　Approximately 20 people per 1000 who are 60 years old die annually.

Change range settings to extend the graph to $x = 100$. Why isn't the function a good model for $60 \leqslant x \leqslant 100$? Draw a graph by hand that models death rate in a more accurate manner for $60 \leqslant x \leqslant 100$.

Paul Thek "Dust" 1988, acrylic/gesso on newspaper,
21 1/4 × 27 1/2 in. Photo courtesy Alexander and Bonin.

3 Graph circles and write their equations.

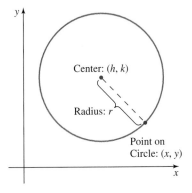

Figure 2.29

A circle centered at (h, k) with radius of length r

Circles

The vertical line test for functions, shown in the preceding box, reveals that the graph of a circle in the Cartesian coordinate plane does not define y as a function of x. However, we can still use the coordinate system to translate the definition of a circle into an algebraic equation involving x and y.

A circle is the set of all points in a plane that are equidistant from a fixed point, called the *center*. The fixed distance from the circle's center to any point on the circle is called the *radius*. Figure 2.29 illustrates a circle with radius of length r and center (h, k) in the Cartesian coordinate plane. The point (x, y) is on this circle if and only if its distance from the center is r. Using the distance formula, we have

$$\sqrt{(x - h)^2 + (y - k)^2} = r.$$

Squaring both sides of this equation yields the *standard form of the equation of a circle.*

Standard form of the equation of a circle

The equation of the circle with the center at (h, k) and radius of length r is

$$(x - h)^2 + (y - k)^2 = r^2.$$

Keith Haring "Monkey Puzzle" 1988, acrylic on canvas, 120 in. diameter.
©The Estate of Keith Haring.

EXAMPLE 4 **Finding the Equation of a Circle and Graphing the Circle**

Find the equation of the circle with a radius of length 2 and the center at $(0, 0)$. Graph the circle.

Figure 2.30

The graph of $x^2 + y^2 = 4$

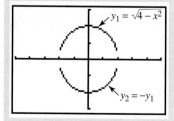

Using technology

To graph a circle with a graphing utility, first solve the equation for y.

$$x^2 + y^2 = 4$$

$$y^2 = 4 - x^2$$

$$y = \pm\sqrt{4 - x^2}$$

Then graph the two functions

$$y_1 = \sqrt{4 - x^2}$$

and

$$y_2 = -\sqrt{4 - x^2}$$

in the same viewing rectangle. Use a $\boxed{\text{ZOOM SQUARE}}$ setting so that the circle looks like a circle. Many graphing utilities have problems connecting the two semicircles because the segments directly across horizontally from the center become nearly vertical.

Solution

Because $(h, k) = (0, 0)$ and $r = 2$, we have

$$(x - h)^2 + (y - k)^2 = r^2 \qquad \text{This is the equation of a circle.}$$

$$(x - 0)^2 + (y - 0)^2 = 2^2 \qquad \text{Substitute 0 for } h, 0 \text{ for } k, \text{ and 2 for } r.$$

$$x^2 + y^2 = 4 \qquad \text{Simplify.}$$

The equation of the circle is $x^2 + y^2 = 4$. Figure 2.30 shows the graph. ∎

Observe from Example 4 that the equation of any circle with the center at the origin $(0, 0)$, and radius r is $x^2 + y^2 = r^2$. The equation $x^2 + y^2 = 4$, graphed in Figure 2.30, does not define y as a function of x. Vertical lines drawn between -2 and 2 intersect the graph more than once.

Example 5 involves finding the equation of a circle whose center is not at the origin.

EXAMPLE 5　**Writing the Equation of a Circle**

Find the equation of the circle with the center at $(-2, 3)$ and a radius of length 4.

Solution

We have $(h, k) = (-2, 3)$ and $r = 4$. Thus,

$$(x - h)^2 + (y - k)^2 = r^2 \qquad \text{This is the equation of a circle.}$$

$$[x - (-2)]^2 + (y - 3)^2 = 4^2 \qquad \text{Substitute } -2 \text{ for } h, 3 \text{ for } k, \text{ and 4 for } r.$$

$$(x + 2)^2 + (y - 3)^2 = 16 \qquad \text{Simplify.}$$

The equation of the circle is $(x + 2)^2 + (y - 3)^2 = 16$. ∎

EXAMPLE 6　**Using the Equation of a Circle to Graph It**

Find the center and radius of the circle whose equation is

$$(x - 2)^2 + (y + 4)^2 = 9$$

and sketch the graph.

Solution

Because $(x - h)^2 + (y - k)^2 = r^2$ and we have $(x - 2)^2 + [(y - (-4)]^2 = 3^2$, the center is at $(2, -4)$ and the radius is 3. Figure 2.31 on page 249 shows the graph. ∎

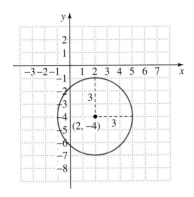

Figure 2.31

The graph of the relation
$(x - 2)^2 + (y + 4)^2 = 9$

If we square $x - 2$ and $y + 4$ in the equation of Example 6, we obtain

$$(x - 2)^2 + (y + 4)^2 = 9 \qquad \text{This is the equation from Example 6.}$$

$$x^2 - 4x + 4 + y^2 + 8y + 16 = 9 \qquad \text{Square } x - 2 \text{ and } y + 4.$$

$$x^2 + y^2 - 4x + 8y + 20 = 9 \qquad \text{Combine numerical terms.}$$

$$x^2 + y^2 - 4x + 8y + 11 = 0 \qquad \text{Subtract 9 from both sides.}$$

This result suggests that an equation in the form $x^2 + y^2 + Dx + Ey + F = 0$ represents a circle. We can return to the familiar form $(x - h)^2 + (y - k)^2 = r^2$ by completing the square on x and y. Let's see how this is done.

EXAMPLE 7 **Completing the Square to Find a Circle's Center and Radius**

Graph: $x^2 + y^2 + 4x - 6y - 23 = 0$

Solution

Because we plan to complete the square on both x and y, let's rearrange terms so that x-terms are arranged in descending order, y-terms are arranged in descending order, and the constant term appears on the right.

$$x^2 + y^2 + 4x - 6y - 23 = 0$$

$$(x^2 + 4x \quad) + (y^2 - 6y \quad) = 23 \qquad \begin{array}{l}\text{Rewrite in anticipation of} \\ \text{completing the square.}\end{array}$$

$$(x^2 + 4x + 4) + (y^2 - 6y + 9) = 23 + 4 + 9 \qquad \begin{array}{l}\text{Complete the square on } x: \\ \frac{1}{2} \cdot 4 = 2 \text{ and } 2^2 = 4, \text{ so add 4} \\ \text{to both sides. Complete the} \\ \text{square on } y: \frac{1}{2}(-6) = -3 \text{ and} \\ (-3)^2 = 9, \text{ so add 9 to both} \\ \text{sides.}\end{array}$$

$$(x + 2)^2 + (y - 3)^2 = 36 \qquad \begin{array}{l}\text{Factor on the left and add on} \\ \text{the right.}\end{array}$$

$$(x + 2)^2 + (y - 3)^2 = 6^2 \qquad \begin{array}{l}\text{This equation is in the form} \\ (x - h)^2 + (y - k)^2 = r^2, \text{ with} \\ \text{the center at } (h, k) \text{ and radius } r.\end{array}$$

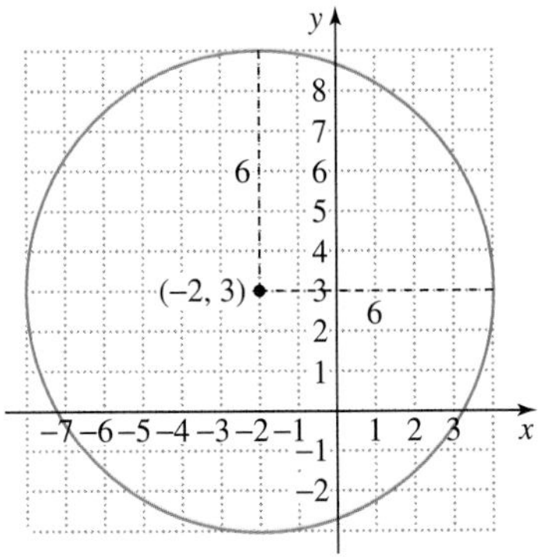

Thus, the center is at $(-2, 3)$, and the radius 6. Figure 2.32 shows the graph. ∎

Figure 2.32

The graph of the relation
$(x + 2)^2 + (y - 3)^2 = 36$

Using technology

To graph $x^2 + y^2 + 4x - 6y - 23 = 0$, rewrite the equation as a quadratic equation in y.

$$y^2 - 6y + (x^2 + 4x - 23) = 0$$

Now solve for y using the quadratic formula, with $a = 1$, $b = -6$, and $c = x^2 + 4x - 23$.

$$y = \frac{-b \pm \sqrt{b^2 - 4ac}}{2a} = \frac{-(-6) \pm \sqrt{(-6)^2 - 4 \cdot 1(x^2 + 4x - 23)}}{2 \cdot 1} = \frac{6 \pm \sqrt{36 - 4(x^2 + 4x - 23)}}{2}$$

Since we will enter these equations, there is no need to simplify. Enter

$$y_1 = \frac{6 + \sqrt{36 - 4(x^2 + 4x - 23)}}{2}$$

and

$$y_2 = \frac{6 - \sqrt{36 - 4(x^2 + 4x - 23)}}{2}.$$

Use a ZOOM SQUARE setting. The graph is shown on the right.

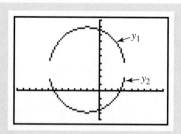

---

4 Graph piecewise functions.

Modeling with Piecewise Functions

Because conditions frequently change over time, functions that model real world phenomena over time must frequently be defined by two or more equations. Therefore, descriptions of the physical world often require piecewise functions. Graphing utilities are helpful in drawing these functions.

EXAMPLE 8   Modeling the Average Miles Driven

The piecewise function

$$f(x) = \begin{cases} 0.0005x^2 + 0.025x + 8.8 & \text{if } 0 \le x \le 30 \\ 0.0202x^2 - 1.58x + 39.22 & \text{if } x > 30 \end{cases}$$

models the average number of miles driven in thousands per car in the United States x years after 1940. Use a graphing utility to draw the graph of the function.

Solution

We will use our graphing utility to graph the functions

$$y_1 = 0.0005x^2 + 0.025x + 8.8$$

and

$$y_2 = 0.0202x^2 - 1.58x + 39.22.$$

If we project the model up to the year 2000, a reasonable range setting for x is

Xmin = 0, Xmax = 60, and Xscl = 2.

Since $f(x)$ represents the average number of miles driven in thousands per car per year, we will use

Ymin = 0, Ymax = 25, and Yscl = 1.

We used the ISECT feature to obtain Figure 2.33, which shows that the graphs intersect at (30, 10). The graph of the first function ends up at the same point where the second one begins. In practical terms, this means that 30 years after 1940, or in 1970, cars in the United States were driven (on the average) 10 thousand miles.

Now we can draw f (by hand) by taking the part of the graph of y_1 to the left of $x = 30$ and combining it with the part of the graph of y_2 to the right of $x = 30$. The resulting graph is shown in Figure 2.34. ∎

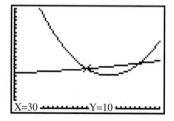

Figure 2.33

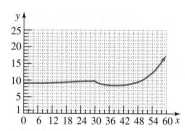

Figure 2.34

A graph showing average miles driven per car in the U.S.

Using technology

The hand-drawn graph in Figure 2.34 can be obtained with a graphing utility. An equation similar to

$$y_1 = (x \le 30)(0.0005x^2 + 0.025x + 8.8) + (x > 30)(0.0202x^2 - 1.58x + 39.22)$$

should give the graph. Consult your manual for the location of the $\boxed{\le}$ and $\boxed{>}$ keys.

A piecewise function can also be drawn by hand by plotting ordered pairs and using techniques for graphing linear functions.

EXAMPLE 9   **Graphing a Piecewise Function**

Sketch by hand the graph of the function f defined by

$$f(x) = \begin{cases} -x + 2 & \text{if } x \le 2 \\ x^2 & \text{if } x > 2 \end{cases}$$

Solution

The graph of $y = -x + 2$ is a line with slope -1 and y-intercept 2, as shown in Figure 2.35

We can graph $y = x^2$ by constructing a table of coordinates. The graph is shown in Figure 2.36.

x	-2	-1	0	1	2	3
$y = x^2$	4	1	0	1	4	9

To graph f (Figure 2.37), we graph the part of the line that lies to the left of $x = 2$. The solid dot at $(2, 0)$ shows that this point is included on the graph. The graph of $y = x^2$ is shown only for $x > 2$; the open dot at $(2, 4)$ indicates that this point is not part of the graph.   ■

Some piecewise functions have graphs that consist of a collection of horizontal line segments, or *steps*, with a jump occurring at each integer.

EXAMPLE 10   **Modeling the Cost of a Long Distance Phone Call**

The cost of a daytime long distance phone call from Miami to Talahassee is $0.50 for the first minute and $0.25 for each additional minute (or portion of a minute). Draw a graph of the cost (C, in dollars) of the phone call as a function of time (t, in minutes).

Solution

Let $C(t)$ be the cost for t minutes. For calls up to 1 minute, the cost is $0.50. For calls between 1 and 2 minutes, the cost is $0.75; between 2 and 3 minutes, the cost is $1.00, and so on.

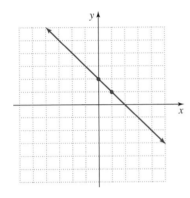

Figure 2.35

The graph of $y = -x + 2$

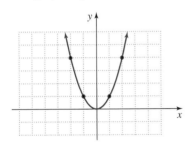

Figure 2.36

The graph of $y = x^2$

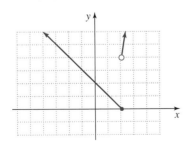

Figure 2.37

The graph of $y = -x + 2$ for $x \le 2$ and $y = x^2$ for $x > 2$

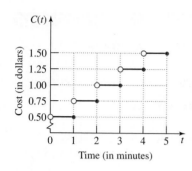

Figure 2.38

Cost of a long distance call

t: Length of call	$0 < t \leqslant 1$	$1 < t \leqslant 2$	$2 < t \leqslant 3$	$3 < t \leqslant 4$
$C(t)$: Cost of call	0.50	0.75	1.00	1.25

Using these values, we obtain the graph of the cost function shown in Figure 2.38. Since $C(t)$ is constant in each interval, the graph consists of a different horizontal line in each interval.   ■

A general step function, called the *greatest integer function,* is denoted by $[[x]]$.

Definition of the greatest integer function

$f(x) = [[x]] = $ the greatest integer less than or equal to x.

Some values of the greatest integer function are as follows:

If $-1 \leqslant x < 0$,

$$[[-1]] = -1, \quad [[-0.9]] = -1, \quad [[-0.5]] = -1, \quad [[-0.001]] = -1$$

Greatest integer less than or equal to -1 Greater integer less than or equal to -0.9 Greater integer less than or equal to -0.5 Greater integer less than or equal to -0.001

Thus, if $-1 \leqslant x < 0$, then $[[x]] = -1$.

If $0 \leqslant x < 1$,

$$[[0]] = 0, \quad [[0.3]] = 0, \quad [[\tfrac{1}{2}]] = 0, \quad [[0.9]] = 0.$$

The greatest integer less than or equal to $0, 0.3, \tfrac{1}{2}$, and 0.9 is 0.

Thus, if $0 \leqslant x < 1$, then $[[x]] = 0$.

If $1 \leqslant x < 2$,

$$[[1]] = 1, \quad [[1.3]] = 1, \quad [[1\tfrac{1}{2}]] = 1, \quad [[1.9]] = 1.$$

The greatest integer less than or equal to $1, 1.3, 1\tfrac{1}{2}$, and 1.9 is 1.

Thus, if $1 \leqslant x < 2$ then $[[x]] = 1$.

In general:

If $n \leqslant x < n + 1$, where n is an integer, then $[[x]] = n$.

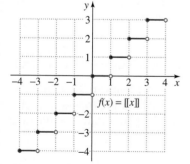

Figure 2.39

The graph of the greatest integer function

The graph of $f(x) = [[x]]$ is shown in Figure 2.39 for x in the interval $[-4, 4)$. The graph of f consists of a collection of horizontal line segments, or steps, with a jump occurring at each integer. The domain is $(-\infty, \infty)$ and the range is the set of integers.

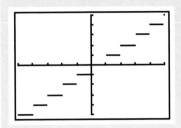

Using technology

You can use a graphing utility to graph the greatest integer function. The function's equation is entered as

$$Y_1 = \boxed{INT}\ X$$

Consult your manual for the location of the $\boxed{INT}$ (greatest integer) key. Without tracing along the graph, it is impossible to tell by looking at the graph that for each step the point on the left is included and the point on the right is not.

PROBLEM SET 2.3

Practice Problems

Graph the relations in Problems 1–30 by either plotting points or, if applicable, using your knowledge about slope and y-intercept for graphing linear functions. Specify the relation's domain and range from the graph.

1. $f(x) = x^2 - 2$ **2.** $f(x) = x^2 - 1$ **3.** $f(x) = x^2 + 2x + 1$ **4.** $f(x) = x^2 - 4x + 4$

5. $f(x) = x^3$ **6.** $f(x) = x^3 + 1$ **7.** $f(x) = x^3 - 2x$ **8.** $f(x) = x^3 - 3x$

9. $f(x) = \sqrt{x} + 1$ **10.** $f(x) = \sqrt{x} + 2$ **11.** $f(x) = \sqrt{x + 1}$ **12.** $f(x) = \sqrt{x + 2}$

13. $f(x) = 2|x|$ **14.** $f(x) = 3|x|$ **15.** $f(x) = |x| - 1$ **16.** $f(x) = |x| - 2$

17. $f(x) = |x - 1|$ **18.** $f(x) = |x - 2|$ **19.** $f(x) = |x| + x$ **20.** $f(x) = |x| - x$

21. $f(x) = \dfrac{x^2 - 4}{x - 2}$ **22.** $f(x) = \dfrac{x^2 - 16}{x - 4}$ **23.** $f(x) = \dfrac{x^2 - 1}{x + 1}$ **24.** $f(x) = \dfrac{x^2 - 25}{x + 5}$

25. $x = 4y^2$ **26.** $x = 8y^2$ **27.** $x = -y^2$ **28.** $x = -4y^2$

29. $G(x) = \dfrac{x}{|x|}$ **30.** $H(x) = \dfrac{|x|}{x}$

For Problems 31–42, use the vertical line test to identify graphs in which y is a function of x. Use the graph to give the domain and range of each relation.

31.

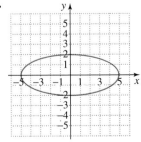

32.

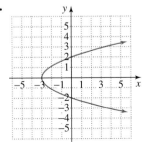

33.

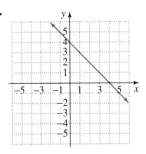

34.

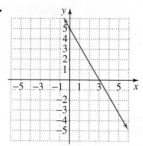

35.

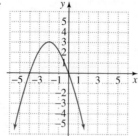

36.

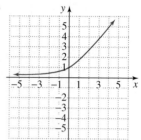

37.

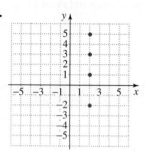

38.

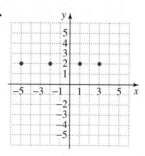

39.

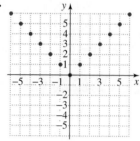

40.

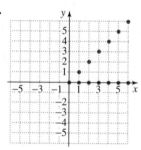

41.

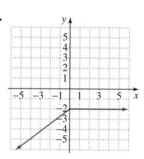

42.

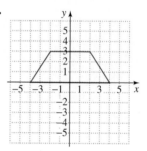

In Problems 43–52, write the standard form of the equation of the circle with the given center and radius.

43. Center $(0, 0)$, $r = 7$

44. Center $(0, 0)$, $r = 8$

45. Center $(3, 2)$, $r = 5$

46. Center $(2, -1)$, $r = 4$

47. Center $(-1, 4)$, $r = 2$

48. Center $(-3, 5)$, $r = 3$

49. Center $(-3, -1)$, $r = \sqrt{3}$

50. Center $(-5, -3)$, $r = \sqrt{5}$

51. Center $(-4, 0)$, $r = 0$

52. Center $(-2, 0)$, $r = 6$

In Problems 53–60, give the center and radius of the circle described by the equation and graph each equation.

53. $x^2 + y^2 = 16$

54. $x^2 + y^2 = 49$

55. $(x - 3)^2 + (y - 1)^2 = 36$

56. $(x - 2)^2 + (y - 3)^2 = 16$

57. $(x + 3)^2 + (y - 2)^2 = 4$

58. $(x + 1)^2 + (y - 4)^2 = 25$

59. $(x + 2)^2 + (y + 2)^2 = 4$

60. $(x + 4)^2 + (y + 5)^2 = 36$

In Problems 61–68, complete the square and then give the center and radius of each circle. Then graph each equation.

61. $x^2 + y^2 + 6x + 2y + 6 = 0$

62. $x^2 + y^2 + 8x + 4y + 16 = 0$

63. $x^2 + y^2 - 10x - 6y - 30 = 0$

64. $x^2 + y^2 - 4x - 12y - 9 = 0$

65. $x^2 + y^2 + 8x - 2y - 8 = 0$

66. $x^2 + y^2 + 12x - 6y - 4 = 0$

67. $x^2 - 2x + y^2 - 15 = 0$

68. $x^2 + y^2 - 6y - 7 = 0$

Graph each piecewise function in Problems 69–78.

69. $f(x) = \begin{cases} 2x + 3 & \text{if } x \leqslant 0 \\ x + 3 & \text{if } x > 0 \end{cases}$

70. $f(x) = \begin{cases} x + 4 & \text{if } x \leqslant 0 \\ 2x + 4 & \text{if } x > 0 \end{cases}$

71. $f(x) = \begin{cases} -x + 1 & \text{if } x \leqslant 1 \\ 2x - 2 & \text{if } x > 1 \end{cases}$

72. $f(x) = \begin{cases} -x + 2 & \text{if } x \leqslant 2 \\ 3x - 6 & \text{if } x > 2 \end{cases}$

73. $f(x) = \begin{cases} 3 - x & \text{if } x < 3 \\ \sqrt{x - 3} & \text{if } x \geqslant 3 \end{cases}$

74. $f(x) = \begin{cases} -x - 2 & \text{if } x < -2 \\ \sqrt{x + 2} & \text{if } x \geqslant -2 \end{cases}$

75. $f(x) = \begin{cases} 2 & \text{if } x \leqslant 0 \\ x^2 & \text{if } x > 0 \end{cases}$

76. $f(x) = \begin{cases} 3 & \text{if } x \leqslant -1 \\ x^2 + 1 & \text{if } x > -1 \end{cases}$

77. $g(x) = \begin{cases} x^2 + 1 & \text{if } x \neq 0 \\ 4 & \text{if } x = 0 \end{cases}$

78. $g(x) = \begin{cases} x^2 - 1 & \text{if } x \neq 0 \\ 2 & \text{if } x = 0 \end{cases}$

Application Problems

79. The concentration of a drug (in parts per million) in a patient's bloodstream t hours after the drug is administered is given by the function $f(t) = -t^4 + 12t^3 - 58t^2 + 132t$. Fill in the following table of coordinates and graph the function. What does the shape of the graph indicate about the drug's concentration over time? After how many hours will the drug be totally eliminated from the bloodstream? How is this indicated by the graph?

t	$f(t) = -t^4 + 12t^3 - 58t^2 + 132t$
0	
1	
2	
3	
4	
5	
6	

80. A cargo service charges a flat fee of $4 plus $1 for each pound or fraction of a pound to mail a package. Let $C(x)$ represent the cost to mail a package that weighs x pounds. Graph the cost function in the interval $(0, 5]$.

81. The cost of a telephone call between two cities is $1 for the first minute and $0.50 for each additional minute or portion of a minute. Draw a graph of the cost (C, in dollars) of the phone call as a function of time (t, in minutes) in the interval $(0, 5]$.

True–False Critical Thinking Problems

82. Which one of the following is true based on the graph of f in the figure?

   a. The domain of f is $[-4, 1) \cup (1, 4]$.

   b. The range of f is $[-2, 2]$.

   c. $f(-1) - f(4) = 2$

   d. $f(0) = 2.1$

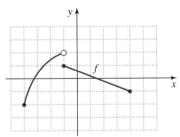

a. There is precisely one value of x for which $g(x) = f(1)$.

b. $f(2) - g(2) > f(1) - g(1)$

c. The sum of $\dfrac{f(x) - f(2)}{x - 2}$ when $x = 3$ and

   $\dfrac{g(x) - g(-2)}{x + 2}$ when $x = -3$ is 1.

d. None of these statements is true.

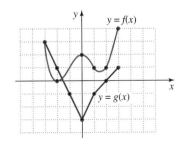

83. Which one of the following is true based on the graphs of f and g in the figure shown at the right?

84. Which one of the following is true?

a. The equation of the circle whose center is at the origin with radius of length 16 is $x^2 + y^2 = 16$.

b. The graph of $(x - 3)^2 + (y + 5)^2 = 36$ is a circle with a radius of 6 centered at $(-3, 5)$.

c. The graph of $(x - 4) + (y + 6) = 25$ is a circle with a radius of 5 centered at $(4, -6)$.

d. None of the above is true.

85. Which one of the following is true?

a. The coordinates of a circle's center satisfy the circle's equation.

b. The radius of the circle whose equation is given by $x^2 - 6x + y^2 = 25$ is 5.

c. If the center of a circle is at $(-1, -2)$ and the circle passes through the origin, then not enough information is provided to write the circle's equation.

d. None of the above is true.

Technology Problems

86. The function

$$C(x) = \begin{cases} -0.35x + 220 & \text{for } 0 \leqslant x \leqslant 20 \\ -0.80x + 229 & \text{for } x > 20 \end{cases}$$

describes the number of milligrams of cholesterol per deciliter of blood for American adults x years after 1960.

a. Graph the function using a graphing utility.

b. What does the graph indicate about cholesterol level from 1960 to the present?

c. Will the goal of lowering cholesterol to a level under 200 be reached by the turn of the century?

Use a graphing utility to graph each relation and function in Problems 87–92.

87. $f(x) = 2[[x+1]]$

88. $g(x) = 4 - [[x]]$

89. $y^2 = x^4 + 3x^2$

90. $f(x) = x^2 + x - 2$ and $g(x) = |x^2 + x - 2|$ (In what ways are these graphs similar and different?)

91. $f(x) = \begin{cases} 3x - x^2 & \text{if } x > 1 \\ (x - 1)^2 & \text{if } x \leqslant 1 \end{cases}$

92. $f(x) = \begin{cases} \sqrt{-x} & \text{if } x < 0 \\ x^2 & \text{if } 0 \leqslant x < 3 \\ \sqrt{x - 3} & \text{if } x \geqslant 3 \end{cases}$

Use a graphing utility to graph each circle whose equation is given in Problems 93–95.

93. $x^2 + y^2 = 25$

94. $(y + 1)^2 = 36 - (x - 3)^2$

95. $x^2 + 10x + y^2 - 4y - 20 = 0$

Writing in Mathematics

96. Discuss one disadvantage to using point plotting as a method for graphing functions.

97. The equation $x^2 + y^2 = 25$ does not define y as a function of x. However, the semicircles whose equations are $y = \sqrt{25 - x^2}$ and $y = -\sqrt{25 - x^2}$ do represent y as a function of x. Does every semicircle formed by $x^2 + y^2 = 25$ define y as a function of x? Explain, illustrating with graphs and equations.

98. Does $(x - 3)^2 + (y - 5)^2 = 0$ represent the equation of a circle? If not, describe the graph of this equation.

99. Does $(x - 3)^2 + (y - 5)^2 = -25$ represent the equation of a circle? What sort of set is the graph of this equation?

Critical Thinking Problems

For Problems 100–103, select the graph from (a)–(h) shown below that best models the given verbal conditions.

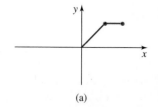

(a)

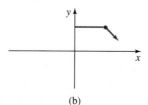

(b)

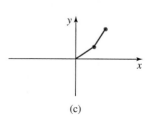

(c)

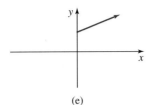

(d)

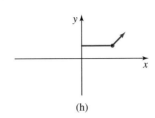

(e)

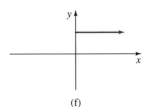

(f)

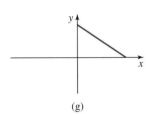

(g)

(h)

100. A utility charges a flat fee of $14 for the first 35 kilo-watt-hours or less, and $0.13 more for each kilowatt-hour above 35. Select the graph that models the relationship between the amount of an electric bill (y) and the total number of kilowatt-hours (x) used by a customer each month.

101. A person walks from home at 1:00 P.M., walking westward at a constant speed of 3 miles per hour for 30 minutes. After 30 minutes, that person meets a friend and stops to talk for 10 minutes. Select the graph that models the relationship between distance from home (y) and time (x).

102. A car travels 60 miles per hour for 2 hours, then slows down to 40 miles per hour for the next 3 hours. Select the graph that models the relationship between the distance that the car travels (y) in x hours.

103. A company cannot sell any radios if the price is $600. However, each price decrease of $45 per radio is accompanied by one additional radio being sold. Select the graph that models the relationship between the price per radio (y) and the number of radios that can be sold (x).

104. Find the area of the region bounded by the graphs of $(x - 2)^2 + (y + 3)^2 = 25$ and $(x - 2)^2 + (y + 3)^2 = 36$.

SECTION 2.4

 Solutions Manual

 Tutorial

 Video 4

Increasing and Decreasing Functions; Extreme Values

Objectives

1 Identify intervals in which a function increases or decreases.

2 Use a verbal model to sketch a function.

3 Find relative extrema.

The graph of a function provides a visual way of showing the relationship between the independent and the dependent variable. Knowing where the graph of a function rises and where it falls provides useful information. Also useful is the ability to identify lowest and highest points on the graph.

Identify intervals in which a function increases or decreases.

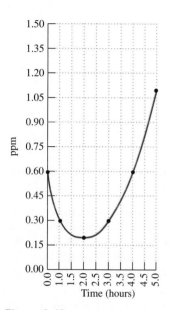

Figure 2.40

Air pollution as a function of time

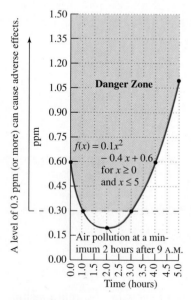

Figure 2.41

Air pollution as a function of time

| **EXAMPLE 1** | **Summer's Air Pollution as a Function of the Time of the Day** |

Although the level of air pollution varies from day to day and from hour to hour, during the summer the level of air pollution is generally a function of the time of the day. The function

$$f(x) = 0.1x^2 - 0.4x + 0.6, \ 0 \leqslant x \leqslant 5$$

describes the level of air pollution (in parts per million, ppm), where x corresponds to the number of hours after 9 A.M. The graph of this function is shown in Figure 2.40.

a. Verify the point $(1, 0.3)$ and describe what this means in practical terms.
b. For what values of x is the graph of the function falling? What does this mean?
c. For what values of x is the graph of the function rising? What does this mean?
d. For what value of x does the lowest point on the graph of f occur? What is the minimum value of y? Describe the practical significance of this minimum value.
e. Researchers have determined that a level of 0.3 ppm of pollutants in the air can be hazardous to your health. Based on the graph, at what time of day should runners exercise to be at a safe level?

Solution

a. We can verify the point $(1, 0.3)$ by evaluating the function at $x = 1$.

$$f(x) = 0.1x^2 - 0.4x + 0.6 \qquad \text{This is the given function.}$$

$$f(1) = 0.1(1)^2 - 0.4(1) + 0.6 \qquad \text{Find } f \text{ at 1.}$$

$$= 0.3$$

This means that one hour after 9 A.M., or at 10 A.M., the level of air pollution is 0.3 ppm.

b. The graph in Figure 2.40 falls and then rises as we move from left to right. It is falling between $x = 0$ and $x = 2$. We can say that the function f is *decreasing* on the interval $(0, 2)$. This means that between 0 and 2 hours after 9 A.M., or between 9 A.M. and 11 A.M., the level of air pollution is decreasing.

c. The graph of the function f is rising between $x = 2$ and $x = 5$. We can say that the function f is *increasing* on the interval $(2, 5)$. This means that between 2 and 5 hours after 9 A.M., or between 11 A.M. and 2 P.M., the level of air pollution is increasing.

d. The lowest point on the graph of f occurs when $x = 2$. The minimum value of y can be found by evaluating the function at $x = 2$.

$$f(x) = 0.1x^2 - 0.4x + 0.6 \qquad \text{This is the given function.}$$

$$f(2) = 0.1(2)^2 - 0.4(2) + 0.6 \qquad \text{Find } f \text{ at 2.}$$

$$= 0.2$$

This means that air pollution reaches its lowest level 2 hours after 9 A.M., or at 11 A.M., and that the minimum level is 0.2 ppm.

e. The graph in Figure 2.41 indicates that the level of air pollution decreases to a minimum at 11 A.M. and then increases above safe levels. A runner should

Discover for yourself

The lowest point on the graph of

$$f(x) = 0.1x^2 - 0.4x + 0.6$$

is $(2, 0.2)$. Provide numerical support to this statement by evaluating $f(1.9)$ and $f(2.1)$.

M.C. Escher (1898–1972) "Other World" © 1997 Cordon Art – Baarn – Holland. All rights reserved.

exercise sometime between 1 and 3 hours after 9 A.M. (between 10 A.M. and noon), with 11 A.M. being the ideal time of day.   ∎

Example 1 illustrates that a function f is increasing when its graph rises and decreasing when its graph falls. It is also possible for a function to remain constant, neither rising nor falling. We provide a formal definition of these concepts as follows.

Increasing, decreasing, and constant functions

1. A function is *increasing* on an interval if for any x_1 and x_2 in the interval, where $x_1 < x_2$, then $f(x_1) < f(x_2)$.
2. A function is *decreasing* on an interval if for any x_1 and x_2 in the interval, where $x_1 < x_2$, then $f(x_1) > f(x_2)$.
3. A function is *constant* on an interval if for any x_1 and x_2 in the interval, where $x_1 < x_2$, then $f(x_1) = f(x_2)$.

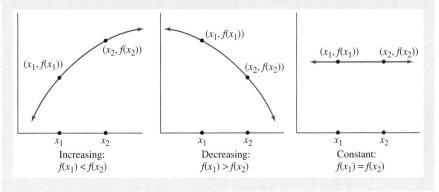

Increasing:
$f(x_1) < f(x_2)$

Decreasing:
$f(x_1) > f(x_2)$

Constant:
$f(x_1) = f(x_2)$

EXAMPLE 2   **Intervals on Which a Function Increases or Decreases**

When a person receives a drug injected into a muscle, the concentration of the drug in the blood (measured in milligrams per 100 milliliters) is a function of the time elapsed since the injection (measured in hours). Figure 2.42 shows the graph of such a function, where

$x =$ hours since the injection

and

$f(x) =$ drug concentration at time x.

State the intervals on which the function is increasing and decreasing and describe what this means in terms of the variables modeled in the graph.

Solution

The function is increasing on the interval $(0, 3)$, that is, between $x = 0$ and $x = 3$. It is decreasing on the interval $(3, 13)$, or between $x = 3$ and $x = 13$. This means that once the drug is injected into the muscle, the drug spreads into the blood and its concentration increases up to 3 hours after the injection. (A maximum concentration of 0.05 milligram per 100 milliliters is reached after 3

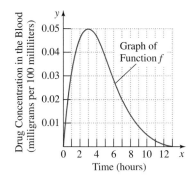

Figure 2.42

Concentration of a drug as a function of time

hours.) Between 3 and 13 hours after the injection, the drug concentration in the blood decreases. Since $f(13) = 0$, at the end of 13 hours there is no longer any drug in the blood.   ■

2 Use a verbal model to sketch a function.

Hannah Wilke "Why Not Sneeze" 1992, wire bird cage, plastic medicine bottles and syringes, $7 \times 9 \times 6\ 7/8$ in. Courtesy Ronald Feldman Fine Arts, NY. Photo by Dennis Cowley.

EXAMPLE 3   Sketching a Function Using Given Conditions

A person who is ill with the flu discovers that body temperature, $f(x)$, is a function of time x, where x represents the number of hours after 8 A.M. Sketch the function based on the following properties.

a. $f(0) = 101$ (Temperature at 8 A.M. is 101°.)
b. Temperature decreases on the interval $(0, 3)$, reaching normal (98.6°) by 11 A.M.
c. Temperature increases on the interval $(3, 5)$.
d. $f(5) = 100$ (Temperature at 1 P.M. is 100°.)
e. Temperature remains constant at 100° on the interval $(5, 7)$. (The last time temperature is taken is at 3 P.M., where $f(7) = 100$.)

Solution

Figure 2.43 shows the graph of body temperature as a function of time.

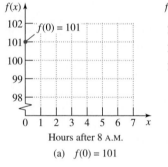

(a) $f(0) = 101$

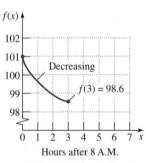

(b)   Temperature decreases in $(0, 3)$, reaching 98.6° by 11 A.M.

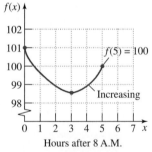

(c)   Temperature increases in $(3, 5)$.
(d) $f(5) = 100$

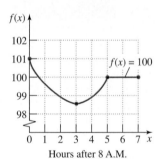

(e)   Temperature remains constant at 100° in $(5, 7)$.

Figure 2.43
Body temperature as a function of time   ■

Verbal descriptions can often be translated into graphs, allowing us to visualize a given situation.

EXAMPLE 4   A Graph That Models a Verbal Description

Represent the following description by a possible graph that models distance from home as a function of time.

I left home for school driving at a constant rate. I briefly stopped at a friend's house when I remembered I had forgotten my homework. I returned home, still driving at a constant rate, although faster than I had originally driven. At that point, I left home again and continued to speed up as I became concerned about being late for my first class.

Solution

Figure 2.44 shows a possible graph.

1. This piece of the graph shows the distance from home increasing at a constant rate, since the narrator drove at a constant speed.
2. This piece of the graph shows the narrator's stop at a friend's house. During the duration of the stop, the distance from home remains constant.
3. This piece of the graph represents the return home, steadily decreasing until the distance from home is 0. Notice that (3) falls more rapidly than (1) rises because the return home was at a faster rate than originally driven.
4. A brief period of time is spent at home to find the forgotten homework. The distance from home remains constant at 0.
5. The increasing steepness of this part of the graph shows distance from home increasing at a faster and faster rate as the narrator continued to speed up. The *y*-coordinate of the point on the right represents the distance from home to school.   ■

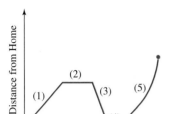

Figure 2.44

Modeling a verbal description with a graph

If a function's equation is known, a graphing utility can be used to determine intervals on which the function increases and decreases.

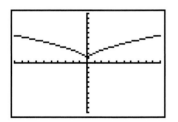

Figure 2.45

The graph of $y = x^{2/3} + 1$

tudy tip

Some graphing utilities do not evaluate $x^{2/3}$ (entered as x ⌃ $(2/3)$) when x is negative. If you do not obtain the graph in Figure 2.45, enter the function's equation as

$$y_1 = (x \; \boxed{\land} \; (1/3)) \; \boxed{\land} \; 2 + 1.$$

EXAMPLE 5   **Using a Graphing Utility**

a. Use a graphing utility to graph the function $f(x) = x^{2/3} + 1$.
b. Find the domain and range of the function.
c. State the intervals on which f increases and decreases.

Solution

a. Using a graphing utility, we graph $f(x) = x^{2/3} + 1$ as shown in Figure 2.45.
b. From the graph, we can see that the domain of f is $(-\infty, \infty)$ and the range is $[1, \infty)$.
c. The graph indicates that f is decreasing on the interval $(-\infty, 0)$ and increasing on the interval $(0, \infty)$.   ■

Look again at the graph of the function in Figure 2.45. The function is decreasing on the interval $(-\infty, 0)$ and increasing on the interval $(0, \infty)$. At the point $(0, 1)$, the function changes from decreasing to increasing, and 1 is a *minimum* value of the function, with this minimum occurring at $x = 0$.

Lowest and highest points on functions are extremely important in applications. These extreme values of a function are the lowest and largest value of the function on some interval and can be formally defined as follows.

3 Find relative extrema.

Definitions of relative maximum and relative minimum

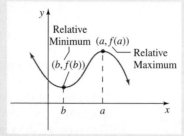

1. A function value $f(a)$ is a *relative maximum* of f if there exists an open interval about a such that $f(a) \geq f(x)$ for all x in the open interval.
2. A function value $f(b)$ is a *relative minimum* of f if there exists an open interval about b such that $f(b) \leq f(x)$ for all x in the open interval.

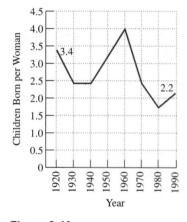

Average Number of Children Born to Each U.S. Woman

Figure 2.46

Source: U.S. Bureau of the Census

EXAMPLE 6 **U.S. Women and Childbearing**

The average number of children born to each U.S. woman from 1920 to 1990 is shown in Figure 2.46. Approximate the maximum number of children born per woman. When did this occur?

Solution

From the graph, we can see that the maximum number of children born per woman is approximately 4. This occurred in 1960. ∎

Techniques for finding *relative extrema* (relative maximum and/or relative minimum) are studied extensively in calculus. However, a graphing utility with FMIN (function minimum) and FMAX (function maximum), or similar command, will approximate relative extrema in an interval.

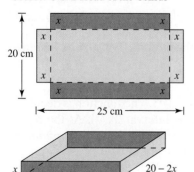

Figure 2.47

Constructing an open-top box

EXAMPLE 7 **An Application of Extrema**

From each of the four corners of a rectangular piece of cardboard measuring 20 by 25 centimeters, a square piece x centimeters on a side is cut out. The flaps are then turned up to form an open-top box (Figure 2.47)

a. Express the volume V of the open-top box as a function of x.
b. Use a graphing utility to graph the function, drawing a relatively complete graph for meaningful values of x.
c. Use the graph and the utility's maximum function FMAX feature to determine x so that the resulting box has maximum possible volume. What is the maximum possible volume?

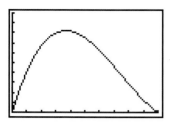

Figure 2.48

The graph of volume as a function of the size of squares cut from corners

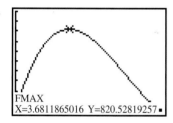

Figure 2.49

Squares whose sides measure 3.68 centimeters give a maximum volume of 820.5 cubic centimeters.

Solution

a. We begin with the volume of the open-top box, or the volume of a rectangular solid.

Volume	=	Length	times	Width	times	Height
V	=	$(25 - 2x)$	·	$(20 - 2x)$	·	x

$$V(x) = (25 - x)(20 - 2x)x \qquad \text{This is the volume } V \text{ of the box as a function of } x.$$

b. The width of the cardboard is 20 centimeters. At most, 10 centimeters can be cut from each end of this line segment and at least, 0 centimeters can be cut (no cutting). This suggests the range setting Xmin = 0, Xmax = 10, and Xscl = 1. Volume must be positive, so we can begin with Ymin = 0. Substituting a few values of x into $V(x)$ suggests a fairly large setting for Ymax. We've used Ymax = 1000 and Yscl = 100 to graph the volume function as shown in Figure 2.48.

c. We now use the maximum function feature of our graphing utility. Although we can also use the ⎡TRACE⎤ and ⎡ZOOM⎤ capabilities to identify coordinates of the graph's peak, we chose the ⎡FMAX⎤ feature, as shown in Figure 2.49. The maximum possible volume results when the square cut from each corner measures 3.68 centimeters. The maximum volume of the box is 820.5 cubic centimeters. ■

PROBLEM SET 2.4

Practice Problems

1. The graph of a function f is shown in the figure.

  a. State the values of $f(-6), f(0),$ and $f(3)$.

  b. State the intervals on which f is increasing and on which f is decreasing.

  c. State the minimum value of f. At what value of x does this occur?

  d. If the domain of f is $(-\infty, \infty)$, what is the function's range?

  a. State the values of $g(0)$ and $g(1)$.

  b. State the intervals on which g is increasing and on which g is decreasing.

  c. State the minimum value of g. At what value of x does this occur?

  d. State the maximum value of g. At what value of x does this occur?

  e. If the domain of g is $[-4, 4]$, what is the function's range?

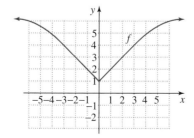

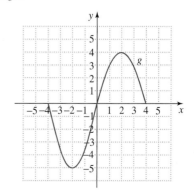

2. The graph of a function g is shown in the figure on the right.

3. The graph of a function f is shown in the figure.

    a. State the values of $f(-2)$, $f(-1)$, and $f(5)$.

    b. State the intervals on which f is increasing, decreasing, and constant.

    c. If the domain of f is $[-5, 6]$, what is the function's range?

4. The graph of a function g is shown in the figure.

    a. State the values of $g(-1)$ and $g(2)$.

    b. State the intervals on which g is increasing and decreasing.

    c. If the domain of g is $[-6, 6]$, what is the function's range?

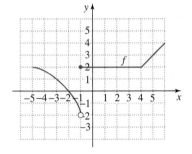

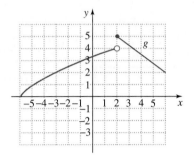

Sketch each function in Problems 5–14 using the given properties. (More than one correct graph is possible.)

5. The domain of f is $[0, 4]$ and the range of f is $[-1, 1]$. $f(0) = 0, f(1) = 1, f(2) = 0, f(3) = -1$, and $f(4) = 0$.

6. The domain of f is $[0, 4]$ and the range of f is $[0, -3]$. -3 is a relative minimum and occurs at $x = 2$.

7. The domain of f is $[-3, \infty)$. $f(-3) = 2, f(0) = 5$, and f is increasing on the interval $(-3, \infty)$.

8. f is increasing on the intervals $(-\infty, 0)$ and $(4, \infty)$, decreasing on the interval $(0, 4)$, $f(-3) = f(3) = 0$, and $f(0) = 4$.

9. f is a piecewise function that is linear and decreasing on $(-\infty, 2)$, $f(2) = 0$, f is nonlinear and increasing on $(2, \infty)$, and the range of f is $[0, \infty)$.

10. f is a piecewise function that is linear and decreasing on $(-\infty, 4)$, $f(4) = 0$, f is nonlinear and increasing on $(4, \infty)$, and the range of f is $[0, \infty)$.

11. The domain of f is $(-\infty, \infty)$, $f(8) = 6$, $-\frac{1}{2}$ is a relative minimum that occurs at $x = \frac{1}{2}$, and $f(0) = 0$.

12. The domain of f is $(-\infty, \infty)$, $f(0) = 0$, and -3 is a relative minimum that occurs at $x = 1$.

13. The domain of f is $(-\infty, \infty)$, 2 is a relative maximum that occurs at $x = 0$, and -1 is a relative minimum that occurs at $x = 3.5$.

14. f is increasing on $(-\infty, 0)$, constant on $(0, 2)$, and decreasing on $(2, \infty)$; $f(-1) = f(3) = 0$; and $f(1.7) = 1$.

Application Problems

15. The graph in the top left column on page 265 shows the percent of females and males who have dropped out of high school in the United States from 1965 through 1990.

    a. During what time intervals is the dropout rate for U.S. males decreasing?

    b. During what time interval is the dropout rate for U.S. males increasing?

    c. For what year is the dropout rate for males at a minimum? What percent of males dropped out of high school in that year? (You'll need to approximate your answers in this part.)

    d. In approximately what year did males and females have the same dropout rate? Approximately what percent of the females and males dropped out of high school in that year?

High School Dropout Rates, by Gender, 1965–1990

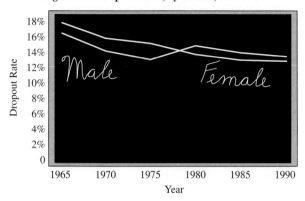

Source: U.S. Department of Education

16. The graph below shows the number of fires in the United States from 1985 through 1994.

   a. During what time periods is the number of fires on the increase?

   b. During what time periods is the number of fires on the decrease?

   c. Between what two consecutive years did the number of fires remain constant?

   d. In what year is the number of fires at a minimum? Approximately how many fires were there in that year?

   e. In what year is the number of fires at a maximum? Approximately how many fires were there in that year?

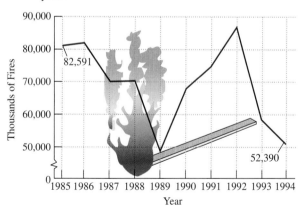

Source: National Interagency Fire Center, National Oceanic and Atmospheric Administration

17. The function $f(x) = 0.4x^2 - 36x + 1000$ describes the number of accidents per 50 million miles driven as a function of age (x, in years), where $16 \leqslant x \leqslant 74$. The graph of f is shown in the figure at the top of the next column.

a. State the intervals on which the function is increasing and decreasing and describe what this means in terms of the variables modeled by the function.

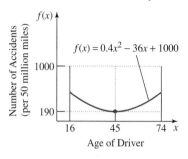

b. For what value of x does the function have a relative minimum? What is the minimum value of y? Describe the practical significance of this minimum value.

c. The domain of f is $[16, 74]$. Use the function's equation to determine the range of f. What is the practical significance of this range in terms of the meaning of $f(x)$ in the given model?

18. Based on a study by Vance Tucker (*Scientific American,* May 1969) the power expenditure of migratory birds in flight is a function of their flying speed (x, in miles per hour), approximated by $f(x) = 0.67x^2 - 27.74x + 387$. Power expenditure, $f(x)$, is measured in calories, and migratory birds generally fly between 12 and 30 miles per hour. The graph of f is shown in the figure, with a domain of $[12, 30]$.

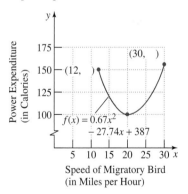

a. State the approximate intervals on which the function is increasing and decreasing and describe what this means in terms of the variables modeled by the function.

b. For what approximate value of x does the function have a relative minimum? What is the minimum value of y? Describe the practical significance of this minimum value.

c. The domain of f is $[12, 30]$. Use the function's equation to find the range of f. What is the practical significance of this range in terms of the meaning of $f(x)$ in the given model?

In Problems 19–20, use the given description to sketch each function.

19. According to the U.S. Center for Disease Control, in 1960 there were approximately 50,000 cases of tuberculosis in the United States. For the function that models these variables, let x represent the number of years after 1960 and let $f(x)$ represent the number of tuberculosis cases. The number of cases steadily increased, reaching a maximum of 60,000 in 1965. Then there was a decrease, with 21,000 cases reported in 1987. From 1987 to 1991, the number of cases continued to increase, reaching 25,500 in 1991. This number remained relatively constant until 1995.

20. According to the College Board, in 1967 the average math SAT score was 490. For the function that models these variables, let x represent the number of years after 1967 and let $f(x)$ represent the average SAT score. Scores decreased, reaching a minimum of 465 in 1981. Scores generally increased from 1981 through 1993, although there was a slight bit of fluctuation, reaching an average of 478 in 1993.

For Problems 21–22, select the graph from (a)–(e) that best models the given situation.

(a)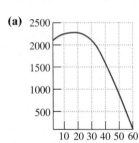

(b)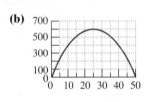

21. The perimeter of a rectangle is 100 feet. Express the area A of the rectangle as a function of x, one of the rectangle's dimensions. Then select the graph that most closely represents the graph of the area function.

(c)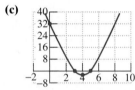

(d)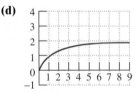

22. A small cruising ship that has a capacity of 50 passengers provides two-day excursions to groups of 35 or more. The cost is $60 per person for groups of 35 people, $59 per person for groups of 36 people, $58 per person for groups of 37 people, with everyone's cost reduced by $1 for each person in excess of 35. If x represents the number of people in excess of 35 who take the cruise, express the ship's revenue R as a function of x. Then select the graph that most closely represents the graph of the revenue function.

(e)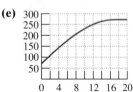

True–False Critical Thinking Problems

23. Which one of the following is true based on the graph of f in the figure shown?

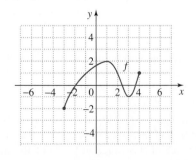

a. The domain of f is $[-2, 2]$.

b. The function has a relative maximum of 2 that occurs at $x = 2$.

c. f is decreasing on the interval $(-1, 2)$.

d. $f(3) - f(-3) = 1$

24. Which one of the following is true based on the graph of f in the figure shown?

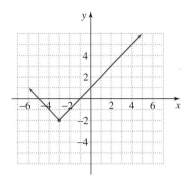

    a. The domain and range of f is $(-\infty, \infty)$.

    b. This piecewise function can be defined by three different equations, all in the form $y = mx + b$.

    c. f is decreasing on the interval $(-\infty, -2)$.

    d. The function has a minimum value of -2 that occurs at $x = -3$.

Technology Problems

In Problems 25–32, use a graphing utility to graph each function. Approximate the intervals over which the function is increasing, decreasing, or constant. Also approximate (to two decimal places) any relative maximum or minimum values of the function.

25. $f(x) = x^3 - 6x^2 + 9x + 1$

26. $f(x) = x\sqrt{4 - x}$

27. $g(x) = |4 - x^2|$

28. $g(x) = |x - 2| + |x + 2|$

29. $h(x) = x^{1/3}(x - 4)$

30. $h(x) = \begin{cases} 3 & \text{if } x < 0 \\ \sqrt{x} + 3 & \text{if } x \geqslant 0 \end{cases}$

31. $f(x) = x^{2/3}$

32. $f(x) = 2 - x^{2/5}$

33. a. Graph the functions $f(x) = x^n$ for $n = 2, 4,$ and 6 in the viewing rectangle $[-2, 2]$ by $[-1, 3]$.

    b. Graph the functions $f(x) = x^n$ for $n = 1, 3,$ and 5 in the viewing rectangle $[-2, 2]$ by $[-2, 2]$.

    c. If n is even, where is the graph of $f(x) = x^n$ increasing and where is it decreasing?

    d. If n is odd, what can you conclude about the graph of $f(x) = x^n$ in terms of increasing or decreasing behavior?

    e. Graph all six functions in a $[-1, 3]$ by $[-1, 3]$ viewing rectangle. What do you observe about the graphs in terms of how flat or how steep they are?

34. a. Use a graphing utility to graph both $f(x) = x^3 - 3x^2 - 9x + 1$ and $f'(x) = 3x^2 - 6x - 9$. (We read $f'(x)$ as "f prime of x.")

    b. Use the graphs to describe the relationship between intervals where f is increasing and those in which f' is positive.

    c. Use the graphs to describe the relationship between intervals where f is decreasing and those in which f' is negative.

    d. Describe what happens to the graph of f' when f has a relative maximum or minimum.

Problems 35–37 involve relative extrema. Use a graphing utility to graph all functions.

35. Two positive numbers have a sum of 9. One of the numbers is represented by x.

    a. Express the product (P) of the numbers as a function of x.

    b. Graph the function, drawing a relatively complete graph for meaningful values of x.

    c. Use the graph to determine x so that the two numbers have a maximum product. What is the maximum product possible?

36. From each of the four corners of a rectangular piece of cardboard measuring 5 by 8 inches, a square piece x inches on a side is cut out. The flaps are then turned up to form an open-top box.

    a. Express the volume V of the open-top box as a function of x.

    b. Graph the function, using an appropriate range setting to draw a relatively complete graph for meaningful values of x.

c. Use the graph to determine x so that the resulting box has a maximum possible volume. What is the maximum possible volume?

37. A manufacturer can produce and sell x items a week for a revenue r given by $r(x) = 200x - 0.01x^2$ cents. The manufacturer has fixed weekly costs of $200 and variable costs of 50¢ to produce each item.

a. Express the weekly profit (P) as a function of x.

b. Graph the function, using an appropriate range setting to draw a relatively complete graph for meaningful values of x.

c. Use the graph to determine the number of items that should be produced and sold to maximize weekly profit. What is the maximum weekly profit?

Writing in Mathematics

For Problems 38–41, create a verbal problem or situation that can be modeled by the graph of each function shown.

38.

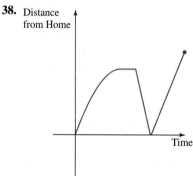

39.

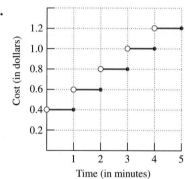

40.

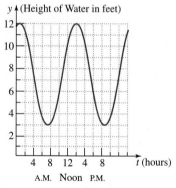

41.

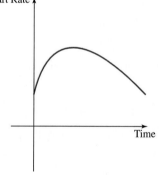

Critical Thinking Problem

42. Define a piecewise function on the intervals $(-\infty, 2]$, $(2, 5)$, and $[5, \infty)$ that does not "jump" at 2 or 5 such that one piece is a constant function, another piece is a linear function that is increasing, and another piece is a nonlinear function that is decreasing.

Group Activity Problem

43. Group members should consult almanacs and appropriate references to find examples of variables that increased, decreased, and remained constant from 1985 through 1995. Once each member finds an example of each, these examples should be presented to other people in the group. Group members should identify increasing or decreasing phenomena whose data points appear to lie on a line. Use a graphing utility with a linear regression option to model the data. Try using this option to model data that clearly do not represent linear growth or decline. What value of the correlation coefficient does the graphing utility give? What does this mean?

Solutions Manual

Tutorial

Video 5

Transformations of Functions

Objectives

1 Use vertical and horizontal shifts to graph functions.
2 Use reflections to graph functions.
3 Use stretching and shrinking to graph functions.
4 Graph functions involving a sequence of transformations.
5 Identify even and odd functions.

In Section 2.3, we focused on the graphs of relations and functions. In this section, we continue our study of graphing techniques by using graphs of basic functions and transforming them into the graphs of new functions.

Algebra's Common Graphs

Figure 2.50 shows the graphs of six commonly used functions in algebra. You should become familiar with these graphs for your work in this section.

Hans Bellmer "Peg-Top" c. 1937–52, oil on canvas, 648 × 648 mm. Tate Gallery, London. Photo: John Webb. Art Resource, NY.

Identity Function $f(x) = x$

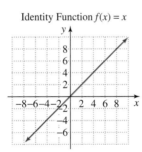

Standard Quadratic Function $f(x) = x^2$

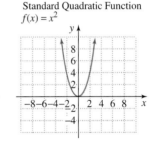

Standard Cubic Function $f(x) = x^3$

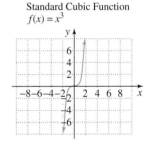

Absolute Value Function $f(x) = |x|$

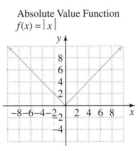

Square Root Function $f(x) = \sqrt{x}$

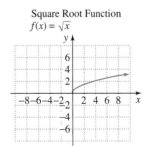

Cube Root Function $f(x) = \sqrt[3]{x}$

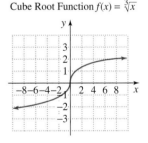

Figure 2.50
Algebra's common graphs

Discover for yourself

Use a graphing utility to verify the six graphs shown in Figure 2.50.

1 Use vertical and horizontal shifts to graph functions.

Vertical and Horizontal Shifts

The study of how transformations of a function can affect its graph can be previewed and explored with a graphing utility. Use your graphing utility and work the Discover for yourself exercises as you read this section.

tudy tip

If you are not using a graphing utility in this course, work the first 3 steps in all the Discover for yourself exercises by graphing the functions in each exercise by hand. Use a table of coordinates and point plotting.

Study tip

To work the first three parts of the Discover for yourself exercise without a graphing utility, begin by filling in this table of coordinates. Then graph f, g, and h in the same rectangular coordinate system.

x	$f(x) = x^2$	$g(x) = x^2 + 2$	$h(x) = x^2 - 3$
-2			
-1			
0			
1			
2			

Discover for yourself

1. Use your graphing utility to graph the functions $f(x) = x^2$, $g(x) = x^2 + 2$, and $h(x) = x^2 - 3$ in the same viewing rectangle.

2. Describe the relationship among the graphs of f, g, and h.

3. Generalize by describing the relationship among the graphs of f, g, and h, if $g(x) = f(x) + c$ and $h(x) = f(x) - c$, with c a positive constant.

4. Reinforce the generalization in part 3 by using a graphing utility, determining a particular function f and a particular value for c. Graph $f(x)$ and $g(x) = f(x) + c$.

In the Discover for yourself, did you observe that the graph of $g(x) = x^2 + 2$ shifts the graph of $f(x) = x^2$ two units upward? Similarly, the graph of $g(x) = x^2 - 3$ shifts the graph of $f(x) = x^2$ three units downward. These relationships are shown in Figure 2.51 at the top of page 271. In general, if c is positive, $y = f(x) + c$ shifts the graph of f upward c units and $y = f(x) - c$ shifts the graph of f downward c units. These are called *vertical shifts* of the graph of f.

Variations in the equation of a function also enable us to shift the function's graph horizontally.

Study tip

To work the first three parts of the Discover for yourself exercise without a graphing utility, begin by filling in these tables of coordinates. Then graph f, g, and h in the same rectangular coordinate system.

x	$f(x) = x^2$	x	$g(x) = (x-5)^2$	x	$h(x) = (x+5)^2$
-2		3		-7	
-1		4		-6	
0		5		-5	
1		6		-4	
2		7		-3	

Discover for yourself

1. Use your graphing utility to graph the functions $f(x) = x^2$, $g(x) = (x - 5)^2$, and $h(x) = (x + 5)^2$ in the same viewing rectangle.

2. Describe the relationship among the graphs of f, g, and h.

3. Generalize by describing the relationship among the graphs of f, g, and h if $g(x) = f(x - c)$ and $h(x) = f(x + c)$, with c a positive real number.

4. Reinforce the generalization in part 3 by using a graphing utility, determining a particular function f and a particular value for c. Graph $f(x)$ and $g(x) = f(x + c)$.

In the Discover for yourself, did you observe that the graph of $g(x) = (x - 5)^2$ shifts the graph of $f(x) = x^2$ five units to the right? Similarly, the graph of $h(x) = (x + 5)^2$ shifts the graph of $f(x) = x^2$ five units to the left. These relationships are shown in Figure 2.52 on page 271. In general, if c is positive, $y = f(x - c)$ shifts the graph of f to the right c units and $y = f(x + c)$ shifts the graph of f to the left c units. These are called *horizontal shifts* of the graph of f.

Vertical and horizontal shifts, also called *translations*, are summarized in the box on page 271.

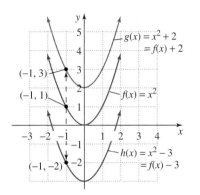

Figure 2.51

Shifting $f(x) = x^2$ two units upward and three units downward

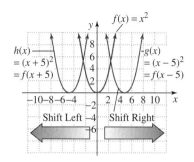

Figure 2.52

Shifting $f(x) = x^2$ five units to the right and five units to the left

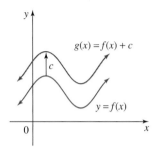

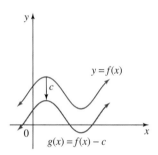

Figure 2.53

Upward and downward shifts

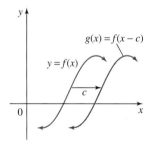

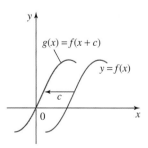

Figure 2.54

Shifting to the right and the left

Vertical and horizontal shifts

Let c represent a positive real number. *Vertical* and *horizontal shifts* in the graph of $y = f(x)$ can be obtained as follows.

Function	How to Obtain the Graph of g from the Graph of f
$g(x) = f(x) + c$	Shift f *upward* c units.
$g(x) = f(x) - c$	Shift f *downward* c units.

These shifts are illustrated in Figure 2.53.

$g(x) = f(x - c)$	Shift f horizontally to the *right* c units.
$g(x) = f(x + c)$	Shift f horizontally to the *left* c units.

These shifts are illustrated in Figure 2.54.

EXAMPLE I Graphing Translated Functions

Use the graph of $f(x) = \sqrt{x}$ to sketch the graph of:

a. $g(x) = \sqrt{x} - 5$ **b.** $g(x) = \sqrt{x + 5}$ **c.** $g(x) = \sqrt{x - 1} - 2$

Solution

(The three graphs described below appear on page 272, so be sure to turn the page and look at the graphs as you read each part of the solution.)

a. We obtain the graph of $g(x) = \sqrt{x} - 5$ by shifting the graph of $f(x) = \sqrt{x}$ downward five units. As shown in Figure 2.55, every point in the graph of g is exactly five units below the corresponding point on the graph of f.

b. We obtain the graph of $g(x) = \sqrt{x + 5}$ by shifting the graph of $f(x) = \sqrt{x}$ horizontally to the left five units. As shown in Figure 2.56, every point in the graph of g is exactly five units to the left of a corresponding point on the graph of f.

c. The graph of $g(x) = \sqrt{x - 1} - 2$ involves two shifts in terms of the graph of $f(x) = \sqrt{x}$. First we shift f one unit to the right ($\sqrt{x - 1}$) and then shift the resulting graph two units downward ($\sqrt{x - 1} - 2$). The graph of g is shown in Figure 2.57. ∎

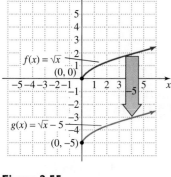

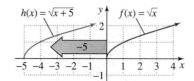

Figure 2.56

Shifting $f(x) = \sqrt{x}$ five units left

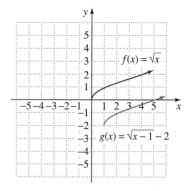

Figure 2.57

Shifting $f(x) = \sqrt{x}$ one unit to the right and two units downward

Figure 2.55

Shifting $f(x) = \sqrt{x}$ five units downward

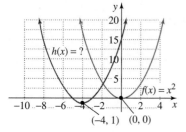

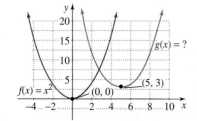

Figure 2.58

EXAMPLE 2 **Finding Equations from Graphs**

Find an equation for each function shown in Figure 2.58.

Solution

a. The graph of g is both a horizontal shift of five units to the right of the graph of f ($y = (x - 5)^2$) and a vertical shift of three units upward of the graph of f. Thus, the equation of g is $g(x) = (x - 5)^2 + 3$.

b. The graph of h is both a horizontal shift of four units to the left of the graph of f ($y = (x + 4)^2$) and a vertical shift of one unit downward of the graph of f. Thus, the equation of h is $h(x) = (x + 4)^2 - 1$. ■

Reflections

2 Use reflections to graph functions.

A second type of transformation is called a *reflection*. Once again, a graphing utility is an ideal tool for exploring this transformation.

Study tip

To work the first three parts of the Discover for yourself exercise without a graphing utility, begin by filling in these tables of coordinates. Then graph f, g, and h in the same rectangular coordinate system.

x	$f(x) = \sqrt{x}$	x	$g(x) = -\sqrt{x}$	x	$h(x) = \sqrt{-x}$
0		0		0	
1		1		-1	
4		4		-4	
9		9		-9	

Discover for yourself

1. Use your graphing utility to graph the functions $f(x) = \sqrt{x}$, $g(x) = -\sqrt{x}$, and $h(x) = \sqrt{-x}$ in the same viewing rectangle.
2. Describe the relationship among the graphs of f, g, and h.
3. Generalize by describing the relationship among the graphs of f, g, and h if $g(x) = -f(x)$ and $h(x) = f(-x)$.
4. Reinforce the generalization in part 3 by using a graphing utility and determining a particular function f. Graph $f(x)$, $-f(x)$, and $f(-x)$.

Figure 2.59

Reflecting $f(x) = \sqrt{x}$

In the Discover for yourself, did you observe that the graph of $g(x) = -\sqrt{x}$ reflects the graph of $f(x) = \sqrt{x}$ in the x-axis? Similarly, the graph of $h(x) = \sqrt{-x}$ reflects the graph of $f(x) = \sqrt{x}$ in the y-axis. These relationships are shown in Figure 2.59. In general, the graph of $y = -f(x)$ reflects the graph of f in the x-axis and the graph of $y = f(-x)$ reflects the graph of f in the y-axis.

These observations can be summarized as follows.

Reflections in the rectangular coordinate system

Reflections of the graph of $y = f(x)$ can be obtained as follows.

Function	How to Obtain the Graph of g from the Graph of f
$g(x) = -f(x)$	Reflect f in the x-axis (Figure 2.60)
$g(x) = f(-x)$	Reflect f in the y-axis (Figure 2.61)

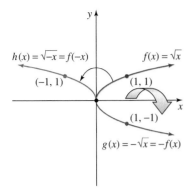

Werner Forman Archive, Tishman Collection, New York. Art Resource, New York.

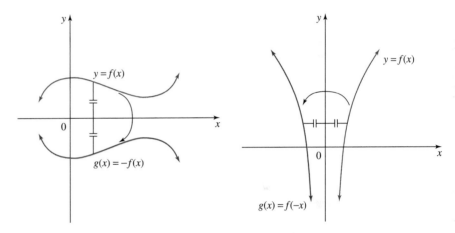

Figure 2.60

Reflecting f in the x-axis

Figure 2.61

Reflecting f in the y-axis

Figure 2.62 illustrates reflections of $f(x) = x^2 - 2x$ in the x-axis and the y-axis.

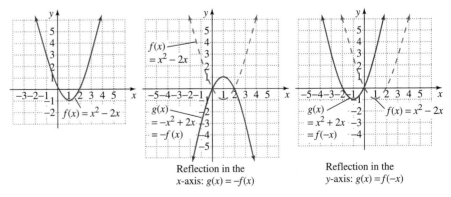

Reflection in the
x-axis: $g(x) = -f(x)$

Reflection in the
y-axis: $g(x) = f(-x)$

Figure 2.62

EXAMPLE 3 **Graphing Reflected Functions**

As shown in Figure 2.63, the graph of f is a line segment joining the points $(-3, 2)$ and $(4, 5)$. Use the graph of f to graph:

a. $y = -f(x)$ **b.** $y = f(-x)$ **c.** $y = -f(-x)$

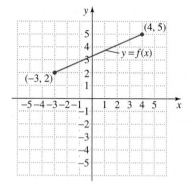

Figure 2.63

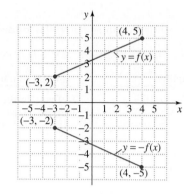

Figure 2.64

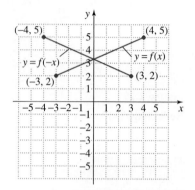

Figure 2.65

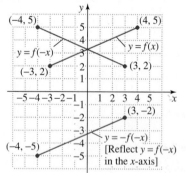

Figure 2.66

Solution

a. Figure 2.64 shows the graph of $y = -f(x)$, obtained by reflecting the graph of f in the x-axis. Observe that points on the reflected function have the same x-coordinates as corresponding points on the given function and opposite y-coordinates.

b. Figure 2.65 shows the graph of $y = f(-x)$, obtained by reflecting the graph of f in the y-axis. In this situation, points on the reflected function have opposite x-coordinates as corresponding points on the given function and the same y-coordinates.

c. Figure 2.66 shows the graph of $y = -f(-x)$, obtained by reflecting the graph of $y = f(-x)$ in the x-axis. ∎

Discover for yourself

Describe how the graph of $y = -f(-x)$ can be obtained from the graph of $y = -f(x)$.

EXAMPLE 4 **Reflections and Shifts**

Compare the graph of each function with the graph $f(x) = |x|$:

a. $g(x) = -|x|$ **b.** $h(x) = -|x - 2|$ **c.** $k(x) = -|x - 2| + 3$

Solution

a. The graph of g is a reflection of the graph of $f(x) = |x|$ in the x-axis because

$$g(x) = -|x| = -f(x).$$

(Figure 2.67)

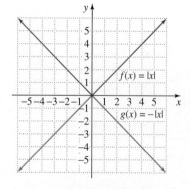

Figure 2.67

b. From the equation

$$h(x) = -|x - 2| = -f(x - 2)$$

we see that the graph of h is a right shift of the graph of f two units ($f(x - 2)$), followed by a reflection in the x-axis ($-f(x - 2)$). (Figure 2.68)

c. From the equation

$$k(x) = -|x - 2| + 3 = -f(x - 2) + 3$$

we see that the graph of k is a right shift of the graph of f two units ($f(x - 2)$), followed by a reflection in the x-axis ($-f(x - 2)$), followed by an upward shift of three units ($-f(x - 2) + 3$). (Figure 2.69) We also could obtain the graph of k by taking the graph of h in Figure 2.68 and shifting it upward three units. ∎

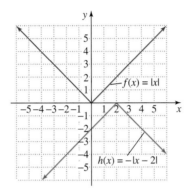

Figure 2.68

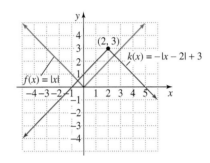

Figure 2.69

3 Use stretching and shrinking to graph functions.

Nonrigid Transformations: Stretching and Shrinking

Up to this point, we have studied horizontal shifts, vertical shifts, and reflections. These transformations are called *rigid* because they do not change the basic shape of the graph.

Let's contrast these rigid transformations with transformations that cause a distortion or a change in shape in a function's graph. These distortions involve stretching or shrinking the graph.

Study tip

To work the first three parts of the Discover for yourself exercise without a graphing utility, begin by filling in this table of coordinates. Then graph f, g, and h in the same coordinate system.

x	$f(x) = x^2$	$g(x) = 2x^2$	$h(x) = \frac{1}{2}x^2$
-2			
-1			
0			
1			
2			

Discover for yourself

1. Use your graphing utility to graph the functions $f(x) = x^2$, $g(x) = 2x^2$, and $h(x) = \frac{1}{2}x^2$ in the same viewing rectangle.

2. Describe the relationship among the graphs of f, g, and h.

3. Generalize by describing the relationship among the graphs of f, g, and h if $g(x) = cf(x)$ for $c > 1$, and $h(x) = cf(x)$ for $0 < c < 1$.

4. Reinforce the generalization in part 3 by using a graphing utility, determining a particular function f, and choosing two values for c, one where $c > 1$ and one where $0 < c < 1$. Graph $f(x)$ and $cf(x)$ for $c > 1$, and $cf(x)$ for $0 < c < 1$.

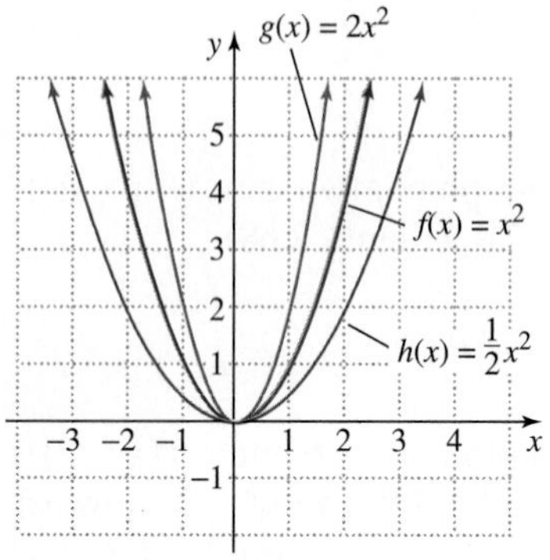

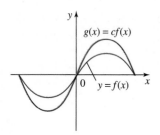

Figure 2.70

Stretching and shrinking
$f(x) = x^2$

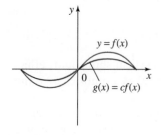

Figure 2.71

$c > 1$: Stretching

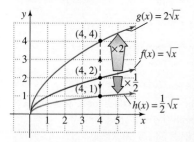

Figure 2.72

$0 < c < 1$: Shrinking

4 Graph functions involving a sequence of transformations.

In the Discover for yourself, did you observe that the graph of $g(x) = 2x^2$ or $g(x) = 2f(x)$, has y-coordinates that are twice as large as the corresponding y-coordinates on f? Figure 2.70 shows that the graph of g is obtained by *stretching* the graph of f. The result is a narrower graph. On the other hand, the graph of $h(x) = \frac{1}{2}x^2$, or $h(x) = \frac{1}{2}f(x)$, has y-coordinates that are one-half as large as the corresponding y-coordinates on f. We say that the graph of h is obtained by *shrinking* the graph of f. The result is a wider graph.

These observations can be summarized as follows.

Nonrigid transformations: stretching and shrinking

Let c represent a positive real number. Stretching and shrinking the graph of $y = f(x)$ can be obtained as follows.

Function	How to Obtain the Graph of g from the Graph of f
$g(x) = cf(x)$, where $c > 1$	Stretch the graph of f, multiplying each of its y-coordinates by c. (Figure 2.71)
$g(x) = cf(x)$, where $0 < c < 1$	Shrink the graph of f, multiplying each of its y-coordinates by c. (Figure 2.72)

EXAMPLE 5   Nonrigid Transformations

Compare the graph of each function in Figure 2.73 at the bottom of this page with the graph of $f(x) = \sqrt{x}$.

Solution

a. Relative to the graph $f(x) = \sqrt{x}$, the graph of

$$g(x) = 2\sqrt{x} = 2f(x)$$

is a vertical stretch (multiply each y-value by 2) of the graph of f.

b. Relative to the graph $f(x) = \sqrt{x}$, the graph of

$$h(x) = \frac{1}{2}\sqrt{x} = \frac{1}{2}f(x)$$

is a vertical shrink (multiply each y-value by $\frac{1}{2}$) of the graph of f. ∎

Sequences of Transformations

If a function involves more than one transformation and is to be graphed without a graphing utility, use the following procedure.

Graphing involving a sequence of transformations

A function involving more than one transformation can be graphed in the following order:

1. Horizontal translation
2. Stretching or shrinking
3. Reflecting
4. Vertical translation

Figure 2.73

EXAMPLE 6 **Graphing Using a Sequence of Transformations**

Use the graph of $f(x) = \sqrt{x}$ to graph $g(x) = 5 - 2\sqrt{x+1}$.

Solution

The following sequence of steps is illustrated below in Figure 2.74. We begin with the graph of $f(x) = \sqrt{x}$.

1. *Horizontal translation:* The graph of $y = \sqrt{x+1} = f(x+1)$ is obtained from the graph of f by shifting the graph of f horizontally one unit to the left.
2. *Stretching:* The graph of $y = 2\sqrt{x+1}$ is obtained from $y = \sqrt{x+1}$ by stretching the graph by a factor of 2. Each y-coordinate is multiplied by 2.
3. *Reflecting:* The graph of $y = -2\sqrt{x+1}$ is obtained by reflecting the graph of $y = 2\sqrt{x+1}$ in the x-axis.
4. *Vertical translation:* The graph of $g(x) = 5 - 2\sqrt{x+1}$ is obtained from the graph of $y = -2\sqrt{x+1}$ by shifting the graph five units upward.

Begin with the graph of $f(x) = \sqrt{x}$

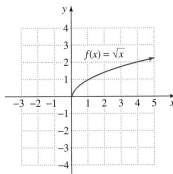

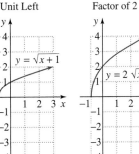

1. Translate 1 Unit Left

2. Stretch by a Factor of 2

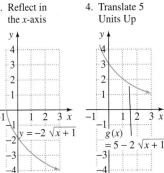

3. Reflect in the x-axis

4. Translate 5 Units Up

Figure 2.74

Transformations of $f(x) = \sqrt{x}$

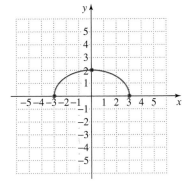

Figure 2.75

EXAMPLE 7 **Graphing Using a Sequence of Transformations**

Use the graph of f shown in Figure 2.75 to graph $g(x) = -3 - \frac{1}{2}f(x-2)$.

Solution

The following sequence of steps is illustrated in Figure 2.76 on page 278. We begin with the graph of f.

1. *Horizontal translation:* Shift f two units to the right. $(f(x-2))$
2. *Shrinking:* Shrink the graph in step 2 by a factor of $\frac{1}{2}$
   $$\left(\tfrac{1}{2}f(x-2)\right)$$
3. *Reflecting:* Reflect the graph in step 3 in the x-axis.
   $$\left(y = -\tfrac{1}{2}f(x-2)\right)$$
4. *Vertical translation:* Shift the graph in step 4 three units downward.
   $$\left(g(x) = -3 - \tfrac{1}{2}f(x-2)\right)$$

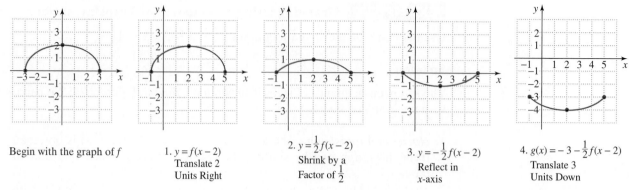

Begin with the graph of f

1. $y = f(x - 2)$
Translate 2
Units Right

2. $y = \frac{1}{2}f(x - 2)$
Shrink by a
Factor of $\frac{1}{2}$

3. $y = -\frac{1}{2}f(x - 2)$
Reflect in
x-axis

4. $g(x) = -3 - \frac{1}{2}f(x - 2)$
Translate 3
Units Down

Figure 2.76

Transformations of f described in Example 7

EXAMPLE 8 **Describing a Sequence of Transformations**

Use a graphing utility to graph the two functions $f(x) = x^3$ and $g(x) = 4 - 3(x - 2)^3$ on the same screen. Describe how the graph of g is obtained from the graph of f as a sequence of transformations.

Solution

Enter

$$Y_1 \boxed{=} X \boxed{\wedge} 3 \text{ and } Y_2 \boxed{=} 4 - 3(X - 2) \boxed{\wedge} 3$$

with the following range setting

$$\text{Xmin} = -5, \text{Xmax} = 5, \text{Xscl} = 1,$$

$$\text{Ymin} = -10, \text{Ymax} = 10, \text{Yscl} = 1$$

The graphs are shown in Figure 2.77. The graph of g can be obtained from the graph of f through the following transformations:

1. *Horizontal translation:* Shift the graph of $f(x) = x^3$ two units to the right.

$$(y = (x - 2)^3)$$

2. *Stretching:* Stretch the graph of $y = (x - 2)^3$ by a factor of 3.

$$(y = 3(x - 2)^3)$$

3. *Reflecting:* Reflect the graph of $y = 3(x - 2)^3$ in the x-axis.

$$(y = -3(x - 2)^3)$$

4. *Vertical translation:* Shift the graph of $y = -3(x - 2)^3$ four units upward.

$$(g(x) = 4 - 3(x - 2)^3) \qquad \blacksquare$$

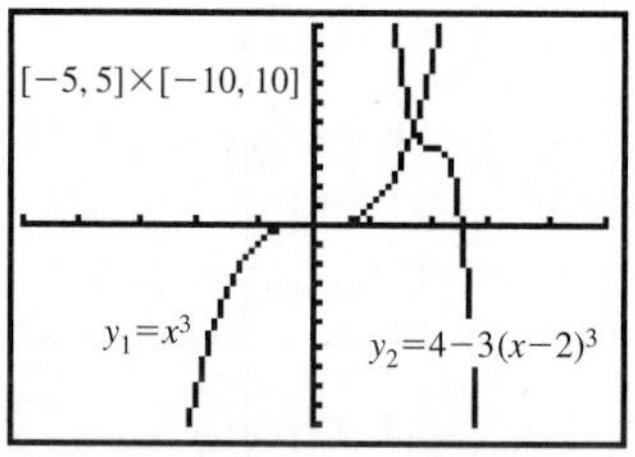

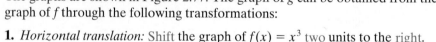

Figure 2.77

The graphs of $y_1 = x^3$ and $y_2 = 4 - 3(x - 2)^3$

Discover for yourself

Use your graphing utility to graph the functions given in each of the four steps on the right. The resulting graphs visually show each of the transformations.

5 Identify even and odd functions.

Even and Odd Functions

We know that $y = f(-x)$ reflects the graph of $y = f(x)$ in the y-axis. For some functions, $f(-x) = f(x)$ for every number x in its domain. Under these conditions, we say that f is an *even function*. For example, the function $f(x) = x^2 - 4$ is even because

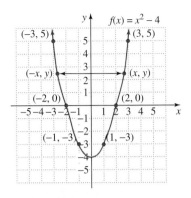

Figure 2.78

y-axis symmetry with
$f(-x) = f(x)$

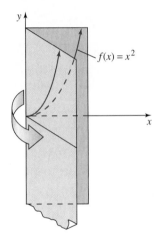

$$f(-x) = (-x)^2 - 4 = x^2 - 4 = f(x).$$

The graph of $f(x) = x^2 - 4$ is shown in Figure 2.78.

Since $f(-x) = f(x)$, this causes the graph to be *symmetric with respect to the y-axis*. Each half of the curve is a mirror image of the other half about the y-axis. If we were to fold the paper along the y-axis, the two halves of the graph would coincide.

y-Axis symmetry and even functions

The graph of f is *symmetric with respect to the y-axis* if $f(-x) = f(x)$. The graph of this *even function* can be obtained by graphing f for $x \geqslant 0$ and then reflecting this portion in the y-axis.

Some functions satisfy $f(-x) = -f(x)$ for all numbers x in their domains. Under these conditions, we say that f is an *odd function*. For example, the function $f(x) = x^3$ is odd because

$$f(-x) = (-x)^3 = -x^3 = -f(x).$$

The graph of $f(x) = x^3$ is shown in Figure 2.79 at the bottom of the page.

Since $f(-x) = -f(x)$, this causes the graph to be *symmetric with respect to the origin*. Points such as $(2, 8)$ and $(-2, -8)$ are reflections of one another through the origin, meaning that they are the same distance from the origin and lie on a line through the origin. For each point (x, y) on the graph, the point $(-x, -y)$ is also on the graph, causing the first- and third-quadrant portions of $f(x) = x^3$ to be reflections of each other about the origin.

Origin symmetry and odd functions

The graph of f is *symmetric with respect to the origin* if $f(-x) = -f(x)$. The graph of this *odd function* can be obtained by graphing f for $x \geqslant 0$ and then rotating this portion through $180°$ about the origin.

Figure 2.79

Origin symmetry with
$f(-x) = -f(x)$

EXAMPLE 9   **Determining Even and Odd Functions**

Determine whether the following functions are even, odd, or neither.

a. $f(x) = x^3 - x$ **b.** $f(x) = x^4 - 4x^2$ **c.** $f(x) = x^4 - 6x^3 + 8x^2$

Solution

In each case, we must find $f(-x)$. If $f(-x) = f(x)$, then f is even and its graph has y-axis symmetry. If $f(-x) = -f(x)$, then f is odd and its graph has origin symmetry. If neither of these things occurs, the graph has neither kind of symmetry and f is neither even nor odd.

a. $f(x) = x^3 - x$ This is the given polynomial function.

   $f(-x) = (-x)^3 - (-x)$ Replace x with $-x$.

   $\quad\quad = -x^3 + x$ Simplify.

ENRICHMENT ESSAY

Shape and Symmetry

Origin symmetry and y-axis symmetry are just two examples of a powerful concept whose workings appear in art, nature, the sciences, poetry, and architecture. The two halves of a bridge span, the wings of a bird or an aircraft, the blades of a propeller—all have symmetry. Although we take symmetry for granted, a form's lack of symmetry may be the quality that makes it interesting and appealing.

Pablo Picasso "Three Musicians," Fontainebleau, Summer 1921, oil on canvas, 6 ft. 7 in. × 7 ft. 3 3/4 in. The Museum of Modern Art, New York. Mrs. Simon Guggenheim Fund. Photograph © 1997 The Museum of Modern Art, New York. © 1998 Estate of Pablo Picasso / Artists Rights Society (ARS), New York.

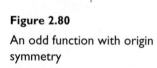

Figure 2.80

An odd function with origin symmetry

Because $f(-x) = -f(x)$, the graph of f is symmetric with respect to the origin (Figure 2.80) and f is an odd function.

b. $f(x) = x^4 - 4x^2$ This is the given polynomial function.

$f(-x) = (-x)^4 - 4(-x)^2$ Replace x with $-x$.

$\quad\quad\;\; = x^4 - 4x^2$ Simplify.

Because $f(-x) = f(x)$, the graph of f is symmetric with respect to the y-axis (Figure 2.81 on page 281) and f is an even function.

c. $f(x) = x^4 - 6x^3 + 8x^2$ This is the given polynomial function.

$f(-x) = (-x)^4 - 6(-x)^3 + 8(-x)^2$ Replace x with $-x$.

$\quad\quad\;\; = x^4 - 6(-x^3) + 8x^2$ Simplify.

$\quad\quad\;\; = x^4 + 6x^3 + 8x^2$

Because $f(-x) \neq f(x)$ and $f(-x) \neq -f(x)$, the graph has neither y-axis nor origin symmetry (Figure 2.82 on page 281) and f is neither even nor odd. ∎

 **iscover for yourself**

How can you tell whether a polynomial function is even, odd, or neither simply by looking at its equation?

In the Discover for yourself, did you observe that polynomial functions with only even powers are even functions and those with only odd powers are odd functions? If the function's equation contains both even and odd powers, it is neither even nor odd.

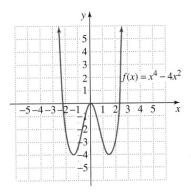

Figure 2.81

An even function with y-axis symmetry

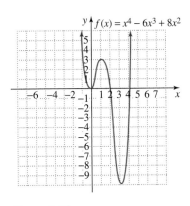

Figure 2.82

Neither y-axis nor origin symmetry

PROBLEM SET 2.5

Practice Problems

In Problems 1–26, sketch the graphs of the three given functions by hand on the same Cartesian coordinate plane. Verify the result with a graphing utility.

1. $f(x) = x^2$
$g(x) = x^2 - 3$
$h(x) = x^2 + 2$

2. $f(x) = x^2$
$g(x) = x^2 - 2$
$h(x) = x^2 + 3$

3. $f(x) = x^2$
$g(x) = (x - 3)^2$
$h(x) = (x + 2)^2$

4. $f(x) = x^2$
$g(x) = (x - 2)^2$
$h(x) = (x + 3)^2$

5. $f(x) = \sqrt{x}$
$g(x) = \sqrt{x} - 3$
$h(x) = \sqrt{x} + 2$

6. $f(x) = \sqrt{x}$
$g(x) = \sqrt{x} - 2$
$h(x) = \sqrt{x} + 3$

7. $f(x) = \sqrt{x}$
$g(x) = \sqrt{x - 3}$
$h(x) = \sqrt{x + 2}$

8. $f(x) = \sqrt{x}$
$g(x) = \sqrt{x - 2}$
$h(x) = \sqrt{x + 3}$

9. $f(x) = \sqrt{x}$
$g(x) = \sqrt{x - 3} + 2$
$h(x) = \sqrt{x + 2} - 4$

10. $f(x) = \sqrt{x}$
$g(x) = \sqrt{x - 2} + 3$
$h(x) = \sqrt{x + 3} - 5$

11. $f(x) = |x|$
$g(x) = |x - 2| + 3$
$h(x) = -|x - 2|$

12. $f(x) = |x|$
$g(x) = |x - 3| + 2$
$h(x) = -|x - 3|$

13. $f(x) = |x|$
$g(x) = 2|x - 3|$
$h(x) = -2|x - 3|$

14. $f(x) = |x|$
$g(x) = \frac{1}{2}|x - 4|$
$h(x) = -\frac{1}{2}|x - 4|$

15. $f(x) = x^2$
$g(x) = (x - 2)^2 + 1$
$h(x) = (x + 2)^2 - 1$

16. $f(x) = x^2$
$g(x) = (x - 1)^2 + 2$
$h(x) = (x + 1)^2 - 2$

17. $f(x) = x^3$
$g(x) = (x - 2)^3 + 1$
$h(x) = -x^3$

18. $f(x) = x^3$
$g(x) = (x - 1)^3 + 2$
$h(x) = -x^3$

19. $f(x) = |x|$
$g(x) = |x - 4|$
$h(x) = -|x - 4| + 2$

20. $f(x) = |x|$
$g(x) = |x - 3|$
$h(x) = -|x + 3| + 4$

21. $f(x) = \sqrt[3]{x}$
$g(x) = \sqrt[3]{x + 1}$
$h(x) = -2 + \sqrt[3]{x + 1}$

22. $f(x) = \sqrt[3]{x}$
$g(x) = \sqrt[3]{x + 2}$
$h(x) = -3 + \sqrt[3]{x + 2}$

23. $f(x) = \sqrt[3]{x}$
$g(x) = -\sqrt[3]{x + 1}$
$h(x) = \frac{1}{2}\sqrt[3]{x}$

24. $f(x) = \sqrt[3]{x}$
$g(x) = -\sqrt[3]{x + 2}$
$h(x) = 2\sqrt[3]{x}$

25. $f(x) = |x|$
$g(x) = -\frac{1}{2}|x| + 4$
$h(x) = -\frac{1}{2}(|x| + 4)$

26. $f(x) = |x|$
$g(x) = 3|x| - 2$
$h(x) = 3(|x| - 2)$

In Problems 27–34, functions f and g are graphed on the same Cartesian coordinate plane. If g is obtained from f through a sequence of transformations, find an equation for g.

27.

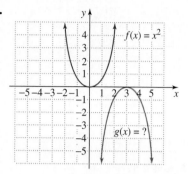

28.

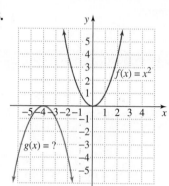

29.

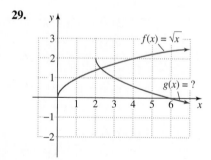

30.

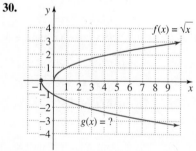

31.

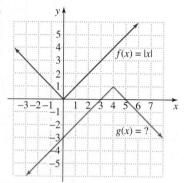

32.

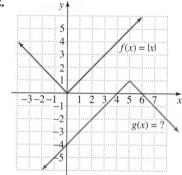

33.

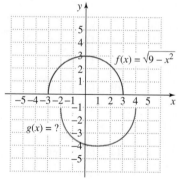

34.

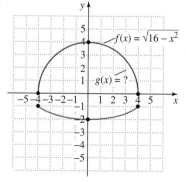

35. Use the graph of $f(x) = \sqrt{x}$ to graph
$g(x) = 4 - 2\sqrt{x - 1}$.

36. Use the graph of $f(x) = \sqrt{x}$ to graph
$g(x) = 3 - \frac{1}{2}\sqrt{x + 2}$.

37. Use the graph of $f(x) = x^2$ to graph
$g(x) = -\frac{1}{2}(x + 2)^2 + 3$.

38. Use the graph of $f(x) = x^2$ to graph $g(x) = -2(x - 1)^2 + 3$.

39. Use the graph of $f(x) = |x|$ to graph
$g(x) = -2|x + 3| - 4$.

40. Use the graph of $f(x) = |x|$ to graph
$g(x) = -\frac{1}{2}|x - 2| - 3$.

41. Use the graph of $f(x) = \sqrt[3]{x}$ to graph
$g(x) = 3 - 4\sqrt[3]{x - 1}$.

42. Use the graph of $f(x) = \sqrt[3]{x}$ to graph
$g(x) = 2 - 2\sqrt[3]{x - 2}$.

43. Use the graph of $f(x) = [[x]]$ to graph $g(x) = f(x) - 4$.

44. Use the graph of $f(x) = [[x]]$ to graph $g(x) = f(x + 2)$.

In Problems 45–52, use the graph of the function f to sketch the graph of the function g.

45.

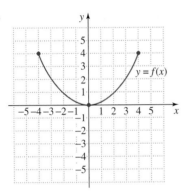

a. $g(x) = f(x - 3)$
b. $g(x) = f(x) + 2$
c. $g(x) = f(x + 1) - 2$
d. $g(x) = f(-x)$
e. $g(x) = -f(x) + 3$
f. $g(x) = -f(x - 1)$

46.

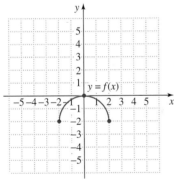

a. $g(x) = f(x - 2)$
b. $g(x) = f(x + 2) - 3$
c. $g(x) = f(-x)$
d. $g(x) = -f(x) + 4$
e. $g(x) = -f(x - 1)$
f. $g(x) = -1 - f(x + 1)$

47.

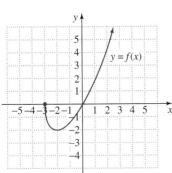

a. $g(x) = f(x - 1)$
b. $g(x) = -f(x + 1)$
c. $g(x) = f(-x) - 2$
d. $g(x) = f(x - 2) + 3$

48.

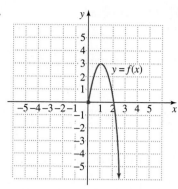

a. $g(x) = f(x - 1)$

b. $g(x) = -f(x + 1)$

c. $g(x) = f(-x) - 2$

d. $g(x) = f(x - 2) + 3$

49.

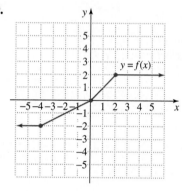

a. $g(x) = f(x - 2)$

b. $g(x) = f(x) + 2$

c. $g(x) = -f(x)$

d. $g(x) = f(-x)$

e. $g(x) = 2f(x)$

f. $g(x) = \frac{1}{2}f(x)$

50.

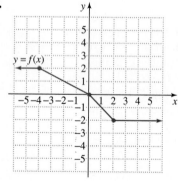

a. $g(x) = f(x - 2)$

b. $g(x) = f(x) + 2$

c. $g(x) = -f(x)$

d. $g(x) = f(-x)$

e. $g(x) = 3f(x)$

f. $g(x) = \frac{1}{2}f(x)$

51.

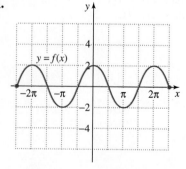

a. $g(x) = 2f(x)$

b. $g(x) = \frac{1}{2}f(x)$

c. $g(x) = f(x + \pi)$

d. $g(x) = -f(x)$

52.

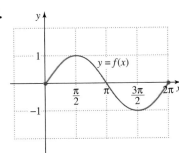

a. $g(x) = 2f(x)$
b. $g(x) = \frac{1}{2}f(x)$
c. $g(x) = f(x + \frac{\pi}{2})$
d. $g(x) = -f(x - \pi) + 1$

In Problems 53–62, determine whether each function is even, odd, or neither.

53. $f(x) = \frac{1}{5}x^6 - 3x^2$

54. $f(x) = x^3 - 6x$

55. $f(x) = x\sqrt{1 - x^2}$

56. $f(x) = x^2\sqrt{1 - x^2}$

57. $f(x) = x^4 + 3x - 5$

58. $f(x) = x^2 + 5x - 2$

59. $f(x) = 3x^{2/3}$

60. $f(x) = 5x^{2/5}$

61. $f(x) = 4x^{2n} + 6x^{2n-2} + 5$; n represents an integer.

62. $f(x) = 5x^{2n+1} + 3x^{2n-1} + 7$; n represents an integer.

Graph each function in Problems 63–67 using the given properties. (More than one correct graph is possible.)

63. The domain of f is $(-\infty, \infty)$, $f(4) = 5$, -1 is a relative minimum that occurs at $x = 1$, $f(0) = 0$, and f is an odd function.

64. The domain of f is $(-\infty, \infty)$, $f(0) = 0$, -3 is a relative minimum that occurs at $x = 1$, and f is an even function.

65. The domain of f is $(-\infty, \infty)$, $f(0) = 1$, 0 is a relative minimum that occurs at $x = 1$, and f is an even function.

66. The domain of f is $(-\infty, \infty)$, 2 is a relative maximum that occurs at $x = 0$, -1 is a relative minimum that occurs at $x = 3.5$, and f is neither even nor odd.

67. The domain of f is $(-\infty, \infty)$, 4 is a relative maximum that occurs at $x = 0$, -2 is a relative minimum that occurs at $x = 3.5$, and f is neither even nor odd.

True–False Critical Thinking Problems

68. Which one of the following is true?

a. If $f(x) = |x|$ and $g(x) = |x + 3| + 3$, then the graph of g is a translation of three units to the right and three units upward of the graph of f.

b. If $f(x) = -\sqrt{x}$ and $g(x) = \sqrt{-x}$, then f and g have identical graphs.

c. If $f(x) = x^2$ and $g(x) = 5(x^2 - 2)$, then the graph of g can be obtained from the graph of f by stretching f five units followed by a downward shift of two units.

d. If $f(x) = x^3$ and $g(x) = -(x - 3)^3 - 4$, then the graph of g can be obtained from the graph of f by moving f three units to the right, reflecting in the x-axis, and then moving the resulting graph down four units.

69. Which one of the following is true?

a. If $f(x) = |x|$ and $g(x) = |x - 3| - 3$, then the graph of g is a translation of three units to the left and three units downward of the graph of f.

b. If $f(x) = x^2$ and $g(x) = -(x - 2)^2 - 4$, then the graph of g can be obtained from the graph of f by moving f two units to the right and four units downward, and then reflecting in the x-axis.

c. If $f(x) = x^2$ and $g(x) = 5x^2 - 2$, then the graph of g is a vertical stretch of five units and a downward shift of two units of the graph of f.

d. If $f(x) = \sqrt{x}$ and $g(x) = -\sqrt{-x}$, then f and g have identical graphs.

Technology Problems

In Problems 70–75, use a graphing utility to graph the three given functions on the same Cartesian coordinate plane. Describe the graphs of g and h relative to the graph of f.

70. $f(x) = x^2 - 2x$
$g(x) = f(x - 1)$
$h(x) = -f(x)$

71. $f(x) = x^2 - 3x - 4$
$g(x) = f(x - 1)$
$h(x) = -f(x)$

72. $f(x) = x^3 - 2x^2$
$g(x) = -\frac{1}{2}f(x)$
$h(x) = f(-x)$

73. $f(x) = x^3 - 3x^2$
$g(x) = -\frac{1}{2}f(x)$
$h(x) = f(-x)$

74. $f(x) = \sqrt{x}$ (Xmin $= -6$, Xmax $= 6$, Xscl $= 1$, Ymin $= -4$, Ymax $= 4$, Yscl $= 1$)

$g(x) = f(x + 3)$　　　　$h(x) = f(x + 3) - 3$

75. $f(x) = \sqrt{2x - x^2}$ (Xmin $= -5$, Xmax $= 5$, Xscl $= 1$, Ymin $= -4$, Ymax $= 4$, Yscl $= 1$)

$g(x) = 2f(x - 1)$　　　　$h(x) = -2f(-x)$

76. a. Use a graphing utility to graph $f(x) = x^2 + 1$.

　　b. Graph $f(x) = x^2 + 1$, $g(x) = f(2x)$, $h(x) = f(3x)$, and $k(x) = f(4x)$ on the same viewing rectangle.

　　c. Describe the relationship among the graphs of f, g, h, and k with emphasis on different values of x for points on all four graphs that give the same y-coordinate.

　　d. Generalize by describing the relationship between the graph of f and the graph of g, where $g(x) = f(cx)$ for $c > 1$.

　　e. Try out your generalization by sketching the graphs of $f(cx)$ for $c = 1$, $c = 2$, $c = 3$, and $c = 4$ for a function of your choice.

77. a. Use a graphing utility to graph $f(x) = x^2 + 1$.

　　b. Graph $f(x) = x^2 + 1$, and $g(x) = f(\frac{1}{2}x)$, and $h(x) = f(\frac{1}{4}x)$ on the same viewing rectangle.

　　c. Describe the relationship among the graphs of f, g, and h with emphasis on different values of x for points on all three graphs that give the same y-coordinate.

　　d. Generalize by describing the relationship between the graph of f and the graph of g, where $g(x) = f(cx)$ for $0 < c < 1$.

　　e. Try out your generalization by sketching the graphs of $f(cx)$ for $c = 1$, and $c = \frac{1}{2}$, and $c = \frac{1}{4}$ for a function of your choice.

78. a. Use a graphing utility to graph f and g on the same viewing rectangle, where $f(x) = x^2$ and $g(x) = f(x - 1) - 4$.

　　b. Use the graph of g to find all x such that $g(x) < 0$.

　　c. What inequality did you solve in part (c)?

Writing in Mathematics

79. Describe the sequence of transformations that yield the graph of $g(x) = 4(x^2 - 3)$ and $h(x) = 4x^2 - 3$ from the graph of $f(x) = x^2$. Contrast the order of applying transformations to obtain the graphs of g and h.

80. A company's profit is given by the function $y = P(x)$, where x represents the amount spent on advertising and P represents weekly profits, both expressed in hundreds of dollars.

　　a. Describe a situation that might occur in the company that would result in the graph of its profit function undergoing a vertical shift.

　　b. Now consider the function $y = D(x)$, where x represents the amount spent on advertising and D represents weekly profits, both expressed in dollars rather than hundreds of dollars. If D and P are both graphed on the same axes, describe the relationship between the two graphs.

81. Many people now study college algebra assisted by a graphing utility. With the use of a graphing utility, what purpose, if any, is there in studying translations, reflections, stretching, and shrinking?

82. Excluding $f(x) = 0$, can the graph of a function be symmetric with respect to the x-axis? Explain.

83. Excluding $f(x) = 0$, can a function be both even and odd? Explain.

Critical Thinking Problems

For Problems 84–90, assume that (a, b) is a point on the graph of f. What is the corresponding point on the graph of each of the following functions?

84. $y = f(-x)$　　　　**85.** $y = 2f(x)$　　　　**86.** $y = f(x - 3)$　　　　**87.** $y = f(x) - 3$

88. $y = f(x + 2) - 4$　　**89.** $y = -f(x - 5)$　　**90.** $y = -f(x + 3) + 2$

For Problems 91–93, assume that f is decreasing on the interval $(-\infty, 1)$, increasing on the interval $(1, 6)$, and decreasing on the interval $(6, \infty)$. Describe the interval(s) on which each function is increasing and decreasing.

91. $y = f(-x)$　　　　　　**92.** $y = -f(x)$

93. $y = -f(-x)$

94. Suppose that $h(x) = \dfrac{f(x)}{g(x)}$. The function f can be even, odd, or neither. The same is true for the function g.

　　a. Under what conditions is h definitely an even function?

　　b. Under what conditions is h definitely an odd function?

95. A line whose equation is $y = mx + b$ is shifted vertically so that it passes through, (x_1, y_1). What is the general form of the transformed line's equation?

S E C T I O N   2 . 6

Solutions Manual

Tutorial

Video 5

Combinations of Functions

Objectives

1 Combine functions arithmetically, specifying domains.

2 Form composite functions.

3 Write functions as compositions.

4 Model with combinations of functions.

5 Find the iterates of a function.

6 Determine whether a number belongs to the Mandelbrot set.

In this section, we study ways of combining functions to form new functions. One such combination leads to a discussion of continuously inputting a function back into itself, hinting at a new branch of mathematics called chaos.

1 Combine functions arithmetically, specifying domains.

Arithmetic Operations with Functions

Functions can be combined using addition, subtraction, multiplication, and division to create new functions. For example, the functions

$$f(x) = 2x + 1 \quad \text{and} \quad g(x) = x^2 - 4$$

can be combined to form the sum, difference, product, and quotient of f and g. Here's how it's done.

$$f(x) + g(x) = (2x + 1) + (x^2 - 4) = x^2 + 2x - 3 \qquad \text{Sum: } f + g$$

$$f(x) - g(x) = (2x + 1) - (x^2 - 4) = -x^2 + 2x + 5 \qquad \text{Difference: } f - g$$

$$f(x)g(x) = (2x + 1)(x^2 - 4) = 2x^3 + x^2 - 8x - 4 \qquad \text{Product: } fg$$

$$\frac{f(x)}{g(x)} = \frac{2x + 1}{x^2 - 4}, \qquad x \neq \pm 2 \qquad \text{Quotient: } \frac{f}{g}$$

The domain of an arithmetic combination of functions f and g consists of all real numbers that are common to the domains of f and g. In the case of the quotient function $\dfrac{f(x)}{g(x)}$, we must remember not to divide by 0, so we add the further restriction that $g(x) \neq 0$.

We define arithmetic operations with functions using these ideas.

S
tudy tip

In the definition

$(f + g)(x) = f(x) + g(x)$

the name of the new function is $f + g$. The $+$ sign represents function addition. The $+$ sign on the right in the definition represents ordinary numerical addition.

Definitions: sum, difference, product, and quotient of functions

Let f and g be two functions. The *sum $f + g$*, the *difference $f - g$*, the *product fg*, and the *quotient $\dfrac{f}{g}$* are functions whose domains are the set of all real numbers common to the domains of f and g, defined as follows:

1. Sum: $(f + g)(x) = f(x) + g(x)$

2. Difference: $(f - g)(x) = f(x) - g(x)$

3. Product: $(fg)(x) = f(x) \cdot g(x)$

4. Quotient: $\left(\dfrac{f}{g}\right)(x) = \dfrac{f(x)}{g(x)}$, provided $g(x) \neq 0$

ENRICHMENT ESSAY

Graphs and the Sum of Functions

Graphs that appear in magazines and newspapers are often based on the sum of functions. Shown here is

$$F(x) = f(x) + g(x)$$

where $F(x)$ denotes total world population over time, x.

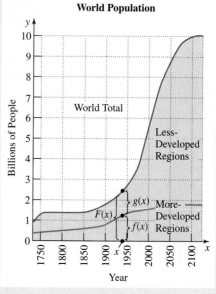

World Population

Billions of People

Year

Source: Population Reference Bureau

Using technology

You can graph the array of functions in Example 1 using a graphing utility and entering

$Y_1 = \sqrt{\ }\ (X + 3)$,

$Y_2 = \sqrt{\ }\ (X - 2)$,

$Y_3 = Y_1 + Y_2$,

$Y_4 = Y_1 - Y_2$,

$Y_5 = Y_1\ Y_2$,

$Y_6 = Y_1/Y_2$

A possible range setting is

Xmin $= -4$, Xmax $= 10$,

Xscl $= 1$, Ymin $= 0$,

Ymax $= 6$, Yscl $= 1$.

The figure below shows all six graphs on the same screen. However, you can deselect any of y_3 through y_6 and study one combined function at a time. The graph shows that the domain for $f + g$, $f - g$, and fg is $[2, \infty)$, but even using a draw dot format it's hard to see that the domain for $\dfrac{f}{g}$ excludes 2.

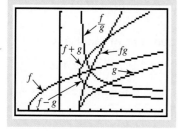

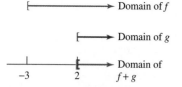

Figure 2.83

EXAMPLE 1 **Combining Functions**

Let $f(x) = \sqrt{x + 3}$ and $g(x) = \sqrt{x - 2}$. Find the following functions, and the domain of each function.

a. $(f + g)(x)$ **b.** $(f - g)(x)$ **c.** $(fg)(x)$ **d.** $\left(\dfrac{f}{g}\right)(x)$

Solution

a. $(f + g)(x) = f(x) + g(x)$ This is the definition $(f + g)(x)$.

$\qquad\qquad\quad = \sqrt{x + 3} + \sqrt{x - 2}$ Substitute the given functions.

The domain of $f + g$ is all x common to the domains of f and g. Since $f(x) = \sqrt{x + 3}$, the domain of f is all x such that $x + 3 \geqslant 0$ or, equivalently, $x \geqslant -3$. The domain of f is $[-3, \infty)$. Similarly, since $g(x) = \sqrt{x - 2}$, its domain is all x such that $x - 2 \geqslant 0$ or, equivalently, $x \geqslant 2$. The domain of g is $[2, \infty)$. Figure 2.83 shows the set of all real numbers common to the domains of f and g. Shown on a number line, we see that the numbers in the two domains intersect at $x \geqslant 2$. Thus, the domain of $f + g$ is $[2, \infty)$.

b. $(f - g)(x) = f(x) - g(x)$ This is the definition of $(f - g)(x)$.

$\qquad\qquad\quad = \sqrt{x + 3} - \sqrt{x - 2}$ Substitute the given functions.

The domain of $f - g$ is all x common to the domains of f and g, so $(f - g)(x) = \sqrt{x + 3} - \sqrt{x - 2}$ has a domain of $[2, \infty)$, the same as the domain of $f + g$.

c. $(fg)(x) = f(x)g(x)$ This is the definition of $(fg)(x)$.

$\qquad\quad = \sqrt{x + 3}\ \sqrt{x - 2}$ Substitute the given functions.

We see that $(fg)(x) = \sqrt{x+3}\,\sqrt{x-2}$ or, equivalently, $(fg)(x) = \sqrt{x^2+x-6}$. The domain of fg is the same as the domains of $f+g$ and $f-g$, namely, $[2, \infty)$.

d. $\left(\dfrac{f}{g}\right)(x) = \dfrac{f(x)}{g(x)}$ This is the definition of $\left(\dfrac{f}{g}\right)(x)$.

$= \dfrac{\sqrt{x+3}}{\sqrt{x-2}}$ Substitute the given functions.

We see that $\left(\dfrac{f}{g}\right)(x) = \dfrac{\sqrt{x+3}}{\sqrt{x-2}}$ or, equivalently, $\left(\dfrac{f}{g}\right)(x) = \sqrt{\dfrac{x+3}{x-2}}$. Since

the denominator of $\dfrac{f}{g}$ cannot equal zero, we must exclude 2 from the domain. Unlike $f+g, f-g$, and fg, whose domain was $[2, \infty)$, the domain of $\dfrac{f}{g}$ is $(2, \infty)$.   ■

2 Form composite functions.

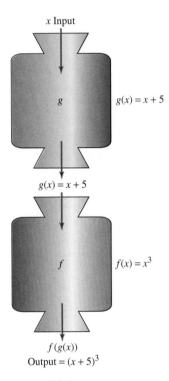

Figure 2.84

Inputting one function into a second function

Composition of Functions

Another method for combining functions is to form the *composition* of one with the other. For example, if $f(x) = x^3$ and $g(x) = x + 5$, then we can form the composition of f with g by taking the output of g, namely, $g(x) = x + 5$, and using it as the input of f.

$f(g(x))$ We must substitute $g(x)$ for x in function f.

$= [g(x)]^3$ Use $f(x)$ and replace x by $g(x)$.

$= (x+5)^3$ Replace $g(x)$ by $x + 5$.

Thus, $f(g(x)) = (x+5)^3$. Observe how this composition, represented by $f \circ g$, is illustrated in Figure 2.84.

> **Definition: composition function $f \circ g$**
>
> Given two functions f and g, the *composition* of f and g, denoted by $f \circ g$, is defined by
>
> $$(f \circ g)(x) = f(g(x)).$$ We read $f(g(x))$ as "f of g of x."
>
> The domain of $f \circ g$ is the set of all x in the domain of g such that $g(x)$ is in the domain of f.

The composition of f and g, $f \circ g$, is pictured with the arrow diagram in Figure 2.85. The diagram indicates that the output of g, or $g(x)$, becomes the input for "machine" f. If $g(x)$ is not in the domain of f, it cannot be inputted into machine f, and so $g(x)$ must be discarded.

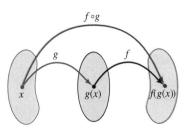

Figure 2.85

The composition of f and g: Inputting $g(x)$ into f

EXAMPLE 2   **Forming the Composition of f with g**

Let $f(x) = \sqrt{x}$ and $g(x) = 2x - 1$. Find:

a. $(f \circ g)(x)$ **b.** The domain of $f \circ g$ **c.** $(f \circ g)(5)$ and $(f \circ g)(0)$

Using technology

The graphs of f, g, and $f \circ g$ can be obtained on a graphing utility by entering

$$y_1 = \sqrt{x}$$

$$y_2 = 2x - 1$$

$$y_3 = \sqrt{y_2}$$

Using a $[-2, 8]$ by $[-1, 6]$ viewing rectangle, the graph indicates that the domain of $f \circ g$ is $[\frac{1}{2}, \infty)$.

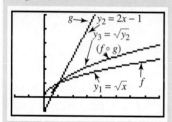

Solution

a. $(f \circ g)(x) = f(g(x))$ This is the definition of $f \circ g$.

$\qquad\qquad = \sqrt{g(x)}$ Use f and replace x with $g(x)$.

$\qquad\qquad = \sqrt{2x - 1}$ Replace $g(x)$ with $2x - 1$.

b. The domain of f is $[0, \infty)$. For $\sqrt{2x - 1}$ to be defined, we must have $2x - 1 \geqslant 0$, so $x \geqslant \frac{1}{2}$. The domain of $f \circ g$ is $[\frac{1}{2}, \infty)$.

c. $(f \circ g)(x) = \sqrt{2x - 1}$ This is the function from part (a).

$\quad (f \circ g)(5) = \sqrt{2(5) - 1}$ To find $(f \circ g)(5)$, replace x with 5.

$\qquad\qquad\quad = 3$ An x-value of 5 in $f \circ g$ has a corresponding y-value of 3.

We can find $(f \circ g)(5)$ because 5 is in the domain of $f \circ g$, namely, $[\frac{1}{2}, \infty)$. However, $(f \circ g)(0)$ is undefined because 0 is not in the domain. ∎

Study tip

An alternative method for Example 2c is to use the given functions

$$f(x) = \sqrt{x} \quad \text{and} \quad g(x) = 2x - 1.$$

We have

$$(f \circ g)(5) = f(g(5)) = f(9) = \sqrt{9} = 3 \quad \text{Note that } g(5) = 2 \cdot 5 - 1 = 9.$$

$$(f \circ g)(0) = f(g(0)) = f(-1) = \sqrt{-1} \quad \sqrt{-1} \text{ is not a real number.}$$

The composition of f with g, $f \circ g$, is generally not the same function as the composition of g with f, $g \circ f$. Furthermore, the functions may not have the same domain.

EXAMPLE 3   Forming the Composition of g with f

Let $f(x) = \sqrt{x}$ and $g(x) = 2x - 1$. Find:

a. $(g \circ f)(x)$ **b.** The domain of $g \circ f$

Solution

a. $(g \circ f)(x) = g(f(x))$ This is the definition of $g \circ f$.

$\qquad\qquad = 2f(x) - 1$ Use g and replace x with $f(x)$.

$\qquad\qquad = 2\sqrt{x} - 1$ Replace $f(x)$ with $\sqrt{x}$

b. The domain of g is $(-\infty, \infty)$. Since $f(x)$ must be in this interval, then $\sqrt{x}$ must be a real number, so $x \geqslant 0$. The domain of $g \circ f$ is $[0, \infty)$. ∎

Using technology

The graphs of $f \circ g$ and $g \circ f$ can be obtained on a graphing utility by entering

$Y_1 = \boxed{\sqrt{}}\, X$

$Y_2 = 2X - 1$

$Y_3 = \boxed{\sqrt{}}\, Y_2$ This is $f \circ g$.

$Y_4 = 2Y_1 - 1$ This is $g \circ f$.

Deselect y_1 and y_2 and use

$Xmin = -2$, $Xmax = 4$,

$Xscl = 1$, $Ymin = -1$,

$Ymax = 3$, $Yscl = 1$

The graph of y_3 ($f \circ g$) has a domain of $[\frac{1}{2}, \infty)$ and the graph of y_4 ($g \circ f$) has a domain of $[0, \infty)$.

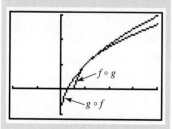

Study tip

The graphs in the Using Technology box show that in general, $f \circ g \neq g \circ f$. It is helpful to remember that the notation $f \circ g$ means that the function g is applied first and then f is applied second.

We need to be careful in describing the domain of a composite function, as we will see in the next example.

Using technology

We can use a graphing utility to find the domain of $f \circ g$. Enter

$Y_1 \;\boxed{=}\; X \;\boxed{\wedge}\; 2 - 4$

This is f.

$Y_2 \;\boxed{=}\; \boxed{\sqrt{}}\;(4 - X \;\boxed{\wedge}\; 2)$

This is g.

$Y_3 \;\boxed{=}\; Y_2 \;\boxed{\wedge}\; 2 - 4$

This $f(g(x))$.

Use a $[-4, 4]$ by $[-8, 2]$ viewing rectangle. The graph of $f \circ g$ does not extend to the left of $x = -2$ or to the right of $x = 2$, reinforcing the fact that the domain of $f \circ g$ is $[-2, 2]$.

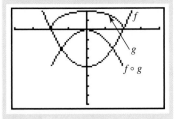

3 Write functions as compositions.

Discover for yourself

Consider

$h(x) = \sqrt[3]{x^2 + 1}$

the function in Example 5. Let

$f(x) = \sqrt[3]{x + 1}$

and

$g(x) = x^2$.

Show that

$f(g(x)) = h(x)$.

What can you conclude about writing a function as a composition?

EXAMPLE 4 **Finding the Domain of a Composite Function**

If $f(x) = x^2 - 4$ and $g(x) = \sqrt{4 - x^2}$ find:

a. $(f \circ g)(x)$ and   **b.** The domain of $f \circ g$

Solution

a. $(f \circ g)(x) = f(g(x))$ This is the definition of $f \circ g$.

$\qquad\qquad = [g(x)]^2 - 4$ Use f and replace x by $g(x)$.

$\qquad\qquad = (\sqrt{4 - x^2})^2 - 4$ Replace $g(x)$ by $\sqrt{4 - x^2}$.

$\qquad\qquad = 4 - x^2 - 4$

$\qquad\qquad = -x^2$ Simplify.

b. We see that $(f \circ g)(x) = -x^2$. You may initially think that the domain of $f \circ g$ is the set of all real numbers, since any real number can be used as an input in the expression $-x^2$. However, the expression $(\sqrt{4 - x^2})^2 - 4$, which simplifies to $-x^2$, is defined when $4 - x^2 \geq 0$.

$\qquad 4 - x^2 \geq 0$

$\qquad\qquad x^2 \leq 4$

$\qquad -2 \leq x \leq 2$

Thus, the domain of $f \circ g$ is $[-2, 2]$.   ■

Up to this point, our examples have called for the formation of the composition of two functions. Often it is useful to recognize the two (or more) simpler functions that make up a given complicated composite function. For example, the function h defined by

$$h(x) = (3x^2 - 4x + 1)^5$$

is the composition of f with g, where $f(x) = x^5$ and $g(x) = 3x^2 - 4x + 1$. That is,

$$h(x) = (3x^2 - 4x + 1)^5 = (g(x))^5 = f(g(x)).$$

Note that in expressing h as $f \circ g$, we choose g to be the "inner" function, that is, the expression inside parentheses: $g(x) = 3x^2 - 4x + 1$. Similarly, $f(x) = x^5$ is the "outer" function.

EXAMPLE 5 **Writing a Function as a Composition**

Express as a composition of two functions: $h(x) = \sqrt[3]{x^2 + 1}$.

Solution

We let g be the inner function; that is, let $g(x)$ be the binomial under the radical.

$\qquad g(x) = x^2 + 1$

Then we let f be the outer function.

$\qquad f(x) = \sqrt[3]{x}$

Check

These choices for f and g are appropriate if $f(g(x)) = h(x)$.

$f(g(x))$ We chose $f(x) = \sqrt[3]{x}$ and $g(x) = x^2 + 1$.

$= \sqrt[3]{g(x)}$ Replace x in f with $g(x)$.

$= \sqrt[3]{x^2 + 1}$ Replace $g(x)$ with $x^2 + 1$.

$= h(x)$ This is the definition of the given function h.   ■

Study tip

If $h(x) = (\text{polynomial in } x)^p$, where p represents a real number, then one way to express h as a composition of two functions is to select

$g(x) = \text{polynomial in } x$

and

$f(x) = x^p$.

With these choices,

$f(g(x)) = h(x)$.

Modeling with Combinations of Functions

Combinations of functions can be used to describe the physical world. For example, Figure 2.86 indicates that math SAT scores for males and females are a function of time. The function f represents the scores for males over time and the function g the scores of females over time. Total scores over time can be obtained by taking the average of the scores for males and females, represented by the function

$$y = \frac{(f + g)(x)}{2}.$$

The graph of $\dfrac{f + g}{2}$ can be obtained by adding the corresponding y-coordinates of f and g and dividing this sum by 2.

Composite functions often model real world phenomena, as illustrated in Example 6.

Math SAT Scores, 1967–1993

Math SAT Scores vs. Year, with curves $y = f(x)$ (Males), $y = \dfrac{(f+g)(x)}{2}$ (Total), and $y = g(x)$ (Females).

Figure 2.86

Source: The College Board

4 Model with combinations of functions

April Gornick "Two Fires" 1983, oil on canvas, 74 × 96 in. Courtesy Edward Thorp Gallery, New York.

EXAMPLE 6   **An Environmental Application**

The relationship between average global temperature increase (G, in degrees Fahrenheit) and carbon dioxide (CO_2) concentration (C, in parts per million, ppm) is modeled by the function

$$G(C) = 0.0193C - 5.4.$$

One possible model for CO_2 concentration (C, in parts per million) t years after 1939 is

$$C(t) = 1.44t + 280.$$

Find the following:

a. The composite function $G(C(t))$. What does this function represent? What information is conveyed by its slope?

b. $G(C(61))$. What information is conveyed by the resulting function value?

Study tip

Reminder: slope represents average rate of change.

Solution

Take a moment to examine the significance of the slope of the given functions.

$$G(C) = 0.0193C - 5.4$$

There is a 0.0193°F increase in global temperature for each ppm increase in CO_2 concentration.

Slope = 0.0193

$$C(t) = 1.44t + 280$$

There is an increase in CO_2 concentration of 1.44 ppm each year.

Slope = 1.44

Discover for yourself

Environmentalists have found that a temperature increase of 1.8°F corresponds to a one-foot rise in ocean level. Use the result of part (a) to find by how much the oceans will rise each year.

a.
$$\begin{aligned} G(C(t)) &= 0.0193C(t) - 5.4 & &\text{Use } G \text{ and replace } C \text{ with } C(t).\\ &= 0.0193(1.44t + 280) - 5.4 & &\text{Replace } C(t) \text{ with } 1.44t + 280.\\ &= 0.027792t + 5.404 - 5.4 & &\text{Apply the distributive property.}\\ &= 0.027792t + 0.004 & &\text{Simplify.} \end{aligned}$$

Slope = 0.027792

This composite function represents average global temperature increase as a function of time. The slope, 0.027792, indicates that average global temperature will increase 0.027792°F each year.

b.
$$\begin{aligned} G(C(t)) &= 0.027792t + 0.004 & &\text{This is the composite function derived in part (a).}\\ G(C(61)) &= 0.027792(61) + 0.004 & &\text{Replace } t \text{ with } 61.\\ &= 1.695712 \end{aligned}$$

This result indicates that we can expect average global temperature increase of approximately 1.7°F 61 years after 1939, in the year 2000. ∎

Modeling the Mandelbrot Set

By opening up our discussion to include the complex plane and by composing a function with itself, we can construct dramatic visual images.

Recall from our prerequisites chapter that the complex number $a + bi$ can be represented as a point in the complex plane. In this plane, the horizontal axis represents the real part of the complex number and the vertical axis the imaginary part. Figure 2.87 shows the graph of the two points corresponding to $3 + 2i$ and $-2 - 4i$.

Now let's see what is meant by composing a function with itself.

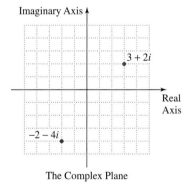

The Complex Plane

Figure 2.87

Plotting complex numbers

5 Find the iterates of a function.

The iterates of a function

$$f^2(x) = f(f(x))$$
$$f^3(x) = f(f(f(x))) = f(f^2(x))$$
$$\vdots$$

$f^n(x)$, called the nth iterate of a function, is found by composing f with itself n times.

For example, consider a function that adds a constant c to the square of each input: $f(x) = x^2 + c$. The second and third iterates of this function are:

$$f^2(x) = f(f(x)) = (f(x))^2 + c = (x^2 + c)^2 + c$$
$$f^3(x) = f(f^2(x)) = (f^2(x))^2 + c = [(x^2 + c)^2 + c]^2 + c$$

If we evaluate the given function and its first two iterates at 0, we obtain

$$f(0) = 0^2 + c = c$$
$$f^2(0) = (0^2 + c)^2 + c = c^2 + c$$
$$f^3(0) = [(0^2 + c)^2 + c]^2 + c = (c^2 + c)^2 + c.$$

Since each successive iterate is obtained by adding c to the square of the previous result, the next iterate is

$$[(c^2 + c)^2 + c]^2 + c.$$

This sequence of iterates remains small for some values of c and is said to be bounded. For other values of c, the successive numbers in the sequence become larger and larger, and the sequence is unbounded. The set of all complex numbers for which the sequence is bounded is called the *Mandelbrot set*.

6 Determine whether a number belongs to the Mandelbrot set.

EXAMPLE 7 The Mandelbrot Set

Determine which of the following numbers belong to the Mandelbrot set: $1, -1, i, 1 + i$.

Solution

Complex Number c	c	$(c^2 + c)$	$(c^2 + c)^2 + c$	$[(c^2 + c)^2 + c]^2 + c$	Keep squaring the previous result and add c
$c = 1$	1	$1^2 + 1 = 2$	$2^2 + 1 = 5$	$5^2 + 1 = 26$	$26^2 + 1 = 677$

The numbers $1, 2, 5, 26, 677, \ldots$ are getting larger and larger. Thus, 1 is not a member of the Mandelbrot set.

$c = -1$	-1	$(-1)^2 + (-1) = 0$	$0^2 + (-1) = -1$	$(-1)^2 + (-1) = 0$	$0^2 + (-1) = -1$

The sequence of numbers $-1, 0, -1, 0, -1, \ldots$ remains small. Thus, -1 is a member of the Mandelbrot set.

$c = i$	i	$i^2 + i$ $= -1 + i$	$(-1 + i)^2 + i$ $= 1 - 2i + i^2 + i$ $= -i$	$(-i)^2 + i$ $= -1 + i$	$(-1 + i)^2 + i$ $= -i$

The next number in the sequence is $-1 + i$, and the sequence $i, -1 + i, -i, -1 + i, -i, -1 + i, \ldots$ is bounded. Thus, i is a member of the Mandelbrot set.

$c = 1 + i$	$1 + i$	$(1 + i)^2 + 1 + i$ $= 1 + 3i$	$(1 + 3i)^2 + 1 + i$ $= -7 + 7i$	$(-7 + 7i)^2 + 1 + i$ $= 1 - 97i$	$(1 - 97i)^2 + 1 + i$ $= -9407 - 193i$

The sequence of numbers $1 + i, 1 + 3i, -7 + 7i, 1 - 97i, -9407 - 193i \ldots$ does not remain small and is not bounded. Thus, $1 + i$ is not a member of the Mandelbrot set.

The numbers -1 and i are members of the Mandelbrot set. ∎

A black-and-white rendering of the Mandelbrot set is shown in Figure 2.88 on page 295. Points that are in the set consist of the number for which the sequence of iterates is bounded. The resulting graph assumes an image that is "buglike" in shape.

Mathematicians and computer artists have added colors to the graph of the boundary of the Mandelbrot set (see the image in the chapter introduction).

ENRICHMENT ESSAY

Recursive Processes; The Sierpinski Gasket

So Nat'ralists observe, A Flea
Hath Smaller Fleas that on him prey
and these have smaller Fleas to bite'em
And so proceed, ad infinitum
　　　　　　　　—Jonathan Swift

A specified sequence of computational steps is called an algorithm. If an algorithm proceeds by repeating a sequence of steps, using the result from the previous step as the input for the next step, it is called *recursive*. Each repetition is an iteration.

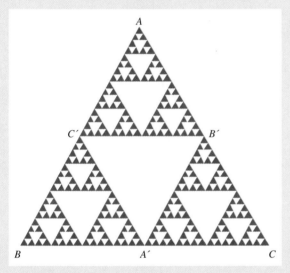

Shown here is an iteration where triangles are repeatedly subtracted from within triangles. From the midpoints of the sides of ΔABC, remove

$\Delta A'B'C'$. Repeat the process for the three remaining triangles, then with each of the triangles left after that, and so on ad infinitum.

After infinitely many steps of this recursive process, an unusual "Swiss cheese"-like triangle called the Sierpinski gasket, named after mathematician Waclaw Sierpinski (1822–1910), emerges. The Sierpinski iteration, shown below, was generated on a computer by scientists at the University of Regina in Canada by subtracting pyramids from within pyramids.

The word *fractal* (from the Latin word *fractus,* meaning "broken up, fragmented") was first used by mathematician Benoit Mandelbrot to describe objects like the Sierpinski gasket.

Daryl H. Hepting and Allan N. Snider "Desktop Tetrahedron" 1990. Computer generated at the University of Regina in Canada.

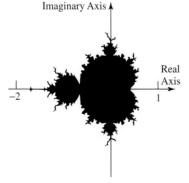

Figure 2.88

The Mandelbrot set

Color choices depend on how quickly the numbers in the boundary approach infinity. The magnified boundary yields repetition of the original overall bug-like structure as well as new and interesting patterns. With each new level of magnification, repetition and unpredictable formations interact to create what has been called the most complicated mathematical object ever known.

The Mandelbrot set is a mathematical *fractal*, a set of points that exhibits increasing detail with increasing magnification. Fractals are as complex in their details as in their overall form. The blending of infinite repetition and infinite variety is one of the hallmarks of the Mandelbrot set in particular and mathematics in general.

PROBLEM SET 2.6

Practice Problems

In Problems 1–8, find $f + g, f - g, fg,$ and $\dfrac{f}{g}$ and their domains.

1. $f(x) = 2x^2 - x - 3, \quad g(x) = x + 1$

2. $f(x) = 6x^2 - x - 1, \quad g(x) = 2x - 1$

3. $f(x) = \sqrt{x + 4}, g(x) = \sqrt{x - 1}$

4. $f(x) = \sqrt{x + 6}, g(x) = \sqrt{x - 3}$

5. $f(x) = \dfrac{1}{x}, g(x) = \dfrac{1}{x^3}$

6. $f(x) = \dfrac{1}{x - 2}, g(x) = \dfrac{1}{(x - 2)^2}$

7. $f(x) = \sqrt{2x - 6}, g(x) = \dfrac{1}{x}$

8. $f(x) = \sqrt{3x - 12}, g(x) = \dfrac{1}{x}$

In Problems 9–12, graph the functions $f, g,$ and $f + g$ on the same set of coordinate axes.

9. $f(x) = 3x, g(x) = 2x - 4$

10. $f(x) = 2x, g(x) = x + 5$

11. $f(x) = 2x, g(x) = -2x + 3$

12. $f(x) = 4x, g(x) = -4x + 1$

13. If $f(x) = 2x^2 - x - 10$ and $g(x) = x + 2$, find $\left(\dfrac{f}{g}\right)(x)$, simplify and graph the function.

14. If $f(x) = x^2 + 8x + 15$ and $g(x) = x + 3$, find $\left(\dfrac{f}{g}\right)(x)$, simplify and graph the function.

In Problems 15–20, find **a.** $f \circ g$, **b.** $g \circ f$, and **c.** $f \circ f$.

15. $f(x) = x^2, g(x) = x + 2$

16. $f(x) = x^2, g(x) = x - 3$

17. $f(x) = 2x + 3, g(x) = 4 - x$

18. $f(x) = 4x - 1, g(x) = 3 - x$

19. $f(x) = 4x^3 - 1, g(x) = \sqrt[3]{\dfrac{x + 1}{4}}$

20. $f(x) = x^3 - 2, g(x) = \sqrt[3]{x + 2}$

In Problems 21–34, find **a.** $f \circ g$ and **b.** $g \circ f$. Also give the domain for each function.

21. $f(x) = \sqrt{x}, g(x) = 2x + 1$

22. $f(x) = \sqrt{x}, g(x) = 3x - 1$

23. $f(x) = \dfrac{1}{x}, g(x) = \dfrac{1}{2x}$

24. $f(x) = \dfrac{1}{x}, g(x) = \dfrac{1}{3x}$

25. $f(x) = x^2 - 9, g(x) = \sqrt{9 - x^2}$

26. $f(x) = x^2 - 1, g(x) = \sqrt{1 - x^2}$

27. $f(x) = x^{1/3}, g(x) = x^6$

28. $f(x) = x^{2/3}, g(x) = x^3$

29. $f(x) = |x|, g(x) = 3x - 1$

30. $f(x) = |x|, g(x) = 2x + 3$

31. $f(x) = \dfrac{3x + 1}{x - 1}, g(x) = \dfrac{x + 1}{x - 3}$

32. $f(x) = \dfrac{x + 1}{1 - x}, g(x) = \dfrac{x - 1}{x + 1}$

33. $f(x) = \sqrt[5]{\dfrac{x^3 + 1}{2}}, g(x) = \sqrt[3]{2x^5 - 1}$

34. $f(x) = (x + 4)^{5/3}, g(x) = x^{3/5} - 4$

In Problems 35–44, use the graphs of f and g to evaluate the expression.

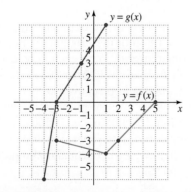

35. $(f \circ g)(1)$

36. $(g \circ f)(1)$

37. $(g \circ g)(-3)$

38. $(f \circ f)(1)$

39. $(g \circ f)(5)$

40. $(f \circ g)(-1)$

41. $(f + g)(1)$

42. $(f + g)(-3)$

43. $(f - g)(-3)$

44. $(f - g)(1)$

In Problems 45–56, find two functions f and g so that h(x) = (f ∘ g)(x). (More than one correct answer is possible for each problem.)

45. $h(x) = (3x - 1)^4$

46. $h(x) = (2x - 5)^3$

47. $h(x) = \sqrt[3]{x^2 - 9}$

48. $h(x) = \sqrt{5x^2 + 3}$

49. $h(x) = |2x - 5|$

50. $h(x) = |3x - 4|$

51. $h(x) = \dfrac{1}{2x - 3}$

52. $h(x) = \dfrac{3}{(5x - 1)^2}$

53. $h(x) = (2x^2 - 5x + 1)^{2/3}$

54. $h(x) = (3x^2 - 7x + 4)^{3/4}$

55. $h(x) = 3(x - 1)^2 + 5(x - 1)$

56. $h(x) = 4(x - 3)^2 + 6(x - 3)$

In Problems 57–62, let f(x) = 3x − 1, g(x) = x², and h(x) = |x|. Write each function as a composition of functions chosen from f, g, and h.

57. $k(x) = (3x - 1)^2$

58. $k(x) = 3x^2 - 1$

59. $k(x) = 3|x| - 1$

60. $k(x) = |3x - 1|$

61. $k(x) = (3|x| - 1)^2$

62. $k(x) = |3x - 1|^2$

In Problems 63–68, find f³(x), the third iterate of the given function f.

63. $f(x) = 2x - 3$

64. $f(x) = 3x + 2$

65. $f(x) = x^2 + 1$

66. $f(x) = 2x^2 - 1$

67. $f(x) = x^3$

68. $f(x) = -2x^3$

69. If $f(x) = 1 + \dfrac{1}{x}$, find $f^4(x)$ and predict the result for $f^n(x)$ as n continues to get larger.

Application Problems

70. Consider two functions M and F that represent the number of male and female members of the House of Representatives for the years 1977, 1981, 1991, and 1994. Sketch the graphs of M and F in the same Cartesian coordinate plane, using the data in the graphs. Each graph should consist of four points whose first coordinates are the years and second coordinates are the numbers of representatives, male or female. Now add to the graphs in your coordinate plane the graph of $M + F$. What constant function do you obtain? What is the significance of this constant?

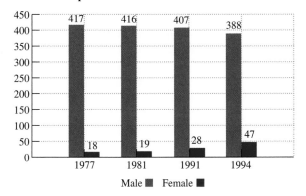

Gender Breakdown of the House of Representatives for Four Selected Years

417   416   407   388

18   19   28   47

1977   1981   1991   1994

Male ■   Female ■

71. A department store has two locations in a city. From 1994 through 1998, the profits for each of the store's two branches are modeled by the functions $f(x) = -0.44x + 13.62$ and $g(x) = 0.51 x + 11.14$. In each model, x represents the number of years after 1994 and f and g represent the profit in millions of dollars.

a. What is the slope for f? Describe what this means.

b. What is the slope for g? Describe what this means.

c. Find $f + g$. What is the slope for this function? What does this mean?

72. Skin temperature (T, in degrees Celsius) is a function of the Celsius temperature C of the environment, modeled by $T(C) = 0.27C + 27.4$. Celsius temperature C is a function of Fahrenheit temperature F, described by $C(F) = \frac{5}{9}(F - 32)$.

a. Find and interpret $T(C(F))$.

b. Find and interpret $T(C(41))$.

73. A stone is dropped into a calm lake, creating a circular ripple that travels outward at a speed of 0.4 feet per second. Thus, $r(t) = 0.4t$, where t represents time in seconds. If the area of the circle is described by $A(r) = \pi r^2$, find $(A \circ r)(t)$ and interpret the resulting composite function.

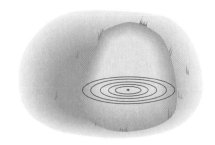

74. A spherical ballon is inflated in such a way that its radius is increasing at a rate of 2 centimeters per second. Thus, $r(t) = 2t$, where t represents time in seconds. If the volume of a sphere is described by $V(r) = \frac{4}{3}\pi r^3$, find $(V \circ r)(t)$ and interpret the resulting composite function.

75. An automobile sells for x dollars.

a. If $f(x) = x - 2000$, describe f using the word *rebate*.

b. If $g(x) = 0.85x$, describe the cost of the car in terms of a percent discount of the car's regular price.

c. Find $(f \circ g)(x)$ and describe what this means in terms of the regular price of the car.

d. Repeat part (c) for $(g \circ f)(x)$

e. Use the results of parts (c) and (d) to compare the car's discounted price.

76. Suppose that deforestation of a rain forest occurs at a rate of 6% per year. If x represents the area of the rain forest at the start of 1997, then $f(x) = 0.94x$ is the area of the rain forest at the start of 1998. Find $f^3(x)$ and interpret this third iterate of f.

In Problems 77–82, use the given value of c to find the following four iterates: $c, c^2 + c, (c^2 + c)^2 + c, [(c^2 + c)^2 + c]^2 + c$ Use the pattern to determine if c is a member of the Mandelbrot set. Check your answer using Figure 2.88 on page 295.

77. 0 **78.** 2 **79.** $-i$ **80.** $\dfrac{i}{2}$

81. $1 - i$ **82.** $2 - i$

True–False Critical Thinking Problems

83. Which one of the following is true?

a. If $f(x) = x^2 - 4$ and $g(x) = \sqrt{x^2 - 4}$, then $(f \circ g)(x) = -x^2$, so that $(f \circ g)(5) = -25$.

b. There can never be two functions f and g, where $f \neq g$, for which $(f \circ g)(x) = (g \circ f)(x)$.

c. If $f(7) = 5$ and $g(4) = 7$, then $(f \circ g)(4) = 5$.

d. If $f(x) = \sqrt{x}$ and $g(x) = 2x - 1$, then $(f \circ g)(5) \neq g(2)$.

84. Which one of the following is true?

a. If $f(x) = x^2 - 9$ and $g(x) = \sqrt{9 - x^2}$, then $(f \circ g)(x) = -x^2$, so that $(f \circ g)(4) = -16$.

b. If $f(x) = 3x - 1$, $g(x) = x^2 + 1$, and $h(x) = \dfrac{x + 1}{3}$, then $(f \circ g)(-1) = (f \circ h)(5)$.

c. If $f(x) = x^4$, then $(f \circ f)(x) = x^8$.

d. If f and g are both quadratic functions (in the form $y = ax^2 + bx + c$, $a \neq 0$), then $f \circ g$ is also a quadratic function.

Technology Problems

For Problems 85–90, use a graphing utility to graph f, g, and $f \circ g$ on the same viewing screen. Use the graphs to describe the domain for all three functions.

85. $f(x) = x^2 - 4$, $g(x) = \sqrt{4 - x^2}$

86. $f(x) = \sqrt{x - 1}$, $g(x) = x^2$

87. $f(x) = \sqrt{x^2 - 1}$, $g(x) = \sqrt{1 - x}$

88. $f(x) = \dfrac{1}{\sqrt{x}}$, $g(x) = x^2 - 4$

89. $f(x) = \sqrt{x - 1}$, $g(x) = \sqrt{1 - x}$

90. $f(x) = \sqrt{1 - x^2}$, $g(x) = \sqrt{x^2 - 1}$

91. a. Graph $f(x) = x^2$, $g(x) = x + 3$, and $h(x) = f(x) + g(x)$ using the following range setting:

Xmin $= -3$, Xmax $= 4$, Xscl $= 1$,

Ymin $= 0$, Ymax $= 15$, and Yscl $= 1$

b. Use the graphs to determine the relationship among the y-coordinates of f, g, and $f + g$.

c. Use the relationship in part (b) to describe a process for obtaining the graph of $f + g$ from the graphs of f and g.

d. Use the process obtained in part (c) to graph $f + g$ for f and g in the figure.

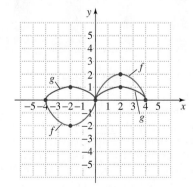

92. The radius of a circle (r) is increasing according to the function $r(t) = 5t + 3$, where r is expressed in inches, t in seconds, and $0 \leq t \leq 8$. The area of the circle is described by $A(r) = \pi r^2$.

  a. Find $A(r(t))$ and graph the function.

  b. Use the graph, along with the $\boxed{\text{TRACE}}$ and $\boxed{\text{ZOOM}}$ features, to determine how many seconds it will take for the circle to have an area of 324π square inches.

93. If $f(x) = x^2$, then $f^n(x_0)$ can be found with a graphing utility by entering x_0 and pressing the $\boxed{x^2}$ key n times. Find $f^{12}(0.96)$ and $f^{12}(1.04)$. Describe what happens to $f^n(0.96)$ and $f^n(1.04)$ as n continues to get larger.

Writing in Mathematics

94. **a.** Describe the process illustrated by the lines with the arrows in the figure.

  b. Describe what happens to $f^n(0.2)$ as n continues to get larger.

  c. Use a calculator to confirm your conclusion in part (b).

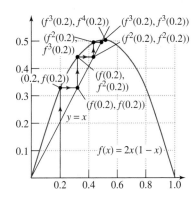

Critical Thinking Problems

95. If $f(x) = x^2 + 5x - 3$ and $h(x) = 16x^2 + 12x - 7$, find a function g such that $f \circ g = h$.

96. If $f \circ g \circ h$ is defined by $(f \circ g \circ h)(x) = f(g(h(x)))$, find $f \circ g \circ h$ if $f(x) = x^2$, $g(x) = x + 2$, and $h(x) = \dfrac{x}{3}$.

97. Define two functions f and g so that $f \circ g = g \circ f$.

98. If $f(x) = x^2$ and $g(x) = 2x + 1$, find
$$\frac{(f \circ g)(x) - (f \circ g)(a)}{x - a}.$$

99. Prove that if f and g are even functions, then fg is also an even function.

100. If f and g are both odd functions, is $\dfrac{f}{g}$ even or odd? Prove your conjecture.

101. If f is an even function and g an odd function, is fg even or odd? Prove your conjecture.

102. If f and g are even functions, is $f \circ g$ even or odd? Prove your conjecture.

103. If f and g are odd functions, is $f \circ g$ even or odd? Prove your conjecture.

104. If f represents any function and $g(x) = \dfrac{f(x) - f(-x)}{2}$, is g even or odd? Prove your conjecture.

105. If $f(x) = -2x + 1$ and g is a linear function, find the slope and y-intercept of g if $f(g(x)) = x$.

Group Activity Problems

106. Use a computer with appropriate software to generate images of fractals. Include an image of the Sierpinski gasket described on page 295.

107. Each group member should be assigned one of the following topics on fractals: algorithms, iterations, itera-tion number, chaos, and fractals in nature. Consult appropriate references, with particular emphasis on ideas and visual images that you believe will interest other group members. After each member researches the assigned topic, the group should come together to share their research with one another.

SECTION 2.7 Inverse Functions

Solutions Manual   Tutorial   Video 5

Objectives

1 Apply the horizontal line test to identify one-to-one functions.
2 Find f^{-1} for specified values.
3 Verify that two functions are inverses of each other.
4 Find the equation of the inverse.
5 Verify inverses with a graphing utility.
6 Solve applied modeling problems using inverses.

The idea behind an inverse operation is to "undo" a previously performed operation. Functions can also undo one another. For example, cubing undoes the operation of taking a cube root, and the functions $f(x) = \sqrt[3]{x}$ and $g(x) = x^3$ are called *inverses*. In this section, we study such functions in detail.

Apply the horizontal line test to identify one-to-one functions.

One-to-One Functions

The graph in Figure 2.89 indicates that the over-85 population in the United States is a function of time. We call this function f, and we observe that the number of over-85 population is different for every year. If we are told that the over-85 population is 8 million, the graph uniquely reveals that this is projected to occur in the year 2020.

Contrast this situation with the one in Figure 2.90 on page 301 which shows U.S. gas consumption. We call this function g, and we notice that the gas consumption is not different for every year. If we are told that the consumption is 120 billion gallons, we cannot uniquely determine the year; this occurred in 1977, 1979, and 1984.

The graph of g in Figure 2.90 reveals that

$$g(1977) = g(1979) = g(1984).$$

In other words, for three different years—x_1, x_2, and x_3—the function values $g(x_1)$, $g(x_2)$, and $g(x_3)$ are the same. On the other hand, for the graph of f in Figure 2.89 below, if the years are different ($x_1 \neq x_2$), then the function values

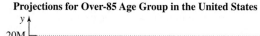

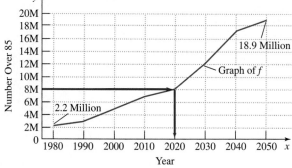

Figure 2.89

Source: U.S. Bureau of the Census

U. S. Gas Consumption
(billions of gallons per year)

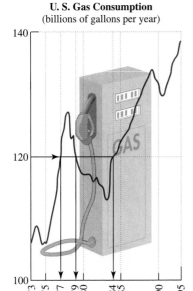

Figure 2.90

Source: U.S. Department of Energy

$f(x_1)$ and $f(x_2)$ are also different. Functions that have the property that $f(x_1) \neq f(x_2)$ whenever $x_1 \neq x_2$ are called *one-to-one functions*.

Definition of a one-to-one function

A *one-to-one function* is one in which no two different ordered pairs have the same second component. This means that

$$f(x_1) \neq f(x_2) \quad \text{whenever} \quad x_1 \neq x_2.$$

How can you tell from a function's graph if it is one-to-one? Compare the horizontal lines drawn in Figures 2.89 and 2.90. In Figure 2.89, the horizontal line intersects the graph only once. In Figure 2.90, the horizontal line intersects the graph in three points.

The horizontal line test

If no horizontal line intersects the graph of a function more than once, then the function is one-to-one.

In Figure 2.91, all the graphs are functions since they all pass the vertical line test. However, the graphs in parts (b), (c), and (e) are not one-to-one because horizontal lines can be drawn that intersect the graphs more than once.

Figure 2.91 illustrates that functions whose graphs are increasing on their entire domain (parts (a), (d), (f), and (g)) all pass the horizontal line test. The same is true if a function is decreasing on its entire domain.

Increasing and decreasing functions

1. If f is increasing on its entire domain, then f is one-to-one.
2. If f is decreasing on its entire domain, then f is one-to-one.

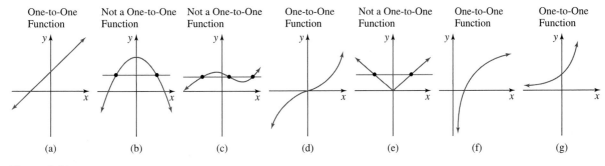

Figure 2.91

Using the horizontal line test to identify one-to-one functions

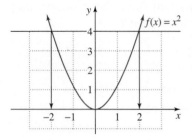

Figure 2.92

$f(x) = x^2$ is not-one-to-one.

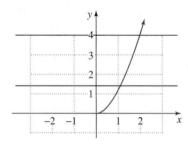

Figure 2.93

$g(x) = x^2 \ (x \geq 0)$ is one-to-one.

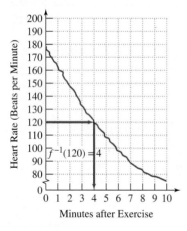

Minutes after Exercise

Figure 2.94

Heart rate as a function of time after exercising

2 Find f^{-1} for specified values.

Study tip

The notation f^{-1} represents the inverse function of f. The -1 is not an exponent. The notation f^{-1} does not mean $\dfrac{1}{f}$. The reciprocal $\dfrac{1}{f(x)}$ is expressed as $[f(x)]^{-1}$.

EXAMPLE 1 **Restricting a Function's Domain**

a. Explain why the function $f(x) = x^2$ is not one-to-one.

b. Suggest a restriction on the domain of f so that the resulting function is one-to-one.

Solution

a. The graph of $f(x) = x^2$ is intersected twice by the horizontal line in Figure 2.92. Since $f(2) = 4$ and $f(-2) = 4$, both 2 and -2 have the same function value, so the function is not one-to-one.

b. If we restrict the domain of f so that $x \geq 0$, the resulting function (call it g)

$$g(x) = x^2, \quad x \geq 0$$

is increasing and passes the horizontal line test. (Figure 2.93)   ■

Discover for yourself

Suggest another restriction on the domain of f so that the resulting function is one-to-one.

Inverse Functions

One-to-one functions are important because they have an inverse function. Consider the graph in Figure 2.94. The graph shows that the rate at which a person's heart beats (in beats per minute) is a function of the number of minutes that elapse after exercising. For example, if 4 minutes elapse, then the rate is 120 beats per minute. If we call this function f, we can write that $f(4) = 120$.

The function f passes the horizontal line test and is one-to-one. The inverse of this one-to-one function is the function that shows the time elapsed after exercising as a function of a person's heartbeat. For example, if a person's heart is beating at a rate of 120 beats per minute, then 4 minutes have elapsed. We express this by writing $f^{-1}(120) = 4$. The inverse function, denoted by f^{-1}, reverses the coordinates of x and y. Since $f(4) = 120$, then $f^{-1}(120) = 4$.

In general, a function f has an inverse function if it is one-to-one. If f takes x to y, that is, $f(x) = y$, then f^{-1} reverses this direction and takes y back to x: $f^{-1}(y) = x$.

Definition of an inverse function

Let f be a one-to-one function with domain A and range B such that

$$f(x) = y.$$

Its inverse function f^{-1} has domain B and range A, where

$$f^{-1}(y) = x$$

for any y in B.

EXAMPLE 2 **Finding f^{-1} for Specific Values**

The function

$$f(x) = 0.06x^2 + 0.8x$$

TABLE 2.2	Speed and Stopping Distance	
Speed: x	Stopping Distance: $f(x)$	Function Value
20	40	$f(20) = 40$
30	78	$f(30) = 78$
40	128	$f(40) = 128$
50	190	$f(50) = 190$
60	264	$f(60) = 264$

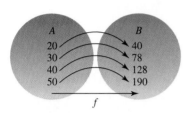

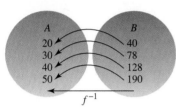

Figure 2.95

f and f^{-1}

3 Verify that two functions are inverses of each other.

is a model for the distance ($f(x)$, in feet) required to stop a car whose speed is x miles per hour. Table 2.2 shows some ordered pairs that express stopping distance as a function of speed. Use the values in the table to find $f^{-1}(40), f^{-1}(78), f^{-1}(128), f^{-1}(190)$, and $f^{-1}(264)$.

Solution

An inverse function f^{-1} reverses the effect of f. Thus,

> If $f(20) = 40$, then $f^{-1}(40) = 20$.
> If $f(30) = 78$, then $f^{-1}(78) = 30$.
> If $f(40) = 128$, then $f^{-1}(128) = 40$.
> If $f(50) = 190$, then $f^{-1}(190) = 50$.
> If $f(60) = 264$, then $f^{-1}(264) = 60$.

The diagram in Figure 2.95 shows how f^{-1} reverses the effects of f for four ordered pairs. There are situations in which it makes sense to reverse the ordered pairs of f, expressing speed as a function of distance. For example, skid marks at the scene of an accident indicate stopping distance and could provide information about the velocity of an automobile just prior to the impact. ∎

Properties of Inverse Functions

The inverse function f^{-1} reverses the effects of f. If we start with x and apply f, we get $f(x)$. If we then apply f^{-1}, we arrive back at x, where we began. That is,

$$f^{-1}(f(x)) = x.$$

In the same way, f undoes what f^{-1} does.

This mutual undoing process is described next. These properties show that f and f^{-1} are inverses of each other.

Cancellation properties of inverse functions

If f is a one-to-one function, the inverse function f^{-1} satisfies the following cancellation properties:

$$f^{-1}(f(x)) = x \quad \text{for every } x \text{ in the domain of } f$$

and

$$f(f^{-1}(x)) = x \quad \text{for every } x \text{ in the domain of } f^{-1}.$$

Any function f^{-1} satisfying these two cancellation properties must be the inverse of f.

EXAMPLE 3 **Verifying Inverse Functions**

Show that each function is an inverse of the other.

$$f(x) = 3x + 2 \quad \text{and} \quad g(x) = \frac{x - 2}{3}$$

Solution

To show that f and g are inverses of each other, we must show that $f(g(x)) = x$ and $g(f(x)) = x$.

$$f(g(x)) = 3[g(x)] + 2$$

Put $g(x)$ into f.

$$= 3\left(\frac{x-2}{3}\right) + 2$$

$$= x - 2 + 2$$

$$= x$$

$$g(f(x)) = \frac{f(x) - 2}{3}$$

Put $f(x)$ into g.

$$= \frac{3x + 2 - 2}{3}$$

$$= \frac{3x}{3}$$

$$= x$$

Because g is the inverse of f (and vice versa), we can use the notation

$$f(x) = 3x + 2 \quad \text{and} \quad f^{-1}(x) = \frac{x-2}{3}.$$

Notice how f^{-1} undoes the changes produced by f. f changes x by *multiplying* by 3 and *adding* 2. f^{-1} undoes this by *substracting* 2 and *dividing* by 3. This "undoing" process is illustrated in Figure 2.96.

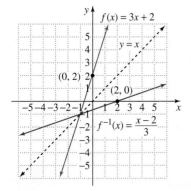

Figure 2.97

Symmetry about $y = x$

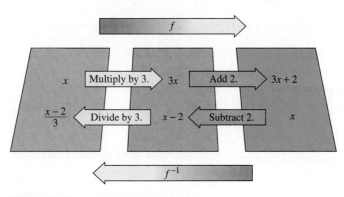

Figure 2.96

f^{-1} undoes the changes produced by f. ∎

Graphs of f and f^{-1}

Figure 2.97 shows the graphs of $f(x) = 3x + 2$ and $f^{-1}(x) = \dfrac{x-2}{3}$. The graphs are mirror images of each other with respect to the line $y = x$.

In general, if f is a one-to-one function and $f(a) = b$, then $f^{-1}(b) = a$, because the inverse function reverses the coordinates of the function. This means that the point (a, b) belongs the the graph of f, and the point (b, a) belongs to the graph of f^{-1}. Since the points (a, b) and (b, a) are symmetric with respect to the line $y = x$, the graph of f^{-1} is reflection of the graph of f with respect to the line $y = x$. This is illustrated in Figure 2.98.

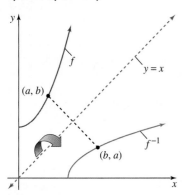

Figure 2.98

The graph of f^{-1} as a reflection of f in $y = x$

4 Find the equation of the inverse.

Finding Formulas for Inverses

If a function is one-to-one, its inverse function may be obtained by reversing the first and second coordinates in each ordered pair of the function. In this reversal, the domain of the function becomes the range of the inverse function, and the range of the function becomes the domain of the inverse function. This means that for each ordered pair (x, y) in the function, the ordered pair (y, x) will be in the inverse function.

In short, if a one-to-one function is defined by an equation, we can obtain the equation for f^{-1}, the inverse of f, by interchanging the role of x and y in the equation for f.

© Scott Kim. From Inversions (W. H. Freeman, 1989).

> ### Finding the inverse of a one-to-one function
>
> The formula for the inverse of a one-to-one function f can be found as follows.
>
> 1. Replace $f(x)$ by y in the equation for $f(x)$.
> 2. Interchange x and y.
> 3. Solve for y.
> 4. Replace y by $f^{-1}(x)$.
>
> This process can be checked by showing that $f(f^{-1}(x)) = x$ and $f^{-1}(f(x)) = x$.

EXAMPLE 4 **Finding the Inverse of a Function**

Find the inverse of $f(x) = 7x - 5$.

Solution

Since the graph of $f(x) = 7x - 5$ is a non-horizontal line, this function is one-to-one.

> ***Discover for yourself***
>
> In Example 4, we found that if $f(x) = 7x - 5$, then
> $$f^{-1}(x) = \frac{x + 5}{7}.$$
>
> 1. Verify this result by showing that
> $$f(f^{-1}(x)) = x$$
> and
> $$f^{-1}(f(x)) = x.$$
>
> 2. Find $f(0), f^{-1}(-5), f(1),$ $f^{-1}(2), f(3),$ and $f^{-1}(16).$ What is f^{-1} doing to the ordered pairs of f?

Step 1. Replace $f(x)$ by y:
$$y = 7x - 5$$

Step 2. Interchange x and y:
$$x = 7y - 5 \qquad \text{This is the inverse function.}$$

Step 3. Solve for y:
$$x + 5 = 7y \qquad \text{Add 5 to both sides.}$$
$$\frac{x + 5}{7} = y \qquad \text{Divide both sides by 7.}$$

Step 4. Replace y by $f^{-1}(x)$:
$$f^{-1}(x) = \frac{x + 5}{7} \qquad \text{The equation is written with } f^{-1} \text{ on the left.}$$

Thus, $f(x) = 7x - 5$ and $f^{-1}(x) = \dfrac{x + 5}{7}$.

Once again, f^{-1} undoes the changes produced by f. f changes x by multiplying by 7 and subtracting 5. f^{-1} undoes this by adding 5 and dividing by 7. ■

EXAMPLE 5 **Finding the Equation of the Inverse**

a. Find the inverse of $f(x) = 4x^3 - 1$.
b. Verify that $f(f^{-1}(x)) = x$ and $f^{-1}(f(x)) = x$.

Solution

The graph of $f(x) = 4x^3 - 1$, shown in the Using Technology, passes the horizontal line test and is one-to-one.

a. **Step 1.** Replace $f(x)$ by y: $\qquad\qquad y = 4x^3 - 1$

Step 2. Interchange x and y: $\qquad\quad x = 4y^3 - 1$

Step 3. Solve for y: $\qquad\qquad\quad x + 1 = 4y^3$

$$\frac{x+1}{4} = y^3$$

$$\sqrt[3]{\frac{x+1}{4}} = y$$

Step 4. Replace y by $f^{-1}(x)$: $\qquad f^{-1}(x) = \sqrt[3]{\frac{x+1}{4}}$

Thus, the inverse of $f(x) = 4x^3 - 1$ is

$$f^{-1}(x) = \sqrt[3]{\frac{x+1}{4}}.$$

b. To verify algebraically that $f(x) = 4x^3 - 1$ and $f^{-1}(x) = \sqrt[3]{\frac{x+1}{4}}$ are inverses of each other, we must show that

$$f(f^{-1}(x)) = x \qquad\qquad \text{and} \quad f^{-1}(f(x)) = x.$$

$$f(f^{-1}(x)) = 4[f^{-1}(x)]^3 - 1 \qquad\qquad f^{-1}(f(x)) = \sqrt[3]{\frac{f(x)+1}{4}}$$

Put $f^{-1}(x)$ into f. $\quad = 4\left(\sqrt[3]{\frac{x+1}{4}}\right)^3 - 1 \qquad$ Put $f(x)$ into f^{-1}. $\quad = \sqrt[3]{\frac{4x^3 - 1 + 1}{4}}$

$$= 4\left(\frac{x+1}{4}\right) - 1 \qquad\qquad = \sqrt[3]{\frac{4x^3}{4}}$$

$$= x + 1 - 1 \qquad\qquad\qquad = \sqrt[3]{x^3}$$

$$= x \qquad\qquad\qquad\qquad\quad = x \qquad\qquad \blacksquare$$

Using technology

The graphs of

$$y_1 = 4x^3 - 1$$

$$y_2 = \sqrt[3]{\frac{x+1}{4}}$$

and

$$y_3 = x$$

were obtained with a graphing utility. The graphs of f and f^{-1} indicate that the graph of f^{-1} is the reflection of the graph of f in the line $y = x$.

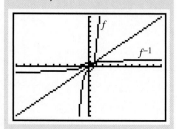

5 Verify inverses with a graphing utility.

EXAMPLE 6 **Using a Graphing Utility to Verify Inverse Functions**

Use a graphing utility to determine whether $f(x) = x^3 + 2$ and $g(x) = \sqrt[3]{x} - 2$ are inverse functions.

Solution

Enter y_1 through y_4 as follows:

$Y_1 \boxed{=} X \boxed{\wedge} 3 + 2$ $\qquad$ This is f.

$Y_2 \boxed{=} \boxed{\sqrt[3]{}} (X - 2)$ $\qquad$ This is g.

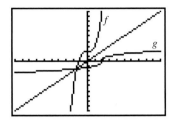

Figure 2.99

The graphs of the inverse functions $f(x) = x^3 + 2$ and $g(x) = \sqrt[3]{x - 2}$

6 Solve applied modeling problems using inverses.

Giacomo Balla "Abstract Speed, The Car Has Passed" 1913. Tate Gallery, London. Photo: John Webb. Art Resource, NY.

$Y_3 \boxed{=} Y_2 \boxed{\wedge} 3 + 2$ This is $f \circ g$.

$Y_4 \boxed{=} \boxed{\sqrt[3]{\ }} (Y_1 - 2)$ This is $g \circ f$.

Figure 2.99 shows the resulting graphs. The functions f and g are inverses if $f(g(x)) = x$ and $g(f(x)) = x$, so that the graphs of y_3 and y_4 should be the line $y = x$. This is indeed the case, verified using the $\boxed{\text{TRACE}}$ feature along the line and observing that the x- and y-coordinates are equal. Notice the symmetry of the graphs of f and g about the line $y = x$.   ■

Modeling with Inverses

We began our discussion of inverse functions with a graph that described heart rate as a function of time after exercising. We then considered time after exercising as a function of heart rate. We now do the same thing—that is, switch the roles of independent and dependent variables—with a mathematical model for speed and stopping distance.

EXAMPLE 7   Speed and Stopping Distance

The function $d = f(v) = 0.06v^2 + 0.8v$ expresses the stopping distance of a car (d, in feet) as a function of the car's speed (v, in miles per hour). Find a formula for the function giving the car's speed as a function of its stopping distance.

Solution

We are given that $d = 0.06v^2 + 0.8v$, a model expressing d in terms of v. We want v in terms of d, so we must solve for v.

$d = 0.06v^2 + 0.8v$ This is the given function.

$0 = 0.06v^2 + 0.8v - d$ To solve for v, set the quadratic equation equal to 0.

$v = \dfrac{-b \pm \sqrt{b^2 - 4ac}}{2a}$ Use the quadratic formula.

$= \dfrac{-0.8 \pm \sqrt{(0.8)^2 - 4(0.06)(-d)}}{2(0.06)}$ Substitute $a = 0.06$, $b = 0.8$, and $c = -d$.

$= \dfrac{-0.8 \pm \sqrt{0.64 + 0.24d}}{0.12}$ Perform the indicated operations.

$= \dfrac{-0.8 \pm \sqrt{0.04(16 + 6d)}}{0.12}$ Simplify.

$= \dfrac{-0.8 \pm 0.2\sqrt{16 + 6d}}{0.12}$

$= \dfrac{-20 \pm 5\sqrt{16 + 6d}}{3}$ Divide the numerator and denominator by 0.04.

$v = \dfrac{-20 + 5\sqrt{16 + 6d}}{3}$ Reject $\dfrac{-20 - 5\sqrt{16 + 6d}}{3}$ because v, the car's speed, must be positive.

We now have the car's speed in terms of its stopping distance. Since we called the given function f and reversed the roles of the variables, we can use f^{-1} to represent speed as a function of stopping distance:

$$f^{-1}(d) = \frac{-20 + 5\sqrt{16 + 6d}}{3}.$$

Let's find just one function value for f^{-1}, say, $f^{-1}(40)$.

$$f^{-1}(40) = \frac{-20 + 5\sqrt{16 + 6(40)}}{3} = \frac{-20 + 5(16)}{3} = 20$$

With $f^{-1}(40) = 20$, a police officer investigating an accident in which skid marks measured 40 feet can determine that the car in question was traveling at 20 miles per hour at the moment the brakes were applied.   ∎

Andy Warhol "White Car Crash 19 Times" 1963, synthetic polymer paint and silkscreen ink on canvas, 145 × 83 1/4 in. © The Andy Warhol Foundation for the Visual Arts/ARS, New York. Art Resource, NY.

PROBLEM SET 2.7

Practice Problems

Problems 1–10 show graphs of relations. Which of these relations are functions? Which of these functions are one-to-one?

1.

2.

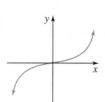

3.

4.

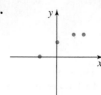

5.

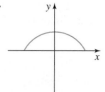

6.

7.

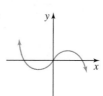

8.

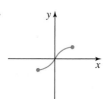

9.

10.

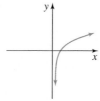

Explain why the functions whose graphs are shown in Problems 11–14 are not one-to-one. Then suggest a restriction on each function's domain so that the resulting function is one-to-one. More than one restriction is possible.

11.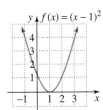

12.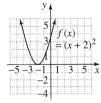

13.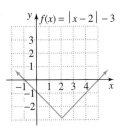

14.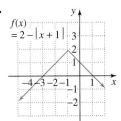

In Problems 15–18, assume that all functions are one-to-one.

15. a. If $f(13) = 19$, find $f^{-1}(19)$.

    b. If $f^{-1}(7) = -4$, find $f(-4)$.

16. a. If $f\left(\dfrac{\pi}{2}\right) = 1$, find $f^{-1}(1)$.

    b. If $f^{-1}\left(\dfrac{\sqrt{3}}{2}\right) = \dfrac{\pi}{3}$, find $f\left(\dfrac{\pi}{3}\right)$.

17. If $f(x) = 3x - 6$, find $f^{-1}(9)$.

18. If $f(x) = 8 - 3x$, find $f^{-1}(14)$.

Problems 19–22 show the graphs of one-to-one functions. Three points are clearly shown on each graph. Reverse the coordinates of these points to find points on the graph of the inverse function. Then use these three points and the fact that the graph of the inverse function is a reflection of the graph of the given function about the line $y = x$ to graph the inverse function on the same set of coordinate axes as the function.

19.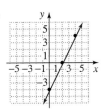

20.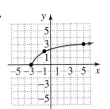

21.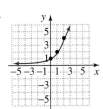

22.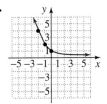

In Problems 23–28, use the cancellation properties of inverse functions to show that f and g are inverses of each other. In particular, show that $f(g(x)) = x$ and $g(f(x)) = x$.

23. $f(x) = 3x - 7; g(x) = \dfrac{x + 7}{3}$

24. $f(x) = \dfrac{5 - x}{9}; g(x) = 5 - 9x$

25. $f(x) = x^2 - 9, x \geqslant 0; g(x) = \sqrt{x + 9}, x \geqslant -9$

26. $f(x) = x^3 + 4; g(x) = \sqrt[3]{x - 4}$

27. $f(x) = \dfrac{1}{x - 2}, x \neq 2; g(x) = \dfrac{1}{x} + 2, x \neq 0$

28. $f(x) = \sqrt{9 - x^2}, 0 \leqslant x \leqslant 3; g(x) = \sqrt{9 - x^2}, 0 \leqslant x \leqslant 3$

For the functions in Problems 29–48, **a.** *find $f^{-1}(x)$ and* **b.** *verify that $f(f^{-1}(x)) = x$ and $f^{-1}(f(x)) = x$.*

29. $f(x) = 3x + 7$

30. $f(x) = 2x - 5$

31. $f(x) = -\dfrac{1}{2}x + 3$

32. $f(x) = -\dfrac{1}{3}x - 4$

33. $f(x) = \sqrt{2x + 7}$

34. $f(x) = \sqrt{3x + 5}$

35. $f(x) = \sqrt[3]{2x - 3}$

36. $f(x) = \sqrt[3]{4x - 5}$

37. $f(x) = 5x^3 - 7$

38. $f(x) = 3x^3 - 4$

39. $f(x) = \dfrac{2x + 1}{x - 3}$

40. $f(x) = \dfrac{2x - 3}{x + 1}$

41. $f(x) = (x-3)^2$ for $x \geq 3$ **42.** $f(x) = (x-1)^2$ for $x \geq 1$ **43.** $f(x) = \sqrt[3]{x-4} + 3$ **44.** $f(x) = \sqrt[3]{x-7} - 4$

45. $f(x) = \dfrac{1}{x} + 3$ **46.** $f(x) = \dfrac{1}{x+3}$ **47.** $f(x) = x^{3/5}$ **48.** $f(x) = x^{5/7} - 1$

In Problems 49–56, use the graph of f to graph each function.

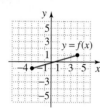

49. $y = f^{-1}(x)$ **50.** $y = f^{-1}(x) + 1$ **51.** $y = f^{-1}(x-2)$

52. $y = f^{-1}(x+1)$ **53.** $y = f^{-1}(x) - 2$ **54.** $y = f^{-1}(-x)$

55. $y = -f^{-1}(x)$ **56.** $y = f^{-1}(x+1) + 1$

In Problems 57–62, use the graph of g to graph each function.

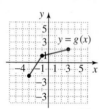

57. $y = g^{-1}(x)$ **58.** $y = g^{-1}(-x)$ **59.** $y = g^{-1}(x) + 1$

60. $y = g^{-1}(x+1)$ **61.** $y = -g^{-1}(x)$ **62.** $y = -g^{-1}(-x)$

In Problems 63–66, use the functions $f(x) = x + 3$ and $g(x) = 2x - 1$ to find the specified function.

63. $f^{-1} \circ g^{-1}$ **64.** $g^{-1} \circ f^{-1}$ **65.** $(f \circ g)^{-1}$ **66.** $(g \circ f)^{-1}$

Application Problems

67. The graph represents the probability of two people in the same room sharing a birthday as a function of the number of people in the room. Call the function f.

   a. Explain why f has an inverse that is a function.

   b. Describe in practical terms the meaning of $f^{-1}(0.25)$, $f^{-1}(0.5)$, and $f^{-1}(0.7)$.

68. A river that contains 20 parts of DDT per million at the beginning of a study has this concentration decreasing as shown in the figure. The graph indicates that DDT concentration is a function of time. Call the function f.

   a. Explain why f has an inverse that is a function.

   b. Describe in practical terms the meaning of $f^{-1}(14)$.

   c. If a safe concentration for swimming is 4 parts of DDT per million, after how many years will this occur?

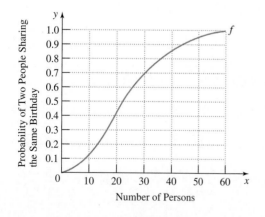

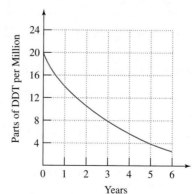

69. The federal government plays a role in the patronage of the arts. The graph in this figure shows appropriations to the National Endowment for the Arts under seven presidential administrations.

    a. Does the graph indicate that arts endowment is a function of time for the period 1966 through 1995? Explain your answer.

    b. Does the graph represent a one-to-one function? What does this mean in terms of time and arts endowment?

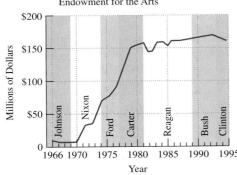

Arts Endowment
Appropriations to the National
Endowment for the Arts

70. The regular price of a pair of jeans is x dollars. Let $f(x) = x - 5$ and $g(x) = x - 0.4x$.

    a. Describe what functions f and g model in terms of the price of the jeans.

    b. Find $(f \circ g)(x)$ and describe what this models in terms of the price of the jeans.

    c. Repeat part (b) for $(g \circ f)(x)$.

    d. Which composite function models the greater discount on the jeans, $f \circ g$ or $g \circ f$? Explain.

    e. Find f^{-1} and describe what this models in terms of the price of the jeans.

71. The linear model $D = 0.2F - 1$ expresses age-adjusted death rate D from breast cancer per 100,000 women as a function of daily fat intake F in grams. Find a formula expressing F as a function of D.

72. The function $s = f(t) = -16t^2 + 192t$ describes the height s in feet of a ball above the ground after t seconds. The model indicates that the ball was thrown with an initial velocity of 192 feet per second.

    a. Find a formula that expresses t in terms of s.

    b. Does the formula in part (a) define t as a function of s? Explain.

    c. Substitute 432 for s in the formula in part (a). Find the corresponding value(s) for t and explain what this means in terms of the ball's height above the ground.

73. Conversions between Celsius and Fahrenheit temperatures are given by $F = \frac{9}{5}C + 32$ and $C = \frac{5}{9}(F - 32)$. Show that these functions are inverses of each other.

True–False Critical Thinking Problems

74. Which one of the following is true?

    a. Every even function is one-to-one.

    b. If $f(x) = x^n$ and n is odd, then f^{-1} exists.

    c. The inverse of the function $\{(3, 4), (6, 6)\}$ is $\{(6, 3), (6, 4)\}$.

    d. A function cannot be its own inverse.

75. Which one of the following is true?

    a. If a function f is one-to-one, then the x-intercept of f is the y-intercept of f^{-1}.

    b. If $f(x) = x^{n+1}$ and n is odd, then f^{-1} exists.

    c. If $f(x) = |x + 4|$, then $f^{-1}(x) = |x| - 4$.

    d. If $f(x) = x^2$, then $f^{-1}(x) = \sqrt{x}$.

Technology Problems

In Problems 76–82, find an equation for f^{-1}. Then use a graphing utility to graph f and f^{-1} in the same viewing rectangle.

76. $f(x) = 3x - 4$
    77. $f(x) = x^3 + 2$
    78. $f(x) = x^2, x \leqslant 0$
    79. $f(x) = \sqrt{9 - x^2}, \ 0 \leqslant x \leqslant 3$

80. $f(x) = \dfrac{1}{x}$
    81. $f(x) = \sqrt[3]{x - 2}$
    82. $f(x) = x^{3/5}$

In Problems 83–91, use a graphing utility to graph the function. Use the graph to determine whether the function has an inverse that is a function (that is, whether the function is one-to-one).

83. $f(x) = 3$

84. $f(x) = [[x - 2]]$

85. $f(x) = |x - 2|$

86. $f(x) = |x + 3| - |x - 3|$

87. $f(x) = (x - 1)^3$

88. $f(x) = \frac{1}{4}(x + 1)^2 - 2$

89. $f(x) = -\sqrt{16 - x^2}$

90. $f(x) = -x\sqrt{9 - x^2}$

91. $f(x) = x^3 + x + 1$

In Problems 92–96, use a graphing utility to graph f, g, f ∘ g, and g ∘ f in the same viewing rectangle. Use the graphs of f ∘ g and g ∘ f to determine whether f and g are inverses of each other. (If the viewing rectangle is too cluttered, deselect the graphs of f and g, focusing on the graphs of f ∘ g, and g ∘ f.)

92. $f(x) = 4x + 4, g(x) = 0.25x - 1$

93. $f(x) = \frac{1}{x} + 2, g(x) = \frac{1}{x - 2}$

94. $f(x) = 2 - x^2 \ (x \geq 0), g(x) = \sqrt{2 - x}$

95. $f(x) = |x + 1| \ (x \geq -1), g(x) = x - 1 \ (x \leq 0)$

96. $f(x) = \sqrt[3]{x} - 2, g(x) = (x + 2)^3$

97. The study of trigonometry focuses on six functions. One of these functions, called the sine function, is represented by $y = \sin x$ or $f(x) = \sin x$.

a. Use a graphing utility to graph $y = \sin x$, selecting ZTRIG (zoom trig).

b. Does $f(x) = \sin x$ have an inverse that is a function? If not, use the graph and the TRACE feature to suggest a domain restriction for $f(x) = \sin x$ so that its inverse will be a function.

c. What real world phenomena might be modeled by $f(x) = \sin x$?

Writing in Mathematics

98. Consider a position function $s(t)$ for *Freedom 7*, the spacecraft that carried the first American into space in 1961. Total flight time was 15 minutes, and the spacecraft reached a maximum altitude of 116 miles. Is the function $s(t)$, expressing height in miles as a function of time (t, in minutes) one-to-one? Explain your answer in a sentence or two.

99. Discuss similarities between f and f^{-1} and real numbers and their multiplicative inverses.

100. Describe how to draw a picture (similar to the one shown here) in which the function is not one-to-one.

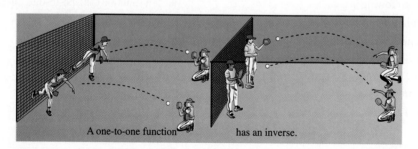

A one-to-one function has an inverse.

Critical Thinking Problems

101. If $f(x) = \dfrac{Ax + B}{Cx - A}$, show that f is its own inverse. $(A^2 + BC \neq 0)$

102. If $f(2) = 6$, find x satisfying $8 + f^{-1}(x - 1) = 10$.

103. If $f(5) = 19$, find x satisfying $1 + f^{-1}(2x + 3) = 6$.

104. The one-to-one property for functions can be stated as follows: If $f(a) = f(b)$, then $a = b$. Use this condition to show that

$$f(x) = \frac{5}{3x - 2}$$

is one-to-one, letting $f(a) = f(b)$ and showing that $a = b$.

105. Prove that if f is an increasing function on its entire domain, then f is one-to-one.

CHAPTER PROJECT
The Game of Life

The Game of Life was invented in 1970 by the mathematician John Conway. Conway created a game where the simplest of rules would lead to a series of interesting and complex events. Life is usually played on a computer, but it is not a game played against the computer, like chess, or a game played against other "opponents," like a video game. For the Game of Life, we set up the game and watch the computer play. We can picture Life being played on an enormous checkerboard, where some of the squares (cells) have small tokens on them and some of the squares are empty. The original version of Life, played by Conway and his colleagues at Cambridge University, was actually played on boards with the same simple rules that would be used on a computer.

The Rules for Life

- If a token is in a cell, the cell is alive. Cells with no tokens are considered dead.
- Each cell has eight neighbors

```
1  2  3
4  *  5
6  7  8
```

four on the sides, and four on the corners.
- *Survival:* If a cell is alive and has two or three neighbors, it will stay alive.
- *Death:* If a cell is alive and has less than two neighbors, it will die (of loneliness). If a cell is alive and has more than three neighbors, it will die (of starvation).
- *Birth:* If a cell is dead and has exactly three neighbors, it will be reborn (come alive).

1. Use a sheet of graph paper and some small objects that can be placed on the squares to start your own game. Begin in the center of the graph paper with the following group:

The next "generation" will look like:

Continue producing generations, keeping a record of the pattern for each one, until you observe a pattern repeating. What is the repeating pattern?

Conway and his colleagues observed some patterns occurring quite often as they played the game and named them after the shapes they suggested, such as loaf, pond, boat, and beehive. Some shapes, called *oscillators*, moved in periodic patterns, the same shape showing up over and over in different generations. All of these were referred to as "life-forms." When the game was introduced to the public in 1970 in the "Mathematical Games" column of *Scientific American*, people began playing it with a passion and soon discovered how easy it was to program a computer to play the game. Using a computer, we can set up a particular pattern and watch thousands of generations unfold from any starting pattern we choose. The computer adds animation, showing us how some of the life-forms created by the game appear to move on forever, drifting away, while others explode into new forms or give birth and die in complex patterns.

2. Set up the following pattern on your grid and create at least 15 generations by moving your tokens. What do you observe? The pattern is called a glider. Do you see why it was so named?

	*	
		*
*	*	*

Although we can predict every move on a cell basis, we find it extremely difficult to see the end result of thousands of generations. The game evolves and changes, shifting into new forms rapidly, sometimes creating moving and self-sustaining forms as other forms expire. In short, Life behaves very much like life. The Game of Life is an example of a cellular automaton: a collection of cells, spread out in space, with states, in our case live and dead, that can be changed according to a set of rules. The entire process has a very machine-like look, resembling a small army of robots programmed to certain tasks. Cellular automata may be "simple," but they are capable of displaying a surprising array of complex behavior and they play a central role in a field of study called *Artificial Life*.

Cellular automata may take forms which are more restrictive than the free-flowing movement in the Game of Life. For example, we may begin with a row of cells across the top of a page and use rules to generate the row below it, moving only downward as we progress through the generations. Displays of this type have shown some remarkable applications, appearing to mimic the flow of oil or water as it trickles through different kinds of soil, the outbreak of forest fires, and the dispersion of an oil spill, to name just a few. We may even see fractal forms, such as the Sierpinski Triangle, shown on page 295.

3. A variation of the Game of Life may be played on a finite field wrapped around a cylinder. In this case, the rules remain the same, but we may see very different behavior from the same initial state we would use on a flat plane extending to infinity in all directions. Select your own starting configuration and explore a few generations both on a plane and on a cylinder. If you work in a group, have one person start the configuration and then pass it on to another member, each person moving a generation. Share your initial configuration and discoveries with the class.

4. Set up a table in a corner of your classroom to play a long-running Game of Life. Each day have some students run through a few generations while others look for patterns or record results. If possible, use one of the many free computer programs to explore a similar number of generations. Compare the dynamic nature of the computer simulation to the hands-on approach of the class. Discuss the benefits and drawbacks of each.

Worldwide Web Resources

Go to the Prentice Hall website (http://www.prenhall.com/blitzer) to access other locations on the Internet that will allow you to further explore the concepts presented in this project.

Chapter Review

SUMMARY

1. **Basic Ideas about Functions**
   a. A function is a correspondence between a first set, the domain, and a second set, the range, such that each element in the domain corresponds to exactly one element in the range.
   b. Using function notation $y = f(x)$, the letter f names the function, and $f(x)$, read "f of x" or "f at x" rep-

resents an expression for the value of the function at x. The value y, the dependent variable, is calculated after selecting a value for x, the independent variable.

   c. The domain of a function is the set of all values of the independent variable for which the function is defined. If a function is defined by an algebraic expression, the implied domain is the set of all real numbers excluding those that cause division by zero and those that result in an even root of a negative number.

d. The range of a function is the set of all values taken on by the dependent variable.

2. Slope

a. The slope m of the line through (x_1, y_1) and (x_2, y_2) is $m = \dfrac{y_2 - y_1}{x_2 - x_1} = \dfrac{\Delta y}{\Delta x}$, $\Delta x \neq 0$.

b. Lines with positive slope rise and those with negative slope fall from left to right. Lines with zero slope are horizontal and those with undefined slope are vertical.

c. If y is a function of x, slope is the average change in y per unit of change in x.

d. Two nonvertical lines are parallel if and only if their slopes are equal. Two nonvertical lines are perpendicular if and only if their slopes are negative reciprocals.

3. Equations of Lines

a. General Form: $Ax + By + C = 0$

b. Vertical line: $x = a$

c. Horizontal line: $y = b$

d. Point-slope form: $y - y_1 = m(x - x_1)$

e. Slope-intercept form: $y = mx + b$

f. Linear Function: $f(x) = mx + b$

4. Graphs of Relations and Functions

a. The graph of a relation is the graph of its ordered pairs. The graph of a function f is the set of points $(x, f(x))$ such that x is in the domain of f.

b. The Vertical Line Test for Functions: If any vertical line intercepts the graph of a relation in more than one point, the relation does not define y as a function of x.

5. The Circle

a. A circle is the set of all points in a plane that are equidistant from a fixed point called the center. The fixed distance from the circle's center to any point on the circle is called the radius.

b. The equation of the circle with center at (h, k) and radius r is given by $(x - h)^2 + (y - k)^2 = r^2$.

6. Increasing and Decreasing Functions, Extreme Values

a. *Increasing, Decreasing, Constant Functions*

If x_1 and x_2 are any values for x in an interval, where $x_1 < x_2$, then:

1. f is increasing on the interval if $f(x_1) < f(x_2)$.

2. f is decreasing on the interval if $f(x_1) > f(x_2)$.

3. f is constant on the interval if $f(x_1) = f(x_2)$.

b. *Relative Extrema*

1. A function value $f(a)$ is a relative maximum of f if there exists an open interval about a such that $f(a) \geq f(x)$ for all x in the open interval.

2. A function value $f(b)$ is a relative minimum of f if there exists an open interval about b such that $f(b) \leq f(x)$ for all x in the open interval.

7. Algebra's Common Graphs

Identity Function $f(x) = x$

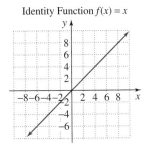

Standard Quadratic Function $f(x) = x^2$

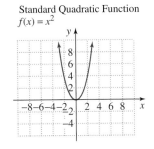

Standard Cubic Function $f(x) = x^3$

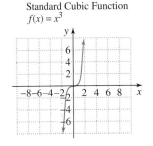

Absolute Value Function $f(x) = |x|$

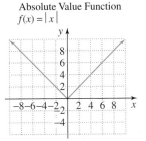

Square Root Function $f(x) = \sqrt{x}$

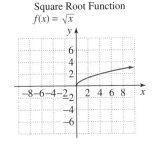

Cube Root Function $f(x) = \sqrt[3]{x}$

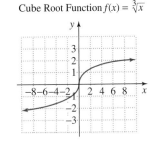

Greatest Integer Function $f(x) = [[x]]$

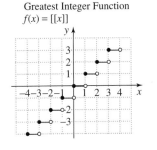

8. Techniques in Graphing; Transforming the Graph of $y = f(x)$

a. In each case, c represents a positive real number.

Function		How to obtain the graph of g
Vertical translations	$\begin{cases} g(x) = f(x) + c \\ g(x) = f(x) - c \end{cases}$	Shift f upward c units. Shift f downward c units.
Horizontal translations	$\begin{cases} g(x) = f(x - c) \\ g(x) = f(x + c) \end{cases}$	Shift f to the right c units. Shift f to the left c units.
Reflections	$\begin{cases} g(x) = -f(x) \\ g(x) = f(-x) \end{cases}$	Reflect f in the x-axis. Reflect f in the y-axis.
Stretching or Shrinking	$\begin{cases} g(x) = cf(x); c > 1 \\ g(x) = cf(x); 0 < x < 1 \end{cases}$	Stretch f, multiplying each of its y-values by c. Shrink f, multiplying each of its y-values by c.

b. *Order of Transformations:*
   1. Horizontal translation
   2. Stretching or shrinking
   3. Reflecting
   4. Vertical translation

c. *Symmetry; Even and Odd Functions*

Find $f(-x)$ and test for symmetry. If $f(-x) = f(x)$, then f is an even function and its graph has y-axis symmetry. If $f(-x) = -f(x)$, then f is an odd function and its graph has origin symmetry.

9. Operations with Functions

a. *Sum:* $(f + g)(x) = f(x) + g(x)$

b. *Difference:* $(f - g)(x) = f(x) - g(x)$

c. *Product:* $(fg)(x) = f(x) \cdot g(x)$

d. *Quotient:* $\left(\dfrac{f}{g}\right) = \dfrac{f(x)}{g(x)}$, provided $g(x) \neq 0$

The domains are the set of all real numbers common to the domains of f and g.

10. Composition Functions

a. $(f \circ g)(x) = f(g(x))$; $f \circ g$ is the composition of f and g. We read $(f \circ g)(x)$ as "f of g of x."

b. The domain of $f \circ g$ is the set of all x in the domain of g such that $g(x)$ is in the domain of f.

c. Iterates of f: $f^2(x) = f(f(x))$, $f^3(x) = f(f^2(x))$, etc.

11. Inverse Functions

a. A function is one-to-one if $f(x_1) \neq f(x_2)$ whenever $x_1 \neq x_2$.

b. *The Horizontal Line Test:* If no horizontal line intersects the graph of a function more than once, then the function is one-to-one.

c. *Increasing and Decreasing Functions*
   1. If f is increasing on its entire domain, then f is one-to-one.
   2. If f is decreasing on its entire domain, then f is one-to-one.

d. If f is one-to-one such that $f(x) = y$, then its inverse function reverses this direction so that $f^{-1}(y) = x$.

e. *Cancellation Properties:* $f^{-1}(f(x)) = x$ and $f(f^{-1}(x)) = x$.

f. The graphs of f and f^{-1} are symmetric with respect to the line $y = x$.

g. If f is one-to-one, the formula for f^{-1} is found by:
   1. Replacing $f(x)$ by y in the equation for $f(x)$.
   2. Interchanging x and y.
   3. Solving for y.
   4. Replacing y by $f^{-1}(x)$.

REVIEW PROBLEMS

1. The function $L(v) = 900\sqrt{1 - \left(\dfrac{v}{186,000}\right)^2}$ describes the length of a 900-meter tall starship moving at velocity v (in miles per second) from the perspective of an observer at rest. Find $L(74,400)$ and describe in practical terms what the function value means.

2. The function
$$f(x) = \begin{cases} 0.0005x^2 + 0.025x + 8.8 & \text{if } 0 \leq x < 30 \\ 0.0202x^2 - 1.58x + 39.2 & \text{if } x \geq 30 \end{cases}$$
describes the average number of miles (in thousands) that each automobile in the United States was driven x years after 1940. Find and interpret $f(45) - f(10)$.

In Problems 3–4, find: **a.** $f(0)$ **b.** $f(-2)$ **c.** $f(a)$ **d.** $f(a + h)$

e. $\dfrac{f(a + h) - f(a)}{h}$ **f.** $f(a) + f(h)$ **g.** $f(2a)$

3. $f(x) = 7x - 3$

4. $f(x) = 4x^2 - 5x + 11$

In Problems 5–10, find the domain of each function. Check your result with a graphing utility.

5. $f(x) = \sqrt{8 - 2x}$

6. $f(x) = \sqrt[3]{8 - 2x}$

7. $f(x) = \sqrt{x^2 - 5x + 4}$

8. $f(x) = \dfrac{2}{\sqrt{x^2 - 5x + 4}}$

9. $f(x) = \sqrt{4 - x^2}$

10. $f(x) = \dfrac{\sqrt{x - 2}}{x - 5}$

In Problems 11–14, use the given conditions to write the equation of each line in point-slope form, slope-intercept form, and general form.

11. Passing through $(-3, 2)$ with a slope of -6

12. Passing through $(1, 6)$ and $(-1, 2)$

13. Passing through $(4, -7)$ and parallel to the line whose equation is $3x + y - 9 = 0$

14. Passing through $(-3, 6)$ and perpendicular to the line whose equation is $x - 3y - 5 = 0$

15. The graph below shows the number of victims of violent crime in the United States for the period 1975 through 1990.

   a. Find a reasonable estimate for the number of victims (in millions) for 1980. Then use this estimate to compute the slope of the line drawn through 1975 and 1980. Interpret this value in terms of average rate of change.

   b. Find an estimate for the slope of the line drawn through 1980 and 1985. Interpret this value in terms of average rate of change.

Violent Crime in the United States, 1975–1990

Source: Bureau of Justice Statistics

16. The graph in the next column shows the number of registered nurses in the United States per 100,000 population. Find a reasonable estimate for the number of nurses per 100,000 population for the years 1983 and 1993. Write the equation that models the data for the line drawn through these years. Then extrapolate from the data and predict the number of nurses per 100,000 population for the year 2000.

Number of Registered Nurses
(per 100,000 Population)

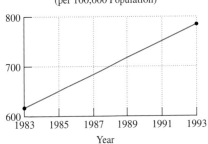

Year

Source: American Hospital Association

17. The data points $(3, 546)$ and $(5, 666.2)$ shown in the table fall on a straight line.

x (Number of Years after 1985)	y (Total of All Health-Care Expenditures in the United States, in Billions of Dollars)
3	546.0
5	666.2

   a. Write the equation of the line on which these measurements fall in point-slope form and slope-intercept form.

   b. Extrapolate from the data to find U.S. health-care expenditures for 1987.

   c. Extrapolate from the data to find U.S. health-care expenditures for the year 2000.

18. Sketch the graphs of $y = \frac{3}{4}x - 2$, $y = -\frac{4}{3}x + 3$, $y = 2x + 1$, and $4x - 2y - 8 = 0$ on the same set of coordinates axes. What two ideas about slope are illustrated?

19. The graph on page 318 shows the relationship between the volume of a room and the power needed to achieve a peak volume level of 106 decibels for a pair of acoustic loudspeakers. Three different kinds of rooms—dead, average, and live—require different power to achieve loud peak volume levels. Find the slope for each kind of room (you'll need to approximate) and interpret the slope for each room in terms of power and volume.

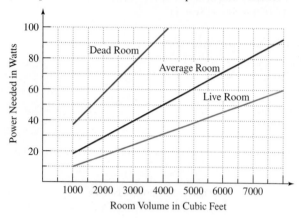

Power Needed in Watts / Room Volume in Cubic Feet

20. Drive-On Rental Company offers cars for $30 a day and 15¢ a mile. Hit the Road Rental charges $40 a day and 10¢ a mile.

    a. Write equations for each company that model daily cost of renting as a function of miles traveled.

    b. Graph both linear functions on the same set of coordinate axes.

    c. What do the graphs indicate in terms of which company has the better offer?

21. A state has an income tax that is directly proportional to a resident's gross income. A person working in that state with a gross yearly income of $23,000 paid $400 in state taxes.

    a. Write an equation that models income tax as a function of gross income.

    b. Use the model to determine state taxes for a yearly gross income of $35,000.

Graph each relation in Problems 22–29 by plotting ordered pairs or, if applicable, using your knowledge about slope and y-intercept for graphing linear functions. Specify the relation's domain and range from the graph. Verify your hand-drawn graph with a graphing utility.

22. $f(x) = x^2 - 5x + 6$

23. $f(x) = x^3 - 5x$

24. $f(x) = \dfrac{x^2 - 25}{x - 5}$

25. $g(x) = |2 - x|$

26. $H(x) = \sqrt{x + 3}$

27. $x = -y^2$

28. $g(x) = \begin{cases} 2 - x & \text{if } x < 0 \\ x^2 & \text{if } x \geqslant 0 \end{cases}$

29. $f(x) = \begin{cases} \sqrt{x - 4} & \text{if } x \geqslant 4 \\ 4 - x & \text{if } x < 4 \end{cases}$

30. The graph of function f is shown in the figure.

    a. What is the domain of f?

    b. What is the range of f?

    c. Find $|f(3) - f(0)|$.

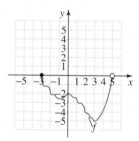

In Problems 31–32, give the center and radius of the circle described by each equation and then graph the equation.

31. $(x - 2)^2 + (y + 1)^2 = 9$

32. $x^2 + 2x + y^2 - 4y - 4 = 0$

33. The graph of a function f is shown in the figure.

    a. State the values of $f(-6), f(-2),$ and $f(6)$.

    b. State the intervals on which f is increasing and on which f is decreasing.

    c. State the relative maximum and minimum values of the function and the values of x at which they occur.

    d. Is f a one-to-one function? If not, suggest a restriction on its domain so that the resulting function is one-to-one.

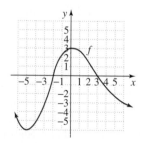

34. The graph shows the height (in meters) of a vulture as a function of its time (in seconds) in flight.

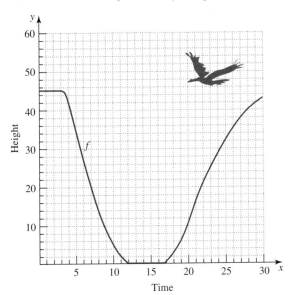

a. How does the graph indicate that height is a function of time?

b. In what interval is the function decreasing? Describe what this means in practical terms.

c. In what intervals is the function constant? What does this mean for each of these intervals?

d. In what interval is the function increasing? What does this mean?

e. Is the function one-to-one? Why is it incorrect to write $f^{-1}(30)$?

35. The model $f(x) = -0.02x^2 + x + 1$ describes the number of inches $f(x)$ that a young redwood tree grows per year as a function of annual rainfall (x, in inches).

a. For what value of x does the function have a maximum value? What is the maximum value of y? Describe the practical significance of this maximum value.

b. The domain of f is [0, 50]. Use the function's equation to find the range of f. What is the practical significance of this range in terms of the meaning of $f(x)$ in the given model?

Sketch each function in Problems 36–37 using the given properties. (More than one graph is possible.)

36. f is constant on the interval $(-\infty, -1)$, increasing on the intervals $(-1, 0)$ and $(1, \infty)$, and decreasing on the interval $(0, 1)$. f has a y-intercept equal to 2.

37. The domain of f is $[-2\pi, 2\pi]$ and its range is $[-1, 1]$. f, an even function, is decreasing on $(0, \pi)$ and increasing on $(\pi, 2\pi)$.

38. For a function g, 2 and 3 are the only relative maxima, occurring at -2 and 3.5, respectively. Furthermore, -1 is the only relative minimum occurring at 2.

a. Sketch g. (More than one graph is possible.)

b. How many x-intercepts does g have?

c. Where is g increasing?

d. Where is g decreasing?

39. Model the following narrative by a possible graph that shows distance from home as a function of time.

I left for school concerned about being late, so I drove quite rapidly at first but then slowed down. Realizing that I felt sluggish, I stopped briefly at a diner for a cup of coffee. I then continued my drive to school at a constant rate of travel.

40. Model the following narrative by a possible graph that shows income as a function of rooms rented.

A hotel gives special rates to groups that reserve between 30 and 60 rooms, decreasing the charge per room by a fixed amount times the number of rooms over 30.

41. A cargo service charges a flat fee of $5 plus $1.50 for each pound or fraction of a pound. Graph shipping cost (in dollars) as a function of weight (x, in pounds) for $0 < x \leq 5$.

Write a narrative that can be modeled by each graph shown in Problems 42–43.

42.

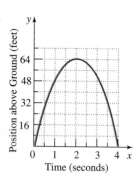

43.

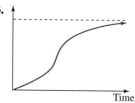

*In Problems 44–47, use a graphing utility to graph the function and use the graph to **a.** determine (to two decimals places where an approximation is needed) intervals in which the function is increasing, decreasing, or constant; **b.** determine any relative maximum or minimum values of the function (use an approximation if necessary); and **c.** state whether the function is even, odd, or neither.*

44. $f(x) = |x + 1| - |x - 1|$

45. $f(x) = 2x^3 - x^2$

46. $g(x) = (x^2 + 1)^2$

47. $h(x) = \sqrt[3]{x}(x - 2)^2$

48. From each of the four corners of a square piece of cardboard measuring 18 inches on a side, a square piece x inches on a side is cut out. The flaps are then turned up to form an open-top box.

   a. Express the volume of the box as a function of x.

   b. Use a graphing utility to graph the function, drawing a relatively complete graph for meaningful values of x.

   c. Use the graph to determine x so that the resulting box has maximum possible volume. What is the maximum volume?

49. A tour operator uses a 400-seat airplane for trips to London. If the plane is full, the airfare is $200 per passenger. If the group has fewer than 400 people, the airfare rises $25 per person for each unoccupied seat.

   a. If x represents the number of passengers under 400, write a model expressing airfare income as a function of x.

   b. Use a graphing utility to graph the function, drawing a relatively complete graph for meaningful values of x.

   c. Use the graph to determine the number of people on the airplane that will generate maximum airfare income. What is the maximum income?

50. A child's playpen can be attached to a wall and opened as shown in the figure. Each side of the playpen is 4 feet in length.

   a. Write a formula in terms of x that models the rectangular area of the playpen.

   b. Write a formula in terms of x that models the triangular area of the playpen.

   c. Add the formulas from parts (a) and (b) and write a function f that models the total area enclosed by the playpen.

   d. Use a graphing utility to graph the area function and determine an approximate value for x that will maximize the playpen's area. Approximately what is the maximum area that can be enclosed?

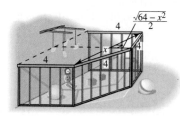

51. Suppose that the graph of f is given. Describe how the graph of each of the following functions can be obtained using the graph of f.

   a. $y = f(x) + 3$ **b.** $y = f(x + 3)$

   c. $y = 2f(x)$ **d.** $y = f(x - 1) - 3$

   e. $y = f(-x)$ **f.** $y = -f(x)$

   g. $y = -f(-x)$ **h.** $y = f^{-1}(x)$

In Problems 52–58, sketch the graphs of the three given functions by hand on the same set of coordinate axes. Verify the result with a graphing utility.

52. $f(x) = \sqrt{x}$
$g(x) = \sqrt{x} - 2$
$h(x) = \sqrt{x - 2}$

53. $f(x) = \sqrt{x}$
$g(x) = \sqrt{x - 3} + 1$
$h(x) = \sqrt{x + 2} - 1$

54. $f(x) = |x|$
$g(x) = |x - 1| + 2$
$h(x) = -2|x - 1|$

55. $f(x) = x^2$
$g(x) = (x - 3)^2 + 1$
$h(x) = -2x^2 - 1$

56. $f(x) = x^3$
$g(x) = (x - 2)^3 - 1$
$h(x) = -x^3$

57. $f(x) = \sqrt[3]{x}$
$g(x) = \sqrt[3]{x + 2}$
$h(x) = -1 + \sqrt[3]{x + 2}$

58. $f(x) = \sqrt[3]{x}$
$g(x) = -\frac{1}{2}\sqrt[3]{x}$
$h(x) = -\sqrt[3]{x - 1}$

59. The graphs in the figure were obtained by transformations of the graph of $f(x) = |x|$. Write equations for g and h and state the domain and range for g and h.

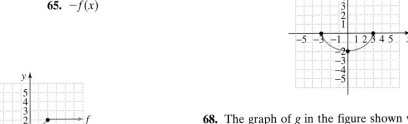

In Problems 60–66, use the graph of f shown below to sketch the graph of each function.

60. $f(x - 1)$

61. $f(x) - 1$

62. $2f(x)$

63. $\frac{1}{2}f(x)$

64. $f(-x)$

65. $-f(x)$

66. $-f(-x)$

67. Use the graph of f to graph $g(x) = -2 - \frac{1}{2}f(x + 2)$.

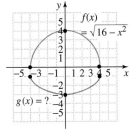

68. The graph of g in the figure shown was obtained from the graph of f through a sequence of transformations. Find an equation for g.

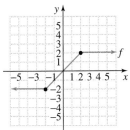

In Problems 69–71, use algebraic methods to determine whether each function is even, odd, or neither. State each function's symmetry. Verify your results by using a graphing utility to graph each function.

69. $f(x) = x^3 - 5x$

70. $f(x) = x^4 - 2x^2 + 1$

71. $f(x) = 2x\sqrt{1 - x^2}$

72. If $f(x) = x^2 - 7x + 8$ and $g(x) = 5 - 2x$, find each of the following functions.

a. $f + g$
b. $f - g$
c. fg
d. $\frac{f}{g}$
e. $f \circ g$
f. $g \circ f$

73. Let $f(x) = \sqrt{x + 4}$ and $g(x) = \sqrt{x - 1}$. Find each function and state its domain.

a. $f + g$ **b.** $f - g$

c. fg **d.** $\frac{f}{g}$

Verify the domains by graphing $f + g$, $f - g$, fg, and $\frac{f}{g}$ with a graphing utility.

74. If $f(x) = \sqrt{x}$ and $g(x) = 2x + 1$, find:

a. $f \circ g$ **b.** $g \circ f$

c. $(f \circ g)(4)$ **d.** $(g \circ f)(25)$

e. State the domain for each composite function in parts (a) and (b).

75. If $f(x) = x^2 - 25$ and $g(x) = \sqrt{25 - x^2}$, find $(f \circ g)(x)$ and the domain of $f \circ g$. Check your result with a graphing utility.

76. The function $L(p) = 0.7\sqrt{p^2 + 3}$ models the level L of carbon monoxide in the air (in ppm) as a function of population (p, in hundred-thousands). The population p of a hypothetical city is described by $p(t) = 1 + 0.01t^3$, where t is time from 1990 and p is population in hundred-thousands.

a. Find $L \circ p$ and describe in practical terms what the function models.

b. Find $(L \circ p)(10)$. What information is conveyed by 10 and the resulting function value?

In Problems 77–79, find two functions f and g so that $h(x) = (f \circ g)(x)$. (More than one correct answer is possible for each problem.)

77. $h(x) = (2x + 7)^4$ **78.** $h(x) = \sqrt{3x^2 - 7}$ **79.** $h(x) = 3(x - 4)^2 + 5(x - 4)$

In Problems 80–81, find $f^3(x)$, the third iterate of the given function f.

80. $f(x) = 3x - 2$

81. $f(x) = x^2 - 1$

82. Suppose that deforestation of a rain forest occurs at a rate of 5% per year. If x represents the area of the rain

forest at the start of 1997, then $f(x) = 0.95x$ models the area of the rain forest at the start of 1998. Find $f^3(x)$ and interpret the third iterate in practical terms.

Problems 83–86 show graphs of relations. Which of these relations are functions? Which of these functions are one-to-one?

83.

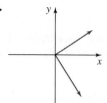

84.

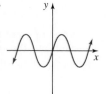

85.

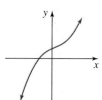

86.

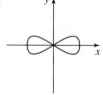

*In Problems 87–92, **a.** find $f^{-1}(x)$; **b.** verify algebraically that $f(f^{-1}(x)) = x$ and $f^{-1}(f(x)) = x$; **c.** use a graphing utility to verify part (b).*

87. $f(x) = \frac{1}{2}x - 4$ **88.** $f(x) = 4x - 3$

89. $f(x) = \sqrt{x + 2}$ **90.** $f(x) = 8x^3 + 1$

91. $f(x) = x^2 - 4, x \geq 0$ **92.** $f(x) = \sqrt[3]{2x - 1}$

93. If $10,000 is invested in an account paying 12% yearly with compounding N times a year, then the balance in the account at the end of the year is a function of N, the number of compounding periods. The graph of this function f is shown in the figure. The figure indicates that the graph approaches but never reaches the constant function $g(x) = 11,274.97$.

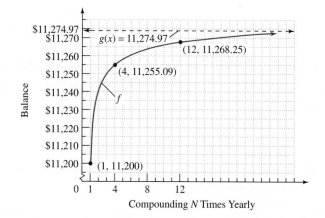

a. Is f a one-to-one function? Explain.

b. What is the practical meaning of $f^{-1}(11, 250)$?

c. An investor stated that by investing $10,000 with more and more compounding periods, there would be no limit to the balance in the account at the end of the year. Critique this statement using the graph of f. In your critique, describe the practical meaning of the range of f.

For Problems 94–96:

a. *Use a graphing utility to graph f.*

b. *Use the graph of f to restrict its domain to an interval where f is increasing.*

c. *Use algebraic methods to find f^{-1} on the restricted domain.*

d. *Use the graphing utility to graph f^{-1} in the same viewing rectangle as f.*

94. $f(x) = (x - 2)^2$

95. $f(x) = |x + 3|$

96. $f(x) = \sqrt{x^2 - 9}$

In Problems 97–100, use the graph of f to sketch the graph of each function.

97. $f^{-1}(x)$

98. $f^{-1}(x + 1)$

99. $f^{-1}(x) + 1$

100. $-f^{-1}(x)$

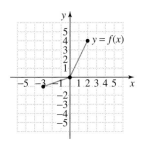

101. The snow cricket chirps in such a way that the chirp rate is a linear function of temperature. At a temperature of 50°F, the cricket chirps 40 times per minute and at a temperature of 70°F, the cricket chirps 120 times per minute.

a. Model this situation with a linear function expressing chirp rate (C) as a function of Fahrenheit temperature (T) $(C = f(T))$.

b. Find $f(90)$ and describe what this means in practical terms.

c. The domain of the model consists of all temperatures for which the predicted chirp rate is positive, up to 136°F. What is the domain of f?

d. What is the range of f?

e. Find a formula expressing temperature as a function of chirp rate.

102. Complete the table so that:

a. f has origin symmetry.   **b.** g has y-axis symmetry.

c. $h = f \circ g$

x	$f(x)$	$g(x)$	$h(x)$
-3	0	0	
-2	5	5	
-1	5	5	
0	0	0	
1			
2			
3			

CHAPTER 2 TEST

1. The function $f(x) = 0.002x^2 + 0.41x + 7.34$ models the percent of the population in the United States $(f(x))$ that graduated from college x years after 1960. Find and interpret $f(10)$.

2. If $f(x) = 7x^2 - 9x + 13$, find $\dfrac{f(a + h) - f(a)}{h}$.

3. If $f(x) = \sqrt{12 - 3x}$, find the domain of f.

Use the conditions given in Problems 4–5 to write the equation of each line in point-slope form, slope-intercept form, and general form.

4. Passing through $(2, 1)$ and $(-1, -8)$

5. Passing through $(-4, 6)$ and perpendicular to the line whose equation is $y = -\frac{1}{4}x + 5$

6. The data points $(4, 401.1)$ and $(9, 475.6)$, shown and described in the table, fall on a straight line.

x (Number of Years after 1985)	y (Average Weekly Earnings of U.S. Workers)
4	401.1
9	475.6

  a. Write the equation of the line on which these measurements fall in point-slope form and slope-intercept form.

  b. Extrapolate from the data to predict the average weekly earnings for U.S. workers for the year 2005.

Graph each function in Problems 7–8

7. $f(x) = \dfrac{x^2 - 36}{x - 6}$

8. $f(x) = \begin{cases} \sqrt{x - 3} & \text{if } x \geqslant 3 \\ 3 - x & \text{if } x < 3 \end{cases}$

9. Use the graph of the function f shown below to answer the following questions.

  a. What is $f(4) - f(-3)$?

  b. On what interval or intervals is f increasing?

  c. On what interval or intervals is f decreasing?

  d. What is the relative maximum value of the function and where does this occur?

  e. What is the relative minimum value of the function and where does this occur?

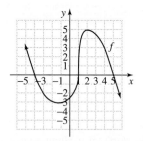

10. From each of the four corners of a square piece of cardboard 12 inches on a side, a square piece x inches on a side is cut out and the flaps are turned up to form an open-top box. The volume of the box, $f(x)$ in cubic inches, is a function of the size of the square cut from each corner, given by $f(x) = 4x^3 - 48x^2 + 144x$. Use the graph of f in the figure shown to find the maximum possible volume of the box.

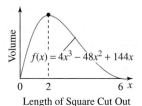

Length of Square Cut Out

11. Sketch the graph of a function f using the following properties: f is decreasing on $(-\infty, -2)$, increasing on $(-2, \infty)$, has a y-intercept of 3, and a minimum y-value of -1. (More than one correct graph is possible.)

12. The graph of g shown in the figure below is a transformation of the graph of $f(x) = x^3$. Write an equation for g.

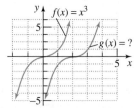

13. The figure shows how the graph of $h(x) = -2(x - 3)^2$ is obtained from the graph of $f(x) = x^2$. Describe this process, using the graph of g in your description.

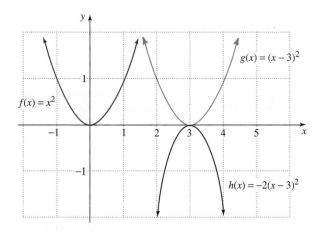

14. Determine whether $f(x) = x^4 - x^2$ even or odd. Then use your answer to explain why the graph in the figure shown cannot be the graph of f.

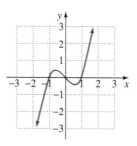

If $f(x) = x^2 + 3x - 4$ and $g(x) = 5x - 2$, find each function or function value in Problems 15–19.

15. $f - g$

16. $f \circ g$

17. $g \circ f$

18. $f(g(2))$

19. $g(f(2))$

20. If $f(x) = \sqrt{x - 2}$, find $f^{-1}(x)$. Then verify either algebraically or graphically that the two functions are inverses.

21. A function f defines a man's recommended weight (in pounds) as a function of his height (in inches). The graph of f is shown in the figure.

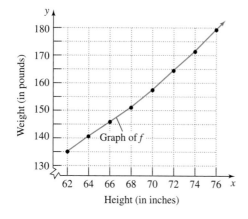

Graph of f

a. Can we also say that a man's height is a function of his recommended weight? Explain your answer using the graph of f and the horizontal line test.

b. Describe in practical terms the meaning of $f^{-1}(170)$.

22. Use a graphing utility to graph $f(x) = x^4 - 6x^3 + 8x^2$ in a $[-3, 5]$ by $[-10, 6]$ viewing rectangle. Now use the graph to answer the following.

a. Is f one-to-one? Explain.

b. Is f even, odd, or neither? Explain.

c. Find the relative maximum and minimum values of f and the values of x at which they occur.

d. What is the range of f?

e. Find the interval or intervals on which f is increasing.

f. Find the interval or intervals on which f is decreasing.

CUMULATIVE REVIEW PROBLEMS (CHAPTERS P–2)

Simplify Problems 1–2.

1. $\left(\dfrac{4x^2 y}{2x^5 y^{-3}} \right)^{-3}$

2. $-6\sqrt{32y} + 10\sqrt{2y}$

3. Factor: $x^2 - 2x + 1 - y^2$.

In Problems 4–5, perform the operations and simplify.

4. $\dfrac{x^2 - 2x}{x^2 - 9} \div \dfrac{x^4 - 4x^3 + 4x^2}{x^2 + 3x}$

5. $\dfrac{x - 1}{\sqrt{x + 1} - \sqrt{2}}$

Solve Problems 6–10.

6. $(12x - 5)(x - 1) = 10$

7. $\dfrac{2x}{x - 1} - \dfrac{4}{x^2 - x} = \dfrac{x - 1}{x}$

8. $\sqrt{x + 12} - \sqrt{x} = 2$

9. $x^{2/3} - x^{1/3} - 6 = 0$

10. $-5 < 8 - 2(x + 3) \leqslant 13$

11. Write the point-slope form, the slope-intercept form, and the general form of the line passing through $(-2, 5)$ and perpendicular to the line whose equation is $x - 2y - 3 = 0$.

12. Graph $f(x) = \sqrt{x}$ and then use horizontal and vertical shifts to graph $g(x) = \sqrt{x - 3} + 4$ on the same rectangular coordinate system.

13. Find the inverse function of $f(x) = \dfrac{x^5 - 3}{2}$.

14. The graph of $f(x) = x^2 - 2x - 3$ is shown in the figure. Describe one way the domain of f can be restricted so that the resulting function is one-to-one.

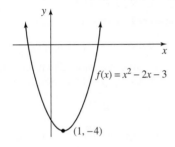

$f(x) = x^2 - 2x - 3$

$(1, -4)$

15. If $f(x) = mx + b$, find $\dfrac{f(a + h) - f(a)}{h}$ and describe what this represents in terms of the given function f.

16. Solve for r: $G = \dfrac{a}{1 - r}$.

17. The number of diagonals D of a polygon of n sides is given by

$$D = \frac{n^2 - 3n}{2}.$$

How many sides does a polygon with 90 diagonals have?

18. In 1970, the U.S. gross federal debt was approximately $380 billion, increasing at about $43 billion each year. In what year was the gross federal debt $1369 billion?

19. The length of a rectangular garden is 2 feet more than twice its width. If 22 feet of planking is needed to enclose the garden, what are its dimensions?

20. A bridge toll currently charges $2.50 per car. The average number of cars crossing per day is 200. However, this number is expected to decrease by one car per day for every 5¢ increase over $2.50. If the bridge needs to bring in an income of $630 per day, what should the toll be?

Modeling with Polynomial and Rational Functions

Henry Small "Friends in the Florida Keys," acrylic on wood, 3 ft. × 4 ft., 1985.

In this century, human impacts on populations and ecosystems have caused hundreds of species of plants and animals to become extinct each year. If present trends continue, we may destroy thousands of species in the next few decades.

Large-scale public projects aimed at reversing this trend and reducing environmental pollution are subject to cost-benefit analyses. The costs of a project must be compared with the benefits that will be achieved. One type of function we will be studying in this chapter is used to model the costs of a given project, illustrating how algebra contributes to decision making on environmental policies.

SECTION 3.1

Solutions Manual   Tutorial   Video 6

Quadratic Functions

Objectives

1 Graph quadratic functions.
2 Solve optimization problems using given models.
3 Solve optimization problems by modeling verbal conditions.

In the prerequisites chapter, we encountered the graph of $y = ax^2 + bx + c$ and called it a *parabola*. We have seen that the real solutions of $ax^2 + bx + c = 0$ can be found by determining where the parabola crosses the x-axis. The quadratic models in Chapter 1 ($y = ax^2 + bx + c$) were viewed as quadratic functions ($f(x) = ax^2 + bx + c$) in Chapter 2. In this section, we continue our ongoing study of quadratic functions.

▍ Graph quadratic functions.

Quadratic Functions as Second-Degree Polynomial Functions

An important class of functions in algebra is defined using polynomials; such functions are called *polynomial functions*.

> **Definition of a polynomial function**
>
> Let n be a nonnegative integer and let $a_n, a_{n-1}, \ldots, a_2, a_1, a_0$, be real numbers with $a_n \neq 0$. The function defined by
>
> $$f(x) = a_n x^n + a_{n-1} x^{n-1} + \cdots + a_2 x^2 + a_1 x + a_0$$
>
> is called a *polynomial function of x of degree n*. The number a_n, the coefficient of the highest power, is called the *leading coefficient*.

A constant function $f(x) = a$, where $a \neq 0$, is a polynomial function of degree 0. A linear function $f(x) = ax + b$, where $a \neq 0$, is a polynomial function of degree 1. Our focus in this section is on second-degree polynomial functions, or *quadratic functions*.

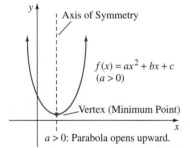

> **Definition of a quadratic function**
>
> A *quadratic function* is any function of the form
>
> $$f(x) = ax^2 + bx + c$$
>
> where a, b, and c are real numbers and $a \neq 0$. The graph of a quadratic function is called a *parabola*.

Characteristics of Parabolas

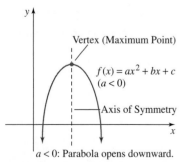

Figure 3.1

Characteristics of parabolas

As shown in Figure 3.1, parabolas are symmetric to a line called the *axis of symmetry*. The turning point on the parabola, the point where the axis of symmetry intersects the parabola, is called the *vertex*. If $a > 0$, the graph of $f(x) = ax^2 + bx + c$ opens upward and the vertex is the *minimum point* on the graph. If $a < 0$, then the parabola opens downward and the vertex is the *maximum point* on the graph.

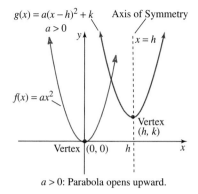

$g(x) = a(x - h)^2 + k$ Axis of Symmetry
$a > 0$

$f(x) = ax^2$

Vertex $(0, 0)$ h

Vertex (h, k)

$x = h$

$a > 0$: Parabola opens upward.

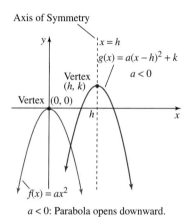

Axis of Symmetry

$x = h$

$g(x) = a(x - h)^2 + k$

$a < 0$

Vertex (h, k)

Vertex $(0, 0)$

h

$f(x) = ax^2$

$a < 0$: Parabola opens downward.

Figure 3.2

Transformations of $f(x) = ax^2$

Using Transformations to Graph Parabolas

In Section 2.5, we applied a series of transformations to the graph of the quadratic function $f(x) = x^2$. Figure 3.2 compares the graphs of $f(x) = ax^2$ and $g(x) = a(x - h)^2 + k$. Observe that h determines the horizontal shift and k determines the vertical shift of the graph of $f(x) = ax^2$. Consequently, the vertex $(0, 0)$ on the graph of $f(x) = ax^2$ moves to the point (h, k) on the graph of $g(x) = a(x - h)^2 + k$. The axis of symmetry is the vertical line whose equation is $x = h$.

The form of the expression for g is convenient since it immediately identifies the vertex of the parabola as (h, k). This is the *standard form* of a quadratic function.

> ### The standard form of a quadratic function
>
> The quadratic function
>
> $$f(x) = a(x - h)^2 + k, \qquad a \neq 0$$
>
> is in *standard form*. The graph of f is a parabola whose vertex is the point (h, k). The parabola is symmetric to the line $x = h$. If $a > 0$, the parabola opens upward; if $a < 0$, the parabola opens downward.

EXAMPLE 1   Graphing a Parabola

Use the vertex and intercepts to sketch the graph of:

a. $f(x) = -2(x - 3)^2 + 2$ **b.** $g(x) = (x + 3)^2 + 1$

Solution

a. The equation for f is expressed in standard form.

$$f(x) = -2(x - 3)^2 + 2 \qquad \text{The parabola has a vertex at } (h, k) = (3, 2).$$

$$f(x) = a(x - h)^2 + k \qquad \text{Because } a = -2, a < 0 \text{ and the parabola opens downward.}$$

Let's find the x-intercepts by replacing $f(x)$ in the given equation by 0.

$$0 = -2(x - 3)^2 + 2 \qquad \text{Find } x\text{-intercepts, letting } f(x) \text{ (or } y\text{) equal 0.}$$

$$2(x - 3)^2 = 2 \qquad \text{Solve for } x. \text{ Add } 2(x - 3)^2 \text{ to both sides of the equation.}$$

$$(x - 3)^2 = 1 \qquad \text{Divide both sides by 2.}$$

$$(x - 3) = \pm 1 \qquad \text{Apply the square root principle. If } (x - c)^2 = d, \text{ then } x - c = \pm \sqrt{d}.$$

$$x = 2 \quad \text{or} \quad x = 4 \qquad \text{The } x\text{-intercepts are 2 and 4.}$$

Finally, we find the y-intercept by replacing x in the given equation with 0.

$$f(0) = -2(0 - 3)^2 + 2 \qquad f(0) \text{ is the value that gives the } y\text{-intercept.}$$

$$f(0) = -16 \qquad \text{The } y\text{-intercept is } -16.$$

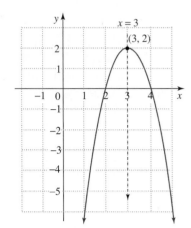

Figure 3.3

The graph of $f(x) = -2(x - 3)^2 + 2$

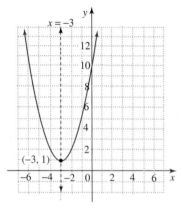

Figure 3.4

The graph of $g(x) = (x + 3)^2 + 1$

iscover for yourself

Use a graphing utility to verify the hand-drawn graphs in Figures 3.3 and 3.4. Trace along each graph and verify the vertex.

With a vertex at $(3, 2)$, x-intercepts at 2 and 4, and a y-intercept at -16, the graph of f is shown in Figure 3.3. (Because of the scale chosen on the axes, the y-intercept is not shown.) The axis of symmetry is the vertical line whose equation is $x = 3$.

b. The equation for g is also expressed in standard form.

$$g(x) = (x + 3)^2 + 1$$

$$g(x) = 1[x - (-3)]^2 + 1 \qquad \text{The parabola has a vertex at } (h, k) = (-3, 1).$$
$$\qquad\qquad\qquad\qquad\qquad\qquad \text{Because } a = 1, a > 0 \text{ and the parabola opens upward.}$$

$$g(x) = a(x - h)^2 + k$$

With a vertex at $(-3, 1)$ and opening upward, the parabola appears to have no x-intercepts (see Figure 3.4). We can verify this observation algebraically.

$$0 = (x + 3)^2 + 1 \qquad \text{Find possible } x\text{-intercepts letting } g(x) \text{ (or } y\text{) equal 0.}$$

$$-1 = (x + 3)^2 \qquad \text{Solve for } x. \text{ Subtract 1 from both sides.}$$

$$x + 3 = \pm\sqrt{-1} \qquad \text{Apply the square root principle.}$$

$$x + 3 = \pm i \qquad \text{Recall that } \sqrt{-1} = i, \text{ an imaginary number.}$$

$$x = -3 \pm i \qquad \text{The equation has no real solution, so there are no } x\text{-intercepts.}$$

The y-intercept is found by replacing x in the given equation with 0.

$$g(0) = (0 + 3)^2 + 1 \qquad \text{The } y\text{-intercept is } g(0).$$

$$g(0) = 10 \qquad \text{The } y\text{-intercept is 10.}$$

With a vertex at $(-3, 1)$, no x-intercepts, and a y-intercept at 10, the graph of g is shown in Figure 3.4. The axis of symmetry is the vertical line whose equation is $x = -3$. ∎

Quadratic Functions in the Form $f(x) = ax^2 + bx + c$

Quadratic functions are frequently expressed in the form

$$f(x) = ax^2 + bx + c.$$

We can complete the square to express the function's equation in standard form. Let's see how this is done.

EXAMPLE 2 **Writing a Quadratic Function in Standard Form**

Rewrite in standard form and sketch the graph: $f(x) = -2x^2 + 4x - 5$

Solution

We begin by completing the square. The first step is to factor out any coefficient of x^2 from the x-terms that differs from 1.

$$f(x) = -2x^2 + 4x - 5 \qquad\qquad \text{This is the given function.}$$

$$= -2(x^2 - 2x) - 5 \qquad\qquad \text{Factor } -2 \text{ out of the } x\text{-terms.}$$

$$= -2(x^2 - 2x + 1 - 1) - 5$$

Complete the square for $x^2 - 2x$ by taking half the x-coefficient and squaring. The resulting number (1) is added and subtracted within the parentheses.

$$\frac{1}{2}(-2) = -1$$
and
$$(-1)^2 = 1$$

$$= -2(x^2 - 2x + 1) + (-2)(-1) - 5$$

Regroup terms by removing the -1 from the parentheses.

$$= -2(x - 1)^2 - 3$$

Factor $x^2 - 2x + 1$ using $A^2 - 2AB + B^2 = (A - B)^2$. Simplify numerical terms.

The function is now expressed in standard form.

$$f(x) = -2(x - 1)^2 - 3$$

The parabola has a vertex at $(h, k) = (1, -3)$ and opens downward.

$$f(x) = a(x - h)^2 + k$$

The parabola has no x-intercepts. By letting $x = 0$ in either the given form of the function or the standard form, we find that the y-intercept is -5. The graph of f is shown in Figure 3.5. ∎

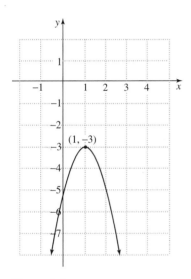

Figure 3.5

The graph of $f(x) = -2x^2 + 4x - 5$

Because quadratic functions are frequently given in the form $f(x) = ax^2 + bx + c$, we would like to immediately recognize the parabola's vertex when its equation is in this form. To do this, let's complete the square on the general quadratic function $f(x) = ax^2 + bx + c$ to obtain its standard form.

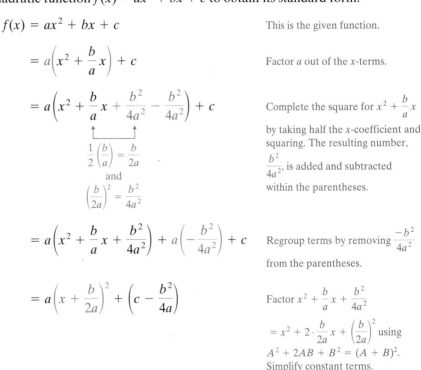

$$f(x) = ax^2 + bx + c$$

This is the given function.

$$= a\left(x^2 + \frac{b}{a}x\right) + c$$

Factor a out of the x-terms.

$$= a\left(x^2 + \frac{b}{a}x + \frac{b^2}{4a^2} - \frac{b^2}{4a^2}\right) + c$$

Complete the square for $x^2 + \frac{b}{a}x$ by taking half the x-coefficient and squaring. The resulting number, $\frac{b^2}{4a^2}$, is added and subtracted within the parentheses.

$$\frac{1}{2}\left(\frac{b}{a}\right) = \frac{b}{2a}$$
and
$$\left(\frac{b}{2a}\right)^2 = \frac{b^2}{4a^2}$$

$$= a\left(x^2 + \frac{b}{a}x + \frac{b^2}{4a^2}\right) + a\left(-\frac{b^2}{4a^2}\right) + c$$

Regroup terms by removing $\frac{-b^2}{4a^2}$ from the parentheses.

$$= a\left(x + \frac{b}{2a}\right)^2 + \left(c - \frac{b^2}{4a}\right)$$

Factor $x^2 + \frac{b}{a}x + \frac{b^2}{4a^2}$

$$= x^2 + 2 \cdot \frac{b}{2a}x + \left(\frac{b}{2a}\right)^2 \text{ using}$$
$$A^2 + 2AB + B^2 = (A + B)^2.$$
Simplify constant terms.

By writing $f(x) = ax^2 + bx + c$ in standard form,

$$f(x) = a\left(x + \frac{b}{2a}\right)^2 + \left(c - \frac{b^2}{4a}\right)$$

we see that $h = -\dfrac{b}{2a}$ and $k = c - \dfrac{b^2}{4a}$. The important part of this result is that

the x-coordinate of the vertex is $-\dfrac{b}{2a}$. The y-coordinate can be found by eval-

uating the function at $-\dfrac{b}{2a}$.

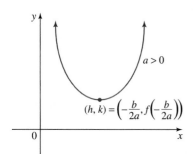

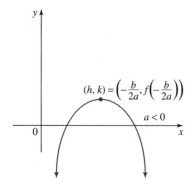

Figure 3.6

Graphs of $f(x) = a(x - h)^2 + k$
or $f(x) = ax^2 + bx + c$

The vertex of a parabola (see Figure 3.6)

1. Standard form: $f(x) = a(x - h)^2 + k$
   Vertex of the parabola: (h, k)
   If $a > 0$, then the minimum value of f occurs at $x = h$ and this value
   is $f(h) = k$.
   If $a < 0$, then the maximum value of f occurs at $x = h$ and this value
   is $f(h) = k$.
2. The form $f(x) = ax^2 + bx + c$
   Vertex of the parabola: $\left(-\dfrac{b}{2a}, f\left(-\dfrac{b}{2a} \right) \right)$

   If $a > 0$, then the minimum value of f occurs at $x = -\dfrac{b}{2a}$ and this
   value is $f\left(-\dfrac{b}{2a} \right)$.

   If $a < 0$, then the maximum value of f occurs at $x = -\dfrac{b}{2a}$ and this
   value is $f\left(-\dfrac{b}{2a} \right)$.

EXAMPLE 3 **Graphing a Parabola**

Use the vertex and intercepts to sketch the graph of: $f(x) = -2x^2 - x + 5$

Solution

We begin with the vertex. Its x-coordinate is $-\dfrac{b}{2a}$.

$$x = -\frac{b}{2a}$$

$$= \frac{-(-1)}{2(-2)} \qquad \text{We are given } f(x) = -2x^2 - x + 5, \text{ so } a = -2 \text{ and } b = -1.$$

$$= -\frac{1}{4} \qquad \text{The } x\text{-coordinate of the vertex is } -\tfrac{1}{4}.$$

The y-coordinate of the vertex is

$$f\left(-\frac{1}{4} \right) = -2\left(-\frac{1}{4} \right)^2 - \left(-\frac{1}{4} \right) + 5 \qquad \text{Substitute } -\tfrac{1}{4} \text{ for } x \text{ in } -2x^2 - x + 5.$$

$$= 5.125$$

The parabola has a vertex at $(-\tfrac{1}{4}, 5.125)$. Because $a = -2$, the parabola opens
downward. With a vertex in quadrant II, the parabola will have x-intercepts.
We can find them by setting $f(x) = 0$.

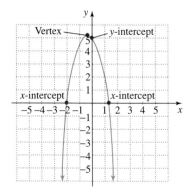

Figure 3.7

The graph of $f(x) = -2x^2 - x + 5$

2 Solve optimization problems using given models.

$$0 = -2x^2 - x + 5 \qquad \text{Set } f(x) \text{ (or } y) \text{ equal to 0.}$$

$$x = \frac{-b \pm \sqrt{b^2 - 4ac}}{2a} = \frac{-(-1) \pm \sqrt{(-1)^2 - 4(-2)(5)}}{2(-2)} = \frac{1 \pm \sqrt{41}}{-4}$$

Use the quadratic formula with $a = -2, b = -1$, and $c = 5$.

With $\sqrt{41} \approx 6.4$, the x-intercepts are approximately -1.9 and 1.4. We find the y-intercept by setting $x = 0$.

$$f(0) = -2 \cdot 0^2 - 0 + 5 \qquad \text{Use the given function } f(x) = -2x^2 - x + 5 \text{ and let } x = 0.$$

$$f(0) = 5 \qquad \text{The } y\text{-intercept is 5.}$$

Summary:
Vertex: $\left(-\frac{1}{4}, 5.125\right)$
x-intercepts: approximately -1.9 and 1.4
y-intercept: 5

The parabola is shown in Figure 3.7.   ■

Optimization: Using Given Mathematical Models

In Section 2.3, we used a graphing utility to find optimum (maximum and minimum) values of a function. In the case of a quadratic function

$$f(x) = ax^2 + bx + c$$

we know that the lowest point on the parabola is the vertex when it opens upward (if $a > 0$) and the highest point on the parabola is the vertex when it opens downward (if $a < 0$) Because the x-coordinate of the vertex is $-\dfrac{b}{2a}$, in either case we can find the minimum or maximum value of f by evaluating the quadratic function at $x = -\dfrac{b}{2a}$.

Minimum and maximum: quadratic functions

Consider $f(x) = ax^2 + bx + c$.

1. If $a > 0$, then f has a minimum that occurs at $x = -\dfrac{b}{2a}$.

2. If $a < 0$, then f has a maximum that occurs at $x = -\dfrac{b}{2a}$.

EXAMPLE 4   **An Application: Alcohol Consumption**

Based on data provided by the U.S. Department of Agriculture, the model

$$f(x) = -0.053x^2 + 1.17x + 35.6$$

describes the average number of gallons of alcohol consumed by each adult in the United States as a function of time, where x represents the number of years after 1970. In what year was alcohol consumption at a maximum? What was the average consumption for that year?

Using technology

The algebraic analysis on the right can be verified with a graphing utility. Enter the given function's equation

$y = -0.053x^2 + 1.17x + 35.6$

with the following range setting:

Xmin = 0, Xmax = 20,

Xscl = 1, Ymin = 0,

Ymax = 50, Yscl = 5

Using the maximum function (FMAX) feature, we obtain the screen below, which verifies that alcohol consumption was at a maximum about 11 years after 1970 and was approximately 42 gallons for each adult.

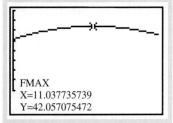

FMAX
X=11.037735739
Y=42.057075472

3 Solve optimization problems by modeling verbal conditions.

Solution

For this quadratic function we have

$$f(x) = -0.053x^2 + 1.17x + 35.6$$
$$f(x) = \quad ax^2 + \quad bx + \quad c$$

which means that $a = -0.053$ and $b = 1.17$. With $a < 0$, the function has a maximum when $x = -\dfrac{b}{2a}$.

$$x = -\frac{b}{2a} = -\frac{1.17}{2(-0.053)} \approx 11.04$$

This means that alcohol consumption was at a maximum approximately 11 years after 1970, or in 1981. The average consumption for that year was

$$f(11) = -0.053(11)^2 + 1.17(11) + 35.6 \approx 42.06$$

or approximately 42 gallons of alcohol by each adult. ■

Optimization: Creating Mathematical Models

Verbal problems involving optimization require us to model functions based on given or implied conditions. If the resulting function is quadratic, we can then use the x-coordinate of the vertex to determine the maximum or minimum value of the function. Here's the general strategy that we will follow in the remainder of our examples.

Strategy for solving optimization problems involving quadratic functions

1. Read the problem carefully and decide which quantity is to be maximized or minimized.
2. Use the conditions of the problem to express the quantity as a function in one variable.
3. Rewrite the function in the form $f(x) = ax^2 + bx + c$.
4. If $a > 0$, f has a minimum at $x = -\dfrac{b}{2a}$. If $a < 0$, f has a maximum at $x = -\dfrac{b}{2a}$.
5. Answer the question posed in the problem.

EXAMPLE 5 Solving a Number Problem

Among all pairs of numbers whose sum is 40, find a pair whose product is as large as possible. What is the maximum product?

Solution

Step 1. Decide what must be maximized or minimized. We must maximize the product of two numbers. Calling the numbers x and y, and calling the product P, we must maximize

$$P = xy.$$

Step 2. Express this quantity as a function in one variable.
Right now, P is expressed in terms of two variables, x and y. However, since the sum of the numbers is 40, we can write

$$x + y = 40.$$

We can solve this equation for y in a terms of x (or vice versa), substitute the result into $P = xy$, and obtain P as a function of one variable. Solving for y, we get

$$y = 40 - x.$$

Now we substitute $40 - x$ for y in $P = xy$.

$$P = xy = x(40 - x).$$

Since P is now a function of x, we can write

$$P(x) = x(40 - x).$$

Step 3. Write the function in the form $f(x) = ax^2 + bx + c$. We apply the distributive property to obtain

$$P(x) = x(40 - x) = 40x - x^2$$

which can be expressed as

$$P(x) = -x^2 + 40x.$$

Step 4. If $a < 0$, the function has a maximum at $x = -\dfrac{b}{2a}$.
Since $P(x) = -x^2 + 40x$, we see that $a = -1$ and $b = 40$. The function P has a maximum at $x = -\dfrac{b}{2a}$.

$$x = -\frac{b}{2a} \qquad \text{Remember that } x \text{ represents one of the numbers.}$$

$$= -\frac{40}{2(-1)} \qquad \text{Since } P(x) = -x^2 + 40x, \text{ substitute } -1 \text{ for } a \text{ and } 40 \text{ for } b.$$

$$= 20 \qquad \text{Simplify.}$$

Step 5. Answer the question posed in the problem. The numbers are $x = 20$ and $y = 40 - x = 40 - 20 = 20$. The number pair whose sum is 40 and whose product is as large as possible is 20, 20. The maximum product is $20(20)$, or 400. ∎

Using technology

The graph of the product function

$$P(x) = x(40 - x)$$

was obtained with a graphing utility using

$$[0, 40] \times [0, 400]$$

$$\text{Xscl} = 4, \text{Yscl} = 20$$

Use the $\boxed{\text{TRACE}}$ or $\boxed{\text{FMAX}}$ feature to verify that a maximum product of 400 occurs when one of the numbers is 20.

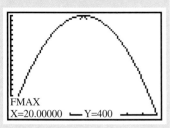

FMAX
X=20.00000 ▭ Y=400

EXAMPLE 6 **Maximizing Income**

A small cruising ship that can hold up to 50 people provides two-day excursions to groups of 34 or more. If the group contains 34 people, each person pays \$60. The cost per person is reduced by \$1 for each person in excess of 34. Find the size of the group that will maximize income for the owners of the ship and determine this maximum income.

Solution

Step 1. Decide what must be maximized or minimized. We must maximize income. What we do not know is how many people in excess of 34 we should have on board to generate maximum income.
Step 2. Express this quantity as a function in one variable.
Let $x =$ the number of people in the group in excess of 34. We write our function based on the verbal model that income equals the number of people in the group times the cost per person.

Income	equals	The number of people in the group			times	The cost per person.		
Income	equals	34 people	plus	The number in excess of 34	times	\$60	minus	\$1 times the number of people in excess of 34.
$I(x)$	$=$	$(\ 34$	$+$	$x\ )$	$\cdot$	$(\ 60$	$-$	$1 \cdot x\ )$

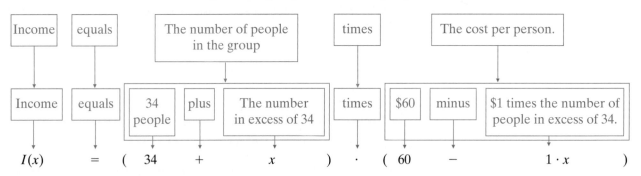

The income function that we must maximize is

$$I(x) = (34 + x)(60 - x).$$

Step 3. Write the function in the form $f(x) = ax^2 + bx + c$.

$I(x) = (34 + x)(60 - x)$ — This is the function from step 2.

$= 2040 + 26x - x^2$ — Multiply using the FOIL method.

$= -x^2 + 26x + 2040$ — Write the function's equation in descending powers of x.

Step 4. Since $a < 0$, the function has a maximum at $x = -\frac{b}{2a}$. Using the formula for $I(x)$, we see that $a = -1$ and $b = 26$.

$x = -\frac{b}{2a} = -\frac{26}{2(-1)} = 13$ — Remember that x represents the number of people in excess of 34.

Step 5. Answer the problem's question. The size of the group that will maximize income for the owners of the ship is $34 + x = 34 + 13$, or 47 people. The maximum income is $I(13)$.

$I(13) = (34 + 13)(60 - 13) = 2209$ — Substitute 13 for x in $I(x) = (34 + x)(60 - x)$.

The maximum income is \$2209. ■

Using technology

The graph of the income function

$I(x) = (34 + x)(60 - x)$

was obtained with a graphing utility using

$[0, 60] \times [0, 2210]$

Xscl = 6, Yscl = 105

Use the TRACE or FMAX feature to verify a maximum income of \$2209 when there are 13 people in excess of 34.

FMAX
X=13
Y=2209

Study tip

In Example 6, the graph of the income function should actually be a collection of discrete points corresponding to the integer values of x, as shown in Figure 3.8. Because

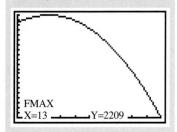

Income

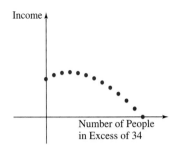

Number of People
in Excess of 34

Figure 3.8

Income Function as a collection
of points

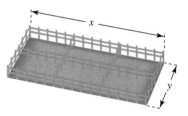

Figure 3.9

Using 1000 feet of fencing to
construct six corrals

x represents the number of people in excess of 34, we cannot permit values of x such as $x = 12.57$. The function $I(x) = -x^2 + 26x + 2040$ is a continuous model whose graph connects the points in Figure 3.8. In Example 6, the solution to the continuous function also gave the solution for the discrete function.

In our next problem, although the function we need to maximize is continuous, the challenge will be to represent this function in one variable.

EXAMPLE 7   **Maximizing Area**

A rancher has 1000 feet of fencing to construct six corrals, as shown in Figure 3.9. Find the dimensions that maximize the enclosed area. What is the maximum area?

Solution

Step 1. Decide what must be maximized or minimized. We must maximize area. What we do not know are the rectangle's dimensions, x and y.

Step 2. Express this quantity as a function in one variable.

Since we must maximize area, we have

$$A = xy.$$

Once again, A is represented in terms of two variables. However, we are given that the rancher has 1000 feet of fencing. Referring to Figure 3.9, this means that

$$3x + 4y = 1000$$

Fencing is placed along three lengths, each of dimension x, and four widths, each of dimension y.

$$y = \frac{1000 - 3x}{4}$$

Solve for y.

$$A = xy$$

This is our expression for the area.

$$A(x) = x\left(\frac{1000 - 3x}{4}\right)$$

Substitute $\dfrac{1000 - 3x}{4}$ for y. This is the function implied by the problem's conditions.

Step 3. Write the function in the form $f(x) = ax^2 + bx + c$. We apply the distributive property.

$$A(x) = x\left(\frac{1000 - 3x}{4}\right) = \frac{1000x - 3x^2}{4} = -\frac{3}{4}x^2 + 250x$$

Step 4. Since $a < 0$ the function has a maximum at $x = -\dfrac{b}{2a}$. With $A(x) = -\dfrac{3}{4}x^2 + 250x$, we see that $a = -\dfrac{3}{4}$ and $b = 250$.

$$x = -\frac{b}{2a}$$

Remember that x represents the length of the structure.

ENRICHMENT ESSAY

The Road to Nowhere

The polynomial function $M(t) = 0.0075t^2 - 0.2676t + 14.8$ models fuel efficiency of passenger cars (M, measured in miles per gallon) t years after 1940. The vertex of the graph is approximately $(18, 12.4)$, indicating that fuel efficiency was poorest in 1958 $(1940 + 18)$, a dismal 12.4 miles per gallon. Huge cars with hanging dice and big fins dominated the roadways in the late 1950s.

By 1989, the miles-per-gallon average for American cars had increased to 20.5. However, with increasingly efficient automobile engines, Americans collectively logged more highway miles in the 1990s (62% more miles in 1992 than in 1978) and purchased 75% more vehicles. In spite of appeals to save gasoline, the falling cost of driving a mile has encouraged more driving. The plan for increased gas savings due to the increased fuel efficiency predicted by the quadratic model backfired; increased consumption was instead the result.

Alexis Smith "Pair O'Dice" 1990, mixed media collage, 85 1/2 × 67 × 3 in. Photo: Douglas M. Parker Studio, Los Angeles. Courtesy of Margo Leavin Gallery, Los Angeles.

Using technology

The graph of the area function

$A(x) = -0.75x^2 + 250x$

was obtained with a graphing utility using

$[0, 250] \times [0, 22{,}000]$

Xscl = 25, Yscl = 1200

Using the trace or maximum function feature, the graph verifies a maximum area of $20{,}833\frac{1}{3}$ square feet when one dimension is $166\frac{2}{3}$ feet.

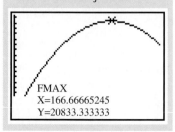

FMAX
X=166.66665245
Y=20833.333333

$$= -\frac{250}{2\left(-\dfrac{3}{4}\right)} \qquad \text{Since } A(x) = -\tfrac{3}{4}x^2 + 250x, a = -\tfrac{3}{4} \text{ and } b = 250.$$

$$= 166\frac{2}{3} \qquad \text{This is the length of the divided structure that maximizes its area.}$$

Step 5. Answer the problem's question. To find the width, we substitute $166\frac{2}{3}$ into the equation for y.

$$y = \frac{1000 - 3x}{4} = \frac{1000 - 3(166\frac{2}{3})}{4} = 125 \qquad y \text{ represents the structure's width.}$$

The dimensions that maximize area are $166\frac{2}{3}$ feet by 125 feet. The maximum area is $(166\frac{2}{3})(125)$, or $20{,}833\frac{1}{3}$ square feet. ∎

Let's conclude with an example that calls for finding a minimum value.

EXAMPLE 8 Minimizing Distance

Find a point on the graph of $y = \sqrt{x - 2} + 1$ that is closest to the point $(4, 1)$.

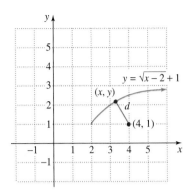

Figure 3.10

Minimizing the distance d from a point on $y = \sqrt{x-2} + 1$ to $(4, 1)$

Solution

Step 1. Decide what must be maximized or minimized.

The situation is illustrated in Figure 3.10. We must find (x, y) on $y = \sqrt{x-2} + 1$ that minimizes the distance d to $(4, 1)$.

Step 2. Express this quantity as a function in one variable.

We'll use the distance formula to write the distance between the point (x, y) on the graph and the point $(4, 1)$. We must find x and y.

$$d = \sqrt{(x-4)^2 + (y-1)^2} \qquad \text{The distance between } (x_1, y_1) \text{ and } (x_2, y_2) \text{ is } \sqrt{(x_2 - x_1)^2 + (y_2 - y_1)^2}.$$

This expresses distance in terms of two variables. To express d in terms of x alone, we use the given equation $y = \sqrt{x-2} + 1$ and substitute $\sqrt{x-2} + 1$ for y in the formula for d.

We obtain

$$d(x) = \sqrt{(x-4)^2 + (\sqrt{x-2} + 1 - 1)^2}$$

$$= \sqrt{(x-4)^2 + x - 2} \qquad (\sqrt{x-2} + 1 - 1)^2 = (\sqrt{x-2})^2 \text{ and } (\sqrt{x-2})^2 = x - 2.$$

$$= \sqrt{x^2 - 7x + 14} \qquad \text{Square } x - 4 \text{ and combine like terms.}$$

Step 3. Write the function in the form $f(x) = ax^2 + bx + c$. At this point, the problem is that $d(x)$ is not a quadratic function. However, we can minimize $d(x)$ finding the x-value that minimizes the algebraic expression under the square root, namely, $x^2 - 7x + 14$. Let's call this expression $f(x)$.

$$f(x) = x^2 - 7x + 14 \qquad \text{Minimize } \sqrt{x^2 - 7x + 14} \text{ by minimizing } x^2 - 7x + 14.$$

Step 4. Since $a > 0$, the function $f(x)$ has a minimum at $x = -\dfrac{b}{2a}$. We are working with $f(x) = x^2 - 7x + 14$, so $a = 1$ and $b = -7$.

$$x = -\frac{b}{2a} = -\frac{(-7)}{2 \cdot 1} = \frac{7}{2}$$

Step 5. Answer the problem's question.

We now have the x-coordinate of the point on $y = \sqrt{x-2} + 1$ closest to $(4, 1)$. We are ready to find the y-coordinate.

$$y = \sqrt{x-2} + 1 \qquad \text{The point is on this function.}$$

$$= \sqrt{\frac{7}{2} - 2} + 1 \qquad \text{Substitute } \tfrac{7}{2} \text{ for } x.$$

$$= \sqrt{\frac{3}{2}} + 1$$

$$= \frac{\sqrt{6}}{2} + \frac{2}{2} \qquad \frac{\sqrt{3}}{\sqrt{2}} = \frac{\sqrt{3}}{\sqrt{2}} \cdot \frac{\sqrt{2}}{\sqrt{2}} = \frac{\sqrt{6}}{2}$$

$$= \frac{2 + \sqrt{6}}{2}$$

The point on the graph of $y = \sqrt{x-2} + 1$ that is closest to $(4, 1)$ is $\left(\dfrac{7}{2}, \dfrac{2 + \sqrt{6}}{2} \right)$. ∎

Discover for yourself

Use a graphing utility to verify the result of Example 8. Is it easier to graph $y = \sqrt{x^2 - 7x + 14}$ or $y = x^2 - 7x + 14$? Which one of these functions will give the minimum distance (a value that we did not compute in the solution process)?

PROBLEM SET 3.1

Practice Problems

In Problems 1–18, use the vertex and intercepts to sketch the graph of each quadratic function by hand. Give the equation for the parabola's axis of symmetry. Then use a graphing utility to verify that your graph is correct.

1. $f(x) = (x - 4)^2 - 1$

2. $f(x) = (x - 1)^2 - 2$

3. $f(x) = (x - 1)^2 - 2$

4. $f(x) = (x - 3)^2 + 2$

5. $y - 1 = (x - 3)^2$

6. $y - 3 = (x - 1)^2$

7. $y = 2(x + 2)^2 - 1$

8. $y = \frac{5}{4} - (x - \frac{1}{2})^2$

9. $f(x) = 4 - (x - 1)^2$

10. $f(x) = 1 - (x - 3)^2$

11. $f(x) = x^2 - 2x - 3$

12. $f(x) = x^2 - 2x - 15$

13. $f(x) = x^2 + 3x - 10$

14. $f(x) = 2x^2 - 7x - 4$

15. $y = 2x - x^2 + 3$

16. $y = 5 - 4x - x^2$

17. $y = 2x - x^2 - 2$

18. $y = 6 - 4x + x^2$

Application Problems

19. The U.S. Center for Disease Control modeled the average annual per capita consumption C of cigarettes by Americans 18 and older as a function of time. The function is $C(t) = -3.1t^2 + 51.4t + 4024.5$, where t represents years after 1960. According to this model, in what year did cigarette consumption per capita reach a maximum?

20. The function $R(x) = -0.0065x^2 + 0.23x + 8.47$ models the American marriage rate R (the number of marriages per 1000 population) x years after 1960. According to this model, in what year was the marriage rate the highest?

Johnson Antonio "Smoking Break" 1986, 28 1/2 × 5 × 6 in. Collection of Chuck and Jan Rosenak. Photo Credit: Lynn Lown.

Solve Problems 21–35, which involve optimization of quadratic functions, without the use of a graphing utility. Then use a graphing utility to verify your result.

21. Among all pairs of numbers whose sum is 16, find a pair whose product is as large as possible. What is the maximum product?

22. Among all pairs of numbers whose sum is 9, find a pair whose product is large as possible. What is the maximum product?

23. Among all rectangles that have a perimeter of 20 yards, find the dimensions of the one with the largest area.

24. Among all rectangles that have a perimeter of 80 feet, find the dimensions of the one with the largest area.

25. A rancher has 1000 feet of fencing to put around a rectangular field and then subdivide the field into three identical smaller rectangular plots by placing two fences parallel to one of the field's shorter sides. Find the dimensions that maximize the enclosed area. What is the maximum area?

26. A rancher has 600 yards of fencing to put around a rectangular field and then subdivide the field into two identical plots by placing a fence parallel to one of the field's shorter sides. Find the dimensions that maximize the enclosed area. What is the maximum area?

27. Find the point on the graph of $y = \sqrt{x}$ that is closest to the point $(1, 0)$.

28. Find the point on the graph of $2x + y = 6$ that is closest to the origin.

29. Find the number that exceeds its square by the greatest amount.

30. Find two numbers whose difference is 50 with the smallest possible product.

31. Find two positive numbers whose sum is 40 and the sum of whose squares is as small as possible.

32. Use the position function $s(t) = -16t^2 + v_0t + s_0$, in which s_0 is the initial height of an object that is thrown directly upward and v_0 is the initial velocity with which the object is thrown to answer this problem. The bridge over the Royal Gorge in Colorado is the highest bridge in the world, and is 1053 feet above the Arkansas River. A rock is thrown vertically upward from the bridge with an initial velocity of 64 feet per second.

    a. Write a function that models the height of the rock above the water as a function of time.

    b. After how many seconds does the rock reach its maximum height? What is the maximum height?

33. A theater that has 300 seats presents revivals of classic American musicals to groups of 100 people or more. The theater charges $6 per person to groups of 100, reducing the price by 20¢ per ticket for each 10 people in excess of 100. What size group will maximize the theater's income? What is the maximum income?

34. A watermelon grower can now ship 6 tons at a profit of $2 per ton. For each week that the grower waits, 3 tons can be added to the shipment, but the profit will be reduced by $\frac{1}{3}$ dollar per ton per week. How many weeks should the grower wait to maximize profit? What is the maximum profit?

35. A rain gutter is made from sheets of aluminum that are 20 inches wide. As shown in the figure at the top of the next column, the edges are turned up to form right angles. Determine the depth of the gutter that will maximize its cross-sectional area and allow the greatest amount of water to flow.

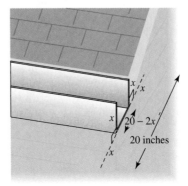

36. The figure shows a Norman window that has the shape of a rectangle with a semicircle attached at the top. The diameter of the circle is equal to the width of the rectangle. If the window has a perimeter of 12 feet, find the dimensions of h and r that allow the maximum amount of light to enter. (*Hint:* Maximize the window's area. You'll need to use the formulas for a circle's circumference and area: Circumference $= 2\pi r$; Area $= \pi r^2$).

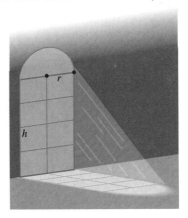

37. An outdoor athletic field is to be constructed in the shape of a rectangle with semicircles at each end. The perimeter of the entire field is to be 440 yards. Find the dimensions of x and r that maximize the area of the rectangular portion of the field.

True–False Critical Thinking Problems

38. Which one of the following is true?

a. No quadratic functions have a range of $(-\infty, \infty)$.

b. The vertex of the parabola described by $f(x) = 2(x - 5)^2 - 1$ is at $(5, 1)$.

c. The graph of $f(x) = -2(x + 4)^2 - 8$ has one y-intercept and two x-intercepts.

d. The maximum value of y for the quadratic function $f(x) = -x^2 + x + 1$ is 1.

39. Which one of the following is true?

a. The minimum value of y in the function $f(x) = 3(x + 2)^2 - 5$ is -2.

b. The graphs of some quadratic functions have two y-intercepts.

c. The function $f(x) = -2(x + \frac{5}{4})^2 + \frac{33}{8}$ is increasing on the interval $(-\infty, -\frac{5}{4})$.

d. Some quadratic functions have a range of $(-\infty, \infty)$.

Technology Problems

In this section we've focused on quadratic functions. In the next section we'll be studying polynomial functions of higher degree. Problems 40–46 will give you the opportunity to explore the ideas that will be discussed in Section 3.2.

40. Graph $y = 2x^2$, $y = x^4$, and $y = 0.5x^6$ using a graphing utility. Which one of the following seems to be true about the graph of $y = a_n x^n$ if n is even and $a_n > 0$?

  a. The graph rises to the left and falls to the right.

  b. The graph rises to the left and right.

  c. The graph falls to the left and rises to the right.

  d. The graph falls to the left and right.

41. Use a graphing utility to graph $y = -2x^2$, $y = -x^4$, and $y = -0.5x^6$. Which statement listed in Problem 40 seems to be true about the graph of $y = a_n x^n$ if n is even and $a_n < 0$?

42. Use a graphing utility to graph $y = 2x^3$, $y = x^5$, and $y = 0.5x^7$. Which statement in Problem 40 seems to be true about the graph of $y = a_n x^n$ if n is odd and $a_n > 0$?

43. Use a graphing utility to graph $y = -2x^3$, $y = -x^5$, and $y = -0.5x^7$. Which statement listed in Problem 40 appears to be true about the graph of $y = a_n x^n$ if n is odd and $a_n < 0$?

44. Graph both $f(x) = 2x^3 - 7x^2 - 8x + 16$ and $g(x) = 2x^3$ using the following range setting

  $\text{Xmin} = -8, \text{Xmax} = 8, \text{Xscl} = 1,$

  $\text{Ymin} = -100, \text{Ymax} = 100, \text{Yscl} = 10.$

Then zoom out using the following range setting:

  $\text{Xmin} = -40, \text{Xmax} = 40, \text{Xscl} = 10,$

  $\text{Ymin} = -10,000, \text{Ymax} = 10,000, \text{Yscl} = 1000.$

Notice that $2x^3$ models the function $f(x) = 2x^3 - 7x^2 - 8x + 16$ when $|x|$ is large.

45. Repeat Problem 44 for each of the following functions, showing that the behavior of f is the same as that of g when $|x|$ is large.

  a. $f(x) = x^4 - 3x^2 + 5x^2 - 6x + 4; g(x) = x^4$

  b. $f(x) = -2x^3 + 7x^2 + 8x - 16; g(x) = -2x^3$

  c. $f(x) = -2x^4 - x^3 + 6x^2 - x - 2; g(x) = -2x^4$

46. Use your observations in Problems 44–45 to select the statement listed in Problem 40 that applies to the graph of

$$f(x) = a_n x^n + a_{n-1} x^{n-1} + \cdots + a_1 x + a_0$$

for each of the following cases.

  a. n is odd; $a_n > 0$

  b. n is odd; $a_n < 0$

  c. n is even; $a_n > 0$

  d. n is even; $a_n < 0$

  e. Experiment with additional polynomial functions and their graphs to reinforce the results obtained in parts (a–d).

47. The number of AIDS cases reported from 1983 through 1990 can be modeled by $f(x) = -143x^3 + 1810x^2 - 187x + 2331$, where x represents the number of years after 1983. Graph the polynomial function using the following range setting:

  $\text{Xmin} = 0, \text{Xmax} = 15, \text{Xscl} = 1,$

  $\text{Ymin} = 0, \text{Ymax} = 50,000, \text{Yscl} = 5000.$

How does the graph indicate that the polynomial model does not describe the spread of AIDS over an extended period of time? Is this true in general for polynomial functions modeling nonnegative phenomena over time?

Writing in Mathematics

48. What explanations can you offer for your answer to Problem 19? Use a graphing utility to graph C. Do you agree with the long-term predictions made by the graph? Explain.

49. Discuss the relationship between the midpoint of the two x-intercepts for $f(x) = ax^2 + bx + c$ and the x-coordinate of the vertex. Assume that $b^2 - 4ac$ is positive. Why is this assumption necessary?

Critical Thinking Problems

In Problems 50–51, find the axis of symmetry for each parabola whose equation is given. Use the axis of symmetry to find a second point on the parabola whose y-coordinate is the same as the given point.

50. $f(x) = 3(x + 2)^2 - 5; (-1, -2)$

51. $f(x) = (x - 3)^2 + 2; (6, 11)$

In Problems 52–53, write an equation of the quadratic function whose graph satisfies the given conditions. Check your result with a graphing utility.

52. Vertex: $(-1, 5)$; point on graph: $(-2, 3)$

53. Vertex: $(3, 1)$; x-intercept: 2

54. If $f(x) = x - 2$ and $g(x) = x^2 - 3x + 1$, find the absolute value of the sum of the minimum values for $f \circ g$ and $g \circ f$.

55. Prove that of all rectangles with a given perimeter P, the square has a maximum area.

56. Prove that of all rectangles with a given perimeter P, the one with the shortest possible diagonal is a square.

57. A right triangle has legs of lengths 3 and 4 feet. What is the area of the largest rectangle that can be inscribed in the right triangle?

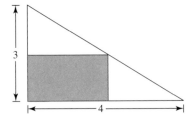

58. For what value of c will the minimum of the function $f(x) = 3x^2 - 2x + c$ be $-\frac{19}{2}$?

59. Find the vertex of the quadratic function $f(x) = x^2 + 2(d + e)x + 2(d^2 + e^2)$. Can the graph of f have two x-intercepts? Explain.

60. Find the equation of the circle that passes through the origin and whose center is at the vertex of the parabola described by $y - 2 = (x - 3)^2$.

61. A quadratic function is described by $f(x) = dx^2 + dx + e$.

    a. Prove that if the vertex of the parabola whose equation is f lies on the x-axis, then $d = 4e$.

    b. Prove the converse of part (a). (The converse of "if p then q" is "if q then p.")

62. If $f(x) = x^2 + kx + 1$, find k that minimizes the distance from the parabola's vertex to the origin.

| Solutions Manual | Tutorial | Video 6 |

SECTION 3.2

Polynomial Functions of Higher Degree and Polynomial Variation

Objectives

1 Determine the end behavior of a polynomial function.

2 Use factoring to find zeros of polynomial functions.

3 Model using polynomial functions.

4 Model using polynomial variation.

1 Determine the end behavior of a polynomial function.

Discover for yourself

You can use your graphing utility to discover all aspects of this discussion on end behavior and the Leading Coefficient Test by working Problems 40–46 in Problem Set 3.1.

Now that we've studied polynomial functions of degree 2, we turn our attention to those of degree 3 or higher. The graphs of these functions exhibit the same kinds of behavior that we found in parabolas.

End Behavior and the Leading Coefficient Test

Polynomial functions of degree 2 or less have graphs that are either parabolas or lines and can be graphed by plotting points. Although polynomial functions of degree 3 or higher also can be graphed by plotting numerous points, the process is rather tedious. Thus, recognizing some of the basic features of the

Bridget Riley "Drift No. 2" 1966, acrylic on canvas, p 1 1/2 × 89 1/2 in. Albright-Knox Art Gallery, Gift of Seymour H. Knox, 1967.

graphs of polynomial functions is important for graphing by hand and in choosing an appropriate viewing rectangle for a graphing utility.

Like quadratic functions, all polynomial functions have graphs that are continuous curves with no breaks (such as those found in the graph of the greatest integer function). Polynomial functions have graphs with smooth, rounded turns (unlike the graph of $f(x) = |x|$, which has a sharp corner at $(0, 0)$).

The behavior of the graph of a function to the far right (as $x \to \infty$; that is, as x goes to infinity) and to the far left (as $x \to -\infty$; that is, as x goes to negative infinity) is called the *end behavior* of a function. All polynomial functions have graphs that eventually rise or fall without bound as $x \to \infty$ or $x \to -\infty$. We can express this symbolically in a number of ways.

The symbolism of end behavior

Symbolic Statement	What It Means	Example
$f(x) \to \infty$ as $x \to \infty$	$f(x)$ increases without bound as x moves to the right without bound.	$f(x) = x^3 - 3x + 2$
$f(x) \to -\infty$ as $x \to -\infty$	$f(x)$ decreases without bound as x moves to the left without bound.	
$f(x) \to \infty$ as $x \to -\infty$	$f(x)$ increases without bound as x moves to the left without bound.	$f(x) = -x^3 + 4x$
$f(x) \to -\infty$ as $x \to \infty$	$f(x)$ decreases without bound as x moves to the right without bound.	

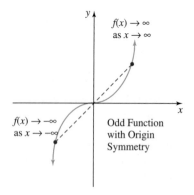

Figure 3.11

If n is odd and $a_n > 0$, the graph of $y = a_n x^n$ falls to the left and rises to the right

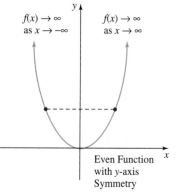

Figure 3.12

If n is even and $a_n > 0$, the graph of $y = a_n x^n$ rises to the left and to the right

There is a relationship between a polynomial function's end behavior and its term of highest degree. The end behavior is actually the same as that of $y = a_n x^n$, the term of highest degree, so let's take a moment to examine the end behavior of this term.

If a_n is positive and n is odd, the function $y = a_n x^n$ is an odd function with origin symmetry. Figure 3.11 shows that its graph falls to the left and rises to the right. On the other hand, if a_n is positive and n is even, the function $y = a_n x^n$ is an even function with y-axis symmetry. Figure 3.12 shows that its graph rises to the left and to the right.

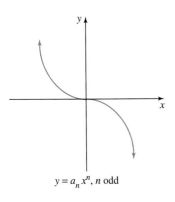

$y = a_n x^n$, n odd

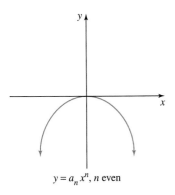

$y = a_n x^n$, n even

Figure 3.13

If a_n is negative, the end behavior shown in Figures 3.11 and 3.12 is reversed.

If a_n is negative, the graphs in Figures 3.11 and 3.12 are reflected in the x-axis. Figure 3.13 illustrates how this reflection reverses the end behavior.

As we noted on the previous page, the end behavior of

$$f(x) = a_n x^n + a_{n-1} x^{n-1} + \cdots + a_1 x + a_0$$

is the same as that of $y = a_n x^n$. In terms of end behavior, only the term of highest degree counts. This observation is summarized by the *Leading Coefficient Test*.

The Leading Coefficient Test

As x increases ($x \to \infty$) or decreases ($x \to -\infty$) without bound, the graph of the polynomial function $f(x) = a_n x^n + a_{n-1} x^{n-1} + \cdots + a_1 x + a_0$ eventually rises ($f(x) \to \infty$) or falls ($f(x) \to -\infty$). In particular:

1. For n odd and $a_n > 0$:

The graph falls to the left and rises to the right.

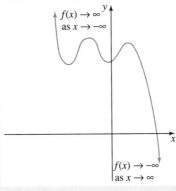

2. For n odd and $a_n < 0$:

The graph rises to the left and falls to the right.

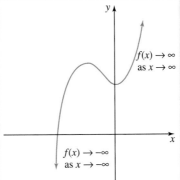

3. For n even and $a_n > 0$:

The graph rises to the left and to the right.

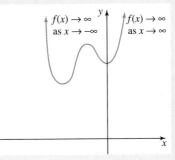

Study tip

The Leading Coefficient Test states that the end behavior of the polynomial f is determined by the term that contains the highest power of the variable. This is the case because when $|x|$ is large, most of the value of $f(x)$ is determined by the term of highest power.

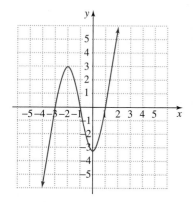

Figure 3.14

The graph of $f(x) = x^3 + 3x^2 - x - 3$, rising to the right and falling to the left

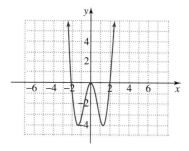

Figure 3.15

The graph of $f(x) = x^4 - 4x^2$, rising to the left and right

2 Use factoring to find zeros of polynomial functions.

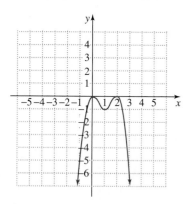

Figure 3.16

The graph of $f(x) = -x^4 + 4x^3 - 4x^2$, falling to the left and right

4. For n even and $a_n < 0$:   The graph falls to the left and to the right.

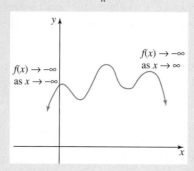

EXAMPLE 1   Using the Leading Coefficient Test

Use the Leading Coefficient Test to determine the end behavior of the graph of:

a. $f(x) = x^3 + 3x^2 - x - 3$ **b.** $f(x) = x^4 - 4x^2$ **c.** $f(x) = -x^4 + 4x^3 - 4x^2$

Solution

a. $f(x) = x^3 + 3x^2 - x - 3$
   Because the degree is odd ($n = 3$) and the leading coefficient is positive, the graph rises to the right and falls to the left, as shown in Figure 3.14.

b. $f(x) = x^4 - 4x^2$
   Because the degree is even ($n = 4$) and the leading coefficient is positive, the graph rises to the left and right, as shown in Figure 3.15.

c. $f(x) = -x^4 + 4x^3 - 4x^2$
   Because the degree is even ($n = 4$) and the leading coefficient is negative, the graph falls to the left and right, as shown in Figure 3.16.   ∎

Zeros of Polynomial Functions

There is an important relationship between the graph of $f(x) = a_n x^n + a_{n-1}x^{n-1} + \cdots + a_1 x + a_0$ and the real solutions to the polynomial equation $a_n x^n + a_{n-1}x^{n-1} + \cdots + a_1 x + a_0 = 0$. Let's explore this relationship for the three polynomial functions in Example 1.

EXAMPLE 2   Solving a Polynomial Equation

Solve: $x^3 + 3x^2 - x - 3 = 0$

Solution

$$x^3 + 3x^2 - x - 3 = 0$$ This is the given polynomial equation.

$$x^2(x + 3) - 1(x + 3) = 0$$ Factor x^2 from the first two terms and -1 from the last two terms.

$$(x + 3)(x^2 - 1) = 0$$ A common factor of $x + 3$ is factored from the expression on the left.

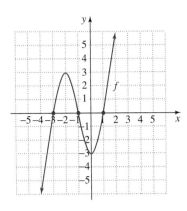

Figure 3.17

The solutions of $x^3 + 3x^2 - x - 3 = 0$ are the x-intercepts for $f(x) = x^3 + 3x^2 - x - 3$.

tudy tip

Like end behavior, zeros of a polynomial function are another important property that determines the shape of the graph.

$$x + 3 = 0 \quad \text{or} \quad x^2 - 1 = 0 \qquad \text{Set each factor equal to 0.}$$

$$x = -3 \qquad\qquad x^2 = 1 \qquad \text{Solve for } x.$$

$$x = \pm 1 \qquad \text{Remember that if } x^2 = d, \text{ then } x = \pm\sqrt{d}.$$

The solution set is $\{-3, -1, 1\}$. Look at Figure 3.17 and observe that the graph crosses the x-axis at -3, -1, and 1 Because $f(-3) = 0$, $f(-1) = 0$, and $f(1) = 0$, we call the solutions, or roots, of the polynomial equation (that is, -3, -1, and 1) the *zeros* of the polynomial function. ∎

Example 2 demonstrates the interrelationship among solving polynomial equations, finding zeros of polynomial functions, and finding x-intercepts. These relationships are summarized as follows.

Real zeros of polynomial functions

If $f(x) = a_n x^n + a_{n-1} x^{n-1} + \cdots + a_1 x + a_0$ and $f(a) = 0$ for any real number a,

1. a is *zero* of the function f.
2. a is a *solution* or *root* of the polynomial equation $f(x) = 0$.
3. a is an *x-intercept* of the graph f.
4. $x - a$ is a *factor* of the polynomial $f(x)$.

EXAMPLE 3 **Finding Zeros of a Fourth-Degree Polynomial Function**

Find all real zeros of: $f(x) = x^4 - 4x^2$

Solution

To find the zeros of the function, we solve the equation $f(x) = 0$.

$$x^4 - 4x^2 = 0 \qquad \text{Set } f(x) \text{ equal to 0.}$$

$$x^2(x^2 - 4) = 0 \qquad \text{Factor.}$$

$$x^2 = 0 \quad \text{or} \quad x^2 - 4 = 0 \qquad \text{Set each factor equal to 0.}$$

$$x = 0 \qquad\qquad x^2 = 4 \qquad \text{Solve for } x.$$

$$x = \pm 2$$

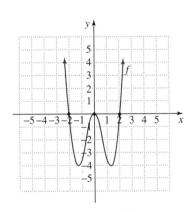

Figure 3.18

The real zeros of $f(x) = x^4 - 4x^2$ are the x-intercepts for the graph of f.

The real zeros of $f(x) = x^2 - 4x^2$ are $-2, 0,$ and 2. The graph of f, shown in Figure 3.18, has x-intercepts at $-2, 0,$ and 2. ∎

EXAMPLE 4 **Finding Zeros of a Polynomial Function**

Find all real zeros of: $f(x) = -x^4 + 4x^3 - 4x^2$

Figure 3.19

The real zeros of $f(x) = -x^4 + 4x^3 - 4x^2$, namely 0 and 2, are the *x*-intercepts for the graph of *f*.

Solution

We find the zeros of *f* by setting $f(x)$ equal to 0.

$-x^4 + 4x^3 - 4x^2 = 0$	We now have a polynomial equation.
$x^4 - 4x^3 + 4x^2 = 0$	Multiply both sides by −1. This step is optional.
$x^2(x^2 - 4x + 4) = 0$	Factor out x^2.
$x^2(x - 2)^2 = 0$	Factor completely.
$x^2 = 0 \quad \text{or} \quad (x - 2)^2 = 0$	Set each factor equal to 0.
$x = 0 \qquad\qquad x = 2$	Solve for *x*.

The real zeros of $f(x) = -x^4 + 4x^3 - 4x^2$ are 0 and 2. The graph of *f*, shown in Figure 3.19 has *x*-intercepts at 0 and 2. ∎

Multiple Roots and x-Intercepts

In Example 3, the real zero arising from $x^2 = 0$ is a repeated zero. In Example 4, both real zeros arising from $x^2 = 0$ and $(x - 2)^2 = 0$ are repeated zeros. The graph in Example 4 touched, but did not cross, the *x*-axis at 0 and 2. In general, a factor of $(x - a)^k$ yields the repeated zero *a* of *multiplicity k*.

Multiplicity and x-intercepts

1. If the factor $x - a$ occurs *k* times ($k \geqslant 2$) in the complete factorization of the polynomial $f(x)$, then *a* is called a repeated zero with multiplicity *k*.
2. If *k* is even, then the graph touches but does not cross the *x*-axis at $x = a$.
3. If *k* is odd, the graph crosses the *x*-axis at $x = a$.
4. Regardless of whether *k* is even or odd, if *a* is a repeated zero, the graph flattens out at $(a, 0)$.

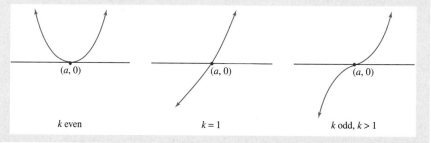

| *k* even | *k* = 1 | *k* odd, *k* > 1 |

Discover for yourself

There is a relationship between the degree of a polynomial and the number of real zeros. Another relationship exists between the polynomial's degree and the number of turning points on its graph. Study the results of Examples 2–4, summarized below, and see if you can state these relationships for an *n*th-degree polynomial.

Polynomial Function	Degree	Real Zeros	Number of Real Zeros	Number of Turning Points on the Graph
$f(x) = x^3 + 3x^2 - x - 3$	3	$-3, -1, 1$	Three	Two
$f(x) = x^4 - 4x^2$	4	$-2, 0, 2$	Three	Three
$f(x) = -x^4 + 4x^3 - 4x^2$	4	$0, 2$	Two	Three

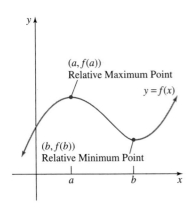

In the Discover for yourself, did you observe the following relationships?

Degrees, real zeros, and turning points

1. An nth-degree polynomial function can have *at most n* real zeros.
2. An nth-degree polynomial function has a graph with *at most $n - 1$* turning points. Equivalently, the graph has at most $n - 1$ relative extrema (relative minimums or maximums).

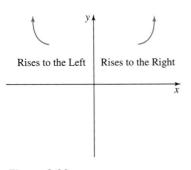

Rises to the Left   Rises to the Right

Figure 3.20

End behavior of the graph of $f(x) = 2x^4 - 4x^3$

EXAMPLE 5 **Sketching the Graph of a Polynomial Function**

Sketch the graph of $f(x) = 2x^4 - 4x^3$ without the use of a graphing utility.

Solution

The end behavior of f is the same as that of $y = 2x^4$. Because the leading coefficient is positive and the degree is even, the graph rises to the left and right, as shown in Figure 3.20.

Let's find the zeros of f, thereby determining the graph's x-intercepts.

$$2x^4 - 4x^3 = 0 \qquad \text{Set } f(x) \text{ equal to 0.}$$

$$2x^3(x - 2) = 0 \qquad \text{Factor.}$$

$$2x^3 = 0 \quad \text{or} \quad x - 2 = 0 \qquad \text{Set each factor equal to 0.}$$

$$x = 0 \qquad\qquad x = 2 \qquad \text{Solve for } x.$$

Since 0 and 2 are of odd multiplicity, the graph of f crosses the x-axis at both of these intercepts. At $x = 0$, because the multiplicity is 3, the graph will flatten out.

We'll plot a few additional points as shown in the accompanying table.

x	-1	0.5	1	1.5
$f(x) = 2x^4 - 4x^3$	6	-0.375	-2	-3.375

The graph of $f(x) = 2x^4 - 4x^3$ appears in Figure 3.21. ∎

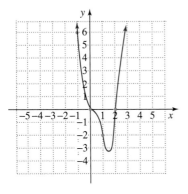

Figure 3.21

The graph of $f(x) = 2x^4 - 4x^3$

A Strategy for Graphing Polynomial Functions

Here's a general strategy for graphing a polynomial function. A graphing utility is a valuable complement to this strategy. Some of the steps listed below will help you to select a viewing rectangle that shows the important parts of the graph.

Graphing a polynomial function

$$f(x) = a_n x^n + a_{n-1} x^{n-1} + a_{n-2} x^{n-2} + \cdots + a_1 x + a_0$$

1. Use the Leading Coefficient Test to determine the graph's end behavior.

2. Find x-intercepts by setting $f(x) = 0$ and solving the resulting polynomial equation. If there is an x-intercept at a as a result of $(x - a)^k$ in the complete factorization of $f(x)$, then:
  a. If k is even, the graph touches the x-axis at a, but does not cross the x-axis.
  b. If k is odd, the graph crosses the x-axis at a.
  c. If $k > 1$, the graph flattens out at $(a, 0)$.
3. Find the y-intercept by setting x equal to 0 and computing $f(0)$.
4. Use symmetry, if applicable, to help draw the graph:
  a. y-axis symmetry: $f(-x) = f(x)$
  b. Origin symmetry: $f(-x) = -f(x)$
5. Use the fact that the maximum number of turning points of the graph is $n - 1$ to check whether it is drawn correctly.

3 Model using polynomial functions.

Modeling with Polynomial Functions

Polynomial models are often useful for describing nonnegative real world phenomena over short periods of time. However, based on end behavior, accurate modeling for extended time periods is a problem. Let's see why.

EXAMPLE 6   AIDS Cases

The number of AIDS cases for the years 1983 through 1991 is approximated by the polynomial function

$$f(x) = -143x^3 + 1810x^2 - 187x + 2331$$

where x represents years after 1983. Use end behavior to explain why the model is only valid for a limited time period.

Solution

Because the leading coefficient (-143) is negative and the degree (3) is odd, the graph falls to the right. This indicates at some point the number of AIDS cases will be negative, an impossibility. No function with a graph that decreases without bound as x (time) increases can model nonnegative real world phenomena over a long period.

Figure 3.22 shows the graph of f from 1983 through 1991 using the range setting:

Xmin $= 0$, Xmax $= 8$, Xscl $= 1$,

Ymin $= 0$, Ymax $= 50,000$, and Yscl $= 5000$.

Look what happens to the graph when we change the range setting to

Xmin $= 0$, Xmax $= 15$, Xscl $= 1$,

Ymin $= -5000$, Ymax $= 50,000$, and Yscl $= 5000$.

(Figure 3.23). By year 13 (1996), the values of y are negative and the function no longer models AIDS cases. ∎

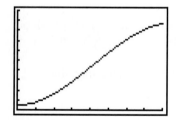

Figure 3.22

The graph of a function modeling the number of AIDS cases from 1983 through 1991

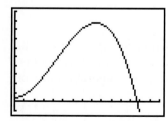

Figure 3.23

By extending the range setting, y is negative and the function no longer models the number of AIDS cases.

ENRICHMENT ESSAY

The Hot Zone

As the tropical forests of the world are destroyed, viruses that have lived undetected in the wilderness are entering human populations. The appearance of AIDS is part of a pattern that, according to author Richard Preston (*The Hot Zone*) involves viruses of extraordinary virulence capable of being spread by airborne mucus or casual contact, bringing grotesque, agonizing, and near-certain death within weeks after infection. The ongoing tragedy of AIDS and the outbreak of the Ebola virus in Zaire in 1995 confirm the reality of modern-day plagues reminiscent of the black death of times gone by.

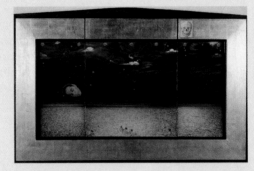

Juan Gonzalez "Mar de Lagrimas" 1987–88, w/c and gouache on paper, 27 1/2 × 43 1/4 in. Courtesy Nancy Hoffman Gallery.

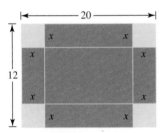

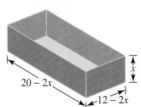

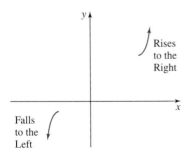

Figure 3.24

Polynomial models are often useful for describing real-world phenomena over short periods of time. They also model many geometric situations, as illustrated in Example 7.

EXAMPLE 7 A Geometric Application

A rectangular piece of cardboard measures 12 inches by 20 inches. A box with an open top is formed by cutting identical squares that measure x by x from each corner and then folding up the sides, as shown in Figure 3.24.

a. Express the volume V of the box as a polynomial function of x.
b. Graph the function.

Solution

a. The lengths of the sides of the open-top box are x, $12 - 2x$, and $20 - 2x$ (see Figure 3.24). The volume of the box is the product of its length, width, and height, and is given by

$$V(x) = x(12 - 2x)(20 - 2x).$$

By multiplying on the right, we can express the volume as

$$V(x) = 4x^3 - 64x^2 + 240x.$$

b. Now we are ready to graph the function that models the volume of the box.
    1. Determine end behavior. Since n is odd and the leading coefficient (4) is positive, the graph falls to the left and rises to the right (see Figure 3.25).
    2. Find x-intercepts by setting $V(x) = 0$.

$$4x^3 - 64x^2 + 240x = 0$$

$$x(12 - 2x)(20 - 2x) = 0 \qquad \text{Use the factored form of the function.}$$

Rises to the Right

Falls to the Left

Figure 3.25

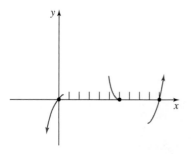

Figure 3.26

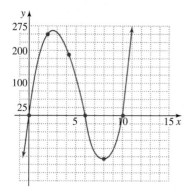

Figure 3.27

The graph of $V(x) = 4x^3 - 64x^2 + 240x$, with uncertainty as to the coordinates of the turning points

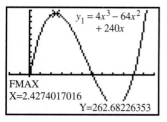

Figure 3.28

Cutting 2.4-inch squares gives a maximum volume of 263 cubic inches.

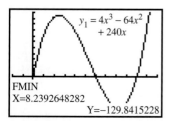

Figure 3.29

Coordinates of the second turning point

$$x = 0 \quad \text{or} \quad 12 - 2x = 0 \quad \text{or} \quad 20 - 2x = 0 \qquad \text{Set each factor equal to 0.}$$

$$x = 0 \qquad\qquad x = 6 \qquad\qquad x = 10 \quad \text{Solve for } x.$$

The multiplicity of each factor is 1, so the graph crosses the x-axis (and does not flatten out) at 0, 6, and 10 (see Figure 3.26).

3. Since the graph passes through $(0, 0)$, we have also found the y-intercept.

4. It seems unlikely that the graph has either y-axis symmetry or origin symmetry. We confirm this by finding $V(-x)$:

$$V(-x) = 4(-x)^3 - 64(-x)^2 + 240(-x) \qquad \text{Replace } x \text{ with } -x \text{ in the formula for } V(x).$$

$$= -4x^3 - 64x^2 - 240x$$

Because $V(-x) \neq V(x)$ and $V(-x) \neq -V(x)$, the graph has neither y-axis nor origin symmetry.

5. Plot a few additional points as shown in the table.

x	2	4	8
$V(x) = 4x^3 - 64x^2 + 240x$	256	192	-128

Figure 3.27 shows the graph of $V(x) = 4x^3 - 64x^2 + 240x$. The maximum number of turning points is $n - 1$ or $3 - 1 = 2$. Since we have 2 turning points, this does not violate the maximum number possible. Although we do not know the precise coordinates of these turning points, we can use point plotting to further refine the graph. ∎

Discover for yourself

Since the original piece of cardboard measured 12 inches by 20 inches, what is the largest possible square that can be cut from each corner? How much "volume" would be attained by cutting a square this large? Trace over the part of the graph in Figure 3.27 that is significant in terms of serving as a realistic model for the volume for the open-top box. For what values of x does the function model the problem?

Using technology

We can use a graphing utility to refine the hand-drawn graph in Figure 3.27 and determine the coordinates of the two turning points. Figure 3.28 was obtained using the FMAX feature. We can see that one turning point for the function occurs at approximately (2.4, 263). This indicates that a box of maximum volume (263 cubic inches) can be obtained by cutting squares that measure 2.4 inches from each corner. Figure 3.29 was obtained using the FMIN feature. Here we see that the other turning point for the function occurs at approximately (8.2, −130).

4 Model using polynomial variation.

Modeling Using Polynomial Variation

In Section 2.2, we modeled the statement "*y* varies directly as *x*" with the linear equation $y = mx$. Variation can also be modeled by higher-order polynomial equations.

Direct variation with powers

y varies directly as the nth power of x (or *y is proportional to the nth power of x*) if there exists some nonzero constant *k* such that

$$y = kx^n.$$

The number *k* is called the *constant of variation* or the *constant of proportionality*.

EXAMPLE 8   **Solving a Direct Variation Problem**

The distance (*d*) that a body falls from rest varies directly as the square of the time (*t*) of the fall. If a body falls 64 feet in 2 seconds, how far will it fall in 4.5 seconds?

Solution

H. C. Westermann "High Swan Dive: The Sea of Cortez" 1973. Courtesy of Allan Franklin, NY. © Estate of H. C. Westermann / Licensed by VAGA, New York, 1998.

$d = kt^2$	Translate "Distance (*d*) varies directly as the square of time (*t*)."
$64 = k \cdot 2^2$	Find *k*. Because a body falls 64 feet in 2 seconds, when $t = 2$, $d = 64$.
$64 = 4k$	
$16 = k$	Solve for *k*, dividing both sides by 4.
$d = 16t^2$	Substitute the value for *k* into the original equation.
$\quad = 16(4.5)^2$	Find *d* when $t = 4.5$.
$\quad = 16(20.25)$	
$\quad = 324$	

Thus, in 4.5 seconds, a body will fall 324 feet.   ■

It is easy to determine whether a data set can be modeled by polynomial variation.

$y = kx^n$	This is the polynomial model.
$\dfrac{y}{x^n} = k$	Solve for *k*.

This last form of the variation model tells us that if a data set is to be modeled by polynomial variation, the ratio $\dfrac{y}{x^n}$ must remain constant.

Let's apply this idea to the model in Example 8. Expressing the model as

$$y = 16x^2$$

we obtain the data points for the function shown in the margin. The graph of $f(x) = 16x^2$ appears below the data points on page 354 (Figure 3.30).

Given only the data points (the ordered pairs in the third column of the table of coordinates), the ratio $\dfrac{y}{x^2}$ remains constant.

x	$y = 16x^2$	Ordered Pair
0	$16 \cdot 0^2 = 0$	$(0, 0)$
1	$16 \cdot 1^2 = 16$	$(1, 16)$
2	$16 \cdot 2^2 = 64$	$(2, 64)$
3	$16 \cdot 3^2 = 144$	$(3, 144)$
4	$16 \cdot 4^2 = 256$	$(4, 256)$
5	$16 \cdot 5^2 = 400$	$(5, 400)$

$(1, 16)$: $\quad \dfrac{y}{x^2} = \dfrac{16}{1^2} = 16$

$(2, 64)$: $\quad \dfrac{y}{x^2} = \dfrac{64}{2^2} = 16$

$(3, 144)$: $\quad \dfrac{y}{x^2} = \dfrac{144}{3^2} = 16$

$(4, 256)$: $\quad \dfrac{y}{x^2} = \dfrac{256}{4^2} = 16$

$(5, 400)$: $\quad \dfrac{y}{x^2} = \dfrac{400}{5^2} = 16$

Discover for yourself

Why isn't the ordered pair $(0, 0)$ considered in these ratios?

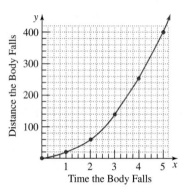

Figure 3.30

The graph of $f(x) = 16x^2$

Determining whether data can be modeled by polynomial variation

A data set with ordered pairs in the form (x, y) can be modeled by polynomial variation if the ratio

$$\frac{y}{x^n}$$

remains constant for some positive integer n.

PROBLEM SET 3.2

Practice Problems

In Problems 1–6, use end behavior to match each polynomial function with its graph (The graphs are labeled (a) through (f).)

1. $f(x) = x^3 - 4x^2$ **2.** $f(x) = -x^4 + x^2$ **3.** $f(x) = -x^3 - x^2 + 5x - 3$ **4.** $f(x) = (x - 3)^2$

5. $f(x) = (x + 1)^2(x - 1)^2$ **6.** $f(x) = x - 3$

a.

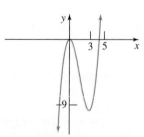

b.

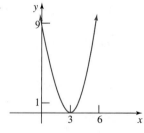

c.

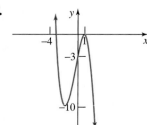

d.

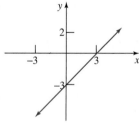

e.

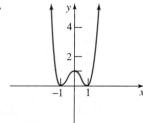

f.

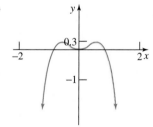

In Problems 7–22:

a. *Use the Leading Coefficient Test to determine the right-hand* $(x \to \infty)$ *and left-hand* $(x \to -\infty)$ *end behavior of the polynomial function.*

b. *Find all real zeros of the polynomial function.*

c. *Graph the function without the use of a graphing utility.*

d. *Verify your hand-drawn graph with a graphing utility.*

7. $f(x) = x^3 + x^2 - 4x - 4$

8. $f(x) = x^3 + 2x^2 - x - 2$

9. $f(x) = x^4 - x^2$

10. $f(x) = x^4 - 9x^2$

11. $f(x) = -x^4 + 4x^2$

12. $f(x) = -x^4 + 16x^2$

13. $f(x) = x^4 - 6x^3 + 9x^2$

14. $f(x) = x^4 - 2x^3 + x^2$

15. $f(x) = -2x^4 + 2x^3$

16. $f(x) = -2x^4 + 4x^3$

17. $f(x) = 6x - x^3 - x^5$

18. $f(x) = 6x^3 - 9x - x^5$

19. $f(x) = \frac{1}{2} - \frac{1}{2}x^4$

20. $f(x) = 3x^2 - x^3$

21. $f(x) = -2(x - 4)^2(x^2 - 25)$

22. $f(x) = -3(x - 1)^2(x^2 - 4)$

Application Problems _____

23. The table below shows the accumulated total number of AIDS cases in the United States from 1983 through 1995.

Year	1983	1984	1985	1986	1987	1988	1989	1990	1991	1992	1993	1994	1995
Number of AIDS Cases	4589	10,750	22,399	41,256	69,592	104,644	146,574	193,878	251,638	326,648	399,613	457,280	496,896

Source: Center for Disease Control

This data can be modeled by the cubic polynomial function

$$f(x) = -68.8x^3 + 4781.7x^2 - 2666.2x + 7094.5$$

where $f(x)$ is the number of AIDS cases and x is the number of years after 1983. The graph of this function is shown in the figure superimposed over actual data points.

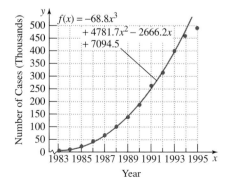

a. How well does the graph indicate that the cubic polynomial models the data?

b. Select a year where the polynomial gives nearly the exact number of AIDS cases. Verify this observation by evaluating the polynomial function for this year and comparing your answer to the number of AIDS cases in the table.

c. Select a year where the polynomial either underestimates or overestimates the actual number of AIDS cases. Verify this observation by evaluating the polynomial function and comparing your answer to the number of AIDS cases given in the table.

d. Why will this cubic polynomial function not model the number of AIDS cases over an extended period of time?

24. A herd of 100 elk is introduced to a small island. The number of elk $N(t)$ after t years is described by the polynomial function $N(t) = -t^4 + 21t^2 + 100$.

a. Use the Leading Coefficient Test to determine the graph's end behavior to the right. What does this mean about what will eventually happen to the elk population?

b. Graph the function.

c. Graph only the portion of the function that serves as a realistic model for the elk population over time. When does the population die off?

25. A rectangular piece of cardboard measures 12 inches by 18 inches. A box with an open top is formed by cutting identical squares that measure x by x from each corner and then folding up the sides.

a. Express the volume (V) of the box as a polynomial function of x.

b. Graph the function.

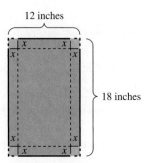

12 inches

18 inches

26. The distance required to stop a car varies directly as the square of its speed. If 200 feet are required to stop a car traveling 60 miles per hour, how many feet are required to stop a car traveling 100 miles per hour.

27. The distance that a body falls varies directly as the square of time in which it falls. A body falls 64 feet in 2 seconds. Predict the distance the same body will fall in 10 seconds.

In Problems 28–29, show that each data set can be modeled by polynomial variation for the given value of n. Then write a polynomial function that models the data.

28.

x	$y\,(n = 3)$
2	$\dfrac{32\pi}{3}$
3	36π
5	$\dfrac{500\pi}{3}$
9	972π

(What familiar formula do you obtain? What do x and y represent?)

29.

x	$y\,(n = 5)$
1	0.1
2	3.2
3	24.3
5	312.5

True–False Critical Thinking Problems

30. Which one of the following is true?

a. If $f(x) = -x^3 + 4x$, then the graph of f falls to the left and to the right.

b. If $f(x) = x^4 - 5x^2 + 3$, then $f(x) \to \infty$ as $x \to -\infty$.

c. A mathematical model that is a polynomial of degree n whose leading term is $a_n x^n$, n odd and $a_n < 0$, is ideally suited to describe nonnegative phenomena over long periods of time.

d. There is only one fourth-degree polynomial with four known x-intercepts.

31. Which one of the following is true?

a. If $f(x) = x^5 - 3x$, the graph of f falls to the left and to the right.

b. If $f(x) = -3x^4 + 3x^2$, then $f(x) \to -\infty$ as $x \to \infty$.

c. There is only one third-degree polynomial with three known x-intercepts.

d. The graph of a function with origin symmetry can rise to the left and to the right.

Technology Problems _____

Write a polynomial function that imitates the end behavior of each graph in Problems 32–35. Then use your graphing utility to graph the polynomial function and verify your result.

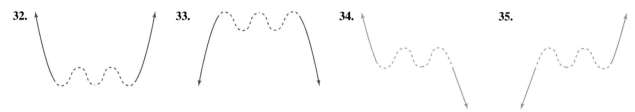

32. **33.** **34.** **35.**

For Problems 36–39, describe the end behavior of the graph of each polynomial function using the Leading Coefficient Test. After describing what happens as $x \to \infty$ and $x \to -\infty$, use a graphing utility to obtain a complete graph of the function to verify your description.

36. $f(x) = -2x^3 + 6x^2 + 3x - 1$

37. $g(x) = -\frac{1}{2}(3x^4 - 2x + 5)$

38. $h(x) = 10^4 + x^3$

39. $h(t) = -3t^2 + 4t^4$

For Problems 40–43:

a. *Use a graphing utility to graph each function in a standard viewing rectangle.*

b. *Based on what you know about the end behavior of the graph, explain why the graph is not complete.*

c. *Modify the range setting to obtain a complete graph of the function showing its end behavior.*

40. $f(x) = x^3 + 13x^2 + 10x - 4$

41. $f(x) = -x^4 + 8x^3 + 4x^2 + 2$

42. $f(x) = (x + 2)(x + 1)(x - 1)(x - 3)(x - 5)$

43. $f(x) = -x^5 + 5x^4 - 6x^3 + 2x + 20$

For Problems 44–47, use a graphing utility to graph f and g in the same viewing rectangle. Then use the $\boxed{\text{ZOOM OUT}}$ *feature to show that the end behavior of f and g (as $x \to \infty$ and $x \to -\infty$) is identical.*

44. $f(x) = x^3 - 6x + 1, g(x) = x^3$

45. $f(x) = 2x^4 - 3x^2, g(x) = 2x^4$

46. $f(x) = -\frac{1}{2}x^3 + x - 4, g(x) = -\frac{1}{2}x^3$

47. $f(x) = -x^4 + 2x^3 - 6x, g(x) = -x^4$

48. Is $|a - b - c| = |a| - |b| - |c|$? Answer the question by graphing $f(x) = |x^3 - 4x^2 - x|$ and $g(x) = |x^3| - |4x^2| - |x|$ in the same viewing rectangle.

49. In South America and Africa, trees are being cut at the rate of 30 acres a minute, day and night. An area of forest nearly twice as large as the state of New York is destroyed every year. This alarming rate of destruction of tropical rainforests is a major factor in the overall increase of atmospheric CO_2 concentration. A short-term polynomial model for the years 1980 through 1982 is $f(x) = 36x^4 - 142x^3 + 175x^2 - 67x + 340$, where x denotes the year ($x = 0$ represents April 1980) and $f(x)$ approximates CO_2 concentration (in ppm). Use your graphing utility to graph the function. What do you observe about the CO_2 concentration during the course of each year?

Fanny Brennan "Forest" 1987, 2 1/8 × 3 in. Courtesy of Salander–O'Reilly Galleries.

Writing in Mathematics

50. Explain how you can determine whether the graph shown below represents $f(x) = (x - 2)^2, f(x) = (x - 2)^3,$ or $f(x) = (x - 2)^4.$

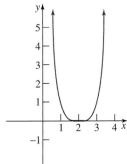

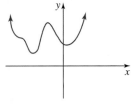

52. What is the smallest possible degree of the polynomial function in the graph? How do you know?

51. Explain why the graph at the top of the next column cannot represent a fifth-degree polynomial function.

Group Activity Problems

Explore the capability of a graphing utility to fit a polynomial function to a data set. This is usually found in the statistics menu. Look for the polynomial regression feature. There are characteristics of a data set that indicate it can be modeled by a polynomial function. For example, take the function $f(x) = x^3$ and consider first-, second-, and third-order differences:

x	$f(x) = x^3$	First-order Differences	Second-order Differences	Third-order Differences
-3	-27			
		$-8 - (-27)$ 19		
-2	-8		$7 - 19$ -12	
		$-1 - (-8)$ 7		$-6 - (-12)$ 6
-1	-1		$1 - 7$ -6	
		$0 - (-1)$ 1		$0 - (-6)$ 6
0	0		$1 - 1$ 0	
		$1 - 0$ 1		$6 - 0$ 6
1	1		$7 - 1$ 6	
		$8 - 1$ 7		
2	8			

Observe that for equally spaced x-values the third-order differences are constant. This observation can be generalized as follows:

> *If the nth-order differences of y-values of a data set are constant for equally spaced x-values, the data can be modeled by a polynomial function of degree n.*

53. Once members of the group have learned to use technology to model data, work with the data given on the next page related to the common cold. The common cold is caused by a rhinovirus. The virus enters our bodies, multiplies, and although there is no cure, the virus begins to die at a certain point. After t days there are N billion viral particles in our bodies, as shown in the table.

t (Days)	N (Viral Particles, in billions)
0	5
1	7.25
2	17
3	22.25
4	5

a. Model the data with a fourth-degree polynomial that describes the number of viral particles as a function of time.

b. Use a graphing utility to draw a complete graph of the function. When (to the nearest hour) does the model indicate that we feel sickest, that is, when is the number of particles at a maximum?

54. Consult the Internet or the research department of your library to find data of interest to the group. Look for data that can be modeled by a polynomial function— that is, data with relatively constant nth-order differences. Use the computer or a graphing calculator to fit a polynomial function to the data, and then describe how well the function predicts the actual data values.

SECTION 3.3 Zeros of Polynomial Functions

Solutions Manual   Tutorial   Video 6

Objectives

1 Use long division to divide polynomials.
2 Use synthetic division to divide polynomials.
3 Evaluate a polynomial using the Remainder Theorem.
4 Use the Factor Theorem to solve a polynomial equation.
5 Use the Rational Zero Theorem to find possible rational zeros.
6 Use technology to eliminate possible rational zeros.
7 Use Descartes' Rule of Signs.
8 Find bounds for the roots of a polynomial equation.
9 Use a strategy to solve polynomial equations.

We have seen that the zeros of the polynomial function

$$f(x) = a_n x^n + a_{n-1} x^{n-1} + \cdots + a_1 x + a_0$$

are the solutions of the polynomial equation

$$a_n x^n + a_{n-1} x^{n-1} + \cdots + a_1 x + a_0 = 0.$$

Furthermore, if a is a real zero, then a is an x-intercept of the graph of f. In this section, we focus in greater detail on these ideas.

The Division Algorithm: Long Division and Synthetic Division

Consider the polynomial function

$$f(x) = 15x^3 + 14x^2 - 3x - 2.$$

Suppose we know that a zero of f occurs at -1. This means that $x + 1$ is a factor of $f(x)$ and that $f(x)$ can be expressed as

$$f(x) = (x + 1)q(x)$$

where $q(x)$ is a second-degree polynomial. We find $q(x)$ using long division of polynomials.

Use long division to divide polynomials.

EXAMPLE I   Reviewing Long Division of Polynomials

Divide $f(x) = 15x^3 + 14x^2 - 3x - 2$ by $x + 1$. Use the result to find all real zeros of f.

Solution

$$
\begin{array}{r}
15x^2 \\
x + 1 \overline{) 15x^3 + 14x^2 - 3x - 2} \\
15x^3 + 15x^2 \\
\hline
-x^2 - 3x
\end{array}
$$

Divide: $\dfrac{15x^3}{x} = 15x^2$

Multiply: $15x^2(x + 1) = 15x^3 + 15x^2$

Subtract $15x^3 + 15x^2$ from $15x^3 + 14x^2$ and bring down $-3x$.

Although the process of division, multiplication, subtraction, and bringing down terms is continued without interruption, we'll rewrite what we have done so far to make the algorithm easier to follow.

$$
\begin{array}{r}
15x^2 - x \\
x + 1 \overline{) 15x^3 + 14x^2 - 3x - 2} \\
15x^3 + 15x^2 \\
\hline
-x^2 - 3x \\
-x^2 - x \\
\hline
-2x - 2
\end{array}
$$

Divide: $-\dfrac{x^2}{x} = -x$; the first term of $-x^2 - 3x$ is divided by the first term of the divisor $x + 1$.

Multiply: $-x(x + 1) = -x^2 - x$

Subtract $-x^2 - x$ from $-x^2 - 3x$ and bring down -2.

Again we will rewrite what we have done so far to make the process easier to follow.

$$
\begin{array}{r}
15x^2 - x - 2 \\
x + 1 \overline{) 15x^3 + 14x^2 - 3x - 2} \\
15x^3 + 15x^2 \\
\hline
-x^2 - 3x \\
-x^2 - x \\
\hline
-2x - 2 \\
-2x - 2 \\
\hline
0
\end{array}
$$

Divide: $\dfrac{-2x}{x} = -2$; the first term of $-2x - 2$ is divided by the first term of the divisor $x + 1$.

Multiply: $-2(x + 1) = -2x - 2$

Subtract $-2x - 2$ from $-2x - 2$, obtaining a remainder of 0.

The completed division shows that

$$f(x) = 15x^3 + 14x^2 - 3x - 2 = (x + 1)(15x^2 - x - 2).$$

We can use this result to find all zeros of f.

$$15x^3 + 14x^2 - 3x - 2 = 0$$ — Set $f(x)$ equal to 0.

$$(x + 1)(15x^2 - x - 2) = 0$$ — Factor using the result from the division.

$$(x + 1)(5x - 2)(3x + 1) = 0$$ — Factor the trinomial.

$$3x + 1 = 0 \quad x + 1 = 0 \quad \text{or} \quad 5x - 2 = 0 \quad \text{or}$$ — Set each factor equal to 0.

$$x = -\frac{1}{3} \quad x = -1 \qquad\qquad x = \frac{2}{5}$$ — Solve for x.

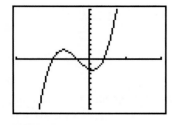

Figure 3.31

The graph of $f(x) = 15x^3 + 14x^2 - 3x - 2$ with zeros of -1, $-\frac{1}{3}$, and $\frac{2}{5}$.

The real zeros of $f(x) = 15x^3 + 14x^2 - 3x - 2$ are -1, $-\frac{1}{3}$, and $\frac{2}{5}$. This can be seen from the graph of f (Figure 3.31), where the x-intercepts occur at -1, $-\frac{1}{3}$, and $\frac{2}{5}$. ∎

2 Use synthetic division to divide polynomials.

There is a faster method, called *synthetic division,* for long division by polynomials of the form $x - c$. Compare the two methods showing $15x^3 + 14x^2 - 3x - 2$ divided by $x + 1$.

Long Division of Polynomials

$$
\begin{array}{r}
15x^2 - \ \ x - 2 \quad \leftarrow \text{Quotient} \\
x + 1 \ \overline{\smash{\big)}\ 15x^3 + 14x^2 - 3x - 2} \quad \leftarrow \text{Dividend} \\
\underline{15x^3 + 15x^2} \quad\quad\quad\quad\quad \\
-\ \ x^2 - 3x \quad\quad\quad \\
\underline{-\ \ x^2 - \ \ x} \quad\quad\quad \\
-\ 2x - 2 \\
\underline{-\ 2x - 2} \\
0 \quad \leftarrow \text{Remainder}
\end{array}
$$

Divisor $\rightarrow$

Synthetic Division

$$
\begin{array}{r|rrrr}
-1 & 15 & 14 & -3 & -2 \\
 & \downarrow & -15 & 1 & 2 \\
\hline
 & 15 & -1 & -2 & 0
\end{array}
$$

Discover for yourself

Study the two methods and answer these questions.

1. How is the dividend represented in the synthetic division process?
2. How is the divisor represented in the synthetic division process? If the divisor is $x + c$, what number would appear to the left of the coefficients of the dividend?
3. How is the quotient represented in the synthetic division process?
4. Where is the remainder in the synthetic division process?
5. The red arrow indicates that the 15 is brought down to the last row. Describe how you might obtain the other three numbers in this row (-1, -2, and 0) using a series of multiplications and additions.

Synthetic division works only for divisors of the form $x - c$. The process of synthetic division is a series of multiplications and additions, shown in the following pattern. How many questions in the Discover for yourself were you able to answer?

Details Behind the Synthetic Division

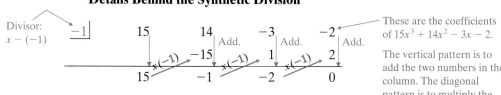

Divisor:
$x - (-1)$

These are the coefficients of $15x^3 + 14x^2 - 3x - 2$.

The vertical pattern is to add the two numbers in the column. The diagonal pattern is to multiply the resulting sum by -1.

Using long division, we found that

$$
\frac{15x^3 + 14x^2 - 3x - 2}{x + 1} = 15x^2 - x - 2.
$$

The first three numbers in the bottom row of the synthetic division, 15, -1, and -2, are the coefficients of the quotient $15x^2 - x - 2$. The last number, 0, is the remainder.

EXAMPLE 2   **Using Synthetic Division**

Use synthetic division to divide: $x^4 - 6x^3 + 2$ by $x + 4$

Solution

Synthetic Division
To divide a polynomial by $x - c$:

1. Arrange polynomials in descending powers, with a 0 coefficient for any missing term.

$$x - (-4) \, \overline{)\, x^4 - 6x^3 + 0x^2 + 0x + 2}$$

2. Write c for the divisor, $x - c$. To the right, write the coefficients of the dividend.

$$-4 \, \rvert \quad 1 \;\; -6 \;\; 0 \;\; 0 \;\; 2$$

3. Write the leading coefficient of the dividend on the bottom row.

$$
\begin{array}{r|rrrrr}
-4 & 1 & -6 & 0 & 0 & 2 \\
& \downarrow & & & & \\
\hline
& 1 & & & &
\end{array}
$$

4. Multiply c (in this case, -4) times the value just written on the bottom row. Write the product in the next column in the second row.

$$
\begin{array}{r|rrrrr}
-4 & 1 & -6 & 0 & 0 & 2 \\
\text{multiply} & & -4 & & & \\
& 1 & & & &
\end{array}
$$

5. Add the values in this new column, writing the sum in the bottom row.

$$
\begin{array}{r|rrrrr}
-4 & 1 & -6 & 0 & 0 & 2 \\
& & -4 & & & \text{Add.} \\
\hline
& 1 & -10 & & &
\end{array}
$$

6. Repeat this series of multiplications and additions until all columns are filled in.

$$
\begin{array}{r|rrrrr}
-4 & 1 & -6 & 0 & 0 & 2 \\
\text{multiply} & & -4 & 40 & & \text{Add.} \\
\hline
& 1 & -10 & 40 & &
\end{array}
$$

$$
\begin{array}{r|rrrrr}
-4 & 1 & -6 & 0 & 0 & 2 \\
& \downarrow & -4 & 40 & -160 & 640 \\
\hline
& 1 & -10 & 40 & -160 & 642
\end{array}
$$

7. Use the last row to write the quotient. The final value in this row is the remainder.

$$\frac{x^4 - 6x^3 + 2}{x + 4} = x^3 - 10x^2 + 40x - 160 + \frac{642}{x + 4}$$

The degree of the first term of the quotient is one less than the degree of the first term of the dividend. ■

The result of Example 2 can be interpreted in two ways:

$$\underbrace{\frac{x^4 - 6x^3 + 2}{\underbrace{x + 4}_{\text{Divisor: } d(x)}}}_{\text{Dividend: } f(x)} = \underbrace{x^3 - 10x^2 + 40x - 160}_{\text{Quotient: } q(x)} + \underbrace{\frac{642}{\underbrace{x + 4}_{\text{Divisor: } d(x)}}}_{\text{Remainder: } r(x)}$$

or

$$\underbrace{x^4 - 6x^3 + 2}_{\text{Dividend: } f(x)} = \underbrace{(x + 4)}_{\text{Divisor: } d(x)}\underbrace{(x^3 - 10x^2 + 40x - 160)}_{\text{Quotient: } q(x)} + \underbrace{642}_{\text{Remainder: } r(x)}$$

Paul Klee (1879–1940) "Woman in Peasant Dress" (Frau in Tracht) 1940.254, paste colour on paper, 48 × 31.3 cm. Kunstmuseum Bern, Paul–Klee–Stiftung, Bern. © ARS, New York.

Notice that the remainder must be a constant when the divisor is in the form $x - c$. We can generalize the last line of this result and write

$$f(x) = d(x)q(x) + r.$$

Dividend Divisor Quotient Remainder

This is known as the *Division Algorithm*.

The Division Algorithm

If $f(x)$ and $d(x)$ are polynomials, with $d(x) \neq 0$, and the degree of $d(x)$ is less than or equal to the degree of $f(x)$, then there exist unique polynomials $q(x)$ and $r(x)$ such that

$$f(x) = d(x) \cdot q(x) + r(x).$$

Dividend   Divisor · Quotient + Remainder

The remainder $r(x)$ equals 0 or it is of degree less than the degree of $d(x)$. If $r(x) = 0$, we say that $d(x)$ *divides evenly* into $f(x)$ and that $d(x)$ and $q(x)$ are *factors* of $f(x)$.

The Remainder and Factor Theorems

If the divisor in the Division Algorithm is of the form $x - c$, then the remainder must be a constant because its degree is less than one, the degree of $x - c$.

$$f(x) = d(x)q(x) + r(x)$$ This is the Division Algorithm.

Dividend Divisor Quotient Remainder

$$f(x) = (x - c)q(x) + r$$ The divisor is $x - c$. Call the constant remainder r.

$$f(c) = (c - c)q(c) + r$$ Find $f(c)$, setting $x = c$. This will give an expression for r.

$$f(c) = 0 \cdot q(c) + r$$ $c - c = 0$ and $0 \cdot q(c) = 0$.

$$f(c) = r$$ This final result gives the remainder when $f(x)$ is divided by $x - c$.

This result is called the *Remainder Theorem*.

3 Evaluate a polynomial using the Remainder Theorem.

The Remainder Theorem

If the polynomial $f(x)$ is divided by $x - c$, then the remainder is $f(c)$.

Example 3 shows how we can use the Remainder Theorem to evaluate a polynomial.

EXAMPLE 3 **Using the Remainder Theorem to Evaluate a Polynomial Function**

Given $f(x) = x^3 - 4x^2 + 5x + 3$, use the Remainder Theorem to find $f(2)$.

Using technology

Since $f(2) = 5$, this means $(2, 5)$ is a point on the graph of $f(x) = x^3 - 4x^2 + 5x + 3$. You can check by substituting 2 for x in $f(x)$.

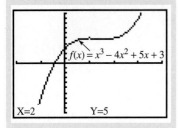

Solution

By the Remainder Theorem, if $f(x)$ is divided by $x - 2$, then the remainder is $f(2)$. We'll use synthetic division to divide.

$$\begin{array}{r|rrrr} 2 & 1 & -4 & 5 & 3 \\ & & 2 & -4 & 2 \\ \hline & 1 & -2 & 1 & 5 \end{array}$$

Remainder

$$x^2 - 2x + 1 + \dfrac{5}{x-2}$$
$$x - 2 \,\overline{)\, x^3 - 4x^2 + 5x + 3}$$

The remainder, 5, is the value of $f(2)$. Thus, $f(2) = 5$. ∎

Let's look again at the Division Algorithm when the divisor is of the form $x - c$.

$$f(x) = (x - c)q(x) + r$$

Dividend   Divisor   Quotient   Constant remainder

By the Remainder Theorem, the remainder r is $f(c)$, so we can substitute $f(c)$ for r:

$$f(x) = (x - c)q(x) + f(c).$$

Notice that if $f(c) = 0$, then

$$f(x) = (x - c)q(x)$$

so that $x - c$ is a factor of $f(x)$. This means that for the polynomial function $f(x)$, if $f(c) = 0$ then $x - c$ is a factor of $f(x)$.

Let's reverse directions and see what happens if $x - c$ is a factor of $f(x)$. This means that

$$f(x) = (x - c)q(x).$$

If we replace x with c we obtain

$$f(c) = (c - c)q(c) = 0.$$

Thus, if $x - c$ is a factor of $f(x)$, then $f(c) = 0$.

We have proved a result known as the *Factor Theorem*.

The Factor Theorem

Let $f(x)$ be a polynomial.

a. If $f(c) = 0$, then $x - c$ is a factor of $f(x)$.
b. If $x - c$ is a factor of $f(x)$, then $f(c) = 0$.

---

4 Use the Factor Theorem to solve a polynomial equation.

The examples that follow show how the Factor Theorem can be used to solve polynomial equations.

EXAMPLE 4 **Using the Factor Theorem**

Solve the equation $2x^3 - 3x^2 - 11x + 6 = 0$ given that 3 is a zero of $f(x) = 2x^3 - 3x^2 - 11x + 6$.

Solution

Since $f(3) = 0$, the Factor Theorem tells us that $x - 3$ is a factor of $f(x)$. We'll use the synthetic division to divide $f(x)$ by $x - 3$.

$$\begin{array}{r|rrrr} 3 & 2 & -3 & -11 & 6 \\ & & 6 & 9 & -6 \\ \hline & 2 & 3 & -2 & 0 \end{array}$$

$$\begin{array}{r} 2x^2 + 3x - 2 \\ x - 3 \overline{\smash{\big)}\, 2x^3 - 3x^2 - 11x + 6} \end{array}$$

Equivalently:

$$2x^3 - 3x^2 - 11x + 6 = (x - 3)(2x^2 + 3x - 2)$$

Now we can solve the polynomial equation.

$2x^3 - 3x^2 - 11x + 6 = 0$	This is the given equation.
$(x - 3)(2x^2 + 3x - 2) = 0$	Factor using the result from the synthetic division.
$(x - 3)(2x - 1)(x + 2) = 0$	Factor the trinomial.
$x - 3 = 0$ or $2x - 1 = 0$ or $x + 2 = 0$	Set each factor equal to 0.
$x = 3$ $\qquad x = \frac{1}{2}$ $\qquad x = -2$	Solve for x.

The solution set is $\{-2, \frac{1}{2}, 3\}$. ∎

Since the solution set of

$$2x^3 - 3x^2 - 11x + 6 = 0$$

is $\{-2, \frac{1}{2}, 3\}$, this implies that the polynomial function

$$f(x) = 2x^3 - 3x^2 - 11x + 6$$

has x-intercepts (or zeros) at $-2, \frac{1}{2}$, and 3. This is verified by the graph of f.

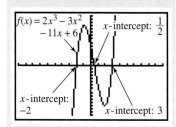

S tudy tip

Synthetic Division and Remainders of Zero

In Example 4 we used synthetic division to divide $f(x) = 2x^3 - 3x^2 - 11x + 6$ by $x - 3$. We obtained a remainder of 0. This indicated that 3 was a solution of the polynomial equation $f(x) = 0$, namely, $2x^3 - 3x^2 - 11x + 6 = 0$. By the Factor Theorem, the following statements are useful in solving polynomial equations.

1. If $f(x)$ is divided by $x - c$ and the remainder is zero, then c is a zero of f and c is a solution of the polynomial equation $f(x) = 0$.
2. If $f(x)$ is divided by $x - c$ and the remainder is 0, then $x - c$ is a factor of $f(x)$.

EXAMPLE 5 **Solving a Polynomial Equation**

Solve the equation $3x^4 - 11x^3 - x^2 + 19x + 6 = 0$ given that -1 and 2 are roots.

Solution

Related to the given equation is the polynomial function

$$f(x) = 3x^4 - 11x^3 - x^2 + 19x + 6.$$

Since -1 is given to be a root, then $f(-1) = 0$. The Factor Theorem tells us that $x - (-1)$, or $x + 1$, is a factor of $f(x)$. We'll use synthetic division to divide $f(x)$ by $x + 1$.

$$\underline{-1}\begin{array}{rrrrr} 3 & -11 & -1 & 19 & 6 \\ & -3 & 14 & -13 & -6 \\ \hline 3 & -14 & 13 & 6 & 0 \end{array}$$

$$\begin{array}{r} 3x^3 - 14x^2 + 13x + 6 \\ x + 1 \overline{)\,3x^4 - 11x^3 - x^2 + 19x + 6} \end{array}$$

Equivalently:

$$3x^4 - 11x^3 - x^2 + 19x + 6$$
$$= (x + 1)(3x^3 - 14x^2 + 13x + 6)$$

Let's see what we have so far.

$$3x^4 - 11x^3 - x^2 + 19x + 6 = 0 \qquad \text{This is the given equation.}$$

$$(x + 1)(3x^3 - 14x^2 + 13x + 6) = 0 \qquad \text{Factor using the result from the synthetic division.}$$

We were also told that 2 is a root of this equation. Since 2 is not a root of $x + 1 = 0$, it must be a solution of the *reduced equation* $3x^3 - 14x^2 + 13x + 6 = 0$. By the Factor Theorem, $x - 2$ must be a factor of $3x^3 - 14x^2 + 13x + 6$. Now we can use synthetic division to find the other factor.

$$\underline{2}\begin{array}{rrrr} 3 & -14 & 13 & 6 \\ & 6 & -16 & -6 \\ \hline 3 & -8 & -3 & 0 \end{array}$$

$$\begin{array}{r} 3x^2 - 8x - 3 \\ x - 2 \overline{)\,3x^3 - 14x^2 + 13x + 6} \end{array}$$

Equivalently:

$$3x^3 - 14x^2 + 13x + 6$$
$$= (x - 2)(3x^2 - 8x - 3)$$

Let's put this all together and see what we have.

$$3x^4 - 11x^3 - x^2 + 19x + 6 = 0 \qquad \text{Once again, this is the given equation.}$$

$$(x + 1)(3x^3 - 14x^2 + 13x + 6) = 0 \qquad \text{This was the result of the first synthetic division.}$$

$$(x + 1)(x - 2)(3x^2 - 8x - 3) = 0 \qquad \text{Factor the cubic factor, using the result from the second synthetic division.}$$

$$(x + 1)(x - 2)(3x + 1)(x - 3) = 0 \qquad \text{Factor the trinomial.}$$

$$x + 1 = 0 \text{ or } x - 2 = 0 \text{ or } 3x + 1 = 0 \text{ or } x - 3 = 0 \qquad \text{Set each factor equal to 0.}$$

$$x = -1 \qquad x = 2 \qquad x = -\tfrac{1}{3} \qquad x = 3 \qquad \text{Solve for } x.$$

Thus, the solution set of $3x^4 - 11x^3 - x^2 + 19x + 6 = 0$ is $\{-1, -\tfrac{1}{3}, 2, 3\}$. ■

Using technology

Since the solution set of

$$3x^4 - 11x^3 - x^2 + 19x + 6 = 0$$

is $\{-1, -\tfrac{1}{3}, 2, 3\}$, this implies that the polynomial function

$$f(x) = 3x^4 - 11x^3 - x^2 + 19x + 6$$

has *x*-intercepts (or zeros) at $-1, -\tfrac{1}{3}, 2,$ and 3. This is verified by the graph of f.

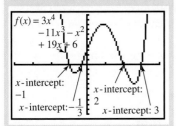

$[-4, 4] \times [-20, 20]$ Yscl = 2

5 Use the Rational Zero Theorem to find possible rational zeros.

The Rational Zero Theorem

In Examples 4 and 5, we were given a root of the polynomial equation $f(x) = 0$. We used the fact that if c is a root of $f(x) = 0$, then $x - c$ is a factor of $f(x)$. But how do we find c? Once we obtain this first zero, the search for roots is simplified by working with the lower-degree polynomial factor obtained by synthetic division.

The best that mathematicians can provide is a theorem that gives a list of *possible* rational roots of the polynomial equation $f(x) = 0$. Not every number in the list will be a zero, but every rational zero of the polynomial equation will appear somewhere in the list.

The Rational Zero Theorem

If $f(x) = a_n x^n + a_{n-1} x^{n-1} + \cdots + a_1 x + a_0$ has *integer* coefficients and $\dfrac{p}{q}$ (where $\dfrac{p}{q}$ is reduced) is a rational zero, then p is a factor of the constant term a_0 and q is a factor of the leading coefficient a_n.

The "why" behind the Rational Zero Theorem is discussed in Problem 64 of Problem Set 3.3. For now, let's see how we can use this theorem to find possible rational zeros.

EXAMPLE 6   Using the Rational Zero Theorem

List all possible rational zeros of $f(x) = 15x^3 + 14x^2 - 3x - 2$.

Solution

$$\text{Possible rational zeros} = \frac{\text{Factors of the constant term}}{\text{Factors of the leading coefficient}}$$

p (factors of the constant term, -2): $\pm 1, \pm 2$
q (factors of the leading coefficient, 15): $\pm 1, \pm 3, \pm 5, \pm 15$

Possible rational zeros are in the form $\dfrac{p}{q}$:

$\pm 1, \pm 2$	Divide each p by $q = \pm 1$.
$\pm \frac{1}{3}, \pm \frac{2}{3},$	Divide each p by $q = \pm 3$.
$\pm \frac{1}{5}, \pm \frac{2}{5}$	Divide each p by $q = \pm 5$.
$\pm \frac{1}{15}, \pm \frac{2}{15},$	Divide each p by $q = \pm 15$.

There are 16 possible rational zeros. The actual solution set to $15x^3 + 14x^2 - 3x - 2 = 0$ is $\{-1, -\frac{1}{3}, \frac{2}{5}\}$, which contains 3 of the 16 possible roots. ∎

6 Use technology to eliminate possible rational zeros.

Using Technology to Narrow Down the List of Possible Rational Zeros

Example 6 illustrates that a list of possible rational zeros may be quite large. Although we can use synthetic division to check some of the easier values, such as 1, −1, 2, and −2, there are ways of ruling out many of the possible rational roots generated by the Rational Zero Theorem.

A graphing utility can help guide us in determining which possible rational zeros to test. Let's see how this works.

EXAMPLE 7   Eliminating Possible Rational Zeros with a Graphing Utility

Find all real zeros of $f(x) = 4x^3 + x^2 - 20x - 5$.

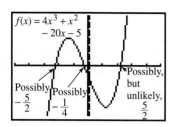

Figure 3.32

The graph of $f(x) = 4x^3 + x^2 - 20x - 5$

Solution

We begin by listing all possible rational zeros.

p (factors of the constant term, -5): $\pm 1, \pm 5$
q (factors of the leading coefficient, 4): $\pm 1, \pm 2, \pm 4$

Possible rational zeros $\left(\dfrac{p}{q}\right)$: $\pm 1, \pm 5, \pm\frac{1}{2}, \pm\frac{5}{2}, \pm\frac{1}{4}, \pm\frac{5}{4}$

With 12 possible rational zeros, we'll graph f using a graphing utility. From Figure 3.32, it looks like possible choices occur at $-\frac{5}{2}, -\frac{1}{4}$, and $\frac{5}{2}$. The most reasonable of these options seems to be at $-\frac{1}{4}$, so let's test this value by synthetic division.

$$
\begin{array}{r|rrrr}
-\frac{1}{4} & 4 & 1 & -20 & -5 \\
 & & -1 & 0 & 5 \\
\hline
 & 4 & 0 & -20 & 0
\end{array}
$$

$$
\begin{array}{r}
4x^2 \qquad\quad -20 \\
x + \tfrac{1}{4}\,\overline{\smash{)}\,4x^3 + x^2 - 20x - 5}
\end{array}
$$

Equivalently:

$$4x^3 + x^2 - 20x - 5 = \left(x + \tfrac{1}{4}\right)(4x^2 - 20)$$

The zero remainder tells us that $-\frac{1}{4}$ is a solution of $4x^3 + x^2 - 20x - 5 = 0$, so $x + \frac{1}{4}$ is a factor of the polynomial.

$$4x^3 + x^2 - 20x - 5 = 0$$
Finding zeros of $f(x) = 4x^3 + x^2 - 20x - 5$ is the same as finding solutions of this polynomial equation.

$$\left(x + \tfrac{1}{4}\right)(4x^2 - 20) = 0$$
Factor using the result from the synthetic division.

$$x + \tfrac{1}{4} = 0 \quad \text{or} \quad 4x^2 - 20 = 0$$
Set each factor equal to 0.

$$x = -\tfrac{1}{4} \qquad\qquad 4x^2 = 20$$
Solve for x.

$$x^2 = 5$$

$$x = \pm\sqrt{5}$$

The solution set is: $\{-\sqrt{5}, -\frac{1}{4}, \sqrt{5}\}$. The real zeros of f are $-\sqrt{5}, -\frac{1}{4}$, and $\sqrt{5}$, and these are the x-intercepts shown in Figure 3.32. ∎

Using Algebraic Methods to Narrow Down the List of Possible Rational Zeros

In addition to the graphing utility method, there are two algebraic methods that are useful for eliminating possible rational zeros. One is *Descartes' Rule of Signs* and the other is the *Upper and Lower Bound Theorem*.

7 Use Descartes' Rule of Signs.

Descartes' Rule of Signs

In Section 3.2, we noted that an *n*th-degree polynomial function can have at most *n* real zeros. *Descartes' Rule of Signs* provides even more specific information about the number of real zeros that a polynomial can have. The rule is based on considering *variations in sign* between consecutive coefficients. For example, the function

$$f(x) = 3x^7 - 2x^5 - x^4 + 7x^2 + x - 3$$

has three sign changes.

An equation can have as many true [positive] roots as it contains changes of sign, from plus to minus or from minus to plus. ... René Descartes (1596–1650) in *La Géométrie* (1637)
Library of Congress

The real zeros of a polynomial function f with rational coefficients are the same as the real zeros of the function obtained by multiplying each term of f by a nonzero constant that clears the fractional coefficients of f. This is necessary to apply the Rational Zero Theorem.

Descartes' Rule of Signs

Let $f(x) = a_n x^n + a_{n-1} x^{n-1} + \cdots + a_2 x^2 + a_1 x + a_0$ be a polynomial with real coefficients and $a_0 \neq 0$.
1. The number of *positive real zeros* of f is either equal to the number of sign changes of $f(x)$ or is less than that number by an even integer. If there is only one variation in sign, there is exactly one positive real zero.
2. The number of *negative real zeros* of f is either equal to the number of sign changes of $f(-x)$ or is less than that number by an even integer. If $f(-x)$ has only one variation in sign, then f has exactly one negative real zero.

EXAMPLE 8 **Eliminating Possible Rational Zeros Using Descartes' Rule of Signs**

Solve: $\dfrac{1}{12}x^4 - \dfrac{7}{6}x^3 + \dfrac{71}{12}x^2 - \dfrac{77}{6}x + 10 = 0$

Solution

$\dfrac{1}{12}x^4 - \dfrac{7}{6}x^3 + \dfrac{71}{12}x^2 - \dfrac{77}{6}x + 10 = 0$ This is the given equation.

$12\left(\dfrac{1}{12}x^4 - \dfrac{7}{6}x^3 + \dfrac{71}{12}x^2 - \dfrac{77}{6}x + 10\right) = 12(0)$ Clear fractions, multiplying both sides by 12.

$x^4 - 14x^3 + 71x^2 - 154x + 120 = 0$ Simplify. With integer coefficients, we can apply the Rational Zero Theorem.

We begin by listing possible rational roots dividing factors of the constant term by factors of the leading coefficient.

$$\frac{\text{Factors of 120 (the constant term)}}{\text{Factors of 1 (the leading coefficient)}}$$

$$= \pm 1, \pm 2, \pm 3, \pm 4, \pm 5, \pm 6, \pm 8, \pm 10, \pm 12, \pm 15, \pm 20,$$

$$\pm 24, \pm 30, \pm 40, \pm 60, \pm 120$$

Now let's apply Descartes' Rule of Signs to determine the possible number of positive and negative real roots of $x^4 - 14x^3 + 71x^2 - 154x + 120 = 0$.

$$f(x) = x^4 - 14x^3 + 71x^2 - 154x + 120$$ There are four variations in sign.
$$\quad\quad\quad 1 \quad\quad 2 \quad\quad 3 \quad\quad 4$$

This means that the equation has either four, two, or zero positive real roots.
To find possibilities for negative roots, replace x with $-x$ in the polynomial function $f(x) = x^4 - 14x^3 + 71x^2 - 154x + 120$.

$$f(-x) = (-x)^4 - 14(-x)^3 + 71(-x)^2 - 154(-x) + 120$$ Replace x with $-x$.
$$= x^4 + 14x^3 + 71x^2 + 154x + 120$$ There are no variations of sign.

This means that the equation has no negative real roots.

At this point, we have the possible rational roots:

$$1, 2, 3, 4, 5, 6, 8, 10, 12, 15, 20, 24, 30, 40, 60, 120$$

Let's test the two smallest using synthetic division.

Test 1:

$$
\begin{array}{r|rrrrr}
1 & 1 & -14 & 71 & -154 & 120 \\
  &   & 1   & -13 & 58   & -96 \\
\hline
  & 1 & -13 & 58 & -96 & 24
\end{array}
$$

Is 1 a solution of $x^4 - 14x^3 + 71x^2 - 154x + 120$?

← The remainder is not 0, so 1 is not a solution.

Test 2:

$$
\begin{array}{r|rrrrr}
2 & 1 & -14 & 71 & -154 & 120 \\
  &   & 2   & -24 & 94   & -120 \\
\hline
  & 1 & -12 & 47 & -60 & 0
\end{array}
$$

Is 2 a solution of $x^4 - 14x^3 + 71x^2 - 154x + 120$?

← The remainder is 0, so 2 is a solution.

$x^4 - 14x^3 + 71x^2 - 154x + 120 = 0$	This is the simplified equation.
$(x - 2)(x^3 - 12x^2 + 47x - 60) = 0$	Factor using the result from the second synthetic division.
$x - 2 = 0$ or $x^3 - 12x^2 + 47x - 60 = 0$	Set each factor equal to 0.

Now we focus our attention on the equation

$$x^3 - 12x^2 + 47x - 60 = 0.$$

$$\text{Possible rational roots} = \frac{\text{Factors of } -60}{\text{Factors of } 1}$$

$$= 1, 2, 3, 4, 5, 6, 10, 12, 15, 20, 30, 60$$

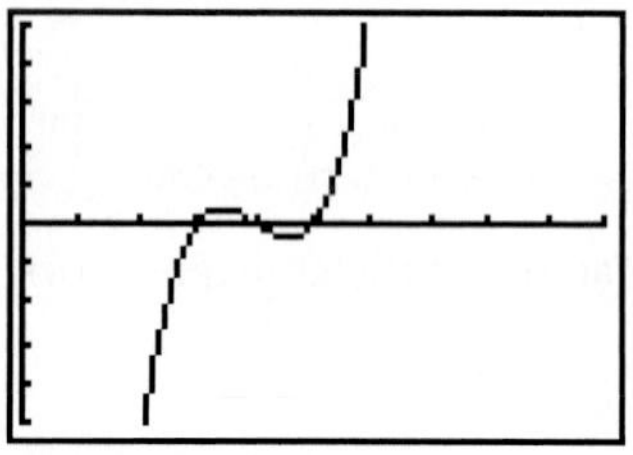

Notice that we omitted possible negative rational roots because we have determined that the equation $x^4 - 14x^3 + 71x^2 - 154x + 120 = 0$ has no negative real roots, and $x^3 - 12x^2 + 47x - 60$ is a factor of $x^4 - 143x^3 + 71x^2 - 154x + 120$.

The graph of $f(x) = x^3 - 12x^2 + 47x - 60$, shown in Figure 3.33 for $x \geqslant 0$, suggests that 3 is a zero of the cubic polynomial. Let's verify this using synthetic division.

Figure 3.33

The graph of $f(x) = x^3 - 12x^2 + 47x - 60$ suggests that 3 is a zero.

Test 3:

$$
\begin{array}{r|rrrr}
3 & 1 & -12 & 47 & -60 \\
  &   & 3   & -27 & 60 \\
\hline
  & 1 & -9 & 20 & 0
\end{array}
$$

Is 3 a solution of $x^3 - 12x^2 + 47x - 60 = 0$?

← The remainder is 0, so 3 is a solution.

Putting this all together, here's what we have:

$x^4 - 14x^3 + 71x^2 - 154x + 120 = 0$	This is the simplified equation.
$(x - 2)(x^3 - 12x^2 + 47x - 60) = 0$	This is the result of the first synthetic division.
$(x - 2)(x - 3)(x^2 - 9x + 20) = 0$	This is the result of the synthetic division shown immediately above.
$(x - 2)(x - 3)(x - 4)(x - 5) = 0$	Factor completely.

$x - 2 = 0$ or $x - 3 = 0$ or $x - 4 = 0$ or $x - 5 = 0$ Set each factor equal to 0.

$x = 2$ $x = 3$ $x = 4$ $x = 5$ Solve for x.

The solution set is $\{2, 3, 4, 5\}$.   ■

⑧ Find bounds for the roots of a polynomial equation.

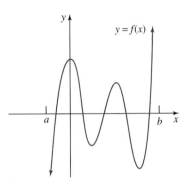

Figure 3.34

b is an upper bound and a is a lower bound for the set of zeros.

Barry Flanagan "Six Foot Leaping Hare on Steel Pyramid" 1990, bronze and steel, 94 × 74 × 20 in. (239 × 188 × 50.8 cm). © Barry Flanagan, courtesy Waddington Galleries, London.

Upper and Lower Bounds for Roots

The *Upper and Lower Bound Theorem* also helps us rule out many of a polynomial equation's possible rational roots. Figure 3.34 illustrates that a is a *lower bound* and b is an *upper bound* for the roots of $f(x) = 0$ because every real root c of the equation satisfies $a \leqslant c \leqslant b$.

The Upper and Lower Bound Theorem

Let $f(x)$ be a polynomial with real coefficients and a positive leading coefficient, and let a and b be nonzero real numbers.

1. Divide $f(x)$ by $x - b$ (where $b > 0$) using synthetic division. If the last row containing the quotient and remainder has no negative numbers, then b is an *upper bound* for the real roots of $f(x) = 0$.
2. Divide $f(x)$ by $x - a$ (where $a < 0$) using synthetic division. If the last row containing the quotient and remainder has numbers that alternate in sign (zero entries count as positive or negative), then a is a *lower bound* for the real roots of $f(x) = 0$.

EXAMPLE 9   Finding Bounds for the Roots

Show that all the real roots of the equation $8x^3 + 10x^2 - 39x + 9 = 0$ lie between -3 and 2.

Solution

We divide the polynomial by $x - 2$ and $x + 3$, respectively.

$$
\begin{array}{r|rrrr}
2 & 8 & 10 & -39 & 9 \\
  &   & 16 & 52 & 26 \\
\hline
  & 8 & 26 & 13 & 35
\end{array}
$$
← All numbers in this row are nonnegative.

By the Upper and Lower Bound Theorem, the nonnegative entries in the last row indicate that 2 is an upper bound for the roots.

$$
\begin{array}{r|rrrr}
-3 & 8 & 10 & -39 & 9 \\
   &   & -24 & 42 & -9 \\
\hline
   & 8 & -14 & 3 & 0
\end{array}
$$
← Counting 0 as negative, the signs alternate: $+, -, +, -$.

By the Upper and Lower Bound Theorem, the alternating signs in the last row indicate that -3 is a lower bound for the roots (and is also a root).   ■

9 Use a strategy to solve polynomial equations.

A General Strategy for Solving Polynomial Equations

Let's tie together the information presented in this section and establish a step-by-step procedure for finding rational roots of a polynomial equation.

Finding rational roots of a polynomial equation

1. Use the Rational Zero Theorem to list all possible rational roots (in order).
2. Use Descartes' Rule of Signs to determine how many positive and negative real roots the equation may have. (In some instances, possible positive or negative rational roots may be eliminated in this step.)
3. Use synthetic division to check the possible positive and negative rational roots. A graphing utility, showing the graph of the polynomial function, might help narrow down the choices. Stop when you find a root, (remainder $= 0$), reach an upper or lower bound, or have found the maximum number of positive or negative roots determined by Descartes' Rule of Signs.
4. If c is a root, then one factor is $x - c$. The second factor is the quotient obtained from the synthetic division.
5. Set each factor equal to zero.
6. **a.** Solve the equation obtained by setting the linear factor equal to zero.
   b. If the second factor is quadratic, solve the resulting quadratic equation by factoring (if possible), by the quadratic formula, or by the square root method.
   c. If the second factor is of degree 3 or greater, repeat steps 1 through 5. It is not necessary to check possible roots that have not worked earlier.

EXAMPLE 10 **Solving a Polynomial Equation**

Find all real zeros of: $f(x) = x^4 + 3x^3 - 27x^2 + 3x - 28$

Solution

The possible rational zeros of $f(x)$ are

$$\pm 1, \pm 2, \pm 4, \pm 7, \pm 14, \pm 28.$$ p (the divisors of -28) are $\pm 1, \pm 2, \pm 4, \pm 7, \pm 14, \pm 28$ and q (the divisors of 1) are ± 1. We've listed all possibilities for $\frac{p}{q}$.

Now we apply Descartes' Rule of Signs.

$$f(x) = x^4 + 3x^3 - 27x^2 + 3x - 28$$ There are three real positive zeros or there is one real positive zero.

$$f(-x) = x^4 - 3x^3 - 27x^2 - 3x - 28$$ There is one real negative zero.

Although Descartes' Rule of Signs gives us an idea of what to expect, it does not permit us to eliminate possible rational roots from our list of 12 candidates.

ENRICHMENT ESSAY

A Brief History of Polynomial Equations

- The Babylonians (approximately 2000 B.C.) solved quadratic equations by completing the square.
- Euclid (approximately 300 B.C.) a Greek mathematician, presented a geometric method for finding the positive solution of $x^2 + bx + c = 0$.
- Omar Khayyam (1050–1122), a Persian poet, used intersecting conic sections to find positive solutions to third-degree polynomial equations. Khayyam incorrectly stated that formulas cannot be used to solve cubic equations.
- Nicolo Tartaglia (1499–1557) derived a formula for finding one of the solutions to the cubic equation $x^3 + mx = n$.
- Lodovico Ferrari (1522–1565) derived a formula for the solution of quartic (fourth-degree) polynomial equations about 1545.
- Leonhard Euler (1707–1783), a Swiss mathe-

matician, presented in 1732 the first complete solution to the cubic equation $x^3 + mx = n$.
- Carl Friedrich Gauss (1777–1855), a German mathematician, proved in his doctoral dissertation that (counting multiplicities) a polynomial equation of degree n has n roots. Derived in 1779, the result is known as the fundamental theorem of algebra because that is what Gauss called it.
- Niels Henrik Abel (1802–1829), a Norwegian mathematician, proved in 1823 that fifth-degree polynomial equations cannot be solved with a formula.
- Evariste Galois (1811–1832), a French mathematician, proved that there is no general formula for solving polynomial equations of degree 5 or greater.

Since we know $f(x)$ has positive zeros, we can use synthetic division to check the positive candidates first, beginning with the smallest.

$$\underline{1}\ \begin{array}{rrrrr} 1 & 3 & -27 & 3 & -28 \\ & 1 & 4 & -23 & -20 \\ \hline 1 & 4 & -23 & -20 & -48 \end{array} \qquad \underline{2}\ \begin{array}{rrrrr} 1 & 3 & -27 & 3 & -28 \\ & 2 & 10 & -34 & -62 \\ \hline 1 & 5 & -17 & -31 & -90 \end{array}$$

$$\underline{4}\ \begin{array}{rrrrr} 1 & 3 & -27 & 3 & -28 \\ & 4 & 28 & 4 & 28 \\ \hline 1 & 7 & 1 & 7 & 0 \end{array}\ \leftarrow\ 4 \text{ is a zero of } f(x) \text{ because the remainder is } 0.$$

$$\underbrace{\qquad\qquad\qquad}_{\text{Nonnegative numbers}}\ \leftarrow\ 4 \text{ is an upper bound for the zeros of } f(x).$$

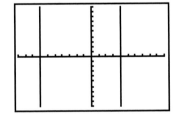

Figure 3.35

An incomplete graph of $f(x) = x^4 + 3x^3 - 27x^2 + 3x - 28$

Because 4 is both a zero and an upper bound for the zeros, we can eliminate 7, 14, and 28 as zeros since they are all greater than 4.

Figure 3.35 shows an incomplete graph of $f(x) = x^4 + 3x^3 - 27x^2 + 3x - 28$, obtained using a graphing utility and a standard range setting. Observe that 4 is a zero, although based on the incomplete graph we cannot be sure that 4 is the upper bound for the zeros.

Let's see what we have so far.

$$x^4 + 3x^3 - 27x^2 + 3x - 28 = 0 \qquad \begin{array}{l}\text{Finding zeros for } f(x) = x^4 + 3x^3 - 27x^2 \\ - 28 \text{ is equivalent to solving the equation} \\ f(x) = 0.\end{array}$$

$$(x - 4)(x^3 + 7x^2 + x + 7) = 0 \qquad \text{Use the result of the last synthetic division.}$$

Now let's reapply what we have done up to this point to the polynomial function

$$g(x) = x^3 + 7x^2 + x + 7.$$

The possible rational zeros of $g(x)$ are

$\pm 1, \pm 7$ p (the divisors of 7) are $\pm 1, \pm 7$ and q (the divisors of 1) are ± 1. We've listed all possibilities for $\frac{p}{q}$.

Next we apply Descartes' Rule of Signs.

$$g(x) = x^3 + 7x^2 + x + 7$$ With no sign changes, g has no positive real zeros.

(No sign variations)

$$g(-x) = -x^3 + 7x^2 - x + 7$$ With 3 sign changes, g has three or one negative real zero(s).

$\quad\quad\quad\quad\quad$ 1 $\quad$ 2 $\quad$ 3

Descartes' Rule of Signs narrows our list of possible rational zeros to -1 and -7. However, we have already found that the given function has only one negative real zero, so function g will have only one also. We have the option of consulting Figure 3.35 on the previous page, where the incomplete graph indicates that -7 is a zero, or checking the negative candidates using synthetic division.

$$
\begin{array}{r|rrrr}
-1 & 1 & 7 & 1 & 7 \\
   &   & -1 & -6 & 5 \\
\hline
   & 1 & 6 & -5 & 12
\end{array}
\qquad
\begin{array}{r|rrrr}
-7 & 1 & 7 & 1 & 7 \\
   &   & -7 & 0 & -7 \\
\hline
   & 1 & 0 & 1 & 0
\end{array}
$$

The final row of the second synthetic division indicates that -7 is a zero of $g(x) = x^3 + 7x^2 + x + 7$. The solution process now moves quite quickly.

$$x^4 + 3x^3 - 27x^2 + 3x - 28 = 0$$ This is the original equation.

$$(x - 4)(x^3 + 7x^2 + x + 7) = 0$$ This result was previously obtained.

$$(x - 4)(x + 7)(x^2 + 1) = 0$$ Factor using the result from the last synthetic division.

$$x - 4 = 0 \ \text{ or } \ x + 7 = 0 \ \text{ or } \ x^2 + 1 = 0$$ Set each factor equal to 0.

$$x^2 = -1$$

$$x = 4 \qquad\qquad x = -7 \qquad\qquad x = \pm i \quad \text{Solve for } x.$$

The solution set for $x^4 + 3x^3 - 27x^2 + 3x - 28 = 0$ is $\{-7, 4, -i, i\}$. The function $f(x) = x^4 + 3x^3 - 27x^2 + 3x - 28$ has four zeros and two of them are real. The real zeros are -7 and 4.   ∎

We can use our work in Example 10 to obtain a more complete graph of

$$f(x) = x^4 + 3x^3 - 27x^2 + 3x - 28$$

Figure 3.36

The end behavior of $f(x) = x^4 + 3x^3 - 27x^2 + 3x - 28$

than the one shown in Figure 3.35 on page 373. Only the real zeros, -7 and 4, appear as x-intercepts. Because n is even $(n = 4)$ and $a_n > 0$ $(a_n = 1)$, the graph of f rises to the left and right, as indicated in Figure 3.36. A range setting for x such as Xmin $= -10$, Xmax $= 10$, and Xscl $= 1$ seems reasonable, but how can we determine Ymin? Synthetic division might help, finding $f(-6)$ and $f(-5)$.

$$
\begin{array}{r|rrrrr}
-6 & 1 & 3 & -27 & 3 & -28 \\
   &   & -6 & 18 & 54 & -342 \\
\hline
   & 1 & -3 & -9 & 57 & -370 \\
\end{array}
\qquad
\begin{array}{r|rrrrr}
-5 & 1 & 3 & -27 & 3 & -28 \\
   &   & -5 & 10 & 85 & -440 \\
\hline
   & 1 & -2 & -17 & 88 & -468 \\
\end{array}
$$

$$f(-6) = -370 \qquad\qquad f(-5) = -468$$

Thus we will use Ymin $= -500$.

Figure 3.37 shows a more complete graph of f using the following range setting:

Xmin $= -10$, Xmax $= 10$, Xscl $= 1$, Ymin $= -500$,

Ymax $= 100$, Yscl $= 20$

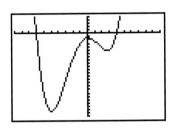

Figure 3.37

The graph of $f(x) = x^4 + 3x^3 - 27x^2 + 3x - 28$

> **Study tip**
>
> Our work in obtaining the graph in Figure 3.37 is an excellent illustration of the fact that technology is complemented by human knowledge and is not intended to replace it.

PROBLEM SET 3.3

Practice Problems

1. Use synthetic division to divide $f(x) = x^3 - 2x^2 - x + 2$ by $x + 1$. Use the result to find all zeros of f.
2. Use synthetic division to divide $f(x) = x^3 - 4x^2 + x + 6$ by $x + 1$. Use the result to find all zeros of f.
3. Solve the equation $2x^3 - 3x^2 - 11x + 6 = 0$ given that -2 is a zero of $f(x) = 2x^3 - 3x^2 - 11x + 6$.
4. Solve the equation $2x^3 - 5x^2 + x + 2 = 0$ given that 2 is a zero of $f(x) = 2x^3 - 5x^2 + x + 2$.
5. Solve the equation $3x^3 + 7x^2 - 22x - 8 = 0$ given that $-\frac{1}{3}$ is a solution.
6. Solve the equation $12x^3 + 16x^2 - 5x - 3 = 0$ given that $-\frac{3}{2}$ is a solution.

In Problems 7–10, use the Rational Zero Theorem to list all possible rational zeros of each equation.

7. $3x^4 - 11x^3 - x^2 + 19x + 6 = 0$
8. $2x^4 + 3x^3 - 11x^2 - 9x + 15 = 0$
9. $4x^4 - x^3 + 5x^2 - 2x - 6 = 0$
10. $3x^4 - 11x^3 - 3x^2 - 6x + 8 = 0$

In Problems 11–16, use Descartes' Rule of Signs to determine the possible number of positive and negative solutions of each polynomial equation.

11. $x^3 + 2x^2 + 5x + 4 = 0$
12. $5x^3 - 3x^2 + 3x - 1 = 0$
13. $-2x^3 + x^2 - x + 7 = 0$
14. $2x^4 - 5x^3 - x^2 - 6x + 4 = 0$
15. $2x^4 - 5x + 2 = 0$
16. $4x^3 - 3x^2 - 3 = 0$

In Problems 17–20:

a. *List all possible rational zeros of the function f.*
b. *Test the possible rational zeros using synthetic division until an actual zero is determined.*
c. *Use the zero from part (b) to find all real zeros of f.*

17. $f(x) = 10x^3 - 15x^2 - 16x + 12$
18. $f(x) = 3x^3 - 13x^2 + 7x + 15$
19. $f(x) = 4x^3 - 12x^2 + 11x - 3$
20. $f(x) = 2x^3 - 3x^2 - 32x - 15$
21. Show that all the real roots of the equation $x^4 - 5x^3 + 11x^2 + 33x - 18 = 0$ lie between -4 and 7.
22. Show that all the real roots of the equation $x^4 + 11x^3 - 12x^2 + 6 = 0$ lie between -13 and 1.
23. Consider the equation $x^4 + 3x^3 + 2x^2 - 5x + 12 = 0$.
    a. List all possible rational roots.

b. Determine whether 1 is a root using synthetic division. What two conclusions can you draw?

c. Based on part (b), what possible rational roots can you eliminate?

d. Determine whether -3 is a root using synthetic division. What two conclusions can you draw?

e. Based on part (d), what possible rational roots can you eliminate?

24. Consider the equation $2x^5 + 5x^4 - 8x^3 - 14x^2 + 6x + 9 = 0$.

a. List all possible rational roots.

b. Determine whether $\frac{3}{2}$ is a root using synthetic division. What two conclusions can you draw?

c. Based on part (b), what possible rational roots can you eliminate?

d. Determine whether -3 is a root using synthetic division. What two conclusions can you draw?

e. Based on part (d), what possible rational roots can you eliminate?

Use the general strategy for finding rational roots of a polynomial equation (page 372) to solve each equation in Problems 25–34.

25. $x^3 - 14x^2 + 25x - 12 = 0$

26. $x^3 + x^2 - 5x + 3 = 0$

27. $12x^3 + 28x^2 - 23x + 3 = 0$

28. $6x^3 + 7x^2 - 16x - 12 = 0$

29. $x^4 + 6x^3 + 7x^2 - 6x - 8 = 0$

30. $x^4 + 3x^3 - 19x^2 - 3x + 18 = 0$

31. $3x^4 + 5x^3 - x^2 - 5x - 2 = 0$

32. $3x^4 + 10x^3 - 14x^2 - 20x + 16 = 0$

33. $5x^5 - x^4 - 25x^3 + 5x^2 + 20x - 4 = 0$

34. $8x^5 - 44x^4 + 86x^3 - 73x^2 + 28x - 4 = 0$

Use the same general strategy as above to find all real zeros of each function in Problems 35–40.

35. $f(x) = 3x^4 + 5x^3 - 11x^2 - 15x + 6$

36. $f(x) = 2x^4 - 5x^3 - 14x^2 + 5x + 12$

37. $f(x) = 4x^5 - 24x^4 + 25x^3 + 39x^2 - 38x - 24$

38. $f(x) = 3x^5 - 9x^4 - 28x^3 + 84x^2 + 9x - 27$

39. $f(x) = x^3 - \frac{5}{2}x^2 - 23x + 12$

40. $f(x) = x^3 + 2x^2 + \frac{11}{9}x + \frac{2}{9}$

Application Problems

41. A box with an open top is to be formed by cutting squares out of the corners of a rectangular piece of cardboard 10 inches by 5 inches and then folding up the sides. What size square must be cut from each corner if the volume of the box is to be 24 cubic inches?

42. An open-top box is to have a square base and a volume of 10 cubic feet. If the material for the bottom costs 15 cents per square foot and that for the sides cost 6 cents per square foot, what dimensions will result in a cost of three dollars?

43. If the volume of the solid shown in the figure is 208 cubic inches, find the value of x.

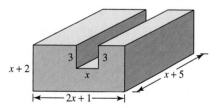

44. An automobile manufacturer can produce x cars per hour. Hourly costs are $5200, plus $2000 per car. The manufacturer can sell x cars per hour at a price of $7070 - 0.4x^2$ per car. How many cars should be produced and sold per hour to generate a profit of $214,500?

45. The common cold is caused by a rhinovirus. The number of viral invaders present on a given day is estimated by the function $N(t) = -\frac{3}{4}t^4 + 3t^3 + 5$, where N is in billions. What is the fewest number of days such that there are 17 billion viral particles present?

46. The function $S_n(n) = \dfrac{n(n + 1)(2n + 1)}{6}$ describes the sum of the squares of the first n natural numbers. How many natural numbers must be added to produce a sum of 1015?

True–False Critical Thinking Problems

47. Which one of the following is true?

a. The equation $x^3 + 5x^2 + 6x + 1 = 0$ has one positive real solution.

b. Descartes' Rule of Signs gives the exact number of positive and negative roots for a polynomial equation.

c. If $x^3 - 5x^2 + 3x - 2 = 0$, we must replace x with $-x$ to find possibilities for negative real solutions, obtaining $-x^3 + 5x^2 - 3x - 2 = 0$.

d. All real roots of the equation $x^4 - 3x^2 + 2x - 5 = 0$ must lie between -3 and 2.

48. Which one of the following is true?

a. Descartes' Rule of Signs never gives the exact number of positive real roots to a polynomial equation.

b. All real roots of the equation $x^4 - 2x^3 - 9x^2 + 2x + 8 = 0$ must lie between -3 and 5.

c. The equation $x^3 + 8x - 30 = 0$ cannot have a solution between 2 and 3.

d. By graphing $f(x) = 3x^4 + 4x^3 - 7x^2 - 2x - 3$ in a $[-3, 2]$ by $[-20, 20]$ viewing rectangle, it is obvious that the real zeros are rational.

Technology Problems

The equations in Problems 49–52 have real solutions that are rational. Use the Rational Zero Theorem to list all possible rational roots. Then graph the polynomial in the given viewing rectangle to determine which possible rational roots are actual solutions of the equation.

49. $2x^3 - 15x^2 + 22x + 15 = 0$; $[-1, 6]$ by $[-50, 50]$
50. $6x^3 - 19x^2 + 16x - 4 = 0$; $[0, 2]$ by $[-3, 2]$
51. $2x^4 + 7x^3 - 4x^2 - 27x - 18 = 0$; $[-4, 3]$ by $[-45, 45]$
52. $4x^4 + 4x^3 + 7x^2 - x - 2 = 0$; $[-2, 2]$ by $[-5, -5]$
53. Use Descartes' Rule of Signs to determine the possible number of positive and negative real zeros of $f(x) = 3x^4 + 5x^2 + 2$. What does this mean in terms of the graph of f? Verify your result by using a graphing utility to graph f.
54. Use Descartes' Rule of Signs to determine the possible number of positive and negative real zeros of $f(x) = x^5$
$- x^4 + x^3 - x^2 + x - 8$. Verify your result by using a graphing utility to graph f.

55. Make up a number of polynomial functions of odd degree and graph each function. Is it possible for the graph to have no real zeros? Explain. Try doing the same thing for polynomial functions of even degree. Now is it possible to have no real zeros?

56. Show that -1 is a lower bound of $f(x) = x^3 - 53x^2 + 103x - 51$. Show that 60 is an upper bound. Use this information and a graphing utility to draw a relatively complete graph of f.

For Problems 57–58, use a graphing utility to determine upper and lower bounds for the zeros of f. Verify your observations using synthetic division.

57. $f(x) = 2x^3 + x^2 - 14x - 7$

58. $f(x) = 2x^4 - 7x^3 - 5x^2 + 28x - 12$

Writing in Mathematics

59. Give an example of a polynomial function for which Descartes' Rule of Signs eliminates all possible negative rational zeros but retains the possibility of positive rational zeros. Describe the process by which you generated the function.

60. Describe the relationship between the zeros of $f(x) = \frac{1}{4}x^3 + x^2 + 2x - 3$ and $g(x) = x^3 + 4x^2 + 8x - 12$. Are f and g the same function? Explain.

61. How many real solutions does $x^5 + x + 1 = 0$ have? How do you know?

62. Give an example of a polynomial equation that has no real solutions. Describe how you obtained the equation.

63. Most graphing utilities have the ability to find solutions to polynomial equations. The polynomial root-finding

capabilities are accessed using a sequence similar to $\boxed{\text{2nd}}$ $\boxed{\text{POLY}}$. Once the POLY order screen appears, enter an integer representing the order of the equation and then enter its coefficients. Try doing this for the following polynomial equations whose solution sets are given.

$$2x^4 + 7x^3 - 4x^2 - 27x - 18 = 0 \quad \{-3, -\tfrac{3}{2}, -1, 2\}$$

$$6x^4 - 19x^3 + 21x^2 - 19x + 15 = 0 \quad \{\tfrac{3}{2}, \tfrac{5}{3}, \pm i\}$$

Describe what happens. Should the techniques for solving polynomial equations discussed in this section be eliminated, using only calculator key sequences to obtain zeros? Defend your position on this issue.

Critical Thinking Problems

64. In this problem, we lead you through the steps involved in the proof of the Rational Zero Theorem. Consider the polynomial equation

$$a_n x^n + a_{n-1} x^{n-1} + a_{n-2} x^{n-2} + \cdots + a_1 x + a_0 = 0$$

where $\dfrac{p}{q}$ is a rational solution reduced to lowest terms.

a. Substitute $\dfrac{p}{q}$ for x in the equation and show that the equation can be written as

$$a_n p^n + a_{n-1} p^{n-1} q + a_{n-2} p^{n-2} q^2 + \cdots + a_1 p q^{n-1} = -a_0 q^n$$

b. Why is p a factor of the left side of the equation?
c. Because p divides the left side, it must also divide the right side. However, because $\dfrac{p}{q}$ is reduced to

lowest terms, p cannot divide q. Thus, p and q have no common factors other than -1 and 1. Because p does divide the right side and it is not a factor of q^n, what can you conclude?

d. Rewrite the equation from part (a) with all terms containing q on the left and the term that does not have a factor of q on the right. Use an argument that parallels parts (b) and (c) to conclude that q is a factor of a_n.

65. Consider the polynomial equation $x^2 - 3 = 0$. Use the Rational Root Theorem to show that $\sqrt{3}$ is not a rational number.

66. Suppose the equation $x^3 + x^2 + x - p = 0$ (p is a prime number) has one rational root. Find value(s) of p. For each value of p, find the solution set of the equation.

SECTION 3.4

Solutions   Tutorial   Video
Manual   6

More on Zeros of Polynomial Functions

Objectives

1 Approximate irrational zeros.
2 Factor polynomials over the complex nonreals.
3 Find polynomials with given zeros.
4 Solve a polynomial equation with a given nonreal complex root.

In this section, we continue our study of zeros of polynomial functions. We'll be focusing more closely on both irrational and nonreal complex zeros.

1 Approximating irrational zeros.

Approximating Irrational Zeros

A graphing utility provides an effective method for finding decimal approximations to irrational roots of polynomial equations. There is also an effective numerical method for obtaining these approximations.

Let's consider a specific example. The graph of $f(x) = x^3 + x - 1$, shown in Figure 3.38, was obtained with a graphing utility using a standard range setting. It appears that f has a real zero somewhere between 0.5 and 1. This seems plausible because the graph of f lies below the x-axis to the left of this zero and above the x-axis to the right of this zero.

We can provide numerical support for our observation that there may be a real zero between 0.5 and 1 by evaluating the function at these values:

$$f(0.5) = -0.375 < 0 \quad \text{and} \quad f(1) = 1 > 0.$$

Since the function value is negative at 0.5 and positive at 1, its graph will have to cross the x-axis somewhere between 0.5 and 1.

Figure 3.39 once again shows the graph of f, but this time in a $[-1, 1]$ by $[-2, 2]$ viewing rectangle. It now appears that there is a real zero somewhere between 0.6 and 0.8. Once again, the graph lies below the x-axis prior to this value and above the x-axis after this value. This is reinforced by evaluating the function at each value:

$$f(0.6) = -0.184 < 0 \quad \text{and} \quad f(0.8) = 0.312 > 0.$$

Figure 3.40 shows an approximation for the function's zero. Correct to the nearest hundredth, the zero is 0.68.

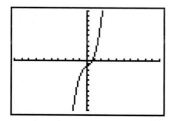

Figure 3.38

The graph of $f(x) = x^3 + x - 1$ with a real zero between 0.5 and 1

Discover for yourself

Prove that the real zero for $f(x) = x^3 + x - 1$, shown in Figure 3.38, is not rational. Use the Rational Zero Theorem and show that the function does not have any possible rational zeros between 0 and 1.

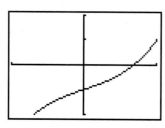

Figure 3.39

The graph of $f(x) = x^3 + x - 1$

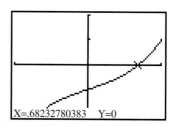

X=.68232780383   Y=0

Figure 3.40

The zero is approximately 0.68.

Using technology

The graph in Figure 3.40 was obtained by entering

$$y_1 = x^3 + x - 1$$

and

$$y_2 = 0$$

and then using the $\boxed{\text{ISECT}}$ feature. Another option is to enter only y_1, changing the scale on the x-axis to, say, 0.001, and zooming in several times.

A nonvisual approach is to use the utility's polynomial equation-solving feature as shown in the figures. To solve $x^3 + 0x^2 + x - 1 = 0$, enter the order (3), followed by the coefficients (1, 0, 1, and -1). Then press $\boxed{\text{SOLVE}}$. The third screen on the right verifies that the real zero (x_3) is approximately 0.68.

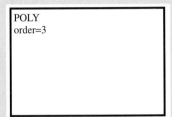

The order (degree) of $x^3 + x - 1 = 0$ is 3.

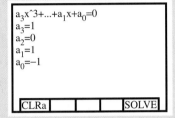

The coefficients of $x^3 + 0x^2 + x - 1 = 0$ are 1, 0, 1, -1.

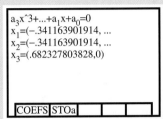

The real zero (x_3) is approximately 0.68.

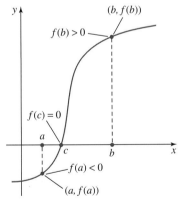

Figure 3.41

The graph must cross the x-axis at some value between a and b.

Discover for yourself

Use the Rational Zero Theorem to show that

$$f(x) = x^3 - 2x - 5$$

does not have any possible rational zeros between 2 and 3. Thus, the zero between 2 and 3 must be irrational.

The numerical support that we provided in narrowing down the location of a real zero is generalized in Figure 3.41. The figure shows that if $(a, f(a))$ lies below the x-axis and $(b, f(b))$ lies above the x-axis, the smooth, continuous graph of a polynomial function f must cross the x-axis at some value c between a and b. This value is a real zero for the function.

These observations are summarized in the *Intermediate Value Theorem*.

The Intermediate Value Theorem for polynomials

Let $f(x)$ be a polynomial with real coefficients. If $f(a)$ and $f(b)$ have opposite signs, then there is at least one value of c between a and b for which $f(c) = 0$. Equivalently, the equation $f(x) = 0$ has at least one real root between a and b.

EXAMPLE 1 Approximating a Real Zero

a. Show that the polynomial $f(x) = x^3 - 2x - 5$ has a zero between 2 and 3.
b. Use the Intermediate Value Theorem to find an approximation for this zero to the nearest tenth.
c. Use a graphing utility to verify part (b) and find the function's real zero correct to the nearest thousandth.

Solution

a. Since $f(2) = -1 < 0$ and $f(3) = 16 > 0$, the Intermediate Value Theorem states that there is a number c between 2 and 3 for which $f(c) = 0$. In other words, the polynomial has a zero between 2 and 3.

b. A numerical approach is to evaluate f at successive tenths between 2 and 3, looking for a sign change. This sign change will place the real zero between the successive tenths.

x	$f(x) = x^3 - 2x - 5$	
2	$f(2) = 2^3 - 2(2) - 5 =$ $\qquad -1$	⎤ sign
2.1	$f(2.1) = (2.1)^3 - 2(2.1) - 5 = 0.061$	⎦ change

The sign change indicates that f has a real zero between 2 and 2.1. We now follow a similar procedure to locate the real zero between successive hundredths. We divide the interval $[2, 2.1]$ into ten equal subintervals, evaluate f at each endpoint, and look for a sign change.

$f(2.00) = -1$ $f(2.06) = -0.378184$

$f(2.01) = -0.899399$ $f(2.07) = -0.270257$

$f(2.02) = -0.797592$ $f(2.08) = -0.161088$

$f(2.03) = -0.694573$ $f(2.09) = -0.050671$ ⎤ sign

$f(2.04) = -0.590336$ $f(2.1) = 0.061$ ⎦ change

$f(2.05) = -0.484875$

The sign change indicates that f has a real zero between 2.09 and 2.1. Correct to the nearest tenth, the zero is 2.1.

c. The graph in Figure 3.42 was obtained by entering $Y_1 = X \boxed{\wedge} 3 - 2X - 5$ and $Y_2 = 0$ and using the $\boxed{\text{ISECT}}$ feature in a $[-3, 3]$ by $[-10, 10]$ viewing rectangle. You can also change the scale on the x-axis to 0.001 and zoom in several times. Correct to the nearest thousandth, the function's real zero is 2.095. ∎

Discover for yourself

Spreadsheet programs or the table function can be used to make tables and perform calculations on a graphing utility. Use these features to find the values for

$f(x) = x^3 - 2x - 5$

for values of the domain shown on the right. You can use the values of $f(x)$ to determine where zeros occur. Now modify this procedure to get closer approximations of the zero.

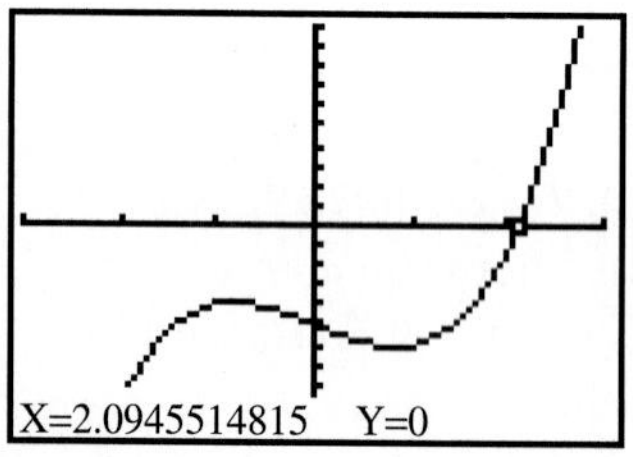

X=2.0945514815   Y=0

Figure 3.42
The real zero for $f(x) = x^3 - 2x - 5$ is approximately 2.095.

Nonreal Complex Zeros

Up to this point, our focus has been on real zeros of polynomials. Let's now look at an example of a polynomial equation that has two roots that are complex and nonreal.

EXAMPLE 2   **Solving a Polynomial Equation**

Solve: $x^4 - 6x^2 - 8x + 24 = 0$

Solution

First we list all possible rational solutions.

p (factors of the constant term, 24): $\pm 1, \pm 2, \pm 3, \pm 4, \pm 6, \pm 8, \pm 12, \pm 24$

q (factors of the leading coefficient, 1): ± 1

Possible rational $\left(\frac{p}{q}\right)$ solutions are $\pm 1, \pm 2, \pm 3, \pm 4, \pm 6, \pm 8, \pm 12, \pm 24$.

Using synthetic division, we can show that 2 is a solution.

$$\underline{2}\,\begin{array}{rrrrr} 1 & 0 & -6 & -8 & 24 \\ & 2 & 4 & -4 & -24 \\ \hline 1 & 2 & -2 & -12 & 0 \end{array}$$ ← Coefficients of $x^4 + 0x^3 - 6x^2 - 8x + 24$

← The zero remainder indicates that 2 is a solution.

Now we can rewrite the given equation in factored form.

$x^4 - 6x^2 - 8x + 24 = 0$ This is the given equation.

$(x - 2)(x^3 + 2x^2 - 2x - 12) = 0$ This is the result obtained from the synthetic division.

$(x - 2)(x - 2)(x^2 + 4x + 6) = 0$ Reapply the Rational Zero Theorem and synthetic division to

$(x - 2)^2(x^2 + 4x + 6) = 0$ $x^3 + 2x^2 - 2x - 12 = 0$.
2 is a root:

$$\underline{2}\,\begin{array}{rrrr} 1 & 2 & -2 & -12 \\ & 2 & 8 & 12 \\ \hline 1 & 4 & 6 & 0 \end{array}$$

$(x - 2)^2 = 0$ or $x^2 + 4x + 6 = 0$ Set each factor equal to 0.

$x = 2$ Solve for x.

We can use the quadratic formula to solve $x^2 + 4x + 6 = 0$.

$$x = \frac{-4 \pm \sqrt{4^2 - 4(1)(6)}}{2(1)} = \frac{-4 \pm \sqrt{-8}}{2}$$ $a = 1, b = 4$, and $c = 6$. Substitute in $\dfrac{-b \pm \sqrt{b^2 - 4ac}}{2a}$.

$$= \frac{-4 \pm 2i\sqrt{2}}{2}$$ $\sqrt{-8} = \sqrt{4 \cdot 2}\,i = 2i\sqrt{2}$

$$= -2 \pm i\sqrt{2}$$ Divide numerator and denominator by 2.

The solution set of $x^4 - 6x^2 - 8x + 24 = 0$ is $\{2, -2 - i\sqrt{2}, -2 + i\sqrt{2}\}$. ■

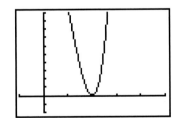

Figure 3.43

The graph of $f(x) = x^4 - 6x^2 - 8x + 24$ in a $[-1, 5]$ by $[-2, 10]$ viewing rectangle

In Example 2, the factor $x - 2$ occurred twice in the factorization. The factor $(x - 2)^2$ yielded the repeated root 2 of multiplicity 2. Notice how the graph of $f(x) = x^4 - 6x^2 - 8x + 24 = 0$ in Figure 3.43 flattens out at $(2, 0)$. The end behavior (rising to the left and to the right) suggests that we have a relatively complete picture of the function. The graph illustrates that nonreal complex zeros do not appear as x-intercepts.

We can generalize our work in Example 2 and discover two useful properties of polynomial equations.

Discover for yourself

Shown on page 382 are four polynomial equations and their solutions. Study the given information and:
1. Write a statement about the number of solutions for a polynomial equation of degree n if multiplicities are counted.
2. Write a statement about the relationship between the two nonreal complex solutions.

Equation	Degree	Solutions	Number of Solutions (Counting Multiplicities)
$x^4 - 6x^2 - 8x + 24 = 0$ (Example 2)	4	2(Multiplicity two) $-2 + i\sqrt{2}$ $-2 - i\sqrt{2}$	Four
$2x^4 - 3x^3 + 12x^2 + 22x - 60 = 0$	4	$-2, \dfrac{3}{2},$ $1 + 3i, 1 - 3i$	Four
$x^5 - 5x^4 + 12x^3 - 24x^2 + 32x - 16 = 0$	5	1, 2, (Multiplicity two), $2i$ (or $0 + 2i$), $-2i$ (or $0 - 2i$)	Five
$3x^3 + 13x^2 - 16 = 0$	3	$-4, -\frac{4}{3}, 1$	Three

In the Discover for yourself, were you able to write the following relationships?

> **Properties of polynomial equations**
>
> **1.** If a polynomial equation is of degree n, then counting multiplicities, the equation has exactly n solutions.
> **2.** If $a + bi$ is a solution of a polynomial equation ($b \neq 0$), then the nonreal complex number $a - bi$ is also a solution. Nonreal complex roots, if they exist, occur in conjugate pairs.

Let's develop these ideas in more detail.

The Fundamental Theorem of Algebra

In 1799, a 22-year-old student named Carl Friedrich Gauss proved in his doctoral dissertation that every polynomial equation of degree n, where $n \geqslant 1$, has at least one complex solution. His result is called the *Fundamental Theorem of Algebra*.

> **The Fundamental Theorem of Algebra**
>
> If $f(x)$ is a polynomial of degree n, where $n \geqslant 1$, then the equation $f(x) = 0$ has at least one complex root.

As an immediate consequence of the Fundamental Theorem, we can prove the following theorem.

> **The Linear Factorization Theorem**
>
> If $f(x) = a_n x^n + a_{n-1} x^{n-1} + \cdots + a_1 x + a_0$, where $n \geqslant 1$ and $a_n \neq 0$, then
>
> $$f(x) = a_n(x - c_1)(x - c_2) \cdots (x - c_n)$$

2 Factor polynomials over the complex nonreals.

where $c_1, c_2, \ldots, c_n$ are complex numbers (possibly real and not necessarily distinct).

In words: An nth-degree polynomial can be expressed as the product of n linear factors.

Proof of the Linear Factorizaton Theorem

To prove the Linear Factorization Theorem, we'll begin with a polynomial equation of degree n:

$$f(x) = 0.$$

By the Fundamental Theorem of Algebra, we know that this equation has at least one complex root; we'll call it c_1. By the Factor Theorem, we know that $x - c_1$ is a factor of $f(x)$. Therefore, we obtain

$$(x - c_1)q_1(x) = 0 \qquad \text{The degree of the polynomial } q_1(x) \text{ is } n - 1. \text{ (Since we divided by } x - c_1, q_1(x) \text{ has } a_n \text{ as its leading coefficient).}$$

$$x - c_1 = 0 \quad \text{or} \quad q_1(x) = 0 \qquad \text{Set each factor equal to 0.}$$

If the degree of $q_1(x)$ is at least 1, by the Fundamental Theorem the equation $q_1(x) = 0$ has at least one complex root. We'll call it c_2. The Factor Theorem gives us

$$q_1(x) = 0 \qquad \text{The degree of } q_1(x) \text{ is } n - 1.$$

$$(x - c_2)q_2(x) = 0 \qquad \text{The degree of } q_2(x) \text{ is } n - 2.$$

$$x - c_2 = 0 \quad \text{or} \quad q_2(x) = 0 \qquad \text{Set each factor equal to 0.}$$

Let's see what we have up to this point, and then continue the process.

$$f(x) = 0 \qquad \text{This is the original polynomial equation of degree } n.$$

$$(x - c_1)q_1(x) = 0 \qquad \text{This is the result from our first application of the Fundamental Theorem.}$$

$$(x - c_1)(x - c_2)q_2(x) = 0 \qquad \text{This is the result from our second application of the Fundamental Theorem.}$$
$$\vdots$$

$$(x - c_1)(x - c_2)(x - c_3) \cdots (x - c_n)a_n = 0 \qquad \text{Continue the process } n \text{ times until obtaining the quotient } q_n(x) = a_n.$$

Equating the first and last expressions that are both equal to 0, we obtain

$$f(x) = a_n(x - c_1)(x - c_2) \cdots (x - c_n)$$

as we wished to show.

In Example 2, we found that $x^4 - 6x^2 - 8x + 24 = 0$ has $\{2, -2 - i\sqrt{2}, -2 + i\sqrt{2}\}$ as a solution set, where 2 is a root of multiplicity 2. With these zeros, we can write the polynomial $f(x) = x^4 - 6x^2 - 8x + 24$ as a product of linear factors.

$$x^4 - 6x^2 - 8x + 24 = (x - 2)(x - 2)[x - (-2 - i\sqrt{2})][x - (-2 + i\sqrt{2})]$$
$$= (x - 2)^2(x + 2 + i\sqrt{2})(x + 2 - i\sqrt{2})$$

Study tip

The Linear Factorization Theorem tells us that an nth-degree polynomial can be expressed as the product of n linear factors. However, it does not tell us how to find the factors. Finding the factors involves the techniques for solving polynomial equations illustrated in Example 2.

Notice that the Linear Factorization Theorem tells us that an nth-degree polynomial has n linear factors. Equivalently, an nth-degree polynomial equation has precisely n solutions. Keep in mind that these solutions can be real or nonreal complex, and they may be repeated.

> **The Numbers of Roots Theorem**
>
> If $f(x)$ is a polynomial of degree at least 1, then $f(x) = 0$ has exactly n roots, where roots are counted according to their multiplicity.

The Linear Factorization Theorem involves factors somewhat different than you are used to seeing. For example, the polynomial $x^2 - 3$, although irreducible over the rational numbers, can be factored over the reals as follows:

$$x^2 - 3 = (x + \sqrt{3})(x - \sqrt{3}).$$ Use $a^2 - b^2 = (a + b)(a - b)$ with $a = x$ and $b = \sqrt{3}$.

However, the polynomial $x^2 + 1$ is irreducible over the reals, but reducible over the complex nonreals.

$$x^2 + 1 = (x + i)(x - i)$$

Study tip

The sum of squares, irreducible over the reals, can be factored over the complex nonreals as

$$a^2 + b^2 = (a + bi)(a - bi)$$

EXAMPLE 3 Factoring a Polynomial

Factor $x^4 - 3x^2 - 28$:

a. As the product of factors that are irreducible over the rationals.
b. As the product of factors that are irreducible over the reals.
c. In completely factored form involving complex nonreals.

Solution

a. $x^4 - 3x^2 - 28 = (x^2 - 7)(x^2 + 4)$

Both quadratic factors are irreducible over the rationals.

b. $= (x + \sqrt{7})(x - \sqrt{7})(x^2 + 4)$

The third factor is still irreducible over the reals.

c. $= (x + \sqrt{7})(x - \sqrt{7})(x + 2i)(x - 2i)$

This is the completely factored form using complex nonreals. ■

Conjugate Pairs

Another result that is sometimes helpful in finding roots and in understanding graphing patterns for polynomial functions is the Conjugate Roots Theorem.

> **The Conjugate Roots Theorem**
>
> Let $f(x)$ be a polynomial with real coefficients. If the complex number $a + bi$ is a root of $f(x) = 0$, then $a - bi$ is also a root (a and b are real numbers).
>
> In words: Complex nonreal zeros occur in conjugate pairs.

Barry Flanagan "Large Mirror Nijinski" 1992, bronze, 121 3/4 × 89 × 36 in. (309.2 × 226 × 91.4 cm). © Barry Flanagan, courtesy Waddington Galleries, London.

3 Find polynomials with given zeros.

EXAMPLE 4   **Finding a Polynomial Satisfying a Given Description**

Find a fourth-degree polynomial function $f(x)$ with real coefficients that has $-2, 2,$ and i as zeros and such that $f(3) = -150$.

Solution

Because i is a zero and the polynomial has real coefficients, the conjugate $-i$ must also be a zero. We can now use the Linear Factorization Theorem.

$$f(x) = a_n(x - c_1)(x - c_2)(x - c_3)(x - c_4)$$ This is the linear factorization for a fourth-degree polynomial.

$$= a_n(x + 2)(x - 2)(x - i)(x + i)$$ Use the given zeros: $c_1 = -2, c_2 = 2, c_3 = i,$ and, from above, $c_4 = -i$.

$$= a_n(x^2 - 4)(x^2 + 1)$$ Multiply.

$$= a_n(x^4 - 3x^2 - 4)$$ Complete the multiplication.

$$f(3) = a_n(3^4 - 3 \cdot 3^2 - 4) = -150$$ To find a_n, use the fact that $f(3) = -150$.

$$50a_n = -150$$ Solve for a_n.

$$a_n = -3$$

Substituting -3 for a_n in the formula for $f(x)$, we obtain

$$f(x) = -3(x^4 - 3x^2 - 4).$$

Equivalently,

$$f(x) = -3x^4 + 9x^2 + 12.$$ ∎

Using technology

In Example 4, we found that

$$f(x) = -3x^4 + 9x^2 + 12$$

is the function that has $-2, 2,$ and i as zeros and such that $f(3) = -150$. Now let's use a graphing utility to check that this function really does satisfy the four given conditions.

Since we are checking real zeros of -2 and 2, a reasonable range setting for x is

Xmin $= -3$, Xmax $= 3$, Xscl $= 1$.

To establish a range setting for y, notice that the leading coefficient (-3) is negative and the degree (4) is even, so the graph falls to the left and to the right. The y-intercept is

$$f(0) = -3(0)^4 + 9(0)^2 + 12, \text{ or } 12$$

and we also know that $f(3) = -150$. Using this information, we've chosen a range setting for y as

Ymin $= -200$, Ymax $= 20$, Yscl $= 20$.

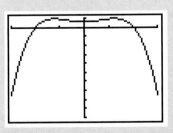

Figure 3.44

The graph of $f(x) = -3x^4 + 9x^2 + 12$, verifying real zeros of -2 and 2

The graph of $f(x) = -3x^4 + 9x^2 + 12$ is shown in Figure 3.44, verifying real zeros of -2 and 2. The graph crosses the x-axis at each zero without flattening out, a convincing visual representation that each zero has multiplicity 1. Trace along the curve and verify that $f(3) = -150$.

EXAMPLE 5 **Finding a Polynomial with Given Zeros**

Find a third-degree polynomial function $f(x)$ with real coefficients that has 3 and $1 - 2i$ as zeros and such that $f(2) = 20$.

Using technology

The graph of

$f(x) = -4x^3 + 20x^2 - 44x + 60$

was obtained using

Xmin = 0, Xmax = 5, Xscl = 1,

Ymin = -80, Ymax = 80,

Yscl = 20

The graph verifies a real zero at 3, and even without using the TRACE feature, we can see that $f(2) = 20$ since (2, 20) appears to be a point on the graph.

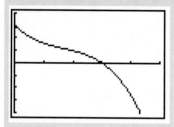

Solution

Because $1 - 2i$ is a zero of f, so is $1 + 2i$.

$f(x) = a_n(x - c_1)(x - c_2)(x - c_3)$ — This is the linear factorization for a cubic (third-degree) polynomial.

$= a_n(x - 3)[x - (1 - 2i)][x - (1 + 2i)]$ — Use the zeros: $c_1 = 3, c_2 = 1 - 2i,$ and $c_3 = 1 + 2i$.

$$\overset{\text{F}}{}\overset{\text{O}}{}\overset{\text{I}}{}\overset{\text{L}}{}$$
$= a_n(x - 3)[x^2 - x(1 + 2i) - x(1 - 2i) + (1 - 4i^2)]$ — Multiply the last two factors using the FOIL pattern for binomial multiplication.

$= a_n(x - 3)(x^2 - 2x + 5)$ — Simplify. Use the fact that $i^2 = -1$.

$= a_n(x^3 - 5x^2 + 11x - 15)$ — Complete the multiplication.

$f(2) = a_n(2^3 - 5 \cdot 2^2 + 11 \cdot 2 - 15) = 20$ — Use the fact that $f(2) = 20$ to find the value of a_n.

$-5a_n = 20$

$a_n = -4$

Substituting -4 for a_n in the formula for $f(x)$, we obtain

$f(x) = -4(x^3 - 5x^2 + 11x - 15) = -4x^3 + 20x^2 - 44x + 60$ ∎

4 Solve a polynomial equation with a given nonreal complex root.

EXAMPLE 6 **Using the Conjugate Roots Theorem to Solve a Polynomial Equation**

Solve $x^4 - 4x^3 + 3x^2 + 8x - 10 = 0$ given that $2 + i$ is a root.

Solution

Because complex nonreal roots come in conjugate pairs, we know that $2 - i$ is also a root. By the Factor Theorem, both

$[x - (2 + i)]$ and $[x - (2 - i)]$

are factors of the given polynomial. We multiply these known factors.

$[x - (2 + i)][x - (2 - i)]$

$$\overset{\text{F}}{}\overset{\text{O}}{}\overset{\text{I}}{}\overset{\text{L}}{}$$
$= x^2 - x(2 - i) - x(2 + i) + (2 + i)(2 - i)$ — Multiply using the FOIL method.

$= x^2 - 2x + ix - 2x - ix + (4 - i^2)$ — Continue multiplying.

$= x^2 - 4x + 5$ — Simplify, using $i^2 = -1$.

We can find the other factor(s) by dividing $x^2 - 4x + 5$ into the polynomial on the left side of the given equation.

$$
\begin{array}{r}
x^2 \qquad\quad - 2 \\
x^2 - 4x + 5 \overline{)\ x^4 - 4x^3 + 3x^2 + 8x - 10} \\
\underline{x^4 - 4x^3 + 5x^2} \\
-2x^2 + 8x - 10 \\
\underline{-2x^2 + 8x - 10} \\
0
\end{array}
$$

We can now solve the given equation.

$x^4 - 4x^3 + 3x^2 + 8x - 10 = 0$	This is the original equation.
$(x^2 - 4x + 5)(x^2 - 2) = 0$	Factor using the result of the polynomial long division.
$x^2 - 4x + 5 = 0 \quad \text{or} \quad x^2 - 2 = 0$	Set each factor equal to 0.
$x = 2 \pm i \qquad\qquad x = \pm\sqrt{2}$	Solve for x. We know the roots for the first equation by our previous analysis.

The solution set is $\{2 \pm i, \pm\sqrt{2}\}$. ■

Using technology

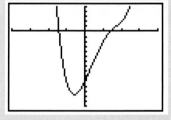

Figure 3.45

The graph of $f(x) = x^4 - 4x^3 + 3x^2 + 8x - 10$ with real zeros approximately equal to -1.4 and 1.4

We can use the results of Example 6 to obtain a relatively complete graph of

$$f(x) = x^4 - 4x^3 + 3x^2 + 8x - 10.$$

The real zeros $\pm\sqrt{2}$ correspond to x-intercepts at approximately ±1.4, so we'll set Xmin $= -4$ and Xmax $= 4$, with Xscl $= 1$. Since $n = 4$ and $a_n = 1 > 0$, the graph rises to the left and right.

The graph of f shown in Figure 3.45 was obtained using the following range setting:

Xmin $= -4$, Xmax $= 4$, Xscl $= 1$, Ymin $= -15$, Ymax $= 5$, Yscl $= 1$

The graph confirms that the real zeros of f are $\pm\sqrt{2}$, or approximately ±1.4.

Study tip

You can use the graph of a polynomial function to convince yourself that a polynomial of degree n has n roots, and that nonreal complex zeros come in pairs. Let's do this for the third-degree polynomial function

$$f(x) = a_3 x^3 + a_2 x^2 + a_1 x + a_0.$$

If $a_3 > 0$, we know that the graph of f falls to the left and rises to the right. As we see in Figure 3.46 on page 388, the first graph of f has three distinct real zeros, with each real zero corresponding to an x-intercept. By sliding the graph up and down, nonreal complex zeros are gained in pairs.

This also occurs in the case of multiple zeros, shown in Figure 3.47 on page 388. (Remember that a graph crosses the x-axis at a zero of odd multiplicity.)

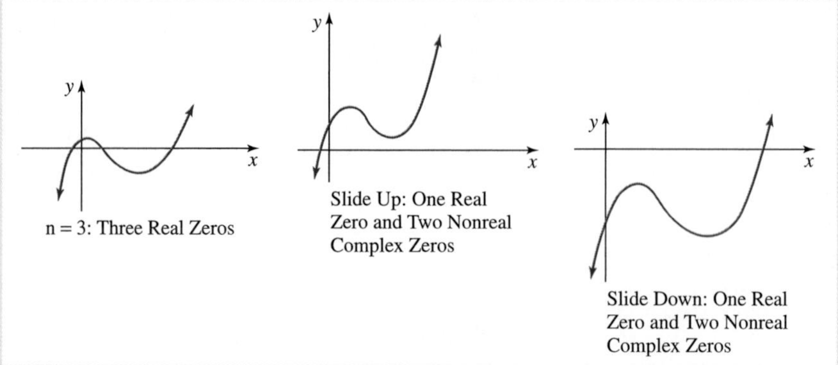

Figure 3.46

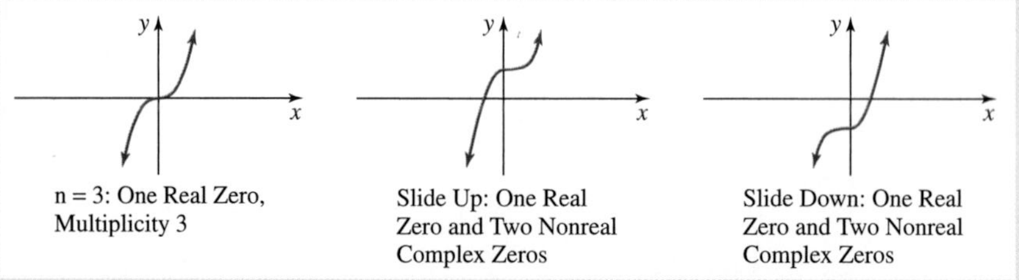

Figure 3.47

PROBLEM SET 3.4

Practice Problems_____

In Problems 1–8, show that each polynomial has a zero between the given integers. Then use the Intermediate Value Theorem to find an approximation for this zero to the nearest tenth. Use a graphing utility to verify your result.

1. $f(x) = x^3 - x - 1$; between 1 and 2

2. $f(x) = x^3 - 4x^2 + 2$; between 0 and 1

3. $f(x) = 2x^4 - 4x^2 + 1$; between -1 and 0

4. $f(x) = x^4 + 6x^3 - 18x^2$; between 2 and 3

5. $f(x) = x^3 + x^2 - 2x + 1$; between -3 and -2

6. $f(x) = x^5 - x^3 - 1$; between 1 and 2

7. $f(x) = 3x^3 - 10x + 9$; between -3 and -2

8. $f(x) = 3x^3 - 8x^2 + x + 2$; between 2 and 3

In Problems 9–12, find an nth-degree polynominal function with real coefficients satisfying the given conditions. Use a graphing utility to graph the function and verify your answers.

9. $n = 3$; -5 and $4 + 3i$ are zeros; $f(2) = 91$

10. $n = 3$; $-\frac{3}{4}$ and $1 + \frac{1}{2}i$ are zeros; $f(1) = 7$

11. $n = 4$; i and $3i$ are zeros; $f(-1) = 20$

12. $n = 4$; -2, $-\frac{1}{2}$, and i are zeros; $f(1) = 18$

In Problems 13–24, use the given root to find the solution set of the polynomial equation.

13. $x^3 - 2x^2 + 4x - 8 = 0$; $-2i$

14. $x^4 + 13x^2 + 36 = 0$; $3i$

15. $3x^3 - 7x^2 + 8x - 2 = 0$; $1 + i$

16. $x^3 - 7x^2 + 16x - 10 = 0$; $3 + i$

17. $x^4 - 6x^2 + 25 = 0$; $2 - i$

18. $x^4 - x^3 - 9x^2 + 29x - 60 = 0$; $1 + 2i$

19. $3x^4 + 2x^3 + 13x^2 + 22x - 6 = 0$; $1 - i$

20. $4x^4 - 28x^3 + 129x^2 - 130x + 125 = 0$; $3 - 4i$

21. $x^4 - 12x^3 + 56x^2 - 120x + 96 = 0; 3 - i\sqrt{3}$

22. $9x^4 + 18x^3 + 20x^2 - 32x - 64 = 0; -1 - i\sqrt{3}$

23. $x^5 - 2x^4 + 6x^3 + 24x^2 + 5x + 26 = 0; 2 - 3i$

24. $x^5 + x^4 + 7x^3 - x^2 + 12x - 20 = 0; -1 + 2i$

In Problems 25–32, find all the zeros of the function and write the polynomial as a product of linear factors.

25. $f(x) = x^3 - x^2 + 25x - 25$

26. $g(x) = x^3 - 10x^2 + 33x - 34$

27. $p(x) = x^4 + 37x^2 + 36$

28. $p(x) = x^4 + 8x^3 + 9x^2 - 10x + 100$

29. $p(x) = 16x^4 + 36x^3 + 16x^2 + x - 30$

30. $p(x) = 2x^4 - x^3 + 7x^2 - 4x - 4$

31. $g(r) = 3r^6 + 5r^5 + r^4 + 5r^3 + r^2 + 5r - 2$

32. $g(x) = x^5 + x^4 - 14x^3 - 14x^2 + 49x + 49$

In Problems 33–34, use the Rational Zero Theorem to show that the equation has no rational roots. Approximate any irrational root to the nearest tenth, checking your answer using the polynomial root-finding capability of your graphing utility.

33. $x^3 - x^2 - 2 = 0$

34. $x^4 + 2x - 2 = 0$

Application Problems

35. The model $f(x) = -0.0013x^3 + 0.078x^2 - 1.43x + 18.1$ describes the percent of U.S. families below the poverty level x years after 1960, where $0 \le x \le 30$.

   a. What percent of U.S. families were below the poverty level in 1970?

   b. In what other years were the percent of U.S. families below the poverty level the same as that of 1970?

36. One of Egypt's largest pyramids was built by Khufu, a king of the fourth dynasty. The pyramid has a square base and a height that is 8 dekameters less than the length of a side of the square base. If the pyramid's volume is 2645 cubic dekameters, what are its dimensions? (The volume of a pyramid is the product of $\frac{1}{3}$, the area of the base, and the height.)

Romilly Lockyer / The Image Bank

37. A slice 2 centimeters thick is cut from one face of a cube. If the volume of the resulting solid is 567 cubic centimeters, find the dimensions of the cube.

38. A hollow cubical box has a uniform thickness of 1 inch. The space inside the box is $\frac{1}{8}$ the volume of the material used in its construction. Find the dimensions of the box to the nearest tenth of an inch.

39. A person standing on the roof of a hotel 30 feet above the ground observes a bungee jumper leaping from a platform 70 feet above the roof. The bungee jumper passes the roof level after 3 seconds, again after 7 seconds on the way back up, and after 10 seconds on the way back down. Find a cubic polynomial function that models the height of the bungee jumper relative to the hotel's roof level.

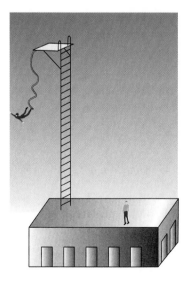

True–False Critical Thinking Problems

40. Which one of the following is true?

a. Some polynomials of odd degree have no x-intercepts.

b. The graph of $f(x) = x^5 + 2x^4 - 6x^3 + 2x - 3$ shown in the figure is drawn completely.

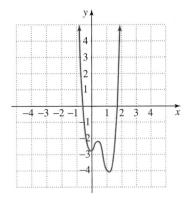

c. If $c > 0$, $d > 0$, and $e > 0$, then the solution set of $x^4 + cx^2 + dx - e = 0$ contains one positive number, one negative number, and two nonreal complex numbers.

d. Fourth-degree polynomial functions have graphs with at least one x-intercept.

41. Which one of the following is true?

a. The function whose graph is shown in the figure at the top of the next column cannot be a fourth-degree polynomial.

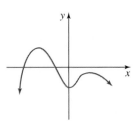

b. The equation $x^n - 1 = 0$ has $n - 2$ nonreal complex roots if n is even. ($n \neq 0$)

c. The graph of $f(x) = (x - 2)^3(x + 1)^2$ shown in the figure below is not drawn completely.

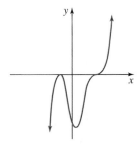

d. A polynomial function of degree n has a graph with precisely $n - 1$ turning points.

Technology Problems

42. The function $f(x) = -0.00002x^3 + 0.008x^2 - 0.3x + 6.95$ models the number of annual physician visits f by a person of age x.

a. Graph the function for meaningful values of x and discuss what the graph reveals in terms of the variables described by the model.

b. Use the polynomial root-finding capability of your graphing utility to find the age (to the nearest year)

for the group that averages 13.43 annual physician visits.

c. Verify part (b) using the graph of f.

43. Graph the model in Problem 39 and determine the relative minimum and maximum height of the bungee jumper relative to the hotel's roof level.

Use a graphing utility to obtain a complete graph for each polynomial function in Problems 44–51. Then determine the number of real zeros and the number of nonreal complex zeros for each function.

44. $f(x) = x^3 - 6x - 9$

45. $f(x) = x^5 + x^3 + 2x^2 - 12x + 8$

46. $f(x) = 3x^5 - 2x^4 + 6x^3 - 4x^2 - 24x + 16$

47. $f(x) = 3x^4 + 4x^3 - 7x^2 - 2x - 3$

48. $f(x) = x^6 - 64$

49. $f(x) = 3x^5 + 24x^3 + 48x$

50. $f(x) = x^4 - 3x^3 - 4x^2$

51. $f(x) = x^5 - 1.5x^4 - 3.2x^3 + 1.3x^2 + 1.5x - 0.6$

Writing in Mathematics

In Problems 52–55, what is the smallest degree that each polynomial could have? Explain how you arrived at your answer.

52.

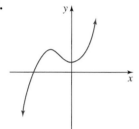

53.

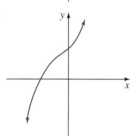

54.

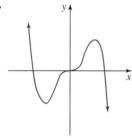

55.

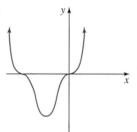

56. Explain why nonreal complex zeros are gained or lost in pairs in terms of graphs of polynomial functions.

57. Explain why a polynomial function of degree 20 cannot cross the x-axis exactly once.

58. Explain how you can find a polynomial of least degree with real coefficients that has zeros of 3, $-1 + 2i$, and $1 - 2i$.

59. Is it possible for a third-degree polynomial function to have exactly one relative maximum? Explain.

Critical Thinking Problems

60. Draw a polynomial function with a positive leading coefficient whose degree is 4 and with four distinct real zeros. Then use the shape of this graph and slide it upward (or downward) to graph a fourth-degree polynomial function having:
   a. Three distinct real zeros, one with multiplicity 2
   b. Two distinct real zeros
   c. One real zero of multiplicity 2
   d. No real zeros

61. Draw a polynomial function with a positive leading coefficient whose degree is 5 and with five distinct real zeros. Then use the shape of this graph and slide it upward (or downward) to graph a fifth-degree polynomial function having:

   a. Four distinct real zeros, one with multiplicity 2
   b. Three distinct real zeros
   c. Two distinct real zeros, one with multiplicity 2
   d. One real zero
   e. Is it possible to have no real zeros? Explain.

Solve each polynomial inequality in Problems 62–64.

62. $x^5 - x^3 \geq 0$
63. $x^3 - 4x^2 - x + 4 \leq 0$
64. $x^3 - 4x + 1 < 0$
65. **a.** Show that $2i$ is a solution of $x^3 - 2ix^2 - 4x + 8i = 0$.
   b. Show that $-2i$ is *not* a solution of the equation in part (a).
   c. Since $0 + 2i$ is a solution of the given equation, shouldn't $0 - 2i$ also be a solution? Explain why this is *not* a contradiction of the Conjugate Roots Theorem.

66. Give an example of a function that is not subject to the Intermediate Value Theorem.

Group Activity Problem _____

67. Use the Intermediate Value Theorem to write a computer or graphing calculator program for estimating a positive irrational zero of a polynomial function to any specified degree of accuracy.

| **S E C T I O N   3 . 5** | **Graphing Rational Functions** |

Solutions Manual   **Tutorial**   **Video 7**

Objectives

1 Graph the reciprocal function.
2 Graph transformations of the reciprocal function.
3 Give the equation of a rational function's horizontal asymptote.
4 Determine symmetry for a rational function's graph.
5 Graph rational functions.
6 Graph rational functions with slant asymptotes.

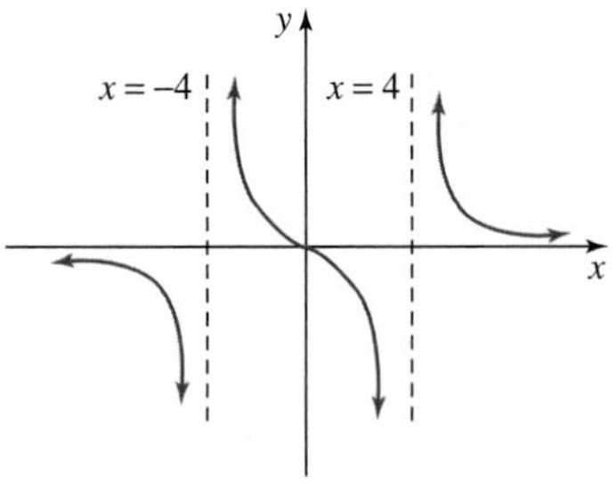

Rational functions such as

$$f(x) = \frac{x}{x^2 - 16} \quad \text{and} \quad g(x) = \frac{2x^2 + 3x - 5}{x - 2}$$

consist of the quotient of two polynomial functions. Unlike polynomial functions, many rational functions have graphs wtith breaks in their curves (see Figure 3.48). Asymptotes—lines that curves approach—are helpful in graphing rational functions.

Figure 3.48

The graph of $f(x) = \dfrac{x}{x^2 - 16}$ has breaks at -4 and 4 but approaches $x = -4$ and $x = 4$.

Rational Functions

Rational functions are quotients of polynomial functions.

> **Rational functions and their domains**
>
> A *rational function* is one that can be written in the form
>
> $$f(x) = \frac{p(x)}{q(x)}$$
>
> where $p(x)$ and $q(x)$ are polynomial functions and $q(x) \neq 0$. The *domain* of a rational function is the set of all real numbers except the x-values that make the denominator $q(x)$ equal zero.

▌ Graph the reciprocal function.

The most basic rational function is the *reciprocal function*, defined by

$$f(x) = \frac{1}{x}.$$

EXAMPLE 1   **The Reciprocal Function**

Find the domain of $f(x) = \dfrac{1}{x}$ and graph the function.

Solution

The denominator of the reciprocal function is zero when $x = 0$, so the domain of f is the set of all real numbers except for 0.

To examine the behavior of f near the excluded value 0, we evaluate $f(x)$ to the left and right of 0.

x approaches 0 from the left.

x	-1	-0.5	-0.1	-0.01	-0.001
$f(x) = \dfrac{1}{x}$	-1	-2	-10	-100	-1000

x approaches 0 from the right.

x	0.001	0.01	0.1	0.5	1
$f(x) = \dfrac{1}{x}$	1000	100	10	2	1

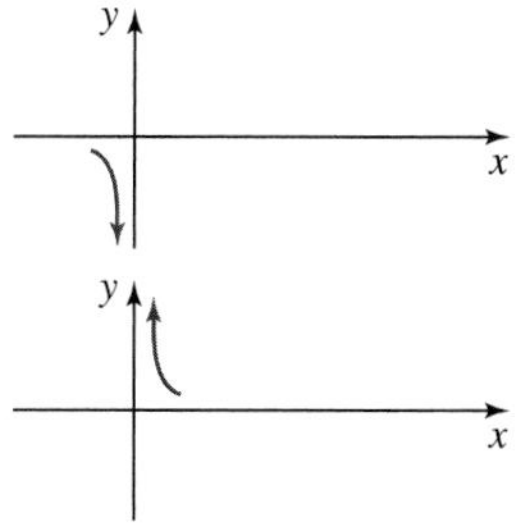

Figure 3.49 illustrates the behavior of $f(x) = \dfrac{1}{x}$ near 0, decreasing without bound as x approaches 0 from the left and increasing without bound as x approaches 0 from the right. We use our special arrow notation to describe this situation symbolically:

$$f(x) \to -\infty \quad \text{as} \quad x \to 0^- \qquad \text{and} \qquad f(x) \to \infty \quad \text{as} \quad x \to 0^+$$

$f(x)$ approaches negative infinity (that is, the graph falls) as x approaches 0 from the left.

$f(x)$ approaches infinity (that is, the graph rises) as x approaches 0 from the right.

Figure 3.49 illustrates that the function is getting closer and closer to the y-axis. A line that the graph of a function gets closer and closer to is called an *asymptote*. Thus, the line whose equation is $x = 0$ (that is, the y-axis) is a *vertical asymptote* of the graph of $f(x) = \dfrac{1}{x}$.

Now let's see what happens for values of x that get farther away from the origin. We create tables for x as $|x|$ increases without bound.

x increases without bound:

x	1	10	100	1000
$f(x) = \dfrac{1}{x}$	1	0.1	0.01	0.001

x decreases without bound:

x	-1	-10	-100	-1000
$f(x) = \dfrac{1}{x}$	-1	-0.1	-0.01	-0.001

Figure 3.50 illustrates the behavior of $f(x) = \dfrac{1}{x}$ as x increases or decreases without bound. The values of y are getting progressively closer to 0. This means that the graph of f is approaching the horizontal line $y = 0$ (that is, the x-axis) as x increases or decreases without bound. We use the arrow notation to describe this situation.

$$f(x) \to 0 \quad \text{as} \quad x \to \infty \qquad \text{and} \qquad f(x) \to 0 \quad \text{as} \quad x \to -\infty$$

$f(x)$ approaches 0 as x increases without bound.

$f(x)$ approaches 0 as x decreases without bound.

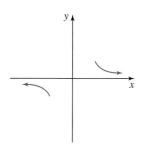

Figure 3.49

Study tip

The minus $(-)$ superscript on the 0 $(x \to 0^-)$ is read "from the left;" the plus $(+)$ superscript $(x \to 0^+)$ is read "from the right."

Figure 3.50

$f(x)$ approaches 0 as x increases or decreases without bound.

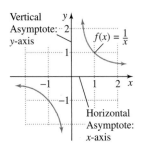

Figure 3.51

The graph of $f(x) = \dfrac{1}{x}$

Since the graph of f is approaching the horizontal line $y = 0$ as $x \to \infty$ and as $x \to -\infty$, the line whose equation is $y = 0$ (the x-axis) is a *horizontal asymptote* of the graph of $f(x) = \dfrac{1}{x}$. The completed graph of f is shown in Figure 3.51. ■

The arrow notation used throughout Example 1 is summarized in the following box.

Arrow notation

Symbol	Meaning
$x \to a^+$	x approaches a from the right.
$x \to a^-$	x approaches a from the left.
$x \to \infty$	x approaches infinity; that is, x increases without bound.
$x \to -\infty$	x approaches negative infinity; that is, x decreases without bound.

Horizontal and Vertical Asymptotes

The graph of the rational function $f(x) = \dfrac{1}{x}$ differs from the graph of a polynomial function in two ways. First, the graph has a break in it and is composed of two distinct branches. Second, the curve has *linear asymptotes*, lines that the graph approaches arbitrarily closely. Many, but not all, rational functions have vertical and horizontal asymptotes.

Study tip

The zeros of the denominator of a rational function will often give the vertical asymptotes.

Vertical and horizontal asymptotes

1. The line $x = a$ is a *vertical asymptote* of a function f if $f(x)$ increases or decreases without bound as x approaches a.

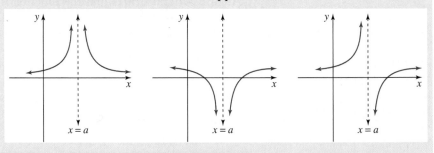

Thus, $x = a$ is a vertical asymptote of f if $f(x) \to \infty$ or $f(x) \to -\infty$ as $x \to a^+$ or $x \to a^-$.

2. The line $y = b$ is a *horizontal asymptote* of a function f if $f(x)$ approaches b as x increases or decreases without bound.

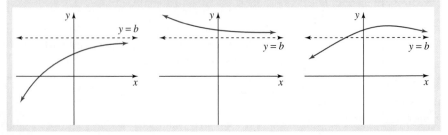

Study tip

A curve can never cross a vertical asymptote. However, it can intersect a horizontal asymptote.

Thus, $y = b$ is a horizontal asymptote of f if $f(x) \to b$ as $x \to \infty$ or $x \to -\infty$.

2 Graph transformations of the reciprocal function.

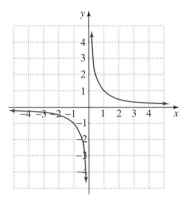

Figure 3.52

The graph of the reciprocal function $f(x) = \dfrac{1}{x}$

Some rational functions can be graphed using transformations of the graph of $f(x) = \dfrac{1}{x}$.

EXAMPLE 2 **Transformations of the Reciprocal Function**

Use the graph of the reciprocal function $f(x) = \dfrac{1}{x}$, shown again in Figure 3.52, to graph each function. Give equations for the vertical and horizontal asymptotes of each function.

a. $y = \dfrac{1}{x - 3}$ **b.** $y = \dfrac{1}{x} - 2$ **c.** $y = \dfrac{2}{x}$

d. $y = -\dfrac{1}{x}$ **e.** $y = -\dfrac{3}{x - 2} - 1$

Solution

Equation and Transformation	Graph	Vertical Asymptote	Horizontal Asymptote
a. $y = \dfrac{1}{x-3}$ Shift the graph of $y = \dfrac{1}{x}$ to the right three units.	Shift right 3 units.	$x = 3$ $y \to -\infty$ as $x \to 3^-$ and $y \to \infty$ as $x \to 3^+$	$y = 0$ (x-axis) $y \to 0$ as $x \to -\infty$ and $y \to 0$ as $x \to \infty$
b. $y = \dfrac{1}{x} - 2$ Shift the graph of $y = \dfrac{1}{x}$ downward two units.	Shift down 2 units.	$x = 0$ (y-axis) $y \to -\infty$ as $x \to 0^-$ and $y \to \infty$ as $x \to 0^+$	$y = -2$ $y \to -2$ as $x \to -\infty$ and $y \to -2$ as $x \to \infty$
c. $y = \dfrac{2}{x}$ Multiply each y-value of $y = \dfrac{1}{x}$ by 2. Each point on the graph is twice as far from the x-axis.		$x = 0$ (y-axis) $y \to -\infty$ as $x \to 0^-$ and $y \to \infty$ as $x \to 0^+$	$y = 0$ (x-axis) $y \to 0$ as $x \to -\infty$ and $y \to 0$ as $x \to \infty$

Equation and Transformation	Graph	Vertical Asymptote	Horizontal Asymptote
d. $y = -\dfrac{1}{x}$ Reflect the graph of $y = \dfrac{1}{x}$ in the x-axis.	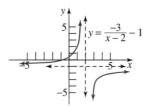	$x = 0$ (y-axis) $y \to \infty$ as $x \to 0^-$ and $y \to -\infty$ as $x \to 0^+$	$y = 0$ (x-axis) $y \to 0$ as $x \to -\infty$ and $y \to 0$ as $x \to \infty$

e. $y = -\dfrac{3}{x-2} - 1$
1. Shift the graph of
   $y = \dfrac{1}{x}$ to the right two
   units. $\left(y = \dfrac{1}{x-2}\right)$
2. Stretch, multiplying
   y-values by 3. $\left(y = \dfrac{3}{x-2}\right)$
3. Reflect in the
   x-axis. $\left(y = -\dfrac{3}{x-2}\right)$
4. Shift downword one unit.
   $\left(y = -\dfrac{3}{x-2} - 1\right)$
	$y = \dfrac{-3}{x-2} - 1$	$x = 2$
		$y \to \infty$ as $x \to 2^-$
		and
		$y \to -\infty$ as $x \to 2^+$
			$y = -1$
			$y \to -1$ as $x \to -\infty$
			and
			$y \to -1$ as $x \to \infty$

■

In Example 2, we can obtain the equation for each rational function's vertical asymptote by setting the denominator equal to 0. For example, the rational function in part (a)

$$y = \frac{1}{x - 3}$$

has a vertical asymptote when $x - 3 = 0$. The equation of the vertical asymptote is $x = 3$. The zeros of the denominator of a rational function will often give the vertical asymptotes.

There is also a rapid way to determine the horizontal asymptotes, if there are any. Before we state this procedure, take a few minutes to discover it for yourself with the aid of your graphing utility. (Use the Study tip on page 397 to work the Discover for yourself without a graphing utility.)

M. C. Escher (1898–1972) "Bond of Union" © 1997 Cordon Art – Baarn – Holland. All rights reserved.

Discover for yourself

1. Use a graphing utility to graph f and g in the same viewing rectangle. Begin with a standard viewing rectangle. Then zoom out to a viewing rectangle such as $[-100, 100]$ by $[-1000, 1000]$. In each case, show that g is a horizontal asymptote for f.

$$f(x) = \frac{4x}{2x^2 + 1} \quad \text{and} \quad g(x) = 0$$

$$f(x) = \frac{-6x}{2x^2 + 1} \quad \text{and} \quad g(x) = 0$$

If the highest power in the numerator of a rational function's equation is less than the highest power in the denominator, generalize from these examples and write the

Study tip

You can work the first two parts of the Discover for yourself without a graphing utility. Simply generalize from the equation of the function and the horizontal asymptote, using the pattern to write the horizontal asymptote's equation based on the highest powers in the numerator and denominator of the given rational function.

equation of the function's horizontal asymptote. Use your graphing utility to verify this observation for a rational function of your choice.

2. Now repeat the exercise with your graphing utility, again showing that g is a horizontal asymptote for f in the following cases:

$$f(x) = \frac{4x^2}{2x^2 + 1} \quad \text{and} \quad g(x) = 2$$

$$f(x) = \frac{-6x^2}{2x^2 + 1} \quad \text{and} \quad g(x) = -3$$

If a rational function has the same highest power in the numerator and denominator, generalize from these examples and describe how you can use the leading coefficients to write the equation of the function's horizontal asymptote. Use your graphing utility to verify this observation for a function of your choice.

3. Now graph the rational functions

$$f(x) = \frac{4x^3}{2x^2 + 1} \quad \text{and} \quad f(x) = \frac{4x^2}{2x + 1}$$

where the highest power in the numerator is greater than the highest power in the denominator. Begin with a standard viewing rectangle, zooming out to $[-100, 100]$ by $[-1000, 1000]$. What can you conclude about horizontal asymptotes when the highest power in the numerator is greater than the highest power in the denominator? Use your graphing utility to verify this observation for a function of your choice.

In the Discover for yourself, did you observe that the equation for the horizontal asymptote, if there is one, depends on the highest power in the numerator and denominator of the rational function's equation? In particular, were you able to discover the following rules for obtaining horizontal asymptotes for rational functions?

3 Give the equation of a rational function's horizontal asymptote.

Study tip

For the rational function whose equation appears in the box on the right, recall that n, the highest power in the numerator, is called the *degree* of the numerator. Similarly, m, the highest power in the denominator, is called the *degree* of the denominator.

Horizontal asymptote of a rational function

Let f be the rational function given by

$$f(x) = \frac{a_n x^n + a_{n-1} x^{n-1} + \cdots + a_1 x + a_0}{b_m x^m + b_{m-1} x^{m-1} + \cdots + b_1 x + b_0}, \ a_n \neq 0, b_m \neq 0.$$

1. If $n < m$, the x-axis is a horizontal asymptote.

2. If $n = m$, the line $y = \dfrac{a_n}{b_m}$ is a horizontal asymptote.

3. If $n > m$, there is no horizontal asymptote.

EXAMPLE 3 **Horizontal Asymptotes of Rational Functions**

Give the equation for the horizontal asymptote, if there is one, for each rational function:

a. $f(x) = \dfrac{4x}{2x^2 + 1}$ **b.** $g(x) = \dfrac{4x^2}{2x^2 + 1}$ **c.** $h(x) = \dfrac{4x^3}{2x^2 + 1}$

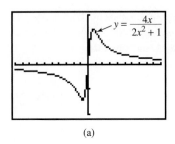

(a)

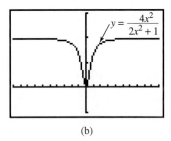

(b)

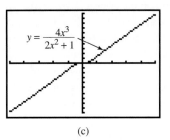

(c)

Figure 3.53

4 Determine symmetry for a rational function's graph.

Solution

a. $f(x) = \dfrac{4x}{2x^2 + 1}$

The degree of the numerator, 1, is less than the degree of the denominator, 2. Thus, the graph of f has the x-axis as a horizontal asymptote (see Figure 3.53(a)). The equation of the horizontal asymptote is $y = 0$.

b. $g(x) = \dfrac{4x^2}{2x^2 + 1}$

The degree of the numerator, 2, is equal to the degree of the denominator, 2. The leading coefficients of the numerator and denominator, 4 and 2, are used to obtain the equation of the horizontal asymptote. The equation of the horizontal asymptote is $y = \frac{4}{2}$ or $y = 2$ (see Figure 3.53(b)).

c. $h(x) = \dfrac{4x^3}{2x^2 + 1}$

The degree of the numerator, 3, is greater than the degree of the denominator, 2. Thus, the graph of h has no horizontal asymptote (see Figure 3.53(c)). ∎

Symmetry

The presence of symmetry is helpful in graphing rational functions. The tests of symmetry are the same as those used for polynomial functions and are restated below.

> **y-axis and origin symmetry**
>
> The graph of a rational function f is symmetric with respect to the y-axis if $f(-x) = f(x)$. The graph is symmetric with respect to the origin if $f(-x) = -f(x)$.

EXAMPLE 4 **Determining Symmetry for Graphs of Rational Functions**

Discuss the symmetry of the graph of each rational function.

a. $f(x) = \dfrac{4x}{2x^2 + 1}$ **b.** $g(x) = \dfrac{4x^2}{2x^2 + 1}$

Solution

In each case, replace x with $-x$.

a. $f(x) = \dfrac{4x}{2x^2 + 1}$

$f(-x) = \dfrac{4(-x)}{2(-x)^2 + 1} = \dfrac{-4x}{2x^2 + 1} = -f(x)$

Since $f(-x) = -f(x)$, the graph is symmetric with respect to the origin (see Figure 3.54(a)).

b. $g(x) = \dfrac{4x^2}{2x^2 + 1}$

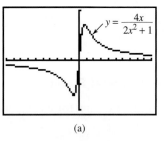

(a)

Figure 3.54(a)

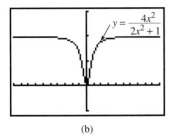

(b)

Figure 3.54(b)

$$g(-x) = \frac{4(-x)^2}{2(-x)^2 + 1} = \frac{4x^2}{2x^2 + 1} = g(x)$$

Since $g(-x) = g(x)$, the graph is symmetric with respect to the y-axis (see Figure 3.54 (b)). ∎

 iscover for yourself

Give an example of a rational function that has neither y-axis nor origin symmetry. Then verify this fact by graphing the function with a graphing utility.

5 Graph rational functions.

Graphing Rational Functions

Here are some suggestions for graphing rational functions.

Strategy for graphing a rational function

Suppose that

$$f(x) = \frac{p(x)}{q(x)}$$

where $p(x)$ and $q(x)$ are polynomial functions with no common factors.
1. Determine whether the graph of f has symmetry.

$f(-x) = f(x)$: y-axis symmetry

$f(-x) = -f(x)$: origin symmetry

2. Find and plot the y-intercept (if there is one) by evaluating $f(0)$.
3. Find and plot the x-intercepts (if there are any) by solving the equation $p(x) = 0$.
4. Find any vertical asymptote(s) by solving the equation $q(x) = 0$. Sketch the vertical asymptotes.
5. Find and sketch the horizontal asymptote (if there is one) using the rule for determining the horizontal asymptote of a rational function.
6. Plot at least one point between and beyond each x-intercept and vertical asymptote.
7. Use smooth curves to graph the function between and beyond the vertical asymptotes.
8. Use a graphing utility in the DOT mode to verify that the graph has been drawn correctly.

Study tip

Values where the denominator of a rational function are zero do not necessarily result in vertical asymptotes. There is a hole and not a vertical asymptote in the graph if there is a reduced form of the function's equation that does not make the denominator zero. Here's an example that we graphed in Chapter 2:

$$f(x) = \frac{x^2 - 9}{x - 3}$$

$$= \frac{(x + 3)(x - 3)}{x - 3}$$

$$= x + 3 \quad (x \neq 3)$$

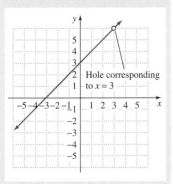

EXAMPLE 5 **Graphing a Rational Function**

Graph: $f(x) = \dfrac{2x}{x - 1}$

Solution

Step 1. Determine symmetry.

$$f(-x) = \frac{2(-x)}{-x - 1} = \frac{-2x}{-x - 1} = \frac{2x}{x + 1}$$

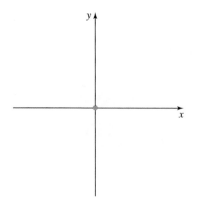

Figure 3.55

The y-intercept is 0.

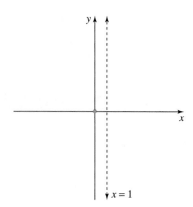

Figure 3.56

The vertical asymptote is $x = 1$.

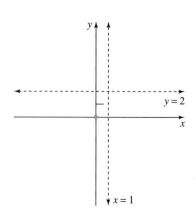

Figure 3.57

The horizontal asymptote is $y = 2$.

Since $f(-x)$ does not equal $f(x)$ or $-f(x)$, the graph has neither y-axis nor origin symmetry.

Step 2. Find the y-intercept by evaluating $f(0)$.

$$f(0) = \frac{2 \cdot 0}{0 - 1} = \frac{0}{-1} = 0$$

The y-intercept is 0, and so the graph passes through the origin (Figure 3.55).

Step 3. Find x-intercept(s) by solving $p(x) = 0$.

$2x = 0$ Set the numerator equal to 0.

$x = 0$

There is only one x-intercept. Again, the graph passes through the origin.

Step 4. Find the vertical asymptote(s) by solving $q(x) = 0$.

$x - 1 = 0$ Set the denominator equal to 0.

$x = 1$

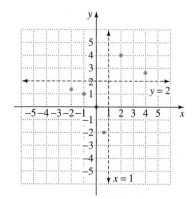

Figure 3.58

The equation of the vertical asymptote is $x = 1$ (Figure 3.56).

Step 5. Find the horizontal asymptote. Since the numerator and denominator have the same degree, the equation of the horizontal asymptote is

$$y = \frac{2}{1} = 2.$$

The graph of $y = 2$ is shown in Figure 3.57.

Step 6. Plot points between and beyond each x-intercept and vertical asymptote.

x	-2	-1	$\frac{1}{2}$	2	4
$f(x) = \dfrac{2x}{x-1}$	$\dfrac{4}{3}$	1	-2	4	$\dfrac{8}{3}$

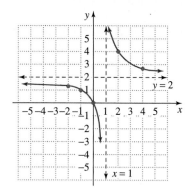

Figure 3.59

The graph of $f(x) = \dfrac{2x}{x - 1}$

Figure 3.58 shows these points, the y-intercept, and the asymptotes.

Step 7. Use a smooth curve to graph the function. The graph of $f(x) = \dfrac{2x}{x - 1}$ is shown in Figure 3.59.

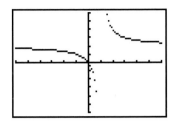

Figure 3.60

The graph of $y = \dfrac{2x}{x-1}$ generated by a graphing utility

Step 8. Although most graphing utilities do not produce good graphs of rational functions, the graph obtained in Figure 3.60 (using the DOT mode) verifies that our hand-drawn graph is correct.   ■

EXAMPLE 6   Graphing a Rational Function

Graph: $f(x) = \dfrac{3x}{x^2 - 1}$

Solution

Step 1. Determine symmetry.

$$f(-x) = \frac{3(-x)}{(-x)^2 - 1} = \frac{-3x}{x^2 - 1} = -f(x)$$

The graph is symmetric with respect to the origin. This should become more obvious as we continue with our graphing strategy.

Step 2. Find the y-intercept by evaluating $f(0)$.

$$f(0) = \frac{3 \cdot 0}{0^2 - 1} = \frac{0}{-1} = 0$$

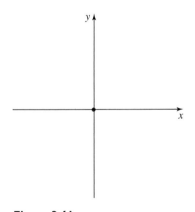

Figure 3.61

The y-intercept is 0.

The y-intercept is 0, and so the graph passes through the origin (Figure 3.61).

Step 3. Find the x-intercept(s) by solving $p(x) = 0$.

$3x = 0$ Set the numerator equal to 0.

$x = 0$

There is only one x-intercept. Again, the graph passes through the origin.

Step 4. Find the vertical asymptote(s) by solving $q(x) = 0$.

$x^2 - 1 = 0$ Set the denominator equal to 0.

$x^2 = 1$

$x = \pm 1$

The equations of the vertical asymptotes are $x = -1$ and $x = 1$.

Step 5. Find the horizontal asymptote. Since the degree of the numerator, 1, is less than the degree of the denominator, 2, the x-axis is a horizontal asymptote. The equation of the horizontal asymptote is $y = 0$. Figure 3.62 shows the intercept and the asymptotes.

Step 6. Plot points between and beyond the x-intercept and the vertical asymptotes.

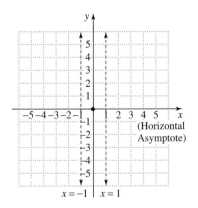

Figure 3.62

The function's intercept and asymptotes

x	-3	-2	-0.5	0.5	2	3
$f(x) = \dfrac{3x}{x^2 - 1}$	$-\dfrac{9}{8}$	-2	2	-2	2	$\dfrac{9}{8}$

Figure 3.63 on page 402 shows these points, the y-intercept, and the asymptotes.

Step 7. Use a smooth curve to graph the function. The graph of $f(x) = \dfrac{3x}{x^2 - 1}$ is shown on page 402 in Figure 3.64. The origin symmetry is now obvious.

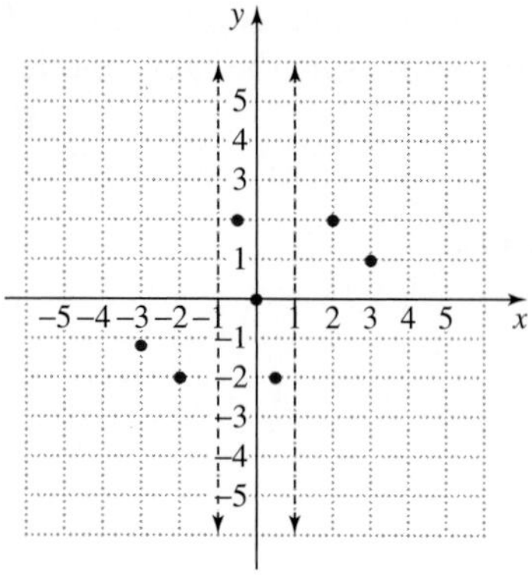

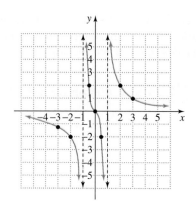

Figure 3.63

Figure 3.64

The graph of $f(x) = \dfrac{3x}{x^2 - 1}$

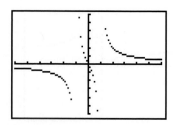

Figure 3.65

The graph of $y = \dfrac{3x}{x^2 - 1}$

generated by a graphing utility

Step 8. The graph shown in Figure 3.65, obtained with a graphing utility, verifies that our hand-drawn graph is correct. ∎

EXAMPLE 7 **Graphing a Rational Function**

Graph: $f(x) = \dfrac{3x^2}{x^2 - 4}$

Solution

Step 1. Symmetry: $f(-x) = \dfrac{3(-x)^2}{(-x)^2 - 4} = \dfrac{3x^2}{x^2 + 4} = f(x)$: Symmetric with respect to the y-axis.

Step 2. y-intercept: $f(0) = \dfrac{3 \cdot 0^2}{0^2 - 4} = \dfrac{0}{-4} = 0$: y-intercept is 0.

Step 3. x-intercept: $3x^2 = 0$, so $x = 0$: x-intercept is 0.

Step 4. Vertical asymptotes: Set $q(x) = 0$.

$$x^2 - 4 = 0 \qquad \text{Set the denominator equal to 0.}$$

$$x^2 = 4$$

$$x = \pm 2$$

Vertical asymtotes: $x = -2$ and $x = 2$

Step 5. Horizontal asymptote: $y = \dfrac{3}{1} = 3$

Step 6. Additional points:

x		-3	-1	1	3	4
$f(x) = \dfrac{3x^2}{x^2 - 4}$		$\dfrac{27}{5}$	-1	-1	$\dfrac{27}{5}$	4

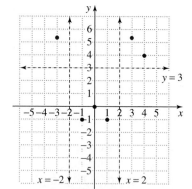

Figure 3.66

Figure 3.66 shows these points, the y-intercept, and the asymptotes.

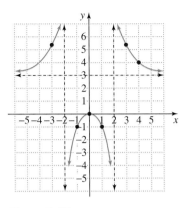

Figure 3.67

The graph of $f(x) = \dfrac{3x^2}{x^2 - 4}$

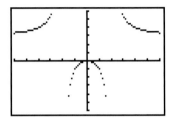

Figure 3.68

The graph of $y = \dfrac{3x^2}{x^2 - 4}$

generated by a graphing utility

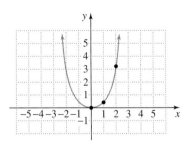

Figure 3.69

The graph of $f(x) = \dfrac{x^4}{x^2 + 1}$

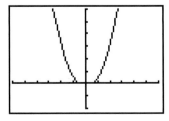

Figure 3.70

The graph of $y = \dfrac{x^4}{x^2 + 1}$,

generated by a graphing utility

Step 7. Use a smooth curve to graph the function. The graph of $f(x) = \dfrac{3x^2}{x^2 - 4}$ is shown in Figure 3.67. The y-axis symmetry is now obvious.

Step 8. The graph shown in Figure 3.68, obtained with a graphing utility, verifies that our hand-drawn graph is correct. ■

Example 8 illustrates that not every rational function has vertical and horizontal asymptotes.

EXAMPLE 8 **Graphing a Rational Function**

Graph: $f(x) = \dfrac{x^4}{x^2 + 1}$

Solution

Step 1. Symmetry: $f(-x) = \dfrac{(-x)^4}{(-x)^2 + 1} = \dfrac{x^4}{x^2 + 1}$: Symmetric with respect to the y-axis

Step 2. y-intercept: $f(0) = \dfrac{0^4}{0^2 + 1} = \dfrac{0}{1} = 0$: y-intercept is 0.

Step 3. x-intercept: also 0

Step 4. Vertical asymptote: Set $q(x) = 0$.

$$x^2 + 1 = 0 \qquad \text{Set the denominator equal to 0.}$$
$$x^2 = -1$$

Although this equation has imaginary solutions ($x = \pm i$), there are no real solutions. Thus, there is no vertical asymptote.

Step 5. Horizontal asymptote: Since the degree of the numerator, 4, is greater than the degree of the denominator, 2, there is no horizontal asymptote.

Step 6. Additional points:

x	0	1	2	3
$f(x) = \dfrac{x^4}{x^2 + 1}$	0	$\dfrac{1}{2}$	$\dfrac{16}{5}$	$\dfrac{81}{10}$

Step 7. Figure 3.69 shows the graph of f, using three points obtained from the table and y-axis symmetry. Notice that as x gets infinitely large or infinitely small, the function is getting larger without bound.

Step 8. The graph shown in Figure 3.70, obtained with a graphing utility, verifies that our hand-drawn graph is correct. The graph has no asymptotes, so we did not use the DOT mode. ■

6 Graph rational functions with slant asymptotes.

Slant Asymptotes

The graph of

$$f(x) = \frac{x^2 + 1}{x - 1}$$

Degree = 2
Degree = 1

has no horizontal asymptote because the degree of the numerator is greater than the degree of the denominator. However, in this situation, there is a *slant asymptote,* as shown in Figure 3.71.

The equation of the slant asymptote can be found by division.

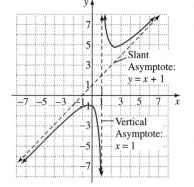

$$\begin{array}{r|rrr} 1 & 1 & 0 & 1 \\ & & 1 & 1 \\ \hline & 1 & 1 & 2 \end{array}$$

$$x + 1 + \dfrac{2}{x + 1}$$
$$x - 1 \overline{)\, x^2 + 0x + 1}$$

Since $f(x) = \dfrac{x^2 + 1}{x - 1}$, divide

$x - 1$ into $x^2 + 1$.

Observe that

$$f(x) = \frac{x^2 + 1}{x - 1} = \underbrace{x + 1}_{\substack{\text{Slant asymptote:} \\ y = x + 1}} + \frac{2}{x - 1}$$

Figure 3.71

The graph of $f(x) = \dfrac{x^2 + 1}{x - 1}$

with a slant asymptote

If $|x| \to \infty$, the value of $\dfrac{2}{x - 1}$ is approximately 0. Thus, when $|x|$ is large, the function is very close to $y = x + 1 + 0$. This means that as $x \to \infty$ or as $x \to -\infty$, the graph of f gets closer and closer to the line whose equation is $y = x + 1$. The line is a slant asymptote of the graph.

We can generalize from these observations and obtain a procedure for finding slant asymptotes, sometimes called *oblique asymptotes.*

Finding slant (or oblique) asymptotes

If $f(x) = \dfrac{p(x)}{q(x)}$ and the degree of p is one greater than the degree of q, the graph of f has a slant asymptote. Find the slant asymptote by dividing $q(x)$ into $p(x)$. The division will take the form

$$\frac{p(x)}{q(x)} = \underbrace{mx + b}_{\substack{\text{Slant asymptote:} \\ y = mx + b}} + \frac{R(x)}{q(x)}$$

As $|x| \to \infty$, the term $\dfrac{R(x)}{q(x)}$ is close to 0. Consequently, $y = mx + b$ is the equation of the slant or oblique asymptote.

EXAMPLE 9   **A Rational Function with a Slant Asymptote**

Graph the function: $f(x) = \dfrac{x^2 - 4x - 5}{x - 3}$

Solution

Notice that the degree of the numerator is one more than the degree of the denominator, so the function will have a slant asymptote. We can express the function in two ways:

$$f(x) = \frac{x^2 - 4x - 5}{x - 3} = \frac{(x - 5)(x + 1)}{x - 3}$$

In this form, we see that there are x-intercepts at 5 and -1, the values of x where the numerator is 0.

$$f(x) = \frac{x^2 - 4x - 5}{x - 3} = \underbrace{x - 1} - \frac{8}{x - 3}$$

Using this form, $y = x - 1$ is a slant asymptote.

Use synthetic division.

$$\begin{array}{r|rrr} 3 & 1 & -4 & -5 \\ & & 3 & -3 \\ \hline & 1 & -1 & -8 \end{array}$$

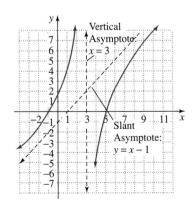

> **Study tip**
>
> The equation of the slant asymptote is obtained by setting y equal to the partial quotient without the remainder.

Now we continue by using our strategy for graphing rational functions.

Step 1. Symmetry: $f(-x) = \dfrac{(-x)^2 - 4(-x) - 5}{-x - 3} = \dfrac{x^2 + 4x - 5}{-x - 3}$

Since $f(-x)$ is not equal to $f(x)$ or $-f(x)$, we do not have symmetry with respect to the y-axis or the origin.

Step 2. y-intercept: $f(0) = \dfrac{0^2 - 4 \cdot 0 - 5}{0 - 3} = \dfrac{5}{3}$: y-intercept is $1\dfrac{2}{3}$.

Step 3. x-intercepts: Set the numerator equal to 0.

$$(x - 5)(x + 1) = 0$$

$$x = 5 \quad \text{or} \quad x = -1 \qquad \text{There are } x\text{-intercepts at 5 and } -1.$$

Step 4. Vertical asymptote: Set $q(x) = 0$.

$$x - 3 = 0 \qquad \text{Set the denominator equal to 0.}$$

$$x = 3 \qquad \text{The equation of the vertical asymptote is } x = 3.$$

Step 5. Horizontal asymptote: Since the degree of the numerator is greater than the degree of the denominator, there are no horizontal asymptotes.

Step 6. Additional points:

x		-3	1	4	7
$f(x) = $	$\dfrac{x^2 - 4x - 5}{x - 3}$	$-2\dfrac{2}{3}$	4	-5	4

Figure 3.72

The graph of

$f(x) = \dfrac{x^2 - 4x - 5}{x - 3}$

Step 7. Remembering that the equation of the slant asymptote is $y = x - 1$, we draw the graph of f shown in Figure 3.72.

Step 8. See the Discover for yourself that follows. ∎

Discover for yourself

Use a graphing utility to graph

$$f(x) = \frac{x^2 - 4x - 5}{x - 3}$$

and $y = x - 1$ in the same viewing rectangle. Use the $\boxed{\text{ZOOM OUT}}$ feature to verify that $y = x - 1$ is indeed a slant asymptote.

PROBLEM SET 3.5

Practice Problems

In Problems 1–18, use the graph of $f(x) = \dfrac{1}{x}$ to graph each function. Graph f and g on the same axes. Identify the horizontal and vertical asymptotes of g.

1. $g(x) = \dfrac{1}{x - 2}$

2. $g(x) = \dfrac{1}{x - 1}$

3. $g(x) = \dfrac{1}{x + 2}$

4. $g(x) = \dfrac{1}{x + 1}$

5. $g(x) = \dfrac{1}{x} - 1$

6. $g(x) = \dfrac{1}{x} - 3$

7. $g(x) = \dfrac{1}{x} + 2$

8. $g(x) = \dfrac{1}{x} + 1$

9. $g(x) = \dfrac{0.5}{x}$

10. $g(x) = \dfrac{0.25}{x}$

11. $g(x) = -\dfrac{2}{x}$

12. $g(x) = -\dfrac{3}{x}$

13. $g(x) = \dfrac{1}{x + 2} - 1$

14. $g(x) = \dfrac{1}{x + 1} - 2$

15. $g(x) = -\dfrac{2}{x + 2} - 1$

16. $g(x) = -\dfrac{2}{x + 1} - 2$

17. $g(x) = \dfrac{2}{x - 2} + 1$

18. $g(x) = \dfrac{2}{x - 1} + 2$

In Problems 19–50, use the strategy for graphing rational functions outlined on page 399 to graph each function. Do not use a graphing utility until your hand-drawn graph is complete. Then verify your graph with a graphing utility.

19. $f(x) = \dfrac{2x}{x^2 - 4}$

20. $f(x) = \dfrac{3x}{x^2 - 9}$

21. $g(x) = \dfrac{3x - 6}{x + 4}$

22. $g(x) = \dfrac{4x - 8}{x + 2}$

23. $h(x) = \dfrac{2x^2}{x^2 - 1}$

24. $h(x) = \dfrac{4x^2}{x^2 - 9}$

25. $r(x) = \dfrac{-x}{x + 1}$

26. $r(x) = \dfrac{-3x}{x + 2}$

27. $f(x) = -\dfrac{1}{x^2 - 4}$

28. $f(x) = -\dfrac{2}{x^2 - 1}$

29. $g(x) = \dfrac{4x}{x^2 - 2x + 1}$

30. $g(x) = \dfrac{2x}{x^2 + 6x + 9}$

31. $h(x) = \dfrac{x-1}{x^2 - x - 6}$

32. $h(x) = \dfrac{x+1}{x^2 - x - 6}$

33. $r(x) = \dfrac{3x^2 + x - 4}{2x^2 - 5x}$

34. $r(x) = \dfrac{x^2 - 4x + 3}{x^2 - 2x}$

35. $f(t) = \dfrac{t^2 - 2t - 8}{t^2 - 4t + 3}$

36. $f(t) = \dfrac{t^2 - 4t + 3}{(t+1)^2}$

37. $h(v) = \dfrac{2-v}{3-v}$

38. $h(v) = \dfrac{3+v}{1-v}$

39. $f(x) = \dfrac{x^2 - 1}{x}$

40. $f(x) = \dfrac{x^2 - 4}{x}$

41. $f(x) = \dfrac{x^2 + 1}{x}$

42. $f(x) = \dfrac{x^2 + 4}{x}$

43. $g(x) = \dfrac{x^2 + x - 6}{x - 3}$

44. $g(x) = \dfrac{x^2 - x + 1}{x - 1}$

45. $r(x) = \dfrac{6x^2 - 6x - 12}{3x^2 + 4x + 5}$

46. $r(x) = \dfrac{4x^2 + 8x - 12}{2x^2 + 3x + 5}$

47. $h(v) = \dfrac{3v^3 - 2v^2 + 3v - 2}{v^2 + 3}$

48. $h(v) = \dfrac{v^3 + v^2 + 2v + 2}{v^2 + 9}$

49. $f(x) = \dfrac{x^2 - 4}{x^2 + 1}$

50. $f(x) = \dfrac{x^2 - 1}{x^2 + 4}$

True–False Critical Thinking Problems

51. Which one of the following is true?

a. The function $f(x) = \dfrac{1}{\sqrt{x-3}}$ is a rational function.

b. The domain of $f(x) = \dfrac{x+5}{x-5}$ is $(-\infty, -5) \cup (-5, 5)$ $\cup (5, \infty)$.

c. The graph of a rational function cannot intersect any of its asymptotes.

d. None of the above is true.

52. Which one of the following is true?

a. The x-axis is a horizontal asymptote for the graph of $f(x) = \dfrac{4x - 1}{x + 3}$.

b. The graph shown in the figure on the right cannot possibly represent the rational function $f(x) = \dfrac{-x^3 + 1}{x^2}$.

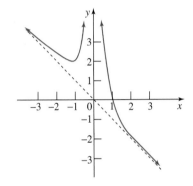

c. The graph of $f(x) = \dfrac{2x + 4}{x^2 - 4}$ has only one vertical asymptote.

d. The line whose equation is $y = x - 1$ is an asymptote for the graph of $f(x) = x - 4 + \dfrac{1}{x - 1}$.

Technology Problems

53. Use a graphing utility to graph $y = \dfrac{1}{x}$, $y = \dfrac{1}{x^3}$, and $\dfrac{1}{x^5}$ in the same viewing rectangle. For odd values of n, how does changing n affect the graph?

54. Use a graphing utility to graph $y = \dfrac{1}{x^2}$, $y = \dfrac{1}{x^4}$, and $y = \dfrac{1}{x^6}$ in the same viewing rectangle. For even values of n, how does changing n affect the graph?

55. The model $f(x) = \dfrac{72{,}900}{100x^2 + 729}$ describes the 1992 unemployment rate (as a percent) in the United States as a function of years of education (x).

a. Use a graphing utility to graph the model for $x \geq 0$.

b. What does the shape of the graph indicate about increasing levels of education?

c. Explain why f is one-to-one. What is $f^{-1}(3.6)$ and what practical information does this provide?

d. Is there an education level that leads to guaranteed employment? How is your answer indicated on the graph?

Throughout Section 3.5 we assumed that the rational function $f(x) = \dfrac{p(x)}{q(x)}$ was written so that $p(x)$ and $q(x)$ contained no common factors. Problems 56–59 explore the possibility of p and q sharing common factors.

56. Use a graphing utility to graph $f(x) = \dfrac{x^2 - 4}{x - 2}$. Does your graphing utility show that f has a "hole" at $x = 2$?

(Note: $f(x) = \dfrac{(x + 2)(x - 2)}{x - 2} = x + 2; x \ne 2$.)

57. Use a graphing utility to graph $f(x) = \dfrac{x^2 - 2x}{x}$. Does your graphing utility show that f has a "hole" at $x = 0$?

(Note: $f(x) = \dfrac{x^2 - 2x}{x} = \dfrac{x(x - 2)}{x} = x - 2; x \ne 0$.)

What does your graphing utility indicate about the domain of f?

58. a. Use a graphing utility to graph $f(x) = \dfrac{x^3 + 4x^2 + 7x + 6}{x + 2}$. Does your graphing utility show that f has an asymptote, a "hole," or neither at $x = -2$?

b. Divide $x^3 + 4x^2 + 7x + 6$ by $x + 2$. What is the relationship between the quotient and the graph of f in part (a)?

59. If $f(x) = \dfrac{p(x)}{q(x)}$ and $q(a) = 0$, use Problems 57 and 58 to generalize when the graph of f has a "hole" at $x = a$ and when the graph of f has a vertical asymptote whose equation is $x = a$.

60. Use a graphing utility to graph $f(x) = \dfrac{2|x - 1|}{x + 2}$. How does the situation with horizontal asymptotes differ from the other functions graphed throughout this section?

61. Use a graphing utility to graph $f(x) = \dfrac{x^2 - 4x + 3}{x - 2}$ and $g(x) = \dfrac{x^2 - 5x + 6}{x - 2}$. What differences do you observe between the graphs of f and g? How do you account for these differences?

62. a. Graph $f(x) = \dfrac{2x^4 + 7x^3 + 7x^2 + 2x}{x^3 - x + 50}$ using the following range setting:

$\text{Xmin} = -20, \text{Xmax} = 20, \text{Xscl} = 10,$

$\text{Ymin} = -20, \text{Ymax} = 20, \text{Yscl} = 10$

b. Now change the range setting to

$\text{Xmin} = -2.1, \text{Xmax} = 1.1, \text{Xscl} = 0.1,$

$\text{Ymin} = -0.04, \text{Ymax} = 0.04, \text{Yscl} = 0.01$

What behavior do you now observe that was not obvious from the graph in part (a)?

c. What theory would you need to know to predict the hidden behavior that becomes more obvious in the graph of part (b)? (This theory is studied in calculus.)

63. Some rational functions have parabolic asymptotes. For example, if $f(x) = \dfrac{x^3 + 4}{x}$, using the technique to find slant asymptotes, we can divide and obtain

$$f(x) = \frac{x^3 + 4}{x} = \underbrace{x^2}_{\substack{y = x^2 \text{ is} \\ \text{a parabolic} \\ \text{asymptote.}}} + \frac{4}{x}$$

a. Use a graphing utility to graph f and $y = x^2$ in the same viewing rectangle. Use the $\boxed{\text{ZOOM OUT}}$ feature to convince yourself that $y = x^2$ is a parabolic asymptote.

b. In general, if $f(x) = \dfrac{p(x)}{q(x)}$, under what conditions will f have a parabolic asymptote?

c. Give an example of another rational function with a parabolic asymptote, where the equation of the asymptote must be found by long division of polynomials. Verify your result with a graphing utility.

64. Explain how you can use a graphing utility to solve $\dfrac{2x^2 + 6x - 8}{2x^2 + 5x - 3} - 1 < 0$. Then use your graphing utility to show that the solution set is $(-\infty, -3) \cup \left(\frac{1}{2}, 5\right)$.

Writing in Mathematics _____

65. The figure on page 409 shows the graphs of $f(x) = \dfrac{1}{x - 1}$ and $g(x) = \dfrac{1}{(x - 1)^2}$, but not necessarily in that order. The graphs are only shown for $x > 1$. Describe how you can determine which is the graph of f and which is the graph of g.

66. Explain why $f(x) = \dfrac{x^6 + 1}{x^4 + 8x^2 + 12}$ has no vertical, horizontal, or slant asymptotes, and no x-intercepts. Verify by graphing f using a graphing utility.

67. Is every rational function a polynomial function? Why or why not? Does a true statement result if the two adjectives *rational* and *polynomial* are reversed? Explain.

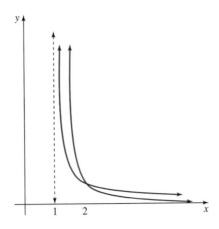

Critical Thinking Problems

68. If $c \neq 0$ and $d \neq 0$, describe the domain and range of $f(x) = \dfrac{ax + b}{cx + d}$.

In Problems 69–72, write the equation of a rational function $f(x) = \dfrac{p(x)}{q(x)}$ having the indicated properties, in which the degrees of p and q are as small as possible. More than one correct function may be possible. Graph your function using a graphing utility to verify that it has the required properties.

69. f has a vertical asymptote given by $x = 3$, a horizontal asymptote $y = 0$, y-intercept $= -1$, and no x-intercept.

70. f has vertical asymptotes given by $x = -2$ and $x = 2$, a horizontal asymptote $y = 2$, y-intercept $= \dfrac{9}{2}$, x-intercepts of -3 and 3, and y-axis symmetry.

71. f has a vertical asymptote given by $x = 1$, a slant asymptote whose equation is $y = x$, y-intercept $= 2$, and x-intercepts of -1 and 2.

72. f has no vertical, horizontal, or slant asymptotes, and no x-intercepts.

| SECTION 3.6 | Modeling with Rational Functions and Inverse Variation |

Solutions Manual   **Tutorial**   **Video 7**

Objectives

1 Solve problems using given rational functions.
2 Solve problems by creating rational models.
3 Model data using inverse variation.

There are numerous examples of asymptotic behavior in functions that model real world phenomena.

| Solve problems using given rational functions.

EXAMPLE 1   **Modeling Average Cost of Producing a Product**

A company that manufactures running shoes has costs given by the function

$$C(x) = 40x + 50{,}000$$

where x is the number of pairs of shoes produced per week and $C(x)$ is measured in dollars. The average cost per pair for the company is given by

tudy tip

The cost function

$C(x) = 40x + 50,000$

implies that the company has fixed costs of $50,000 and variable costs of $40 for each pair of running shoes manufactured.

$$\overline{C}(x) = \frac{40x + 50,000}{x}.$$

a. Find and interpret $\overline{C}(1000)$, $\overline{C}(10,000)$, and $\overline{C}(100,000)$.
b. What is the horizontal asymptote for this function? Describe what this represents for the company.

Solution

a. $\overline{C}(1000) = \dfrac{40(1000) + 50,000}{1000} = 90$

The average cost per pair of shoes of producing 1000 pairs per week is $90.00.

$$\overline{C}(10,000) = \frac{40(10,000) + 50,000}{10,000} = 45$$

The average cost per pair of shoes of producing 10,000 pairs per week is $45.00.

$$\overline{C}(100,000) = \frac{40(100,000) + 50,000}{100,000} = 40.5$$

The average cost per pair of shoes of producing 100,000 pairs per week is $40.50. Notice that with higher production levels, the cost of producing each unit of the product decreases.

b. We are given the average cost function

$$\overline{C}(x) = \frac{40x + 50,000}{x}$$

in which the degree of the numerator, 1, is equal to the degree of the denominator, 1. The leading coefficients of the numerator and denominator, 40 and 1, are used to obtain the equation of the horizontal asymptote. The equation of the horizontal asymptote is

$$y = \frac{40}{1} \quad \text{or} \quad y = 40.$$

The horizontal asymptote is shown in Figure 3.73. This means that the more pairs of running shoes produced per week, the closer the average cost per pair for the company comes to $40. The least possible cost per pair is approaching $40.00. Competitively low prices take place with high production levels, posing a major problem for small businesses. ∎

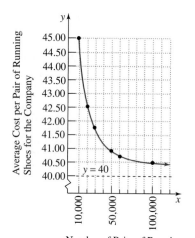

Number of Pairs of Running Shoes Produced per Week

Figure 3.73

As production level increases, the average cost per unit approaches $40.00.

An economic tool involving a *cost-benefit model* is often applied when making decisions on public policies dealing with the environment. For example, a graph of reducing air pollution in Tarnobrzeg, Poland, is shown in Figure 3.74. The graph indicates that removing the highest 40 to 50 percent of particles of pollution costs far less than removing 70 percent. Costs seem to be spiraling upward as the percent of removed pollutants increases. There is also a vertical asymptote associated with the graph of the cost function. Example 2 shows how these ideas can be expressed using a rational function.

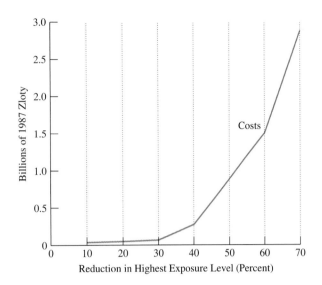

Figure 3.74

Source: J. Cofala, T. Lis, and H. Balandynowicz, *Cost-Benefit Analysis of Regional Air Pollution Control: Case Study for Tarnobrzeg*, 1991, Polish Academy of Sciences, Warsaw.

Jim Richardson / Richardson Photography

EXAMPLE 2   **A Cost-Benefit Model**

Cost-benefit models involving environmental pollution express cost as a function of the percentage of pollutants removed from the environment. For example, the rational function

$$C(x) = \frac{250x}{100 - x}$$

models the cost ($C(x)$, in millions of dollars) to remove x percent of the pollutants that are discharged into a river.

a. Find and interpret $C(90) - C(40)$.
b. What is the equation of the vertical asymptote for the cost-benefit model? What does this mean in terms of cleaning up the river?

Solution

a. First we will find and interpret $C(40)$ and $C(90)$.

$$C(40) = \frac{250(40)}{100 - 40} \approx 166.67$$

The cost to remove 40% of the river's pollutants is approximately $166.67 million.

$$C(90) = \frac{250(90)}{100 - 90} = 2250$$

The cost to remove 90% of the river's pollutants is $2250 million. Thus,

$$C(90) - C(40) \approx 2250 - 166.67 = 2083.33.$$

The difference in cost between removing 90% of the river's pollutants and 40% of its pollutants is approximately $2083.33 million.

b. The vertical asymptote for

$$C(x) = \frac{250x}{100 - x}$$

is $x = 100$, the zero of the denominator. As x approaches 100, $C(x)$ increases without bound (see Figure 3.75). This means that no amount of money can possibly clean up 100% of the river's pollutants. How much would it cost to remove 99% of the pollutants?

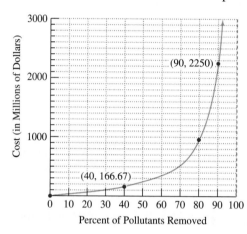

Figure 3.75

The graph of $C(x) = \dfrac{250x}{100 - x}$ ∎

2 Solve problems by creating rational models.

In Example 3, we use a rational function to model uniform motion. The example is based on the formula

$$RT = D$$

which states that the product of rate and time equals distance. Equivalently,

$$T = \frac{D}{R}$$

so that time is the quotient of distance and rate.

EXAMPLE 3   Modeling Involving Uniform Motion

Two commuters drove to work a distance of 40 miles and then returned again on the same route. The average rate on the return trip was 30 miles per hour faster than the average rate on the outgoing trip.

a. Find a rational function that models the total time T required to complete the round trip as a function of the rate x on the outgoing trip.
b. Use a graphing utility to graph the function for meaningful values of x.
c. If the commuters would like to complete the round trip in 2 hours, use the graph to determine what the average rate on the outgoing trip should be.

Solution

Roy Lichtenstein "In the Car" 1963. The Scottish National Gallery of Modern Art. © Roy Lichtenstein.

a. Let

$$x = \text{average rate on the outgoing trip}$$
$$x + 30 = \text{average rate on the return trip}$$

The critical sentence that we must model is:

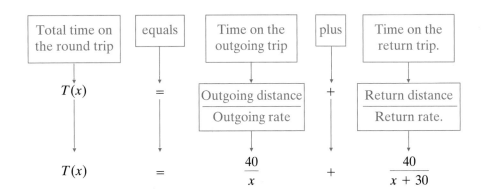

$$T(x) = \frac{40}{x} + \frac{40}{x + 30} \qquad \text{This is the function that models total time } T.$$

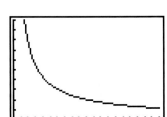

Figure 3.76

The graph of
$T(x) = \dfrac{40}{x} + \dfrac{40}{x + 30}$. As the
rate increases, time decreases.

b. Let's graph the function for $0 \leqslant x \leqslant 60$, since it seem unlikely that an average outgoing rate exceeds 60 miles per hour with a return rate that is 30 miles per hour faster.

Figure 3.76 shows the graph of $T(x) = \dfrac{40}{x} + \dfrac{40}{x + 30}$, obtained on a graphing utility using the following range setting:

Xmin $= 0$, Xmax $= 60$, Xscl $= 3$,

Ymin $= 0$, Ymax $= 10$, Yscl $= 1$.

The graph shows decreasing times with increasing rates. The vertical asymptote, $x = 0$, indicates that close to an outgoing rate of zero miles per hour, the round trip will take nearly forever.

c. The commuters have 2 hours, so we trace along the curve until $y = 2$, remembering that y represents total time. Figure 3.77 shows that $x = 30$, meaning that the commuters must average 30 miles per hour on the outgoing trip to complete the round trip in 2 hours. For this one-to-one function, $T^{-1}(2) = 30$. ■

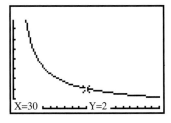

Figure 3.77

An average speed of 30 miles per hour on the outgoing trip results in 2 hours for the round trip.

3 Model data using inverse variation.

Modeling Using Inverse Variation

It is possible to look at a data set and notice patterns to suggest modeling with rational functions in the form

$$y = \frac{k}{x^n}, \quad \text{where } n \geqslant 1.$$

If x and y are related by this equation, we say that *y varies inversely as the nth power of x*.

Inverse variation

y varies inversely as the nth power of x (or *y is inversely proportional to the nth power of x*) if there exists some nonzero constant *k* such that

$$y = \frac{k}{x^n}.$$

k is called the *constant of variation* or the *constant of proportionality*.

EXAMPLE 4   **Solving an Inverse Variation Problem**

The number of pens sold varies inversely as the price per pen. If 4000 pens are sold at a price of $1.50 per pen, write a function *f* that models the number of pens sold and the price per pen. Then graph the function. Is the function one-to-one? What is the practical meaning of $f^{-1}(2000)$?

Solution

$$y = \frac{k}{x}$$ Translate "Number sold (*y*) varies inversely as price per pen (*x*)."

$$4000 = \frac{k}{1.5}$$ Find *k*. Because 4000 pens are sold at $1.50 per pen, when *x* = 1.5, then *y* = 4000.

$$6000 = k$$ Multiply both sides of the equation by 1.5 and solve for *k*.

$$y = \frac{6000}{x}$$ Substitute the value for *k* into the original equation.

The function that models the number of pens sold and the price per pen is

$$y = \frac{6000}{x} \quad \text{or} \quad f(x) = \frac{6000}{x}.$$

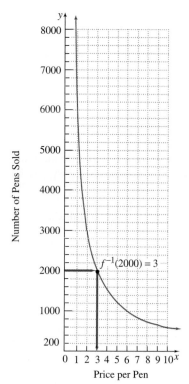

Figure 3.78

The graph of $f(x) = \dfrac{6000}{x}$

We can graph the function by multiplying each *y*-value of the graph of $y = \frac{1}{x}$ by 6000. The graph of $f(x) = \frac{6000}{x}$ appears in Figure 3.78. The graph is drawn only for meaningful values of *x*, namely, *x* > 0. The curve indicates that as the price per pen increases, the number of pens sold decreases quite rapidly. Since every horizontal line that can be drawn intersects the graph no more than once, the function is one-to-one—that is, its inverse is a function. The graph indicates the practical meaning of $f^{-1}(2000)$: If 2000 pens are sold, then the price per pen is $3.00.   ∎

Because $y = \dfrac{k}{x^n}$ means *y* varies inversely as the *n*th power of *x*, we can tell if a given data set is consistent with an inverse *n*th power variation by multiplying both sides of the equation by x^n. We obtain

$$yx^n = k$$

indicating that the product of x^n and *y* remains constant.

> ### Modeling with inverse variation
>
> If a data set is given such that the product $x^n y$ remains constant (k), then the data can be modeled by the function $y = \dfrac{k}{x^n}$.

EXAMPLE 5 | Modeling Gravitational Force

Use the data shown below to write a rational function that models gravitational force and distance.

x (Distance of Object from Earth's Center)	y (Gravitational Force with which Earth Attracts the Object)
1000 miles	2560 pounds
2000 miles	640 pounds
4000 miles	160 pounds
10,000 miles	25.6 pounds

Solution

If we are to model the data using inverse variation, then $x^n y = k$.

x	y	If $n = 1$	If $n = 2$
1000	2560	$xy = 2{,}560{,}000$	$x^2 y = (1000)^2(2560) = 2{,}560{,}000{,}000$
2000	640	$xy = 1{,}280{,}000$	$x^2 y = (2000)^2(640) = 2{,}560{,}000{,}000$
4000	160	Products are not constant.	$x^2 y = (4000)^2(160) = 2{,}560{,}000{,}000$
10,000	25.6	$y = \dfrac{k}{x}$ does not model the data.	$x^2 y = (10{,}000)^2(25.6) = 2{,}560{,}000{,}000$

Since the products in the last column are constant, $x^2 y = 2{,}560{,}000{,}000$. Equivalently, the rational function that models the data is

$$ y = \frac{2{,}560{,}000{,}000}{x^2} \quad \text{or} \quad f(x) = \frac{2{,}560{,}000{,}000}{x^2}. $$

In practical terms, the model states: The gravitational force with which Earth attracts an object varies inversely with the square of the distance the object is from the center of Earth. Thus, the force of gravity falls off rapidly as an object gets farther from Earth's center. ■

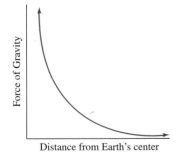

Force of Gravity

Distance from Earth's center

The force of gravity falls off rapidly as objects get farther apart from Earth's center.

Isaac Newton's model for gravitation involves *combined variation,* where gravitational force depends on several other variables.

Jane Lewis "The Illusionist" 1995, oil on canvas. Original paintings available: Portal Gallery, London, England.

David Madison / Duomo Photography

Basic variation models

y varies directly as x^n: $y = kx^n$

y varies inversely as x^n: $y = \dfrac{k}{x^n}$

z varies jointly as x^n and y^m: $z = kx^n y^m$

Newton's theory of gravitation is modeled by

$$F = G\frac{m_1 m_2}{d^2}$$

which states that the force of gravitation (F) between two bodies varies jointly as the product of their masses (m_1 and m_2) and inversely as the square of the distance between them (d^2). (G is the gravitational constant.) The formula indicates that gravitational force exists between any two objects in the universe, increasing as the distance between the bodies decreases. One practical result is that the pull of the moon on the oceans is greater on the side of Earth closer to the moon. This gravitational imbalance is what produces tides.

EXAMPLE 6 Combined Variation: Modeling Centrifugal Force

The centrifugal force (C) of a body moving in a circle varies jointly with the radius of the circular path (r) and the body's mass (m), and inversely with the square of the time (t) it takes to move about one full circle. A 6-gram body moving in a circle with radius 100 centimeters at a rate of 1 revolution in 2 seconds has a centrifugal force of 6000 dynes. Find the centrifugal force of an 18-gram body moving in a circle with radius 100 centimeters at a rate of 1 revolution in 3 seconds.

Solution

$C = \dfrac{krm}{t^2}$ Translate "Centrifugal force (C) varies jointly with radius (r) and mass (m) and inversely with the square of time (t)."

$6000 = \dfrac{k(100)(6)}{2^2}$ If $r = 100$, $m = 6$, and $t = 2$, then $C = 6000$.

$40 = k$ Solve for k.

$C = \dfrac{40rm}{t^2}$ Substitute 40 for k in the model for centrifugal force.

$\quad = \dfrac{40(100)(18)}{3^2}$ Find C when $r = 100$, $m = 18$, and $t = 3$.

$\quad = 8000$

The centrifugal force is 8000 dynes. ∎

ENRICHMENT ESSAY

Atonal Music for the Planetary Spheres

Chloe: The future is all programmed like a computer—that's a proper theory, isn't it?
Valentine: The deterministic universe, yes ... There was someone, forgot his name, 1820s, who pointed out that from Newton's laws you could predict everything to come...
Chloe: But it doesn't work, does it?
Valentine: No. It turns out the mathematics is different.

Tom Stoppard, *Arcadia*

Kingsley Parker "Primitive Orrery #1"
1993. Courtesy of Kingsley Parker.

PROBLEM SET 3.6

Practice and Application Problems

1. A company that manufactures small canoes has costs given by the function $C(x) = 20x + 20{,}000$, where x is the number of canoes manufactured and $C(x)$ is measured in dollars. The average cost to manufacture each canoe is given by

$$\overline{C}(x) = \frac{20x + 20{,}000}{x}.$$

   a. Find the average cost per canoe when $x = 100$, 1000, $10{,}000$, and $100{,}000$.

   b. What is the horizontal asymptote for this function, and what does it represent?

2. A bicycle manufacturing business has fixed costs of $100,000 and variable costs of $100 per bicycle.

   a. Find a function $\overline{C}$ that models the average cost per bicycle for x bicycles. (*Hint:* Costs, $C(x)$, are given by 100 times the number of bicycles manufactured plus fixed costs. The average cost to manufacture each bicycle is the quotient of $C(x)$ and the number of bicycles manufactured.)

   b. Find and interpret $\overline{C}(500)$, $\overline{C}(1000)$, $\overline{C}(2000)$, and $\overline{C}(4000)$.

   c. What is the horizontal asymptote for the function $\overline{C}$? Describe what this means in practical terms.

3. The cost in dollars of removing p percent of the air pollutants in the stack emission of a utility company that burns coal to generate electricity is given by

$$C(p) = \frac{60{,}000p}{100 - p}.$$

   a. Current law requires that the company remove 80% of the pollutants from their smokestack emissions. A new law before the legislature would require increasing this amount by 5%. How much will it cost to remove another 5% of the pollutants?

   b. Does this model indicate the possibility of removing 100% of the pollutants? Explain.

4. The rational function

$$C(x) = \frac{130x}{100 - x} \quad 0 \leqslant x < 100,$$

   models the cost (C, in millions of dollars) to inoculate $x\%$ of the population against a particular strain of flu.

   a. Find and interpret $C(80) - C(40)$.

   b. Graph the function.

   c. Describe the practical meaning of the observation that $x = 100$ is an asymptote.

5. The temperature (F, in degrees Fahrenheit) of a dessert placed in an icebox for t hours is modeled by

$$f(t) = \frac{80}{t^2 + 4t + 1}.$$

   a. Find and interpret $F(0)$.

   b. Find the temperature of the dessert after 1 hour, 2 hours, 3 hours, 4 hours, and 5 hours.

   c. What is the equation of the horizontal asymptote associated with this function? Describe what this means in terms of the dessert's temperature over time.

   d. Graph the function.

6. The electrical resistance of a wire varies directly as its length and inversely as the square of its diameter. A wire of 720 feet with a $\frac{1}{4}$-inch diameter has a resistance of 1.5 ohms. Predict the resistance for 960 feet of the same kind of wire if its diameter is doubled.

7. The force of attraction between two bodies varies jointly as the product of their masses and inversely as the square of the distance between them. Two 1-kilogram masses separated by 1 meter exert a force of attraction of 6.67×10^{-11} newton. What is the gravita-

tional force exerted by Earth on a 1000-kilogram satellite orbiting at an altitude of 300 kilometers? (The mass of Earth is 5.98×10^{24} kilograms and its radius is 6400 kilometers.)

8. Use the results of Example 5 on page 415 to answer this question: If a person weighs 160 pounds on Earth's surface and if Earth's radius is 6370 kilometers, how much does that person weigh on a jet that is flying at an altitude of 8 kilometers above the surface?

9. Use the data shown below to write a rational function of the form $y = \dfrac{k}{x^n}$ that models intensity of illumination on a surface (y) and the distance of the light source from the surface (x). Describe what the model states in practical terms.

x (in feet)	y (in foot-candles)
2	100
4	25
10	4
20	1

In Problems 10–12, determine whether the data can be modeled by the rational function $y = \dfrac{k}{x^n}$ for the given value of n. If it can, write the function.

10.

x	$y\,(n = 2)$
4	100
5	64
8	25
10	16

11.

x	$y\,(n = 3)$
3	8000
4	3375
5	1728
10	216

12.

x	$y\,(n = 3)$
2	25
4	$3\frac{1}{8}$
5	$1\frac{3}{5}$
10	$\frac{1}{5}$

Technology Problems

13. A person drove from home to a vacation resort 600 miles away. The return trip was on the same highway. The average rate on the return trip was 10 miles per hour slower than the average rate on the outgoing trip.

   a. Find a rational function that models the total time T required to complete the round trip as a function of the rate x on the outgoing trip.

   b. Use a graphing utility to graph the function for meaningful values of x.

   c. Use the graph to find $T^{-1}(22)$ and describe what this means in practical terms.

   d. Modify the given word problem by writing a problem whose solution can be found by graphing $y = 20$

in the same viewing rectangle as the graph of T, and then finding positive values of x for which the graph for T lies below the graph of $y = 20$.

14. A contractor is constructing the house shown in the figure on page 419. The cross section up to the roof is in the shape of a rectangle with an area of 2500 square feet. The width of the rectangle is represented by x.

   a. Write a rational function that models the perimeter of the rectangle.

   b. A minimum perimeter will reduce construction costs for the house. Use a graphing utility to graph the perimeter function and find the dimensions of the rectangle with the least possible perimeter.

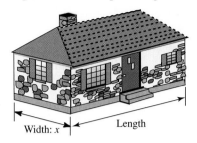

Width: x Length

15. A grocery store sells 4000 cases of canned soup per year. By averaging costs and storage costs, the owner has determined that if x cases are ordered at a time, the inventory cost will be

$$C(x) = \frac{10,000}{x} + 3x.$$

a. Use a graphing utility to graph the inventory function.

b. Use the [ZOOM] and [TRACE] features or the minimum function feature of your graphing utility to approximate the number of cases that should be ordered to minimize inventory cost. What is the minimum cost?

Writing in Mathematics

16. You have been hired as vice-president in charge of advertising for the Worldwide Widget Corporation. You were told that the demand for widgets varies directly as the amount spent on advertising and inversely as the price of the product. However, as more money is spent on advertising, the price of a widget rises. Under what conditions would you recommend an increased expense in advertising? Once you've determined what a "widget" is, write mathematical models for the given conditions and experiment with hypothetical numbers. What other factors might you take into consideration in terms of your recommendation?

17. Suppose you are a sales analyst for a home video game company. It has been determined that the rational function

$$f(x) = \frac{200x}{x^2 + 100}$$

models the monthly sales (in thousands of games) of a new video game as a function of the number of months (x) after the game is introduced. Use a graphing utility to graph the function. What are your recommendations to the company in terms of how long the video game should be on the market before another new video game is introduced? What other factors might you want to take into account in terms of your recommendations? What will eventually happen to sales, and how is this indicated by the graph? What does this have to do with a horizontal asymptote? What could the company do to change the behavior of this function and continue generating sales? Would this be cost effective?

Critical Thinking Problems

18. The heat generated by a stove element varies directly as the square of the voltage and inversely as the resistance. If the voltage remains constant, what needs to be done to triple the amount of heat generated?

19. Galileo's telescope brought about revolutionary changes in astronomy. A comparable leap in our ability to observe the universe is now taking place as a result of the Hubble Space Telescope. The space telescope can see stars and galaxies whose brightness is $\frac{1}{50}$ of the faintest objects now observable using ground-based telescopes. Use the fact that the brightness of a point source, such as a star, varies inversely as the square of its distance from an observer to show that the space telescope can see about seven times farther than a ground-based telescope.

Group Activity Problem

20. A cost-benefit analysis compares the estimated costs of a project with the benefits that will be achieved. Costs and benefits are given monetary values and compared by using a benefit-cost ratio. As shown in the figure on page 420, a favorable ratio for a project means that the benefits outweigh the costs, and the project is cost-effective. As a group, select an environmental project that has been implemented in your area of the country.

Research the cost and benefit graphs that resulted in the implementation of this project. How were the benefits converted into monetary terms? Is there an equation for either the cost model or the benefit model? Group members may need to interview members of environmental groups and businesses that were part of the project. You may also wish to consult an environmental science textbook to find out more about cost-benefit analyses. After doing your research, the group should write or present a report explaining why the cost-benefit analysis resulted in the project's implementation.

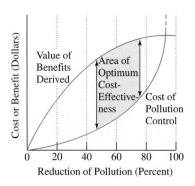

CHAPTER PROJECT
Women in Mathematics

Glancing at most mathematics textbooks, it is obvious that the majority of people mentioned are men. For example, look at the essay on page 373 entitled "A Brief History of Polynomial Equations" and read the list of names contributing to the study of polynomials. No women appear on the list which covers almost 2000 years. If we looked through a similar list of notable names for the same time period, roughly 200 B.C. to the mid 1800s, in almost any area of science and mathematics, we would see very few women mentioned.

1. People have argued that the absence of women in mathematics for the time period discussed above "proves" women cannot do math. Compare the following two arguments and discuss their validity:
   * Before the mid 1900s, there were no African-Americans playing major league baseball. Therefore, African-Americans do not play baseball very well.
   * Before the mid 1800s, there were almost no women working in mathematics. Therefore, women do not do mathematics very well.
   Consider as part of your argument the fact that in 1991, 47 percent of the Bachelor's degrees in mathematics from American schools went to women.

As the above problem illustrates, if we are looking for the contributions of women in mathematics, we must keep in mind the historical period we are studying. Historically, how did women obtain an education? For many years women were denied a formal education, precisely the kind of education required for work in advanced mathematics. While men worked with mentors or tutors, and subsequently may have progressed on to universities, women had virtually no contact with this style of learning. In addition, most advanced learning in the sciences or mathematics required a high standard of living. In many cases, only the affluent would have the large blocks of time needed to devote to study. Some women, who did contribute to mathematics in the 1800s, had access to learning because they came from wealthy families and were able to use private libraries.

2. Working with a group, select a woman mathematician from the list below. Prepare a report about women and learning in the time period in which she lived. Please note, this is not necessarily a report on the mathematics, it is a discussion of how any subject in general was available for study to women. As we move closer to our time period, the mathematics discussed may be beyond the scope of this course, but the human aspect of learning remains.

Theano (wife of Pythagoras, around 5th Century B.C.)	Emilie du Chatelet (1706–1749)
Hypatia (3__?–415)	Maria Gaetana Agnesi (1718–1799)
Elena Lucrezia Cornaro Piscopia (first woman to receive a doctoral degree, 1646–1684)	Caroline Herschel (1750–1848)
	Sophie Germain (1776–1831)
	Mary Fairfax Somerville (1780–1872)

3. The first PhD in mathematics was awarded to Winifred Edgerton Merrill (1862–1951) by Columbia University in 1886. Her son later wrote that she "said that a condition of her admission was to dust the astro-

nomical [instruments] and so comport herself as not to disturb the male students." The separate treatment of men and women, even when allowing women access to the same universities, was not unusual. In a small group, use resources in your library or on the Worldwide Web to find a man in the sciences or mathematics who lived and worked in approximately the same time frame as one of the women below. Compare and contrast the opportunities given for education and the lifestyles of each. Share your results with the class.

Ada Byron Lovelace (1815–1852)
Florence Nightingale (1820–1910)
Mary Everest Boole (1832–1916)
Christine Ladd-Franklin (1847–1930)
Sonya Kovalevskaya (1850–1891)
Charlotte Angas Scott (1858–1931)
Charlotte Barnum (1860–1934)
Winifred Edgerton Merrill (1862–1951)

Clara Eliza Smith (1865–1943)
Grace Chisholm Young (1868–1944)
Ada Isabel Maddison (1869–1950)
Mary Frances Winston Newson (1869–1959)
Agnes Baxter (1870–1917)
Emmy Noether (1882–1935)
Anna Pell Wheeler (1883–1966)

Worldwide Web Resources

Go to the Prentice Hall website (http://www.prenhall.com/blitzer) to access other locations on the Internet that will allow you to further explore the concepts presented in this project.

Chapter Review

SUMMARY

1. Quadratic Functions

a. Standard form: $f(x) = a(x - h)^2 + k$
Vertex of the graph, which is a parabola: (h, k)

b. The form $f(x) = ax^2 + bx + c$
Vertex of the parabola: $\left(-\dfrac{b}{2a}, f\left(-\dfrac{b}{2a}\right)\right)$

c. Optimization

1. If $a > 0$, the minimum value of f occurs at $x = h$ or $x = -\dfrac{b}{2a}$. This minimum value is $f(h) = k$ or $f\left(-\dfrac{b}{2a}\right)$.

2. If $a < 0$, the maximum value of f occurs at $x = h$ or $x = -\dfrac{b}{2a}$. This maximum value is $f(h) = k$ or $f\left(-\dfrac{b}{2a}\right)$.

2. Graphing a Polynomial Function
$f(x) = a_n x^n + a_{n-1}x^{n-1} + a_{n-2}x^{n-2} + \cdots + a_1 x + a_0$

a. Use the Leading Coefficient Test to determine the graph's end behavior.

1. If n is odd and $a_n > 0$, the graph falls to the left and rises to the right.

2. If n is odd and $a_n < 0$, the graph rises to the left and falls to the right.

3. If n is even and $a_n > 0$, the graph rises to the left and to the right.

4. If n is even and $a_n < 0$, the graph falls to the left and to the right.

b. Find x-intercepts by setting $f(x) = 0$ and solving the resulting polynomial equation. If there is an x-intercept at a as a result of $(x - a)^k$ in the complete factorization of $f(x)$, then:

1. If k is even, the graph touches the x-axis at a but does not cross the x-axis.

2. If k is odd, the graph crosses the x-axis at a.

3. If $k > 1$, the graph flattens out at $(a, 0)$.

c. Find the y-intercept by setting x equal to 0 and calculating $f(0)$.

d. Use symmetry, if applicable, to help draw the graph:

1. y-axis symmetry: $f(-x) = f(x)$

2. Origin symmetry: $f(-x) = -f(x)$

e. Use the fact that the maximum number of turning points of the graph is $n - 1$ to check whether the graph is drawn correctly.

3. **Real Zeros of Polynomial Functions**

   If f is a polynomial function and $f(a) = 0$ for any real number a:

   a. a is a zero of f.

   b. a is a solution or root of $f(x) = 0$.

   c. a is an x-intercept of the graph of f.

   d. $x - a$ is a factor of the polynomial f.

4. **Solving Polynomial Equations; Finding Zeros of Polynomial Functions**

   $$a_n x^n + a_{n-1} x^{n-1} + a_{n-2} x^{n-2} + \cdots + a_1 x + a_0 = 0$$

   a. Use the Rational Zero Theorem to list all possible rational roots, $\dfrac{p}{q}$. p is a factor of the constant term, a_0, and q is a factor of the leading coefficient, a_n.

   b. Use Descartes' Rule of Signs to determine the possibilities for positive and negative roots. The number of positive roots is equal to the number of variations in sign in the equation or less than the number of variations by a positive even integer. The number of negative real roots applies to a similar statement when x is replaced by $-x$ in the given equation.

   c. Use synthetic division to check the positive and negative candidates for rational zeros listed in step (a). A graphing utility, showing the graph of the polynomial function and its zeros, might be helpful. Stop when you find a root (remainder $= 0$).

   d. If step (c) indicates that c is a root, then one factor is $x - c$. The second factor is the quotient obtained from the synthetic division.

   e. Set each factor equal to zero.

   f.
   1. Solve the equation obtained by setting the linear factor equal to zero.

   2. If the second factor is quadratic, solve the resulting quadratic equation by factoring (if possible), by the quadratic formula, or by the square root method.

   3. If the second factor is of degree 3 or greater, repeat steps (a) through (e) on the resulting polynomial equation.

5. **Theorems that Help to Find Zeros of Polynomial Functions**

   a. *The Factor Theorem:* If $f(c) = 0$, then $x - c$ is a factor of $f(x)$. If $x - c$ is a factor of $f(x)$, then $f(c) = 0$.

   b. *The Rational Zero Theorem:* If $f(x) = a_n x^n + a_{n-1} x^{n-1} + \cdots + a_1 x + a_0$ has integer coefficients and $\dfrac{p}{q}$ (where $\dfrac{p}{q}$ is reduced) is a rational zero, then p divides a_0 and q divides a_n.

   c. *Descartes' Rule of Signs:* The number of positive real zeros of f equals the sign changes of $f(x)$ or is less than that number by an even integer. The number of negative real zeros of f applies a similar statement to $f(-x)$.

   d. *The Upper and Lower Bound Theorem:* The number $b > 0$ is an upper bound for the real roots of $f(x) =$ 0 if synthetic division of $f(x)$ by $x - b$ results in no negative numbers. The number a ($a < 0$) is a lower bound if similar synthetic division results in numbers that alternate in sign, counting zero entries as positive or negative.

   e. *The Linear Factorization Theorem:* If

   $$f(x) = a_n x^n + a_{n-1} x^{n-1} + \cdots + a_1 x + a_0$$
   $$(n \geq 1, a_n \neq 0),$$

   then

   $$f(x) = a_n (x - c_1)(x - c_2) \cdots (x - c_n)$$

   where $c_1, c_2, \ldots, c_n$ are complex numbers (possibly real and not necessarily distinct).

   f. *The Number of Roots Theorem:* If $f(x)$ is a polynomial of degree $n \geq 1$, then $f(x) = 0$ has exactly n roots, where roots are counted according to their multiplicity.

   g. *The Conjugate Roots Theorem:* If $a + bi$ is a root of $f(x) = 0$, then $a - bi$ is also a root.

   h. *The Intermediate Value Theorem:* If $f(a)$ and $f(b)$ have opposite signs, there is at least one value of c between a and b for which $f(c) = 0$.

6. **Rational Functions**

   a. A rational function is one that can be written in the form $f(x) = \dfrac{p(x)}{q(x)}$, where $p(x)$ and $q(x)$ are polynomial functions, and $q(x) \neq 0$.

   b. The domain of a rational function is the set of all real numbers except those for which $q(x) = 0$.

   c. *Vertical Asymptotes:* The line $x = a$ is a vertical asymptote of f if $f(x) \to \infty$ or $f(x) \to -\infty$ as $x \to a^+$ or $x \to a^-$.

   d. *Horizontal Asymptotes:* The line $y = b$ is a horizontal asymptote of f if $f(x) \to b$ as $x \to \infty$ or $x \to -\infty$.

   e. *The Reciprocal Function:* $f(x) = \dfrac{1}{x}$

   Horizontal asymptote: $y = 0$ (the x-axis)

   Vertical asymptote: $x = 0$ (the y-axis)

   Some rational functions can be graphed using transformations of the reciprocal function.

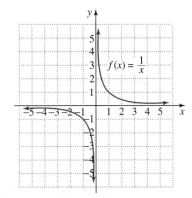

7. **Graphing Rational Functions**

$f(x) = \dfrac{p(x)}{q(x)}$: $p(x)$ and $q(x)$ are polynomial functions

with no common factors.

  a. Determine whether the graph of f has symmetry. (See 2(d).)

  b. Find and plot the y-intercept (if there is one) by evaluating $f(0)$.

  c. Find and plot the x-intercepts (if there are any) by solving the equation $p(x) = 0$.

  d. Find any vertical asymptote(s) by solving the equation $q(x) = 0$. Sketch the vertical asymptotes.

  e. Find and sketch the horizontal asymptote (if there is one).

  1. If the degree of p is less than the degree of q, then $y = 0$ (the x-axis) is a horizontal asymptote.

  2. If the degree of p is equal to the degree of q, the horizontal asymptote is found by taking the quotient of the leading coefficient of p and q. Setting y equal to this quotient gives the equation of the horizontal asymptote.

  3. If the degree of p is greater than the degree of q, there is no horizontal asymptote.

  f. Find and sketch the slant asymptote (if there is one). If the degree of p is one greater than the degree of q, divide $q(x)$ into $p(x)$. If $\dfrac{p(x)}{q(x)} = mx + b + \dfrac{R(x)}{q(x)}$, then $y = mx + b$ describes the slant asymptote.

  g. Plot at least one point between and beyond each x-intercept and vertical asymptote or between and beyond each x-intercept and slant asymptote.

  h. Use smooth curves to graph the function by having the curve pass through the indicated points and then having it approach but never touch the indicated asymptotes.

  i. Use a graping utility in the DOT mode to verity that the graph has been drawn correctly.

8. **Modeling with Variation**

  a. *Direct Variation: y* varies directly as the nth power of x (or y is proportional to the nth power of x) if there exists $k > 0$ such that $y = kx^n$. A data set in the form (x, y) can be modeled by direct variation if the ratio $\dfrac{y}{x^n}$ remains constant for some positive integer n.

  b. *Joint Variation: z* varies jointly as x^n and y^m is modeled by $z = kx^n y^m$.

  c. *Inverse Variation: y* varies inversely as the nth power of x (or y is inversely proportional to the nth power of x) if there exists $k > 0$ such that $y = \dfrac{k}{x^n}$. A data set in the form (x, y) can be modeled by inverse variation if the product $x^n y$ remains constant for some positive integer n.

REVIEW PROBLEMS

In Problems 1–4, use the vertex and the intercepts to sketch the graph of each quadratic function by hand. Give the equation for the parabola's axis of symmetry. Use a graphing utility to verify that your graph is correct.

1. $f(x) = -2(x - 1)^2 + 3$

2. $g(x) = (x + 4)^2 - 2$

3. $r(x) = -x^2 + 2x + 3$

4. $h(x) = 2x^2 - 4x - 6$

5. The function $p(x) = 0.0014x^2 - 0.1529x + 5.855$ models the percent (p) of income that Americans contribute to charities in terms of annual household income (x), where x is expressed in thousands of dollars, where $5 \leqslant x \leqslant 100$. What income level corresponds to the minimum percentage given to charities?

6. The function $M(t) = 0.0075t^2 - 0.2676t + 14.8$ describes the fuel efficiency of passengers cars (M, measured in miles per gallon) t years after 1940. What is the vertex for the graph of this quadratic function? Describe the significance of the vertex in practical terms.

7. A person standing close to the edge on the top of an 80-foot building throws a ball vertically upward with an initial velocity of 64 feet per second.

  a. Use the position function $s(t) = -16t^2 + v_0 t + s_0$, in which v_0 represents initial velocity and s_0 represents initial position to write the function modeling the height of the ball above the ground as a function of time.

  b. After how many seconds does the ball reach its maximum height? What is the maximum height?

  c. Sketch the graph of the function in part (a) by hand, verifying your result with a graphing utility.

8. A field bordering a straight stream is to be enclosed. The side bordering the stream is not to be fenced. If 1000 yards of fencing material are to be used, what are the dimensions of the largest rectangular field that can be fenced? What is the maximum area?

9. An owner of a large diving boat that can hold as many as 70 people charges $10 per passenger to groups between 15 and 20 people. If more than 20 divers charter the boat, the fee per passenger is decreased by 10¢ times the number of people in excess of 20. Find the size of the group that will maximize the boat owner's income. What is the maximum income?

10. Find the point on the line whose equation is $2x + y = 2$ that is closest to the origin.

In Problems 11–14:

a. Use the Leading Coefficient Test to determine the right-hand ($x \to \infty$) and left-hand ($x \to -\infty$) end behavior of the polynomial function.

b. Use factoring to find all zeros of the polynomial function.

c. Graph the function by hand.

d. Use a graphing utility to verify that your graph is correct.

11. $f(x) = x^3 - 5x^2 - x + 5$

12. $g(x) = -x^4 + 25x^2$

13. $h(x) = -x^4 + 6x^3 - 9x^2$

14. $r(x) = 3x^4 - 15x^3$

15. The function $f(x) = -0.0013x^3 + 0.078x^2 - 1.43x + 18.1$ describes the percentage of U.S. families below the poverty level x years after 1960. Use end behavior to explain why the model is valid only for a limited period of time.

In Problems 16–19, use end behavior and, if necessary, zeros to match each polynomial function with its graph.

a.

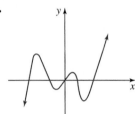

b.

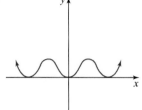

c.

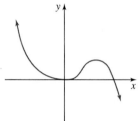

d.

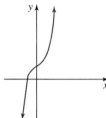

16. $f(x) = -x^3 + 12x^2 - x$

17. $g(x) = x^6 - 6x^4 + 9x^2$

18. $h(x) = x^5 - 5x^3 + 4x$

19. $r(x) = x^3 + 1$

20. Can the data shown on the right be modeled by a direct variation equation in which $n = 4$? If so, write a polynomial function that models the data.

x	y
0.2	48
0.3	243
0.5	1875
0.8	12,288
1.0	30,000

In Problems 21–22, use Descartes' Rule of Signs to determine the possible number of positive and negative solutions of the polynomial equation.

21. $3x^4 - 2x^3 - 8x + 5 = 0$

22. $2x^5 - 3x^3 - 5x^2 + 3x - 1 = 0$

23. Explain why $2x^4 + 6x^2 + 8 = 0$ has no real roots.

For Problems 24–28:

a. List all possible rational solutions.

b. Use Descartes' Rule of Signs to determine the possible number of positive and negative real solutions.

c. (Optional) If there is a fairly long list of possible rational zeros for f (where $f(x) = 0$), use a graphing utility to graph f and eliminate some of the possible rational zeros.

d. Find the solution set.

24. $8x^3 - 36x^2 + 46x - 15 = 0$

25. $x^4 - x^3 - 7x^2 + x + 6 = 0$

26. $2x^4 - 5x^3 - 8x^2 + 25x - 10 = 0$

27. $4x^4 - 8x^3 - 43x^2 + 29x + 60 = 0$

28. $3x^5 - 2x^4 - 15x^3 + 10x^2 + 12x - 8 = 0$

For Problems 29–31:

a. List all possible rational zeros.

b. Use Descartes' Rule of Signs to determine the possible number of positive and negative real zeros.

c. (Optional) If there is a fairly long list of possible rational zeros, use a graphing utility to graph the function and eliminate some of the possible rational zeros.

d. Find all zeros of the function.

29. $f(x) = 6x^3 + x^2 - 4x + 1$

30. $g(x) = 2x^4 + x^3 - 2x^2 - 4x - 3$

31. $h(x) = 3x^4 - 7x^3 + 6x^2 - 28x - 24$

32. Consider the equation $2x^4 - x^3 - 5x^2 + 10x + 12 = 0$.

   a. List all possible rational roots.

   b. Determine whether 2 is a root using synthetic division. In terms of bounds, what can you conclude?

   c. Determine whether -2 is a root using synthetic division. In terms of bounds, what can you conclude?

   d. Use the results of parts (b) and (c) to discard some of the possible rational roots from part (a). Now what are the possible rational roots?

33. Show that all real solutions of the equation $2x^4 - 7x^3 - 5x^2 + 28x - 12 = 0$ lie between -2 and 6. Use this result to list all possible rational roots.

34. The function

$$S_n(n) = \frac{n(n+1)(2n+1)}{6}$$

describes the sum of the squares of the first n natural numbers. How many natural numbers must be added to produce a sum of 140?

35. The number of eggs (N) in a female moth is a function of her abdominal width (W, in millimeters), modeled by $N = 14W^3 - 17W^2 - 16W + 34$, for $1.5 \leqslant W \leqslant 3.5$. What is the abdominal width when there are 211 eggs?

36. If the volume of the solid shown in the figure is 164 cubic centimeters, find the value of x.

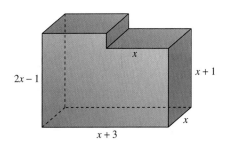

In Problems 37–38, show that the polynomial has a zero between the given integers. Then use the Intermediate Value Theorem to find an approximation for this zero to the nearest tenth. Use a graphing utility to verify your result.

37. $f(x) = x^3 - 2x - 1$; between 1 and 2

38. $f(x) = 3x^3 + 2x^2 - 8x + 7$; between -3 and -2

In Problems 39–41, find an nth-degree polynomial function with real coefficients satisfying the given conditions. Use a graphing utility to graph the function and check your answer.

39. $n = 3$; 2 and $2 - 3i$ are zeros; $f(1) = -10$

40. $n = 4$; i is a zero; -3 is a zero of multiplicity 2; $f(-1) = 16$

41. $n = 4$; -2, 3, and $1 + 3i$ are zeros; $f(2) = -40$

In Problems 42–45, use the given root to find the solution set of each polynomial equation.

42. $4x^3 - 47x^2 + 232x + 61 = 0$; $6 + 5i$

43. $x^4 - 4x^3 + 16x^2 - 24x + 20 = 0$: $1 - 3i$

44. $2x^4 - 17x^3 + 137x^2 - 57x - 65 = 0$; $4 + 7i$

45. $x^5 + 3x^4 + 2x^3 + 14x^2 + 29x + 15 = 0$; $1 + 2i$

In Problems 46–47, find all the zeros of each polynomial function and write the polynomial as a product of linear factors.

46. $f(x) = 2x^4 + 3x^3 + 3x - 2$

47. $g(x) = x^4 - 6x^3 + x^2 + 24x + 16$

In Problems 48–51, graphs of fifth-degree polynomial functions are shown. In each case, specify the number of real zeros and the number of nonreal complex zeros. Indicate whether there are any real zeros with multiplicity other than 1.

48.

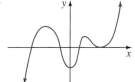

49.

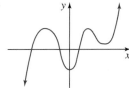

50.

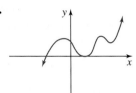

51.

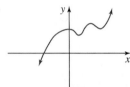

52. Show that $x^4 + 2x - 1 = 0$ has no rational roots. Use a graphing utility to graph $f(x) = x^4 + 2x - 1$. Use the

ZOOM and TRACE features to approximate the irrational roots of $x^4 + 2x - 1 = 0$ to two decimal places.

53. A one-inch thick slice is cut from the edge of a cube. The volume of the remaining solid is 448 cubic inches. Model these conditions with a polynomial equation. Solve the equation with a graphing utility and determine the original length of the cube.

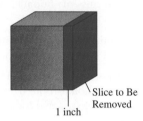

1 inch
Slice to Be Removed

In Problems 54–58, use the graph of $f(x) = \dfrac{1}{x}$ to graph each function. Graph f and g on the same axes. Identify the horizontal and vertical asymptotes of g.

54. $g(x) = \dfrac{1}{x - 4}$

55. $g(x) = \dfrac{1}{x} - 4$

56. $g(x) = \dfrac{1}{x + 3} - 1$

57. $g(x) = -\dfrac{1}{x + 3}$

58. $g(x) = -\dfrac{2}{x - 1} + 1$

In Problems 59–66, use the strategy for graphing rational functions outlined in the chapter summary (item 7) to graph each function. Do not use a graphing utility until your hand-drawn graph is complete. Then verify your graph with a graphing utility.

59. $f(x) = \dfrac{2x}{x^2 - 9}$

60. $g(x) = \dfrac{2x - 4}{x + 3}$

61. $h(x) = \dfrac{x^2 - 3x - 4}{x^2 - x - 6}$

62. $r(x) = \dfrac{x^2 + 4x + 3}{(x + 2)^2}$

63. $y = \dfrac{x^2}{x + 1}$

64. $y = \dfrac{x^2 + 2x - 3}{x - 3}$

65. $f(x) = \dfrac{-2x^3}{x^2 + 1}$

66. $g(x) = \dfrac{4x^2 - 16x + 16}{2x - 3}$

Problems 67–69 involve rational functions that model the given situations. In each case, find the horizontal asymptote as $x \to \infty$ and then describe what this means in practical terms.

67. $N(x) = \dfrac{60(x - 2)}{x}$; the number of words that a person can type per minute (N) after x weeks of practice.

68. $F(x) = \dfrac{30(4 + 5x)}{1 + 0.05x}$; the number of fish $(F,$ in thousands) after x weeks in a lake that was stocked with 120,000 fish.

69. $P(x) = \dfrac{72{,}900}{100x^2 + 729}$; the percentage rate (P) of U.S. unemployment for groups with x years of education.

70. A company that manufactures graphing calculators has fixed costs of \$80,000 and variable costs of \$20 per calculator. The company's total costs are modeled by $C(x)$

$= 20x + 80,000.$ The average cost to manufacture each calculator is given by

$$\overline{C}(x) = \frac{20x + 80,000}{x}.$$

a. Find and interpret $\overline{C}(50), \overline{C}(100), \overline{C}(1000)$, and $\overline{C}(100,000)$.

b. What is the horizontal asymptote? Describe what this means in practical terms.

c. Graph $\overline{C}$.

71. In Silicon Valley, California, a government agency ordered computer-related companies to contribute to a monetary pool to clean up underground water supplies. (The companies had stored toxic chemicals in leaking underground containers.) The rational function

$$C(x) = \frac{200x}{100 - x}$$

models the cost (C, in tens of thousands of dollars) for removing $x\%$ of the contaminants.

a. Find and interpret $C(90) - C(50)$.

b. Graph the function.

c. Describe the practical meaning of the observation that $x = 100$ is an asymptote.

72. In a get-tough drug policy, a politician promises to spend whatever it takes to seize all illegal drugs as they enter the country. If the cost of this venture is

$$C(p) = \frac{Ap}{100 - p}$$

where A is a positive constant, C is expressed in millions of dollars, and p is the percent of illegal drugs seized, use this model to critique the politician's promise.

73. The force of attraction between two bodies varies jointly as the product of their masses and inversely as the square of the distance between them. Two 1-kilogram masses separated by 1 meter exert a force of attraction of 6.67×10^{-11} newton. Earth has a mass of 5.98×10^{24} kilograms and the moon has a mass of 6.7×10^{22} kilograms. If the moon is 385,000 kilometers from Earth, what is the gravitational attraction between the two bodies?

74. Use the data shown below to write a rational function of the form $y = \dfrac{k}{x^n}$ that models intensity of light on a surface (y) and the distance from the light source to the surface (x).

x (in meters)	y (in lumens)
2	300
4	75
10	12
20	3

Describe what the model states in practical terms. Predict the light intensity at a distance of 30 meters. What is the effect on light intensity as the distance increases without limit?

75. A commuter drove to work a distance of 60 miles and then returned again on the same highway. The average rate on the return trip was 30 miles per hour faster than the average rate on the outgoing trip.

a. Find a rational function that models the total time (T) required to complete the round trip as a function of the rate (x) on the outgoing trip.

b. Use a graphing utility to graph the function for meaningful values of x.

c. If the commuter would like to complete the round trip in 3 hours, use the graph to determine what the average rate on the outgoing trip should be.

76. Suppose that x ounces of pure acid is added to a ounces of a $p\%$ acid solution. The concentration of the mixture is given by

$$C = \frac{100x + pa}{x + a}$$

where the mixture is $C\%$ acid.

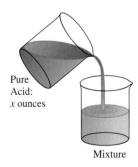

Pure
Acid:
x ounces

Mixture

a. Take the particular case in which x ounces of pure acid is added to 50 ounces of a 35% acid solution. Write the function C for this case. Then graph C using a graphing utility for meaningful values of x.

b. What do you observe about the graph of C in terms of its domain, intercepts, asymptotes, and general shape?

c. Use the graph to determine the amount of pure acid that must be added to produce a 75% acid mixture.

CHAPTER 3 TEST

In Problems 1–2, use the vertex and intercepts to graph each quadratic function. Give the equation of the axis of symmetry.

1. $f(x) = (x - 1)^2 - 4$

2. $f(x) = 2x^2 - 4x - 6$

3. The function $f(x) = -x^2 + 46x - 360$ models the daily profit ($f(x)$, in hundreds of dollars) for a company that manufactures x VCRs daily. How many VCRs should be manufactured each day to maximize profit? What is the maximum daily profit?

4. A rectangular plot of land along a river is to be fenced along 3 sides using 600 feet of fencing. No fencing is to be placed along the river's edge. Find the dimensions that maximize the enclosed area. What is the maximum area?

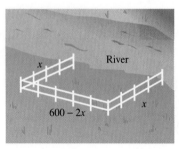

5. Consider the function $f(x) = x^3 - 5x^2 - 4x + 20$.

a. Use factoring to find all zeros of f.

b. Use the Leading Coefficient Test and the zeros of f to graph the function.

6. Use end behavior to explain why the graph at the top of the next column cannot be the graph of $f(x) = x^5 - x$. Then use intercepts to explain why the graph cannot represent $f(x) = x^5 - x$.

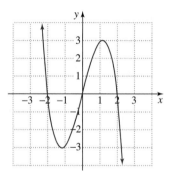

7. The graph of $f(x) = 6x^3 - 19x^2 + 16x - 4$ is shown in the figure.

a. Based on the graph of f, find the integral root of the equation $6x^3 - 19x^2 + 16x - 4 = 0$.

b. Use synthetic division to find the other two roots of $6x^3 - 19x^2 + 16x - 4 = 0$.

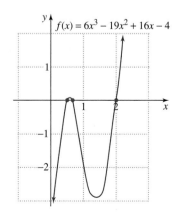

$f(x) = 6x^3 - 19x^2 + 16x - 4$

8. List all possible rational solutions to $2x^3 + 11x^2 - 7x - 6 = 0$.

9. Use Descartes' Rule of Signs to determine the possible number of positive and negative solutions to $3x^5 - 2x^4 - 2x^2 + x - 1 = 0$.

10. Solve: $x^3 + 6x^2 - x - 30 = 0$.

11. Consider the function whose equation is given by $f(x) = 2x^4 - x^3 - 13x^2 + 5x + 15$.

    a. List all possible rational zeros.

    b. Use the graph of f in the figure shown and synthetic division to find all zeros of the function.

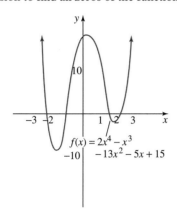

$$f(x) = 2x^4 - x^3 - 13x^2 - 5x + 15$$

12. Use the graph of $f(x) = 3x^4 + 4x^3 - 7x^2 - 2x - 3$ in the figure at the top of the next column to find the smallest positive integer that is an upper bound and the largest negative integer that is a lower bound for the real roots of $3x^4 + 4x^3 - 7x^2 - 2x - 3 = 0$. Then use synthetic division to show that all the real roots of the equation lie between these integers.

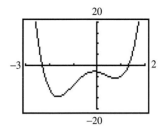

13. Solve $x^4 - 7x^3 + 18x^2 - 22x + 12 = 0$ given that $1 - i$ is a root.

14. Use the graph of $f(x) = x^3 + 3x^2 - 4$ in the figure shown to factor $x^3 + 3x^2 - 4$.

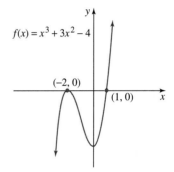

Graph each rational function in Problems 15–18.

15. $f(x) = \dfrac{x}{x^2 - 16}$

16. $f(x) = \dfrac{x^2 - 9}{x - 2}$

17. $f(x) = \dfrac{x + 1}{x^2 + 2x - 3}$

18. $f(x) = \dfrac{4x^2}{x^2 + 3}$

19. The average cost per manual for a company to produce x graphing utility manuals is given by

    $$\overline{C}(x) = \frac{6x + 3000}{x}.$$

    a. How many manuals must be produced so that the average cost per manual is $10?

    b. What is the horizontal asymptote for what this function? Describe what this means for the company.

20. The horsepower required to move a ship varies directly as the cube of its speed. The horsepower required for a speed of 15 knots is 10,125. Find the horsepower required for double this speed.

21. Use a graphing utility to obtain a relatively complete graph of $f(x) = 8x^4 - 8x^2 + 1$. Now use the graph to determine each of the following.

    a. Find all real zeros to two decimal places.

    b. Find the relative maximum and minimum values of f and the values of x at which they occur. Express all answers to two decimal places.

CUMULATIVE REVIEW PROBLEMS (CHAPTERS P–3)

In Problems 1–2, simplify.

1. $\dfrac{1}{2 - \sqrt{3}}$

2. $2\sqrt[3]{54} + 4\sqrt[3]{250}$

3. Factor: $x^3 + x^2y - xy^2 - y^3$.

Solve Problems 4–7.

4. $|x^2 - 1| = 3$

5. $(x^2 + 5x)^2 + 10(x^2 + 5x) + 24 = 0$

6. $3x^2 > 2x + 5$

7. $x^3 + 2x^2 - 5x - 6 = 0$

8. Solve for t: $V = C\left(1 - \dfrac{t}{15}\right)$.

9. If $f(x) = \sqrt{45 - 9x}$, find the domain of f.

10. The graph of g shown in the figure was obtained by a transformation of the graph of $f(x) = \sqrt{x}$. Write an equation for g.

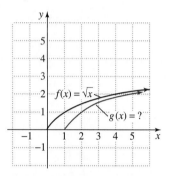

11. If $f(x) = x^3 + 4$, find $f^{-1}(x)$.

12. If $f(x) = x^2 - 5x + 6$ and $g(x) = 7x - 4$, find $f \circ g$ and $g \circ f$.

13. Sketch the graph of a function f having the following properties: f is increasing on $(-\infty, -2)$, decreasing on $(-2, \infty)$, has a y-intercept of -5, and a maximum y-value of 2. (More than one correct graph is possible.)

14. Use the vertex and intercepts to graph the quadratic function whose equation is $f(x) = 4x^2 - 16x + 7$.

15. Consider the function $f(x) = x^3 - 4x^2 - x + 4$.

    a. Use factoring to find all zeros of f.

    b. Use the Leading Coefficient Test and the zeros of f to graph the function.

16. Graph: $f(x) = (x - 1)(x - 2)^2(x - 3)^3$.

17. Graph: $f(x) = \dfrac{x - 1}{x - 2}$.

18. A brick path of uniform width is constructed around the outside of a 20-foot by 30-foot rectangular pool. Find how wide the path should be if there is enough brick to cover 336 square feet.

19. An orchard contains 30 apple trees, each of which yields approximately 400 apples over the growing season. For each new tree added to the orchard, crowding will result in a reduction of 10 apples per tree for all trees in the orchard. How many trees should be added to maximize the total yield of apples? What is the maximum yield?

20. The intensity of illumination on a surface varies inversely as the square of the distance of the light source from the surface. If the illumination from a source is 25 foot-candles when the distance is 4 feet, find the illumination when the distance is 6 feet.

Exponential and Logarithmic Functions

René Magritte "Golconde" 1953. Menil Collection, Houston, TX, USA / Giraudon / Art Resource, NY. © 1998 C. Herscovici, Brussels / Artists Rights Society (ARS), New York.

In a report entitled *Resources and Man*, the U.S. National Academy of Sciences concluded that a world population of 10 billion "is close to (if not above) the maximum that an intensely managed world might hope to support with some degree of comfort and individual choice." For this reason, 10 billion is called the carrying capacity of Earth. How long will it take for this carrying capacity to be reached? Exponential and logarithmic functions provide a frightening answer to this question (see Problem 39 in Problem Set 4.1), raising serious concerns about the kind of world we are leaving for our children. Modeling with these remarkable functions helps us to both predict the future and rediscover the past.

S E C T I O N   4 . 1

**Solutions
Manual**

Tutorial

**Video
8**

Exponential Functions

Objectives

1 Evaluate exponential functions.
2 Graph exponential functions.
3 Use compound interest formulas.
4 Solve applied problems using exponential functions.

Throughout this book, we have seen examples of polynomial functions such as $f(x) = x^2$ and $g(x) = x^3$. These functions involve a variable raised to a constant power. We will now interchange the variable and the constant, obtaining functions such as $F(x) = 2^x$ and $G(x) = 3^x$. These functions, with constant bases and variable exponents, are called *exponential functions*.

Exponential functions are used to model many situations. For example, we can use an exponential function to model the accumulated total number of AIDS cases in the United States. Table 4.1 shows the accumulated number of cases from 1983 through 1995.

TABLE 4.1	**Accumulated Number of AIDS Cases in the United States**												
Year	1983	1984	1985	1986	1987	1988	1989	1990	1991	1992	1993	1994	1995
Number of AIDS Cases	4589	10,750	22,399	41,256	69,592	104,644	146,574	193,878	251,638	326,648	399,613	457,280	496,896

Source: Center for Disease Control

This data can be modeled by the function

$$f(t) = 9420.9(1.492)^t$$

where $f(t)$ is the number of AIDS cases and t is the number of years after 1983. Notice that the variable, t, is in the exponent, making this an example of an exponential function with base 1.492. All exponential functions contain a constant base and a variable exponent.

Definition of the exponential function

The exponential function f with base b is defined by

$$f(x) = b^x \quad \text{or} \quad y = b^x$$

where b is a positive constant other than 1 $(b > 0$ and $b \neq 1)$ and x is any real number.

$\mathcal{S}$tudy tip

Notice the difference between polynomial and exponential functions.

Polynomial Functions	**Exponential Functions**
Constant exponent	Variable exponent
$y = x^2, y = x^{10}$	$y = 2^x, y = 10^x$
Variable base	Constant base

1 Evaluate exponential functions.

Pablo Picasso "Weeping Woman," 1937. Tate Gallery, London / Art Resource, NY. © 1998 Estate of Pablo Picasso / Artists Rights Society (ARS), New York.

EXAMPLE 1 **Evaluating an Exponential Model**

The exponential function $f(t) = 9420.9(1.492)^t$ describes the accumulated total number of AIDS cases in the United States t years after 1983.

a. According to the model, how many cases were there in 1983? How does this compare with the actual number?

b. How well does the function describe the actual number of AIDS cases for the years 1985, 1989, and 1992?

Solution

a. Since 1983 is zero years after 1983, we substitute 0 for t.

$$f(t) = 9420.9(1.492)^t \quad \text{This is the given function.}$$

$$f(0) = 9420.9(1.492)^0 \quad \text{Substitute 0 for } t.$$

$$= 9420.9(1) \quad b^0 = 1. \text{ A calculator is not necessary.}$$

$$= 9420.9$$

According to the model, there were approximately 9421 cases, which over-estimates the actual number of cases (4589) given in Table 4.1.

b. We'll begin with the year 1985. Since 1985 is 2 years after 1983, we substitute 2 for t.

$$f(t) = 9420.9(1.492)^t \quad \text{This is the given function.}$$

$$f(2) = 9420.9(1.492)^2 \quad \text{Substitute 2 for } t.$$

$$\approx 20{,}972 \quad \text{Use a calculator: } 9420.9 \boxed{\times} 1.492 \boxed{\wedge} 2 \boxed{\text{ENTER}}$$

The model predicts 20,972 cases, and the actual number from Table 4.1 is 22,399. The function slightly underestimates the actual number.

We now follow the same procedure for the years 1989 and 1992.

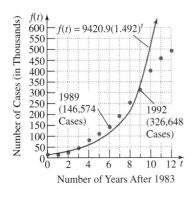

Figure 4.1

Comparing a model for the number of AIDS cases in the United States with the actual numbers

| **1989:** | **1992:** |
| **6 years after 1983** | **9 years after 1983** |

$$f(6) = 9420.9(1.492)^6 \qquad f(9) = 9420.9(1.492)^9$$
$$\approx 103{,}921 \qquad\qquad \approx 345{,}153$$

Actual number = 146,574 Actual number = 326,648
The model underestimates The model overestimates
the actual number. the actual number.

The graph of the exponential function is superimposed over the actual data points in Figure 4.1. Notice how the curve lies below the data point for 1989, underestimating the actual number, and above the data point for 1992, overestimating the actual number. The function does not appear to fit the data too well beyond 1992. ■

2 Graph exponential functions.

Graphing Exponential Functions

We are familiar with expressions involving b^x where x is a rational number. For example:

$$b^{1.7} = b^{17/10} = \sqrt[10]{b^{17}} \quad \text{and} \quad b^{1.73} = b^{173/100} = \sqrt[100]{b^{173}}.$$

Using technology

Use a calculator to find $2^{1.7}$, $2^{1.73}$, $2^{1.732}$, $2^{1.73205}$, and $2^{1.7320508}$. Now find $2^{\sqrt{3}}$. What do you observe?

However, because the definition $f(x) = b^x$ includes all real numbers for x, you may wonder what b^x means when x is an irrational number, such as $b^{\sqrt{3}}$ or b^{π}. A precise definition for $b^{\sqrt{3}}$ is really impossible without a background in calculus. However, using the nonrepeating and nonterminating approximation 1.73205 for $\sqrt{3}$ we can think of $b^{\sqrt{3}}$ as the value that has the successively closer approximations

$$b^{1.7}, b^{1.73}, b^{1.732}, b^{1.73205}, \dots .$$

In this way, we will be able to graph the exponential function with no holes or points of discontinuity at the irrational domain values. Let's consider the graphs of some exponential functions.

EXAMPLE 2 **Graphing an Exponential Function**

Graph: $f(x) = 2^x$

Solution

We begin by setting up a table of coordinates.

x	$f(x) = 2^x$
-3	$f(-3) = 2^{-3} = \frac{1}{8}$
-2	$f(-2) = 2^{-2} = \frac{1}{4}$
-1	$f(-1) = 2^{-1} = \frac{1}{2}$
0	$f(0) = 2^0 = 1$
1	$f(1) = 2^1 = 2$
2	$f(2) = 2^2 = 4$
3	$f(3) = 2^3 = 8$

Study tip

Since $2^x > 0$, the graph of $f(x) = 2^x$ comes close to the x-axis but does not touch or cross it. The x-axis is a horizontal asymptote.

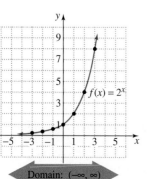

Figure 4.2

The graph of $f(x) = 2^x$

We plot these points, connecting them with a continuous curve. Figure 4.2 shows the graph of $f(x) = 2^x$. Observe that the graph approaches but never touches the negative portion of the x-axis, indicating a range of all positive real numbers. There are no values of x that will result in 0 or negative numbers in the range of $f(x) = 2^x$. Furthermore, we chose integers for x in our table of coordinates. However, you can use a calculator to find additional coordinates. For example, $f(0.3) = 2^{0.3} \approx 1.231$, $f(0.95) = 2^{0.95} \approx 1.932$, and the points $(0.3, 1.231)$ and $(0.95, 1.932)$ really do fit the graph. ∎

EXAMPLE 3 **Graphing an Exponential Function**

Graph: $f(x) = \left(\dfrac{1}{3}\right)^x$

Solution

We again begin with a table of coordinates.

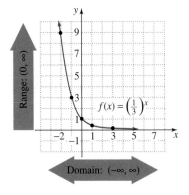

Figure 4.3

The graph of $f(x) = \left(\frac{1}{3}\right)^x$

x	$f(x) = \left(\dfrac{1}{3}\right)^x$
-2	$f(-2) = \left(\dfrac{1}{3}\right)^{-2} = \dfrac{1}{\left(\dfrac{1}{3}\right)^2} = \dfrac{1}{\dfrac{1}{9}} = 9$
-1	$f(-1) = \left(\dfrac{1}{3}\right)^{-1} = \dfrac{1}{\left(\dfrac{1}{3}\right)} = 3$
0	$f(0) = \left(\dfrac{1}{3}\right)^0 = 1$
1	$f(1) = \left(\dfrac{1}{3}\right)^1 = \dfrac{1}{3}$
1.2	$f(1.2) = \left(\dfrac{1}{3}\right)^{1.2} \approx 0.268$ ⟵ Using a calculator: $\boxed{(}\,1\,\boxed{\div}\,3\,\boxed{)}\,\boxed{\wedge}\,1.2\,\boxed{\text{ENTER}}$
2	$f(2) = \left(\dfrac{1}{3}\right)^2 = \dfrac{1}{9}$

We plot these points and connect them with a smooth continuous curve. Figure 4.3 shows the graph. ∎

Four exponential functions have been graphed in Figure 4.4. Compare the graphs of functions where $b > 1$ to those where $b < 1$. When $b > 1$, the value of y increases as the value of x increases. When $b < 1$, the value of y decreases as the value of x-increases.

These graphs illustrate the following general characteristics of exponential functions.

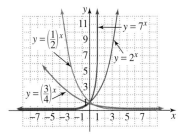

Figure 4.4

Graphs of four exponential functions

Study tip

Since $\left(\frac{1}{3}\right)^x = (3^{-1})^x = 3^{-x}$, the graphs of $f(x) = \left(\frac{1}{3}\right)^x$ and $g(x) = 3^{-x}$ are identical.

Characteristics of exponential functions

1. The domain of $f(x) = b^x$ consists of all real numbers. The range of $f(x) = b^x$ consists of all positive real numbers.
2. The graphs of all exponential functions pass through the point $(0, 1)$ because $f(0) = b^0 = 1\ (b \neq 0)$.
3. If $b > 1$, $f(x) = b^x$ has a graph that goes up to the right and is an increasing function.
4. If $0 < b < 1$, $f(x) = b^x$ has a graph that goes down to the right and is a decreasing function.
5. $f(x) = b^x$ is one-to-one and has an inverse that is a function.
6. The graph of $f(x) = b^x$ approaches but does not cross the x-axis. The x-axis is a horizontal asymptote.

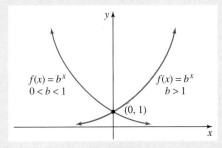

The graphs of exponential functions can be translated vertically or horizontally, reflected, stretched, or shrunk using the ideas of Section 2.5, as summarized in Table 4.2.

TABLE 4.2 Transformations Involving Exponential Functions

Transformation	Equation	Description
Horizontal translation	$y = b^{x+c}$	Shifts the graph of $y = b^x$ to the left c units if $c > 0$ and to the right c units if $c < 0$.
Stretching or shrinking	$y = cb^x$	Stretches the graph of $y = b^x$ for $c > 1$ and shrinks the graph of $y = b^x$ for $0 < c < 1$, multiplying y-coordinates of $y = b^x$ by c.
Reflecting	$y = -b^x$	Reflects the graph of $y = b^x$ in the x-axis.
	$y = b^{-x}$	Reflects the graph of $y = b^x$ in the y-axis.
Vertical translation	$y = b^x + c$	Shifts the graph of $y = b^x$ upward c units if $c > 0$ and downward c units if $c < 0$.

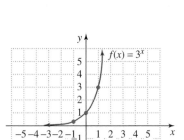

Red Grooms "Subway" (detail) from Ruckus Manhattan, 1976. Mixed media, 9 ft. × 18 ft. 7 in. × 37 ft. 2 in. Photo courtesy Marlborough Gallery, New York. © 1998 Red Grooms / Artists Rights Society (ARS), New York.

By realizing that a function is a transformation of the graph of $f(x) = b^x$, you should have a good idea of what its graph looks like. Using this information and a table of coordinates will result in relatively accurate graphs that can be verified using a graphing utility. This idea is illustrated in the next example.

EXAMPLE 4 **Transformations Involving Exponential Functions**

Sketch the graph of each function by using transformations of the graph of $f(x) = 3^x$. State each function's domain and range.

a. $g(x) = 3^{x+1}$ **b.** $h(x) = 3^{x+2} - 4$
c. $k(x) = 2 \cdot 3^{x+1} - 5$ **d.** $r(x) = -3^{-x} + 1$

Solution

We begin with the graph of $f(x) = 3^x$. Using a table of coordinates, we obtain the graph shown in Figure 4.5.

Figure 4.5

The graph of $f(x) = 3^x$

x	$f(x) = 3^x$
-2	$f(-2) = 3^{-2} = \frac{1}{9}$
-1	$f(-1) = 3^{-1} = \frac{1}{3}$
0	$f(0) = 3^0 \quad = 1$
1	$f(1) = 3^1 \quad = 3$
2	$f(2) = 3^2 \quad = 9$

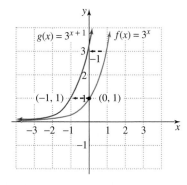

Figure 4.6

The graph of $g(x) = 3^{x+1}$ shifts the graph of $f(x) = 3^x$ to the left one unit.

a. To graph $g(x) = 3^{x+1}$, we shift the graph of $f(x) = 3^x$ one unit to the left, as shown in Figure 4.6. This is the case because $g(x) = 3^{x+1} = f(x + 1)$. As a check, we can find a few ordered pairs satisfying $g(x) = 3^{x+1}$.

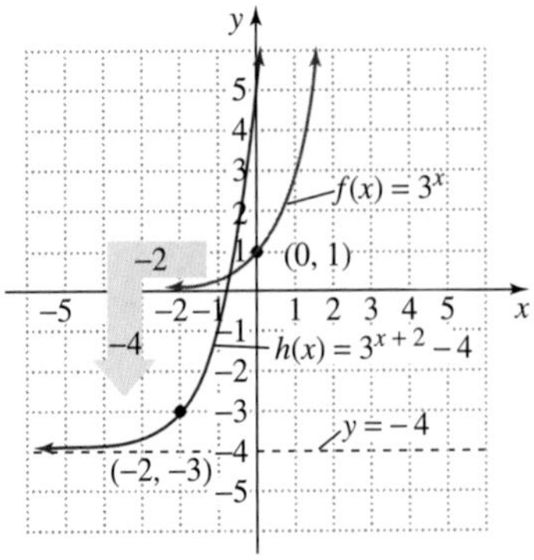

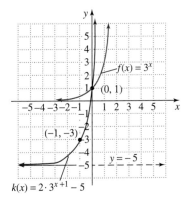

Figure 4.7

The graph of $h(x) = 3^{x+2} - 4$ shifts the graph of $f(x) = 3^x$ to the left two units and downward four units.

x	-3	-2	-1	0
$g(x) = 3^{x+1}$	$3^{-2} = \frac{1}{9}$	$3^{-1} = \frac{1}{3}$	$3^0 = 1$	$3^1 = 3$

The domain of $g(x) = 3^{x+1}$ is $(-\infty, \infty)$, and the range is $(0, \infty)$.

b. To graph $h(x) = 3^{x+2} - 4$, we first shift the graph of $f(x) = 3^x$ to the left two units ($y = 3^{x+2}$) and then shift the graph of $y = 3^{x+2}$ downward four units, as shown in Figure 4.7. This graph can be verified with a graphing utility or with a few ordered pairs.

x	-2	-1	0
$h(x) = 3^{x+2} - 4$	$3^0 - 4 = -3$	$3^1 - 4 = -1$	$3^2 - 4 = 5$

The domain of $h(x) = 3^{x+2} - 4$ is $(-\infty, \infty)$, and the range is $(-4, \infty)$.

> **Study tip**
>
> If an exponential function is translated upward or downward, the horizontal asymptote is shifted by the amount of the vertical shift. This is illustrated in Figure 4.7.

c. We obtain the graph of $k(x) = 2 \cdot 3^{x+1} - 5$ from the graph of $f(x) = 3^x$ using the following transformations.
  1. Translate to the left one unit, obtaining $y = 3^{x+1}$.
  2. Stretch by a factor of 2, obtaining $y = 2 \cdot 3^{x+1}$.
  3. Translate downward five units, obtaining $y = 2 \cdot 3^{x+1} - 5$.
The graph of $k(x) = 2 \cdot 3^{x+1} - 5$ is shown in Figure 4.8. Once again, we can verify the graph with a graphing utility or with a few ordered pairs.

Figure 4.8

The graph of $k(x) = 2 \cdot 3^{x+1} - 5$ translates and stretches the graph of $f(x) = 3^x$.

x	-2	-1	0
$k(x) = 2 \cdot 3^{x+1} - 5$	$2 \cdot 3^{-1} - 5 = -4\frac{1}{3}$	$2 \cdot 3^0 - 5 = -3$	$2 \cdot 3 - 5 = 1$

The domain of $k(x) = 2 \cdot 3^{x+1} - 5$ is $(-\infty, \infty)$, and the range is $(-5, \infty)$.

d. We obtain the graph of $r(x) = -3^{-x} + 1$ from the graph of $f(x) = 3^x$ using three transformations at the top of the next page (see Figure 4.9).

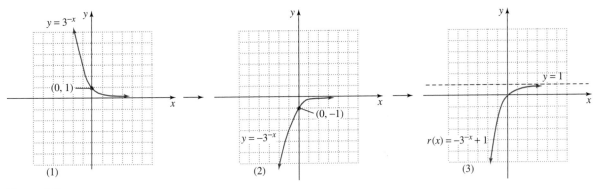

Figure 4.9

Using transformations to graph $r(x) = -3^{-x} + 1$

1. Reflect f in the y-axis, obtaining $y = 3^{-x}$.
2. Reflect $y = 3^{-x}$ in the x-axis, obtaining $y = -3^{-x}$.
3. Translate $y = -3^{-x}$ upward one unit, obtaining $r(x) = -3^{-x} + 1$. The domain of $r(x) = -3^{-x} + 1$ is $(-\infty, \infty)$, and the range is $(-\infty, 1)$. ∎

3 Use compound interest formulas.

Compound Interest and the Irrational Number e

Compound interest is another situation that can be modeled by exponential functions. Suppose a sum of money P (called the principal) is invested at an annual percentage rate r (in decimal form), compounded once a year. Because the interest is added to the principal at year's end, the accumulated value (A) is

$$A = P + Pr = P(1 + r).$$

The accumulated amount of money follows this pattern of multiplying the previous principal by $(1 + r)$ for each successive year, as indicated in Table 4.3.

TABLE 4.3

Time in Years	Accumulated Value after Each Compounding
0	$A = P$
1	$A = P(1 + r)$
2	$A = P(1 + r)(1 + r) = P(1 + r)^2$
3	$A = P(1 + r)^2(1 + r) = P(1 + r)^3$
4	$A = P(1 + r)^3(1 + r) = P(1 + r)^4$
⋮	⋮
t	$A = P(1 + r)^t$

If money invested at a specified rate of interest is compounded more than once a year, then the model $A = P(1 + r)^t$ can be adjusted to take into account the number of compounding periods in a year. If n represents the number of compounding periods in a year, the mathematical model becomes

$$A = P\left(1 + \frac{r}{n}\right)^{nt}.$$

Compounded Annually

Compounded Quarterly

Compounded Monthly

Compounded Daily

$0 $500 $1000 $1500 $2000 $2500 $3000

Future value of $1000 invested at 10% interest in 10 years; the increasing bar lengths show greater accumulated values as the number of compounding periods in a year increases.

Using technology

Here's the graphing calculator keystroke sequence for the expression $10,000\left(1 + \dfrac{0.08}{4}\right)^{4 \cdot 5}$:

10,000 ⎡(1 + .08 ÷ 4)⎤ ⎡^⎤
⎡(4 × 5)⎤ ⎡ENTER⎤

EXAMPLE 5   Using the Compound Interest Model

A sum of $10,000 is invested at an annual interest rate of 8%, compounded quarterly. Find the balance in the account after 5 years.

Solution

$$A = P\left(1 + \frac{r}{n}\right)^{nt} \qquad \text{Use the compound interest model.}$$

We are given that P (the principal) = $10,000, r (the interest rate) = 8% = 0.08, n (the number of compounding periods with quarterly compounding) = 4, and t (the time, in years) = 5. We substitute these values into the model.

$$A = 10,000\left(1 + \frac{0.08}{4}\right)^{4 \cdot 5} \approx 14,859.47$$

After 5 years, the balance in the account is $14,859.47. ∎

Barton Lidice Benes "Dustpan" 1990,
mixed media with currency,
12 × 14 1/2 × 2 in.

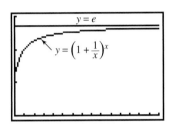

Figure 4.10

As *x* increases to the right,
$(1 + \frac{1}{x})^x$ gets closer to e.

The Irrational Number e

Suppose that \$1 is invested at an interest rate of 100% for 1 year, and the compounding periods increase infinitely (compounding interest every trillionth of a second, every quadrillionth of a second, etc.). As the compounding periods increase, the amount you would receive at the end of the year gets closer and closer to \$2.718281828459045.... Mathematicians represent this irrational number by the symbol *e*, where *e* is approximately equal to 2.72 ($e \approx 2.72$).

Let's rewrite our compound interest model so that we can visualize what all this means.

$$A = P\left(1 + \frac{r}{n}\right)^{nt} \qquad \text{This is the compound interest model.}$$

$$= 1\left(1 + \frac{1}{n}\right)^{n \cdot 1} \qquad \begin{array}{l} P \text{ (the principal)} = \$1, r \text{ (the interest rate)} = 100\% = 1, \\ \text{and } t \text{ (time)} = 1 \text{ year.} \end{array}$$

$$= \left(1 + \frac{1}{n}\right)^{n} \qquad \text{Simplify.}$$

Remembering that *A* represents the balance in the account, we've used a graphing utility to graph $A = (1 + \frac{1}{n})^n$ and $A = e$ (approximately 2.72) in the same viewing rectangle. Figure 4.10 illustrates that as *n* (the number of compounding periods) increases, the balance in the account gets closer and closer to *e*, or approximately \$2.72.

The irrational number *e* is the base for the *natural exponential function*.

The natural exponential function

The function

$$f(x) = e^x$$

is called the *natural exponential function*. The irrational number *e* is called the *natural base*, where $e \approx 2.7183$.

Study tip

Remember that e is the constant number 2.71828... and x is a variable.

You will need to use a calculator to evaluate the natural exponential function. Here are some examples obtained using the natural exponential key $\boxed{e^x}$.

$f(x) = e^x$		Approximate
Function Value	**Keystrokes**	**Display**
$f(2) = e^2$	$\boxed{e^x}$ 2 $\boxed{\text{ENTER}}$	7.389056
$f(-3) = e^{-3}$	$\boxed{e^x}$ $\boxed{(-)}$ 3 $\boxed{\text{ENTER}}$	0.049787
$f(4.6) = e^{4.6}$	$\boxed{e^x}$ 4.6 $\boxed{\text{ENTER}}$	99.4843156

Because $2 < e < 3$, the graph of $y = e^x$ is between the graphs of $y = 2^x$ and $y = 3^x$ as shown in Figure 4.11 at the top of the next page.

Discover for yourself

Use a graphing utility to reproduce the graphs shown in Figure 4.11.

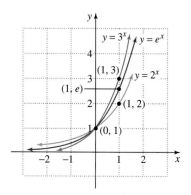

Figure 4.11

Graphs of three exponential functions

Since the number e is related to compounding periods increasing infinitely, or *continuous compounding*, the number appears in the formula for the balance in an account subject to this type of compounding. The formulas for both types of compounding are given as follows.

Formulas for compound interest

After t years, the balance A in an account with principal P and annual interest rate r (in decimal form) is given by the following mathematical models.

1. For n compoundings per year: $A = P\left(1 + \dfrac{r}{n}\right)^{nt}$

2. For continuous compounding: $A = Pe^{rt}$

EXAMPLE 6 **Choosing Between Investments**

You want to invest $8000 for 6 years, and you have a choice between two accounts. The first pays 7% per year, compounded monthly. The second pays 6.85% per year, compounded continuously. Which is the better investment?

Solution

The better investment is the one with the greater balance in the account after 6 years. Let's begin with the account with monthly compounding. We use the compound interest model with $P = 8000$, $r = 0.07$, $n = 12$ (monthly compounding, means 12 compoundings per year), and $t = 6$.

$$A = P\left(1 + \frac{r}{n}\right)^{nt} = 8000\left(1 + \frac{0.07}{12}\right)^{12 \cdot 6} \approx 12{,}160.84$$

The balance in this account after 6 years is $12,160.84. For the second investment option, we use the model for continuous compounding with $P = 8000$, $r = 6.85\% = 0.0685$, and $t = 6$.

$$A = Pe^{rt} = 8000\, e^{0.0685(6)} \approx 12{,}066.60$$

The balance in this account after 6 years is $12,066.60, slightly less than the previous amount. Thus, the better investment is the 7% monthly compounding option. ∎

sing technology

Here's the graphing calculator keystroke sequence for the expression $8000e^{0.0685(6)}$:

8000 [e^] [(] .0685 [×] 6 [)]

[ENTER]

Discover for yourself

Write an example similar to Example 6 in which continuous compounding at a slightly lower yearly interest rate is a better investment than compounding n times per year.

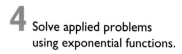

4 Solve applied problems using exponential functions.

Applications of Exponential Functions

Exponential functions involving base e describe a wide variety of situations in addition to continuous compounding. Frequently encountered are *growth* and *decay models*.

Exponential growth and decay models

The mathematical model for exponential growth or decay is given by

$$f(t) = A_0 e^{kt} \quad \text{or} \quad A = A_0 e^{kt}.$$

Within this model, t represents time, A_0 is the original amount (or size) of the growing (or decaying) quantity, and A or $f(t)$ represents the amount (or size) of the quantity at time t. The number k represents a constant that depends on the rate of growth or decay. If $k > 0$, the formula represents *exponential growth*; if $k < 0$, the formula represents *exponential decay*.

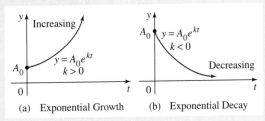

(a) Exponential Growth   (b) Exponential Decay

Study tip

With exponential growth and decay, quantities grow or decay at a rate directly proportional to their size. Populations that are growing exponentially grow extremely rapidly as they get larger because there are more adults to have offspring.

Study tip

Living beings continually receive carbon-14 that balances the carbon-14 that decays. Thus, the level of carbon-14 is constant in living organisms.

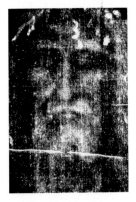

After generations of controversy, the shroud of Turin, long believed to be the burial cloth of Jesus Christ, was dated by carbon-14 techniques to centuries after the death of Christ.

Gianni Tortoli / Science Source / Photo Researchers, Inc.

EXAMPLE 7 **Carbon-14 Dating**

Archaeologists, anthropologists, and geologists use a technique called *radiocarbon dating* to estimate when a particular organism died. Carbon-14 is a radioactive substance that is absorbed by an organism while it is alive. Once an organism dies, carbon-14 is no longer replaced in its tissues, but continues to decay. The mathematical model

$$A = A_0 e^{-0.000121t}$$

describes the amount A of carbon-14 present after t years, where A_0 is the amount present at time $t = 0$. If an artifact originally had 16 grams of carbon-14 present, how many grams will be present in:

a. 5715 years?   **b.** 11,430 years?   **c.** 17,145 years?

Solution

a. $A = A_0 e^{-0.000121t}$ This is the given model.

$\quad = 16 e^{-0.000121(5715)}$ A_0 (the initial amount) = 16 grams and $t = 5715$ years.

$\quad \approx 8$ Use a calculator.

After 5715 years, approximately 8 grams of carbon-14 will be present.

b. $A = A_0 e^{-0.000121t}$ This is the given model.

$\quad = 16 e^{-0.000121(11,430)}$ $A_0 = 16$ grams and $t = 11,430$ years.

$\quad \approx 4$ Use a calculator.

After 11,430 years, approximately 4 grams of carbon-14 will be present.

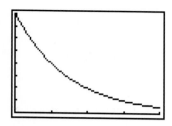

Figure 4.12

The graph of the exponential decay model for carbon-14

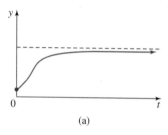

(a)

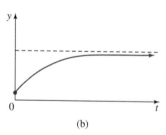

(b)

Figure 4.13

Limits to growth: The learning curve (b) illustrates that when we learn new tasks, learning is initially quite rapid but tends to taper off as time increases.

c. $A = A_0 e^{-0.000121t}$ We'll repeat the same procedure.

$\quad = 16e^{-0.000121(17,145)}$ $A_0 = 16$ grams and $t = 17{,}145$ years.

$\quad \approx 2$ Use a calculator.

After 17,145 years, approximately 2 grams of carbon-14 will be present.   ■

The *half-life* of a substance that is decaying exponentially is the time required for half of the atoms in the substance to disintegrate. As we saw in Example 7, every 5715 years half of the carbon-14 decays. Thus, carbon-14 decays exponentially with a half-life of approximately 5715 years.

As always, an understanding of a function's behavior enables us to choose realistic range settings on a graphing utility. The graph of $A = 16e^{-0.000121t}$ or, equivalently, $y = 16e^{-0.000121x}$ is shown in Figure 4.12. We've selected Xmin = 0, Xmax = 22,860 (four multiples of 5715, the half-life), and Xscl = 5715. We chose Ymin = 0, Ymax = 16 (the original amount of carbon-14 present), and Yscl = 2 as range settings for y. If you know the amount of carbon-14 present, how might you use the graph to determine an artifact's age? We'll discuss this in more detail in Section 4.5.

Exponential functions can be used to model long-term population growth, the spread of an epidemic, the sales of new products, or learning. However, nothing on Earth (population growth or the spread of an epidemic) can grow exponentially indefinitely. There are always limits to growth, shown in Figure 4.13 by the horizontal asymptote.

EXAMPLE 8   **Spread of an Epidemic**

The function

$$N(t) = \frac{30{,}000}{1 + 20e^{-1.5t}}$$

describes the number of people $N(t)$ who have become ill with influenza t weeks after its initial outbreak in a town with 30,000 inhabitants.

a. How many people became ill with the flu when the epidemic began?
b. How many people were ill by the end of the fourth week?
c. What is the horizontal asymptote for this function? Describe what this means in practical terms.
d. Use a graphing utility to graph the function.

Solution

a. $N(t) = \dfrac{30{,}000}{1 + 20e^{-1.5t}}$ This is the given function.

$\quad N(0) = \dfrac{30{,}000}{1 + 20e^{-1.5(0)}}$ When the epidemic began, $t = 0$.

$\quad\quad\quad = \dfrac{30{,}000}{1 + 20}$ $e^{-1.5(0)} = e^0 = 1$

$\quad\quad\quad \approx 1429$

Approximately 1429 people were ill when the epidemic began.

b. $N(t) = \dfrac{30{,}000}{1 + 20e^{-1.5t}}$ Use the given function.

ENRICHMENT ESSAY

Exponential Growth of Carbon Dioxide; The Nuclear Option

The amount of carbon dioxide (CO_2) in the atmosphere is an environmental issue that we have looked at throughout this book. The present concentration is about 352 parts per million and continues to increase exponentially at 0.4% per year. Using the exponential growth model $A = 352e^{0.004t}$ that describes the concentration of CO_2 (in parts per million) t years after 1990, climatologists forecast a warming trend of about 1°F per decade.

The major sources of carbon dioxide are energy production and deforestation. Today's American household uses an average of 3200 kilowatts-hours of electricity a year, accounting for emissions of nearly three tons of CO_2, the leading greenhouse gas, and 60 pounds of sulfur dioxide, a major cause of acid rain.

One solution might be nuclear power plants, which do not generate greenhouse gases during operation. However, numerous problems stand in the way of nuclear power expansion, not the least of which is what to do with nuclear waste. By mid-1990 more than 23,000 tons of intensely radioactive nuclear waste had accumulated around the 111 operating nuclear power plants in the United States.

Bill Woodrow "Self Portrait in the Nuclear Age" 1986, shelving unit, wall map, coat, globe, acrylic paint, 205 × 145 × 265 cm. Saatchi Collection, London. Photo credit: Bill Woodrow.

$$N(4) = \frac{30,000}{1 + 20e^{-1.5(4)}}$$ To find the number of people ill by the end of week four, let $t = 4$.

$$\approx 28,583$$ Use a calculator.

Approximately 28,583 people were ill by the end of the fourth week. Compared with the number of people who were ill initially, this illustrates the virulence of the epidemic.

c. The horizontal asymptote is determined by finding out what happens to $N(t)$ as t increases without bound (as $t \to \infty$).

$$N(t) = \frac{30,000}{1 + 20e^{-1.5t}}$$ Again, use the given function.

$$= \frac{30,000}{1 + \dfrac{20}{e^{1.5t}}}$$ $e^{-1.5t} = \dfrac{1}{e^{1.5t}}$, by the definition of a negative exponent.

As $t \to \infty$ (time increases), $\dfrac{20}{e^{1.5t}}$ gets smaller and smaller $\left(\dfrac{20}{e^{1.5t}} \to 0\right)$ because the numerator (20) is staying the same size and the denominator ($e^{1.5t}$) is getting larger without bound. Consequently, we see that $\dfrac{20}{e^{1.5t}}$ is approximately equal to 0 as t grows increasingly larger. Thus, the function gets

Discover for yourself

Show that

$$\frac{20}{e^{1.5t}}$$

approaches 0 as t gets larger by using your calculator to evaluate the expression for $t = 10, 50, 100,$ and $1000.$

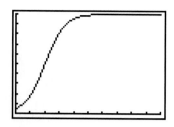

Figure 4.14

The graph of a function modeling the spread of an epidemic

closer and closer to $\dfrac{30,000}{1+0}$, indicating that $y = 30,000$ is the function's horizontal asymptote. In practical terms, this indicates that 30,000 is the limiting size of the population that becomes ill.

d. The graph of $Y = 30,000/(1 + 20 \boxed{e\wedge}(-1.5X))$, obtained with a graphing utility, is shown in Figure 4.14. Based on our previous results, we used the following range settings: Xmin = 0, Xmax = 10 (This takes us to week 10, and by week 4 approximately 28,583 people were ill), Xscl = 1. We also know that 30,000 is the limiting size, so we select Ymin = 0, Ymax = 30,000 (the limiting size), and Yscl = 3000. As always, other range settings are possible.   ■

PROBLEM SET 4.1

Practice Problems

In Problems 1–8, sketch by hand the graph of each function by making a table of coordinates. Then use a graphing utility to confirm your results.

1. $f(x) = 4^x$ **2.** $g(x) = 5^x$ **3.** $h(x) = \left(\frac{1}{3}\right)^x$ **4.** $f(x) = \left(\frac{1}{4}\right)^x$
5. $y = \left(\frac{3}{2}\right)^x$ **6.** $y = \left(\frac{4}{3}\right)^x$ **7.** $g(x) = (1.2)^x$ **8.** $h(x) = (0.6)^x$

In Problems 9–20, sketch by hand the graph of each function by using transformations of the graph of $f(x) = 2^x$ and a table of coordinates. State each function's domain and range. Then use a graphing utility to confirm your results.

9. $g(x) = 2^{x-1}$ **10.** $h(x) = 2^{x-2}$ **11.** $y = 2^x + 3$ **12.** $y = 2^x - 1$
13. $j(x) = 2^{x+1} - 1$ **14.** $k(x) = 2^{x+2} - 1$ **15.** $r(x) = 3 \cdot 2^{x+1} - 2$ **16.** $v(x) = 3 \cdot 2^{x+2} - 2$
17. $y = -2^x$ **18.** $y = -2^{x-2}$ **19.** $y = 2^{-x} + 3$ **20.** $y = 2^{-x} + 4$

In Problems 21–30, use transformations to graph each function. Then use a graphing utility to confirm your graph.

21. $f(x) = 3^{x-1}$ **22.** $g(x) = \frac{1}{2} \cdot 3^{x+1}$ **23.** $y = 3^{x+2} - 1$ **24.** $y = 4^{x+1} - 3$
25. $h(x) = \left(\frac{1}{2}\right)^{x+1}$ **26.** $j(x) = \left(\frac{1}{2}\right)^{x+2}$ **27.** $k(x) = \left(\frac{3}{2}\right)^{x-1}$ **28.** $r(x) = \left(\frac{3}{2}\right)^{x-2}$
29. $y = 1 - 5^x$ **30.** $y = -1 - 5^{-x}$

Application Problems

31. The exponential function $f(t) = 67.38(1.026)^t$ describes the population (in millions) of Mexico t years after 1980.
   a. What was the 1980 population?
   b. Predict Mexico's population in 2007, 2034, and 2061.
   c. How many years does it take for Mexico's population to double?

32. The model $N(t) = 2000(2)^{\frac{t}{20}}$ describes the growth of the bacteria Escherichi coli often found in the human bladder. The formula indicates that 2000 of the bacteria are in the bladder at time $t = 0$. If t is measured in minutes, sketch by hand the function for $0 \leqslant t \leqslant 80$.

Use the compound interest models

$$A = P\left(1 + \frac{r}{n}\right)^{nt} \quad and \quad A = Pe^{rt}$$

to answer Problems 33–37.

33. In 1626 Peter Minuit convinced the Wappinger Indians to sell him Manhattan Island for $24. If the Native Americans had put the $24 into a bank account paying 5% interest, how much would the investment be worth in the year 2000 if interest were compounded
   a. Monthly?   **b.** Daily?   **c.** Continuously?

Dan V. Lomahaftewa "Spring Arrival" Collagraph, 42 × 30.

34. Suppose that you have $8000 to invest. What investment yields the greater return over 3 years: 7% compounded monthly or 6.85% compounded continuously?

35. Suppose that you have $7000 to invest. What investment yields the greater return over 5 years: 8.25% compounded quarterly or 8.3% compounded semiannually?

36. Complete the following table, which indicates what happens to $5000 invested at 5.5% for different time periods and different compounding periods. (Round each entry to the nearest dollar.)

	1 year	5 years	10 years	20 years
Compounded Semiannually				
Compounded Monthly				
Compounded 365 Times per Year				
Compounded Continuously				

37. Repeat Problem 36 for 30,000 invested at 6.5%.

38. The exponential growth model $A = A_0 e^{kt}$ $(k > 0)$ describes the number of AIDS cases in the United States among intravenous drug users. In 1989, there were 24,000 cases of AIDS among intravenous drug users $(A_0 = 24,000)$ with a growth rate of 21% $(k = 0.21)$. If this growth rate doesn't change, how many cases of AIDS will be expected by the end of

a. 2000? **b.** 2020?

39. Take a moment to reread the chapter introduction. At the time *Resources and Man* was issued in 1969, the world population was approximately 3.6 billion, with a relative growth rate of 2% per year. If exponential growth continues at this rate, use the model $A = 3.6e^{0.02t}$, where A is the world population in billions t years after 1969, to show that Earth's carrying capacity will be reached in the year 2020.

40. Polonium-210 has a half-life of 140 days. This means that the rate of decay is proportional to the amount present, and half of any given quantity will disintegrate in 140 days. The mathematical model $A = A_0 e^{-0.004951t}$ describes the amount A of polonium-210 present after t days, where A_0 is the amount present at time $t = 0$.

a. A sample has a mass of 100 milligrams. How many milligrams of polonium-210 will be present in 140 days? 210 days? 280 days? 420 days? 630 days?

b. Display your computations in part (a) graphically, with time (in days) along the horizontal axis, and the amount of polonium-210 present (in milligrams) along the vertical axis. Connect the data points with a smooth curve. Write the equation of the function that you have graphed.

The model $A = A_0 \left(\frac{1}{2}\right)^{\frac{t}{h}}$ is also used to describe exponential decay. If there is an initial amount, A_0, of a radioactive substance with a half-life of h, the formula describes the amount of substance remaining, A, after a period of time t. Units of measure for t and h should be the same. Use this formula to answer Problem 41.

41. The half-life of iodine-131 is 8 years. An island is contaminated with debris from illegal above-ground nuclear testing. The debris contains 3000 grams of iodine-131, which is considered 30,000 times above a safe level for a team of scientists to visit the island and possibly identify the country violating the ban. Assuming that they must wait until there is only 0.1 gram of iodine-131, can the scientists safely visit the island after 80 years?

42. In a learning theory project, psychologists discovered that

$$f(x) = \frac{0.8}{1 + e^{-0.2x}}$$

is a model for describing the *proportion* of correct responses after x learning trials.

a. Find the proportion of correct responses prior to learning trials taking place.

b. Find the proportion of correct responses after 10 learning trials.

c. What proportion of correct responses will be approached as continued learning trials take place?

43. The function

$$f(x) = \frac{100}{1 + 100{,}000e^{-0.4x}}$$

describes the percentage of pilots who suffer from nitrogen bubbling out of the blood (a disorder called "the bends") at an altitude of x thousand feet.

a. What percentage of pilots suffer from the bends at an altitude of 10,000 feet?

b. What is the horizontal asymptote for this function? Describe what this means in practical terms.

44. In college, we study large volumes of information— information that, unfortunately, is often not retained for a very long period of time. The Ebbinghaus model for human memory,

$$f(x) = (100 - a)e^{-bx} + a$$

describes the percentage of information that we retain after x weeks. The letters a and b represent constants that vary from person to person. Suppose that for a particular person $a = 20$ and $b = 0.5$. For that person,

$$f(x) = 80e^{-0.5x} + 20.$$

a. What percentage of material is retained after 1 week?

b. What percentage of material is retained after 4 weeks?

c. What percentage of material is ultimately retained?

d. What percentage of material is retained immediately after it is learned?

e. Sketch the graph of $f(x)$.

45. The formula $S = C(1 + r)^t$ models inflation, with $C =$ the value today, $r =$ the annual inflation rate, and $S =$ the inflated value t years from now. If there is an inflation rate of 6%, what is the value of an $85,000 house in 9 years?

True–False Critical Thinking Problems

46. Which one of the following is true?

a. As the number of compounding periods increases on a fixed investment, the amount of money in the account over a fixed interval of time will increase without bound.

b. The functions $f(x) = 3^{-x}$ and $g(x) = -3^x$ have the same graph.

c. $e = 2.718$.

d. The functions $f(x) = (\frac{1}{3})^x$ and $g(x) = 3^{-x}$ have the same graph.

47. Which one of the following is true?

a. The function $f(x) = 1^x$ is an exponential function.

b. If the population of a country is increasing steadily from year to year, the population over time can be modeled by an exponential function.

c. The number e is not a real number.

d. After t months of training, the function $N(t) = 800 - 5000e^{-0.5t}$ describes the number of letters per hour $(N(t))$ that a postal clerk can sort. Based on the model, the production of 800 letters per hour is the level of peak efficiency that is being approached.

Technology Problems

Use a graphing utility to sketch the graph of each function in Problems 48–52. Use the graph to determine intervals in which the function is increasing and decreasing. Approximate any relative maximum or minimum values for the function.

48. $f(x) = x^2 e^x$

49. $f(x) = 3^{-x^2}$

50. $f(x) = \dfrac{e^x + e^{-x}}{2}$

51. $f(x) = \dfrac{e^x - e^{-x}}{2}$

52. $f(x) = \dfrac{2^{|x|}}{1 + 2^x}$

53. a. Use a graphing utility to graph $y = b^x$ with $b = 2, 3$, and 5.

b. For what values of x is it true that $2^x > 3^x > 5^x$?

c. For what values of x is it true that $2^x < 3^x < 5^x$?

d. Is there a value of x for which $2^x = 3^x = 5^x$?

e. Generalize by making a statement in terms of which graph is on the top and which is on the bottom in the intervals $(-\infty, 0)$ and $(0, \infty)$ for $y = b^x$, with $b > 1$.

f. Repeat parts (a)–(e) for $b = \frac{1}{2}, b = \frac{1}{3}$, and $b = \frac{1}{5}$.

54. The space shuttle *Challenger* exploded approximately 73 seconds into flight on January 28, 1986. The tragedy involved O-rings used to seal the connections between different sections of the shuttle engines. The number of O-rings damaged increases dramatically with falling Fahrenheit temperature t, modeled by $N = 13.49(0.967)^t - 1$.

NASA Headquarters

a. On the morning the *Challenger* was launched, the temperature was 31°, colder than any previous experience. Find the number of O-rings expected to fail at this temperature.

b. Use a graphing utility to graph the function. If NASA engineers had used this function and its graph, is it likely they would have allowed the *Challenger* to be launched? Explain.

55. An important function used in statistical decision making is the *normal distribution probability density function*, a model developed by German mathematician Carl Gauss. The model

$$f(x) = \frac{1}{\sqrt{2\pi}} e^{-x^2/2}$$

has a graph that is the familiar bell-shaped curve. Measurements of data collected in nature often follow the pattern determined by this curve. The heights and weights of large populations of human beings are examples of distributions that are approximately normal. In these distributions, data items tend to cluster around the mean (the average) and become more spread out as they differ from the mean.

a. Show that f has y-axis symmetry.

b. What is the horizontal asymptote for this function? Describe what this means in practical terms.

c. Verify parts (a) and (b) by graphing the normal distribution function on a graphing utility. Use the range setting

Xmin = −4, Xmax = 4, Xscl = 1,

Ymin = 0, Ymax = 0.5, Yscl = 0.05

56. A free-hanging power line between two vertical supporting towers appears to be a parabola, but is actually a curve called a *catenary*. A catenary has an equation given by

$$y = \frac{a}{2}(e^{x/a} + e^{-x/a})$$

where a is a constant. Graph the catenary for $a = 1$, $a = 2$, $a = 3$, $a = -1$, and $a = -2$. Describe the effect that a has on each of the graphs.

57. You have $10,000 to invest. One bank pays 5% interest compounded quarterly and the other pays 4.5% interest compounded monthly.

a. Use the formula for compound interest to write a model for the balance in each account at any time t.

b. Use a graphing utility to graph both models on an appropriate viewing rectangle. Based on the graphs, which bank offers the better return on your money?

58. The function

$$N(t) = \frac{100,000}{1 + 5000e^{-t}}$$

describes the number of people who have become ill with influenza t weeks after its initial outbreak in a particular community.

a. How many people became ill with the flu when the epidemic began?

b. How many people were ill by the end of the fourth week?

c. What is the horizontal asymptote for this function? Describe what this means in practical terms.

d. Use a graphing utility to graph the function.

e. Describe the behavior of the epidemic's growth.

59. Graph $y = (1 + \frac{1}{x})^x$ using a graphing utility with $0 \leq x \leq 1000$ and $0 \leq y \leq 10$. Use the $\boxed{\text{TRACE}}$ feature to trace along the curve as x continues to get larger. What value is y approaching? Expand the viewing rectangle for Xmax, tracing along the curve until $y = 2.7183$. Explain what is happening in terms of the definition of the irrational number e.

60. a. Use a graphing utility to graph $y = \dfrac{e^x - 1}{x}$.

   b. Use the graph to determine the value that y approaches $(y \to ?)$ as x approaches 0 $(x \to 0)$.

61. The function $S(t) = 0.32e^{0.11t}$ describes the size of a tumor in a mouse $(S(t)$, measured in cubic centimeters) over time $(t$, in days). Use a graphing utility to graph the function for $0 \le t \le 20$ and $0 \le S(t) \le 3$. Then use the TRACE feature to determine the tumor's size after ten days and after twenty days.

62. Read the description of exponential decay just prior to Problem 41. The half-life of polonium-210 is 140 days. If there are 200 milligrams initially:

   a. Use a graphing utility to graph the decay model.

   b. Use the graph to determine when the mass will decay to 50 milligrams.

63. a. Graph $y = e^x$ and $y = 1 + x + \dfrac{x^2}{2}$ in the same viewing rectangle.

   b. Graph $y = e^x$ and $y = 1 + x + \dfrac{x^2}{2} + \dfrac{x^3}{6}$ in the same viewing rectangle.

   c. Graph $y = e^x$ and $y = 1 + x + \dfrac{x^2}{2} + \dfrac{x^3}{6} + \dfrac{x^4}{24}$ in the same viewing rectangle.

   d. Describe what you observe in parts (a)–(c). Try generalizing this observation.

64. Use a graphing utility to graph $y = e^x$ and $y = e^{x-2} + 2$ in the same viewing rectangle. Describe the transformations of the first function's graph that result in the graph of $y = e^{x-2} + 2$.

Writing in Mathematics

65. The exponential function $y = b^x$ is one-to-one and has an inverse that is a function. Try finding the inverse of $y = 2^x$ by interchanging x and y and solving for y. Describe the difficulty that you encounter in this process. What is needed to overcome this problem?

66. Describe how you could use the graph of $y = 2^x$ to obtain a decimal approximation for $\sqrt{2}$.

67. The graphs labeled (a)–(d) in the figure represent $y = 3^x, y = 5^x, y = (\frac{1}{3})^x$, and $y = (\frac{1}{5})^x$, but not necessarily in that order. Which is which? Describe the process that enables you to make this decision.

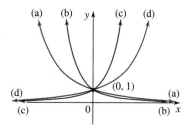

68. Each function in the table is increasing. The graphs labeled (a)–(c) in the figure represent these functions. Which is which? Describe the process that enables you to make this decision.

x	$f(x)$	$g(x)$	$h(x)$
1	23	15	2.7
2	25	25	3.1
3	28	34	3.5
4	32	42	3.9
5	37	49	4.3
6	43	55	4.7

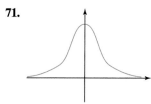

For Problems 69–71, create a verbal problem or situation that can be modeled by the graph of the function shown.

69.

70.

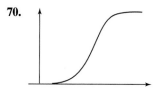

71.

Critical Thinking Problems

72. If $f(x) = b^x + 1$, show that $\dfrac{1}{f(x)} + \dfrac{1}{f(-x)} = 1$.

73. If $f(x) = 3^x$, show that $\dfrac{f(x + h) - f(x)}{h} = 3^x\left(\dfrac{3^h - 1}{h}\right)$.

74. The half-life of a substance that is decaying exponentially is the time required for half of the substance to disintegrate. Use the figure to estimate the half-life of each substance whose exponential decay curve is shown.

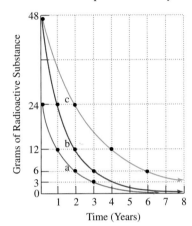

75. The graph shows the exponential growth of a population over time. Approximately how long will it take for the population to double in size?

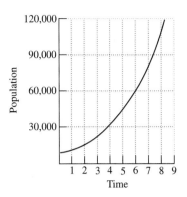

76. The hyperbolic cosine and hyperbolic sine functions are defined by

$$\cosh x = \frac{e^x + e^{-x}}{2} \quad \text{and} \quad \sinh x = \frac{e^x - e^{-x}}{2}.$$

a. Prove that $(\cosh x)^2 - (\sinh x)^2 = 1$.

b. Prove that the hyperbolic cosine is an even function.

c. Prove that the hyperbolic sine is an odd function.

d. Prove that $(\cosh x)^2 = \dfrac{1 + \cosh 2x}{2}$.

e. Prove that $\sinh(x + y) = \sinh x \cosh y + \cosh x \sinh y$.

S E C T I O N 4 . 2 Logarithmic Functions

Solutions Manual   **Tutorial**   **Video 8**

Objectives

1 Evaluate logarithms.

2 Graph logarithmic functions.

3 Evaluate common and natural logarithms.

4 Use logarithmic functions in applications.

In this section our focus is on the inverse of the exponential function, called the logarithmic function. The logarithmic function serves as a model for understanding phenomena as seemingly unrelated as human memory, loudness of sound, earthquake magnitude, and an object's cooling time.

The Definition of Logarithmic Functions

We begin with the problem of finding the inverse of exponential functions. The graphs of exponential functions $f(x) = b^x$ all pass through $(0, 1)$ and have

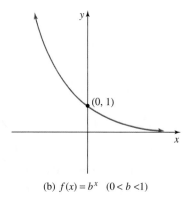

(a) $f(x) = b^x$ $(b > 1)$

(b) $f(x) = b^x$ $(0 < b < 1)$

Figure 4.15

Exponential functions have inverses that are functions.

Evaluate logarithms.

the general shapes shown in Figure 4.15. Because no horizontal line intersects the graph of $f(x) = b^x$ in more than one point, the exponential function is one-to-one. Thus, $f(x) = b^x$ has an inverse that is also a function. We can use the procedure of Section 2.7 to find a formula for f^{-1}.

$$f(x) = b^x$$

1. Replace $f(x)$ by y: $y = b^x$
2. Interchange x and y: $x = b^y$
3. Solve for y: ?

The question mark indicates that we need some sort of notation to solve $x = b^y$ for y. In words, we can write

$y =$ the power on b that gives x.

We solve for y with the following definition.

Definition of the logarithmic function

For $x > 0$ and $b > 0, b \neq 1$,

$$y = \log_b x \text{ is equivalent to } b^y = x.$$

The function $f(x) = \log_b x$ is the *logarithmic function with base b*.

Some logarithms can be evaluated by inspection.

EXAMPLE I   Evaluating Logarithms

Evaluate:

a. $\log_2 16$ **b.** $\log_3 9$ **c.** $\log_{25} 5$
d. $\log_{10} \frac{1}{1000}$ **e.** $\log_7 1$ **f.** $\log_{\frac{1}{3}} 9$ **g.** $\log_3 3$

Solution

Study tip

A logarithm is an exponent. The logarithm of x with base b,

$\log_b x$

is the exponent to which b must be raised to get x. Thus, $\log_2 32 = 5$ because 2 must be raised to the fifth power to get 32. ($2^5 = 32$).

Logarithmic Expression	Question Needed for Evaluation	Logarithmic Expression Evaluated
a. $\log_2 16$	2 to what power gives 16?	$\log_2 16 = 4$ because $2^4 = 16$.
b. $\log_3 9$	3 to what power gives 9?	$\log_3 9 = 2$ because $3^2 = 9$.
c. $\log_{25} 5$	25 to what power gives 5?	$\log_{25} 5 = \frac{1}{2}$ because $25^{1/2} = \sqrt{25} = 5$.
d. $\log_{10} \frac{1}{1000}$	10 to what power gives $\frac{1}{1000}$?	$\log_{10} \frac{1}{1000} = -3$ because $10^{-3} = \frac{1}{10^3} = \frac{1}{1000}$.
e. $\log_7 1$	7 to what power gives 1?	$\log_7 1 = 0$ because $7^0 = 1$.
f. $\log_{\frac{1}{3}} 9$	$\frac{1}{3}$ to what power gives 9?	$\log_{\frac{1}{3}} 9 = -2$ because $\left(\frac{1}{3}\right)^{-2} = \frac{1}{\left(\frac{1}{3}\right)^2} = \frac{1}{\left(\frac{1}{9}\right)} = 9$.
g. $\log_3 3$	3 to what power gives 3?	$\log_3 3 = 1$ because $3^1 = 3$.

We can generalize from Example 1 and obtain the following logarithmic properties.

Basic logarithmic properties

1. $\log_b b = 1$ because 1 is the exponent to which b must be raised to obtain b. $(b^1 = b)$

2. $\log_b 1 = 0$ because 0 is the exponent to which b must be raised to obtain 1. $(b^0 = 1)$

3. $\log_b b^x = x$ because x is the exponent to which b must be raised to obtain b^x. $(b^x = b^x)$

---

2 Graph logarithmic functions.

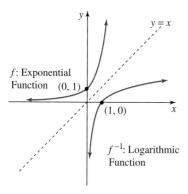

Figure 4.16

An exponential function and its inverse logarithmic function are reflections in the line $y = x$.

Graphs of Logarithmic Functions

Figure 4.16 illustrates the fact that the graphs of $y = \log_b x$ and $y = b^x$ $(b > 0)$ are reflections of each other in the line $y = x$ because $y = \log_b x$ is the inverse of $y = b^x$. Let's see specifically what this means for $b = 2$.

EXAMPLE 2   **Graphing an Exponential Function and Its Inverse Logarithmic Function**

Graph $y = 2^x$ [or $f(x) = 2^x$] and $y = \log_2 x$ [or $f^{-1}(x) = \log_2 x$] on the same set of axes.

Solution

We first set up a table of coordinates for $f(x) = 2^x$. Reversing these coordinates gives the coordinates for the inverse function $f^{-1}(x) = \log_2 x$.

$f(x) = 2^x$ or $y = 2^x$						
x	-2	-1	0	1	2	3
y	$\frac{1}{4}$	$\frac{1}{2}$	1	2	4	8
(x, y)	$(-2, \frac{1}{4})$	$(-1, \frac{1}{2})$	$(0, 1)$	$(1, 2)$	$(2, 4)$	$(3, 8)$

Reverse coordinates

$f^{-1}(x) = \log_2 x$ or $y = \log_2 x$						
x	$\frac{1}{4}$	$\frac{1}{2}$	1	2	4	8
y	-2	-1	0	1	2	3
(x, y)	$(\frac{1}{4}, -2)$	$(\frac{1}{2}, -1)$	$(1, 0)$	$(2, 1)$	$(4, 2)$	$(8, 3)$

We now plot the ordered pairs in both tables, connecting them with smooth curves. Figure 4.17 at the top of the next page shows the graphs of $f(x) = 2^x$ and its inverse function $f^{-1}(x) = \log_2 x$. The graph of the inverse can also be drawn by reflecting the graph of $f(x) = 2^x$ in the line $y = x$. ∎

If we generalize the results of this example from base 2 to base b (with $b > 1$) we see that the inverse of $f(x) = b^x$ is $f^{-1}(x) = \log_b x$, as shown in Figure 4.18 on page 452. These graphs illustrate characteristics of $y = b^x$ and $y = \log_b x$ that are listed in the box on page 452.

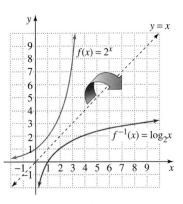

Figure 4.17

The graphs of $f(x) = 2^x$ and its inverse function

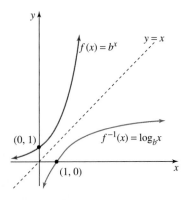

Figure 4.18

Shapes for the graphs of exponential and logarithmic functions, $b > 1$

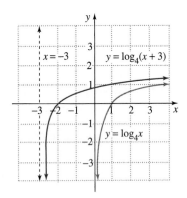

Figure 4.19

A horizontal translation

Characteristics of exponential and logarithmic functions

1. The domain of $y = b^x$ is the set of all real numbers, and the range is the set of all real numbers > 0. Domain $= (-\infty, \infty)$ and range $= (0, \infty)$.
2. The domain of $y = \log_b x$ is the set of all real numbers > 0, and the range of $y = \log_b x$ is the set of all real numbers. Domain $= (0, \infty)$ and range $= (-\infty, \infty)$.
3. The graph of $y = b^x$ passes through $(0, 1)$, and the graph of $y = \log_b x$ passes through $(1, 0)$.
4. The x-axis is a horizontal asymptote of the graph of $y = b^x$ for $b > 1$. ($b^x \to 0$ as $x \to -\infty$)
5. The y-axis is a vertical asymptote of the graph of $y = \log_b x$ for $b > 1$. ($\log_b x \to -\infty$ as $x \to 0^+$)

The graph of $y = \log_b x$ can be transformed through translations, reflections, stretchings, or shrinkings. For example, the graph of $y = \log_4 (x + 3)$ lies three units to the left of the graph of $y = \log_4 x$ (Figure 4.19). The graph of $y = \log_4 x + 3$ lies three units above the graph of $y = \log_4 x$ (Figure 4.20.)

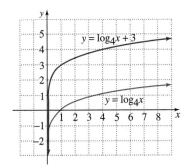

Figure 4.20

A vertical translation

Study tip

The graph of $y = \log_b (x + a) + c$ with $a > 0$ and $c > 0$, lies a units to the left and c units above the graph of $y = \log_b x$. The domain of $f(x) = \log_b (x + a) + c$ consists of all x for which $x + a > 0$. The vertical asymptote is $x = -a$.

EXAMPLE 3   **Transformations of Logarithmic Functions**

Sketch the graph of each function using transformations of the graph of $f(x) = \log_2 x$. State each function's domain and range.

Study tip

If a logarithmic function is translated horizontally, its vertical asymptote will change. Figure 4.21 shows that the vertical asymptote for

$$g(x) = \log_2(x + 2) - 3$$

is $x = -2$.

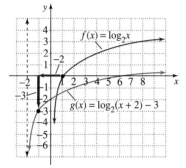

Figure 4.21

The graph of g shifts the graph of f to the left two units and downward three units.

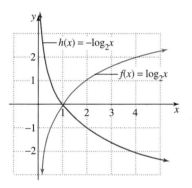

Figure 4.22

The graph of h reflects the graph of f in the x-axis. The line $x = 0$ is the vertical asymptote of h.

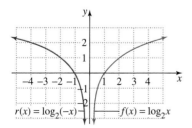

Figure 4.23

The graph of r reflects the graph of f in the y-axis. The y-axis ($x = 0$) is a vertical asymptote of f and r.

a. $g(x) = \log_2(x + 2) - 3$
b. $h(x) = -\log_2 x$
c. $r(x) = \log_2(-x)$

Solution

a. The graph of $g(x) = \log_2(x + 2) - 3$ is obtained from the graph of $f(x) = \log_2 x$ using the following transformations.
  1. A horizontal translation two units to the left: $y = \log_2(x + 2)$.
  2. A vertical translation three units downward: $y = \log_2(x + 2) - 3$.
  The graph of $g(x) = \log_2(x + 2) - 3$ is shown in Figure 4.21. As a check, we can find a few ordered pairs satisfying $g(x) = \log_2(x + 2) - 3$.

x	$g(x) = \log_2(x + 2) - 3$	$(x, g(x))$
-1	$g(-1) = \log_2(-1 + 2) - 3$ $= \log_2 1 - 3 = 0 - 3 = -3$	$(-1, -3)$
2	$g(2) = \log_2(2 + 2) - 3$ $= \log_2 4 - 3 = 2 - 3 = -1$	$(2, -1)$
6	$g(6) = \log_2(6 + 2) - 3$ $= \log_2 8 - 3 = 3 - 3 = 0$	$(6, 0)$

The domain of g consists of all x such that $x + 2 > 0$. Thus, $x > -2$ and the domain is $(-2, \infty)$. The range of g is $(-\infty, \infty)$.

b. Since $h(x) = -\log_2 x = -f(x)$, the graph of h can be obtained from the graph of f by reflecting in the x-axis, as shown in Figure 4.22. The domain of h is $(0, \infty)$ and the range is $(-\infty, \infty)$.

Discover for yourself

Check the graph of h in Figure 4.22 by finding three ordered pairs satisfying

$$h(x) = -\log_2 x.$$

In particular, let $x = 1, 2,$ and 4, and find corresponding values for y. Locate these points on the graph.

c. Since $r(x) = \log_2(-x) = f(-x)$, the graph of r can be obtained from the graph of f by reflecting in the y-axis, as shown in Figure 4.23. The domain of r is $(-\infty, 0)$ and the range is $(-\infty, \infty)$.

Discover for yourself

Check the graph of r in Figure 4.23 by finding three ordered pairs satisfying

$$r(x) = \log_2(-x).$$

In particular, let $x = -4, -2,$ and -1, and find corresponding values for y. Locate these points on the graph.

3 Evaluate common and natural logarithms.

Common Logarithms and Natural Logarithms

The bases that most frequently appear on logarithmic functions are e and 10. If $f(x) = 10^x$, then $f^{-1}(x) = \log_{10}x$. Similarly, if $f(x) = e^x$, then $f^{-1}(x) = \log_e x$.

> **Notations for common and natural logarithms**
>
> We write $y = \log_{10}x$ as $y = \log x$. Logarithms to the base 10 are called *common logarithms*. We write $y = \log_e x$ as $y = \ln x$. Logarithms to the base e are called *natural logarithms*. The symbol $\ln x$ is read as "el en of x."

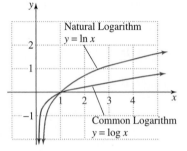

Figure 4.24

Graphs of $y = \log x$ and $y = \ln x$

Figure 4.24 shows the graphs of $y = \log x$ and $y = \ln x$.

You will need to use a calculator to evaluate most common and natural logarithms. Here are some examples obtained by using the common logarithm key $\boxed{\log}$ and the natural logarithm key $\boxed{\ln}$.

Logarithm	Keystrokes	Approximate Display
$\log 1.75$	$\boxed{\log}$ 1.75 $\boxed{\text{ENTER}}$	0.24304
$\ln 1.6$	$\boxed{\ln}$ 1.6 $\boxed{\text{ENTER}}$	0.47000
$\ln(-3)$	$\boxed{\ln}$ $\boxed{(-)}$ 3 $\boxed{\text{ENTER}}$	ERROR

Using technology

Verify the graphs in Figure 4.24 with a graphing utility, using the keys labeled $\boxed{\log}$ and $\boxed{\ln}$.

Although we write $\log 1.75 = 0.24304$, it turns out that 0.24304 is only an *approximate* value. We can reinforce this fact by writing

$$\log 1.75 \approx 0.24304.$$

Similarly, although our calculator tells us that $\ln 1.75 \approx 0.5596$, a more accurate statement is $\ln 1.75 \approx 0.5596$.

We can use a calculator to find, say, $\log 0.192$ and $\ln 2.4$ and then locate points on the graph corresponding to these values.

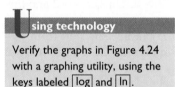

Figure 4.25

$\log 0.192 \approx -0.7167$ and the point $(0.2, -0.7)$ can be approximately located on the graph of $y = \log x$. (See Figure 4.25.)

$\ln 1.6 \approx 0.47$ and the point $(1.6, 0.47)$ can be approximately located on the graph of $y = \ln x$. (See Figure 4.25.)

You can also use a graphing utility's $\boxed{\text{TRACE}}$ feature with the graphs of $y = \log x$ and $y = \ln x$ to verify these values.

Notice that $\ln(-3)$ results in an error message in a calculator's display. Remember that the domain of $y = \log_b x$ is the set of positive real numbers, which means that $\ln(-3)$ is undefined. There is no graph for $y = \ln x$ at $x = -3$ (see Figure 4.25). Since -3 is not in the domain of the natural log function, there is no corresponding range value.

The basic properties of logarithms can be applied to common and natural logarithms.

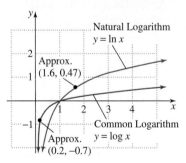

Study tip

A logarithm can be negative or zero. However, the logarithm of a negative number or zero is undefined.

Properties of common and natural logarithms

	General Properties	Common Logarithms	Natural Logarithms
1.	$\log_b 1 = 0$	$\log 1 = 0$	$\ln 1 = 0$
2.	$\log_b b = 1$	$\log 10 = 1$	$\ln e = 1$
3.	$\log_b b^x = x$	$\log 10^x = x$	$\ln e^x = x$

We can use the property $\ln e^x = x$ to identify points on the graph of the natural logarithmic function.

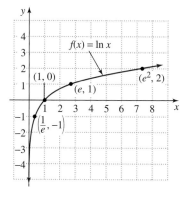

Figure 4.26

Identifying points on the graph of the natural logarithm function

x	$f(x) = \ln x$	$(x, f(x))$	Conclusion (see Figure 4.26)
$\dfrac{1}{e}$	$f\!\left(\dfrac{1}{e}\right) = \ln \dfrac{1}{e} = \ln e^{-1} = -1$	$\left(\dfrac{1}{e}, -1\right)$	$\left(\dfrac{1}{e}, -1\right) \approx (0.37, -1)$ lies on the graph of $f(x) = \ln x$.
e	$f(e) = \ln e = \ln e^1 = 1$	$(e, 1)$	$(e, 1) \approx (2.72, 1)$ lies on the graph of $f(x) = \ln x$.
e^2	$f(e^2) = \ln e^2 = 2$	$(e^2, 2)$	$(e^2, 2) \approx (7.4, 2)$ lies on the graph of $f(x) = \ln x$.

EXAMPLE 4 **Graphing Utilities and Natural Logarithmic Functions**

Find the domain for each function and verify the domain by graphing the function with a graphing utility.

a. $f(x) = \ln (x + 3)$ **b.** $g(x) = \ln (3 - x)$ **c.** $r(x) = \ln (x - 3)^2$

Solution

a. We know that $\ln (x + 3)$ is defined only if $x + 3 > 0$. The domain of f is $(-3, \infty)$ verified by the graph in Figure 4.27.

b. Since $\ln (3 - x)$ is defined only if $3 - x > 0$, the domain of g is $(-\infty, 3)$. This is verified by the graph in Figure 4.28.

c. We know that $\ln (x - 3)^2$ is defined only if $(x - 3)^2 > 0$. The domain of r is all real numbers except $x = 3$, represented by $(-\infty, 3) \cup (3, \infty)$. This is shown by the graph in Figure 4.29. If it is not obvious that 3 is excluded from the domain, try using a $\boxed{\text{DOT}}$ format.

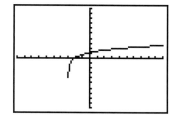

Figure 4.27

The domain of $f(x) = \ln (x + 3)$ is $(-3, \infty)$.

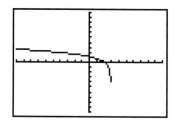

Figure 4.28

The domain of $g(x) = \ln (3 - x)$ is $(-\infty, 3)$.

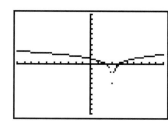

Figure 4.29

The domain of $r(x) = \ln (x - 3)^2$ is $(-\infty, 3) \cup (3, \infty)$.   ∎

4 Use logarithmic functions in applications.

Modeling with Logarithmic Functions

Numerous mathematical models contain natural and common logarithmic functions, illustrated in Table 4.4.

TABLE 4.4   Mathematical Models Containing Logarithms

Area Being Described	Model	Variables in Model	
Walking speed and city population	$W = 0.35 \ln P + 2.74$	P:	population of a city (in thousands)
		W:	average walking speed (in feet per second) for a resident of the city
Cooling time	$t = k \ln \left(\dfrac{T_0 - C}{T - C} \right)$ called Newton's Law of Cooling	t:	time it takes for an object to cool from an initial temperature T_0 to a temperature T
		C:	temperature of the surrounding air
		k:	a constant
The pH of a solution (An acid solution has a pH < 7; those with pH > 7 are called basic.)	$\text{pH} = -\log[\text{H}^+]$	$[\text{H}^+]$:	concentration of the hydrogen ion in moles per liter
		pH:	pH of the solution
Human memory	$f(t) = A - B \ln(t + 1)$	$f(t)$:	Average score on a test for a group of students who take the test t months after learning the material
		A:	average test score on the original test (when $t = 0$)
		B:	a constant
Earthquake magnitude	$R = \log \dfrac{I}{I_0}$	I:	intensity of the earthquake
		I_0:	intensity of a barely felt zero-level earthquake
		R:	magnitude of the earthquake on the Richter scale

Magnitude on Richter scale	Destructive power
$R < 4.5$	Small
$4.5 \leqslant R < 5.5$	Moderate
$5.5 \leqslant R < 6.5$	Large
$6.5 \leqslant R < 7.5$	Major
$7.5 \leqslant R$	Greatest

EXAMPLE 5   Walking Speed and City Population

a. Find the average walking speed of people living in the cities shown in the table.

City	Population (in thousands)
Tulsa, Oklahoma	367
San Diego, California	1111
New York, New York	7323

Jeff Wall "The Stumbling Block" Digitally manipulated cibachrome transparency, 229 × 337 cm. Courtesy Ydessa Hendeles Art Foundation, Toronto.

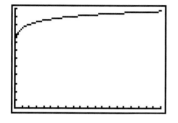

Figure 4.30

The graph of $W = 0.35 \ln P + 2.74$, a model for walking speed as a function of a city's population

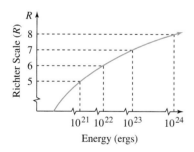

Roger Brown "San Andreas Fault Line" 1995. Photo courtesy Phyllis Kind Gallery, New York & Chicago.

b. Choose an appropriate viewing rectangle and graph the model on a graphing utility.

Solution

$W = 0.35 \ln P + 2.74$ This is the model from Table 4.4.

a. $W = 0.35 \ln(367) + 2.74$ Tulsa: $P = 367$ (thousand)

≈ 4.8 Average walking speed is 4.8 feet per second.

$W = 0.35 \ln(1111) + 2.74$ San Diego: $P = 1111$ (thousand)

≈ 5.2 Average walking speed is 5.2 feet per second.

$W = 0.35 \ln(7323) + 2.74$ New York: $P = 7323$ (thousand)

≈ 5.9 Average walking speed is 5.9 feet per second.

b. The graph of $y = 0.35 \ln x + 2.74$, shown in Figure 4.30, was obtained using the following range setting (suggested by our prior computations):

Xmin = 0, Xmax = 7500, Xscl = 375, Ymin = 0,

Ymax = 6, Yscl = 0.5   ∎

EXAMPLE 6   **Earthquake Intensity**

The Richter scale, used to measure magnitude of earthquakes, is named for U.S. seismologist Charles Richter (1900–1985). The scale is a common logarithmic scale: For each increase of one unit on the Richter scale, there is a tenfold increase in the energy in an earthquake, as shown in the figure on the left.

The magnitude R on the Richter scale of an earthquake of intensity I is given by

$$R = \log \frac{I}{I_0}$$

where I_0 is the intensity of a barely felt zero-level earthquake.

a. Compare the magnitudes on the Richter scale of earthquakes that are 10 times, 100 times, 1000 times, and 10,000 times as intense as a zero-level earthquake.

b. Northern California's 1989 earthquake was $10^{7.1}$ times as intense as a zero-level earthquake. What was its magnitude on the Richter scale?

Solution

a. $R = \log \dfrac{I}{I_0}$ This is the given model.

$= \log \dfrac{10I_0}{I_0}$ $I = 10I_0$ for an earthquake 10 times as intense as a zero-level quake. I_0 is the intensity of the weakest earthquake that can be recorded on a seismograph.

$= \log 10$ Simplify.

$= 1$ $\log 10 = 1$ because the power to which 10 must be raised to get 10 is 1. You can also use a calculator.

Continue using the given model:

$$R = \log \frac{I}{I_0}.$$

100 times as intense as a zero-level quake:	**1000 times as intense as a zero-level quake:**	**10,000 times as intense as a zero-level quake:**
$I = 100I_0$, so	$I = 1000I_0$, so	$I = 10{,}000I_0$, so
$R = \log \dfrac{100I_0}{I_0}$	$R = \log \dfrac{1000I_0}{I_0}$	$R = \dfrac{10{,}000I_0}{I_0}$
$= \log 100$	$= \log 1000$	$= \log 10{,}000$
$= 2$	$= 3$	$= 4$

Thus, for earthquakes that are 10 times, 100 times, 1000 times, and 10,000 times as intense as a zero-level earthquake, the magnitudes on the Richter scale are 1, 2, 3, and 4, respectively.

b. $R = \log \dfrac{I}{I_0}$ This is the given model.

$= \log \dfrac{10^{7.1}I_0}{I_0}$ Northern California's earthquarke was $10^{7.1}$ times the intensity of I_0, so $I = 10^{7.1}I_0$.

$= \log 10^{7.1}$ Simplify.

$= 7.1$ Use a calculator or the property $\log 10^x = x$.

Northern California's 1989 earthquake registered 7.1 on the Richter scale. ■

Sound intensity is described by a model that looks just like the model for earthquake intensity. In particular, the loudness level of a sound D, in decibels, is given by

$$D = 10 \log \frac{I}{I_0}$$

where I is the intensity of the sound and I_0 is the intensity of sound that is barely audible to the human ear. Table 4.5 on the next page indicates the intensity and loudness level of some sounds.

tudy tip

Equal differences in loudness level are equivalent to equal ratios of sound intensity. If the loudness level increases by 10 decibels, the sound intensity increases by a factor of 10.

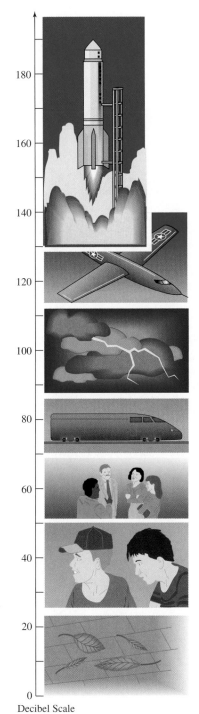

Decibel Scale

TABLE 4.5   Intensity and Loudness Level of Various Sounds

Sound	I: Intensity (in watts per meter^2)	D: Loudness Level (in decibels)
Threshold of hearing	10^{-12}	0
Normal breathing	10^{-11}	10
Rustling leaves	10^{-10}	20
Whisper	10^{-9}	30
Quiet conversation	10^{-8}	40
Quiet television	10^{-7}	50
Normal conversation	10^{-6}	60
Ordinary traffic	10^{-5}	70
Heavy traffic	10^{-4}	80
Niagara Falls	10^{-3}	90
Subway	10^{-2}	100
Loud thunder	10^{-1}	110
Rock concert (2 m from speakers)	10^{0} or 1	120
Jackhammer	10	130
Jet takeoff (40 m away)	10^{2}	140
Sound resulting in a ruptured eardrum	10^{4}	160

EXAMPLE 7   Sound Intensity

a. Use the model

$$D = 10 \log \frac{I}{I_0}$$

to verify the decibel level for jet takeoff in Table 4.5.

b. The cry of a blue whale can be heard nearly 500 miles away, reaching an intensity of 6.3×10^6 watts per meter^2. Determine the decibel level of this sound.

Solution

a. $D = 10 \log \dfrac{I}{I_0}$ This is the logarithmic model for measuring sound intensity.

$\quad D = 10 \log \dfrac{10^2}{10^{-12}}$ $I_0 = 10^{-12}$ (the threshold of hearing) and $I = 10^2$ (for jet takeoff)

$\quad = 10 \log 10^{14}$ $\dfrac{10^2}{10^{-12}} = 10^{2-(-12)} = 10^{14}$

$\quad = 10(14)$ $\log 10^{14} = 14$ because $\log 10^x = x$.

$\quad = 140$ This verifies the decibel level given in Table 4.5 for jet takeoff.

b. $D = 10 \log \dfrac{I}{I_0}$ This is the given model.

$\quad = 10 \log \dfrac{6.3 \times 10^6}{10^{-12}}$ $I_0 = 10^{-12}$ and I (for the blue whale) $= 6.3 \times 10^6$.

$\quad = 10 \log 6.3 \times 10^{18}$ $\dfrac{10^6}{10^{-12}} = 10^{6-(-12)} = 10^{18}$

$\quad \approx 188$ Use a calculator.

With a loudness level that can reach 188 decibels, at close range the cry of a blue whale can rupture the human eardrum.   ∎

ENRICHMENT ESSAY

Earthquakes and the Richter scale

Earthquakes release an enormous amount of energy. Northern California's 1989 earthquake, measuring 7.1 on the Richter scale, released as much energy as 30 million tons of high explosives, nearly ten times the explosive power of all bombs used in World War II (including the two atomic bombs). Magnitude and energy released by an earthquake are not the same thing. The energy released by an earthquake's motion is approximately 30 times greater for each increase of 1 on the Richter scale.

San Francisco's Marina District, built on landfill, was seriously damaged in California's 1989 earthquake. Gordon L. Kallio/The Image Bank

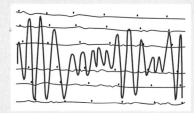

Magnitude	Energy Released (million of ergs)	Earthquake Magnitude
9	20,000,000,000,000,000,000	8.9 Japan 1933
8	600,000,000,000,000,000	7.9 San Francisco 1906
7	20,000,000,000,000,000	6.9 Armenia 1988
6	600,000,000,000,000	
5	20,000,000,000,000	
4	600,000,000,000	
3	20,000,000,000	3.0 Smallest earth-
2	600,000,000	quakes commonly
1	20,000,000	felt
0	600,000	

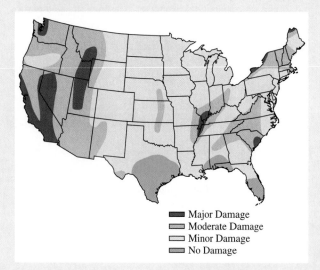

■ Major Damage
■ Moderate Damage
□ Minor Damage
▨ No Damage

This map shows potential earthquake damage for the 48 contiguous states.

PROBLEM SET 4.2

Practice Problems _____

In Problems 1–32, evaluate the expression without the use of a calculator.

1. $\log_3 9$

2. $\log_9 81$

3. $\log_2 64$

4. $\log_3 27$

5. $\log_7 \sqrt{7}$

6. $\log_6 \sqrt{6}$

7. $\log_2 \frac{1}{32}$

8. $\log_3 \frac{1}{81}$

9. $\log_{36} 6$

10. $\log_{81} 9$

11. $\log_{12} 12$

12. $\log_7 7$

13. $\log_4 1$

14. $\log_5 1$

15. $\log_{16} 8$

16. $\log_{125} 5$

17. $\log 10{,}000$

18. $\log 100{,}000$

19. $\log 0.01$

20. $\log 0.001$

21. $\ln e^7$

22. $\ln e^{13}$

23. $\ln \sqrt[3]{e}$

24. $\ln \sqrt[5]{e}$

25. $\log_5 5^7$

26. $\log_4 4^6$

27. $\log 10^{19}$

28. $\log 10^{34}$

29. $\log_b b^7$

30. $\log_b \dfrac{1}{b^2}$

31. $\log_3 (\log_7 7)$

32. $\log_5 (\log_2 32)$

33. Graph $y = 3^x$ [or $f(x) = 3^x$] and $y = \log_3 x$ [or $f^{-1}(x) = \log_3 x$] on the same set of axes.

34. Graph $y = \left(\frac{1}{2}\right)^x$ [or $f(x) = \left(\frac{1}{2}\right)^x$] and $y = \log_{\frac{1}{2}} x$ [or $f^{-1}(x) = \log_{\frac{1}{2}} x$] on the same set of axes.

In Problems 35–44, sketch the graph of each function by using transformations of the graph of $f(x) = \log_2 x$. State each function's domain and range.

35. $g(x) = \log_2 x + 3$

36. $g(x) = \log_2 x - 1$

37. $g(x) = \log_2 (x + 3)$

38. $g(x) = \log_2 (x - 1)$

39. $g(x) = \log_2 (x - 1) + 3$

40. $g(x) = \log_2 (x - 2) - 1$

41. $h(x) = -\log_2 (x + 1)$

42. $h(x) = -\log_2 (x + 2)$

43. $r(x) = -\frac{1}{2} \log_2 (-x)$

44. $r(x) = -2 \log_2 (-x)$

In Problems 45–48, use the graph of $y = \ln x$ to match the function with the appropriate graph.

45. $f(x) = 2 - \ln x$

46. $f(x) = \frac{1}{2} \ln x$

47. $f(x) = 4 \ln x$

48. $f(x) = \ln (x - 2)$

a.

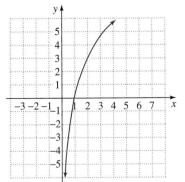

b.

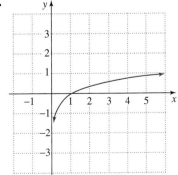

c.

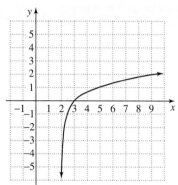

d.

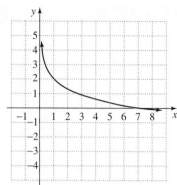

In Problems 49–52, match the function with the appropriate graph.

49. $f(x) = \log_2 x + 1$ **50.** $f(x) = \log_2(x + 1)$ **51.** $f(x) = \log_2 x - 2$ **52.** $f(x) = \log_2(x - 2)$

a.

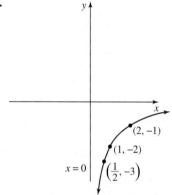

b.

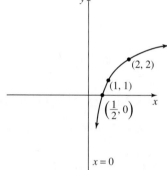

c.

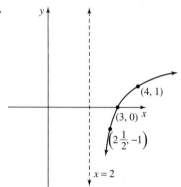

d.

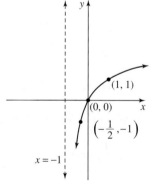

In Problems 53–58, use the property $\ln e^x = x$ to simplify each expression.

53. $\ln\left(\frac{1}{e}\right)$ **54.** $\ln e$ **55.** $\ln \sqrt{e}$

56. $\sqrt{\ln e}$ **57.** $\ln e^{-3}$ **58.** $(\ln e)^{-3}$

In Problems 59–66, find the domain, vertical asymptote, and x-intercept for each logarithmic function. Verify your results by graphing each function with a graphing utility.

59. $f(x) = \ln(x + 2)$ **60.** $f(x) = \ln(x - 1)$ **61.** $f(x) = \ln(2 - x)$ **62.** $f(x) = \ln(1 - x)$

63. $f(x) = \ln x^2$ **64.** $f(x) = \ln(x + 1)^2$ **65.** $f(x) = -\log(x - 2)$ **66.** $f(x) = -\log(x + 3)$

Application Problems

Use the appropriate model from Table 4.4 on page 456 to answer Problems 67–74. In Problems 69–70, assume cooling time is measured in hours.

67. How much faster do residents of Chicago (population: 2,784,000) walk than those of Madison, Wisconsin (population: 191,000)?

68. What is the difference in walking speed for residents of Buffalo (population: 328,000) and Philadelphia (population: 1,586,000)?

69. A bottle of wine at a room temperature of 72° Fahrenheit is placed into a refrigerator. If the temperature of the refrigerator is 40° Fahrenheit and $k = 0.4$, how long will it take the wine to cool to 50° Fahrenheit?

70. The police discover the body of a murder victim at 1:00 P.M. The victim's body temperature is 81.2° Fahrenheit. If the victim is in a room with a temperature of 75° Fahrenheit and $k = 10$, at what time was the murder committed? (Assume temperature at the time of death to be 98.6°.)

71. Find the pH of the following substances, classifying each as acid or basic. (See the figure in the next column.)

    a. Milk: $[H^+] = 3.97 \times 10^{-7}$ mole per liter

    b. Orange juice: $[H^+] = 2.8 \times 10^{-4}$ mole per liter

72. Find the pH of the following substances, classifying each as acid or basic. (See the figure in the next column.)

    a. Beer: $[H^+] = 3.16 \times 10^{-5}$ mole per liter

    b. Vinegar: $[H^+] = 9.32 \times 10^{-4}$ mole per liter

Use the model for sound intensity

$$D = 10 \log \frac{I}{I_0}$$

to answer Problems 75–76.

75. What is the decibel level of a normal conversation, 3.2×10^{-6} watt per meter^2? ($I_0 = 10^{-12}$)

76. People should not be exposed to sound intensities of 3.16×10^{-4} watt per meter^2 for extended periods without ear protection. What decibel level does this represent? ($I_0 = 10^{-12}$)

77. Students in a psychology class took a final examination. As part of an experiment to see how much of the course content they remembered over time, they took equivalent forms of the exam in monthly intervals thereafter. The average score for the group, $f(t)$, after t months was given by the human memory model

$$f(t) = 88 - 15 \ln (t + 1), \qquad 0 \leqslant t \leqslant 12$$

    a. What was the average score on the original exam?

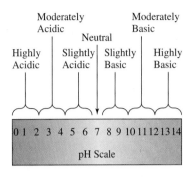

73. Find the Richter scale ratings of earthquakes having the following intensities:

    a. $100,000I_0$ **b.** $1,000,000I_0$

    c. $10,000,000I_0$ **d.** $100,000,000I_0$

74. Compute the magnitude on the Richter scale for the following earthquakes:

    a. Iran, 1978: $10^{7.7}$ times as intense as a zero-level earthquake

    b. San Francisco, 1906: $10^{8.25}$ times as intense as a zero-level earthquake

    c. An earthquake that is 10^k times as intense as a zero-level earthquake

    b. What was the average score after 2 months? 4 months? 6 months? 8 months? 10 months? one year?

    c. Sketch the graph of f (either by hand or with a graphing utility). Describe what the graph indicates in terms of the material retained by the students.

78. Recall that a prime number is a natural number greater than 1 divisible only by itself and 1. The number of prime numbers less than or equal to a given number x can be approximated by the expressions

$$\frac{x}{\ln x} \quad \text{and} \quad \frac{x}{\ln x - 1.08366}$$

when x is large. Fill in the last two columns of the table on page 464, rounding results to the nearest whole number. Then answer the following questions:

    Which of the two expressions provides a better approximation for the numbers in the second column?

What happens to each expression, compared with the numbers in the second column, as x gets larger?

x	Number of Prime Numbers That Are Less Than x	$\dfrac{x}{\ln x}$	$\dfrac{x}{\ln x - 1.083\ 66}$
100	25		
10,000	1229		
10^6	78,498		
10^8	5,761,455		
10^9	50,847,534		
10^{10}	455,052,512		

79. The number $n!$ (n factorial) is defined by

$$n! = n(n-1)(n-2) \cdot \cdots \cdot 3 \cdot 2 \cdot 1$$

where n is a positive integer. For example,

$$6! = 6 \cdot 5 \cdot 4 \cdot 3 \cdot 2 \cdot 1 = 720.$$

Very large factorials can be approximated by $n! \approx \left(\frac{n}{e}\right)^n \sqrt{2\pi n}$, called Stirling's formula. Use the formula to approximate $12!$ and compare it to the actual value.

True–False Critical Thinking Problems

80. Which one of the following is true?

   a. If (a, b) satisfies $y = 2^x$, then (a, b) satisfies $y = \log_2 x$.

   b. $\log(-100) = -2$.

   c. The domain of $f(x) = \log_2 x$ is $(-\infty, \infty)$.

   d. $\log_b x$ is the exponent to which b must be raised to obtain x.

81. Which one of the following is true?

   a. $\log_{36} 6 = 2$

   b. There is no relationship between the graphs of $f(x) = 3^x$ and $g(x) = \log_3 x$.

   c. $\dfrac{\log_2 8}{\log_2 4} = \dfrac{8}{4}$

   d. $\log_{1/2} 32 = -5$

82. Which one of the following is true?

   a. $\log_b 0 = 0$, for $b > 0$ and $b \neq 1$.

   b. If $\log_b 9 = 2$, then $b = 3$ or $b = -3$.

   c. $\log_5 5^7 = 7$

   d. The inverse of $y = 4^x$ is $y = x^4$.

Technology Problems

In Problems 83–87, graph f and g in the same viewing rectangle. Then describe the transformations needed on the graph of f to obtain the graph of g.

83. $f(x) = \ln x$ $g(x) = \ln(x + 3)$
85. $f(x) = \ln x$ $g(x) = \ln(-x) - 2$
87. $f(x) = \ln x$ $g(x) = \ln(x - 2) + 1$

84. $f(x) = \ln x$ $g(x) = \ln x + 3$
86. $f(x) = \ln x$ $g(x) = -\ln(x + 1)$

In Problems 88–93, use a graphing utility to graph each function. Use the graph to determine the function's domain and range, as well as intervals in which the function is increasing and decreasing. Where applicable, approximate any relative maximum or minimum values for the function.

88. $f(x) = x \ln x$

89. $f(x) = \dfrac{\ln x}{x}$

90. $f(x) = \log(x^2 + 1)$

91. $f(x) = \log(\log x)$

92. $f(x) = \sqrt{\ln x}$

93. $f(x) = \ln|x|$

94. Students in a mathematics class took a final examination. As part of an experiment to see how much of the course content they remembered over time, they took equivalent forms of the exam in monthly intervals thereafter. The average score for the group, $f(t)$, after t months was given by the human memory model

$$f(t) = 75 - 10 \log(t + 1), \qquad 0 \le t \le 12.$$

Use a graphing utility to graph the model. Then determine how many months will elapse before the average score falls below 65.

95. Graph f and g in the same viewing rectangle.

   a. $f(x) = \ln(3x)$ $g(x) = \ln 3 + \ln x$

   b. $f(x) = \log(5x^2)$ $g(x) = \log 5 + \log x^2$

c. $f(x) = \ln(2x^3)$ $g(x) = \ln 2 + \ln x^3$

d. Describe what you observe in parts (a)–(c). Generalize this observation by writing an equivalent expression for $\log_b(MN)$, where $M > 0$ and $N > 0$.

e. Complete this statement: The log of a product is equal to _____ .

96. Graph f and g in the same viewing rectangle.

a. $f(x) = \ln\dfrac{x}{2}$ $g(x) = \ln x - \ln 2$

b. $f(x) = \log\dfrac{x}{5}$ $g(x) = \log x - \log 5$

c. $f(x) = \ln\dfrac{x^2}{3}$ $g(x) = \ln x^2 - \ln 3$

d. Describe what you observe in parts (a)–(c). Generalize this observation by writing an equivalent expression for $\log_b\left(\dfrac{M}{N}\right)$, where $M > 0$ and $N > 0$.

e. Complete this statement: The log of a quotient is equal to _____ .

97. Graph f and g in the same viewing rectangle.

a. $f(x) = \ln x^2$ $g(x) = 2 \ln x$

b. $f(x) = \log x^3$ $g(x) = 3 \log x$

c. $f(x) = \ln \sqrt{x} = \ln x^{1/2}$ $g(x) = \frac{1}{2}\ln x$

d. Describe what you observe in parts (a)–(c). Generalize this observation by writing an equivalent expression for $\log_b M^p$, where $M > 0$.

e. Complete this statement: The log of a number with an exponent is equal to _____ .

In Problems 98–106, graph the functions on each side of the = sign in the same viewing rectangle to determine whether each statement is true or false.

98. $\ln(x + 3) = \ln x + \ln 3$ (Graph $y = \ln(x + 3)$ and $y = \ln x + \ln 3$.)

99. $\log\dfrac{x}{2} = \dfrac{\log x}{\log 2}$

100. $\log\dfrac{x}{2} = \log x - \log 2$

101. $\ln x^3 = (\ln x)^3$

102. $x + \ln x^{-1} = x - \ln x$

103. $x^x = e^{x \ln x}$

104. $\ln e^x = x$

105. $(\log x)(\log 2) = \log 2x$

106. $\log x + \log 2 = \log 2x$

107. Graph each of the following pairs of functions in the same viewing rectangle, $-1 \le x \le 3;\ -2 \le y \le 2$.

a. $y = \ln(1 + x)$ and $y = x - \dfrac{x^2}{2}$

b. $y = \ln(1 + x)$ and $y = x - \dfrac{x^2}{2} + \dfrac{x^3}{3}$

c. $y = \ln(1 + x)$ and $y = x - \dfrac{x^2}{2} + \dfrac{x^3}{3} - \dfrac{x^4}{4}$

d. Describe the emerging pattern.

108. Graph each of the following functions in the same viewing rectangle and then place the functions in order from the one that increases most slowly to the one that increases most rapidly.

$$y = x,\ y = \sqrt{x},\ y = e^x,\ y = \ln x,\ y = x^x,\ y = x^2$$

Writing in Mathematics

109. Suppose that we use the model $W = 0.35 \ln P + 2.74$ relating population (P) of a city and average walking speed (W) of the city's residents. If we know the average walking speed is 2.4 feet per second, what difficulty will we encounter by attempting to approximate the city's population? What skill is needed to overcome this difficulty?

110. Select a mathematical model from Table 4.4 and write a verbal problem that can be solved using the formula. Solve the problem.

111. Without using a calculator, how can you determine the greater number: $\log_4 60$ or $\log_3 40$?

Critical Thinking Problems

112. Without using a calculator, find the exact value of

$$\dfrac{\log_3 81 - \log_\pi 1}{\log_{2\sqrt{2}} 8 - \log 0.001}.$$

113. Without using a calculator, find the exact value of

$$\log \sqrt[3]{10 \sqrt{1000}} - \log \dfrac{1}{100} \sqrt{\dfrac{1}{10} \sqrt[3]{100}}.$$

114. Solve for x: $\log_4[\log_3(\log_2 x)] = 0$.

115. Write the point-slope equation of the line passing through $(3 \ln e + \ln(\frac{1}{e}),\ \ln(\ln e))$, and $(3 \ln(e \ln e),\ \ln e^2)$.

116. If $f(x) = \ln(\ln x)$, find $f^{-1}(x)$.

S E C T I O N   4 . 3

Solutions   **Tutorial**   **Video**
Manual   **8**

Properties of Logarithmic Functions

Objectives

1 Expand logarithmic expressions.
2 Condense logarithmic expressions.
3 Use the inverse properties of logarithms.
4 Use the change-of-base property.

In this section, we consider six important properties of logarithms. These properties can be used to answer questions about variables that appear in logarithmic models.

Three Basic Logarithmic Properties

iscover for yourself

You can discover three basic logarithmic properties on your own with your graphing utility. To do so work Problems 95–97 in Problem Set 4.2 before continuing.

There are three properties of exponents that have corresponding logarithmic properties:

1. $b^M \cdot b^N = b^{M+N}$

2. $\dfrac{b^M}{b^N} = b^{M-N}$

3. $(b^M)^p = b^{Mp}$

These identities, coupled with an awareness that a logarithm is an exponent, suggest the following properties.

Properties of logarithms

Let b, M, and N be positive real numbers with $b \neq 1$, and let p be any real number.

Property	The Property in Words	Example
The Product Rule $\log_b(MN) = \log_b M + \log_b N$	The log of a product is the sum of the logs.	$\ln(4x) = \ln 4 + \ln x$
The Quotient Rule $\log_b\left(\dfrac{M}{N}\right) = \log_b M - \log_b N$	The log of a quotient is the difference of the logs.	$\log \dfrac{x}{2} = \log x - \log 2$
The Power Rule $\log_b M^p = p \log_b M$	The log of a number raised to the power p is the product of p and the log of that number.	$\ln \sqrt{x} = \ln x^{1/2} = \dfrac{1}{2}\ln x$

---

sing technology

The relationships in the examples given in the preceding box can be confirmed using a graphing utility. The graphs on each side of the three equalities are the same, verifying these special cases of the product, quotient, and power rules.

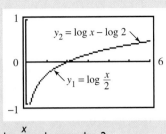

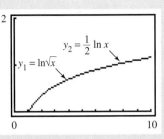

$\ln(4x) = \ln 4 + \ln x$ $\log \dfrac{x}{2} = \log x - \log 2$ $\ln x^{1/2} = \dfrac{1}{2}\ln x$

Proving the Logarithmic Properties

All three of the properties of logarithms can be proved in the same way, Let's look at the proof for the product rule

$$\log_b(MN) = \log_b M + \log_b N.$$

We begin by letting $\log_b M = R$ and $\log_b N = S$.

Now we write each logarithm in exponential form.

$$\log_b M = R \quad \text{means} \quad b^R = M$$
$$\log_b N = S \quad \text{means} \quad b^S = N$$

By substituting and using a property of exponents, we see that

$$MN = b^R b^S = b^{R+S}.$$

Now we change $MN = b^{R+S}$ to logarithmic form.

$$MN = b^{R+S} \quad \text{means} \quad \log_b(MN) = R + S.$$

Finally, substituting $\log_b M$ for R and $\log_b N$ for S gives us

$$\log_b(MN) = \log_b M + \log_b N$$

the property that we wanted to prove.

The proofs of the quotient and power rules are nearly identical and are left for the problem set.

Using the Properties Together to Expand Logarithmic Expressions

When we expand a logarithmic expression, it is possible to change the domain of a function. For example, the power rule tells us that

$$\log_b x^2 = 2\log_b x.$$

If $f(x) = \log_b x^2$, then the domain of f is the set of all nonzero real numbers. However, if $g(x) = 2\log_b x$, the domain is the set of positive real numbers. Consequently, we should write

$$\log_b x^2 = 2\log_b x, \quad \text{if } x > 0.$$

In short, when expanding or condensing a logarithmic expression, you might want to observe whether the rewriting has changed the domain of the expression.

Example 1 involves expanding logarithmic expressions. In summary, most of our work will be based on the following three properties.

sing technology

The graphs of $y = \ln x^2$ and $y = 2 \ln x$ have different domains.

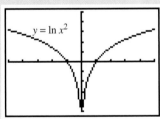

Domain: $(-\infty, 0) \cup (0, \infty)$

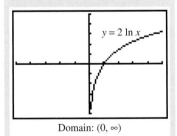

Domain: $(0, \infty)$

Notice that the graphs are only the same if $x > 0$. Thus, we should write

$\ln x^2 = 2 \ln x$ for $x > 0$.

Expand logarithmic
expressions.

Properties for expanding logarithmic expressions

1. $\log_b(MN) = \log_b M + \log_b N$ Product rule

2. $\log_b\left(\dfrac{M}{N}\right) = \log_b M - \log_b N$ Quotient rule

3. $\log_b M^p = p \log_b M$ Power rule

In all cases, $M > 0$ and $N > 0$.

EXAMPLE 1   Expanding Logarithmic Expressions

Use logarithmic properties to expand each expression as much as possible.
Expanded forms should involve logarithms of x, y, and z.

a. $\log_b x^2 \sqrt{y}$ **b.** $\log_6\left(\dfrac{\sqrt[3]{x}}{36y^4}\right)$ **c.** $\log \sqrt[5]{\dfrac{1000x}{y^3z^2}}$

Solution

We will have to use two or more of the properties of logarithms in each part of
this example.

a. $\log_b x^2 \sqrt{y} = \log_b x^2 y^{1/2}$ Use exponential notation.

$\qquad = \log_b x^2 + \log_b y^{1/2}$ Use the product rule.

$\qquad = 2 \log_b x + \dfrac{1}{2} \log_b y$ Use the power rule.

b. $\log_6\left(\dfrac{\sqrt[3]{x}}{36y^4}\right) = \log_6 \dfrac{x^{1/3}}{36y^4}$ Use exponential notation.

$\qquad = \log_6 x^{1/3} - \log_6 36y^4$ Use the quotient rule.

$\qquad = \log_6 x^{1/3} - (\log_6 36 + \log_6 y^4)$ Use the product rule on $\log_6 36y^4$.

$\qquad = \dfrac{1}{3} \log_6 x - (\log_6 36 + 4 \log_6 y)$ Use the power rule.

$\qquad = \dfrac{1}{3} \log_6 x - \log_6 36 - 4 \log_6 y$ Apply the distributive property.

$\qquad = \dfrac{1}{3} \log_6 x - 2 - 4 \log_6 y$ $\log_6 36 = 2$ because 2 is the power to which we must raise 6 to get 36. $(6^2 = 36)$

c. $\log \sqrt[5]{\dfrac{1000x}{y^3z^2}} = \log\left(\dfrac{1000x}{y^3z^2}\right)^{\frac{1}{5}}$ Use exponential notation.

$\qquad = \dfrac{1}{5} \log\left(\dfrac{1000x}{y^3z^2}\right)$ Use the power rule.

$\qquad = \dfrac{1}{5}(\log 1000x - \log y^3z^2)$ Use the quotient rule.

$\qquad = \dfrac{1}{5}[\log 1000 + \log x - (\log y^3 + \log z^2)]$ Use the product rule.

$\qquad = \dfrac{1}{5}(\log 1000 + \log x - \log y^3 - \log z^2)$ Apply the distributive property.

$\qquad = \dfrac{1}{5}(\log 1000 + \log x - 3\log y - 2\log z)$ Use the power rule.

$$= \frac{1}{5}(3 + \log x - 3\log y - 2\log z)$$

log 1000 means $\log_{10}1000$, which is 3, because $10^3 = 1000$.

■

2 Condense logarithmic expressions.

Using the Properties Together to Condense Logarithmic Expressions

In Example 1, we wrote each expression as the sum and the difference of simpler logarithmic expressions. We can also use the logarithmic properties to reverse the direction of this procedure, writing the sum and/or difference of logarithms as a single logarithm. Most of our work will use the following three properties.

Properties for condensing logarithmic expressions

1. $\log_b M + \log_b N = \log_b(MN)$ Product rule

2. $\log_b M - \log_b N = \log_b\left(\dfrac{M}{N}\right)$ Quotient rule

3. $p\log_b M = \log_b M^p$ Power rule

In all cases, $M > 0$ and $N > 0$.

EXAMPLE 2   Condensing Logarithmic Expressions

Write as a single logarithm:

a. $\log_4 2 + \log_4 32$ **b.** $\ln 5 - 5\ln x$ **c.** $4\log_b x - 2\log_b 6 + \frac{1}{2}\log_b y$

Solution

a. $\log_4 2 + \log_4 32 = \log_4(2 \cdot 32)$ Use the product rule.

$\qquad\qquad\qquad\quad = \log_4 64$ We now have a single logarithm. However, we can simplify.

$\qquad\qquad\qquad\quad = 3$ $\log_4 64 = 3$ because $4^3 = 64$.

b. $\ln 5 - 5\ln x = \ln 5 - \ln x^5$ Use the power rule.

$\qquad\qquad\qquad = \ln \dfrac{5}{x^5}$ Use the quotient rule.

Study tip

Coefficients of logarithms must be 1 before you can condense them using the product and quotient rules. Thus, we must write 5 ln x as ln x^5.

c. $4\log_b x - 2\log_b 6 + \dfrac{1}{2}\log_b y$

$= \log_b x^4 - \log_b 6^2 + \log_b y^{1/2}$ Use the power rule so that log coefficients are all 1.

$= (\log_b x^4 - \log_b 36) + \log_b y^{1/2}$ This optional step emphasizes the order of operations.

$= \log_b \dfrac{x^4}{36} + \log_b y^{1/2}$ Use the quotient rule.

$= \log_b \dfrac{x^4}{36} \cdot y^{1/2}$ or $\log_b \dfrac{x^4\sqrt{y}}{36}$ Use the product rule.   ■

Study tip

Shown on the next page are the graphs of $y_1 = \ln(x + 3)$ and $y_2 = \ln x + \ln 3$. Our graphing utility indicates that the graphs are *not the same*. The graph of $y_1 = \ln(x + 3)$ is

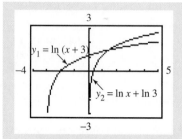

actually the graph of the natural log function shifted 3 units to the left. The graph of $y_2 = \ln x + \ln 3$ is the graph of the natural log function shifted upward by ln 3, which is an approximate vertical shift of 1.1 units.

We see, then, that $\ln(x + 3) \neq \ln x + \ln 3$. In general,

$$\log_b(M + N) \neq \log_b M + \log_b N.$$

Try to avoid each of the following incorrect procedures.

INCORRECT

$$\log_b(M + N) = \log_b M + \log_b N \qquad \log_b(M - N) = \log_b M - \log_b N \qquad \log_b(M \cdot N) = (\log_b M)(\log_b N)$$

$$\log_b \frac{M}{N} = \frac{\log_b M}{\log_b N} \qquad \frac{\log_b M}{\log_b N} = \log_b M - \log_b N$$

Use your graphing utility, replacing $\log_b$ by ln, M by x, and N by 3 to show that all five of these equations are not true in general.

3 Use the inverse properties of logarithms.

"True–False" 1992. Rimma Gerlovina and Valeriy Gerlovina

The Inverse Properties

The inverse properties are a result of the fact that the exponential and logarithmic functions are inverses.

Discover for yourself

We previously learned that $f(f^{-1}(x)) = x$ and $f^{-1}(f(x)) = x$. Apply these relationships to $f(x) = b^x$ and $f^{-1}(x) = \log_b x$ and write two inverse properties involving logarithms.

In the Discover for yourself, were you able to write the following properties?

Inverse properties of logarithms

For $b > 0, b \neq 1$:

$$b^{\log_b x} = x \quad \text{and} \quad \log_b b^x = x \qquad (x > 0)$$

We have already introduced the second of these properties, stating that $\log_b b^x = x$ because x is the exponent to which b must be raised to obtain b^x. To prove the first part of this property, let

$$y = \log_b x.$$

$$b^y = x \qquad \text{Rewrite the logarithmic expression in its equivalent exponential form.}$$

$$b^{\log_b x} = x \qquad \text{Substitute } \log_b x \text{ for } y \text{ (from the first equation) into the equation in the previous step. This is precisely what we wanted to prove.}$$

The inverse properties for logarithms can be applied to common and natural logarithms, as shown in Table 4.6.

TABLE 4.6	Inverse Properties Applied to Common and Natural Logarithms	
Inverse Property	**Base 10**	**Base e**
$b^{\log_b x} = x$	$10^{\log x} = x$	$e^{\ln x} = x$
$\log_b b^x = x$	$\log 10^x = x$	$\ln e^x = x$

EXAMPLE 3 **Expanding Logarithmic Expressions**

Use the properties of logarithms to expand each expression, simplifying terms where possible.

a. $\log(A \cdot 10^{7t})$ **b.** $\ln\left(\dfrac{B}{e^{7a}}\right)$ $(A > 0, B > 0)$

Solution

a. $\log(A \cdot 10^{7t}) = \log A + \log 10^{7t}$ $\log_b(MN) = \log_b M + \log_b N$
$\qquad\qquad\qquad\; = \log A + 7t$ $\log 10^x = x$

b. $\qquad \ln\left(\dfrac{B}{e^{7a}}\right) = \ln B - \ln e^{7a}$ $\log_b \dfrac{M}{N} = \log_b M - \log_b N$

$\qquad\qquad\qquad\qquad = \ln B - 7a$ $\ln e^x = x$ ∎

EXAMPLE 4 **Condensing Logarithmic Expressions**

Use the properties of logarithms to write each expression as a single term that does not contain a logarithm.

a. $10^{\log 4x + \log 3x}$ **b.** $e^{\ln 8x^5 - \ln 2x^2}$

Solution

a. $10^{\log 4x + \log 3x} = 10^{\log[(4x)(3x)]}$ Condense the log expression in the exponent using
$\qquad\qquad\qquad\qquad\qquad\qquad$ $\log_b M + \log_b N = \log_b(MN)$.

$\qquad\qquad\qquad\quad = 10^{\log 12x^2}$ Multiply.

$\qquad\qquad\qquad\quad = 12x^2$ $10^{\log x} = x$

b. $e^{\ln 8x^5 - \ln 2x^2} = e^{\ln(8x^5/2x^2)}$ Condense the log expression in exponent using

$\qquad\qquad\qquad\qquad\qquad\qquad$ $\log_b M - \log_b N = \log_b \dfrac{M}{N}$.

$\qquad\qquad\qquad\quad = e^{\ln 4x^3}$ Divide.

$\qquad\qquad\qquad\quad = 4x^3$ $e^{\ln x} = x$ ∎

In the next section we will use the inverse properties to solve exponential equations, gaining further insight into the mathematical models that we have already discussed.

4 Use the change-of-base property.

The Change-of-Base Property

We have seen that calculators give the values of both common logarithms (base 10) and natural logarithms (base e). To find a logarithm with any other base, we can use the following change-of-base property.

Study tip

The change-of-base property is used to write a logarithm in terms of quantities that can be evaluated with a calculator.

The change-of-base property

For any logarithmic bases a and b, and any positive number M,

$$\log_b M = \frac{\log_a M}{\log_a b}.$$

Base a is a new base that we introduce. The logarithm on the left side has a base of b, allowing us to change from base b to any other base a, as long as the newly introduced base is a positive number not equal to 1.

To prove the change-of-base property, we let x equal the logarithm on the left side:

$$\log_b M = x.$$

Now we rewrite this logarithm in exponential form.

$$\log_b M = x \quad \text{means} \quad b^x = M.$$

Since b^x and M are equal, the logarithms with base a for each of these expressions must be equal. This means that

$$\log_a b^x = \log_a M$$

$$x \log_a b = \log_a M \qquad \text{Apply the power rule for logarithms on the left side.}$$

$$x = \frac{\log_a M}{\log_a b} \qquad \text{Solve for } x \text{ dividing both sides by } \log_a b.$$

In our first step we let x equal $\log_b M$. Replacing x on the left side by $\log_b M$ gives us

$$\log_b M = \frac{\log_a M}{\log_a b}$$

which is the change-of-base property.

Because calculators contain keys for common and natural logarithms, we will frequently introduce base 10 or base e.

EXAMPLE 5   **Changing Base to Common Logarithms**

Use common logarithms to evaluate: $\log_5 140$

Discover for yourself

Let $\log_5 140 = y$. Since this means that $5^y = 140$, what is a reasonable estimate for y, to the nearest whole number? Now compare this estimate to the value given in the solution.

Study tip

Since the change-of-base property usually involves introducing base 10 or base e, here's what the property looks like in each of these situations.

Introducing Common Logarithms (Base 10)

$$\log_b M = \frac{\log M}{\log b}$$

Introducing Natural Logarithms (Base e)

$$\log_b M = \frac{\ln M}{\ln b}$$

Solution

$$\log_5 140 = \frac{\log_{10} 140}{\log_{10} 5} \qquad \text{Use } \log_b M = \frac{\log_a M}{\log_a b}. \text{ The newly introduced base is 10.}$$

$$= \frac{\log 140}{\log 5} \qquad \text{We omit the 10 with common logs, writing } \log_{10} \text{ as log.}$$

$$\approx 3.07 \qquad \text{Use a calculator: } \boxed{\log} \; 140 \; \boxed{\div} \; \boxed{\log} \; 5 \; \boxed{\text{ENTER}}$$

This means that $\log_7 50 \approx 3.07$.   ■

EXAMPLE 6   **Changing Base to Natural Logarithms**

Use natural logarithms to evaluate: $\log_5 140$

Solution

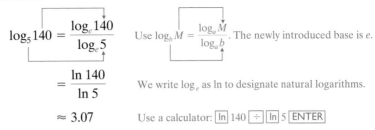

$$\log_5 140 = \frac{\log_e 140}{\log_e 5} \qquad \text{Use } \log_b M = \frac{\log_a M}{\log_a b}. \text{ The newly introduced base is } e.$$

$$= \frac{\ln 140}{\ln 5} \qquad \text{We write } \log_e \text{ as ln to designate natural logarithms.}$$

$$\approx 3.07 \qquad \text{Use a calculator: } \boxed{\text{ln}}\ 140\ \boxed{\div}\ \boxed{\text{ln}}\ 5\ \boxed{\text{ENTER}}$$

We have again shown that $\log_5 140 \approx 3.07$. ∎

We can use the change-of-base property to determine a relationship between $y = \ln x$ and $y = \log x$, whose graphs are shown partially in Figure 4.31. We begin with

$$\log_{10} x = \frac{\log_e x}{\log_e 10} \qquad \text{We've introduced base } e.$$

Using the agreed-upon notation for common and natural logarithms, we can write

$$\log x = \frac{\ln x}{\ln 10}.$$

Multiplying both sides by $\ln 10$ results in

$$(\log x)(\ln 10) = \ln x.$$

Because $\ln 10$ is approximately 2.3026, this equation gives us a relationship between the natural and common logarithmic functions:

$$\ln x \approx (\log x)(2.3026).$$

In terms of the graphs in Figure 4.31, the natural logarithm of any (positive) number is approximately equal to the common logarithm of that number multiplied by 2.3026. Thus, the graph of $y = \ln x$ stretches the graph of $y = \log x$ by a factor of approximately 2.3.

The change-of-base property is useful when a graphing utility is used to graph logarithmic functions with bases other than 10 or e.

EXAMPLE 7 **Using a Graphing Utility to Graph Logarithmic Functions**

Graph $y = \log_2 x$ and $y = \log_{20} x$ in the same viewing rectangle.

Solution

Since $\log_2 x = \dfrac{\ln x}{\ln 2}$ and $\log_{20} x = \dfrac{\ln x}{\ln 20}$, the functions are entered as

$$Y_1 \boxed{=} \text{lnX/ln2} \text{ and } Y_2 \boxed{=} \text{lnX/ln20}$$

using

Xmin = 0, Xmax = 10, Xscl = 1, Ymin = −3, Ymax = 3, and Yscl = 1.

The graphs are shown in Figure 4.32. ∎

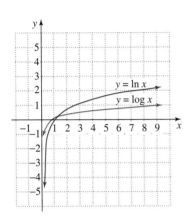

Figure 4.31

Partial graphs of natural and common logarithmic functions

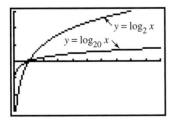

Figure 4.32

Using the change-of-base property to graph logarithmic functions

ENRICHMENT ESSAY

The Curious Number e

- The number e was named by the Swiss mathematician Leonhard Euler (1707–1783), who proved that it is the limit as N tends to infinity of $\left(1 + \dfrac{1}{N}\right)^N$.

- e features in Euler's remarkable relationship $e^{i\pi} = -1$.

- The first few decimal places of e are fairly easy to remember: $e = 2.7\ 1828\ 1828\ 45\ 90\ 45\ \ldots$.

- The best approximation of e using numbers less than 1000 is also easy to remember: $e \approx \dfrac{878}{323} \approx$ 2.71826 ….

- Newton showed that $e^x = 1 + x + \dfrac{x^2}{2!} + \dfrac{x^3}{3!} + \dfrac{x^4}{4!} + \cdots$, from which we obtain $e = 1 + 1 + \dfrac{1}{2!} + \dfrac{1}{3!} + \dfrac{1}{4!} \cdots$, an infinite sum suitable for calculation because its terms decrease so rapidly. (*Note: n*! (*n* factorial) is the product of all the consecutive integers from n down to 1: $n! = n(n-1)(n-2)(n-3) \cdot \cdots \cdot 3 \cdot 2 \cdot 1$.)

- The area of the region bounded by $y = \dfrac{1}{x}$, the x-axis, $x = 1$ and $x = t$ (shown in orange in Figure 4.33) is a function of t, designated by $A(t)$. Grégoire de Saint-Vincent, a Belgian Jesuit (1584–1667), spent his entire professional life attempting to find a formula for $A(t)$. With his student, he showed that $A(t) = \ln t$, becoming one of the first mathematicians to make use of the logarithmic function for something other than a computational device.

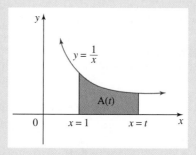

Figure 4.33

EXAMPLE 8   **Logarithmic Properties and Transformations**

Describe the relationship between the graphs of $y = \log_3 x$ and $y = \log_3 \dfrac{1}{(x+1)^3}$. Then use a graphing utility to verify your description.

Solution

We begin by using the power rule to rewrite the second equation.

$$y = \log_3 \frac{1}{(x+1)^3}$$

$$= \log_3 (x+1)^{-3} \qquad \frac{1}{b^3} = b^{-3}$$

$$= -3 \log_3 (x+1) \qquad \log_b M^p = p \log_b M$$

We can obtain the graph of $y = -3 \log_3 (x + 1)$ from the graph of $y = \log_3 x$ by translating one unit to the left ($y = \log_3 (x + 1)$), stretching by multiplying by a factor of 3 ($y = 3 \log_3 (x + 1)$), and reflecting in the x-axis ($y = -3 \log_3 (x + 1)$). Using a $[-2, 5]$ by $[-4, 4]$ range setting and entering

$$y_1 = \log_3 x \quad \text{as} \quad Y_1 = \ln X / \ln 3$$

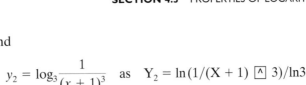

and

$$y_2 = \log_3 \frac{1}{(x+1)^3} \quad \text{as} \quad Y_2 = \ln(1/(X+1) \boxed{\wedge} 3)/\ln 3$$

we obtain the graphs shown in Figure 4.34. The graphs verify our description of the transformations. Observe that the domain of $y = \log_3 x$ is $(0, \infty)$ and the domain of $y = \log_3 \dfrac{1}{(x+1)^3}$ is $(-1, \infty)$. Both functions have a range of $(-\infty, \infty)$. ∎

Figure 4.34

The graph of y_2 moves the graph of y_1 to the left one unit, stretches it by a factor of 3, and reflects it in the *x*-axis.

PROBLEM SET 4.3

Practice Problems

In Problems 1–18, use logarithmic properties to expand each expression as much as possible. Where possible, evaluate logarithmic expressions without the use of a calculator.

1. $\log_7(7x)$

2. $\log(100x)$

3. $\ln(x^3 y^2)$

4. $\ln(x^5 y^4)$

5. $\log_4\left(\dfrac{\sqrt{x}}{64}\right)$

6. $\log_5\left(\dfrac{\sqrt[3]{x}}{25}\right)$

7. $\log_6\left(\dfrac{36}{\sqrt{3x-2}}\right)$

8. $\log_8\left(\dfrac{64}{\sqrt{x^2+1}}\right)$

9. $\ln\left(\dfrac{9}{e^4}\right)$

10. $\ln\left(\dfrac{23}{e^7}\right)$

11. $\log\sqrt[3]{\dfrac{x}{y}}$

12. $\log\sqrt[5]{\dfrac{x}{y}}$

13. $\log_b\left(\dfrac{\sqrt{x}y^3}{z^3}\right)$

14. $\log_b\left(\dfrac{\sqrt[3]{x}y^4}{z^5}\right)$

15. $\log_5\sqrt[3]{\dfrac{x^2 y}{25}}$

16. $\log_2\sqrt[5]{\dfrac{xy^4}{16}}$

17. $\ln\left(\sqrt{\sqrt{x}y^3}\right)$

18. $\log_4(2\sqrt[3]{x}\sqrt{y})$

In Problems 19–38, write each expression as a single logarithm whose coefficient is 1.

19. $\ln x + \ln 7$

20. $\ln y + \ln 3$

21. $\log_5 x - \log_5 y$

22. $\log_4 7 - \log_4 y$

23. $2\log_3(x+1)$

24. $2\log_3(x-1)$

25. $\frac{1}{2}\log_2(5x)$

26. $\frac{1}{3}\log_2(7x)$

27. $\log_3(x^2-9) - \log_3(x-3)$

28. $\log_5(x^3+8) - \log_5(x+2)$

29. $\frac{1}{2}\log x + 3\log y$

30. $\frac{1}{3}\log x + 2\log y$

31. $\frac{1}{3}\log_4 x + 2\log_4(3x+2)$

32. $\frac{1}{4}\log_6 x + 2\log_6(x-7)$

33. $3\ln x + 5\ln y - 6\ln z$

34. $4\ln x + 7\ln y - 3\ln z$

35. $\frac{1}{2}(\log x + \log y)$

36. $\frac{1}{3}(\log_4 x - \log_4 y)$

37. $\frac{1}{2}(\log_5 x + \log_5 y) - 2\log_5(x+1)$

38. $\frac{1}{3}(\log_4 x - \log_4 y) + 2\log_4(x+1)$

In Problems 39–44, use properties of logarithms to expand each expression, simplifying terms where possible.

39. $\log(c \cdot 10^{9t})$

40. $\log\left(\dfrac{B}{10^{14a}}\right)$

41. $\ln(A \cdot e^{7x^3})$

42. $\ln\left(\dfrac{10^3}{e^7}\right)$

43. $\ln(A\sqrt[4]{e^3})$

44. $\ln\left(\dfrac{\sqrt[7]{e^4}}{B}\right)$

In Problems 45–50, use properties of logarithms to write each expression as a single term that does not contain a logarithm.

45. $10^{\log 6x^2 + \log 3x^4}$

46. $e^{\ln 7x^3 + \ln 2x^5}$

47. $10^{\log 18y^7 - \log 2y^4}$

48. $e^{\ln 25x^5 y^3 - \ln 5x^2 y}$

49. $e^{-2 + \ln y^2}$

50. $e^{5 + \ln y^4}$

In Problems 51–58, evaluate each logarithm using the change-of-base property. Work each problem twice, first using common logarithms and then using natural logarithms. Round each expression to three decimal places.

51. $\log_3 17$ **52.** $\log_7 3$ **53.** $\log_{\frac{1}{2}} 6$ **54.** $\log_4 0.65$

55. $\log_6 0.8$ **56.** $\log_{30} 1365$ **57.** $\log_{15} 195$ **58.** $\log_{\frac{1}{4}} 0.025$

Application Problem

59. The pH of a solution is a measure of its acidity. A low pH indicates an acidic solution, and a high pH indicates a basic solution. Neutral water has a pH of 7. Most people have a blood pH of approximately 7.4. The precise pH of a person's blood is modeled by the Henderson–Hasselbach formula, pH $= 6.1 + \log B - \log C$. In the model, B represents the concentration of bicarbonate, which is a base, and C represents the concentration of carbonic acid, which is acidic.

   a. Rewrite the model so that the common logarithm appears only once.

   b. Use the form of the model in part (a) and a calculator to answer this question: If the concentration of bicarbonate is 25 and the concentration of carbonic acid is 2, what is the pH for this person's blood?

True–False Critical Thinking Problems

60. Which of the following is true?

   a. $\dfrac{\log_8 64}{\log_8 8} = \log_8 64 - \log_8 8$

   b. $(\log 10)^3 = 3 \log 10$

   c. $\ln (7e^{10}) = 10 + \ln 7$

   d. $\frac{1}{5} \log_5 x = \sqrt[x]{x}, x > 0$

61. Which one of the following is true?

   a. Because $\ln x^2 = 2 \ln x$, we can obtain a complete graph of $y = \ln x^2$ by doubling all the values on the graph of $y = \ln x$.

   b. There is a value of x for which $(\log x)^3 = 3 \log x$.

   c. There is no value of x for which $\log (x + 3) = \log x + \log 3$.

   d. None of the above is true.

Technology Problems

62. a. Use a graphing utility (and the change-of-base property) to graph $y = \log_3 x$.

   b. Graph $y = (\log_3 x) + 2$, $y = \log_3 (x + 2)$, and $y = -\log_3 x$ in the same viewing rectangle as $y = \log_3 x$. Then describe the transformation(s) of the graph of $y = \log_3 x$ needed to obtain each of these three graphs.

63. a. Use a graphing utility (and the change-of-base property) to graph $y = \log_{\frac{1}{2}} x$.

   b. Graph $y = (\log_{\frac{1}{2}} x) - 1$, $y = \log_{\frac{1}{2}} (x - 1)$, and $y = -\log_{\frac{1}{2}} (x - 1) + 1$ in the same viewing rectangle as $y = \log_{\frac{1}{2}} x$. Then describe the transformations of the graph of $y = \log_{\frac{1}{2}} x$ needed to obtain each of these three graphs.

64. Graph $y = \log x$, $y = \log (10x)$, and $y = \log (0.1x)$ in the same viewing rectangle. Describe the relationship among the three graphs. What logarithmic property accounts for this relationship?

65. a. Use a graphing utility and the change-of-base property to graph $y = \log_b x$ with $b = 3, 25$, and 100 in the same viewing rectangle.

   b. Which graph is on the top in the interval $(0, 1)$? Which is on the bottom?

   c. Which graph is on the top in the interval $(1, \infty)$? Which is on the bottom?

   d. Repeat parts (a)–(c) with $b = \frac{1}{2}$, $b = \frac{1}{4}$, and $b = \frac{1}{10}$.

   e. Generalize by writing a statement about which graph is on top, which is on the bottom, and in which intervals for $y = \log_b x$ for $b > 1$ and $b < 1$.

66. a. Use the power rule to rewrite $\log_4 \dfrac{1}{(x - 2)^3}$.

   b. Describe the relationship between the graphs of $y = \log_4 x$ and $y = \log_4 \dfrac{1}{(x - 2)^3}$.

   c. Use a graphing utility to verify your description.

67. a. Use a graphing utility to graph $y = \log_2 x$ and $y = \log_2 \sqrt{\dfrac{x}{8}}$ in the same viewing rectangle.

   b. Show that $\log_2 \sqrt{\dfrac{x}{8}} = \dfrac{1}{2} \log_2 \left(\dfrac{x}{8}\right)$. Describe the transformations of the graph of $y = \log_2 x$ needed to obtain the graph of $y = \frac{1}{2} \log_2 (\frac{x}{8})$. Visually reinforce these transformations with the two graphs that appear in the viewing rectangle.

68. By letting y equal a positive constant of your choice, use a graphing utility to *disprove* each of the following statements.

   a. $\log_3 (xy) = (\log_3 x)(\log_3 y)$ (*Hint:* If $y = 2$, first graph $y_1 = \log_3 (2x)$. Then graph $y_2 = (\log_3 x)(\log_3 2)$, the right side of the given statement, and show that y_1 and y_2 do not have the same graphs.)

   b. $\log_5 \dfrac{x}{y} = \dfrac{\log_5 x}{\log_5 y}$

   c. $\dfrac{\log_4 x}{\log_4 y} = \log_4 x - \log_4 y$

69. Graph $y = \ln (e^2 x)$. Now graph $y = k + \ln x$ so that the two graphs are identical. (First you will need to determine a value for k.)

Writing in Mathematics

70. In Problem 68a, you disproved the property $\log_3(xy) = (\log_3 x)(\log_3 y)$ using a graphing utility. (Take a moment to reread the problem.) If the property $\log_3(xy) = \log_3 x + \log_3 y$ can be visually verified with a graphing utility for, say, fifty different choices for y (where $y = a$ positive constant), is this a *proof* that $\log_3(xy) = \log_3 x + \log_3 y$? Explain. How does this compare with the proof given on page 467?

71. Why is there no logarithmic property for $\log_b(M + N)$?

72. Describe the error in the following argument:

$$4 > 2$$

$4 \log \frac{1}{3} > 2 \log \frac{1}{3}$ Multiply both sides by $\log \frac{1}{3}$.

$\log\left(\frac{1}{3}\right)^4 > \log\left(\frac{1}{3}\right)^2$ $p \log_b M = \log_b M^p$

$\left(\frac{1}{3}\right)^4 > \left(\frac{1}{3}\right)^2$

$\frac{1}{81} > \frac{1}{9}$

$1 > 9$ Multiply both sides by 81.

Critical Thinking Problems

73. Use the change-of-base property to prove that
$$\log e = \frac{1}{\ln 10}.$$

74. If $f(x) = e^x$, find the value of $f(\ln 4) + f(\ln 3^2) - f(\ln \frac{1}{e})$.

75. Prove that
$$\log_b\left(\frac{M}{N}\right) = \log_b M - \log_b N$$
by letting $R = \log_b M$ and $S = \log_b N$ and following the procedure used to prove logarithmic property 1 on page 467.

76. Prove that $\log_b M^p = p \log_b M$.

77. Logarithms were conceived by John Napier (1550–1617), a Scots mathematician who was searching for easier methods of computation with large numbers. Napier's logarithms were neither common nor natural, defined instead by Nap $\log x = 10^7 \log_{\frac{1}{e}}\left(\frac{x}{10^7}\right)$. Use logarithmic properties to prove that
$$\text{Nap } \log x = 10^7 \left(7 \ln 10 - \ln x\right).$$

78. Simplify: $\log_b x + \log_{\frac{1}{b}} x$.

79. Show that $\ln \dfrac{\sqrt{3} + \sqrt{2}}{\sqrt{3} - \sqrt{2}} = 2 \ln\left(\sqrt{3} + \sqrt{2}\right)$. Do not use a calculator.

S E C T I O N   4 . 4

Solutions Manual

Tutorial

Video 8

Exponential and Logarithmic Equations

Objectives

1 Solve exponential equations.

2 Solve applied problems involving exponential equations.

3 Solve logarithmic equations.

4 Solve applied problems involving logarithmic equations.

In Section 4.1, we considered exponential models that described the number of AIDS cases in the United States, the accumulated value of an interest-bearing account, and the age of an artifact. These models contained variable exponents. Techniques for solving equations with variable exponents will enable us to work in more meaningful ways with exponential models.

1 Solve exponential equations.

Exponential Equations

An exponential equation is an equation containing a variable in an exponent. Examples of exponential equations include

$$10^x = 5.71, \quad e^{7-4x} - 3 = 10, \quad \text{and} \quad 2^{x+2} = 3^{2x+1}.$$

These equations can be solved by isolating the exponential expression, taking the natural logarithm on both sides, simplifying, and then solving for the variable.

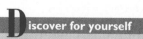

iscover for yourself

Solve

$$10^x = 5.71$$

by taking the common logarithm on both sides. Use the change-of-base property to show that the answer is the same as the one in Example 1.

EXAMPLE 1 Solving an Exponential Equation

Solve: $10^x = 5.71$

Solution

$$10^x = 5.71 \qquad \text{This is the given equation.}$$

$$\ln 10^x = \ln 5.71 \qquad \text{Take the natural logarithm on both sides.}$$

$$x \ln 10 = \ln 5.71 \qquad \text{Use the power rule for logarithms.}$$

$$x = \frac{\ln 5.71}{\ln 10} \qquad \text{Solve for } x \text{ by dividing both sides by } \ln 10.$$

We now have an exact value for x. We can obtain a decimal approximation using a calculator.

$$x \approx 0.757 \qquad \text{Calculator keystroke sequence:} \boxed{\text{ln}}\ 5.71\ \boxed{\div}\ \boxed{\text{ln}}\ 10\ \boxed{\text{ENTER}}$$

Since $10^0 = 1$ and $10^1 = 10$, it seems reasonable that the solution to $10^x = 5.71$ is approximately 0.757. The equation's solution set is $\left\{ \dfrac{\ln 5.71}{\ln 10} \right\}$. We use the exact value for x in expressing the solution set.

We can verify this solution with a graphing utility. Using methods discussed in Section 1.1, enter

$$Y_1 = 10\ \boxed{\wedge}\ X$$

$$Y_2 = 5.71$$

$$Y_3\ \boxed{=}\ Y_1 - Y_2.$$

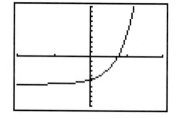

Figure 4.35

The x-intercept for $y_3 = 10^x - 5.71$ is approximately 0.757. This value is the approximate solution for $10^x = 5.71$.

Deselect y_1 and y_2. The graph of y_3 is shown in Figure 4.35. It appears that the x-intercept for y_3 is approximately 0.7. The more accurate approximation of 0.757 can be obtained using the $\boxed{\text{TRACE}}$ and $\boxed{\text{ZOOM}}$ features. ∎

EXAMPLE 2 Solving Exponential Equations

Solve:

a. $7e^{3x} = 312$ **b.** $e^{7-4x} - 3 = 10$ **c.** $2^{x+2} = 3^{2x+1}$

iscover for yourself

Since e is approximately equal to 2.72, what is a reasonable estimate (to the nearest whole number) of the solution to $7e^{3x} = 312$? Experiment with a few integers for x. When the integer is multiplied by 3 and e is raised to this power, multiplication of this result by 7 should give 312. Now use this approach to find a reasonable estimate (to the nearest whole number) of the solution to the equation in Example 2b.

Solution

Study tip

Once you have taken the natural logarithm on both sides of an exponential equation, you will simplify using one of the following properties:

$\ln b^x = x \ln b$ or $\ln e^x = x$.

a. We begin by dividing both sides by 7 to isolate the exponential expression.

$$7e^{3x} = 312 \qquad \text{This is the given equation.}$$

$$e^{3x} = \frac{312}{7} \qquad \text{Divide both sides by 7.}$$

$$\ln e^{3x} = \ln\left(\frac{312}{7}\right) \qquad \text{Take the natural logarithm on both sides.}$$

$$3x = \ln\left(\frac{312}{7}\right) \qquad \text{Use the inverse property: } \ln e^{\square} = \square$$

$$x = \frac{\ln\left(\frac{312}{7}\right)}{3} \qquad \text{Divide both sides by 3.}$$

$$\approx 1.266 \qquad \text{Use a calculator.}$$

The solution set is $\left\{ \dfrac{\ln\left(\dfrac{312}{7}\right)}{3} \right\}$.

b. We begin by adding 3 on both sides to isolate the exponential expression.

$$e^{7-4x} - 3 = 10 \qquad \text{This is the given equation.}$$

$$e^{7-4x} = 13 \qquad \text{Add 3 on both sides.}$$

$$\ln e^{7-4x} = \ln 13 \qquad \text{Take the natural logarithm on both sides.}$$

$$7 - 4x = \ln 13 \qquad \text{Use the inverse property: } \ln e^{\square} = \square$$

$$4x = 7 - \ln 13$$

$$x = \frac{7 - \ln 13}{4} \qquad \text{Divide both sides by 4.}$$

$$\approx 1.109 \qquad \text{Use a calculator.}$$

The solution set is $\left\{ \dfrac{7 - \ln 13}{4} \right\}$.

c.

$$2^{x+2} = 3^{2x+1} \qquad \text{This is the given equation.}$$

$$\ln 2^{x+2} = \ln 3^{2x+1} \qquad \text{Take the natural log on both sides: If } M = N, \text{ then } \ln M = \ln N.$$

$$(x + 2)\ln 2 = (2x + 1)\ln 3 \qquad \ln b^x = x \ln b$$

$$x \ln 2 + 2 \ln 2 = 2x \ln 3 + \ln 3 \qquad \text{Apply the distributive property.}$$

$$2 \ln 2 - \ln 3 = 2x \ln 3 - x \ln 2 \qquad \text{Isolate terms with the variable on one side of the equation.}$$

$$2 \ln 2 - \ln 3 = x(2 \ln 3 - \ln 2) \qquad \text{Factor } x \text{ on the right.}$$

$$\frac{2 \ln 2 - \ln 3}{2 \ln 3 - \ln 2} = x \qquad \text{Solve for } x.$$

The solution set is $\left\{ \dfrac{2 \ln 2 - \ln 3}{2 \ln 3 - \ln 2} \right\}$, approximately 0.191.

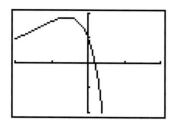

Figure 4.36

The *x*-intercept for $y_3 = 2^{x+2} - 3^{2x+1}$ is approximately 0.191. This value is the approximate solution for $2^{x+2} = 3^{2x+1}$.

sing technology

Shown below is the graph of $y = e^{2x} - 4e^x + 3$. There are two *x*-intercepts, one at 0 and one (using the ⟨TRACE⟩ and ⟨ZOOM⟩ features) approximately at 1.099, verifying our algebraic solution.

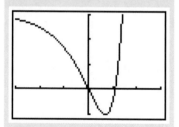

2 Solve applied problems involving exponential equations.

Barton Lidice Benes "Penny dreaming of becoming a dollar." Photo by Karen Furth.

Figure 4.36 shows the graph of y_3, where $y_1 = 2^{x+2}$, $y_2 = 3^{2x+1}$, and y_3 ⟨=⟩ $y_1 - y_2$.

Use the ⟨TRACE⟩ and ⟨ZOOM⟩ features to approximate the *x*-intercept as 0.191, verifying the algebraic solution. ∎

EXAMPLE 3   Solving an Exponential Equation

Solve: $e^{2x} - 4e^x + 3 = 0$

Solution

The given equation is quadratic in form. If $t = e^x$, the equation can be expressed as $t^2 - 4t + 3 = 0$. Since this equation can be solved by factoring, we factor to isolate the exponential term.

$e^{2x} - 4e^x + 3 = 0$	This is the given equation.
$(e^x - 3)(e^x - 1) = 0$	Factor on the left. Notice that if $t = e^x$, $t^2 - 4t + 3 = (t - 3)(t - 1)$.
$e^x - 3 = 0$ or $e^x - 1 = 0$	Set each factor equal to 0.
$e^x = 3$ $e^x = 1$	Solve for e^x.
$\ln e^x = \ln 3$ $x = 0$	Take the natural log on both sides of the first equation. The second equation can be solved by inspection.
$x = \ln 3$ $x = 0$	$\ln e^x = x$

The solution set is $\{0, \ln 3\}$. The solutions are 0 and (approximately) 1.099. ∎

Modeling Applications

Knowing how to solve exponential equations enables us to work with exponential models and solve applied problems.

EXAMPLE 4   Revisiting the Continuous Compounding Model

The model $A = Pe^{rt}$ describes the accumulated value A of a sum of money P after t years at an annual percentage rate r (in decimal form) compounded continuously. Suppose that \$1250 is invested at 5.5% compounded continuously.

a. How long will it take the investment to triple in value?
b. What interest rate is required for the investment to double in 10 years time?

Solution

a.
$A = Pe^{rt}$	Use the model for continuous compounding.
$3750 = 1250\, e^{0.055t}$	$P = 1250$ (the amount invested), $A = 3750$ (accumulated amount is the triple of $1250 = 3 \cdot 1250 = 3750$), and r (the interest rate) $= 0.055$.
$3 = e^{0.055t}$	Divide both sides of the equation by 1250, isolating the exponential term.

$$\ln 3 = \ln e^{0.055t} \qquad \text{Take the natural logarithm on both sides.}$$

$$\ln 3 = 0.055t \qquad \ln e^x = x$$

$$t = \frac{\ln 3}{0.055} \qquad \text{Solve for } t.$$

$$\approx 20 \qquad \text{Use a calculator.}$$

We see that the investment will triple in value in approximately 20 years.

b. $\quad A = Pe^{rt} \qquad$ Once again, use the model for continuous compounding.

$$2500 = 1250\, e^{r(10)} \qquad P = 1250 \text{ (the amount invested) } A = 2500 \text{ (we want the investment to double), and } t = 10 \text{ (the doubling is to happen in 10 years).}$$

$$2 = e^{10r} \qquad \text{Isolate the exponential term.}$$

$$\ln 2 = \ln e^{10r} \qquad \text{Take the natural logarithm on both sides.}$$

$$\ln 2 = 10r \qquad \ln e^x = x$$

$$r = \frac{\ln 2}{10} \qquad \text{Solve for } r.$$

$$\approx 0.069 \qquad \text{Use a calculator.}$$

An interest rate of approximately 6.9% is required for the investment to double over a period of 10 years. ∎

EXAMPLE 5   Revisiting the Formula for Compound Interest

The model

$$A = P\left(1 + \frac{r}{n}\right)^{nt}$$

describes the accumulated value A of a sum of money P, the principal, after t years at annual percentage rate r (in decimal form), compounded n times a year (rather than compounded continuously). How long will it take \$1000 to grow to \$3600 at 8% annual interest compounded quarterly?

Solution

$$A = P\left(1 + \frac{r}{n}\right)^{nt} \qquad \text{This is the given model.}$$

$$3600 = 1000\left(1 + \frac{0.08}{4}\right)^{4t} \qquad A \text{ (the desired accumulated value)} = \$3600, \\ P\,(\text{the principal}) = \$1000,\, r\,(\text{the interest rate}) = 8\% = \\ 0.08, \text{ and } n = 4 \text{ (quarterly compounding).}$$

Our goal is to solve for t. Let's first simplify within parentheses and reverse the two sides of the equation.

$$1000(1.02)^{4t} = 3600$$

$$(1.02)^{4t} = 3.6 \qquad \text{Isolate the exponential expression by dividing both sides by 1000.}$$

Using technology

The model

$$y = 1000 \, (1.02)^{4x}$$

representing the balance in the account, y, after x years is graphed below. The point with approximate coordinates (16.2, 3600) verifies a balance of $3600 after 16.2 years.

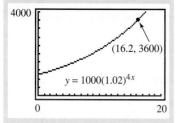

$$\ln (1.02)^{4t} = \ln 3.6 \qquad \text{Take the natural logarithm on both sides.}$$

$$4t \ln 1.02 = \ln 3.6 \qquad \text{Use the power rule for logarithms.}$$

$$t = \frac{\ln 3.6}{4 \ln 1.02} \qquad \text{Solve for } t, \text{ dividing both sides by } 4 \ln 1.02.$$

$$\approx 16.2 \qquad \text{Use a calculator.}$$

After (approximately) 16.2 years, the $1000 will grow to an accumulated value of $3600. ∎

One aim of algebra is to use mathematical models to predict the behavior of variables. The model in the next example, taken from a report to former president Jimmy Carter, involves a prediction that is cause for alarm.

EXAMPLE 6   Depletion of Nonrenewable Energy Resources

The model

$$A = \frac{A_0}{k} \left(e^{kt} - 1 \right)$$

describes the amount of nonrenewable energy resources A (oil, natural gas, coal, uranium) from time $t = 0$ to time t, where A_0 is the amount of the resource consumed during the year $t = 0$ and k is the relative growth rate of annual consumption. In the Global 2000 Report to former president Jimmy Carter, the 1976 worldwide consumption of oil was 21.7 billion barrels, and the predicted growth rate for oil consumption was 3% per year ($k = 0.03$). At that time, the total known remaining oil resources ultimately available were 1661 billion barrels. Using these figures, by what year will the planet be depleted of its oil resources that were known at the time?

Solution

$$A = \frac{A_0}{k} \left(e^{kt} - 1 \right) \qquad \text{This is the given model.}$$

$$1661 = \frac{21.7}{0.03} \left(e^{0.03t} - 1 \right) \qquad \begin{array}{l} A_0 = 21.7 \text{ (billion barrels) in } t = 0 \\ (1976). \ k = 0.03. \text{ We must find } t \\ \text{corresponding to } A = 1661 \\ \text{(billion barrels).} \end{array}$$

$$1661 \left(\frac{0.03}{21.7} \right) = e^{0.03t} - 1 \qquad \text{Our goal is to isolate } e^{0.03t}.$$

$$1661 \left(\frac{0.03}{21.7} \right) + 1 = e^{0.03t} \qquad \text{Add 1 to both sides.}$$

$$\ln \left[1661 \left(\frac{0.03}{21.7} \right) + 1 \right] = \ln e^{0.03t} \qquad \text{Take the natural logarithm on both sides.}$$

$$\ln \left[1661 \left(\frac{0.03}{21.7} \right) + 1 \right] = 0.03t \qquad \text{On the right: } \ln e^x = x$$

I am directing the Council on Environmental Quality and the Department of State, working in cooperation with … other appropriate agencies, to make a one-year study of the probable changes in the world's population, natural resources, and environment through the end of the century.

President Jimmy Carter, May 23, 1977

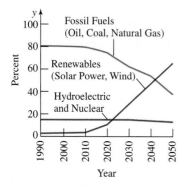

Fossil fuels supply about 80% of the energy used in industrialized countries. Supplies are diminishing, and they could be replaced by cleaner renewable resources such as wind and solar power.

Source: World Bank

$$\frac{\ln\left[1661\left(\frac{0.03}{21.7}\right) + 1\right]}{0.03} = t \qquad \text{Solve for } t.$$

Using a calculator, we find $t \approx 40$. The planet will be depleted of its then-known oil resources 40 years after 1976, in 2016. ∎

Logarithmic Equations

Just as exponential equations contain a variable in an exponent, logarithmic equations contain a variable in a logarithmic expression. Examples of logarithmic equations include

$$\log_4(x + 3) = 2 \quad \text{and} \quad \log(2x - 1) = \log(4x - 3) - \log x.$$

The first equation contains a logarithmic expression and a constant. The second equation contains only logarithmic expressions. We now consider methods for solving both kinds of logarithmic equations. These methods will enable us to work in more meaningful ways with logarithmic models.

3 Solve logarithmic equations.

Logarithmic Equations Containing Logarithmic Expressions and Constants

The equation

$$\log_4(x + 3) = 2$$

contains the logarithmic expression $\log_4(x + 3)$ and the constant 2. We can solve this equation by rewriting it as an equivalent exponential equation. Example 7 illustrates how this is done.

EXAMPLE 7 Solving a Logarithmic Equation

Solve: $\log_4(x + 3) = 2$

Solution

We first rewrite the equation as an equivalent exponential equation.

$$\log_b M = N \qquad \text{means} \qquad b^N = M$$
$$\log_4(x + 3) = 2 \quad \text{means} \quad 4^2 = x + 3$$

Now we solve the equivalent exponential equation for x.

$$4^2 = x + 3$$

$$16 = x + 3$$

$$13 = x$$

sing technology

The graphs of

$y_1 = \log_4(x + 3)$ and $y_2 = 2$

have an intersection point whose x-coordinate is 13. This verifies that $\{13\}$ is the solution set for $\log_4(x + 3) = 2$. *Note:* Since

$$\log_b x = \frac{\ln x}{\ln b}$$

(change-of-base property),

we entered y_1 using

$$y_1 = \log_4(x + 3) = \frac{\ln(x + 3)}{\ln 4}.$$

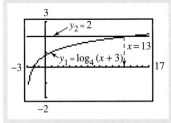

Check

$$\log_4(x + 3) = 2 \qquad \text{This is the given logarithmic equation.}$$

$$\log_4(13 + 3) \stackrel{?}{=} 2 \qquad \text{Substitute 13 for } x.$$

$$\log_4 16 \stackrel{?}{=} 2 \qquad \text{Substitute 13 for } x.$$

$$2 = 2 \checkmark \qquad \log_4 16 = 2 \text{ because } 4^2 = 16.$$

This true statement indicates that the solution set is $\{13\}$. ∎

The solution method used in Example 7 works when one side of the equation contains a single logarithm and the other side contains a constant. Once the equation is in the form $\log_b M = N$, we rewrite it in the exponential form $b^N = M$. In the next example we use the product rule for logarithms to obtain a single logarithmic expression on the left side.

EXAMPLE 8 **Using the Product Rule to Solve a Logarithmic Equation**

Solve: $\log_2 x + \log_2(x - 7) = 3$

Solution

$\log_2 x + \log_2(x - 7) = 3$	This is the given equation.
$\log_2 x(x - 7) = 3$	Use the product rule to obtain a single logarithm.
$2^3 = x(x - 7)$	If $\log_b M = N$, then $b^N = M$.
$8 = x^2 - 7x$	Apply the distributive property on the right.
$0 = x^2 - 7x - 8$	Set the equation equal to 0.
$0 = (x - 8)(x + 1)$	Factor.
$x - 8 = 0$ or $x + 1 = 0$	Set each factor equal to 0.
$x = 8 \qquad\qquad x = -1$	Solve for x.

Check

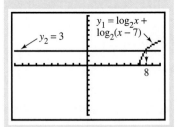sing technology

The graphs of

$y_1 = \log_2 x + \log_2(x - 7),$

entered as

$$y_1 = \frac{\ln x}{\ln 2} + \frac{\ln(x - 7)}{\ln 2}$$

and

$$y_2 = 3$$

have only one intersection point whose x-coordinate is 8. This verifies that $\{8\}$ is the solution set for $\log_2 x + \log_2(x - 7) = 3$.

Checking 8:

$$\log_2 x + \log_2(x - 7) = 3$$

$$\log_2 8 + \log_2(8 - 7) \stackrel{?}{=} 3$$

$$\log_2 8 + \log_2 1 \stackrel{?}{=} 3$$

$$3 + 0 \stackrel{?}{=} 3$$

$$3 = 3 \checkmark$$

Checking -1:

$$\log_2 x + \log_2(x - 7) = 3$$

$$\log_2(-1) + \log_2(-1 - 7) \stackrel{?}{=} 3$$

The number -1 does not check. Negative numbers do not have logarithms.

The solution set is $\{8\}$. ∎

Solving an equation containing logarithmic expressions and constants

1. Combine all logarithmic expressions on one side and all constants on the other side of the equation.
2. Use logarithmic properties to rewrite the logarithmic expression as a single logarithm whose coefficient is 1. The form of the equation is $\log_b M = N$.
3. Rewrite step 2 in exponential form $b^N = M$ and solve the resulting equation for the variable.
4. Check each solution in the original equation, rejecting values that produce the logarithm of a negative number or the logarithm of 0.

Logarithmic Equations Containing Only Logarithmic Expressions

We continue our discussion with the solution of equations such as

$$\log(2x - 1) = \log(4x - 3) - \log x.$$

Notice that every term contains a logarithmic expression. The fact that the logarithmic function is one-to-one is very important in solving such equations. Stated symbolically, this means that

> If $\log_b M = \log_b N$ $(M > 0, N > 0)$, then $M = N$.

Consequently, we can solve these equations by using logarithmic properties to rewrite each side as the logarithm of a single quantity. We then set the quantities equal to each other and solve for the variable. Example 9 illustrates this process.

Using technology

Using

$$y_1 = \log(2x - 1)$$

$$y_2 = \log(4x - 3) - \log x$$

and

$$y_3 \boxed{=} y_1 - y_2$$

we obtain the graph of y_3 shown below. The range setting

Xmin = 0, Xmax = 2, Xscl = 1, Ymin = −0.2, Ymax = 0.2, Yscl = 1

makes it fairly easy to see that y_3 has x-intercepts at 1 and $\frac{3}{2}$, verifying the solution set.

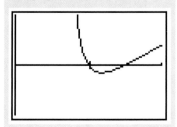

EXAMPLE 9 **A Logarithmic Equation Containing Only Logarithmic Expressions**

Solve: $\log(2x - 1) = \log(4x - 3) - \log x$

Solution

$$\log(2x - 1) = \log(4x - 3) - \log x \qquad \text{This is the given equation.}$$

$$\log(2x - 1) = \log\left(\frac{4x - 3}{x}\right) \qquad \log_b M - \log_b N = \log_b \frac{M}{N}$$

$$2x - 1 = \frac{4x - 3}{x} \qquad \text{If } \log_b M = \log_b N, \text{ then } M = N.$$

$$(2x - 1)x = 4x - 3 \qquad \text{Clear fractions by multiplying both sides by } x.$$

$$2x^2 - 5x + 3 = 0 \qquad \text{Write the quadratic equation in standard form.}$$

$$(2x - 3)(x - 1) = 0 \qquad \text{Factor.}$$

$$2x - 3 = 0 \quad \text{or} \quad x - 1 = 0 \qquad \text{Set each factor equal to 0.}$$

$$x = \frac{3}{2} \qquad\qquad x = 1 \qquad\qquad \text{Solve for } x.$$

Both proposed solutions are valid, since neither causes the logarithm of 0 or the logarithm of a negative number in the given equation. The solution set is $\{1, \frac{3}{2}\}$. ∎

---

4 Solve applied problems involving logarithmic equations.

Using technology

The graph of

$$y = \frac{1}{0.03} \ln\left(\frac{65}{65 - x}\right)$$

continues increasing at a faster and faster rate. This shows that as the number of signs to be learned increases, more and more time is needed to achieve the maximum mastery of 65 signs.

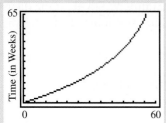

Number of signs to be mastered
$[0, 60] \times [0, 65]$
Xscl = 5 Yscl = 5

The utility's TRACE feature confirms that approximately 30 signs can be mastered in 21 weeks.

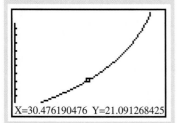

X=30.476190476 Y=21.091268425

Solving an equation containing only logarithmic expressions

1. Use properties of logarithms to rewrite each side of the equation as the logarithm of a single quantity. (Be sure that the coefficient of each logarithm is 1.)
2. Set the quantities equal to each other and solve for the variable.
3. Check each solution in the original equation, rejecting values that produce the logarithm of a negative number or the logarithm of 0.

Modeling Applications

Solving logarithmic equations enables us to work with logarithmic models and solve applied problems.

EXAMPLE 10   **A Learning Model**

The model

$$t = \frac{1}{c} \ln\left(\frac{A}{A - N}\right)$$

describes the time (t, in weeks) that it takes to achieve mastery of a portion of a task, in which

A = Maximum learning possible

N = Portion of the learning that is to be achieved

c = Constant used to measure an individual's learning style

The formula is also used to determine how long it will take chimpanzees and apes to master a task. For example, a typical chimpanzee learning sign language can master a maximum of 65 signs. How many signs can a chimpanzee master in 21 weeks if c for that chimp is 0.03?

Solution

$$t = \frac{1}{c} \ln\left(\frac{A}{A - N}\right) \qquad \text{This is the given model.}$$

$$21 = \frac{1}{0.03} \ln\left(\frac{65}{65 - N}\right) \qquad \begin{array}{l} t\,(\text{time}) = 21 \text{ weeks}, c = 0.03, \text{ and } A \\ (\text{maximum learning possible}) = 65. \text{ We} \\ \text{must solve for } N, \text{ the portion (or the} \\ \text{number of signs) to be learned.} \end{array}$$

$$0.63 = \ln\left(\frac{65}{65 - N}\right)$$

Multiply both sides of the equation by 0.03.

$$e^{0.63} = \frac{65}{65 - N}$$

Since $\log_b M = N$ is equivalent to $b^N = M$, then $\ln M = N$ is equivalent to $e^N = M$.

$$e^{0.63}(65 - N) = 65$$

Multiply both sides of the equation by $65 - N$.

$$e^{0.63}(65) - e^{0.63}N = 65$$

Apply the distributive property on the left.

$$e^{0.63}(65) - 65 = e^{0.63}N$$

Collect the term containing N on the right and all other terms on the left.

$$\frac{e^{0.63}(65) - 65}{e^{0.63}} = N$$

Solve for N by dividing both sides of the equation by $e^{0.63}$.

$$N \approx 30$$

Use a calculator.

The chimpanzee can master approximately 30 signs in 21 weeks.   ■

PROBLEM SET 4.4

Practice Problems

Solve each equation in Problems 1–26 by taking the natural logarithm on both sides. Express the answer in terms of natural logarithms. Then use a calculator to obtain a decimal approximation, correct to the nearest thousandth, for the solution.

1. $10^x = 3.91$

2. $10^x = 8.07$

3. $e^x = 5.7$

4. $e^x = 0.83$

5. $5^x = 17$

6. $19^x = 143$

7. $5e^x = 23$

8. $9e^x = 107$

9. $3e^{5x} = 1977$

10. $4e^{7x} = 10{,}273$

11. $e^{1-5x} = 793$

12. $e^{1-8x} = 7957$

13. $e^{5x-3} - 2 = 10{,}476$

14. $e^{4x-5} - 7 = 11{,}243$

15. $7^{x+2} = 410$

16. $5^{x-3} = 137$

17. $7^{0.3x} = 813$

18. $3^{x/7} = 0.2$

19. $5^x = 3^{x+1}$

20. $11^{1-x} = 4^x$

21. $5^{2x+3} = 3^{x-1}$

22. $7^{2x+1} = 3^{x+2}$

23. $e^{2x} - 3e^x + 2 = 0$

24. $e^{2x} - 2e^x - 3 = 0$

25. $e^{4x} + 5e^{2x} - 24 = 0$

26. $e^{4x} - 3e^{2x} - 18 = 0$

In Problems 27–56, solve each logarithmic equation algebraically. Check each proposed solution by direct substitution or with a graphing utility, eliminating values that produce the logarithm of a negative number or the logarithm of 0.

Problems 27–40 involve logarithmic equations containing logarithmic expressions and constants.

27. $\log_3 x = 4$

28. $\log_5 x = 3$

29. $\log_4(x + 5) = 3$

30. $\log_5(x - 7) = 2$

31. $\log_3(x - 4) = -3$

32. $\log_7(x + 2) = -2$

33. $\log_4(3x + 2) = 3$

34. $\log_2(4x + 1) = 5$

35. $\log_5 x + \log_5(4x - 1) = 1$

36. $\log_6(x + 5) + \log_6 x = 2$

37. $\log_3(x - 5) + \log_3(x + 3) = 2$

38. $\log_2(x - 1) + \log_2(x + 1) = 3$

39. $\log_2(x + 2) - \log_2(x - 5) = 3$

40. $\log_4(x + 2) - \log_4(x - 1) = 1$

Problems 41–56 involve logarithmic equations containing only logarithmic expressions.

41. $\log(x + 2) = \log(3x - 4)$

42. $\log\left(\dfrac{3x - 9}{3}\right) = \log\left(\dfrac{2x - 6}{6}\right)$

43. $\log_3 x + \log_3 2 = \log_3 6$

44. $\log_6 3x + \log_6 4 = \log_6 24$

45. $\log(9x + 2) - \log(3x - 1) = \log 4$

46. $\log_2(x + 5) - \log_2 x = \log_2 4$

47. $\log x + \log(x + 3) = \log 4$

48. $\log_3 x + \log_3(x + 6) = \log_3 27$

49. $\log_2(x + 1) + \log_2(x - 2) = \log_2(x + 6)$

50. $\ln(x + 2) + \ln(x + 6) = \ln(x + 20)$

51. $\ln(7x + 12) - \ln(x + 3) = \ln x$

52. $\ln(12 - x) - \ln(x + 2) = \ln 2x$

53. $2 \log x - \log 4 = \log 100$

54. $2 \log x - \log 5 = \log 1000$

55. $\ln(2x - 3) = 2 \ln x - \ln(x - 2)$

56. $2 \log x = \log(2 - x) + \log(4 - x)$

Application Problems

In Problems 57–60, complete the table for a savings account subjected to continuous compounding ($A = Pe^{rt}$).

	Amount Invested	Annual Interest Rate	Accumulated Amount	Time t in Years
57.	$8000	8%	Double the amount invested	
58.	$8000		$12,000	2
59.	$2350		Triple the amount invested	7
60.	$17,425	4.25%	$25,000	

In Problems 61–64, complete the table for a savings account subjected to n compoundings yearly $\left(A = P\left(1 + \dfrac{r}{n}\right)^{nt}\right)$.

	Amount Invested	Number of Compounding Periods	Annual Interest Rate	Accumulated Amount	Time t in Years
61.	$12,500	4	5.75%	$20,000	
62.	$7250	12	6.5%	$15,000	
63.	$1000	360		$1400	2
64.	$5000	360		$9000	4

65. In the Global 2000 Report to former president Jimmy Carter, the 1976 worldwide consumption of natural gas was 50 trillion cubic feet, and the growth rate for natural gas consumption was 2% per year ($k = 0.02$). At that time, the remaining known natural gas resources ultimately available were 8493 trillion cubic feet. Using these figures and the model in Example 6 (page 482), find the year the planet will be depleted of its then-known natural gas resources.

66. In the Global 2000 Report to former president Jimmy Carter, the 1976 consumption of nonrenewable energy resources (petroleum, natural gas, coal, shale oil, and uranium) was 250 quadrillion Btu, and the predicted growth rate for consumption was 2% per year ($k = 0.02$). At that time, the total remaining known resources ultimately available were 161,241 quadrillion Btu. Using

these figures and the model in Example 6 (page 482), find the year the planet will be depleted of its then-known nonrenewable energy resources.

Use the exponential growth model $A = A_0 e^{kt}$ to solve Problems 67–69.

67. These statistics were issued in 1995 by the Population Reference Bureau in Washington, D.C.:

Region	1995 Population (in billions)	Percentage of Population in 1995	Growth Rate (k)
World	5.702	100	1.5% = 0.015
More developed regions	1.169	20.5	0.2% = 0.002
Less developed regions	4.533	79.5	1.9% = 0.019

Let $t = 0$ correspond to 1995 and assume that the indicated growth rates are valid for at least 15 years. By what year (that is, how many years after 1995) will the world population be 6.37 billion? At that time, what will be the percentage of the world's population for less developed regions?

68. These statistics were issued in 1995 by the Population Reference Bureau in Washington, D.C.:

Country	1995 Population (in millions)	Growth Rate (k)
United Kingdom	58.6	0.2% = 0.002
Iraq	20.6	3.7% = 0.037

The given growth rates (percentage per year) indicate that the United Kingdom has one of the lowest growth

rates in the world and Iraq has one of the highest. Let $t = 0$ correspond to 1995 and assume that the indicated growth rates are valid for at least 15 years. By what year (that is, how many years after 1995) will the United Kingdom have a population of 59.2 million? What will be Iraq's population at that time?

69. In a report entitled *Resources and Man*, the U.S. National Academy of Sciences concluded that a world population of 10 billion "is close to (if not above) the maximum that an intensely managed world might hope to support with some degree of comfort and individual choice." For this reason, 10 billion is called the carrying capacity of Earth. In the 1969 report, world population figures for 1975 were projected at 4.043 billion (low) and 4.134 billion (high), with a relative growth rate of 1.5% per year (low) and 2% per year (high).

a. Use the low figures and the exponential growth model to estimate the year in which the carrying capacity might be reached.

b. Repeat part (a) using the high figures.

70. The model $V(t) = C\left(1 - \dfrac{2}{n}\right)^t$ describes the value of a piece of equipment after t years, where C is the original cost and n is the life expectancy in years. How long will it take a piece of equipment that costs $50,000 and has an expected life of 20 years to be worth $20,000?

Chemists have defined the pH (hydrogen potential) of a solution by $pH = -\log[H^+]$, where $[H^+]$ represents the concentration of the hydrogen ion in moles per liter. The pH scale ranges from 0 to 14, depending on a solution's acidity or alkalinity. Values below 7 indicate progressively greater acidity, while those above 7 are progressively more alkaline. Use this model to answer Problems 71–72.

71. Normal, unpolluted rain has a pH of about 5.6. An environmental concern involves the destructive effects of acid rain, which is caused primarily by sulfur dioxide emissions. The most acidic rainfall ever had a pH of 2.4. What was the hydrogen ion concentration of that rainfall?

72. The pH of human blood ranges between 7.37 and 7.44. What are the corresponding bounds for hydrogen ion concentration?

73. We have seen that the magnitude R on the Richter scale of an earthquake of intensity I is given by

$$R = \log\frac{I}{I_0}$$

where I_0 is the intensity of a barely felt zero-level earthquake.

a. Northern California's 1989 earthquake measured 7.1 on the Richter scale. How many times more intense was this earthquake than a zero-level earthquake?

b. One of the largest earthquakes ever recorded was the 8.9 magnitude quake that struck Japan in 1933. How many times more intense was the earthquake than a zero-level earthquake?

c. How many times greater in intensity was Japan's 1933 earthquake than California's 1989 earthquake?

d. Earthquakes release an enormous amount of energy. The model $\log E = 11.4 + 1.5R$ describes the relationship between the energy released by an earthquake (measured in ergs) and the earthquake's magnitude on the Richter scale. How many ergs of energy were released by California's 1989 earthquake and Japan's 1933 earthquake?

74. The total TNT equivalent of all nuclear weapons is 25,000 megatons. Given that 1 megaton = 1 million tons and one ton of TNT releases 4.2×10^6 ergs of energy, use the formula in Problem 73 (d) to find the Richter scale reading for the simultaneous release of all nuclear weapons.

75. The brightness of stars, as seen by the naked eye and measured in magnitudes M, is given by

$$M = 6 - 2.5 \log \frac{I}{I_0}$$

where

I = the intensity of light from the star

and

I_0 = the intensity of light from a just-visible star

The ancient Greek astronomer Ptolemy used six categories for the magnitude M of a star: 6 for the dimmest stars and 1 for the brightest stars.

a. How many times more intense is the light from a star of magnitude 1 than the light from a just-visible star?

b. How many times more intense is the light from a star of magnitude 4 than the light from a just-visible star?

c. How many times more intense is the light from a star of magnitude 1 than the light from a star of magnitude 4?

Ptolemy's six categories for the brightness of stars has been expanded to negative magnitudes for bright bodies such as the sun and to positive magnitudes for stars visible only with telescopes. Two stars of magnitudes M_1 and M_2 and intensity I_1 and I_2 are related by

$$M_2 - M_1 = 2.5 \log \frac{I_1}{I_2}.$$

Use this model to answer Problems 76–77.

76. How many times more intense is the light from Sirius (magnitude = −1.5) than the light from Polaris (magnitude = 2)?

77. How many times more intense is the light from the sun (magnitude = −26) than the light from the full moon (magnitude = −13)?

True–False Critical Thinking Problems

78. Which one of the following is true?

a. The solution of the equation $3^{x+2} = 5$ must be a positive number.

b. A reasonable estimate for the solution of the equation $8^{0.4x} = 5$ is 2.

c. A reasonable estimate for the solution of the equation $e^{1-4x} = 2$ is 10.

d. Examples of exponential equations include $10^x = 5.71$, $e^x = 0.72$, and $x^{10} = 5.71$.

79. Which one of the following is true?

a. If $\log(x + 3) = 2$, then $e^2 = x + 3$.

b. If $\log(7x + 3) - \log(2x + 5) = 4$, then in exponential form $10^4 = (7x + 3) - (2x + 5)$.

c. If $x = \dfrac{1}{k} \ln y$, then $y = e^{kx}$.

d. An earthquake measuring 8.6 on the Richter scale is twice as intense as one measuring 4.3 on the Richter scale.

80. Which one of the following is true?

a. Models involving logarithmic functions can be expressed in exponential form and vice versa.

b. If $\log(x + 2) + \log(x + 6) = \log(x + 20)$, then $(x + 2) + (x + 6) = x + 20$.

c. $\log(-4) + \log(-5) = \log 20$

d. If $\log 4 + \log(x + 1) = 2 \log x + \log 3$, then $4(x + 1) = (2x)(3)$.

Technology Problems

In Problems 81–84, use a graphing utility to find the solution set for each exponential equation, expressing solutions to the nearest tenth.

81. $\dfrac{e^x + e^{-x}}{2} = 2$

82. $\dfrac{3^x - 3^{-x}}{2} = 5$

83. $3^x = 2x + 3$

84. $5^x = 3x + 4$

85. The model $W = 2600(1 - 0.51e^{-0.075t})^3$ describes the weight (in kilograms) of a female African elephant at age t (in years). Use a graphing utility to graph the weight function. Then trace along the curve to estimate the age of an adult female elephant weighing 1800 kilograms.

Renée Young / Photo Researchers, Inc.

86. The model $h = 79.041 + 6.39x - e^{3.261-0.993x}$ is used to predict the height h in centimeters for a preschooler who is x years old, $\frac{1}{4} \le x \le 6$. Graph the model and trace along the graph to predict the age for a preschooler with a height of 75.8 centimeters. What difficulties would you encounter if you attempted to make this prediction using algebraic methods?

87. The model $P = 145e^{-0.092t}$ describes a runner's pulse rate (P, in beats per minute) t minutes after a race ($0 \le t \le 15$). Graph the equation with a graphing utility. Trace along the graph and determine after how many minutes the runner's pulse rate will be 70 beats per minute.

Use a graphing utility to solve each logarithmic equation in Problems 88–91. Round your solutions to two decimal places.

88. $\ln(2x + 1) - \ln(x - 3) = 2\ln 3$

89. $\log(x - 1) = x - 5$

90. $(\log x)^2 - 3x = 1$

91. $\log_3 x + 7 = 4 - \log_5 x$

92. The model $P(t) = 95 - 30 \log_2 t$ is a hypothetical formula describing the percentage of students ($P(t)$) who could recall the important features of a lecture as a function of time, where t represents the number of days that have elapsed since the lecture was given.

  a. Enter the function on your graphing utility as $y_1 = 95 - 30 \dfrac{\ln x}{\ln 2}$, using the change-of-base property, with

Xmin = 0, Xmax = 10, Xscl = 1,

Ymin = 0, Ymax = 100, Yscl = 10

  b. Use the graph to solve the equation

$95 - 30 \log_2 t = 0.$

Describe what the solution means in practical terms.

  c. Use the graph to solve the equation

$95 - 30 \log_2 t = 50.$

Describe what the solution means in practical terms.

Writing in Mathematics

93. The exponential equation $2^{4-5x} = 8$ can be solved by taking the natural logarithm on both sides. Since $2^3 = 8$, the equation can also be solved by writing each side with a common base, $2^{4-5x} = 2^3$, and then equating exponents, resulting in $4 - 5x = 3$. Solve the equation using both methods and then compare and contrast the two methods. Explain why we did not use the second method in solving the exponential equations that appeared in this section.

94. Describe similarities and differences in the procedures needed to solve $\log_2 x + \log_2(x - 7) = 3$ and $\log_2 x + \log_2(x - 7) = \log_2 8$.

95. In Example 6 we used the model

$$A = \frac{A_0}{k}(e^{kt} - 1)$$

to describe the amount of nonrenewable energy resources from time $t = 0$ to time t, where A_0 is the amount of the resource consumed during the year $t = 0$ and k is the relative growth rate of annual consumption. Using principles for solving logarithmic equations, this model can be solved for t, resulting in

$$t = \frac{\ln\left(\frac{kA}{A_0} + 1\right)}{k}.$$

Under what applied conditions would it be easier to work with the logarithmic form of the model? Which form of the model would be easier to work with to solve Example 6?

Critical Thinking Problems

96. If $4000 is deposited into an account paying 3% interest compounded annually and at the same time $2000 is deposited into an account paying 5% interest com- pounded annually, after how long will the two accounts have the same balance?

Some populations exhibit an increasing rate of growth, followed by a declining rate as the population levels off and approaches the maximum level that can be supported by the habitat. Illustrated by the growth in the figure shown, this growth is described by the logistic model

$$N = \frac{N_0 N_{max}}{N_0 + (N_{max} - N_0)e^{-kt}}$$

where

$N_0 =$ the initial population size

$N_{max} =$ the maximum population that can be supported by the habitat

$N =$ the population at time t

$k =$ the population rate of growth

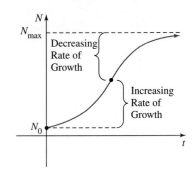

Use this model to solve Problem 97.

97. A student carrying a contagious influenza virus enrolls at a school with 6000 students. The spread of the virus follows the logistics growth model with $k = 0.3$ and time t measured in days. In how many days will half of the students be infected?

98. Use the model $A = \frac{A_0}{k}(e^{kt} - 1)$ and show that

$$t = \frac{\ln\left(\frac{kA}{A_0} + 1\right)}{k}.$$

99. We have seen that the perceived loudness of a sound (D, in decibels) at the eardrum is described by

$$D = 10 \log \frac{I}{I_0}$$

where

$I =$ the intensity of the sound (in watts per meter^2)

$I_0 =$ the intensity of a sound barely audible to the human ear

A worker uses a machine that produces 90 decibels. A second machine in the same area produces 80 decibels. Union regulations prohibit exposure to sound levels greater than 100 decibels. What is the combined decibel level of the two machines? Does this situation violate union regulations? (Hint: In measuring sound, we add intensities but not decibels.)

Solve the logarithmic equations in Problems 100–103 algebraically. Check each proposed solution by direct substitution or with a graphing utility.

100. $(\ln x)^2 = \ln x^2$

101. $(\ln x)^2 - 3 \ln x + 1 = 0$

102. $\log x(2 \log x + 1) = 6$

103. $\ln (\ln x) = 0$

Group Activity Problem

104. This problem is appropriate for discussion in a small group.

India is currently one of the world's fastest-growing countries, with a growth rate of about 2.6% annually. Assuming that the relative growth rate is constant, the population will reach 1.5 billion sometime in the year 2010. Treating the year 1974 as year 0, the growth model is $y = 574,220,000e^{0.026t}$.

One problem with this model is that nothing can grow exponentially indefinitely. Although India grew at an average rate of 2.6% per year from 1974 to 1984, the model is based on the assumption that India's growth rate will remain the same at least until the year 2010.

Is this a likely assumption? Can you suggest factors that influence population growth rates for India and other countries? What factors might limit the size of a population? Can you suggest factors that might lead to a

greater population growth rate than the one predicted by an exponential growth model? How might these factors be accounted for in a mathematical model predicting population growth or in the model's graph? (In answering this last question, consider the model

$$y = \frac{5000}{1 + e^{-t/6}}$$

which describes the population y of a certain species t years after it is introduced into a new habitat.)

If the assumption of an indefinite growth at a constant rate is unrealistic, why do you think models based on this assumption are widely used? What does this tell you about making predictions based on mathematical models?

SECTION 4.5 Modeling with Exponential and Logarithmic Functions

Solutions Manual   Tutorial   Video 8

Objectives

1 Model exponential growth and decay.
2 Model cooling and heating times.
3 Fit exponential functions to data.
4 Fit logarithmic functions to data.

In this section, we continue our work with applications of exponential and logarithmic functions. All examples and problems involve modeling actual real world situations.

Growth and Decay Models

I Model exponential growth and decay.

In Section 4.1, we introduced the exponential growth and decay models

$$A = A_0 e^{kt}$$

where A_0 is the original amount (or size) of a growing (or decaying) entity at $t = 0$, A is the amount at time t, and k is a growth ($k > 0$) or decay ($k < 0$) constant. Some situations require us to use given data in growth and decay models to determine k. Once this value is computed, the model $A = A_0 e^{kt}$ can be used to solve problems and make predictions. This idea is illustrated in our first two examples.

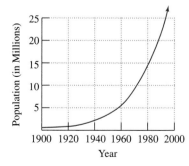

Figure 4.37

Mexico City's population has grown exponentially.

EXAMPLE I Modeling Mexico City's Growth

The graph in Figure 4.37 shows the growth of the Mexico City metropolitan area from 1900 through 2000. In 1970, the population of Mexico City was 9.4 million and by 1990 it had grown to 20.2 million.

a. Find the exponential growth function that models the data.
b. By what year will the population reach 100 million?

Solution

a. We use the exponential growth model

$$A = A_0 e^{kt}$$

in which t is the number of years since 1970. This means that 1970 corresponds to $t = 0$. At that time there were 9.4 million inhabitants, so we substitute 9.4 for A_0 in the growth model.

$$A = 9.4 e^{kt}$$

We are given that there were 20.2 million inhabitants in 1990. Since 1990 is 20 years after 1970, when $t = 20$ the value of A is 20.2. Substituting these numbers into the growth model will enable us to find k, the growth constant or the growth rate.

$A = 9.4 e^{kt}$	Use the growth model with $A_0 = 9.4$
$20.2 = 9.4 e^{k \cdot 20}$	When $t = 20$, $A = 20.2$. Substitute these numbers into the model.
$e^{20k} = \dfrac{20.2}{9.4}$	Isolate the exponential term by dividing both sides by 9.4. We also reversed the sides.
$\ln e^{20k} = \ln \dfrac{20.2}{9.4}$	Take the natural logarithm on both sides.
$20k = \ln \dfrac{20.2}{9.4}$	Simplify the left side using $\ln e^x = x$
$k = \dfrac{\ln \dfrac{20.2}{9.4}}{20} \approx 0.038$	Divide both sides by 20 and solve for k.

The exponential growth function is

$$A = 9.4 e^{0.038t}$$

where t is measured in years since 1970.

b. To find the year in which the population will reach 100 million, we substitute 100 for A in the model from part (a) and solve for t.

$A = 9.4 e^{0.038t}$	This is the model from part (a).
$100 = 9.4 e^{0.038t}$	Substitute 100 for A.
$e^{0.038t} = \dfrac{100}{9.4}$	Divide both sides by 9.4.
$\ln e^{0.038t} = \ln \dfrac{100}{9.4}$	Take the natural logarithm on both sides.
$0.038t = \ln \dfrac{100}{9.4}$	Simplify on the left using $\ln e^x = x$.
$t = \dfrac{\ln \dfrac{100}{9.4}}{0.038} \approx 62$	Solve for t by dividing both sides by 0.038.

George Segal "Rush Hour" 1983, plaster, acrylic paint, 72 × 96 × 96 in. (182.9 × 243.8 × 243.8 cm). Courtesy Sidney Janis Gallery, New York. © George Segal / Licensed by VAGA, New York 1998.

Since 62 is the number of years after 1970, the model indicates that the population of Mexico City will reach 100 million by 1970 + 62, or in the year 2032. ∎

EXAMPLE 2   Carbon-14 Dating: The Dead Sea Scrolls

Carbon-14 (C-14) decays exponentially with a half-life of approximately 5715 years, meaning that after 5715 years a given amount of C-14 will have decayed to half the original amount.

a. Find the exponential decay model for C-14.
b. In 1947 earthenware jars containing what are known as the Dead Sea Scrolls were found by an Arab Bedouin herdsman. Analysis indicated that the scroll wrappings contained 76% of their original C-14. Estimate the age of the Dead Sea Scrolls.

Study tip

The half-life of a radioactive substance is the time required for half of a given sample to disintegrate. Half-life is a property of the substance and does not depend on its size.

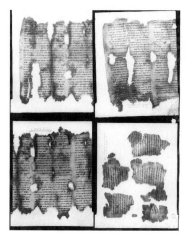

Shrine of the Book, a portion of the Dead Sea Scrolls. Robert Slosser / Israel Museum

Solution

a. We use the exponential decay model and the information about the half-life of C-14 to find k.

$$A = A_0 e^{kt}$$ This is the exponential decay model.

$$\frac{A_0}{2} = A_0 e^{k5715}$$ After 5715 years ($t = 5715$), $A = \dfrac{A_0}{2}$ (since the amount present (A) is half the original amount (A_0).)

$$\frac{1}{2} = e^{5715k}$$ Divide both sides of the equation by A_0.

$$\ln \frac{1}{2} = \ln e^{5715k}$$ Take the natural logarithm of both sides.

$$\ln \frac{1}{2} = 5715k$$ $\ln e^x = x$

$$k = \frac{\ln \dfrac{1}{2}}{5715} \approx -0.000121$$ Solve for k.

Substituting for k in the decay model, the model for C-14 is $A = A_0 e^{-0.000121t}$.

b.
$$A = A_0 e^{-0.000121t}$$ This is the decay model for C-14.

$$0.76 A_0 = A_0 e^{-0.000121t}$$ A, the amount present, is 76% of the original amount, so $A = 0.76 A_0$.

$$0.76 = e^{-0.000121t}$$ Divide both sides of the equation by A_0.

$$\ln 0.76 = \ln e^{-0.000121t}$$ Take the natural logarithm on both sides.

$$\ln 0.76 = -0.000121t$$ $\ln e^x = x$

$$t = \frac{\ln 0.76}{-0.000121} \approx 2268$$ Solve for t.

The Dead Sea Scrolls are approximately 2268 years old plus the number of years between 1947 and the current year. ∎

ENRICHMENT ESSAY

Carbon Dating and Artistic Development

Carbon dating is a method for estimating the age of organic material. Until recently, the oldest prehistoric cave paintings known were from the Lascaux cave in France. Charcoal from the site was analyzed, and the paintings were estimated to be 15,505 years old.

The artistic community was electrified by the discovery in 1995 of spectacular cave paintings in a limestone cavern near Lascaux. Carbon dating showed that the images, created by artists of remarkable talent, were 30,000 years old, making them the oldest cave paintings ever found. The artists seemed to have used the cavern's natural contours to heighten a sense of perspective. The quality of the painting suggests that the art of early humans did not mature steadily from primitive to sophisticated in any simple linear fashion.

Jean Clottes, France's foremost expert on prehistoric cave art, was moved by the sophisticated imagery of the 30,000-year-old work. "I remember standing in front of the paintings and being profoundly moved by the artistry," he exclaimed. "Tears were running down my cheeks. I was witnessing one of the world's great masterpieces."

Jean-Marie Chauvet / Sygma

2 Model cooling and heating times.

Alexis Rockman "Thaw" 1996. Envirotex, soil, oil & spray paint, glaciated gravel, modelling paste, clay, plaster bones, digitized photograph, plastic plants, plexiglas, kosher salt on two wood panels. 56 in. H × 88 in. W × 3.5 in. Courtesy of Jay Gorney Modern Art, New York. Photographer: Oren Slor, New York.

Modeling Cooling & Heating Times

Over a period of time, a cup of hot coffee cools to the temperature of the surrounding air. The rate of cooling of an object is proportional to the difference between the object's temperature and the temperature of the surrounding medium. The difference between the two temperatures decreases exponentially, given by

$$T = C + (T_0 - C)e^{-kt}$$

where t is the time it takes for an object to cool from an initial temperature T_0 to a temperature T, C is the temperature of the surrounding medium, and k is a positive constant that is associated with the cooling object. This model, called Newton's Law of Cooling, applies to warming as well. Earlier we encountered Newton's Law in which the formula was solved for t.

EXAMPLE 3 Using Newton's Law of Cooling

A cake removed from the oven has a temperature of 210°F and is left to cool in a room that has a temperature of 70°F. After 30 minutes the temperature of the cake is 140°F.

a. What is the temperature of the cake after 40 minutes?
b. When will the temperature of the cake be 90°F?
c. Illustrate the problem by using a graphing utility to sketch the temperature function.

Solution

a.

$$T = C + (T_0 - C)e^{-kt}$$

Use Newton's cooling model. Our first step will be to determine k.

$$140 = 70 + (210 - 70)e^{-k(30)}$$

Since the cake's temperature after 30 minutes is 140°F, $T = 140$, $C = 70$ (room temperature), $T_0 = 210$ (initial temperature), and $t = 30$ (30 minutes to cool to 140°F).

$$70 = 140e^{-30k}$$

First isolate the exponential term by substracting 70 from both sides.

$$0.5 = e^{-30k}$$

Now divide both sides by 140. The exponential term is isolated.

$$\ln 0.5 = \ln e^{-30k}$$

Take the natural logarithm on both sides.

$$\ln 0.5 = -30k$$

$\ln e^x = x$

$$\frac{\ln 0.5}{-30} = k$$

Solve for k.

Now we substitute this value for k into Newton's formula.

$$T = C + (T_0 - C)e^{-kt}$$

Once again, this is Newton's cooling model.

$$T = 70 + (210 - 70)e^{-\left(\frac{\ln 0.5}{-30}\right)t}$$

$C = 70$ (room temperature), $T_0 = 210$ (cake's initial temperature), and $k = \dfrac{\ln 0.5}{-30}$ (from above).

$$= 70 + (210 - 70)e^{\frac{\ln 0.5}{30}t}$$

We can now answer the question and determine the cake's temperature after 40 minutes.

$$T = 70 + (210 - 70)e^{\frac{\ln 0.5}{30}(40)}$$

Substitute 40 for t.

$$\approx 126$$

Use a calculator.

After 40 minutes the temperature of the cake will be approximately 126°F.

b. The temperature of the cake will be 90°F when

$$70 + (210 - 70)e^{\frac{\ln 0.5}{30}t} = 90$$

Substitute 90 for T in the model and solve for t (time).

$$140e^{\frac{\ln 0.5}{30}t} = 20$$

Once again, isolate the exponential term. Subtract 70 from both sides.

$$e^{\frac{\ln 0.5}{30}t} = \frac{1}{7}$$

Divide both sides by 140. The exponential term is isolated.

$$\ln e^{\frac{\ln 0.5}{30}t} = \ln \frac{1}{7}$$

Take the natural logarithm on both sides.

$$\frac{\ln 0.5}{30}t = \ln \frac{1}{7}$$

$\ln e^x = x$

$$t = \frac{\ln \dfrac{1}{7}}{\dfrac{\ln 0.5}{30}}$$

Divide both sides by $\dfrac{\ln 0.5}{30}$.

$$\approx 84$$

Use a calculator.

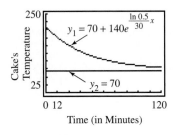

Figure 4.38

The graph of the cake's cooling model using a [0, 120] by [0, 250] viewing rectangle and Xscl = 12, Yscl = 25

3 Fit exponential functions to data.

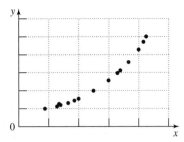

Figure 4.39

Data points suggest an exponential model.

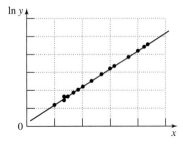

Figure 4.40

Data points suggest a linear model.

The temperature of the cake will be 90°F after approximately 84 minutes.

c. Figure 4.38 shows the graph of the cooling model, entered as

$$Y_1 = 70 + 140e^{\wedge}((\ln 0.5) X/30).$$

Also graphed is $y_2 = 70$, which serves as a horizontal asymptote. Why is $y = 70$ the horizontal asymptote?   ■

Fitting Exponential Functions to Data

Throughout this chapter, we have been working with models that were given. However, we can create models that fit data by observing patterns in the graph of the data points.

For example, the data in Figure 4.39 suggest that we model it with an exponential function. Since $\ln e^y = y$, if ordered pairs (x, y) lie on an exponential curve, then the ordered pairs $(x, \ln y)$ will lie on a straight line. This is shown in Figure 4.40, where we transformed the data in Figure 4.39 by taking the natural logarithm of each y value.

These observations give us a procedure for fitting exponential functions to data such as the points shown in Figure 4.39.

Modeling with exponential functions

1. Represent all data points using (x, y).
2. Take the natural logarithm of each y-coordinate.
3. Write an equation of the line that best fits the data points $(x, \ln y)$. The equation for this line, called the regression line, can be obtained with a graphing utility.
4. Since exponential and logarithmic functions are inverses, then $e^{\ln y} = y$. Use this relationship to find y by taking the equation in step 3 and writing each side as an exponent with base e.

EXAMPLE 4   Fitting an Exponential Model to Data

The data below indicate world population (in billions) for five years.

Year	World Population (in billions)
1950	2.6
1960	3.1
1970	3.7
1980	4.5
1989	5.3

Find a model for the data. Then use the model to predict world population in the year 2020.

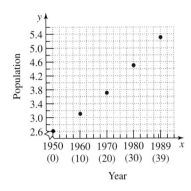

Figure 4.41

The graph of world population over time suggests an exponential model.

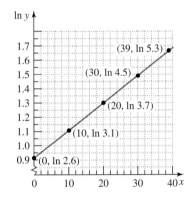

Figure 4.42

Data points suggest a linear model.

Solution

The graph of the five data points in Figure 4.41 has a shape that suggests modeling with an exponential function.

Letting $x = 0$ represent 1950, we transform the points by taking the natural logarithm of each y-coordinate.

Year (x)	World Population y	In y
1950 (0)	2.6	0.9555
1960 (10)	3.1	1.1314
1970 (20)	3.7	1.3083
1980 (30)	4.5	1.5041
1989 (39)	5.3	1.6677

The five points for $(x, \ln y)$ suggest using a linear model, as shown in Figure 4.42. Using the linear regression option on a graphing utility, the equation of the line that best fits the $(x, \ln y)$ transformed data is

$$\ln y = 0.0183x + 0.9504.$$

Since we are interested in a model with y on the left, we use the property for inverse functions

$$e^{\ln y} = y$$

and write both sides of the equation as an exponent on base e.

$$e^{\ln y} = e^{0.0183x + 0.9504}$$

$$e^{\ln y} = e^{0.0183x} \cdot e^{0.9504} \qquad b^{M+N} = b^M \cdot b^N$$

$$y = e^{0.0183x} \cdot 2.5867 \qquad e^{\ln y} = y. \text{ Use a calculator to approximate } e^{0.9504}.$$

$$= 2.5867e^{0.0183x} \qquad \text{Apply the commutative property on the right.}$$

The exponential model that fits the given data is

$$y = 2.5867e^{0.0183x}.$$

Since we often use t as the variable standing for time, we can equivalently write the model as

$$y = 2.5867e^{0.0183t} \quad \text{or} \quad f(t) = 2.5867e^{0.0183t}.$$

We are now ready to use the model to predict population in the year 2020.

$$f(t) = 2.5867e^{0.0183t} \qquad \text{This is one of the forms of the model obtained from above.}$$

$$f(70) = 2.5867e^{0.0183(70)} \qquad \text{The year 2020 is 70 years after 1950, so substitute 70 for } t.$$

$$\approx 9.3 \qquad \text{Use a calculator.}$$

Using the model that we constructed from the data, predicted world population in the year 2020 is approximately 9.3 billion. ∎

4 Fit logarithmic functions to data.

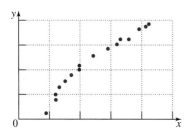

Figure 4.43

Data points suggest a logarithmic model.

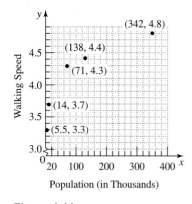

Figure 4.44

The graph of walking speed as a function of population suggests a logarithmic model.

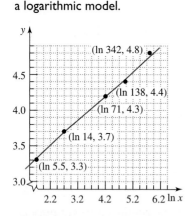

Figure 4.45

Data points suggest a linear model.

Fitting Logarithmic Functions to Data

Suppose that the data we are attempting to model falls into a pattern like the one in Figure 4.43. The graph is increasing, although increasing at a slower and slower rate, but does not seem to be leveling out to approach a horizontal asymptote. The shape suggests a logarithmic function, so we will attempt to model the data with such a curve. We can fit a logarithmic model to a set of points of the form (x, y) by fitting a linear model to points of the form $(\ln x, y)$. This idea is illustrated in our next example.

EXAMPLE 5 **Fitting a Logarithmic Model to Data**

Earlier in the chapter we encountered a model relating the size of a city with the average walking speed, in feet per second, of pedestrians. A study collected the following data.

Population (in thousands)	Walking Speed (in feet per second)
5.5	3.3
14	3.7
71	4.3
138	4.4
342	4.8

Source: Mark and Helen Bornstein, "The Pace of Life," *Nature*, 259 (19 Feb. 1976) 557–559

Find a model that fits the given data.

Solution

The graph of the five data points in Figure 4.44 has a shape that suggests modeling the data with a logarithmic function. Represent each data value by (x, y) and transform the points by taking the natural logarithm of each x coordinate.

x	$\ln x$	y
5.5	1.7047	3.3
14	2.6391	3.7
71	4.2627	4.3
138	4.9273	4.4
342	5.8348	4.8

The five points for $(\ln x, y)$ are graphed in Figure 4.45. The points approximately fall along a line, suggesting a linear model.

Using the linear regression option on a graphing utility, the equation of the regression line that models the transformed $(\ln x, y)$ data is

$$y = 0.3525 \ln x + 2.735.$$

This is the logarithmic model that fits the data. Compare this model with the one in Example 5 on pages 456–457. ∎

tudy tip

If any data values fall approximately along a line, the equation of the regression line (or the line of best fit) can be obtained without a graphing utility using the following formulas.

Regression Line: $y = mx + b$

$$m = \frac{n(\Sigma xy) - (\Sigma x)(\Sigma y)}{n(\Sigma x^2) - (\Sigma x)^2}$$

$$b = \frac{(\Sigma y)(\Sigma x^2) - (\Sigma x)(\Sigma xy)}{n(\Sigma x^2) - (\Sigma x)^2}$$

where

n = the number of data points (x, y)

Σx = the sum of the x-values

Σy = the sum of the y-values

$\Sigma (xy)$ = the sum of the product of x and y in each pair

Σx^2 = the sum of the squares of the x-values

$(\Sigma x)^2$ = the square of the sum of the x-values

Summary of curve-fitting procedures

1. If a graph of the data points appears to follow an exponential pattern, fit an exponential model to the set of points of the form (x, y) by fitting a *linear* model to points of the form $(x, \ln y)$. Write the equation of the regression line for $(x, \ln y)$ and then exponentiate both sides to find the model for the data points (x, y).

2. If a graph of the data points appears to follow a logarithmic model, fit a logarithmic model to the set of points of the form (x, y) by fitting a *linear* model to points of the form $(\ln x, y)$. The model for the data points (x, y) is the equation of the regression line for $(\ln x, y)$.

sing technology

Using the statistical menu of a graphing utility, you can immediately draw data points as coordinates. The resulting drawing is called a *scatter plot* of the data. The shape of the scatter plot determines what sort of function best fits the data. Using a graphing utility, it is not necessary to first linearize the data and write an equation for a regression line. Instead, you can use the utility's EXPonential REGression option ($y = ab^x$) to fit exponential functions to data, or its logarithmic regression (LnReg) option ($y = a + b \ln x$) to fit logarithmic functions to data. By telling your graphing utility the regression option you want to use to model the data, the utility will print the equation of the function for you.

Use your graphing utility's exponential regression option to model the data in Example 4 and its logarithmic regression option to model the data in Example 5.

In this era of technology, since graphing utilities have capabilities that fit linear and nonlinear functions to data, the process of creating mathematical models involves decision-making more than computation.

Study tip

When using a graphing utility to model data, first draw a scatter diagram to obtain a general picture for the shape of the data. If necessary, fit several models to the data. The number r that appears with each model is called the *correlation coefficient* and is a measure of how well the model fits the data. The value of r is such that $0 \le |r| \le 1$. The best model is the one which yields the value r for which $|r|$ is closest to 1.

PROBLEM SET 4.5

Practice and Application Problems

Use the exponential growth and decay model $A = A_0 e^{kt}$, where A_0 is the original amount (or size) of a growing (or decaying) entity at $t = 0$, A is the amount at time t, and k is a growth ($k > 0$) or decay ($k < 0$) constant to answer Problems 1–5.

1. Through the end of 1991, 200,000 cases of AIDS in the United States had been reported to the Center for Disease Control. By the end of 1995, the number had grown to 513,000. Let $t = 0$ correspond to the year 1991.

a. Since 1991 corresponds to $t = 0$, at that time there were 200,000 cases. This means that A_0, the original number in the growth model, is 200,000. Now use the fact that there were 513,000 cases by 1995 (when $t = 4$, $A = 513,000$) to find k, the growth rate or the growth constant.

b. Substitute the values for A_0 and k into the exponential growth model and write the function that models the data.

c. Using this model, in what year will the cumulative number of AIDS cases in the United States reach one million?

d. The model in this problem and your answer in part (c) is based on uninhibited exponential growth. Is this a realistic assumption?

e. The World Health Organization makes predictions about the number of AIDS cases based on a compromise between a conservative linear model and an uninhibited exponential growth model. Use the data in this problem, shown again in the table, to write the point-slope formula and the slope-intercept formula for the line modeling the data. Then use your linear model to predict the year in which the cumulative number of AIDS cases will reach one million.

x (Years after 1991)	y (Cumulative Reported AIDS Cases)
0	200,000
4	513,000

f. What year would be a good compromise between the predictions made by the exponential and the linear models to determine when the number of AIDS cases in the United States will reach one million?

Frank Moore "Hospital" 1992, oil on wood with frame and attachments, 49 × 58 in. (124.5 × 147.3 cm) SW 92637. Private Collection. Courtesy Sperone Westerwater, New York.

2. The 1940 population of New York was 13,479,142 and in 1950 the population was 14,830,192. Let $t = 0$ correspond to the year 1940.

a. Since 1940 corresponds to $t = 0$, at that time there were 13,479,142 people in New York. This means that A_0, the original amount in the growth model, is 13,479,142. Now use the fact that there were 14,830,192 people in 1950 (when $t = 10$, $A = 14,830,192$) to find k, the growth rate for New York during this decade.

b. Substitute the values for A_0 and k into the exponential growth model and write the function that models New York's population t years after 1940.

c. Use the model in part (c) to determine New York's 1994 population ($t = 54$) if the growth rate from 1940 through 1950 had remained in effect from 1940 through 1994. How does this compare with the actual 1994 population of 18,169,000?

d. What are some of the factors that might account for the difference between the number predicted by the exponential growth model and the actual population?

3. Strontium-90 is a waste product from nuclear fission reactors. Its half-life is 28 years.

a. Show that the decay model for strontiun-90 is $A = A_0e^{-0.024755t}$.

b. Strontium-90 in the atmosphere enters the food chain and is absorbed into our bones. As a consequence of fallout from atmospheric nuclear tests, we all have a measurable amount of strontium-90 in our bones. Suppose a nuclear accident occurs and releases 60 grams of strontium-90 into the biosphere. How long will it take for strontium-90 to decay to a level of 10 grams?

c. Graph (by hand or with a graphing utility) the model in part (a) with $A_0 = 60$ for $0 \leqslant t \leqslant 112$ and $0 \leqslant A \leqslant 60$. Show the solution of part (b) on the graph.

4. The August 1978 issue of *National Geographic* described the 1964 find of dinosaur bones of a newly discovered dinosaur weighing 170 pounds, measuring 9 feet, and having a 6-inch claw on one toe of each hind foot. The age of the dinosaur, called *Deinonychus* ("terrible claw"),was estimated using potassium-40 dating of rocks surrounding the bones.

a. Potassium-40 decays exponentially with a half-life of approximately 1.31 billion years, meaning that after 1.31 billion years a given amount of potassium-40 will have decayed to half the original amount. Use this information to show that the decay model for potassium-40 is given by $A = A_0e^{-0.52912t}$, where t is given in billions of years.

b. Analysis of the surrounding rock indicated that 94.5% of the original amount of potassium-40 was still present. Let $A = 0.945A_0$ in the model in part (a) and estimate the age of the bones of *Deinonychus*.

5. A bird species in danger of extinction has a population that is decreasing exponentially $(A = A_0e^{kt})$. Five years ago the population was at 1400 and today only 1000 of the birds are still alive. Once the population drops below 100, the situation will be irreversible. When will this happen?

Use Newton's model $T = C + (T_0 - C)e^{-kt}$, where t is the time it takes for an object to cool from an initial temperature T_0 to a temperature T, C is the temperature of the surrounding medium, and k is a positive constant that is associated with the cooling object, to answer Problems 6–7.

6. A bottle of juice has a temperature of 70°F and is left to cool in a refrigerator that has a temperature of 45°F. After 10 minutes, the temperature of the juice is 55°F.

a. What is the temperature of the juice after 15 minutes?

b. When will the temperature of the juice be 50°F?

c. Illustrate the problem by using a graphing utility to sketch the temperature function. According to the graph, will the temperature of the juice ever reach 45°F? Explain.

7. A pie removed from the oven has a temperature of 375°F and is left to cool in a room that has a temperature of 72°F. After 60 minutes, the temperature of the pie is 75°F.

a. What is the temperature of the pie after 30 minutes?

b. When will the temperature of the pie be 250°F?

c. Illustrate the problem by using a graphing utility to sketch the temperature function. According to the graph, will the temperature of the pie ever reach 72°F? Explain.

8. The intensity of light entering water is reduced according to the model $I = I_0e^{-kd}$, where I is the intensity at a depth of d feet, I_0 is the intensity at the surface, and k is a constant called the *constant of extinction*. In a body of water, half the surface light remained at a depth of 26 feet. Find the depth at which 10% of the surface light remains.

Suppose a set of data in the form (x, y) is graphed as a set of points. The resulting shape of the graph suggests modeling with an exponential function. The natural logarithm of each y-coordinate is taken and an equation of the line that best fits the data *points $(x, \ln y)$ is obtained. The equation of this line is given in Problems 9–10. Write each side of the equation as an exponent with base e, simplify, and find the exponential function that fits the original data.*

9. $\ln y = 0.3782 + 1.0745x$

10. $\ln y = -1.2074 + 0.873x$

11. The amount of carbon dioxide (CO_2) in the atmosphere has increased by over 25% of its original amount since 1900, as a result of the burning of oil and coal. The burning of these fossil fuels results in a buildup of gases and particles that can trap heat and raise the planet's temperature. The resultant gradually increasing temperature is called the greenhouse effect. Carbon dioxide accounts for about half of the warming.

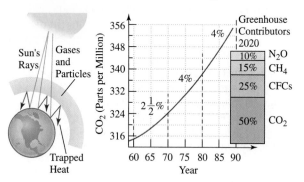

The data below shows the average CO_2 concentration in parts per million (ppm) x years after 1900.

x (Year)	y (Average CO_2 Concentration in ppm)
1900 (0)	280
1950 (50)	310
1975 (75)	331
1980 (80)	338
1988 (88)	351

a. Take the natural logarithm of each y-value.

b. Graph the five data points $(x, \ln y)$. Notice that the points lie nearly on a straight line, suggesting the use of a linear model.

c. Use the linear regression option on a graphing utility or the procedure outlined in the Study Tip on page 501 to write the equation of the line that best fits the $(x, \ln y)$ transformed data.

d. Write both sides of the equation in part (c) as exponents on base e. Simplify and obtain a model for

CO_2 concentration in parts per million x years after 1900.

e. Use your model from part (d) to estimate when the CO_2 concentration might reach 600 ppm. (This would be approximately double the level estimated to exist prior to the Industrial Revolution.)

12. The data below shows the national debt in billions of dollars x years after 1980.

Year x	Debt: y
1980 (0)	907.7
1981 (1)	997.9
1982 (2)	1142.0
1983 (3)	1377.2
1984 (4)	1572.3
1985 (5)	1823.1
1986 (6)	2153.3
1987 (7)	2350.3
1988 (8)	2602.3
1989 (9)	2857.4
1990 (10)	3233.3
1991 (11)	3502.0
1992 (12)	4002.1
1993 (13)	4351.4
1994 (14)	4643.7
1995 (15)	4961.5

a. Take the natural logarithm of each y-value.

b. Graph the 16 data points $(x, \ln y)$. Notice that the points lie nearly on a straight line, suggesting the use of a linear model.

c. Use the linear regression option on a graphing utility or the procedure outlined in the Study Tip on page 501 to write the equation of the line that best fits the $(x, \ln y)$ transformed data.

d. Write both sides of the equation in part (c) as exponents on base e. Simplify and obtain a model for the national debt x years after 1980.

e. Use your model from part (d) to predict the national debt in the year 2000.

13. In a study entitled "Reaction Time and Speed of Movement of Males and Females of Various Ages,"

(*Research Quarterly:* American Assocation of Health and Physical Education and Recreation, 1963), the following data appeared.

x (Age)	19	27	45	60
y (Reflex Time, in seconds)	0.18	0.20	0.23	0.25

a. Graph the four data points. What function might reasonably model the data?

b. Take the natural logarithm of each *x*-value and graph the four data points (ln *x*, *y*). Notice that the points lie nearly on a straight line, suggesting the use of a linear model.

c. Use the linear regression option on a graphing utility or the procedure outlined in the Study Tip on page 501 to find the logarithmic model for the given data.

d. Use the model from part (c) to predict the reflex time for a 75-year-old.

14. The space shuttle *Challenger* exploded approximately 73 seconds into flight on January 28, 1986. The tragedy involved O-rings used to seal the connections between different sections of the shuttle engines. The data below indicate Fahrenheit temperature and the number of damaged O-rings during 26 space shuttle flights.

x Temperature	y (Number of Damaged O-rings)	x Temperature	y (Number of Damaged O-rings)
53	3	70	0
57	1	70	0
58	1	72	0
63	1	73	0
66	0	75	2
67	0	75	0
67	0	76	0
67	0	76	0
68	0	78	0
69	0	79	0
70	1	80	0
70	1	81	0

Find a logarithmic model for the data. Then use the model to predict the number of O-ring problems on the morning the *Challenger* was launched, when the temperature was 31°F.

UPI / Vince Mannino / Corbis–Bettmann

The following table gives the number of deaths from AIDS in the United States for men and for women from 1988 through 1993.

Year	1988	1989	1990	1991	1992	1993
AIDS Deaths among Men	17,947	23,742	26,752	30,725	34,072	35,551
AIDS Deaths among Women	2050	2613	3182	3926	4741	5526

Source: U.S. Center for Disease Control

Use this data to answer Problems 15–16.

15. In order to avoid ln 0, let *x* represent the number of years after 1987 and let *y* represent AIDS deaths among men. Model the data with a linear model, an exponential model, and a logarithmic model. If the actual number of AIDS deaths among men for 1994 was 37,360, which of the three models most accurately makes this prediction?

16. Let *x* represent the number of years after 1987 and let *y* represent AIDS deaths among women. Model the data with a linear model, an exponential model, and a logarithmic model. If the actual number of AIDS deaths among women for 1994 was 6615, which of the three models most accurately makes this prediction?

17. The following table gives per capita health-care expenditures (in dollars) in the United States from 1988 through 1993.

Year	1988	1989	1990	1991	1992	1993
Per Capita Health-care Expenditures	2201	2422	2688	2902	3144	3331

Source: U.S. Health-Care Financing Administration

Let *x* represent the number of years after 1987 and let *y* represent per capita health-care expenditures. Model the data with a linear model, an exponential model, and a logarithmic model. If the actual per capita health-care expenditures for 1994 were $3510, which of the three models most accurately makes this prediction?

We have seen that a logarithmic model can be fitted to a set of data points of the form (x, y) by fitting a linear model to points of the form (ln x, y). However, if x = 0, a problem occurs because ln 0 is undefined. We can avoid this problem by increasing each value of x by 1, fitting a linear model to points of the form (ln (x + 1), y). Use this approach to solve Problem 18.

18. Students in a psychology class were given a final examination. As part of an experiment to see how much of the course content they remembered over time, they took equivalent forms of the exam in monthly intervals thereafter. The average score for the group (*y*) after *x* months is given by the following data.

Find a model for the data in the form

$$y = a + b \ln (x + 1).$$

Use the model to predict the avervage score for months 8 through 12.

x (Month)	0	1	2	3	4	5	6	7
y (Average Score)	88	78	72	67	64	61	59	57

True–False Critical Thinking Problems

19. A hot object is left to cool in a room. The graph of the cooling model is shown in the figure below. Time (*t*) is measured in minutes and temperature (*T*) is measured in degrees Celsius. Which one of the following is true based on the information conveyed by the graph?

  a. The temperature of the hot object was 20° Celsius when it was first left to cool in the room.

  b. After 15 minutes, the object has cooled to approximately 27° Celsius.

  c. The room is kept at 10° Celsius.

  d. The object will cool down to 21° Celsius after about 27 minutes.

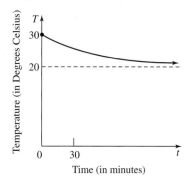

20. Which one of the following is true?

  a. The data points (1, 2), (2, 3), (3, 3.5), (4, 4), (5, 4.1), (6, 4.2), and (7, 4.4) should be modeled using an exponential function.

  b. The data points (1, 4.4), (1.5, 4.7), (2, 5.5), (4, 9.9), (6, 18.1), and (8, 34) should be modeled using a logarithmic function.

  c. The points shown in the figure can be modeled by an exponential function, taking the set of points of the form (*x*, *y*) and fitting a linear model to points of the form (ln *x*, ln *y*).

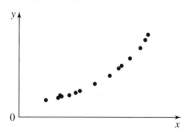

  d. The points shown in the figure can be modeled by a logarithmic function, taking the set of points of the form (*x*, *y*) and fitting a linear model to points of the form (ln *x*, *y*).

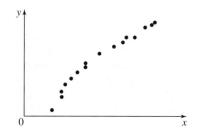

Technology Problems

21. Use a graphing utility to solve Problem 11. In particular:
  a. Draw a scatter diagram for the data.
  b. Use the exponential regression option to fit an exponential model to the data.
  c. Use the utility to graph the model from part (b).
  d. Trace along the curve and estimate when the concentration might reach 600 ppm.

22. Use a graphing utility to solve Problem 12. In particular:
  a. Draw a scatter diagram for the data.
  b. Use the exponential regression option to fit an exponential model to the data.
  c. Use the utility to graph the model from part (b).
  d. Trace along the curve and estimate the national debt in the year 2000.

23. Use a graphing utility to solve Problem 13. In particular:
  a. Draw a scatter diagram for the data.
  b. Use the logarithmic regression option to fit a logarithmic model to the data.
  c. Use the utility to graph the model from part (b).
  d. Trace along the curve and estimate the reflex time for a 75-year-old.

24. Use a graphing utility to solve Problem 14. In particular:
  a. Draw a scatter diagram for the data.
  b. Use the logarithmic regression option to fit a logarithmic model to the data.
  c. Use the utility to graph the model from part (b).
  d. Trace along the curve and predict the number of damaged O-rings on the morning the *Challenger* was launched, when the temperature was 31°F.

25. Use a graphing utility to solve Problem 15. In particular:
  a. Draw a scatter diagram for the data.
  b. Use the linear, exponential, and logarithmic regression options to fit a linear model, an exponential model, and a logarithmic model to the data.
  c. The number r that appears measures how good the fit is. The closer $|r|$ is to 1, the better the fit. What is the best model for AIDS deaths among men over time? How closely does this model predict the actual number of 37,360 deaths in 1994?.

26. Use a graphing utility to solve Problem 16. In particular:
  a. Draw a scatter diagram for the data.
  b. Use the linear, exponential, and logarithmic regression options to fit a linear model, an exponential model, and a logarithmic model to the data.
  c. The number r that appears measures how good the fit is. The closer $|r|$ is to 1, the better the fit. What is the best model for AIDS deaths among women over time? How closely does this model predict the actual number of 6615 deaths in 1994?

27. Use a graphing utility to solve Problem 17. In particular:
  a. Draw a scatter diagram for the data.
  b. Use the linear, exponential, and logarithmic regression options to fit a linear model, an exponential model, and a logarithmic model to the data.
  c. The number r that appears measures how good the fit is. The closer $|r|$ is to 1, the better the fit. What is the best model for per capita health-care expenditures over time? How closely does this model predict the per capita health-care expenditures for 1994, which was $3510?

28. The following table shows the gross federal debt (money borrowed by the Treasury and by various federal agencies) from 1980 through 1995.

Year	Gross Federal Debt (in millions)
1980	909,050
1981	994,845
1982	1,137,345
1983	1,371,710
1984	1,564,657
1985	1,817,521
1986	2,120,629
1987	2,346,125
1988	2,601,307
1989	2,868,039
1990	3,206,564
1991	3,598,498
1992	4,002,136
1993	4,351,416
1994	4,643,711
1995	4,921,025

Source: U.S. Office of Management and Budget

  a. Use a graphing utility to draw a scatter diagram for the data, with values of x representing the number of years after 1979 and values of y representing federal debt.

  b. Use the linear, exponential, and logarithmic regression options to fit a linear model, an exponential model, and a logarithmic model to the data.

  c. Use the model with the value of r closest to 1 as the model of best fit. How well does this model predict the 1996 debt of $5,207,298 million?

Writing in Mathematics

29. Describe some of the decisions that must be made in modeling data.

30. Suppose that (60, 250) and (70, 300) are two points for a set of data that can be modeled exponentially. Explain why (65, 400) cannot be a data point in the scatter plot.

31. What does the model in Example 5 on page 500 imply about urban growth and stress?

32. The strength of a habit H is a function of the number of times the habit is repeated N, modeled by $H = 1000(1 - e^{-kN})$. Suppose that an ordered pair (H, N) is known and that you want to predict habit strength for another value of N. Describe how you would do this using the given model.

33. Describe how you would create a mathematical model to fit the data in the graph.

Group Activity Problem

34. Group members should consult the current edition of *The World Almanac* or an equivalent reference. Select data that is of interest to the group that can be modeled with an exponential or logarithmic function. Model the data. Each group member should make one prediction based on the model, and then discuss a consequence of this prediction. What factors might change the accuracy of each prediction?

CHAPTER PROJECT
Helices and Spirals

Although there are numerous types of spirals mathematicians study, we commonly find the Archimedian spiral and the logarithmic spiral around us. An Archimedian spiral resembles the shape of a tightly coiled rope neatly laid out on a dock. We find the distinctive shape of a logarithmic spiral in nature. This is the spiral we see in the cut-a-way view of a nautilus shell. If we pull a spiral up, out of a two-dimensional flat plane and into three dimensions, we have a helix. We see the rising form of a helix in the double-helix of a DNA molecule and the swirl of an antelope's horns. Mathematically speaking, the twisting staircase known as a "spiral" staircase is actually a helix.

To understand why a spiral is referred to as "logarithmic" we need to see the equation of the spiral written in a *polar* coordinate system. In this coordinate system, the equation would be written as

$$\ln r = a\theta$$

where r is a radial length measured out from the origin (pole), a is a constant, and θ is an angle measure. The angle needs to be measured not in degrees, but in *radians*, a real number.

One radian is approximately 57 degrees. The logarithmic spiral is also referred to as the *equiangular* spiral because a line radiating outward from the origin of the spiral always intersects the spiral at the same angle.

1. In modern times, we usually solve for r and rewrite the logarithmic equation $\ln r = a\theta$ in exponential form. What is the equation in exponential form?

2. One simple way to see the difference between an Archimedian and logarithmic spiral is to create a pair of them by folding paper. Take two long strips of paper and mark off measurements in the following way:
- For the Archimedian spiral, mark off a distance measuring one inch in length. Place the next mark after measuring two inches in length, then mark off three inches, and so on.

- For the logarithmic spiral, mark off inches measured by using the sequence of numbers 1, 1, 2, 3, 5, 8, 13, 21. Thus, the first mark is made at one inch, the second mark is placed after measuring another inch, the third mark is made after measuring two inches further, and so on.
- Bend each strip of paper in a right angle at the mark and study the curves.

3. As an experiment, draw some spirals on a piece of paper, as if you were doodling. Switch hands and draw a few more. How do the two sets of spirals compare? Are they turning in the same direction? How does the distance between the coils compare? Which style of spiral discussed do your sketches resemble?

4. Use a cardboard cylinder from a paper towel roll to investigate a helical curve on the surface of a cylinder. Cut open the cylinder, and lay it out flat. Using a protractor, measure the angle made by the spiral on different levels of the flattened cylinder. What do you observe? Wrap a piece of wire around the cylinder, following the original outline of the helix. Take the wire off and press it flat to the table. What kind of spiral do you see?

5. You can reverse the procedure in the above problem to create the double helix of a DNA molecule. Begin with the flattened cylinder and draw another set of lines duplicating the lines already on the cylinder, but slanted in the opposite direction. When the cylinder is closed, you will see the distinct form of the double helix.

Compare the spiral on the paper core to the spirals found on different styles of common screws. What do you observe? Notice that some screws are basically cylindrical and others are conical in shape. How does this affect the helix you observe? Make your own versions by wrapping wire around two forms, one cone and one cylinder, and compare the shapes. Florist shops and craft stores sell many varieties of forms and wire suitable for this experiment. Using resources in the library or on the Worldwide Web, find out what is meant by an *Archimedian water screw*.

7. The logarithmic spiral may be drawn by geometric means using a Golden Rectangle. This rectangle is discussed in Problem 43 on page 157. Using resources in the library or on the Worldwide Web, discover how to create a logarithmic spiral by geometric construction. After completing the drawing, use a protractor to check for the equal angle property of the spiral. What effect does the initial size of the Golden Rectangle have on the logarithmic spiral?

Worldwide Web Resources

Go to the Prentice Hall website (http://www.prenhall.com/blitzer) to access other locations on the Internet that will allow you to further explore the concepts presented in this project.

Chapter Review

SUMMARY

1. **Exponential Functions**

   a. The exponential function f with base b is defined by $f(x) = b^x$, where $b > 0$ and $b \neq 1$.

   b. Graphs for $f(x) = b^x$ are shown on the right and on the next page.

   Graph of $f(x) = b^x$, $b > 1$

   - Domain: $(-\infty, \infty)$
   - Range: $(0, \infty)$
   - y-Intercept: 1

- Increasing
- Horizontal asymptote: x-axis

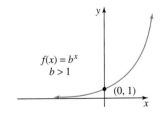

Graph of $f(x) = b^x$, $0 < b < 1$

- Domain: $(-\infty, \infty)$

- Range: $(0, \infty)$

- y-Intercept: 1

- Decreasing

- Horizontal asymptote: x-axis

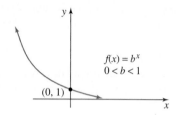

c. Transformations Involving Exponential Functions

   1. *Horizontal Translation:* $y = b^{x+c}$ shifts $y = b^x$ to the left c units if $c > 0$ and to the right c units if $c < 0$.

   2. *Stretching or Shrinking:* $y = cb^x$ stretches $y = b^x$ for $c > 1$ and shrinks $y = b^x$ for $0 < c < 1$.

   3. *Reflecting:* $y = -b^x$ reflects $y = b^x$ in the x-axis and $y = b^{-x}$ reflects $y = b^x$ in the y-axis.

   4. *Vertical Translation:* $y = b^x + c$ shifts $y = b^x$ upward c units if $c > 0$ and downward c units if $c < 0$.

d. The Natural Exponential Function: $f(x) = e^x$
   The irrational number e is called the natural base, where $e \approx 2.7183$.

2. Formulas for Compound Interest
After t years, the balance A in an account with principal P and annual interest rate r (in decimal form) is given by the following mathematical models.

a. For n compoundings per year: $A = P\left(1 + \frac{r}{n}\right)^{nt}$

b. For continuous compounding: $A = Pe^{rt}$

3. Growth and Decay Models

a. *Exponential Growth:* $A = A_0 e^{kt}, k > 0$

b. *Exponential Decay:* $A = A_0 e^{kt}, k < 0$. In both models, t represents time, A_0 is the amount present at $t = 0$, and A is the amount present at time t.

4. Logarithmic Functions

a. $y = \log_b x$ is equivalent to $b^y = x$ where $x > 0, b > 0$, and $b \neq 1$. The function $f(x) = \log_b x$ is the logarith-mic function with base b. Thus, a logarithm is an exponent.

b. $y = b^x$ and $y = \log_b x$ are inverses. Thus if $f(x) = b^x$, then $f^{-1}(x) = \log_b x$.

c. The graphs of $f(x) = b^x$ and $f^{-1}(x) = \log_b x$ are shown below (with $b > 1$).

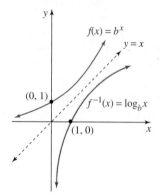

For $f(x) = b^x$:

   Domain $= (-\infty, \infty)$ and range $= (0, \infty)$

   As $x \to -\infty$, $b^x \to 0$.

   The x-axis is a horizontal asymptote.

For $f^{-1}(x) = \log_b x$:

   Domain $= (0, \infty)$ and range $= (-\infty, \infty)$

   As $x \to 0^+$, $\log_b x \to -\infty$.

   The y-axis is a vertical asymptote.

d. *Transformations Involving Logarithmic Functions:* The graph of $y = \log_b(x + a) + c$, with $a > 0$ and $c > 0$, lies a units to the left and c units above the graph of $y = \log_b x$. The domain of $f(x) = \log_b(x + a) + c$ consists of all x for which $x + a > 0$. The vertical asymptote is $x = -a$.

e. *Common and Natural Logarithms:* $y = \log x$ means $y = \log_{10} x$ and is called the common logarithmic function. $y = \ln x$ means $y = \log_e x$ and is called the natural logarithmic function. If $f(x) = 10^x$, then $f^{-1}(x) = \log x$. If $f(x) = e^x$, then $f^{-1}(x) = \ln x$.

5. Properties of Logarithmic Functions

a. *Basic Properties:*

Base b ($b > 0, b \neq 1$)	Base 10 (Common Logarithms)	Base e (Natural Logarithms)
$\log_b 1 = 0$	$\log 1 = 0$	$\ln 1 = 0$
$\log_b b = 1$	$\log 10 = 1$	$\ln e = 1$
$\log_b b^x = x$	$\log 10^x = x$	$\ln e^x = x$

b. *The Product Rule:* $\log_b(MN) = \log_b M + \log_b N$

c. *The Quotient Rule:* $\log_b\left(\dfrac{M}{N}\right) = \log_b M - \log_b N$

d. *The Power Rule:* $\log_b M^p = p \log_b M$

e. *The Inverse Properties:*

Base b	Base 10	Base e
$b^{\log_b x} = x$	$10^{\log x} = x$	$e^{\ln x} = x$
$\log_b b^x = x$	$\log 10^x = x$	$\ln e^x = x$

f. *The Change-of Base Property:*

The General Property	Introducing Common Logarithms	Introducing Natural Logarithms
$\log_b M = \dfrac{\log_a M}{\log_a b}$	$\log_b M = \dfrac{\log M}{\log b}$	$\log_b M = \dfrac{\ln M}{\ln b}$

6. Exponential Equations

a. Exponential equations contain a variable in an exponent.

b. To solve an exponential equation:

   1. Isolate the exponential expression.

   2. Take the natural logarithm on both sides of the equation.

   3. Simplify using one of the following properties:

$$\ln b^x = x \ln b \quad \text{or} \quad \ln e^x = x$$

   4. Solve for the variable.

7. Logarithmic Equations

a. Logarithmic equations contain a variable in a logarithmic expression.

b. Equations containing logarithmic expressions and nonlogarithmic constants can be solved by using properties of logarithms to write the equation in the form $\log_b M = N$ and then solving $b^N = M$ for the desired variable.

c. Equations containing only logarithmic expressions can be solved by using properties of logarithms to write the equation in the form $\log_b M = \log_b N$ and then setting $M = N$.

d. Be sure to reject values that produce the logarithm of a negative number or the logarithm of 0 in the original equation.

8. Modeling with Exponential and Logarithmic Functions

a. If the graph of the data points appears to follow an exponential pattern, fit a linear model to $(x, \ln y)$ and then exponentiate both sides to find the model for the data points (x, y). Or use a graphing utility's exponential regression option ($y = ab^x$).

b. If the graph of the data points appears to follow a logarithmic pattern, fit a linear model to $(\ln x, y)$. The model for the data points (x, y) is the equation of the regression line for $(\ln x, y)$. Or use a graphing utility's logarithmic regression option ($y = a + b \ln x$).

c. When using a graphing utility to model data, first draw a scatter diagram to obtain a general picture for the shape of the data. If necessary, fit several models to the data. The model that best fits the data is the one with the value of r closest to 1.

REVIEW PROBLEMS

In Problems 1–8, match each function with its graph.

1. $f(x) = 4^x$

2. $f(x) = 4^{-x}$

3. $f(x) = -4^{-x}$

4. $f(x) = -4^{-x} + 3$

5. $f(x) = \log x$

6. $f(x) = \log(-x)$

7. $f(x) = \log(2 - x)$

8. $f(x) = 1 + \log(2 - x)$

a.

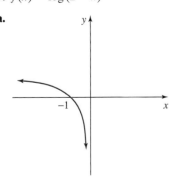

b.

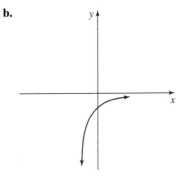

c.

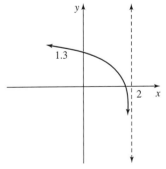

d.

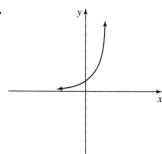

e.

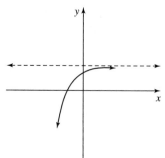

f.

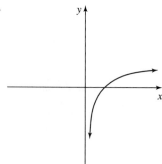

g.

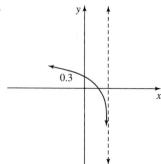

h.

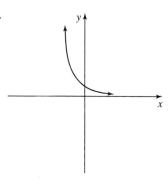

In Problems 9–15, sketch by hand the graphs of the two functions in the same rectangular coordinate system. Use a table of coordinates to sketch the first given function and transformations of the first function plus a table of coordinates to graph the second function. State each function's domain and range. Then use a graphing utility to confirm your results.

9. $y = 2^x$ and $y = 2^{x+1}$

10. $f(x) = 3^x$ and $g(x) = 3^{x-1} - 2$

11. $f(x) = 2^x$ and $y = 2^{-x}$

12. $y = 3^x$ and $y = -3^{-x} + 2$

13. $y = \left(\frac{1}{2}\right)^x$ and $y = \left(\frac{1}{2}\right)^{-x} - 1$

14. $f(x) = \log_2 x$ and $g(x) = \log_2(x - 2) + 3$

15. $y = \log_2 x$ and $y = -\log_2(-x)$

Use the compound interest models

$$A = P\left(1 + \frac{r}{n}\right)^{nt} \quad \text{and} \quad A = Pe^{rt}$$

to answer Problems 16–17.

16. Suppose that you have $5000 to invest. What investment yields the greater return over 5 years: 5.5% compounded semiannually or 5.25% compounded monthly?

17. Suppose that you have $14,000 to invest. What investment yields the greater return over 10 years: 7% compounded monthly or 6.85% compounded continuously?

18. Use the exponential growth model $A = A_0 e^{kt}$ (where $k > 0$) to answer this question. In 1990 China had a population of 1119.9 million with a growth rate of 1.3% per year ($k = 0.013$). India's 1990 population was 853.4 million with a growth rate of 2.1% per year. In their 1990 *World Population Profile*, the United States Bureau of the Census stated that "The latest projections suggest that India's population may surpass China's in less than 60 years." Estimate the population of the two countries in the year 2050. Do these estimations reinforce the prediction by the Bureau of the Census?

19. The function

$$N(t) = \frac{200{,}000}{1 + 1999e^{-0.06t}}$$

describes the number of people $N(t)$ who become ill with an infectious disease t days after its initial outbreak.

a. How many people became ill with the disease when the outbreak began?

b. How many people were ill by the end of the fourth day?

c. What is the horizontal asymptote for this function? Describe what this means in practical terms.

d. Use a graphing utility to graph the function.

In Problems 20–22, evaluate each expression without the use of a calculator.

20. $\log_3 9 + \log_2 32$

21. $\log_3 \frac{1}{27} - \ln e^5$

22. $\log_3(\log_8 8) + \log_3(\log_2 8)$

In Problems 23–25, find the domain, the vertical asymptote, and the x-intercept for each function. Verify your results by graphing the function with a graphing utility.

23. $f(x) = \ln(x - 2)$

24. $g(x) = \ln(4 - x)$

25. $r(x) = \ln(x - 2)^2$

26. The model

$$t = \frac{1}{c} \ln\left(\frac{A}{A - N}\right)$$

describes the time (t, in weeks) that it takes to achieve mastery of a portion of a task. In the model, A represents maximum learning possible, N is the portion of the learning that is to be achieved, and c is a constant used to measure an individual's learning style. A 50-year-old man decides to start running as a way to maintain good health. He feels that the maximum rate he could ever hope to achieve is 12 miles per hour. How many weeks will it take before the man can run 5 miles in 1 hour if $c = 0.06$ for this person?

In Problems 27–31, use logarithmic properties to expand each expression as much as possible. Where possible, evaluate or simplify logarithmic expressions without the use of a calculator.

27. $\log_7\left(\dfrac{\sqrt{x}}{49y^2}\right)$

28. $\log \dfrac{10{,}000}{\sqrt[3]{3x + 1}}$

29. $\ln \sqrt{\dfrac{x^3 y}{e^4}}$

30. $\ln(9e^{5t})$

31. $\log_b(\sqrt{x^3}\sqrt{y})^{-1/5}$

In Problems 32–34, write the expression as a single logarithm whose coefficient is 1.

32. $5\log(x - 4) + \log 7 - 3\log x$

33. $\frac{1}{2}\ln x - \frac{1}{7}\ln y$

34. $\ln(x^2 - 9) - \ln(x + 3) + \frac{2}{3}\ln(x - 3)$

35. Rewrite as a single term that does not contain a logarithm: $10^{\log 5x^2 + \log 3x} + e^{\ln 16x^7 - \ln 2x^4}$

In Problems 36–37, evaluate the logarithm using the change-of-base property. Work each problem twice, first using common logarithms and then using natural logarithms. Round each approximation to three decimal places.

36. $\log_5 17$

37. $\log_{15} 3000$

On the Richter scale, the magnitude R of an earthquake of intensity I is given by

$$R = \log \frac{I}{I_0}$$

where I_0 is the intensity of a barely felt zero-level earthquake. Use this model to answer Problems 38–39.

38. What is the magnitude on the Richter scale of an earthquake that is $10^{6.7}$ times as intense as a zero-level earthquake?

39. Compare the intensity of Colombia's 1906 earthquake measuring 8.6 on the Richter scale to Northern California's 1989 earthquake that measured 7.1. Calculate the ratio of their intensities and determine how many times greater in intensity was Colombia's 1906 earthquake than California's 1989 earthquake.

40. Another form of Richter's equation is

$$R = 0.67 \log E - 2.9,$$

where E is the energy in joules of an earthquake and R is its magnitude on the Richter scale. What is the difference between the energy released in the two earthquakes described in Problem 39?

41. Describe the relationship between the graphs of $y = \log_4 x$ and $y = \log_4 \dfrac{1}{(x - 1)^2}$. Then use a graphing utility to verify your description.

42. The relationship between the number of decibels (D) and the intensity of a sound (I) in watts per meter^2 is given by

$$D = 10 \log \frac{I}{I_0}$$

where I_0 is the intensity of sound barely audible to the human ear. If damage to the average ear occurs at 90 decibels or greater, will a jet aircraft from 500 feet with an intensity of 1×10^2 watts per meter^2 cause ear damage? (Assume I_0, the threshold of hearing, to be 10^{-12} watt per meter^2.)

Solve the exponential equations in Problems 43–47 by taking the natural logarithm of both sides. Express the answer in terms of natural logarithms. Then use a calculator to obtain a decimal approximation, correct to the nearest thousandth, for the solution.

43. $8^x = 12,143$

44. $9e^{5x} = 1268$

45. $e^{12-5x} - 7 = 123$

46. $7^{x-3} = 5^{4x+2}$

47. $e^{2x} - e^x - 6 = 0$

48. The model $A = Pe^{rt}$ describes the accumulated value (A) of a sum of money (P) after t years at an annual percentage rate (r, in decimal form) compounded continuously. Suppose that $5500 is invested at 4.5% compounded continuously.

    a. How long will it take the investment to triple in value?

    b. What interest rate is required for the investment to double in 8 years time?

49. The model $A = P(1 + \frac{r}{n})^{nt}$ describes the accumulated value (A) of a sum of money (P) after t years at an annual percentage rate (r, in decimal form) compounded n times a year. How long will it take $12,500 to grow to $20,000 at 6.5% annual interest compounded quarterly?

50. The model

$$A = \frac{A_0}{k}(e^{kt} - 1)$$

describes the amount of nonrenewable energy resources (A) from time $t = 0$ to time t, where A_0 is the amount of the resource consumed during the year $t = 0$ and k is the relative growth rate of annual consumption. In 1990, the total remaining known oil resources ultimately available were 983.4 billion barrels. There were 21.3 billion barrels consumed that year. With a rate of growth in oil consumption of 2.5%, by what year will the planet be depleted of its known oil resources?

51. Use the exponential growth model $A = A_0 e^{kt}$ (where $k > 0$) to solve this problem. In 1995 the United States had a population of 263.2 million and a relative growth rate of 0.7% per year. Brazil's 1995 population was 157 million with a growth rate of 1.9% per year. In what year will the two countries have the same population?

In Problems 52–56, solve the logarithmic equation algebraically. Check each proposed solution by direct substitution or with a graphing utility, eliminating values that produce the logarithm of a negative number or the logarithm of 0 from the equation's solution set.

52. $\log_4(3x - 5) = 3$

53. $\log_2(x + 3) + \log_2(x - 3) = 4$

54. $\log_3(x - 1) = 2 + \log_3(x + 2)$

55. $\ln x + \ln(x - 4) = \ln(2x - 5)$

56. $\log(x + 7) - \log(x + 2) = \log(x + 1)$

57. The formula $W = 0.35 \ln P + 2.74$ is a model for the average walking speed (W, in feet per second) for a resident of a city whose population is P thousand. What is New York City's population (to the nearest thousand) if the average walking speed is 5.9 feet per second?

58. The brightness of stars (or planets), measured in magnitudes (M), is given by

$$M = 6 - 2.5 \log \frac{I}{I_0}$$

where

    I = the intensity of light from the star

    I_0 = the intensity of light from a just-visible star

How many times more intense is the light from the planet Jupiter (magnitude $= -2.9$) than the light from Polaris, the north star (magnitude $= 2$)?

Use the exponential growth and decay models

$$A = A_0 e^{kt} \quad \text{or} \quad f(t) = A_0 e^{kt}$$

where A_0 is the original amount (or size) of a growing (or decaying) entity at $t = 0$, A is the amount at time t, and k is a growth ($k > 0$) or decay ($k < 0$) constant, to answer Problems 59–60.

59. According to the U.S. Bureau of the Census, in 1980 there were 14,609 thousand residents of Hispanic origin living in the United States. By 1994, the number had increased to 26,077 thousand.

    a. Let $t = 0$ correspond to 1980. This means that A_0, the original amount in the growth model, is 14,609.

Now use the fact that in 1994 the number was 26,077 (when $t = 14$, $A = 26,077$) to find k, the growth rate.

    b. Substitute the values for A_0 and k into the exponential growth model and write the function that models the Hispanic resident population of the United States t years after 1980.

c. Use the model in part (c) to project the Hispanic resident population (in thousands) for the years 2005, 2010, and 2025.

60. Carbon-14 decays exponentially with a half-life of approximately 5715 years, meaning that after 5715 years a given amount of carbon-14 will have decayed to half the original amount. Prehistoric cave paintings were discovered in the Lascaux cave in France. The paint contained 15% of the original Carbon-14. Estimate the age of the paintings.

61. A cup of hot coffee is taken out of a microwave oven and placed in a room. The temperature (T, in degrees Fahrenheit) after t minutes is modeled by the function $T = 70 + 130e^{-0.04855t}$. The graph of the function is shown in the figure below. Use the graph to answer each of the following questions.

a. What was the temperature of the coffee when it was first taken out of the microwave?

b. What is a reasonable estimate of the temperature of the coffee after 20 minutes? Use your calculator to verify this estimate.

c. What is the limit of the temperature to which the coffee will cool? What does this tell you about the temperature of the room?

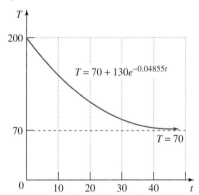

62. Use Newton's Law of Cooling, $T = C + (T_0 - C)e^{-kt}$, where t is the time it takes for an object to cool from an initial temperature T_0 to a temperature T, C is the temperature of the surrounding medium, and k is a positive constant associated with the cooling object, to answer this question. At 9:00 A.M., a coroner arrived at the home of a person who had died during the night. The temperature of the room was 70°F, and at the time of death the person had a body temperature of 98.6°F. The coroner took the body's temperature at 9:30 A.M., at which time it was 85.6°F, and again at 10:00 A.M., when it was 82.7°F. At what time did the person die?

63. The figure at the top of the next column shows world population projections through the year 2150. The data is from the United Nations Family Planning Program and is based on optimistic or pessimistic expectations for suc-

cessful control of human population growth. Suppose that you are interested in modeling this data using exponential, linear, and quadratic functions. Which one of these functions would you use to model each of the projections? Explain your choices. For the choice corresponding to a quadratic model, would your formula involve one with a positive or negative leading coefficient? Explain.

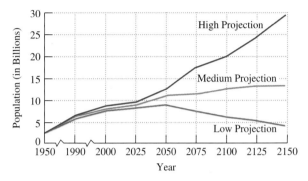

64. The figure shows the number of people in the United States age 65 and over, with projected figures for the year 2000 and beyond.

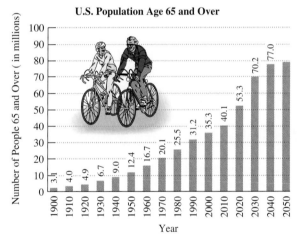

Source: U.S. Bureau of the Census

a. Let x represent the number of decades after 1900, and let y represent the number of people 65 and over. Use the exponential regression option of a graphing utility to model the data from 1900 through 2040, or follow these steps:

1. Take the natural logarithm of each y-value.

2. Graph the 15 data points $(x, \ln y)$. Notice that the points lie nearly on a straight line, suggesting the use of a linear model.

3. Use the linear regression option on a graphing utility or the procedure outlined in the Study Tip on page 501 to write the equation of the line that best fits the $(x, \ln y)$ transformed data.

4. Write both sides of the equation from step 3 as exponents on base e. Simplify and obtain a model for U.S. population 65 and over x decades after 1900.

b. Use the model to predict the U.S. population age 65 and over for the year 2050.

c. In 1990 the United States had a population of 263.2 million. The projected population for the year 2050 is 392.7 million. What is the percentage of U.S. population 65 and over for each of these years?

65. The data below indicate the amount of rain forest cleared for cattle ranching in the indicated year.

x Year	y **Rain Forest Cleared (in millions of hectares) (1 hectare = 247 acres)**
1960 (1)	0.30
1970 (11)	0.95
1980 (21)	1.36
1988 (29)	1.40

a. Graph the four data points. What function might reasonably model the data?

b. Use the appropriate regression option on a graphing utility to find a function that models the data, or follow these steps:

1. Take the natural logarithm of each x-value and graph the four data points ($\ln x$, y). Notice that the points lie nearly on a straight line, suggesting the use of a linear model.

2. Use the linear regression option on a graphing utility or the procedure outlined in the Study Tip on page 501 to find the logarithmic model for the given data.

c. Use the model from part (b) to determine the year in which 1.5 million hectares will be cleared for cattle ranching.

66. The data below show consumer expenditures on books in the United States.

x (Year)	y **(Consumer Expenditures on Books, in millions of dollars)**
1982	9869
1985	12,611
1990	19,043
1992	21,244
1993	22,635
1994	23,798

a. Use a graphing utility to draw a scatter diagram of the data.

b. Use a graphing utility to model the data with a linear model, an exponential model, and a logarithmic model. Which model best describes book expenditures as a function of time?

c. Use the best of the three models to predict the amount that consumers in the United States will spend on books in the year 2005.

CHAPTER 4 TEST

1. Graph $y = 2^x$ and $y = 2^{x-1} + 1$ in the same rectangular coordinate system.

2. Use the compound interest models

$$A = P\left(1 + \frac{r}{n}\right)^{nt} \quad \text{and} \quad A = Pe^{rt}$$

to answer this question. Suppose you have \$9000 to invest. What investment yields the greater return over 10 years: 6.5% compounded semiannually or 6% compounded continuously? How much more, to the nearest dollar, is yielded by the better investment?

3. A cup of coffee at 180° Fahrenheit is cooling in a room at a rate of 5°F per minute. The function $f(t) = 72 +$

$112e^{-0.05t}$ describes the coffee's temperature ($f(t)$) after t minutes.

a. What is the temperature, to the nearest degree, after 10 minutes?

b. What is the horizontal asymptote for this function? Describe what this means in practical terms.

4. Evaluate without using a calculator: $\log_2 16 - \log_{1/3} 9$

5. Graph $y = \log_2 x$ and $y = \log_2(x - 1)$ in the same rectangular coordinate system.

6. Find the domain of $f(x) = \ln(3 - x)$.

Use logarithmic properties to expand the expressions in Problems 7–8 as much as possible. Where possible, evaluate logarithmic expressions.

7. $\log_5\left(\dfrac{\sqrt[3]{x}}{125y^2}\right)$

8. $\ln(10e^{7t})$

Write each expression in Problems 9–10 as a single logarithm.

9. $6 \log x + 2 \log y$

10. $\ln 7 - 3 \ln x$

Solve each exponential equation in Problems 11–12. Express your answers in terms of natural logarithms. Then use a calculator to find a decimal approximation of each answer, correct to the nearest thousandth.

11. $e^{9-3x} - 17 = 25$

12. $e^{2x} - 6e^x + 5 = 0$

13. Use the compound interest model $A = Pe^{rt}$ to answer this question. If \$4000 is invested at 6% annually, compounded continuously, how long will it take (to the nearest tenth of a year) for the investment to double in value?

14. Use the exponential decay model $A = A_0 e^{kt}$ to answer this question. A radioactive substance that initially contains 200 grams decays at the rate of 8% per year. How long will it take until 100 grams of the substance remain?

Solve each logarithmic equation in Problems 15–16.

15. $\log_5 (x - 2) = \log_5 x - \log_5 8$

16. $\log_{16}(10x + 3) + \log_{16} x = \frac{1}{2}$

17. The perceived loudness of a sound (D, in decibels) at the eardrum is described by

$$D = 10 \log \frac{I}{I_0}$$

where

$I =$ the intensity of the sound (in watts per meter^2)

$I_0 =$ the intensity of a sound barely audible to the human ear

An intensity level of 120 decibels is potentially damaging to the ear. How many times louder than I_0 is a level of 120 decibels?

18. Use the exponential growth model $A = A_0 e^{kt}$ to answer this problem.

  a. In 1992, world population was approximately 5.5 billion and in 1995 it was approximately 5.8 billion. Find the exponential growth function that models the data t years after 1992.

  b. In what year will world population reach 10 billion?

19. The model $T = C + (T_0 - C)e^{-kt}$ describes the time t it takes for an object to cool from an initial temperature T_0 to a temperature T when the surrounding temperature is C.

  a. A house whose temperature is 68° Fahrenheit loses power because of an ice storm. The temperature outside the house is 28°F and the temperature of the house drops from 68° to 64° in an hour. Find k and write the model for the temperature T of the house after t hours.

  b. After how long will the temperature in the house be 50°F?

20. Use a graphing utility to graph $f(x) = 5 \ln x - \frac{x}{2}$ in an appropriate viewing rectangle. Use the graph to answer each of the following.

  a. What is the vertical asymptote of f?

  b. Find, correct to two decimal places, the maximum value of f and the value of x at which this occurs.

  c. Find the range of f.

  d. Solve the equation $5 \ln x - \frac{1}{2}x = 0$. Express each solution correct to two decimal places.

CUMULATIVE REVIEW PROBLEMS (CHAPTERS P–4)

In Problems 1–2, simplify.

1. $\dfrac{3}{4x^2 + 4x + 1} + \dfrac{x + 3}{2x^2 - x - 1} - \dfrac{2}{x - 1}$

2. $\dfrac{1 - \dfrac{2}{x}}{1 - \dfrac{3}{x} + \dfrac{2}{x^2}}$

Solve Problems 3–7.

3. $|3x - 4| = 2$

4. $\sqrt{2x - 5} - \sqrt{x - 3} = 1$

5. $x^4 + x^3 - 3x^2 - x + 2 = 0$

6. $e^{11-5x} - 32 = 96$

7. $\log_2(x + 5) + \log_2(x - 1) = 4$

8. Multiply: $[3y - (2x + 5)][3y + (2x - 5)]$.

9. Graph: $f(x) = \begin{cases} \sqrt{x - 2} & \text{if } x \geq 2 \\ 2 - x & \text{if } x < 2. \end{cases}$

10. Use the graph of $f(x) = x^2$ to graph $g(x) = -(x + 2)^2 + 4$.

11. If $2 - i$ is a solution of $x^4 - 4x^3 + 6x^2 - 4x + 5 = 0$, find the complete solution set.

12. Graph: $f(x) = \dfrac{x^2 - 4}{x - 1}$.

13. The model $F = 1 - k \ln(t + 1)$ describes the fraction of people (F) who remember all the words in a list of nonsense words t hours after memorizing the list. After 3 hours only half the people could remember all the words. Determine the value of k and then predict the fraction of people in the group who will remember all the words after 6 hours.

14. Expand and evaluate logarithms where possible: $\log_3(81 \sqrt[3]{x})$.

15. Factor completely: $x^3 - 5x^2 - 25x + 125$.

16. Solve for n: $I = \dfrac{nE}{R + nr}$.

17. The life expectancy for women born in the United States in 1950 is 71 years. This amount has increased by about 0.2 years for each year of birth after 1950. What is the year of birth for which a woman in the United States can expect to live 81.8 years?

18. A water tank has the shape of an inverted right circular cone of altitude 8 feet and base radius 4 feet. How much water is in the tank when the water is 3 feet deep?

19. What number exceeds its square by a maximum amount?

20. The cost (C) of removing $p\%$ of a river's pollutants is given by the model

$$C = \frac{60{,}000p}{100 - p}.$$

What is the vertical asymptote for this function? What does this mean in practical terms?

Matrices and Linear Systems

The Berlin Airlift (1948–1949) was an operation put into effect by the United States and Great Britain after the former Soviet Union closed all roads and rail lines between West Germany and Berlin, cutting off supply routes to the city. The Allies used a technique called linear programming to break the blockade. The eleven-month airlift, in 272,264 flights, provided basic necessities to blockaded Berlin, saving one of the world's greatest cities.

Mathematical models often have thousands of equations, sometimes a million variables. Problems ranging from saving a blockaded Berlin to routing phone calls over our nation's communication network require solutions in a matter of moments. In this chapter, we study methods for modeling and solving problems with several variables.

Grisha Bruskin "Birth of a Hero, Figure C" 1990, stainless steel with painted aluminum (work in progress – partially painted), 73 1/2 × 37 × 25 1/2 in. (186.7 × 94 × 64.8 cm). 1997 © Grisha Bruskin / Licensed by VAGA, New York, NY / Courtesy, Marlborough Gallery, NY.

SECTION 5.1

Solving Linear Systems Using Substitution and Addition

Solutions Manual

Tutorial

Video 9

Objectives

1 Solve linear systems in two variables by substitution.
2 Solve linear systems in two variables by addition.
3 Identify inconsistent and dependent systems.
4 Solve linear systems in three variables by addition.
5 Solve problems using linear systems.

Linear Systems in Two Variables

In Chapter 2, we studied the linear equation in two variables

$$Ax + By = C$$

whose graph is a straight line. Many applied problems are modeled by two or more linear equations, such as

$$2x - 3y = -4$$
$$2x + y = 4.$$

We call these equations *simultaneous linear equations* or a *system of linear equations.* A *solution* of such a system is an ordered pair of real numbers that makes *both* equations true.

> **Solution set of a system**
>
> The *solution set* of a system of equations in two variables is the set of all ordered pairs of values (a, b) that satisfy every equation in the system.

EXAMPLE 1 **Determining Whether an Ordered Pair Is a Solution of a System**

Determine whether $(1, 2)$ is a solution of the system:

$$2x - 3y = -4$$
$$2x + y = 4$$

Solution

Since 1 is the *x*-coordinate and 2 is the *y*-coordinate of $(1, 2)$, we replace *x* with 1 and *y* with 2 in both equations.

$$2x - 3y = -4 \qquad 2x + y = 4$$
$$2(1) - 3(2) \overset{?}{=} -4 \qquad 2(1) + 2 \overset{?}{=} 4$$
$$2 - 6 \overset{?}{=} -4 \qquad 2 + 2 \overset{?}{=} 4$$
$$-4 = -4 \checkmark \qquad 4 = 4 \checkmark$$

The ordered pair $(1, 2)$ satisfies both equations, so it is a solution of the system. Although such a solution can be described by saying that $x = 1$ and $y = 2$, we

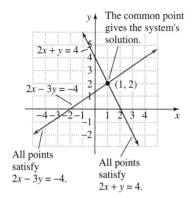

Figure 5.1

A linear system's solution

Solve linear systems in two variables by substitution.

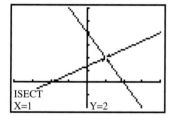

Figure 5.2

Using a graphing utility to visualize a system's solution

will use set-builder notation. The solution set to the system is $\{(1, 2)\}$—that is, the set consisting of the ordered pair $(1, 2)$. Figure 5.1 illustrates that the solution of the system corresponds to the point of intersection of the graphs.   ■

We can use a graphing utility to verify the graphs in Figure 5.1 and thereby find a system's solution set. We first solve each equation in the system for y:

$$2x - 3y = -4 \qquad\qquad 2x + y = 4$$
$$-3y = -2x - 4 \qquad\qquad y = -2x + 4$$
$$y = \tfrac{2}{3}x + \tfrac{4}{3}$$

Now we enter the equations and graph them in the same viewing rectangle. We can use the utility's $\boxed{\text{TRACE}}$ or $\boxed{\text{ISECT}}$ feature to find the intersection point, illustrated in Figure 5.2.

Eliminating a Variable Using the Substitution Method

Finding a linear system's solution by graphing lines can be awkward. For example, a solution of $\left(-\tfrac{2}{3}, \tfrac{157}{29}\right)$ would be difficult to "see" as an intersection point on a graph.

Therefore we will now consider a method that does not depend on finding a system's solution visually—the substitution method. This method involves converting the system to one equation in one variable by an appropriate substitution.

EXAMPLE 2 **Solving a System by the Substitution Method**

Solve by the substitution method:

$$3x - y = -5$$
$$5x - 2y = -7$$

Solution

The substitution method relies on having one variable isolated. By solving for y in the first equation, which has a coefficient of -1, we can avoid fractions.

$3x - y = -5$	This is the first given equation.
$-y = -3x - 5$	Subtract $3x$ from both sides.
$y = 3x + 5$	Solve for y by multiplying both sides by -1.

Since y is now isolated, we can substitute the expression for y in the second equation.

$$y = 3x + 5 \qquad 5x - 2y = -7$$

Here are the details:

$5x - 2y = -7$	This is the second equation in the given system.
$5x - 2(3x + 5) = -7$	Substitute $3x + 5$ from the first equation for y. We now have one equation in one variable.
$5x - 6x - 10 = -7$	Apply the distributive property.

$$-x - 10 = -7 \qquad \text{Combine like terms.}$$

$$-x = 3 \qquad \text{Add 10 to both sides.}$$

$$x = -3 \qquad \text{Multiply both sides by } -1.$$

tudy tip

Back-substitute means to, after finding the value for a variable, substitute that value *back* into one of the system's equations to find the value of the other variable.

We now know that the *x*-coordinate of the solution is -3. To find the *y*-coordinate, we *back-substitute* the *x*-value into either one of the given equations. It is easiest to use the first equation, $3x - y = -5$, in the form $y = 3x + 5$, solved for *y*. (This is the form of the equation that we obtained at the beginning of the solution when we isolated *y*.)

$$y = 3x + 5 \qquad \text{Use the form of the first equation obtained at the start of the solution process.}$$

$$y = 3(-3) + 5 \qquad \text{Substitute } -3 \text{ for } x.$$

$$y = -4 \qquad \text{Simplify.}$$

sing technology

Check Example 2 with a graphing utility. Solve each equation for *y*. Enter

$$y_1 = 3x + 5 \qquad y_2 = \frac{5x + 7}{2}.$$

Graph both equations in the same viewing rectangle. The solution, shown below, is $(-3, -4)$.

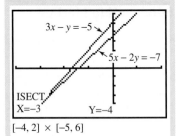

$[-4, 2] \times [-5, 6]$

Since $x = -3$ and $y = -4$, we now check the proposed solution $(-3, -4)$ in both of the system's given equations.

$3x - y = -5$	$5x - 2y = -7$
$3(-3) - (-4) \overset{?}{=} -5$	$5(-3) - 2(-4) \overset{?}{=} -7$
$-9 + 4 \overset{?}{=} -5$	$-15 + 8 \overset{?}{=} -7$
$-5 = -5 \checkmark$	$-7 = -7 \checkmark$

The proposed solution, $(-3, -4)$, satisfies both of the system's equations, so the system's solution set is $\{(-3, -4)\}$. ∎

tudy tip

Get into the habit of checking ordered-pair solutions in *both* equations of a system in two variables.

Before considering additional examples, let's summarize the steps used in the substitution method.

tudy tip

In step 1, if possible, solve for the variable whose coefficient is 1 or −1 to avoid working with fractions.

Solving linear systems by the substitution method

1. Solve one of the equations for one variable in terms of the other. (If one of the equations is already in this form, you can skip this step.)
2. Substitute the expression found in step 1 into the other equation. This will result in an equation in one variable.
3. Solve the equation obtained in step 2.
4. Back-substitute the value found from step 3 into the equation from step 1 to find the value of the remaining variable.
5. Check the proposed solution in both of the system's given equations.

Since the graph of a linear equation in two variables is a straight line, there are three possibilities for the solution of a system of two linear equations.

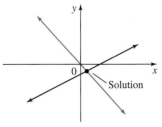

Exactly One Solution

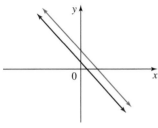

No Solution (Parallel Lines)

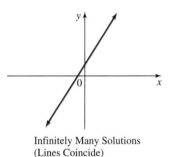

Infinitely Many Solutions
(Lines Coincide)

Figure 5.3

Possibilities for two linear
equations in two variables

Step 1. Solve one of the equations for
one variable in terms of the other.

Step 2. Substitute the expression from
step 1 into the other equation.

Step 3. Attempt to solve the
resulting equation.

Steps 4–5 do not apply. The equation
in one variable has no solution, so
there is no value to back-substitute.

**The number of solutions to a system
of two linear equations (see Figure 5.3)**

The number of solutions for a system of two linear equations in two
variables is given by one of the following.

Number of Solutions	What This Means Graphically
Exactly one ordered pair solution	The two lines intersect at one point.
No solution	The two lines are parallel.
Infinitely many solutions	The two lines are identical.

In Example 2, we solved a system with one solution using the substitution
method. Now let's see what occurs when we attempt to solve a system with no
solution using this method.

EXAMPLE 3   **The Substitution Method: No-Solution Case**

Solve by the substitution method:

$$3x - 3y = -2$$

$$x - y = 5$$

Solution

Begin by isolating one of the variables in either of the equations. By solving for
x in the second equation, which has a coefficient of 1, we can avoid fractions.

$x - y = 5$ This is the second given equation.

$x = y + 5$ Solve for x by adding y to each side.

Since x is now isolated, we can substitute the expression for x in the first equation.

$$x = y + 5 \qquad 3x - 3y = -2$$

Here are the details:

$3x - 3y = -2$ This is the first equation in the given system.

$3(y + 5) - 3y = -2$ Substitute $y + 5$ for x.

$3y + 15 - 3y = -2$ Apply the distributive property.

$15 = -2$ Simplify. There are no values of x and y for which $15 = -2$.

The false statement $15 = -2$ indicates that the system has no solution. The
contradiction $15 = -2$ tells us that the solution set for the system is the empty
set, $\emptyset$.

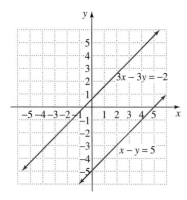

Figure 5.4

A system with no solution

2 Solve linear systems in two variables by addition.

The lines corresponding to the two equations in the given system are shown in Figure 5.4. The lines are parallel and have no point of intersection. ∎

Discover for yourself

Show that the graphs of $3x - 3y = -2$ and $x - y = 5$ must be parallel lines by solving each equation for y. What is the slope and y-intercept for each line? What does this mean? If a linear system has no solution, what must be true about the slopes and y-intercepts for the system's graphs?

Eliminating a Variable Using the Addition Method

The substitution method is most useful if one of the given equations has either an isolated variable or a variable with a coefficient of 1 or -1. Such a variable can easily be isolated without introducing fractions. Another, and frequently the easiest, method for solving a linear system is the addition method. As with the substitution method, the addition method involves eliminating a variable and ultimately solving an equation containing only one variable. However, this time the elimination process is achieved by adding the equations.

The object of the addition method is to obtain two equations whose sum will be an equation containing only one variable. This occurs when the coefficients of one of the variables, x or y, are opposites (additive inverses) of each other. It is often necessary to multiply one or both equations by some nonzero number so that the coefficients of one of the variables become additive inverses. Only then do we add the equations. Let's see exactly what this means by looking at an example.

EXAMPLE 4   **Solving a System by the Addition Method**

Solve by the addition method:

$$5x + 6y = 2$$
$$3x - 3y = 10$$

Solution

Discover for yourself

Solve the system in Example 4 by eliminating x. Multiply the first equation by 3 and the second equation by -5. Compare your solution with the one shown on the right.

We must rewrite one or both equations in equivalent forms so that the coefficients of the same variable (either x or y) will be opposites of one another. We can accomplish this in a number of ways. Let's consider the terms in y in each equation, that is, $6y$ and $-3y$. To eliminate y, we can multiply each term of the second equation by 2 and then add the equations.

$$
\begin{array}{lll}
5x + 6y = 2 & \xrightarrow{\text{No change}} & 5x + 6y = 2 \\
3x - 3y = 10 & \xrightarrow{\text{Multiply by 2.}} & 6x - 6y = 20 \\
\hline
& \text{Add: } 11x & = 22 \\
& x = 2 & \text{Solve for } x, \text{ dividing both sides by 11.}
\end{array}
$$

Now we back-substitute 2 for x into either one of the given equations and solve for y. We will use the first equation.

Using technology

Check Example 4 with a graphing utility. Solve each equation for y. Enter

$$y_1 = \frac{-5x + 2}{6}$$ and

$$y_2 = \frac{3x - 10}{3}.$$

The intersection point verifies that the system's solution is $(2, -\frac{4}{3})$.

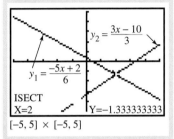

ISECT
X=2   Y=-1.333333333
$[-5, 5] \times [-5, 5]$

$$5x + 6y = 2 \qquad \text{This is the first equation in the given system.}$$

$$5 \cdot 2 + 6y = 2 \qquad \text{Substitute 2 for } x.$$

$$10 + 6y = 2 \qquad \text{Multiply.}$$

$$6y = -8 \qquad \text{Subtract 10 from both sides.}$$

$$y = -\frac{8}{6} = -\frac{4}{3} \qquad \text{Divide both sides by 6.}$$

We have found that $x = 2$ and $y = -\frac{4}{3}$. Now we must verify that the ordered pair $(2, -\frac{4}{3})$ satisfies both equations in the system.

$$5x + 6y = 2 \qquad\qquad 3x - 3y = 10 \qquad \text{These are the given equations.}$$

$$5(2) + 6(-\tfrac{4}{3}) \overset{?}{=} 2 \qquad 3(2) - 3(-\tfrac{4}{3}) \overset{?}{=} 10 \qquad \text{Replace } x \text{ with 2 and } y \text{ with } -\tfrac{4}{3}.$$

$$10 + (-8) \overset{?}{=} 2 \qquad\qquad 6 + 4 \overset{?}{=} 10$$

$$2 = 2 \checkmark \qquad\qquad\qquad 10 = 10 \checkmark$$

Since $(2, -\frac{4}{3})$ checks, the system's solution set is $\{(2, -\frac{4}{3})\}$. ∎

Before considering additional examples, let's summarize the steps involved in the solution of a system of two equations in two variables by the addition method.

Solving linear systems by the addition method

1. If necessary, rewrite both equations in the form $Ax + By = C$.
2. If necessary, multiply either equation or both equations by appropriate numbers so that the coefficients of x or y will be opposites with a sum of 0.
3. Add the equations in step 2. The sum is an equation in one variable.
4. Solve the equation in one variable from step 3.
5. Back-substitute the value obtained in step 4 into either of the given equations and solve for the other variable.
6. Check the solution in both of the original equations.

We mentioned earlier that a linear system in two variables can have one solution, no solution, or infinitely many solutions. Let's see what happens when we apply the addition method to a system with infinitely many solutions.

EXAMPLE 5 **The Addition Method: Infinitely Many-Solutions Case**

Solve by the addition method:

$$8x = 2y + 8$$

$$3y = 12x - 12$$

Solution

We must first arrange the system so that variable terms appear on the left and constants appear on the right. Subtracting $2y$ from both sides of the first equation and $12x$ from both sides of the second equation, we obtain

$$8x - 2y = 8$$
$$-12x + 3y = -12.$$

We can eliminate x or y. Let's eliminate y by multiplying the first equation by 3 and the second equation by 2.

$$8x - 2y = 8 \xrightarrow{\text{Multiply by 3.}} 24x - 6y = 24$$
$$-12x + 3y = -12 \xrightarrow{\text{Multiply by 2.}} -24x + 6y = -24$$

Add: $\quad 0 = 0 \quad$ This statement is true for all values of x and y.

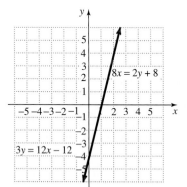

Figure 5.5

The equations in the graph are two different ways of writing $y = 4x - 4$.

In our final step, both variables have been eliminated, and the resulting statement $0 = 0$ is true. The equations in the given system represent two different ways of writing the equation of the same line, illustrated in Figure 5.5. All points on the line have coordinates that satisfy the system. We can use either equation to obtain the line's slope-intercept form by solving for y.

Equation I	**Equation 2**
$8x = 2y + 8$	$3y = 12x - 12$
$8x - 8 = 2y$ Subtract 8 from both sides.	$y = 4x - 4$ Divide both sides by 3.
$4x - 4 = y$ Divide both sides by 2.	

The line's equation in slope-intercept form is $y = 4x - 4$. This means that any ordered pair of the form $(x, 4x - 4)$ is a solution of the system. The system's infinite solution set can be expressed as $\{(x, 4x - 4)\}$. ∎

Discover for yourself

The solution set for the system

$$8x = 2y + 8$$
$$3y = 12x - 12$$

can be expressed as

$$\left\{\left(\frac{t+4}{4}, t\right)\right\}.$$

Select three values for t and write the resulting three ordered pairs. Then show that each of these ordered pairs satisfies the two given equations in the system.

Study tip

The solution set for a linear system with infinitely many solutions is usually expressed in terms of one of the variables. In Example 5, since the equations in the system are two forms of the equation $y = 4x - 4$, we express the solution set as $\{(x, 4x - 4)\}$. We could also solve for x and express the solution set in terms of y.

$$y = 4x - 4$$
$$y + 4 = 4x \qquad \text{Add 4 to both sides.}$$
$$\frac{y + 4}{4} = x \qquad \text{Divide both sides by 4.}$$

The solution set can be expressed as

$$\left\{\left(\frac{y+4}{4}, y\right)\right\}.$$

Letting $y = t$ (or any letter of our choice), the solutions to the system are all of the form

$$x = \frac{t + 4}{4}, y = t,$$

where t is any real number. This means that a third way to express the system's solution set is

$$\left\{ \left(\frac{t + 4}{4}, t \right) \right\}.$$

We will use this latter notation when we study linear systems with infinitely many solutions in more detail in Section 5.3.

3 Identify inconsistent and dependent systems.

Raoul Hausmann (b. Vienna 1886) "The Spirit of Our Times" 1919, wood with leather box, stamp, ruler, cogs, eyeglass and tape measure, h. 32 cm. Photo: Georges Meguerditchian. © Centre Georges Pompidou, Paris. Musée National d'Art Moderne.

Identifying Inconsistent and Dependent Systems

A linear system with no solution is called an *inconsistent system.* When solving such a system by either substitution or addition, both variables will be eliminated and the resulting statement will be a *contradiction.* In Example 3, we applied the substitution method to an inconsistent system and obtained the contradiction $15 = -2$. The two equations represented parallel lines.

A linear system with infinitely many solutions is called a *dependent system.* When solving such a system by either substitution or addition, both variables will be eliminated, and the resulting statement will be true. In Example 5, we applied the addition method to a dependent system and obtained the true statement $0 = 0$. The two equations represented the same line.

These observations are summarized as follows.

Inconsistent and dependent systems of linear equations

If both variables are eliminated when a system of linear equations is solved by substitution or addition:

1. There is no solution if the resulting statement is false. The system is inconsistent.
2. There are infinitely many solutions if the resulting statement is true. The system is dependent.

Any system that has at least one solution is called a *consistent system.* Thus, a linear system with one solution and a linear system with infinitely many solutions are both said to be consistent.

Linear Systems in Three Variables

An equation such as $5x - 2y - 4z = 3$ is called a *linear equation in three variables.* In general, any equation of the form

$$Ax + By + Cz = D$$

where A, B, C, and D are real numbers such that A, B, and C are not all 0, is a linear equation in the variables x, y, and z. The graph of this linear equation in three variables is a plane in three-dimensional space.

The process of solving a system of three linear equations in three variables is geometrically equivalent to finding the point of intersection (assuming that there is one) of three planes in space (see Figure 5.6). The addition method can be used to find the three coordinates of this intersection point.

4 Solve linear systems in three variables by addition.

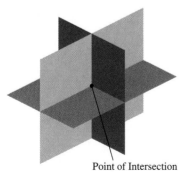

Point of Intersection

Figure 5.6

Solution set of a system

A *solution* of a system of three linear equations in three variables is an ordered triple of real numbers that satisfies all equations of the system. The *solution set* of the system is the set of all its solutions.

The method for solving a system of linear equations in three variables is similar to that used on systems of linear equations in two variables. We use addition to eliminate any variable, reducing the system to two equations in two variables. Once we obtain a system of two equations in two variables, we use the addition method again to solve the system.

EXAMPLE 6   **Solving a System in Three Variables**

Solve the system:

$$5x - 2y - 4z = 3 \qquad \text{Equation 1}$$
$$3x + 3y + 2z = -3 \qquad \text{Equation 2}$$
$$-2x + 5y + 3z = 3 \qquad \text{Equation 3}$$

Solution

There are many ways to proceed. Because our initial goal is to reduce the system to two equations in two variables, *the central idea is to take two different pairs of equations and eliminate the same variable from each pair.*

Step 1. Reduce the system to two equations in two variables.

First we choose any two equations and use the addition method to eliminate a variable. Let's eliminate z from Equations 1 and 2 by first multiplying Equation 2 by 2. Then we'll add equations.

$$5x - 2y - 4z = 3 \qquad \text{Equation 1}$$
$$6x + 6y + 4z = -6 \qquad \text{Twice Equation 2}$$

Add: $11x + 4y \qquad\quad = -3 \qquad \text{Equation 4}$

Thus, we use Equation 4 in place of Equations 1 and 2.

Now we must eliminate the *same* variable from another pair of equations. We can eliminate z from Equations 2 and 3 by multiplying Equation 2 by -3, multiplying Equation 3 by 2, and then adding equations.

(Equation 2) $3x + 3y + 2z = -3$ $\xrightarrow{\text{Multiply by } -3.}$ $-9x - 9y - 6z = 9$

(Equation 3) $-2x + 5y + 3z = 3$ $\xrightarrow{\text{Multiply by 2.}}$ $-4x + 10y + 6z = 6$

Add: $-13x + \quad y \qquad\quad = 15$

Equation 5

Thus, we use Equation 5 in place of Equations 2 and 3.

Step 2. Solve the resulting system of two equations in two variables.

Now we'll solve the resulting system of Equations 4 and 5 for x and y. We multiply Equation 5 on both sides by -4 and add this equation to Equation 4.

$$
\begin{array}{ll}
\text{(Equation 4)} \quad 11x + 4y = -3 & \xrightarrow{\text{No change}} \quad 11x + 4y = -3 \\
\text{(Equation 5)} \quad -13x + y = 15 & \xrightarrow{\text{Multiply by } -4.} \quad 52x - 4y = -60 \\
& \qquad\qquad \text{Add: } \overline{63x \qquad\quad = -63} \\
& \qquad\qquad\qquad\qquad x = -1 \\
& \qquad\qquad\qquad \text{Divide both sides by 63.}
\end{array}
$$

Step 3. Use back-substitution in one of the equations in two variables to find the value of the second variable.

We back-substitute -1 for x in either Equation 4 or 5 to find the value of y.

$$
\begin{array}{ll}
-13x + y = 15 & \text{Equation 5} \\
-13(-1) + y = 15 & \text{Substitute } -1 \text{ for } x. \\
13 + y = 15 & \text{Multiply.} \\
y = 2 & \text{Subtract 13 from both sides.}
\end{array}
$$

Step 4. Back-substitute the values found for two variables into one of the original equations to find the value of the third variable.

We can now use any one of the original equations and back-substitute the values of x and y to find the value for z. We will use Equation 2.

$$
\begin{array}{ll}
3x + 3y + 2z = -3 & \text{Equation 2} \\
3(-1) + 3(2) + 2z = -3 & \text{Substitute } -1 \text{ for } x \text{ and 2 for } y. \\
3 + 2z = -3 & \text{Multiply and then add.} \\
2z = -6 & \text{Subtract 3 from both sides.} \\
z = -3 & \text{Divide both sides by 2.}
\end{array}
$$

With $x = -1$, $y = 2$, and $z = -3$, the proposed solution is the ordered triple $(-1, 2, -3)$.

Step 5. Check the proposed solution in each of the original equations.

We now check the proposed solution in each of the system's given equations. We must show that the numbers in $(-1, 2, -3)$ satisfy all three equations when used as replacements for x, y, and z, respectively.

$$
\begin{array}{lll}
5x - 2y - 4z = 3 & 3x + 3y + 2z = -3 & -2x + 5y + 3z = 3 \\
5(-1) - 2(2) - 4(-3) \overset{?}{=} 3 & 3(-1) + 3(2) + 2(-3) \overset{?}{=} -3 & -2(-1) + 5(2) + 3(-3) \overset{?}{=} 3 \\
-5 - 4 + 12 \overset{?}{=} 3 & -3 + 6 - 6 \overset{?}{=} -3 & 2 + 10 - 9 \overset{?}{=} 3 \\
3 = 3 \ \checkmark & -3 = -3 \ \checkmark & 3 = 3 \ \checkmark
\end{array}
$$

The proposed solution $(-1, 2, -3)$ satisfies the system's equations, so the system's solution set is $\{(-1, 2, -3)\}$. ∎

A number of approaches can be used to solve systems involving three variables. We are first faced with three options regarding which variable to eliminate. Then we must choose which equations to use to eliminate the desired variable. Keep in mind that the initial goal is to reduce the original system to one involving two equations in two variables.

In summary, a system of three linear equations in three variables can be solved by the addition method as follows.

Solving linear systems in three variables by the addition method

1. Reduce the system to two equations in two variables. This is usually accomplished by taking two different pairs of equations and using the addition method to eliminate the same variable from each pair.

2. Solve the resulting system of two equations in two variables using addition or substitution. The result is an equation in one variable that gives the value of that variable.
3. Back-substitute the value of the variable found in step 2 into either of the equations in two variables to find the value of the second variable.
4. Use the values of the two variables from steps 2 and 3 to find the value of the third variable by back-substituting into one of the original equations.
5. Check the solution in each of the original equations.

5 Solve problems using linear systems.

Problem Solving Using Linear Systems

Our strategy for solving problems using linear systems is similar to the problem-solving strategy for one equation in one variable. Here are some general steps we will follow in solving problems.

Strategy for solving problems using linear systems

1. Read the problem and assign letters to represent the unknown quantities.
2. Write a linear system of equations that models the verbal conditions of the problem.
3. Solve the system and answer the problem's question.
4. Check the answers in the original wording of the problem.

Width: W

Length: L

Figure 5.7
A rectangular waterfront lot

EXAMPLE 7 **Solving a Fencing Problem**

A rectangular waterfront lot has a perimeter of 1000 feet. The lot's owner decides to fence along three sides, excluding the side that fronts the water (see Figure 5.7). An expensive fencing along the lot's front length costs $25 per foot, and an inexpensive fencing along the two side widths costs only $5 per foot. The total cost of the fencing along all three sides comes to $9500.

a. What are the lot's dimensions?
b. The owner is considering using the expensive fencing on all three sides of the lot, but is limited by a $16,000 budget. Can this be done given the budget constraints?

Solution

Step 1. Use variables to represent unknown quantities.

Step 2. Write a linear system of equations modeling the problem's conditions.

a. Let L represent the lot's length and let W represent its width.
The lot has a perimeter of 1000 feet.

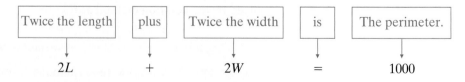

Twice the length	plus	Twice the width	is	The perimeter.
$2L$	$+$	$2W$	$=$	1000

The cost of fencing three sides of the lot is $9500.

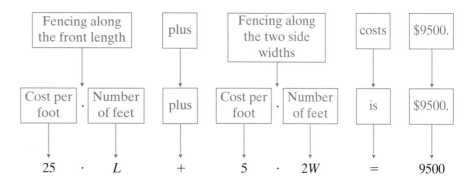

25	·	L	+	5	·	$2W$	=	9500

Step 3. Solve the system and answer the problem's question.

The system

$$2L + 2W = 1000$$

$$25L + 10W = 9500$$

can be solved by addition. We'll multiply the first equation by -5 to eliminate W.

$$2L + 2W = 1000 \xrightarrow{\text{Multiply by } -5.} -10L - 10W = -5000$$

$$25L + 10W = 9500 \xrightarrow{\text{No change}} \underline{25L + 10W = 9500}$$

$$\text{Add:} \quad 15L \quad\quad = 4500$$

$$L \quad\quad = \frac{4500}{15} = 300$$

Now let's find the value of W. We must back-substitute 300 for L in either of the system's equations.

$$2L + 2W = 1000 \quad \text{We'll use the first equation.}$$

$$2(300) + 2W = 1000 \quad \text{Back-substitute 300 for } L.$$

$$600 + 2W = 1000 \quad \text{Multiply.}$$

$$2W = 400 \quad \text{Subtract 600 from both sides.}$$

$$W = 200 \quad \text{Divide both sides by 2.}$$

The lot's dimensions are 300 feet by 200 feet.

Step 4. Check the proposed answers in the original wording of the problem.

The lot's perimeter is

$$2(300) + 2(200) = 1000 \text{ feet}$$

which checks with the given conditions.

The cost of fencing along the front (the length) and the two sides (the two widths) is

$$300(\$25) + 400(\$5) = \$9500$$

which also checks with the given conditions.

b. The owner plans to fence $300 + 2(200)$ or 700 feet. With the expensive fencing, this will cost $700(\$25) = \$17,500$. Limited by a \$16,000 budget, the owner cannot use the expensive fencing on three sides of the lot. ■

EXAMPLE 8 **Solving a Packaging Problem**

A certain brand of razor blades comes in packages of 6, 12, and 24 blades, costing \$2, \$3, and \$4 per package, respectively. A store sold 12 packages containing a total of 162 razor blades and took in \$35. How many packages of each type were sold?

Step 1. Use variables to represent unknown quantities.

Solution

Let

$x = $ the number of \$2 packages sold

$y = $ the number of \$3 packages sold

$z = $ the number of \$4 packages sold.

Step 2. Write a linear system of equations modeling the problem's conditions.

The store sold 12 packages.

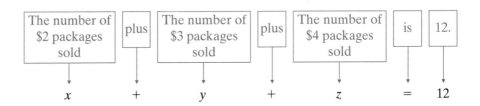

The store sold 162 razor blades.

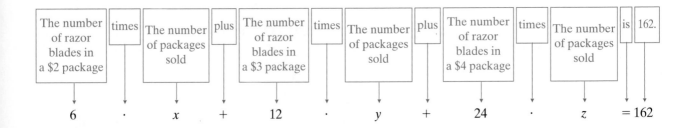

The store took in $35.

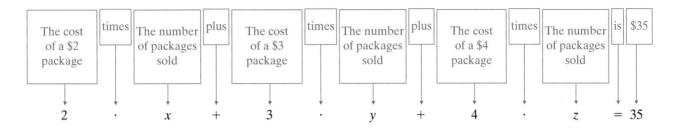

$$2 \cdot x + 3 \cdot y + 4 \cdot z = 35$$

The second equation in the system, $6x + 12y + 24z = 162$, can be simplified by dividing both sides by 6.

$$\frac{6x}{6} + \frac{12y}{6} + \frac{24z}{6} = \frac{162}{6}$$

This results in

$$x + 2y + 4z = 27.$$

The system that models the problem's conditions is

$$x + y + z = 12 \qquad \text{Equation 1}$$

$$x + 2y + 4z = 27 \qquad \text{Equation 2}$$

$$2x + 3y + 4z = 35 \qquad \text{Equation 3}$$

Step 3. Solve the system and answer the problem's question.

This system can be solved by addition. We'll eliminate x and obtain two equations in y and z. We can begin by multiplying Equation 1 by -1 and adding the resulting equation to Equation 2.

(Equation 1) $x + y + z = 12 \xrightarrow{\text{Multiply by } -1.} -x - y - z = -12$

(Equation 2) $x + 2y + 4z = 27 \xrightarrow{\text{No change}} x + 2y + 4z = 27$

Add: $y + 3z = 15$
Equation 4

Now we eliminate the same variable, x, from another pair of equations. One option is to multiply Equation 2 by -2 and add the resulting equation to Equation 3.

(Equation 2) $x + 2y + 4z = 27 \xrightarrow{\text{Multiply by } -2.} -2x - 4y - 8z = -54$

(Equation 3) $2x + 3y + 4z = 35 \xrightarrow{\text{No change}} 2x + 3y + 4z = 35$

Add: $-y - 4z = -19$
Equation 5

The next step is to solve the resulting system of Equations 4 and 5 for y and z. Since the coefficients of y are opposites, we can immediately add these equations and eliminate y.

$$\text{(Equation 4)} \quad y + 3z = 15$$

$$\text{(Equation 5)} \quad -y - 4z = -19$$

$$\text{Add:} \quad -z = -4$$

$$z = 4 \qquad \text{Multiply both sides by } -1.$$

Using back-substitution, we now will find the values for y and x. We back-substitute 4 for z in either Equation 4 or 5 to find the value of y.

$$y + 3z = 15 \qquad \text{Equation 4.}$$

$$y + 3(4) = 15 \qquad \text{Substitute 4 for } z.$$

$$y + 12 = 15 \qquad \text{Multiply.}$$

$$y = 3 \qquad \text{Subtract 12 from both sides.}$$

Now we can use any one of the original equations and back-substitute the values of y and z to find the value of x. We will use Equation 1.

$$x + y + z = 12 \qquad \text{Equation 1}$$

$$x + 3 + 4 = 12 \qquad \text{Substitute 3 for } y \text{ and 4 for } z.$$

$$x + 7 = 12 \qquad \text{Add.}$$

$$x = 5 \qquad \text{Subtract 7 from both sides.}$$

With $x = 5$, $y = 3$, and $z = 4$, there were five \$2 packages, three \$3 packages, and four \$4 packages sold.

Step 4. Check the proposed answers in the original wording of the problem.

Now let's see if the problem's conditions are satisfied. The store sold $5 + 3 + 4$ or 12 packages, which checks with the first condition. Remembering that the packages contain 6, 12, and 24 razor blades, respectively, the store sold $5(6) + 3(12) + 4(24) = 30 + 36 + 96$ or 162 blades, which checks with the second condition. Finally, the store's income is $5(\$2) + 3(\$3) + 4(\$4) = \$10 + \$9 + \$16$ or \$35, which checks with the third condition. ∎

PROBLEM SET 5.1

Practice Problems

Solve each system in Problems 1–8 by the substitution method. Verify solutions by using a graphing utility to graph each system's equations to find intersection points.

1. $\begin{aligned} 2x - y &= 3 \\ y &= 4x + 5 \end{aligned}$

2. $\begin{aligned} 2x - 3y &= 2 \\ y &= 8x + 3 \end{aligned}$

3. $\begin{aligned} 2x - 3y &= -8 \\ x + 3y &= 5 \end{aligned}$

4. $\begin{aligned} -3x + y &= 7 \\ 2x - 5y &= 4 \end{aligned}$

5. $\begin{aligned} 3x + 5y &= -11 \\ x - 2y &= 11 \end{aligned}$

6. $\begin{aligned} 2x + 3y &= -5 \\ x - y &= -10 \end{aligned}$

7. $\begin{aligned} 5x - 4y &= -7 \\ x - \tfrac{3}{5}y &= -2 \end{aligned}$

8. $\begin{aligned} 2x + y &= 4 \\ \tfrac{2}{3}x + \tfrac{1}{4}y &= 2 \end{aligned}$

Solve each system in Problems 9–18 by the addition method. Use a graphing utility to verify solutions as on page 534.

9. $3x - 4y = 1$
$3x + 2y = 13$

10. $3x + 2y = -4$
$3x - y = 11$

11. $3x - y = 13$
$5x + 2y = 7$

12. $3x - 2y = -1$
$2x + y = 4$

13. $6x - y = 14$
$3x - 10y = 45$

14. $4x - y = 4$
$3x + 3y = 18$

15. $3x + 8y = 16$
$2x + 5y = 11$

16. $5x + 3y = 17$
$2x - 5y = 13$

17. $2x - 5y = -14$
$4x - 3y = 8$

18. $2x - 5y = -3$
$3x - 4y = 5$

Solve each system in Problems 19–38 by the method of your choice. Explain why you chose the method you used. Identify inconsistent and dependent systems, expressing the solution set for a dependent system in terms of one of the variables.

19. $x = 2y - 7$
$3x + 2y = 15$

20. $y = 3x + 5$
$5x - 2y = -7$

21. $2x + 5y = -1$
$3x + 4y = 2$

22. $3x + 4y = 2$
$2x + 3y = 5$

23. $x - 3y = 6$
$y = \dfrac{x}{3} - 2$

24. $y = 2x - 4$
$x - \dfrac{y}{2} = 2$

25. $3(x - y) + y = -1$
$2(2x + y) + 2x = y + 8$

26. $4(x + 3y) - 5(y - 2) = -2x$
$3(x - y) - 5(x + y) = -8$

27. $8x - 5y = -23$
$4x + 9y = 0$

28. $3x + 10y = 2$
$x + 3y = 0$

29. $2x - 3y = 6$
$6x - 9y = 36$

30. $5x + 10y = 10$
$4x + 8y = -20$

31. $9x - 2y = 12$
$5x = 14 + 8y$

32. $30x + 14y = -70$
$15x = -7y - 35$

33. $9x - 9y = -45$
$y = 2x + 2$

34. $x = 2y + 1$
$9x - 9y = 36$

35. $2x - y = 1$
$\dfrac{4x}{5} = \dfrac{3y}{2} + \dfrac{1}{5}$

36. $4x + y = 2$
$\dfrac{3x}{2} = \dfrac{2y}{7} - 1$

37. $y = 3x$
$\dfrac{x}{2} + \dfrac{y}{3} = 3$

38. $y - 3x = 2$
$x = \dfrac{1}{4}y$

Solve each system in Problems 39–54.

39. $x + y + 3z = 14$
$x + y + 2z = 11$
$x + 2y - z = 5$

40. $x + y + 3z = 3$
$x + 2y + 4z = 7$
$x + y + 6z = 3$

41. $x + y + z = 6$
$x + 2y - 3z = -11$
$2x - y + z = 11$

42. $x + y + z = 7$
$2x - y + z = 7$
$5x - 3y - 2z = -1$

43. $x - 4y - z = 6$
$2x - y + 3z = 0$
$-3x + 2y - z = -4$

44. $x - 2y + 2z = 4$
$2x + y - 3z = 5$
$-3x + y - 4z = -4$

45. $x + z = 3$
$x + 2y - z = 1$
$2x - y + z = 3$

46. $2x + y = 2$
$x + y - z = 4$
$3x + 2y + z = 0$

47. $x + 3y + 5z = 20$
$y - 4z = -16$
$3x - 2y + 9z = 36$

48. $x + y = -4$
$y - z = 1$
$2x + y + 3z = -21$

49. $x + y = 11$
$y + 2z = 5$
$x - 2z = 4$

50. $x + y = 4$
$x + z = 4$
$y + z = 4$

51. $2x + y + 2z = 1$
$3x - y + z = 2$
$x - 2y - z = 0$

52. $3x + 4y + 5z = 8$
$x - 2y + 3z = -6$
$2x - 4y + 6z = 8$

53. $\dfrac{x}{2} - \dfrac{y}{2} + \dfrac{z}{4} = 1$
$\dfrac{x}{2} + \dfrac{y}{3} - \dfrac{z}{4} = 2$
$\dfrac{x}{4} - \dfrac{y}{2} + \dfrac{z}{2} = 2$

54. $\dfrac{x}{2} + y + z = \dfrac{5}{2}$
$\dfrac{x}{4} - \dfrac{y}{4} + \dfrac{z}{4} = \dfrac{3}{2}$
$\dfrac{2x}{3} + y - \dfrac{z}{3} = \dfrac{1}{3}$

Application Problems

Use a linear system in two variables to solve Problems 55–64.

55. The graph shows the number of executions in the United States from 1976 through 1994. (The death penalty was reinstated in 1977.) The difference between the number of executions in 1993 and 1994 was half the number for 1991. The total number of executions in 1993 and 1994 was triple the number for 1990. How many executions were there in 1993 and 1994?

**Prisoners Executed in the
United States, 1976 –1994**

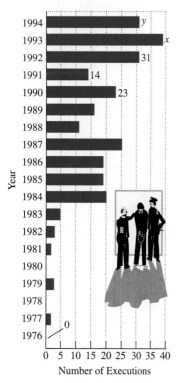

Source: U.S. Bureau of the
Census, Statistical Abstract of the
United States

56. The graph in the next column makes Super Bowl Sunday look like a day of snack food binging in the United States. The number of pounds of guacamole consumed is ten times the difference between the number of pounds of potato and tortilla chips eaten on the same day. On Super Bowl Sunday Americans also wolf down a total quantity of potato and tortilla chips that exceeds popcorn consumption by 7.3 million pounds. How many millions of pounds of potato chips and tortilla chips are consumed on Super Bowl Sunday?

**Millions of Pounds of Snack Food
Consumed on Super Bowl Sunday**

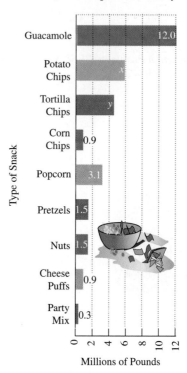

Source: Association of American
Snackfoods. Adapted from *The
Macmillan Visual Almanac*, Simon &
Schuster, 1996

57. A rectangular lot has a perimeter of 320 feet. An expensive fencing along the lot's length costs $16 per foot, and an inexpensive fencing along the two side widths costs only $5 per foot. The total cost of the fencing along all three sides comes to $2140. What are the lot's dimensions?

58. A rectangular lot has a perimeter of 360 feet. An expensive fencing along the lot's length costs $20 per foot, and an inexpensive fencing along the two side widths costs only $8 per foot. The total cost of the fencing along all three sides comes to $3280. What are the lot's dimensions?

59. The verdict is in: after years of research, the nation's health experts agree that high cholesterol in the blood is a major contributor to heart disease. Cholesterol intake should be limited to 300 mg or less each day. Fast foods provide a cholesterol carnival. Two McDonald's

Quarter Pounders and three Burger King Whoppers with cheese contain 520 mg of cholesterol. Three Quarter Pounders and one Whopper with cheese exceed the suggested daily cholesterol intake by 53 mg. Determine the cholesterol content in each item.

Claes Oldenburg (American, 1929) "Floor Burger" 1962, canvas filled with foam rubber and cardboard boxes, painted with latex paint, 132.1 × 213.4 cm. Art Gallery of Ontario.

60. Cake can vary in cholesterol content. Four slices of sponge cake and 2 slices of pound cake contain 784 mg of cholesterol. One slice of sponge cake and 3 slices of pound cake contain 366 mg of cholesterol. Find the cholesterol content in each item.

61. A heat-loss survey by an electric company indicated that a wall of a house containing 40 square feet of glass

and 60 square feet of plaster lost 1920 Btu (British thermal units) of heat. A second wall containing 10 square feet of glass and 100 square feet of plaster lost 1160 Btu. Determine the heat lost per square foot for the glass and for the plaster.

62. A football team, as part of a special promotion at one of its games, will give a team athletic bag or a jacket to the first 750 fans who arrive. Each athletic bag costs $4.50 and each jacket costs $8.25, and the team's promotion manager has a budget of $4500. How many team bags and how many jackets can be given away at the promotion?

63. Find x and y in the figure below.

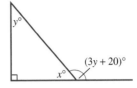

64. Find x and y in the figure below.

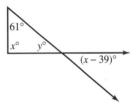

Use a linear system in three variables to solve Problems 65–71.

65. The graph shows a low savings rate in the United States compared to that of many industrialized countries. The combined rate for Japan, Germany, and France is 45%. The savings rate in Japan exceeds that for Germany by 1% and is 12% less than twice that for France. Find the savings rate for Japan, Germany, and France.

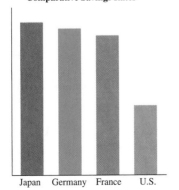

Comparative Savings Rates

Source: Office of Management and Budget

66. The largest angle in a triangle measures 93° more than the smallest angle. The measure of the largest angle exceeds three times the measure of the remaining angle by 7°. Find the measure of the triangle's three interior angles.

67. In a triangle, the largest angle is 80° greater than the smallest angle. The largest angle is also 20° less than twice as large as the remaining angle. Find the measure of each angle.

68. At a college production of *Evita,* 400 tickets were sold. The ticket prices were $8, $10, and $12, and the total income from ticket sales was $3700. How many tickets of each type were sold if the combined number of $8 and $10 tickets sold was 7 times the number of $12 tickets sold?

69. The figure at the top of page 538 represents three circular pulleys that are mutually tangent. The center-to-center distances AB, AC, and BC are 8 inches, 11 inches, and 9 inches, respectively. Find the radius of each pulley.

70. An individual invests $10,000 in three stocks that pay dividends of 6%, 8%, and 10%. The amount invested at 10% is twice the amount invested at 6%, and the return

on the three investments combined is $860. What is the amount invested in each stock?

71. The equation $y = \frac{1}{2}Ax^2 + Bx + C$ gives the relationship between the number of feet a car travels once the brakes are applied (y) and the number of seconds the car is in motion after the brakes are applied (x). A research firm discovered that when a car was in motion for 1 second after the brakes were applied, the car traveled 46 feet. (When $x = 1$, $y = 46$.) Similarly, it was found that when x was 2, y was 84, and when x was 3, y was 114. Use these values to find the constants A, B, and C in the equation. What is the value for y when $x = 6$? Describe what this means.

True–False Critical Thinking Problems

72. Which one of the following is true?

a. The addition method cannot be used to eliminate either variable in a system of two equations in two variables.

b. The solution set to the system

$$5x - y = 1$$
$$10x - 2y = 2$$

is $\{(2, 9)\}$.

c. A linear system of equations can have a solution set consisting of precisely two ordered pairs.

d. If two equations in a system are $x + y - z = 5$ and $2x + 2y - 2z = 7$, then the system cannot have a solution.

73. Which one of the following is true?

a. A linear system in three variables cannot have $(0, 0, 0)$ as a solution.

b. Because no variable has a coefficient of 1 or -1, the substitution method cannot be used to solve the system

$$3x + 5y = 19$$
$$2x + 7y = 23.$$

c. The equation $x - y - z = -6$ is satisfied by $(2, -3, 5)$.

d. The addition method can be used to solve an equation containing irrational coefficients and constants such as

$$\sqrt{2}x + \sqrt{3}y = 5$$
$$\sqrt{8}x + 2\sqrt{6}y = \sqrt{5}.$$

Technology Problems

74. Some graphing utilities can give the solution to a linear system of equations. (Consult your manual for details.) This capability is usually accessed with the $\boxed{\text{SIMULT}}$ (simultaneous equations) feature. First, you will enter 2, for two equations in two variables. With each equation in standard form, you will then enter the coefficients for x and y and the constant term, one equation at a time. After entering all six numbers, press $\boxed{\text{SOLVE}}$. The solution will be displayed on the screen. (The x-coordinate may be displayed as $x_1 =$ and the y-value as $x_2 =$.) Use this capability to verify the solution to some of the problems you solved from Problems 1–38 in this problem set.

75. Use the $\boxed{\text{SIMULT}}$ feature of your graphing utility to verify the solution to some of the problems from Problems 39–54 in this problem set. On most utilities, you will enter 3 (for three equations in three variables), and then enter the coefficients for x, y, z, and the constant term, one equation at a time. After entering all nine numbers, press $\boxed{\text{SOLVE}}$. The solution will be displayed on the

screen. (The x-, y-, and z-coordinates may be displayed as x_1, x_2, and x_3.) Consult your manual for details.

76. Some graphing utilities will do three-dimensional graphing. For example, on the TI-92, press $\boxed{\text{MODE}}$, go to $\boxed{\text{GRAPH}}$, press the arrow to the right, select $\boxed{\text{3D}}$, then $\boxed{\text{ENTER}}$. When you display the $\boxed{\text{Y} =}$ screen, you will see the equations are functions of x and y. Thus, you must solve each of a linear system's equations for z before entering the equation. For example,

$$x + y + z = 19$$

is solved for z, giving

$$z = 19 - x - y.$$

(Consult your manual.) If your utility does three-dimensional graphing, graph some of the systems in Problems 39–54 and trace along the planes to find their common point of intersection.

Writing in Mathematics

77. The daily cost (y) for the Worldwide Widget Corporation to produce x widgets is given by the linear function $y = 8x + 200$. The daily revenue (y) generated by the sale of x widgets is described by another linear function, $y = 16x$. Use the graphs in the figure to write a short paragraph describing how the company breaks even and under what conditions there will be either profit or loss.

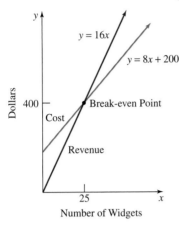

78. When is it easier to use the addition method rather than the substitution method for solving a linear system of equations in two variables?

79. When using the addition method to solve a linear system of three equations in three variables, a variable is eliminated from any two of the given equations. Why is it necessary to eliminate the *same* variable from any other two equations?

80. Describe how the system

$$x + y - z - 2w = -8$$

$$x - 2y + 3z + w = 18$$

$$2x + 2y + 2z - 2w = 10$$

$$2x + y - z + w = 3$$

could be solved. Is it likely that in the near future a graphing utility will be available to provide a geometric solution (using intersecting graphs) to this system? Explain.

Critical Thinking Problems

81. Write a system of equations having $\{(-2, 7)\}$ as a solution set. (More than one system is possible.)

82. Write the point-slope form, the slope-intercept form, and the standard form of the line passing through the intersection of $x + 2y = 4$ and $x - y = -5$ and perpendicular to the line whose equation is $3x - 9y = 10$.

83. Solve the system for x and y in terms of a_1, b_1, c_1, a_2, b_2, and c_2:

$$a_1 x + b_1 y = c_1$$

$$a_2 x + b_2 y = c_2.$$

84. Solve:

$$\frac{5}{x} + \frac{6}{y} = \frac{19}{6}$$

$$\frac{3}{x} + \frac{4}{y} = 2.$$

(*Hint:* Let $a = \dfrac{1}{x}$ and $b = \dfrac{1}{y}$. Substitute these expressions into the system to find a and b. Then find x and y.)

85. A high school has 900 students, 40% of whom are Hispanic. A second high school has a 75% Hispanic population. The school board plans to merge the two schools into one school that will then have a 52.5% Hispanic student body. How many students are at the second school?

86. On a particular train route, adult tickets cost $20, tickets for an adult with one child cost $30, and tickets for an adult with two children cost $35. The train's conductor collected $415 in fares for 28 passengers, 15 of whom were adults. How many tickets of each type were sold?

87. Solve for x, y, and z in terms of the constants a, b, and c:

$$x + y = a$$

$$y + z = b$$

$$x + z = c.$$

Solve each system in Problems 88–89 for x and y in terms of A and B. What assumptions must be made about A and B so that the system has a solution?

88. $Ax + A^2y = 1$
$Bx + B^2y = 1$

89. $(A + B)x + (A^2 + B^2)y = A^3 + B^3$
$(A - B)x + (A^2 - B^2)y = A^3 - B^3$

Group Activity Problem _____

90. In your group, write three problems that are similar to application Problems 65–71 in this problem set. Each problem should give rise to three equations in three variables. If you write problems like Problem 65, consult the reference section of your library for appropriate graphs. Then solve each of the three problems.

S E C T I O N   5 . 2 | **Matrix Solutions to Linear Systems**

Solutions Manual | **Tutorial** | **Video 9**

Objectives

1 Write the augmented matrix for a system.
2 Use matrices to solve systems.
3 Use matrices to model data and solve problems.

In this section, we look at a rather rapid technique for solving linear systems of equations. Since systems involving two equations in two variables can easily be solved by substitution or addition, our focus will be on linear systems in three or more variables. The solution method for these systems, using rectangular arrays of numbers called *matrices,* is essentially a streamlined way of using the addition method to eliminate variables.

> **Definition of a matrix**
>
> A *matrix* (plural: *matrices*) is a rectangular array of numbers arranged in rows and columns and placed in brackets. Each number in the matrix is an *element* or *entry.*

Here are three examples of matrices.

$$\begin{bmatrix} 7 & -3 & 0 \\ 5 & 2 & -11 \end{bmatrix}$$

This is a 2 × 3 (2 by 3) matrix with two rows and three columns.

$$\begin{bmatrix} 7 & -4 & -5 \\ -2 & 0 & -6 \\ -5 & 0 & 0.25 \\ \pi & \sqrt{3} & -\frac{1}{3} \end{bmatrix}$$

This is a 4 × 3 (4 by 3) matrix with four rows and three columns.

$$\begin{bmatrix} -3 & 0 \\ 7 & -5 \end{bmatrix}$$

This is a 2 × 2 (2 by 2) matrix with two rows and two columns. This is also a *square matrix* because it has the same number of rows as columns.

Alexandra Exter (Russian 1884–1949) "Robot" 1926, cardboard, fabric, wood, glass and string, ht.: 50.8 cm. Block, Serlin, and Saidenberg Fund; through prior gifts of Mr. and Mrs. Carter H. Harrison, Arthur Keeting, and Albert Kunstadter Family Foundation, 1990.133. Photograph by Robert Hashimoto. Photograph © 1997, The Art Institute of Chicago, All Rights Reserved.

Augmented Matrices

1 Write the augmented matrix for a system.

A matrix derived from a linear system of equations, each in standard form, is called the *augmented matrix* of the system. An augmented matrix has a vertical bar separating the columns of the matrix into two groups. The coefficients of each variable in a linear system are placed to the left of the vertical line, and the constants are placed to the right.

The augmented matrix for a linear system in three variables

<table>
<tr><th>System of Linear Equations</th><th>Augmented Matrix</th></tr>
<tr><td>
$$a_1x + b_1y + c_1z = d_1$$
$$a_2x + b_2y + c_2z = d_2$$
$$a_3x + b_3y + c_3z = d_3$$
</td><td>
$$\begin{bmatrix} a_1 & b_1 & c_1 & \vert & d_1 \\ a_2 & b_2 & c_2 & \vert & d_2 \\ a_3 & b_3 & c_3 & \vert & d_3 \end{bmatrix}$$
</td></tr>
</table>

S tudy tip

The augmented matrix for a linear system is a matrix with the system's variables and equal signs eliminated.

Here are some examples of augmented matrices.

<table>
<tr><th>System</th><th>Augmented Matrix</th></tr>
<tr><td>
$$3x + y + 2z = 31$$
$$x + y + 2z = 19$$
$$x + 3y + 2z = 25$$
</td><td>
$$\begin{bmatrix} 3 & 1 & 2 & \vert & 31 \\ 1 & 1 & 2 & \vert & 19 \\ 1 & 3 & 2 & \vert & 25 \end{bmatrix}$$
</td></tr>
<tr><td>
$$x + 2y - 5z = -19$$
$$y + 3z = 9$$
$$z = 4$$
</td><td>
$$\begin{bmatrix} 1 & 2 & -5 & \vert & -19 \\ 0 & 1 & 3 & \vert & 9 \\ 0 & 0 & 1 & \vert & 4 \end{bmatrix}$$
</td></tr>
</table>

Notice how the second matrix contains 1s down the diagonal from upper left to lower right and 0s below the 1s. This arrangement makes it easy to find the solution of the system of equations, as Example 1 shows.

EXAMPLE 1 Solving a System Using a Matrix

Write the solution set for a system of equations represented by the matrix:

$$\begin{bmatrix} 1 & 2 & -5 & \vert & -19 \\ 0 & 1 & 3 & \vert & 9 \\ 0 & 0 & 1 & \vert & 4 \end{bmatrix}$$

Solution

The system represented by the given matrix is

$$\begin{bmatrix} 1 & 2 & -5 & \vert & -19 \\ 0 & 1 & 3 & \vert & 9 \\ 0 & 0 & 1 & \vert & 4 \end{bmatrix} \longrightarrow \begin{array}{l} 1x + 2y - 5z = -19 \\ 0x + 1y + 3z = 9 \\ 0x + 0y + 1z = 4 \end{array}$$

This system can be simplified as follows.

$$x + 2y - 5z = -19 \quad \text{Equation 1}$$

$$y + 3z = 9 \quad \text{Equation 2}$$

$$z = 4 \quad \text{Equation 3}$$

The value of z is known. We can find y by back-substitution.

$$y + 3z = 9 \quad \text{Equation 2}$$

$$y + 3(4) = 9 \qquad \text{Substitute 4 for } z.$$
$$y + 12 = 9 \qquad \text{Multiply.}$$
$$y = -3 \qquad \text{Subtract 12 from both sides.}$$

With values for y and z, we can now use back-substitution to find x.

$$x + 2y - 5z = -19 \qquad \text{Equation 1}$$
$$x + 2(-3) - 5(4) = -19 \qquad \text{Substitute } -3 \text{ for } y \text{ and 4 for } z.$$
$$x - 6 - 20 = -19 \qquad \text{Multiply.}$$
$$x - 26 = -19 \qquad \text{Add.}$$
$$x = 7 \qquad \text{Add 26 to both sides.}$$

We see that $x = 7$, $y = -3$, and $z = 4$. The solution set for the system is $\{(7, -3, 4)\}$. ■

If all augmented matrices could be converted to the form of Example 1, with 1s down the main diagonal and 0s below the 1s, a system's solution could easily be obtained from the matrix. As it turns out, there are ways of rewriting matrices to convert them to this more useable form.

Our goal in solving a linear system in three variables is to produce a matrix similar to the one in Example 1. In general, the matrix will be of the form

$$\left[\begin{array}{ccc|c} 1 & A & B & C \\ 0 & 1 & D & E \\ 0 & 0 & 1 & F \end{array}\right].$$

The system represented by this matrix is

$$x + Ay + Bz = C$$
$$y + Dz = E$$
$$z = F$$

Study tip

In triangular form, the second equation does not have an x in it, and the third equation has neither x nor y.

This system is said to be in *triangular form*. Example 1 illustrated that it is quite easy to solve a system in this form.

We wish to solve a system of linear equations by using the augmented matrix for that system. We use *row operations* on the augmented matrix. The row operations may change the numbers in the matrix, but they produce new matrices representing systems of equations with the same solution set as that of the original system.

Matrix row operations

Three row operations produce matrices that lead to systems with the same solution set as the original system.

Description of the operation	Symbol and meaning	Example
Interchange two rows of the matrix.	$R_i \leftrightarrow R_j$ Interchange the entries in ith and jth rows.	$\left[\begin{array}{cc\|c} 1 & 2 & -1 \\ 4 & -3 & -15 \end{array}\right]$ $R_1 \leftrightarrow R_2$: Interchange rows 1 and 2. $\left[\begin{array}{cc\|c} 4 & -3 & -15 \\ 1 & 2 & -1 \end{array}\right]$

Multiply the entries in any row by the same nonzero real number.	kR_i Multiply each entry in the ith row by k.	$\begin{bmatrix} 1 & 2 & 3 \\ 0 & 6 & 4 \end{bmatrix}$ $\frac{1}{6}R_2$: Multiply row 2 by $\frac{1}{6}$. $\begin{bmatrix} 1 & 2 & 3 \\ 0 & 1 & \frac{2}{3} \end{bmatrix}$
Add to the entries of any row a multiple of the corresponding entries of another row.	$kR_i + R_j$ Add k times the entries in row i to the corresponding entries in row j.	$\begin{bmatrix} 1 & 2 & -1 \\ 4 & -3 & -15 \end{bmatrix}$ $-4R_1 + R_2$: Multiply row 1 by -4 and add to row 2. $\begin{bmatrix} 1 & 2 & -1 \\ (-4)(1)+4 & (-4)(2)+(-3) & (-4)(-1)+(-15) \end{bmatrix}$ which simplifies to $\begin{bmatrix} 1 & 2 & -1 \\ 0 & -11 & -11 \end{bmatrix}$

Two matrices are *row-equivalent* if one can be obtained from the other by a sequence of row operations.

EXAMPLE 2 Performing Matrix Row Operations

Perform the indicated row operation:

a. $\begin{bmatrix} 3 & 1 & -5 & 2 \\ -1 & 4 & 2 & 1 \\ 1 & 2 & -1 & 3 \end{bmatrix}$, $R_1 \leftrightarrow R_3$ **b.** $\begin{bmatrix} 1 & 2 & -1 & 3 \\ 0 & 6 & 1 & 4 \\ 3 & 1 & -5 & 2 \end{bmatrix}$, $-3R_1 + R_3$

c. $\begin{bmatrix} 1 & 2 & -1 & 3 \\ 0 & 6 & 1 & 4 \\ 0 & -5 & -2 & -7 \end{bmatrix}$, $\frac{1}{6}R_2$

Discover for yourself

The system for the augmented matrix in part (a) is

$3x + y - 5z = 2$

$-x + 4y + 2z = 1$

$x + 2y - z = 3$.

This system has the same solution as

$x + 2y - z = 3$

$-x + 4y + 2z = 1$

$3x + y - 5z = 2$

obtained by interchanging the first and third equations. Now write the systems for the augmented matrices in parts (b) and (c). Then write the system corresponding to the matrices once we have performed the indicated row operations. Explain how the resulting systems can be obtained without using matrices.

Solution

a. Interchanging rows 1 and 3 $(R_1 \leftrightarrow R_3)$ gives us

$$\begin{bmatrix} 1 & 2 & -1 & 3 \\ -1 & 4 & 2 & 1 \\ 3 & 1 & -5 & 2 \end{bmatrix}.$$

b. Multiplying the first row by -3 and adding to the third row $(-3R_1 + R_3)$ gives us

$$\begin{bmatrix} 1 & 2 & -1 & 3 \\ 0 & 6 & 1 & 4 \\ 3+(-3)\cdot 1 & 1+(-3)\cdot 2 & -5+(-3)\cdot(-1) & 2+(-3)\cdot(3) \end{bmatrix}$$

↑	↑	↑	↑	↑	↑	↑	↑
Original number from row 3	−3 times the number from row 1	Original number from row 3	−3 times the number from row 1	Original number from row 3	−3 times the number from row 1	Original number from row 3	−3 times the number from row 1

$$= \begin{bmatrix} 1 & 2 & -1 & 3 \\ 0 & 6 & 1 & 4 \\ 0 & -5 & -2 & -7 \end{bmatrix}.$$

c. Multiplying the second row by $\frac{1}{6}$ $\left(\frac{1}{6}R_2\right)$ gives us

$$\begin{bmatrix} 1 & 2 & -1 & 3 \\ \frac{1}{6}\cdot 0 & \frac{1}{6}\cdot 6 & \frac{1}{6}\cdot 1 & \frac{1}{6}\cdot 4 \\ 0 & -5 & -2 & -7 \end{bmatrix} = \begin{bmatrix} 1 & 2 & -1 & 3 \\ 0 & 1 & \frac{1}{6} & \frac{2}{3} \\ 0 & -5 & -2 & -7 \end{bmatrix}. \qquad \blacksquare$$

② Use matrices to solve systems.

Using Matrices and Row Operations to Solve Linear Systems

To solve a linear system using matrix row operations, we begin with the augmented matrix. Then we use row operations to obtain a row-equivalent matrix with 1s down the diagonal from left to right and 0s below each 1.

> **Study tip**
>
> Begin with the augmented matrix. Then proceed column by column from left to right. In each column, obtain a 1 in the diagonal position. Then clear out the entries below the 1 and obtain 0s, one by one, using row operations.

EXAMPLE 3 **Using Row Operations to Solve a System**

Use matrices to solve:

$$3x + y + 2z = 31$$

$$x + y + 2z = 19$$

$$x + 3y + 2z = 25$$

Solution

We begin by writing the system in terms of an augmented matrix:

$$\begin{bmatrix} \boxed{3} & 1 & 2 & 31 \\ 1 & 1 & 2 & 19 \\ 1 & 3 & 2 & 25 \end{bmatrix}$$ Our first goal is to get a 1 where the 3 is in the upper-left cell.

To get a 1 in the top position of the first column, we interchange rows 1 and 2 (or rows 1 and 3).

$$\begin{bmatrix} 1 & 1 & 2 & 19 \\ \boxed{3} & 1 & 2 & 31 \\ \boxed{1} & 3 & 2 & 25 \end{bmatrix}$$ $R_1 \leftrightarrow R_2$
Now the boxed entries should be 0s.

Now we want 0s in the second two positions of column 1. The first 0 (where there is now 3) can be obtained by adding to row 2 the results of multiplying each number in row 1 by -3. The second 0 (where there is now 1) can be obtained by adding to row 3 the results of multiplying each number in row 1 by -1.

$$\begin{bmatrix} 1 & 1 & 2 & 19 \\ -3\cdot 1 + 3 & -3\cdot 1 + 1 & -3\cdot 2 + 2 & -3\cdot 19 + 31 \\ -1\cdot 1 + 1 & -1\cdot 1 + 3 & -1\cdot 2 + 2 & -1\cdot 19 + 25 \end{bmatrix}$$ $\leftarrow -3R_1 + R_2$
$\leftarrow -1R_1 + R_3$

This matrix can be simplified.

$$\begin{bmatrix} 1 & 1 & 2 & | & 19 \\ 0 & \boxed{-2} & -4 & | & -26 \\ 0 & 2 & 0 & | & 6 \end{bmatrix}$$
Because we want 1s down the main diagonal, the boxed entry should be 1.

We can obtain 1 in row 2, column 2, by multiplying each number in row 2 by $-\frac{1}{2}$.

$$\begin{bmatrix} 1 & 1 & 2 & | & 19 \\ 0 & 1 & 2 & | & 13 \\ 0 & \boxed{2} & 0 & | & 6 \end{bmatrix} \leftarrow -\frac{1}{2}R_2 \qquad \text{Now the boxed entry should be 0.}$$

Now we want 0 below the 1 instead of 2. We get 0 in row 3, column 2, by adding to row 3 the results of multiplying each number in row 2 by -2.

$$\begin{bmatrix} 1 & 1 & 2 & | & 19 \\ 0 & 1 & 2 & | & 13 \\ -2 \cdot 0 + 0 & -2 \cdot 1 + 2 & -2 \cdot 2 + 0 & | & -2 \cdot 13 + 6 \end{bmatrix} \leftarrow -2R_2 + R_3$$

Now we simplify the entries.

$$\begin{bmatrix} 1 & 1 & 2 & | & 19 \\ 0 & 1 & 2 & | & 13 \\ 0 & 0 & \boxed{-4} & | & -20 \end{bmatrix}$$
Remember that we want 1s down the main diagonal. The boxed entry should be 1.

The last step involves obtaining 1 in row 3, column 3. We multiply each number in row 3 by $-\frac{1}{4}$.

$$\begin{bmatrix} 1 & 1 & 2 & | & 19 \\ 0 & 1 & 2 & | & 13 \\ 0 & 0 & 1 & | & 5 \end{bmatrix} \leftarrow -\frac{1}{4}R_3$$

Converting back to a system of equations, we obtain the *triangular system*

$$x + y + 2z = 19$$
$$y + 2z = 13$$
$$z = 5.$$

Alexander Calder (American 1898–1976) "Obus" 1972, steel, painted, 3.618 × 3.859 × 2.276 (142 1/2 × 152 × 89 5/8). Collection of Mr. and Mrs. Paul Mellon, © 1997 Board of Trustees, National Gallery of Art, Washington, D.C. Photo by: Philip A. Charles. © 1998 Estate of Alexander Calder / Artists Rights Society (ARS), New York.

To find y, we back-substitute 5 for z in the second equation:

$$y + 2z = 13 \qquad \text{Equation 2}$$
$$y + 2(5) = 13 \qquad \text{Substitute 5 for } z.$$
$$y = 3 \qquad \text{Solve for } y.$$

Finally, back-substitute 3 for y and 5 for z in the first equation:

$$x + y + 2z = 19 \quad \text{Equation 1}$$

$$x + 3 + 2(5) = 19 \quad \text{Substitute 3 for } y \text{ and 5 for } z.$$

$$x + 13 = 19 \quad \text{Multiply and add.}$$

$$x = 6 \quad \text{Subtract 13 from both sides.}$$

The solution set for the original system is $\{(6, 3, 5)\}$. ∎

The process that we used to solve the system in Example 3 is called *Gaussian elimination,* after the German mathematician Carl Friedrich Gauss (1777–1855). Here's a summary of the steps used in Gaussian elimination.

> **Solving linear systems using Gaussian elimination**
>
> 1. Write the augmented matrix for the system.
> 2. Use matrix row operations to simplify the matrix to one with 1s down the diagonal from upper left to lower right, and 0s below the 1s.
> 3. Write the system of linear equations corresponding to the matrix in step 2, and use back-substitution to find the system's solution.

Example 4 illustrates Gaussian elimination for a linear system involving four equations in four variables.

Study tip

Begin with the augmented matrix. The final matrix should be of the form

$$\begin{bmatrix} 1 & - & - & - & A \\ 0 & 1 & - & - & B \\ 0 & 0 & 1 & - & C \\ 0 & 0 & 0 & 1 & D \end{bmatrix}$$

From this matrix we can conclude that $w = D$. We then can use back-substitution to find the values for the other three variables.

EXAMPLE 4 **Gaussian Elimination with Back-Substitution**

Use matrices to solve the system:

$$2x + y + 3z - w = 6$$

$$x - y + 2z - 2w = -1$$

$$x - y - z + w = -4$$

$$-x + 2y - 2z - w = -7$$

Solution

We begin by writing the system in terms of an augmented matrix.

$$\begin{bmatrix} \boxed{2} & 1 & 3 & -1 & 6 \\ 1 & -1 & 2 & -2 & -1 \\ 1 & -1 & -1 & 1 & -4 \\ -1 & 2 & -2 & -1 & -7 \end{bmatrix}$$

Our first goal is to get a 1 where the 2 is in the upper-left cell.

ENRICHMENT ESSAY

Disasters!

Greg O'Halloran "Natural Disasters: Tornado #1" 1985–86, powdered graphite, watercolor, watercolor pencil, wood and compo, 56 × 56 in.

Meteorologists modeling atmospheric conditions surrounding tornados and hurricanes must solve huge systems rapidly and efficiently. Such solutions can result in effective warning times that mean survival for those in the path of nature's destructive forces.

Otto Freundlich "Composition" 1930. Donation Freundlich. Musée de Pontoise, France.

To get a 1 in the top position of the first column, we interchange rows 1 and 2.

$$\left[\begin{array}{cccc|c} 1 & -1 & 2 & -2 & -1 \\ \boxed{2} & 1 & 3 & -1 & 6 \\ \boxed{1} & -1 & -1 & 1 & -4 \\ \boxed{-1} & 2 & -2 & -1 & -7 \end{array}\right] \quad R_1 \leftrightarrow R_2$$

Now the boxed entries should be 0s.

Now we want to get 0s below the first 1 on the diagonal. First, to get a 0 in the first position of the second row, we multiply the first row by -2 and add the result to the second row. Second, to get a 0 in the first position of the third row, we multiply the first row by -1 and add the result to the third row. Finally, to get a 0 in the first position of the last row, we add the corresponding entries of rows 1 and 4.

$$\left[\begin{array}{cccc|c} 1 & -1 & 2 & -2 & -1 \\ 0 & \boxed{3} & -1 & 3 & 8 \\ 0 & 0 & -3 & 3 & -3 \\ 0 & 1 & 0 & -3 & -8 \end{array}\right] \begin{array}{l} \leftarrow -2R_1 + R_2 \\ \leftarrow -1R_1 + R_3 \\ \leftarrow 1R_1 + R_4 \end{array}$$

Because we want 1s down the main diagonal, the boxed entry should be 1.

We can obtain 1 in row 2, column 2 by multiplying each entry in row 2 by $\frac{1}{3}$.

$$\left[\begin{array}{cccc|c} 1 & -1 & 2 & -2 & -1 \\ 0 & 1 & -\frac{1}{3} & 1 & \frac{8}{3} \\ 0 & \boxed{0} & -3 & 3 & -3 \\ 0 & \boxed{1} & 0 & -3 & -8 \end{array}\right] \leftarrow \frac{1}{3}R_2$$

Now the boxed entries should be 0s.

Notice that we already have a 0 in row 3, column 2. To get a 0 in the final position of column 2, we multiply the second row by -1 and add the result to the corresponding entries of the last row. (What would happen if we added rows 1 and 4?)

$$\left[\begin{array}{cccc|c} 1 & -1 & 2 & -2 & -1 \\ 0 & 1 & -\frac{1}{3} & 1 & \frac{8}{3} \\ 0 & 0 & \boxed{-3} & 3 & -3 \\ 0 & 0 & \frac{1}{3} & -4 & -\frac{32}{3} \end{array}\right] \leftarrow -1R_2 + R_4$$

Now the boxed entry should be 1.

We can obtain 1 in row 3, column 3 by multiplying each entry in row 3 by $-\frac{1}{3}$.

$$\left[\begin{array}{cccc|c} 1 & -1 & 2 & -2 & -1 \\ 0 & 1 & -\frac{1}{3} & 1 & \frac{8}{3} \\ 0 & 0 & 1 & -1 & 1 \\ 0 & 0 & \boxed{\frac{1}{3}} & -4 & -\frac{32}{3} \end{array}\right] \leftarrow -\frac{1}{3}R_3$$

Now the boxed entry should be 0.

Remember that we want to obtain 1s down the main diagonal and 0s below the 1s. To get a 0 in row 4, column 3, we multiply the third row by $-\frac{1}{3}$ and add the result to the corresponding entry in row 4.

$$\left[\begin{array}{cccc|c} 1 & -1 & 2 & -2 & -1 \\ 0 & 1 & -\frac{1}{3} & 1 & \frac{8}{3} \\ 0 & 0 & 1 & -1 & 1 \\ 0 & 0 & 0 & \boxed{-\frac{11}{3}} & -11 \end{array}\right] \leftarrow -\frac{1}{3}R_3 + R_4$$

Because we want 1s down the main diagonal, the boxed entry should be 1.

The last step involves obtaining 1 in row 4, column 4. We multiply each entry in row 4 by $-\frac{3}{11}$.

$$\left[\begin{array}{cccc|c} 1 & -1 & 2 & -2 & -1 \\ 0 & 1 & -\frac{1}{3} & 1 & \frac{8}{3} \\ 0 & 0 & 1 & -1 & 1 \\ 0 & 0 & 0 & 1 & 3 \end{array}\right] \begin{array}{l} \\ \\ \\ \leftarrow -\frac{3}{11}R_4 \end{array}$$

We now have 1s down the main diagonal and 0s below the 1s.

Converting back to a system of equations, we get

$$x - y + 2z - 2w = -1$$
$$y - \frac{1}{3}z + w = \frac{8}{3}$$
$$z - w = 1$$
$$w = 3$$

This system in triangular form is obtained by attaching the variables x, y, z, and w to the coefficients of the last augmented matrix.

We can now use back-substitution to find the values for z, y, and x.

$$w = 3 \qquad z - w = 1 \qquad y - \frac{1}{3}z + w = \frac{8}{3} \qquad x - y + 2z - 2w = -1$$

$$z - 3 = 1 \qquad y - \frac{1}{3}(4) + 3 = \frac{8}{3} \qquad x - 1 + 2(4) - 2(3) = -1$$

$$z = 4 \qquad y + \frac{5}{3} = \frac{8}{3} \qquad x - 1 + 8 - 6 = -1$$

$$y = 1 \qquad x + 1 = -1$$

$$x = -2$$

Let's agree to write the solution set for the system in the order in which the variables for the given system appeared from left to right, namely (x, y, z, w). Thus, the solution set is $\{(-2, 1, 4, 3)\}$, which can be verified by substitution in the original system of equations. ◼

Gauss-Jordan Elimination

Using Gaussian elimination, matrix row operations are applied to an augmented matrix to obtain a row-equivalent matrix corresponding to a linear system in triangular form. A second method, called *Gauss-Jordan elimination*, after Carl Friedrich Gauss and Wilhelm Jordan (1842–1899), continues the process until a matrix with 1s down the main diagonal from left to right and 0s in every position *above and below* each 1 is found. For three linear equations in three variables, x, y, and z, we try to get the augmented matrix into form

$$\left[\begin{array}{ccc|c} 1 & 0 & 0 & A \\ 0 & 1 & 0 & B \\ 0 & 0 & 1 & C \end{array}\right]$$

from which we conclude that $x = A$, $y = B$, and $z = C$.

EXAMPLE 5 Using Gauss-Jordan Elimination

Use Gauss-Jordan elimination to solve the system:

$$3x + y + 2z = 31$$

$$x + y + 2z = 19$$

$$x + 3y + 2z = 25$$

Solution

In Example 3 we used Gaussian elimination to obtain the following matrix:

$$\begin{bmatrix} 1 & \boxed{1} & \boxed{2} & | & 19 \\ 0 & 1 & \boxed{2} & | & 13 \\ 0 & 0 & 1 & | & 5 \end{bmatrix}$$
Using Gauss-Jordan elimination, the boxed entries should be 0s.

Rather than using back-substitution, we now will apply matrix row operations to get 0s *above* the 1s in the main diagonal. To obtain 0 in row 1, column 2 (where there is now a 1), we multiply each entry in row 2 by -1 and add this result to the corresponding entry in row 1.

$$\begin{bmatrix} 1 & 0 & \boxed{0} & | & 6 \\ 0 & 1 & \boxed{2} & | & 13 \\ 0 & 0 & 1 & | & 5 \end{bmatrix} \leftarrow -1R_2 + R_1$$
The boxed entries should be 0s.

Since we already have a 0 in row 1, column 3, we need only concentrate on obtaining a 0 in row 2, column 3 (where there is now a 2). To obtain 0, we multiply row 3 by -2 and add the resulting products to row 2.

$$\begin{bmatrix} 1 & 0 & 0 & | & 6 \\ 0 & 1 & 0 & | & 3 \\ 0 & 0 & 1 & | & 5 \end{bmatrix} \leftarrow -2R_3 + R_2$$

This last matrix corresponds to

$$x = 6, y = 3, z = 5.$$

As we found in Example 3, the solution set is $\{(6, 3, 5)\}$. ∎

Study tip

The advantage to Gauss-Jordan elimination is that from the augmented matrix we can simply read the solution. The disadvantage is that we must continue row operations in the augmented matrix from the Gaussian elimination process, and it's fairly easy to make computational errors.

3 Use matrices to model data and solve problems.

Modeling Data and Solving Problems Using Matrices

A frequently encountered function is called the *position function*. The height of a free-falling object above the ground is a function of the time that the object is falling. The position function has the form

$$s = \frac{1}{2}at^2 + v_0t + s_0 \quad \text{or} \quad s(t) = \frac{1}{2}at^2 + v_0t + s_0$$

in which

t = the time, in seconds, the object is falling

a = the object's acceleration, measured in feet per second squared

s (or $s(t)$) = the position (or height), in feet, of the object above the ground

Jordan Massengale "The Jump" 1992, mixed med., 50×68 in.

v_0 (called *initial velocity*) = the velocity when $t = 0$

s_0 (called *initial height*) = the object's height above the ground when $t = 0$

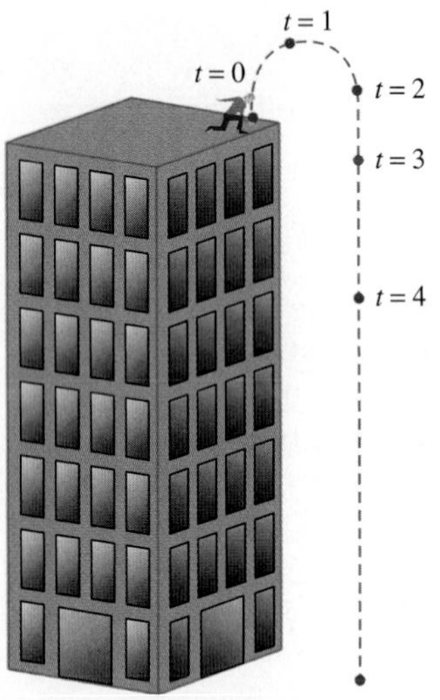

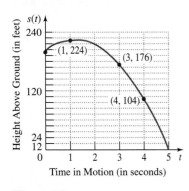

Figure 5.8

Throwing a ball directly upward

Figure 5.9

The ball's height above the ground is a function of its time in motion.

EXAMPLE 6   Modeling Position

As shown in Figure 5.8, a ball is thrown directly upward from the top of a building. The graph in Figure 5.9 shows the ball's height above the ground, $s(t)$, as a function of its time in motion. Use the three data points labeled on the graph to model the height of the ball.

Solution

We substitute the three values of t and s into the position function

$$s(t) = \frac{1}{2} at^2 + v_0 t + s_0.$$

Our aim is to determine values for a, v_0, and s_0.

$$s(t) = \frac{1}{2} at^2 + v_0 t + s_0 \qquad \text{Use the position function.}$$

$$s(1) = \frac{1}{2} a \cdot 1^2 + v_0 \cdot 1 + s_0 = 224 \qquad \text{When } t = 1, s = 224.$$

$$s(3) = \frac{1}{2} a \cdot 3^2 + v_0 \cdot 3 + s_0 = 176 \qquad \text{When } t = 3, s = 176.$$

$$s(4) = \frac{1}{2} a \cdot 4^2 + v_0 \cdot 4 + s_0 = 104 \qquad \text{When } t = 4, s = 104.$$

Before solving the system, we clear fractions by multiplying both sides of the first and second equation by 2. We obtain the system

$$a + 2v_0 + 2s_0 = 448$$

$$9a + 6v_0 + 2s_0 = 352$$

$$8a + 4v_0 + s_0 = 104.$$

We now solve the system using Gaussian elimination.

$$\begin{bmatrix} 1 & 2 & 2 & | & 448 \\ 9 & 6 & 2 & | & 352 \\ 8 & 4 & 1 & | & 104 \end{bmatrix} \xrightarrow[-8R_1 + R_3]{-9R_1 + R_2} \begin{bmatrix} 1 & 2 & 2 & | & 448 \\ 0 & -12 & -16 & | & -3680 \\ 0 & -12 & -15 & | & -3480 \end{bmatrix}$$

$$\xrightarrow{-\frac{1}{12}R_2} \begin{bmatrix} 1 & 2 & 2 & | & 448 \\ 0 & 1 & \frac{4}{3} & | & \frac{920}{3} \\ 0 & -12 & -15 & | & -3480 \end{bmatrix} \xrightarrow{12\,R_2+R_3} \begin{bmatrix} 1 & 2 & 2 & | & 448 \\ 0 & 1 & \frac{4}{3} & | & \frac{920}{3} \\ 0 & 0 & 1 & | & 200 \end{bmatrix}$$

Our final augmented matrix results in the linear system (in triangular form)

$$a + 2v_0 + 2s_0 = 448$$
$$v_0 + \frac{4}{3}s_0 = \frac{920}{3}$$
$$s_0 = 200$$

We can use back-substitution to find the values for v_0 and a.

$$s_0 = 200 \qquad v_0 + \frac{4}{3}s_0 = \frac{920}{3} \qquad\qquad a + 2v_0 + 2s_0 = 448$$

$$v_0 + \frac{4}{3}(200) = \frac{920}{3} \qquad\qquad a + 2\cdot40 + 2\cdot200 = 448$$

$$v_0 = \frac{120}{3} = 40 \qquad\qquad a + 480 = 448$$

$$a = -32$$

Now we substitute these values into the position function.

$$s(t) = \frac{1}{2}at^2 + v_0 t + s_0 \qquad \text{Use the position function.}$$

$$s(t) = \frac{1}{2}(-32)t^2 + 40t + 200 \qquad \text{Substitute the three obtained values.}$$

The position function for the ball is

$$s(t) = -16t^2 + 40t + 200.$$

Using this function, we can find the ball's height at any time t. The function is evaluated for four values of t in the following table and is illustrated in Figure 5.10.

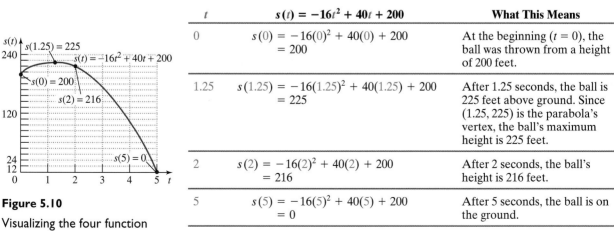

Figure 5.10

Visualizing the four function values in the table

t	$s(t) = -16t^2 + 40t + 200$	What This Means
0	$s(0) = -16(0)^2 + 40(0) + 200$ $= 200$	At the beginning ($t = 0$), the ball was thrown from a height of 200 feet.
1.25	$s(1.25) = -16(1.25)^2 + 40(1.25) + 200$ $= 225$	After 1.25 seconds, the ball is 225 feet above ground. Since (1.25, 225) is the parabola's vertex, the ball's maximum height is 225 feet.
2	$s(2) = -16(2)^2 + 40(2) + 200$ $= 216$	After 2 seconds, the ball's height is 216 feet.
5	$s(5) = -16(5)^2 + 40(5) + 200$ $= 0$	After 5 seconds, the ball is on the ground.

iscover for yourself

According to Example 6 and the Study tip on the right, every object at the surface of the Earth, once released, should accelerate downward at the same rate. If a ball and a feather are dropped from the top of a building, the ball will hit the ground well before the feather. How can this be explained if acceleration is constant?

Joseph Cornell "Untitled (Pharmacy)" 1942. Peggy Guggenheim Collection, The Solomon R. Guggenheim Museum, New York. Photograph by David Heald. © The Solomon R. Guggenheim Foundation, New York. 9FN 76.2553 PG128.

Step 1. Use variables to represent unknown quantities.

Step 2. Write a linear system of equations modeling the problem's conditions.

tudy tip

Any object dropped near the Earth's surface, no matter how heavy or light, falls with exactly the same constant acceleration. By modeling the data for the falling ball in Example 6, we see that a, the acceleration due to gravity, is -32 feet per second squared. The ball underwent a constant downward acceleration. A falling object accelerates from a stationary position to a velocity of 32 feet per second (about 22 miles per hour) after 1 second. After 2 seconds, the velocity doubles to 64 feet per second; after 3 seconds, it triples to 96 feet per second; and so on.

In Example 6, we modeled the given data. In Example 7, we model verbal conditions and use our strategy for solving problems with linear systems.

EXAMPLE 7 Providing Nutritional Needs

A nutritionist in a hospital is arranging special diets that consist of a combination of three basic foods. It is important that patients on this diet consume exactly 310 units of calcium, 190 units of iron, and 250 units of vitamin A each day. The amounts of these nutrients in one ounce of food are given in the following table.

| | Units per Ounce | | |
	Calcium	Iron	Vitamin A
Food A	30	10	10
Food B	10	10	30
Food C	20	20	20

How many ounces of each food must be used to satisfy the nutrient requirements exactly?

Solution

Let

x = the number of ounces of Food A

y = the number of ounces of Food B

z = the number of ounces of Food C

The patients must consume 310 units of calcium daily. Using the data in the calcium column, we get

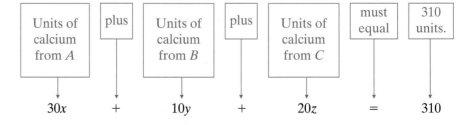

| Units of calcium from A | plus | Units of calcium from B | plus | Units of calcium from C | must equal | 310 units. |

$$30x \qquad + \qquad 10y \qquad + \qquad 20z \qquad = \qquad 310$$

The patients must consume 190 units of iron daily. Using the data in the iron column, we get

ENRICHMENT ESSAY

Using Matrices to Present Data

Matrices are extremely useful for presenting rectangular arrays of numbers, as shown in the following example.

Organ Transplants

	Number of people waiting, 1995	1-year survival rates, 1994 (percent)
Transplant:		
Heart	3468	84.3
Liver	5691	82.9
Kidney	31,045	95.0
Heart-lung	208	75.8
Lung	1923	(NA)
Pancreas/Islet cell	285	(NA)

Source: U.S. Bureau of the Census, Statistical Abstract of the United States

Alfredo Castañeda "The Lady of the Mirror" 1986, oil on canvas, 39 1/2 × 39 1/2 in. Courtesy of Mary-Anne Martin / Fine Art, New York.

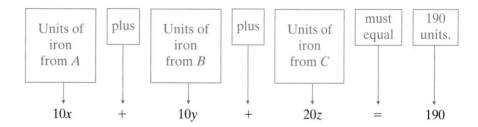

Units of iron from A	plus	Units of iron from B	plus	Units of iron from C	must equal	190 units.
$10x$	$+$	$10y$	$+$	$20z$	$=$	190

The patients must consume 250 units of vitamin A daily. Using the data in the vitamin A column, we get

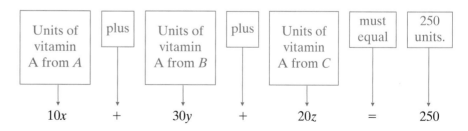

Units of vitamin A from A	plus	Units of vitamin A from B	plus	Units of vitamin A from C	must equal	250 units.
$10x$	$+$	$30y$	$+$	$20z$	$=$	250

We obtain the following linear system, which can be simplified as shown.

$$30x + 10y + 20z = 310 \quad \xrightarrow{\text{Divide by 10.}} \quad 3x + y + 2z = 31$$
$$10x + 10y + 20z = 190 \quad \xrightarrow{\text{Divide by 10.}} \quad x + y + 2z = 19$$
$$10x + 30y + 20z = 250 \quad \xrightarrow{\text{Divide by 10.}} \quad x + 3y + 2z = 25$$

Step 3. Solve the system and answer the problem's question.

The simplified system on the right at the bottom of page 553 is the system that we solved in Example 3 using Gaussian elimination. (We solved the system again in Example 5 using Gauss-Jordan elimination.) We found that $x = 6$, $y = 3$, and $z = 5$.

Patients on the special diet should be fed 6 ounces of Food A, 3 ounces of Food B, and 5 ounces of Food C every day. Take a few minutes to check this solution in terms of calcium, iron, and vitamin A requirements, respectively. ∎

PROBLEM SET 5.2

Practice Problems

Write the system of equations corresponding to each matrix in Problems 1–6. Once the system is written, use back-substitution to find its solution.

1. $\begin{bmatrix} 1 & 0 & -4 & | & 5 \\ 0 & 1 & -12 & | & 13 \\ 0 & 0 & 1 & | & -\frac{1}{2} \end{bmatrix}$

2. $\begin{bmatrix} 1 & 2 & 1 & | & 0 \\ 0 & 1 & 0 & | & -2 \\ 0 & 0 & 1 & | & 3 \end{bmatrix}$

3. $\begin{bmatrix} 1 & \frac{1}{2} & 1 & | & \frac{11}{2} \\ 0 & 1 & \frac{3}{2} & | & 7 \\ 0 & 0 & 1 & | & 4 \end{bmatrix}$

4. $\begin{bmatrix} 1 & 1 & 0 & | & 3 \\ 0 & 1 & \frac{3}{2} & | & -2 \\ 0 & 0 & 1 & | & 0 \end{bmatrix}$

5. $\begin{bmatrix} 1 & -1 & 1 & 1 & | & 3 \\ 0 & 1 & -2 & -1 & | & 0 \\ 0 & 0 & 1 & 6 & | & 17 \\ 0 & 0 & 0 & 1 & | & 3 \end{bmatrix}$

6. $\begin{bmatrix} 1 & 2 & -1 & 0 & | & 2 \\ 0 & 1 & 1 & -2 & | & -3 \\ 0 & 0 & 1 & -1 & | & -2 \\ 0 & 0 & 0 & 1 & | & 3 \end{bmatrix}$

In Problems 7–16, perform the matrix row operation(s) and write the new matrix.

7. $\begin{bmatrix} 2 & -5 & 5 & | & 17 \\ -1 & 3 & 0 & | & -4 \\ 1 & -2 & 3 & | & 9 \end{bmatrix} R_1 \leftrightarrow R_3$

8. $\begin{bmatrix} 5 & -2 & 1 & | & 2 \\ 1 & 3 & -3 & | & 0 \\ 2 & -4 & 6 & | & -2 \end{bmatrix} R_1 \leftrightarrow R_2$

9. $\begin{bmatrix} 10 & 16 & 2 & | & 100 \\ 0 & -18 & 15 & | & 150 \\ 9 & 10 & 17 & | & 23 \end{bmatrix} \begin{matrix} \frac{1}{2}R_1 \\ -\frac{1}{3}R_2 \end{matrix}$

10. $\begin{bmatrix} 12 & 14 & 6 & | & 80 \\ 0 & -25 & 50 & | & 15 \\ 7 & 3 & 9 & | & 11 \end{bmatrix} \begin{matrix} \frac{1}{2}R_1 \\ -\frac{1}{5}R_2 \end{matrix}$

11. $\begin{bmatrix} 1 & -3 & 2 & | & 0 \\ 3 & 1 & -1 & | & 7 \\ 2 & -2 & 1 & | & 3 \end{bmatrix} \begin{matrix} -3R_1 + R_2 \\ -2R_1 + R_3 \end{matrix}$

12. $\begin{bmatrix} 1 & -1 & 5 & | & -6 \\ 3 & 3 & -1 & | & 10 \\ 1 & 3 & 2 & | & 5 \end{bmatrix} \begin{matrix} -3R_1 + R_2 \\ -1R_1 + R_3 \end{matrix}$

13. $\begin{bmatrix} 1 & 3 & 4 & | & 10 \\ 0 & -5 & -15 & | & -38 \\ 4 & 8 & 4 & | & 9 \end{bmatrix} \begin{matrix} -5R_1 + R_2 \\ -4R_1 + R_3 \end{matrix}$

14. $\begin{bmatrix} 1 & 1 & -1 & | & 6 \\ 2 & -1 & 1 & | & -3 \\ 3 & -1 & -1 & | & 4 \end{bmatrix} \begin{matrix} -2R_1 + R_2 \\ -3R_1 + R_3 \end{matrix}$

15. $\begin{bmatrix} 1 & -1 & 1 & 1 & | & 3 \\ 0 & 1 & -2 & -1 & | & 0 \\ 2 & 0 & 3 & 4 & | & 11 \\ 5 & 1 & 2 & 4 & | & 6 \end{bmatrix} \begin{matrix} -2R_1 + R_3 \\ -5R_1 + R_4 \end{matrix}$

16. $\begin{bmatrix} 1 & -5 & 2 & -2 & | & 4 \\ 0 & 1 & -3 & -1 & | & 0 \\ 3 & 0 & 2 & -1 & | & 6 \\ -4 & 1 & 4 & 2 & | & -3 \end{bmatrix} \begin{matrix} -3R_1 + R_3 \\ 4R_1 + R_4 \end{matrix}$

In Problems 17–30, solve each system of equations. Use Gaussian elimination with back-substitution or Gauss-Jordan elimination.

17.
$$x + y - z = -2$$
$$2x - y + z = 5$$
$$-x + 2y + 2z = 1$$

18.
$$x - 2y - z = 2$$
$$2x - y + z = 4$$
$$-x + y - 2z = -4$$

19.
$$x + 3y = 0$$
$$x + y + z = 1$$
$$3x - y - z = 11$$

20.
$$3y - z = -1$$
$$x + 5y - z = -4$$
$$-3x + 6y + 2z = 11$$

21. $2x + 2y + 7z = -1$
$$2x + y + 2z = 2$$
$$4x + 6y + z = 15$$

22. $3x + 2y + 3z = 3$
$$4x - 5y + 7z = 1$$
$$2x + 3y - 2z = 6$$

23.
$$x + y + z + w = 4$$
$$2x + y - 2z - w = 0$$
$$x - 2y - z - 2w = -2$$
$$3x + 2y + z + 3w = 4$$

24.
$$x + y + z + w = 5$$
$$x + 2y - z - 2w = -1$$
$$x - 3y - 3z - w = -1$$
$$2x - y + 2z - w = -2$$

25.
$$3x - 4y + z + w = 9$$
$$x + y - z - w = 0$$
$$2x + y + 4z - 2w = 3$$
$$-x + 2y + z - 3w = 3$$

26.
$$2x + z - 3w = 8$$
$$x - y + 4w = -10$$
$$3x + 5y - z - w = 20$$
$$x + y - z - w = 6$$

27.
$$2x + 3y - z - w = -3$$
$$2x - y - 3z + 2w = -5$$
$$x - y + z - w = -4$$
$$3x - 2y + z + w = 0$$

28.
$$2x - y - z + w = 4$$
$$x + 3y - 2z - 3w = 6$$
$$x - y + z - w = 2$$
$$-x + 2y - z - w = -1$$

29.
$$2x_1 - 2x_2 + 3x_3 - x_4 = 12$$
$$x_1 + 2x_2 - x_3 + 2x_4 - x_5 = -7$$
$$x_1 + x_3 - x_4 - 5x_5 = 5$$
$$-x_1 + x_2 - x_3 - 2x_4 - 3x_5 = 0$$
$$x_1 - x_2 - x_4 + x_5 = 4$$

30.
$$2x - 2z + 4w - 4v = -6$$
$$-x - y - z - w - u - v = -12$$
$$x + y - z - w = -2$$
$$y - z + u - v = -1$$
$$x - y + z - w + u - v = 0$$
$$3y - z + v = 4$$

Application Problems

31. A football is launched straight upward. The graph shows the ball's height above the ground (s) as a function of its time (t) in motion.

  a. Use the position function $s(t) = \frac{1}{2}at^2 + v_0t + s_0$ and the three data points labeled on the graph to model the height of the football.

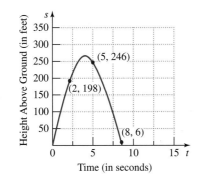

  b. Find and interpret $s(0)$, $s(4)$, and $s(7)$.

  c. Complete the following statement: The ball was launched straight upward at ____ feet per second from a height of ____ feet.

32. A ball is thrown vertically upward, as shown in the figure. The graph shows the ball's height (in feet) above the ground (s) as a function of its time (t) in motion (in seconds).

  a. Use the position function $s(t) = \frac{1}{2}at^2 + v_0t + s_0$ and the three data points labeled on the graph to model the height of the ball. Use matrices to solve the system to determine the values for a, v_0, and s_0.

  b. Use a graphing utility to graph the parabola. Approximate the maximum height of the ball and the time at which the ball strikes the ground.

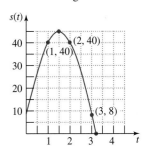

33. The table below shows the number of sides of a polygon and the total number of its sides and diagonals.

x (Number of Sides)	y (Number of Sides and Diagonals)
3	3
4	6
5	10

Use the function $y = ax^2 + bx + c$ to model the data. Then use the data given in the figure for the hexagon to verify the model.

$x = 3, y = 3$

$x = 4, y = 6$

$x = 5, y = 10$

$x = 6, y = 15$

34. In a study relating sleep and death rate, the following data were obtained.

x (Average Number of Hours of Sleep)	y or f(x) (Death Rate per Year per 100,000 Males)
4	1682
7	626
9	967

Model the data using a quadratic function, $f(x) = ax^2 + bx + c$, and then find the death rate of males who sleep 5 hours, 6 hours, and 11 hours.

Write a linear system in three variables to solve Problems 35–38. Then use matrices to solve the system.

35. In the figure, triangle ABC is isosceles, with $AB = CB$. Point D is located midway between B and C. The perimeters of $\triangle ABC$, $\triangle ADB$, and $\triangle ACD$ are 80 feet, 70 feet, and 48 feet, respectively. Find the lengths of AB, AD, and AC.

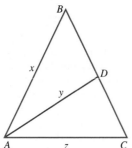

36. Three foods have the following nutritional content per ounce.

	Calories	Protein (in grams)	Vitamin C (in milligrams)
Food A	40	5	30
Food B	200	2	10
Food C	400	4	300

If a meal consisting of the three foods allows exactly 660 calories, 25 grams of protein, and 425 milligrams of vitamin C, how many ounces of each kind of food should be used?

37. Three diet foods contain grams of fat, carbohydrate, and protein as shown in the table in the next column.

	Fat (in grams)	Carbohydrate (in grams)	Protein (in grams)
Serving of Food A	2	1	3
Serving of Food B	1	2	4
Serving of Food C	2	4	3

If a diet allows exactly 16 grams of fat, 23 grams of carbohydrate, and 29 grams of protein, how many servings of each kind of food can be eaten?

38. A furniture company produces three types of desks: a children's model, an office model, and a deluxe model. Each desk is manufactured in three stages: cutting, construction, and finishing. The time requirements for each model and manufacturing stage are given in the following table.

	Children's model	Office model	Deluxe model
Cutting	2 hr	3 hr	2 hr
Construction	2 hr	1 hr	3 hr
Finishing	1 hr	1 hr	2 hr

Each week the company has available a maximum of 100 hours for cutting, 100 hours for construction, and 65 hours for finishing. If all available time must be used, how many of each type of desk should be produced each week? Use matrices to solve the system.

True–False Critical Thinking Problems _____

39. Which one of the following is true?

a. A row-equivalent matrix will not be obtained if the numbers in any row are multiplied by a negative improper fraction.

b. The augmented matrix for the system

$$x - 3y = 5$$
$$y - 2z = 7 \quad \text{is} \quad \begin{bmatrix} 1 & -3 & 5 \\ 1 & -2 & 7 \\ 2 & 1 & 4 \end{bmatrix}.$$
$$2x + z = 4$$

c. In solving a linear system of three equations in three variables, we begin with the augmented matrix and use row operations to obtain a row-equivalent matrix with 0s down the diagonal from left to right and 1s below each 0.

d. The augmented matrix for a system of three linear equations in three variables is always a matrix with three rows and four columns, described as a 3×4 matrix.

40. Which one of the following is true?

a. The augmented matrix for a system of three linear equations in three variables contains three rows and three columns.

b. The notation $-2R_1 + R_3$ means to replace all entries in the third row of a matrix by -2 times the entries in the first row.

c. A system whose augmented matrix is given by

$$\begin{bmatrix} 1 & -2 & 3 & | & 9 \\ 0 & 1 & 3 & | & 5 \\ 0 & 0 & 1 & | & 2 \end{bmatrix}$$

has $\{(1, -1, 2)\}$ as its solution set.

d. The matrices

$$\begin{bmatrix} 1 & 2 & -3 & | & 15 \\ -2 & -3 & 1 & | & -15 \\ 4 & 9 & -4 & | & 49 \end{bmatrix} \quad \text{and}$$

$$\begin{bmatrix} 1 & 2 & -3 & | & 15 \\ 0 & 1 & -5 & | & 15 \\ 0 & 1 & -16 & | & -11 \end{bmatrix}$$

are row-equivalent.

Technology Problems

41. Most graphing utilities can perform row operations on matrices. Consult the owner's manual for your graphing utility to learn proper keystrokes for performing these operations. Then duplicate the row operations of any three problems that you solved from Problems 7–16.

42. The final augmented matrix that we obtain when using Gaussian elimination is said to be in *row-echelon form*. For systems of linear equations with unique solutions, this form results when each entry in the main diagonal is 1 and all entries below the main diagonal are 0s. Some graphing utilities can transform a matrix to row-echelon form. Consult the owner's manual for your graphing utility. If your utility has this capability, obtain the final matrices of Example 3 on page 544, and Example 4 on page 546. Then use this capability to solve some of the systems in Problems 17–30.

43. The final augmented matrix that we obtain when using Gauss-Jordan elimination is said to be in *reduced row-echelon form*. For systems of linear equations with unique solutions, this form results when each entry on the main diagonal is 1 and all entries below and above that main diagonal are 0s. Some graphing utilities can transform a matrix to reduced row-echelon form. Consult the owner's manual for your graphing utility. If your utility has this capability, obtain the final matrix of Example 5 on page 549 beginning with the matrix for Example 3 on page 544. Then use this capability to solve some of the systems in Problems 17–30.

44. a. The function $y = ax^3 + bx^2 + cx + d$ can be used to estimate the number of bachelor's degrees (*y*) conferred in mathematics *x* years after 1970, where $0 \le x \le 20$. Use the four data values shown in the bar graph, namely, (5, 14,685), (10, 13,140), (15, 15,095), and (20, 15,150) to create a polynomial

model that fits the data. You'll need to substitute the given data points into the function, one at a time, and solve the resulting linear system of four equations for *a, b, c,* and *d*. Solve using the simultaneous equation feature of your graphing utility.

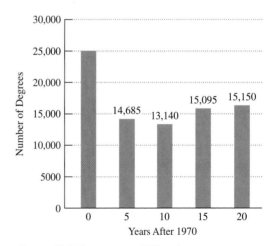

Mathematics Degrees, 1970–1990

Source: U.S. Department of Education

b. Use a graphing utility to graph the function with the following range setting:

Xmin = 0, Xmax = 30, Xscl = 3, Ymin = 0,

Ymax = 25,000, Yscl = 5000

Trace along the curve and verify the four given data points. Explain, based on the graph's shape, why it is not a good model for $x > 20$.

Writing in Mathematics

45. Describe similarities between Gaussian elimination as a method for solving linear systems of equations and synthetic division as a method for dividing polynomials.

46. Would it make sense to study Gaussian elimination before covering the addition and substitution methods? Explain. What does this tell you about learning mathematics?

47. Many graphing utilities are now capable of displaying the final matrix for solving a system of linear equations when the augmented matrix is entered. Put in another way, the solution set of a system of linear equations can be found using an appropriate keystroke sequence on a graphing utility. Given this reality, do you think it would be a good idea not to teach the addition method, the substitution method, and the matrix methods for solving linear systems and focus strictly on the keystroke sequences needed to find a solution set? Justify your opinion.

Critical Thinking Problems

48. In the following augmented matrix, explain how a 1 can be obtained in row 1, column 1, without introducing fractions.

$$\begin{bmatrix} 2 & 3 & 4 & | & 1 \\ 5 & 1 & -1 & | & 4 \\ 3 & 7 & 2 & | & 0 \end{bmatrix}$$

49. The parabola that is the graph of $y = ax^2 + bx + c$ passes through the points $(-1, -11)$, $(1, 1)$, and $(2, 4)$. What are the coordinates of the parabola's vertex?

50. If $A > 0$, solve the following system.

$$e^x + 2e^y - 4e^z = 3A$$

$$e^x + e^y - 2e^z = 2A$$

$$e^x - 6e^y + 2e^z = -10A$$

Group Activity Problem

51. Members of the group should research a realistic application involving foods and nutrients or any other area of interest that gives rise to a fairly large linear system. Use matrices, a graphing utility, or a computer to solve the system.

SECTION 5.3

Solutions Manual   **Tutorial**   **Video 9**

Inconsistent and Dependent Systems and Their Applications

Objectives

1 Apply Gaussian elimination to inconsistent and dependent systems.

2 Apply Gaussian elimination to systems with differing numbers of variables and equations.

3 Solve problems involving inconsistent and dependent systems.

In Section 5.1, we saw that linear systems can have one solution for each variable, no solutions, or infinitely many solutions. We can use Gaussian elimina-

René Magritte "La Bonne Parole" (The Good Word), 1939, gouache on paper. © 1998 C. Herscovici, Brussels / Artists Rights Society (ARS), New York. Art Resource, NY.

 Apply Gaussian elimination to inconsistent and dependent systems.

Discover for yourself

Use the addition method to solve Example 1. Describe what happens. Why does this mean that there is no solution? Since the graph of each equation is a plane, what is true about the three planes in this system? Now read the solution to Example 1 and compare what you discovered using addition to the matrix approach.

tion on systems with three or more variables to determine how many solutions such systems may have. In the case of systems with no solutions or infinitely many solutions, it is impossible to rewrite the augmented matrix in the desired form with 1s down the main diagonal and 0s below the 1s. Let's see what this means by looking at a system that has no solutions.

EXAMPLE 1 A System With No Solutions

Use Gaussian elimination to solve the system:

$$x - y - 2z = 2$$
$$2x - 3y + 6z = 5$$
$$3x - 4y + 4z = 12$$

Solution

We start with the augmented matrix.

$$\begin{bmatrix} 1 & -1 & -2 & | & 2 \\ \boxed{2} & -3 & 6 & | & 5 \\ \boxed{3} & -4 & 4 & | & 12 \end{bmatrix}$$ The boxed entries should be 0s.

To get 0 in the first position of R_2, we multiply R_1 by -2, and add the result to R_2. To get 0 in the first position of R_3, we multiply R_1 by -3, and add the result to R_3.

$$\begin{bmatrix} 1 & -1 & -2 & | & 2 \\ 2 & -3 & 6 & | & 5 \\ 3 & -4 & 4 & | & 12 \end{bmatrix} \xrightarrow[-3R_1 + R_3]{-2R_1 + R_2} \begin{bmatrix} 1 & -1 & -2 & | & 2 \\ 0 & \boxed{-1} & 10 & | & 1 \\ 0 & -1 & 10 & | & 6 \end{bmatrix}$$ Because we want 1s down the main diagonal, the boxed entry should be 1.

To get 1 in row 2, column 2, we multiply row 2 by -1.

$$\begin{bmatrix} 1 & -1 & -2 & | & 2 \\ 0 & -1 & 10 & | & 1 \\ 0 & -1 & 10 & | & 6 \end{bmatrix} \xrightarrow{-1R_2} \begin{bmatrix} 1 & -1 & -2 & | & 2 \\ 0 & 1 & -10 & | & -1 \\ 0 & \boxed{-1} & 10 & | & 6 \end{bmatrix}$$ The boxed entry should be 0.

To get 0 in row 3, column 2, we add row 2 to row 3.

$$\begin{bmatrix} 1 & -1 & -2 & | & 2 \\ 0 & 1 & -10 & | & -1 \\ 0 & -1 & 10 & | & 6 \end{bmatrix} \xrightarrow{1R_2 + R_3} \begin{bmatrix} 1 & -1 & -2 & | & 2 \\ 0 & 1 & -10 & | & -1 \\ 0 & 0 & 0 & | & 5 \end{bmatrix}$$

It is impossible to convert this last matrix to the desired form of 1s down the main diagonal. If we translate the last row back into equation form, we get

$$0x + 0y + 0z = 5$$

which is false. Regardless of what values we select for x, y, and z, the last equation can never be a true statement. Consequently, the system has no solution. The solution set is $\emptyset$, the empty set.   ∎

Three parallel planes with no
common intersection point

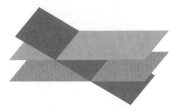

Two planes are parallel with
no common intersection point

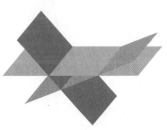

Planes intersect two at a time with
no common intersection point

Figure 5.11

Three planes may have no
common point of intersection.

Recall that the graph of a system of three linear equations in three variables consists of three planes. When these planes intersect in a single point, the system has precisely one ordered-triple solution. When the planes have no point in common, the system has no solution, like the one in Example 1. Figure 5.11 illustrates some of the geometric possibilities for these inconsistent systems.

Now let's see what happens when we apply Gaussian elimination to a system with infinitely many solutions.

EXAMPLE 2   **A System with an Infinite Number of Solutions**

Use Gaussian elimination to solve the following system:

$$3x - 4y + 4z = 7$$

$$x - y - 2z = 2$$

$$2x - 3y + 6z = 5$$

Discover for yourself

Use the addition method to solve this system. What happens in the solution process? Why does this mean that there are infinitely many solutions? What does this mean in terms of the three planes that make up the graph of this system? Now read the following solution and compare what you discovered to the matrix approach.

Solution

As always, we start with the augmented matrix.

$$\begin{bmatrix} 3 & -4 & 4 & | & 7 \\ 1 & -1 & -2 & | & 2 \\ 2 & -3 & 6 & | & 5 \end{bmatrix} \xrightarrow{R_1 \leftrightarrow R_2} \begin{bmatrix} 1 & -1 & -2 & | & 2 \\ 3 & -4 & 4 & | & 7 \\ 2 & -3 & 6 & | & 5 \end{bmatrix}$$

$$\xrightarrow[-2R_1 + R_3]{-3R_1 + R_2} \begin{bmatrix} 1 & -1 & -2 & | & 2 \\ 0 & -1 & 10 & | & 1 \\ 0 & -1 & 10 & | & 1 \end{bmatrix} \xrightarrow{-1R_2} \begin{bmatrix} 1 & -1 & -2 & | & 2 \\ 0 & 1 & -10 & | & -1 \\ 0 & -1 & 10 & | & 1 \end{bmatrix}$$

$$\xrightarrow{1R_2 + R_3} \begin{bmatrix} 1 & -1 & -2 & | & 2 \\ 0 & 1 & -10 & | & -1 \\ 0 & 0 & 0 & | & 0 \end{bmatrix}$$

If we translate row 3 of the matrix into equation form, we obtain

$$0x + 0y + 0z = 0$$

or

$$0 = 0.$$

Study tip

Identifying a dependent system
using matrices is not too
difficult. However, representing
the solution set for such
systems is a bit tricky.
Remember from our work in
Section 5.1 that the solutions
are always expressed in terms of
one of the variables.

This equation results in a true statement regardless of what values we select for
x, y, and z. Consequently, the equation $0x + 0y + 0z = 0$ is *dependent* on the
other two equations in the system in the sense that it adds no new information
about the variables. Thus, we can drop it from the system, which can now be
expressed in the form

$$\begin{bmatrix} 1 & -1 & -2 & 2 \\ 0 & 1 & -10 & -1 \end{bmatrix}$$

The original system is equivalent to the system

$$x - y - 2z = 2$$
$$y - 10z = -1$$

Although neither of these equations gives a value for z, we can use them to
express x and y in terms of z. From the last equation we obtain

$$y = 10z - 1.$$

Back-substituting for y into the previous equation, we can find x in terms of z.

$x - y - 2z = 2$	This is the first equation obtained from the final matrix.
$x - (10z - 1) - 2z = 2$	Since $y = 10z - 1$, substitute $10z - 1$ for y.
$x - 10z + 1 - 2z = 2$	Apply the distributive property.
$x - 12z + 1 = 2$	Combine like terms.
$x = 12z + 1$	Solve for x in terms of z.

Because no value is determined for z, we can find a solution to the system by
letting z equal any real number and then using the above equations to obtain x
and y. For example, if $z = 1$, then

$$x = 12z + 1 = 12(1) + 1 = 13 \text{ and}$$
$$y = 10z - 1 = 10(1) - 1 = 9.$$

Consequently, $(13, 9, 1)$ is a solution to the system. On the other hand, if we let
$z = -1$, then

$$x = 12z + 1 = 12(-1) + 1 = -11 \text{ and}$$
$$y = 10z - 1 = 10(-1) - 1 = -11.$$

Thus, $(-11, -11, -1)$ is another solution to the system. Finally, letting $z = t$
(or any letter of our choice), the solutions to the system are all of the form

$$x = 12t + 1, \quad y = 10t - 1, \quad z = t$$

where t is a real number. Therefore, every ordered triple that is of the form
$(12t + 1, 10t - 1, t)$, where t is a real number, is a solution of the system. The
solution set of the dependent system can be written as $\{(12t + 1, 10t - 1, t)\}$. ■

Figure 5.12

Three planes may intersect at
infinitely many points.

We have seen that when three planes have no point in common, the corresponding system has no solution. When the system has infinitely many solutions, like the one in Example 2, the three planes intersect in more than one
point. Figure 5.12 illustrates one geometric possibility for dependent systems.

2 Apply Gaussian elimination to systems with differing numbers of variables and equations.

Nonsquare Systems

Up to this point, we have encountered only *square* systems in which the number of equations is equal to the number of variables. In a nonsquare system, the number of variables differs from the number of equations.

EXAMPLE 3 A System with Fewer Equations Than Variables

Use Gaussian elimination to solve the system:

$$3x + 7y + 6z = 26$$
$$x + 2y + z = 8$$

Solution

$$\begin{bmatrix} 3 & 7 & 6 & | & 26 \\ 1 & 2 & 1 & | & 8 \end{bmatrix} \xrightarrow{R_1 \leftrightarrow R_2} \begin{bmatrix} 1 & 2 & 1 & | & 8 \\ 3 & 7 & 6 & | & 26 \end{bmatrix} \xrightarrow{-3R_1 + R_2} \begin{bmatrix} 1 & 2 & 1 & | & 8 \\ 0 & 1 & 3 & | & 2 \end{bmatrix}$$

Since we now have 1s down the diagonal that begins with the upper-left entry and a 0 below this 1, we translate the matrix back into equation form.

$$x + 2y + z = 8 \quad \text{Equation 1}$$
$$y + 3z = 2 \quad \text{Equation 2}$$

We can let z equal any real number and use back-substitution to express x and y in terms of z.

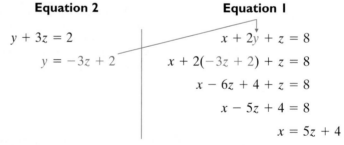

Equation 2	**Equation 1**
$y + 3z = 2$	$x + 2y + z = 8$
$y = -3z + 2$	$x + 2(-3z + 2) + z = 8$
	$x - 6z + 4 + z = 8$
	$x - 5z + 4 = 8$
	$x = 5z + 4$

With $z = t$, the ordered solution (x, y, z) enables us to express the system's solution set as

$$\{(5t + 4, -3t + 2, t)\}$$

where t is any real number. ∎

Applications

Many linear systems that arise in problem-solving situations turn out to be dependent or inconsistent. Our next example, involving a dependent system, hints at a future in which a central computer and computers within cars will be used to manage traffic flow.

EXAMPLE 4 Traffic Control

Figure 5.13 on page 563, shows the intersections of four one-way streets. As you study the figure, notice, for example, that 300 cars per hour want to enter

Red Grooms "Looking Along Broadway Towards Grace Church" 1981, alkyd paint, gator board, celastic, wood, wax foamcore, 71 × 63 3/4 × 28 3/4 in. Photo courtesy of Marlborough Gallery, NY. © 1998 Red Grooms / Artists Rights Society (ARS), New York.

Discover for yourself

Let $t = 1$ for the solution set

$$\{(5t + 4, -3t + 2, t)\}.$$

What solution do you obtain? Substitute these three values in the two equations in Example 3 and show that each equation is satisfied. Repeat this process for another two values for t.

3 Solve problems involving inconsistent and dependent systems.

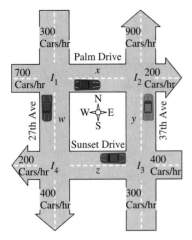

Figure 5.13

The intersections of four one-way streets

intersection I_1 from the north on 27th Avenue, while 200 cars per hour want to head east from intersection I_2 on Palm Drive. The letters x, y, z, and w stand for the number of cars passing between the intersections.

a. If the traffic is to keep moving, at each intersection the number of cars entering per hour must equal the number of cars leaving per hour. Use this idea to set up a linear system of equations involving x, y, z, and w.

b. Use Gaussian elimination to solve the system.

c. If construction on 27th Avenue limits w to 50 cars per hour, how many cars per hour must pass between the other intersections to keep traffic flowing?

Solution

a. Set up the system by considering one intersection at a time, referring to Figure 5.13.

For Intersection I_1: Since $300 + 700 = 1000$ cars enter I_1, and $x + w$ cars leave the intersection, then $x + w = 1000$.

For Intersection I_2: Since $x + y$ cars enter the intersection, and $200 + 900 = 1100$ cars leave I_2, then $x + y = 1100$.

For Intersection I_3: Figure 5.13 indicates that $300 + 400 = 700$ cars enter and $y + z$ leave, so $y + z = 700$.

For Intersection I_4: With $z + w$ cars entering and $200 + 400 = 600$ cars exiting, traffic will keep flowing if $z + w = 600$.

The system of equations that evolves in this situation is given by

$$x + w = 1000$$

$$x + y = 1100$$

$$y + z = 700$$

$$z + w = 600.$$

Study tip

To obtain the augmented matrix, express the given system

$x + w = 1000$

$x + y = 1100$

$y + z = 700$

$z + w = 600$

as

$1x + 0y + 0z + 1w = 1000$

$1x + 1y + 0z + 0w = 1100$

$0x + 1y + 1z + 0w = 700$

$0x + 0y + 1z + 1z = 600$

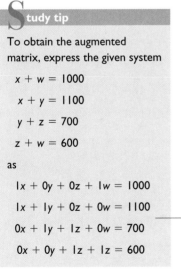

b. The augmented matrix for this system is

$$\begin{bmatrix} 1 & 0 & 0 & 1 & 1000 \\ 1 & 1 & 0 & 0 & 1100 \\ 0 & 1 & 1 & 0 & 700 \\ 0 & 0 & 1 & 1 & 600 \end{bmatrix}$$

We can now use row operations to obtain the equivalent matrix

$$\begin{bmatrix} 1 & 0 & 0 & 1 & 1000 \\ 0 & 1 & 0 & -1 & 100 \\ 0 & 0 & 1 & 1 & 600 \\ 0 & 0 & 0 & 0 & 0 \end{bmatrix}.$$

This matrix has 1s down the diagonal that begins with the upper-left entry and 0s below the 1s. Use row operations to verify the matrix.

The last row of the matrix shows that the system

$$x + w = 1000 \quad \text{Equation 1}$$

$$y - w = 100 \quad \text{Equation 2}$$

$$z + w = 600 \quad \text{Equation 3}$$

is dependent and has no unique solution. We can let w equal any real number and express x, y, and z in terms of w:

Equation 3:	**Equation 2:**	**Equation 1:**
$z + w = 600$	$y - w = 100$	$x + w = 1000$
$z = 600 - w$	$y = 100 + w$	$x = 1000 - w$

With $w = t$, the ordered solution (x, y, z, w) enables us to express the system's solution set as

$$\{(1000 - t, 100 + t, 600 - t, t)\}.$$

c. We are given that construction limits w to 50 cars per hour. Since $w = t$, we replace 50 for t in the system's ordered solution:

$(1000 - t, 100 + t, 600 - t, t)$ Use the system's solution.

$= (1000 - 50, 100 + 50, 600 - 50, 50)$ $t = 50$

$= (950, 150, 550, 50)$

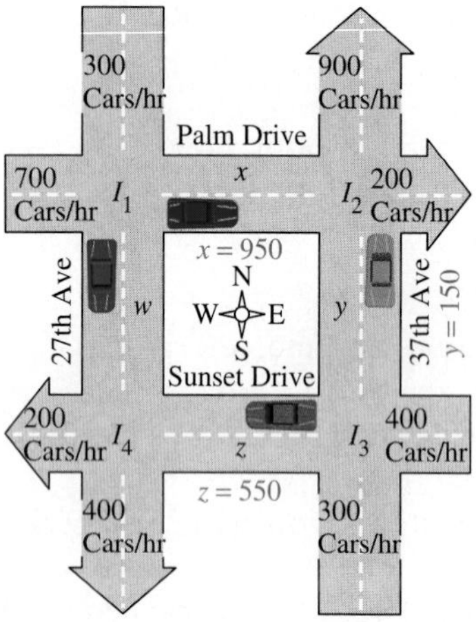

Figure 5.14

With w limited to 50 cars per hour, values for x, y, and z are determined.

Thus, $x = 950$, $y = 150$, and $z = 550$. (See Figure 5.14.) With construction on 27th Avenue, this means that to keep traffic flowing, 950 cars per hour must be routed between I_1 and I_2, 150 per hour between I_3 and I_2, and 550 per hour between I_3 and I_4. ∎

PROBLEM SET 5.3

Practice Problems

In Problems 1–24, use Gaussian elimination to find the complete solution to each system of equations, or show that none exists.

1. $5x + 12y + z = 10$
$2x + 5y + 2z = -1$
$x + 2y - 3z = 5$

2. $2x - 4y + z = 3$
$x - 3y + z = 5$
$3x - 7y + 2z = 12$

3. $5x + 8y - 6z = 14$
$3x + 4y - 2z = 8$
$x + 2y - 2z = 3$

4. $5x - 11y + 6z = 12$
$-x + 3y - 2z = -4$
$3x - 5y + 2z = 4$

5. $3x + 4y + 2z = 3$
$4x - 2y - 8z = -4$
$x + y - z = 3$

6. $2x - y - z = 0$
$x + 2y + z = 3$
$3x + 4y + 2z = 8$

7. $8x + 5y + 11z = 30$
$-x - 4y + 2z = 3$
$2x - y + 5z = 12$

8. $x + y - 10z = -4$
$x - 7z = -5$
$3x + 5y - 36z = -10$

9. $x - 2y - z - 3w = -9$
$x + y - z = 0$
$3x + 4y + w = 6$
$2y - 2z + w = 3$

10. $2x + y - 2z - w = 3$
$x - 2y + z + w = 4$
$-x - 8y + 7z + 5w = 13$
$3x + y - 2z + 2w = 6$

11. $2x + y - z = 3$
$x - 3y + 2z = -4$
$3x + y - 3z + w = 1$
$x + 2y - 4z - w = -2$

12. $2x - y + 3z + w = 0$
$3x + 2y + 4z - w = 0$
$5x - 2y - 2z - w = 0$
$2x + 3y - 7z - 5w = 0$

13. $x - 3y + z - 4w = 4$
$-2x + y + 2z = -2$
$3x - 2y + z - 6w = 2$
$-x + 3y + 2z - w = -6$

14. $3x + 2y - z + 2w = -12$
$4x - y + z + 2w = 1$
$x + y + z + w = -2$
$-2x + 3y + 2z - 3w = 10$

15. $2x + y - z = 2$
$3x + 3y - 2z = 3$

16. $3x + 2y - z = 5$
$x + 2y - z = 1$

17. $x + 2y + 3z = 5$
$y - 5z = 0$

18. $3x - y + 4z = 8$
$y + 2z = 1$

19. $x + y - 2z = 2$
$3x - y - 6z = -7$

20. $-2x - 5y + 10z = 19$
$x + 2y - 4z = 12$

21. $x + y - z + w = -2$
$2x - y + 2z - w = 7$
$-x + 2y + z + 2w = -1$

22. $2x - 3y + 4z + w = 7$
$x - y + 3z - 5w = 10$
$3x + y - 2z - 2w = 6$

23. $x + 2y + 3z - w = 7$
$2y - 3z + w = 4$
$x - 4y + z = 3$

24. $x - y + w = 0$
$x - 4y + z + 2w = 0$
$3x - z + 2w = 0$

Application Problems

25. Three foods have the following nutritional content per ounce.

Units per Ounce

	Vitamin A	Iron	Calcium
Food 1	20	20	10
Food 2	30	10	10
Food 3	10	10	30

a. A diet must consist precisely of 220 units of vitamin A, 180 units of iron, and 340 units of calcium. However, the dietician runs out of Food 1. Use a matrix approach to show that under these conditions the dietary requirements cannot be met.

b. Now suppose that all three foods are available, but due to problems with vitamin A for pregnant women, a hospital dietician no longer wants to include this vitamin in the diet. Use matrices to give two possible ways to meet the iron and calcium requirements with the three foods.

26. The vitamin content per ounce for three foods is given in the following table.

Milligrams per Ounce

	Thiamin	Riboflavin	Niacin
Food A	3	7	1
Food B	1	5	3
Food C	3	8	2

a. Use matrices to show that no combination of these foods can provide exactly 14 mg of thiamin, 32 mg of riboflavin, and 9 mg of niacin.

b. Use matrices to describe in practical terms what happens if the riboflavin requirement is increased by 5 mg and the other requirements stay the same.

27. Use the figure to find values for x, y, z, and w that would realize the desired traffic flow. (All numbers represent cars per hour.)

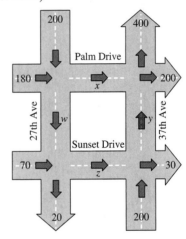

28. A company that manufactures products A, B, and C does both manufacturing and testing. The hours needed to manufacture and test each product are shown in the table.

	Hours Needed Weekly to Manufacture	Hours Needed Weekly to Test
Product A	7	2
Product B	6	2
Product C	3	1

The company has exactly 67 hours per week available for manufacturing and 20 hours per week available for testing. Give two different combinations for the number of products that can be manufactured and tested weekly.

True–False Critical Thinking Problems

29. Which one of the following is true?

  a. The solution set to the system

$$y = 4x - 3$$

$$y = 4x + 5$$

  is the empty set.

  b. The solution set to the system

$$5x - y = 1$$

$$10x - 2y = 2$$

  is $\{(2, 9)\}$.

  c. The system

$$x + y + z = 0$$

$$-x - y + z = 0$$

$$-x + y + z = 0$$

  has $(0, 0, 0)$ as its only solution, and is therefore inconsistent.

  d. The system

$$x - 2y + 3z = 6$$

$$2x - 4y + 6z = 12$$

$$6x - 12y + 18z = 36$$

  is inconsistent.

30. Which one of the following is true?

  a. In a nonsquare system, the number of variables is always equal to the number of equations.

  b. A dependent system in three variables can have a graph consisting of three parallel planes.

  c. A system in three variables that has an augmented matrix given by

$$\begin{bmatrix} 1 & -3 & 4 & | & 16 \\ 0 & 1 & 1 & | & -10 \\ 0 & 0 & 0 & | & 1 \end{bmatrix}$$

  has infinitely many solutions.

  d. One solution to the system in three variables whose augmented matrix is given by

$$\begin{bmatrix} 1 & 1 & -10 & | & -4 \\ 0 & 1 & -3 & | & 1 \\ 0 & 0 & 0 & | & 0 \end{bmatrix}$$

  is $(9, 7, 2)$.

Technology and Group Activity Problem

31. **a.** The figure below shows the intersections of a number of one-way streets. The numbers given represent traffic flow at a peak period (from 4 P.M. to 5:30 P.M.). Use the figure on the right to write a linear system of six equations in seven variables based on the idea that at each intersection the number of cars entering must equal the number of cars leaving.

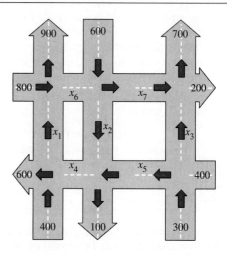

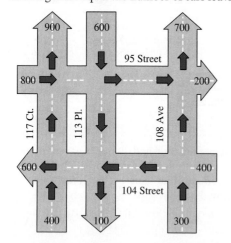

  b. Use a graphing utility with matrix capabilities to find the complete solution to the system.

  c. A political group is planning a demonstration on 95th Street between 113th Place and 117th Court for 5 P.M. Wednesday. The problem becomes one of minimizing traffic flow on 95th Street (between

113th and 117th) without causing traffic tie-ups on other streets. One possible solution is to close off traffic on 95th Street between 113th and 117th (let $x_6 = 0$). What can you conclude about x_7 under these conditions?

d. Choose a value for x_7 (say, $x_7 = 200$) and describe what this means in practical terms.

e. (Group Activity Problem) Notice that working with a matrix allows us to simplify the problem caused by

the political demonstration, but it did not actually solve the problem. There are an infinite number of solutions; each value of x_7 we choose gives us a new picture. We also assumed x_6 was equal to 0; changing that assumption would also lead to different solutions. With your group, design another solution to the traffic flow problem caused by the political demonstration.

Writing in Mathematics

32. Describe what happens when Gaussian elimination is used to solve an inconsistent system.

33. Describe what happens when Gaussian elimination is used to solve a system with dependent equations.

34. In solving a system of dependent equations in three variables, one student simply said that there are infinitely many solutions. A second student expressed the solution set as $\{(4t + 3, 5t - 1, t)\}$. Which is the better form of expressing the solution set and why?

Critical Thinking Problem

35. Consider the linear system

$$x + 3y + z = a^2$$
$$2x + 5y + 2az = 0$$
$$x + y + a^2 z = -9$$

For what values of a will the system be inconsistent?

Group Activity Problem

36. Members of the group should research a realistic application involving foods and nutrients, traffic flow, or any other area of interest that gives rise to a fairly large

dependent or inconsistent system. Discuss how the system's infinitely many solutions or lack of a solution can be used to solve a problem in this specific situation.

SECTION 5.4 **Matrix Operations and Their Applications**

Solutions Manual

Tutorial

Video 9

Objectives

1 Use matrix notation.
2 Determine whether matrices are equal.
3 Perform operations with matrices.
4 Solve matrix equations.
5 Model practical situations with matrix operations.

Up to this point, we have been using matrices to solve linear systems. Matrices, however, have other applications in numerous fields. In this section, we turn our attention to matrix algebra and some of its applications.

Use matrix notation.

Notations for Matrices

We have seen that an array of numbers, arranged in rows and columns and placed in brackets, is called a matrix. Uppercase letters such as A, B, or C are used to name matrices and lowercase letters are used to name their entries.

An $m \times n$ matrix

Let m and n be natural numbers. An $m \times n$ *matrix* A or a matrix of *order $m \times n$* is a rectangular array of numbers with m rows and n columns.

$$
A = \begin{bmatrix}
a_{11} & a_{12} & a_{13} & \cdots & a_{1j} & \cdots & a_{1n} \\
a_{21} & a_{22} & a_{23} & \cdots & a_{2j} & \cdots & a_{2n} \\
a_{31} & a_{32} & a_{33} & \cdots & a_{3j} & \cdots & a_{3n} \\
\vdots & \vdots & \vdots & & \vdots & & \vdots \\
a_{i1} & a_{i2} & a_{i3} & \cdots & \boxed{a_{ij}} & \cdots & a_{in} \\
\vdots & \vdots & \vdots & & \vdots & & \vdots \\
a_{m1} & a_{m2} & a_{m3} & \cdots & a_{mj} & \cdots & a_{mn}
\end{bmatrix}
$$

jth column · ith row · m rows · n columns

The numbers a_{ij} are called the *entries* or *elements* of the matrix. The subscripts indicate the position of each entry: a_{ij} is the entry in the ith row and the jth column. If $m = n$, the matrix is called a *square matrix*.

Although capital letters denote a matrix as a whole, and lowercase letters denote individual entries, a matrix can also be denoted by a representative element enclosed in brackets, such as

$$A = [a_{ij}].$$

The matrix with elements a_{ij}

EXAMPLE 1 Matrix Notation

Let

$$A = \begin{bmatrix} 3 & 2 & 0 \\ -4 & -5 & -\frac{1}{5} \end{bmatrix}$$

a. What is the order of A?

b. If $A = [a_{ij}]$, identify a_{23} and a_{12}.

Solution

a. The matrix has 2 rows and 3 columns, so it is of order 2×3.

b. The entry a_{23} is in the second row and third column. Thus, $a_{23} = -\dfrac{1}{5}$. The entry a_{12} is in the first row and second column, and consequently $a_{12} = 2$.

2 Determine whether matrices are equal.

Equality of Matrices

Study tip

This definition says that equal matrices have the same order and equal corresponding entries.

Definition of equality of matrices

Two matrices A and B are *equal* if and only if they have the same order $m \times n$ and $a_{ij} = b_{ij}$ for $i = 1, 2, \ldots, m$ and $j = 1, 2, \ldots, n$.

EXAMPLE 2 Deciding Whether Two Matrices Are Equal

Determine whether each two matrices are equal:

a. $\begin{bmatrix} \sqrt{9} & 5^2 & 0 & e^0 \\ 0.75 & 3 & \sqrt[3]{-8} & 2-2 \end{bmatrix}$ and $\begin{bmatrix} 3 & 25 & 0 & 1 \\ \frac{3}{4} & 3 & -2 & 0 \end{bmatrix}$

b. $\begin{bmatrix} 1 & 2 & 3 \\ 4 & 5 & 6 \end{bmatrix}$ and $\begin{bmatrix} 1 & 4 \\ 2 & 5 \\ 3 & 6 \end{bmatrix}$

Solution

a. $\begin{bmatrix} \sqrt{9} & 5^2 & 0 & e^0 \\ 0.75 & 3 & \sqrt[3]{-8} & 2-2 \end{bmatrix}$ and $\begin{bmatrix} 3 & 25 & 0 & 1 \\ \frac{3}{4} & 3 & -2 & 0 \end{bmatrix}$ are equal because these matrices have the same order (2×4: 2 rows and 4 columns) and their corresponding entries are equal.

b. $\begin{bmatrix} 1 & 2 & 3 \\ 4 & 5 & 6 \end{bmatrix}$ and $\begin{bmatrix} 1 & 4 \\ 2 & 5 \\ 3 & 6 \end{bmatrix}$ are not equal because they do not have the same order, even though they contain the same numbers. The first matrix has order 2×3 (2 rows and 3 columns) and the second has order 3×2 (3 rows and 2 columns). ∎

3 Perform operations with matrices.

Matrix Addition and Subtraction

Table 5.1 shows that matrices of the same order can be added or subtracted by simply adding or subtracting corresponding entries.

TABLE 5.1 Adding and subtracting matrices (Let $A = [a_{ij}]$ and $B = [b_{ij}]$ be matrices of order $m \times n$.)

Definition	The Definition in Words	Example
Matrix Addition $A + B = [a_{ij} + b_{ij}]$	Matrices of the same order are added by adding the entries in corresponding positions.	$\begin{bmatrix} 1 & -2 \\ 3 & 5 \end{bmatrix} + \begin{bmatrix} -1 & 6 \\ 0 & 4 \end{bmatrix}$ $= \begin{bmatrix} 1+(-1) & -2+6 \\ 3+0 & 5+4 \end{bmatrix} = \begin{bmatrix} 0 & 4 \\ 3 & 9 \end{bmatrix}$
Matrix Subtraction $A - B = [a_{ij} - b_{ij}]$	Matrices of the same order are subtracted by subtracting the entries in corresponding positions.	$\begin{bmatrix} 1 & -2 \\ 3 & 5 \end{bmatrix} - \begin{bmatrix} -1 & 6 \\ 0 & 4 \end{bmatrix}$ $= \begin{bmatrix} 1-(-1) & -2-6 \\ 3-0 & 5-4 \end{bmatrix} = \begin{bmatrix} 2 & -8 \\ 3 & 1 \end{bmatrix}$

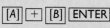
sing technology

Graphing utilities can add and subtract matrices. Enter the matrices and use a keystroke sequence similar to

[A] [+] [B] [ENTER]

[A] [−] [B] [ENTER]

Consult your manual and verify the results in Example 3.

iscover for yourself

If a matrix whose entries are all 0s is added to any matrix, such as

$$\begin{bmatrix} -5 & 2 \\ 3 & 6 \end{bmatrix} + \begin{bmatrix} 0 & 0 \\ 0 & 0 \end{bmatrix},$$

what is the result? How is this similar to the role that 0 plays in ordinary addition?

tudy tip

Matrices of different orders cannot be added or subtracted. For example, both

$$\begin{bmatrix} 0 & 3 \\ 4 & 3 \end{bmatrix} + \begin{bmatrix} 1 & 9 \\ 4 & 5 \\ 2 & 3 \end{bmatrix}$$

and

$$\begin{bmatrix} 0 & 3 \\ 4 & 3 \end{bmatrix} - \begin{bmatrix} 1 & 9 \\ 4 & 5 \\ 2 & 3 \end{bmatrix}$$

are undefined. The first matrix is of order 2×2 and the second is of order 3×2.

EXAMPLE 3 Adding and Subtracting Matrices

Perform the indicated matrix operations:

a. $\begin{bmatrix} 0 & 5 & 3 \\ -2 & 6 & -8 \end{bmatrix} + \begin{bmatrix} -2 & 3 & 5 \\ 7 & -9 & 6 \end{bmatrix}$

b. $\begin{bmatrix} -5 & 2 \\ 3 & 6 \end{bmatrix} + \begin{bmatrix} 5 & -2 \\ -3 & -6 \end{bmatrix}$

c. $\begin{bmatrix} -6 & 7 \\ 2 & -3 \end{bmatrix} - \begin{bmatrix} -5 & 6 \\ 0 & -4 \end{bmatrix}$

Solution

a.
$$\begin{bmatrix} 0 & 5 & 3 \\ -2 & 6 & -8 \end{bmatrix} + \begin{bmatrix} -2 & 3 & 5 \\ 7 & -9 & 6 \end{bmatrix}$$

$$= \begin{bmatrix} 0 + (-2) & 5 + 3 & 3 + 5 \\ -2 + 7 & 6 + (-9) & -8 + 6 \end{bmatrix}$$ Add the corresponding entries in the 2×3 matrices.

$$= \begin{bmatrix} -2 & 8 & 8 \\ 5 & -3 & -2 \end{bmatrix}$$ Simplify.

b.
$$\begin{bmatrix} -5 & 2 \\ 3 & 6 \end{bmatrix} + \begin{bmatrix} 5 & -2 \\ -3 & -6 \end{bmatrix}$$

$$= \begin{bmatrix} -5 + 5 & 2 + (-2) \\ 3 + (-3) & 6 + (-6) \end{bmatrix}$$ Add the corresponding entries in the 2×2 matrices.

$$= \begin{bmatrix} 0 & 0 \\ 0 & 0 \end{bmatrix}$$ A matrix that contains only zero entries is called a *zero matrix*.

c.
$$\begin{bmatrix} -6 & 7 \\ 2 & -3 \end{bmatrix} - \begin{bmatrix} -5 & 6 \\ 0 & -4 \end{bmatrix}$$

$$= \begin{bmatrix} -6 - (-5) & 7 - 6 \\ 2 - 0 & -3 - (-4) \end{bmatrix}$$ Subtract the corresponding entries in the 2×2 matrices.

$$= \begin{bmatrix} -1 & 1 \\ 2 & 1 \end{bmatrix}$$ Simplify. ■

Scalar Multiplication

A matrix of order 1×1, such as [6], contains only one entry. To distinguish this matrix from the number 6, we refer to 6 as a *scalar*. In general, in our work with matrices, we will refer to real numbers as scalars.

To multiply a matrix A by a scalar c, we multiply each entry in A by c. For example,

$$4 \begin{bmatrix} 2 & 5 \\ -3 & 0 \end{bmatrix} = \begin{bmatrix} 4(2) & 4(5) \\ 4(-3) & 4(0) \end{bmatrix} = \begin{bmatrix} 8 & 20 \\ -12 & 0 \end{bmatrix}.$$

Scalar   Matrix

> ### Definition of scalar multiplication
>
> If $A = [a_{ij}]$ is a matrix of order $m \times n$ and c is a scalar, then the matrix cA is the $m \times n$ matrix given by
>
> $$cA = [ca_{ij}]$$
>
> obtained by multiplying each entry of A by the real number c. We call cA a *scalar multiple* of A.

Using technology

Use your graphing utility to verify Example 5. First enter the matrices and then use the following keystrokes:

$\boxed{(-)}\,4\,\boxed{[A]}\,+\,3\,\boxed{[B]}\,\boxed{\text{ENTER}}$

EXAMPLE 4 **Scalar Multiplication and Matrix Addition**

If $A = \begin{bmatrix} 1 & 4 \\ -3 & 0 \\ 2 & 5 \end{bmatrix}$ and $B = \begin{bmatrix} 0 & 0 \\ -4 & -1 \\ 1 & -1 \end{bmatrix}$ find $-4A + 3B$.

Solution

$$-4A + 3B = -4\begin{bmatrix} 1 & 4 \\ -3 & 0 \\ 2 & 5 \end{bmatrix} + 3\begin{bmatrix} 0 & 0 \\ -4 & -1 \\ 1 & -1 \end{bmatrix}$$

$$= \begin{bmatrix} -4(1) & -4(4) \\ -4(-3) & -4(0) \\ -4(2) & -4(5) \end{bmatrix} + \begin{bmatrix} 3(0) & 3(0) \\ 3(-4) & 3(-1) \\ 3(1) & 3(-1) \end{bmatrix}$$

Multiply each entry in the first matrix by -4 and each entry in the second matrix by 3.

$$= \begin{bmatrix} -4 & -16 \\ 12 & 0 \\ -8 & -20 \end{bmatrix} + \begin{bmatrix} 0 & 0 \\ -12 & -3 \\ 3 & -3 \end{bmatrix}$$

Simplify.

$$= \begin{bmatrix} -4 + 0 & -16 + 0 \\ 12 + (-12) & 0 + (-3) \\ -8 + 3 & -20 + (-3) \end{bmatrix}$$

Perform the addition of these 3×2 matrices by adding corresponding entries.

$$= \begin{bmatrix} -4 & -16 \\ 0 & -3 \\ -5 & -23 \end{bmatrix}$$

Simplify.

$\blacksquare$

Matrix Multiplication

Matrix multiplication is *not* defined by multiplying corresponding entries of two matrices. To define a type of multiplication that has meaningful applications, we must begin by thinking of matrix multiplication as *row-by-column* multiplication.

iscover for yourself

You can use your graphing utility to discover a number of important ideas about matrix multiplication before reading about them. Begin by entering the matrices

$$A = \begin{bmatrix} 2 & 3 & 4 \\ -1 & -2 & 0 \end{bmatrix} \quad \text{and} \quad B = \begin{bmatrix} 4 & 2 \\ 6 & 1 \\ 3 & 5 \end{bmatrix}.$$

Use the following keystrokes to find the product of the matrices.

$$\boxed{[A]} \; \boxed{\times} \; \boxed{[B]} \; \boxed{\text{ENTER}}$$

1. What is the order of matrix A? What is the order of matrix B? What is the order of the product AB that appears in the viewing rectangle?

2. Now find the product BA by using these keystrokes.

$$\boxed{[B]} \; \boxed{\times} \; \boxed{[A]} \; \boxed{\text{ENTER}}$$

Is BA the same matrix as AB? What does this mean about matrix multiplication and the commutative property?

3. What is the order of matrix BA?

4. Use your results from (1) and (3) to generalize a pattern. If the order of matrix A is $m \times n$ and the order of matrix B is $n \times p$, then what is the order of the product AB?

Study tip

The product AB has exactly as many rows as A and exactly as many columns as B.

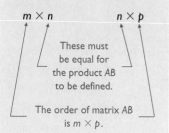

Matrix A: **Matrix B:**

$m \times n$ $n \times p$

These must be equal for the product AB to be defined.

The order of matrix AB is $m \times p$.

Following are two important ideas that eventually we'll formulate into a precise definition of matrix multiplication. Were you able to observe the first of these ideas in the Discover for yourself?

1. If A is an $m \times n$ matrix and B is an $n \times p$ matrix, the matrix product AB is an $m \times p$ matrix. (See the Study Tip.)

2. The entry in the ith row and the jth column of the matrix product AB is found by multiplying each entry in the ith row of A by the corresponding entry in the jth column of B and adding the results. If A and B are matrices such that

$$A = \begin{bmatrix} a_{11} & a_{12} & a_{13} & \cdots & a_{1n} \\ a_{21} & a_{22} & a_{23} & \cdots & a_{2n} \\ \vdots & \vdots & \vdots & & \vdots \\ a_{i1} & a_{i2} & a_{i3} & \cdots & a_{in} \\ \vdots & \vdots & \vdots & & \vdots \\ a_{m1} & a_{m2} & a_{m3} & \cdots & a_{mn} \end{bmatrix} \quad \text{and} \quad B = \begin{bmatrix} b_{11} & b_{12} & \cdots & b_{1j} & \cdots & b_{1p} \\ b_{21} & b_{22} & \cdots & b_{2j} & \cdots & b_{2p} \\ b_{31} & b_{32} & \cdots & b_{3j} & \cdots & b_{3p} \\ \vdots & \vdots & & \vdots & & \vdots \\ b_{n1} & b_{n2} & \cdots & b_{nj} & \cdots & b_{np} \end{bmatrix}$$

then the ith row, jth column element of the matrix AB is

$$a_{i1}b_{1j} + a_{i2}b_{2j} + a_{i3}b_{3j} + \cdots + a_{in}b_{nj}.$$

Let's look at some examples to see if we can unravel what all this means.

EXAMPLE 5   Multiplying Matrices

Matrices A and B are defined as follows.

$$A = \begin{bmatrix} 1 & 2 & 3 \end{bmatrix} \qquad B = \begin{bmatrix} 4 \\ 5 \\ 6 \end{bmatrix}$$

Find: **a.** AB and **b.** BA

Solution

a. Matrix A is a 1×3 matrix and matrix B is a 3×1 matrix. Thus, the product AB is a 1×1 matrix.

$$AB = \begin{bmatrix} 1 & 2 & 3 \end{bmatrix} \begin{bmatrix} 4 \\ 5 \\ 6 \end{bmatrix}$$ We will perform a row-by-column computation.

$$= [(1)(4) + (2)(5) + (3)(6)]$$ Multiply each entry in row 1 of A by each entry in column 1 of B.

$$= [4 + 10 + 18]$$

$$= [32]$$

b. Matrix B is a 3×1 matrix and matrix A is a 1×3 matrix. Thus, the product BA is a 3×3 matrix.

$$BA = \begin{bmatrix} 4 \\ 5 \\ 6 \end{bmatrix} \begin{bmatrix} 1 & 2 & 3 \end{bmatrix}$$ We will perform a row-by-column computation.

$$= \begin{bmatrix} \overset{\text{Row 1} \times \text{Column 1}}{(4)(1)} & \overset{\text{Row 1} \times \text{Column 2}}{(4)(2)} & \overset{\text{Row 1} \times \text{Column 3}}{(4)(3)} \\ \underset{}{(5)(1)} & (5)(2) & (5)(3) \\ (6)(1) & (6)(2) & (6)(3) \end{bmatrix}$$

$$= \begin{bmatrix} 4 & 8 & 12 \\ 5 & 10 & 15 \\ 6 & 12 & 18 \end{bmatrix}$$ Simplify.   ■

EXAMPLE 6   **Multiplying Matrices**

Multiply: $\begin{bmatrix} 2 & 3 & 4 \\ -1 & 2 & 0 \end{bmatrix} \begin{bmatrix} 4 & 2 \\ 6 & 1 \\ 3 & 5 \end{bmatrix}$

Solution

The first matrix is a 2×3 matrix and the second is a 3×2 matrix. The product will be a 2×2 matrix.

$$\begin{bmatrix} 2 & 3 & 4 \\ -1 & 2 & 0 \end{bmatrix} \begin{bmatrix} 4 & 2 \\ 6 & 1 \\ 3 & 5 \end{bmatrix}$$ We will perform a row-by-column computation.

$$= \begin{bmatrix} \overset{\text{Row 1} \times \text{Column 1}}{2(4) + 3(6) + 4(3)} & \overset{\text{Row 1} \times \text{Column 2}}{2(2) + 3(1) + 4(5)} \\ \underset{\text{Row 2} \times \text{Column 1}}{-1(4) + 2(6) + 0(3)} & \underset{\text{Row 2} \times \text{Column 2}}{-1(2) + 2(1) + 0(5)} \end{bmatrix}$$

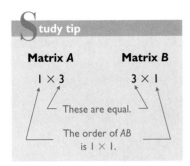

Matrix A   **Matrix B**

1×3 3×1

These are equal.

The order of AB is 1×1.

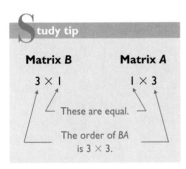

Matrix B   **Matrix A**

3×1 1×3

These are equal.

The order of BA is 3×3.

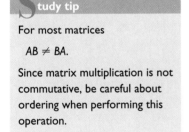

For most matrices

$AB \neq BA$.

Since matrix multiplication is not commutative, be careful about ordering when performing this operation.

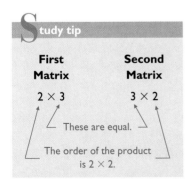

First   **Second**
Matrix   **Matrix**

2×3 3×2

These are equal.

The order of the product is 2×2.

$$= \begin{bmatrix} 8 + 18 + 12 & 4 + 3 + 20 \\ -4 + 12 + 0 & -2 + 2 + 0 \end{bmatrix} = \begin{bmatrix} 38 & 27 \\ 8 & 0 \end{bmatrix} \quad \blacksquare$$

Before considering an additional example, let's look at the formal definition of matrix multiplication.

Definition of matrix multiplication

If $A = [a_{ij}]$ is an $m \times n$ matrix and $B = [b_{ij}]$ is an $n \times p$ matrix, then the product AB is the $m \times p$ matrix

$$AB = [p_{ij}]$$

where $\underbrace{p_{ij}}_{} = \underbrace{a_{i1}b_{1j} + a_{i2}b_{2j} + a_{i3}b_{3j} + \cdots + a_{in}b_{nj}}_{}.$

The entry in   Multiply the entries in the ith
the ith row   row of A by the corresponding
and jth column   entries in the jth column of B.
of the product   Then add the results.

EXAMPLE 7 | Multiplying Matrices

Where possible, find each product:

a. $\begin{bmatrix} 4 & 2 \\ 1 & 3 \end{bmatrix} \begin{bmatrix} 1 & 2 & 3 & 4 \\ 0 & 2 & -1 & 6 \end{bmatrix}$

b. $\begin{bmatrix} 1 & 2 & 3 & 4 \\ 0 & 2 & -1 & 6 \end{bmatrix} \begin{bmatrix} 4 & 2 \\ 1 & 3 \end{bmatrix}$

Solution

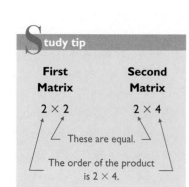

a. The first matrix is a 2×2 matrix and the second is a 2×4 matrix. The product will be a 2×4 matrix.

$$\begin{bmatrix} 4 & 2 \\ 1 & 3 \end{bmatrix} \begin{bmatrix} 1 & 2 & 3 & 4 \\ 0 & 2 & -1 & 6 \end{bmatrix}$$

Once again, we will perform a row-by-column computation.

Row 1   Row 1   Row 1   Row 1
× Column 1   × Column 2   × Column 3   × Column 4

$$= \begin{bmatrix} 4(1) + 2(0) & 4(2) + 2(2) & 4(3) + 2(-1) & 4(4) + 2(6) \\ 1(1) + 3(0) & 1(2) + 3(2) & 1(3) + 3(-1) & 1(4) + 3(6) \end{bmatrix}$$

Row 2   Row 2   Row 2   Row 2
× Column 1   × Column 2   × Column 3   × Column 4

$$= \begin{bmatrix} 4 + 0 & 8 + 4 & 12 - 2 & 16 + 12 \\ 1 + 0 & 2 + 6 & 3 - 3 & 4 + 18 \end{bmatrix}$$

$$= \begin{bmatrix} 4 & 12 & 10 & 28 \\ 1 & 8 & 0 & 22 \end{bmatrix}$$

b. $\begin{bmatrix} 1 & 2 & 3 & 4 \\ 0 & 2 & -1 & 6 \end{bmatrix} \begin{bmatrix} 4 & 2 \\ 1 & 3 \end{bmatrix}$

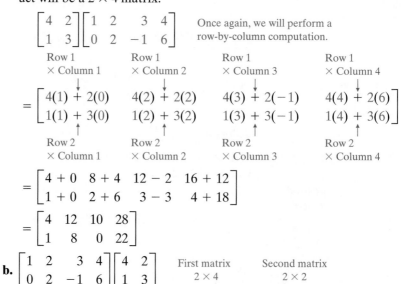

First matrix   Second matrix
2×4 2×2

These numbers must be the same
to multiply the matrices.

Study tip

For the product of two matrices to be defined, the number of columns of the first matrix must equal the number of rows of the second matrix.

The inner two numbers are not the same, meaning that rows and columns will not match up when calculating the product. The product of these two matrices is not defined.   ∎

4 Solve matrix equations.

Matrix Equations

Because of the similarity of the algebra of matrices and the algebra of real numbers, the process of solving a matrix equation parallels the process of solving an equation in which the variable represents a real number. This process is illustrated in Example 8.

Discover for yourself

Try solving Example 8 on your own before reading the solution. First solve

  $6X + A = 2B$

for X. Then find the matrix X by substituting the matrices A and B and performing the resulting operations.

EXAMPLE 8   Solving a Matrix Equation

If $A = \begin{bmatrix} 3 & -4 \\ 0 & 2 \end{bmatrix}$ and $B = \begin{bmatrix} -5 & 2 \\ 3 & 1 \end{bmatrix}$, solve for X in the matrix equation $6X + A = 2B$.

Solution

Let's first solve the given equation for X.

$$6X + A = 2B \qquad \text{This is the given equation.}$$

$$6X = 2B - A \qquad \text{Subtract } A \text{ from both sides.}$$

$$X = \tfrac{1}{6}(2B - A) \qquad \text{Multiply both sides by } \tfrac{1}{6}.$$

This equation tells us exactly what we must do with the given matrices A and B to find the matrix X.

1. Multiply B by 2.
2. Subtract A.
3. Multiply the resulting matrix by $\tfrac{1}{6}$.

We now carry out this three-step process.

Multiply B by 2:

$$2B = 2\begin{bmatrix} -5 & 2 \\ 3 & 1 \end{bmatrix} = \begin{bmatrix} -10 & 4 \\ 6 & 2 \end{bmatrix} \qquad \text{Multiply each entry in } B \text{ by 2.}$$

Subtract A:

$$2B - A = \begin{bmatrix} -10 & 4 \\ 6 & 2 \end{bmatrix} - \begin{bmatrix} 3 & -4 \\ 0 & 2 \end{bmatrix} = \begin{bmatrix} -13 & 8 \\ 6 & 0 \end{bmatrix} \qquad \text{Subtract corresponding entries.}$$

Multiply the resulting matrix by $\tfrac{1}{6}$:

$$\tfrac{1}{6}(2B - A) = \tfrac{1}{6}\begin{bmatrix} -13 & 8 \\ 6 & 0 \end{bmatrix} = \begin{bmatrix} -\tfrac{13}{6} & \tfrac{4}{3} \\ 1 & 0 \end{bmatrix} \qquad \text{Multiply each entry by } \tfrac{1}{6}.$$

We see that $X = \begin{bmatrix} -\tfrac{13}{6} & \tfrac{4}{3} \\ 1 & 0 \end{bmatrix}$.   ∎

iscover for yourself

Check the matrix in the solution to Example 8 by substituting it, as well as A and B, in the given equation

$$6X + A = 2B.$$

You should obtain equal matrices on both sides of the equation.

The Matrix Form of a Linear System

Arthur Cayley

The Granger Collection

Matrices were first studied intensively by the English mathematician Arthur Cayley (1821–1895). Before reaching the age of 25, he published 25 papers, setting a pattern of prolific creativity that lasted throughout his life. Cayley was a lawyer, painter, mountaineer, and Cambridge professor whose greatest invention was that of matrices and matrix theory. Cayley's matrix algebra, especially the noncommutativity of multiplication ($AB \neq BA$) opened up a new area of mathematics called abstract algebra.

Matrix multiplication, not expressed in the anticipated way in which we merely multiply corresponding entries, provides a compact, symbolic notation for representing a system of linear equations.

EXAMPLE 9   **Expressing a Linear System in Matrix Form**

Show that the system of equations

$$a_1 x + b_1 y + c_1 z = d_1$$

$$a_2 x + b_2 y + c_2 z = d_2$$

$$a_3 x + b_3 y + c_3 z = d_3$$

is equivalent to the matrix equation

$$\begin{bmatrix} a_1 & b_1 & c_1 \\ a_2 & b_2 & c_2 \\ a_3 & b_3 & c_3 \end{bmatrix} \begin{bmatrix} x \\ y \\ z \end{bmatrix} = \begin{bmatrix} d_1 \\ d_2 \\ d_3 \end{bmatrix}.$$

Solution

We will work with the matrix equation and obtain the system of equations. Since the first matrix on the left side is a 3×3 matrix and the second matrix is a 3×1 matrix, the order of their product is a 3×1 matrix. We apply the row-column definition of matrix multiplication to obtain this 3×1 matrix.

$$\begin{bmatrix} a_1 & b_1 & c_1 \\ a_2 & b_2 & c_2 \\ a_3 & b_3 & c_3 \end{bmatrix} \begin{bmatrix} x \\ y \\ z \end{bmatrix} = \begin{bmatrix} d_1 \\ d_2 \\ d_3 \end{bmatrix}$$ This is the equation in matrix form. Multiply on the left.

$$\begin{array}{l} \text{Row 1} \times \text{Column 1} \longrightarrow \\ \text{Row 2} \times \text{Column 1} \longrightarrow \\ \text{Row 3} \times \text{Column 1} \longrightarrow \end{array} \begin{bmatrix} a_1 x + b_1 y + c_1 z \\ a_2 x + b_2 y + c_2 z \\ a_3 x + b_3 y + c_3 z \end{bmatrix} = \begin{bmatrix} d_1 \\ d_2 \\ d_3 \end{bmatrix}$$

Since these matrices are equal, their corresponding entries are equal. Equating these entries, we obtain

$$a_1 x + b_1 y + c_1 z = d_1$$

$$a_2 x + b_2 y + c_2 z = d_2$$

$$a_3 x + b_3 y + c_3 z = d_3.$$ ■

The matrix equation

$$\begin{bmatrix} a_1 & b_1 & c_1 \\ a_2 & b_2 & c_2 \\ a_3 & b_3 & c_3 \end{bmatrix} \begin{bmatrix} x \\ y \\ z \end{bmatrix} = \begin{bmatrix} d_1 \\ d_2 \\ d_3 \end{bmatrix}$$

$$\underset{A}{\downarrow} \qquad \underset{X}{\downarrow} \;\underset{=}{\downarrow}\; \underset{B}{\downarrow}$$

is abbreviated as $AX = B$, where A is the *coefficient matrix* of the system, and X and B are matrices containing one column, called *column matrices*. The matrix B is called the *constant matrix*.

EXAMPLE 10 Solving a Matrix Equation

Solve the matrix equation $AX = B$ for X, where

Charles Sheeler "Windows" 1951, oil on canvas, $32 \times 20 \, 1/4$ in. Collection Hirschl & Adler Galleries, New York.

$$\overset{\text{Coefficient matrix}}{A = \begin{bmatrix} 1 & 0 & 1 \\ 2 & 2 & 4 \\ 1 & 6 & 8 \end{bmatrix}} \quad \text{and} \quad \overset{\text{Constant matrix}}{B = \begin{bmatrix} 4 \\ 10 \\ 4 \end{bmatrix}}$$

Solution

Using the definition for matrix multiplication with $X = \begin{bmatrix} x \\ y \\ z \end{bmatrix}$, the system

$$\begin{bmatrix} 1 & 0 & 1 \\ 2 & 2 & 4 \\ 1 & 6 & 8 \end{bmatrix} \begin{bmatrix} x \\ y \\ z \end{bmatrix} = \begin{bmatrix} 4 \\ 10 \\ 4 \end{bmatrix}$$

is equivalent to

$$x + z = 4$$
$$2x + 2y + 4z = 10$$
$$x + 6y + 8z = 4.$$

We begin with the augmented matrix

$$\left[\begin{array}{ccc|c} 1 & 0 & 1 & 4 \\ 2 & 2 & 4 & 10 \\ 1 & 6 & 8 & 4 \end{array} \right]$$

and use Gaussian elimination to obtain

$$\left[\begin{array}{ccc|c} 1 & 0 & 1 & 4 \\ 0 & 1 & 1 & 1 \\ 0 & 0 & 1 & -6 \end{array} \right].$$

The solution of the system of linear equations

$$x + z = 4$$
$$y + z = 1$$
$$z = -6$$

ENRICHMENT ESSAY

Algebra and Poetry

In many ways, algebra exhibits the same elements of beauty that are generally acknowledged to be the essence of poetry. The superiority of poetry over other forms of written communication lies both in its symbolism and in its extreme condensation—its studied economy of words.

In algebra, we do things in much the same way as does the poet. Descriptions appear in compact, symbolic formulas such as

$$\begin{bmatrix} a_1 & b_1 & c_1 \\ a_2 & b_2 & c_2 \\ a_3 & b_3 & c_3 \end{bmatrix} \begin{bmatrix} x \\ y \\ z \end{bmatrix} = \begin{bmatrix} d_1 \\ d_2 \\ d_3 \end{bmatrix}.$$

This symbolic style presents infinitely many different systems of equations in three variables in one compact space.

Giorgio de Chirico (1888–1978) "The Uncertainty of the Poet," 1913, oil on canvas. Tate Gallery, London / Art Resource.

is $z = -6$, $y = 7$ [$y + z = 1$, so $y - 6 = 1$ and $y = 7$], and $x = 10$ [$x + z = 4$, so $x - 6 = 4$ and $x = 10$]. The solution of the matrix equation is

$$X = \begin{bmatrix} x \\ y \\ z \end{bmatrix} = \begin{bmatrix} 10 \\ 7 \\ -6 \end{bmatrix}.$$

∎

Discover for yourself

Substitute the matrix in the solution to Example 10 for X in the given equation. Show that $AX = B$ is a true statement for the given matrices A and B, and the proposed solution matrix X. Perform the multiplication either by hand or with your graphing utility.

5 Model practical situations with matrix operations.

Applications

Matrices are a convenient way to classify data. The matrices contain the essential numbers that describe a given situation, as we see in Example 11.

EXAMPLE 11   Modeling with Matrices

A corporation makes three models of designer telephones that are manufactured in Mexico and assembled in the United States. Manufacturing/assembly and shipping costs for each telephone can be listed in 3×2 matrices.

	Mexico			**United States**	
	Manufacture	**Ship**		**Assemble**	**Ship**
Model 1	60	12	**Model 1**	12	4
Model 2	85	14	**Model 2**	15	6
Model 3	90	16	**Model 3**	20	8

a. Let M be the matrix for Mexico, and let U be the matrix for the United States. Find $M + U$ and describe the significance of this matrix in practical terms.

b. Write the matrix for the situation where the costs in Mexico are cut in half and the costs in the United States are tripled.

Solution

a. $M + U$

$$= \begin{bmatrix} 60 & 12 \\ 85 & 14 \\ 90 & 16 \end{bmatrix} + \begin{bmatrix} 12 & 4 \\ 15 & 6 \\ 20 & 8 \end{bmatrix}$$

$$= \begin{bmatrix} 72 & 16 \\ 100 & 20 \\ 110 & 24 \end{bmatrix} \longleftarrow \text{Add the corresponding entries in the two } 2 \times 3 \text{ matrices.}$$

This matrix describes manufacturing, assembly, and all shipping costs for the three models of telephone in Mexico and the United States. For example, each Model 1 costs $72 to manufacture and assemble, and $16 to ship. In short, the matrix $M + U$ describes total costs.

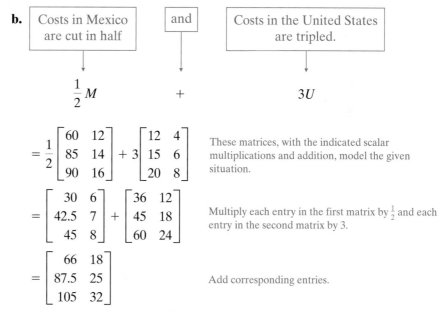

b.

| Costs in Mexico are cut in half | and | Costs in the United States are tripled. |

$$\frac{1}{2}M \qquad + \qquad 3U$$

$$= \frac{1}{2}\begin{bmatrix} 60 & 12 \\ 85 & 14 \\ 90 & 16 \end{bmatrix} + 3\begin{bmatrix} 12 & 4 \\ 15 & 6 \\ 20 & 8 \end{bmatrix}$$

These matrices, with the indicated scalar multiplications and addition, model the given situation.

$$= \begin{bmatrix} 30 & 6 \\ 42.5 & 7 \\ 45 & 8 \end{bmatrix} + \begin{bmatrix} 36 & 12 \\ 45 & 18 \\ 60 & 24 \end{bmatrix}$$

Multiply each entry in the first matrix by $\frac{1}{2}$ and each entry in the second matrix by 3.

$$= \begin{bmatrix} 66 & 18 \\ 87.5 & 25 \\ 105 & 32 \end{bmatrix}$$

Add corresponding entries.

This matrix describes total costs when Mexico's costs are halved and those of the United States are tripled. The first row in this matrix indicates that each Model 1 now costs $66 to manufacture and assemble and $18 to ship. ∎

Since many corporations manufacture far more than three models of a product and costs are generally broken down into many categories, costs are usually modeled by matrices of order far greater than 3×2. Computers are used to store the matrices and perform the matrix operations.

Claes Oldenburg "Soft Pay Telephone" 1963. Solomon R. Guggenheim Museum, New York. Photo by David Heald © The Solomon R. Guggenheim Foundation, New York. FN 80.247.

We have seen how matrix multiplication allows us to express a linear system in a compact, matrix form. Example 12 illustrates how matrix multiplication can be used to model a practical situation.

EXAMPLE 12   Applying Matrix Multiplication

At a certain gas station, the number of gallons of regular, unleaded, and super unleaded gas sold on Monday, Tuesday, and Wednesday of a particular week is given by the following matrix.

	Regular	**Unleaded**	**Super Unleaded**
Monday	240	300	160
Tuesday	200	280	180
Wednesday	260	310	200

$= A$

A second matrix gives the selling price per gallon and the profit per gallon for the three types of gas sold by the station.

	Selling price per gallon	**Profit per gallon**
Regular	1.15	0.15
Unleaded	1.20	0.17
Super Unleaded	1.25	0.19

$= B$

a. Calculate the product AB.
b. Describe what the matrix AB represents and interpret the entries.
c. What is the gas station's profit for Monday through Wednesday?

Solution

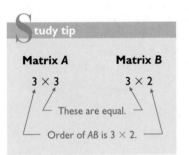

Study tip

Matrix A   **Matrix B**

3×3 3×2

These are equal.

Order of AB is 3×2.

a. $AB = \begin{bmatrix} 240 & 300 & 160 \\ 200 & 280 & 180 \\ 260 & 310 & 200 \end{bmatrix} \begin{bmatrix} 1.15 & 0.15 \\ 1.20 & 0.17 \\ 1.25 & 0.19 \end{bmatrix}$

$= \begin{bmatrix} 240(1.15) + 300(1.20) + 160(1.25) & 240(0.15) + 300(0.17) + 160(0.19) \\ 200(1.15) + 280(1.20) + 180(1.25) & 200(0.15) + 280(0.17) + 180(0.19) \\ 260(1.15) + 310(1.20) + 200(1.25) & 260(0.15) + 310(0.17) + 200(0.19) \end{bmatrix}$

Perform the row-by-column multiplications.

$= \begin{bmatrix} 836 & 117.40 \\ 791 & 111.80 \\ 921 & 129.70 \end{bmatrix}$

Multiply and add as indicated.

b. To describe the significance of matrix AB, let's look at two of the computations. We'll begin with the entry in row 1, column 1:

$$240 \quad (1.15) \; + \; 300 \quad (1.20) \; + \; 160 \quad (1.25)$$

| Regular gallons sold on Monday | Selling price per gallon | Unleaded gallons sold on Monday | Selling price per gallon | Super Unleaded gallons sold on Monday | Selling price per gallon |

This entry represents total sales for Monday. Similarly, the other two entries in the first column represent total sales for Tuesday and Wednesday.

Now let's look at the entry in row 1, column 2:

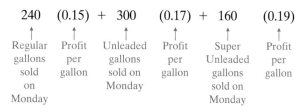

240 (0.15) + 300 (0.17) + 160 (0.19)

Regular gallons sold on Monday — Profit per gallon — Unleaded gallons sold on Monday — Profit per gallon — Super Unleaded gallons sold on Monday — Profit per gallon

This entry represents the gas station's profit for Monday. Similarly, the other two entries in the second column represent the profit for Tuesday and Wednesday.

The matrix AB represents the total sales and the profit for Monday, Tuesday, and Wednesday.

	Total Sales	Profit
Monday	836	117.40
Tuesday	791	111.80
Wednesday	921	129.70

c. The gas station's profit for Monday through Wednesday is \$117.40 + \$111.80 + \$129.70 or \$358.90. ∎

PROBLEM SET 5.4

Practice Problems

In Problems 1–4,

a. *Give the order of each matrix.*

b. *If $A = [a_{ij}]$, identify a_{32} and a_{23} or explain why identification is not possible.*

1. $\begin{bmatrix} 4 & -7 & 5 \\ -6 & 8 & -1 \end{bmatrix}$

2. $\begin{bmatrix} -6 & 4 & -1 \\ -9 & 0 & \frac{1}{2} \end{bmatrix}$

3. $\begin{bmatrix} 1 & -5 & \pi & e \\ 0 & 7 & -6 & -\pi \\ -2 & \frac{1}{2} & 11 & -\frac{1}{5} \end{bmatrix}$

4. $\begin{bmatrix} -4 & 1 & 3 & -5 \\ 2 & -1 & \pi & 0 \\ 1 & 0 & -e & \frac{1}{5} \end{bmatrix}$

In Problems 5–8, find values for the variables that make each matrix equation true.

5. $\begin{bmatrix} 2x & -4 \\ 6 & -3y \end{bmatrix} = \begin{bmatrix} -10 & -4 \\ 6 & 6 \end{bmatrix}$

6. $\begin{bmatrix} -7 & 21 \\ 4y & x-4 \end{bmatrix} = \begin{bmatrix} -7 & 21 \\ 20 & -6 \end{bmatrix}$

7. $\begin{bmatrix} x+3 & 2w-8 \\ y+1 & 4x+6 \\ z-3 & 3z \end{bmatrix} = \begin{bmatrix} 0 & -6 \\ -3 & 2x \\ 2z+4 & -21 \end{bmatrix}$

8. $\begin{bmatrix} x-2 & 8 & -5 \\ 1 & -3y & 2x \\ 1 & -4 & 13 \end{bmatrix} = \begin{bmatrix} 2x+4 & 8 & -5 \\ 1 & -18 & -12 \\ \frac{z}{7} & -4 & 2z-1 \end{bmatrix}$

In Problems 9–14, find:

a. $A + B$ **b.** $A - B$ **c.** $4A$ **d.** $5A - 3B$

9. $A = \begin{bmatrix} 1 & 3 \\ 3 & 4 \\ 5 & 6 \end{bmatrix}$, $B = \begin{bmatrix} 2 & -1 \\ 3 & -2 \\ 0 & 1 \end{bmatrix}$

10. $A = \begin{bmatrix} 3 & 1 & 1 \\ -1 & 2 & 5 \end{bmatrix}$, $B = \begin{bmatrix} 2 & -3 & 6 \\ -3 & 1 & -4 \end{bmatrix}$

11. $A = \begin{bmatrix} 2 \\ -4 \\ 1 \end{bmatrix}$, $B = \begin{bmatrix} -5 \\ 3 \\ -1 \end{bmatrix}$

12. $A = \begin{bmatrix} 6 & 2 & -3 \end{bmatrix}$, $B = \begin{bmatrix} 4 & -2 & 3 \end{bmatrix}$

13. $A = \begin{bmatrix} \sqrt{8} & 1 \\ 2 & 0 \end{bmatrix}$, $B = \begin{bmatrix} \sqrt{2} & -1 \\ -3 & 6 \end{bmatrix}$

14. $A = \begin{bmatrix} 2 & -10 & -2 \\ 14 & 12 & 10 \\ 4 & -2 & 2 \end{bmatrix}$, $B = \begin{bmatrix} 6 & 10 & -2 \\ 0 & -12 & -4 \\ -5 & 2 & -2 \end{bmatrix}$

For Problems 15–24, find (if possible): **a.** *AB and* **b.** *BA*

15. $A = \begin{bmatrix} 1 & 2 & 3 & 4 \end{bmatrix}$, $B = \begin{bmatrix} 1 \\ 2 \\ 3 \\ 4 \end{bmatrix}$

16. $A = \begin{bmatrix} -1 \\ -2 \\ -3 \end{bmatrix}$, $B = \begin{bmatrix} 1 & 2 & 3 \end{bmatrix}$

17. $A = \begin{bmatrix} 1 & 3 \\ 5 & 3 \end{bmatrix}$, $B = \begin{bmatrix} 3 & -2 \\ -1 & 6 \end{bmatrix}$

18. $A = \begin{bmatrix} 3 & -2 \\ 1 & 5 \end{bmatrix}$, $B = \begin{bmatrix} 0 & 0 \\ 5 & -6 \end{bmatrix}$

19. $A = \begin{bmatrix} 1 & -1 & 4 \\ 4 & -1 & 3 \\ 2 & 0 & -2 \end{bmatrix}$, $B = \begin{bmatrix} 1 & 1 & 0 \\ 1 & 2 & 4 \\ 1 & -1 & 3 \end{bmatrix}$

20. $A = \begin{bmatrix} 1 & -1 & 1 \\ 5 & 0 & -2 \\ 3 & -2 & 2 \end{bmatrix}$, $B = \begin{bmatrix} 1 & 1 & 0 \\ 1 & -4 & 5 \\ 3 & -1 & 2 \end{bmatrix}$

21. $A = \begin{bmatrix} 4 & 2 \\ 6 & 1 \\ 3 & 5 \end{bmatrix}$, $B = \begin{bmatrix} 2 & 3 & 4 \\ -1 & -2 & 0 \end{bmatrix}$

22. $A = \begin{bmatrix} 2 & 4 \\ 3 & 1 \\ 4 & 2 \end{bmatrix}$, $B = \begin{bmatrix} 3 & 2 & 0 \\ -1 & -3 & 5 \end{bmatrix}$

23. $A = \begin{bmatrix} 2 & -3 & 1 & -1 \\ 1 & 1 & -2 & 1 \end{bmatrix}$, $B = \begin{bmatrix} 1 & 2 \\ -1 & 1 \\ 5 & 4 \\ 10 & 5 \end{bmatrix}$

24. $A = \begin{bmatrix} 2 & -1 & 3 & 2 \\ 1 & 0 & -2 & 1 \end{bmatrix}$, $B = \begin{bmatrix} -1 & 2 \\ 1 & 1 \\ 3 & -4 \\ 6 & 5 \end{bmatrix}$

In Problems 25–32, perform the indicated matrix operation given that A, B, and C are defined as follows. If an operation is not defined, state the reason.

$$A = \begin{bmatrix} 4 & 0 \\ -3 & 5 \\ 0 & 1 \end{bmatrix} \qquad B = \begin{bmatrix} 5 & 1 \\ -2 & -2 \end{bmatrix} \qquad C = \begin{bmatrix} 1 & -1 \\ -1 & 1 \end{bmatrix}$$

25. $4B - 3C$

26. $5C - 2B$

27. $BC + CB$

28. $A(B + C)$

29. $A - C$

30. $B - A$

31. $A(BC)$

32. $A(CB)$

In Problems 33–38, solve the given equation for X if

$$A = \begin{bmatrix} -4 & -1 \\ 1 & 0 \\ 5 & -6 \end{bmatrix} \text{ and } B = \begin{bmatrix} 0 & -8 \\ 2 & 0 \\ -6 & -2 \end{bmatrix}.$$

33. $2X + A = 4B$

34. $3X - A = 6B$

35. $-4X + 3A = -\frac{1}{2}B$

36. $-6X + 2A = -\frac{1}{2}B$

37. $-2(X + A) = -4X + B$

38. $-3(X - A) = -2X + B$

In Problems 39–42, rewrite each equation as an equivalent equation without using matrices. Then solve for the variable.

39. $\begin{bmatrix} 4 & -7 \\ 2 & -3 \end{bmatrix} \begin{bmatrix} x \\ y \end{bmatrix} = \begin{bmatrix} -3 \\ 1 \end{bmatrix}$

40. $\begin{bmatrix} 3 & 0 \\ -3 & 1 \end{bmatrix} \begin{bmatrix} x \\ y \end{bmatrix} = \begin{bmatrix} 6 \\ -7 \end{bmatrix}$

41. $\begin{bmatrix} 2 & 0 & -1 \\ 0 & 3 & 0 \\ 1 & 1 & 0 \end{bmatrix} \begin{bmatrix} x \\ y \\ z \end{bmatrix} = \begin{bmatrix} 6 \\ 9 \\ 5 \end{bmatrix}$

42. $\begin{bmatrix} -1 & 0 & 1 \\ 0 & -1 & 0 \\ 0 & 1 & 1 \end{bmatrix} \begin{bmatrix} x \\ y \\ z \end{bmatrix} = \begin{bmatrix} -4 \\ 2 \\ 4 \end{bmatrix}$

Application Problems

43. Two salespeople for a company that sells only two models of boats have gross dollar sales for June and July given in the following matrices:

June Sales

	Model 1	Model 2
Person x	$45,000	$62,000
Person y	$30,000	$0

$= A$

July Sales

	Model 1	Model 2
Person x	$64,000	$65,000
Person y	$36,000	$10,000

$= B$

a. Find $A + B$. Describe in a sentence or two what this matrix represents in practical terms.

b. Find $B - A$. Describe in a sentence or two what this matrix represents in practical terms.

c. Each salesperson receives a 6% commission on gross dollar sales. Compute $0.06A$ and describe what this matrix represents.

44. A bank has three branches (B_1, B_2, B_3) in a particular city. Matrix A represents the number of checking (c), savings (s), and money-market (m) accounts at each branch office on January 1.

	c	s	m
B_1	50,002	10,140	480
B_2	16,230	8740	106
B_3	25,614	12,160	86

$= A$

Matrix B shows the number of accounts that were opened at each branch in the first quarter, and matrix C models the number of accounts that were closed during the first quarter.

	c	s	m
B_1	4600	2502	68
B_2	1004	608	18
B_3	1056	780	23

$= B$

	c	s	m
B_1	2400	980	72
B_2	560	180	34
B_3	1000	700	30

$= C$

a. Find $A + B - C$. Describe in a sentence or two what this matrix represents in practical terms.

b. It is anticipated that all accounts will decline by 4% in the second quarter. Construct a matrix that models the anticipated number of accounts at each branch at the end of the second quarter. Round all fractions to the nearest integer.

c. What does matrix $B - C$ describe?

45. In a certain county, the proportion of voters in each age group registered as Republicans, Democrats, or Independents is given by the following matrix, which we'll call A.

	Age		
	18–30	**31–50**	**over 50**
Republicans	0.4	0.30	0.70
Democrats	0.30	0.60	0.25
Independents	0.30	0.10	0.05

The distribution, by age and gender, of this county's voting population is given by the following matrix, which we'll call B.

		Male	Female
	18–30	6000	8000
Age	**31–50**	12,000	14,000
	over 50	14,000	16,000

a. Calculate the product AB.

b. Describe what the matrix AB represents and interpret the entries.

c. How many female Democrats are there in this county?

46. A virus strikes a college campus. Students are either sick, well, or carriers of the virus. The percentages of people in each category are given by the following matrix, which we'll call A.

	Freshman	Sophomore	Junior	Senior
Well	15%	25%	20%	10%
Sick	35%	40%	35%	70%
Carrier	50%	35%	45%	20%

The student population is distributed by class and gender as given by the following matrix, which we'll call B.

	Male	Female
Freshman	820	640
Sophomore	950	1020
Junior	680	720
Senior	930	910

a. Calculate the product AB.

b. Describe what the matrix AB represents and interpret the entries.

c. How many sick females are there? How many male carriers are there?

47. The number of grams per cup of protein, carbohydrate, and fat in skim and low-fat milk is given by the following matrix, which we'll call A.

	Skim	Low-Fat
Protein	9	9
Carbohydrate	13	13
Fat	1	5

The number of cups of milk fed daily by a researcher for a dairy company is given by the following matrix, which we'll call B.

	Children	Adolescents	Adults
Skim	3	1	1
Low-Fat	2	1	0

a. Calculate the product AB.

b. Describe what the matrix AB represents and interpret the entries.

c. How much protein is consumed by children?

48. A company that manufactures inflatable boats has labor-hour and wage requirements given by the following matrices. Call the first matrix A and the second B.

Labor-Hour Requirements for Each Boat

	Cutting	Assembly	Packaging
One-Person Boat	1.2 hours	0.6 hour	0.2 hour
Two-Person Boat	2.0 hours	1.0 hour	0.2 hour
Three-Person Boat	2.4 hours	1.8 hours	0.4 hour

Hourly Wages

	Domestic Factory	Overseas Factory
Cutting Department	$8	$6
Assembly Department	$10	$8
Packaging Department	$6	$4

a. Calculate the product AB.

b. Describe what the matrix AB represents and interpret the entries.

c. What is the total labor cost for a three-person inflatable boat at the overseas factory?

True–False Critical Thinking Problems

49. Which one of the following is true?

a. An $n \times n$ matrix has n^2 entries.

b. If $\begin{bmatrix} x - 2y \\ x + y \end{bmatrix} = \begin{bmatrix} -3 \\ 3 \end{bmatrix}$, then $x = -1$.

c. If A, B, and C represent matrices, if $A = B$ and $B = C$, then A and C are not necessarily equal.

d. $\begin{bmatrix} 1 & 0 & 0 \\ 0 & 2 & 3 \\ 1 & 0 & 4 \end{bmatrix} - \begin{bmatrix} 1 & 0 & 0 \\ 0 & 2 & 3 \\ 1 & 0 & 4 \end{bmatrix} = \begin{bmatrix} 2 & 3 \\ 1 & 2 \end{bmatrix} - \begin{bmatrix} 2 & 3 \\ 1 & 2 \end{bmatrix}$

50. Which one of the following is true?

a. In solving the matrix equation $3X + A = B$, if $A = \begin{bmatrix} 2 & -4 \\ 0 & 3 \end{bmatrix}$ and $B = \begin{bmatrix} -6 & 5 \\ 2 & 7 \end{bmatrix}$ then the matrix X must contain only integers because matrices A and B contain only integers.

b. Addition of more than two matrices is undefined.

c. Every zero matrix contains the same number of entries.

d. The sum of $A = \begin{bmatrix} 1 & 1 \\ 1 & 1 \\ 1 & 1 \end{bmatrix}$ and $B = \begin{bmatrix} 1 & 1 & 1 \\ 1 & 1 & 1 \end{bmatrix}$ is undefined.

51. Which one of the following is true?

a. If A and B are matrices, then if $AB = 0$, either A or B must be zero matrices. (*Hint:* Let $A = \begin{bmatrix} 3 & 3 \\ 4 & 4 \end{bmatrix}$ and $B = \begin{bmatrix} 1 & -1 \\ -1 & 1 \end{bmatrix}$. Find AB.)

b. If $A = \begin{bmatrix} 4 & 2 \\ -8 & -4 \end{bmatrix}$, $B = \begin{bmatrix} -1 & 3 \\ 2 & -6 \end{bmatrix}$, and $C = \begin{bmatrix} 8 & 2 \\ 4 & 6 \end{bmatrix}$, then $C(A + B) = (A + B)C$.

c. Some linear systems cannot be written in matrix form.

d. Some matrices, such as matrices in the form $A = \begin{bmatrix} a & b \\ -b & a \end{bmatrix}$ and $B = \begin{bmatrix} c & d \\ -d & c \end{bmatrix}$, satisfy the commutative property of matrix multiplication.

52. Which one of the following is true?

a. If $A = \begin{bmatrix} 0 & 1 \\ 1 & 0 \end{bmatrix}$, $B = \begin{bmatrix} 0 & 1 \\ -1 & 0 \end{bmatrix}$, and $C = \begin{bmatrix} 1 & 0 \\ 0 & 0 \end{bmatrix}$, then $A(B + C) = AB + CA$.

b. If $AB = -BA$, then A and B are said to be *anticommutative*. Using this definition, the matrices $A = \begin{bmatrix} 0 & -1 \\ 1 & 0 \end{bmatrix}$ and $B = \begin{bmatrix} 1 & 0 \\ 0 & -1 \end{bmatrix}$ are anticommutative.

c. For the product of two matrices to be defined, the number of rows of the first matrix must equal the number of columns of the second matrix.

d. The easiest way to multiply two matrices is to multiply their corresponding entries.

Technology Problems

53. Use the matrix feature of a graphing utility to verify your answers to Problems 25–32.

54. The matrix

$$A = \begin{bmatrix} 0.2 & 0.7 & 0.1 \\ 0.6 & 0.2 & 0.2 \\ 0.2 & 0.1 & 0.7 \end{bmatrix} \begin{matrix} D \\ R \\ I \end{matrix} \Big\} \text{ To}$$

From
D   **R**   **I**

describes the preference in a particular community. Each entry a_{ij}, where $i \neq j$, describes the proportion of registered voters that changes from party i to party j. The entry a_{ij}, where $i = j$ (that is, a_{ii}) gives the proportion of registered voters that stay with the same party from each election to the next. The matrix $A \times A$ (A^2) describes the transition probabilities from the first election in the community to the third.

a. Use a graphing utility to find $A^2, A^3, A^4, A^5, A^6, A^7$, and A^8. Describe what each matrix represents.

b. Describe the pattern as A is raised to higher powers. What does this mean in practical terms?

Writing in Mathematics

55. Reread Problem 48. Describe a more realistic matrix for a small agency that sells a number of different models of, say, Boston Whalers (a popular cruising and fishing boat). What role would an electronic spreadsheet program on a computer play in this situation?

56. Write a word problem similar to Problem 43 or 44 in which matrix addition, matrix subtraction, and scalar multiplication each have practical meaning. The problem should not involve boats or banks. Then solve the problem and describe the practical meaning for each of the three matrix operations.

57. Describe when the multiplication of two matrices is not defined.

58. Describe what must be true about the orders of the matrices A and B if both AB and BA are defined.

59. The final grade in a particular course is determined by grades on the midterm and final. The grades for five students and the two grading systems are modeled by the following matrices. Call the first matrix A and the second B.

	Midterm	Final
Student 1	76	92
Student 2	74	84
Student 3	94	86
Student 4	84	62
Student 5	58	80

	System 1	System 2
Midterm	0.5	0.3
Final	0.5	0.7

a. Describe the grading system that is modeled by matrix B.

b. Compute the matrix AB and assign each of the five students a final course grade first using system 1 and then using system 2. (89.5–100 = A, 79.5–89.4 = B, 69.5–79.4 = C, 59.5–69.4 = D, below 59.5 = F)

c. Choose a course that you might enjoy teaching. You can select from any discipline at any level (elementary–graduate school). Your task is to set up a system for assigning each student in the course a final grade. Factor in as many components as possible (exams, homework, participation, attendance, written work, term papers, etc.) and provide students with a number of options for earning their final grade. Be as fair and creative as possible, fitting your system to the particular course you are teaching. Describe your grading system in words and then write a matrix that models the system.

60. In the matrices described below, a 1 represents a yes, and a 0 represents a no. The first matrix A describes whether or not three colleges in a state university system offer degrees in each program.

	Programs		
	Liberal Arts	Engineering	Education
College 1	1	1	0
College 2	1	1	1
College 3	0	1	0

Each program requires that certain math courses be completed, indicated by the matrix on page 586, called B.

	General College Math	Inter-mediate Algebra	College Algebra	Trigono-metry	Calculus
Liberal Arts	1	1	0	0	0
Engineering	0	0	1	1	1
Education	1	1	1	0	0

Find the product AB. Explain how this helps the college decide which courses to offer.

Critical Thinking Problems

61. The trace of $\begin{bmatrix} a & b \\ c & d \end{bmatrix}$ is defined by $\operatorname{tr}\begin{bmatrix} a & b \\ c & d \end{bmatrix} = a + d$.

If $A = \begin{bmatrix} a & b \\ c & d \end{bmatrix}$ and $B = \begin{bmatrix} e & f \\ g & h \end{bmatrix}$, prove that $\operatorname{tr}(A + B) = \operatorname{tr} A + \operatorname{tr} B$.

62. The transpose of a matrix is obtained by switching its rows and columns. If $A = \begin{bmatrix} a & b \\ c & d \end{bmatrix}$, then the transpose of A is defined by $A^T = \begin{bmatrix} a & c \\ b & d \end{bmatrix}$. If $A = \begin{bmatrix} a & b \\ c & d \end{bmatrix}$ and $B = \begin{bmatrix} e & f \\ g & h \end{bmatrix}$, prove that:

a. $(A + B)^T = A^T + B^T$

b. $(AB)^T = B^T A^T$.

63. Show that the matrix $A = \begin{bmatrix} 0 & 0 \\ 1 & 0 \end{bmatrix}$ satisfies the equation $A^2 = 0$. Describe two kinds of matrices that satisfy the equation $A^2 = 0$.

64. Explain why, in matrix algebra, $(A + B)(A - B) \neq A^2 - B^2$ except in special cases. Devise two 2×2 matrices A and B that illustrate the inequality. Now try finding two 2×2 matrices A and B that illustrate the special case $(A + B)(A - B) = A^2 - B^2$.

65. Consider a square matrix such that each entry that is not on the main diagonal is zero. Experiment with such matrices (call each matrix A) by finding A^2. Then write a sentence or two describing a method for squaring this kind of matrix.

Problems 66–67 involve powers of matrices. Let $A^2 = AA$, $A^3 = AAA$, $A^4 = AAAA$, and so on. In short, powers represent repeated matrix multiplication.

66. If $A = \begin{bmatrix} 1 & 1 \\ 0 & 1 \end{bmatrix}$, find A^2, A^3, A^4. Then generalize and find a formula for A^n.

67. If $A = \begin{bmatrix} 1 & 1 \\ 1 & 1 \end{bmatrix}$, find A^2, A^3, A^4. Then generalize and find a formula for A^n.

68. Find two 2×2 matrices that satisfy the commutative property for matrix multiplication.

Group Activity Problems

69. This question is intended for discussion in a small group. If a graphing utility can perform matrix multiplication, should we learn how to multiply matrices by hand? Answer this question by comparing the number of operations needed to multiply a pair of 2×2 matrices by hand and the number of entries that must be entered into the utility. What if a pair of matrices have order 6×6? Is there any point where using the graphing utility is just not worthwhile?

70. The interesting and useful applications of matrix theory are nearly unlimited. Applications of matrices run the gamut from analyzing marriage rules of various societies to predicting long-range trends in the stock market. Members of the group should research an area describing an application of matrices that members find intriguing. The group should then report to the class about this application.

| Solutions Manual | Tutorial | Video 9 |

Inverses of Matrices

Objectives

1 Find the inverse of a square matrix.
2 Solve linear systems using matrix inverses.
3 Use matrix inverses to encode messages.

Ellsworth Kelly "Red Blue Green Yellow" 1965, 2 panels – top: oil on canvas / bottom: oil on canvas mounted on masonite, 87 1/2 × 54 × 87 1/2 in. Photo: Douglas M. Parker Studio, Los Angeles. Courtesy of Margo Leavin Gallery, Los Angeles.

The Algebra of Matrices

There are many similarities between the algebra of matrices and the algebra of real numbers. For example, just as $a + 0 = a$, we can use a *zero matrix*, whose entries are all zeros, to serve as an additive identity. If A is an $m \times n$ matrix and 0 is an $m \times n$ zero matrix, then

$$A + 0 = A.$$

Furthermore, if A is an $m \times n$ matrix and 0 is an $m \times n$ zero matrix, then

$$A + (-A) = 0$$

so that $-A$ is the additive inverse of A. For example, since

$$\begin{bmatrix} 3 & -7 \\ -2 & 6 \end{bmatrix} + \begin{bmatrix} -3 & 7 \\ 2 & -6 \end{bmatrix} = \begin{bmatrix} 0 & 0 \\ 0 & 0 \end{bmatrix},$$

the additive identity, then $\begin{bmatrix} -3 & 7 \\ 2 & -6 \end{bmatrix}$ is the additive inverse of $\begin{bmatrix} 3 & -7 \\ -2 & 6 \end{bmatrix}$.

Insofar as matrix addition, matrix subtraction, and scalar multiplication are involved, the algebra of matrices is exactly like ordinary algebra of real numbers. The properties of matrix addition and scalar multiplication are summarized as follows:

Discover for yourself

Verify each of the eight properties listed in the box using

$$A = \begin{bmatrix} 2 & -4 \\ -5 & 3 \end{bmatrix},$$

$$B = \begin{bmatrix} 4 & 0 \\ 1 & -6 \end{bmatrix},$$

$$C = \begin{bmatrix} -3 & 1 \\ -2 & 0 \end{bmatrix},$$

$$0 = \begin{bmatrix} 0 & 0 \\ 0 & 0 \end{bmatrix},$$

$c = 4$, and $d = -2$

Properties of matrix addition and scalar multiplication

If A, B, and C are $m \times n$ matrices, 0 is an $m \times n$ zero matrix, and c and d are scalars, then the following properties are true.

1. $A + B = B + A$ Commutative Property of Addition
2. $(A + B) + C = A + (B + C)$ Associative Property of Addition
3. $A + 0 = 0 + A = A$ Additive Identity Property
4. $A + (-A) = (-A) + A = 0$ Additive Inverse Property
5. $(cd)A = c(dA)$ Associative Property of Scalar Multiplication
6. $1A = A$ Scalar Identity Property
7. $c(A + B) = cA + cB$ Distributive Property
8. $(c + d)A = cA + dA$ Distributive Property

Although matrix multiplication is not commutative, it does obey many of the properties of real numbers.

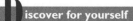iscover for yourself

Verify the properties listed in the box using

$$A = \begin{bmatrix} 3 & 2 \\ -1 & 4 \end{bmatrix}$$

$$B = \begin{bmatrix} 1 & 0 \\ 3 & 2 \end{bmatrix}$$

$$C = \begin{bmatrix} 1 & 2 \\ -1 & 1 \end{bmatrix}$$

and $c = 3$.

Properties of matrix and scalar multiplication

If A, B, and C are matrices and c is a scalar, then the following properties are true. (Assume the order of each matrix is such that all operations in these properties are defined.)

1. $(AB)C = A(BC)$ — Associative Property of Matrix Multiplication

2. $A(B + C) = AB + AC$
$(A + B)C = AC + BC$ — Distributive Properties of Matrix Multiplication

3. $c(AB) = (cA)B = A(cB)$ — Associative Property of Scalar Multiplication

For the real numbers, we know that 1 is the identity of multiplication because

$$a \cdot 1 = 1 \cdot a = a.$$

Is there a similar property for matrix multiplication? That is, is there a matrix I such that

$$AI = A \text{ and } IA = A ?$$

iscover for yourself

Find each of the following products:

$$\begin{bmatrix} 2 & 4 \\ 3 & 5 \end{bmatrix}\begin{bmatrix} 1 & 0 \\ 0 & 1 \end{bmatrix} \qquad \begin{bmatrix} 1 & 2 & 3 \\ 4 & 5 & 6 \\ 7 & 8 & 9 \end{bmatrix}\begin{bmatrix} 1 & 0 & 0 \\ 0 & 1 & 0 \\ 0 & 0 & 1 \end{bmatrix}$$

Describe what happens in each case. Using the pattern in the second matrices of these products, by what matrix can a 4×4 square matrix be multiplied without changing any of its entries?

In the Discover for yourself, did you observe that a square matrix with 1s along the major diagonal and 0s elsewhere does not change the entries in a matrix when it multiplies that matrix? For this reason, such a matrix is called the *identity matrix*.

The identity matrix of order n

The $n \times n$ matrix with 1s along the major diagonal and 0s elsewhere is called the *identity matrix of order n* and is denoted by

$$I_n = \begin{bmatrix} 1 & 0 & 0 & \cdots & 0 \\ 0 & 1 & 0 & \cdots & 0 \\ 0 & 0 & 1 & \cdots & 0 \\ \vdots & \vdots & \vdots & & \vdots \\ 0 & 0 & 0 & \cdots & 1 \end{bmatrix}.$$

Find the inverse of a square
matrix.

The Multiplicative Inverse of a Matrix

The identity matrix I_n plays a major role in the definition of the inverse of a matrix. The definition of a multiplicative inverse of a matrix is similar to the definition of the multiplicative inverse of a nonzero number a. The multiplicative inverse $\frac{1}{a}$ has the property that

$$a \cdot \frac{1}{a} = 1 \quad \text{and} \quad \frac{1}{a} \cdot a = 1.$$

A similar definition exists for the multiplicative inverse of a matrix.

> **Definition of an inverse of a matrix**
>
> Let A be an $n \times n$ matrix. If there exists an $n \times n$ matrix A^{-1} (read: "A inverse") such that
>
> $$AA^{-1} = I_n \text{ and } A^{-1}A = I_n$$
>
> then A^{-1} is the *inverse* of A.

Study tip

Matrix multiplication is not necessarily commutative. Thus, to show that matrix B is the inverse of matrix A, it is necessary to find both AB and BA. Both products should give the identity matrix.

EXAMPLE 1 **The Inverse of a Matrix**

Show that B is the inverse of A, where

$$A = \begin{bmatrix} -1 & 3 \\ 2 & -5 \end{bmatrix} \quad \text{and} \quad B = \begin{bmatrix} 5 & 3 \\ 2 & 1 \end{bmatrix}.$$

Solution

We must find the products AB and BA, showing that $AB = I$ and $BA = I$.

$$AB = \begin{bmatrix} -1 & 3 \\ 2 & -5 \end{bmatrix}\begin{bmatrix} 5 & 3 \\ 2 & 1 \end{bmatrix} = \begin{bmatrix} -1(5) + 3(2) & -1(3) + 3(1) \\ 2(5) + (-5)(2) & 2(3) + (-5)(1) \end{bmatrix} = \begin{bmatrix} 1 & 0 \\ 0 & 1 \end{bmatrix}$$

$$BA = \begin{bmatrix} 5 & 3 \\ 2 & 1 \end{bmatrix}\begin{bmatrix} -1 & 3 \\ 2 & -5 \end{bmatrix} = \begin{bmatrix} 5(-1) + 3(2) & 5(3) + 3(-5) \\ 2(-1) + 1(2) & 2(3) + 1(-5) \end{bmatrix} = \begin{bmatrix} 1 & 0 \\ 0 & 1 \end{bmatrix}$$

Since $AB = I$ and $BA = I$, the matrices A and B are inverses of each other. Equivalently, B is the inverse of A and we can designate B as $A^{-1} = \begin{bmatrix} 5 & 3 \\ 2 & 1 \end{bmatrix}.$ ∎

Example 2 shows one method for finding the inverse of a matrix.

EXAMPLE 2 **Finding the Inverse of a Matrix**

Find the inverse of:

$$A = \begin{bmatrix} 2 & 1 \\ 5 & 3 \end{bmatrix}$$

Solution

Let us denote the inverse by

$$A^{-1} = \begin{bmatrix} x & y \\ z & w \end{bmatrix}.$$

By the definition of a matrix inverse $AA^{-1} = I$, so we set A and A^{-1} equal to the identity matrix.

$$\begin{bmatrix} 2 & 1 \\ 5 & 3 \end{bmatrix} \begin{bmatrix} x & y \\ z & w \end{bmatrix} = \begin{bmatrix} 1 & 0 \\ 0 & 1 \end{bmatrix}$$ Our goal is to find values for the four variables.

$$\begin{bmatrix} 2x + z & 2y + w \\ 5x + 3z & 5y + 3w \end{bmatrix} = \begin{bmatrix} 1 & 0 \\ 0 & 1 \end{bmatrix}$$ Use row-by-column matrix multiplication on the left.

We now equate corresponding entries to obtain the following two systems of linear equations.

$$2x + z = 1 \quad \text{and} \quad 2y + w = 0$$
$$5x + 3z = 0 \qquad\qquad 5y + 3w = 1$$

Each of these systems can be solved fairly rapidly using the addition method.

$$
\begin{array}{l}
2x + z = 1 \xrightarrow[\text{No change}]{\substack{\text{Multiply}\\ \text{by} -3.}} -6x - 3z = -3 \\
5x + 3z = 0 \xrightarrow{\phantom{\text{No change}}} \underline{5x + 3z = 0} \\
 \text{Add:} \quad -x = -3 \\
 x = 3 \\
\end{array}
$$

Use back-substitution. $z = -5$

$$
\begin{array}{l}
2y + w = 0 \xrightarrow[\text{No change}]{\substack{\text{Multiply}\\ \text{by} -3.}} -6y - 3w = 0 \\
5y + 3w = 1 \xrightarrow{\phantom{\text{No change}}} \underline{5y + 3w = 1} \\
 \text{Add:} \quad -y = 1 \\
 y = -1 \\
\end{array}
$$

Use back-substitution. $w = 2$

Using these values, we have

$$A^{-1} = \begin{bmatrix} x & y \\ z & w \end{bmatrix} = \begin{bmatrix} 3 & -1 \\ -5 & 2 \end{bmatrix}. \qquad \blacksquare$$

> ## Discover for yourself
>
> Verify that the inverse matrix found in Example 2 is correct. Use matrix multiplication to show that
>
> $$AA^{-1} = I \quad \text{and} \quad A^{-1}A = I,$$
>
> where
>
> $$I = \begin{bmatrix} 1 & 0 \\ 0 & 1 \end{bmatrix}.$$

Matrices Without Inverses

A nonsquare matrix, one with a different number of rows than columns, cannot have an inverse. If A is an $m \times n$ matrix and B is an $n \times m$ matrix ($n \neq m$), then the products AB and BA are of different orders. This means that they could not be equal to each other, so that AB and BA could not both equal the identity matrix.

Only square matrices of order $n \times n$ possess inverses, but not every square matrix has an inverse. If a square matrix has an inverse, that inverse is unique, meaning that the square matrix has no more than one inverse. If a square matrix has an inverse, it is said to be *invertible*.

Man Ray "Indestructible Object, 1923/1958." Telimage. © 1998 Artists Rights Society (ARS), NY/ADAGP/Man Ray Trust, Paris.

EXAMPLE 3 **A Matrix with No Inverse**

Show that:

$$A = \begin{bmatrix} -6 & 4 \\ -3 & 2 \end{bmatrix}$$

does not have an inverse.

Solution

Letting $A^{-1} = \begin{bmatrix} x & y \\ z & w \end{bmatrix}$, we have

$$\begin{array}{ccc} A & A^{-1} & = & I \end{array}$$

$$\begin{bmatrix} -6 & 4 \\ -3 & 2 \end{bmatrix} \begin{bmatrix} x & y \\ z & w \end{bmatrix} = \begin{bmatrix} 1 & 0 \\ 0 & 1 \end{bmatrix} \qquad \text{Substitute the matrices. Can you anticipate what is going to occur?}$$

$$\begin{bmatrix} -6x + 4z & -6y + 4w \\ -3x + 2z & -3y + 2w \end{bmatrix} = \begin{bmatrix} 1 & 0 \\ 0 & 1 \end{bmatrix} \qquad \text{Use the definition of matrix multiplication on the left.}$$

Now we equate corresponding entries and solve the resulting systems.

$$\begin{array}{ccc} -6x + 4z = 1 & \xrightarrow{\text{No change}} & -6x + 4z = 1 \quad \text{and} \quad -6y + 4w = 0 \\ -3x + 2z = 0 & \xrightarrow{\text{Multiply by } -2.} & \underline{6x - 4z = 0} \qquad\qquad -3y + 2w = 1 \\ & & \text{Add:} \quad 0 = 1 \end{array}$$

The contradiction on the left indicates that this inconsistent system has no solution. Thus, the given matrix does not possess an inverse. ■

A Quick Method for Finding the Inverse of a 2 × 2 Matrix

The following rule enables us to calculate the inverse of a 2 × 2 matrix.

Discover for yourself

Verify that the inverse matrix for A shown in the box is correct. Use matrix multiplication to show that

$$AA^{-1} = I \quad \text{and} \quad A^{-1}A = I.$$

Inverse of a 2 × 2 matrix

If $A = \begin{bmatrix} a & b \\ c & d \end{bmatrix}$, then $A^{-1} = \dfrac{1}{ad - bc} \begin{bmatrix} d & -b \\ -c & a \end{bmatrix}.$

The matrix A is invertible if and only if $ad - bc \neq 0$. If $ad - bc = 0$, then A does not have an inverse.

EXAMPLE 4 **Using the Quick Method to Find Inverses**

Find the inverse of each matrix, if it exists.

a. $A = \begin{bmatrix} -1 & -2 \\ 3 & 4 \end{bmatrix}$ $\qquad B = \begin{bmatrix} 6 & -3 \\ -8 & 4 \end{bmatrix}$

Solution

a. $A = \begin{bmatrix} -1 & -2 \\ 3 & 4 \end{bmatrix}$ (a, b above; c, d below)

This is the given matrix. We've designated the entries a, b, c, and d.

$$A^{-1} = \frac{1}{ad-bc}\begin{bmatrix} d & -b \\ -c & a \end{bmatrix}$$

This is the formula for the inverse of $\begin{bmatrix} a & b \\ c & d \end{bmatrix}$.

$$= \frac{1}{(-1)(4)-(-2)(3)}\begin{bmatrix} 4 & -(-2) \\ -3 & -1 \end{bmatrix}$$

Apply the formula with $a = -1$, $b = -2$, $c = 3$, and $d = 4$.

$$= \frac{1}{2}\begin{bmatrix} 4 & 2 \\ -3 & -1 \end{bmatrix}$$

Simplify.

$$= \begin{bmatrix} 2 & 1 \\ -\dfrac{3}{2} & -\dfrac{1}{2} \end{bmatrix}$$

Perform the scalar multiplication by multiplying each entry in the matrix by $\dfrac{1}{2}$.

The inverse of $A = \begin{bmatrix} -1 & -2 \\ 3 & 4 \end{bmatrix}$ is $A^{-1} = \begin{bmatrix} 2 & 1 \\ -\dfrac{3}{2} & -\dfrac{1}{2} \end{bmatrix}$.

We can verify this result by showing that $AA^{-1} = I$ and $A^{-1}A = I$.

b. $B = \begin{bmatrix} 6 & -3 \\ -8 & 4 \end{bmatrix}$ (a, b above; c, d below)

The entries in the given matrix are designated a, b, c, and d.

$$B^{-1} = \frac{1}{ad-bc}\begin{bmatrix} d & -b \\ -c & a \end{bmatrix}$$

This is the formula for the inverse of $\begin{bmatrix} a & b \\ c & d \end{bmatrix}$.

$$= \frac{1}{6(4)-(-3)(-8)}\begin{bmatrix} 4 & -(-3) \\ -(-8) & 6 \end{bmatrix}$$

Apply the formula with $a = 6$, $b = -3$, $c = -8$, and $d = 4$.

$$= \frac{1}{24-24}\begin{bmatrix} 4 & 3 \\ 8 & 6 \end{bmatrix}$$

$$= \frac{1}{0}\begin{bmatrix} 4 & 3 \\ 8 & 6 \end{bmatrix}$$

Since $ad - bc = 0$, we have an expression that is undefined.

This result indicates that matrix B is not invertible. The matrix does not have an inverse.   ∎

Barbara Hepworth "Hollow Form
with White Interior" 1963, guarea
wood, 39 in. (99 cm) high. BH328.
© Alan Bowness, Hepworth Estate.
Photograph by Ellen Page Wilson,
courtesy PaceWildenstein Gallery.

Inverse of $n \times n$ Matrices

For 3×3 and larger matrices, graphing utilities provide the most efficient method for finding inverses. Matrix operations are performed by most graphing utilities. Use the $\boxed{\text{MATRIX}}$ function, enter the matrix, and then use the inverse key to find the inverse. If the matrix is not invertible, you will be given an error message.

Example 5 illustrates a systematic procedure for finding the inverse of a 3×3 matrix without using a graphing utility.

EXAMPLE 5 **Finding the Inverse of a 3 × 3 Matrix**

Find the inverse of:

$$\begin{bmatrix} -1 & -1 & -1 \\ 4 & 5 & 0 \\ 0 & 1 & -3 \end{bmatrix}$$

Solution

We must find a matrix

$$\begin{bmatrix} x_1 & x_2 & x_3 \\ y_1 & y_2 & y_3 \\ z_1 & z_2 & z_3 \end{bmatrix}$$

such that

$$\begin{bmatrix} -1 & -1 & -1 \\ 4 & 5 & 0 \\ 0 & 1 & -3 \end{bmatrix} \begin{bmatrix} x_1 & x_2 & x_3 \\ y_1 & y_2 & y_3 \\ z_1 & z_2 & z_3 \end{bmatrix} = \begin{bmatrix} 1 & 0 & 0 \\ 0 & 1 & 0 \\ 0 & 0 & 1 \end{bmatrix}.$$

We multiply the matrices on the left, using the row-by-column definition of matrix multiplication.

$$\begin{bmatrix} -x_1 - y_1 - z_1 & -x_2 - y_2 - z_2 & -x_3 - y_3 - z_3 \\ 4x_1 + 5y_1 + 0z_1 & 4x_2 + 5y_2 + 0z_2 & 4x_3 + 5y_3 + 0z_3 \\ 0x_1 + 1y_1 - 3z_1 & 0x_2 + 1y_2 - 3z_2 & 0x_3 + 1y_3 - 3z_3 \end{bmatrix} = \begin{bmatrix} 1 & 0 & 0 \\ 0 & 1 & 0 \\ 0 & 0 & 1 \end{bmatrix}$$

We now equate corresponding entries to obtain the following three systems of linear equations.

$$\begin{aligned} -x_1 - y_1 - z_1 &= 1 \\ 4x_1 + 5y_1 + 0z_1 &= 0 \\ 0x_1 + y_1 - 3z_1 &= 0 \end{aligned} \qquad \begin{aligned} -x_2 - y_2 - z_2 &= 0 \\ 4x_2 + 5y_2 + 0z_2 &= 1 \\ 0x_2 + y_2 - 3z_2 &= 0 \end{aligned} \qquad \begin{aligned} -x_3 - y_3 - z_3 &= 0 \\ 4x_3 + 5y_3 + 0z_3 &= 0 \\ 0x_3 + y_3 - 3z_3 &= 1 \end{aligned}$$

Notice that the variables on the left of the equal sign have the same coefficients in each system. We can use Gauss-Jordan elimination to solve all three systems at once.

$$\left[\begin{array}{ccc|ccc} -1 & -1 & -1 & 1 & 0 & 0 \\ 4 & 5 & 0 & 0 & 1 & 0 \\ 0 & 1 & -3 & 0 & 0 & 1 \end{array} \right]$$

This augmented matrix contains the coefficients of the three systems to the left of the vertical line and the constants for the systems to the right. We want

$$\begin{bmatrix} 1 & 0 & 0 \\ 0 & 1 & 0 \\ 0 & 0 & 1 \end{bmatrix}$$ to the left of the vertical line.

To get 1 in the upper-left corner, we multiply the elements in the first row by -1.

$$\left[\begin{array}{ccc|ccc} 1 & 1 & 1 & -1 & 0 & 0 \\ \boxed{4} & 5 & 0 & 0 & 1 & 0 \\ \boxed{0} & 1 & -3 & 0 & 0 & 1 \end{array}\right] \quad \begin{array}{l} -1 \times R_1. \\ \text{Now the boxed entries should be 0s.} \end{array}$$

To get 0 for the first element in the second row, we multiply the elements of the first row by -4 and add to the second row.

$$\left[\begin{array}{ccc|ccc} 1 & \boxed{1} & 1 & -1 & 0 & 0 \\ 0 & 1 & -4 & 4 & 1 & 0 \\ 0 & \boxed{1} & -3 & 0 & 0 & 1 \end{array}\right] \quad \begin{array}{l} -4R_1 + R_2. \\ \text{Now the boxed entries should be 0s.} \end{array}$$

To get 0 for the second element in the first row, we multiply the elements of the second row by -1 and add to the first row. We do the same thing to get 0 for the second element in the third row.

$$\left[\begin{array}{ccc|ccc} 1 & 0 & \boxed{5} & -5 & -1 & 0 \\ 0 & 1 & \boxed{-4} & 4 & 1 & 0 \\ 0 & 0 & 1 & -4 & -1 & 1 \end{array}\right] \quad \begin{array}{l} -1R_2 + R_1 \\ -1R_2 + R_3 \\ \text{With 1s down the main diagonal, the boxed entries} \\ \text{should be 0s.} \end{array}$$

To get 0 for the third element in the first row, we multiply the elements of the third row by -5 and add to the first row. To get 0 for the third element in the second row, we multiply the elements of the third row by 4 and add to the second row.

$$\left[\begin{array}{ccc|ccc} 1 & 0 & 0 & 15 & 4 & -5 \\ 0 & 1 & 0 & -12 & -3 & 4 \\ 0 & 0 & 1 & -4 & -1 & 1 \end{array}\right] \quad \begin{array}{l} -5R_3 + R_1 \\ 4R_3 + R_2 \end{array}$$

This augmented matrix provides the solutions to the three systems of equations. They are given by

$$\left[\begin{array}{ccc|c} 1 & 0 & 0 & 15 \\ 0 & 1 & 0 & -12 \\ 0 & 0 & 1 & -4 \end{array}\right] \quad \begin{array}{l} x_1 = 15 \\ y_1 = -12 \\ z_1 = -4 \end{array}$$

and

$$\left[\begin{array}{ccc|c} 1 & 0 & 0 & 4 \\ 0 & 1 & 0 & -3 \\ 0 & 0 & 1 & -1 \end{array}\right] \quad \begin{array}{l} x_2 = 4 \\ y_2 = -3 \\ z_2 = -1 \end{array}$$

and

$$\left[\begin{array}{ccc|c} 1 & 0 & 0 & -5 \\ 0 & 1 & 0 & 4 \\ 0 & 0 & 1 & 1 \end{array}\right] \quad \begin{array}{l} x_3 = -5 \\ y_3 = 4 \\ z_3 = 1 \end{array}$$

The inverse matrix is

$$\left[\begin{array}{ccc} x_1 & x_2 & x_3 \\ y_1 & y_2 & y_3 \\ z_1 & z_2 & z_3 \end{array}\right] = \left[\begin{array}{ccc} 15 & 4 & -5 \\ -12 & -3 & 4 \\ -4 & -1 & 1 \end{array}\right].$$

sing technology

You can use a graphing utility to find the inverse of

$$A = \left[\begin{array}{ccc} -1 & -1 & -1 \\ 4 & 5 & 0 \\ 0 & 1 & -3 \end{array}\right].$$

Enter $\boxed{\text{MATRIX}}$ [A] and then use the inverse key. The display should be $[A]^{-1}$

$$\left[\begin{array}{ccc} 15 & 4 & -5 \\ -12 & -3 & 4 \\ -4 & -1 & 1 \end{array}\right]$$

Notice that this is precisely the matrix that appears to the right of the vertical line at the point where we completed Gauss-Jordan elimination. ■

Example 5 leads us to a procedure for finding the inverse of an invertible matrix.

Study tip

Since we have a quick method for finding the inverse of a 2 × 2 matrix, the procedure on the right is recommended for matrices of order 3 × 3 or greater when a graphing utility is not being used.

Procedure for finding the inverse of a matrix

To find A^{-1} for any $n \times n$ matrix A for which A^{-1} exists:

1. Form the augmented matrix $[A\,|\,I]$, where I is the identity matrix of the same order as the given matrix A.
2. Perform row transformations on $[A\,|\,I]$ to obtain a matrix of the form $[I\,|\,B]$. This is equivalent to using Gauss-Jordan elimination to change A into the identity matrix.
3. Matrix B is A^{-1}.
4. Verify the result by showing that $AB = I$ and $BA = I$.

EXAMPLE 6 **Finding the Inverse of a 3 × 3 Matrix**

Find the inverse of:

$$A = \begin{bmatrix} 2 & 7 & -3 \\ 3 & 8 & -4 \\ -10 & 6 & 7 \end{bmatrix}$$

Solution

Step 1. Form the augmented matrix $[A\,|\,I]$.

$$\begin{bmatrix} 2 & 7 & -3 & | & 1 & 0 & 0 \\ 3 & 8 & -4 & | & 0 & 1 & 0 \\ -10 & 6 & 7 & | & 0 & 0 & 1 \end{bmatrix}$$

Remember that I, the identity matrix, has 1s down the main diagonal and 0s elsewhere.

Step 2. Use row transformations to obtain $[I\,|\,B]$. We want 1s down the main diagonal to the left of the vertical dividing line and 0s elsewhere.

$$\begin{bmatrix} 2 & 7 & -3 & | & 1 & 0 & 0 \\ 3 & 8 & -4 & | & 0 & 1 & 0 \\ -10 & 6 & 7 & | & 0 & 0 & 1 \end{bmatrix}$$

$\xrightarrow{\frac{1}{2} \times R_1}$

$$\begin{bmatrix} 1 & \dfrac{7}{2} & -\dfrac{3}{2} & | & \dfrac{1}{2} & 0 & 0 \\ \boxed{3} & 8 & -4 & | & 0 & 1 & 0 \\ \boxed{-10} & 6 & 7 & | & 0 & 0 & 1 \end{bmatrix}$$

The boxed entries should be 0s.

$\xrightarrow{-3R_1 + R_2}$

$\xrightarrow{10R_1 + R_3}$

$$\begin{bmatrix} 1 & \dfrac{7}{2} & -\dfrac{3}{2} & | & \dfrac{1}{2} & 0 & 0 \\ 0 & \boxed{-\dfrac{5}{2}} & \dfrac{1}{2} & | & -\dfrac{3}{2} & 1 & 0 \\ 0 & 41 & -8 & | & 5 & 0 & 1 \end{bmatrix}$$

The boxed entry should be 1.

$$-\frac{2}{5} \times R_2 \longrightarrow \begin{bmatrix} 1 & \boxed{\dfrac{7}{2}} & -\dfrac{3}{2} & \dfrac{1}{2} & 0 & 0 \\ 0 & 1 & -\dfrac{1}{5} & \dfrac{3}{5} & -\dfrac{2}{5} & 0 \\ 0 & \boxed{41} & -8 & 5 & 0 & 1 \end{bmatrix}$$

The boxed entries should be 0s.

$$-\frac{7}{2}R_2 + R_1 \longrightarrow \begin{bmatrix} 1 & 0 & -\dfrac{4}{5} & -\dfrac{8}{5} & \dfrac{7}{5} & 0 \\ 0 & 1 & -\dfrac{1}{5} & \dfrac{3}{5} & -\dfrac{2}{5} & 0 \\ 0 & 0 & \boxed{\dfrac{1}{5}} & -\dfrac{98}{5} & \dfrac{82}{5} & 1 \end{bmatrix}$$

The boxed entry should be 1.

$$-41R_2 + R_3 \longrightarrow$$

$$5 \times R_3 \longrightarrow \begin{bmatrix} 1 & 0 & \boxed{-\dfrac{4}{5}} & -\dfrac{8}{5} & \dfrac{7}{5} & 0 \\ 0 & 1 & \boxed{-\dfrac{1}{5}} & \dfrac{3}{5} & -\dfrac{2}{5} & 0 \\ 0 & 0 & 1 & -98 & 82 & 5 \end{bmatrix}$$

The boxed entries should be 0s.

$$\frac{4}{5}R_3 + R_1 \\ \frac{1}{5}R_3 + R_2 \longrightarrow \begin{bmatrix} 1 & 0 & 0 & -80 & 67 & 4 \\ 0 & 1 & 0 & -19 & 16 & 1 \\ 0 & 0 & 1 & -98 & 82 & 5 \end{bmatrix}$$

iscover for yourself

Verify that the inverse matrix found in Example 7 is correct. Use matrix multiplication to show that

$AA^{-1} = I$ and $A^{-1}A = I.$

Study tip

Not all square matrices have inverses. If row transformations result in all zeros in a row or column to the left of the vertical line, the given matrix does not have an inverse.

This matrix is now in the form $[I\,|\,B]$, where

$$B = \begin{bmatrix} -80 & 67 & 4 \\ -19 & 16 & 1 \\ -98 & 82 & 5 \end{bmatrix}.$$

Step 3. Matrix B is A^{-1}. Thus,

$$A^{-1} = \begin{bmatrix} -80 & 67 & 4 \\ -19 & 16 & 1 \\ -98 & 82 & 5 \end{bmatrix}.$$

Step 4. See the Discover for yourself.   ■

Summary: Finding inverses for invertible matrices

Use a graphing utility with matrix capabilities

or

a. If the matrix is 2×2: The inverse of $A = \begin{bmatrix} a & b \\ c & d \end{bmatrix}$ is

$$A^{-1} = \frac{1}{ad - bc}\begin{bmatrix} d & -b \\ -c & a \end{bmatrix}.$$

b. If the matrix A is $n \times n$ where $n > 2$: Use the procedure on page 595. Form $[A\,|\,I]$ and use row transformations to obtain $[I\,|\,B]$. $A^{-1} = [B]$

ENRICHMENT ESSAY

Matrices and Communication

Figure 5.15 shows a map of the United States with one conceivable network of communication lines connecting major cities. Routing millions of calls over this network requires powerful matrix algorithms and high-speed computers. The diagram in Figure 5.16 shows part of a communication network. Matrices can be used to represent communications between the eight nodes. In the following communication matrix, the element 1 indicates that *direct* communication *to* another point is possible. The element 0 indicates that direct communication to another point is not possible.

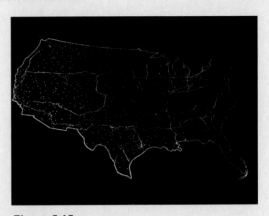

Figure 5.15
Property of AT&T Archives. Reprinted with permission of AT&T.

Communication Matrix

To Node

		1	2	3	4	5	6	7	8
From	1	0	1	0	1	1	0	0	0
Node	2	1	0	1	0	0	1	0	0
	3	0	1	0	0	0	1	0	0
	4	1	0	0	0	0	0	1	0
	5	1	0	0	0	0	1	1	0
	6	0	1	1	0	1	0	0	1
	7	0	0	0	1	1	0	0	1
	8	0	0	0	0	0	1	1	0

Note that the principal diagonal of the matrix consists entirely of zeros, representing the assumption that a member of a communication network does not send a message to himself or herself.
(How can matrix multiplication be used to find all the two-step communication possibilities?)

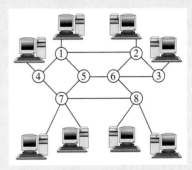

Figure 5.16

2 Solve linear systems using matrix inverses.

Solving Systems of Equations Using Matrix Inverses

In Section 5.4, we saw that a linear system of equations could be expressed in matrix form as

$$AX = B$$

where A is the *coefficient matrix* of the system and B, a matrix with one column, is the *constant matrix*. We can solve this equation as follows:

$$AX = B \qquad \text{This is the given matrix equation. Assume that } A \text{ has an inverse.}$$

$$A^{-1}(AX) = A^{-1}B \qquad \text{Multiply both sides by } A^{-1}.$$

$$(A^{-1}A)X = A^{-1}B$$ Apply the associative property of matrix multiplication on the left.

$$IX = A^{-1}B$$ The multiplicative inverse property tells us that $A^{-1}A = I$.

$$X = A^{-1}B$$ Since I is the multiplicative identity, $IX = X$.

We see that if $AX = B$, then $X = A^{-1}B$.

Solving a system using A^{-1}

If $AX = B$ has a unique solution, then $X = A^{-1}B$. To solve a linear system of equations, multiply A^{-1} and B to find X.

EXAMPLE 7   **Using the Inverse of a Matrix to Solve a System**

Solve the system by using A^{-1}, the inverse of the coefficient matrix.

$$x + 4y + 7z = 3$$

$$2x + 3y + 6z = 0$$

$$5x + y - z = 7$$

Solution

With

$$A = \begin{bmatrix} 1 & 4 & 7 \\ 2 & 3 & 6 \\ 5 & 1 & -1 \end{bmatrix}, \qquad X = \begin{bmatrix} x \\ y \\ z \end{bmatrix}, \quad \text{and} \quad B = \begin{bmatrix} 3 \\ 0 \\ 7 \end{bmatrix}$$

the system can be expressed as $AX = B$. The solution is given by $X = A^{-1}B$. Consequently, we must find A^{-1}.

$$A^{-1} = \begin{bmatrix} -\dfrac{9}{28} & \dfrac{11}{28} & \dfrac{3}{28} \\[6pt] \dfrac{8}{7} & -\dfrac{9}{7} & \dfrac{2}{7} \\[6pt] -\dfrac{13}{28} & \dfrac{19}{28} & -\dfrac{5}{28} \end{bmatrix}$$

Verify this inverse matrix with a graphing utility or by hand using the method for finding the inverse of a 3×3 matrix.

$$X = A^{-1}B = \begin{bmatrix} -\dfrac{9}{28} & \dfrac{11}{28} & \dfrac{3}{28} \\[6pt] \dfrac{8}{7} & -\dfrac{9}{7} & \dfrac{2}{7} \\[6pt] -\dfrac{13}{28} & \dfrac{19}{28} & -\dfrac{5}{28} \end{bmatrix} \begin{bmatrix} 3 \\ 0 \\ 7 \end{bmatrix}$$

If $AX = B$, then $X = A^{-1}B$.

$$= \begin{bmatrix} -\dfrac{3}{14} \\[6pt] \dfrac{38}{7} \\[6pt] -\dfrac{37}{14} \end{bmatrix}$$

Verify this result by performing the matrix multiplication.

Since

$$\begin{bmatrix} x \\ y \\ z \end{bmatrix} = \begin{bmatrix} -\dfrac{3}{14} \\[6pt] \dfrac{38}{7} \\[6pt] -\dfrac{37}{14} \end{bmatrix},$$

the solution to the system is $\{(-\frac{3}{14}, \frac{38}{7}, -\frac{37}{14})\}$. ∎

3 Use matrix inverses to encode messages.

Paul Klee "Broken Key" 1938, Ol auf Jute uber Keilrahman, 50 × 66 cm. Sprengel Museum Hannover. Photo: Michael Herling. © VG Bild–Kunst / ARS, New York.

Application of Matrix Inverses to Coding

A *cryptogram* is a message written so that no one other than the intended recipient can understand it. A matrix can be used to encode the message and its inverse can be used to decode the message.

EXAMPLE 8 Encoding and Decoding a Message

Use matrices to encode and decode the message MATH.

Solution

We begin by assigning a number to each letter in the alphabet: $A = 1$, $B = 2$, $C = 3$, $\cdots$, $Z = 26$, and a space $= 0$.

Encoding:
The numerical equivalent of the word MATH is 13, 1, 20, 8. We list these numbers in two 2×1 matrices.

$$\begin{bmatrix} 13 \\ 1 \end{bmatrix} \quad \begin{bmatrix} 20 \\ 8 \end{bmatrix}$$

Now we multiply the matrices by any coding matrix for which the multiplication is defined. We will use $A = \begin{bmatrix} -2 & -3 \\ 3 & 4 \end{bmatrix}$.

$$\underset{\text{coding matrix}}{\begin{bmatrix} -2 & -3 \\ 3 & 4 \end{bmatrix}} \underset{\text{MA}}{\begin{bmatrix} 13 \\ 1 \end{bmatrix}} = \begin{bmatrix} -29 \\ 43 \end{bmatrix} \quad \text{and} \quad \underset{\text{coding matrix}}{\begin{bmatrix} -2 & -3 \\ 3 & 4 \end{bmatrix}} \underset{\text{TH}}{\begin{bmatrix} 20 \\ 8 \end{bmatrix}} = \begin{bmatrix} -64 \\ 92 \end{bmatrix}$$

The encoded message sent is $-29, 43, -64, 92$.

Decoding:

An authorized receiver who knows the coding matrix A can decode the message $-29, 43, -64, 92$, using A^{-1}.

$$A = \begin{bmatrix} -2 & -3 \\ 3 & 4 \end{bmatrix}$$

Designate entries of the coding matrix a, b, c, and d.

$$A^{-1} = \frac{1}{ad - bc} \begin{bmatrix} d & -b \\ -c & a \end{bmatrix}$$

Use the formula for the inverse of a 2×2 matrix.

$$= \frac{1}{(-2)(4) - (-3)(3)} \begin{bmatrix} 4 & -(-3) \\ -3 & -2 \end{bmatrix}$$

Apply the formula with $a = -2$, $b = -3, c = 3$, and $d = 4$.

$$= \frac{1}{1} \begin{bmatrix} 4 & 3 \\ -3 & 2 \end{bmatrix} = \begin{bmatrix} 4 & 3 \\ -3 & -2 \end{bmatrix}$$

To decode the message, multiply A^{-1} and the encoded column matrices.

$$\begin{bmatrix} 4 & 3 \\ -3 & -2 \end{bmatrix} \begin{bmatrix} -29 \\ 43 \end{bmatrix} = \begin{bmatrix} 4(-29) + & 3(43) \\ -3(-29) + & (-2)(43) \end{bmatrix} = \begin{bmatrix} 13 \\ 1 \end{bmatrix} \quad \begin{matrix} 13 = M \\ 1 = A \end{matrix}$$

Inverse of the coding matrix — First two letters of the encoded message — First two letters of the decoded message

$$\begin{bmatrix} 4 & 3 \\ -3 & -2 \end{bmatrix} \begin{bmatrix} -64 \\ 92 \end{bmatrix} = \begin{bmatrix} 4(-64) + & 3(92) \\ -3(-64) + & (-2)(92) \end{bmatrix} = \begin{bmatrix} 20 \\ 8 \end{bmatrix} \quad \begin{matrix} 20 = T \\ 8 = H \end{matrix}$$

Inverse of the coding matrix — Last two letters of the encoded message — Last two letters of the decoded message

The decoded message is $13, 1, 20, 8$, or MATH. ∎

Encoding and decoding a word or message

1. To encode a message, select an $n \times n$ invertible matrix A and multiply the matrix and the encoded message (as column matrices) to obtain coded column matrices.

2. To decode the message, multiply the inverse of the coding matrix and the coded column matrices to obtain the uncoded message (as column matrices).

Decoding is simple for an authorized receiver who knows the coding matrix A. The receiver need only multiply A^{-1} by the coded column matrices to retrieve the decoded column matrices. Decoding the cryptogram for an unauthorized receiver who does not know A is extremely difficult. Any invertible matrix of any order could be used for the coding scheme.

PROBLEM SET 5.5

Practice Problems

In Problems 1–8, determine whether the matrices in each pair are inverses of each other.

1. $A = \begin{bmatrix} -2 & 1 \\ \frac{3}{2} & -\frac{1}{2} \end{bmatrix}$, $B = \begin{bmatrix} 1 & 2 \\ 3 & 4 \end{bmatrix}$

2. $A = \begin{bmatrix} 4 & 5 \\ 2 & 3 \end{bmatrix}$, $B = \begin{bmatrix} \frac{3}{2} & -\frac{5}{2} \\ -1 & 2 \end{bmatrix}$

3. $A = \begin{bmatrix} -4 & 0 \\ 1 & 3 \end{bmatrix}$, $B = \begin{bmatrix} -2 & 4 \\ 0 & 1 \end{bmatrix}$

4. $A = \begin{bmatrix} -2 & 4 \\ 1 & -2 \end{bmatrix}$, $B = \begin{bmatrix} 1 & 2 \\ -1 & -2 \end{bmatrix}$

5. $A = \begin{bmatrix} 1 & 2 & 3 \\ 1 & 3 & 4 \\ 1 & 4 & 3 \end{bmatrix}$, $B = \begin{bmatrix} \frac{7}{2} & -3 & \frac{1}{2} \\ -\frac{1}{2} & 0 & \frac{1}{2} \\ -\frac{1}{2} & 1 & -\frac{1}{2} \end{bmatrix}$

6. $A = \begin{bmatrix} 0 & 2 & 0 \\ 3 & 3 & 2 \\ 2 & 5 & 1 \end{bmatrix}$ $B = \begin{bmatrix} -3.5 & -1 & 2 \\ 0.5 & 0 & 0 \\ 4.5 & 2 & -3 \end{bmatrix}$

7. $A = \begin{bmatrix} 0 & 0 & -2 & 1 \\ -1 & 0 & 1 & 1 \\ 0 & 1 & -1 & 0 \\ 1 & 0 & 0 & -1 \end{bmatrix}$, $B = \begin{bmatrix} 1 & 2 & 0 & 3 \\ 0 & 1 & 1 & 1 \\ 0 & 1 & 0 & 1 \\ 1 & 2 & 0 & 2 \end{bmatrix}$

8. $A = \begin{bmatrix} 1 & -2 & 1 & 0 \\ 0 & 1 & -2 & 1 \\ 0 & 0 & 1 & -2 \\ 0 & 0 & 0 & 1 \end{bmatrix}$, $B = \begin{bmatrix} 1 & 2 & 3 & 4 \\ 0 & 1 & 2 & 3 \\ 0 & 0 & 1 & 2 \\ 0 & 0 & 0 & 1 \end{bmatrix}$

In Problems 9–14, use the fact that if $A = \begin{bmatrix} a & b \\ c & d \end{bmatrix}$, then $A^{-1} = \frac{1}{ad - bc} \begin{bmatrix} d & -b \\ -c & a \end{bmatrix}$ to find the inverse of each matrix, if possible. Check that $AA^{-1} = I$ and $A^{-1}A = I$. Use a graphing utility to verify the inverse of each matrix that exists.

9. $A = \begin{bmatrix} 2 & 3 \\ -1 & 2 \end{bmatrix}$

10. $A = \begin{bmatrix} 0 & 3 \\ 4 & -2 \end{bmatrix}$

11. $A = \begin{bmatrix} 3 & -1 \\ -4 & 2 \end{bmatrix}$

12. $A = \begin{bmatrix} 2 & -6 \\ 1 & -2 \end{bmatrix}$

13. $A = \begin{bmatrix} 10 & -2 \\ -5 & 1 \end{bmatrix}$

14. $A = \begin{bmatrix} 6 & -3 \\ -2 & 1 \end{bmatrix}$

15. If $A = \begin{bmatrix} 3 & 5 \\ 2 & 4 \end{bmatrix}$, show that $(A^{-1})^{-1} = A$.

16. If $A = \begin{bmatrix} 1 & 4 \\ 2 & 10 \end{bmatrix}$ and $B = \begin{bmatrix} 1 & -3 \\ -2 & 5 \end{bmatrix}$, show that $(AB)^{-1} = B^{-1}A^{-1}$.

In Problems 17–22, find A^{-1} by forming $[A \mid I]$ and then using row transformations to obtain $[I \mid B]$, where $A^{-1} = [B]$. Use a graphing utility to verify the inverse of the matrix. Check that $AA^{-1} = I$ and $A^{-1}A = I$.

17. $A = \begin{bmatrix} 2 & 2 & -1 \\ 0 & 3 & -1 \\ -1 & -2 & 1 \end{bmatrix}$

18. $A = \begin{bmatrix} 1 & -1 & 1 \\ 0 & 2 & -1 \\ 2 & 3 & 0 \end{bmatrix}$

19. $A = \begin{bmatrix} 5 & 0 & 2 \\ 2 & 2 & 1 \\ -3 & 1 & -1 \end{bmatrix}$

20. $A = \begin{bmatrix} 3 & 2 & 6 \\ 1 & 1 & 2 \\ 2 & 2 & 5 \end{bmatrix}$

21. $A = \begin{bmatrix} 1 & 0 & 0 & 0 \\ 0 & -1 & 0 & 0 \\ 0 & 0 & 3 & 0 \\ 1 & 0 & 0 & 1 \end{bmatrix}$

22. $A = \begin{bmatrix} 2 & 0 & 0 & 1 \\ 0 & 1 & 0 & 0 \\ 0 & 0 & -1 & 0 \\ 0 & 0 & 0 & 2 \end{bmatrix}$

In Problems 23–28, solve each system using the inverse of the coefficient matrix.

The inverse of

23. $2x + 6y + 6z = 8$
$2x + 7y + 6z = 10$
$2x + 7y + 7z = 9$

$$\begin{bmatrix} 2 & 6 & 6 \\ 2 & 7 & 6 \\ 2 & 7 & 7 \end{bmatrix} \text{ is } \begin{bmatrix} \frac{7}{2} & 0 & -3 \\ -1 & 1 & 0 \\ 0 & -1 & 1 \end{bmatrix}.$$

The inverse of

24. $x + 2y + 5z = 2$
$2x + 3y + 8z = 3$
$-x + y + 2z = 3$

$$\begin{bmatrix} 1 & 2 & 5 \\ 2 & 3 & 8 \\ -1 & 1 & 2 \end{bmatrix} \text{ is } \begin{bmatrix} 2 & -1 & -1 \\ 12 & -7 & -2 \\ -5 & 3 & 1 \end{bmatrix}.$$

The inverse of

25. $x - y + z = 8$
$2y - z = -7$
$2x + 3y = 1$

$$\begin{bmatrix} 1 & -1 & 1 \\ 0 & 2 & -1 \\ 2 & 3 & 0 \end{bmatrix} \text{ is } \begin{bmatrix} 3 & 3 & -1 \\ -2 & -2 & 1 \\ -4 & -5 & 2 \end{bmatrix}.$$

The inverse of

26. $y - z = -4$
$4x + y = -3$
$3x - y + 3z = 1$

$$\begin{bmatrix} 0 & 1 & -1 \\ 4 & 1 & 0 \\ 3 & -1 & 3 \end{bmatrix} \text{ is } \frac{1}{5}\begin{bmatrix} -3 & 2 & -1 \\ 12 & -3 & 4 \\ 7 & -3 & 4 \end{bmatrix}.$$

The inverse of

27. $x - y + 2z = -3$
$y - z + w = 4$
$-x + y - z + 2w = 2$
$-y + z - 2w = -4$

$$\begin{bmatrix} 1 & -1 & 2 & 0 \\ 0 & 1 & -1 & 1 \\ -1 & 1 & -1 & 2 \\ 0 & -1 & 1 & -2 \end{bmatrix} \text{ is } \begin{bmatrix} 0 & 0 & -1 & -1 \\ 1 & 4 & 1 & 3 \\ 1 & 2 & 1 & 2 \\ 0 & -1 & 0 & -1 \end{bmatrix}.$$

The inverse of

28. $4z + w = -2$
$2x - y + 3w = 5$
$3x + 2y + w = 11$
$2x + 2y + 6z + w = 4$

$$\begin{bmatrix} 0 & 0 & 4 & 1 \\ 2 & -1 & 0 & 3 \\ 3 & 2 & 0 & 1 \\ 2 & 2 & 6 & 1 \end{bmatrix} \text{ is } \begin{bmatrix} -3 & \frac{6}{7} & -\frac{11}{7} & 2 \\ 3 & -1 & 2 & -2 \\ -\frac{1}{2} & \frac{1}{7} & -\frac{3}{7} & \frac{1}{2} \\ 3 & -\frac{4}{7} & \frac{12}{7} & -2 \end{bmatrix}.$$

29. a. Show that $\begin{bmatrix} 1 & -6 & 3 \\ 2 & -7 & 3 \\ 4 & -12 & 5 \end{bmatrix}$ is its own inverse.

b. Use part (a) to solve the following system.

$$x - 6y + 3z = 11$$
$$2x - 7y + 3z = 14$$
$$4x - 12y + 5z = 25$$

In Problems 30–31, find values of a for which the given matrix is not invertible.

30. $\begin{bmatrix} 1 & a+1 \\ a-2 & 4 \end{bmatrix}$

31. $\begin{bmatrix} 3 & a+5 \\ a & -2 \end{bmatrix}$

In Problems 32–35 let

$$A = \begin{bmatrix} 1 & 2 & 3 \\ 3 & 5 & 7 \\ -1 & 2 & 4 \end{bmatrix} \text{ and } B = \begin{bmatrix} 5 & 12 & 6 \\ 3 & 7 & 4 \\ 5 & 13 & 2 \end{bmatrix}.$$

32. Show that $(AB)^{-1} = B^{-1}A^{-1}$.

33. Show that A is not its own inverse.

34. Show that $(A^2)^{-1} = (A^{-1})^2$.

35. Show that $(A + B)^{-1} \neq A^{-1} + B^{-1}$.

Application Problems

In Problems 36–39, use the coding matrix $A = \begin{bmatrix} 4 & -1 \\ -3 & 1 \end{bmatrix}$ *and its inverse* $A^{-1} = \begin{bmatrix} 1 & 1 \\ 3 & 4 \end{bmatrix}$ *to encode and then decode the given message.*

36. HELP

38. S E N D _ C A S H _
19 5 14 4 0 3 1 19 8 0

Use $\begin{bmatrix} 19 \\ 5 \end{bmatrix}, \begin{bmatrix} 14 \\ 4 \end{bmatrix}, \begin{bmatrix} 0 \\ 3 \end{bmatrix}, \begin{bmatrix} 1 \\ 19 \end{bmatrix}$, and $\begin{bmatrix} 8 \\ 0 \end{bmatrix}$.

37. LOVE

39. S T A Y _ W E L L _
19 20 1 25 0 23 5 12 12 0

Use $\begin{bmatrix} 19 \\ 20 \end{bmatrix}, \begin{bmatrix} 1 \\ 25 \end{bmatrix}, \begin{bmatrix} 0 \\ 23 \end{bmatrix}, \begin{bmatrix} 5 \\ 12 \end{bmatrix}$, and $\begin{bmatrix} 12 \\ 0 \end{bmatrix}$.

In Problems 40–41, use the coding matrix $A = \begin{bmatrix} 1 & -1 & 0 \\ 3 & 0 & 2 \\ -1 & 0 & -1 \end{bmatrix}$ *and its inverse* $A^{-1} = \begin{bmatrix} 0 & -1 & 2 \\ -1 & 1 & 2 \\ 0 & -1 & -3 \end{bmatrix}$ *to write a cryptogram for each message. Check your result by decoding the cryptogram.*

40. A R R I V E D _ S A F E L Y _
1 18 18 9 22 5 4 0 19 1 6 5 12 25 0

Use $\begin{bmatrix} 1 \\ 18 \\ 18 \end{bmatrix}, \begin{bmatrix} 9 \\ 22 \\ 5 \end{bmatrix}, \begin{bmatrix} 4 \\ 0 \\ 19 \end{bmatrix}, \begin{bmatrix} 1 \\ 6 \\ 5 \end{bmatrix}$, and $\begin{bmatrix} 12 \\ 25 \\ 0 \end{bmatrix}$.

41. A R T _ E N R I C H E S
1 18 20 0 5 14 18 9 3 8 5 19

Use $\begin{bmatrix} 1 \\ 18 \\ 20 \end{bmatrix}, \begin{bmatrix} 0 \\ 5 \\ 14 \end{bmatrix}, \begin{bmatrix} 18 \\ 9 \\ 3 \end{bmatrix}$, and $\begin{bmatrix} 8 \\ 5 \\ 19 \end{bmatrix}$.

True–False Critical Thinking Problems

42. Which one of the following is true?
  a. Some nonsquare matrices have inverses.
  b. All square 2×2 matrices have inverses since there is a formula for finding these inverses.
  c. Two 2×2 invertible matrices can have a matrix sum that is not invertible.
  d. To solve the matrix equation $AX = B$ for X, multiply A and the inverse of B.

43. Which one of the following is true?
  a. $(AB)^{-1} = A^{-1}B^{-1}$, assuming A, B, and AB are invertible.
  b. $(A + B)^{-1} = A^{-1} + B^{-1}$, assuming A, B, and $A + B$ are invertible.
  c. $\begin{bmatrix} 1 & -3 \\ -1 & 3 \end{bmatrix}$ is an invertible matrix.
  d. None of the above is true.

Technology Problems

In Problems 44–45, use a graphing utility to find the inverse of the matrix (if it exists).

44. $\begin{bmatrix} 7 & -3 & 0 & 2 \\ -2 & 1 & 0 & -1 \\ 4 & 0 & 1 & -2 \\ -1 & 1 & 0 & -1 \end{bmatrix}$

45. $\begin{bmatrix} 1 & 2 & 0 & 0 \\ 0 & 0 & 1 & 0 \\ 1 & 3 & 0 & 1 \\ 4 & 0 & 0 & 2 \end{bmatrix}$

In Problems 46–51, write each system in the form $AX = B$. Then solve the system by entering A and B into your graphing utility and computing $A^{-1}B$.

46. $x - y + z = -6$
    $4x + 2y + z = 9$
    $4x - 2y + z = -3$

47. $3x - 2y + z = -2$
    $4x - 5y + 3z = -9$
    $2x - y + 5z = -5$

48. $y + 2z = 0$
    $-x + y = 1$
    $2x - y + z = -1$

49. $x - y = 1$
    $6x + y + 20z = 14$
    $y + 3z = 1$

50.
$$x - 3z + v = -3$$
$$y + w = -1$$
$$z + v = 7$$
$$x + y - z + 4w = -8$$
$$x + y + z + w + v = 8$$

51.
$$x + y + z + w = 4$$
$$x + 3y - 2z + 2w = 7$$
$$2x + 2y + z + w = 3$$
$$x - y + 2z + 3w = 5$$

52. The matrix T, called a *technology matrix*, is given by

Input		Output Agricul-ture	Output Manufac-turing	Output Transpor-tation
	Agriculture	0.1	0.067	0
	Manufacturing	0.2	0.25	0.25
	Transportation	0.2	0.2	0.167

Call this matrix T. Each column of T gives the fraction of one dollar's worth of input from each sector of the economy needed to produce one dollar's worth of output in the sector represented by that column. For example, to produce one dollar's worth of manufactured goods (column 2) requires about 7 cents' worth of agri-culture, 25 cents' worth of manufacturing, and 20 cents' worth of transportation.

The matrix shown below is called a *production matrix*.

Production (in millions of dollars)

Agriculture	$\begin{bmatrix} 100 \\ 120 \\ 120 \end{bmatrix}$
Manufacturing	
Transportation	

Call this matrix P. The matrix indicates that the econ-omy produces \$100 million worth of agriculture, \$120 million worth of manufacturing, and \$120 million worth of transportation.

Use a graphing utility to answer each of the fol-lowing questions.

a. The matrix TP, called an *internal consumption matrix*, gives the amount of output consumed by an economic system internally. What is TP? Describe the amount of each sector's output needed as input by the other sectors to meet production goals.

b. The *demand matrix, D,* is given by $D = P - TP$. The matrix describes how much output is left to meet the demands of consumers after the input requirements are met. Find D.

Writing in Mathematics

53. List all the methods that we have studied up to this point for solving systems of linear equations. Select the method that you prefer and explain why.

54. Explain why a matrix that does not have the same num-ber of rows and columns cannot have an inverse.

55. Explain what happens if you try to find the inverse of $\begin{bmatrix} 1 & 4 & 2 \\ 0 & 2 & 4 \\ 0 & -3 & -6 \end{bmatrix}$. What part of the process tells you that the matrix is not invertible? Why is this the case?

56. Suppose that A is an invertible matrix. Is it correct to say $(4A)^{-1} = \frac{1}{4}A^{-1}$? Explain.

Critical Thinking Problems

In Problems 57–58, suppose that A and B are matrices and that $A \circ B = AB - BA$.

57. Show that $A \circ I = 0$ and $I \circ A = 0$. Is it possible for A to have an inverse for the operation $\circ$? Explain.

58. If $A = \begin{bmatrix} 1 & 2 \\ 3 & 0 \end{bmatrix}$, $B = \begin{bmatrix} 1 & 0 \\ 3 & 1 \end{bmatrix}$, and $C = \begin{bmatrix} 0 & 1 \\ 1 & 2 \end{bmatrix}$, show that $(A \circ B) \circ C \neq A \circ (B \circ C)$.

59. Give an example of a 2×2 matrix that is its own inverse.

60. Find two 2×2 invertible matrices whose sum is not invertible.

61. A matrix of the form $\begin{bmatrix} 1 & a & b \\ 0 & 1 & c \\ 0 & 0 & 1 \end{bmatrix}$ in which a, b, and c are real numbers is called a Heisenberg matrix. Show that the inverse of a Heisenberg matrix is a Heisenberg matrix.

62. Find A^{-1} if $A = \begin{bmatrix} a & 0 & 0 & 0 \\ 0 & b & 0 & 0 \\ 0 & 0 & c & 0 \\ 0 & 0 & 0 & d \end{bmatrix}$. Assume that $abcd \neq 0$.

63. If $A = \begin{bmatrix} a & 1 & 1 & 1 \\ 0 & b & 0 & 0 \\ 0 & 0 & c & 0 \\ 0 & 0 & 0 & d \end{bmatrix}$, find a formula for A^{-1} and state

the conditions under which A is invertible.

64. If $A = \begin{bmatrix} a & 0 \\ 0 & b \end{bmatrix}$ and $ab \neq 0$, find $(A^n)^{-1}$ for any positive integer n.

65. A 3×3 invertible matrix has six entries that are zeros. Describe the position of the nonzero entries of the matrix.

66. If A and B are invertible $n \times n$ matrices, prove that $(AB)^{-1} = B^{-1}A^{-1}$. Start with the equation $(AB)(AB)^{-1} = I$, multiplying each side first by A^{-1} and then by B^{-1}. State the result strictly in words without the use of any mathematical symbolism.

SECTION 5.6 Determinants and Cramer's Rule

Solutions Manual   Tutorial   Video 10

Objectives

1 Evaluate determinants.

2 Solve linear systems using determinants and Cramer's rule.

In this chapter, we have so far studied substitution, addition, and matrix methods for solving linear systems of equations. We turn now to another method for solving such systems. As with matrix methods, solutions are obtained by writing down the coefficients and constants of a linear system and operating with them. The result is known as Cramer's rule, in honor of the Swiss geometer Gabriel Cramer (1704–1752).

1 Evaluate determinants.

The Determinant of a 2 × 2 Matrix

Associated with every square matrix is a real number called its *determinant*. The determinant for a 2×2 square matrix is defined as follows.

Study tip

To evaluate a determinant, find the difference of the product of the two diagonals.

$$\begin{vmatrix} a_1 & b_1 \\ a_2 & b_2 \end{vmatrix} = a_1 b_2 - a_2 b_1$$

Definition of the determinant of a 2 × 2 matrix

The determinant of the matrix $\begin{bmatrix} a_1 & b_1 \\ a_2 & b_2 \end{bmatrix}$ is denoted by $\begin{vmatrix} a_1 & b_1 \\ a_2 & b_2 \end{vmatrix}$ and is defined by

$$\begin{vmatrix} a_1 & b_1 \\ a_2 & b_2 \end{vmatrix} = a_1 b_2 - a_2 b_1.$$

We also say that the *value* of the *second-order determinant*

$\begin{vmatrix} a_1 & b_1 \\ a_2 & b_2 \end{vmatrix}$ is $a_1 b_2 - a_2 b_1$.

EXAMPLE 1 Evaluating the Determinant of a 2 × 2 Matrix

Evaluate the determinant of:

a. $\begin{bmatrix} 5 & 6 \\ 7 & 3 \end{bmatrix}$ **b.** $\begin{bmatrix} 2 & 4 \\ -3 & -5 \end{bmatrix}$

iscover for yourself

Example I illustrates that the determinant of a matrix may be positive or negative. The determinant can also have 0 as its value. Write and then evaluate three determinants, one whose value is positive, one whose value is negative, and one whose value is 0.

Solution

We multiply and subtract as indicated.

a. $\begin{vmatrix} 5 & 6 \\ 7 & 3 \end{vmatrix} = 5 \cdot 3 - 7 \cdot 6 = 15 - 42 = -27$ The value of the second-order determinant is -27.

b. $\begin{vmatrix} 2 & 4 \\ -3 & -5 \end{vmatrix} = 2(-5) - (-3)(4) = -10 + 12 = 2$ The value of the second-order determinant is 2.

sing technology

You can use your graphing utility to evaluate the determinant of a square matrix. First use the $\boxed{\text{MATRX}}$ feature to enter the matrix as [A]. Then evaluate the determinant using a sequence similar to

$$\det [A] \ \boxed{\text{ENTER}} \ .$$

Use this feature to verify the results in Example I and all subsequent examples in this section. Describe what happens when you attempt to evaluate the determinant of a nonsquare matrix.

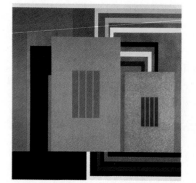
Peter Halley "CUSeeMe" 1995, acryl, Day–Glo, Roll a tex on linen, 274.5 × 280 cm. Sammlung Goetz.

The Determinant of a 3 × 3 Matrix

Associated with every square matrix is a real number called its determinant. The determinant for a 3×3 matrix is defined as follows.

Definition of a third-order determinant

$$\begin{vmatrix} a_1 & b_1 & c_1 \\ a_2 & b_2 & c_2 \\ a_3 & b_3 & c_3 \end{vmatrix} = a_1 b_2 c_3 + b_1 c_2 a_3 + c_1 a_2 b_3 - a_3 b_2 c_1 - b_3 c_2 a_1 - c_3 a_2 b_1$$

The six terms and the three factors in each term in this complicated evaluation formula can be rearranged, and then we can apply the distributive property. We obtain

$$a_1 b_2 c_3 - a_1 b_3 c_2 - a_2 b_1 c_3 + a_2 b_3 c_1 + a_3 b_1 c_2 - a_3 b_2 c_1$$

$$= a_1 (b_2 c_3 - b_3 c_2) - a_2 (b_1 c_3 - b_3 c_1) + a_3 (b_1 c_2 - b_2 c_1)$$

$$= a_1 \begin{vmatrix} b_2 & c_2 \\ b_3 & c_3 \end{vmatrix} - a_2 \begin{vmatrix} b_1 & c_1 \\ b_3 & c_3 \end{vmatrix} + a_3 \begin{vmatrix} b_1 & c_1 \\ b_2 & c_2 \end{vmatrix} .$$

You can evaluate each of the second-order determinants and obtain the three expressions in parentheses in the second step.

In summary, we now have arranged the definition of a third-order determinant as follows.

Definition of the determinant of a 3 × 3 matrix

A third-order determinant is defined by:

Subtract.   Add.

$$\begin{vmatrix} a_1 & b_1 & c_1 \\ a_2 & b_2 & c_2 \\ a_3 & b_3 & c_3 \end{vmatrix} = a_1 \begin{vmatrix} b_2 & c_2 \\ b_3 & c_3 \end{vmatrix} - a_2 \begin{vmatrix} b_1 & c_1 \\ b_3 & c_3 \end{vmatrix} + a_3 \begin{vmatrix} b_1 & c_1 \\ b_2 & c_2 \end{vmatrix}$$

Note that the a's come from the first column.

Study tip

Here are some helpful tips when evaluating a 3 × 3 determinant using the definition given above.

1. Each of the three terms in the definition contains two factors—a number and a second-order determinant.
2. The numerical factor in each term is an element from the first column of the third-order determinant.
3. The minus sign precedes the second term.
4. The second-order determinant that appears in each term is obtained by crossing out the row and the column containing the numerical factor.

$$a_1 \begin{vmatrix} b_2 & c_2 \\ b_3 & c_3 \end{vmatrix} - a_2 \begin{vmatrix} b_1 & c_1 \\ b_3 & c_3 \end{vmatrix} + a_3 \begin{vmatrix} b_1 & c_1 \\ b_2 & c_2 \end{vmatrix}$$

$$\begin{vmatrix} a_1 & b_1 & c_1 \\ a_2 & b_2 & c_2 \\ a_3 & b_3 & c_3 \end{vmatrix} \begin{vmatrix} a_1 & b_1 & c_1 \\ a_2 & b_2 & c_2 \\ a_3 & b_3 & c_3 \end{vmatrix} \begin{vmatrix} a_1 & b_1 & c_1 \\ a_2 & b_2 & c_2 \\ a_3 & b_3 & c_3 \end{vmatrix}$$

The *minor* of an element is precisely the determinant that remains after deleting the row and column of that element. For this reason, we call this method *expansion by minors*.

EXAMPLE 2 **Evaluating the Determinant of a 3 × 3 Matrix**

Evaluate the determinant of:

$$\begin{bmatrix} 1 & -1 & 2 \\ -2 & 1 & 1 \\ 1 & -1 & 3 \end{bmatrix}.$$

Study tip

Here's the crossing out that is needed to obtain the 2 × 2 determinants in the second step of the solution.

$$\begin{vmatrix} 1 & -1 & 2 \\ -2 & 1 & 1 \\ 1 & -1 & 3 \end{vmatrix} \begin{vmatrix} 1 & -1 & 2 \\ -2 & 1 & 1 \\ 1 & -1 & 3 \end{vmatrix}$$

$$\begin{vmatrix} 1 & -1 & 2 \\ -2 & 1 & 1 \\ 1 & -1 & 3 \end{vmatrix}$$

Solution

Don't forget to supply the minus sign.

$$\begin{vmatrix} 1 & -1 & 2 \\ -2 & 1 & 1 \\ 1 & -1 & 3 \end{vmatrix} = 1 \begin{vmatrix} 1 & 1 \\ -1 & 3 \end{vmatrix} - (-2) \begin{vmatrix} -1 & 2 \\ -1 & 3 \end{vmatrix} + 1 \begin{vmatrix} -1 & 2 \\ 1 & 1 \end{vmatrix}$$

$$= 1[1 \cdot 3 - (-1) \cdot 1] + 2[-1 \cdot 3 - (-1) \cdot 2] + 1(-1 \cdot 1 - 1 \cdot 2)$$

$$= 1(3 + 1) + 2(-3 + 2) + 1(-1 - 2) \quad \text{Evaluate the three second-order determinants.}$$

Using technology

Verify the result of Example 2 by using your graphing utility to enter the given matrix as

$$A = \begin{bmatrix} 1 & -1 & 2 \\ -2 & 1 & 1 \\ 1 & -1 & 3 \end{bmatrix}.$$

Then enter

det [A] $\boxed{\text{ENTER}}$.

The result should be -1.

$$= 1(4) + 2(-1) + 1(-3)$$

$$= 4 - 2 - 3$$

$$= -1 \qquad \blacksquare$$

The six terms in the definition of a third-order determinant can be rearranged and factored in a variety of ways, making it possible to expand a determinant by minors about any row or any column. *Minus signs must be supplied preceding any element appearing in a position where the sum of its row and its column is an odd number.* For example, expanding about the elements in column 2 gives us

$$\begin{vmatrix} a_1 & b_1 & c_1 \\ a_2 & b_2 & c_2 \\ a_3 & b_3 & c_3 \end{vmatrix} = -b_1 \begin{vmatrix} a_2 & c_2 \\ a_3 & c_3 \end{vmatrix} + b_2 \begin{vmatrix} a_1 & c_1 \\ a_3 & c_3 \end{vmatrix} - b_3 \begin{vmatrix} a_1 & c_1 \\ a_2 & c_2 \end{vmatrix}.$$

Minus sign is supplied because b_1 appears in row 1 and column 2; $1 + 2 = 3$, an odd number.

Minus sign is supplied because b_3 appears in row 3 and column 2; $3 + 2 = 5$, an odd number.

Discover for yourself

In Example 2, we evaluated the determinant by expanding about column 1. Use a different row or column and evaluate the determinant.

Expanding by minors about column 3, we obtain

$$\begin{vmatrix} a_1 & b_1 & c_1 \\ a_2 & b_2 & c_2 \\ a_3 & b_3 & c_3 \end{vmatrix} = c_1 \begin{vmatrix} a_2 & b_2 \\ a_3 & b_3 \end{vmatrix} - c_2 \begin{vmatrix} a_1 & b_1 \\ a_3 & b_3 \end{vmatrix} + c_3 \begin{vmatrix} a_1 & b_1 \\ a_2 & b_2 \end{vmatrix}$$

Minus sign must be supplied because c_2 appears in row 2 and column 3; $2 + 3 = 5$, an odd number.

Study tip

Here are some helpful tips when evaluating a 3 × 3 determinant using expansion by minors.

1. You can expand the determinant about any row or any column. When possible, it is convenient to expand about the row or column containing one or more 0s, or the row or column containing the smallest numbers in magnitude.
2. It is helpful to remember the following sign array:

$$\begin{vmatrix} + & - & + \\ - & + & - \\ + & - & + \end{vmatrix}.$$

EXAMPLE 3 **Evaluating a Third-Order Determinant**

Evaluate:

$$\begin{vmatrix} 9 & 5 & 0 \\ -2 & -3 & 0 \\ 1 & 4 & 2 \end{vmatrix}$$

Solution

With two 0s in the last column, the deteminant will be expanded about the elements in that column.

$$\begin{vmatrix} 9 & 5 & 0 \\ -2 & -3 & 0 \\ 1 & 4 & 2 \end{vmatrix} = 0\begin{vmatrix} -2 & -3 \\ 1 & 4 \end{vmatrix} - 0\begin{vmatrix} 9 & 5 \\ 1 & 4 \end{vmatrix} + 2\begin{vmatrix} 9 & 5 \\ -2 & -3 \end{vmatrix}$$

$$= 0 - 0 + 2(-27 + 10)$$

$$= 2(-17)$$

$$= -34 \qquad \blacksquare$$

The Determinant of Any $n \times n$ Matrix

A determinant with n rows and n columns is said to be an *nth-order determinant*. The value of an *n*th-order determinant ($n > 2$) can be found in terms of determinants of order $n - 1$. For example, we found the value of a third-order determinant in terms of determinants of order 2.

We can generalize this idea for fourth-order determinants and higher. We have seen that the *minor* of the entry a_{ij} is the determinant obtained by deleting the *i*th row and the *j*th column in the given array of numbers. The *cofactor* of the entry a_{ij} is $(-1)^{i+j}$ times the minor of the a_{ij}th entry. If the sum of the row and column $(i + j)$ is even, the cofactor is the same as the minor. If the sum of the row and column $(i + j)$ is odd, the cofactor is the opposite of the minor.

Let's see what this means in the case of a fourth-order determinant.

EXAMPLE 4   **Evaluating the Determinant of a 4 × 4 Matrix**

Evaluate the determinant of:

$$A = \begin{bmatrix} 1 & -2 & 3 & 0 \\ -1 & 1 & 0 & 2 \\ 0 & 2 & 0 & -3 \\ 2 & 3 & -4 & 1 \end{bmatrix}$$

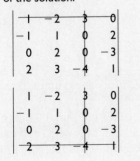

Study tip

Here's the crossing out that is needed to obtain the determinants in the second step of the solution.

Solution

$$|A| = \begin{vmatrix} 1 & -2 & 3 & 0 \\ -1 & 1 & 0 & 2 \\ 0 & 2 & 0 & -3 \\ 2 & 3 & -4 & 1 \end{vmatrix}$$

With two 0s in the third column, we will expand along the third column.

$$= (-1)^{1+3}\,3\begin{vmatrix} -1 & 1 & 2 \\ 0 & 2 & -3 \\ 2 & 3 & 1 \end{vmatrix} + (-1)^{4+3}\,(-4)\begin{vmatrix} 1 & -2 & 0 \\ -1 & 1 & 2 \\ 0 & 2 & -3 \end{vmatrix}$$

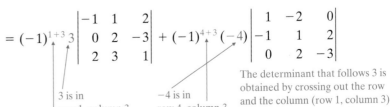

3 is in
row 1, column 3.

−4 is in
row 4, column 3.

The determinant that follows 3 is obtained by crossing out the row and the column (row 1, column 3) in the above step. The minor for −4 is obtained in the same manner. (See the Study tip.)

$$= 3 \begin{vmatrix} -1 & 1 & 2 \\ 0 & 2 & -3 \\ 2 & 3 & 1 \end{vmatrix} + 4 \begin{vmatrix} 1 & -2 & 0 \\ -1 & 1 & 2 \\ 0 & 2 & -3 \end{vmatrix}$$

$(-1)^{1+3}(3) = (-1)^4(3) = 3$ and
$(-1)^{4+3}(-4) = (-1)^7(-4)$
$= (-1)(-4) = 4$

Now we must work with the two third-order determinants.

$$\begin{vmatrix} -1 & 1 & 2 \\ 0 & 2 & -3 \\ 2 & 3 & 1 \end{vmatrix} = -1 \begin{vmatrix} 2 & -3 \\ 3 & 1 \end{vmatrix} + 2 \begin{vmatrix} 1 & 2 \\ 2 & -3 \end{vmatrix} \qquad \text{Expand about column 1.}$$

$$= -1(2 + 9) + 2(-3 - 4)$$

$$= -25$$

$$\begin{vmatrix} 1 & -2 & 0 \\ -1 & 1 & 2 \\ 0 & 2 & -3 \end{vmatrix} = 1 \begin{vmatrix} 1 & 2 \\ 2 & -3 \end{vmatrix} + 1 \begin{vmatrix} -2 & 0 \\ 2 & -3 \end{vmatrix} \qquad \text{Expand about column 1.}$$

$$= 1(-3 - 4) + 1(6 - 0)$$

$$= -1$$

Substituting these values for the two third-order determinants into the evaluation of $|A|$, we obtain

$$|A| = 3(-25) + 4(-1) = -79.$$ ■

Study tip

Here are the sign patterns for the cofactors. Odd positions (where row plus column number is odd) have negative signs. Even positions (where row plus column is even) have positive signs.

$$\begin{vmatrix} + & - & + \\ - & + & - \\ + & - & + \end{vmatrix} \qquad \begin{vmatrix} + & - & + & - \\ - & + & - & + \\ + & - & + & - \\ - & + & - & + \end{vmatrix} \qquad \begin{vmatrix} + & - & + & - & + & \cdots \\ - & + & - & + & - & \cdots \\ + & - & + & - & + & \cdots \\ - & + & - & + & - & \cdots \\ + & - & + & - & + & \cdots \\ \vdots & \vdots & \vdots & \vdots & \vdots & \end{vmatrix}$$

3 × 3 determinant   **4 × 4 determinant**   **n × n determinant**

Properties of Determinants

When evaluating determinants by hand, the three basic properties summarized in Table 5.2 often simplify the arithmetic.

Notice how the third property in Table 5.2 can be used to obtain a 0 entry. The property is just like the one we used to obtain 0s in a matrix when solving systems by Gaussian elimination. Use this property to evaluate a determinant

TABLE 5.2 Properties of Determinants	
Property	**Example**
1. Factoring a common factor from a given row (or column) does not change a determinant's value.	$\begin{vmatrix} 30 & 60 \\ 7 & 5 \end{vmatrix} = 30 \begin{vmatrix} 1 & 2 \\ 7 & 5 \end{vmatrix}$
2. Interchanging two rows (or columns) of a determinant multiplies its value by -1.	$\begin{vmatrix} 3 & 4 \\ 7 & 5 \end{vmatrix} = -\begin{vmatrix} 7 & 5 \\ 3 & 4 \end{vmatrix}$
3. Adding a multiple of one row (or column) to the corresponding entries of another row (or column) does not change a determinant's value.	$\begin{vmatrix} 1 & 2 \\ 8 & 5 \end{vmatrix} = \begin{vmatrix} 1 & 2 \\ 8 + (-8) \cdot 1 & 5 + (-8) \cdot 2 \end{vmatrix} = \begin{vmatrix} 1 & 2 \\ 0 & -11 \end{vmatrix}$ Multiply the first row by -8 and add to the second row.

by obtaining 0s for every entry except for one in any row or column. Then expand about that row or column.

EXAMPLE 5 Using the Properties of Determinants

Evaluate the determinant:

$$\begin{vmatrix} 3 & 1 & 10 \\ 6 & 4 & -15 \\ -12 & 7 & 20 \end{vmatrix}$$

Solution

$$\begin{vmatrix} 3 & 1 & 10 \\ 6 & 4 & -15 \\ -12 & 7 & 20 \end{vmatrix}$$

$$= (3 \times 5) \begin{vmatrix} 1 & 1 & 2 \\ 2 & 4 & -3 \\ -4 & 7 & 4 \end{vmatrix}$$ Use property 1 to factor 3 from the first column and 5 from the third column.

$$= 15 \begin{vmatrix} 1 & 1 & 2 \\ 0 & 2 & -7 \\ 0 & 11 & 12 \end{vmatrix}$$ Let's expand about column 1 by first getting 0s below the 1. Multiply row 1 by -2 and add to row 2. Multiply row 1 by 4 and add to row 3.

$$= 15 \cdot 1 \begin{vmatrix} 2 & -7 \\ 11 & 12 \end{vmatrix}$$ With two 0s in column 1, expand about that column.

$$= 15[24 - (-77)]$$

$$= 1515$$ ∎

2 Solve linear systems using determinants and Cramer's rule.

Determinants and Linear Systems

Determinants can be used to solve linear systems of equations. Let's begin with the easiest case, namely two equations in two variables. In general, such a system appears as

$$a_1 x + b_1 y = c_1$$

$$a_2 x + b_2 y = c_2$$

Let's first solve this system for x using the addition method. We can solve for x by eliminating y from the equations. We multiply the first equation by b_2 and the second equation by $-b_1$. Then we add the two equations.

$$
\begin{array}{ll}
a_1 x + b_1 y = c_1 & \xrightarrow{\text{Multiply by } b_2.} \\
a_2 x + b_2 y = c_2 & \xrightarrow{\text{Multiply by } -b_1.}
\end{array}
\qquad
\begin{array}{rl}
a_1 b_2 x + b_1 b_2 y = & b_2 c_1 \\
-a_2 b_1 x - b_1 b_2 y = & -b_1 c_2 \\
\hline
\text{Add:} \quad (a_1 b_2 - a_2 b_1) x = & b_2 c_1 - b_1 c_2 \\
\\
x = & \dfrac{b_2 c_1 - b_1 c_2}{a_1 b_2 - a_2 b_1}
\end{array}
$$

Because

$$
\begin{vmatrix} c_1 & b_1 \\ c_2 & b_2 \end{vmatrix} = c_1 b_2 - c_2 b_1
\quad \text{and} \quad
\begin{vmatrix} a_1 & b_1 \\ a_2 & b_2 \end{vmatrix} = a_1 b_2 - a_2 b_1
$$

we can express our answer for x as the quotient of two determinants.

$$
x = \frac{\begin{vmatrix} c_1 & b_1 \\ c_2 & b_2 \end{vmatrix}}{\begin{vmatrix} a_1 & b_1 \\ a_2 & b_2 \end{vmatrix}}
$$

In a similar way, we could use the addition method to solve our system for y, again expressing y as the quotient of two determinants. This method of using determinants to solve the linear system is called *Cramer's rule*.

No, not that Kramer!

Michael Richards of "Seinfeld" / Archive Photos.

Solving a system using determinants

Cramer's Rule

If

$$a_1 x + b_1 y = c_1$$

$$a_2 x + b_2 y = c_2$$

then

$$
x = \frac{\begin{vmatrix} c_1 & b_1 \\ c_2 & b_2 \end{vmatrix}}{\begin{vmatrix} a_1 & b_1 \\ a_2 & b_2 \end{vmatrix}}
\quad \text{and} \quad
y = \frac{\begin{vmatrix} a_1 & c_1 \\ a_2 & c_2 \end{vmatrix}}{\begin{vmatrix} a_1 & b_1 \\ a_2 & b_2 \end{vmatrix}}
$$

where

$$
\begin{vmatrix} a_1 & b_1 \\ a_2 & b_2 \end{vmatrix} \neq 0.
$$

S

tudy tip

Here are some helpful tips when solving

$$a_1 x + b_1 y = c_1$$
$$a_2 x + b_2 y = c_2$$

using determinants.

1. Three different determinants are used to find x and y. The determinants in the denominators for x and y are identical. The determinants in the numerators for x and y differ. In abbreviated notation, we write

$$x = \frac{D_x}{D} \quad \text{and} \quad y = \frac{D_y}{D}, \text{ where } D \neq 0.$$

2. The elements of D, the determinant in the denominator, are the coefficients of the variables in the system.

$$D = \begin{vmatrix} a_1 & b_1 \\ a_2 & b_2 \end{vmatrix}$$

3. D_x, the determinant in the numerator of x, is obtained by replacing the x-coefficients a_1, a_2, in D with the constants on the right side of the equations, c_1, c_2.

$$D = \begin{vmatrix} a_1 & b_1 \\ a_2 & b_2 \end{vmatrix} \quad \text{and} \quad D_x = \begin{vmatrix} c_1 & b_1 \\ c_2 & b_2 \end{vmatrix} \qquad \text{Replace the column with } a_1 \text{ and } a_2 \text{ with the constants } c_1 \text{ and } c_2 \text{ to get } D_x.$$

4. D_y, the determinant in the numerator for y, is obtained by replacing the y-coefficients b_1, b_2 in D with the constants on the right side of the equations c_1, c_2.

$$D = \begin{vmatrix} a_1 & b_1 \\ a_2 & b_2 \end{vmatrix} \quad \text{and} \quad D_y = \begin{vmatrix} a_1 & c_1 \\ a_2 & c_2 \end{vmatrix} \qquad \text{Replace the column with } b_1 \text{ and } b_2 \text{ with the constants } c_1 \text{ and } c_2 \text{ to get } D_y.$$

Example 6 illustrates the use of Cramer's rule.

EXAMPLE 6 Using Cramer's Rule to Solve a Linear System

Use Cramer's rule to solve:

$$x + 3y = 14$$
$$5x - 2y = 2$$

Solution

Because

$$x = \frac{D_x}{D} \quad \text{and} \quad y = \frac{D_y}{D},$$

we will set up and evaluate the three determinants D, D_x, and D_y.

1. D, the determinant in both denominators, consists of the x- and y-coefficients.

$$D = \begin{vmatrix} 1 & 3 \\ 5 & -2 \end{vmatrix} = (1)(-2) - (5)(3) = -2 - 15 = -17$$

tudy tip

In order for you not to have to turn back a page, here again is the system.

$x + 3y = 14$

$5x - 2y = 2$

2. D_x, the determinant in the numerator for x, is obtained by replacing the x-coefficients in D, 1 and 5, with the constants on the right side of the equation, 14 and 2.

$$D_x = \begin{vmatrix} 14 & 3 \\ 2 & -2 \end{vmatrix} = (14)(-2) - (2)(3) = -28 - 6 = -34$$

3. D_y, the determinant in the numerator for y, is obtained by replacing the y-coefficients in D, 3 and -2, with the constants on the right side of the equation, 14 and 2.

$$D_y = \begin{vmatrix} 1 & 14 \\ 5 & 2 \end{vmatrix} = (1)(2) - (5)(14) = 2 - 70 = -68$$

4. Thus,

$$x = \frac{D_x}{D} = \frac{-34}{-17} = 2 \quad \text{and} \quad y = \frac{D_y}{D} = \frac{-68}{-17} = 4.$$

As always, the solution (2, 4) can be checked by substituting these values into the original equations. The solution set is $\{(2, 4)\}$. ∎

Cramer's rule can be applied to solving systems of linear equations in three variables. The determinants in the numerator and denominator of all variables are third-order determinants.

Solving three equations in three variables using determinants

Cramer's Rule

If

$$a_1 x + b_1 y + c_1 z = d_1$$

$$a_2 x + b_2 y + c_2 z = d_2$$

$$a_3 x + b_3 y + c_3 z = d_3$$

then

$$x = \frac{D_x}{D}, y = \frac{D_y}{D}, \text{ and } z = \frac{D_z}{D}.$$

These four third-order determinants are given by

$$D = \begin{vmatrix} a_1 & b_1 & c_1 \\ a_2 & b_2 & c_2 \\ a_3 & b_3 & c_3 \end{vmatrix} \quad \text{These are the coefficients of the variables } x, y, \text{ and } z. \, D \neq 0.$$

$$D_x = \begin{vmatrix} d_1 & b_1 & c_1 \\ d_2 & b_2 & c_2 \\ d_3 & b_3 & c_3 \end{vmatrix} \quad \text{Replace } x\text{-coefficients in } D \text{ with the constants at the right of the three equations.}$$

$$D_y = \begin{vmatrix} a_1 & d_1 & c_1 \\ a_2 & d_2 & c_2 \\ a_3 & d_3 & c_3 \end{vmatrix} \quad \text{Replace } y\text{-coefficients in } D \text{ with the constants at the right of the three equations.}$$

$$D_z = \begin{vmatrix} a_1 & b_1 & d_1 \\ a_2 & b_2 & d_2 \\ a_3 & b_3 & d_3 \end{vmatrix}$$

Replace z-coefficients in D with the constants at the right of the three equations.

EXAMPLE 7   Using Cramer's Rule to Solve a Linear System in Three Variables

Use Cramer's rule to solve:

$$x + 2y - z = -4$$
$$x + 4y - 2z = -6$$
$$2x + 3y + z = 3$$

Solution

Because

$$x = \frac{D_x}{D}, \quad y = \frac{D_y}{D}, \quad \text{and} \quad z = \frac{D_z}{D}$$

we need to set up and evaluate four determinants.

Step 1. Set up the determinants.
1. D, the determinant in all three denominators, consists of the x-, y-, and z-coefficients.

$$D = \begin{vmatrix} 1 & 2 & -1 \\ 1 & 4 & -2 \\ 2 & 3 & 1 \end{vmatrix}$$

2. D_x, the determinant in the numerator for x, is obtained by replacing the x-coefficients in D, 1, 1, and 2, with the constants on the right side of the equation, -4, -6, and 3.

$$D_x = \begin{vmatrix} -4 & 2 & -1 \\ -6 & 4 & -2 \\ 3 & 3 & 1 \end{vmatrix}$$

3. D_y, the determinant in the numerator for y, is obtained by replacing the y-coefficients in D, 2, 4, and 3, with the constants on the right side of the equation, -4, -6, and 3.

$$D_y = \begin{vmatrix} 1 & -4 & -1 \\ 1 & -6 & -2 \\ 2 & 3 & 1 \end{vmatrix}$$

4. D_z, the determinant in the numerator for z, is obtained by replacing the z-coefficients in D, -1, -2, and 1, with the constants on the right side of the equation, -4, -6, and 3.

$$D_z = \begin{vmatrix} 1 & 2 & -4 \\ 1 & 4 & -6 \\ 2 & 3 & 3 \end{vmatrix}$$

Step 2. Evaluate the four determinants.

1. $D = \begin{vmatrix} 1 & 2 & -1 \\ 1 & 4 & -2 \\ 2 & 3 & 1 \end{vmatrix} = 1 \begin{vmatrix} 4 & -2 \\ 3 & 1 \end{vmatrix} - 1 \begin{vmatrix} 2 & -1 \\ 3 & 1 \end{vmatrix} + 2 \begin{vmatrix} 2 & -1 \\ 4 & -2 \end{vmatrix}$

$$= 1(4 + 6) - 1(2 + 3) + 2(-4 + 4)$$

$$= 1(10) - 1(5) + 2(0) = 5$$

Using the same technique to evaluate each determinant we obtain:

2. $D_x = -10$

3. $D_y = 5$

4. $D_z = 20$

Step 3. Substitute these four values and solve the system.

$$x = \frac{D_x}{D} = \frac{-10}{5} = -2$$

$$y = \frac{D_y}{D} = \frac{5}{5} = 1$$

$$z = \frac{D_z}{D} = \frac{20}{5} = 4$$

The solution $(-2, 1, 4)$ can be checked by substitution into the original three equations. The solution set is $\{(-2, 1, 4)\}$. ∎

The Idea Behind Cramer's Rule

One way we can convince ourselves that we can rely on Cramer's rule to provide the correct solutions to

$$a_1 x + b_1 y + c_1 z = d_1$$

$$a_2 x + b_2 y + c_2 z = d_2$$

$$a_3 x + b_3 y + c_3 z = d_3$$

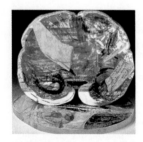

Todd Siler "Mind Icon: A Reverie of Cerebralism" 1991–1992, mixed media on wood, 17 × 19 × 13 in. Courtesy Ronald Feldman Fine Arts, New York. Photo credit: D. James Dee.

is to use Gaussian elimination to find x, y, and z and then show that these values are the same as those given by the quotient of the determinants. A shorter method involves using the properties of determinants. Let's begin with D_x, the determinant in the numerator for x.

$$D_x = \begin{vmatrix} d_1 & b_1 & c_1 \\ d_2 & b_2 & c_2 \\ d_3 & b_3 & c_3 \end{vmatrix}$$

The x-coefficients in D—a_1, a_2, and a_3—are replaced with the constants at the right of the given equations.

$$= \begin{vmatrix} a_1 x + b_1 y + c_1 z & b_1 & c_1 \\ a_2 x + b_2 y + c_2 z & b_2 & c_2 \\ a_3 x + b_3 y + c_3 z & b_3 & c_3 \end{vmatrix}$$

Substitute the left sides of the given equations for d_1, d_2, and d_3, respectively.

$$= \begin{vmatrix} a_1x + b_1y + c_1z - b_1y - c_1z & b_1 & c_1 \\ a_2x + b_2y + c_2z - b_2y - c_2z & b_2 & c_2 \\ a_3x + b_3y + c_3z - b_3y - c_3z & b_3 & c_3 \end{vmatrix}$$

Use the properties of determinants and subtract y times column 2 and z times column 3 from column 1.

$$= \begin{vmatrix} a_1x & b_1 & c_1 \\ a_2x & b_2 & c_2 \\ a_3x & b_3 & c_3 \end{vmatrix}$$

Simplify the entries in the first column.

$$= x \begin{vmatrix} a_1 & b_1 & c_1 \\ a_2 & b_2 & c_2 \\ a_3 & b_3 & c_3 \end{vmatrix}$$

Factor out x from the first column.

$$= xD$$

The determinant consists of the coefficients of x, y, and z which, by definition, is D.

At this point we have

$$D_x = xD.$$

Solving for x, we obtain:

$$x = \frac{D_x}{D}$$

which is precisely the result given by Cramer's rule. We can prove Cramer's rule for y and z in a similar manner.

Cramer's rule can be extended to cover n linear equations in n variables.

Solving n equations in n variables using determinants

Cramer's Rule

If

$$a_{11}x_1 + a_{12}x_2 + a_{13}x_3 + \cdots + a_{1n}x_n = b_1$$
$$a_{21}x_1 + a_{22}x_2 + a_{23}x_3 + \cdots + a_{2n}x_n = b_2$$
$$a_{31}x_1 + a_{32}x_2 + a_{33}x_3 + \cdots + a_{3n}x_n = b_3$$
$$\vdots$$
$$a_{n1}x_1 + a_{n2}x_2 + a_{n3}x_3 + \cdots + a_{nn}x_n = b_n$$

then

$$x_i = \frac{D_i}{D},$$

where D is the determinant of the coefficients of the system $(D \neq 0)$ and D_i is the determinant where the coefficients of x_i have been replaced with the constants $b_1, b_2, b_3, \ldots, b_n$, respectively.

Cramer's Rule with Inconsistent and Dependent Systems

If D, the determinant in the denominator, is 0, the variables described by the quotient of determinants are not real numbers. However, when $D = 0$, this

iscover for yourself

Write a system of two equations that is inconsistent. Now use determinants and the result boxed on the right to verify that this is truly an inconsistent system. Repeat the same process for a system with two dependent equations.

indicates that the system is inconsistent or contains dependent equations. This gives rise to the following two situations.

Determinants: Inconsistent and dependent systems

1. If $D = 0$ and at least one of the determinants in the numerator is not 0, then the system is inconsistent. The solution set is $\emptyset$.
2. If $D = 0$ and all the determinants in the numerators are 0, then the equations in the system are dependent.

Although we have focused on applying determinants to solve linear systems, they have other applications, some of which we consider in the problem set that follows.

PROBLEM SET 5.6

Practice Problems

Evaluate each determinant in Problems 1–18. Use a graphing utility to verify your result.

1. $\begin{vmatrix} 2 & -7 \\ -3 & 4 \end{vmatrix}$

2. $\begin{vmatrix} 5 & -6 \\ -2 & 8 \end{vmatrix}$

3. $\begin{vmatrix} -6 & 7 \\ -1 & 4 \end{vmatrix}$

4. $\begin{vmatrix} -5 & 9 \\ -1 & 8 \end{vmatrix}$

5. $\begin{vmatrix} 3 & 0 & 0 \\ 2 & 1 & -5 \\ 2 & 5 & -1 \end{vmatrix}$

6. $\begin{vmatrix} 4 & 0 & 0 \\ 3 & -1 & 4 \\ 2 & -3 & 5 \end{vmatrix}$

7. $\begin{vmatrix} 3 & 1 & 0 \\ -3 & 4 & 0 \\ -1 & 3 & -5 \end{vmatrix}$

8. $\begin{vmatrix} 2 & -4 & 2 \\ -1 & 0 & 5 \\ 3 & 0 & 4 \end{vmatrix}$

9. $\begin{vmatrix} 1 & 1 & 1 \\ 2 & 2 & 2 \\ -3 & 4 & -5 \end{vmatrix}$

10. $\begin{vmatrix} 1 & 2 & 3 \\ 2 & 2 & -3 \\ 3 & 2 & 1 \end{vmatrix}$

11. $\begin{vmatrix} 4 & 2 & 8 & -7 \\ -2 & 0 & 4 & 1 \\ 5 & 0 & 0 & 5 \\ 4 & 0 & 0 & -1 \end{vmatrix}$

12. $\begin{vmatrix} 3 & -1 & 1 & 2 \\ -2 & 0 & 0 & 0 \\ 2 & -1 & -2 & 3 \\ 1 & 4 & 2 & 3 \end{vmatrix}$

13. $\begin{vmatrix} -2 & -3 & 3 & 5 \\ 1 & -4 & 0 & 0 \\ 1 & 2 & 2 & -3 \\ 2 & 0 & 1 & 1 \end{vmatrix}$

14. $\begin{vmatrix} 1 & -3 & 2 & 0 \\ -3 & -1 & 0 & -2 \\ 2 & 1 & 3 & 1 \\ 2 & 0 & -2 & 0 \end{vmatrix}$

15. $\begin{vmatrix} 2 & 0 & 0 & 0 & 0 \\ 0 & 3 & 0 & 0 & 0 \\ 0 & 0 & 2 & 0 & 0 \\ 0 & 0 & 0 & 1 & 0 \\ 0 & 0 & 0 & 0 & 4 \end{vmatrix}$

16. $\begin{vmatrix} a & a & a & a & a \\ 0 & a & a & a & a \\ 0 & 0 & a & a & a \\ 0 & 0 & 0 & a & a \\ 0 & 0 & 0 & 0 & a \end{vmatrix}$

17. $\begin{vmatrix} 2 & 1 & 0 & 0 & 0 & 0 \\ 1 & 2 & 0 & 0 & 0 & 0 \\ 0 & 0 & 1 & 1 & 0 & 0 \\ 0 & 0 & 2 & 1 & 0 & 0 \\ 0 & 0 & 0 & 0 & 3 & 1 \\ 0 & 0 & 0 & 0 & 1 & 3 \end{vmatrix}$

18. $\begin{vmatrix} \begin{vmatrix} 3 & 1 \\ -2 & 3 \end{vmatrix} & \begin{vmatrix} 7 & 0 \\ 1 & 5 \end{vmatrix} \\ \begin{vmatrix} 3 & 0 \\ 0 & 7 \end{vmatrix} & \begin{vmatrix} 9 & -6 \\ 3 & 5 \end{vmatrix} \end{vmatrix}$

Evaluate the determinants in Problems 19–24 by

a. *Factoring out the greatest common factor from columns or rows.*
b. *Using properties of determinants to write any one row or column with all entries but one as zeros.*
c. *Expanding about that row or column.*

19. $\begin{vmatrix} 4 & 1 & 10 \\ -12 & -2 & -15 \\ 8 & 3 & 20 \end{vmatrix}$

20. $\begin{vmatrix} 5 & 1 & 8 \\ 25 & -3 & 16 \\ -15 & 2 & -24 \end{vmatrix}$

21. $\begin{vmatrix} 25 & 40 & 5 \\ -9 & 0 & 3 \\ 2 & -3 & 5 \end{vmatrix}$

22. $\begin{vmatrix} 18 & 24 & 6 \\ -10 & 0 & 5 \\ 3 & 2 & 5 \end{vmatrix}$

23. $\begin{vmatrix} 5 & 2 & 3 & -3 \\ 25 & 6 & -5 & -9 \\ -30 & 8 & 1 & 12 \\ 15 & 4 & 2 & 9 \end{vmatrix}$

24. $\begin{vmatrix} 5 & -30 & 25 & 15 \\ 2 & 8 & 6 & 4 \\ -5 & 3 & 1 & 2 \\ 12 & -9 & -3 & 6 \end{vmatrix}$

In Problems 25–40, use Cramer's rule to solve the given system. Use a graphing utility to verify the determinants.

25. $x + 2y = 19$
$3x - 7y = -8$

26. $2x - 5y = 13$
$5x + 3y = 17$

27. $2x + 3y = 7$
$5x - 7y = -3$

28. $x - y = 5$
$3x - 2y = 20$

29. $x + y + z = 0$
$2x - y + z = -1$
$-x + 3y - z = -8$

30. $x - y + 2z = 3$
$2x + 3y + z = 9$
$-x - y + 3z = 11$

31. $4x - 5y - 6z = -1$
$x - 2y - 5z = -12$
$2x - y = 7$

32. $x - 3y + z = -2$
$x + 2y = 8$
$2x - y = 1$

33. $x + y + z = 4$
$x - 2y + z = 7$
$x + 3y + 2z = 4$

34. $2x + 2y + 3z = 10$
$4x - y + z = -5$
$5x - 2y + 6z = 1$

35. $x + 2z = 4$
$2y - z = 5$
$2x + 3y = 13$

36. $3x + 2z = 4$
$5x - y = -4$
$4y + 3z = 22$

37. $a - b - 3c - 2d = 2$
$a + 3b - 2c - d = 9$
$3a + b - c + d = 5$
$4a + b + c + 2d = 2$

38. $a - b + c - d = -11$
$2a - 2b - 3c - 3d = 26$
$a + b + c + d = -7$
$3a + 2b + c - d = -9$

39. $a + b + c + d = -1$
$b + c + d = -3$
$c + d = -2$
$a + b + c = 2$

40. $a + b + c + d + e = 1$
$b + c + d + e = 0$
$c + d + e = 1$
$d + e = 0$
$e = 1$

41. Solve for *w:*
$3y - z + 2w = 1$
$x - 2y + 9w = 5$
$5x + y + 3z - w = -4$
$2x + 2z = 3$

42. Solve for *x:*
$4x + y - 3w = 4$
$5x + 2y - 2z + w = 7$
$x - 3y + 2z - 2w = -6$
$3z + 4w = -7$

Application Problems

43. The U.S. Bureau of the Census issued (in 1990) data expressing a relationship between years of education and average yearly earnings.

x (Years of Education)	8	12	16
f (x) (Average Yearly Earnings)	$16,000	$24,000	$36,000

   a. Use the quadratic function $f(x) = ax^2 + bx + c$ to model the data. Use Cramer's rule to determine the values for a, b, and c.

   b. According to the model, how many years of education correspond to the minimum average yearly earnings? What is this minimum? Do you think that the model accurately describes reality?

44. Use the quadratic function $f(x) = ax^2 + bx + c$ to model the following data.

Speed of an Automobile (in miles per hour)	Operating Cost per Mile (in cents)
10	22
20	20
50	20

Use Cramer's rule to determine values for a, b, and c. Then predict the cost of operating the automobile at 75 miles per hour.

Determinants are used to find the area of a triangle whose vertices are given by three points in a rectangular coordinate system. The area of a triangle with vertices (x_1, y_1), (x_2, y_2), and (x_3, y_3) is

$$\text{Area} = \pm\frac{1}{2}\begin{vmatrix} x_1 & y_1 & 1 \\ x_2 & y_2 & 1 \\ x_3 & y_3 & 1 \end{vmatrix}$$

where the symbol $(\pm)$ indicates that the appropriate sign should be chosen to yield a positive area. Use this idea to answer Problems 45–48.

45. a. Use determinants to find the area of the triangle whose vertices are $(3, -5)$, $(2, 6)$, and $(-3, 5)$.

 b. Graph the vertices and triangle in part (a) and then confirm your answer by using the formula for a triangle's area, $A = \frac{1}{2}bh$.

46. Find the area of the triangle whose vertices are $(1, 1)$, $(-2, -3)$, and $(11, -3)$.

47. The point (x_1, y_1) is a vertex of a triangle whose other vertices are $(1, 3)$ and $(5, 1)$. If the area of the triangle is 10 square units, find all points (x_1, y_1) satisfying these conditions.

48. Find the area of a quadrilateral with vertices $(0, 2)$, $(3, 0)$, $(5, 8)$, and $(1, 10)$.

Determinants are used to show that three points lie on the same line (are collinear).

If $\begin{vmatrix} x_1 & y_1 & 1 \\ x_2 & y_2 & 1 \\ x_3 & y_3 & 1 \end{vmatrix} = 0$, then the points (x_1, y_1), (x_2, y_2), and (x_3, y_3) are collinear. If the determinant $\neq 0$, then the points are not collinear. Use this idea to answer Problems 49–50.

49. Are the points $(3, -1)$, $(0, -3)$, and $(12, 5)$ collinear?

50. Are the points $(-4, -6)$, $(1, 0)$, and $(11, 12)$ collinear?

Determinants are used to write an equation of a line passing through two points. An equation of the line passing through the distinct points (x_1, y_1) and (x_2, y_2) is given by

$$\begin{vmatrix} x & y & 1 \\ x_1 & y_1 & 1 \\ x_2 & y_2 & 1 \end{vmatrix} = 0.$$

Use this idea to answer Problems 51–53.

51. Use the determinant to write an equation for the line passing through $(3, -5)$ and $(-2, 6)$. Then expand the determinant, expressing the line's equation in the form $Ax + By + C = 0$.

52. Use the determinant to write an equation for the line passing through $(-1, 3)$ and $(2, 4)$. Then expand the determinant, expressing the line's equation in the form $Ax + By + C = 0$.

53. Verify the determinant form of the line by expanding the determinant and expressing the result in the form

$$y - y_1 = \frac{y_2 - y_1}{x_2 - x_1}(x - x_1).$$

54. Show that the equation

$$\begin{vmatrix} x & y & 1 \\ x_1 & y_1 & 1 \\ 1 & m & 0 \end{vmatrix} = 0$$

represents a line with slope m, passing through (x_1, y_1).

True–False Critical Thinking Problems

55. Which one of the following is true?

a. $\begin{vmatrix} 4a & 4b \\ 4c & 4d \end{vmatrix} = 4 \begin{vmatrix} a & b \\ c & d \end{vmatrix}$

b. $\begin{vmatrix} a & 0 & 0 & b \\ 0 & a & b & 0 \\ 0 & b & a & 0 \\ b & 0 & 0 & a \end{vmatrix} = \begin{vmatrix} a & b \\ b & a \end{vmatrix}^2$

c. Using Cramer's rule, we use $\dfrac{D}{D_z}$ to obtain the value for z.

d. If two rows of a third-order determinant are identical, the value of the determinant is 1.

56. Which one of the following is true?

a. When solving a linear system in two variables using Cramer's rule, there are different determinants in the numerators of x and y. This means that if a system in two variables is solved by Cramer's rule, x and y cannot have the same value.

b. The determinant shown here can be evaluated by taking the product of the entries on the main diagonal.

$\begin{vmatrix} -2 & 0 & 0 & 0 & 0 \\ 0 & 4 & 0 & 0 & 0 \\ 0 & 0 & 1 & 0 & 0 \\ 0 & 0 & 0 & 3 & 0 \\ 0 & 0 & 0 & 0 & -5 \end{vmatrix}$

c. The determinant shown here cannot be evaluated using only one two-by-two determinant.

$\begin{vmatrix} 2 & 3 & -2 \\ 0 & 1 & 3 \\ 0 & 4 & -1 \end{vmatrix}$

d. The inverse of a determinant can be found by taking its matrix.

Technology Problems

In Problems 57–58, use a graphing utility to find the determinant for the given matrix.

57. $\begin{bmatrix} 3 & -2 & -1 & 4 \\ -5 & 1 & 2 & 7 \\ 2 & 4 & 5 & 0 \\ -1 & 3 & -6 & 5 \end{bmatrix}$

58. $\begin{bmatrix} 8 & 2 & 6 & -1 & 0 \\ 2 & 0 & -3 & 4 & 7 \\ 2 & 1 & -3 & 6 & -5 \\ -1 & 2 & 1 & 5 & -1 \\ 4 & 5 & -2 & 3 & -8 \end{bmatrix}$

59. What is the fastest method for solving a linear system with your graphing utility?

Writing in Mathematics

60. A cash prize is divided among 20 people in such a way that the amount that each person receives corresponds to the value of a variable in a linear system of 20 equations in 20 variables. If no technology is available, what method would you use to solve the system? Why? If you are one of the 20 people, what method would you use to determine your portion of the cash prize? Why would you choose this method?

61. If the determinant appearing in the denominator in Cramer's rule is zero, we are not provided with a solution. If this occurs, would it be a good idea to try solving the system using the inverse of the coefficient matrix? Explain your answer.

62. Verify that $(-11, -11, -1)$ is a solution of the system

$$3x - 4y + 4z = 7$$
$$x - y - 2z = 2$$
$$2x - 3y + 6z = 5$$

Explain why this solution cannot be found using Cramer's rule.

63. A student in introductory algebra comes across

$\begin{vmatrix} 5 & 4 \\ -3 & 7 \end{vmatrix}$ and $\begin{bmatrix} 5 & 4 \\ -3 & 7 \end{bmatrix}$

in an article appearing in a professional journal. Explain the differences between these two notations to the student.

64. Suppose that the determinant form for a triangle's area, given just before Problem 45, resulted in zero for three given points. Describe what this means about the points.

65. Write a word problem that can be modeled by a system of at least three linear equations in three variables. Then solve the system using Cramer's rule.

Critical Thinking Problems

66. Solve for y:

$$\begin{vmatrix} y - 4 & 0 & 0 \\ 0 & y + 4 & 0 \\ 0 & 0 & y + 1 \end{vmatrix} = 0$$

67. Solve for y:

$$\begin{vmatrix} y & 0 & 0 & 0 \\ 15 & y - 1 & 0 & 0 \\ -5 & 24 & y + 2 & 0 \\ 24 & -31 & 57 & 2y + 3 \end{vmatrix} = 0$$

68. Show that

$$\begin{vmatrix} a & b & c \\ 0 & d & e \\ 0 & 0 & f \end{vmatrix}$$

is equal to the product of its entries in the main diagonal.

69. Show that

$$\begin{vmatrix} a & 0 & b & 0 \\ 0 & x & 0 & y \\ x & 0 & b & 0 \\ 0 & a & 0 & y \end{vmatrix} = \begin{vmatrix} a & b \\ x & b \end{vmatrix} \cdot \begin{vmatrix} x & y \\ a & y \end{vmatrix}$$

70. A linear system in which all the constant terms are zero is called a *homogeneous system*. What is the solution for an $n \times n$ homogeneous system if Cramer's rule shows that $D \neq 0$? Explain how you arrived at your answer.

71. If

$$\begin{vmatrix} a_1 & b_1 & c_1 \\ a_2 & b_2 & c_2 \\ a_3 & b_3 & c_3 \end{vmatrix} = 12, \text{ what is the value of}$$

$$\begin{vmatrix} a_1 & b_1 & c_1 \\ a_2 - a_1 & b_2 - b_1 & c_2 - c_1 \\ 2a_3 & 2b_3 & 2c_3 \end{vmatrix} ?$$

72. The figure shows that the area A of the triangle with vertices $(0, 0)$, (x_1, y_1) and (x_2, y_2) can be found by subtracting the areas of three right triangles (A_1, A_2, and A_3) from the area of a rectangle. Use this idea to find an equation for A and show that

$$A = \frac{1}{2} \begin{vmatrix} x_1 & y_1 \\ x_2 & y_2 \end{vmatrix}.$$

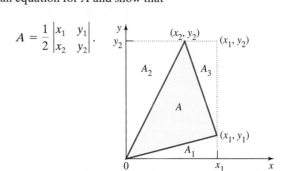

73. The figure represents the area of triangle DBC as A. Use the result of Problem 73 to write the areas of triangles ODB, OCD, and OCB in determinant form. Then show that

$$A = \begin{vmatrix} x_1 & y_1 & 1 \\ x_2 & y_2 & 1 \\ x_3 & y_3 & 1 \end{vmatrix}$$

74. Use an argument like that on pages 616–617 to prove Cramer's rule for three equations in three variables for y and z.

Group Activity Problem

75. We have seen that determinants can be used to solve linear equations, give areas of triangles in rectangular coordinates, and determine equations of lines. Members of the group should research another area in which determinants are used. The group should then report to the class telling about this application.

SECTION 5.7 Partial Fractions

Solutions Manual   **Tutorial**   **Video 10**

Objective

I Find the partial fraction decomposition for a rational expression.

Jiří Kolář "Untitled" 1971. Collection of Meda Mladek, Courtesy of The Czech Museum of Fine Arts.

I Find the partial fraction decomposition for a rational expression.

The theme of reversing a change is found throughout algebra. Changes to a number produced by multiplication can be reversed by division. Raising a number to the nth power can be reversed by taking the nth root of that result. In this section, we study a process for reversing the addition of rational expressions. The process gives rise to linear systems that can be solved by the methods studied throughout the chapter.

Recall that rational expressions are added and subtracted by finding common denominators. For example, consider

$$\frac{3}{x-4} - \frac{2}{x+2}$$

The least common denominator (LCD) is $(x-4)(x+2)$.

$$= \frac{3(x+2)}{(x-4)(x+2)} - \frac{2(x-4)}{(x+2)(x-4)}$$

Rewrite each rational expression with the LCD. Multiply the numerator and denominator by whatever factors are required to form the LCD.

$$= \frac{3(x+2) - 2(x-4)}{(x-4)(x+2)}$$

Subtract the numerators, putting this difference over the LCD.

$$= \frac{x+14}{(x-4)(x+2)}$$

Perform the indicated operations in the numerator.

Since $\dfrac{3}{x-4} - \dfrac{2}{x+2} = \dfrac{x+14}{(x-4)(x+2)}$, it seems reasonable that we can reverse the direction, starting with $\dfrac{x+14}{(x-4)(x+2)}$ and writing the result either as $\dfrac{3}{x-4} - \dfrac{2}{x+2}$ or $\dfrac{3}{x-4} + \dfrac{-2}{x+2}$. Let's see how this is done.

EXAMPLE 1 Reversing the Addition of Rational Expressions

Write as a sum of two rational expressions: $\dfrac{x+14}{(x-4)(x+2)}$.

Solution

$$\frac{x+14}{(x-4)(x+2)} = \frac{A}{x-4} + \frac{B}{x+2}$$

Our goal is to find two numbers A and B so that we can write the given rational expression as a sum.

$$(x-4)(x+2)\frac{x+14}{(x-4)(x+2)} = (x-4)(x+2)\left(\frac{A}{x-4} + \frac{B}{x+2}\right)$$

Clear fractions, multiplying both sides by the LCD.

$$x + 14 = A(x+2) + B(x-4)$$

Multiply as indicated. On the right, distribute $(x-4)(x+2)$ and simplify.

$$x + 14 = Ax + 2A + Bx - 4B$$

Distribute on the right.

$$x + 14 = Ax + Bx + 2A - 4B$$

Let's make both sides look similar by first writing the variable x-terms and then writing the constant terms.

$$1x + 14 = (A + B)x + (2A - 4B)$$

As shown by the arrows, if two polynomials are equal, their coefficients must be equal ($A + B = 1$) and their constant terms must be equal ($2A - 4B = 14$). Consequently, A and B satisfy the following two equations.

$$A + B = 1$$

$$2A - 4B = 14$$

Now we can solve by any method of our choice. We'll use Cramer's rule.

$$A = \frac{\begin{vmatrix} 1 & 1 \\ 14 & -4 \end{vmatrix}}{\begin{vmatrix} 1 & 1 \\ 2 & -4 \end{vmatrix}} = \frac{-4 - 14}{-4 - 2} = \frac{-18}{-6} = 3$$

$$B = \frac{\begin{vmatrix} 1 & 1 \\ 2 & 14 \end{vmatrix}}{\begin{vmatrix} 1 & 1 \\ 2 & -4 \end{vmatrix}} = \frac{14 - 2}{-6} = \frac{12}{-6} = -2$$

Since $A = 3$ and $B = -2$,

$$\frac{x + 14}{(x - 4)(x + 2)} = \frac{A}{x - 4} + \frac{B}{x + 2} = \frac{3}{x - 4} + \frac{-2}{x + 2} \left(\text{or } \frac{3}{x - 4} - \frac{2}{x + 2} \right).$$

Each of the two fractions on the right is called a *partial fraction*. The sum of these fractions is called the *partial fraction decomposition* of the rational expression on the left-hand side. ∎

The same method used on Example 1 will be applied as we break our work into four general cases.

Steps in partial fraction decomposition

1. Set up the partial fraction decomposition with the unknown constants $A, B, C,$ etc.
2. Multiply both sides of the resulting equation by the LCD.
3. Simplify the right-hand side of the equation.
4. With both sides in descending powers, equate coefficients, and equate constant terms.
5. Solve the resulting linear system for $A, B, C,$ etc.
6. Substitute the values for $A, B, C,$ etc., and write the partial fraction decomposition.

Alfredo Castañeda "To Grow" signed and dated 1986, oil on canvas, 31 1/2 × 31 1/2 in. (80 × 80 cm). Mary-Anne Martin / Fine Art, New York.

CASE 1: Decomposing a Rational Expression with Distinct Linear Factors in the Denominator

Consider the rational expression

$$\frac{P(x)}{Q(x)}$$

where the degree of P is less than the degree of Q and $Q(x)$ is the product of distinct linear factors:

$$Q(x) = (a_1 x + b_1)(a_2 x + b_2)(a_3 x + b_3) \cdots (a_n x + b_n).$$

The partial fraction decomposition is of the form

$$\frac{P(x)}{Q(x)} = \frac{A_1}{a_1 x + b_1} + \frac{A_2}{a_2 x + b_2} + \frac{A_3}{a_3 x + b_3} + \cdots + \frac{A_n}{a_n x + b_n}.$$

We can determine the constants $A_1, A_2, A_3, ..., A_n$ exactly as we did in the previous example. Example 2 shows how this is done.

EXAMPLE 2   **Partial Fraction Decomposition with Distinct Linear Factors**

Write the partial fraction decomposition for:

$$\frac{9x^2 - 9x + 6}{2x^3 - x^2 - 8x + 4}$$

Solution

We begin by factoring the denominator.

$$2x^3 - x^2 - 8x + 4 = x^2(2x - 1) - 4(2x - 1)$$ Factor x^2 from the first two terms and -4 from the last two terms.

$$= (2x - 1)(x^2 - 4)$$ Factor $2x - 1$ from the resulting expression.

$$= (2x - 1)(x + 2)(x - 2)$$ Factor $x^2 - 4$, the difference of two squares, using $a^2 - b^2 = (a + b)(a - b)$.

Now we return to the given rational expression, with the denominator factored.

$$\frac{9x^2 - 9x + 6}{(2x - 1)(x + 2)(x - 2)} = \frac{A}{2x - 1} + \frac{B}{x + 2} + \frac{C}{x - 2}$$ A partial fraction is written with a constant numerator for each linear factor in the denominator.

$$(2x - 1)(x + 2)(x - 2)\left[\frac{9x^2 - 9x + 6}{(2x - 1)(x + 2)(x - 2)}\right] = (2x - 1)(x + 2)(x - 2)\left[\frac{A}{2x - 1} + \frac{B}{x + 2} + \frac{C}{x - 2}\right]$$

Multiply both sides by the LCD.

> **S**tudy tip
>
> Take your time with this step. First distribute
>
> $(2x - 1)(x + 2)(x - 2)$
>
> to each of the three terms. Then cancel the common factors in the numerator and denominator in each term.

$$9x^2 - 9x + 6 = A(x + 2)(x - 2) + B(2x - 1)(x - 2) + C(2x - 1)(x + 2)$$

Multiply and simplify.

$$9x^2 - 9x + 6 = A(x^2 - 4) + B(2x^2 - 5x + 2) + C(2x^2 + 3x - 2)$$

Use FOIL to multiply the binomial factors on the right.

$$9x^2 - 9x + 6 = Ax^2 - 4A + 2Bx^2 - 5Bx + 2B + 2Cx^2 + 3Cx - 2C$$

Distribute on the right.

$$9x^2 - 9x + 6 = Ax^2 + 2Bx^2 + 2Cx^2 - 5Bx + 3Cx - 4A + 2B - 2C$$

Write the right side in descending powers of x.

$$9x^2 - 9x + 6 = (A + 2B + 2C)x^2 + (-5B + 3C)x + (-4A + 2B - 2C)$$

Express both sides in the same form.

We are now ready to write a system of equations by equating the coefficients of the like terms. The corresponding coefficients are shown using the same color.

$$A + 2B + 2C = 9$$

$$-5B + 3C = -9$$

$$-4A + 2B - 2C = 6$$

To simplify things, divide by -2.

$$A + 2B + 2C = 9$$

$$-5B + 3C = -9$$

$$2A - B + C = -3$$

We can now solve by any method of our choice. We'll use Gaussian elimination with back-substitution.

$$\begin{bmatrix} 1 & 2 & 2 & | & 9 \\ 0 & -5 & 3 & | & -9 \\ 2 & -1 & 1 & | & -3 \end{bmatrix} \xrightarrow{-2R_1 + R_3} \begin{bmatrix} 1 & 2 & 2 & | & 9 \\ 0 & -5 & 3 & | & -9 \\ 0 & -5 & -3 & | & -21 \end{bmatrix}$$

$$\xrightarrow{-\frac{1}{5}R_2} \begin{bmatrix} 1 & 2 & 2 & | & 9 \\ 0 & 1 & -\frac{3}{5} & | & \frac{9}{5} \\ 0 & -5 & -3 & | & -21 \end{bmatrix} \xrightarrow{5R_2 + R_3} \begin{bmatrix} 1 & 2 & 2 & | & 9 \\ 0 & 1 & -\frac{3}{5} & | & \frac{9}{5} \\ 0 & 0 & -6 & | & -12 \end{bmatrix}$$

$$\xrightarrow{-\frac{1}{6}R_3} \begin{bmatrix} 1 & 2 & 2 & | & 9 \\ 0 & 1 & -\frac{3}{5} & | & \frac{9}{5} \\ 0 & 0 & 1 & | & 2 \end{bmatrix}$$

We obtain

$$A + 2B + 2C = 9$$

$$B - \tfrac{3}{5}C = \tfrac{9}{5}$$

$$C = 2.$$

Using $C = 2$ and back-substitution, we see that $B = 3$, and $A = -1$. Substituting these values into

$$\frac{9x^2 - 9x + 6}{2x^3 - x^2 - 8x + 4} = \frac{A}{2x - 1} + \frac{B}{x + 2} + \frac{C}{x - 2}.$$

We obtain the partial fraction decomposition

$$\frac{9x^2 - 9x + 6}{2x^3 - x^2 - 8x + 4} = \frac{-1}{2x - 1} + \frac{3}{x + 2} + \frac{2}{x - 2}.$$

You can check the decomposition by graphing $y = \dfrac{9x^2 - 9x + 6}{2x^3 - x^2 - 8x + 4}$ and

$y = -\dfrac{1}{2x - 1} + \dfrac{3}{x + 2} + \dfrac{2}{x - 2}$ on the same screen. The graphs should be identical, as shown in Figure 5.17. ■

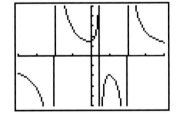

Figure 5.17

The graphs verify the partial fraction decomposition in Example 2.

Discover for yourself

Use an algebraic approach to verify that

$$\frac{9x^2 - 9x + 6}{2x^3 - x^2 - 8x + 4} = -\frac{1}{2x - 1} + \frac{3}{x + 2} + \frac{2}{x - 2}.$$

Combine the three rational expressions on the right using an LCD. You should obtain the expression on the left.

tudy tip

Here's another method for finding the constants in a partial fraction decomposition. This method may involve a bit less work than our approach in Example 2.

When we set up the partial fraction decomposition in Example 2 and multiplied both sides by the LCD, we obtained the equation

$$9x^2 - 9x + 6 = A(x + 2)(x - 2) + B(2x - 1)(x - 2) + C(2x - 1)(x + 2).$$

Since this equation must be true for all x, we can substitute values of x that make the factors $x + 2$, $x - 2$, and $2x - 1$ equal zero.

Set each factor equal to 0.	Solve for x.	Substitute x into $9x^2 - 9x + 6 = A(x + 2)(x - 2) + B(2x - 1)(x - 2) + C(2x - 1)(x + 2)$
$x + 2 = 0$	$x = -2$	$9(-2)^2 - 9(-2) + 6 = A(-2 + 2)(-2 - 2) + B[2(-2) - 1](-2 - 2) + C[2(-2) - 1](-2 + 2)$ $60 = 0A + 20B + 0C$ $3 = B$
$x - 2 = 0$	$x = 2$	$9 \cdot 2^2 - 9 \cdot 2 + 6 = A(2 + 2)(2 - 2) + B(2 \cdot 2 - 1)(2 - 2) + C(2 \cdot 2 - 1)(2 + 2)$ $24 = 0A + 0B + 12C$ $2 = C$
$2x - 1 = 0$	$x = \frac{1}{2}$	$9(\frac{1}{2})^2 - 9(\frac{1}{2}) + 6 = A(\frac{1}{2} + 2)(\frac{1}{2} - 2) + B(2 \cdot \frac{1}{2} - 1)(\frac{1}{2} - 2) + C(2 \cdot \frac{1}{2} - 1)(\frac{1}{2} + 2)$ $\frac{15}{4} = -\frac{15}{4}A + 0B + 0C$ $-1 = A$

We see that $A = -1$, $B = 3$, and $C = 2$. These are precisely the values we found when equating coefficients and solving the resulting linear system in Example 2.

CASE 2: Decomposing a Rational Expression with Linear Factors in the Denominator, Some of Which Are Repeated

Consider, again, the rational expression

$$\frac{P(x)}{Q(x)}.$$

Suppose that $(ax + b)^n$ is a factor of $Q(x)$. This means that the linear factor $ax + b$ is repeated n times. When this occurs, the partial fraction decomposition will contain the following sum of n fractions.

$$\frac{A_1}{ax + b} + \frac{A_2}{(ax + b)^2} + \frac{A_3}{(ax + b)^3} + \cdots + \frac{A_n}{(ax + b)^n}$$

Include one fraction with a constant numerator for each power of $ax + b$.

EXAMPLE 3 **Partial Fraction Decomposition with Repeated Linear Factors**

Write the partial fraction decomposition for:

$$\frac{-10x^2 + 27x - 14}{(x + 2)(x - 1)^3}$$

Solution

Since the factor $x - 1$ is repeated three times, we must include one fraction with a constant numerator for each power of $x - 1$.

$$\frac{-10x^2 + 27x - 14}{(x + 2)(x - 1)^3} = \frac{A}{x + 2} + \frac{B}{x - 1} + \frac{C}{(x - 1)^2} + \frac{D}{(x - 1)^3}$$

Now we clear fractions by multiplying by $(x + 2)(x - 1)^3$, the LCD.

$$(x + 2)(x - 1)^3 \left[\frac{-10x^2 + 27x - 14}{(x + 2)(x - 1)^3} \right] = (x + 2)(x - 1)^3 \left[\frac{A}{x + 2} + \frac{B}{x - 1} + \frac{C}{(x - 1)^2} + \frac{D}{(x - 1)^3} \right]$$

Next we multiply and simplify.

Discover for yourself

We've left out some of the details in this step. Perform the multiplication on the right side and fill in the algebraic details that are not shown.

$$-10x^2 + 27x - 14$$
$$= A(x - 1)^3 + B(x + 2)(x - 1)^2 + C(x + 2)(x - 1) + D(x + 2)$$

Our goal is to write both sides in descending powers of x and then equate coefficients. This means that we must multiply on the right side.

$$-10x^2 + 27x - 14$$
$$= Ax^3 - 3Ax^2 + 3Ax - A + Bx^3 - 3Bx + 2B + Cx^2 + Cx - 2C + Dx + 2D$$

We write the right side in descending powers of x

$$-10x^2 + 27x - 14$$
$$= Ax^3 + Bx^3 - 3Ax^2 + Cx^2 + 3Ax - 3Bx + Cx + Dx - A + 2B - 2C + 2D$$

and express both sides in the same form.

$$0x^3 - 10x^2 + 27x - 14 = (A + B)x^3 + (-3A + C)x^2 + (3A - 3B + C + D)x + (-A + 2B - 2C + 2D)$$

Discover for yourself

Let $x = 1$ and $x = -2$, respectively, in

$-10x^2 + 27x - 14$
$= A(x - 1)^3 + B(x + 2) \cdot$
$(x - 1)^2 + C(x + 2)(x - 1)$
$+ D(x + 2)$.

You should immediately obtain values for D and A. Are there any other values of x that you can substitute to immediately find B and C? If not, how can you use the values for D and A in the system of equations that results when equating coefficients to find values for B and C fairly quickly?

Finally, we equate coefficients.

$$A + B = 0$$
$$-3A + C = -10$$
$$3A - 3B + C + D = 27$$
$$-A + 2B - 2C + 2D = -14$$

Once again, we'll solve this system using Gaussian elimination with back-substitution.

$$\begin{bmatrix} 1 & 1 & 0 & 0 & | & 0 \\ -3 & 0 & 1 & 0 & | & -10 \\ 3 & -3 & 1 & 1 & | & 27 \\ -1 & 2 & -2 & 2 & | & -14 \end{bmatrix} \xrightarrow[\substack{-3R_1 + R_3 \\ 1R_1 + R_4}]{3R_1 + R_2} \begin{bmatrix} 1 & 1 & 0 & 0 & | & 0 \\ 0 & 3 & 1 & 0 & | & -10 \\ 0 & -6 & 1 & 1 & | & 27 \\ 0 & 3 & -2 & 2 & | & -14 \end{bmatrix}$$

$$\xrightarrow{\frac{1}{3}R_2} \begin{bmatrix} 1 & 1 & 0 & 0 & | & 0 \\ 0 & 1 & \frac{1}{3} & 0 & | & -\frac{10}{3} \\ 0 & -6 & 1 & 1 & | & 27 \\ 0 & 3 & -2 & 2 & | & -14 \end{bmatrix}$$

$$\xrightarrow[\substack{-3R_2 + R_4}]{6R_2 + R_3} \begin{bmatrix} 1 & 1 & 0 & 0 & | & 0 \\ 0 & 1 & \frac{1}{3} & 0 & | & -\frac{10}{3} \\ 0 & 0 & 3 & 1 & | & 7 \\ 0 & 0 & -3 & 2 & | & -4 \end{bmatrix}$$

$$\xrightarrow{\frac{1}{3}R_3} \begin{bmatrix} 1 & 1 & 0 & 0 & 0 \\ 0 & 1 & \frac{1}{3} & 0 & -\frac{10}{3} \\ 0 & 0 & 1 & \frac{1}{3} & \frac{7}{3} \\ 0 & 0 & -3 & 2 & -4 \end{bmatrix} \xrightarrow{3R_3 + R_4} \begin{bmatrix} 1 & 1 & 0 & 0 & 0 \\ 0 & 1 & \frac{1}{3} & 0 & -\frac{10}{3} \\ 0 & 0 & 1 & \frac{1}{3} & \frac{7}{3} \\ 0 & 0 & 0 & 3 & 3 \end{bmatrix}$$

$$\xrightarrow{\frac{1}{3}R_4} \begin{bmatrix} 1 & 1 & 0 & 0 & 0 \\ 0 & 1 & \frac{1}{3} & 0 & -\frac{10}{3} \\ 0 & 0 & 1 & \frac{1}{3} & \frac{7}{3} \\ 0 & 0 & 0 & 1 & 1 \end{bmatrix}$$

We obtain

$$A + B = 0$$

$$B + \tfrac{1}{3}C = -\tfrac{10}{3}$$

$$C + \tfrac{1}{3}D = \tfrac{7}{3}$$

$$D = 1$$

Jasper Johns, "Numbers in Color" 1958–59. Encaustic and collage on canvas. 67 × 49 1/2 in. Albright–Knox Art Gallery, Buffalo, New York. Gift of Seymour H. Knox. © Jasper Johns / Licensed by VAGA, New York.

Using $D = 1$ and back-substitution, we see that $C = 2$, $B = -4$, and $A = 4$. The required partial fraction decomposition is

$$\frac{-10x^2 + 27x - 14}{(x + 2)(x - 1)^3} = \frac{A}{x + 2} + \frac{B}{x - 1} + \frac{C}{(x - 1)^2} + \frac{D}{(x - 1)^3}$$

$$= \frac{4}{x + 2} - \frac{4}{x - 1} + \frac{2}{(x - 1)^2} + \frac{1}{(x - 1)^3}$$

Substitute the values for A, B, C, and D. ■

CASE 3: Decomposing a Rational Expression with Irreducible, Nonrepeated Quadratic Factors in the Denominator

Consider, again, the rational expression

$$\frac{P(x)}{Q(x)}.$$

If the complete factorization of $Q(x)$ contains the nonfactorable quadratic $ax^2 + bx + c$, then the corresponding partial fraction decomposition of $\dfrac{P(x)}{Q(x)}$ will have a term with a linear numerator, namely,

$$\frac{Ax + B}{ax^2 + bx + c}.$$

EXAMPLE 4 **Partial Fraction Decomposition with a Quadratic Factor**

Write the partial fraction decomposition for:

$$\frac{3x^2 + 17x + 14}{x^3 - 8}$$

Solution

We begin by factoring the denominator.

$$A^3 - B^3 = (A - B)(A^2 + AB + B^2)$$

$$x^3 - 8 = x^3 - 2^3 = (x - 2)(x^2 + x \cdot 2 + 2^2) = (x - 2)(x^2 + 2x + 4)$$

Now we return to the given rational expression, with the denominator factored.

$$\frac{3x^2 + 17x + 14}{(x - 2)(x^2 + 2x + 4)} = \frac{A}{x - 2} + \frac{Bx + C}{x^2 + 2x + 4}$$

Notice that we put a constant over the linear factor and a linear expression $(Bx + C)$ over the prime quadratic factor.

Next we clear fractions, multiplying both sides by $(x - 2)(x^2 + 2x + 4)$, the LCD.

$$(x - 2)(x^2 + 2x + 4)\left[\frac{3x^2 + 17x + 14}{(x - 2)(x^2 + 2x + 4)}\right] = (x - 2)(x^2 + 2x + 4)\left[\frac{A}{x - 2} + \frac{Bx + C}{x^2 + 2x + 4}\right]$$

Discover for yourself

What value of x can you substitute in

$3x^2 + 17x + 14$
$= A(x^2 + 2x + 4)$
$+ (Bx + C)(x - 2)$

to immediately find the value for A? Do this and determine A. Use the value for A to find values for B and C in the system of equations that results when equating coefficients.

Now we multiply and simplify.

$$3x^2 + 17x + 14 = A(x^2 + 2x + 4) + (Bx + C)(x - 2)$$

As with all partial fraction decompositions, our goal is to write both sides in descending powers of x and then equate coefficients. This means that we must multiply on the right side.

$$3x^2 + 17x + 14 = Ax^2 + 2Ax + 4A + Bx^2 - 2Bx + Cx - 2C$$

We write the right side in descending powers of x

$$3x^2 + 17x + 14 = Ax^2 + Bx^2 + 2Ax - 2Bx + Cx + 4A - 2C$$

and express both sides in the same form.

$$3x^2 + 17x + 14 = (A + B)x^2 + (2A - 2B + C)x + (4A - 2C)$$

Finally, we equate coefficients.

$$A + B = 3$$
$$2A - 2B + C = 17$$
$$4A - 2C = 14$$

Discover for yourself

Take a few minutes to solve the system in Example 4 by the method of your choice, obtaining $A = 5$, $B = -2$, and $C = 3$.

With $A = 5$, $B = -2$, and $C = 3$, the required partial fraction decomposition is

$$\frac{3x^2 + 17x + 14}{x^3 - 8} = \frac{A}{x - 2} + \frac{Bx + C}{x^2 + 2x + 4}$$

$$= \frac{5}{x - 2} + \frac{-2x + 3}{x^2 + 2x + 4} \qquad \text{Substitute the values for } A, B, \text{ and } C.$$

■

CASE 4: Decomposing a Rational Expression with an Irreducible, Repeated Quadratic Factor in the Denominator

Consider once more the rational expression

$$\frac{P(x)}{Q(x)}.$$

Suppose that $(ax^2 + bx + c)^n$ is a factor of $Q(x)$ and that $ax^2 + bx + c$ cannot be factored further. This means that the quadratic factor $ax^2 + bx + c$ is repeated n times. When this occurs, the partial fraction decomposition will contain a linear numerator for each power of $ax^2 + bx + c$.

$$\frac{A_1x + B_1}{ax^2 + bx + c} + \frac{A_2x + B_2}{(ax^2 + bx + c)^2} + \frac{A_3x + B_3}{(ax^2 + bx + c)^3} + \cdots + \frac{A_nx + B_n}{(ax^2 + bx + c)^n}$$

EXAMPLE 5 **Partial Fraction Decomposition with a Repeated Quadratic Factor**

Write the partial fraction decomposition for:

$$\frac{5x^3 - 3x^2 + 7x - 3}{(x^2 + 1)^2}$$

Solution

Since the quadratic factor $x^2 + 1$ is repeated twice, we must include one fraction with a linear numerator for each power of $x^2 + 1$.

$$\frac{5x^3 - 3x^2 + 7x - 3}{(x^2 + 1)^2} = \frac{Ax + B}{x^2 + 1} + \frac{Cx + D}{(x^2 + 1)^2}$$

We clear fractions by multiplying both sides by $(x^2 + 1)^2$, the LCD.

$$(x^2 + 1)^2\left[\frac{5x^3 - 3x^2 + 7x - 3}{(x^2 + 1)^2}\right] = (x^2 + 1)^2\left[\frac{Ax + B}{x^2 + 1} + \frac{Cx + D}{(x^2 + 1)^2}\right]$$

Now we multiply and simplify.

$$5x^3 - 3x^2 + 7x - 3 = (x^2 + 1)(Ax + B) + Cx + D$$

Next we complete the multiplication on the right and express both sides in the same form.

$$5x^3 - 3x^2 + 7x - 3 = Ax^3 + Bx^2 + (A + C)x + (B + D)$$

Finally, we equate coefficients.

$$A = 5$$

$$B = -3$$

$A + C = 7$ With $A = 5$, we immediately obtain $C = 2$.

$B + D = -3$ With $B = -3$, we immediately obtain $D = 0$.

The partial fraction decomposition for the given rational expression is

$$\frac{5x^3 - 3x^2 + 7x - 3}{(x^2 + 1)^2} = \frac{Ax + B}{x^2 + 1} + \frac{Cx + D}{(x^2 + 1)^2}$$

$$= \frac{5x - 3}{x^2 + 1} + \frac{2x}{(x^2 + 1)^2} \qquad \text{Substitute the values for } A, B, C, \text{ and } D.$$ ∎

Up to this point, we have considered situations where the rational expression $\dfrac{P(x)}{Q(x)}$ has a degree of P (the highest power of $P(x)$) less than the degree of Q (the highest power of $Q(x)$). If the degree of Q is not greater than the degree of P, we must first divide to obtain the form

$$\frac{P(x)}{Q(x)} = (\text{polynomial}) + \frac{R(x)}{Q(x)}.$$

We can then decompose $\dfrac{R(x)}{Q(x)}$ into partial fractions. Example 6 illustrates this situation.

EXAMPLE 6   **Partial Fraction Decomposition Where Division Is Necessary**

Write the partial fraction decomposition for:

$$\frac{x^3 + 3x - 2}{x^2 - x}$$

Solution

The degree of the numerator (3) is greater than the degree of the denominator (2), so we begin by dividing.

$$
\begin{array}{r}
x + 1 + \frac{4x - 2}{x^2 - x} \\
x^2 - x \,\overline{)\, x^3 + 0x^2 + 3x - 2} \\
\underline{x^3 - \ x^2} \ \ \ \ \ \ \ \ \ \\
x^2 + 3x \ \ \ \ \\
\underline{x^2 - \ x} \ \ \\
4x - 2
\end{array}
$$

We now apply partial fraction decomposition to $\dfrac{4x - 2}{x^2 - x}$ or $\dfrac{4x - 2}{x(x - 1)}$.

$$\frac{4x - 2}{x(x - 1)} = \frac{A}{x} + \frac{B}{x - 1} \qquad \text{A partial fraction is written with a constant numerator for each linear factor in the denominator.}$$

$$x(x - 1)\left[\frac{4x - 2}{x(x - 1)}\right] = x(x - 1)\left[\frac{A}{x} + \frac{B}{x - 1}\right] \qquad \text{Multiply both sides by the LCD.}$$

$$4x - 2 = A(x - 1) + Bx \qquad \text{Multiply and simplify.}$$

$$4x - 2 = (A + B)x - A$$

Complete the multiplication and express both sides in the same form.

Now we equate coefficients.

$$A + B = 4$$

$$-A = -2$$

We see that $A = 2$ and, using back-substitution, $B = 2$. (Don't you wish that every linear system was as easy as this to solve? By contrast, see Problem 43 in Problem Set 5.7.)

The partial fraction decomposition for the given rational expression is

$$\frac{x^3 + 3x - 2}{x^2 - x} = x + 1 + \frac{4x - 2}{x^2 - x}$$

This was the result of the long division.

$$= x + 1 + \frac{A}{x} + \frac{B}{x - 1}$$

$$= x + 1 + \frac{2}{x} + \frac{2}{x - 1}$$

Substitute the values for A and B. ∎

Here's a summary of the main points to remember for partial fraction decomposition.

Decomposition of $\dfrac{P(x)}{Q(x)}$ into partial fractions

1. If the degree of $P(x)$ is less than the degree of $Q(x)$, factor $Q(x)$ into prime factors that are either linear or quadratic.

  a. Include one partial fraction with a constant numerator for each distinct linear factor.

  b. Include one partial fraction with a constant numerator for each power of a repeated linear factor.

  c. Include one partial fraction with a linear numerator for each distinct quadratic factor.

  d. Include one partial fraction with a linear numerator for each power of a repeated quadratic factor.

2. Set up the partial fraction decomposition with the unknown constants, multiply by the LCD, write both sides of the equation in the same form, and equate coefficients. Solve the resulting linear system for the constants and substitute these values into the partial fraction decomposition.

Dorothea Rockburne "Capernaum Gate" 1984, oil on gessoed linen, 92 in. h. × 85 in. w. × 4 in. d. © 1998 Dorothea Rockburne / Artists Rights Society (ARS), New York.

3. If the degree of $P(x)$ is greater than or equal to the degree of $Q(x)$, use division and express $\dfrac{P(x)}{Q(x)}$ as

$(\text{polynomial}) + \dfrac{R(x)}{Q(x)}$.

Apply partial fraction decomposition to $\dfrac{R(x)}{Q(x)}$.

PROBLEM SET 5.7

Practice Problems

In Problems 1–38, find the partial fraction decomposition for the given rational expression. Check your result using a graphing utility.

1. $\dfrac{4}{x^2 - 4}$

2. $\dfrac{1}{x^2 + x}$

3. $\dfrac{7x - 10}{(x - 2)(x - 1)}$

4. $\dfrac{7x + 2}{(x + 2)(x - 4)}$

5. $\dfrac{5x + 7}{x^2 + 2x - 3}$

6. $\dfrac{3x - 13}{6x^2 - x - 12}$

7. $\dfrac{4x^2 + 13x - 9}{x^3 + 2x^2 - 3x}$

8. $\dfrac{4x^2 - 5x - 15}{x^3 - 4x^2 - 5x}$

9. $\dfrac{6x - 11}{(x - 1)^2}$

10. $\dfrac{x}{(x + 1)^2}$

11. $\dfrac{x^2 - 6x + 3}{(x - 2)^3}$

12. $\dfrac{2x^2 + 8x + 3}{(x + 1)^3}$

13. $\dfrac{x^2 + 2x + 7}{x(x - 1)^2}$

14. $\dfrac{3x^2 + 49}{x(x + 7)^2}$

15. $\dfrac{5x^2 + 21x + 4}{(x + 1)^2(x - 3)}$

16. $\dfrac{x}{(x + 1)(x + 2)^2}$

17. $\dfrac{4x^2 - 7x - 3}{x^3 - x}$

18. $\dfrac{2x^2 - 18x - 12}{x^3 - 4x}$

19. $\dfrac{5x^2 - 6x + 7}{(x - 1)(x^2 + 1)}$

20. $\dfrac{5x^2 - 9x + 19}{(x - 4)(x^2 + 5)}$

21. $\dfrac{6x^2 - x + 1}{x^3 + x^2 + x + 1}$

22. $\dfrac{3x^2 - 2x + 8}{x^3 + 2x^2 + 4x + 8}$

23. $\dfrac{5x^2 + 6x + 3}{(x + 1)(x^2 + 2x + 2)}$

24. $\dfrac{9x + 2}{(x - 2)(x^2 + 2x + 2)}$

25. $\dfrac{x^4 + 2x^2 - x - 1}{x(x^2 + 1)^2}$

26. $\dfrac{3x^4 + 12x^2 - 3x + 9}{x(x^2 + 3)^2}$

27. $\dfrac{x^4 + 4x^2 - x}{(x - 1)(x^2 + 1)^2}$

28. $\dfrac{x^3 + 2x - 1}{(x + 1)(x^2 + 1)^2}$

29. $\dfrac{x^3 - 4x^2 + 9x - 5}{(x^2 - 2x + 3)^2}$

30. $\dfrac{3x^3 - 6x^2 + 7x - 2}{(x^2 - 2x + 2)^2}$

31. $\dfrac{3x^4 - 9x^3 + 14x^2 - 9x + 2}{(x - 1)(x^2 - 2x + 2)^2}$

32. $\dfrac{x^5 - 2x^4 + 2x^3 + x - 2}{x^2(x^2 + 1)^2}$

33. $\dfrac{1}{(x^2 + x + 1)(x^2 + x + 2)}$

34. $\dfrac{x^3 - 2x^2 + 4x - 2}{(x^2 + 4)(x^2 + 1)}$

35. $\dfrac{x^5 - 2x^4 + x^3 + x + 5}{x^3 - 2x^2 + x - 2}$

36. $\dfrac{x^4 + 2x^3 - 4x^2 + x - 3}{x^2 - x - 2}$

37. $\dfrac{x^3 + 8}{x^3 + 4x}$

38. $\dfrac{x^3}{(x - 1)^2}$

Application Problems

39. Find the partial fraction decomposition for $\dfrac{1}{x^2 + x}$ and use the result to find the following sum:

$$\dfrac{1}{1 \cdot 2} + \dfrac{1}{2 \cdot 3} + \dfrac{1}{3 \cdot 4} + \cdots + \dfrac{1}{99 \cdot 100}.$$

40. Find the partial fraction decomposition for $\dfrac{2}{x^2 + 2x}$ and use the result to find the following sum:

$$\dfrac{2}{1 \cdot 3} + \dfrac{2}{3 \cdot 5} + \dfrac{2}{5 \cdot 7} + \cdots + \dfrac{2}{99 \cdot 101}.$$

True–False Critical Thinking Problems

41. Which one of the following correctly gives the first step for finding the partial fraction decomposition?

a. $\dfrac{x}{(x-1)(x^2-x-6)} = \dfrac{A}{x} + \dfrac{Bx+C}{x^2-x-6}$

b. $\dfrac{1}{(x^2+x+5)^3} = \dfrac{Ax+B}{x^2+x+5} + \dfrac{Cx+D}{(x^2+x+5)^2}$

$\quad + \dfrac{Ex+F}{(x^2+x+5)^3}$

c. $\dfrac{3x^2-5x+2}{x^3-8} = \dfrac{A}{x-2} + \dfrac{Bx+C}{x^2+4x+4}$

d. $\dfrac{x^3-7x^2+17x-17}{(x-3)(x-2)} = \dfrac{A}{x-3} + \dfrac{B}{x-2}$

42. Which one of the following is true?

a. When using partial fraction decomposition, if the degree of the denominator is greater than the degree of the numerator, long division must first be used.

b. Partial fraction decomposition reverses the process of multiplying and dividing rational expressions.

c. When using partial fraction decomposition, if a quadratic factor such as $x^2-5x-14$ appears in the denominator, then the corresponding partial fraction decomposition will have a term with a linear numerator, namely,

$\dfrac{Ax+B}{x^2-5x-14}.$

d. There are some rational expressions that cannot be decomposed further using partial fraction decomposition.

Technology Problems

43. Find the partial fraction decomposition for

$$\frac{x^5 - 3x^2 + 12x - 1}{x^3(x^2+x+1)(x^2+2)^3}.$$

Solve the resulting linear systems with your graphing utility.

44. Find the partial fraction decomposition for $\dfrac{1}{ax-bx^2}$.

Then assign values to a and b and verify your result with a graphing utility.

45. Find the partial fraction decomposition for $\dfrac{x}{(ax+b)^2}$.

Then assign values to a and b and verify your result with a graphing utility.

46. We mentioned that you can check decomposition by graphing

y_1 = the given rational expression

and y_2 = its partial fraction decomposition

showing that the two graphs on the same screen of a graphing utility are identical. The problem with this is that it might take you a while to determine the appropriate range setting. Can you suggest another method for showing that $y_1 = y_2$ using your graphing utility? Use this method to check the results on some of the decompositions in Problems 1–38.

Writing in Mathematics

47. How can you verify your result for the partial fraction decomposition for a given rational expression without using a graphing utility?

48. If x^2 is a factor of the denominator, we considered this to be a repeated linear factor, using $\dfrac{A}{x} + \dfrac{B}{x^2}$. However,

x^2 can be viewed as a distinct quadratic factor, leading to the partial fraction $\dfrac{Ax+B}{x^2}$. Show that the two forms are the same. Then describe why the first form is preferable.

Critical Thinking Problems

49. If a, b, and c are constants, find the partial fraction decomposition for

$$\frac{ax+b}{(x-c)^2}.$$

50. Find the partial fraction decomposition for

$$\frac{4x^2 + 5x - 9}{x^3 - 6x - 9}.$$

51. Find the partial fraction decomposition for

$$\frac{\ln x^{11} + 2}{(\ln x + 1)(\ln x^2 - 1)}.$$

Group Activity Problem

52. Three of the options in Problem 41 contain common errors that can occur when applying partial fraction decomposition to a rational expression. In your group, discuss these common errors and then suggest strategies for avoiding them.

S E C T I O N   5 . 8   **Systems of Linear Inequalities and Linear Programming**

Solutions   Tutorial   Video
Manual   10

Objectives

1 Determine whether an ordered pair is a solution of an inequality in two variables.

2 Graph a linear inequality in two variables.

3 Graph a system of linear inequalities.

4 Use linear programming to solve problems.

Much of our work in this chapter has involved solving systems of linear equations. In this section, our focus is on systems containing linear inequalities. Such systems arise in linear programming, a method for solving problems in which a particular quantity that must be maximized or minimized is limited in some ways. Linear programming is one of the most widely used tools in management science, helping businesses allocate resources on hand to manufacture a particular array of products that will maximize profit.

1 Determine whether an ordered pair is a solution of an inequality in two variables.

Solutions of Inequalities in Two Variables

A *linear inequality in two variables* (*x* and *y*) is an inequality that can be written in one of the following forms

$$Ax + By > C, \quad Ax + By \geqslant C, \quad Ax + By < C, \quad \text{or} \quad Ax + By \leqslant C$$

where A, B, and C are real numbers, and A and B are not both zero. Examples of linear inequalities in two variables are $x + y > 4$, $2x - 3y \leqslant 6$, $x \geqslant 3$, and $y < -2$.

An ordered pair (x_1, y_1) is a *solution* to an inequality in two variables if the inequality is true when x_1 is substituted for x and y_1 is substituted for y. Under these conditions, we say that (x_1, y_1) *satisfies* the inequality.

EXAMPLE 1   **Deciding Whether Ordered Pairs Are Solutions of Inequalities**

Determine whether each ordered pair satisfies the inequality $2x - 3y \leqslant 6$.

a. $(2, 3)$ **b.** $(5, -2)$ **c.** $(0, 0)$

Solution

The three ordered pairs are graphed and labeled in Figure 5.18. We can substitute the *x*- and *y*-coordinate of each pair into the given inequality.

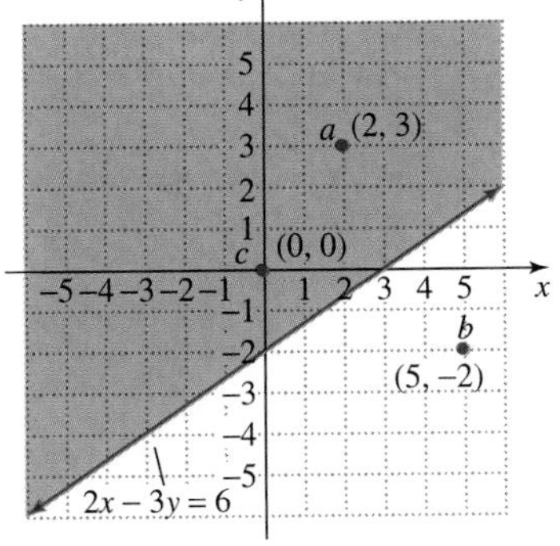

Figure 5.18

Which ordered pairs satisfy $2x - 3y \leqslant 6$?

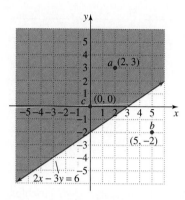

Figure 5.18 (repeated)

Which ordered pairs satisfy $2x - 3y \leq 6$?

a. $(2, 3)$:

$$2x - 3y \leq 6$$

$$2(2) - 3(3) \overset{?}{\leq} 6$$

$$4 - 9 \overset{?}{\leq} 6$$

$$-5 \leq 6 \qquad \text{True}$$

Thus, $(2, 3)$ is a solution.

b. $(5, -2)$:

$$2x - 3y \leq 6$$

$$2(5) - 3(-2) \overset{?}{\leq} 6$$

$$10 + 6 \overset{?}{\leq} 6$$

$$16 \leq 6 \qquad \text{False}$$

Thus, $(5, -2)$ is not a solution.

c. $(0, 0)$:

$$2x - 3y \leq 6$$

$$2(0) - 3(0) \overset{?}{\leq} 6$$

$$0 - 0 \overset{?}{\leq} 6$$

$$0 \leq 6 \qquad \text{True}$$

Thus, $(0, 0)$ is a solution.

The ordered pairs that satisfy the given inequality are those in parts (a) and (c). In Figure 5.18, these ordered pairs all lie in the same *half-plane* that is formed when the line whose equation is $2x - 3y = 6$ divides the plane in two. This observation will be helpful in obtaining the graph of a linear inequality. ∎

2 Graph a linear inequality in two variables.

The Graph of a Linear Inequality in Two Variables

The graph of a linear inequality in two variables is the collection of all points in the rectangular coordinate system whose ordered pairs satisfy the inequality. The graph consists of an entire region rather than a line.

To sketch the graph of a linear inequality such as $2x - 3y \leq 6$ (the inequality of Example 1), we begin by graphing the *corresponding linear equation*

$$2x - 3y = 6.$$

Then we draw a *dashed line* for the inequalities $<$ and $>$ or a *solid line* for the inequalities $\leq$ and $\geq$. The line separates the plane into two half-planes. In each half-plane, one of the following statements is true:

1. All points in the half-plane satisfy the inequality.
2. No points in the half-plane satisfy the inequality.

We can determine whether the points in an entire half-plane are solutions of the inequality by testing just one point in the region. Substitute the coordinates of the test point into the inequality. If a true statement results, shade the half-plane containing this test point. If a false statement results, shade the half-plane not containing this test point.

These ideas are illustrated in the next example.

EXAMPLE 2 Graphing a Linear Inequality in Two Variables

Graph: $2x - 3y \leq 6$

Solution

The graph of the corresponding equation

$$2x - 3y = 6$$

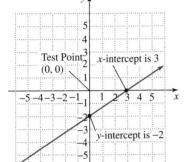

Figure 5.19

The graph of $2x - 3y = 6$

is the line shown in Figure 5.19. Remember that to find the *x*-intercept, we set $y = 0$.

$$2x = 6$$

$$x = 3$$

The *x*-intercept is 3.
To find the *y*-intercept, we set $x = 0$.

$$-3y = 6$$

$$y = -2$$

The *y*-intercept is -2.
The graph is a solid line since equality is included in $2x - 3y \leq 6$. To find which half-plane belongs to the graph, we test a point from either half-plane. The origin, $(0, 0)$, is easiest.

$$2x - 3y \leq 6 \quad \text{This is the given inequality.}$$

$$2 \cdot 0 - 3 \cdot 0 \overset{?}{\leq} 6 \quad \text{Test } (0, 0) \text{ by substituting 0 for } x \text{ and 0 for } y.$$

$$0 \leq 6 \quad \text{The true statement indicates that } (0, 0) \text{ satisfies the inequality.}$$

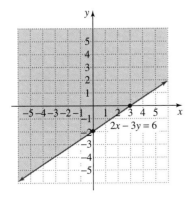

Figure 5.20

The graph of $2x - 3y \leq 6$

Since $(0, 0)$ satisfies the inequality, *all* points in the half-plane above the line also satisfy $2x - 3y \leq 6$. (We checked one of these, $(2, 3)$, in Example 1.) The graph consists of the line and the half-plane lying above the line, shown in Figure 5.20. ∎

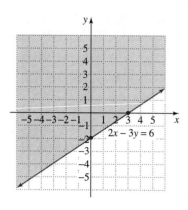

Discover for yourself

Graph $2x - 3y \leq 6$ by first solving for y and writing the inequality in slope-intercept form.

Here's what you should have obtained when working the Discover for yourself exercise and writing the inequality in slope-intercept ($y = mx + b$) form.

$-3y \leq -2x + 6$ Isolate the y-term by subtracting $2x$ from both sides.

$y \geq \frac{2}{3}x - 2$ Solve for y by dividing both sides by -3. (y-intercept $= -2$; slope $= \frac{2}{3}$)

With the symbol $\geq$, you can see that all solution points lie *on* or *above* the line, shown in Figure 5.20 repeated on the left from page 639.

Study tip

If an inequality is in slope-intercept form, it is not necessary to use a test point to obtain the graph.

Inequality	Description of the Graph
$y < mx + b$	Half-plane *below* the line $y = mx + b$
$y \leq mx + b$	Half-plane *below* the line $y = mx + b$ and the boundary line $y = mx + b$
$y > mx + b$	Half-plane *above* the line $y = mx + b$
$y \geq mx + b$	Half-plane *above* the line $y = mx + b$ and the boundary line $y = mx + b$

It is also not necessary to use test points when boundary lines are horizontal or vertical.

Inequality	Graph		Inequality	Graph
$x < a$	The half-plane to the *left* of the line $x = a$		$y > b$	The half-plane *above* the line $y = b$

3 Graph a system of linear inequalities.

Systems of Linear Inequalities in Two Variables

The following is an example of a system of linear inequalities in two variables:

$$x + 2y \leq 8$$
$$3x - 2y \geq 0$$

An ordered pair (a, b) is a solution of the system if a true statement results when a and b are substituted for x and y, respectively, in each inequality in the system.

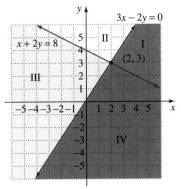

Figure 5.21

The two lines divide the plane into four regions.

The graphs of the corresponding equations for this system, $x + 2y = 8$ and $3x - 2y = 0$, are shown in Figure 5.21. These lines divide the plane into four regions, numbered I, II, III, and IV. All points in region IV are solutions to the system. Two such points are $(4, -2)$ and $(0, -3)$.

System of Inequalities	Solution: $(4, -2)$	Solution: $(0, -3)$
$x + 2y \leq 8$	$4 + 2(-2) \leq 8; 0 \leq 8$ is true.	$0 + 2(-3) \leq 8; -6 \leq 8$ is true.
$3x - 2y \geq 0$	$3(4) - 2(-2) \geq 0; 16 \geq 0$ is true.	$3(0) - 2(-3) \geq 0; 6 \geq 0$ is true.

The solution set to a system of inequalities is the intersection of the solution sets to the individual inequalities. The graphs of the corresponding equations will divide the plane into regions. Either all points in each region are solutions of the system of inequalities or no points in each region are solutions. Consequently, we can determine whether the points in a region satisfy the system by testing one point in the region. This idea is illustrated in our next example.

EXAMPLE 3 **Graphing a System of Inequalities in Two Variables**

Graph the solution set:

$$x + y \geq 1$$
$$x + 2y \leq 10$$
$$2x - y \leq 5$$

Solution

We begin by graphing the corresponding equations:

$x + y = 1$ x-intercept: $x + 0 = 1; x = 1$
              y-intercept: $0 + y = 1; y = 1$

$x + 2y = 10$ x-intercept: $x + 2 \cdot 0 = 10; x = 10$
              y-intercept: $0 + 2y = 10; y = 5$

$2x - y = 5$ x-intercept: $2x - 0 = 5; x = \frac{5}{2}$
              y-intercept: $2 \cdot 0 - y = 5; y = -5$

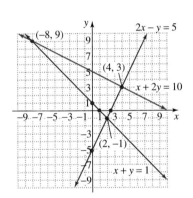

Figure 5.22

Graphs of 3 lines and their intersection points

The graphs of the lines are shown in Figure 5.22.

Also shown are the intersection points, obtained by taking pairs of equations as follows.

$(2, -1)$	$(4, 3)$	$(-8, 9)$
By addition:	**By Cramer's rule:**	**By addition:**

$$\begin{array}{ll} 2x - y = 5 & 2x - y = 5 \\ x + y = 1 & x + 2y = 10 \end{array}$$

Add: $3x = 6$

$x = 2$

$2x - y = 5$
$x + 2y = 10$

$$x = \dfrac{\begin{vmatrix} 5 & -1 \\ 10 & 2 \end{vmatrix}}{\begin{vmatrix} 2 & -1 \\ 1 & 2 \end{vmatrix}} = \dfrac{20}{5} = 4$$

$x + 2y = 10 \xrightarrow{x(-1)} -x - 2y = -10$

$x + y = 1 \xrightarrow{\text{No change}} x + y = 1$

Add: $-y = -9$

$y = 9$

By back-substitution:	**By back-substitution:**	**By back-substitution:**
$y = -1$	$y = 3$	$x = -8$

One way to determine which region belongs to the solution set is to solve each inequality for y.

$x + y \geq 1$	$x + 2y \leq 10$	$2x - y \leq 5$
	$2y \leq -x + 10$	$-y \leq -2x + 5$
$y \geq -x + 1$	$y \leq -\frac{x}{2} + 5$	$y \geq 2x - 5$
Region lies above $x + y = 1$.	Region lies below $x + 2y = 10$.	Region lies above $2x - y = 5$.

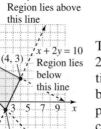

2x − y = 5
Region lies above this line

(−8, 9)

(4, 3)
x + 2y = 10
Region lies below this line

(2, −1)

x + y = 1
Region lies above this line

Figure 5.23

The graph of $x + y \geq 1$ and $x + 2y \leq 10$ and $2x - y \leq 5$

The intersection of the regions above $x + y = 1$, below $x + 2y = 10$, and above $2x - y = 5$ is shown in Figure 5.23. The line segments that belong to the solution set are shown as solid lines. Be sure to change any portion of any solid boundary line not in the intersection of the solutions from solid to dashed. The points $(4, 3)$, $(-8, 9)$ and $(2, -1)$ are the *vertices* of the solution region. Let's check the solution region by taking one point in the region to show that it satisfies all three inequalities. Using $(1, 1)$, we obtain

$x + y \geq 1$	$x + 2y \leq 10$	$2x - y \leq 5$
$1 + 1 \overset{?}{\geq} 1$	$1 + 2(1) \overset{?}{\leq} 10$	$2(1) - 1 \overset{?}{\leq} 5$
$2 \geq 1$ True	$3 \leq 10$ True	$1 \leq 5$ True

■

EXAMPLE 4 Graphing a System of Inequalities in Two Variables

Graph the solution set:

$$x - y < 2$$
$$-2 \leq x < 4$$
$$y < 3$$

Solution

We can begin by graphing the corresponding equations, $x - y = 2$, $x = -2$, $x = 4$, and $y = 3$. For variety, let's consider each of the three inequalities, a technique that you might want to use when the graphs of some of the corresponding equations are horizontal or vertical lines. We'll graph the region associated with each inequality, remembering that the solution set to the system is the intersection of these regions.

We begin by graphing $x - y < 2$, the first given inequality. The corresponding equation, $x - y = 2$, has a graph with an x-intercept $= 2$ and a y-intercept $= -2$. Solving $x - y < 2$ for y gives $y > x - 2$. This means that the graph includes the half-plane above $x - y = 2$, shown in Figure 5.24 as a dashed line.

Now let's consider the second given inequality, $-2 \leq x < 4$. The corresponding equations, $x = -2$ and $x = 4$, are lines parallel to the y-axis. The line of $x = 4$ is not included. The region described by $-2 \leq x < 4$ includes the por-

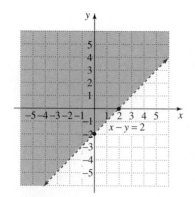

x − y = 2

Figure 5.24

The graph of $x - y < 2$

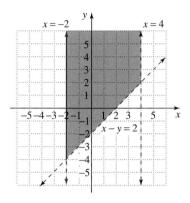

Figure 5.25

The graph of $x - y < 2$ and $-2 \leq x < 4$

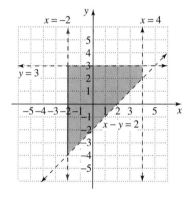

Figure 5.26

The graph of $x - y < 2$ and $-2 \leq x < 4$ and $y < 3$

4 Use linear programming to solve problems.

tion of the plane between the lines of $x = -2$ and $x = 4$. We've added this region to Figure 5.24, intersecting the region between the vertical lines with the region in Figure 5.24. The resulting region is shown in Figure 5.25.

Finally, let's consider the third given inequality, $y < 3$. The corresponding equation, $y = 3$, is a line parallel to the x-axis. The region does not include the line of $y = 3$, but does include the half-plane below the line. The solution is shown by the graph in Figure 5.26, which is the intersection of the half-plane and the region in Figure 5.25. ∎

In Example 4, no point of intersection of the lines is a vertex of the bounded region representing the solution set. This is because the dashed lines are not part of the solution set. This was not the case in Example 3, where the solid lines were part of the solution set so each point of intersection of a pair of boundary lines corresponded to a vertex.

An Application: Linear Programming

Linear programming is a method for solving problems in which a particular quantity that must be maximized or minimized is limited in some ways. For example, suppose that some supplies must be quickly sent to the survivors of a catastrophic earthquake. We want to maximize the amount of bottled water and the number of medical kits that can be shipped to the victims. However, the planes that ship these supplies have a limited capacity. Let's see how systems of inequalities can be used to analyze and solve this problem.

Objective Functions and Constraints

Objective functions and constraints are the crucial elements of linear programming.

> **The vocabulary of linear programming**
>
> *Objective Function:* An algebraic expression in two or more variables modeling a quantity that must be maximized or minimized
> *Constraints:* A system of inequalities that models the limitations in the situation

EXAMPLE 5   An Example of an Objective Function

Bottled water and medical supplies are to be shipped to victims of an earthquake by plane. Each container of bottled water will serve 10 people and each medical kit will aid 6 people. If x represents the number of bottles of water to be shipped and y represents the number of medical kits, write the objective function that models the number of people that can be helped.

ENRICHMENT ESSAY

Cubist Art and Systems of Linear Inequalities

The logic and order underlying graphs of systems of inequalities appears in an artistic style called cubism. Straight lines, a narrow range of color, and slicing of the plane into geometric shapes suggest a highly impersonal and rational approach to the figure. Beyond the superficial world of appearance is the same kind of mathematical order shaping regions graphed in rectangular coordinates.

Juan Gris (Jose Victoriano Gonzalez), (Spanish 1887–1927) "Portrait of Pablo Picasso" 1912, oil on canvas, 74.1 × 93 cm. Gift of Leigh B. Block, 1958.525. Photograph © 1997, The Art Institute of Chicago, All Rights Reserved.

Discover for yourself

Familiarize yourself with the problem's conditions by selecting particular values for the number of water bottles and medical kits.

Water: 10 people per bottle   Medical kits: 6 people per kit

2 bottles of water and 3 medical kits will help $10(2) + 3(6)$ people.
100 bottles of water and 75 medical kits will help ____ people.
3000 bottles of water and 2500 medical kits will help ____ people.

Use the pattern to write an algebraic expression for the number of people helped by x bottles of water and y medical kits.

Solution

Since each bottle of water serves 10 people and each medical kit aids 6 people, we have

The number of people helped	is	10 times the number of bottles of water	plus	6 times the number of medical kits.
	=	$10x$	+	$6y$

Using z to represent the objective function, we have

$$z = 10x + 6y.$$

Unlike the functions that we have seen so far, the objective function has two independent variables.

As you observed in the Discover for yourself, it appears that x and y should be increased without limit until all victims of the earthquake are helped. However, the planes are bound by certain limitations, which are the constraints of this problem. These limitations are considered in Examples 6 and 7.

EXAMPLE 6   An Example of a Constraint

Each plane can carry no more than 80,000 pounds. The bottled water weighs 20 pounds per container and each medical kit weighs 10 pounds. If x represents the number of bottles of water to be shipped and y represents the number of medical kits, write an inequality that models this constraint.

Solution

Since each plane can carry no more than 80,000 pounds, we have

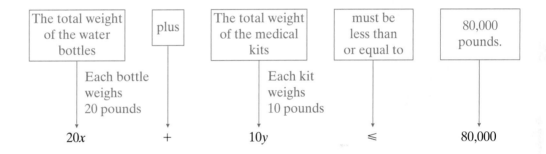

The plane's weight constraint is modeled by the inequality $20x + 10y \leq 80{,}000$. ∎

In addition to a weight constraint on its cargo, each plane has a limited amount of space in which to carry supplies. Example 7 demonstrates how to express this constraint.

EXAMPLE 7   An Example of a Constraint

The planes in our problem can carry a total volume for supplies that does not exceed 6000 cubic feet. Each water bottle is 1 cubic foot, and each medical kit also has a volume of 1 cubic foot. With x still representing the number of water bottles and y the number of medical kits, write an inequality that models this second constraint.

Solution

Since each plane can carry a volume for supplies that does not exceed 6000 cubic feet, we have

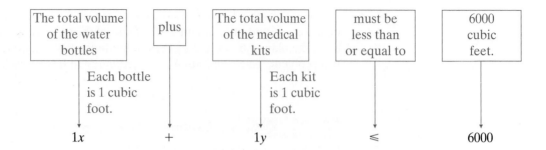

The total volume of the water bottles	plus	The total volume of the medical kits	must be less than or equal to	6000 cubic feet.
Each bottle is 1 cubic foot.		Each kit is 1 cubic foot.		
$1x$	$+$	$1y$	$\leq$	6000

The plane's volume constraint is modeled by the inequality $x + y \leq 6000$. ∎

In summary, here's what we have modeled in this aid-to-earthquake-victims situation:

$$z = 10x + 6y$$

This is the objective function describing the number of people helped with x bottles of water and y medical kits.

$$20x + 10y \leq 80{,}000$$
$$x + y \leq 6000$$

These are the constraints based on each plane's weight and volume limitations.

Linear Programming

The problem in the situation described above is to maximize the number of earthquake victims who can be helped, subject to the planes' weight and volume constraints. The process of solving this problem is called linear programming, based on a theorem that was proven during World War II.

René Magritte "La clef des Champs" (The Key of the Fields), 1948, gouache. © 1998 C. Herscovici, Brussels / Artists Rights Society (ARS), New York. Private Collection. Art Resource, NY.

Solving a linear programming problem

Let $z = ax + by$ be an objective function, and suppose z is subject to some constraints on x and y. If a maximum or minimum value of z exists, it can be determined as follows:

1. Graph the system of inequalities representing the constraints.
2. Find the value of the objective function at each corner, or *vertex*, of the graphed region. The maximum and minimum of the objective function occur at the corner points.

EXAMPLE 8 Solving the Earthquake Victim Problem

Solve the problem of determining how many bottles of water and how many medical kits should be sent on each plane to maximize the number of earthquake victims who can be helped.

Solution

We must maximize $z = 10x + 6y$ subject to the constraints

$$20x + 10y \leq 80{,}000$$
$$x + y \leq 6000.$$

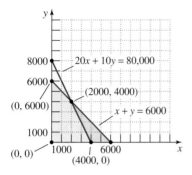

Figure 5.27

The region in quadrant I representing the constraints
$20x + 10y \leqslant 80{,}000$
$x + y \leqslant 6000$

Since x (the number of bottles of water per plane) and y (the number of medical kits per plane) must be positive, we need only graph the system of inequalities in quadrant I. To graph the inequality $20x + 10y \leqslant 80{,}000$, we graph the equation $20x + 10y = 80{,}000$ as a solid line. Setting $y = 0$, we find that the x-intercept is 4000. Setting $x = 0$, we find that the y-intercept is 8000. Using $(0, 0)$ as a test point, we see that the inequality is satisfied, so we graph below the line, shown in Figure 5.27.

Now we graph $x + y \leqslant 6000$ by first graphing $x + y = 6000$ as a solid line. Setting $y = 0$, we find that the x-intercept is 6000. Setting $x = 0$, we find that the y-intercept is 6000. Using $(0, 0)$ as a test point, we see that the inequality is satisfied, so we graph below the line, shown also in Figure 5.27.

Now we'll use the addition method to find where the lines $20x + 10y = 80{,}000$ and $x + y = 6000$ intersect.

$$
\begin{array}{llll}
20x + 10y = 80{,}000 & \xrightarrow{\text{No change}} & 20x + 10y = & 80{,}000 \\
x + y = 6000 & \xrightarrow{\text{Multiply by } -10.} & -10x - 10y = & -60{,}000 \\
& \text{Add:} & \overline{10x = } & 20{,}000 \\
& & x = & 2000
\end{array}
$$

Back-substituting 2000 for x into $x + y = 6000$, we find $y = 4000$, so the intersection point is $(2000, 4000)$.

The region formed by the constraints is shown in green in Figure 5.27. We now evaluate the objective function at the four corners of this region.

Corner (Vertex) (x, y)	Objective Function $z = 10x + 6y$
$(0, 0)$	$z = 10(0) + 6(0) = 0$
$(4000, 0)$	$z = 10(4000) + 6(0) = 40{,}000$
$(2000, 4000)$	$z = 10(2000) + 6(4000) = 44{,}000$ ⟵ maximum
$(0, 6000)$	$z = 10(0) + 6(6000) = 36{,}000$

Thus the maximum value of z is 44,000, and this occurs when $x = 2000$ and $y = 4000$. In practical terms, this means that the maximum number of earthquake victims that can be helped with each plane shipment is 44,000. This can be accomplished by sending 2000 water bottles and 4000 medical kits per plane. ∎

Before considering another example, let's summarize the steps in solving a linear programming problem.

> **Solving a linear programming problem**
>
> **1.** Write the objective function and all necessary constraints.
> **2.** Graph the region for the system of inequalities representing the constraints.
> **3.** Find the value of the objective function at each corner, or vertex, of the graphed region.
> **4.** The maximum or minimum of the objective function occurs at the corner points.

Before considering additional applications, let's use this procedure to solve a linear programming problem in which the objective function and constraints are given.

EXAMPLE 9 Solving a Linear Programming Problem

Find the maximum value of the objective function

$$z = 2x + y$$

subject to the constraints:

$$x \geq 0, y \geq 0$$

$$x + 2y \leq 5$$

$$x - y \leq 2$$

Solution

We begin by graphing the region in quadrant I ($x \geq 0$, $y \geq 0$) formed by the constraints. The graph is shown in Figure 5.28.

Now we evaluate the objective function at the four vertices of this region.

Objective function: $z = 2x + y$

At $(0, 0)$:　$z = 2 \cdot 0 + 0 = 0$

At $(2, 0)$:　$z = 2 \cdot 2 + 0 = 4$

At $(3, 1)$:　$z = 2 \cdot 3 + 1 = 7$　　　Maximum value of z

At $(0, 2.5)$:　$z = 2 \cdot 0 + 2.5 = 2.5$

Thus, the maximum value of z is 7, and this occurs when $x = 3$ and $y = 1$. ∎

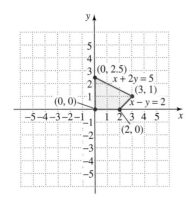

Figure 5.28

The graph of $x + 2y \leq 5$ and $x - y \leq 2$ in quadrant I

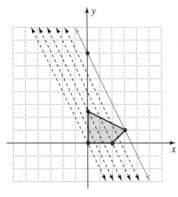

Figure 5.29

The line with slope -2 with the greatest y-intercept that intersects the shaded region passes through one of its vertices.

$\mathcal{S}$**tudy tip**

We can see why the objective function in Example 9 occurs at a vertex by solving the equation for y.

$z = 2x + y$　　　This is the objective function of Example 9.

$y = -2x + z$　　　Solve for y. Recall that the slope-intercept form of a line is $y = mx + b$.

slope $= -2$　　y-intercept $= z$

In this form, z represents the y-intercept of the objective function. The equation describes infinitely many parallel lines, each with a slope of -2. The process in linear programming involves finding the maximum z-value for all lines that intersect the region determined by the constraints. Of all the lines whose slope is -2, we're looking for the one with the greatest y-intercept that intersects the given region. As we see in Figure 5.29, such a line will pass through one (or possibly more) of the vertices of the region.

Linear programming is often used to determine the best allocation of resources available to a company. The word *programming* refers to a "program to allocate resources."

EXAMPLE 10 A Corporate Case Study: Federal Express

In 1978, a ruling by the Civil Aeronautics Board allowed Federal Express to purchase larger aircraft. Federal Express's options included 20 Boeing 727s that United Airlines was retiring and/or the French-built Dassault Fanjet Falcon 20. To aid in their decision, executives at Federal Express analyzed the following data:

	Boeing 727	Falcon 20
Direct Operating Cost	$1400 per hour	$500 per hour
Payload	42,000 pounds	6000 pounds

Federal Express was faced with the following constraints:

1. Hourly operating cost was limited to $35,000.
2. Total payload had to be at least 672,000 pounds.
3. Only twenty 727s were available.

Given the constraints, how many of each kind of aircraft should Federal Express have purchased to maximize the number of aircraft?

Solution

Let

$x =$ the number of 727s

$y =$ the number of Falcon 20s

We first note that the objective function is

$z = x + y.$ This represents the number of the two aircraft to be purchased.

Now let's consider the constraints. Hourly operating cost was limited to $35,000.

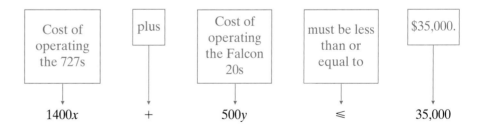

Total payload had to be at least 672,000 pounds.

ENRICHMENT ESSAY

George Dantzig and Linear Programming

In 1947, George Dantzig was a young mathematician looking for methods of solving optimization problems in business and for the military. A new kind of machine called the "computer" led him to success. Since its invention in the 1940s, Dantzig's matrix-computer method, called the simplex method, has provided solutions to problems that have saved industry and the military time and money. Dantzig, now a professor of operations research and computer science at Stanford University, talks about his technique:

> The problems we solve nowadays have thousands of equations, sometimes a million variables. One of the things that still amazes me is to see a program run on the computer—and to see the answer come out. If we think of the number of combinations of different solutions that we're trying to choose the best of, it's akin to the stars in the heavens. Yet we solve them in a matter of moments. This, to me, is staggering. Not that we can solve them—but that we can solve them so rapidly and efficiently.

The search continues for finding faster and faster methods for solving problems. In 1984, Narendra Karmarkar, a research mathematician at Bell Laboratories, invented a powerful new linear programming algorithm that is faster and more efficient than the simplex method. The applications of Karmarkar's algorithm, incorporated into a software product, have a direct impact on the efficiency and profitability of numerous industries, including telephone communications and the airlines.

Narendra Karmarkar. Property of AT&T Archives. Reprinted with permission of AT&T.

Only twenty 727s were available.

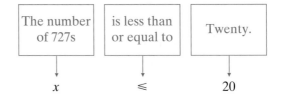

The number of 727s	is less than or equal to	Twenty.
x	$\leq$	20

Thus we have the following constraints:

$$1400x + 500y \leq 35{,}000 \quad \xrightarrow{\text{Simplify, dividing by 100.}} \quad 14x + 5y \leq 350$$

$$42{,}000x + 6000y \geq 672{,}000 \quad \xrightarrow{\text{Simplify, dividing by 6000.}} \quad 7x + y \geq 112$$

$$x \leq 20 \quad \xrightarrow{\text{No simplification is needed.}} \quad x \leq 20$$

$$x \geq 0, y \geq 0 \quad \xrightarrow[\text{the nonnegative constraints.}]{\text{Be sure to include}} \quad x \geq 0, y \geq 0$$

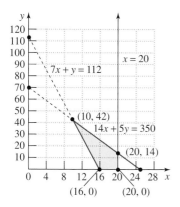

Figure 5.30

The graph of the constraints placed on Federal Express

The graph of the region formed by the constraints is shown in Figure 5.30.

Since the Federal Express purchase is represented by $z = x + y$, evaluate this objective function at the four vertices of the region.

Objective function: $z = x + y$

At $(16, 0)$: $z = 16 + 0 = 16$

At $(20, 0)$: $z = 20 + 0 = 20$

At $(20, 14)$: $z = 20 + 14 = 34$

At $(10, 42)$: $z = 10 + 42 = 52$ Maximum value of z

Thus, Federal Express should have purchased 10 Boeing 727s and 42 Falcon 20s. ■

Example 10 contained two kinds of aircraft, two variables, and five constraints. Problems in linear programming can involve thousands of products subject to thousands of constraints. For real world problems with more than two variables, several nongeometric linear programming methods, from the simplex method to Karmarkar's algorithm, are available on computer programs. These programs are based on matrix methods, helping businesses allocate resources on hand to manufacture a particular array of products that maximizes profit. Linear programming accounts for more than 50% and perhaps as much as 90% of all computing time used for management decisions in business.

PROBLEM SET 5.8

Practice Problems

In Problems 1–12, sketch the graph of each inequality.

1. $x + 2y \leq 8$

2. $3x - 6y \leq 12$

3. $x - 2y > 10$

4. $2x - y > 4$

5. $y < 3$

6. $x < -2$

7. $2 < x \leq 5$

8. $-1 \leq y < 4$

9. $|y| < 4$

10. $|x| > 2$

11. $|3x + 2| \geq 6$

12. $|3y + 2| \leq 6$

In Problems 13–24, sketch the graph of the solution set of each system of inequalities.

13. $x - 2y \leq 12$
$2x + y \geq 14$

14. $x - 5y \leq -5$
$2x + 5y \leq 20$

15. $2x + y \leq 12$
$4x + y \geq 8$

16. $3x - 5y < 15$
$2x + y > 10$

17. $x + y \geq -1$
$3x - 2y < 12$
$x \geq -1$
$y \leq 5$

18. $2x - y > 4$
$x - 5y \leq 2$
$x - y < 4$
$x < 5$

19. $5x - 2y < 10$
$2x - 5y > -10$
$x + y \geq -5$
$x \geq -2$

20. $-2x + y \leq -1$
$x + y \leq 8$
$x \geq 1$
$y \geq -1$

21. $y < x - 3$
$y > -x - 3$
$x + 2y < 6$
$x - 2y < 6$

22. $y > 2x - 3$
$y > -2x - 3$
$x + 4y < 24$
$x - 4y > -24$

23. $x + 5y \geq 5$
$3x + y \leq 15$
$x + y \leq 7$
$2x - y \geq -1$

24. $-x + y \leq 1$
$x + y \geq 1$
$-x + y \geq -1$
$x + y \leq 3$

In Problems 25–28, find the maximum and minimum values of the given objective function in the indicated region.

25. $z = 5x + 6y$

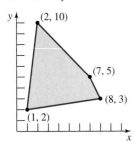

26. $z = 3x + 2y$

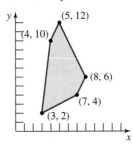

27. $z = 40x + 50y$

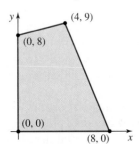

28. $z = 30x + 45y$

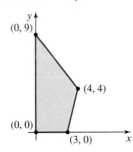

In Problems 29–30, sketch the region determined by the constraints. Then find the maximum and minimum values of the given objective function, subject to the constraints.

29. Objective function:

$$z = 3x - 2y$$

Subject to:

$$1 \leqslant x \leqslant 5$$

$$y \geqslant 2$$

$$y \leqslant x + 3$$

30. Objective function:

$$z = 6x + 10y$$

Subject to:

$$x + y \leqslant 12$$

$$x + 2y \leqslant 20$$

$$x \geqslant 0, y \geqslant 0$$

Application Problems

31. A television manufacturer makes console and wide-screen televisions. The profit per unit is $125 for the console televisions and $200 for the wide-screen televisions.

  a. Let

   x = the number of consoles manufactured in a month and

   y = the number of wide-screens manufactured in a month

   Write the objective function that models the total monthly profit.

  b. The manufacturer is bound by the following constraints:

  1. Equipment in the factory allows for making at most 450 console televisions in one month.

  2. Equipment in the factory allows for making at most 200 wide-screen televisions in one month.

  3. The cost to the manufacturer per unit is $600 for the console televisions and $900 for the wide-screen televisions. Total monthly costs cannot exceed $360,000.

  Write a series of inequalities that models these constraints.

  c. Graph the system of inequalities in part (b). Use only the first quadrant, since x and y must both be positive.

d. Evaluate the objective profit function at each of the five vertices of the graphed region. (The vertices should occur at $(0, 0)$, $(0, 200)$, $(300, 200)$, $(450, 100)$, and $(450, 0)$.)

e. Complete the missing portions of this statement: The television manufacturer will make the greatest profit by manufacturing ___ console televisions each month and ___ wide-screen televisions each month. The maximum monthly profit is $ ___.

Use the four steps given in the box on page 647 to solve each of the following problems.

32. A manufacturer produces two models of bicycles. The times (in hours) required for assembling and painting each model are given in the following table.

	Model *A*	Model *B*
Assembling	5	4
Painting	2	3

The maximum total weekly hours available in the assembly department and the paint department are 200 hours and 108 hours, respectively. The profits per unit are $25 for model *A* and $15 for model *B*. How many of each type should be produced to maximize profit?

33. Food and clothing are shipped to victims of a natural disaster. Each carton of food will feed 5 people, while each carton of clothing will help 6 people. The commercial carriers transporting the food and clothing are bound by the following constraints: Each 30-cubic-foot box of food weighs 50 pounds and each 20-cubic-foot box of clothing weighs 5 pounds. The total weight per carrier cannot exceed 18,000 pounds and the total volume must be less than 12,000 cubic feet. How many cartons of food and clothes should be sent with each shipment to maximize the number of people who can be helped?

34. Suppose that you inherit $10,000 with certain stipulations attached. Some (or all) of the money must be invested in stocks and bonds. The requirements are that at least $3000 be invested in bonds, with expected returns of $0.08 per dollar, and at least $2000 be invested in stocks, with expected returns of $0.12 per dollar. Since the stocks are medium-risk, the final stipulation requires that the investment in bonds should never be less than

the investment in stocks. How should the money be invested so as to maximize your expected returns?

35. On June 24, 1948, the former Soviet Union blocked all land and water routes through East Germany to Berlin. A gigantic airlift was organized using American and British planes to supply food, clothing, and other supplies to the more than 2 million people in West Berlin. The cargo capacity of an American plane was 30,000 cubic feet and that of a British plane was 20,000 cubic feet. The Western Allies wanted to maximize cargo capacity to break the Soviet blockade, but were subject to the following restrictions:

1. No more than 44 planes could be used.

2. The larger American planes required 16 personnel per flight, double the requirement for the British planes, and the total number of personnel could not exceed 512.

3. The cost of an American flight was $9000, and the cost of a British flight was $5000. Total weekly costs could not exceed $300,000.

Find the number of American and British planes that were used to maximize cargo capacity.

36. A theater is presenting a program on drinking and driving for students and their parents. The proceeds will be donated to a local alcohol information center. Admission is $2.00 for parents and $1.00 for students. However, the situation has two constraints: the theater can hold no more than 150 people and every two parents must bring at least one student. How many parents and students should attend to raise the maximum amount of money?

True–False Critical Thinking Problems

37. Which one of the following is true?

a. In linear programming, constraints are limitations placed on problem solvers without access to computer programs.

b. A linear programming problem is solved by evaluating the objective function at the vertices of the region represented by the constraints.

c. The graph of the system

$$x + y \leqslant 4$$

$$2x + y \leqslant 6$$

$$x \geqslant 0, y \geqslant 0$$

does not have a corner, or vertex, at $(2, 2)$.

d. A student who works for x hours at \$15 per hour by tutoring and y hours at \$12 per hour as a teacher's aid has earnings modeled by the objective function $z = 15xy + 12xy$.

38. Which one of the following is true?

a. The point $(-2, 4)$ is in the graph of the system

$$2x + 3y \geqslant 4$$

$$2x - y > -6.$$

b. Any rectangular coordinate system in which nothing at all is graphed is the graph of the system

$$y \geqslant 2x + 3$$

$$y \leqslant 2x - 3.$$

c. In rectangular coordinates, there is no difference between the graphs of

$$x > -2 \quad \text{and} \quad y < 4$$

and

$$x > -2 \quad \text{or} \quad y < 4.$$

d. The solution to a system of linear inequalities is represented by the set of points in rectangular coordinates that satisfies any one of the inequalities in the system.

Technology Problems

39. Read the section of the user's manual for your graphing utility that describes how to shade a region that is above a line, below a line, or between two lines. Once you have learned how to use your graphing utility to graph a linear inequality in two variables, use the graphing utility to verify your results from any six problems that you worked from Problems 1–24.

40. Use a linear programming computer program (such as the simplex method) to solve two of the problems from Problems 31–36. Then consult a textbook on finite mathematics and use the program to solve a problem that cannot be solved by the methods discussed in this section. The problem should contain three or more variables subject to at least five constraints.

Writing in Mathematics

41. Explain how to graph a system of linear inequalities.

42. Choose a particular field that is of interest to you. Research how linear programming is used to solve problems in that field. If possible, investigate the solution of a specific practical problem. Write a report on your findings, including the contributions of George Dantzig, Narendra Karmarkar, and L.G. Khachion to linear programming.

43. Evaluate the objective function $z = 3x + 4y$ at the vertices of the region shown in the figure on the right. What do you observe about the maximum value of z? What is true about the objective function when evaluated at any point on the line segment connecting $(6, 3)$ and $(2, 6)$? Describe why this is so by comparing the slope of the

line through $(6, 3)$ and $(2, 6)$ with the slope of the objective function.

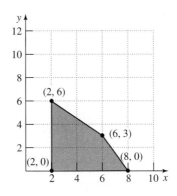

Critical Thinking Problems

44. Sketch the graph of the solution set for the following system of inequalities:

$$|x + y| \leq 3$$
$$|y| \leq 2.$$

45. Sketch the graph of the solution set for the following system of inequalities:

$$y \geq nx + b \quad (n < 0, b > 0)$$
$$y \leq mx + b \quad (m > 0, b > 0)$$

46. Write a system of linear inequalities whose graph consists of all points inside and on the square whose vertices are on the coordinate axes and whose diagonals have length a.

47. Consider the objective function $z = Ax + By$ ($A > 0$ and $B > 0$) subject to the following constraints: $2x + 3y \leq 9$, $x - y \leq 2$, $x \geq 0$, and $y \geq 0$. Prove that the objective function will have the same maximum value at the vertices $(3, 1)$ and $(0, 3)$ if $A = \frac{2}{3}B$.

Group Activity Problem

48. Members of the group should interview a business executive who is in charge of deciding the variety of products for a business. How are production policy decisions made? Are other methods used in conjunction with linear programming? What are these methods? What sort of academic background, particularly in mathematics, does this executive have? Write a group report addressing these questions, emphasizing the role of linear programming for the business.

CHAPTER PROJECT
Cryptography

Cryptography is the study of encoding and decoding messages. One of the oldest examples is called the *Spartan scytale*. Plutarch describes how generals would exchange messages by using long, narrow strips of parchment wound in a spiral around a special staff. The person sending the message would write vertically on the wrapped parchment, unwrap it and send it off to a person who had exactly the same size cylindrical staff. Only by rewrapping the parchment around a staff of precisely the same size would someone be able to see the letters lined up correctly and, thus, read the message.

1. Try your own version of the Spartan scytale. Ask another person to decipher your message without knowing the exact size of the cylinder used. How effective did you find this method?

Another early method of sending secret messages was used by Julius Caesar. He simply replaced each letter of his message by the letter three places after it. So "a" in the original message became "d," and so on. The original message is called *plaintext* and the disguised message is called *ciphertext*. We could describe Caesar's method by showing the two alphabets, one above the other, giving us the substitution to make:

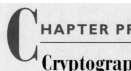

plaintext alphabet: a   b   c   d   e   f   g ...
ciphertext alphabet: d   e   f   g   h   i   j ...

If you intercept the message and you know it was encrypted by shifting letters, you could simply try all 25 letters for the beginning of the shift. This may take some time, but the time spent could possibly be shortened by knowing how frequently certain letters of the alphabet occur. For example, your first guess may be that the most frequently occurring letter in the disguised message corresponds to "e," the most frequently occurring letter in most English text. In order of decreasing frequency, the ten most frequently occurring letters are: e, t, a, o, n, r, i, s, h, and d.

2. Write down a phrase you wish to encrypt and use Caesar's method for the encryption process. You might use the same trick Caesar did to make the message a little harder to decrypt: leave no spaces between the words. Bring your disguised message to class and let your instructor mix all the messages together. Select one, at random, to translate. Compare your speed of decrypting with others in the class.

We might also choose to use different visual aids or symbols to substitute for letters. The most common type of codes used would be *Morse Code*, or the flag-waving nautical codes. One popular style, in use before the Renaissance, arranged letters of the alphabet in boxes and crosses:

A	B	C	J	N	O	P	W
D	E	F	K	Q	R	S	X Y
G	H	I	M	T	U	V	Z

There is a mathematically secure method of encoding called a *One Time Pad* or *Vernam cipher*. As long as this method is, in fact, used only once with any given key, it is secure. Usually the message to be sent is represented with 1s and 0s, in *binary* form, with the following assignment for the individual letters:

A	B	C	D	E	F
00000	00001	00010	00011	00100	00101 ...

For example, the word "bad" would read: 000010000000011

3. Complete the rest of the alphabet with the binary assignments for each letter. The message in 1s and 0s is then combined with a key, also consisting of 1s and 0s, and sent. It is decoded using the same key. The key is usually a very long string of 1s and 0s, randomly generated. The plaintext is combined with the key using a simple rule: matching digits give 0, and a 1 and 0 gives a 1. For example:
Random key string: 00001100010010111001111001010110001

Text:	C	A	L	L	D	A	N
	00010	00000	01011	01011	00011	00000	01101
Key:	00001	10001	00101	11001	11100	10101	10001
Cipher:	00011	10001	01110	10010	11111	10101	11100

Message goes out: 00011100010111010010111111010111100

4. Decode the outgoing message above using the given key. Create your own message and key and exchange it with another member of your class for decoding. Notice that you must transfer the key as well as the message.

There are many systems which encode text messages using numbers. We saw one such method using matrices in Example 8, page 599. Using numbers for the encoding and decoding of messages allows us to take advantage of computers to send messages. Right now, we are in the midst of an explosion of data flowing to and from various places, much of which involves secure transmissions. Each time a bank transfers financial information, the security of the data is crucial. In fact, any online commercial use of credit cards and direct money transfers demands tight security to transfer information. Other styles of data must also remain tamperproof. For example, when filing a federal income tax return via electronic mail, the return must be received on or before April 15th. Thus the time and date stamp for the electronic document must be secure. There is an increasing move towards making monetary transactions via a "smart card" or electronic money transfer card. Not only would this card be transferring "cash" directly between accounts, it would also leave a "trail" behind of where and what was purchased. Although these cards would be treated like cash, they allow a loss of privacy and anonymity that is present in cash transactions.

Worldwide Web Resources

Go to the Prentice Hall website (http://www.prenhall.com/blitzer) to access other locations on the Internet that will allow you to further explore the concepts presented in this project.

Chapter Review

SUMMARY

1. **Linear Systems of Equations in Two Variables**
   a. The solution set is the set of all ordered pairs of values (a, b) that satisfy every equation in the system.
   b. *Solving by Graphing*
      1. Graph each equation in the system.
      2. The solution corresponds to the point of intersection of the graphs.
      3. Inconsistent systems have no solutions. Graphs are parallel lines. Systems with dependent equations have infinitely many solutions. Graphs are the same line.
   c. *Solving by Substitution*
      1. If necessary, solve one of the equations for one variable in terms of the other.
      2. Substitute the expression for that variable into the other equation.
      3. Solve the resulting equation in one variable.
      4. Back-substitute the value from step 3 into the equation in step 1 and find the value of the remaining variable.
      5. Check the solution in both of the system's given equations.
   d. *Solving by Addition*
      1. If necessary, rewrite both equations in the form $Ax + By = C$.
      2. If necessary, multiply either or both equations by appropriate numbers so that coefficients of x or y will have a sum of 0.
      3. Add equations. The sum is an equation in one variable.
      4. Solve the equation in step 3.
      5. Back-substitute the value from step 4 into either of the given equations and solve for the other variable.

   6. Check the solution in both of the system's given equations.
   e. *Inconsistent and Dependent Systems*
      1. If both variables are eliminated and the resulting statement is false, the system is inconsistent.
      2. If both variables are eliminated and the resulting statement is true, the system contains dependent equations.

2. **Solving Linear Systems in Three Variables by Addition**
   Use addition to eliminate any variable, reducing the system to two equations in two variables. Use the addition method again to solve the resulting system in two variables.

3. **Solving Problems Using Linear Systems**
   a. Assign letters to unknown quantities.
   b. Write a linear system that models the problem's conditions.
   c. Solve the system and answer the problem's question.
   d. Check the answers in the problem's wording.

4. **Solving Linear Systems by Gaussian Elimination and Gauss-Jordan Elimination**
   a. Write the augmented matrix for the system whose entries are the coefficients of the variables of the system together with the constants.
   b. For Gaussian elimination with back-substitution, use elementary row operations to simplify the matrix to one with 1s down the diagonal from upper left to lower right, called the main diagonal, and 0s below the 1s.
   c. The three elementary matrix row operations are as follows:
      1. Any two rows of the matrix may be interchanged.
      2. The entries in any row may be multiplied by the same nonzero real number.

3. Any row may be changed by adding to its entries a multiple of the corresponding entries of another row.

d. For Gauss-Jordan elimination, use elementary row operations to simplify the matrix to one with 1s down the main diagonal from upper left to lower right and 0s in every position above and below each 1.

e. If Gaussian elimination results in a matrix with a row containing all 0s to the left of the vertical line and a nonzero number to the right, the system has no solution (is inconsistent). If all 0s result in a row, the system has an infinite number of solutions (is dependent).

5. Matrix Operations

a. *Equal Matrices:* Two matrices are equal if and only if they have the same order and their corresponding entries are equal.

b. *Addition of Matrices:* If A and B are matrices of the same order, then $A + B$ is the matrix formed by adding corresponding entries in A and B.

c. *Scalar Multiplication:* If A is a matrix and c is a scalar, then cA is the matrix formed by multiplying each entry in A by c.

d. *Subtraction of Matrices:* If A and B are matrices of the same order, then $A - B$ can be found by subtracting entries in B from the corresponding entries in A.

e. *Multiplication of Matrices:* If A is an $m \times n$ matrix and B is an $n \times p$ matrix, then AB is an $m \times p$ matrix whose entries are computed as follows: The entry in the ith row and jth column of AB is the sum of the products of the corresponding entries in the ith row of A and the jth column of B. (Remember that, in general, matrix multiplication is not commutative; $AB \neq BA$.)

6. Inverse of a Square Matrix

a. The identity matrix I_n is an $n \times n$ matrix with 1s along the major diagonal and 0s elsewhere. If A is an $n \times n$ matrix, then $AI_n = A$ and $I_n A = A$.

b. Let A be an $n \times n$ square matrix. If there is a square matrix A^{-1} such that $AA^{-1} = I_n$ and $A^{-1}A = I_n$, then A^{-1} is called the inverse of A.

c. *The Inverse of a 2 × 2 matrix:* If $A = \begin{bmatrix} a & b \\ c & d \end{bmatrix}$, then

$$A^{-1} = \frac{1}{ad - bc}\begin{bmatrix} d & -b \\ -c & a \end{bmatrix}.$$

The matrix A is invertible (has an inverse) if and only if $ad - bc \neq 0$.

d. *The Inverse of an Invertible $n \times n$ Matrix:* Form the augmented matrix $[A\,|\,I]$, using row operations to obtain $[I\,|\,B]$. Matrix B is A^{-1}.

7. Determinants and Cramer's Rule

a. *Value of a Second-Order Determinant*

$$\begin{vmatrix} a_1 & b_1 \\ a_2 & b_2 \end{vmatrix} = a_1 b_2 - a_2 b_1$$

b. *Evaluating an nth-order determinant, where $n > 2$*

1. Select a row or column about which to expand.

2. For each entry a_{ij} in the row or column, multiply by $(-1)^{i+j}$ times the determinant obtained by deleting the ith row and the jth column in the given array of numbers.

3. The value of the determinant is the sum of the products found in step 2.

c. *Properties of Determinants*

1. If a common factor is factored from a given row (or column), the value of the determinant is not changed.

2. If two rows (or columns) are interchanged, then the value of the determinant is multiplied by -1.

3. If a multiple of one row (or column) is added to another row (or column), the value of the determinant is not changed.

d. *Cramer's Rule*

If

$$a_{11}x_1 + a_{12}x_2 + a_{13}x_3 + \cdots + a_{1n}x_n = b_1$$
$$a_{21}x_1 + a_{22}x_2 + a_{23}x_3 + \cdots + a_{2n}x_n = b_2$$
$$a_{31}x_1 + a_{32}x_2 + a_{33}x_3 + \cdots + a_{3n}x_n = b_3$$
$$\vdots$$
$$a_{n1}x_1 + a_{n2}x_2 + a_{n3}x_3 + \cdots + a_{nn}x_n = b_n$$

then $x_i = \dfrac{D_i}{D}$, where D is the determinant of the coefficients of the system ($D \neq 0$) and D_i is the determinant where the coefficients of x_i have been replaced by the constants.

8. Decomposition of $\dfrac{P(x)}{Q(x)}$ into Partial Fractions

a. If the degree of $P(x)$ is less than the degree of $Q(x)$, factor $Q(x)$ into prime factors that are either linear or quadratic.

1. Include one partial fraction with a constant numerator for each distinct linear factor.

2. Include one partial fraction with a constant numerator for each power of a repeated linear factor.

3. Include one partial fraction with a linear numerator for each distinct quadratic factor.

4. Include one partial fraction with a linear numerator for each power of a repeated quadratic factor.

b. Multiply by the LCD, write both sides of the equation in the same form, and equate coefficients. Solve the resulting linear system for the constants and substitute these values into the partial fraction decomposition.

c. If the degree of $P(x)$ is greater than or equal to the degree of $Q(x)$, divide and apply partial fraction decomposition to the remainder term.

9. Systems of Linear Inequalities in Two Variables

a. *Graphing a Linear Inequality in Two Variables*

1. Graph the boundary line, using a solid line if equality is included and a dashed line if equality is not included.

2. Determine which open half-plane belongs to the graph by testing one point from a half-plane or by writing the inequality in slope-intercept form. If $y \geq mx + b$, shade the half-plane above the line of $y = mx + b$. If $y \leq mx + b$, shade the half-plane below $y = mx + b$.

b. *Graphing a System of Linear Inequalities in Two Variables*

1. Graph the corresponding equations for the system. Shade the region representing the intersection of all the given inequalities.

or 2. Graph each inequality on the same rectangular coordinate system. Lightly shade each solution set. Then darken the intersection of the lightly shaded regions.

3. In either method 1 or 2, change any portion of any solid boundary line not in the intersection of the solutions from solid to dashed.

10. Linear Programming

If an objective function $z = ax + by$ has an optimal value, it must occur at one of the corners, or vertices, of the region corresponding to the system of constraints.

REVIEW PROBLEMS

For Problems 1–5, use substitution or addition to solve each system. If applicable, state that the system is inconsistent or dependent. Verify solutions graphically.

1. $4x + 6y = 31$
$3x + 4y = 22$

2. $0.05x + 0.06y = 400$
$y = 2x + 1000$

3. $2y - 6x = 7$
$3x - y = 9$

4. $3x - 2y = 12$
$2x - 3y = -2$

5. $4x - 8y = 16$
$3x - 6y = 12$

Solve each system in Problems 6–7 by the addition method.

6. $2x - y + z = 1$
$3x - 3y + 4z = 5$
$4x - 2y + 3z = 4$

7. $x + 2y - z = 5$
$2x - y + 3z = 0$
$2y + z = 1$

8. Health experts agree that cholesterol intake should be limited to 300 mg or less each day. Three ounces of shrimp and 2 ounces of scallops contain 156 mg of cholesterol. Five ounces of shrimp and 3 ounces of scallops contain 45 fewer mg of cholesterol than the suggested maximum daily intake. Determine the cholesterol content in an ounce of each item.

9. The building lot shown in the figure on the right is composed of three rectangles. The lengths of two of the sides, as shown, are designated by A and B. All other sides have lengths that can be expressed in terms of A and B. The perimeter of the lot is 310 feet. For privacy, fencing is to be placed along the three lengths designated by A, B, and A, respectively. Fencing along A costs $30 per foot, and fencing along B costs $8 per foot, with a total cost for the three sides of $2600. Find A and B. Then write a description of the lot's shape for a prospective buyer using these values.

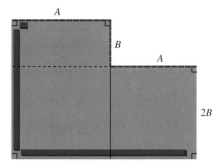

10. New York City, Chicago, and Los Angeles have the largest police forces in the United States, with a total of 50,760 officers. New York City has 22,900 more officers than Los Angeles. The number of officers in New

York City exceeds twice that of Chicago by 5335. What is the size of the three largest police forces in the United States?

Largest Police Forces in the United States

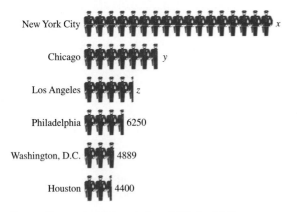

Source: International Association of Chiefs of Police

11. The largest angle in a triangle measures 66° more than the smallest angle. The measure of the largest angle is 17° less than three times the measure of the remaining angle. Find the measure of each angle of the triangle.

12. A dietician wants to prepare a meal that consists of foods A, B, and C. The nutritional component per ounce in these foods is given in the following table.

Nutritional Component per Ounce

	Calories	Protein (in grams)	Vitamin C (in milligrams)
Food A	200	0.2	0
Food B	50	3	10
Food C	10	1	30

The meal with foods A, B, and C must contain exactly 740 calories, 10.6 grams of protein, and 140 milligrams of vitamin C. Model the situation with a system of three linear equations and determine how many ounces of each type of food must be used.

In Problems 13–14, solve the system of equations using either Gaussian elimination with back-substitution or Gauss-Jordan elimination.

13. $x + 2y + 3z = -5$
$2x + y + z = 1$
$x + y - z = 8$

14. $3x_1 + 5x_2 - 8x_3 + 5x_4 = -8$
$x_1 + 2x_2 - 3x_3 + x_4 = -7$
$2x_1 + 3x_2 - 7x_3 + 3x_4 = -11$
$4x_1 + 8x_2 - 10x_3 + 7x_4 = -10$

15. The table on the right shows the pollutants in the air in a city on a typical summer day.

x (Hours after 6 A.M.)	y (Amount of Pollutants in the Air, in parts per million)
2	98
4	138
10	162

a. Use the function $y = ax^2 + bx + c$ to model the data. Use either Gaussian elimination with back-substitution or Gauss-Jordan elimination to find the values for a, b, and c.

b. Use the function to find the time of day at which the city's air pollution level is at a maximum. What is the maximum level?

In Problems 16–19, use Gaussian elimination to find the complete solution to each system of equations, or show that none exists.

16. $2x - 3y + z = 1$
$x - 2y + 3z = 2$
$3x - 4y - z = 1$

17. $x - 3y + z = 1$
$-2x + y + 3z = -7$
$x - 4y + 2z = 0$

18. $x_1 + 4x_2 + 3x_3 - 6x_4 = 5$
$x_1 + 3x_2 + x_3 - 4x_4 = 3$
$2x_1 + 8x_2 + 7x_3 - 5x_4 = 11$
$2x_1 + 5x_2 - 6x_4 = 4$

19. $2x + 3y - 5z = 15$
$x + 2y - z = 4$

20. The figure shows the intersections of three one-way streets. The numbers given represent traffic flow in cars per hour at a peak period (from 4 P.M. to 6 P.M.).

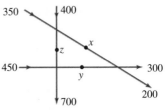

a. Use the idea that the number of cars entering each intersection per hour must equal the number of cars leaving per hour to set up a linear system of equations involving x, y, and z.

b. Use Gaussian elimination to solve the system.

c. If construction limits the value of z to 400, how many cars per hour must pass between the other intersections to keep traffic flowing?

In Problems 21–31, let

$$A = \begin{bmatrix} 2 & -1 & 2 \\ 5 & 3 & -1 \end{bmatrix}, \quad B = \begin{bmatrix} 0 & -2 \\ 3 & 2 \\ 1 & -5 \end{bmatrix}, \quad C = \begin{bmatrix} 1 & 2 & 3 \\ -1 & 1 & 2 \\ -1 & 2 & 1 \end{bmatrix}, \quad \text{and} \quad D = \begin{bmatrix} -2 & 3 & 1 \\ 3 & -2 & 4 \end{bmatrix}.$$

Find the indicated matrix, if possible.

21. $3A + 2D$

22. $-2A - 4D$

23. $-5(A + D)$

24. AB

25. BA

26. BD

27. DB

28. C^2

29. BAC

30. $AB - BA$

31. $(A - D)C$

32. If $A = \begin{bmatrix} 4 & -5 \\ 0 & 2 \end{bmatrix}$ and $B = \begin{bmatrix} -7 & 3 \\ -4 & 5 \end{bmatrix}$, solve fox X in the matrix equation $4X - 3B = 2A$.

33. A company owns two branch stores. The company's costs and retail selling prices (in thousands of dollars) in three divisions (clothing, furniture, and appliances) for each branch store are given by the following matrices.

Branch Store 1

	Clothing	Furniture	Appliances	
Costs	17	37	47	$= A$
Retail Sales	28	96	163	

Branch Store 2

	Clothing	Furniture	Appliances	
Costs	13	21	31	$= B$
Retail Sales	22	79	145	

a. Find $A + B$ and describe the significance of this matrix in practical terms.

b. Find $A - B$. What does this matrix describe?

c. Write a matrix representing the total profits for the two branch stores combined in the three divisions.

34. An automobile dealership sells three models of cars at its three outlets. The inventory of models at each store is given the following matrix.

	Model X	Model Y	Model Z	
Outlet 1	12	7	6	
Outlet 2	20	8	10	$= A$
Outlet 3	7	2	3	

The next matrix gives the wholesale and retail prices for each model.

	Wholesale Price	Retail Price	
Model X	12,000	15,000	
Model Y	8000	11,000	$= B$
Model Z	10,0000	14,500	

a. Calculate the product AB.

b. Describe what the matrix AB represents and interpret the entries.

c. What is the wholesale value of the cars at outlet 1?

d. What is the retail value of the cars at outlet 2?

e. If outlet 3 sells all of the inventory in matrix A, what is the profit for that branch of the dealership?

35. If $A = \begin{bmatrix} 2 & 6 & 1 \\ 1 & 2 & -1 \\ 2 & 4 & -1 \end{bmatrix}$ and $B = \begin{bmatrix} -1 & -5 & 4 \\ \frac{1}{2} & 2 & -\frac{3}{2} \\ 0 & -2 & 1 \end{bmatrix}$, show that B is the inverse of A.

36. If $A = \begin{bmatrix} 2 & 3 \\ -1 & 5 \end{bmatrix}$, find A^{-1}. Verify your result by showing that $AA^{-1} = I$ and $A^{-1}A = I$.

37. If $A = \begin{bmatrix} 1 & 0 & -2 \\ 2 & 1 & 0 \\ 1 & 0 & -3 \end{bmatrix}$, find A^{-1}. Verify your result by showing that $AA^{-1} = I$ and $A^{-1}A = I$.

38. Consider the following linear system:

$$x + y + 2z = 7$$
$$y + 3z = -2$$
$$3x - 2z = 0$$

a. Express the system in the form $AX = B$ where A, X, and B are appropriate matrices.

b. Find A^{-1}, the inverse of the coefficient matrix.

c. Use A^{-1} to solve the given system.

39. Use the procedure of Problem 38 to solve the following system:

$$x + 2y - z = 5$$
$$2x + 3y - z = 8$$
$$3x + 6y - 2z = 14$$

40. Use the coding matrix $A = \begin{bmatrix} 1 & 1 \\ 4 & 5 \end{bmatrix}$ and its inverse to encode and then decode the word BASE.

In Problems 41–43, evaluate each determinant.

41. $\begin{vmatrix} 4 & 5 \\ -6 & -3 \end{vmatrix}$

42. $\begin{vmatrix} 2 & 4 & -3 \\ 1 & -1 & 5 \\ -2 & 4 & 0 \end{vmatrix}$

43. $\begin{vmatrix} 1 & 1 & 0 & 2 \\ 0 & 3 & 2 & 1 \\ 0 & -2 & 4 & 0 \\ 0 & 3 & 0 & 1 \end{vmatrix}$

In Problems 44–46, use Cramer's rule to solve the given system.

44. $x - 2y = 8$
$3x + 2y = -1$

45. $x + 2y + 2z = 5$
$2x + 4y + 7z = 19$
$-2x - 5y - 2z = 8$

46. $x_1 + x_2 = 3$
$x_2 + x_3 + x_4 = 3$
$x_2 + x_4 = 2$
$x_3 + x_4 = 0$

47. Use the quadratic function $f(x) = ax^2 + bx + c$ to model the following data:

(x) (Age of a Driver)	$f(x)$ (Average Number of Automobile Accidents per day in the United States)
20	400
40	150
60	400

Use Cramer's rule to determine values for a, b, and c. Then use the model to write a statement about the average number of automobile accidents in which 30-year-olds and 50-year-olds are involved daily.

In Problems 48–52, find the partial fraction decomposition for the given rational expression. Check your result with a graphing utility.

48. $\dfrac{4x^2 - 3x - 4}{x^3 + x^2 - 2x}$

49. $\dfrac{x^3 - 4x - 1}{x(x - 1)^3}$

50. $\dfrac{x^3 + x^2 + 2x + 3}{x^4 + 5x^2 + 6}$

51. $\dfrac{x^2 + 4}{(x^2 + 1)^2(x^2 + 2)}$

52. $\dfrac{x^3 - x^2 - 11x + 10}{x^3 - 2x + 4}$

In Problems 53–57, sketch the graph of the solution set of the system of inequalities. Use a graphing utility to verify the graph. If you are using a graphing utility with shading capabilities, shade the region representing the solution set.

53. $2x - y \geq -6$
$x \leq 3$
$y \geq 1$

54. $x + 3y \leq 6$
$x - y \leq 2$
$x \geq 0$

55. $2x + y \leq 10$
$4x + y \geq 8$
$y \geq 0$

56. $x + y < 6$
$x - 2y < 3$
$y - x < 2$
$x \geq 0, y \geq 0$

57. $y > x - 2$
$y < -x + 6$
$y < 2x - 3$

In Problems 58–59, sketch the region determined by the constraints. Then find the maximum and minimum values of the given objective function, subject to the constraints.

58. Objective Function:

$$z = 5x + 4y$$

Constraints:

$$5x + 3y \geq 60$$
$$5x + 6y \geq 90$$
$$x \geq 0$$
$$y \geq 0$$

59. Objective Function:

$$z = 12x + y$$

Constraints:

$$3x + 3y \geqslant 6$$

$$2x + y \leqslant 6$$

$$1 \leqslant x \leqslant 2$$

$$y \leqslant 3$$

60. A student earns $10 per hour for tutoring and $7 per hour as a teacher's aid. Although the student wants to maximize weekly earnings, there are two constraints in the situation:

   1. To have enough time for studies, the student can work no more than 20 hours a week.

   2. The tutoring center for which the student works requires that each tutor spend at least 3 hours, but no more than 8 hours, a week tutoring.

How many hours each week should the student work as a tutor and as a teacher's aid to maximize weekly earnings?

61. A manufacturer of lightweight tents makes two models whose specifications are given in the following table.

	Cutting Time per Tent	Assembly Time per Tent
Model A	0.9 hours	0.8 hours
Model B	1.8 hours	1.2 hours

On a monthly basis, the manufacturer has no more than 864 hours of labor available in the cutting department and at most 672 hours in the assembly division. The profits come to $25 per tent for model A and $40 per tent for model B. How many of each should be manufactured monthly to maximize the profit?

CHAPTER 5 TEST

Use addition or substitution to solve each system in Problems 1–2.

1. $2x + 5y = -2$
$3x - 4y = 20$

2. $2x + 3y = 6$
$\quad\quad x = 3y + 8$

3. Solve by the addition method:

$$x + y + z = 6$$

$$3x + 4y - 7z = 1$$

$$2x - y + 3z = 5$$

4. Find x and y in the figure shown, and then find the measure of each of the three angles.

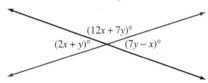

Use matrices to solve each system in Problems 5–6.

5. $\quad\quad x + 2y - z = -3$
$\quad\quad 2x - 4y + z = -7$
$-2x + 2y - 3z = 4$

6. $x - 2y + z = 2$
$2x - y - z = 1$

In Problems 7–10, let

$$A = \begin{bmatrix} 3 & 1 \\ 1 & 0 \\ 2 & 1 \end{bmatrix}, \quad B = \begin{bmatrix} 1 & -1 \\ 2 & 1 \end{bmatrix}, \quad \text{and} \quad C = \begin{bmatrix} 1 & 2 \\ -1 & 3 \end{bmatrix}.$$

Carry out the indicated operations.

7. $2B + 3C$ **8.** AB

9. C^{-1} **10.** $BC - 3B$

11. If $A = \begin{bmatrix} 1 & 2 & 2 \\ 2 & 3 & 3 \\ 1 & -1 & -2 \end{bmatrix}$ and $B = \begin{bmatrix} -3 & 2 & 0 \\ 7 & -4 & 1 \\ -5 & 3 & -1 \end{bmatrix}$,
show that B is the inverse of A.

12. Consider the system

$$3x + 5y = 9$$

$$2x - 3y = -13.$$

   a. Express the system in the form $AX = B$, where A, X, and B are appropriate matrices.

b. Find A^{-1}, the inverse of the coefficient matrix.

c. Use A^{-1} to solve the given system.

13. Evaluate: $\begin{vmatrix} 4 & -1 & 3 \\ 0 & 5 & -1 \\ 5 & 2 & 4 \end{vmatrix}$.

Find the partial fraction decomposition for each rational expression in Problems 15–17.

15. $\dfrac{2x + 1}{x^2 - 4}$

16. $\dfrac{2x^2 - 4x + 5}{(x + 3)(x^2 + 2x + 4)}$

17. $\dfrac{x^3}{(x^2 + 1)^2}$

18. Graph the solution set:

$$2x - y \leqslant 0$$

$$x + 2y \leqslant 4$$

19. The table shows the number of inmates in federal and state prisons in the United States for three selected years.

x (Number of Years after 1980)	1	5	10
y (Number of Inmates, in thousands)	344	480	740

a. Use the quadratic function $y = ax^2 + bx + c$ to model the data.

b. Predict the number of inmates in the year 2010.

c. List one factor that would change the accuracy of this model for the year 2010.

20. A nutritionist is arranging special diets that consist of a combination of three basic foods. The diet must contain exactly 170 units of calcium, 90 units of iron, and 110 units of vitamin A each day. The amounts of these nutrients in one ounce of each food are given in the table in the next column.

14. Solve for x only using Cramer's rule:

$$3x + y - 2z = -3$$

$$2x + 7y + 3z = 9$$

$$4x - 3y - z = 7$$

	Units per Ounce		
	Calcium	**Iron**	**Vitamin A**
Food A	20	10	10
Food B	15	5	10
Food C	5	5	15

How many ounces of each food must be used each day to satisfy the nutrient requirements exactly?

21. In triangle ABC, angle A measures 10° less than the sum of the measures of angles B and C, and angle B measures 50° less than the sum of the measures of angles A and C.

a. Let x represent the measure of angle A, y the measure of angle B, and z the measure of angle C. Write a system of three equations in three variables that models the problem's conditions.

b. Use matrices to solve the system and find the measure of each of the triangle's angles.

22. A paper manufacturing company converts wood pulp to writing paper and newsprint. The profit on a unit of writing paper is $500 and the profit on a unit of newsprint is $350. However, the manufacturer is bound by two constraints: Equipment in the factory allows for making at most 200 units of paper (writing paper and newsprint) in a day. Furthermore, regular customers require at least 10 units of writing paper and at least 80 units of newsprint daily. How many units of writing paper and newsprint should the company produce each day to make the greatest profit? What is the maximum daily profit?

CUMULATIVE REVIEW PROBLEMS (CHAPTERS P–5)

1. Simplify: $\dfrac{2x^2 - x - 10}{2x^2 + 7x + 6}$.

Solve Problems 2–5.

2. $\dfrac{x - 3}{2} \geqslant \dfrac{4x + 1}{5} + 4$

3. $\dfrac{x + 5}{x - 1} > 2$

4. $2x^3 + x^2 - 13x + 6 = 0$

5. $\log (x + 1) + \log (x - 1) = \log (x + 5)$

6. If $f(x) = 11x^2 - 19x + 17$, find $\dfrac{f(a + h) - f(a)}{h}$.

7. Use the graph of f to sketch the graph of $y = 1 + f(x - 1)$.

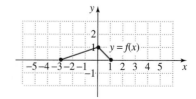

8. Explain why f and g are inverses of each other based on the graphs shown in the figure.

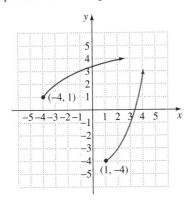

9. The function f is defined by

$$f(x) = \begin{cases} x + 4 & \text{if } x < 0 \\ 4 & \text{if } 0 \le x < 1 \\ 5 - x^2 & \text{if } x \ge 1 \end{cases}$$

Graph f and identify open intervals on which f is increasing, decreasing, and constant.

10. Consider the function $f(x) = x^5 - x$.

  a. Use factoring to find all zeros of f.

  b. Use the zeros of f and the Leading Coefficient Test to graph the function.

11. Show that all the real roots of $16x^3 - 52x^2 - 44x + 105 = 0$ lie between -2 and 4.

12. Graph: $f(x) = \dfrac{x^2 - x - 6}{x + 1}$.

13. Evaluate without a calculator: $\log_2 64 - \log_{25} 5$.

14. Expand and evaluate one of the resulting logarithmic expressions:

$$\ln \frac{27}{e^{5t}}.$$

15. Use the compound interest model $A = Pe^{rt}$ to answer this question:

At what annual rate, compounded continuously, must $6000 be invested to grow to $18,000 in 10 years?

16. Solve using matrices:

$$x_1 + 2x_2 - 3x_3 - 2x_4 = -8$$
$$3x_1 + 4x_2 + 2x_3 + 3x_4 = 10$$
$$-x_1 - 6x_2 + 8x_3 + x_4 = 0$$
$$8x_1 - 7x_2 - x_3 - 2x_4 = -31$$

17. If $A = \begin{bmatrix} 1 & -2 \\ 1 & 0 \end{bmatrix}$ and $B = \begin{bmatrix} -1 & 2 \\ -1 & 1 \end{bmatrix}$, find $AB + BA$.

18. Graph the solution set:

$$2x + 3y \ge 12$$
$$y \le \tfrac{1}{2}x - 4$$

19. A brick path of uniform width is constructed around the outside of a 30 by 40 foot rectangular pool. Find how wide the path should be if there is enough brick to cover 296 square feet.

20. The owner of a barn plans to fence in a rectangular region using 100 feet of fencing. As shown in the figure, no fencing is to be used for the side that is along the barn. Find the dimensions that maximize the enclosed area. What is the maximum area?

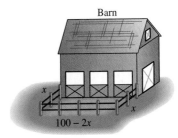

Conic Sections and Nonlinear Systems

Geospace / Science Photo Library / Photo Researchers, Inc.

Intriguing signs show that the world is profoundly mathematical. Over 2000 years ago the ancient Greeks studied ellipses without regard to their immediate usefulness simply because the study elicited ideas that were exciting, challenging, and interesting. Much later, German scientist and mathematician Johannes Kepler (1571–1630) discovered that the ellipse is the path through which Earth, and humanity along with it, journeys through space. The voluntary creation of the human mind called mathematics, invented for aesthetic reasons, gives us an ever-widening grip on the physical world, leading to the philosophic dispute over whether math is discovered or invented. Does it exist out there with Plato's Forms, waiting for the mind to see it? Or do people make it up like, say, a song?

This much is certain: When viewed closely, the familiar face of the ordinary world can be described mathematically. The mathematics of the world is present in the movement of planets, bridge construction, navigational systems used to locate lost ships, manufacture of lenses for telescopes (including the Hubble Space Telescope), and even a procedure for disintegrating kidney stones. The mathematics behind these applications involves conic sections, the focus of this chapter.

S E C T I O N   6 . 1

Solutions   Tutorial   Video
Manual   II

The Ellipse

Objectives

1 Graph ellipses centered at the origin.
2 Write equations of ellipses in standard form.
3 Graph ellipses not centered at the origin.
4 Solve applied problems involving ellipses.

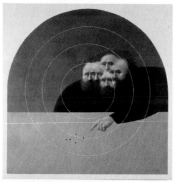

Alfredo Castaneda "Aqui es el
centro" 1984, oil on canvas,
39 3/8 × 39 3/8 in. (100.0 × 100.0 cm).
Courtesy of Mary–Anne Martin /
Fine Art, New York.

Conic sections (or conics, for short) were so named after their historical discovery as intersections of a plane and a right circular cone of two nappes (Figure 6.1). A *circle* results when the cone is cut with a plane that is perpendicular to the axis, but does not pass through the vertex. Incline the plane slightly so that it intersects only one nappe, and the intersected section is an *ellipse*. Incline the plane still more so that it is parallel to a line on the surface of the cone, but intersects only one nappe, and the resulting intersection is a *parabola*. Continue tilting the plane so that it intersects both nappes, but does not contain the vertex, and the resulting intersection is a *hyperbola*, a curve with two branches.

In this section, we study the symmetric oval-shaped curve known as the ellipse (see Figure 6.1). We will use a geometric definition for an ellipse to derive its standard equations. With these equations we can then consider ways in which the ellipse is useful in modeling physical phenomena, including the orbit of planets.

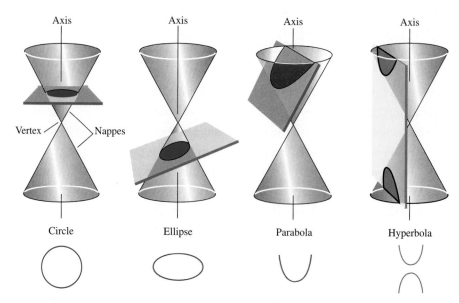

Figure 6.1

Obtaining the conics by intersecting a plane and a cone

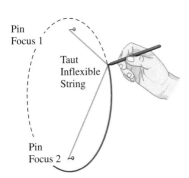

Figure 6.2

Drawing an ellipse

Definition of an Ellipse

Figure 6.2 illustrates how to draw an ellipse. Place two pins at two fixed points, each of which is called a focus (plural: foci). If the ends of a fixed length of string are fastened to the pins and we draw the string taut with a pencil, the path traced by the pencil will be an ellipse. Notice that the sum of the distances of the pencil point from the foci remains constant, because the length of the string is fixed. This procedure for drawing an ellipse illustrates its definition.

Definition of an ellipse

An *ellipse* is the set of all points in a plane the sum of whose distances from two fixed points F_1 and F_2 is constant (see Figure 6.3). These two fixed points are called the *foci* (plural of *focus*). The midpoint of the segment connecting the foci is the *center* of the ellipse.

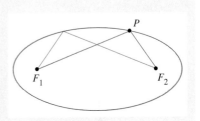

Figure 6.3

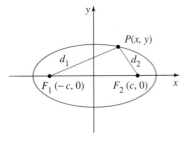

Figure 6.4

By the definition of an ellipse, $d_1 + d_2$ is constant.

The Equation of an Ellipse

Using the rectangular coordinate system, we can obtain equations for ellipses. To derive the simplest equation possible, place the foci on the x-axis at the points $(-c, 0)$ and $(c, 0)$, as in Figure 6.4. In this way the center of the ellipse is at the origin. For any point (x, y) on the ellipse, the sum of the distances to the two foci must be constant, which we will denote by $2a$ for simplicity.

$$d_1 + d_2 = 2a$$

The sum of the distances from any point on the ellipse to the foci is constant. Represent the constant by $2a$.

$$\sqrt{(x + c)^2 + y^2} + \sqrt{(x - c)^2 + y^2} = 2a$$

Use the distance formula. Recall that the distance between (x_1, y_1) and (x_2, y_2) is $\sqrt{(x_1 - x_2)^2 + (y_1 - y_2)^2}$.

$$\sqrt{(x + c)^2 + y^2} = 2a - \sqrt{(x - c)^2 + y^2}$$

Isolate the radical terms.

$$(x + c)^2 + y^2 = 4a^2 - 4a\sqrt{(x - c)^2 + y^2} + (x - c)^2 + y^2$$

Square both sides. On the right side, use $(A - B)^2 = A^2 - 2AB + B^2$ with $A = 2a$ and $B = \sqrt{(x - c)^2 + y^2}$.

$$x^2 + 2cx + c^2 + y^2 = 4a^2 - 4a\sqrt{(x - c)^2 + y^2} + x^2 - 2cx + c^2 + y^2$$

Square $x + c$ and $x - c$.

$$2cx = 4a^2 - 4a\sqrt{(x - c)^2 + y^2} - 2cx$$

Simplify by subtracting $x^2 + c^2 + y^2$ from both sides of the equation.

$$4a\sqrt{(x-c)^2+y^2}=4a^2-4cx$$

Again, isolate the radical term by transposing the radical to the left side and writing all terms without the radical on the right.

$$a\sqrt{(x-c)^2+y^2}=a^2-cx$$

Simplify by dividing both sides by 4.

$$a^2[(x-c)^2+y^2]=(a^2-cx)^2$$

Square both sides.

$$a^2(x^2-2cx+c^2+y^2)=a^4-2a^2cx+c^2x^2$$

Square $x-c$ and a^2-cx using $(A-B)^2 = A^2 - 2AB + B^2$.

$$a^2x^2-2a^2cx+a^2c^2+a^2y^2=a^4-2a^2cx+c^2x^2$$

Apply the distributive property on the left.

$$a^2x^2+a^2c^2+a^2y^2=a^4+c^2x^2$$

Simplify by adding $2a^2cx$ to both sides.

$$a^2x^2-c^2x^2+a^2y^2=a^4-a^2c^2$$

Write all terms involving the variables x^2 and y^2 on the left and all other terms on the right.

$$(a^2-c^2)x^2+a^2y^2=a^2(a^2-c^2)$$

Factor out x^2 from the first two terms on the left. Factor out a^2 on the right.

$$b^2x^2+a^2y^2=a^2b^2$$

From triangle F_1F_2P in Figure 6.4 on page 669, notice that $F_1F_2 < d_1 + d_2$. Equivalently, $2c < 2a$ and $c < a$. Consequently, $a^2 - c^2 > 0$. For convenience, let $b^2 = a^2 - c^2$.

$$\frac{b^2x^2}{a^2b^2}+\frac{a^2y^2}{a^2b^2}=\frac{a^2b^2}{a^2b^2}$$

Divide both sides by a^2b^2.

$$\frac{x^2}{a^2}+\frac{y^2}{b^2}=1$$

Simplify.

This shows that the coordinates of every point (x, y) on the ellipse satisfy the equation

$$\frac{x^2}{a^2}+\frac{y^2}{b^2}=1.$$

By reversing the preceding steps, we can show that if (x, y) is a solution of this equation, then the point (x, y) lies on the ellipse.

The x-intercepts of the ellipse are found by setting $y = 0$ in the equation. Doing so results in $\frac{x^2}{a^2} = 1$, or $x^2 = a^2$, so $x = \pm a$. The corresponding points $V(a, 0)$ and $V'(-a, 0)$ on the graph are called the *vertices* of the ellipse (see Figure 6.5). The line segment VV' that joins the vertices is called the *major axis*. Similarly, to find the y-intercepts, we set $x = 0$ in the equation, obtaining

$$\frac{y^2}{b^2}=1$$

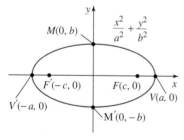

Figure 6.5

The x-intercepts for this ellipse are $-a$ and a. The y-intercepts are $-b$ and b.

or $y^2 = b^2$, so $y = \pm b$. Thus, the y-intercepts are b and $-b$. The line segment between $M'(0, -b)$ and $M(0, b)$ is called the *minor axis* of the ellipse. Since $a > b$, the major axis is always longer than the minor axis.

The equation $\dfrac{x^2}{a^2} + \dfrac{y^2}{b^2} = 1$ is unchanged if x is replaced with $-x$ or if y is replaced with $-y$ or if (x, y) is replaced with $(-x, -y)$. Consequently, the ellipse is symmetric with respect to the y-axis, the x-axis, and the origin.

The equation $\dfrac{x^2}{a^2} + \dfrac{y^2}{b^2} = 1$ is called the *standard form* of the equation of an ellipse. Our discussion is summarized as follows.

Standard form of the equation of an ellipse centered at (0, 0): Foci on x-axis

The graph of the equation

$$\frac{x^2}{a^2} + \frac{y^2}{b^2} = 1$$

for $a^2 > b^2$ is an ellipse with vertices $(a, 0)$ and $(-a, 0)$. (See Figure 6.6.) The endpoints of the minor axis are $(0, -b)$ and $(0, b)$. The foci are $(-c, 0)$ and $(c, 0)$, where $b^2 = a^2 - c^2$.

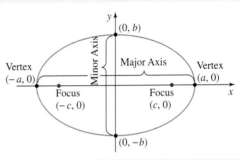

Figure 6.6

| Graph ellipses centered at the origin.

EXAMPLE 1 Graphing an Ellipse Centered at the Origin

Sketch the graph of $4x^2 + 9y^2 = 36$. Find the foci.

Solution

$$4x^2 + 9y^2 = 36 \qquad \text{This is the given equation.}$$

$$\frac{4x^2}{36} + \frac{9y^2}{36} = \frac{36}{36} \qquad \text{Divide both sides by 36 to rewrite the equation in standard form.}$$

$$\frac{x^2}{9} + \frac{y^2}{4} = 1 \qquad \text{This is the standard form } \frac{x^2}{a^2} + \frac{y^2}{b^2} = 1 \text{ with } a^2 = 9 \text{ and } b^2 = 4.$$

$$\underset{a^2 = 9}{\uparrow} \quad \underset{b^2 = 4}{\nwarrow}$$

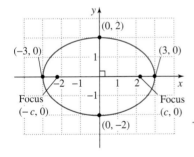

Figure 6.7

The graph of $\dfrac{x^2}{9} + \dfrac{y^2}{4} = 1$

Because $a^2 = 9$, the vertices $[(a, 0)$ and $(-a, 0)]$ are $(3, 0)$ and $(-3, 0)$, as shown in Figure 6.7. With $b^2 = 4$, the endpoints of the minor axis $[(0, -b)$ and $(0, b)]$ are $(0, -2)$ and $(0, 2)$. Because $a^2 = 9$ and $b^2 = 4$, we find the foci using $b^2 = a^2 - c^2$.

$$b^2 = a^2 - c^2 \qquad \text{Remember that the foci are } (-c, 0) \text{ and } (c, 0).$$

$$4 = 9 - c^2 \qquad \text{Substitute 4 for } b^2 \text{ and 9 for } a^2, \text{ obtained from the equation's standard form.}$$

$$c^2 = 5 \qquad \text{Solve for } c^2.$$

Study tip

Although the definition of the ellipse is given in terms of its foci, the foci are not part of the graph of the ellipse. A complete graph can be obtained without graphing the foci.

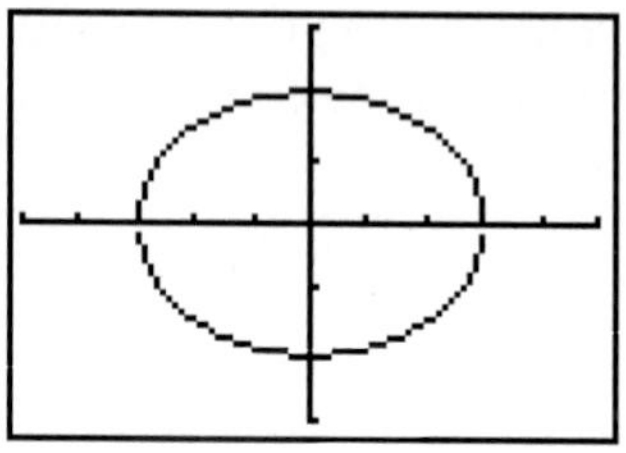

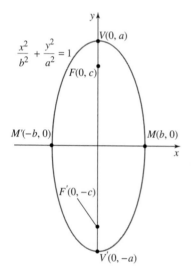

Figure 6.8

The graph of $\dfrac{x^2}{9} + \dfrac{y^2}{4} = 1$ obtained with a graphing utility

Thus, the foci [$(-c, 0)$ and $(c, 0)$] are $(-\sqrt{5}, 0)$ and $(\sqrt{5}, 0)$. The graph is shown in Figure 6.7.   ∎

Using technology

Because the ellipse cannot be represented as a single function, we can graph $4x^2 + 9y^2 = 36$ by solving for y and defining two functions:

$$\frac{x^2}{9} + \frac{y^2}{4} = 1 \qquad \text{Use the standard form of the equation.}$$

$$\frac{y^2}{4} = 1 - \frac{x^2}{9} \qquad \text{We will solve for } y. \text{ Isolate the term containing } y^2.$$

$$y^2 = 4\left(1 - \frac{x^2}{9}\right) \qquad \text{Multiply both sides by 4.}$$

$$y = \pm 2\sqrt{1 - \frac{x^2}{9}} \qquad \text{Apply the square root method.}$$

If you graph the two equations

$$Y_1 \boxed{=} 2\boxed{\sqrt{}}(1 - X \boxed{\wedge} 2/9) \quad \text{and} \quad Y_2 \boxed{=} -2\boxed{\sqrt{}}(1 - X \boxed{\wedge} 2/9) \quad (\text{or } Y_2 = -Y_1)$$

you should obtain the upper and lower portions of the graph shown in Figure 6.8. To see the true shape of the ellipse, use the $\boxed{\text{ZOOM SQUARE}}$ feature so that one unit on the x-axis is the same length as one unit on the y-axis.

It is possible to have an ellipse in standard position with the major axis along the y-axis, as shown in Figure 6.9. With foci $(0, c)$ and $(0, -c)$, then by the same type of derivation used previously, we obtain the following.

Figure 6.9

Ellipses can be elongated vertically, with foci along the y-axis.

Standard form of the equation of an ellipse centered at (0, 0): Foci on y-axis

The graph of the equation

$$\frac{x^2}{b^2} + \frac{y^2}{a^2} = 1$$

for $a^2 > b^2$ is an ellipse with vertices $(0, a)$ and $(0, -a)$. (See Figure 6.10.) The endpoints of the minor axis are $(-b, 0)$ and $(0, b)$. The foci are $(0, -c)$ and $(0, c)$, where $b^2 = a^2 - c^2$.

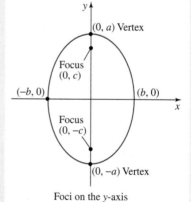

Foci on the y-axis

Figure 6.10

EXAMPLE 2   **Graphing an Ellipse with Foci on the y-Axis**

Sketch the graph of $25x^2 + 16y^2 = 400$. Find the foci.

Solution

$$25x^2 + 16y^2 = 400 \qquad \text{This is the given equation.}$$

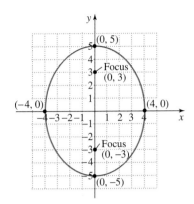

Figure 6.11

The graph of $\dfrac{x^2}{16} + \dfrac{y^2}{25} = 1$

 Write equations of ellipses in standard form.

$$\frac{25x^2}{400} + \frac{16y^2}{400} = \frac{400}{400} \qquad \text{Divide both sides by 400 to rewrite the equation in standard form.}$$

$$\frac{x^2}{16} + \frac{y^2}{25} = 1 \qquad \text{This is the standard form } \frac{x^2}{b^2} + \frac{y^2}{a^2} = 1. \text{ Remember that } a^2 > b^2.$$

$$b^2 = 16 \qquad a^2 = 25$$

Because $a^2 = 25$ and the major axis is vertical, the vertices $[(0, a)$ and $(0, -a)]$ are $(0, 5)$ and $(0, -5)$, as shown in Figure 6.11. With $b^2 = 16$, the endpoints of the minor axis $[(-b, 0)$ and $(b, 0)]$ are $(-4, 0)$ and $(4, 0)$. Because $a^2 = 25$ and $b^2 = 16$, we find the foci $[(0, -c)$ and $(0, c)$ on the y-axis$]$ using $b^2 = a^2 - c^2$. We obtain $16 = 25 - c^2$, or $c^2 = 9$. Thus, the foci are $(0, -3)$ and $(0, 3)$. The graph is shown in Figure 6.11. ∎

Writing Equations of Ellipses

In Examples 1 and 2, we used the equation of an ellipse to find its foci and vertices. In the next example, we reverse this procedure.

EXAMPLE 3 Finding the Equation of an Ellipse from Its Foci and Vertices

Find the standard form of the equation for an ellipse with foci at $(-1, 0)$ and $(1, 0)$ and vertices $(-2, 0)$ and $(2, 0)$.

Solution

The vertices are $(-2, 0)$ and $(2, 0)$, which are on the x-axis. The center of the ellipse is midway between the vertices, located at $(0, 0)$, so the form of the equation is $\dfrac{x^2}{a^2} + \dfrac{y^2}{b^2} = 1$.

$$\frac{x^2}{2^2} + \frac{y^2}{b^2} = 1 \qquad \text{The distance from the center to a vertex is 2, so } a = 2.$$

$$\frac{x^2}{4} + \frac{y^2}{b^2} = 1 \qquad \text{We must still find } b^2. \text{ The distance from the center to a focus is 1, so } c = 1. \text{ Since } b^2 = a^2 - c^2, \text{ we have } b^2 = 2^2 - 1^2 = 3.$$

$$\frac{x^2}{4} + \frac{y^2}{3} = 1 \qquad \text{Substitute 3 for } b^2 \text{ in the standard form of the equation.} \qquad ∎$$

We can also find the equation of an ellipse when only the vertices and one other point are known, as Example 4 illustrates.

EXAMPLE 4 Finding the Equation of an Ellipse

Find the standard form of the equation for an ellipse with vertices at $(0, -6)$ and $(0, 6)$, passing through $(2, -4)$.

Solution

Since the vertices are on the y-axis, so are the foci. The form of the equation is

$$\frac{x^2}{b^2} + \frac{y^2}{a^2} = 1.$$

Discover for yourself

What happens to the equation of an ellipse

$$\frac{x^2}{a^2} + \frac{y^2}{b^2} = 1$$

if $a = b$?

Substitute a for b in the equation and then clear fractions. What conic section does the resulting equation represent?

The vertices $[(0, -a)$ and $(0, a)]$ are $(0, -6)$ and $(0, 6)$, so $a = 6$.

$$\frac{x^2}{b^2} + \frac{y^2}{6^2} = 1$$

$$\frac{x^2}{b^2} + \frac{y^2}{36} = 1$$

Because $(2, -4)$ is a point on the ellipse, its coordinates satisfy the preceding equation.

$$\frac{2^2}{b^2} + \frac{(-4)^2}{36} = 1 \qquad \text{Substitute 2 for } x \text{ and } -4 \text{ for } y.$$

$$\frac{4}{b^2} + \frac{16}{36} = 1$$

$$\frac{4}{b^2} + \frac{4}{9} = 1 \qquad \text{Reduce the fraction.}$$

$$9b^2\left(\frac{4}{b^2} + \frac{4}{9}\right) = 9b^2(1) \qquad \text{Clear fractions by multiplying both sides by } 9b^2.$$

$$36 + 4b^2 = 9b^2 \qquad \text{Apply the distributive property and simplify.}$$

$$36 = 5b^2 \qquad \text{Subtract } 4b^2 \text{ from both sides.}$$

$$\frac{36}{5} = b^2 \qquad \text{Divide both sides by 5.}$$

Substituting this value for b^2 in the preceding equation will give us the standard form for the ellipse's equation.

$$\frac{x^2}{b^2} + \frac{y^2}{36} = 1$$

$$\frac{x^2}{\frac{36}{5}} + \frac{y^2}{36} = 1$$

3 Graph ellipses not centered at the origin.

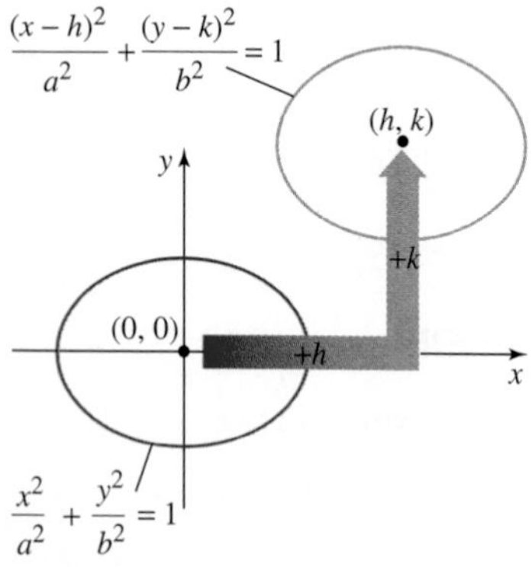

$$\frac{(x-h)^2}{a^2} + \frac{(y-k)^2}{b^2} = 1$$

(h, k)

$(0, 0)$

$+k$

$+h$

$$\frac{x^2}{a^2} + \frac{y^2}{b^2} = 1$$

Figure 6.12

Translating an ellipse's graph

This equation can equivalently be expressed as

$$\frac{5x^2}{36} + \frac{y^2}{36} = 1.$$

∎

Translations of Ellipses

Despite the fact that an ellipse is not the graph of a function, its graph can be translated in the same manner as that of a function. Figure 6.12 illustrates that the graphs of

$$\frac{(x-h)^2}{a^2} + \frac{(y-k)^2}{b^2} = 1 \quad \text{and} \quad \frac{x^2}{a^2} + \frac{y^2}{b^2} = 1$$

have the same size and shape. However, the graph of the first equation is centered at (h, k) rather than at the origin. We can obtain the graph of the first equation by translating the graph of the second one horizontally h units and vertically k units.

EXAMPLE 5   Graphing an Ellipse Centered at (h, k)

Sketch the graph of $\dfrac{(x-1)^2}{4} + \dfrac{(y+2)^2}{9} = 1$. Find the foci.

Solution

We obtain the graph of this equation by translating the graph of

$$\frac{x^2}{4} + \frac{y^2}{9} = 1$$

to the right one unit and down two units, as illustrated in Figure 6.13. The center of the ellipse is $(h, k) = (1, -2)$. Since $a^2 = 9$, $a = 3$ and the vertices lie three units above and below the center. Also, since $b^2 = 4$, $b = 2$ and the endpoints of the minor axis lie two units to the right and left of the center. We categorize these observations as follows:

Center	Vertices	Endpoints of Minor Axis
$(1, -2)$	$(1, -2 + 3) = (1, 1)$	$(1 + 2, -2) = (3, -2)$
	$(1, -2 - 3) = (1, -5)$	$(1 - 2, -2) = (-1, -2)$

Using the center and these four points, we can sketch the ellipse shown in Figure 6.13.

With $b^2 = a^2 - c^2$, we have $4 = 9 - c^2$, and $c^2 = 5$. So the foci are located $\sqrt{5}$ units above and below the center, at $(1, -2 + \sqrt{5})$ and $(1, -2 - \sqrt{5})$. ∎

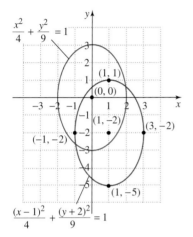

Study tip

$$\frac{(x-h)^2}{b^2} + \frac{(y-k)^2}{a^2} = 1$$

$$\frac{(x-1)^2}{2^2} + \frac{(y-(-2))^2}{3^2} = 1$$

$h = 1, k = -2$

center $= (h, k) = (1, -2)$

$\dfrac{x^2}{4} + \dfrac{y^2}{9} = 1$

$(1, 1)$
$(0, 0)$
$(1, -2)$
$(3, -2)$
$(-1, -2)$
$(1, -5)$

$\dfrac{(x-1)^2}{4} + \dfrac{(y+2)^2}{9} = 1$

Figure 6.13

Using translations to graph an ellipse

Using technology

As in Example 1, we can graph the ellipse in Example 5 by first solving the given equation for y.

$$\frac{(x-1)^2}{4} + \frac{(y+2)^2}{9} = 1 \qquad \text{This is the given equation.}$$

$$\frac{(y+2)^2}{9} = 1 - \frac{(x-1)^2}{4} \qquad \text{To solve for } y, \text{ isolate the term containing } y.$$

$$(y+2)^2 = 9\left[1 - \frac{(x-1)^2}{4}\right] \qquad \text{Multiply both sides by 9.}$$

$$y + 2 = \pm 3\sqrt{1 - \frac{(x-1)^2}{4}} \qquad \text{Apply the square root method.}$$

$$y = -2 \pm 3\sqrt{1 - \frac{(x-1)^2}{4}} \qquad \text{Solve for } y.$$

If you graph the two equations

$$Y_1 \boxed{=} -2 + 3\boxed{\sqrt{}}(1 - (X - 1)\boxed{\wedge}2/4)$$

and

$$Y_2 \boxed{=} -2 - 3\boxed{\sqrt{}}(1 - (X - 1)\boxed{\wedge}2/4)$$

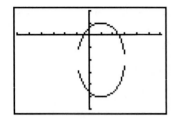

Figure 6.14

The graph of
$$\frac{(x - 1)^2}{4} + \frac{(y + 2)^2}{9} = 1$$
The utility that we used left gaps where the graph is nearly vertical.

you will obtain the upper and lower portions of the graph of the ellipse. The graph in Figure 6.14 was obtained using

    Xmin = −6, Xmax = 6, Xscl = 1, Ymin = −6, Ymax = 2, Yscl = 1

and then using the [ZOOM SQUARE] feature. The graph looks like our hand-drawn graph in Figure 6.13.

We can summarize our discussion of the standard equation of an ellipse, including translations, as follows.

Standard form of the equation of an ellipse centered at (h, k)

The standard form of the equation of an ellipse centered at (h, k), with major and minor axes of lengths $2a$ and $2b$ (where $a > b$) is

$$\frac{(x - h)^2}{a^2} + \frac{(y - k)^2}{b^2} = 1 \qquad \text{Major axis is horizontal. (See Figure 6.15.)}$$

$$\frac{(x - h)^2}{b^2} + \frac{(y - k)^2}{a^2} = 1 \qquad \text{Major axis is vertical. (See Figure 6.16.)}$$

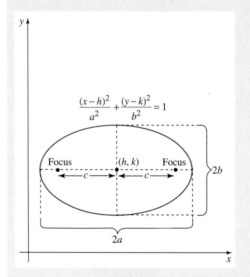

Figure 6.15

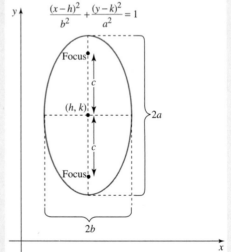

Figure 6.16

The foci lie on the major axis, c units from the center, with $b^2 = a^2 - c^2$ (or, equivalently $c^2 = a^2 - b^2$). The vertices are the endpoints of the major axis.

If the center is at the origin $(0, 0)$, the standard form is

$$\frac{x^2}{a^2} + \frac{y^2}{b^2} = 1 \qquad \text{Major axis is horizontal.}$$

$$\frac{x^2}{b^2} + \frac{y^2}{a^2} = 1 \qquad \text{Major axis is vertical.}$$

We can write an equation of an ellipse in standard form by completing the square on x and y. Let's see how this is done.

EXAMPLE 6 Writing an Equation in Standard Form by Completing the Square

Sketch the graph of the ellipse whose equation is

$$9x^2 + 25y^2 - 54x + 50y - 119 = 0.$$

Solution

Because we plan to complete the square on both x and y, let's rearrange terms so that x-terms are arranged in descending order, y-terms are arranged in descending order, and the constant term appears on the right.

$$9x^2 + 25y^2 - 54x + 50y - 119 = 0$$

This is the given equation.

$$(9x^2 - 54x) + (25y^2 + 50y) = 119$$

Rewrite in anticipation of completing the square.

$$9(x^2 - 6x \quad) + 25(y^2 + 2y \quad) = 119$$

To complete the square, the coefficients of x^2 and y^2 must be 1. Factor 9 and 25, respectively.

We've added 81.   We've added 25.

$$9(x^2 - 6x + 9) + 25(y^2 + 2y + 1) = 119 + 81 + 25$$

Complete the square on x: $\frac{1}{2}(-6) = -3$ and $(-3)^2 = 9$. Do the same on y: $\frac{1}{2}(2) = 1$ and $1^2 = 1$.

$$9(x - 3)^2 + 25(y + 1)^2 = 225$$

Factor.

$$\frac{9(x - 3)^2}{225} + \frac{25(y + 1)^2}{225} = \frac{225}{225}$$

In standard form, the constant term on the right is 1. Divide both sides by 225.

$$\frac{(x - 3)^2}{25} + \frac{(y + 1)^2}{9} = 1$$

Simplify.

This is the standard form

$$\frac{(x - h)^2}{a^2} + \frac{(y - k)^2}{b^2} = 1$$

with $(h, k) = (3, -1)$ and $a^2 = 25$, $b^2 = 9$.

The center of the ellipse is at $(3, -1)$. The ellipse passes through points five units to the right and left of the center and three units above and below the center. Therefore, we categorize these observations as follows:

Study tip

$$\frac{(x - h)^2}{a^2} + \frac{(y - k)^2}{b^2} = 1$$

$$\frac{(x - 3)^2}{5^2} + \frac{(y - (-1))^2}{3^2} = 1$$

$h = 3, k = -1$

center $= (h, k) = (3, -1)$

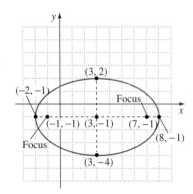

Figure 6.17

The graph of
$$\frac{(x - 3)^2}{25} + \frac{(y + 1)^2}{9} = 1$$

Center	Vertices	Endpoints of Minor Axis
$(3, -1)$	$(3 + 5, -1) = (8, -1)$	$(3, -1 + 3) = (3, 2)$
	$(3 - 5, -1) = (-2, -1)$	$(3, -1 - 3) = (3, -4)$

The graph of the ellipse is shown in Figure 6.17. We can find the foci by determining c using $c^2 = a^2 - b^2$.

$$c^2 = 25 - 9 = 16$$

The foci are four units to the right and left of the center.

Foci: $(3 + 4, -1) = (7, -1)$ and $(3 - 4, -1) = (-1, -1)$ ∎

Using technology

If we use a graphing utility to graph the ellipse in Example 6, we must first solve the given equation for y. To do this, we must use the quadratic formula.

$9x^2 + 25y^2 - 54x + 50y - 119 = 0$	We can use this form, which is the given equation.
$25y^2 + 50y + (9x^2 - 54x - 119) = 0$	Write the equation as a quadratic equation in y.
$y = \dfrac{-b \pm \sqrt{b^2 - 4ac}}{2a}$	Use the quadratic formula to solve for y.
$= \dfrac{-50 \pm \sqrt{50^2 - 4 \cdot 25(9x^2 - 54x - 119)}}{2 \cdot 25}$	$a = 25$, $b = 50$, and $c = 9x^2 - 54x - 119$
$= \dfrac{-50 \pm \sqrt{2500 - 100(9x^2 - 54x - 119)}}{50}$	

We do not have to simplify any further before using a graphing utility. As shown in Figure 6.18, the graph of

$$y = \frac{-50 + \sqrt{2500 - 100(9x^2 - 54x - 119)}}{50}$$

produces the upper half of the graph of the ellipse, and the graph of

$$y = \frac{-50 - \sqrt{2500 - 100(9x^2 - 54x - 119)}}{50}$$

produces the lower half. Using

$$\text{Xmin} = -3, \text{Xmax} = 9, \text{Xscl} = 1, \text{Ymin} = -5, \text{Ymax} = 3, \text{Yscl} = 1$$

and the ZOOM SQUARE feature, we obtain the graph of the ellipse shown in Figure 6.18. Notice that we obtained the graph without having to write the ellipse's equation in standard form.

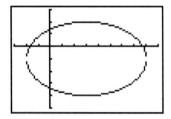

Figure 6.18

The graph of $9x^2 + 25y^2 - 54x + 50y - 119 = 0$, obtained with a graphing utility

Discover for yourself

A circle can be thought of as an ellipse in which the two foci coincide at the center. If the foci coincide, then $c = 0$, so $a = b$ and the ellipse becomes a circle with radius $r = a = b$. Substitute a for b in

$$\frac{(x - h)^2}{a^2} + \frac{(y - k)^2}{b^2} = 1$$

M. C. Escher (1898–1972) "Snakes"
1969. © 1994 M. C. Escher / Cordon
Art – Baarn – Holland. All Rights
Reserved.

and show that the equation of a circle with center at (h, k) and radius a is

$$(x - h)^2 + (y - k)^2 = a^2.$$

Since we can think of a circle as a special case of an ellipse, apply the technique of completing the square to

$$x^2 + y^2 + 6x - 2y + 6 = 0.$$

Determine the center and radius, and sketch the circle's graph.

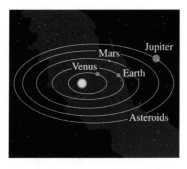

Planets move in elliptical orbits.

Whispering in an elliptical room

Disintegrating kidney stones

Applications

Ellipses have many applications. German scientist Johannes Kepler (1571–1630) showed that the planets in our solar system move in elliptical orbits, with the sun at a focus. Earth satellites also travel in elliptical orbits, with Earth at a focus.

One intriguing aspect of the ellipse is that a line from one focus will be reflected from the side of the ellipse exactly through the other focus. A whispering gallery is an elliptical room with an elliptical, dome-shaped ceiling. People standing at the foci can whisper and hear each other quite clearly, while persons in other locations in the room cannot hear them. Statuary Hall in the U.S. Capitol building is elliptical. President John Quincy Adams, while a member of the House of Representatives, was aware of this acoustical phenomenon. He situated his desk at a focal point of the elliptical ceiling, easily eavesdropping on the private conversations of other House members located near the other focus.

The elliptical reflection principle is used in a procedure for disintegrating kidney stones. The patient is placed within a device that is elliptical in shape. The patient is at one focus, while ultrasound waves from the other focus hit the walls and are reflected to the kidney stone. The convergence of the ultrasound waves at the kidney stone causes vibrations that shatter it into fragments. The small pieces can then be passed painlessly through the patient's system. The patient recovers in days, as opposed to up to six weeks if surgery is used instead.

Ellipses are often used for supporting arches of bridges and in tunnel construction. This application forms the basis of our next example.

4 Solve applied problems involving ellipses.

EXAMPLE 7   An Application Involving an Ellipse

A semielliptical archway over a one-way road has a height of 10 feet and a width of 40 feet (see Figure 6.19). A truck that is 10 feet wide has a height of 9 feet. Will the truck clear the opening of the archway?

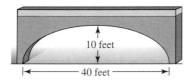

Figure 6.19

A semielliptical archway

Solution

To determine the clearance, we must find the height of the tunnel 5 feet from the center. If that height is 9 feet or less, the truck will not clear the opening.

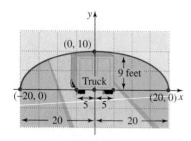

Figure 6.20

In Figure 6.20, we've constructed a coordinate system with the x-axis on the ground and the origin at the center of the archway. Also shown is the truck, whose height is 9 feet.

$$\frac{x^2}{a^2} + \frac{y^2}{b^2} = 1 \qquad \text{Use the equation of an ellipse.}$$

$$\frac{x^2}{20^2} + \frac{y^2}{10^2} = 1 \qquad \text{The } x\text{-intercept is 20 and the } y\text{-intercept is 10.}$$

$$\frac{x^2}{400} + \frac{y^2}{100} = 1$$

The Mormon Tabernacle in Salt Lake City, Utah, has the shape of a whispering room.
UPI / Corbis – Bettmann.

As shown in Figure 6.20, the edge of the 10-foot wide truck corresponds to $x = 5$. We find the height of the archway 5 feet from the center by substituting 5 for x and solving for y.

$$\frac{5^2}{400} + \frac{y^2}{100} = 1 \qquad \text{Substitute 5 for } x.$$

$$\frac{25}{400} + \frac{y^2}{100} = 1$$

$$\frac{1}{16} + \frac{y^2}{100} = 1$$

$$1600\left(\frac{1}{16} + \frac{y^2}{100}\right) = 1600(1) \qquad \text{Clear fractions by multiplying both sides by 1600.}$$

$$100 + 16y^2 = 1600 \qquad \text{Apply the distributive property and simplify.}$$

$$16y^2 = 1500 \qquad \text{Subtract 100 from both sides.}$$

$$y^2 = \frac{1500}{16} \qquad \text{Divide both sides by 16.}$$

$$y = \sqrt{\frac{1500}{16}} \qquad \begin{array}{l}\text{Take only the positive square root. The archway is}\\ \text{above the } x\text{-axis and } y \text{ is nonnegative.}\end{array}$$

$$y \approx 9.68$$

Thus, the height of the opening 5 feet from the center is approximately 9.68 feet. Since the truck's height is 9 feet, this is enough for it to clear the archway. ∎

Copernican Universe, 17th Century.
The Granger Collection, New York.

Although Johannes Kepler showed that the planets in our solar system revolve in elliptical orbits, some of these orbits are very flat, whereas others—like Earth's—are almost circular. To obtain information about the roundness or ovalness of an ellipse, we use its *eccentricity*, defined as follows.

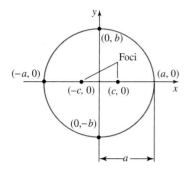

Figure 6.21

Ellipses with almost circular shapes have eccentricities close to 0.

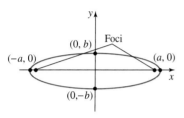

Figure 6.22

Ellipses that are elongated and very flat have eccentricities close to 1.

Definition of eccentricity

The *eccentricity e* of an ellipse is given by the ratio

$$e = \frac{c}{a} = \frac{\sqrt{a^2 - b^2}}{a}.$$

tudy tip

Do not confuse the use of the letter e to describe the eccentricity or shape of an ellipse with the base e of natural logarithms.

Let's see how eccentricity affects the shape of an ellipse by looking at the ellipse whose equation is

$$\frac{x^2}{a^2} + \frac{y^2}{b^2} = 1.$$

Figure 6.21 shows an ellipse that is nearly circular in shape. The foci are close to the center and c is small compared with a. Consequently, the ratio $\frac{c}{a}$ is close to 0. Since $e = \frac{\sqrt{a^2 - b^2}}{a}$, if $e \approx 0$ then $\sqrt{a^2 - b^2} \approx 0$ and $a \approx b$. If a and b are approximately equal, the ellipse is almost circular.

By contrast, Figure 6.22 shows a flattened, or elongated, ellipse. The foci are close to the vertices and c is nearly equal to a. Consequently, the ratio $\frac{c}{a}$ is close to 1. Since $e = \frac{\sqrt{a^2 - b^2}}{a}$, if $e \approx 1$ then $\sqrt{a^2 - b^2} \approx a$ and $b \approx 0$. If $b \approx 0$, the ellipse is very flat.

tudy tip

For every ellipse, its eccentricity e is a number between 0 and 1. Ellipses with almost circular shapes have eccentricities close to 0 and those that are flat have eccentricities close to 1. An ellipse changes from a circle to a line segment (no longer an ellipse) as e increases from 0 to 1.

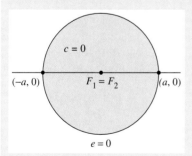

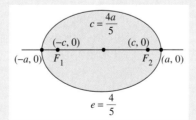

EXAMPLE 8   **Eccentricity of Earth's Orbit**

Aphelion and perihelion are those two points on an elliptic orbit at which a planet or comet is farthest from or closest to the sun, respectively. These points are the vertices of the orbit. Earth's distance from the sun at aphelion is 94.56

Perihelion
91.44
million
miles

Aphelion
94.56
million
miles

Center of
Elliptic
Orbit

Sun

c

a

a

Figure 6.23

Earth at aphelion and perihelion.

ENRICHMENT ESSAY

Modeling Planetary Motion

Polish astronomer Nicolaus Copernicus (1473–1543) postulated a sun-centered universe, although he incorrectly believed that celestial orbits move in perfect circles, calling his system the "ballet of the planets."

The Ptolemaic model

The Copernican model

million miles and its distance at perihelion is 91.44 million miles (see Figure 6.23). What is the eccentricity of Earth's orbit?

Solution

Figure 6.23 shows that the center of the sun is at one focus of Earth's elliptic orbit. The center of the Earth is a vertex at aphelion and at perihelion. The figure shows that

$a + c = 94.56$ Earth's maximum distance from the sun is 94.56 million miles.

$a - c = 91.44$ Earth's minimum distance from the sun is 91.44 million miles.

We can eliminate c from this linear system by adding the equations.

$$a + c = 94.56$$
$$\underline{a - c = 91.44}$$

Add: $2a = 186$

$a = 93$

Since $a + c = 94.56$ and $a = 93$, then $c = 1.56$.

Now we are ready to determine the eccentricity of Earth's orbit.

$e = \dfrac{c}{a}$ Use the formula for eccentricity of an ellipse.

$= \dfrac{1.56}{93}$ Substitute the values for c and a.

≈ 0.017

Earth's orbital eccentricity is approximately 0.017, which is close to zero, indicating a nearly circular orbit. ∎

Table 6.1 indicates that the planets in our solar system have orbits with eccentricities that are much closer to 0 than to 1. Most of these orbits are almost circular, which made it difficult for early astronomers to detect that they are actually ellipses.

TABLE 6.1	Eccentricities of Planetary Orbits		
Mercury	0.2056	Saturn	0.0543
Venus	0.0068	Uranus	0.0460
Earth	0.0167	Neptune	0.0082
Mars	0.0934	Pluto	0.2481
Jupiter	0.0484		

ENRICHMENT ESSAY

Halley's Comet

The time required for a comet to complete an elliptical orbit around the sun depends on the eccentricity of the orbit and ranges from as little as five years to many centuries. Halley's Comet, with an orbital eccentricity of 0.967, is named after the British astronomer Edmond Halley (1656–1742), who calculated the recurrence of its appearance every 76.3 years.

The first recorded sighting of Halley's Comet was in 239 B.C. As shown in the figure, it was last seen in 1986. At that time, spacecraft went close to the comet, measuring its nucleus to be 7 miles long and 4 miles wide. By 2024, Halley's Comet will have reached the farthest point in its elliptical orbit, traveling nearly to Pluto before returning to be next visible from Earth in 2062.

Houghton Cranford Smith "Comet, New Mexico" oil on canvas, 16 3/4 × 30 1/2 in. Courtesy, Richard York Gallery, New York.

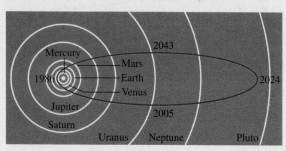

PROBLEM SET 6.1

Practice Problems

In Problems 1–14, sketch the graph of each ellipse by hand and find the foci. Verify your hand-drawn graph using a graphing utility.

1. $4x^2 + 25y^2 = 100$

2. $4x^2 + 9y^2 = 36$

3. $16x^2 + 4y^2 = 64$

4. $12x^2 + 4y^2 = 36$

5. $\dfrac{(x - 2)^2}{9} + \dfrac{(y - 1)^2}{4} = 1$

6. $\dfrac{(x - 1)^2}{16} + \dfrac{(y + 2)^2}{9} = 1$

7. $\dfrac{x^2}{25} + \dfrac{(y - 2)^2}{36} = 1$

8. $\dfrac{(x - 4)^2}{4} + \dfrac{y^2}{25} = 1$

9. $9x^2 + 25y^2 - 36x + 50y - 164 = 0$

10. $4x^2 + 9y^2 - 32x + 36y + 64 = 0$

11. $9x^2 + 16y^2 - 18x + 64y - 71 = 0$

12. $x^2 + 4y^2 + 10x - 8y + 13 = 0$

13. $4x^2 + y^2 + 16x - 6y - 39 = 0$

14. $4x^2 + 25y^2 - 24x + 100y + 36 = 0$

In Problems 15–18, find the standard form of the equation for each ellipse and identify its foci.

15.

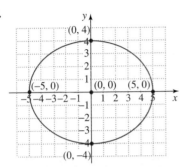

16.

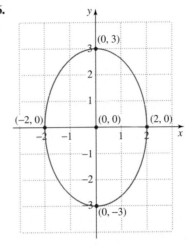

17.

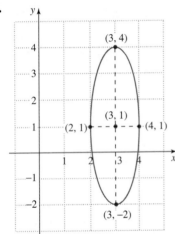

18.

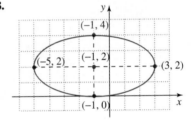

In Problems 19–32, find the standard form of the equation for each ellipse centered at the origin satisfying the given conditions.

19. Foci: $(-5, 0)$, $(5, 0)$; vertices: $(-8, 0)$, $(8, 0)$

20. Foci: $(0, -2)$, $(0, 2)$; vertices: $(0, -6)$, $(0, 6)$

21. Vertices: $(0, -5)$, $(0, 5)$; length of minor axis $= 4$

22. Foci: $(-3, 0)$, $(3, 0)$; length of minor axis $= 2$

23. x-intercepts: ± 4; y-intercepts: $\pm \frac{1}{2}$

24. x-intercepts: $\pm \frac{1}{3}$; y-intercepts: ± 2

25. Vertices: $(-6, 0)$, $(6, 0)$; passes through $(-4, 2)$

26. Vertices: $(0, -5)$, $(0, 5)$; passes through $(4, 2)$

27. Vertices: $(-4, 0)$, $(4, 0)$; $e = \frac{3}{4}$

28. Foci: $(-5, 0)$, $(5, 0)$; $e = \frac{1}{2}$

29. Foci: $(-3, 0)$, $(3, 0)$; Sum of the distances from any point on the ellipse to the foci is 10.

30. Foci: $(-4, 0)$, $(4, 0)$; Sum of the distances from any point on the ellipse to the foci is 9.

31. Foci: $(-6, 0)$, $(6, 0)$; passes through $(6, 5)$

32. Foci: $(-2, 0)$, $(2, 0)$; passes through $(2, 3)$

Application Problems

33. Will a truck that is 8 feet wide carrying a load that reaches 7 feet above the ground clear the semielliptical arch on the one-way road that passes under the bridge shown in the figure?

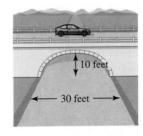

34. A semielliptic archway has a height of 20 feet and a width of 50 feet, as shown in the figure. Can a truck 14 feet high and 10 feet wide drive under the archway without going over the road's center line?

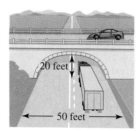

20 feet

50 feet

35. A planet with an elliptical orbit has a major axis of 100 million miles and minor axis of 81 million miles. Write an equation that models the planet's orbit and determine the distance between its foci.

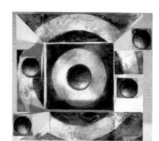

John R. Thompson "Orbital Space" 1992, enamel, oil, varnish, pencil on plywood, 12 1/2 × 13 × 5 3/4 in. Photo courtesy Luise Ross Gallery, New York.

36. The planet Pluto revolves around the sun in an elliptical orbit with the sun at one focus. Pluto's distance from the sun at aphelion is 7.37×10^9 kilometers and its distance at perihelion is 4.43×10^9 kilometers. What is the eccentricity of Pluto's orbit?

37. The planet Saturn moves in an elliptical orbit with the sun at one focus. Saturn's distance from the sun at aphelion is 1.5045×10^9 kilometers and at perihelion is 1.3495×10^9 kilometers. What is the eccentricity of Saturn's orbit?

38. If an elliptical whispering room has a height of 30 feet and a width of 100 feet, where should two people stand if they would like to whisper back and forth and be heard?

39. The figure shows that the orbit of Halley's Comet is an ellipse with eccentricity $e = 0.967$. The sun is at one focus of the orbit and the closest that the comet comes to the sun is 0.587 AU. (One AU, or astronomical unit, is the average distance from Earth to the sun, approximately 93,000,000 miles.) Find the comet's maximum distance from the sun in AU. (*Hint:* Using the figure shown, the minimum distance between the sun and the comet is represented by $a - c$ and the maximum distance is represented by $a + c$.)

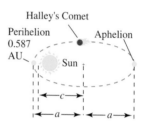

Halley's Comet

Perihelion 0.587 AU

Aphelion

Sun

c

a a

True–False Critical Thinking Problems

40. Which one of the following is true?

a. The three ellipses shown in the figure are arranged in order of decreasing eccentricities.

b. The ellipse whose equation is $\dfrac{x^2}{16} + \dfrac{y^2}{25} = 1$ has a major axis of length 8.

c. The equation of an ellipse centered at the origin with x-intercepts of 4 and -4 and y-intercepts of 9 and -9 is $\dfrac{x^2}{4} + \dfrac{y^2}{9} = 1$.

d. The equation of an ellipse does not define y as a function of x.

41. Which one of the following is true?

a. The circle whose equation is $x^2 + y^2 = 9$ and the ellipse described by $\dfrac{x^2}{25} + \dfrac{y^2}{16} = 1$ intersect twice.

b. The circle whose equation is $x^2 + y^2 = 4$ and the ellipse described by $4x^2 + y^2 = 4$ intersect at $(0, 2)$ and $(0, -2)$.

c. There are infinitely many points satisfying $\dfrac{x^2}{9} + \dfrac{y^2}{36} \leq 1$ and $x \geq 4$.

d. The eccentricity, e, of an ellipse is an irrational number that is approximately equal to 2.718.

Technology Problems

Use a graphing utility to graph each ellipse in Problems 42–45.

42. $\dfrac{x^2}{4} + \dfrac{y^2}{16} = 1$

43. $\dfrac{(x-2)^2}{9} + \dfrac{(y-1)^2}{4} = 1$

44. $16x^2 + 4y^2 + 96x - 8y + 84 = 0$

45. $4x^2 + 13y^2 + 6x + 4y - 10 = 0$

46. Graph $\dfrac{x^2}{2.9} + \dfrac{y^2}{2.1} = 1$ and $\dfrac{x^2}{4.3} + \dfrac{(y-2.1)^2}{4.9} = 1$ in the same viewing rectangle and find intersection points correct to the nearest hundredth.

In Problems 47–48, write each equation as a quadratic equation in y and then use the quadratic formula to express y in terms of x. Graph the resulting two equations using a graphing utility. What effect does the xy-term have on the graph of the resulting ellipse? What problems would you encounter if you attempted to write these equations in standard form by completing the square?

47. $2x^2 - xy + y^2 - 4 = 0$

48. $5x^2 - 8xy + 5y^2 - 9 = 0$

49. Write an equation for the path of the following elliptical orbits. Then use a graphing utility to graph the two ellipses in the same viewing rectangle. Describe what you observe.

Earth's orbit: Length of major axis: 186 million miles
Length of minor axis: 185.8 million miles
Mars's orbit: Length of major axis: 283.5 million miles
Length of minor axis: 278.5 million miles

50. a. Graph $x^2 + y^2 = (3950)^2$ using

$$\text{Xmin} = -4500, \text{Xmax} = 4500,$$

$$\text{Ymin} = -4500, \text{ and Ymax} = 4500.$$

This circular orbit is approximately the same as Earth's.

b. The first artificial satellite to orbit Earth was Sputnik I, launched by the former Soviet Union in 1957. The elliptical orbit of Sputnik I had the equation

$$\frac{(x + 225.5)^2}{4307.5^2} + \frac{y^2}{4301.6^2} = 1$$

where one of the foci was Earth's center. Graph this ellipse in the same viewing rectangle as the graph in part (a). Describe what you observe in terms of the altitude of Sputnik's orbit compared to the size of Earth.

Writing in Mathematics

51. Astronomer Johannes Kepler wrote, "The heavenly motions are nothing but a continuous song for several voices, to be perceived by the intellect, not by the ear." Describe what Kepler meant by this in terms of our work in this section.

Richard E. Prince "The Planets" 1979, copper, brass, sanddollars, rawhide, plexiglass, 36 × 24 × 6 in. Collection: Canada Council Art Bank. Photo: Courtesy of Equinox Gallery, Vancouver, Canada.

52. Are the graphs of $x^2 + 2y^2 + 3 = 0$ and $3x^2 + y^2 = 0$ ellipses? If not, describe the graph of each equation and explain why the equation does or does not represent an ellipse. (Use a graphing utility to verify your answers.)

53. Describe how you might use the method for tracing an ellipse shown in the figure to illustrate the concept of eccentricity.

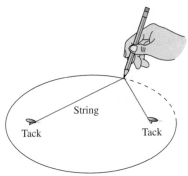

String

Tack   Tack

54. We have seen that the point in a planet's orbit at which the planet is closest to the sun is called *perihelion*, and the point at which it is farthest is called *aphelion*. What do you think is meant by *perilune* and *apolune*? What clue in each word might help determine its meaning? Describe the orbit and what you'd expect to find at the focus when these words are used.

55. An elliptipool is an elliptical pool table with only one pocket. A pool shark places a ball on the table, hits it in what appears to be a random direction, and yet it con-

stantly bounces off the cushion, falling directly into the pocket. Explain why this happens.

Critical Thinking Problems

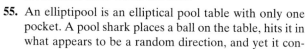

In Problems 56–62, find the standard form of the equation for each ellipse not centered at the origin satisfying the given conditions.

56. Center: (2, 2); one focus: (0, 2); one vertex: (5, 2)

57. Center: (1, −4); foci: (1, −2), (1, −6); length of minor axis = 10

58. Center: (1, 2); length of minor axis = 3; passes through (0, 5)

59. Center: (5, 2); passes through (1, 1) and (3, 0)

60. Foci: (−1, 2), (1, 2); length of major axis = 6

61. Foci: (−4, 3), (2, 3); vertices: (−6, 3), (4, 3)

62. Foci: (−1, 1), (−1, 7); passes through $(\frac{3}{4}, 1)$

63. The Apollo 11 spacecraft was placed in an elliptical orbit about the moon, with the center of the moon at one focus. The point in its orbit nearest the surface of the moon was 110 kilometers and the point farthest from the surface was 314 kilometers. If the radius of the moon is 1728 kilometers, write an equation for the spacecraft's elliptical orbit.

64. A line segment through the focus on an ellipse and perpendicular to the major axis is called a *latus rectum*. Prove that the length of each latus rectum in the figure is $\frac{2b^2}{a}$. Use this information to sketch the graph of $\frac{x^2}{4} + \frac{y^2}{9} = 1$.

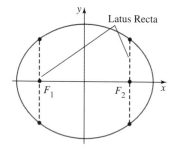

65. a. Show that the equation of an ellipse centered at (h, k) with eccentricity e can be written as

$$\frac{(x - h)^2}{a^2} + \frac{(y - k)^2}{a^2(1 - e^2)} = 1.$$

b. Describe what happens to the equation as e approaches zero.

66. A satellite travels about the Earth in an elliptical orbit with Earth at one focus. The greatest distance from Earth's center is A and the least distance is P. Prove that the eccentricity of the orbit is $e = \frac{A - P}{A + P}$.

67. Write the equation of the ellipse centered at the origin, passing through (2, 1), where the major axis is twice as long as the minor axis. (There are actually two such ellipses. Find the standard form of the equation for the one with the longer major axis.)

68. As shown in the figure, line segment CD has a length of $c + d$. The line segment moves with endpoint D on the y-axis. Prove that the coordinates of point P satisfy the equation of an ellipse.

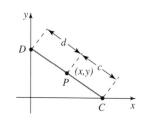

SECTION 6.2

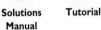

Solutions Manual	Tutorial	Video II

The Hyperbola

Objectives

1 Graph hyperbolas centered at the origin.
2 Write equations of hyperbolas in standard form.
3 Graph hyperbolas not centered at the origin.
4 Solve applied problems involving hyperbolas.

Figure 6.24 shows a cylindrical lampshade casting two shadows on a wall. These shadows indicate the distinguishing feature of hyperbolas: Their graphs contain two disjoint parts called *branches*. Although each branch might look like a parabola, its shape is actually quite different.

   Like ellipses, hyperbolas can be obtained by intersecting a plane and a cone, as shown in Figure 6.25. The definition of a hyperbola is similar to that of an ellipse.

Figure 6.24

Casting hyperbolic shadows

> **Definition of a hyperbola**
>
> A hyperbola is the set of points in a plane the *difference* of whose distances from two fixed points (called *foci*) is a constant. The point midway between the two foci is the *center* of the hyperbola.

Figure 6.25

Intersecting a plane and a cone to obtain a hyperbola

Because of the similarity of this definition to the ellipse, the equations for a hyperbola and an ellipse look very similar. To derive the simplest equation possible, place the foci on the *x*-axis at the points $(-c, 0)$ and $(c, 0)$, as in Figure 6.26 on page 689. In this way the center of the hyperbola is at the origin. For any point (x, y) on the hyperbola, the difference of the distances to the two foci must be constant, which we will denote by $2a$, just as we did for the ellipse.

$$|d_2 - d_1| = 2a$$

$$\left|\sqrt{(x + c)^2 + (y - 0)^2} - \sqrt{(x - c)^2 + (y - 0)^2}\right| = 2a \qquad \text{Use the distance formula.}$$

The procedure for simplifying this expression is identical to that used to derive an equation for an ellipse.

$$\frac{x^2}{a^2} - \frac{y^2}{c^2 - a^2} = 1 \qquad \text{Upon simplification, we can write the preceding equation in this form.}$$

$$\frac{x^2}{a^2} - \frac{y^2}{b^2} = 1 \qquad \text{Introduce the substitution } b^2 = c^2 - a^2.$$

This shows that the coordinates of every point (x, y) on the hyperbola satisfy the equation

$$\frac{x^2}{a^2} - \frac{y^2}{b^2} = 1.$$

Conversely, if (x, y) is a solution of this equation, then the point (x, y) lies on the hyperbola.

Study tip

There are two things you should remember about deriving the hyperbola's equation:

1. The substitution $b^2 = c^2 - a^2$ (or, equivalently, $a^2 + b^2 = c^2$) is needed to locate the foci when a and b are known.

2. The expression $2a$ represents the constant difference of the distances to the foci. We will use this later in applications.

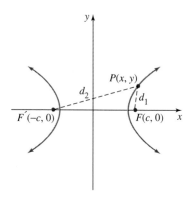

Figure 6.26

By the definition of a hyperbola, $|d_2 - d_1|$ is constant.

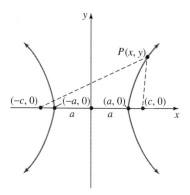

Figure 6.27

$\dfrac{x^2}{a^2} - \dfrac{y^2}{b^2} = 1$ has vertices at $(-a, 0)$ and $(a, 0)$.

The x-intercepts of the hyperbola are found by setting $y = 0$ in the equation. Doing so results in $\dfrac{x^2}{a^2} = 1$, or $x^2 = a^2$, so $x = \pm a$. The corresponding points $(a, 0)$ and $(-a, 0)$ are the *vertices* of the hyperbola (see Figure 6.27).

As with ellipses, hyperbolas centered at the origin can have foci on the x-axis or on the y-axis, illustrated in Figure 6.28. The intercepts of a hyperbola are called its *vertices*. The vertices and the foci are both located on either the x-axis or the y-axis. The line segment connecting the vertices is the *transverse axis*. Figure 6.28 illustrates that the midpoint of the transverse axis is the hyperbola's center.

Standard form of the equation of a hyperbola centered at (0, 0)

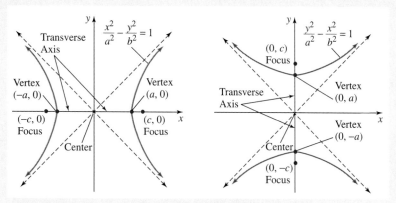

Figure 6.28

Foci can be located on the x-axis or on the y-axis. The value of c in both cases is found using $b^2 = c^2 - a^2$.

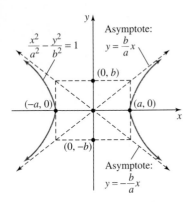

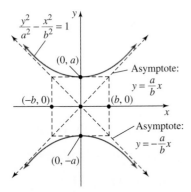

Figure 6.29

Hyperbolas centered at the origin and equations of asymptotes

The Asymptotes of a Hyperbola

As x and y get larger and larger, Figure 6.28 in the box on page 689 indicates that in both cases the two branches of the graph of a hyperbola approach a pair of intersecting straight lines called *asymptotes*. The asymptotes pass through the center of the hyperbola and are helpful in graphing a hyperbola.

In Figure 6.29, the equations of the asymptotes have been added to the two basic graphs of hyperbolas that are centered at the origin. The figure indicates that the asymptotes pass through the corners of a rectangle of dimensions $2a$ by $2b$. The line segment of length $2b$ is the *conjugate axis* of the hyperbola and is perpendicular to the transverse axis.

Let's take a moment to verify the equations of the asymptotes for the first case of a hyperbola, that is, the case where the transverse axis is horizontal. We do this by solving the hyperbola's equation for y.

$$\frac{x^2}{a^2} - \frac{y^2}{b^2} = 1 \qquad \text{This is the standard form equation of a hyperbola.}$$

$$\frac{y^2}{b^2} = \frac{x^2}{a^2} - 1 \qquad \text{We isolate the term involving } y^2 \text{ to solve for } y.$$

$$y^2 = \frac{b^2x^2}{a^2} - b^2 \qquad \text{Multiply both sides by } b^2.$$

$$y^2 = \frac{b^2x^2}{a^2}\left(1 - \frac{a^2}{x^2}\right) \qquad \text{Factor out } \frac{b^2x^2}{a^2} \text{ on the right. Verify that this result is correct by multiplying using the distributive property and obtaining the previous step.}$$

$$y = \pm\sqrt{\frac{b^2x^2}{a^2}\left(1 - \frac{a^2}{x^2}\right)} \qquad \text{Solve for } y \text{ using the square root method: If } u^2 = d, \text{ then } u = \pm\sqrt{d}.$$

$$y = \pm\frac{b}{a}x\sqrt{1 - \frac{a^2}{x^2}} \qquad \text{Simplify.}$$

As $|x| \to \infty$, the value of $\frac{a^2}{x^2}$ approaches 0. Consequently, the value of y can be approximated by

$$y = \pm\frac{b}{a}x.$$

In practical terms, this means that the lines $y = \frac{b}{a}x$ and $y = -\frac{b}{a}x$ are indeed asymptotes for the graph of the hyperbola.

| Graph hyperbolas centered at the origin.

Graphing Hyperbolas Centered at the Origin

Hyperbolas are graphed using vertices and asymptotes.

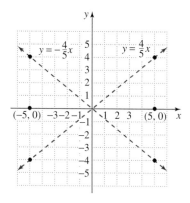

Figure 6.30

The asymptotes for
$$\frac{x^2}{25} - \frac{y^2}{16} = 1$$

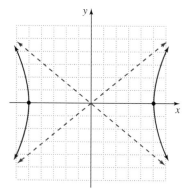

Figure 6.31

The graph of $\dfrac{x^2}{25} - \dfrac{y^2}{16} = 1$

Using technology

Graph $\dfrac{x^2}{25} - \dfrac{y^2}{16} = 1$ by solving

for y:

$$y_1 = \frac{\sqrt{16x^2 - 400}}{5}$$

$$y_2 = -\frac{\sqrt{16x^2 - 400}}{5}$$

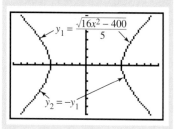

EXAMPLE I Graphing a Hyperbola

Sketch the graph of $\dfrac{x^2}{25} - \dfrac{y^2}{16} = 1$. Find the foci.

Solution

We can express the given equation as

$$\frac{x^2}{5^2} - \frac{y^2}{4^2} = 1.$$ The equation is in the form $\dfrac{x^2}{a^2} - \dfrac{y^2}{b^2} = 1$. Notice that $a = 5$ and $b = 4$.

This means that the x-intercepts are -5 and 5. The hyperbola's branches pass through $(-5, 0)$ and $(5, 0)$, which are its vertices.

Using $a = 5$ and $b = 4$, we find that the asymptotes are

$$y = \frac{b}{a}x \quad \text{and} \quad y = -\frac{b}{a}x$$

or

$$y = \frac{4}{5}x \quad \text{and} \quad y = -\frac{4}{5}x.$$

We sketch the asymptotes, as shown in Figure 6.30.

Now that we have the vertices and the asymptotes, through each vertex we draw a smooth curve that approaches the asymptotes closely. The graph of the hyperbola is shown in Figure 6.31. The asymptotes are drawn using dashed lines because they are not part of the hyperbola.

We find the foci using $b^2 = c^2 - a^2$.

$$b^2 = c^2 - a^2 \quad \text{Remember that the foci are } (-c, 0) \text{ and } (c, 0).$$

$$16 = c^2 - 25 \quad \text{Substitute 25 for } a^2 \text{ and 16 for } b^2.$$

$$c^2 = 41 \quad \text{Solve for } c^2.$$

$$c = \sqrt{41} \quad \text{We need only the positive square root.}$$

The foci are $(-\sqrt{41}, 0)$ and $(\sqrt{41}, 0)$, approximately $(-6.4, 0)$ and $(6.4, 0)$. ■

Study tip

Asymptotes can be drawn quickly by sketching a rectangle to use as a guide. To obtain the rectangle for

$$\frac{x^2}{5^2} - \frac{y^2}{4^2} = 1$$

use 5 and -5 on the x-axis (the intercepts) and 4 and -4 on the y-axis. The rectangle passes through these four points, shown in the figure on the right. Now draw dashed lines through the opposite corners of the rectangle to obtain the graph of the asymptotes.

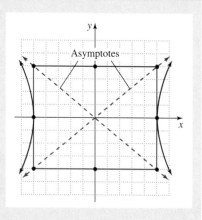

Discover for yourself

Work with $9y^2 - 4x^2 = 36$ and show that the y-intercepts are -2 and 2 by setting x equal to 0 and solving for y. Show that there are no x-intercepts by setting y equal to 0. Describe what happens.

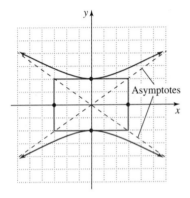

Figure 6.32

The graph of $9y^2 - 4x^2 = 36$

EXAMPLE 2   **Graphing a Hyperbola with No *x*-Intercepts**

Sketch the graph of $9y^2 - 4x^2 = 36$. Find the foci.

Solution

We begin by writing the equation in standard form. The right side should be 1, so we divide both sides by 36.

$$\frac{9y^2}{36} - \frac{4x^2}{36} = \frac{36}{36}$$

$$\frac{y^2}{4} - \frac{x^2}{9} = 1 \qquad \text{Simplify. The right side is now 1.}$$

We can express this equation as

$$\frac{y^2}{2^2} - \frac{x^2}{3^2} = 1. \qquad \text{The equation is in the form } \frac{y^2}{a^2} - \frac{x^2}{b^2} = 1 \text{ with } a = 2 \text{ and } b = 3.$$

We find the intercepts using the term without the minus sign. Thus, the intercepts are on the y-axis. In particular, the y-intercepts are -2 and 2. The vertices are $(0, -2)$ and $(0, 2)$.

We can construct a rectangle to find the asymptotes, using 2 and -2 on the y-axis (the intercepts) and 3 and -3 on the x-axis. The rectangle passes through these four points. We draw dashed lines through the opposite corners to show the asymptotes. The equations for these asymptotes are

$$y = \pm \frac{a}{b}x \quad \text{or} \quad y = \pm \frac{2}{3}x.$$

Finally, through each intercept we draw a smooth curve that approaches the asymptotes. The graph of the hyperbola is shown in Figure 6.32.

We again find the foci using $b^2 = c^2 - a^2$.

$$b^2 = c^2 - a^2 \qquad \text{This time the foci are on the } y\text{-axis and are } (0, -c) \text{ and } (0, c).$$

$$9 = c^2 - 4 \qquad \text{Substitute 4 for } a^2 \text{ and 9 for } b^2.$$

$$c^2 = 13 \qquad \text{Solve for } c^2.$$

$$c = \sqrt{13}$$

The foci are $(0, -\sqrt{13})$ and $(0, \sqrt{13})$, approximately $(0, -3.6)$ and $(0, 3.6)$. ∎

The following procedure is useful when graphing hyperbolas by hand.

Graphing Hyperbolas

(See Figure 6.33 on page 693)

1. Locate the vertices.
2. Draw the rectangle centered at the origin with sides parallel to the axes, crossing one axis at $\pm a$ and the other at $\pm b$.
3. Draw the diagonals for this rectangle and extend them to obtain the asymptotes.
4. Draw the two branches of the hyperbola by starting at each vertex and approaching the asymptotes.

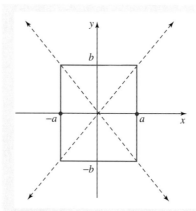 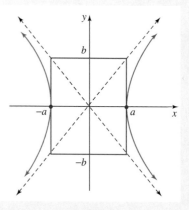

Figure 6.33

Graphing the hyperbola $\dfrac{x^2}{a^2} - \dfrac{y^2}{b^2} = 1$

Study tip

Here are some helpful similarities and differences between equations of ellipses and hyperbolas centered at the origin.

1. In standard form, equations of ellipses and hyperbolas both have a 1 on the right.
2. The equation for an ellipse has a plus sign between terms. The equation for a hyperbola has a minus sign between terms.
3. When the equation is expressed with a 1 on the right, for the ellipse a^2 is the larger number and b^2 is the smaller number. For the hyperbola, a^2 is the number under the variable whose term is preceded by a $+$ sign and b^2 is the number under the variable whose term is preceded by a $-$ sign.

    Ellipse: **Hyperbola:**

$$\dfrac{x^2}{4} + \dfrac{y^2}{9} = 1 \qquad\qquad \dfrac{y^2}{4} - \dfrac{x^2}{9} = 1$$

$$b^2 = 4 \quad a^2 = 9 \qquad\qquad a^2 = 4 \quad b^2 = 9$$

4. Foci are determined using a and b. For the ellipse, the equation is $b^2 = a^2 - c^2$, and for the hyperbola, the equation is $b^2 = c^2 - a^2$.
5. Foci are on the axis corresponding to the variable that appears over a^2.

    Ellipse: **Hyperbola:**

$$\dfrac{x^2}{4} + \dfrac{y^2}{9} = 1 \quad\text{Foci on } y\text{-axis.} \qquad \dfrac{y^2}{4} - \dfrac{x^2}{9} = 1 \quad\text{Foci on } y\text{-axis.}$$

$$a^2 \qquad\qquad\qquad\qquad a^2$$

---

 2 Write equations of hyperbolas in standard form.

Writing Equations of Hyperbolas

In Examples 1 and 2, we used the equation of a hyperbola to find its vertices and foci. In Example 3, we reverse this procedure.

EXAMPLE 3   **Finding the Equation of a Hyperbola from Its Vertices and Foci**

Find the equation of the hyperbola with vertices $(\pm 4, 0)$ and foci $(\pm 5, 0)$. Sketch the graph.

Solution

Since the vertices are on the x-axis, we use the form of the equation in which a^2 is written under x^2. The hyperbola has a horizontal transverse axis and its equation is of the form

$$\frac{x^2}{a^2} - \frac{y^2}{b^2} = 1.$$

With vertices $(\pm 4, 0)$, $a = 4$, so the equation becomes

$$\frac{x^2}{4^2} - \frac{y^2}{b^2} = 1 \quad \text{or} \quad \frac{x^2}{16} - \frac{y^2}{b^2} = 1.$$

We now determine the value of b using $b^2 = c^2 - a^2$. Since the foci are $(\pm 5, 0)$, $c = 5$.

$$b^2 = c^2 - a^2 = 5^2 - 4^2 = 25 - 16 = 9$$

Thus, the equation of the hyperbola is

$$\frac{x^2}{16} - \frac{y^2}{9} = 1.$$

The graph is shown in Figure 6.34.   ■

We can also find the equation of a hyperbola when only its vertices and asymptotes are known, as Example 4 illustrates.

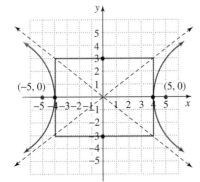

Figure 6.34

The graph of $\dfrac{x^2}{16} - \dfrac{y^2}{9} = 1$

EXAMPLE 4   **Finding the Equation of a Hyperbola from Its Vertices and Asymptotes**

Find the equation and the foci of the hyperbola with vertices $(0, \pm 3)$ and asymptotes $y = \pm \frac{3}{2}x$. Sketch the graph.

Solution

Since the vertices are on the y-axis, we use the form of the equation in which a^2 is written under y^2. The hyperbola has a vertical transverse axis and opens upward and downward. Its equation is of the form

$$\frac{y^2}{a^2} - \frac{x^2}{b^2} = 1.$$

With vertices $(0, \pm 3)$, $a = 3$, so the equation becomes

$$\frac{y^2}{3^2} - \frac{x^2}{b^2} = 1 \quad \text{or} \quad \frac{y^2}{9} - \frac{x^2}{b^2} = 1.$$

The equations of the asymptotes, $y = \pm \frac{3}{2}x$, are of the form $y = \pm \frac{a}{b}x$. This means that

$$\frac{a}{b} = \frac{3}{2}$$

$$\frac{3}{b} = \frac{3}{2} \qquad \text{We are given that } a = 3.$$

$$b = 2$$

Thus, the equation of the hyperbola is

$$\frac{y^2}{9} - \frac{x^2}{4} = 1. \qquad \text{Substitute 2 for } b \text{ in } \frac{y^2}{9} - \frac{x^2}{b^2} = 1.$$

We find the foci using $b^2 = c^2 - a^2$.

$$b^2 = c^2 - a^2 \qquad \text{In this case, the foci are } (0, \pm c), \text{ on the } y\text{-axis.}$$

$$2^2 = c^2 - 3^2 \qquad \text{Substitute 3 for } a \text{ and 2 for } b.$$

$$4 = c^2 - 9$$

$$13 = c^2 \qquad \text{Solve for } c.$$

$$c = \sqrt{13}$$

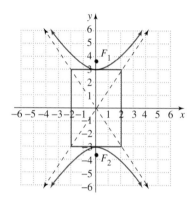

Figure 6.35

The graph of $\dfrac{y^2}{9} - \dfrac{x^2}{4} = 1$

The foci are $(0, \pm\sqrt{13})$, approximately $(0, \pm 3.6)$. The graph is shown in Figure 6.35. ∎

Translations of Hyperbolas

The graph of a hyperbola can be centered at (h, k) rather than at the origin. Horizontal and vertical translations are accomplished by replacing x with $x - h$ and y with $y - k$ in the standard form of the hyperbola's equation.

Standard form of the equation of a hyperbola centered at (h, k)

The standard form of the equation of a hyperbola with center at (h, k) is

$$\frac{(x - h)^2}{a^2} - \frac{(y - k)^2}{b^2} = 1 \qquad \text{Transverse axis is horizontal.}$$

$$\frac{(y - k)^2}{a^2} - \frac{(x - h)^2}{b^2} = 1 \qquad \text{Transverse axis is vertical.}$$

See Figure 6.36. The vertices are a units from the center and the foci are c units from the center, where $b^2 = c^2 - a^2$.

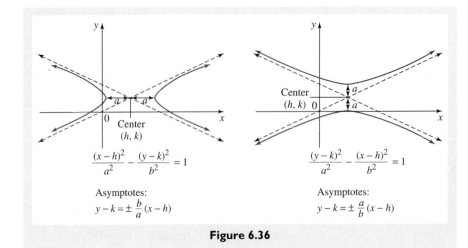

Figure 6.36

3 Graph hyperbolas not centered at the origin.

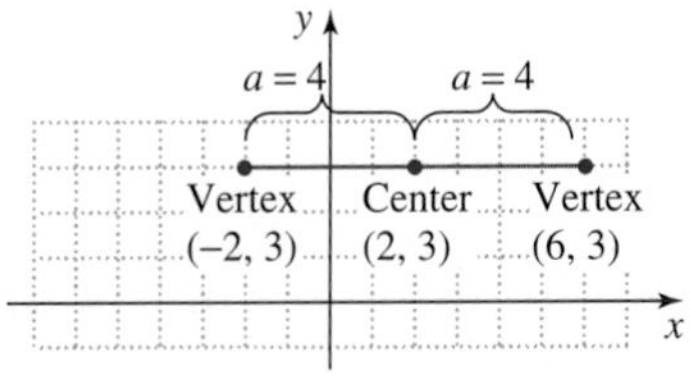

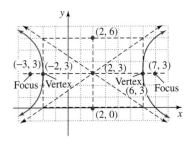

Figure 6.37

Locating the center and vertices of $\dfrac{(x-2)^2}{16} - \dfrac{(y-3)^2}{9} = 1$

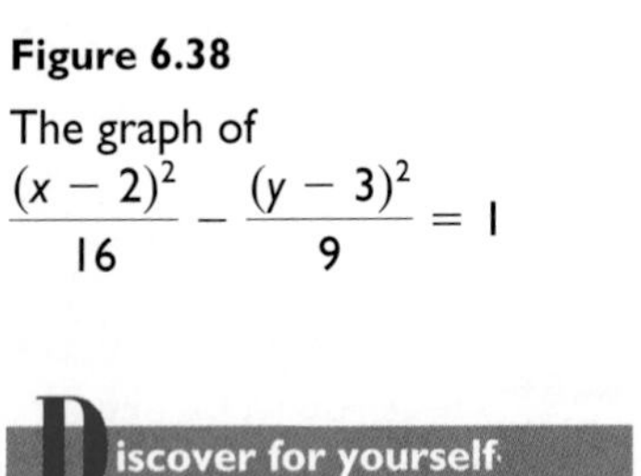

Figure 6.38

The graph of
$\dfrac{(x-2)^2}{16} - \dfrac{(y-3)^2}{9} = 1$

Discover for yourself

Use the graph in Figure 6.38 and the point-slope form of the line to obtain the equations of the asymptotes.

EXAMPLE 5   **Graphing a Hyperbola Centered at (h, k)**

Sketch the graph of $\dfrac{(x-2)^2}{16} - \dfrac{(y-3)^2}{9} = 1$. Find the foci.

Solution

The graph that we want is the graph of $\dfrac{x^2}{16} - \dfrac{y^2}{9} = 1$ translated so that its center is $(2, 3)$. Because $a^2 = 16$, then $a = 4$, meaning that the vertices are four units to the right and left of the center. The vertices are located at $(2 + 4, 3)$ or $(6, 3)$ and $(2 - 4, 3)$ or $(-2, 3)$, as shown in Figure 6.37.

Because $b^2 = 9$ and $b = 3$, the fundamental rectangle passes through the vertices and points that are three units above and below $(2, 3)$—namely, $(2, 6)$ and $(2, 0)$. We construct the fundamental rectangle and extend its diagonals to get the asymptotes. Next we draw the hyperbola with branches opening to the left and right, as shown in Figure 6.38.

With $a^2 = 16$ and $b^2 = 9$, we find the foci using $b^2 = c^2 - a^2$.

$b^2 = c^2 - a^2$ Remember that the foci are c units to the left and right of the center.

$c^2 = a^2 + b^2$ Solve for c^2.

$c^2 = 16 + 9$ Substitute 16 for a^2 and 9 for b^2.

$\quad = 25$

Thus, $c = 5$. The foci are on the transverse axis, five units from the center, located at $(2 + 5, 3)$ or $(7, 3)$ and $(2 - 5, 3)$ or $(-3, 3)$ as shown in Figure 6.38.

The asymptotes of the unshifted hyperbola

$$\underset{a^2}{\underbrace{\dfrac{x^2}{16}}} - \underset{b^2}{\underbrace{\dfrac{y^2}{9}}} = 1$$

are

$$y = \pm \frac{b}{a} x = \pm \frac{3}{4} x.$$

Thus, the asymptotes for the hyperbola that is shifted two units to the right and three units up, namely

$$\frac{(x-2)^2}{16} - \frac{(y-3)^2}{9} = 1$$

have equations that can be expressed as

$$y - 3 = \pm \frac{3}{4}(x - 2). \qquad ■$$

Using technology

We can graph the hyperbola in Example 5 by first solving the given equation for y.

$$\frac{(x-2)^2}{16} - \frac{(y-3)^2}{9} = 1 \qquad \text{This is the given equation.}$$

$$\frac{(y-3)^2}{9} = \frac{(x-2)^2}{16} - 1 \qquad \begin{array}{l}\text{To solve for } y\text{, isolate the term} \\ \text{containing } y.\end{array}$$

$$(y-3)^2 = 9\left[\frac{(x-2)^2}{16} - 1\right] \qquad \text{Multiply both sides by 9.}$$

$$y - 3 = \pm 3 \sqrt{\frac{(x-2)^2}{16} - 1} \qquad \text{Apply the square root method.}$$

$$y = 3 \pm 3 \sqrt{\frac{(x-2)^2}{16} - 1} \qquad \text{Solve for } y.$$

By entering

$$\text{Y}_1 \boxed{=} 3 + 3 \boxed{\sqrt{\ }} ((\text{X} - 2) \boxed{\wedge} 2/16 - 1)$$

and

$$\text{Y}_2 \boxed{=} 3 - 3 \boxed{\sqrt{\ }} ((\text{X} - 2) \boxed{\wedge} 2/16 - 1)$$

using

$$\text{Xmin} = -4, \text{Xmax} = 8, \text{Xscl} = 1,$$

$$\text{Ymin} = -4, \text{Ymax} = 8, \text{Yscl} = 1$$

we obtain the graph in Figure 6.39. The x-intercepts are shown more accurately than in Figure 6.38, our hand-drawn graph on page 696.

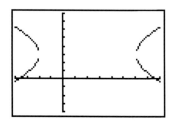

Figure 6.39

The graph of
$$\frac{(x-2)^2}{16} - \frac{(y-3)^2}{9} = 1$$
obtained with a graphing utility

As with the ellipse, we can write an equation of a hyperbola in standard form by completing the square on x and y. Let's see how this is done.

EXAMPLE 6 **Writing a Hyperbola's Equation in Standard Form by Completing the Square**

Sketch the graph of $4x^2 - 24x - 25y^2 + 250y - 489 = 0$. Find the foci.

Solution

$$4x^2 - 24x - 25y^2 + 250y - 489 = 0$$
This is the given equation.

$$(4x^2 - 24x) + (-25y^2 + 250y) = 489$$
Rewrite in anticipation of completing the square.

$$4(x^2 - 6x) - 25(y^2 - 10y) = 489$$
Factor 4 and -25, so that coefficients of x^2 and y^2 are 1.

We've added 36.   We've added -625

$$4(x^2 - 6x + 9) - 25(y^2 - 10y + 25) = 489 + 36 + (-625)$$
Complete the square on x and y.
For x: $\frac{1}{2}(-6) = -3$ and $(-3)^2 = 9$.
For y: $\frac{1}{2}(-10) = -5$ and $(-5)^2 = 25$.

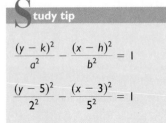

$$\frac{(y-k)^2}{a^2} - \frac{(x-h)^2}{b^2} = 1$$	**Study tip**

$$\frac{(y-5)^2}{2^2} - \frac{(x-3)^2}{5^2} = 1$$

$h = 3, k = 5$

center $= (h, k) = (3, 5)$

$$4(x-3)^2 - 25(y-5)^2 = -100$$
Factor.

$$\frac{4(x-3)^2}{-100} - \frac{25(y-5)^2}{-100} = \frac{-100}{-100}$$
In standard form, the constant term is 1, so divide both sides by -100.

$$\frac{(x-3)^2}{-25} + \frac{(y-5)^2}{4} = 1$$
Simplify.

$$\frac{(y-5)^2}{4} - \frac{(x-3)^2}{25} = 1$$
Write the equation in the standard form $\frac{(y-k)^2}{a^2} - \frac{(x-h)^2}{b^2} = 1$.
$$\underset{a^2 = 4}{\uparrow} \qquad \underset{b^2 = 25}{\nwarrow}$$
Notice that $h = 3$ and $k = 5$, so $(h, k) = (3, 5)$.

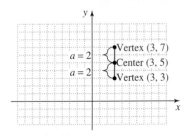

Figure 6.40

Locating the center and vertices of $\dfrac{(y-5)^2}{4} - \dfrac{(x-3)^2}{25} = 1$

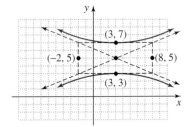

Figure 6.41

The graph of $\dfrac{(y-5)^2}{4} - \dfrac{(x-3)^2}{25} = 1$

The graph we want is the graph of $\dfrac{y^2}{4} - \dfrac{x^2}{25} = 1$ translated so that its center is $(3, 5)$. Since the x^2-term is being subtracted, the transverse axis is vertical and the hyperbola opens upward and downward. Because $a^2 = 4$, then $a = 2$, meaning that the vertices are two units above and below the center. The vertices are located at $(3, 5 + 2)$ or $(3, 7)$, and $(3, 5 - 2)$ or $(3, 3)$, as shown in Figure 6.40. Because $b^2 = 25$ and $b = 5$, the fundamental rectangle passes through points that are five units to the left and right of $(3, 5)$—namely, $(-2, 5)$ and $(8, 5)$. We construct the fundamental rectangle through these points and the vertices and extend its diagonals to get the asymptotes. Then we draw the hyperbola with branches opening upward and downward, as shown in Figure 6.41.

With $a^2 = 4$ and $b^2 = 25$, we find the foci using $b^2 = c^2 - a^2$.

$$b^2 = c^2 - a^2$$
Remember that in this situation the foci are c units above and below the center.

$$c^2 = a^2 + b^2$$
Solve for c^2.

$$= 4 + 25$$
Substitute 4 for a^2 and 25 for b^2.

$$= 29$$

Thus, $c = \sqrt{29}$. The foci are $\sqrt{29}$ units from the center, located at $(3, 5 + \sqrt{29})$ and $(3, 5 - \sqrt{29})$.

The asymptotes of the unshifted hyperbola

$$\frac{y^2}{4} - \frac{x^2}{25} = 1$$
$$\underset{a^2}{\uparrow} \qquad \underset{b^2}{\nwarrow}$$

are

$$y = \pm \frac{a}{b} x = \pm \frac{2}{5} x.$$

Thus, the asymptotes for the hyperbola that is shifted three units to the right and five units up, namely

$$\frac{(y-5)^2}{4} - \frac{(x-3)^2}{25} = 1$$

have equations that can be expressed as

$$y - 5 = \pm \frac{2}{5}(x-3). \qquad \blacksquare$$

Using technology

If we use a graphing utility to graph the hyperbola in Example 6, we must first solve the given equation for y. To do this, we must use the quadratic formula.

$4x^2 - 24x - 25y^2 + 250y - 489 = 0$	We can use this form, which is the given equation.
$-25y^2 + 250y + (4x^2 - 24x - 489) = 0$	Write the equation as a quadratic equation in y.
$y = \dfrac{-b \pm \sqrt{b^2 - 4ac}}{2a}$	Use the quadratic formula to solve for y.
$= \dfrac{-250 \pm \sqrt{250^2 - 4(-25)(4x^2 - 24x - 489)}}{2(-25)}$	$a = -25, b = 250,$ and $c = 4x^2 - 24x - 489.$
$= \dfrac{-250 \pm \sqrt{62{,}500 + 100(4x^2 - 24x - 489)}}{-50}$	

We do not have to simplify any further before using a graphing utility. As shown in Figure 6.42, the graph of

$$y_1 = \frac{-250 + \sqrt{62{,}500 + 100(4x^2 - 24x - 489)}}{-50}$$

produces the lower branch of the hyperbola and

$$y_2 = \frac{-250 - \sqrt{62{,}500 + 100(4x^2 - 24x - 489)}}{-50}$$

produces the upper branch.

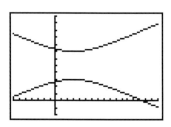

Figure 6.42

The graph of $4x^2 - 24x - 25y^2 + 250y - 489 = 0$, obtained with a graphing utility

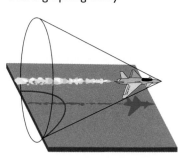

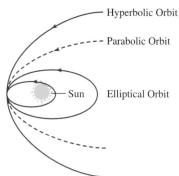

— Hyperbolic Orbit

-- Parabolic Orbit

— Sun   Elliptical Orbit

Applications

Hyperbolas have many applications. When a jet flies at a speed greater than the speed of sound, the shock wave that is created is heard as a sonic boom. The wave has the shape of a cone. The shape formed as the cone hits the ground is one branch of a hyperbola.

Halley's Comet, a permanent part of our solar system, travels around the sun in an elliptical orbit. Other comets pass through the solar system only once, following a hyperbolic path with the sun as a focus.

ENRICHMENT ESSAY

Hyperbolas and the Atomic Nucleus

Ernest Rutherford.
Corbis–Bettmann.

As a result of Rutherford's work, the atom was conceptualized as a small, dense, positively charged nucleus sitting at the center, with light, negatively charged electrons circling it, like planets orbiting the sun. Although Rutherford's model has been modified, his scattering experiments replaced the old image of an atom in which the negatively charged electrons were thought of as being distributed throughout a thin, positively charged material.

Scattering experiments, in which moving particles are deflected by various forces, led to the concept of the nucleus of an atom. In 1911, the physicist Ernest Rutherford (1871–1937) used radioactive material as sources of tiny subatomic "bullets" called alpha particles. Most of the particles either passed through the gold foil at which they were aimed or were scattered through very small angles. However, Rutherford discovered that one particle in a thousand directed toward the nuclei of gold atoms was scattered back, deflected along a hyperbolic path. Rutherford concluded that a large part of each atom's mass is located in a very small, compact object at the center—what he called the nucleus. When the alpha particle got close to the nucleus, it was bounced back along a hyperbolic path.

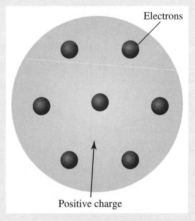

This is the model of the atom prior to Rutherford's work.

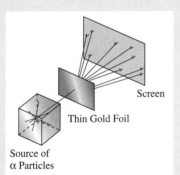

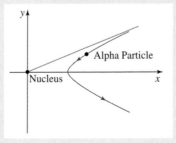

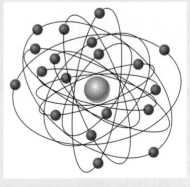

Following Rutherford's discovery, the atom was pictured with its mass concentrated at the nucleus, with electrons orbiting around it.

4 Solve applied problems involving hyperbolas.

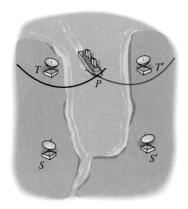

Figure 6.43

Using hyperbolas to locate a ship

Hyperbolas are of practical importance in fields ranging from architecture to navigation. Cooling towers used in the design for nuclear power plants have cross sections that are both ellipses and hyperbolas. Three-dimensional solids whose cross sections are hyperbolas are used in some rather unique architectural creations, including the TWA building at Kennedy Airport and the St. Louis Science Center Planetarium.

The hyperbola is the basis for the navigational system LORAN (for long-range navigation). As shown in Figure 6.43, signals sent out by radio transmitters at T and T' reach a radio receiver in a ship located at point P. LORAN is based on the time difference between the reception of signals sent simultaneously from stations T and T'. The difference in times of arrival of the signals is used to determine that the ship lies on one branch of a hyperbola. The process is then repeated for radio transmitters at S and S'. The point of intersection of the two hyperbolas is the location of the ship.

LORAN is used to construct highly accurate navigational maps. One of these may have been used by the crew of the Enola Gay when it dropped the first atomic bomb over Hiroshima.

EXAMPLE 7 An Application Involving Hyperbolas

An explosion is recorded by two microphones that are 2 miles apart. Microphone M_1 received the sound 4 seconds before microphone M_2. Assuming sound travels at 1100 feet per second, determine the possible locations of the explosion relative to the location of the microphones.

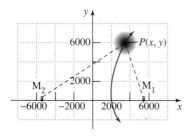

Figure 6.44

Locating an explosion on the branch of a hyperbola

Solution

We begin by putting the microphones in a coordinate system. Since 1 mile = 5280 feet, we place M_1 5280 feet on a horizontal axis to the right of the origin and M_2 5280 feet on a horizontal axis to the left of the origin. Figure 6.44 illustrates that the two microphones are 2 miles apart.

We know that M_2 received the sound 4 seconds after M_1. Because sound travels at 1100 feet per second, the difference between the distance from P to M_1 and the distance from P to M_2 is 4400 feet. The set of all points P (or locations of the explosion) satisfying these conditions fits the definition of a hyperbola, with microphones M_1 and M_2 at the foci.

$$\frac{x^2}{a^2} - \frac{y^2}{b^2} = 1$$

Use the standard form of the hyperbola's equation. $P(x, y)$, the explosion point, lies on this hyperbola. We must find a^2 and b^2.

The difference between the distances, represented by $2a$ in the derivation of the hyperbola's equation, is 4400 feet. Thus, $2a = 4400$ and $a = 2200$.

$$\frac{x^2}{(2200)^2} - \frac{y^2}{b^2} = 1$$

Substitute 2200 for a.

Since $c = 5280$ and $a = 2200$, then $b^2 = c^2 - a^2 = 5280^2 - 2200^2 = 23{,}038{,}400$.

$$\frac{x^2}{4,840,000} - \frac{y^2}{23,038,400} = 1 \quad \text{Substitute 23,038,400 for } b^2.$$

We can conclude that the explosion occurred somewhere on the right branch (the branch closest to M_1) of the hyperbola given by

$$\frac{x^2}{4,840,000} - \frac{y^2}{23,038,400} = 1.$$ ∎

In Example 7, we determined that the explosion occurred somewhere along one branch of a hyperbola, but not exactly where on the hyperbola. If, however, we had received the sound from another pair of microphones, we could locate the sound along a branch of another hyperbola. The exact location of the explosion would be the point where the two hyperbolas intersect.

PROBLEM SET 6.2

Practice Problems

In Problems 1–20, sketch the graph of each hyperbola by hand and find the foci. Verify your graph using a graphing utility.

1. $\dfrac{x^2}{4} - \dfrac{y^2}{9} = 1$

2. $\dfrac{x^2}{25} - \dfrac{y^2}{4} = 1$

3. $9y^2 - 25x^2 = 225$

4. $16y^2 - 9x^2 = 144$

5. $y^2 = 1 + x^2$

6. $x^2 = 1 + y^2$

7. $\dfrac{(x+4)^2}{9} - \dfrac{(y+3)^2}{16} = 1$

8. $\dfrac{(x+2)^2}{9} - \dfrac{(y-1)^2}{25} = 1$

9. $\dfrac{(x+3)^2}{25} - \dfrac{y^2}{16} = 1$

10. $\dfrac{(x+2)^2}{9} - \dfrac{y^2}{25} = 1$

11. $\dfrac{(y+2)^2}{4} - \dfrac{(x-1)^2}{16} = 1$

12. $\dfrac{(y-2)^2}{36} - \dfrac{(x+1)^2}{49} = 1$

13. $x^2 - y^2 - 2x - 4y - 4 = 0$

14. $4x^2 - y^2 + 32x + 6y + 39 = 0$

15. $16x^2 - y^2 + 64x - 2y + 67 = 0$

16. $9y^2 - 4x^2 - 18y + 24x - 63 = 0$

17. $4x^2 - 9y^2 - 16x + 54y - 101 = 0$

18. $4x^2 - 9y^2 + 8x - 18y - 6 = 0$

19. $4x^2 - 25y^2 - 32x + 164 = 0$

20. $9x^2 - 16y^2 - 36x - 64y + 116 = 0$

In Problems 21–24, find the standard form of the equation for each hyperbola and identify its foci.

21.

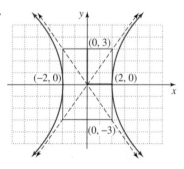

22.

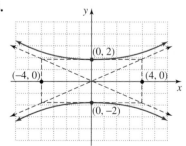

23.

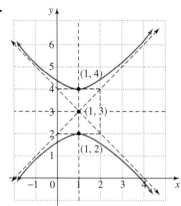

24.

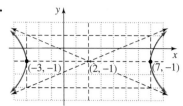

In Problems 25–34, find the standard form of the equation for each hyperbola centered at the origin satisfying the given conditions.

25. Foci: $(-4, 0), (4, 0)$; vertices: $(-3, 0), (3, 0)$

26. Foci: $(0, -3), (0, 3)$; vertices: $(0, -1), (0, 1)$

27. Vertices: $(-1, 0), (1, 0)$; asymptotes: $y = \pm 3x$

28. Vertices: $(0, -1), (0, 1)$; asymptotes: $y = \pm 2x$

29. Vertices of fundamental rectangle: $(0, -3), (0, 3),$ $(-6, 0), (6, 0)$; opens left and right

30. Vertices of fundamental rectangle: $(0, -3), (0, 3),$ $(-2, 0), (2, 0)$; opens upward and downward

31. Vertices: $(0, -3), (0, 3)$; passes through $(-2, 5)$

32. Vertices: $(-2, 0), (2, 0)$; passes through $(3, \sqrt{3})$

33. Length of transverse axis = 8; length of conjugate axis = 24; opens to the left and right

34. Length of transverse axis = 4; length of conjugate axis = 2; opens upward and downward

Application Problems

35. An explosion is recorded by two microphones that are 1 mile apart. Microphone M_1 received the sound 2 seconds before microphone M_2. Assuming sound travels at 1100 feet per second, determine the possible locations of the explosion relative to the location of the microphones.

36. Radio towers A and B, 200 kilometers apart, are situated along the coast, with A located due west of B. Simultaneous radio signals are sent from each tower to a ship, with the signal from B received 500 microseconds before the signal from A.

   a. Assuming that the radio signals travel 300 meters per microsecond, determine the equation of the hyperbola on which the ship is located.

   b. If the ship lies due north of tower B, how far out at sea is it?

37. The figure shows a scattering experiment in which an alpha particle aimed at the nucleus of a gold atom is deflected along a hyperbolic path. If the particle gets as close as three units to the nucleus, what is the equation of its path?

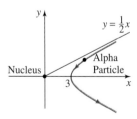

38. An architect designs two houses that are shaped and positioned like a part of the branches of the hyperbola whose equation is $625y^2 - 400x^2 = 250,000$, where x

and y are in yards. How far apart are the houses at their closest point?

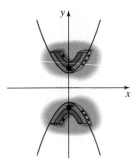

39. When an airplane flies at a speed greater than the speed of sound (1100 feet per second), the shock wave that is created is heard as a sonic boom. The shock wave is in the shape of a cone. The shape formed as the cone hits the ground is a hyperbola, as shown in the figure at the top right. Find the equation of this hyperbola for the Concorde, traveling at 2200 feet per second at an altitude of 65,000 feet. (Under these conditions, the vertex of the hyperbola is 24 miles from the point on the ground beneath the plane's nose and the equations for the asymptotes of the hyperbolic boundary are $y = \pm\frac{1}{2}x$.) How wide is the region affected by the sonic boom when $x = -50$?

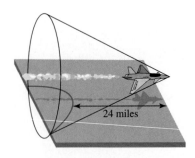

24 miles

40. Hyperbolas have a reflective property that is important in the construction of telescopes, glasses, and cameras. As shown in the figure, a light ray aimed at focus F_1 will be reflected to the other focus F_2. If a ray of light strikes the mirror at the point $(5, 3)$, what is the equation for the reflected ray?

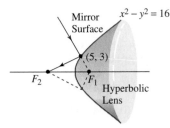

Mirror Surface
$x^2 - y^2 = 16$
$(5, 3)$
F_2
F_1
Hyperbolic Lens

True–False Critical Thinking Problems

41. Which one of the following is true?

  a. The transverse axis of a hyperbola can be shorter than the conjugate axis.

  b. The graph of $\dfrac{x^2}{9} - \dfrac{y}{25} = 1$ is a hyperbola.

  c. The graph of $\dfrac{x^2}{9} - \dfrac{y^2}{16} = 1$ has vertices at $(0, 4)$ and $(0, -4)$.

  d. All points on $y = \pm\dfrac{b}{a}x$, the asymptotes for the hyperbola $\dfrac{x^2}{a^2} - \dfrac{y^2}{b^2} = 1$, also satisfy the hyperbola's equation.

42. Which one of the following is true?

  a. The hyperbola whose equation is $4y^2 - 9x^2 = 36$ has asymptotes whose equations are $y = \dfrac{2}{3}x$ and $y = -\dfrac{2}{3}x$.

  b. The graph of $y^2 = 4 - x^2$ is a hyperbola centered at the origin.

  c. The foci for $\dfrac{x^2}{25} - \dfrac{y^2}{9} = 1$ are $(4, 0)$ and $(-4, 0)$.

  d. The graph of $\dfrac{x^2}{9} - \dfrac{y^2}{4} = 1$ does not intersect the line $y = -\dfrac{2}{3}x$.

Technology Problems

Use a graphing utility to graph each hyperbola in Problems 43–46.

43. $\dfrac{x^2}{16} - \dfrac{y^2}{4} = 1$

44. $\dfrac{(x + 3)^2}{8} - \dfrac{(y - 2)^2}{2} = 1$

45. $x^2 - y^2 - 2x - 4y = 4$

46. $3y^2 - 5x^2 = 13$

47. Graph $x^2 - y^2 = 16$ and $2y^2 - x^2 = 8$ in the same viewing rectangle and find intersection points correct to the nearest tenth.

48. Write $4x^2 - 6xy + 2y^2 - 3x + 10y - 6 = 0$ as a quadratic equation in y and then use the quadratic formula to express y in terms of x. Graph the resulting two equations using a graphing utility. What effect does the xy-term have on the graph of the resulting hyperbola? What problems would you encounter if you attempted to write the given equation in standard form by completing the square?

49. Two orbiting bodies move on paths given by the equations $2x^2 - 2xy + y^2 = 2$ and $3x^2 + 2xy - y^2 = 3$. Use the suggestion in Problem 48 to graph both equations in the same viewing rectangle and find the four ordered pairs where the paths will intersect.

50. Use a graphing utility to graph $\dfrac{x^2}{4} - \dfrac{y^2}{9} = 0$. Is the graph a hyperbola? In general, what is the graph of $\dfrac{x^2}{a^2} - \dfrac{y^2}{b^2} = 0$?

51. Graph $\dfrac{x^2}{a^2} - \dfrac{y^2}{b^2} = 1$ and $\dfrac{x^2}{a^2} - \dfrac{y^2}{b^2} = -1$ in the same viewing rectangle for values of a^2 and b^2 of your choice. Describe the relationship between the two graphs.

52. Graph $\dfrac{x^2}{16} - \dfrac{y^2}{9} = 1$ and $\dfrac{x|x|}{16} - \dfrac{y|y|}{9} = 1$ in the same viewing rectangle. Explain why the graphs are not the same.

Writing in Mathematics _____

53. Describe why the graph in the figure must be a hyperbola. How would you find its equation?

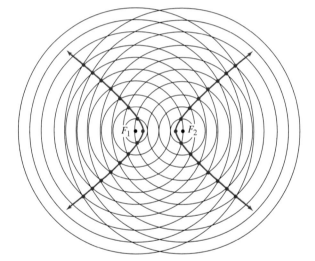

54. Describe what happens to the shape of a hyperbola if the center is fixed but the foci move in toward the center.

55. Consider two strings attached to the tip of a pen, shown in the figure. Turn the pen so that the string winds around the tip. Explain why, as the lengths of string shorten, the pen's tip will trace out a hyperbolic path.

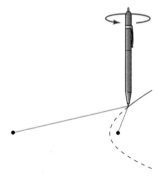

Critical Thinking Problems _____

56. In this problem, we use algebra to explore the two-branch distinguishing feature of hyperbolas.

a. Consider the hyperbola given by
$$\frac{x^2}{a^2} - \frac{y^2}{b^2} = 1.$$
Solve for $\dfrac{x^2}{a^2}$ and explain why $\dfrac{x^2}{a^2}$ must be greater than or equal to 1.

b. Use your result from part (a),
$$\frac{x^2}{a^2} \geq 1$$
to solve for x^2 and then for x.

c. What do your solutions for x in part (b) mean in terms of the hyperbola's graph?

In Problems 57–60, we explore a hyperbola's shape in terms of its eccentricity. Asymptotes are important in determining the proper shape or eccentricity of a hyperbola. As with ellipses, the eccentricity is defined by
$$e = \frac{c}{a}.$$

57. We know that the eccentricity for an ellipse lies between 0 and 1. Based on the relationship between c and a for the hyperbola, what must be true about its eccentricity? Express this idea as an inequality.

58. What happens to the shape of a hyperbola as its eccentricity gets larger and larger? What happens to the shape as the eccentricity is close to the smallest permissible value?

59. Write the equation in standard form for a hyperbola centered at the origin with one focus $(6, 0)$ and $e = \dfrac{3}{2}$.

60. Describe what is incorrect about these directions: Write the standard form of the equation of a hyperbola centered at $(-3, -3)$, $e = 2$, and length of conjugate axis $= 6$.

In Problems 61–65, find the standard form of the equation for each hyperbola not centered at the origin satisfying the given conditions.

61. Center: $(3, -1)$; length of transverse axis $= 11$; length of conjugate axis $= 6$; opens upward and downward

62. Foci: $(1, -1)$, $(7, -1)$; length of transverse axis $= 2$

63. Vertices: $(5, -6)$, $(5, 6)$; passes through $(0, 9)$

64. Foci: $(-4, -3)$, $(4, -3)$; difference of distances from any point on the hyperbola to the foci is 6

65. Foci: $(2, 2)$, $(6, 2)$; asymptotes: $y = x - 2$, $y = 6 - x$

66. Suppose that the eccentricity of the hyperbolas described by $\dfrac{x^2}{a^2} - \dfrac{y^2}{b^2} = 1$ and $\dfrac{y^2}{b^2} - \dfrac{x^2}{a^2} = 1$ are represented by e_1 and e_2, respectively. Prove that $e_1^2 e_2^2 = e_1^2 + e_2^2$.

67. If $P(x, y)$ is a point on the hyperbola whose equation is $\dfrac{x^2}{a^2} - \dfrac{y^2}{b^2} = 1$, prove that the distance to the focus divided by the distance to the line whose equation is $x = \dfrac{a^2}{c}$ equals the eccentricity.

68. Prove that if the lengths of the transverse axis and conjugate axis are both multiplied by the same constant, the eccentricity and asymptotes of a hyperbola do not change.

69. The graph shown in the figure can be described from two perspectives. From the point of view of the $x'y'$ system, the equation is

$$\frac{x'^2}{2} - \frac{y'^2}{2} = 1.$$

From the point of view of the xy system, the equation is

$$xy = 1.$$

The $x'y'$ system is obtained by rotating the xy system through $45°$.

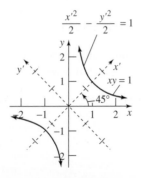

a. Use the equation

$$\frac{x'^2}{2} - \frac{y'^2}{2} = 1$$

to find the vertices and asymptotes for the hyperbola with respect to the $x'y'$ system.

b. Use the graph of the equation

$$xy = 1$$

to find the vertices and asymptotes for the hyperbola with respect to the xy system.

c. In general, describe the graph of $xy = c$ if $c > 0$. What can you say about the x- and y-intercepts? What are the asymptotes? Reinforce these observations by graphing $xy = 4$ and $xy = 6$.

d. Describe the graph of $xy = c$ if $c < 0$. What similarities and differences are there with respect to the graph of $xy = c$ if $c > 0$? Reinforce these observations by graphing $xy = -4$ and $xy = -6$.

70. Find the equation of a hyperbola whose asymptotes are perpendicular. (Many equations are possible.) If the asymptotes of $\dfrac{x^2}{a^2} - \dfrac{y^2}{b^2} = 1$ are perpendicular, what is the relationship between a and b? Prove that this relationship is correct.

71. Given the hyperbola described by $\dfrac{x^2}{a^2} - \dfrac{y^2}{b^2} = 1$, prove that the length of the diagonal of the fundamental rectangle is $2c$.

72. A line segment through a focus perpendicular to the transverse axis and cut off by the hyperbola is called the *focal chord*. Show that for $\dfrac{x^2}{a^2} - \dfrac{y^2}{b^2} = 1$ the focal chord has length $\dfrac{2b^2}{a}$.

SECTION 6.3

Solutions Manual

Tutorial

Video II

The Parabola

Objectives

1 Graph parabolas with vertices at the origin.

2 Write equations of parabolas in standard form.

3 Graph parabolas with vertices not at the origin.

4 Solve applied problems involving parabolas.

5 Recognize equations of conic sections.

In Chapter 3, we studied parabolas, viewing them as graphs of quadratic functions expressed as

$$f(x) = ax^2 + bx + c \quad \text{or} \quad f(x) = a(x - h)^2 + k.$$

Like the other conic sections, parabolas can be obtained by intersecting a plane and a cone, shown in Figure 6.45. In this section, we will use a geometric definition for a parabola to derive its standard equation. We will also consider applications of parabolas, including parabolic shapes that gather distant rays of light and focus them into spectacular images.

Figure 6.45

Intersecting a plane and a cone to obtain a parabola

Definition of a Parabola

We have defined each of the conics (the circle, ellipse, and hyperbola) as the set of points satisfying certain geometric conditions. As in the preceding sections, a geometric definition for the parabola is our starting point.

Definition of a parabola

A *parabola* is the set of all points in a plane that are equidistant from a fixed line (the *directrix*) and a fixed point (the *focus*) that is not on the line (see Figure 6.46).

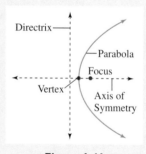

Figure 6.46

As shown in Figure 6.46, the midpoint between the focus and the directrix is the *vertex* and the line passing through the focus and the vertex is the *axis of symmetry* of the parabola.

The Equation of a Parabola

Using the rectangular coordinate system, we can obtain equations for parabolas. To derive a simple equation, place the focus on the x-axis at the point $(p, 0)$, as in Figure 6.47. The directrix has an equation given by $x = -p$ and the vertex, midway between the focus and the directrix, is located at the origin.

If (x, y) is any point on the parabola, the distance d_1 to the directrix is equal to the distance d_2 to the focus.

Figure 6.47

By the definition of a parabola, $d_1 = d_2$.

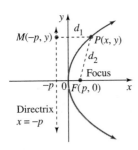

Figure 6.47

By the definition of a parabola, $d_1 = d_2$.

$$d_1 = d_2$$

By definition, all points on the parabola are equidistant from the directrix and the focus.

$$\sqrt{(x + p)^2 + (y - y)^2} = \sqrt{(x - p)^2 + (y - 0)^2}$$

Refer to Figure 6.47 (shown again in the margin) and use the distance formula. Recall that the distance between (x_1, y_1) and (x_2, y_2) is $\sqrt{(x_1 - x_2)^2 + (y_1 - y_2)^2}$.

$$(x + p)^2 = (x - p)^2 + y^2$$

Square both sides of the equation.

$$x^2 + 2px + p^2 = x^2 - 2px + p^2 + y^2$$

Square $x + p$ and $x - p$.

$$2px = -2px + y^2$$

Subtract $x^2 + p^2$ from both sides of the equation.

$$y^2 = 4px$$

Solve for y^2.

The equation $y^2 = 4px$ is the *standard form* of a parabola with directrix $x = -p$ and focus at $(p, 0)$.

■ Graph parabolas with vertices at the origin.

EXAMPLE 1 **Finding the Focus and Directrix of a Parabola**

Find the focus and directrix of the parabola given by $y^2 = 12x$. Then graph the parabola.

Solution

$$y^2 = 12x$$ 　This is the given equation. The standard form is $y^2 = 4px$, so $4p = 12$.

$$4p = 12$$ 　Remember that the focus is at $(p, 0)$ and the directrix is given by $x = -p$.

$$p = 3$$ 　Divide both sides by 4.

Using this value for p, we obtain

Focus: $(p, 0) = (3, 0)$

Directrix: $x = -p; x = -3$

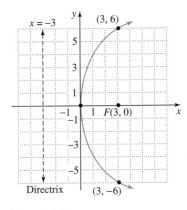

Figure 6.48

The graph of $y^2 = 12x$

Observe that the vertex, midway between the focus and the directrix, is at the origin. To graph $y^2 = 12x$, we assign x a value that makes the right side a perfect square. If $x = 3$, then $y^2 = 12(3)$ or $y^2 = 36$. Since $y = \pm 6$, the parabola passes through the points $(3, 6)$ and $(3, -6)$. The graph is sketched in Figure 6.48. ■

Parabolas with vertices at the origin can open to the right, left, upward, or downward. These cases and the corresponding standard equations are summarized on the next page.

Standard form of the equations of a parabola with vertex at (0, 0)

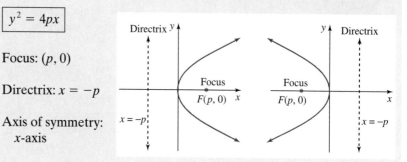

$$\boxed{y^2 = 4px}$$

Focus: $(p, 0)$

Directrix: $x = -p$

Axis of symmetry:
x-axis

$$\boxed{x^2 = 4py}$$

Focus: $(0, p)$

Directrix: $y = -p$

Axis of symmetry:
y-axis

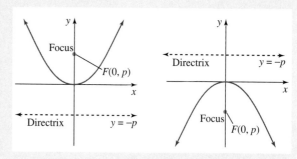

Study tip

The focus is located on the axis corresponding to the variable that is not squared:

$y^2 = 4px$ — Focus on x-axis

$x^2 = 4py$ — Focus on y-axis

EXAMPLE 2 **Using the Standard Equation of a Parabola**

Find the focus and directrix of the parabola given by $x^2 = -8y$, and sketch its graph.

Solution

$x^2 = -8y$ — This is the given equation. The standard form is $x^2 = 4py$, so $4p = -8$.

$4p = -8$ — The focus, on the y-axis, is at $(0, p)$.

$p = -2$ — The directrix is given by $y = -p$.

Using this value for p, we obtain

Focus: $(0, p) = (0, -2)$

Directrix: $y = -p;\ y = 2$

To graph $x^2 = -8y$, we assign y a value that makes the right side a perfect square. If $y = -2$, then $x^2 = -8(-2) = 16$, so $x = \pm 4$. The parabola passes through the points $(4, -2)$ and $(-4, -2)$. The graph is sketched in Figure 6.49.

∎

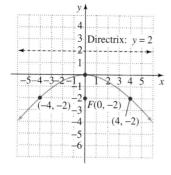

Figure 6.49

The graph of $x^2 = -8y$

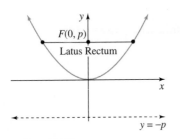

Figure 6.50

The latus rectum for a parabola that opens upward

The Latus Rectum as a Graphing Aid

A line segment that passes through a parabola's focus and is parallel to its directrix is called its *latus rectum*, shown in Figure 6.50. The latus rectum is helpful when graphing a parabola.

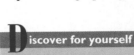

Shown in Figure 6.51 are the two parabolas from Examples 1 and 2 and their latera recta (plural of latus rectum). Study the figure and find a formula for the length of the latus rectum for either $y^2 = 4px$ or $x^2 = 4py$. How can this length be used to graph a parabola?

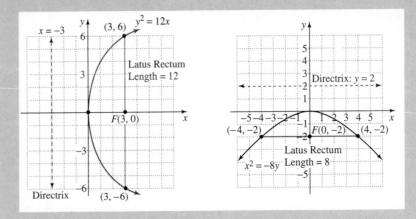

Figure 6.51
The latus rectum for $y^2 = 12x$ has length 12 and the latus rectum for $x^2 = -8y$ has length 8.

In the Discover for yourself did you observe that the length of the latus rectum is $|4p|$? You can find two points on the graph by counting half this length above and below the focus or to the right and the left of the focus.

EXAMPLE 3 Using the Latus Rectum to Sketch a Parabola

Find the focus, directrix, and latus rectum length of the parabola $y = \frac{1}{10}x^2$, and sketch its graph.

Solution

We begin by expressing the equation in the form $x^2 = 4py$.

$$y = \frac{1}{10}x^2 \qquad \text{This is the given equation.}$$

$$x^2 = 10y \qquad \text{Multiply both sides by 10.}$$

Using this equation, we see that $4p = 10$ and $p = 2.5$. With this value for p, we obtain

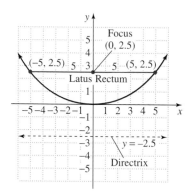

Figure 6.52

Using the latus rectum to graph
$y = \frac{1}{10}x^2$

2 Write equations of parabolas in standard form.

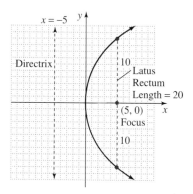

Figure 6.53

The graph of $y^2 = 20x$

Focus: $(0, p) = (0, 2.5)$

Directrix: $y = -p$; $y = -2.5$

Length of latus rectum $= |4p| = |4(2.5)| = 10$

This means that the length of the latus rectum extends five units to the left and five units to the right of the focus. This enables us to sketch the parabola, shown in Figure 6.52. ∎

Writing Equations of Parabolas

Up to this point, we have used the equation of a parabola to find its focus and directrix. In the next example, we reverse this procedure.

EXAMPLE 4 **Finding the Equation of a Parabola from Its Focus and Directrix**

Find the equation of the parabola with focus $(5, 0)$ and directrix $x = -5$. Sketch the graph.

Solution

Because the focus is $(5, 0)$, which is on the *x*-axis, we use $y^2 = 4px$, the standard form of the equation in which the variable *x* is not squared. With the focus $(5, 0)$ (since the focus is given by $(p, 0)$), we see that $p = 5$. We can also use the directrix $x = -5$ ($x = -p$) to see that $p = 5$. The equation becomes

$$y^2 = 4 \cdot 5x \quad \text{or} \quad y^2 = 20x.$$

The length of the latus rectum is $|4p| = |4 \cdot 5| = 20$, extending 10 units below and 10 units above the focus. This enables us to sketch the parabola, shown in Figure 6.53. ∎

We can also find the equation of a parabola when only its vertex, axis of symmetry, and one other point are known. Let's see how this is done.

EXAMPLE 5 **Finding the Equation of a Parabola**

Find the equation of the parabola with vertex at the origin, the *y*-axis as the axis of symmetry, and passing through $(-10, -5)$. Sketch the graph.

Solution

Since the *y*-axis is the axis of symmetry, the focus is on the *y*-axis. We use $x^2 = 4py$, the standard form of the equation in which the variable *y* is not squared.

Because $(-10, -5)$ is a point on the parabola, its coordinates satisfy the preceding equation.

$$(-10)^2 = 4p(-5)$$
$$100 = -20p$$
$$-5 = p$$

Substituting -5 for p in $x^2 = 4py$, we obtain the equation

$$x^2 = 4(-5)y \quad \text{or} \quad x^2 = -20y.$$

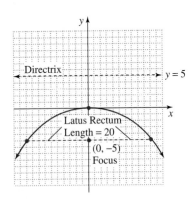

Figure 6.54

The graph of $x^2 = -20y$

Using either this equation or the value of p, we obtain

Focus: $(0, p) = (0, -5)$

Directrix: $y = -p$; $y = 5$

Length of latus rectum $= |4p| = |4(-5)| = 20$

As in Example 4, the length of the latus rectum is 20, extending 10 units to the left and 10 units to the right of the focus. The parabola is sketched in Figure 6.54. ■

iscover for yourself

Look at the graphs of $y^2 = 20x$ and $x^2 = -20y$, shown in Figures 6.53 and 6.54, respectively. Describe two ways in which the graphs are similar and two ways in which they differ. To obtain a complete "family" of parabolas, we must add the graphs of $y^2 = -20x$ and $x^2 = 20y$. Why do you think these four graphs are called a family of parabolas?

3 Graph parabolas with vertices not at the origin.

Translations of Parabolas

The graph of a parabola can have its vertex at (h, k) rather than at the origin. Horizontal and vertical translations are accomplished by replacing x with $x - h$ and y with $y - k$ in the standard form of the parabola's equation.

Standard form of the equation of a parabola with vertex at (h, k)

The standard form of the equation of a parabola with vertex at (h, k) is

$(y - k)^2 = 4p(x - h)$ Focus: $(h + p, k)$
Directrix: $x = h - p$
Axis of symmetry is horizontal.

$(x - h)^2 = 4p(y - k)$ Focus: $(h, k + p)$
Directrix: $y = k - p$
Axis of symmetry is vertical.

$\boxed{(y - k)^2 = 4p(x - h)}$ $\boxed{(x - h)^2 = 4p(y - k)}$

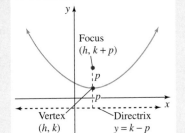

If $p < 0$, the parabola opens to the left. If $p < 0$, the parabola opens downward.

EXAMPLE 6   Graphing a Parabola with Vertex at (h, k)

Sketch the graph of the parabola given by $(x - 3)^2 = 8(y + 1)$. Find the vertex, focus, and directrix.

Solution

Let's begin by identifying the vertex.

$$(x - 3)^2 = 8(y + 1)$$ The given equation is in the form $(x - h)^2 = 4p(y - k)$, so $h = 3$ and $k = -1$. The vertex (h, k) is $(3, -1)$.

The graph that we want is the graph of $x^2 = 8y$ translated so that its vertex is $(3, -1)$. Because $4p = 8$ and $p = 2$, the focus is two units above the vertex, located at $(3, -1 + 2)$ or at $(3, 1)$, as shown in Figure 6.55. Also shown is the directrix, located two units below the vertex. Its equation is $y = -1 - 2$, or $y = -3$.

We can use the length of the latus rectum, $|4p|$, to find two additional points on the parabola. Since $p = 2$, the latus rectum has length $|4 \cdot 2|$ or 8 units. We count four units to the left and the right of the focus $(3, 1)$. The parabola passes through $(3 - 4, 1)$ or $(-1, 1)$ and $(3 + 4, 1)$ or $(7, 1)$. Passing a smooth curve through the vertex and these two points, we sketch the parabola shown in Figure 6.56.   ■

As with the ellipse and the hyperbola, we can write an equation of a parabola in standard form by completing the square. Let's see how this is done.

EXAMPLE 7   Writing a Parabola's Equation in Standard Form by Completing the Square

Sketch the graph of the parabola given by $y^2 + 2y + 12x - 23 = 0$. Find the vertex, focus, and directrix.

Solution

$$y^2 + 2y + 12x - 23 = 0$$ This is the given equation. Since there is a y^2-term, the parabola has a horizontal axis of symmetry.

$$y^2 + 2y = -12x + 23$$ Rewrite in anticipation of completing the square. Since there is no x^2-term, we will complete the square only on the y-terms.

$$y^2 + 2y + 1 = -12x + 23 + 1$$ Complete the square on y: $\frac{1}{2}(2) = 1$ and $1^2 = 1$, so add 1 to both sides.

$$(y + 1)^2 = -12x + 24$$ Factor on the left.

$$(y + 1)^2 = -12(x - 2)$$ To write the equation in the form $(y - k)^2 = 4p(x - h)$, factor -12 on the right. Thus, $h = 2$ and $k = -1$. The vertex (h, k) is $(2, -1)$.

The graph that we want is the graph of $y^2 = -12x$ translated so that its vertex is $(2, -1)$. Because $4p = -12$ and $p = -3$, the focus is three units to the left of

Study tip

$$(x - h)^2 = 4p(y - k)$$
$$\downarrow \quad \downarrow \quad \downarrow$$
$$(x - 3)^2 = 8(y - (-1))$$

$h = 3, k = -1$

Vertex $= (h, k) = (3, -1)$

$4p = 8$, so $p = 2$

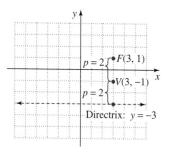

Figure 6.55

Identifying the vertex, focus, and directrix of $(x - 3)^2 = 8(y + 1)$

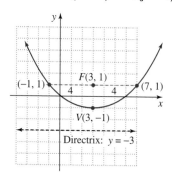

Figure 6.56

The graph of $(x - 3)^2 = 8(y + 1)$

Study tip

$$(y - k)^2 = 4p(x - h)$$
$$\searrow \quad \searrow \quad \searrow$$
$$(y - (-1))^2 = -12(x - 2)$$

$h = 2, k = -1$

Vertex $(h, k) = (2, -1)$

$4p = -12$, so $p = -3$

ENRICHMENT ESSAY

The Parabola's Rival

If a cable hung between two structures supports its own weight (such as a telephone line), it does not form a parabola, but rather a curve called a catenary (after the Latin word for *chain*). The Gateway Arch in St. Louis, Missouri, is in the shape of a cate-

nary. The accompanying figure illustrates that visually it is impossible to distinguish a catenary from a parabola. However, near the vertex of the parabola, the catenary lies above the parabola; elsewhere, it lies below the parabola.

Bachmann / Photri, Inc.

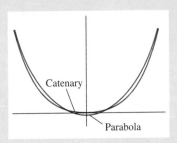

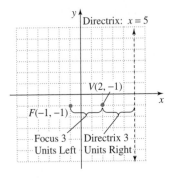

Figure 6.57

Identifying the vertex, focus, and directrix of $(y + 1)^2 = -12(x - 2)$

the vertex (why to the left and not below?), located at $(2 - 3, -1)$ or at $(-1, -1)$, as shown in Figure 6.57. The directrix, also shown in Figure 6.57, is three units to the right of the vertex, and its equation is $x = 2 + 3$, or $x = 5$.

Once again, we use the length of the latus rectum, $|4p|$, to find two additional points on the parabola. Since $p = -3$, the latus rectum has length $|4(-3)| = 12$ units. The two points on the parabola are found by counting six units below and above the focus $(-1, -1)$. They are $(-1, -1 - 6)$ or $(-1, -7)$ and $(-1, -1 + 6)$ or $(-1, 5)$. Passing a smooth curve through the vertex and these two points, we sketch the parabola shown in Figure 6.58. ■

Applications

Parabolas have many applications. Cables hung between structures to form suspension bridges form parabolas. Arches constructed of steel and concrete, whose main purpose is strength, are usually parabolic in shape.

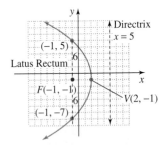

Figure 6.58

The graph of $(y + 1)^2 = -12(x - 2)$

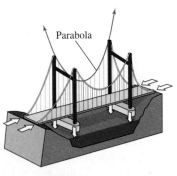

Suspension bridge

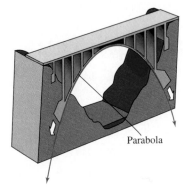

Arch bridge

4 Solve applied problems involving parabolas.

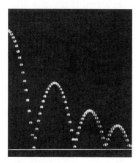

Figure 6.59

Multiflash photo showing the parabolic path of a ball thrown into the air.

Berenice Abbott / Commerce Graphics Ltd., Inc.

We have seen that comets in our solar system travel in orbits that are ellipses and hyperbolas. Some comets also follow parabolic paths. Only comets with elliptical orbits, such as Halley's Comet, return to our part of the galaxy.

A projectile, such as a baseball thrown directly upward, moves along a parabolic path, illustrated in Figure 6.59.

If a parabola is rotated about its axis of symmetry, a parabolic surface is formed. Figure 6.60 illustrates that a light source originating at the focus will reflect from the surface parallel to the axis of symmetry. This property is used in the design of searchlights, automobile headlights, and parabolic microphones.

The same principle is used in reverse in reflecting telescopes, radar, and TV satellite dishes. Figure 6.61 illustrates that light rays striking the surface parallel to the axis of symmetry are reflected to the focus. Reflecting telescopes, from the 3-inch type used at home to the 200-inch instrument on Mount Palomar in California magnify the light from distant stars by reflecting the parallel light rays from these bodies to the focus of a parabolic mirror. Figure 6.61 shows that the light waves travel the length of the telescope tube and are reflected to the focus of the parabola where the eyepiece of the telescope is located. Our next examples utilize the reflection property of the parabola.

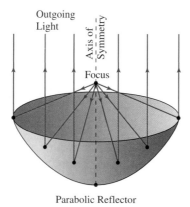

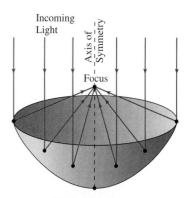

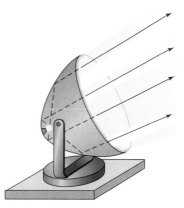

Figure 6.60

Light from the focus is reflected parallel to the axis of symmetry.

Figure 6.61

Incoming light rays are reflected to the focus.

| EXAMPLE 8 | **Using the Reflection Property of Parabolas** |

An engineer is designing a flashlight using a parabolic reflecting mirror and a light source, shown in Figure 6.62. The casting has a diameter of 4 inches and a depth of 2 inches. What is the equation for the parabola used to shape the mirror? At what point should the light source be placed relative to the mirror's vertex?

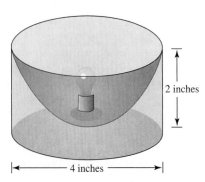

Figure 6.62

Designing a flashlight

Figure 6.63

Solution

We position the parabola with its vertex at the origin and opening upward (Figure 6.63). Thus its equation is of the form $x^2 = 4py$.

$x^2 = 4py$ This is the standard equation of a parabola with vertex at $(0, 0)$ and opening upward.

$2^2 = 4p \cdot 2$ We must find p. Since $(2, 2)$ lies on the parabola, let $x = 2$ and $y = 2$.

$4 = 8p$ Simplify and solve for p.

$p = \dfrac{1}{2}$ Divide both sides by 8.

$x^2 = 4 \cdot \dfrac{1}{2} y$ Substitute $\frac{1}{2}$ for p in the equation $x^2 = 4py$.

$x^2 = 2y$

The equation for the parabola used to shape the mirror is $x^2 = 2y$. The light source should be placed at the focus $(0, p)$. Since $p = \frac{1}{2}$, the light should be placed at $(0, \frac{1}{2})$, or $\frac{1}{2}$ inch above the vertex. ■

David Wojnarowicz "Something From Sleep III" 1988–89, acrylic on canvas, 48 1/2 × 38 1/2 in. Courtesy of PPOW.

| EXAMPLE 9 | **The Hale Telescope** |

The Hale Telescope on Palomar Mountain in California was for many years the world's largest and was unrivaled in its light-collecting power.

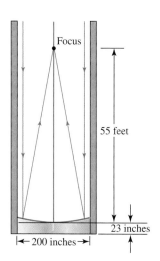

Figure 6.64

The Hale Telescope

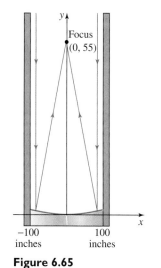

Figure 6.65

Positioning the Hale Telescope in an *xy* system; Remember that 55 is in feet.

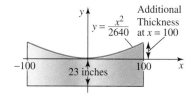

Figure 6.66

The thickness at the outside edge is 23 inches plus the additional thickness at *x* = 100.

The telescope uses a Pyrex glass mirror with a parabolic cross section and a diameter of 200 inches, as shown in Figure 6.64. (For clarity in visualizing this situation, the figure is not drawn to scale.)

a. If the focus of the parabola is 55 feet from the vertex, what is the equation for the parabola that was used to shape the mirror?
b. If the mirror is 23 inches thick at the center, what is the thickness of the glass on the outside edge?

Solution

a. We position the parabola with its vertex at the origin and opening upward (Figure 6.65). Its focus is at (0, 55 feet), and its equation is of the form $x^2 = 4py$.

$$x^2 = 4py$$
This is the standard equation of a parabola with vertex at $(0,0)$, focus at $(0, p)$, and opening upward.

$$x^2 = 4(660)y$$
Since $(0, p) = (0, 55)$, then $p = 55$ feet or, equivalently, $p = 55(12) = 660$ inches.

$$x^2 = 2640y$$
This is the equation for the parabola that was used to shape the Pyrex glass mirror.

b. We are now ready to find the thickness of the glass on the outside edge. Figure 6.66 shows that its thickness is

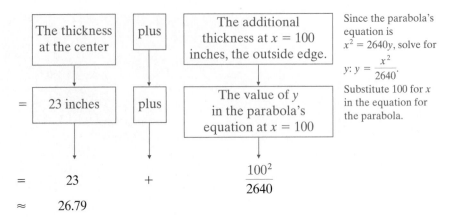

The thickness of the glass mirror on the outside edge is approximately 26.79 inches. ∎

5 Recognize equations of conic sections.

Summary of Conic Sections

The following box contains a summary of the equations and graphs of the conic sections.

Conic sections

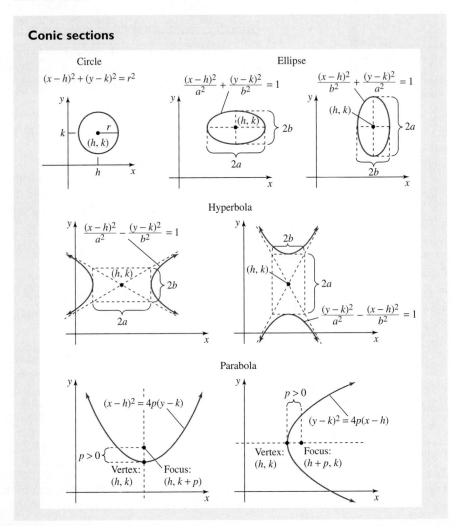

ENRICHMENT ESSAY

The Hubble Space Telescope

Astronomers' vision of an observatory above the atmosphere providing an unobscured view of the universe came true with the 1990 launching of the Hubble Space Telescope. Its main reflector, shown in the figure, is parabolic, 94.5 inches in diameter. The telescope also contains a smaller hyperbolic mirror positioned so that its focus coincides with that of the parabolic mirror. Notice that the eyepiece of the telescope is located at the other focus of the hyperbolic reflector.

profound mysteries of the cosmos: How big and how old is the universe? What is it made of? How did the galaxies come to exist? Do other Earth-like planets orbit other sun-like stars?

The Hubble Space Telescope.
Space Telescope Science Institute.

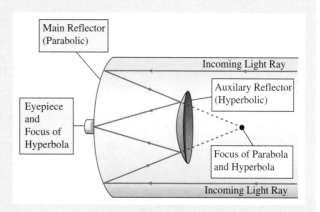

The Hubble telescope initially had blurred vision due to problems with its parabolic mirror that had been ground two millionths of a meter smaller than design specifications. (See Problem 53 in Problem Set 6.3.) In 1993 astronauts from the Space Shuttle *Endeavor* equipped the telescope with optics to correct the blurred vision and new solar panels. "A small change for a mirror, a giant leap for astronomy," Christopher J. Burrows of the Space Telescope Science Institute said when the first Hubble images were presented to the public in January, 1993.

The Hubble Space Telescope has captured images from the ends of the universe, more than 400 million times as far away as the sun, showing infant star systems the size of our solar system emerging from the gas and dust that shrouded their creation. With still at least a decade of useful life, the Hubble is expected to answer many of the most

M100 Galactic Nucleus

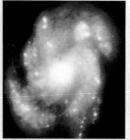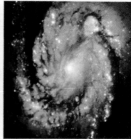

Comparison of images of spiral galaxy M100 in the Virgo cluster of galaxies taken by the Hubble Space Telescope before (left) and after (right) repairs were made. Space Telescope Science Institute.

We opened the chapter by noting that conic sections were named after their historical discovery as intersections of a plane and a right circular cone of two nappes. However, these intersections might not result in a conic. Three degenerate cases occur when the cutting plane passes through the vertex. These *degenerate conic sections* are a point, a line, and a pair of intersecting lines, illustrated in Figure 6.67.

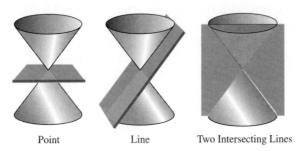

Point   Line   Two Intersecting Lines

Figure 6.67

Degenerate conics

Conic sections can be represented both geometrically (intersecting planes and cones) and algebraically. Their equations can all be expressed in the form

$$Ax^2 + Cy^2 + Dx + Ey + F = 0$$

where A and C are not both zero. This equation is called the *general form* for the equation of a conic section.

Study tip

A nondegenerate conic section of the form

$$Ax^2 + Cy^2 + Dx + Ey + F = 0$$

can be classified by inspecting the coefficients of x^2 and y^2.

Condition	Description	Resulting Conic	Example
$A = C$	The squared terms have equal coefficients.	Circle	$x^2 + y^2 - 6x + y + 3 = 0$ $A = 1 \quad C = 1 \qquad A = C$
$AC > 0$	The coefficients of the squared terms have the same sign.	Ellipse	$3x^2 + 2y^2 + 12x - 4y + 2 = 0$ $A = 3 \quad C = 2 \qquad A$ and C have the same sign. $A \ne C$.
$AC < 0$	The coefficients of the squared terms have opposite signs.	Hyperbola	$9x^2 - 16y^2 - 90x + 64y + 17 = 0$ $A = 9 \quad C = -16 \quad A$ and C have opposite signs.
$AC = 0$	$A = 0$ or $C = 0$ (but not both). There is only one squared term.	Parabola	$y^2 - 12x - 4y + 52 = 0$ or $0x^2 + y^2 - 12x - 4y + 52 = 0$ $A = 0 \quad C = 1$

Richard E. Prince "The Cone of Apollonius" (detail), fiberglass, steel, paint, graphite, 51 × 18 × 14 in. Private collection, Vancouver. Photo courtesy of Equinox Gallery, Vancouver, Canada.

EXAMPLE 10   Recognizing Equations of Conic Sections

Identify the graph of the following nondegenerate conics as a circle, an ellipse, a hyperbola, or a parabola.

a. $4x^2 - 25y^2 - 24x + 250y - 489 = 0$
b. $x^2 + y^2 + 6x - 2y + 6 = 0$
c. $y^2 + 12x + 2y - 23 = 0$
d. $9x^2 + 25y^2 - 54x + 50y - 119 = 0$

Solution

These are some of the equations from examples throughout the chapter, which we solved by completing the square.

a. The graph of $4x^2 - 25y^2 - 24x + 250y - 489 = 0$, with $A = 4$ and $C = -25$, is a hyperbola because A and C have opposite signs. (See Example 6, page 697.)
b. The graph of $x^2 + y^2 + 6x - 2y + 6 = 0$, with $A = 1$ and $C = 1$, is a circle because $A = C$. (See the Discover for yourself, page 678.)
c. The graph of $y^2 + 12x + 2y - 23 = 0$, with $A = 0$ and $C = 1$, is a parabola because there is only one squared term. (See Example 7, page 713.)
d. The graph of $9x^2 + 25y^2 - 54x + 50y - 119 = 0$, with $A = 9$ and $C = 25$, is an ellipse because $AC = (9)(25) > 0$. (See Example 6, page 677.) ■

PROBLEM SET 6.3

Practice Problems

In Problems 1–18, sketch the graph of each parabola by hand. Identify the vertex, focus, and directrix. Verify your graph using a graphing utility by solving for y in terms of x and entering the resulting equation(s).

1. $y^2 = 4x$
2. $y^2 = -8x$
3. $x^2 = -12y$
4. $x^2 = 16y$
5. $(x - 2)^2 = 8(y - 1)$
6. $(x + 2)^2 = 4(y + 1)$
7. $(y + 3)^2 = -16(x + 1)$
8. $(y - 2)^2 = -12(x - 1)$
9. $(x + 1)^2 = -8y$
10. $(x - 2\sqrt{2})^2 = y + \sqrt{2}$
11. $y^2 + 6y + 8x + 25 = 0$
12. $y^2 - 2y + 12x - 35 = 0$
13. $x^2 - 2x - 4y + 9 = 0$
14. $x^2 + 6x + 8y + 1 = 0$
15. $2x^2 + 8x - 3y + 4 = 0$
16. $2x^2 + 6x + 5y - 20 = 0$
17. $4y^2 - 12y + 9x = 0$
18. $4y^2 + 4y + 4x + 13 = 0$

In Problems 19–20, the parabola whose vertex is at the origin is translated to a new position, indicated in the graph. Write the standard form of the equation for each translated parabola.

19.

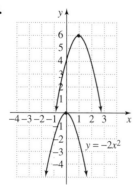

20.

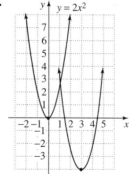

In Problems 21–24, find the standard form of the equation for each parabola and identify its directrix.

21.

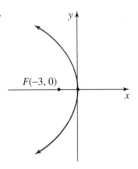

22.

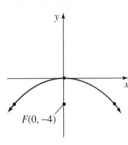

23.

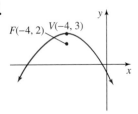

24.

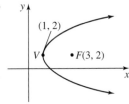

In Problems 25–42, find the standard form of the equation for each parabola satisfying the given conditions.

25. Focus: $(3, 0)$; directrix: $x = -3$

26. Focus: $(0, -4)$; directrix: $y = 4$

27. Vertex: origin; axis of symmetry: y-axis; passes through $(-1, 2)$

28. Vertex: origin; axis of symmetry: x-axis; passes through $(4, 3)$

29. Vertex: origin; focus on y-axis; passes through $(7, -10)$

30. Vertex: origin; focus: smaller of the two x-intercepts of the circle whose equation is $x^2 - 8x + y^2 - 6y + 9 = 0$

31. Vertex: origin; focus on x-axis; length of latus rectum $= 8$; parabola opens to the right.

32. Vertex: origin; focus on y-axis; length of latus rectum $= 6$; parabola opens upward.

33. Vertex: $(2, 3)$; focus: $(0, 3)$

34. Vertex: $(5, 4)$; focus: $(2, 4)$

35. Focus: $(-4, 1)$; directrix: $y = 5$

36. Focus: $(2, 1)$; directrix $x = -4$

37. Vertex: $(-3, 2)$; vertical axis of symmetry; passes through $(-2, -1)$

38. Vertex: $(3, -5)$; horizontal axis of symmetry; passes through $(4, 3)$

39. Set of points equidistant from $y = 4$ and $(-1, 0)$

40. Set of points equidistant from $y = 5$ and $(1, 2)$

41. Focus: $(4, -2)$; directrix parallel to y-axis; length of latus rectum $= 6$; parabola opens to the right.

42. Focus: $(-3, -3)$; directrix parallel to y-axis; length of latus rectum $= 8$; parabola opens to the left.

In Problems 43–50, classify the graph of each equation as a circle, an ellipse, a hyperbola, or a parabola. Verify your result using a graphing utility to graph the equation.

43. $4x^2 - 9y^2 - 8x - 36y - 68 = 0$

44. $9x^2 + 25y^2 - 54x - 200y + 256 = 0$

45. $y^2 + 8x + 6y + 25 = 0$

46. $4x^2 + 4y^2 + 12x + 4y + 1 = 0$

47. $9x^2 - 36x + 31 = -4y^2 - 8y$

48. $100x^2 + 90y = 7y^2 + 368$

49. $y^2 + 2y = 4x - 21$

50. $y^2 - 4y = 4x$

Application Problems

51. Parabolic microphones are often used in sporting events. The device consists of a parabolic dish to reflect sound to a receiver located at the focus. A parabolic microphone is formed by revolving the portion of the parabola $15x = y^2$ between $y = -9$ and $y = 9$ about its axis of symmetry. Where should the receiver be placed?

52. A parabolic headlight is formed by revolving the portion of the parabola $x^2 = 12y$ between the lines $x = -4$ and $x = 4$ about its axis of symmetry. Where should the light be placed for maximum illumination?

53. Galileo's telescope brought about revolutionary changes in astronomy. A comparable leap in our ability to observe the universe is now taking place as a result of the Hubble Space Telescope. The Hubble telescope can see stars and galaxies whose brightness is $\frac{1}{50}$ of the faintest objects now observable using ground-based telescopes. Like the Hale telescope (Example 9), the Hubble telescope uses a glass mirror, but its parabolic cross section is 2.4 meters in diameter. The focus of the parabola is 57.6 meters from the vertex. Initially the Hubble telescope had problems with its glass mirror, which had been ground to be 0.006248 meter thicker at the outside edge than at the center. Find the equation of the parabola that was used to shape the mirror. How much thicker, to six decimal places, should the mirror have been at the edge than at its center? What does this say in practical terms about the glass mirror used in a properly working telescope?

54. A TV satellite dish consists of a parabolic dish with the receiver placed at its focus. (See the figure.) Why must the receiver for a shallow dish be farther from the vertex than for a deeper dish of the same diameter? If the dish is to have a diameter of 20 feet and the receiver is to be placed 6 feet from the vertex, how deep should the dish be?

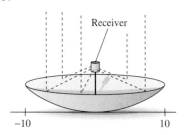

55. The towers of the Golden Gate Bridge in San Francisco are 1280 meters apart and rise 160 meters above the road. The cable between the towers has the shape of a parabola, and the cable just touches the sides of the road midway between the towers. What is the height of the cable 200 meters from a tower?

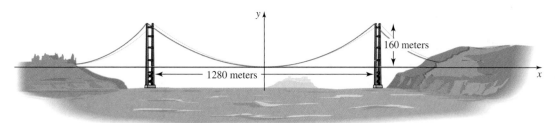

56. The parabolic arch shown in the figure is 50 feet above the water at the center and 200 feet wide at the base. Will a boat that is 30 feet tall clear the arch 30 feet from the center?

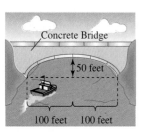

True–False Critical Thinking Problems

57. Which one of the following is true?
  a. The parabola whose equation is $x = 2y - y^2 + 5$ opens to the right.
  b. If the parabola whose equation is $x = ay^2 + by + c$ has its vertex at $(3, 2)$ and $a > 0$, then it has no y-intercepts.
  c. Some parabolas that open to the right have equations that define y as a function of x.
  d. The graph of $x = a(y - k) + h$ is a parabola with vertex at (h, k).

58. Which one of the following is true?

  a. The graph of $x^2 - 4y^2 - 8x + 16 = 0$ is a hyperbola.
  b. If the coefficients of the nonlinear terms in any equation involving x and y have opposite signs, such as $9x^3 - 16y^3 - 90x + 64y + 17 = 0$, then the equation's graph is a hyperbola.
  c. For any parabola, the distance from any point to the focus equals the distance from the focus to the directrix.
  d. The number -2 appears in the description of both the focus and the directrix for the parabola whose equation is $x^2 + 4x - 4y = 0$.

Technology Problems

Use a graphing utility to graph each parabola in Problems 59–62. Identify the coordinates of each vertex.

59. $x^2 - 6x - 12y + 9 = 0$

60. $y^2 - 6x + 2y + 13 = 0$

61. $y^2 + 10y - x + 25 = 0$

62. $12x^2 - x + 3y - 14 = 0$

In Problems 63–64, write each equation as a quadratic equation in y and then use the quadratic formula to express y in terms of x. Graph the resulting two equations using a graphing utility. What effect does the xy-term have on the graph of the resulting parabola?

63. $16x^2 - 24xy + 9y^2 - 60x - 80y + 100 = 0$

64. $x^2 + 2\sqrt{3}\,xy + 3y^2 + 8\sqrt{3}\,x - 8y + 32 = 0$

Each equation in Problems 65–69 represents a degenerate conic. Describe what you would expect each equation to represent based on the values of A and C, the coefficients of the squared terms. Then use a graphing utility to graph the equation. Which graphs clearly indicate degenerate conics?

65. $x^2 - 4y^2 - 6x + 8y + 5 = 0$

66. $9x^2 - 4y^2 - 18x + 9 = 0$

67. $3x^2 + y^2 - 6x + 6y + 13 = 0$

68. $x^2 + 4y^2 - 24y + 36 = 0$

69. $x^2 + y^2 + 4x - 6y + 13 = 0$

70. Graph $x = y^2$ using a graphing utility. In the same viewing rectangle, make it appear that a particle coming in horizontally from the right toward $(1, 1)$ is reflected off the parabola and moves toward the focus $\left(\frac{1}{4}, 0\right)$. What property of the parabola is illustrated?

71. As shown in the figure on the right, the line that touches the parabola at the point (x_0, y_0), called the *tangent line*, has a y-intercept of $-y_0$. Find the equation of the tan-

gent line to $x^2 = 4y$ (in this case, $p = 1$) at the point $(2, 1)$. Then use a graphing calculator to graph the parabola and the tangent line in the same viewing rectangle. Repeat this process for a few other points on the parabola. In what way might the slope of these tangent lines be useful in describing the parabola's changing steepness?

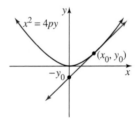

Writing in Mathematics

72. In Problems 37 and 38 you were asked to determine the equation of a parabola given its vertex, direction of the axis of symmetry, and one other point. If you were given only two of these things, how many parabolas would be possible? Discuss what this means in terms of the number of conditions needed to determine a unique parabola.

73. Explain why the 13 points shown in the figure on the right must lie along a parabola.

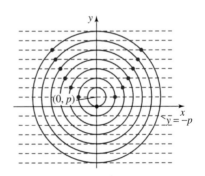

74. For the parabola shown in the figure, is enough information given to determine its equation? Explain. If you were asked to match the graph with one of the following equations, describe the process that would enable you to determine the correct match.

$$y = (x - 1)^2 - 1; \qquad x = (y - 1)^2 - 1;$$

$$x = (y - 1)^2; \qquad y = (x - 1)^2$$

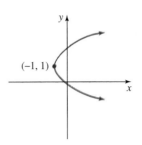

Critical Thinking Problems

75. Use the figure to write a formula for the area of the triangle in terms of A and B if it is known that C is the midpoint of line segment AB.

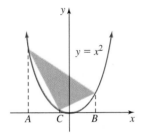

76. Use the figure to prove that the sum of the lengths of line segments FR and RG is $2p$, where R is any point on the parabola.

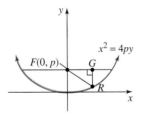

77. Complete the square on the equation

$$Ax^2 + Cy^2 + Dx + Ey + F = 0$$

and describe, using the sign of A and C and the expression $\dfrac{D^2}{4A} + \dfrac{E^2}{4C} - F$, when the conic section degenerates into a point and into pairs of intersecting lines.

78. Identify the vertex, focus, and directrix of a parabola given by $y = ax^2 + bx + c$, where $a \neq 0$.

79. Suppose that line segment PQ is the latus rectum of the parabola given by $x^2 = 4py$. If the coordinates of P are (x_1, y_1), prove that Q has coordinates given by

$$\left(\frac{-4p^2}{x_1}, \frac{p^2}{y_1} \right).$$

Group Activity Problem

80. Consult the research department of your library or the Internet to find an example of architecture that incorporates one or more conic sections in its design. Share this example with other group members. Explain precisely how conic sections are used. Do conic sections enhance the appeal of the architecture? In what ways?

Solutions Manual

Tutorial

Video II

SECTION 6.4 Nonlinear Systems

Objectives

1 Solve nonlinear systems of equations.
2 Solve problems that can be modeled by nonlinear systems of equations.
3 Solve nonlinear systems of inequalities.

An equation in which one or more terms have a variable of degree 2 or higher, such as the equation $x^2 = 2y + 10$, is called a *nonlinear equation*. All conic sections are graphs of nonlinear equations. A *nonlinear system* of equations contains at least one nonlinear equation.

1 Solve nonlinear systems of equations.

Systems Containing One Nonlinear Equation

A system consisting of one linear equation and one nonlinear equation whose graph is a conic section can have no real solutions, one real solution, or two real solutions. Real solutions correspond to intersection points of the graphs of the equations in a system. Figure 6.68 illustrates possibilities for intersection points when one graph is a line.

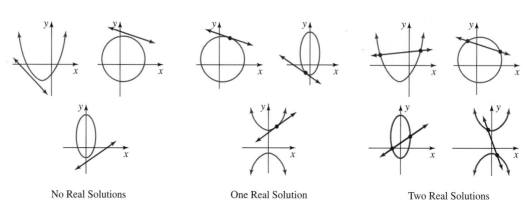

No Real Solutions   One Real Solution   Two Real Solutions

Figure 6.68
Possibilities for nonlinear systems with one linear equation

When solving a system in which one equation is linear, it is usually easiest to use the substitution method. Before considering some examples, let's summarize the steps that we used to solve linear systems by the substitution method—this same procedure will be applied to nonlinear systems.

> **The substitution method for a system involving two equations in two variables**
>
> 1. Solve one equation for one variable in terms of the other.
> 2. Substitute this expression for that variable into the other equation.
> 3. Solve the resulting equation in one variable.
> 4. Back-substitute the solution you get into the equation in step 1 to find the value of the other variable.
> 5. Check the solution by substituting into the given equations or using a graphing analysis to find intersection points for the system's graphs.

When solving a system in which one of the equations is quadratic, the substitution method often leads to a quadratic equation in one variable. Example 1 illustrates this.

EXAMPLE 1 **Solving a Nonlinear System by the Substitution Method**

Solve by the substitution method:

$$2x + 3y = 12 \qquad \text{Equation 1. The graph is a line.}$$

$$4x^2 + 9y^2 = 144 \qquad \text{Equation 2. The graph is an ellipse.}$$

Solution

Study tip

If a system contains a linear equation and a nonlinear equation, always solve the linear equation and substitute the resulting expression into the nonlinear equation.

1. We can solve for either x or y in the linear equation (Equation 1). We will solve for y.

$$2x + 3y = 12 \qquad \text{This is the given linear equation.}$$

$$3y = 12 - 2x \qquad \text{Subtract } 2x \text{ from both sides.}$$

$$y = \frac{12 - 2x}{3} \qquad \text{Divide both sides by 3.}$$

2. Substitute $\dfrac{12 - 2x}{3}$ for y in Equation 2.

$$4x^2 + 9y^2 = 144 \qquad \text{Equation 2}$$

$$4x^2 + 9\left(\frac{12 - 2x}{3}\right)^2 = 144 \qquad \text{Replace } y \text{ with } \frac{12 - 2x}{3}.$$

3. Solve this equation in one variable.

$$4x^2 + 9\left(\frac{12 - 2x}{3}\right)^2 = 144 \qquad \text{Solve for } x.$$

$$4x^2 + 9\left(\frac{144 - 48x + 4x^2}{9}\right) = 144 \qquad \begin{array}{l}\text{Square the numerator and the}\\\text{denominator of the rational expression.}\end{array}$$

$$4x^2 + \cancel{9}\left(\frac{144 - 48x + 4x^2}{\cancel{9}}\right) = 144 \qquad \begin{array}{l}\text{Divide the numerator and denominator}\\\text{by 9.}\end{array}$$

$$8x^2 - 48x + 144 = 144 \qquad \text{Combine like terms on the left.}$$

$$8x^2 - 48x = 0 \qquad \begin{array}{l}\text{Set the quadratic equation equal to 0 by}\\\text{subtracting 144 from both sides.}\end{array}$$

$$8x(x - 6) = 0 \qquad \text{Factor.}$$

$$8x = 0 \quad \text{or} \quad x - 6 = 0 \qquad \text{Set each factor equal to 0.}$$

$$x = 0 \qquad\qquad x = 6 \qquad \text{Solve for } x.$$

4. Back-substitute these values for x into $y = \dfrac{12 - 2x}{3}$, the form of Equation 1 obtained in step 1.

If $x = 0$, $y = \dfrac{12 - 2(0)}{3} = \dfrac{12}{3} = 4$, so $(0, 4)$ is a solution.

If $x = 6$, $y = \dfrac{12 - 2(6)}{3} = \dfrac{0}{3} = 0$, so $(6, 0)$ is a solution.

5. Check by showing that both ordered pairs satisfy both equations.

Check (0, 4)

$$2x + 3y = 12 \qquad 4x^2 + 9y^2 = 144$$
$$2 \cdot 0 + 3 \cdot 4 \overset{?}{=} 12 \qquad 4 \cdot 0^2 + 9 \cdot 4^2 \overset{?}{=} 144$$
$$12 = 12 \checkmark \qquad 144 = 144 \checkmark$$

Check (6, 0)

$$2x + 3y = 12 \qquad 4x^2 + 9y^2 = 144$$
$$2 \cdot 6 + 3 \cdot 0 \overset{?}{=} 12 \qquad 4 \cdot 6^2 + 9 \cdot 0^2 \overset{?}{=} 144$$
$$12 = 12 \checkmark \qquad 144 = 144 \checkmark$$

The solution set of the given system is $\{(0, 4), (6, 0)\}$. Figure 6.69 shows the graphs of the solutions. ■

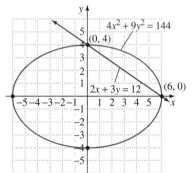

Figure 6.69

Points of intersection illustrate the nonlinear system's solutions.

Systems Containing Two Nonlinear Equations

We now consider systems consisting of two equations whose graphs are both conic sections. The following figure shows that such systems may have no real solutions, or they may have one, two, three, or four real solutions.

No Real
Solutions
One Real
Solution
Two Real
Solutions
Three Real
Solutions
Four Real
Solutions

EXAMPLE 2 Solving a Nonlinear System by the Substitution Method

Solve by the substitution method:

$$y^2 - x + 3 = 0 \qquad \text{Equation 1. The graph is a parabola.}$$
$$y^2 + 2x - 3y - 12 = 0 \qquad \text{Equation 2. The graph is a parabola.}$$

Discover for yourself

Graph the equations in Example 2. How many solutions do you expect for this nonlinear system?

Solution

1. We will solve for x in Equation 1.

$$y^2 - x + 3 = 0 \qquad \text{This is Equation 1.}$$
$$y^2 + 3 = x \qquad \text{Add } x \text{ to both sides.}$$

2. Substitute $(y^2 + 3)$ for x in Equation 2.

$$y^2 + 2x - 3y - 12 = 0 \qquad \text{Equation 2}$$
$$y^2 + 2(y^2 + 3) - 3y - 12 = 0 \qquad \text{Replace } x \text{ with } y^2 + 3.$$

3. Solve this equation in one variable.

$$y^2 + 2y^2 + 6 - 3y - 12 = 0 \qquad \text{Apply the distributive property.}$$

$$3y^2 - 3y - 6 = 0 \qquad \text{Combine like terms.}$$
$$y^2 - y - 2 = 0 \qquad \text{Simplify by dividing both sides of the equation by 3.}$$
$$(y - 2)(y + 1) = 0 \qquad \text{Factor.}$$
$$y - 2 = 0 \quad \text{or} \quad y + 1 = 0 \qquad \text{Set each factor equal to 0.}$$
$$y = 2 \qquad\qquad y = -1 \qquad \text{Solve for } y.$$

4. Back-substitute these values for y into $x = y^2 + 3$, the form of Equation 1 obtained in step 1.

If y is 2, $x = 2^2 + 3 = 7$, so $(7, 2)$ is a solution.

If y is -1, $x = (-1)^2 + 3 = 4$, so $(4, -1)$ is a solution.

5. Let's use a graphing utility to check the solution. Both equations in the system contain only one squared term, so their graphs are parabolas. To use a graphing utility, we must first solve each equation for y.

Equation 1

$$y^2 - x + 3 = 0$$

$$y^2 = x - 3$$

$$y = \pm\sqrt{x - 3}$$

Equation 2

$$y^2 - 3y + (2x - 12) = 0$$

$$y = \frac{-b \pm \sqrt{b^2 - 4ac}}{2a}$$

$$= \frac{3 \pm \sqrt{9 - 4(2x - 12)}}{2} \qquad a = 1, b = -3, \text{ and } c = 2x - 12$$

We now enter four equations:

$$y_1 = \sqrt{x - 3}$$
$$y_2 = -y_1$$
$$y_3 = \frac{3 + \sqrt{9 - 4(2x - 12)}}{2}$$
$$y_4 = \frac{3 - \sqrt{9 - 4(2x - 12)}}{2}.$$

The system is graphed in Figure 6.70. Intersection points occur at $(7, 2)$ and $(4, -1)$, verifying that the solution set for the given system is $\{(7, 2), (4, -1)\}$. ∎

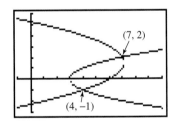

Figure 6.70

A system with $(4, -1)$ and $(7, 2)$ as real solutions

EXAMPLE 3   **Solving a Nonlinear System by the Substitution Method**

Solve by the substitution method:

$$xy = 4 \qquad \text{Equation 1. The graph is a hyperbola. (See the Study Tip on page 730.)}$$
$$x^2 + 4y^2 = 20 \qquad \text{Equation 2. The graph is an ellipse.}$$

Solution

1. We will solve for y in Equation 1.

$$xy = 4 \qquad \text{Equation 1}$$

$$y = \frac{4}{x} \qquad \text{Divide both sides by } x.$$

tudy tip

Another form of the hyperbola is

$$xy = c$$

where c is a nonzero constant. These hyperbolas have the x- and y-axes as asymptotes. You can explore this idea in more depth by working Problem 69 in Problem Set 6.2.

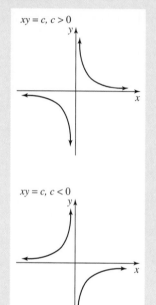

$xy = c, c > 0$

$xy = c, c < 0$

2. Substitute $\dfrac{4}{x}$ for y in Equation 2.

$$x^2 + 4y^2 = 20$$

$$x^2 + 4\left(\frac{4}{x}\right)^2 = 20$$

3. Solve this equation in one variable.

$$x^2 + \frac{64}{x^2} = 20$$
Square $\dfrac{4}{x}$, squaring the numerator and the denominator. Then multiply by 4.

$$x^2\left(x^2 + \frac{64}{x^2}\right) = x^2 \cdot 20$$
Multiply both sides by x^2, clearing the equation of fractions ($x \neq 0$).

$$x^4 + 64 = 20x^2$$
Apply the distributive property and simplify.

$$x^4 - 20x^2 + 64 = 0$$
Set the equation equal to 0.

$$(x^2 - 16)(x^2 - 4) = 0$$
Factor.

$$x^2 - 16 = 0 \quad \text{or} \quad x^2 - 4 = 0$$
Set each factor equal to 0.

$$x^2 = 16 \qquad\qquad x^2 = 4$$
Solve for x.

$$x = \pm\sqrt{16} \qquad\quad x = \pm\sqrt{4}$$
If $x^2 = a$, then $x = \pm\sqrt{a}$.

$$x = 4 \ \text{ or } \ x = -4 \quad x = 2 \ \text{ or } \ x = -2$$
Simplify.

4. Back-substitute these values for x into $y = \dfrac{4}{x}$, the form of Equation 1 obtained in step 1.

If $x = 4$, $y = \dfrac{4}{4} = 1$. $(4, 1)$ is a solution.

If $x = -4$, $y = \dfrac{4}{-4} = -1$. $(-4, -1)$ is a solution.

If $x = 2$, $y = \dfrac{4}{2} = 2$. $(2, 2)$ is a solution.

If $x = -2$, $y = \dfrac{4}{-2} = -2$. $(-2, -2)$ is a solution.

5. Let's use a graphing utility to check the solution. First we solve each equation in the system for y.

Equation 1	**Equation 2**

$$x^2 + 4y^2 = 20$$

Equation 1	**Equation 2**
$xy = 4$	$4y^2 = 20 - x^2$
$y = \dfrac{4}{x}$	$y^2 = \dfrac{20 - x^2}{4}$
	$y = \pm\dfrac{\sqrt{20 - x^2}}{2}$

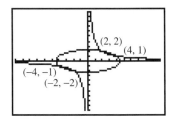

Figure 6.71

A system with four real
solutions

We now enter three equations:

$$y_1 = \frac{4}{x} \qquad y_2 = \frac{\sqrt{20 - x^2}}{2} \qquad y_3 = -y_2$$

The system is graphed in Figure 6.71. The four intersection points verify
that the solution set is $\{(4, 1), (-4, -1), (2, 2), (-2, -2)\}$. ■

In Chapter 5, we saw that linear systems in two variables can be solved
using substitution and the addition method. The addition method works par-
ticularly well on nonlinear systems when each equation is in the form $Ax^2 + By^2 = C$. If necessary we will multiply either equation or both equations by
appropriate numbers so that the coefficients of x^2 or y^2 will have a sum of 0.
We then add the equations. The sum will be an equation in one variable.

EXAMPLE 4 **Solving a Nonlinear System by the Addition Method**

Solve by the addition method:

$$4x^2 + y^2 = 13 \qquad \text{Equation 1. The graph is an ellipse.}$$

$$x^2 + y^2 = 10 \qquad \text{Equation 2. The graph is a circle.}$$

Solution

1. We can eliminate y^2 by multiplying Equation 2 by -1 and adding equations.

$$
\begin{array}{ll}
4x^2 + y^2 = 13 & \xrightarrow{\text{No change}} \\
x^2 + y^2 = 10 & \xrightarrow{\text{Multiply by } -1.}
\end{array}
\quad
\begin{array}{rl}
4x^2 + y^2 = & 13 \\
-x^2 - y^2 = & -10 \\
\hline
\text{Add: } 3x^2 \qquad = & 3 \\
x^2 = & 1 \\
x = & \pm 1
\end{array}
$$

2. Now we back-substitute these values for x into either one of the original
equations. Let's use $x^2 + y^2 = 10$, Equation 2. If $x = 1$,

$$1^2 + y^2 = 10 \qquad \text{Replace } x \text{ with 1 in Equation 2.}$$

$$y^2 = 9 \qquad \text{Subtract 1 from both sides.}$$

$$y = \pm 3 \qquad \text{Apply the square root method.}$$

$(1, 3)$ and $(1, -3)$ are solutions. If $x = -1$,

$$(-1)^2 + y^2 = 10 \qquad \text{Replace } x \text{ with } -1 \text{ in Equation 2.}$$

$$y^2 = 9 \qquad \text{The steps are the same as above.}$$

$$y = \pm 3$$

$(-1, 3)$ and $(-1, -3)$ are solutions.

Each ordered pair can be checked and shown to satisfy both equations, indicating a solution set of $\{(1, 3), (1, -3), (-1, 3), (-1, -3)\}$ (see
Figure 6.72). ■

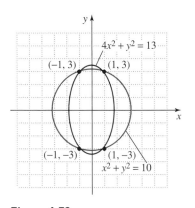

Figure 6.72

A system with four solutions

EXAMPLE 5	**Solving a Nonlinear System by the Addition Method**

Solve by the addition method:

$$16x^2 + y^2 - 24y + 80 = 0 \qquad \text{Equation 1. The graph is an ellipse.}$$

$$16x^2 + 25y^2 - 400 = 0 \qquad \text{Equation 2. The graph is an ellipse.}$$

Solution

1. The addition method is effective on the given system because we can eliminate the x^2-term by multiplying either equation by -1, adding equations, and eliminating all occurrences of the variable x.

$$
\begin{array}{l}
16x^2 + y^2 - 24y + 80 = 0 \xrightarrow{\text{Multiply by } -1.} -16x^2 - y^2 + 24y - 80 = 0 \\
16x^2 + 25y^2 - 400 = 0 \xrightarrow{\text{No change}} 16x^2 + 25y^2 - 400 = 0 \\
\hline
\text{Add:} 24y^2 + 24y - 480 = 0
\end{array}
$$

We will now work with the resulting quadratic equation, solving for y.

$$24y^2 + 24y - 480 = 0 \qquad \text{This the equation obtained above.}$$

$$y^2 + y - 20 = 0 \qquad \text{Divide both sides of the equation by 24.}$$

$$(y - 4)(y + 5) = 0 \qquad \text{Factor.}$$

$$y - 4 = 0 \quad \text{or} \quad y + 5 = 0 \qquad \text{Set each factor equal to 0.}$$

$$y = 4 \qquad\qquad y = -5 \qquad \text{Solve for } y.$$

2. Now we back-substitute these values for y into either of the original equations. We'll use Equation 2 because it is a bit less complicated than Equation 1.

$$16x^2 + 25y^2 - 400 = 0 \qquad \text{Equation 2}$$

Let $y = 4$:	**Let $y = -5$:**
$16x^2 + 25 \cdot 4^2 - 400 = 0$	$16x^2 + 25(-5)^2 - 400 = 0$
$16x^2 + 400 - 400 = 0$	$16x^2 + 225 = 0$
$16x^2 = 0$	$x^2 = -\dfrac{225}{16}$
$x = 0$	$x = \pm\sqrt{-\dfrac{225}{16}}$
	$x = \pm\dfrac{15}{4}i$

The system's solution set is $\left\{ (0, 4), \left(\dfrac{15}{4}i, -5\right), \left(-\dfrac{15}{4}i, -5\right) \right\}$. Let's see what this means by graphing the system. In standard form, the first equation can be expressed as

$$\frac{x^2}{4} + \frac{(y - 12)^2}{64} = 1$$

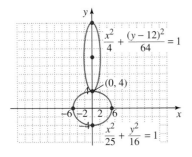

Figure 6.73

A system of intersecting ellipses with (0, 4) as a real solution

Jacques Villon "L'Espace" 1932, oil on canvas, 116 × 89 cm. Collection Galerie Louis Carré & Cie, Paris (France). © 1998 Artists Rights Society (ARS), NY / ADAGP, Paris.

and the second equation as

$$\frac{x^2}{25} + \frac{y^2}{16} = 1.$$

The equations are graphed in Figure 6.73. The ellipses intersect at (0, 4). The solutions with imaginary coordinates do not appear as points of intersection. ∎

Study tip

When an imaginary number appears as either the x-coordinate or y-coordinate of a system's solution, this solution is not a point of intersection for the graphs of the system's equations.

Example 6 illustrates a system of second-degree equations that requires both the addition and substitution methods to solve.

EXAMPLE 6 **Solving a Nonlinear System by a Combination of Methods**

Solve the system:

$$2x^2 - xy + 2y^2 = 32 \qquad \text{Equation 1. It's not obvious what the graph is.}$$

$$x^2 + y^2 = 13 \qquad \text{Equation 2. The graph is a circle.}$$

Solution

We can begin by eliminating the squared terms by multiplying both sides of Equation 2 by -2 and then adding the result to Equation 1.

$$
\begin{array}{llll}
2x^2 - xy + 2y^2 = 32 & \xrightarrow{\text{No change}} & 2x^2 - xy + 2y^2 = & 32 \\
x^2 + \ y^2 = 13 & \xrightarrow{\text{Multiply by } -2.} & \underline{-2x^2 \qquad -\ 2y^2 = -26} \\
& \text{Add:} & -xy \qquad \quad = & 6
\end{array}
$$

Next we solve $-xy = 6$ for either x or y. Let's solve for y.

$$-xy = 6$$

$$y = -\frac{6}{x} \qquad \text{Divide both sides by } -x.$$

Now we back-substitute $-\dfrac{6}{x}$ for y in one of the original equations. We'll work with Equation 2 because it is less complicated.

$$x^2 + y^2 = 13 \qquad \text{Equation 2}$$

$$x^2 + \left(-\frac{6}{x}\right)^2 = 13 \qquad \text{Because } y = -\frac{6}{x},$$

$$\text{substitute } -\frac{6}{x} \text{ for } y.$$

$$x^2 + \frac{36}{x^2} = 13 \qquad \text{Square } \left(-\frac{6}{x}\right), \text{ squaring}$$

$$\text{the numerator and the denominator.}$$

$$x^2\left(x^2 + \frac{36}{x^2}\right) = x^2 \cdot 13$$

Multiply both sides by x^2, clearing the equation of fractions ($x \neq 0$).

$$x^4 + 36 = 13x^2$$

Apply the distributive property and simplify.

$$x^4 - 13x^2 + 36 = 0$$

Set the equation equal to 0.

$$(x^2 - 4)(x^2 - 9) = 0$$

Factor.

$$x^2 - 4 = 0 \quad \text{or} \quad x^2 - 9 = 0$$

Set each factor equal to 0.

$$x^2 = 4 \qquad\qquad x^2 = 9$$

Solve for x^2.

$$x = 2 \quad \text{or} \quad x = -2 \qquad x = 3 \quad \text{or} \quad x = -3$$

If $x^2 = a$, then $x = \sqrt{a}$ or $x = -\sqrt{a}$.

Discover for yourself

Substitute the x-values 2, -2, 3, and -3 into Equation 2 ($x^2 + y^2 = 13$). What additional ordered pairs do you obtain that are not listed among the four pairs on the right? Do these additional ordered pairs satisfy Equation 1? Based on the graph of the system shown in Figure 6.75, can they possibly be solutions?

We can find the corresponding values for y by back-substituting the four values for x into $y = -\dfrac{6}{x}$:

If $x = 2$, $y = -\dfrac{6}{2} = -3$. $(2, -3)$ is a solution.

If $x = -2$, $y = -\dfrac{6}{-2} = 3$. $(-2, 3)$ is a solution.

If $x = 3$, $y = -\dfrac{6}{3} = -2$. $(3, -2)$ is a solution.

If $x = -3$, $y = -\dfrac{6}{-3} = 2$. $(-3, 2)$ is a solution.

Let's use a graphing utility to visualize what these solutions mean. The graph of Equation 1, $2x^2 - xy + 2y^2 = 32$, is not obvious because of the xy-term. To obtain the graph, let's first express y in terms of x.

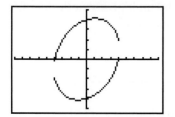

Figure 6.74

The graph of $2x^2 - xy + 2y^2 = 32$, a tilted or rotated ellipse

$$2x^2 - xy + 2y^2 = 32$$

Equation 1.

$$2y^2 - xy + (2x^2 - 32) = 0$$

Write the equation as a quadratic in y.

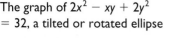

$a = 2 \quad b = -x \quad c = 2x^2 - 32$

$$y = \frac{-b \pm \sqrt{b^2 - 4ac}}{2a}$$

Use the quadratic formula.

$$y = \frac{x \pm \sqrt{x^2 - 8(2x^2 - 32)}}{4}$$

Let $a = 2$, $b = -x$, and $c = 2x^2 - 32$.

We now enter two equations:

$$y_1 = \frac{x + \sqrt{x^2 - 8(2x^2 - 32)}}{4}$$

$$y_2 = \frac{x - \sqrt{x^2 - 8(2x^2 - 32)}}{4}$$

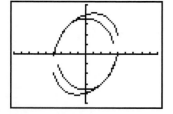

Figure 6.75

Intersection points are difficult to see, but they can certainly be $(2, -3)$, $(-2, 3)$, $(3, -2)$ and $(-3, 2)$, the solutions obtained algebraically.

We obtain what appears to be a tilted or rotated ellipse, shown in Figure 6.74. In Figure 6.75 we added the graph of $x^2 + y^2 = 13$, a circle, to the viewing rectangle. We obtained that graph by solving for y and entering

$$y_3 = \sqrt{13 - x^2}$$

and

$$y_4 = -\sqrt{13 - x^2} \quad (\text{or } y_4 = -y_3).$$

It's a bit difficult to tell where the tilted ellipse and the circle intersect. We can use the $\boxed{\text{TRACE}}$ and $\boxed{\text{ZOOM}}$ features to determine the ordered pairs for the intersection points, verifying that the solution set is $\{(2, -3), (-2, 3), (3, -2), (-3, 2)\}$. ■

Study tip

1. In the Discover for yourself, did you observe that it is possible to obtain ordered pairs that satisfy one, but not both, equations in a nonlinear system and that consequently are not solutions? Always check proposed solutions with real coordinates by direct substitution in each equation in the system or by graphing the system and using the graphs to eliminate ordered pairs that are not intersection points.

2. In trigonometry, it is shown that the graph of

   $$Ax^2 + Bxy + Cy^2 + Dx + Ey + F = 0, B \neq 0,$$

   is a *rotated* or *tilted conic* section. You can identify the graph by computing $B^2 - 4AC$. In particular:
   a. If $B^2 - 4AC < 0$, the graph is a tilted ellipse.
   b. If $B^2 - 4AC > 0$, the graph is a tilted hyperbola.
   c. If $B^2 - 4AC = 0$, the graph is a tilted parabola.

---

2 Solve problems that can be modeled by nonlinear systems of equations.

Problem Solving and Modeling Using Nonlinear Systems

Many geometric problems can be modeled and solved by the use of nonlinear systems of equations.

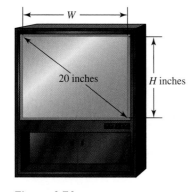

Figure 6.76

The screen's area is 192 square inches.

Step 1. Use variables to represent unknown quantities.

EXAMPLE 7   An Application of a Nonlinear System

A 20-inch diagonal television that is wider than it is high has a viewing area of 192 square inches. The selling price of the television is cheaper than a smaller television that is only 10 inches high. A consumer has already built a shelf for the television that can accommodate a height of 11 inches. The smaller, more expensive television will fit with no problem onto the shelf. Will the shelf's height accommodate the larger television? Answer the question by computing the width and height of the screen.

Solution

Let

$W =$ the screen's width

$H =$ the screen's height

(See Figure 6.76.)

We can use the formula for the area of a rectangle and the Pythagorean Theorem to obtain a system of equations.

Step 2. Write a system of equations modeling the problem's conditions.

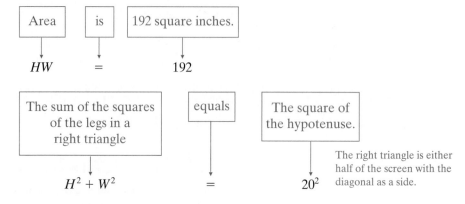

Step 3. Solve the system and answer the problem's question.

We must solve the system

$$HW = 192$$

$$H^2 + W^2 = 400.$$

We solve the first equation for H.

$$H = \frac{192}{W}$$

Now we substitute $\dfrac{192}{W}$ for H in the second equation and solve for W.

$$\left(\frac{192}{W}\right)^2 + W^2 = 400$$

$\dfrac{36{,}864}{W^2} + W^2 = 400$ Square the numerator and the denominator.

$36{,}864 + W^4 = 400W^2$ Clear fractions by multiplying both sides by W^2.

$W^4 - 400W^2 + 36{,}864 = 0$ Set the equation equal to 0.

$(W^2 - 256)(W^2 - 144) = 0$ Factor.

$W^2 - 256 = 0$ or $W^2 - 144 = 0$ Set each factor equal to 0.

$\qquad W^2 = 256 \qquad\qquad W^2 = 144$

$\qquad\quad W = \pm 16 \qquad\qquad\;\, W = \pm 12$

Since W cannot be negative (because it represents the screen's width), W is either 16 or 12.

We back-substitute these values into $H = \dfrac{192}{W}$.

If $W = 16$, then $H = \dfrac{192}{16} = 12.$

If $W = 12$, then $H = \dfrac{192}{12} = 16.$

The screen is wider than it is high, so its width is 16 inches and its height is 12 inches. The shelf's height of 11 inches will not accommodate this television.

Step 4. Check the proposed answers in the original wording of the problem.

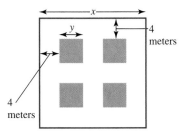

Figure 6.77

The floor plan of a library. Books are to be shelved in each of the four shaded square areas

Step 1. Use variables to represent unknown quantities.

Step 2. Write a system of equations describing the problem's conditions.

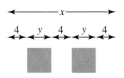

Figure 6.78

Step 3. Solve the system and answer the problem's question.

With the shelf already built, the consumer will have to purchase the smaller, more expensive model.

Take a moment to check the solutions in each of the two sentences on page 736 that are boxed and that we used to obtain our equations.   ∎

EXAMPLE 8   **Using a Geometric Figure to Set Up a Nonlinear System**

An architect is designing the floor plan for one room of a library. Figure 6.77 shows that the room is to have a square floor plan and that books are to be shelved in each of the four smaller shaded square areas. The border around each of the smaller squares is to uniformly measure 4 meters so that people have ample space to browse. The area of the larger square, the room's floor, excluding the four squares for shelving books in its interior, is to be 720 square meters. Find the dimensions of the larger square, that is the library's floor, and the smaller squares where the books are to be shelved.

Solution

Let

x = the length of each side of the larger square

y = the length of each side of the smaller squares

These variables appear in Figure 6.77.

We must refer to the picture to set up a system of equations. Because the border around the smaller squares uniformly measures 4 meters, let's consider the portion of the diagram shown in Figure 6.78.

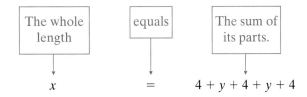

We simplify this equation to obtain

$x = 2y + 12.$

We obtain our second equation by translating the following sentence.

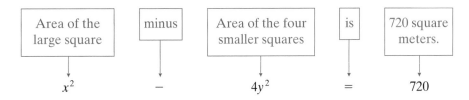

Now the system

$$x = 2y + 12$$
$$x^2 - 4y^2 = 720$$

can be solved by the substitution method. We substitute $2y + 12$ for x in the second equation.

$$(2y + 12)^2 - 4y^2 = 720$$

$$4y^2 + 48y + 144 - 4y^2 = 720 \quad \text{Square the binomial using } (A + B)^2 = A^2 + 2AB + B^2.$$

$$48y + 144 = 720 \quad \text{Simplify.}$$

$$48y = 576 \quad \text{Subtract 144 from both sides.}$$

$$y = 12 \quad \text{Divide both sides by 48.}$$

Now we back-substitute 12 for y into $x = 2y + 12$.

$$x = 2(12) + 12 = 36$$

Step 4. Check the proposed answers in the original wording of the problem.

We see that $x = 36$ and $y = 12$. This means that the larger square—the library's floor—measures 36 meters by 36 meters, and the smaller squares—the book display areas—each measure 12 meters by 12 meters.

Take a moment to check the solutions in each of the two sentences that are boxed and that we used to obtain our equations. ∎

3 Solve nonlinear systems of inequalities.

Wassily Kandinsky "White Figure" Jan. 1915, oil on board, 22 5/8 × 16 1/2 in. Solomon R. Guggenheim Museum, New York. Photo by David Heald. © The Solomon R. Guggenheim Foundation, New York. FN 47.1140.

Nonlinear Systems of Inequalities

In Section 5.8, we graphed linear inequalities by first graphing the boundary and then choosing a test point not on the boundary to determine the region used for the solution set. A *nonlinear inequality in two variables* (x and y) has a corresponding equation that is nonlinear. For example,

$$4x^2 + y^2 \leq 36$$

is a nonlinear inequality because $4x^2 + y^2 = 36$ is a nonlinear equation. Second-degree inequalities like this are graphed exactly like linear inequalities. The boundary of the inequality is the graph of the equation

$$4x^2 + y^2 = 36.$$

Writing this equation in standard form by dividing both sides by 36, we obtain

$$\frac{4x^2}{36} + \frac{y^2}{36} = \frac{36}{36} \quad \text{In standard form, we want 1 on the right.}$$

$$\frac{x^2}{9} + \frac{y^2}{36} = 1$$

The graph of this equation is an ellipse with x-intercepts at -3 and 3 and y-intercepts at -6 and 6. The vertices are $(0, -6)$ and $(0, 6)$. The graph of $4x^2 + y^2 \leq 36$ will also include either all points inside the ellipse or all points outside the ellipse. We can determine which region to shade by substituting any point not on the ellipse, such as $(0, 0)$, into $4x^2 + y^2 \leq 36$. Because

$$4 \cdot 0^2 + 0^2 \leq 36$$

$$0 \leq 36$$

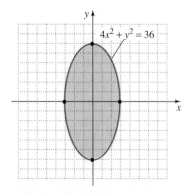

Figure 6.79

The graph of $4x^2 + y^2 \leq 36$

is a true statement, the graph of $4x^2 + y^2 \leq 36$ includes the ellipse whose equation is $4x^2 + y^2 = 36$ and all points inside the ellipse. The graph is shown in Figure 6.79.

ENRICHMENT ESSAY

Space Debris: Asteroids and Comets

L. Brower Hatcher "Atlas"
1992 (52 – 9 × 9 squares),
117 × 36.

An asteroid is a tiny planet that revolves around the sun. Most of them can be found in an asteroid belt between the orbits of Mars and Jupiter, as shown in the figure below. In this ring of rocky debris, the largest—Ceres—is 620 miles across. But most of the 5000 or so known pieces are only a few miles across.

At the end of the known solar system, there is more debris in the form of comets. Comets are believed to be leftover particles from the beginning of the solar system when most of these particles collided with each other to form the planets. The solid part of a comet, made up chiefly of dirt and icy materials, is only a few miles across, but the cloud and gas extending away from the center can stretch millions of miles.

Over 650 comets have been identified in our solar system, and each has a path that is elliptical, parabolic, or hyperbolic, with the sun at one focus. A comet's orbit is determined by its velocity; those with open parabolic and hyperbolic orbits pass by the Earth only once, so it is probable that many have not been seen or identified.

Scientists debate the probability that a "doomsday rock" will collide with Earth. It has been estimated that an asteroid crashes into Earth about once every 250,000 years, and that such a collision would have disastrous results. In 1908 a small fragment, possibly from the comet Encke, struck Siberia, leveling thousands of acres of trees. One theory about the extinction of dinosaurs 65 million years ago involves Earth's collision with a large asteroid and the resulting drastic changes in Earth's climate.

Understanding the path of a comet, using an analysis of conic sections and their properties, is essential in establishing some sort of detection system for troublesome asteroids and comets. In 1992 a NASA team began a project called Spaceguard Survey, calling for an international watch for threatening space debris. The danger of a collision is greatest for comets with open parabolic and hyperbolic orbits, seen only once and consequently difficult to detect.

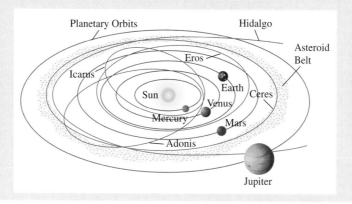

In Example 9, we will be graphing the solution set of two inequalities. If two or more inequalities are considered and at least one of them is nonlinear, we have a *nonlinear system of inequalities*. We can graph the solution set for the nonlinear system by finding the intersection of the graphs of each inequality in the system. Let's see exactly what this means.

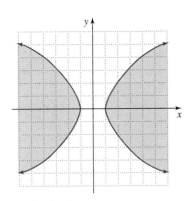

Figure 6.80

The graph of $x^2 - y^2 \geq 1$

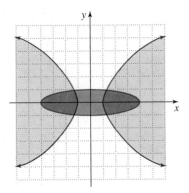

Figure 6.81

Adding $\dfrac{x^2}{16} + y^2 \leq 1$ to

$x^2 - y^2 \geq 1$

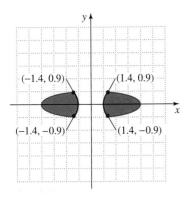

Figure 6.82

The graph of $x^2 - y^2 \geq 1$ and

$\dfrac{x^2}{16} + y^2 \leq 1$

 EXAMPLE 9   **Graphing a Nonlinear System of Inequalities**

Graph the solution set of:

$$x^2 - y^2 \geq 1$$

$$\frac{x^2}{16} + y^2 \leq 1$$

Solution

We begin by graphing the solutions of $x^2 - y^2 \geq 1$. The boundary is the graph of

$$\frac{x^2}{1} - \frac{y^2}{1} = 1$$

a hyperbola with vertices at $(1, 0)$ and $(-1, 0)$. The graph of $x^2 - y^2 \geq 1$ will include the hyperbola and either the region between its branches or the region to the left of the left branch and to the right of the right branch.
We'll use $(0, 0)$ as a test point.

$$x^2 - y^2 \geq 1$$

$$0^2 - 0^2 \overset{?}{\geq} 1$$

$$0 \geq 1 \quad \text{False}$$

The region between the branches is not included; Figure 6.80 shows the graph of $x^2 - y^2 \geq 1$.
On the same axes, we graph

$$\frac{x^2}{16} + y^2 \leq 1.$$

The boundary is the graph of

$$\frac{x^2}{16} + \frac{y^2}{1} = 1$$

an ellipse with x-intercepts at 4 and -4 and y-intercepts at 1 and -1. We'll use $(0, 0)$ as a test point.

$$\frac{0^2}{16} + 0^2 \overset{?}{\leq} 1$$

$$0 \leq 1 \quad \text{True}$$

The graph of $\left(\dfrac{x^2}{16}\right) + y^2 \leq 1$ includes both the ellipse and its interior (Figure 6.81).
The graph of the solution set (Figure 6.82) is the intersection of the graphs of the two inequalities. All points in these shaded regions have coordinates that satisfy the two given inequalities. The intersection points shown in the figure can be obtained using the addition method.

$$x^2 - y^2 = 1$$

$$\frac{x^2}{16} + y^2 = 1$$

Add: $\dfrac{17x^2}{16} = 2$

$$17x^2 = 32 \qquad \text{Multiply both sides by 16.}$$

$$x^2 = \frac{32}{17} \qquad \text{Divide both sides by 17.}$$

$$x = \pm\sqrt{\frac{32}{17}} = \pm\frac{4\sqrt{2}}{\sqrt{17}} \approx \pm 1.4 \qquad \text{Apply the square root method.}$$

Using back-substitution and $x^2 - y^2 = 1$, we can find y.

$$x^2 - y^2 = 1$$

$$\left(\pm\sqrt{\frac{32}{17}}\right)^2 - y^2 = 1 \qquad \text{Substitute the values for } x.$$

$$\frac{32}{17} - 1 = y^2 \qquad \text{Simplify.}$$

$$\frac{15}{17} = y^2 \qquad\qquad \frac{32}{17} - 1 = \frac{32}{17} - \frac{17}{17} = \frac{15}{17}$$

$$y = \pm\sqrt{\frac{15}{17}} \approx \pm 0.9 \qquad \text{Apply the square root method.}$$

Thus, the ellipse and the hyperbola intersect at approximately $(1.4, 0.9)$, $(1.4, -0.9)$, $(-1.4, 0.9)$, and $(-1.4, -0.9)$. ■

PROBLEM SET 6.4

Practice Problems

Solve each nonlinear system in Problems 1–16 by the substitution method. Confirm real solutions by direct substitution or by using graphs.

1. $x + y = 2$
$y = x^2 - 4$

2. $x + y = 3$
$x^2 + y^2 = 9$

3. $y = x^2 - 1$
$4x + y + 5 = 0$

4. $\dfrac{x^2}{9} + \dfrac{y^2}{4} = 1$
$2x + 3y - 6 = 0$

5. $x^2 + y^2 = 40$
$3x - y + 20 = 0$

6. $x^2 + 2x + 4y - 11 = 0$
$y = x + 5$

7. $xy = 6$
$2x - y = 1$

8. $xy = -12$
$x + 14 = 2y$

9. $x^2 + y^2 = 25$
$x - y = 1$

10. $y = x^2$
$2x - y = -3$

11. $x^2 + 2y = 19$
$2x - y = 1$

12. $x^2 - 2y^2 + 3x - 4y = -2$
$2x - y = 1$

13. $x - y = 2$
$x^2 - 3y^2 = 8$

14. $y - 4x = 0$
$y = x^2 + 5$

15. $xy = 4$
$2x^2 + y^2 = 18$

16. $xy = 4$
$x^2 + y^2 = 10$

Solve each nonlinear system in Problems 17–24 by the addition method. Confirm real solutions by direct substitution or by using graphs.

17. $4x^2 - y^2 = 4$
$4x^2 + y^2 = 4$

18. $x^2 + y^2 = 13$
$x^2 - y^2 = 5$

19. $3x^2 + 4y^2 = 16$
$2x^2 - 3y^2 = 5$

20. $x^2 + y^2 = 1$
$\dfrac{x^2}{4} + \dfrac{y^2}{16} = 1$

21. $x^2 + y^2 - 13 = 0$
$x^2 - y - 7 = 0$

22. $x^2 + 2y^2 - 11 = 0$
$3x^2 + 4y - 23 = 0$

23. $y = x^2 - 4$
$x^2 + y^2 = 10$

24. $y = x^2 + 5$
$x^2 + y^2 = 25$

Solve each nonlinear system in Problems 25–42 by the method of your choice. In some of the problems you will need to use a combination of methods. Confirm real solutions by direct substitution or by using graphs.

25. $x^2 + 4y^2 = 20$
$xy = 4$

26. $x^2 - 2y^2 = 2$
$xy = 2$

27. $y = x^2 + 4$
$x^2 + y^2 = 16$

28. $y = 3x^2 - 3$
$9x^2 + y^2 = 27$

29. $4x^2 - y = 3$
$8x^2 - y^2 = -9$

30. $x^2 + (y - 2)^2 = 4$
$x^2 - 2y = 0$

31. $y = (x + 3)^2$
$x + 2y = -2$

32. $x^2 + y^2 = 4$
$\dfrac{x^2}{4} - \dfrac{y^2}{8} = 1$

33. $x^2 + y^2 = 25$
$\dfrac{x^2}{18} + \dfrac{y^2}{32} = 1$

34. $x^2 - y^2 - 4x + 6y - 4 = 0$
$x^2 + y^2 - 4x - 6y + 12 = 0$

35. $x^2 + y^2 - 16x + 39 = 0$
$x^2 - y^2 - 9 = 0$

36. $x^2 + y^2 + 3y - 22 = 0$
$2x + y + 1 = 0$

37. $x^2 + 2xy - y^2 = 14$
$x^2 - y^2 = -16$

38. $x^2 - 2xy + y^2 = 1$
$x^2 + y^2 = 5$

39. $x^2 - xy + y^2 = 7$
$x^2 + y^2 = 5$

40. $5x^2 - xy + 5y^2 = 89$
$x^2 + y^2 = 17$

41. $x^2 + y^2 + 6y + 5 = 0$
$x^2 + y^2 - 2x - 8 = 0$

42. $x^2 + y^2 = 25$
$x^2 + y^2 - 2x - 4y = 5$

Graph each nonlinear system of inequalities in Problems 43–66.

43. $x^2 + y^2 \leqslant 9$
$\dfrac{x^2}{4} + \dfrac{y^2}{25} \geqslant 1$

44. $x^2 + y^2 \leqslant 25$
$\dfrac{x^2}{4} + \dfrac{y^2}{16} \geqslant 1$

45. $\dfrac{x^2}{16} + \dfrac{y^2}{9} \leqslant 1$
$\dfrac{x^2}{9} + \dfrac{y^2}{16} \leqslant 1$

46. $\dfrac{x^2}{16} + \dfrac{y^2}{9} \leqslant 1$
$\dfrac{x^2}{9} + \dfrac{y^2}{16} \geqslant 1$

47. $\dfrac{x^2}{4} + y^2 < 1$
$y \leqslant 0$

48. $x^2 + \dfrac{y^2}{4} < 1$
$x \leqslant 0$

49. $9x^2 + 25y^2 \leqslant 225$
$x^2 + 4y^2 \geqslant 16$

50. $9x^2 + 16y^2 > 144$
$25x^2 + 16y^2 < 400$

51. $\dfrac{x^2}{4} - \dfrac{y^2}{9} \geqslant 1$
$\dfrac{x^2}{9} + \dfrac{y^2}{25} < 1$

52. $\dfrac{x^2}{9} - \dfrac{y^2}{25} \leqslant 1$
$\dfrac{x^2}{4} + \dfrac{y^2}{16} > 1$

53. $4y^2 - 9x^2 < 36$
$3x + y \leqslant 1$

54. $x^2 - y^2 < 1$
$y \leqslant 0$

55. $x^2 - y^2 \leqslant 1$
$x^2 + y^2 < 4$

56. $x^2 - y^2 > 1$
$\dfrac{x^2}{9} + \dfrac{y^2}{4} \leqslant 1$

57. $-x^2 + \dfrac{y^2}{4} \geqslant 1$
$|y| < 4$

58. $-9x^2 + 4y^2 < 36$
$x^2 + y^2 \leqslant 16$

59. $4x^2 - y^2 \leqslant 4$
$x^2 + 4y^2 \geqslant 4$

60. $x^2 - 4y^2 \leqslant 4$
$x^2 + 4y^2 \geqslant 4$

61. $x \geqslant y^2 - 6y + 5$

$x^2 + y^2 \leqslant 4$

62. $x \leqslant -y^2 - 2y + 3$

$x^2 + y^2 \leqslant 9$

63. $y > -(x - 3)^2 + 4$

$\dfrac{x^2}{9} + \dfrac{y^2}{4} \leqslant 1$

64. $y > -(x + 2)^2 + 1$

$x^2 + \dfrac{y^2}{4} \leqslant 1$

65. $x \leqslant (y + 2)^2 - 1$

$(x - 2)^2 + (y - 2)^2 \geqslant 1$

66. $9x^2 + y^2 \leqslant 9$

$x \leqslant y^2 - 4y$

Application Problems

67. A LORAN system indicates that a ship is on the hyperbola given by $16y^2 - x^2 = 16$. The process is repeated and the ship is found to lie on the hyperbola given by $9x^2 - 4y^2 = 36$. If it is known that the ship is located in the first quadrant of the coordinate system, determine its exact location.

68. An orbiting body follows the elliptic path

$$\frac{x^2}{4} + \frac{y^2}{16} = 1.$$

A comet follows the parabolic path $y = x^2 - 4$. Assume that all units are in 100,000 kilometers. Where might the comet intersect the orbiting body? Illustrate the situation graphically.

69. Thirty-six meters of fencing is to be used to build the enclosure shown in the figure. Some of this fencing is to be used to build an internal divider. Find x and y so that the area of the enclosure is 54 square meters.

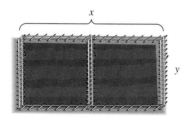

70. The perimeter of the larger rectangle in the figure shown is 58 meters. Find x and y so that the area of the smaller, green rectangle is 29.25 square meters.

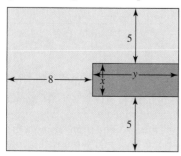

71. The area of a rug is 108 square feet, and the length of its diagonal is 15 feet. Use the formula for the area of a rec-

tangle and the Pythagorean Theorem to find the dimensions of the rug.

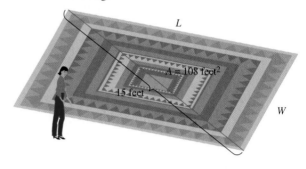

72. The perimeter of a rug is 70 feet and its diagonal length is 25 feet. Find the dimensions of the rug.

73. If $CD = 8$ inches, find the length of the sides of the right triangle shown in the figure where variables now appear.

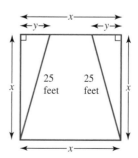

74. Use the Pythagorean Theorem and the fact that the perimeter of the trapezoid in this figure is 84 feet to find x and y.

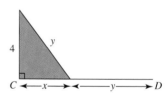

75. The figure in the left column on page 744 shows a square floor plan with a smaller square area that will accommodate a combination fountain and pool. The floor with the fountain-pool area removed has an area of 21 square meters and a perimeter of 24 meters. Find the dimensions of the floor and the dimensions of the square that will accommodate the pool.

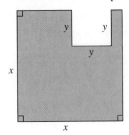

76. The tank shown in the figure has a height of 10 feet, a volume of 3000 cubic feet, and a surface area of 1340 square feet. Find the length and width of the tank.

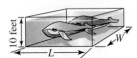

77. Find the radius of each circle in the figure if the total area enclosed by the two circles is 12.5π square inches.

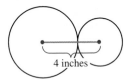

78. Find the lengths of the sides in which variables appear in the figure.

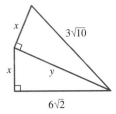

True–False Critical Thinking Problems

79. Which one of the following is true?

a. The system

$$y = x^2 - 1$$
$$x + y = 1$$

contains one linear equation and so it cannot be solved by addition.

b. By visualizing the graphs of the following system, it is immediately obvious that the system has no real solutions.

$$x^2 + y^2 = 100$$
$$\frac{x^2}{9} + \frac{y^2}{4} = 1$$

c. When solving a system with a linear equation and a nonlinear equation by substitution, it is easiest to solve the nonlinear equation and substitute the resulting expression back into the same equation.

d. By visualizing the graphs of the following system, it is immediately obvious that the system has no real solutions.

$$y = x - 1$$
$$(x - 3)^2 + (y + 1)^2 = 9$$

80. Which one of the following is true?

a. A system of two equations in two variables whose graphs represent a circle and a line can have four real solutions.

b. If a nonlinear system has nonintersecting graphs, there are no real number ordered-pair solutions.

c. The graphs of a nonlinear system cannot be used to determine the number of real solutions.

d. A nonlinear system whose graphs represent a circle and a hyperbola cannot have three real solutions.

Technology Problems

In Problems 81–82, use a graphing utility to find all real solutions.

81. $x^2 + y^2 = 17$
        $x^2 - 2x + y^2 = 13$

82. $3x^2 + 2y^2 + 15x = 0$
        $xy + y^2 = 0$

83. The orbit of Earth around the sun is described by

$\dfrac{x^2}{1.03} + \dfrac{y^2}{0.97} = 1$, where distances are measured in astronomical units (1 AU ≈ 93,000,000 miles). The orbit of a comet is described by $x^2 - 2xy + y^2 - 2x - 2y + 1 = 0$. Graph the two equations using a graphing utility and determine how many times the comet will cross Earth's orbit. At what point(s) will this occur?

Writing in Mathematics

84. Explain why the system

$$x^2 + y^2 = 9$$

$$\frac{x^2}{25} + \frac{y^2}{16} = 1$$

has no ordered pairs that are real numbers in its solution set.

85. Write a system of equations, one equation whose graph is a line and the other whose graph is an ellipse, that has no ordered pairs that are real numbers in its solution set. Describe how you determined the two equations.

86. Describe the process needed to find the equation of the line in the figure shown.

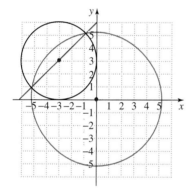

87. Describe how the solution of nonlinear systems of equations can be used by three rescue teams searching a wooded area for a lost hiker. Assume that the teams hear the hiker's cry for help a few seconds apart. Try making up a specific problem that fits these conditions. (*Useful information:* Sound travels at 1100 feet per second.)

88. The orbit of a comet is described by $x^2 - xy + y^2 - 2 = 0$. How can you determine if the orbit is open or closed? Refer to the Study Tip, number 2, on page 735.

Critical Thinking Problems

89. The points of intersection of the graphs of $xy = 20$ and $x^2 + y^2 = 41$ are joined to form a rectangle. Find the area of the rectangle.

90. For what values of a will the graphs of $x^2 + y^2 = 25$ and $\frac{x^2}{a^2} + \frac{y^2}{25} = 1$ intersect twice?

91. Prove that the graphs of $y = mx$ and $\frac{x^2}{a^2} - \frac{y^2}{b^2} = 1$ intersect if and only if $|m| < \left|\frac{a}{b}\right|$.

92. Prove that the hyperbola given by $\frac{x^2}{a^2} - \frac{y^2}{b^2} = 1$ does not intersect its asymptotes.

93. The radio transmitters at A, B, C, and D in the figure shown are attempting to use signals to locate a ship that is lost at sea. Assuming that all units are given in miles, it is determined that the lost ship is 4 miles closer to B than to A and 2 miles closer to C than to D. What is the ship's exact location?

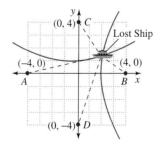

Group Activity Problem

94. As we know, the planets in our solar system revolve around the sun in elliptical orbits with the sun at one focus. The eccentricities and the lengths of the major axes for these elliptical orbits are given in the following table.

Planet	Eccentricity	Length of Major Axis (in millions of kilometers)
Mercury	0.206	116
Venus	0.007	216
Earth	0.017	299
Mars	0.093	456
Jupiter	0.048	1557
Saturn	0.056	2854
Uranus	0.047	5738
Neptune	0.008	8996
Pluto	0.249	11,800

Ptolemaic Universe, 16th Century. Italy 16th c., Portulan of Admiral Coligny: Zodiaque, Ms. 700/1602 fol. 4v and 5r. Chantilly, Musée Conde. Giraudon / Art Resource, NY.

Find an equation for the orbit of each of the nine planets around the sun. Then consider the equations in pairs and solve the resulting systems to determine whether or not any two of the planets in our solar system could collide.

CHAPTER PROJECT

Minimal Surfaces

In the essay on page 714, we introduced a curve called a *catenary*. This curve is observed when a free-hanging cable (such as a telephone or power line) is suspended between two points. We can certainly use two nails and a piece of chain to see this curve hanging in space, or we can use a different approach and discover a three-dimensional version of this curve. We will make a surface called a *catenoid*, by using a soap film. The catenoid is in the class of objects called *minimal surfaces;* surfaces characterized by having zero curvature at every point. Zero curvature will certainly be found on a perfectly flat surface, like a plane. However, a distinctive shape, like a saddle, where the surface curves both away from and towards the point by the same amount, also has a curvature of zero.

We can create the catenoid simply by dipping two circular loops of wire into a soap solution and slowly pulling the circles apart. The graceful, arching film that appears between the two frames is a catenoid. We can study this curved surface better if we use larger rings and a heavier solution. There are many mixtures for bubble and soap film solutions, but to create the longest-lasting creations you will need to add glycerin, available at pharmacies or from your chemistry department. For a large amount of solution, you might try 2/3 of a cup of good quality dishwashing soap, added to 1 gallon of water, with 3 tablespoons of glycerin. For a very heavy mixture, use 1/3 cup good quality dishwashing soap, 1/3 cup water (preferably distilled, but it is not crucial), and 1/3 cup glycerin. The mixtures seem to get better with age, so keep them in containers that can be used for several days.

1. Work in small groups to create your soap solution and a pair of rings to use for dipping. You can create frames out of wire coat hangers, pipe cleaners, or heavy wire. Experiment until you discover the style you prefer. To create a catenary, dip two wire frame circles into the soap solution and pull them apart. Another method is to tie string to points around the edge of the circle frame and pull the string up to knot together above the center of the circle, forming what looks like the outline of a cone. You can lower the circle into a round dish filled with soap solution and slowly pull it upwards, watching for the catenoid.

Scientists, engineers, and architects have studied the graceful curves of soap films stretched across wire frames long before the introduction of computer modeling techniques. One of the most famous architectural examples of a structure built by studying soap films are the roofs over the Munich Olympic park, used for the 1972 Olympics. An important step in the design of cable-net and membrane structures is the study of the geometry of the surface. The architect Otto Frei made numerous wire-frame and thread models, each of them increasingly large and detailed, and dipped them in a soap solution to study the tent-like forms he wished to create.

2. Create a wire frame of a rectangle or square and then twist the figure up out of a planar shape. Make a handle for your creation, and dip it into the soap solution. The film clinging to the frame is a minimal surface for the boundary you have created.
3. Using resources in your library or on the Worldwide Web, research the geometric forms known as the Platonic Solids. Create a wire-frame model of one of the Platonic Solids and dip it into your solution. What do you observe? Make a sketch of your observations and compare your results with others in the class.
4. Create a helix for a wire-frame by using a coiled wire around a long stick. Make the coils large enough so that they stand out in space well away from the central stick. Dip the frame into your soap solution and observe the results.
5. In a small group, design a wire and string model to be dipped into soap solution to create your own roof "tent." Make a sketch of your result or bring your frame to class and perform the experiment.

A soap bubble is an example of a form which has the minimum possible surface area for the volume it encloses. Soap bubbles may also be used to model surface interactions in crystalline forms by observing what happens when two or more bubbles touch.

6. Work with others to blow two soap bubbles, then touch them together and observe the interface. Try the same thing with three bubbles. Three bubbles will always meet at the same angle. What is it?
7. Microscopic marine organisms called *Radiolarians* have skeletons with the same basic shape as soap bubbles trapped in frameworks. Using resources in your library or on the Worldwide Web, discover what some of the Radiolarian skeletons look like and make your own frame to recreate the structure with soap bubbles. Share your results with the class.
8. Dip your hands into the soap solution to study the forms you can create to compress soap bubbles. Try to capture two or three soap bubbles in a string. You may also push your hand inside large bubbles to study the surface created. Work with another person to make sketches of your experiments.

Worldwide Web Resources

Go to the Prentice Hall website (http://www.prenhall.com/blitzer) to access other locations on the Internet that will allow you to further explore the concepts presented in this project.

Chapter Review

SUMMARY

1. The Ellipse

a. An ellipse is the set of all points in a plane the sum of whose distances from two fixed points (the foci) is a constant.

b. *Standard Equations of an Ellipse Centered at (0, 0)*

Foci on *x*-axis:

$$\frac{x^2}{a^2} + \frac{y^2}{b^2} = 1$$

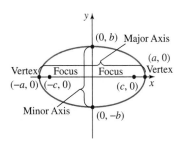

Foci on *y*-axis:

$$\frac{x^2}{b^2} + \frac{y^2}{a^2} = 1$$

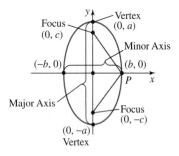

In both cases $a^2 > b^2$ and $b^2 = a^2 - c^2$.

c. *Standard Equations of an Ellipse Centered at (h, k)*

Major axis horizontal:

$$\frac{(x-h)^2}{a^2} + \frac{(y-k)^2}{b^2} = 1$$

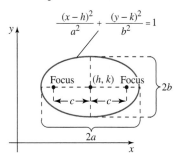

Major axis vertical:

$$\frac{(x-h)^2}{b^2} + \frac{(y-k)^2}{a^2} = 1$$

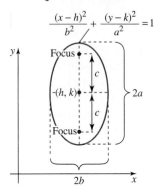

In both cases $a^2 > b^2$, the foci lie on the major axis c units from the center, and $b^2 = a^2 - c^2$.

d. *Eccentricity*

$$e = \frac{c}{a} = \frac{\sqrt{a^2 - b^2}}{a}: \text{For every ellipse, } 0 \le e < 1.$$

An ellipse is circular in shape if $e = 0$ and becomes very flat as e approaches 1.

2. The Hyperbola

a. A hyperbola is the set of all points in a plane the difference of whose distances from two fixed points (the foci) is constant.

b. *Standard Equations of a Hyperbola Centered at (0, 0)*

Foci on *x*-axis:

$$\frac{x^2}{a^2} - \frac{y^2}{b^2} = 1$$

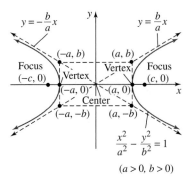

$$\frac{x^2}{a^2} - \frac{y^2}{b^2} = 1$$

$(a > 0, b > 0)$

Foci on *y*-axis:

$$\frac{y^2}{a^2} - \frac{x^2}{b^2} = 1$$

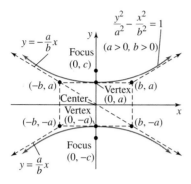

In both cases, $b^2 = c^2 - a^2$.

c. *Standard Equations of a Hyperbola Centered at (h, k)*

Horizontal Transverse Axis:

$$\frac{(x-h)^2}{a^2} - \frac{(y-k)^2}{b^2} = 1$$

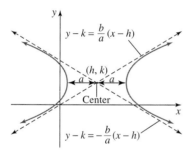

Vertical Transverse Axis:

$$\frac{(y-k)^2}{a^2} - \frac{(x-h)}{b^2} = 1$$

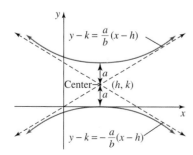

In both cases the vertices are *a* units from the center, the foci are *c* units from the center, and $b^2 = c^2 - a^2$.

3. The Parabola

a. A parabola is the set of all points in a plane that are equidistant from a fixed line (the directrix) and a fixed point (the focus) that is not on the line.

b. *Standard Equations of a Parabola with Vertex at (0,0)*

Axis of symmetry: *x*-axis

$$y^2 = 4px$$

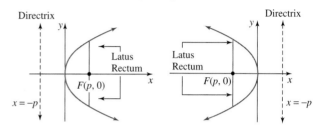

Axis of symmetry: *y*-axis

$$x^2 = 4py$$

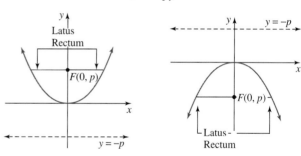

In all cases, the line segment with endpoints on the parabola that passes through the focus and is parallel to the directrix is called the *latus rectum*. Its length is $|4p|$.

c. *Standard Equations of a Parabola with Vertex at (h, k)*

Horizontal Axis of Symmetry:

$$(y-k)^2 = 4p(x-h)$$

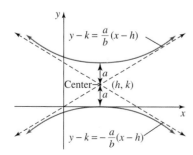

(If $p < 0$, the parabola opens to the left.)

Vertical Axis of Symmetry:

$$(x - h)^2 = 4p(y - k)$$

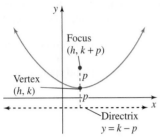

(If $p < 0$, the parabola opens down.)

In both cases, the focus and directrix are p units from the vertex.

4. Identifying Nondegenerate Conics
The graph of $Ax^2 + Cy^2 + Dx + Ey + F = 0$ is a circle if $A = C$; an ellipse if $AC > 0$; a hyperbola if $AC < 0$; a parabola if $AC = 0$.

5. Nonlinear Systems

a. Nonlinear systems of equations can be solved algebraically by eliminating all occurrences of one of the variables by the substitution and addition methods. A graphical analysis, using the system's graphs to determine intersection points, is useful in both predicting and verifying real solutions (if there are any).

b. The solution set for nonlinear systems of inequalities can be graphed by finding the intersection of the graphs of each inequality in the system.

REVIEW PROBLEMS

In Problems 1–13, identify each conic and sketch its graph by hand. For ellipses and hyperbolas, find the foci. For parabolas, identify the vertex, focus, and directrix. Use a graphing utility to verify all graphs.

1. $y^2 + 8x = 0$

2. $4x^2 + y^2 = 16$

3. $9x^2 - 16y^2 - 144 = 0$

4. $x^2 + 16y = 0$

5. $(y - 2)^2 = -16x$

6. $\dfrac{(x - 1)^2}{16} + \dfrac{(y + 2)^2}{9} = 1$

7. $\dfrac{(x - 2)^2}{25} - \dfrac{(y + 3)^2}{16} = 1$

8. $(x - 4)^2 = 4(y + 1)$

9. $4x^2 - y^2 - 8x - 4y - 16 = 0$

10. $4x^2 - 40x - y + 102 = 0$

11. $4x^2 + 9y^2 + 24x - 36y + 36 = 0$

12. $y^2 - 4x - 10y + 21 = 0$

13. $4x^2 - y^2 + 8x + 4y + 4 = 0$

For Problems 14–17, assume that each graph represents an ellipse, hyperbola, or parabola. Find the standard form of the equation for each graph.

14.

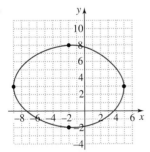

15.

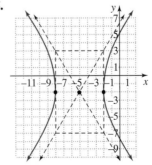

16.

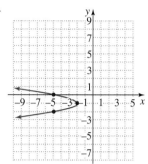

17.

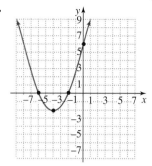

In Problems 18–20, find the standard form of the equation for each ellipse satisfying the given conditions.

18. Foci: $(0, -4)$, $(0, 4)$; vertices: $(0, -5)$, $(0, 5)$

19. Vertices: $(5, 0)$, $(-5, 0)$; passes through $(3, -4)$

20. Foci: $(-6, 0)$, $(6, 0)$; $e = \frac{3}{5}$

In Problems 21–23, find the standard form of the equation for each hyperbola satisfying the given conditions.

21. Foci: $(0, 5)$, $(0, -5)$; vertices: $(0, -3)$, $(0, 3)$

22. Foci: $(0, -9)$, $(0, 9)$; asymptotes: $y = \pm\frac{3}{4}x$

23. Foci: $(-3, 0)$, $(3, 0)$; passes through $(4, 1)$

In Problems 24–26, find the standard form of the equation for the parabola satisfying the given conditions.

24. Focus: $(-2, 0)$; directrix: $x = 2$

25. Vertex: origin; axis of symmetry: x-axis; passes through $(2, 1)$

26. Vertex: $(4, 2)$; focus $(4, 0)$

27. A semielliptic archway has a height of 15 feet at the center and a width of 50 feet, as shown in the figure. The 50-foot width consists of a two-lane road. Can a truck that is 12 feet high and 14 feet wide drive under the archway without going over the road's center line?

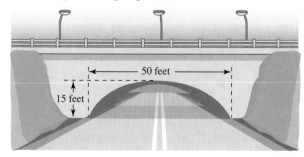

28. The eccentricity of Earth's orbit about the sun is approximately 0.017. The sun is at one focus of the orbit and the closest that Earth comes to the sun is 91.44 million miles. Find the maximum distance between Earth and the sun, represented by $a + c$ in the figure shown.

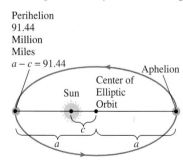

29. Radio tower M_2 is located 200 miles due west of radio tower M_1. The situation is illustrated in the figure in the left column on page 752, where a coordinate system has been superimposed. Simultaneous radio signals are sent from each tower to a ship, with the signal from M_2 received 500 microseconds before the signal from M_1.

a. Assuming that radio signals travel at 0.186 miles per microsecond, determine the equation of the hyperbola on which the ship is located.

b. If the ship is traveling a course that is 60 miles north and parallel to the relatively straight east–west shoreline, determine its approximate location using an ordered pair and the rectangular system shown in the figure.

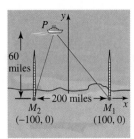

30. An engineer is designing headlight units for automobiles. The unit has a parabolic cross section with a diameter of 12 inches and a depth of 3 inches. The situation is illustrated in the figure, where a coordinate system has been superimposed. What is the equation of the parabola in this system? Where should the light source be placed? Describe this placement relative to the vertex.

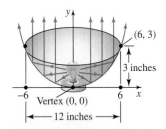

David Woods / The Stock Market

31. A telescope uses a glass mirror with a parabolic cross section and a diameter of 30 inches, as shown in the figure at the top of the next column.

a. If the focus of the parabola is 10 inches from the vertex, what is the equation for the parabola that was used to shape the mirror?

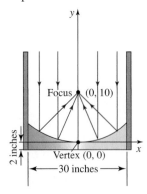

b. If the mirror is 2 inches thick at the center, what is the thickness of the glass on the outside edge?

32. The George Washington Bridge spans the Hudson River from New York to New Jersey. Its two towers are 3500 feet apart and rise 316 feet above the road. The cable between the towers has the shape of a parabola, and the cable just touches the sides of the road midway between the towers. What is the height of the cable 1000 feet from a tower?

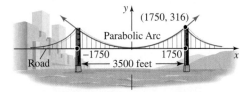

The George Washington Bridge.
RAGA / The Stock Market

In Problems 33–41, solve each system by the addition method, the substitution method, or possibly a combination of methods. Verify real number solutions by graphing the equations in the system either by hand or with a graphing utility.

33. $x - y = -3$
$y = x^2 + 2x + 1$

34. $x + y = 4$
$x^2 - y^2 = 4$

35. $2x^2 + 3y^2 = 21$
$3x^2 - 4y^2 = 23$

36. $x + y = 1$
$xy = -12$

37. $x^2 + 2x - y + 1 = 0$
$x + y - 1 = 0$

38. $xy = 4$
$x^2 + y^2 = 8$

39. $y = 12 - x^2$
$x^2 + (y + 2)^2 = 14$

40. $x^2 + x - y - 4 = 0$
$x^2 - 2xy + y^2 - 9 = 0$

41. $x^2 + 2xy - y^2 = 14$
$x^2 - y^2 = -16$

42. The plot plan of a waterfront property is shown in the figure. The plot has an area of 12,500 square feet and a perimeter of 500 feet. The plot is advertised as having 110 feet that front the water. Find x and y and determine if this advertisement is accurate.

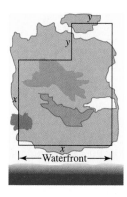

43. A television screen has a 26-inch diagonal and an area of 240 square inches. The television is wider than it is high, and the console extends 2 inches below and 2 inches above the screen. Find the dimensions of the screen. Will the television fit onto a shelf that can accommodate a height of 13 inches?

44. Find a and b, the lengths of the legs of the right triangle shown in the figure.

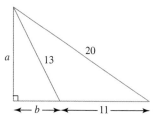

45. Find the coordinates of all points (x, y) that lie on the line whose equation is $2x + y = 8$, so that the area of the rectangle in the figure shown is 6 square units.

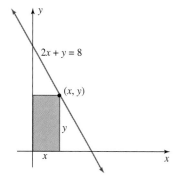

46. Neptune and Pluto revolve around the sun in elliptical orbits with the sun at one focus. The eccentricity of Neptune's orbit is 0.008 and the length of its major axis is 8996 million kilometers. The eccentricity of Pluto's orbit is 0.249 and the length of its major axis is 11,800 million kilometers. Is it possible that Neptune and Pluto could collide? Answer the question using a system of equations.

In Problems 47–52, graph the solution to each system of inequalities.

47. $x^2 + y^2 < 9$
$2x + 3y > 6$

48. $4x^2 + 9y^2 \leq 36$
$-2 \leq x \leq 2$

49. $\dfrac{x^2}{9} + y^2 < 1$
$x^2 - \dfrac{y^2}{4} \geq 1$

50. $x^2 + y^2 \leq 9$
$25x^2 + 4y^2 \geq 100$

51. $x^2 + y^2 \leq 4$
$x^2 - y^2 < 1$

52. $9x^2 + y^2 \leq 9$
$x \leq y^2 - 4y$

53. Explain why it is not possible for a hyperbola to have foci at $(0, -2)$ and $(0, 2)$ and vertices at $(0, -3)$ and $(0, 3)$.

54. An elliptical pool table has a ball placed at each focus. If one ball is hit toward the side of the table, explain what will occur.

CHAPTER 6 TEST

In Problems 1–5, identify the conic section and sketch its graph. For ellipses and hyperbolas, find the foci. For parabolas, identify the vertex, focus, and directrix.

1. $9x^2 - 4y^2 = 36$

2. $x^2 = -8y$

3. $\dfrac{(x + 2)^2}{25} + \dfrac{(y - 5)^2}{9} = 1$

4. $4x^2 - y^2 + 8x + 2y + 7 = 0$

5. $(x + 5)^2 = 8(y - 1)$

In Problems 6–8, find the standard form of the equation for each conic section with the given conditions.

6. Ellipse: foci $(0, -\sqrt{5})$, $(0, \sqrt{5})$; vertices $(0, -3)$, $(0, 3)$

7. Hyperbola: vertices $(-2, 0)$, $(2, 0)$; asymptotes $y = \pm\frac{5}{2}x$

8. Parabola: focus $(\frac{3}{2}, 0)$; directrix $x = -\frac{3}{2}$

9. The figure shows a semielliptic archway with a height of 20 feet at the center and a width of 60 feet. Can a truck that is 15 feet high and 10 feet wide drive under the archway without going over the road's center line?

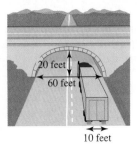

20 feet
60 feet
10 feet

10. An engineer is designing headlight units for cars. The unit shown in the figure has a parabolic cross section with a diameter of 6 inches and a depth of 3 inches.

a. Using the coordinate system that has been positioned on the unit, find the parabola's equation.

b. If the light source is located at the focus, describe its placement relative to the vertex.

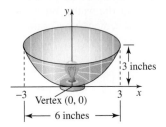

3 inches
−3   3 x
Vertex (0, 0)
6 inches

Solve each nonlinear system in Problems 11–13.

11.
$$y = x^2 - 4$$
$$x - 2y + 2 = 0$$

12. $x^2 + 2y^2 = 8$
$2x^2 - y^2 = 6$

13.
$$x^2 + y^2 = 25$$
$$x^2 - xy + y^2 = 13$$

14. Graph the solution to this system of inequalities:

$$x^2 + y^2 \leq 16$$
$$9x^2 + 16y^2 \geq 144$$

15. Find the dimensions of a rectangle whose perimeter is 33 feet and whose area is 65 square feet.

16. A rectangle has a diagonal of 15 feet and a perimeter of 42 feet. Find the rectangle's dimensions.

17. Use a graphing utility to graph the conics $x^2 + y^2 = 9$ and $\frac{y^2}{6} - \frac{x^2}{3} = 1$. Then solve the system consisting of these two equations, expressing coordinates of all intersection points correct to two decimal places.

CUMULATIVE REVIEW PROBLEMS (CHAPTERS P–6)

1. Simplify: $\dfrac{\dfrac{x}{x+3} + \dfrac{x}{x^2-9}}{\dfrac{1}{x-3} + 1}$.

Solve the equations, inequalities, and linear systems in Problems 2–7.

2. $\dfrac{1}{x} + \dfrac{1}{x+3} = 2$

3. $4x^2 - 4x + 2 < (2x + 3)(2x - 2)$

4. $\sqrt{2x + 4} - \sqrt{x + 3} - 1 = 0$

5. $3x^3 + 8x^2 - 15x + 4 = 0$

6. $e^{2x} - 14e^x + 45 = 0$

7. $\log(x - 2) + \log x = \log(x + 4)$

8. Use matrices to solve this system.
$$x - y + z = 17$$
$$2x + 3y + z = 8$$
$$-4x + y + 5z = -2$$

9. Solve for y using only Cramer's rule.
$$x - 2y + z = 7$$
$$2x + y - z = 0$$
$$3x + 2y - 2z = -2$$

10. Write the general form for the equation of the line that is perpendicular to the line whose equation is $4x + 6y + 5 = 0$ and passes through the origin.

11. The graph of g shown in the figure was obtained by transforming the graph of $f(x) = \sqrt{x}$. Write an equation for the function g.

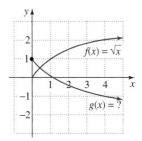

12. If $f(x) = \sqrt{4x - 7}$, find $f^{-1}(x)$.

13. Graph: $f(x) = \dfrac{x}{x^2 - 16}$.

14. Use the graph of $f(x) = 4x^4 - 4x^3 - 25x^2 + x + 6$ shown in the figure to factor the polynomial completely.

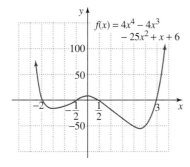

15. Graph $y = \log_2 x$ and $y = \log_2(x + 1)$ in the same rectangular coordinate system.

16. The magnitude R on the Richter scale of an earthquake of intensity I is given by

$$R = \log \frac{I}{I_0}$$

where I_0 is the intensity of a barely felt zero-level earthquake. How much more intense is a 7.2 earthquake than a 3.6 earthquake?

17. Use the exponential decay model $A = A_0 e^{kt}$ to solve this problem. A radioactive substance has a half-life of 40 days. There are initially 900 grams of the substance.

   a. Find the decay model for this substance.

   b. How much of the substance will remain after 10 days?

18. Multiply the matrices: $\begin{bmatrix} 1 & -1 & 0 \\ 2 & 1 & 3 \end{bmatrix} \begin{bmatrix} 4 & -1 \\ 2 & 0 \\ 1 & 1 \end{bmatrix}.$

19. Find the partial fraction decomposition for

$$\frac{2x^3 - 3x + 4}{x(x^2 + 1)^2}.$$

20. Identify and graph the conic section whose equation is $2x^2 - 3y^2 - 4x + 12y - 28 = 0$.

Sequences, Series, and Probability

P. J. Crook "Time and Time Again" 1981. Courtesy Montpelier Sandelson Gallery, London.

Infinity has been the object of speculation for millenia. Because it is so difficult to use human intuition to grasp infinite processes, one can ask the following question: Is infinity something that reason can investigate?

In this chapter, we will discover that infinite processes can be examined systematically. We will see how infinite sums can be used to model numerous situations, including tax rebates. The ideas used to analyze patterns in nonending sums such as

$$\frac{1}{2} + \frac{1}{4} + \frac{1}{8} + \frac{1}{16} + \frac{1}{32} + \frac{1}{64} + \frac{1}{128} + \frac{1}{256} + \cdots$$

led to the development of calculus, without which modern technology would not exist.

Solutions Manual

Tutorial

Video 12

SECTION 7.1 — Sequences and Series

Objectives

1 Find particular terms of a sequence given the general term.
2 Use factorial notation.
3 Use recursion formulas.
4 Find the general term of a sequence.
5 Find partial sums.
6 Expand and evaluate sums in summation notation.
7 Write sums in summation notation.

Sequences

Many creations in nature involve intricate mathematical designs, including a variety of spirals. For example, the arrangement of the individual florets in the head of a sunflower form spirals. In some species, there are 21 spirals in the clockwise direction and 34 in the counterclockwise direction. The precise numbers depend on the species of sunflower: 21 and 34, or 34 and 55, or 55 and 89, or even 89 and 144.

Dick Morton

This observation becomes even more interesting when we consider a sequence of numbers investigated by Leonardo of Pisa, also known as Fibonacci, an Italian mathematician of the 13th century. The *Fibonacci sequence* of numbers is an infinite sequence that begins as follows:

1, 1, 2, 3, 5, 8, 13, 21, 34, 55, 89, 144, 233, ...

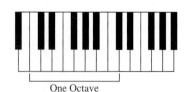

One Octave

Figure 7.1

Fibonacci numbers occur in an octave on the piano keyboard.

The first two terms are 1. Every term thereafter is the sum of the two preceding terms. The number of spirals in a daisy or a sunflower (21 and 34) corresponds to two consecutive Fibonacci numbers, as do the number in a pine cone (8 and 13) and a pineapple (8 and 13). Furthermore, numbers in the Fibonacci sequence can be found in an octave on the piano keyboard. The octave contains 2 black keys in one cluster, 3 black keys in another cluster, 5 black keys, 8 white keys, and a total of 13 keys altogether (Figure 7.1). Observe that 2, 3, 5, 8, and 13 are the third through seventh terms of the Fibonacci sequence.

We can think of the Fibonacci sequence as a function. The terms of the sequence

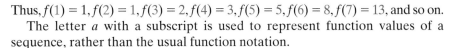

$$1, 1, 2, 3, 5, 8, 13, 21, 34, 55, 89, 144, 233, \ldots$$

are the range values for a function whose domain is the set of positive integers.

Domain: $1, 2, 3, 4, 5, 6, \ 7, \ldots$
$\downarrow \downarrow \downarrow \downarrow \downarrow \downarrow \ \downarrow$
Range: $1, 1, 2, 3, 5, 8, 13, \ldots$

Thus, $f(1) = 1, f(2) = 1, f(3) = 2, f(4) = 3, f(5) = 5, f(6) = 8, f(7) = 13$, and so on.

The letter a with a subscript is used to represent function values of a sequence, rather than the usual function notation.

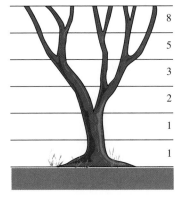

As this tree branches, the number of branches forms the Fibonacci sequence.

Term of a Sequence	Term of the Fibonacci Sequence $1, 1, 2, 3, 5, 8, 13, \ldots$
a_1 represents the first term of a sequence.	$a_1 = 1$
a_2 represents the second term of a sequence.	$a_2 = 1$
a_3 represents the third term of a sequence.	$a_3 = 2$
$\vdots$	$\vdots$
a_n represents the nth term of a sequence.	$a_n = $ sum of previous two terms

The subscripts make up the domain of the sequence, and they identify the location of a term. Thus a_3 is the third term of the sequence, and a_7 is the seventh term of the sequence. In general, a_n represents the nth term, or general term, of the sequence. The notation $\{a_n\}$ represents the entire sequence.

Definition of a sequence

An *infinite sequence* $\{a_n\}$ is a function whose domain is the set of positive integers. The function values, or terms, of the sequence are represented by

$$a_1, a_2, a_3, a_4, \ldots, a_n, \ldots.$$

Sequences whose domains consist only of the first n positive integers are called *finite sequences*.

Find particular terms of a sequence given the general term.

EXAMPLE I   **Finding the Terms of a Sequence from the General Term**

Write the first five terms for the sequence whose general term is defined as:

a. $a_n = 3n + 4$ **b.** $a_n = \dfrac{n}{n + 2}$ **c.** $a_n = \dfrac{(-1)^n}{3^n - 1}$

Solution

a. The first five terms of the sequence whose nth term, or general term, is $a_n = 3n + 4$ are

$$a_1 = 3(1) + 4 = 7$$

$$a_2 = 3(2) + 4 = 10$$

$$a_3 = 3(3) + 4 = 13$$

$$a_4 = 3(4) + 4 = 16$$

$$a_5 = 3(5) + 4 = 19$$

The sequence defined by $a_n = 3n + 4$ can be written as

$$7, 10, 13, \ldots, 3n + 4, \ldots.$$

iscover for yourself

Write the general term of a decreasing sequence.

Because each term in the sequence is greater than the preceding term, this is an example of an *increasing sequence*.

b. The first five terms of the sequence whose nth term, or general term, is $a_n = \dfrac{n}{n + 2}$ are:

$$a_1 = \frac{1}{1 + 2} = \frac{1}{3}$$

$$a_2 = \frac{2}{2 + 2} = \frac{2}{4} = \frac{1}{2}$$

$$a_3 = \frac{3}{3 + 2} = \frac{3}{5}$$

$$a_4 = \frac{4}{4 + 2} = \frac{4}{6} = \frac{2}{3}$$

$$a_5 = \frac{5}{5 + 2} = \frac{5}{7}$$

The sequence defined by $a_n = \dfrac{n}{n + 2}$ can be written as

$$\frac{1}{3}, \frac{1}{2}, \frac{3}{5}, \ldots, \frac{n}{n + 2}, \ldots.$$

tudy tip

The factor $(-1)^n$ in the general term of a sequence causes the signs of the terms to alternate between positive and negative, depending on whether n is even or odd.

c. The first five terms of the sequence whose nth term, or general term, is $a_n = \dfrac{(-1)^n}{3^n - 1}$ are:

$$a_1 = \frac{(-1)^1}{3^1 - 1} = -\frac{1}{2}$$

$$a_2 = \frac{(-1)^2}{3^2 - 1} = \frac{1}{8}$$

$$a_3 = \frac{(-1)^3}{3^3 - 1} = -\frac{1}{26}$$

Discover for yourself

Write the general term for this sequence:

$$\frac{1}{2}, \ -\frac{1}{8}, \ \frac{1}{26}, \ -\frac{1}{80}, \ \frac{1}{242}, \cdots$$

This is almost the same as the sequence in Example 1c, but the signs of the terms alternate differently.

$$a_4 = \frac{(-1)^4}{3^4 - 1} = \frac{1}{80}$$

$$a_5 = \frac{(-1)^5}{3^5 - 1} = -\frac{1}{242}$$

The sequence defined by $a_n = \dfrac{(-1)^n}{3^n - 1}$ can be written as

$$-\frac{1}{2}, \ \frac{1}{8}, \ -\frac{1}{26}, \cdots, \frac{(-1)^n}{3^n - 1}, \cdots$$ ∎

2   Use factorial notation.

Factorial Notation

Products of consecutive positive integers occur quite often in sequences. These products can be expressed in a special notation, called *factorial notation*.

Factorial notation

If n is a positive integer, the notation $n!$ (read "n factorial") is the product of consecutive integers from 1 to n.

$$n! = 1 \cdot 2 \cdot 3 \cdot 4 \cdot \cdots \cdot (n - 1)n$$

$0!$, by definition, is 1.

$$0! = 1$$

Using technology

Graphing utilities have factorial keys. To find 5 factorial, enter

5 [!] [ENTER] .

Since $n!$ becomes quite large as n increases, your utility will display these larger values in scientific notation.

The values of $n!$ for the first six positive integers are

$$1! = 1$$

$$2! = 1 \cdot 2 = 2$$

$$3! = 1 \cdot 2 \cdot 3 = 6$$

$$4! = 1 \cdot 2 \cdot 3 \cdot 4 = 24$$

$$5! = 1 \cdot 2 \cdot 3 \cdot 4 \cdot 5 = 120$$

$$6! = 1 \cdot 2 \cdot 3 \cdot 4 \cdot 5 \cdot 6 = 720$$

Study tip

Factorials affect only the number or variable that they follow unless grouping symbols appear.

$$2 \cdot 3! = 2(1 \cdot 2 \cdot 3)$$

$$= 2 \cdot 6 = 12$$

$$(2 \cdot 3)! = 6! = 1 \cdot 2 \cdot 3 \cdot 4 \cdot 5 \cdot 6$$

$$= 720$$

In this sense, factorials are similar to exponents.

EXAMPLE 2   **Finding Terms of a Sequence Involving Factorials**

Write the first five terms of the infinite sequence whose nth term is:

$$a_n = \frac{2^n}{(n - 1)!}$$

Using technology

Some graphing utilities can create tables. (Consult your manual.) You can use this capability to create a table that shows the terms of

$$a_n = \frac{2^n}{(n-1)!}$$

using steps similar to these.

1. Enter the sequence in $\boxed{Y=}$ as

   $Y_1 \boxed{=} 2 \boxed{\wedge} X \boxed{\div} (X - 1)!$

2. With the TblSet menu, set TblStart = 1 or TblMin = 1 and ΔTbl = 1.

3. View the terms of the sequence using the keystroke $\boxed{\text{TABLE}}$.

Use the cursor to look at additional terms.

3 Use recursion formulas.

Solution

$$a_1 = \frac{2^1}{(1-1)!} = \frac{2}{0!} = \frac{2}{1} = 2$$

To write the first term, let $n = 1$ in the given formula for the general term.

$$a_2 = \frac{2^2}{(2-1)!} = \frac{4}{1!} = \frac{4}{1} = 4$$

To write the second term, let $n = 2$ in the given formula.

$$a_3 = \frac{2^3}{(3-1)!} = \frac{8}{2!} = \frac{8}{2} = 4$$

Now let $n = 3$.

$$a_4 = \frac{2^4}{(4-1)!} = \frac{16}{3!} = \frac{16}{6} = \frac{8}{3}$$

Let $n = 4$.

$$a_5 = \frac{2^5}{(5-1)!} = \frac{32}{4!} = \frac{32}{24} = \frac{4}{3}$$

Finally, let $n = 5$.

Using the first five terms, we can express the sequence as

$$2, 4, 4, \frac{8}{3}, \frac{4}{3}, \ldots, \frac{2^n}{(n-1)!}, \ldots.$$

Recursion Formulas

In Examples 1 and 2, the formulas used for the nth term of a sequence expressed the term as a function of n, the number of the term. Sequences can also be defined using *recursion formulas*. A recursion formula defines the nth term of a sequence as a function of the previous term. Our next example illustrates that if the first term of a sequence is known, then the recursion formula can be used to determine the remaining terms.

EXAMPLE 3 **Using a Recursion Formula**

Find the first four terms of the infinite sequence in which $a_1 = 5$ and $a_n = 3a_{n-1} + 2$ for $n \geq 2$.

Solution

$$a_1 = 5$$

This is the given first term.

$$a_2 = 3a_1 + 2$$

Use $a_n = 3a_{n-1} + 2$, with $n = 2$. Thus, $a_2 = 3a_{2-1} + 2 = 3a_1 + 2$.

$$= 3(5) + 2 = 17$$

Substitute 5 for a_1.

$$a_3 = 3a_2 + 2$$

Again use $a_n = 3a_{n-1} + 2$, with $n = 3$.

$$= 3(17) + 2 = 53$$

Substitute 17 for a_2.

$$a_4 = 3a_3 + 2$$

Notice that a_4 is defined in terms of a_3. We used $a_n = 3a_{n-1} + 2$, with $n = 4$.

$$= 3(53) + 2 = 161$$

Use the value of a_3, the third term, obtained from above.

The first four terms are 5, 17, 53, and 161.

4 Find the general term of a sequence.

Josef Albers "Homage to the Square: Star Blue" 1957, oil on board, 75.9 × 75.9 cm. © The Cleveland Museum of Art, Contemporary Collection of the Cleveland Museum of Art, 1965.1. © 1998 The Josef and Anni Albers Foundation / Artists Rights Society (ARS), New York.

Finding the nth Term of a Sequence

So far we have listed terms in a sequence with the general term given. In contrast, listing the first few terms of a sequence is not enough to say for certain what the general term is. However, we can make a prediction by looking for a pattern. For example, the sequence,

$$1, 4, 9, 16, \ldots$$

has terms that are squares of consecutive even integers. Thus, a possible formula for the general term is

$$a_n = n^2.$$

Discover for yourself

Write the first five terms for the following sequences whose general term is given.

$$a_n = n^2$$

$$a_n = n^2 + (n - 1)(n - 2)(n - 3)(n - 4)\left(\frac{\pi - 25}{1 \cdot 2 \cdot 3 \cdot 4}\right)$$

What do you observe? Can you see why listing the first few terms of a sequence is not enough to define the general term in only one way?

If we are given the first few terms of a sequence, the best we can do is to find one possible general term, while realizing that more than one formula for a_n may be correct. Let's try it in Example 4.

Study tip

In Example 4a, when we replace n with successive positive integers, $2n + 1$ generates the odd numbers. Similarly, $2n$ generates the even numbers.

Study tip

In Example 4b, if you start with $n = 1$ and use consecutive integers, the factor

$$(-1)^n$$

will result in terms with the alternating signs

$$- + - + - + \text{ etc.}$$

The factor

$$(-1)^{n+1}$$

will result in terms with the alternating signs

$$+ - + - + - \text{ etc.}$$

EXAMPLE 4 **Finding a Possible General Term of a Sequence**

Find a possible formula for the general term of the sequence:

a. $3, 5, 7, 9, 11, \ldots$
b. $-1, 4, -9, 16, -25, \ldots$
c. $3, 12, 27, 48, 75, \ldots$

Solution

a. n:

1,	2,	3,	4,	5,	...

Terms:

3,	5,	7,	9,	11,	...
$2 \cdot 1 + 1,$	$2 \cdot 2 + 1,$	$2 \cdot 3 + 1,$	$2 \cdot 4 + 1,$	$2 \cdot 5 + 1,$	...

One pattern is that each term is 1 more than twice n. A possible formula for the general term is

$$a_n = 2n + 1.$$

b. n:

1,	2,	3,	4,	5,	...

Terms:

$-1,$	4,	$-9,$	16,	$-25,$	...
$(-1)^1 \cdot 1^2,$	$(-1)^2 \cdot 2^2,$	$(-1)^3 \cdot 3^2,$	$(-1)^4 \cdot 4^2,$	$(-1)^5 \cdot 5^2,$	...

The terms of the sequence are the squares of consecutive positive integers with alternating signs. A possible formula for the general term is

$$a_n = (-1)^n n^2.$$

c. *n:*   1,   2,   3,   4,   5,   ...

| | | | | | |

Terms: 3,   12,   27,   48,   75,   ...

| | | | | | |

$3 \cdot 1^2$, $3 \cdot 2^2$, $3 \cdot 3^2$, $3 \cdot 4^2$, $3 \cdot 5^2$,   ...

One pattern is that each term is 3 times the square of *n*. A possible formula for the general term is

$$a_n = 3n^2.$$ ∎

5 Find partial sums.

Series and Summation Notation

Now let's focus on the *sum of the terms* of a sequence. Consider the sequence given by

$$\frac{1}{2}, \frac{1}{4}, \frac{1}{8}, \frac{1}{16}, \frac{1}{32}, \ldots, \left(\frac{1}{2}\right)^n, \ldots.$$

The figure below shows how we can model this sequence using parts of a square.

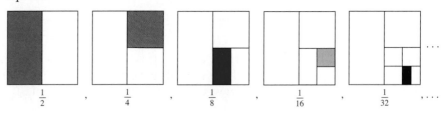

$$\frac{1}{2}, \qquad \frac{1}{4}, \qquad \frac{1}{8}, \qquad \frac{1}{16}, \qquad \frac{1}{32}, \ldots$$

The sum of a number of terms in a sequence is called a finite series. The sum of the first *n* terms is denoted by S_n, and is called the *nth partial sum*. The figures show the first five partial sums for the sequence

$$\frac{1}{2}, \frac{1}{4}, \frac{1}{8}, \frac{1}{16}, \frac{1}{32}, \ldots, \left(\frac{1}{2}\right)^n, \ldots$$

with geometric models.

S_1: First Partial Sum S_2: Second Partial Sum S_3: Third Partial Sum S_4: Fourth Partial Sum S_5: Fifth Partial Sum

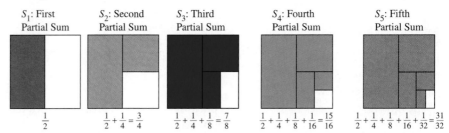

$$\frac{1}{2} \qquad \frac{1}{2}+\frac{1}{4}=\frac{3}{4} \qquad \frac{1}{2}+\frac{1}{4}+\frac{1}{8}=\frac{7}{8} \qquad \frac{1}{2}+\frac{1}{4}+\frac{1}{8}+\frac{1}{16}=\frac{15}{16} \qquad \frac{1}{2}+\frac{1}{4}+\frac{1}{8}+\frac{1}{16}+\frac{1}{32}=\frac{31}{32}$$

As we continue to add terms, the sum is getting closer to modeling the area of one whole square. Thus, these partial sums are getting closer and closer to 1. Although the number 1 will never be reached, we can say that the *infinite series*

$$\frac{1}{2} + \frac{1}{4} + \frac{1}{8} + \frac{1}{16} + \frac{1}{32} + \cdots + \left(\frac{1}{2}\right)^n + \cdots$$

has a sum that is equal to 1.

This discussion can be generalized with the following definitions.

Partial sums and infinite series

Consider the infinite sequence

$$a_1, a_2, a_3, \ldots, a_n, \ldots.$$

The finite series S_n is the sum of the first n terms of the sequence, and is called the *nth partial sum*.

$$S_n = a_1 + a_2 + a_3 + \cdots + a_n$$

The sum of the terms

$$a_1 + a_2 + a_3 + \cdots + a_n + \cdots$$

is called an *infinite series*.

0!	1
1!	1
2!	2
3!	6
4!	24
5!	120
6!	720
7!	5040
8!	40,320
9!	362,880
10!	3,628,800
11!	39,916,800
12!	479,001,600
13!	6,227,020,800
14!	87,178,291,200
15!	1,307,674,368,000
16!	20,922,789,888,000
17!	355,687,428,096,000
18!	6,402,373,705,728,000
19!	121,645,100,408,832,000
20!	2,432,902,008,176,640,000

Factorials from 0 through 20

Discover for yourself

Compare the sequence in Example 5 to the one in Example 4a on page 763. What do you observe? What does this mean about the formula for a_n for the two sequences?

$$a_n = 2n + 1 \text{ (Example 4a)}$$

$$a_n = \frac{(2n + 1)!}{(2n)!} \text{ (Example 5)}$$

Use the factorial definition to prove your observation about the formula for a_n.

EXAMPLE 5 **Finding Partial Sums**

Consider the sequence whose general term is given by

$$a_n = \frac{(2n + 1)!}{(2n)!}$$

Find:

a. S_3, the third partial sum
b. S_5, the fifth partial sum

Solution

To begin, we replace n with 1, 2, 3, 4, and 5 to find the first five terms of the sequence.

$$a_1 = \frac{(2 \cdot 1 + 1)!}{(2 \cdot 1)!} = \frac{3!}{2!} = \frac{1 \cdot 2 \cdot 3}{1 \cdot 2} = 3$$

$$a_2 = \frac{(2 \cdot 2 + 1)!}{(2 \cdot 2)!} = \frac{5!}{4!} = \frac{1 \cdot 2 \cdot 3 \cdot 4 \cdot 5}{1 \cdot 2 \cdot 3 \cdot 4} = 5$$

$$a_3 = \frac{(2 \cdot 3 + 1)!}{(2 \cdot 3)!} = \frac{7!}{6!} = \frac{1 \cdot 2 \cdot 3 \cdot 4 \cdot 5 \cdot 6 \cdot 7}{1 \cdot 2 \cdot 3 \cdot 4 \cdot 5 \cdot 6} = 7$$

$$a_4 = \frac{(2 \cdot 4 + 1)!}{(2 \cdot 4)!} = \frac{9!}{8!} = \frac{1 \cdot 2 \cdot 3 \cdot 4 \cdot 5 \cdot 6 \cdot 7 \cdot 8 \cdot 9}{1 \cdot 2 \cdot 3 \cdot 4 \cdot 5 \cdot 6 \cdot 7 \cdot 8} = 9$$

$$a_5 = \frac{(2 \cdot 5 + 1)!}{(2 \cdot 5)!} = \frac{11!}{10!} = \frac{1 \cdot 2 \cdot 3 \cdot 4 \cdot 5 \cdot 6 \cdot 7 \cdot 8 \cdot 9 \cdot 10 \cdot 11}{1 \cdot 2 \cdot 3 \cdot 4 \cdot 5 \cdot 6 \cdot 7 \cdot 8 \cdot 9 \cdot 10} = 11$$

a. S_3, the third partial sum, is found by adding the first three terms of the sequence:

$$S_3 = a_1 + a_2 + a_3 = 3 + 5 + 7 = 15.$$

b. S_5, the fifth partial sum, is found by adding the first five terms of the sequence:

$$S_5 = a_1 + a_2 + a_3 + a_4 + a_5 = 3 + 5 + 7 + 9 + 11 = 35.$$ ■

In the Discover for yourself on page 765, did you observe that the formula for the general term in Example 5 can be simplified?

$$\frac{(2n + 1)!}{(2n)!} = \frac{1 \cdot 2 \cdot 3 \cdots 2n \cdot (2n + 1)}{1 \cdot 2 \cdot 3 \cdots 2n} = 2n + 1$$

Consequently,

$$a_n = \frac{(2n + 1)!}{(2n)!} \quad \text{and} \quad a_n = 2n + 1$$

are equal and represent the general term for the same sequence. Observe that it would have been easier to first reduce the fraction in the formula for a_n rather than to perform this simplification for each of the first five terms.

> **S**tudy tip
>
> Quotients of factorials, such as
>
> $$\frac{(2n + 1)!}{(2n)!}$$
>
> can often be reduced and rewritten as equivalent expressions without factorials.

6 Expand and evaluate sums in summation notation.

Summation Notation

Mathematicians have a compact notation for dealing with series, called *summation notation* or *sigma notation*. The name comes from the use of a stylized version of the capital Greek letter sigma, denoted by Σ. An example of this notation is the fourth partial sum

$$\sum_{i=1}^{4} 16(2i - 1).$$

To find this partial sum, we replace the letter i in $16(2i - 1)$ with $1, 2, 3,$ and 4, as follows:

$$\sum_{i=1}^{4} 16(2i - 1) = 16(2 \cdot 1 - 1) + 16(2 \cdot 2 - 1) + 16(2 \cdot 3 - 1) + 16(2 \cdot 4 - 1)$$

$$= 16(1) + 16(3) + 16(5) + 16(7)$$

$$= 16 + 48 + 80 + 112$$

$$= 256$$

We read

$$\sum_{i=1}^{4} 16(2i - 1)$$

as "the sum as i goes from 1 to 4 of $16(2i - 1$.)" The letter i is called the *index of summation* and is not related to the use of i to represent $\sqrt{-1}$. Any letter can be used for the index of summation. Furthermore, the index of summation can start at a number other than 1.

EXAMPLE 6   Using Summation Notation

Expand and evaluate the sum:

a. $\displaystyle\sum_{i=1}^{6} (i^2 + 1)$ **b.** $\displaystyle\sum_{k=4}^{7} [(-2)^k - 5]$ **c.** $\displaystyle\sum_{i=1}^{5} 3$

Discover for yourself

Expand and evaluate each of the following:

$$\sum_{i=0}^{5} [(i + 1)^2 + 1]$$

$$\sum_{i=2}^{7} [(i - 1)^2 + 1]$$

Compare these results with the evaluation shown in Example 6a. Describe what this means about writing a sum in summation notation.

Using technology

Graphing utilities can calculate the sum of a sequence. For example, to find the sum of the sequence in Example 6a, enter

$\boxed{\text{sumseq}}\,(I \boxed{\wedge} 2 \boxed{+} 1, I, 1, 6, 1)$.

Then press $\boxed{\text{ENTER}}$; 97 should be displayed. Use this capability to verify Example 6b.

Solution

a. We must replace i in the expression $i^2 + 1$ with all consecutive integers from 1 to 6 inclusively. Then we add.

$$\sum_{i=1}^{6} (i^2 + 1) = (1^2 + 1) + (2^2 + 1) + (3^2 + 1) + (4^2 + 1)$$
$$+ (5^2 + 1) + (6^2 + 1)$$
$$= 2 + 5 + 10 + 17 + 26 + 37$$
$$= 97$$

b. This time the index of summation is k. First we evaluate $(-2)^k - 5$ for all integers from 4 through 7. Then we add.

$$\sum_{k=4}^{7} [(-2)^k - 5] = [(-2)^4 - 5] + [(-2)^5 - 5]$$
$$+ [(-2)^6 - 5] + [(-2)^7 - 5]$$
$$= (16 - 5) + (-32 - 5) + (64 - 5) + (-128 - 5)$$
$$= 11 + (-37) + 59 + (-133)$$
$$= -100$$

c. To find $\sum_{i=1}^{5} 3$, we observe that every term of the series is 3. The notation $i = 1$ through 5 indicates that we must add the first five 3s from a sequence in which every term is 3.

$$\sum_{i=1}^{5} 3 = 3 + 3 + 3 + 3 + 3 = 15 \qquad \blacksquare$$

In the Discover for yourself at the top of the page, did you observe that the summation notation for a series can be written so that the index begins at any number? Varying both limits of summation can produce different-looking summation notations for the same sum. Furthermore, any letter can be used. For example, the sum of the squares of the first four integers can be expressed in a number of equivalent ways, including

$$\sum_{i=1}^{4} i^2 = 1^2 + 2^2 + 3^2 + 4^2 = 30$$

$$\sum_{i=0}^{3} (i + 1)^2 = (0 + 1)^2 + (1 + 1)^2 + (2 + 1)^2 + (3 + 1)^2$$
$$= 1^2 + 2^2 + 3^2 + 4^2 = 30$$

$$\sum_{i=2}^{5} (i - 1)^2 = (2 - 1)^2 + (3 - 1)^2 + (4 - 1)^2 + (5 - 1)^2$$
$$= 1^2 + 2^2 + 3^2 + 4^2 = 30$$

$$\sum_{k=1}^{4} k^2 = 1^2 + 2^2 + 3^2 + 4^2 = 30$$

The ability to recognize patterns plays an important role in mathematics. Patterns with partial sums are sometimes helpful in finding the sum of a relatively large number of terms.

Discover for yourself

Express the sum of the squares of the first four integers with an index of summation that starts at 4. Do this again, with an index of summation starting at −2.

John Okulick "Wraparound" 1990,
wood and mixed media,
$84 \times 76 \times 9$ in. Photo courtesy Nancy
Hoffman Gallery.

EXAMPLE 7 **Using Partial Sum Patterns to Find the Sum of a Series**

Find the sum of the series: $\displaystyle\sum_{i=1}^{500} \left(\frac{1}{i} - \frac{1}{i+1} \right)$

Solution

Let's see if there is a pattern for the partial sums.

$$S_1 = \left(1 - \frac{1}{2}\right) \qquad\qquad = 1 - \frac{1}{2}$$

$$S_2 = \left(1 - \frac{1}{2}\right) + \left(\frac{1}{2} - \frac{1}{3}\right) \qquad = 1 - \frac{1}{3}$$

$$S_3 = \left(1 - \frac{1}{2}\right) + \left(\frac{1}{2} - \frac{1}{3}\right) + \left(\frac{1}{3} - \frac{1}{4}\right) \qquad = 1 - \frac{1}{4}$$

$$S_4 = \left(1 - \frac{1}{2}\right) + \left(\frac{1}{2} - \frac{1}{3}\right) + \left(\frac{1}{3} - \frac{1}{4}\right) + \left(\frac{1}{4} - \frac{1}{5}\right) = 1 - \frac{1}{5}$$

Discover for yourself

Find S_5 and S_6. What is the general pattern for S_n? Express this as a formula. Now use the formula to find S_{500}, the sum of the first 500 terms.

In the Discover for yourself, did you observe that

$$S_n = 1 - \frac{1}{n+1}?$$ The denominator in the second term is one more than the subscript on S.

We now substitute 500 for n in this formula to find the sum of the first 500 terms.

$$S_{500} = 1 - \frac{1}{500 + 1} = 1 - \frac{1}{501} = \frac{501}{501} - \frac{1}{501} = \frac{500}{501}$$

Thus,

$$\sum_{i=1}^{500} \left(\frac{1}{i} - \frac{1}{i+1} \right) = \frac{500}{501}. \qquad\blacksquare$$

We could use the formula from Example 7 for the sum of the first n terms to find the sum of the first million terms if we wished to do so. The sum of the terms of a huge finite sequence is always a finite number. In Section 7.3, we will see that some infinite sequences also have finite sums.

EXAMPLE 8 **Finding the Sum of an Infinite Sequence**

Find the sum of the infinite sequence:

$$\frac{6}{10}, \frac{6}{10^2}, \frac{6}{10^3}, \frac{6}{10^4}, \frac{6}{10^5}, \dots, \frac{6}{10^n}, \dots$$

ENRICHMENT ESSAY

Golden Ratios and the Fibonacci Sequence

The golden ratio appears in the Fibonacci sequence

$$1, 1, 2, 3, 5, 8, 13, 21, 34, 55, 89, 144, 233, \ldots$$

As the terms progress, the ratio of each number to its predecessor keeps getting closer to the golden ratio,

$$\frac{1 + \sqrt{5}}{2} \text{ to } 1$$

or approximately 1.618 to 1.

$$\frac{a_6}{a_5} = \frac{8}{5} = 1.6 \text{ to } 1$$

$$\frac{a_7}{a_6} = \frac{13}{8} = 1.625 \text{ to } 1$$

$$\frac{a_8}{a_7} = \frac{21}{13} = 1.\overline{615384} \text{ to } 1$$

$$\frac{a_9}{a_8} = \frac{34}{21} \approx 1.619048 \text{ to } 1$$

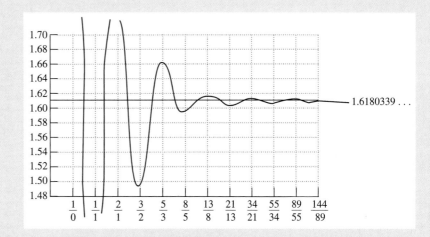

$$\vdots$$

$$\frac{a_{99}}{a_{98}} = \frac{218922995834555169026}{135301852344706746049} \approx 1.6180339887498948482045868343656381177203 \text{ to } 1$$

$$\frac{a_{100}}{a_{99}} = \frac{354224848179261915075}{218922995834555169026} \approx 1.6180339887498948482045868343656381177202 \text{ to } 1$$

Shown below is a compositional analysis of Grant Wood's *American Gothic*. Using the figure on the right, how many rectangles in the 1.618 to 1 ratio can you find?

Grant Wood, American (1891–1942), "American Gothic" 1930, oil on beaver board, 29 7/8 × 24 7/8 in. (74.3 × 62.4 cm). Friends of American Art Collection, All rights reserved by The Art Institute of Chicago and VAGA, New York, NY, 1930.934.

Solution

$$\sum_{n=1}^{\infty} \frac{6}{10^n} = \frac{6}{10^1} + \frac{6}{10^2} + \frac{6}{10^3} + \frac{6}{10^4} + \frac{6}{10^5} + \cdots$$

$$= 0.6 + 0.06 + 0.006 + 0.0006 + 0.00006 + \cdots \qquad \text{Write each fraction in decimal form.}$$

$$= 0.66666 \ldots \qquad \text{This can be expressed as } 0.\overline{6}.$$

$$\approx \frac{2}{3} \qquad \text{The previous repeating decimal is the decimal representation for } \frac{2}{3}.$$

Because mathematicians express their ideas in compact, symbolic notation, a sum is frequently written in compact form using summation notation.

7 Write sums in summation notation.

EXAMPLE 9 **Writing Sums in Summation Notation**

Write the sum using summation notation:

a. $1 + 8 + 27 + 64 + 125 + 216 + 343$
b. $-1 + 8 - 27 + 64 - 125 + 216 - 343$
c. $3 + 9 + 27 + 81 + \cdots$

Solution

a. $1 + 8 + 27 + 64 + 125 + 216 + 343$
This is the sum of cubes, $1^3 + 2^3 + 3^3 + 4^3 + 5^3 + 6^3 + 7^3$. Using i for the index of summation, a possible general term is i^3. Thus,

$$1 + 8 + 27 + 64 + 125 + 216 + 343 = \sum_{i=1}^{7} i^3.$$

b. $-1 + 8 - 27 + 64 - 125 + 216 - 343$
This is almost like the series in part (a), except that the signs alternate. Since $(-1)^i = 1$ when i is even and $(-1)^i = -1$ when i is odd, we write the series in summation notation as follows:

$$-1 + 8 - 27 + 64 - 125 + 216 - 343 = \sum_{i=1}^{7} (-1)^i i^3.$$

c. $3 + 9 + 27 + 81 + \cdots$
This infinite series consists of powers of 3. We use the symbol ∞ to represent infinity and write the series in summation notation as follows:

$$3 + 9 + 27 + 81 + \cdots = \sum_{i=1}^{\infty} 3^i.$$ ■

Table 7.1 contains some important properties of sums expressed in summation notation.

iscover for yourself

Each of the properties in Table 7.1 follows directly from three basic properties of the real numbers. What are these three properties? Use these properties to prove each of the properties of sums in the table.

TABLE 7.1   Properties of Sums

Property	Example
1. $\displaystyle\sum_{i=1}^{n} ca_i = c \sum_{i=1}^{n} a_i$, c any real number	$\displaystyle\sum_{i=1}^{4} 3i^2 = 3 \cdot 1^2 + 3 \cdot 2^2 + 3 \cdot 3^2 + 3 \cdot 4^2$
	$\displaystyle 3\sum_{i=1}^{4} i^2 = 3(1^2 + 2^2 + 3^2 + 4^2) = 3 \cdot 1^2 + 3 \cdot 2^2 + 3 \cdot 3^2 + 3 \cdot 4^2$
	Conclusion: $\displaystyle\sum_{i=1}^{4} 3i^2 = 3\sum_{i=1}^{4} i^2$
2. $\displaystyle\sum_{i=1}^{n} (a_i + b_i) = \sum_{i=1}^{n} a_i + \sum_{i=1}^{n} b_i$	$\displaystyle\sum_{i=1}^{4} (i + i^2) = (1 + 1^2) + (2 + 2^2) + (3 + 3^2) + (4 + 4^2)$
	$\displaystyle\sum_{i=1}^{4} i + \sum_{i=1}^{4} i^2 = (1 + 2 + 3 + 4) + (1^2 + 2^2 + 3^2 + 4^2)$
	$= (1 + 1^2) + (2 + 2^2) + (3 + 3^2) + (4 + 4^2)$
	Conclusion: $\displaystyle\sum_{i=1}^{4} (i + i^2) = \sum_{i=1}^{4} i + \sum_{i=1}^{4} i^2$
3. $\displaystyle\sum_{i=1}^{n} (a_i - b_i) = \sum_{i=1}^{n} a_i - \sum_{i=1}^{n} b_i$	$\displaystyle\sum_{i=3}^{5} (i^2 - i^3) = (3^2 - 3^3) + (4^2 - 4^3) + (5^2 - 5^3)$
	$\displaystyle\sum_{i=3}^{5} i^2 - \sum_{i=3}^{5} i^3 = (3^2 + 4^2 + 5^2) - (3^3 + 4^3 + 5^3)$
	$= (3^2 - 3^3) + (4^2 - 4^3) + (5^2 - 5^3)$
	Conclusion: $\displaystyle\sum_{i=3}^{5} (i^2 - i^3) = \sum_{i=3}^{5} i^2 - \sum_{i=3}^{5} i^3$

PROBLEM SET 7.1

Practice Problems

In Problems 1–18, write the first five terms of each sequence whose general term is given.

1. $a_n = 3n + 2$

2. $a_n = 4n - 1$

3. $a_n = 3^n$

4. $a_n = \left(\dfrac{1}{3}\right)^n$

5. $a_n = (-3)^n$

6. $a_n = \left(-\dfrac{1}{3}\right)^n$

7. $a_n = (-1)^n(n + 3)$

8. $a_n = (-1)^{n+1}(n + 4)$

9. $a_n = \dfrac{2n}{n + 4}$

10. $a_n = \dfrac{3n}{n + 5}$

11. $a_n = \dfrac{(-1)^{n+1}}{2^n - 1}$

12. $a_n = \dfrac{(-1)^{n+1}}{2^n + 1}$

13. $a_n = \dfrac{n^2}{n!}$

14. $a_n = \dfrac{(n + 1)!}{n^2}$

15. $a_n = 2(n + 1)!$

16. $a_n = -2(n - 1)!$

17. $a_n = \dfrac{(n + 1)!}{n!}$

18. $a_n = \dfrac{(2n - 1)!}{(2n + 1)!}$

The sequences in Problems 19–24 are defined using recursion formulas. Write the first five terms for each sequence.

19. $a_1 = 4$ and $a_n = 2a_{n-1} + 3$

20. $a_1 = -5$ and $a_n = 3a_{n-1} - 1$

21. $a_1 = 2$ and $a_n = (a_{n-1})^2 - 4$

22. $a_1 = -11$ and $a_n = (a_{n-1})^2 - 5$

23. $a_1 = 1, a_2 = 1,$ and $a_n = a_{n-2} + a_{n-1}$

24. $a_1 = 2, a_2 = -1,$ and $a_n = a_{n-2} - a_{n-1}$

For Problems 25–34, write an expression for the general, or nth, term of each sequence. More than one expression may be possible.

25. $1, 3, 5, 7, 9, \ldots$

26. $1, 4, 7, 10, 13, \ldots$

27. $\dfrac{2}{1}, \dfrac{3}{2}, \dfrac{4}{3}, \dfrac{5}{4}, \dfrac{6}{5}, \ldots$

28. $\dfrac{1}{2}, \dfrac{1}{4}, \dfrac{1}{8}, \dfrac{1}{16}, \dfrac{1}{32}, \ldots$

29. $-1, 1, -1, 1, -1, \ldots$

30. $1, -1, 1, -1, 1, \ldots$

31. $1, -\dfrac{1}{4}, \dfrac{1}{9}, -\dfrac{1}{16}, \dfrac{1}{25}, \ldots$

32. $2, -4, 6, -8, 10, \ldots$

33. $1, 2, 6, 24, 120, \ldots$

34. $1, \dfrac{1}{2}, \dfrac{1}{6}, \dfrac{1}{24}, \dfrac{1}{120}, \ldots$

In Problems 35–46, find the indicated partial sum for each given sequence.

35. $5, 10, 15, 20, \ldots; S_6$

36. $7, 14, 21, 28, \ldots; S_6$

37. $1, \dfrac{1}{2}, \dfrac{1}{4}, \dfrac{1}{8}, \ldots; S_5$

38. $\dfrac{1}{3}, \dfrac{2}{3}, \dfrac{3}{3}, \dfrac{4}{3}, \ldots; S_8$

39. $a_n = \dfrac{n+1}{n}; S_3$

40. $a_n = \dfrac{n}{n+1}; S_3$

41. $a_n = (-1)^{n+1}(2n+1); S_7$

42. $a_n = (-1)^{n+1}(3n-2); S_7$

43. $a_n = (-1)^n n!; S_4$

44. $a_n = (-1)^n(n+1)!; S_4$

45. $a_n = \dfrac{(n+1)!}{n!}; S_6$

46. $a_n = \dfrac{(n+2)!}{n!}; S_6$

In Problems 47–60, find each indicated sum.

47. $\displaystyle\sum_{i=1}^{6} 5i$

48. $\displaystyle\sum_{i=1}^{6} 7i$

49. $\displaystyle\sum_{i=1}^{4} 2i^2$

50. $\displaystyle\sum_{i=1}^{5} i^3$

51. $\displaystyle\sum_{k=1}^{5} k(k+4)$

52. $\displaystyle\sum_{k=1}^{4} (k-3)(k+2)$

53. $\displaystyle\sum_{i=1}^{4} \left(-\dfrac{1}{2}\right)^i$

54. $\displaystyle\sum_{i=2}^{4} \left(-\dfrac{1}{3}\right)^i$

55. $\displaystyle\sum_{i=5}^{9} 11$

56. $\displaystyle\sum_{i=3}^{7} 12$

57. $\displaystyle\sum_{i=0}^{4} \dfrac{(-1)^i}{i!}$

58. $\displaystyle\sum_{i=0}^{4} \dfrac{(-1)^{i+1}}{(i+1)!}$

59. $\displaystyle\sum_{i=1}^{5} \dfrac{i!}{(i-1)!}$

60. $\displaystyle\sum_{i=1}^{5} \dfrac{(i+2)!}{i!}$

In Problems 61–66, find a formula for the nth partial sum S_n of each series. Then find the indicated sum.

61. $\displaystyle\sum_{i=1}^{1000} \left(\dfrac{1}{i+1} - \dfrac{1}{i+2}\right)$

62. $\displaystyle\sum_{i=1}^{1000} \left(\dfrac{1}{i+3} - \dfrac{1}{i+4}\right)$

63. $\displaystyle\sum_{i=1}^{50} \left(\dfrac{1}{2i-1} - \dfrac{1}{2i+1}\right)$

64. $\displaystyle\sum_{i=1}^{20} (2^i - 2^{i+1})$

65. $\displaystyle\sum_{i=1}^{1000} (a_i - a_{i+1})$

66. $\displaystyle\sum_{i=1}^{9999} [\ln i - \ln(i+1)]$

In Problems 67–68, write the first five terms for each infinite series, expressing the terms in decimal form. Then use your knowledge of repeating decimals to find the sum of the infinite series.

67. $\displaystyle\sum_{i=1}^{\infty} \dfrac{3}{10^i}$

68. $\displaystyle\sum_{i=1}^{\infty} \dfrac{9}{100^i}$

In Problems 69–80, write each series in summation notation. Use the index i and let i begin at 1.

69. $1 + 2 + 3 + 4 + 5$

70. $2 + 4 + 6 + 8 + 10$

71. $\dfrac{1}{2} + \dfrac{1}{2^2} + \dfrac{1}{2^3} + \dfrac{1}{2^4} + \dfrac{1}{2^5} + \dfrac{1}{2^6}$

72. $\dfrac{1}{1^3} + \dfrac{1}{2^3} + \dfrac{1}{3^3} + \dfrac{1}{4^3} + \dfrac{1}{5^3} + \dfrac{1}{6^3} + \dfrac{1}{7^3}$

73. $1 + 4 + 9 + 16 + 25 + 36$

74. $1 + 8 + 27 + 64 + 125 + 216$

75. $1 - \dfrac{1}{2} + \dfrac{1}{4} - \dfrac{1}{8} + \cdots - \dfrac{1}{128}$

76. $\dfrac{1}{1 \cdot 3} + \dfrac{1}{2 \cdot 4} + \dfrac{1}{3 \cdot 5} + \cdots + \dfrac{1}{11 \cdot 13}$

77. $\dfrac{x}{x + 1} + \dfrac{x}{x + 2} + \dfrac{x}{x + 3} + \dfrac{x}{x + 4}$

78. $x + \dfrac{x^3}{3} + \dfrac{x^5}{5} + \dfrac{x^7}{7} + \dfrac{x^9}{9}$

79. $8 + 16 + 24 + 32 + 40 + \cdots$

80. $6 + 12 + 18 + 24 + 30 + \cdots$

Application Problems

81. The finite sequence whose general term is

$$a_n = 0.16n^2 - 1.04n + 7.39$$

where $n = 1, 2, 3, \ldots, 8$ models the total number of dollars (in billions) that Americans spent on recreational boating from 1991 through 1998. Find and interpret S_5, the fifth partial sum.

82. The finite sequence whose general term is

$$a_n = 14.9n + 267$$

where $n = 5, 6, 7, \ldots, 15$, models the average weekly earnings for U.S. workers from 1985 through 1995. Find and interpret $a_{15} - a_5$.

83. The finite sequence whose general term is

$$a_n = 0.255n^3 - 4.096n^2 + 1570.417$$

where $n = 1, 2, 3, \ldots, 15$ models the per capita federal debt in the United States from 1981 through 1995. Find and interpret $a_{15} - a_1$.

84. The number of AIDS cases reported in the United States for 1984 through 1990 is approximated by the model

$$a_n = -143n^3 + 1810n^2 - 187n + 2331$$

where $n = 1$ corresponds to 1984, $n = 2$ to 1985, and so on. Find and interpret S_3, the third partial sum.

True–False Critical Thinking Problems

85. Which one of the following is true?

a. $\dfrac{n!}{(n - 1)!} = \dfrac{1}{n - 1}$

b. The Fibonacci sequence

$$1, 1, 2, 3, 5, 8, 13, 21, 34, 55, 89, 144, \ldots$$

can be defined recursively using

$$a_0 = 1, a_1 = 1, a_n = a_{n-2} + a_{n-1}, \text{ where } n \geq 2.$$

c. The same sum cannot be expressed by varying the limits of summation.

d. As more terms are added, the sequence whose general term is given by $a_n = \dfrac{1}{n!}$ has a sum that is getting farther and farther from e.

86. Which one of the following is true?

a. $\displaystyle\sum_{i=1}^{3} (2i + 5) = \left(\sum_{i=1}^{3} 2i \right) + 5$

b. $\displaystyle\sum_{i=1}^{5} 4 = 4$

c. The series

$$\sum_{i=0}^{6} (-1)^i (i + 1)^2 \quad \text{and} \quad \sum_{j=1}^{7} (-1)^j j^2$$

have the same sum.

d. $\displaystyle\sum_{i=1}^{5} 6i = 6 \sum_{i=1}^{5} i$

87. Which one of the following is true?

a. A sequence is the indicated sum of the terms of a series.

b. There is only one way to write a given sum in summation notation.

c. $\displaystyle\sum_{i=1}^{4} 3i + \sum_{i=1}^{4} 4i = \sum_{i=1}^{4} 7i$

d. $\displaystyle\sum_{i=1}^{2} (-1)^i 2^i = 0$

Technology Problems

Most graphing utilities have a sequence graphing mode that plots the terms of a sequence as points on a rectangular coordinate system. Use this capability to graph each of the sequences in Problems 88–91. Try to duplicate the graph of the sequence shown in the indicated figure.

88. $a_n = \dfrac{n}{n + 1}$

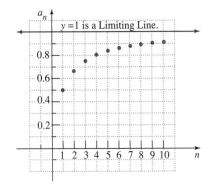

89. $a_n = \dfrac{100}{n}$

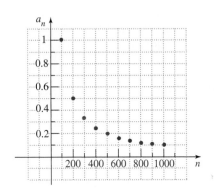

90. $a_n = \dfrac{2n^2 + 5n - 7}{n^3}$

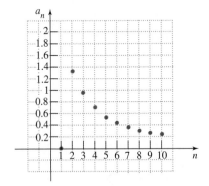

91. $a_n = \dfrac{3n^4 + n - 1}{5n^4 + 2n^2 + 1}$

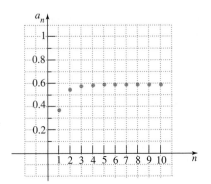

A sequence $a_1, a_2, a_3, \ldots, a_n, \ldots$ has a limit L if the terms a_n approach L $(a_n \to L)$ as n increases without bound $(n \to \infty)$. For example, if you continue writing the terms of $a_n = \dfrac{n}{n + 1}$ as $n \to \infty$, the sequence gets closer and closer to 1. Thus, as

$n \to \infty, \dfrac{n}{n + 1} \to 1$. *Of course, not all sequences have a limit. For example, the terms of the sequence 3, 7, 11, 15, ...*

increase without bound and do not approach any number. The terms of the sequence 2, −2, 2, −2, ... alternate between 2 and −2 and do not approach a single number. The terms of a sequence must approach a unique real number in order to have a limit.

   In Problems 92–97, use a graphing utility to graph each sequence and determine whether the sequence has a limit. If so, what is the limit? If the sequence has no limit, explain why it does not.

92. $a_n = \dfrac{1}{n}$

93. $a_n = \dfrac{2n}{n + 1}$

94. $a_n = \dfrac{3n^2 + 4n}{2n + 1}$

95. $a_n = (-1)^n$

96. $a_n = (1 + n)^{1/n}$

97. $a_n = \dfrac{2n + 1}{2n^2 + 4n}$

Writing in Mathematics

98. Describe the difference between a sequence and a series.

99. If $f(x) = \sum_{i=0}^{6} \frac{x^i}{i!}$, find $f(1)$, $f(2)$, and $f(3)$ to four decimal places. Compare these values to e, e^2, and e^3, respectively. Describe what you observe.

100. Newton's formula for approximating $\sqrt{c}\ (c \geq 0)$ is
$$\sqrt{c} \approx a_{n+1} = \frac{1}{2}\left(a_n + \frac{c}{a_n}\right),$$ where $n = 0, 1, 2, 3, \ldots$. The number a_0 is a first guess, and the sequence $a_1, a_2, a_3, \ldots$ generates successively closer approximations for $\sqrt{c}$.

a. Find $\sqrt{10}$ correct to four decimal places, starting with $a_0 = 3$.

b. Find an approximation for $\sqrt{10}$ using your graphing utility. Describe why Newton's formula provides successively closer approximations to $\sqrt{10}$.

Critical Thinking Problems

In Problems 101–102, write the first five terms of each sequence whose general term is given.

101. $a_n = \dfrac{3 \cdot 5 \cdots (2n+1)}{2 \cdot 4 \cdots (2n)}$

102. $a_{n+1} = \begin{cases} \dfrac{a_n}{2} & \text{if } a_n \text{ is even} \\[2mm] 2a_n + 5 & \text{if } a_n \text{ is odd and } a_1 = 9 \end{cases}$

103. Write the first eight terms of the sequence whose general term is given by
$$a_n = \frac{1 + (-1)^{n+1}}{2i^{n-1}} \qquad (i = \sqrt{-1})$$

104. The first seven terms of a sequence are 2, 3, 5, 7, 11, 13, and 17. Describe at least one pattern for the terms of this sequence and use the pattern to write the next ten terms.

Group Activity Problem

105. Enough curiosities involving the Fibonacci sequence exist to warrant a flourishing Fibonacci Association that publishes a quarterly journal. Do some research on the Fibonacci sequence by consulting the research department of your library or the Internet and find one property that interests you. After doing this research, group members should share these intriguing properties with each other. Depending on the size of the group, it will be interesting to see whether duplication occurs or if each member researches a unique property for this fascinating sequence.

S E C T I O N 7 . 2

Solutions Manual

Tutorial

Video
12

Arithmetic Sequences and Series

Objectives

1 Find the common difference for an arithmetic sequence.

2 Find a term for an arithmetic sequence.

3 Find partial sums of arithmetic sequences.

4 Use sums of arithmetic series to solve applied problems.

Arithmetic Sequences

A mathematical model for the average annual salaries of major league baseball players generates the following data.

Year	1990	1991	1992	1993	1994	1995	1996
Salary	710,000	801,000	892,000	983,000	1,074,000	1,165,000	1,256,000

Jacob J. Kass "Picking A Team" 1985, oil and acrylic magna on steel, 7 × 27.5 in. Photo courtesy Nancy Hoffman Gallery, NY.

The sequence consisting of annual salaries generated by the model indicates a yearly salary increase of $91,000. This sequence is called an *arithmetic sequence, or arithmetic progression*, because each term after the first (the 1990 salary) differs from the preceding term by a constant amount.

> **Definition of an arithmetic sequence**
>
> An *arithmetic sequence*, or an *arithmetic progression*, is a sequence in which each term after the first differs from the preceding term by a constant amount. The difference between consecutive terms is called the *common difference* of the sequence.

Find the common difference for an arithmetic sequence.

EXAMPLE 1 **Finding the Common Difference of Arithmetic Sequences**

For each arithmetic sequence, identify the common difference.

a. $3, 9, 15, 21, \ldots$
b. $-2, -7, -12, -17, \ldots$
c. $a_1, a_1 - 4d, a_1 - 8d, a_1 - 12d, \ldots$

Solution

The common difference is obtained by subtracting from any term its predecessor. We will subtract the second term from the first term.

Discover for yourself

Confirm each common difference shown on the right by computing

third term − second term

and

fourth term − third term

for each of the three arithmetic sequences.

Sequence	Second Term − First Term	Common Difference
a. $3, 9, 15, 21, \ldots$	$9 - 3 = 6$	6
b. $-2, -7, -12, -17, \ldots$	$-7 - (-2) = -5$	-5
c. $a_1, a_1 - 4d, a_1 - 8d, a_1 - 12d, \ldots$	$(a_1 - 4d) - a_1 = -4d$	$-4d$

■

EXAMPLE 2 **Writing the Terms of an Arithmetic Sequence Using the First Term and the Common Difference**

Find the first five terms of the arithmetic sequence in which $a_1 = 624$ and $a_{n+1} = a_n - 24$.

Solution

The recursion formula $a_{n+1} = a_n - 24$ indicates that each term after the first is obtained by adding -24 to the previous term.

$a_1 = 624$ — This is given.

$a_2 = a_1 - 24 = 624 - 24 = 600$ — Use $a_{n+1} = a_n - 24$ with $n = 1$.

$a_3 = a_2 - 24 = 600 - 24 = 576$ — Use $a_{n+1} = a_n - 24$ with $n = 2$.

$a_4 = a_3 - 24 = 576 - 24 = 552$ — Use $a_{n+1} = a_n - 24$ with $n = 3$.

$a_5 = a_4 - 24 = 552 - 24 = 528$ — Use $a_{n+1} = a_n - 24$ with $n = 4$.

The first five terms are

624, 600, 576, 552, and 528.

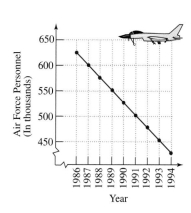

Figure 7.2

Active U.S. Air Force personnel

The recursion formula in Example 2 is actually a model for the total number (in thousands) of Air Force personnel on active duty for each year from 1986 through 1990. The model can be extended up through 1994 (see Figure 7.2) and can be expressed using

$$a_{n+1} = a_n - 24, a_1 = 624$$

or

$$a_n = -24n + 624.$$

To use the recursion formula, we must be given the first term.

Let's now consider generalizing from Example 2 for an arithmetic sequence whose first term is a_1 and whose common difference is d. We start with the first term and add d to each successive term as follows:

2 Find a term for an arithmetic sequence.

a_1: first term $= a_1$
a_2: second term $= a_1 + d$
a_3: third term $= a_2 + d = (a_1 + d) + d = a_1 + 2d$
a_4: fourth term $= a_3 + d = (a_1 + 2d) + d = a_1 + 3d$
a_5: fifth term $= a_4 + d = (a_1 + 3d) + d = a_1 + 4d$
a_6: sixth term $= a_5 + d = (a_1 + 4d) + d = a_1 + 5d$
⋮   ⋮
a_n: nth term $= ?$

Discover for yourself

Look at the pattern for the first six terms of an arithmetic sequence whose first term is a_1 and whose common difference is d. What relationship do you observe between the coefficient of d and the subscript of a denoting the term number? Use this relationship to write the missing formula for the nth term.

In the Discover for yourself, did you notice that the coefficient of d is 1 less than the subscript of a denoting the term number? Thus, the formula for the nth term is

$$a_n\text{: }n\text{th term} = a_1 + (n - 1)d.$$

1 less than the
subscript of a

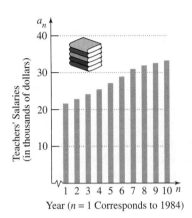

Figure 7.3
Teachers' salaries
Source: N.E.A.

Using technology

The formula for the nth term of
21,700, 23,172, 24,644, ... is

$a_n = a_1 + (n - 1)d$

$\quad = 21,700 + (n - 1)1472$

or

$a_n = 1472n + 20,228.$

Consult the manual to see if
your graphing utility is capable of
generating tables. If so, use the
TABLE function and the
formula for the nth term of the
sequence to display the first 40
terms.

Honore Desmond Sharrer "Tribute
to the American People." National
Museum of American Art,
Washington, D.C. / Art Resource,
NY.

> **General term of an arithmetic sequence**
>
> The nth term (the general term) of an arithmetic sequence with first
> term a_1 and common difference d is
>
> $$a_n = a_1 + (n - 1)d.$$

Because the formula contains four variables (a_n, a_1, n, and d), if any three of
these are known, we can find the value of the fourth variable. This idea is illustrated in Examples 3 and 4.

EXAMPLE 3 **Finding a Specified Term of an Arithmetic Sequence**

In 1984, teachers in the United States earned an average of $21,700 per year.
This amount has increased by about $1472 yearly, generating the arithmetic
sequence

$\qquad$ 21,700, 23,172, 24,644, 26,116, ...

(see Figure 7.3). Find a_{30}, the 30th term, and describe what this means in practical terms.

Solution

The first term of this sequence is 21,700, so $a_1 = 21,700$. The common difference can be obtained from the problem's wording or by subtracting any two
adjacent terms. We'll use the first two terms. The common difference is

$\qquad$ $23,172 - 21,700 = 1472$

so $d = 1472$. (This represents the yearly increase in salary.) Using the formula
for the general term of an arithmetic sequence, we obtain

$$a_n = a_1 + (n - 1)d$$

$$a_{30} = 21,700 + (30 - 1)1472 = 21,700 + (29)1472 = 64,388$$

The 30th term is 64,388. Since $n = 1$ corresponds to the year 1984, this means
that 30 years after 1983—in the year 2013—U.S. teachers are predicted to earn
$64,388. ∎

EXAMPLE 4 **Using the Formula for the General Term of an Arithmetic Sequence**

In the sequence in Example 3, which term is 314,628?

Solution

We must find n so that $a_n = 314,628$. In terms of the variables being modeled,
we want to know when teachers will be earning $314,628. We substitute 21,700
for a_1 (the first term), 1472 for d (the common difference) and 314,628 for a_n
into the formula for the general term of an arithmetic sequence. Our goal is to
solve for n.

$$a_n = a_1 + (n - 1)d$$ 　　This is the formula for the nth term of an arithmetic sequence.

$$314{,}628 = 21{,}700 + (n - 1)1472$$ 　　Substitute the given values.

$$314{,}628 = 21{,}700 + 1472n - 1472$$ 　　Apply the distributive property.

$$314{,}628 = 20{,}228 + 1472n$$ 　　Combine numerical terms.

$$294{,}400 = 1472n$$ 　　Subtract 20,228 from both sides.

$$200 = n$$ 　　Divide both sides by 1472.

The term 314,628 is the 200th term of the sequence. Since $n = 1$ corresponds to 1984, this means that 200 years after 1983—in the year 2183—U.S. teachers are predicted to earn $314,628. 　■

iscover for yourself

Although the answer in Example 4 is theoretically correct, problems arise when the model for teachers' salaries is extended 200 years after 1983. What are two possible circumstances that would render the predicted salary incorrect?

3 Find partial sums of arithmetic sequences.

Arithmetic Series

Annual teachers' salaries in the United States can be modeled using

$$a_n = 1472n + 20{,}228$$

where $n = 1$ corresponds to 1984. Using this formula, we can compute partial sums such as S_3. This third partial sum describes the total three-year salaries for 1984, 1985, and 1986. Instead of adding the three values, the partial sums of arithmetic sequences can be calculated using a quicker method.

Let's consider the sum of the first n terms of an arithmetic sequence. This sum, denoted by S_n, is called an *arithmetic series*. Recall that the first n terms of the arithmetic sequence are

$$a_1, a_1 + d, a_1 + 2d, \ldots, a_1 + (n - 1)d.$$

We proceed as follows:

$$S_n = a_1 + (a_1 + d) + (a_1 + 2d) + \cdots + [a_1 + (n - 1)d]$$ 　　S_n is the sum of the first n terms of the sequence.

$$S_n = a_n + (a_n - d) + (a_n - 2d) + \cdots + [a_n - (n - 1)d]$$ 　　The same series can be obtained by beginning with the last term, a_n, and subtracting d.

$$2S_n = (a_1 + a_n) + (a_1 + a_n) + (a_1 + a_n) + \cdots + (a_1 + a_n)$$ 　　Add the two equations.

$$2S_n = n(a_1 + a_n)$$ 　　Simplify the right side using the fact that there are n sums of $(a_1 + a_n)$.

$$S_n = \frac{n}{2}(a_1 + a_n)$$ 　　Solve for S_n.

tudy tip

The terms involving d cancel when adding the equations in the first two lines:

$$d - d = 0$$
$$2d - 2d = 0$$
$$(n - 1)d - (n - 1)d = 0.$$

Study tip

The formula in the box shows that to find the sum of the terms of an arithmetic sequence, we need to know the first term, a_1, the last term, a_n, and the number of terms, n.

We have proved the following result.

The nth partial sum of an arithmetic sequence

The sum, S_n, of the first n terms of an arithmetic sequence is given by

$$S_n = \frac{n}{2}(a_1 + a_n)$$

in which a_1 is the first term and a_n is the nth term.

The following examples illustrate how to use this formula.

EXAMPLE 5   **Finding the Sum of an Arithmetic Sequence**

Find the sum of the first 100 terms of the arithmetic sequence: $1, 3, 5, 7, \ldots$

Solution

We are actually finding the sum of the first 100 odd numbers. Before finding this sum, we need to know the last term, the 100th term. The common difference is 2.

$$a_n = a_1 + (n - 1)d$$
This is the formula for the nth term of an arithmetic sequence. Use it to find the 100th term.

$$a_{100} = 1 + (100 - 1) \cdot 2$$
Substitute 100 for n, 2 for d, and 1 (the first term) for a_1.

$$= 1 + 99 \cdot 2$$

$$= 199$$

Now we are ready to find the sum of the first 100 terms of $1, 3, 5, 7, \ldots, 199$.

$$S_n = \frac{n}{2}(a_1 + a_n)$$
Use the formula for the sum of the first n terms of an arithmetic sequence. Let $n = 100$, $a_1 = 1$, and $a_{100} = 199$.

$$S_{100} = \frac{100}{2}(1 + 199) = 50(200) = 10,000$$

The sum of the first 100 odd numbers is 10,000.   ◼

Discover for yourself

Find

$$\sum_{i=1}^{25}(5i - 9)$$

by expanding the summation and adding the resulting 25 terms. What is the advantage to the method in Example 6, where a formula is used? Are there any disadvantages to using the formula for S_n to evaluate the summation?

EXAMPLE 6   **Using S_n to Evaluate a Summation**

Find the following sum: $\displaystyle\sum_{i=1}^{25}(5i - 9)$

Solution

$$\sum_{i=1}^{25}(5i - 9) = (5 \cdot 1 - 9) + (5 \cdot 2 - 9) + (5 \cdot 3 - 9) + \cdots + (5 \cdot 25 - 9)$$

$$= -4 \quad\ + 1 \quad\ + 6 \quad\ + \cdots + 116$$

By evaluating the first three terms and the last term, we see that $a_1 = -4$; d, the common difference, is $1 - (-4)$ or 5; and a_{25}, the last term, is 116.

$$S_n = \frac{n}{2}(a_1 + a_n)$$

> Use the formula for the sum of the first n terms of an arithmetic sequence. Let $n = 25$, $a_1 = -4$, and $a_{25} = 116$.

$$S_{25} = \frac{25}{2}(-4 + 116) = \frac{25}{2}(112) = 1400$$

Thus,

$$\sum_{i=1}^{25}(5i - 9) = 1400.$$ ■

4 Use sums of arithmetic series to solve applied problems.

Paul Klee "Portrait of Mrs. P in the South" 1924. Peggy Guggenheim collection, The Solomon R. Guggenheim Museum, New York. Photograph by David Heald. © The Solomon R. Guggenheim Foundation, New York (FN 76.2553 PG 89). © ARS, New York.

tudy tip

Be careful when selecting a value of n that corresponds to a particular year. The value is usually 1 more than what you expect. Can you explain why?

Modeling Using Partial Sums of Arithmetic Sequences

Partial sums of arithmetic sequences can be used to solve many types of applied problems. For example, if we are interested in total cost over a period of years, we can use the formula for the sum of the terms of the arithmetic sequence that describes yearly costs.

EXAMPLE 7 **Modeling Total Nursing Home Costs over a Six-Year Period**

Your grandmother and her financial counselor are looking at options in case nursing home care is needed in the future. One possibility involves immediate nursing home care for a six-year period beginning in 1998. Your grandmother has saved $250,000. Using the model

$$a_n = 28{,}130 + 1800n$$

which describes yearly nursing home costs with $n = 1$ corresponding to 1990, does your grandmother have enough in savings to pay for the facility?

Solution

We must find the sum of an arithmetic sequence whose first term corresponds to nursing home costs in the year 1998 and whose last term corresponds to nursing home costs in the year 2003. Since $n = 1$ describes the year 1990, then $n = 9$ describes the year 1998 and $n = 14$ describes the year 2003.

$$a_n = 28{,}130 + 1800n$$

$$a_9 = 28{,}130 + 1800 \cdot 9 = 44{,}330$$

$$a_{14} = 28{,}130 + 1800 \cdot 14 = 53{,}330$$

The first year the facility will cost $44,330 and by year six the facility will cost $53,330. Now we want the sum of these costs for all six years. Our focus is on the sum of the first six terms of the arithmetic series

$$44{,}330, 46{,}130, \ldots, 53{,}330.$$

We can call a_1 (the first term) 44,330 and a_6 (the last term) 53,330.

$$S_n = \frac{n}{2}(a_1 + a_n)$$

$$S_6 = \frac{6}{2}(44{,}330 + 53{,}330) = 3(97{,}660) = 292{,}980$$

Total nursing home costs for your grandmother are predicted to be $292,980 for the six-year period. Since she has saved $250,000, her savings are not enough to pay for the facility. ■

PROBLEM SET 7.2

Practice Problems

Write the first six terms of each arithmetic sequence in Problems 1–6.

1. First term = 11, common difference = −5

2. First term = 17, common difference = −6

3. $a_{n+1} = a_n + 6, a_1 = -9$

4. $a_{n+1} = a_n + 4, a_1 = -7$

5. $a_{n+1} = a_n - 0.4, a_1 = -1.6$

6. $a_{n+1} = a_n - 0.3, a_1 = -1.7$

In Problems 7–20, write a formula for the nth term of each arithmetic sequence. Do not use a recursion formula. Then use the formula for a_n to find a_{20}, the twentieth term of the sequence.

7. 1, 5, 9, 13, …

8. 2, 7, 12, 17, …

9. 7, 3, −1, −5, …

10. 6, 1, −4, −9, …

11. 3.15, 3.10, 3.05, 3.00, …

12. 3, 3.75, 4.5, 5.25, …

13. $\frac{e}{6}, \frac{e}{3}, \frac{e}{2}, \frac{2e}{3}, …$

14. $\frac{\pi}{12}, \frac{\pi}{6}, \frac{\pi}{4}, \frac{\pi}{3}, …$

15. $a_1 = 9, d = 2$

16. $a_1 = -\frac{1}{3}, d = \frac{1}{3}$

17. $a_1 = 4, a_{n+1} = a_n + 3$

18. $a_1 = 6, a_{n+1} = a_n - 2$

19. $a_1 = -\frac{1}{3}, a_{n+1} = a_1 + \frac{1}{3}$

20. $a_1 = 5\frac{1}{2}, a_{n+1} = a_n - \frac{1}{4}$

21. In the arithmetic sequence 5, 8, 11, 14, …, which term is 32?

22. In the arithmetic sequence −8, −13, −18, −23, …, which term is −53?

23. In the sequence 4, 1, −2, −5, −8, …, which term is −281?

24. In the sequence $\frac{5}{4}, \frac{1}{2}, -\frac{1}{4}, -1, …$, which term is $-\frac{55}{4}$?

25. In the sequence $-\frac{5}{3}, -2, -\frac{7}{3}, -\frac{8}{3}, …$, which term is −9?

26. In the sequence 10, 14.5, 19, 23.5, …, which term is 203.5?

27. Find the sum of the first 30 terms of the arithmetic sequence 7, 19, 31, 43, ….

28. Find the sum of the first 30 terms of the arithmetic sequence 4, 10, 16, 22, ….

29. Find the sum of the first 46 terms of the arithmetic sequence −12, −3, 6, 15, ….

30. Find the sum of the first 46 terms of the arithmetic sequence −15, −7, 1, 9, ….

For Problems 31–40, write out the first three terms and the last term. Then use the formula for the sum of the first n terms of an arithmetic sequence to find the sum indicated.

31. $\displaystyle\sum_{i=1}^{60} (6i - 4)$

32. $\displaystyle\sum_{i=1}^{60} (5i + 3)$

33. $\displaystyle\sum_{i=1}^{40} (-2i + 7)$

34. $\displaystyle\sum_{i=1}^{40} (-3i + 14)$

35. $\displaystyle\sum_{i=1}^{100} 5i$

36. $\displaystyle\sum_{i=1}^{100} 3i$

37. $\displaystyle\sum_{i=1}^{50} -2i$

38. $\displaystyle\sum_{i=1}^{50} -4i$

39. $\displaystyle\sum_{i=3}^{12} (-0.1i + 1)$

40. $\displaystyle\sum_{i=4}^{20} (-0.4i + 2)$

41. Find the sum of the first 60 positive even integers.

42. Find the sum of the first 80 positive even integers.

43. Find the sum of the first 48 positive odd integers.

44. Find the sum of the first 42 positive odd integers.

45. Find the sum of the even integers between 21 and 479.

46. Find the sum of the odd integers between 30 and 210.

Application Problems

47. According to the U.S. Bureau of Economic Analysis, in 1984 U.S. travelers spent $22,208 million in other countries. This amount has increased by approximately $2350 million each year.

   a. Write the general term for the arithmetic sequence modeling the amount spent by U.S. visitors abroad (in millions of dollars), where $n = 1$ corresponds to 1984.

   b. Use the model to predict the amount that U.S. travelers will spend in other countries in the year 2000.

   c. In what year will U.S. travelers spend $97,408 million abroad?

48. According to the U.S. Bureau of Labor Statistics, in 1990 there were 126,424 thousand employees in the United States. This number has increased by approximately 1265 thousand employees each year.

   a. Write the general term for the arithmetic sequence modeling the number of employees (in thousands) in the United States, where $n = 1$ corresponds to 1990.

   b. Use the model to predict the number of employees in the United States in the year 2000.

   c. In what year will there be 152,989 thousand employees?

49. The starting salary for a teacher's assistant is $12,000 with yearly increments of $1150.

   a. Write the general term for the arithmetic sequence modeling the assistant's salary during year n.

   b. How many years will it take to reach a salary of $28,100?

50. An orange tree produces 19 oranges the first year and is expected to produce 14 more oranges each year than it had the preceding year.

   a. Write the general term for the arithmetic sequence modeling the number of oranges produced during year n.

   b. How many oranges can the tree be expected to produce during the 12th year?

51. According to the Environmental Protection Agency, in 1960 the United States recovered 3.78 million tons of solid waste. Due primarily to recycling programs, this amount has increased by approximately 0.576 million tons each year.

   a. Write the general term for the arithmetic sequence modeling the amount of solid waste recovered in the United States, where $n = 1$ corresponds to 1960.

   b. What is the total amount of solid waste recovered from 1960 through 1998?

52. According to the Environmental Protection Agency, in 1960 the United States generated 87.1 million tons of solid waste. This amount has increased by approximately 3.14 million tons each year.

   a. Write the general term for the arithmetic sequence modeling the amount of solid waste generated in the United States, where $n = 1$ corresponds to 1960.

b. What is the total amount of solid waste generated from 1960 through 1998?

53. A company offers a starting yearly salary of $15,000 with raises of $500 per year. Find the total salary over a 10-year period.

54. Suppose you are considering two job offers. You are told that company A will start you at $15,000 a year and guarantee you a raise of $500 each year and company B will start you at $16,000 a year but will only guarantee you a raise of $400 each year. Over a 10-year period, which company will pay you the greater total amount?

55. A *degree-day* is a unit used to measure the fuel requirements of buildings. By definition, each degree that the average daily temperature is below 65°F is 1 degree-day. For example, a temperature of 42°F constitutes 23 degree-days. If the average temperature on January 1 was 42°F and fell 2°F for each subsequent day up to and including January 10, how many degree-days are included from January 1 to January 10?

56. The model $a_n = 69.3n + 580$ describes the annual expenditures in billion of dollars of the federal government beginning with the year 1980, where $n = 0$ corresponds to 1980. Find $\sum_{n=0}^{20} a_n$ and describe what this number represents in practical terms.

True–False Critical Thinking Problems

57. Which one of the following is true?

a. The sequence given by $\log 2$, $\log 4$, $\log 8$, $\log 16$, $\log 32$, ... is arithmetic.

b. Two different arithmetic sequences cannot have the same fifth term.

c. The sum of an arithmetic sequence can never be negative.

d. It is not possible to find the 51st term of an arithmetic sequence without knowing all 50 of the preceding terms.

58. Which one of the following is true?

a. The sequence 3, 6, 3, 0, 3, 6, 3, 0, ... is an arithmetic sequence.

b. A sequence that follows a pattern involving addition is an arithmetic sequence.

c. If the first term of an arithmetic sequence is -3 and the third term is -2, then the fourth term is -1.

d. In any arithmetic sequence, we can write as many terms as we want by adding the same number repeatedly.

59. Which one of the following is true?

a. An arithmetic sequence is the indicated sum of an arithmetic series.

b. Any series that can be expressed in sigma notation, such as $\sum_{i=1}^{5} (2i + 3)$, must be an arithmetic series.

c. The sum of the even integers from 10 through 30 inclusively is $\frac{20}{2}(10 + 30)$.

d. The formula $\frac{n(n + 1)}{2}$ describes the sum of the first n natural numbers.

60. Which one of the following is true?

a. The common difference for the arithmetic sequence given by 1, -1, -3, -5, ... is 2.

b. The sequence 1, 4, 8, 13, 19, 26, ... is an arithmetic sequence.

c. The nth term of an arithmetic sequence whose first term is a_1 and whose common difference is d is $a_n = a_1 + nd$.

d. If the first term of an arithmetic sequence is 5 and the third term is -3, then the fourth term is -7.

Technology Problems

61. One way to graph a sequence using a graphing utility is to set the mode to the DOT mode rather than the connected mode and use an integer viewing rectangle. For example, the arithmetic sequence

    10, 14.5, 19, 23.5, ...

whose general term is $a_n = 4.5n + 5.5$ can be graphed by entering $y_1 = 4.5x + 5.5$ in the DOT mode. You could then use the ⊞TRACE⊞ feature to find a particular term. For example, the ⊞TRACE⊞ feature indicates that if $x = 26$, then $y = 122.5$, so the 26th term of the sequence is 122.5. Use this technique to graph the arithmetic sequences whose general term you found in Problems 7–20. Then trace along the graph of the sequence and verify the indicated term that you found.

62. Use the capability of a graphing utility to calculate the sum of a sequence to verify your answers to Problems 27–40.

Writing in Mathematics

63. Give examples of two different arithmetic sequences with $a_4 = 10$. Describe how you obtained each sequence.

64. Consider the following finite arithmetic sequence:

1, 2, 3, 4, 5, 6, 7, 8, 9.

Underline some of the terms, and leave the rest not underlined. (Leave at least four terms that are not underlined.)

a. Consider for example, 1, $\underline{2}$, 3, $\underline{4}$, 5, 6, $\underline{7}$, 8, 9.

Can you find three of the not-underlined values that will form a finite arithmetic sequence?

b. Repeat part (a) for 1, $\underline{2}$, 3, $\underline{4}$, $\underline{5}$, 6, $\underline{7}$, 8, 9.

c. What always seems to be true for three of the underlined or three of the not-underlined values in the sequence 1, 2, 3, ... , 9? Create further examples of this observation. (For more details on this sequence, see "Ramsey Theory" by Graham and Spencer in the July 1990 issue of *Scientific American*.)

Critical Thinking Problems

65. Find the sum of the first 20 terms of an arithmetic sequence whose first term is 4 and whose tenth term is 31.

66. A person wants to save $40,000 over a 16-year period. If $1600 is saved the first year and each year the amount saved is to be increased by a fixed sum, what should this fixed sum be?

67. The sum of the first n terms of an arithmetic sequence is $n^2 + 3n$. Find the 50th term of the sequence.

68. Twelve consecutive integers are added. Find the remainder when the sum is divided by 4.

69. Prove that the sum of the first n consecutive positive integers plus n^2 is equal to the sum of the next n consecutive integers.

Group Activity Problem

70. Members of your group have been hired by the Environmental Protection Agency to write a report on whether we are making significant progress in recovering solid waste. Use the models from Problems 51 and 52 as the basis for your report. A graph of each model from 1960 through 2000 would be helpful. What percent of solid waste generated is actually recovered on a year-to-year basis? Be as creative as you want in your report and then draw conclusions. The group should actually write up the report and perhaps even include suggestions as to how we might improve recycling progress.

SECTION 7.3 Geometric Sequences and Series

Solutions   Tutorial   Video
Manual   12

Objectives

1 Find the common ratio for a geometric sequence.
2 Find a specified term for a geometric sequence.
3 Use geometric sequences to model applied situations.
4 Find partial sums of geometric sequences.
5 Use sums of geometric sequences to solve applied problems.
6 Find sums of certain infinite geometric series.

Geometric Sequences

Table 7.2 on page 786 contains two sequences. The sequence for total food production capacity is an arithmetic sequence; each term is 50,000 more than the previous term. However, the sequence for total population does not appear to be arithmetic. Each year the population is increasing by 7% of what it had been during the previous year. The sequence in the table representing total population is an example of a geometric sequence. In such a sequence, each term after the first is obtained by *multiplying* by a constant, in this case 1.07.

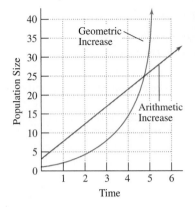

Figure 7.4

Arithmetic and geometric growth curves. The geometric or exponential curve may begin slowly, but will overtake the steadily increasing arithmetic curve.

TABLE 7.2 Population and Food Production Sequences		
Time	**Total Population**	**Total Food Production Capacity**
Beginning year	100,000	100,000
1	107,000	150,000
2	114,490	200,000
3	122,504	250,000
13	240,984	750,000
25	542,743	1,350,000
40	1,497,446	2,100,000
46	2,247,262	2,400,000
47	2,404,571	2,450,000
48	2,572,891	2,500,000

Economist Thomas Malthus (1766–1834) predicted that population growth increases in a geometric sequence and food production increases only in an arithmetic sequence. He concluded that eventually a point would be reached where the world's population would exceed the world's food production. If two sequences, one geometric and one arithmetic, are increasing, the geometric sequence will eventually overtake the arithmetic sequence, regardless of the head start that the arithmetic sequence might initially have (see Figure 7.4). In short, according to Malthus, unchecked population growth will eventually lead to mass starvation. The numbers in Table 7.2 indicate that this would occur during the 48th year.

The sequence for total food production capacity in Table 7.2 is arithmetic because each term after the first is obtained by adding a constant amount—50,000—to the preceding term. Let's contrast this with the definition of a geometric sequence.

Study tip

Each term after the first in a geometric sequence is obtained by multiplying the preceding term by a nonzero constant.

Definition of a geometric sequence

A sequence $a_1, a_2, a_3, a_4, \ldots, a_n, \ldots$ is a *geometric sequence* or *geometric progression* if there is a number $r \neq 0$, called the *common ratio*, such that

$$\frac{a_2}{a_1} = r, \frac{a_3}{a_2} = r, \frac{a_4}{a_3} = r, \cdots, \frac{a_{n+1}}{a_n} = r$$

and so on.

1 Find the common ratio for a geometric sequence.

EXAMPLE 1 Finding the Common Ratio of Geometric Sequences

Identify the common ratio for the geometric sequence:

a. $1, 3, 9, 27, \ldots$
b. $-64, 16, -4, 1, \ldots$

iscover for yourself

Confirm each common ratio shown on the right by computing

Third term
Second term

and

Fourth term
Third term

for each of the three geometric sequences.

tudy tip

When the common ratio of a geometric sequence is negative ($r < 0$), the signs of the terms alternate.

Solution

The common ratio can be obtained by dividing any term by its predecessor. We will divide the second term by the first term.

Sequence	Second term / First term	Common Ratio
a. $1, 3, 9, 27, \ldots$	$\dfrac{3}{1} = 3$	3
b. $-64, 16, -4, 1, \ldots$	$\dfrac{16}{-64} = -\dfrac{1}{4}$	$-\dfrac{1}{4}$

■

We can reverse the direction of Example 1 and write the first four terms of a geometric sequence using the first term and the common ratio. We start with the first term and multiply each successive term by the common ratio. Thus, the first four terms of the geometric sequence with first term 12 and common ratio $-\dfrac{1}{3}$ are

$$12, \quad 12\left(-\frac{1}{3}\right) \text{ or } -4, \quad -4\left(-\frac{1}{3}\right) \text{ or } \frac{4}{3}, \quad \frac{4}{3}\left(-\frac{1}{3}\right) \text{ or } -\frac{4}{9}.$$

Just as we can express the general term (or the nth term), a_n, of an arithmetic sequence in terms of the first term a_1 and the common difference d, the general term of a geometric sequence can be written in terms of the first term a_1 and the common ratio, designated by r $(r \neq 0)$. We start with the first term and multiply each successive term by r as follows:

$$\underset{\substack{a_1 \\ \text{First} \\ \text{term}}}{a_1}, \quad \underset{\substack{a_2 \\ \text{Second} \\ \text{term}}}{a_1 r}, \quad \underset{\substack{a_3 \\ \text{Third} \\ \text{term}}}{a_1 r^2}, \quad \underset{\substack{a_4 \\ \text{Fourth} \\ \text{term}}}{a_1 r^3}, \quad \underset{\substack{a_5 \\ \text{Fifth} \\ \text{term}}}{a_1 r^4}, \quad \underset{\substack{a_6 \\ \text{Sixth} \\ \text{term}}}{a_1 r^5}, \quad \ldots, \quad \underset{\substack{a_n \\ \text{General} \\ (n\text{th}) \text{ term}}}{?}$$

iscover for yourself

Look at the pattern for the first six terms of a geometric sequence whose first term is a_1 and whose common ratio is r. What relationship do you observe between the exponent on r and the subscript of a denoting the term number? Use this relationship to write the missing formula for the nth term.

In the Discover for yourself, did you notice that the exponent on r is 1 less than the subscript of a denoting the term number? Thus, the formula for the nth term is

$$a_n = a_1 r^{n-1}.$$

1 less than the
subscript of a

> **General term of a geometric sequence**
>
> The nth term (the general term) of a geometric sequence with first term a_1 and common ratio r is
>
> $$a_n = a_1 r^{n-1}.$$

Examples 2 through 4 illustrate the use of this formula.

2 Find a specified term for a geometric sequence.

EXAMPLE 2 **Finding a Specified Term of a Geometric Sequence**

Find the general term and the eighth term, a_8, of the geometric sequence:

$24, -36, 54, -81, \dots$.

Solution

To find the general term of a geometric sequence, we need the first term and the common ratio. The first term, a_1, is 24. The common ratio, r, is found by dividing any term by its predecessor. Thus,

$$r = \frac{-36}{24} = -\frac{3}{2}.$$

Now we use the formula for the general term of a geometric sequence.

$$a_n = a_1 r^{n-1}$$

Substituting 24 for a_1 and $-\frac{3}{2}$ for r gives the general term of the sequence.

$$a_n = 24\left(-\frac{3}{2}\right)^{n-1}$$

The eighth term of the sequence is found by substituting 8 for n.

$$a_8 = 24\left(-\frac{3}{2}\right)^{8-1} = 24\left(-\frac{3}{2}\right)^7 = 24\left(-\frac{2187}{128}\right) = -\frac{6561}{16}$$

The eighth term of the geometric sequence is $-\frac{6561}{16}$ (or $-410\frac{1}{16}$). ∎

Example 3 shows us that the first term of a geometric sequence can be determined if other terms in the sequence are known.

EXAMPLE 3 **Finding the First Term of a Geometric Sequence Using Two Known Terms**

If the third term of a geometric sequence is 27 and the fifth term is 243, find the first term of the sequence.

Solution

$$a_n = a_1 r^{n-1} \qquad a_n = a_1 r^{n-1} \qquad \text{Use the formula for the } n\text{th term}$$
$$\text{of a geometric sequence twice to find } r.$$

$$a_3 = a_1 r^{3-1} \qquad a_5 = a_1 r^{5-1} \qquad \text{Let } a_3 = 27 \text{ and } a_5 = 243.$$

$$27 = a_1 r^2 \qquad 243 = a_1 r^4$$

S tudy tip

Be careful with the order of operations when evaluating

$a_1 r^{n-1}$.

First find r^{n-1}. Then multiply the result by a_1.

$$\frac{a_1 r^4}{a_1 r^2} = \frac{243}{27}$$ 〔Solve for r by dividing the two equations.〕

$$r^2 = 9$$ 〔Divide the numerator and denominator by a_1.〕

$$r = \pm 3$$

$$27 = a_1 r^2$$ 〔Solve for a_1 using $27 = a_1 r^2$ (or $243 = a_1 r^4$).〕

$$27 = a_1 (\pm 3)^2$$ 〔Substitute the values that we obtained for r.〕

$$27 = 9a_1$$ 〔$(\pm 3)^2 = 9$〕

$$a_1 = 3$$ 〔Solve for a_1.〕

Thus, the first term of the geometric sequence is 3. Because $r = \pm 3$, two sequences are possible: 3, 9, 27, 81, 243, ... and 3, −9, 27, −81, 243, ... In both sequences, $a_3 = 27$ and $a_5 = 243$. ∎

3 Use geometric sequences to model applied situations.

Sandy Skoglund "The Green House" © 1990 Sandy Skoglund.

EXAMPLE 4 Modeling Population Growth

In 1798, English economist Thomas Malthus claimed that populations increase in a geometric sequence. The population of Florida from 1980 through 1987 is shown in the following table.

Year	1980	1981	1982	1983	1984	1985	1986	1987
Population (millions)	9.75	10.03	10.32	10.62	10.93	11.25	11.59	11.91

a. Show that the population is increasing geometrically.
b. Write the general term for the geometric sequence modeling population growth for Florida, where $n = 1$ corresponds to 1980.
c. Estimate Florida's population (in millions) for the year 2000.

Solution

a. First, we divide the population for each year by the population in the preceding year.

$$\frac{10.03}{9.75} \approx 1.029, \quad \frac{10.32}{10.03} \approx 1.029, \quad \frac{10.62}{10.32} \approx 1.029$$

Continuing in this manner, we will keep getting approximately 1.029, which means that the population is increasing geometrically with $r \approx 1.029$. In this situation, the common ratio is the growth rate, indicating that the population of Florida in any year shown in the table is approximately 1.029 times the population the year before.

b. Since the population is increasing geometrically, we can use the formula for the general term of a geometric sequence to model the data. Using $a_1 = 9.75$ (the 1980 population) and $r = 1.029$, we have

$$a_n = a_1 r^{n-1}$$ 〔Use the formula for the nth term of a geometric sequence.〕

$$a_n = 9.75(1.029)^{n-1}$$ 〔Substitute the given values.〕

The general term for the geometric sequence modeling Florida's population growth is

Discover for yourself

Use Florida's population model to find the state's population one year ago. Now consult the research department of your library or the Internet to find the actual population. How accurate is the model?

$$a_n = 9.75(1.029)^{n-1}$$

where $n = 1$ corresponds to 1980.

c. We can use the model from part (b) to predict Florida's population (in millions) for the year 2000. We substitute 21 for n.

$$a_{21} = 9.75(1.029)^{21-1} \approx 17.27$$

The model predicts a population of 17.27 million in the year 2000. ■

4 Find partial sums of geometric sequences.

Finite Geometric Series

Now let's consider the sum of the first n terms of a geometric sequence. This sum, denoted by S_n, is called a *geometric series*. Recall that the first n terms of a geometric sequence are

$$a_1, a_1 r, a_1 r^2, \ldots, a_1 r^{n-2}, a_1 r^{n-1}.$$

We proceed as follows:

$$S_n = a_1 + a_1 r + a_1 r^2 + \cdots + a_1 r^{n-2} + a_1 r^{n-1} \qquad \text{S_n is the sum of the first n terms of the sequence.}$$

$$rS_n = a_1 r + a_1 r^2 + a_1 r^3 + \cdots + a_1 r^{n-1} + a_1 r^n \qquad \text{Multiply both sides of the equation by r.}$$

$$S_n - rS_n = a_1 - a_1 r^n \qquad \text{Subtract the second equation from the first equation.}$$

$$S_n(1 - r) = a_1(1 - r^n) \qquad \text{Factor out S_n on the left and a_1 on the right.}$$

$$S_n = \frac{a_1(1 - r^n)}{1 - r} \qquad \text{Solve for S_n by dividing both sides by $1 - r$ (assuming that $r \neq 1$).}$$

We have proved the following result.

study tip

If the common ratio is 1, the geometric sequence is

$$a_1, a_1, a_1, a_1, \ldots$$

The sum of the first n terms of this sequence is na_1:

$$S_n = \underbrace{a_1 + a_1 + a_1 + \cdots + a_1}_{\text{There are n terms.}}$$

$$= na_1.$$

The nth partial sum of a geometric sequence

The sum, S_n, of the first n terms of a geometric sequence is given by

$$S_n = \frac{a_1(1 - r^n)}{1 - r}$$

in which a_1 is the first term and r is the common ratio ($r \neq 1$).

The following examples illustrate the application of this formula.

EXAMPLE 5 **Finding the 18th Partial Sum of a Geometric Sequence**

Find the 18th partial sum of the geometric sequence:

$$50, 50(1.04), 50(1.04)^2, 50(1.04)^3, \ldots$$

Solution

Before finding the 18th partial sum of this sequence, we must find the common ratio.

$$r = \frac{a_2}{a_1} = \frac{50(1.04)}{50} = 1.04$$

Now we are ready to find the sum of the first 18 terms.

$$S_n = \frac{a_1(1 - r^n)}{1 - r}$$ Use the formula for the sum of the first n terms of a geometric sequence.

$$S_{18} = \frac{50[1 - (1.04)^{18}]}{1 - 1.04}$$ a_1 (the first term) $= 50$, $r = 1.04$, and $n = 18$ because we want the sum of the first 18 terms.

$$\approx 1282.27$$ Use a calculator.

To the nearest hundredth, the sum of the first 18 terms is 1282.27. ∎

EXAMPLE 6 **Using S_n to Evaluate a Summation**

Evaluate: $\displaystyle\sum_{i=3}^{10} 6\left(-\frac{1}{2}\right)^i$

Solution

It might help to write out a few terms. We have

$$\sum_{i=3}^{10} 6\left(-\frac{1}{2}\right)^i = 6\left(-\frac{1}{2}\right)^3 + 6\left(-\frac{1}{2}\right)^4 + 6\left(-\frac{1}{2}\right)^5 + \cdots + 6\left(-\frac{1}{2}\right)^{10}.$$

The first term, a_1, is $6\left(-\frac{1}{2}\right)^3$. The common ratio, r, is $-\frac{1}{2}$. Notice that by starting with $i = 3$ and ending with $i = 10$, there are *eight* terms in the sum, so $n = 8$. Thus,

$$\sum_{i=3}^{10} 6\left(-\frac{1}{2}\right)^i = \frac{a_1(1 - r^n)}{1 - r}$$ Use the formula for the sum of the first n terms of a geometric series.

$$= \frac{6\left(-\frac{1}{2}\right)^3\left[1 - \left(-\frac{1}{2}\right)^8\right]}{1 - \left(-\frac{1}{2}\right)}$$ a_1 (the first term) $= 6\left(-\frac{1}{2}\right)^3$, r (the common ratio) $= -\frac{1}{2}$, and $n = 8$ because we are adding eight terms.

$$\approx -0.498$$

Thus,

$$\sum_{i=3}^{10} 6\left(-\frac{1}{2}\right)^i \approx -0.498.$$ ∎

5 Use sums of geometric sequences to solve applied problems.

Modeling Using Partial Sums of Geometric Sequences

Some of the problems in the previous problem set involved situations in which salaries increase by a fixed amount each year. A more realistic situation is one in which salary raises amount to a certain percent each year. Example 7 shows how such a situation can be modeled using a geometric series.

U.S. Bureau of Engraving and Printing

EXAMPLE 7   **Computing a Lifetime Salary**

A union contract specifies that workers shall receive an 8% pay increase each year for the next 30 years. Workers are paid $20,000 the first year. What is the total lifetime salary over a 30-year period?

Solution

The salary for the first year is $20,000. With an 8% raise, the second-year salary is computed as follows:

Salary for year 2 = 20,000 + 20,000(0.08) = 20,000(1.08)

Each year, the salary is 1.08 times what it was in the previous year. Thus, the salary for year 3 is 1.08 times 20,000(1.08), or $20,000(1.08)^2$. The salaries for the first five years are given in the table.

Yearly Salaries

Year 1	Year 2	Year 3	Year 4	Year 5	
20,000	20,000(1.08)	$20,000(1.08)^2$	$20,000(1.08)^3$	$20,000(1.08)^4$	...

The numbers in the second row form a geometric sequence with $a_1 = 20,000$ and $r = 1.08$. To find the total salary over 30 years, we use the formula for the nth partial sum of a geometric sequence, with $n = 30$.

$$S_n = \frac{a_1(1 - r^n)}{1 - r}$$

$$\text{Total salary over 30 years} = \frac{20,000(1 - (1.08)^{30})}{1 - 1.08}$$

$$= \frac{20,000(1 - (1.08)^{30})}{-0.08}$$

$$\approx 2,265,664 \qquad \text{Use a calculator.}$$

The total salary over the 30-year period is approximately $2,265,664.   ∎

An important application of the formula for the nth partial sum of a geometric sequence is in projecting the value of an annuity. An *annuity* is a sequence of equal payments made at equal time periods. The payments can be investments or loan payments.

EXAMPLE 8   **Value of an Annuity**

A deposit of $1000 is made the first day of each month in an account that pays 6% annual interest compounded monthly. What is the value of this annuity at the end of three years?

Solution

To solve this problem, we must use the formula for compound interest from Chapter 5.

$$A = P\left(1 + \frac{r}{n}\right)^{nt}$$

A = the balance in the account
P = the initial deposit ($1000)
r = the annual interest rate (0.06)
n = the number of compounding periods per year
(12, with monthly compounding)
t = the time in years

Since $t = 3$ years $= 36$ months, we can consider each of the 36 deposits separately. The first deposit of $1000 will earn interest for 36 months, so its balance will be

$$A_{36} = 1000\left(1 + \frac{0.06}{12}\right)^{12 \cdot 3}$$

$$= 1000(1.005)^{36}$$

The second deposit of $1000 will earn interest for 35 months, and amounts to

$$A_{35} = 1000\left(1 + \frac{0.06}{12}\right)^{35}$$

$$= 1000(1.005)^{35}$$

Continuing in this manner, the last (36th) deposit will earn interest for only one month, and its balance will be

$$A_1 = 1000\left(1 + \frac{0.06}{12}\right)$$

$$= 1000(1.005)$$

The total balance in the account at the end of three years is the sum of the balances of the 36 deposits.

$$S_{36} = A_1 + A_2 + A_3 + \cdots + A_{35} + A_{36}$$

$$= 1000(1.005) + 1000(1.005)^2 + 1000(1.005)^3 + \cdots$$

$$+ 1000(1.005)^{35} + 1000(1.005)^{36}$$

$$= \frac{a_1(1 - r^n)}{1 - r}$$

The series is geometric with $r = 1.005$, so use the formula for the sum of the first n terms of a geometric series.

$$= \frac{1000(1.005)(1 - 1.005^{36})}{1 - 1.005}$$

$a_1 = 1000(1.005)$, $r = 1.005$, and $n = 36$ because we are adding 36 terms.

$$\approx 39,533$$

The value of the annuity at the end of three years is approximately $39,533. ■

Barton Benes "Stretched Dollar Bill." Photo by Karen Furth.

6 Find sums of certain infinite geometric series.

Infinite Geometric Series

In some cases, it is possible to find the sum of an infinite number of terms of a geometric sequence. Let's see how by investigating the sum

$$\frac{1}{2} + \frac{1}{4} + \frac{1}{8} + \frac{1}{16} + \frac{1}{32} + \frac{1}{64} + \cdots$$

Here are a few partial sums:

$$\frac{1}{2} + \frac{1}{4} = \frac{3}{4} = 0.75$$

$$\frac{1}{2} + \frac{1}{4} + \frac{1}{8} = \frac{7}{8} = 0.875$$

$$\frac{1}{2} + \frac{1}{4} + \frac{1}{8} + \frac{1}{16} = \frac{15}{16} = 0.9375$$

$$\frac{1}{2} + \frac{1}{4} + \frac{1}{8} + \frac{1}{16} + \frac{1}{32} = \frac{31}{32} = 0.96875$$

$$\frac{1}{2} + \frac{1}{4} + \frac{1}{8} + \frac{1}{16} + \frac{1}{32} + \frac{1}{64} = \frac{63}{64} = 0.984375$$

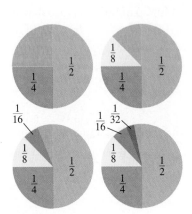

Figure 7.5

The sum
$\frac{1}{2} + \frac{1}{4} + \frac{1}{8} + \frac{1}{16} + \frac{1}{32} + \cdots$
is approaching 1.

It appears that these partial sums, or the *sequence of partial sums*,

$$0.75, 0.875, 0.9375, 0.96875, 0.984375, \ldots$$

is approaching 1. Notice how this is modeled in Figure 7.5. As more terms are included, the sum

$$\frac{1}{2} + \frac{1}{4} + \frac{1}{8} + \frac{1}{16} + \frac{1}{32} + \cdots$$

is approaching the area of one complete circle.

One possible way to algebraically investigate this infinite sum is to write a formula for the sum of the first n terms and then consider what will happen to this expression as n gets larger and larger.

$$S_n = \frac{a_1(1 - r^n)}{1 - r} \qquad \text{Use the formula for the } n\text{th partial sum.}$$

$$= \frac{\frac{1}{2}\left(1 - \left(\frac{1}{2}\right)^n\right)}{1 - \frac{1}{2}} \qquad \text{For the sum } \frac{1}{2} + \frac{1}{4} + \frac{1}{8} + \frac{1}{16} + \cdots, a_1 = \frac{1}{2} \text{ and } r, \text{ the common ratio, is also } \frac{1}{2}.$$

$$= \frac{\frac{1}{2}\left(1 - \left(\frac{1}{2}\right)^n\right)}{\frac{1}{2}} \qquad \text{Subtract in the denominator.}$$

$$= 1 - \left(\frac{1}{2}\right)^n \qquad \text{Cancel factors of } \frac{1}{2} \text{ in the numerator and the denominator.}$$

iscover for yourself

What happens to $\left(\frac{1}{2}\right)^n$ as n gets larger and larger? Use your calculator and find $\left(\frac{1}{2}\right)^5$, $\left(\frac{1}{2}\right)^{100}$, $\left(\frac{1}{2}\right)^{1000}$, and so on. What does this mean in terms of the expression for S_n as n gets larger and larger?

In the Discover for yourself, did you observe that as n gets larger, the term $\left(\frac{1}{2}\right)^n$ gets closer to 0? The expression for S_n is getting closer and closer to 1.

$$S_n = 1 - \left(\frac{1}{2}\right)^n \qquad \text{Approximately 0} \qquad \text{For a very large } n, \left(\frac{1}{2}\right)^n \text{ is approximately 0.}$$

$$\approx 1 \qquad\qquad\qquad \text{As we continue to add more terms, our sum approaches 1.}$$

Consequently, we define 1 to be the sum of the infinite geometric series. Thus,

$$\frac{1}{2} + \frac{1}{4} + \frac{1}{8} + \frac{1}{16} + \frac{1}{32} + \frac{1}{64} + \cdots = 1.$$

There are, of course, some infinite geometric series to which we cannot assign a sum, such as

$$2 + 4 + 8 + 16 + 32 + 64 + \cdots .$$

As more and more terms are added, the sum is getting larger and larger without bound.

An infinite series of the form

$$a_1 + a_1 r + a_1 r^2 + a_1 r^3 + a_1 r^4 + \cdots + a_1 r^{n-1} + \cdots$$

is an *infinite geometric series*. We can determine which infinite geometric series have sums and which do not by looking at what happens to r^n as n gets larger in the formula for the nth partial sum of this series, namely

$$S_n = \frac{a_1(1 - r^n)}{1 - r}.$$

iscover for yourself

Write the second, third, fourth, and fifth partial sums of

$$2 + 4 + 8 + 32 + 64 + \cdots .$$

Write these partial sums as a sequence. State how this sequence of partial sums differs from the sequence

0.75, 0.875, 0.9375, 0.96875, 0.984375, …

representing the partial sums of

$$\frac{1}{2} + \frac{1}{4} + \frac{1}{8} + \frac{1}{32} + \frac{1}{64} + \cdots .$$

iscover for yourself

If r is any number between -1 and 1, what happens to r^n as n gets larger and larger? Use your calculator to answer this question by selecting values between -1 and 1 and find larger and larger powers.

In the Discover for yourself, did you see that if $-1 < r < 1$ (equivalently, $|r| < 1$) and n is large, r^n is close to 0? Thus, for the sum of an infinite geometric series, the expression r^n approaches 0 when $|r| < 1$, and S_n approaches $\dfrac{a_1}{1 - r}$.

Formula for the sum of an infinite geometric series

If $|r| < 1$, then the sum of the infinite geometric series

$$a_1 + a_1 r + a_1 r^2 + a_1 r^3 + \cdots$$

Study tip

Here's another way to express the information in the box on the right:

$$\sum_{n=1}^{\infty} a_1 r^{n-1} = \frac{a_1}{1-r}, |r| < 1.$$

in which a_1 is the first term and r is the common ratio is given by

$$S_\infty = \frac{a_1}{1-r}.$$

If $|r| \geqslant 1$, the infinite series does not have a sum.

EXAMPLE 9   **Finding the Sum of an Infinite Geometric Series**

Find the sum of the infinite series:

$$\sum_{n=1}^{\infty} 3\left(\frac{1}{2}\right)^{n-1} = 3 + \frac{3}{2} + \frac{3}{2^2} + \frac{3}{2^3} + \frac{3}{2^4} + \cdots$$

Discover for yourself

Before reading the algebraic solution, investigate this sum numerically with the help of your calculator. Find each of the following partial sums:

$$3 + \frac{3}{2}, 3 + \frac{3}{2} + \frac{3}{2^2}, 3 + \frac{3}{2} + \frac{3}{2^2} + \frac{3}{2^3}, 3 + \frac{3}{2} + \frac{3}{2^2} + \frac{3}{2^3} + \frac{3}{2^4}, 3 + \frac{3}{2} + \frac{3}{2^2} + \frac{3}{2^3} + \frac{3}{2^4} + \frac{3}{2^5}$$

Continue computing more partial sums until it is clear what

$$3 + \frac{3}{2} + \frac{3}{2^2} + \frac{3}{2^3} + \frac{3}{2^4} + \frac{3}{2^5} + \cdots$$

is approaching.

Study tip

Some problems, such as finding the sum

$$3 + \frac{3}{2} + \frac{3}{2^2} + \frac{3}{2^3} + \frac{3}{2^4} + \cdots$$

can be investigated numerically (see the Discover for yourself), algebraically (see the solution on the right) and geometrically (see the discussion that follows this solution).

Solution

We can find the sum algebraically by using the formula for the sum of an infinite geometric series.

$$S_\infty = \sum_{n=1}^{\infty} 3\left(\frac{1}{2}\right)^{n-1}$$

$$= \frac{a_1}{1-r} \qquad \text{Use the formula for the sum of an infinite geometric series.}$$

$$= \frac{3}{1 - \frac{1}{2}} \qquad a_1 \text{ (the first term)} = 3 \text{ and } r \text{ (the common ratio)} = \frac{1}{2}.$$

$$= 6$$

Thus, the sum of the given infinite geometric series is 6. Put in an informal way, as we continue adding more and more terms, the sum is approximately 6.

∎

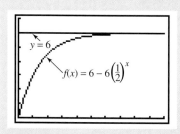

Figure 7.6

The function f representing partial sums approaches 6, the sum of the infinite series.

We can visualize

$$3 + \frac{3}{2} + \frac{3}{2^2} + \frac{3}{2^3} + \frac{3}{2^4} + \cdots = 6$$

by writing the formula for the partial sum as a function of x, namely

$$f(x) = \frac{a_1(1 - r^x)}{1 - r} = \frac{a_1}{1 - r} - \frac{a_1 r^x}{1 - r}.$$

With $a_1 = 3$ and $r = \frac{1}{2}$, the partial sum function is

$$f(x) = \frac{3}{1 - \frac{1}{2}} - \frac{3\left(\frac{1}{2}\right)^x}{1 - \frac{1}{2}} = 6 - 6\left(\frac{1}{2}\right)^x.$$

Figure 7.6 shows the graph of $f(x)$ and $y = 6$. The line $y = 6$ is a horizontal asymptote for the graph f. Since the graph of S_n consists of points on the graph of f, we see that $S_n \to 6$ as $n \to \infty$. Therefore, the sum of the series in Example 9 is 6.

We can use the formula for the sum of an infinite series to express a repeating decimal as a fraction in lowest terms.

EXAMPLE 10 **Writing a Repeating Decimal as a Fraction**

Express as a fraction in lowest terms: $3.2\overline{47}$

Solution

We observe that $3.2\overline{47} = 3.247474747\ldots$, so we begin by separating the repeating part of the decimal from the nonrepeating part.

$$3.247474747\ldots = 3.2 + 0.047 + 0.00047 + 0.0000047 + 0.000000047 + \cdots$$

$$= \frac{32}{10} + \frac{47}{1000} + \frac{47}{100,000} + \frac{47}{10,000,000} + \frac{47}{1,000,000,000} + \cdots$$

$$= \frac{32}{10} + \frac{\dfrac{47}{1000}}{1 - \dfrac{1}{100}} \qquad \text{The terms of the series after the first term form an infinite geometric series with } a_1 = \frac{47}{1000} \text{ and } r = \frac{1}{100}. \text{ Use } S_\infty = \frac{a_1}{1 - r}.$$

$$= \frac{32}{10} + \frac{47}{1000} \cdot \frac{100}{99}$$

$$= \frac{32}{10} + \frac{47}{990} = \frac{3168}{990} + \frac{47}{990} = \frac{3215}{990}$$

An equivalent fraction for $3.2\overline{47}$ is $\frac{3215}{990}$. ∎

ENRICHMENT ESSAY

Off to Infinity

The sum of the infinite series

$$\frac{1}{2} + \frac{1}{4} + \frac{1}{8} + \frac{1}{16} + \frac{1}{32} + \frac{1}{64} + \cdots$$

in which $a_1 = \frac{1}{2}$ (the first term is $\frac{1}{2}$) and $r = \frac{1}{2}$ (the common ratio is $\frac{1}{2}$) is

$$S_\infty = \frac{a_1}{1-r} = \frac{\frac{1}{2}}{1-\frac{1}{2}} = \frac{\frac{1}{2}}{\frac{1}{2}} = 1.$$

As more terms are calculated, the sum gets closer and closer to 1.

Over the course of centuries, mathematicians have tried to use infinite sums to calculate special numbers, including π, e, and $\log 2$. These include:

$$1 - \frac{1}{3} + \frac{1}{5} - \frac{1}{7} + \frac{1}{9} - \frac{1}{11} + - \cdots = \frac{\pi}{4}$$

$$\frac{1}{1^2} + \frac{1}{2^2} + \frac{1}{3^2} + \frac{1}{4^2} + \frac{1}{5^2} + \frac{1}{6^2} + \cdots = \frac{\pi^2}{6}$$

$$\frac{1}{1!} + \frac{1}{2!} + \frac{1}{3!} + \frac{1}{4!} + \frac{1}{5!} + \frac{1}{6!} + \cdots = e$$

$$1 - \frac{1}{2} + \frac{1}{3} - \frac{1}{4} + \frac{1}{5} - \frac{1}{6} + - \cdots = \log 2$$

Check these equations by calculating the value of the expressions on the left for increasing numbers of terms, verifying that the sums get closer and closer to the special numbers on the right.

Richard Megna /
Fundamental Photographs

Modeling with Infinite Geometric Series

Infinite sums can model numerous situations, as illustrated in Example 11.

EXAMPLE 11 Tax Rebates and the Multiplier Effect

A tax rebate that returns a certain amount of money to taxpayers can have a total effect on the economy that is many times this amount. In economics this phenomenon is called the *multiplier effect*. Suppose, for example, that the government reduces taxes so that each consumer has $2000 more income. The government feels that each person will spend 70% of this (= $1400). The individuals and businesses receiving this $1400 in turn spend 70% of it (= $980), creating extra income for yet other people to spend, and so on. Determine the total amount spent on consumer goods from the initial $2000 tax rebate.

$1400

70%
is spent.

$980

70%
is spent.

$686

Solution

The total amount spent is given by the infinite geometric series

$$1400 + 980 + 686 + \cdots$$

in which $a_1 = 1400$ and $r = 0.7$. Using our formula for the sum of an infinite geometric series, we obtain

$$S_\infty = \frac{a_1}{1 - r} = \frac{1400}{1 - 0.7} \approx 4667.$$

This means that the total amount spent on consumer goods from the initial $2000 rebate is approximately $4667. ∎

PROBLEM SET 7.3

Practice Problems

In Problems 1–6, write the first five terms of each geometric sequence with the given first term and common ratio.

1. $a_1 = 20$, $r = \dfrac{1}{2}$

2. $a_1 = 24$, $r = \dfrac{1}{3}$

3. $a_1 = -\dfrac{1}{3}$, $r = -3$

4. $a_1 = -\dfrac{1}{20}$, $r = -5$

5. $a_1 = \dfrac{x^2}{y}$, $r = \dfrac{2y}{x}$

6. $a_1 = c$, $r = \dfrac{b}{c}$

In Problems 7–20, find the general term (the nth term) and the seventh term for each geometric sequence.

7. $-3, -12, -48, -192, \ldots$

8. $18, -6, 2, -\dfrac{2}{3}, \ldots$

9. $1.5, -3, 6, -12, \ldots$

10. $5, -1, \dfrac{1}{5}, -\dfrac{1}{25}, \ldots$

11. $-2, 2\sqrt{3}, -6, 6\sqrt{3}, \ldots$

12. $-4, 2\sqrt{2}, -2, \sqrt{2}, \ldots$

13. $0.0004, -0.004, 0.04, -0.4, \ldots$

14. $222\dfrac{2}{9}, 22\dfrac{2}{9}, 2\dfrac{2}{9}, \ldots$

15. $a^6, a^5b, a^4b^2, \ldots$

16. $a^7b^6, a^6b^4, a^5b^2, \ldots$

17. $3, 3^{d+1}, 3^{2d+1}, 3^{3d+1}, \ldots$

18. $1, a^{2/7}, a^{4/7}, a^{6/7}, \ldots$

19. $a_1 = -3, r = -2$

20. $a_1 = -4, r = -2$

21. If the third term of a geometric sequence is 28 and the fifth term is 112, find the first term of the sequence.

22. If the third term of a geometric sequence is 54 and the sixth term is -2, find the first term of the sequence.

23. If the third term of a geometric sequence is 4 and the sixth term is $\frac{1}{2}$, find the first term of the sequence.

24. If the fourth term of a geometric sequence is -3 and the sixth term is $-\frac{1}{3}$, find the two possibilities for the first term of the sequence.

Use the formula for the nth partial sum of a geometric sequence to solve Problems 25–30.

25. Find the sum of the first 12 terms of the geometric sequence $2, 6, 18, 54, \ldots$.

26. Find the sum of the first 12 terms of the geometric sequence $3, 6, 12, 24, \ldots$.

27. Find the sum of the first 11 terms of the geometric sequence $3, -6, 12, -24, \ldots$.

28. Find the sum of the first 11 terms of the geometric sequence $4, -12, 36, -108, \ldots$.

29. Find the sum of the first 14 terms of the geometric sequence $-\frac{3}{2}, 3, -6, 12, \ldots$.

30. Find the sum of the first 14 terms of the geometric sequence $-\frac{1}{24}, \frac{1}{12}, -\frac{1}{6}, \frac{1}{3}, \ldots$.

In Problems 31–40, find the indicated sum.

31. $\displaystyle\sum_{i=1}^{8} 3^{i-1}$

32. $\displaystyle\sum_{i=1}^{8} (-2)^{i-1}$

33. $\displaystyle\sum_{n=0}^{6} 64\left(\frac{1}{4}\right)^{n-1}$

34. $\displaystyle\sum_{n=0}^{6} 128\left(-\frac{1}{2}\right)^{n-1}$

35. $\displaystyle\sum_{i=3}^{9} 8\left(-\frac{1}{4}\right)^{i-1}$

36. $\displaystyle\sum_{i=3}^{6} 6\left(-\frac{1}{3}\right)^{i-1}$

37. $\displaystyle\sum_{n=5}^{16} 3\left(\frac{3}{2}\right)^{n}$

38. $\displaystyle\sum_{n=5}^{15} 2\left(\frac{3}{4}\right)^{n}$

39. $\displaystyle\sum_{n=0}^{6} 200(1.08)^{n}$

40. $\displaystyle\sum_{n=0}^{6} 300(1.02)^{n}$

In Problems 41–50, find the sum of each infinite geometric series.

41. $\displaystyle\sum_{n=0}^{\infty} \left(\frac{1}{3}\right)^{n} = 1 + \frac{1}{3} + \frac{1}{9} + \frac{1}{27} + \cdots$

42. $\displaystyle\sum_{n=0}^{\infty} \left(\frac{1}{4}\right)^{n} = 1 + \frac{1}{4} + \frac{1}{16} + \frac{1}{64} + \cdots$

43. $\displaystyle\sum_{n=0}^{\infty} \left(-\frac{1}{2}\right)^{n} = 1 - \frac{1}{2} + \frac{1}{4} - \frac{1}{8} + \cdots$

44. $\displaystyle\sum_{n=0}^{\infty} \left(-\frac{2}{3}\right)^{n} = 1 - \frac{2}{3} + \frac{4}{9} - \frac{8}{27} + \cdots$

45. $\displaystyle\sum_{n=1}^{\infty} 3\left(\frac{1}{4}\right)^{n-1} = 3 + \frac{3}{4} + \frac{3}{4^2} + \frac{3}{4^3} + \cdots$

46. $\displaystyle\sum_{n=1}^{\infty} 5\left(\frac{1}{6}\right)^{n-1} = 5 + \frac{5}{6} + \frac{5}{6^2} + \frac{5}{6^3} + \cdots$

47. $\displaystyle\sum_{i=0}^{\infty} 8(-0.3)^{i}$

48. $\displaystyle\sum_{i=0}^{\infty} 12(-0.7)^{i}$

49. $3 - 1 + \dfrac{1}{3} - \dfrac{1}{9} + \cdots$

50. $4 - 2 + 1 - \dfrac{1}{2} + \cdots$

Express each repeating decimal in Problems 51–62 as a fraction in lowest terms.

51. $0.\overline{2}$

52. $0.\overline{7}$

53. $0.\overline{47}$

54. $0.\overline{83}$

55. $0.\overline{347}$

56. $0.\overline{659}$

57. $3.\overline{72}$

58. $4.\overline{731}$

59. $3.2\overline{53}$

60. $4.1\overline{27}$

61. $1.3\overline{517}$

62. $3.27\overline{15}$

Application Problems

Use the formula for the general term (the nth term) of a geometric sequence to answer Problems 63–66.

63. In 1798, English economist Thomas Malthus argued that populations grow geometrically. Britain's population based on predictions of the Malthusian model is shown in the table. At the time, the population of Britain was 7 million, denoted in the table as year 0.

   a. Show that Britain's population is increasing geometrically every five years.

   b. Write the general term for the geometric sequence modeling Britain's population growth, where $n = 1$ corresponds to 1798, $n = 2$ corresponds to 1803, $n = 3$ corresponds to 1808, and so on. (This is precisely the model that Malthus used.)

   c. What does the Malthusian model predict for Britain's population in 1998? Consult an appropriate reference to find the actual population.

How well does the Malthusian model work when extended this far from 1798?

Number of Years after 1798	Britain's Population in Millions
0	7.00
5	8.04
10	9.23
15	10.59
20	12.16
25	13.96
30	16.03
35	18.40
40	21.13
45	24.25
50	27.85
100	110.77

64. A machine that cost a company $40,000 depreciates 10% each year. Find a formula that models the value of the machine after n years. What is its value after 8 years?

65. *E. coli* is a rod-shaped bacterium approximately 10^{-6} meters long found in the intestinal tracts of humans. The cells of *E. coli* reproduce quite rapidly, as shown in the table.

   a. Write the general term for the geometric sequence modeling the growth of the bacteria, where $n = 1$ corresponds to the first 20-minute time period, $n = 2$ corresponds to the second 20-minute time period, and so on.

   b. Use the model to predict the number of bacteria after 8 hours, which is the 24th time period.

Initial population = 100 bacteria	
20-Minute Time Interval	Number of E. coli Bacteria
1	200
2	400
3	800
4	1600
5	3200
6	6400

66. A person is investigating two employment opportunities. They both have a beginning salary of $30,000 per year. Company A offers an increase of $1400 per year. Company B offers guaranteed annual increases of 3% per year. Which company will pay the greater salary after 20 years?

Use the formula for the nth partial sum of a geometric sequence to answer Problems 67–76.

67. A job pays a salary of $24,000 the first year. During the next 19 years, the salary increases by 5% each year. What is the total lifetime salary over the 20-year period?

68. A person is investigating two employment opportunities. Company A offers $30,000 the first year. During the next four years, the salary is guaranteed to increase by 6% per year. Company B offers $32,000 the first year, with guaranteed annual increases of 3% per year after that. Which company offers the better total salary for a five-year contract?

69. A deposit of $200 is made at the beginning of each month for six years in an account that pays 9% annual interest compounded monthly. What is the balance in the account at the end of six years?

$$A = 200\left(1 + \frac{0.09}{12}\right) + 200\left(1 + \frac{0.09}{12}\right)^2$$
$$+ \cdots + 200\left(1 + \frac{0.09}{12}\right)^{72}$$

70. A deposit of $300 is made at the beginning of each month for six years in an account that pays 8% interest compounded monthly. What is the balance in the account at the end of six years?

$$A = 300\left(1 + \frac{0.08}{12}\right) + 300\left(1 + \frac{0.08}{12}\right)^2$$
$$+ \cdots + 300\left(1 + \frac{0.08}{12}\right)^{72}$$

71. A deposit of $2000 is made the first day of each year in an account that pays 5% interest compounded annually. What is the balance in the account at the end of six years?

72. A deposit of $4000 is made the first day of each year in an account that pays 9% interest compounded annually. What is the balance in the account at the end of ten years?

73. The future value of an annuity with periodic payments R made at the *end* of each of n periods with interest rate i per period is given by $S = R + R(1 + i) + R(1 + i)^2 + \cdots + R(1 + i)^{n-1}$.

   a. Use the formula for the sum of a finite geometric series to show that

   $$S = R\frac{(1 + i)^n - 1}{i}.$$

   b. Suppose that $2000 is deposited at the end of each year into an account that earns 6% compounded annually. Use the formula in part (a) to find the future value of the annuity.

74. Suppose a person saves $1 the first day of a month, $2 the second day, $4 the third day, and continues to double the savings each day. What will be the total savings for the first 15 days?

75. A pendulum swings through an arc of 16 inches. On each successive swing, the length of the arc is 96% of the previous length.

16,	0.96(16),	$(0.96)^2(16)$,	$(0.96)^3(16)$,	$\cdots$
1st swing	2nd swing	3rd swing	4th swing	

After 10 swings, what is the total length the pendulum has swung?

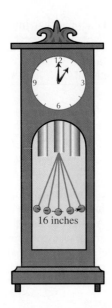

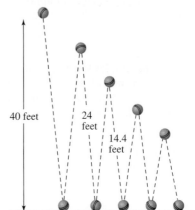

height. Find the total distance traveled by the ball by the time it strikes the ground for the fifth time.

$$40 + 2(40)(0.6) + 2(40)(0.6)^2 + \cdots + ?$$

40 feet   24 feet   14.4 feet

76. A ball is dropped from a height of 40 feet. Each time the ball strikes the ground, it bounces up 60% of its previous

Use the formula for the sum of an infinite geometric series to answer Problems 77–85.

77. If the shading process shown in the figure is continued indefinitely, what fractional part of the largest square is eventually shaded?

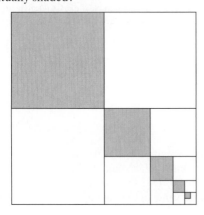

78. A square piece of paper whose sides measure 40 inches is cut into four smaller squares, each with 20 inches on a side. One of these squares is cut into four smaller squares, each with 10 inches on a side. One of these squares is cut into four smaller squares, each with 5 inches on a side. Assuming that this process is continued indefinitely, find the sum of the perimeters of all the squares, including the original square.

79. A new factory in a small town has an annual payroll of $6 million. It is expected that 60% of this money will be spent in the town by factory personnel. The people in the town who receive this money are expected to spend 60% of what they receive in the town, and so on. What is the total of all this spending, called the total economic impact of the factory, on the town each year?

80. How much additional spending will be generated by a $10 billion tax rebate if 60% of all income is spent?

81. The basic attitude of the ancient Greeks was to avoid the infinite. One extremely strong reason why they avoided it was the irritating paradox, attributed to the Greek philosopher Zeno (495–435 B.C.), called "Achilles and the Tortoise." Zeno argued that Achilles, who can run 10 meters per second, can never pass a tortoise, who can run 1 meter per second, if the tortoise is given a 10-meter head start (see the figure). Zeno's proof was as follows.

To pass the tortoise, Achilles must get to where the tortoise was at the start of the race. With the tortoise's 10-meter head start, this will take Achilles 1 second. However, during this 1 second, the tortoise has moved ahead 1 meter. To pass the tortoise, Achilles must move this 1 meter, which takes him $\frac{1}{10}$ of a second. But during this $\frac{1}{10}$ of a second, the tortoise has again moved ahead $\frac{1}{10}$ of a meter. To pass the tortoise, Achilles must move $\frac{1}{10}$ of a meter, which takes him $\frac{1}{100}$ of a second.

The argument continues infinitely. To pass the tortoise, Achilles is going to have to run an infinite number of distances $(10 + 1 + \frac{1}{10} + \frac{1}{100} + \cdots)$ in a finite amount of time. Unravel the paradox by finding the sum of the infinite geometric series $10 + 1 + \frac{1}{10} + \frac{1}{100} + \cdots$.

How far must Achilles run to overtake the tortoise? How long will this take?

82. A hot-air balloon rises 60 feet in the first minute of its flight. In each succeeding minute the balloon rises 80% as far as in the previous minute. What will be its maximum altitude if it is allowed to rise without limit?

83. A ball is dropped from a height of 8 feet and begins bouncing. The height of each bounce is half the height of the previous bounce. The total vertical distance traveled by the ball is given by

$$8 + 8\left(\frac{1}{2}\right) + 8\left(\frac{1}{2}\right) + 8\left(\frac{1}{2}\right)\left(\frac{1}{2}\right) + 8\left(\frac{1}{2}\right)\left(\frac{1}{2}\right) + \cdots$$

$$\underbrace{\qquad}_{\text{Up}} \underbrace{\qquad}_{\text{Down}} \underbrace{\qquad}_{\text{Up}} \underbrace{\qquad}_{\text{Down}}$$

$$16\left(\frac{1}{2}\right) \qquad\qquad 16\left(\frac{1}{2}\right)^2$$

$$= 8 + 16\left(\frac{1}{2}\right) + 16\left(\frac{1}{2}\right)^2 + 16\left(\frac{1}{2}\right)^3 + \cdots$$

What is the total vertical distance traveled by the ball?

Richard Megna / Fundamental Photographs

84. A ball is dropped from a height of a feet and begins bouncing. The height of each bounce is r times the height of the previous bounce, where r is positive and less than 1. Generalize from Problem 83 and show that the total vertical distance traveled by the ball is $a\left(\dfrac{1 + r}{1 - r}\right)$.

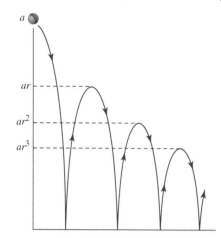

85. In a pest-eradication program, sterilized male flies are released into the general population each day. Ninety percent of those flies will survive a given day. How many flies should be released each day if the long-range goal of the program is to keep 20,000 sterilized flies in the population?

True–False Critical Thinking Problems

86. Which one of the following is true?

  a. The sequence $2, 6, 24, 120, \ldots$ is an example of a geometric sequence.

  b. The sum of the geometric series $\frac{1}{2} + \frac{1}{4} + \frac{1}{8} + \cdots + \frac{1}{512}$ can only be estimated without knowing precisely what terms occur between $\frac{1}{8}$ and $\frac{1}{512}$.

  c. $10 - 5 + \frac{5}{2} - \frac{5}{4} + \cdots = \dfrac{10}{1 - \frac{1}{2}}$

  d. If the nth term of a geometric sequence is $a_n = 3(0.5)^{n-1}$, the common ratio is $\frac{1}{2}$.

87. Which one of the following is true?

  a. Not every geometric sequence is a function.

  b. The only way to find $\displaystyle\sum_{i=1}^{\infty} (0.1)^i$ is to write out all terms and then find their sum.

  c. The sum of the first seven terms of $\sqrt{3}, \sqrt{6}, 2\sqrt{3}, 2\sqrt{6}, \ldots$ contains two radical terms, one of which is $15\sqrt{3}$.

  d. $\displaystyle\sum_{i=0}^{\infty} 3(0.6)^{i+1} = \dfrac{1.8(1 - 0.6)^8}{1 - 0.6}$

Technology Problems

88. **a.** Find a complete graph of the function $f(x) = \dfrac{1 - r^x}{1 - r}$ for the following values of r: $r = 0.5$, $r = 0.9$, $r = 1.1$, $r = 3$.

  b. For what values of r does the complete graph of f have a horizontal asymptote?

  c. What does part (b) indicate about when the infinite series $1 + r + r^2 + r^3 + r^4 + \cdots$ has a sum?

In Problems 89–90, use a graphing utility to graph the function. Determine the horizontal asymptote for the graph of f and discuss its relationship to the sum of the given series.

Function	Series

89. $f(x) = \dfrac{2\left[1 - \left(\frac{1}{3}\right)^x\right]}{1 - \dfrac{1}{3}}$ $\displaystyle\sum_{n=0}^{\infty} 2\left(\frac{1}{3}\right)^n$

90. $f(x) = \dfrac{4[1 - (0.6)^x]}{1 - 0.6}$ $\displaystyle\sum_{n=0}^{\infty} 4(0.6)^n$

91. Graph $f(x) = \dfrac{4(1 - 3^x)}{1 - 3}$ in the viewing rectangle [0, 10] by [0, 1000]. Does the graph of f have a horizontal asymptote? What does the complete graph of f indicate about the sum of the infinite series $\displaystyle\sum_{n=1}^{\infty} 4 \cdot 3^{n-1}$?

Writing in Mathematics

92. Would you rather have $10,000,000 and a new BMW or 1¢ today, 2¢ tomorrow, 4¢ on day 3, 8¢ on day 4, 16¢ on day 5, and so on for 30 days? Explain.

93. Infinite series often occur in applications in which we want to estimate the long-term behavior of a process that changes at regularly spaced intervals. Describe a situation that can be modeled by an infinite geometric series.

94. What happens to the series $\displaystyle\sum_{n=1}^{N} 2^n$ as N increases? Why does the formula $S = \dfrac{a_1}{1 - r}$ not apply to $\displaystyle\sum_{n=1}^{\infty} 2^n$?

Critical Thinking Problems

95. Give an example of two different geometric series that have the same sum, the same first term, and the same number of terms.

96. A piece of equipment worth $60,000 depreciates 5% of its present value each year. During the first year, $60,000(0.05) = $3000 is depreciated. During the second year, $57,000(0.05) = $2850 is depreciated, and so on. If this depreciation continues indefinitely, what is the total depreciation?

97. Find the indicated sum: $\displaystyle\sum_{n=3}^{\infty} \left[\left(\frac{1}{4}\right)^n + \left(\frac{3}{4}\right)^n\right]$.

98. Find the indicated sum: $\displaystyle\sum_{i=1}^{3}\left(\sum_{n=1}^{\infty}\left(\frac{i}{4}\right)^{n-1}\right)$.

99. The Cantor set, named after the German mathematician Georg Cantor (1845–1918), is formed as follows.

Begin with the closed interval [0, 1] and remove the middle one-third, namely, $(\frac{1}{3}, \frac{2}{3})$. This leaves the two intervals $[0, \frac{1}{3}]$ and $[\frac{2}{3}, 1]$. Again, remove the middle one-third of each of them. Continue indefinitely by removing the middle one-third of each interval that remains from the preceding step. The Cantor set consists of the numbers that remain in [0, 1] after all those intervals have been removed. Find the sum of the lengths of all the intervals that have been removed.

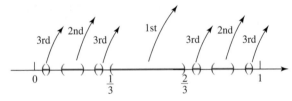

Group Activity Problem

100. Problem 81 presented a paradox posed by the Greek philosopher Zeno. Paradoxes lead to contradictions, such as this famous paradox presented by the twentieth-century philosopher and mathematician Bertrand Russell: "*In a town a male barber shaves all and only those men who do not shave themselves. Does the barber shave himself?*" Why is this a paradox? Group members should now try forming an original paradox that leads to a contradiction.

Solutions Manual

Tutorial

Video 12

SECTION 7.4 Mathematical Induction

Objectives

1 Use mathematical induction to prove statements.

2 Verify patterns using mathematical induction.

In this section, our focus is on a method for proving that certain statements involving the positive integers (1, 2, 3, 4, ...) are true for all positive integers. The form of proof used in these situations is called the principle of *mathematical induction*.

Let's begin with a statement that appears to give a correct formula for the sum of the first n positive integers.

1 Use mathematical induction to prove statements.

Robert Hudson "Double Time" 1963, painted metal, 58.75 × 50 in. Collection of the Oakland Museum of California, Gift of the Women's Board, Oakland Museum Association.

$$S_n: 1 + 2 + 3 + \cdots + n = \frac{n(n + 1)}{2}$$

We can verify this statement for, say, the first four positive integers.

If $n = 1$, the statement S_1 is

$$\left[1 = \frac{1(1 + 1)}{2} \right] \text{ a true statement, because } 1 = 1.$$

If $n = 2$, the statement S_2 is

$$\left[1 + 2 = \frac{2(2 + 1)}{2} \right] \text{ a true statement, because } 3 = 3.$$

If $n = 3$, the statement S_3 is

$$\left[1 + 2 + 3 = \frac{3(3 + 1)}{2} \right] \text{ a true statement, because } 6 = 6.$$

Finally, if $n = 4$, the statement S_4 is

$$\left[1 + 2 + 3 + 4 = \frac{4(4 + 1)}{2} \right] \text{ a true statement, because } 10 = 10.$$

This approach does not prove that the given statement S_n is true for every positive integer n. The fact that the formula produces true statements for $n = 1, 2, 3,$ and 4 does not guarantee that it is valid for all positive integers n. Thus, we need to be able to verify the truth of S_n without one verification after another.

A legitimate proof of the given statement S_n involves a technique called *mathematical induction*.

The principle of mathematical induction

Let S_n be a statement involving the positive integer n. If

1. S_1 is true, and
2. the truth of the statement S_k implies the truth of the statement S_{k+1}, for every positive integer k,

then the statement S_n is true for all positive integers n.

Figure 7.7

Falling dominoes illustrate the principle of mathematical induction.

The principle of mathematical induction can be illustrated using an unending line of dominoes, as shown in Figure 7.7. If the first domino is pushed over, it knocks down the next, which knocks down the next, and so on, in a chain reaction. To topple all the dominoes in the infinite sequence, two conditions must be satisfied:

1. The first domino must be knocked down.
2. If the domino in position k is knocked down, then the domino in position $k + 1$ must be knocked down.

If the second condition is not satisfied (suppose the dominoes are spaced far enough apart so that a falling domino does not push over the next domino in the line), it does not follow that all the dominoes will topple.

The domino analogy provides the two steps that are required in a proof by mathematical induction.

The steps in a proof by mathematical induction

Let S_n be a statement involving the positive integer n. To prove that S_n is true for all positive integers n requires two steps.

> **Step 1.** Show that S_1 is true.
> **Step 2.** Show that if S_k is assumed to be true, then S_{k+1} is also true, for every positive integer k.

Notice that to prove S_n, we work only with the statements S_1, S_k, and S_{k+1}. Our first example provides practice in writing these statements.

EXAMPLE 1   Writing S_1, S_k, and S_{k+1}

For the given statement S_n, write the three statements S_1, S_k, and S_{k+1}.

a. S_n: $1 + 2 + 3 + \cdots + n = \dfrac{n(n + 1)}{2}$

b. S_n: $1^2 + 2^2 + 3^2 + \cdots + n^2 = \dfrac{n(n + 1)(2n + 1)}{6}$

Solution

a. S_n: $1 + 2 + 3 + \cdots + n = \dfrac{n(n + 1)}{2}$

This is the given statement S_n.

S_1: $1 = \dfrac{1(1 + 1)}{2}$

Write S_1 by replacing n with 1 in the statement S_n.

S_k: $1 + 2 + 3 + \cdots + k = \dfrac{k(k + 1)}{2}$

Write S_k by replacing n with k in the statement S_n.

S_{k+1}: $1 + 2 + 3 + \cdots + (k + 1) = \dfrac{(k + 1)[(k + 1) + 1]}{2}$

Write S_{k+1} by replacing n with $k + 1$ in the statement S_n.

S_{k+1}: $1 + 2 + 3 + \cdots + (k + 1) = \dfrac{(k + 1)(k + 2)}{2}$

Simplify on the right.

b. $S_n: 1^2 + 2^2 + 3^2 + \cdots + n^2 = \dfrac{n(n+1)(2n+1)}{6}$

This is the given statement S_n.

$S_1: 1^2 = \dfrac{1(1+1)(2 \cdot 1 + 1)}{6}$

Write S_1 by replacing n with 1 in the statement S_n.

$S_k: 1^2 + 2^2 + 3^2 + \cdots + k^2 = \dfrac{k(k+1)(2k+1)}{6}$

Write S_k by replacing n by k in the statement S_n.

$S_{k+1}: 1^2 + 2^2 + 3^2 + \cdots + (k+1)^2 = \dfrac{(k+1)[(k+1)+1][2(k+1)+1]}{6}$

Write S_{k+1} by replacing n by $k+1$ in the statement S_n.

$S_{k+1}: 1^2 + 2^2 + 3^2 + \cdots + (k+1)^2 = \dfrac{(k+1)(k+2)(2k+3)}{6}$

Simplify on the right.   ■

ENRICHMENT ESSAY

Visualizing Summation Formulas

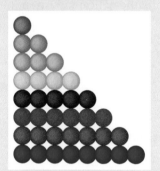

Modeling the sum of consecutive positive integers leads to *triangular numbers* of the form $\dfrac{n(n+1)}{2}$.

$\dfrac{n(n+1)}{2}$ $\dfrac{n(n+1)}{2}$ $\dfrac{n(n+1)}{2}$ $\dfrac{n(n+1)}{2}$

$n = 1$: $n = 2$: $n = 3$: $n = 4$:
   1   3   6   10

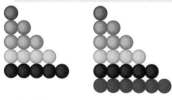

$\dfrac{n(n+1)}{2}$ $\dfrac{n(n+1)}{2}$

$n = 5$: $n = 6$:
   15   21

In Examples 2 and 3, we will use the statements S_1, S_k, and S_{k+1} to prove each of the statements S_n of Example 1.

EXAMPLE 2 **Proving a Formula by Mathematical Induction**

Use mathematical induction to prove that

$$1 + 2 + 3 + \cdots + n = \dfrac{n(n+1)}{2}$$

for all positive integers n.

Solution

Step 1. Show that S_1 is true. If $n = 1$, the statement S_1 is $1 = \dfrac{1(1+1)}{2}$, a true statement, since $1 = 1$.

Step 2. Show that if S_k is true, then S_{k+1} is true. Using our results from Example 1, show that the truth of S_k,

$$1 + 2 + 3 + \cdots + k = \dfrac{k(k+1)}{2}$$

implies the truth of S_{k+1},

$$1 + 2 + 3 + \cdots + (k+1) = \dfrac{(k+1)(k+2)}{2}.$$

We will work with S_k. Since S_k is true, we add the next consecutive integer after k—namely, $k+1$—to both sides.

$$1 + 2 + 3 + \cdots + k = \dfrac{k(k+1)}{2}$$

This is S_k, which we assume is true.

$$1 + 2 + 3 + \cdots + k + (k+1) = \dfrac{k(k+1)}{2} + (k+1)$$

Add $k+1$ to both sides of the equation.

$$1 + 2 + 3 + \cdots + (k + 1) = \frac{k(k + 1)}{2} + \frac{2(k + 1)}{2}$$

We do not write k on the left-hand side because k is understood to be the integer that precedes $k + 1$. Write the right-hand side with a common denominator of 2.

ENRICHMENT ESSAY

Visualizing Summation Formulas

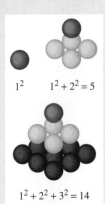

1^2 $1^2 + 2^2 = 5$

$1^2 + 2^2 + 3^2 = 14$

$1^2 + 2^2 + 3^2 + 4^2 = 30$

$$1 + 2 + 3 + \cdots + (k + 1) = \frac{(k + 1)}{2}(k + 2)$$

Factor out the common factor $\dfrac{k + 1}{2}$ on the right.

$$1 + 2 + 3 + \cdots + (k + 1) = \frac{(k + 1)(k + 2)}{2}$$

This final result is the statement for $n = k + 1$.

We have shown that by assuming that S_k is true (and adding $k + 1$ to both sides), then S_{k+1} is also true. By the principle of mathematical induction, the statement S_n, namely,

$$1 + 2 + 3 + \cdots + n = \frac{n(n + 1)}{2}$$

is true for every positive integer n. ∎

EXAMPLE 3   Proving a Formula by Mathematical Induction

Use mathematical induction to prove that

$$1^2 + 2^2 + 3^2 + \cdots + n^2 = \frac{n(n + 1)(2n + 1)}{6}$$

for all positive integers n.

Solution

Step 1. Show that S_1 is true. If $n = 1$, the statement S_1 is

$$1^2 = \frac{1(1 + 1)(2 \cdot 1 + 1)}{6}$$

$$1 = \frac{1 \cdot 2 \cdot 3}{6} \qquad \text{Simplify, leading to the true statement } 1 = 1.$$

Step 2. Show that if S_k is true, then S_{k+1} is true. From Example 1, show that the truth of

$$S_k : 1^2 + 2^2 + 3^2 + \cdots + k^2 = \frac{k(k + 1)(2k + 1)}{6}$$

implies the truth of

$$S_{k+1} : 1^2 + 2^2 + 3^2 + \cdots + (k + 1)^2 = \frac{(k + 1)(k + 2)(2k + 3)}{6}.$$

$1^2 + 2^2 + 3^2 + 4^2 + 5^2 = 55$

Modeling the sum of the squares of consecutive integers leads to *square pyramid numbers* of the form $\dfrac{n(n + 1)(2n + 1)}{6}$.
Cannonballs in old war memorials are often stacked in this manner.

Since S_k is assumed true, we add the square of the next consecutive integer after k^2—namely, $(k + 1)^2$—to both sides.

$$1^2 + 2^2 + 3^2 + \cdots + k^2 = \frac{k(k + 1)(2k + 1)}{6}$$

This is S_k, assumed to be true. We must work with this and show S_{k+1} is true.

$$1^2 + 2^2 + 3^2 + \cdots + k^2 + (k + 1)^2 = \frac{k(k + 1)(2k + 1)}{6} + (k + 1)^2$$

Add $(k + 1)^2$ to both sides.

$$1^2 + 2^2 + 3^2 + \cdots + (k + 1)^2 = \frac{k(k + 1)(2k + 1)}{6} + \frac{6(k + 1)^2}{6}$$

It is not necessary to write k^2 on the left. Express the right side with the LCD.

$$= \frac{(k + 1)}{6}[k(2k + 1) + 6(k + 1)]$$

Factor out the common factor $\frac{k + 1}{6}$.

$$= \frac{(k + 1)}{6}(2k^2 + 7k + 6)$$

Multiply and combine like terms.

$$= \frac{(k + 1)}{6}(k + 2)(2k + 3)$$

Factor $2k^2 + 7k + 6$.

$$= \frac{(k + 1)(k + 2)(2k + 3)}{6}$$

This final statement is S_{k+1}.

We have shown that by assuming that S_k is true (and adding $(k + 1)^2$ to both sides), then S_{k+1} is also true. By the principle of mathematical induction, the statement S_n, namely,

$$1^2 + 2^2 + 3^2 + \cdots + n^2 = \frac{n(n + 1)(2n + 1)}{6}$$

is true for every positive integer n. ∎

2 Verify patterns using mathematical induction.

Sums of Powers of Positive Integers: Recognizing Patterns

A more difficult problem involves trying to find a formula for the sum of the first n terms of a series. Once we have a pattern that might work, we can then use mathematical induction to prove the conjectured formula.

EXAMPLE 4   **Verifying Patterns Using Mathematical Induction**

a. Find a formula for the following sum.

$$S_n: \frac{1}{1\cdot 2} + \frac{1}{2\cdot 3} + \frac{1}{3\cdot 4} + \cdots + \frac{1}{n(n+1)}$$

b. Verify the result using mathematical induction.

Solution

a. Let's find S_1 through S_4 and then use the pattern to make a conjecture about S_n.

$$S_1 = \frac{1}{1\cdot 2} = \frac{1}{2}$$

$$S_2 = \frac{1}{1\cdot 2} + \frac{1}{2\cdot 3} = \frac{1}{2} + \frac{1}{6} = \frac{3}{6} + \frac{1}{6} = \frac{4}{6} = \frac{2}{3}$$

$$S_3 = \frac{1}{1\cdot 2} + \frac{1}{2\cdot 3} + \frac{1}{3\cdot 4} = \frac{1}{2} + \frac{1}{6} + \frac{1}{12} = \frac{9}{12} = \frac{3}{4}$$

$$S_4 = \frac{1}{1\cdot 2} + \frac{1}{2\cdot 3} + \frac{1}{3\cdot 4} + \frac{1}{4\cdot 5} = \frac{1}{2} + \frac{1}{6} + \frac{1}{12} + \frac{1}{20} = \frac{48}{60} = \frac{4}{5}$$

Discover for yourself

Look at the formulas for S_1 through S_4 shown on the right. What do you observe about the numerator of each fraction in relationship to the subscript on S? What do you notice about the denominator of each fraction compared to the numerator? Using these patterns, what is the formula for S_n?

In the Discover for yourself, did you notice that the numerator of each partial sum was the same as the subscript and that the denominator was one greater than the numerator?

$$S_1 = \frac{1}{2}, S_2 = \frac{2}{3}, S_3 = \frac{3}{4}, \text{ and } S_4 = \frac{4}{5}.$$

Using the emerging pattern, it appears that

$$S_n = \frac{n}{n+1}.$$

Our conjecture is that

$$S_n: \frac{1}{1\cdot 2} + \frac{1}{2\cdot 3} + \frac{1}{3\cdot 4} + \cdots + \frac{1}{n(n+1)} = \frac{n}{n+1}.$$

b. Now we will prove the conjectured formula for S_n using mathematical induction.

Step 1. Show that S_1 is true. If $n = 1$, the statement S_1 is $\frac{1}{1\cdot 2} = \frac{1}{1+1}$, a true statement since $\frac{1}{2} = \frac{1}{2}$.

Step 2. Show that if S_k is true, then S_{k+1} is true.

$$S_k: \frac{1}{1\cdot 2} + \frac{1}{2\cdot 3} + \frac{1}{3\cdot 4} + \cdots + \frac{1}{k(k+1)} = \frac{k}{k+1}$$ Replace n with k in S_n.

$$S_{k+1}: \frac{1}{1\cdot 2} + \frac{1}{2\cdot 3} + \frac{1}{3\cdot 4} + \cdots + \frac{1}{(k+1)[(k+1)+1]} = \frac{k+1}{(k+1)+1}$$ Replace n with $k+1$ in S_n.

$$\frac{1}{1\cdot 2} + \frac{1}{2\cdot 3} + \frac{1}{3\cdot 4} + \cdots + \frac{1}{(k+1)(k+2)} = \frac{k+1}{k+2}$$ Simplify S_{k+1}.

We now work with S_k. We add to both sides the term after $\dfrac{1}{k(k+1)}$. Since each factor in the denominator increases by 1 with each successive term, we will add $\dfrac{1}{(k+1)(k+2)}$ to both sides.

$$\frac{1}{1\cdot 2} + \frac{1}{2\cdot 3} + \frac{1}{3\cdot 4} + \cdots + \frac{1}{k(k+1)} = \frac{k}{k+1}$$

This is S_k.

$$\frac{1}{1\cdot 2} + \frac{1}{2\cdot 3} + \frac{1}{3\cdot 4} + \cdots + \frac{1}{k(k+1)} + \frac{1}{(k+1)(k+2)} = \frac{k}{k+1} + \frac{1}{(k+1)(k+2)}$$

Add $\dfrac{1}{(k+1)(k+2)}$ to both sides.

$$\frac{1}{1\cdot 2} + \frac{1}{2\cdot 3} + \frac{1}{3\cdot 4} + \cdots + \frac{1}{(k+1)(k+2)} = \frac{k(k+2)}{(k+1)(k+2)} + \frac{1}{(k+1)(k+2)}$$

Express the right side with the LCD, namely, $(k+1)(k+2)$. What happened to $\dfrac{1}{k(k+1)}$ on the left?

$$= \frac{k^2 + 2k + 1}{(k+1)(k+2)}$$

Add numerators, putting this sum over the LCD.

$$= \frac{(k+1)(k+1)}{(k+1)(k+2)}$$

Factor $k^2 + 2k + 1$.

$$= \frac{(k+1)}{(k+2)}$$

Cancel identical factors of $k+1$ in the numerator and denominator on the right. This final statement is the statement S_{k+1}.

We have shown that by assuming S_k is true and adding $\dfrac{1}{(k+1)(k+2)}$ to both sides, then S_{k+1} is also true. This verifies the formula that we conjectured for S_n. ■

In our final example, we use mathematical induction to prove a statement involving an inequality.

Jack Youngerman
"Black/Red/Yellow" 1964, oil on
canvas, 54 × 42 in. © Jack
Youngerman / Licensed by VAGA,
New York, NY.

EXAMPLE 5 **Proving an Inequality Using Mathematical Induction**

Prove that $3^n < (n+2)!$ for all positive integers n.

Solution

Step 1. For $n = 1$, the formula is true because

$$3^1 < (1+2)!$$ $(1+2)! = 3! = 1\cdot 2\cdot 3 = 6$

$$3 < 6$$ A true inequality results when $n = 1$.

Step 2. Assuming that

$$3^k < (k+2)!$$

we must prove that

$$3^{k+1} < (k+3)!$$ Replace n with $k+1$ in the given inequality.

$$3^k < (k+2)!$$ This is S_k. To prove S_{k+1}, with 3^{k+1} on the left, we will multiply both sides of S_k by 3.

$$3 \cdot 3^k < 3(k+2)!$$ Multiply both sides by 3.

$$3^{k+1} < 3(k+2)!$$ $3 \cdot 3^k = 3^1 \cdot 3^k = 3^{k+1}$

$$3^{k+1} < (k+3)(k+2)!$$ If $k \geq 1$, $3 < k+3$.

$$3^{k+1} < (k+3)!$$ $(k+3)(k+2)! = (k+3)(k+2)(k+1) \cdot \cdots \cdot 1$
   $= (k+3)!$

By assuming that S_k is true and multiplying both sides by 3, then S_{k+1} is also true. This verifies that the statement S_n, namely,

$$3^n < (n+2)!$$

is true for every positive integer n.   ∎

PROBLEM SET 7.4

Practice Problems

In Problems 1–14, use mathematical induction to prove that each given formula is true for every positive integer n.

1. $2 + 4 + 6 + \cdots + 2n = n(n+1)$

2. $3 + 6 + 9 + \cdots + 3n = \dfrac{3n(n+1)}{2}$

3. $6 + 10 + 14 + \cdots + (4n+2) = 2n(n+2)$

4. $3 + 7 + 11 + \cdots + (4n-1) = n(2n+1)$

5. $2 + 4 + 8 + \cdots + 2^n = 2^{n+1} - 2$

6. $\dfrac{1}{2} + \dfrac{1}{4} + \dfrac{1}{8} + \cdots + \dfrac{1}{2^n} = 1 - \dfrac{1}{2^n}$

7. $1^3 + 2^3 + 3^3 + \cdots + n^3 = \dfrac{n^2(n+1)^2}{4}$

8. $2^2 + 4^2 + 6^2 + \cdots + (2n)^2 = \dfrac{2n(n+1)(2n+1)}{3}$

9. $1 \cdot 2 + 2 \cdot 3 + 3 \cdot 4 + \cdots + n(n+1) = \dfrac{n(n+1)(n+2)}{3}$

10. $1 \cdot 3 + 2 \cdot 4 + 3 \cdot 5 + \cdots + n(n+2) = \dfrac{n(n+1)(2n+7)}{6}$

11. $\dfrac{1}{1 \cdot 2 \cdot 3} + \dfrac{1}{2 \cdot 3 \cdot 4} + \dfrac{1}{3 \cdot 4 \cdot 5} + \cdots + \dfrac{1}{n(n+1)(n+2)} = \dfrac{n(n+3)}{4(n+1)(n+2)}$

12. $1 + \dfrac{3}{2} + \dfrac{5}{2^2} + \cdots + \dfrac{2n-1}{2^{n-1}} = 6 - \dfrac{2n+3}{2^{n-1}}$

13. $\displaystyle\sum_{i=1}^{n} i^4 = \dfrac{n(n+1)(2n+1)(3n^2 + 3n - 1)}{30}$

14. $\displaystyle\sum_{i=1}^{n} i^5 = \dfrac{n^2(n+1)^2(2n^2 + 2n - 1)}{12}$

In Problems 15–16 and 18–19, find S_1 through S_5 and then use the pattern to make a conjecture about S_n. In Problem 17, find S_2 through S_6 and make a conjecture about S_n. Prove the conjectured formula for S_n by mathematical induction.

15. $S_n: \dfrac{1}{2 \cdot 3} + \dfrac{1}{3 \cdot 4} + \dfrac{1}{4 \cdot 5} + \cdots + \dfrac{1}{(n+1)(n+2)}$

16. $S_n: \dfrac{1}{1 \cdot 3} + \dfrac{1}{3 \cdot 5} + \dfrac{1}{5 \cdot 7} + \cdots + \dfrac{1}{(2n-1)(2n+1)}$

17. $S_n: \dfrac{1}{4} + \dfrac{1}{12} + \dfrac{1}{24} + \cdots + \dfrac{1}{2n(n-1)}$

18. $S_n: \left(1 + \dfrac{1}{1}\right)\left(1 + \dfrac{1}{2}\right)\left(1 + \dfrac{1}{3}\right) \cdots \left(1 + \dfrac{1}{n}\right)$

19. $S_n: \left(1 - \dfrac{1}{2}\right)\left(1 - \dfrac{1}{3}\right)\left(1 - \dfrac{1}{4}\right) \cdots \left(1 - \dfrac{1}{n+1}\right)$

20. **a.** Find each of the following sums:

$1 + 2 + 1 =$

$1 + 2 + 3 + 2 + 1 =$

$1 + 2 + 3 + 4 + 3 + 2 + 1 =$

$1 + 2 + 3 + 4 + 5 + 4 + 3 + 2 + 1 =$

$1 + 2 + 3 + 4 + 5 + 6 + 5 + 4 + 3 + 2 + 1 =$

(How is this illustrated in the figure on the right?)

b. Use pattern recognition to make a conjecture about

$1 + 2 + 3 + \cdots + (n-1) + n + (n-1)$

$+ \cdots + 3 + 2 + 1 =$

c. Prove the conjectured formula in part (b) using mathematical induction.

In Problems 21–30, use mathematical induction to prove that the given statement is true for every positive integer n.

21. $n < n + 1$

22. $3 \le 3^n$

23. $2^n > n$

24. $1 + 2n \le 3^n$

25. $2^{n-1} \le n!$

26. $e^n < n!$ (Assume $n \ge 6$.)

27. $(ab)^n = a^n b^n$

28. $\left(\dfrac{a}{b}\right)^n = \dfrac{a^n}{b^n}$

29. $n^2 + n$ is divisible by 2.

30. $n^3 - n$ is divisible by 3.

Application Problems _____

31. The Tower of Hanoi is a game that has three pegs on a board. On one peg are n disks, each smaller than the one on which it rests. The game's object is to move all the disks to another peg. Only one disk may be moved at a time and a larger disk may not be placed on a smaller disk.

a. Determine the least number of moves it takes to move 2 disks; 3 disks; 4 disks; 1 disk. Use the pattern to state a conjectured formula for the least number of moves it takes to move n disks.

b. Prove the conjectured formula in part (a) by mathematical induction.

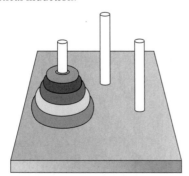

32. The fractal constructed in the figure on the right starts with an equilateral triangle having sides of length 1.

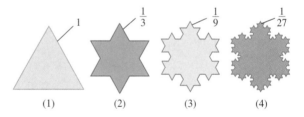

At each step in an infinite sequence of steps, we build on the preceding figure. Starting with the equilateral triangle of length 1, add to each side another equilateral triangle with side of length $\frac{1}{3}$. Continue in this manner by adding equilateral triangles whose sides are $\frac{1}{3}$ the length of the preceding triangle to each side.

a. Complete the following table.

Figure	Number of Sides	Perimeter
(1)	3	$3 \cdot 1$
(2)		
(3)		
(4)		
$\vdots$ (n)	$\vdots$	$\vdots$

b. Using mathematical induction, prove each of the conjectured formulas for the number of sides of the nth figure and the perimeter of the nth figure.

True–False Critical Thinking Problems

33. Which one of the following is true?

a. Mathematical induction is used to prove statements true for all real numbers.

b. Since $(a + b)^n = a^n + b^n$ for $n = 1$, we can use mathematical induction to prove that the statement is true for every positive integer n.

c. If S_n is the statement

$$1^3 + 2^3 + 3^3 + \cdots + n^3 = \frac{n^2(n + 1)^2}{4}$$

then S_{k+1} is the statement

$$1^3 + 2^3 + 3^3 + \cdots + (k + 1)^3 = \frac{(k + 1)^2(k + 2)^2}{5}.$$

d. When using mathematical induction to prove a summation formula, we work with S_k and add a_{k+1} to both sides, where a_{k+1} is the $k + 1$ term of the original sum.

34. Which one of the following is true?

a. If S_n is the statement

$$1^2 + 2^2 + 3^2 + \cdots + n^2 = \frac{n(n + 1)(2n + 1)}{6}$$

then S_{k+1} is the statement

$$1^2 + 2^2 + 3^2 + \cdots + (k + 1)^2$$
$$= \frac{(k + 1)(k + 1 + 1)(2(k + 1) + 1)}{7}$$

which can be simplified as

$$1^2 + 2^2 + 3^2 + \cdots + (k + 1)^2$$
$$= \frac{(k + 1)(k + 2)(2k + 3)}{7}.$$

b. To prove that

$$1^3 + 2^3 + 3^3 + \cdots + n^3 = \frac{n^2(n + 1)^2}{4}$$

we work with S_k and add $(k + 1)^3$ to each side of S_k. This gives

$$1^3 + 2^3 + 3^3 + \cdots + k^3 + (k + 1)^3$$
$$= \frac{k^2(k + 1)^2}{4} + (k + 1)^3.$$

However, the next step

$$1^3 + 2^3 + 3^3 + \cdots + (k + 1)^3$$
$$= \frac{k^2(k + 1)^2}{4} + \frac{4(k + 1)^3}{4}$$

is incorrect because you can't just drop the term k^3.

c. If a statement about the positive integers can be verified for the first one million positive integers, then because we have done this for one million numbers and climbed a huge ladder, this is essentially the same as proving the statement true using mathematical induction.

d. To prove that there are no positive integers that satisfy $a^n + b^n = c^n$ for $n > 2$, it is not enough to show that no positive integers satisfy this equation from $n = 3$ through $n = 3,000,000,000,000$.

Writing in Mathematics

35. Describe the principle of mathematical induction using the concept of climbing up an infinite ladder.

36. Consider the statement S_n given by

$$n^2 - n + 41 \text{ is prime}$$

Although $S_1, S_2, \ldots, S_{40}$ are true, S_{41} is false. Describe how this is illustrated by the dominoes in the figure. What does this tell you about a pattern, or formula, that seems to work for several values of n?

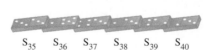

37. What is meant by inductive reasoning? What is the connection between inductive reasoning and mathematical induction?

Critical Thinking Problems

The Fibonacci sequence is defined by $a_{n+1} = a_n + a_{n-1}$ with $a_1 = a_2 = 1$. Use mathematical induction to prove each of the following properties of the Fibonacci sequence.

38. $a_1 + a_2 + a_3 + \cdots + a_n = a_{n+2} - 1$ **39.** $a_1^2 + a_2^2 + a_3^2 + \cdots + a_n^2 = a_n a_{n+1}$ **40.** $a_2 + a_4 + a_6 + \cdots + a_{2n} = a_{2n+1} - 1$

Some statements are false for the first few positive integers, but true for some positive integer on. In these instances, you can prove S_n for $n \ge k$ by showing that S_k is true and that S_k implies S_{k+1}. Use this extended principle of mathematical induction to prove that each statement in Problems 41–44 is true.

41. Prove that for every natural number $n \ge 2$,
$$\left(1 - \frac{1}{2^2}\right)\left(1 - \frac{1}{3^2}\right)\left(1 - \frac{1}{4^2}\right) \cdots \left(1 - \frac{1}{n^2}\right) = \frac{n+1}{2n}.$$
Show that the statement is true for $n = 2$ and then use step 2 of mathematical induction.

42. Prove that $4n < 2^n$ for $n \ge 5$. Show that the statement is true for $n = 5$ and then use step 2 of mathematical induction.

43. Prove that $n + 12 \le n^2$ for $n \ge 5$.

44. Prove that for every natural number $n \ge 2$,
$$\frac{1}{\sqrt{1}} + \frac{1}{\sqrt{2}} + \frac{1}{\sqrt{3}} + \cdots + \frac{1}{\sqrt{n}} > \sqrt{n}.$$

SECTION 7.5 The Binomial Theorem

Solutions Manual   Tutorial   Video 13

Objectives

1 Evaluate a binomial coefficient.

2 Expand a binomial raised to a power.

3 Find a specified term in a binomial's expansion.

Throughout algebra, we used the FOIL method to square a binomial such as $(x + 3)^2$. In this section, we study higher powers of binomials.

Patterns in Binomial Expansions

When we write out the *binomial expression* $(a + b)^n$, where n is a positive integer, a number of patterns begin to appear.

$$(a + b)^1 = a + b$$

$$(a + b)^2 = a^2 + 2ab + b^2$$

$$(a + b)^3 = a^3 + 3a^2b + 3ab^2 + b^3$$

$$(a + b)^4 = a^4 + 4a^3b + 6a^2b^2 + 4ab^3 + b^4$$

$$(a + b)^5 = a^5 + 5a^4b + 10a^3b^2 + 10a^2b^3 + 5ab^4 + b^5$$

P. S. Gordon "The Captain's All
Vegged Out" 1982, w/c and gouache,
46×54 in. © 1982 by P. S. Gordon.

Discover for yourself

Each expanded form of the binomial expression is a polynomial. Study the five polynomials on the bottom of page 815 and answer the following questions.

1. For each polynomial, describe the pattern for the exponents on a. What is the largest exponent of a? What happens to the exponent on a from term to term?
2. Describe the pattern for the exponents on b. What is the exponent on b in the first term? What is the exponent on b in the second term? What happens to the exponent on b from term to term?
3. Find the sum of the exponents on the variables in each term for the polynomials in the five rows. Describe the pattern.
4. How many terms are there in the polynomials on the right in relation to the power of the binomial?

In the Discover for yourself, how many of the following patterns were you able to discover?

1. The first term is a^n. The exponent on a decreases by 1 in each successive term.
2. The exponents on b increase by 1 in each successive term. In the first term, the exponent on b is 0. (Because $b^0 = 1$, b is not shown in the first term.) The last term is b^n.
3. The sum of the exponents on the variables in any term is equal to the exponent on $(a + b)^n$.
4. There is one more term in the polynomial expansion than there is in the power of the binomial, n. There are $n + 1$ terms in the expanded form of $(a + b)^n$.

Using these observations, the variable parts of the expansion of $(a + b)^6$ are

$$a^6, a^5b, a^4b^2, a^3b^3, a^2b^4, ab^5, b^6.$$

The first term is a^6, with the exponent on a decreasing by 1 in each successive term. The exponents on b increase from 0 to 6, with the last term being b^6. The sum of the exponents in each term is equal to 6.

Generalizing from these observations, the variable parts of the expansion of $(a + b)^n$ are

$$a^n, a^{n-1}b, a^{n-2}b^2, a^{n-3}b^3, \ldots, ab^{n-1}, b^n.$$

Let's now establish a pattern for the coefficients of the terms in the binomial expansion. Notice that each row in the following figure begins and ends with 1. Any other number in the row can be obtained by adding the two numbers immediately above it.

Study tip

We have not shown the number in the top row of Pascal's triangle on the right. The top row is *row zero* because it corresponds to $(a + b)^0 = 1$. With row zero, the triangle appears as:

```
            1
         1     1
      1     2     1
   1     3     3     1
 1     4     6     4     1
            etc.
```

Coefficients for $(a + b)^1$.					1		1			
Coefficients for $(a + b)^2$.				1		2		1		
Coefficients for $(a + b)^3$.			1		3		3		1	
Coefficients for $(a + b)^4$.		1		4		6		4		1
Coefficients for $(a + b)^5$.	1		5		10		10		5		1

The following triangular array of coefficients is called *Pascal's triangle*. If we continue with the sixth row, the first and last numbers are 1. Each of the other numbers is obtained by finding the sum of the two closest numbers above it in the fifth row.

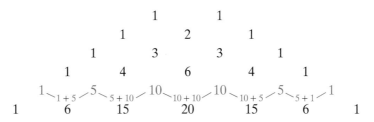

Using the numbers in the sixth row and the variable parts we found, the expansion for $(a + b)^6$ is

$$(a + b)^6 = a^6 + 6a^5b + 15a^4b^2 + 20a^3b^3 + 15a^2b^4 + 6ab^5 + b^6.$$

Binomial Coefficients

Pascal's triangle becomes cumbersome when a binomial contains a relatively large power. Therefore, the coefficients in a binomial expansion are instead given in terms of factorials. The coefficients are written in a special notation, which we define next.

Discover for yourself

What problems are there in using Pascal's triangle when n is large, such as in trying to expand $(a + b)^{23}$ or $(a + b)^{143}$?

1 Evaluate a binomial coefficient.

Study tip

The symbol

$$_nC_r$$

is often used in place of

$$\binom{n}{r}$$

to denote binomial coefficients.

Definition of a binomial coefficient $\binom{n}{r}$

For nonnegative integers n and r, with $n \geqslant r$, the expression $\binom{n}{r}$ (read "n above r") is called a binomial coefficient and is defined by

$$\binom{n}{r} = \frac{n!}{r!(n - r)!}.$$

Using technology

Graphing utilities have the capability of computing binomial coefficients. For example, to find $\binom{6}{2}$, many utilities require the sequence

6 [nCr] 2 [ENTER]

and 15 is displayed. Consult your manual and verify the other evaluations in Example 1.

EXAMPLE 1 **Evaluating Binomial Coefficients**

Evaluate: **a.** $\binom{6}{2}$ **b.** $\binom{3}{0}$ **c.** $\binom{9}{3}$ **d.** $\binom{4}{4}$

Solution

In each case, we apply the definition for the binomial coefficient.

a. $\binom{6}{2} = \frac{6!}{2!(6 - 2)!} = \frac{6!}{2!4!} = \frac{6 \cdot 5 \cdot 4 \cdot 3 \cdot 2 \cdot 1}{(2 \cdot 1)(4 \cdot 3 \cdot 2 \cdot 1)} = 15$

b. $\binom{3}{0} = \frac{3!}{0!(3 - 0)!} = \frac{3!}{0!3!} = \frac{3 \cdot 2 \cdot 1}{(1)(3 \cdot 2 \cdot 1)} = 1$

c. $\dbinom{9}{3} = \dfrac{9!}{3!(9-3)!} = \dfrac{9!}{3!6!} = \dfrac{9 \cdot 8 \cdot 7 \cdot 6 \cdot 5 \cdot 4 \cdot 3 \cdot 2 \cdot 1}{(3 \cdot 2 \cdot 1)(6 \cdot 5 \cdot 4 \cdot 3 \cdot 2 \cdot 1)} = 84$

d. $\dbinom{4}{4} = \dfrac{4!}{4!(4-4)!} = \dfrac{4!}{4!0!} = \dfrac{4 \cdot 3 \cdot 2 \cdot 1}{(4 \cdot 3 \cdot 2 \cdot 1)(1)} = 1$ ∎

EXAMPLE 2   Showing That Binomial Coefficients Are Equal

Show that $\dbinom{n}{r} = \dbinom{n}{n-r}$

Solution

$$\dbinom{n}{r} = \dfrac{n!}{r!(n-r)!}$$

Use the definition of a binomial coefficient on the two given binomial coefficients.

$$\dbinom{n}{n-r} = \dfrac{n!}{(n-r)![n-(n-r)]!}$$

$$= \dfrac{n!}{(n-r)!r!}$$

Simplify within brackets.

$$= \dfrac{n!}{r!(n-r)!}$$

Apply the commutative property.

$$= \dbinom{n}{r}$$

By definition, the above expression is equal to $\dbinom{n}{r}$.   ∎

Discover for yourself

Describe how the equation

$$\dbinom{n}{r} = \dbinom{n}{n-r}$$

is modeled by Pascal's triangle by looking at the pairs of equal numbers in each row. (The evaluation of each binomial coefficient is shown by the small red number at its top right.)

$$\dbinom{0}{0}^1$$
$$\dbinom{1}{0}^1 \quad \dbinom{1}{1}^1$$
$$\dbinom{2}{0}^1 \quad \dbinom{2}{1}^2 \quad \dbinom{2}{2}^1$$
$$\dbinom{3}{0}^1 \quad \dbinom{3}{1}^3 \quad \dbinom{3}{2}^3 \quad \dbinom{3}{3}^1$$
$$\dbinom{4}{0}^1 \quad \dbinom{4}{1}^4 \quad \dbinom{4}{2}^6 \quad \dbinom{4}{3}^4 \quad \dbinom{4}{4}^1$$
$$\dbinom{5}{0}^1 \quad \dbinom{5}{1}^5 \quad \dbinom{5}{2}^{10} \quad \dbinom{5}{3}^{10} \quad \dbinom{5}{4}^5 \quad \dbinom{5}{5}^1$$
⋮

---

2 Expand a binomial raised to a power.

The Binomial Theorem

If we use binomial coefficients and the pattern for the variable part of each term, a formula called the *binomial theorem* or the *binomial expansion* can be written for any natural number power of a binomial.

Discover for yourself

Describe how you would use mathematical induction to prove the Binomial Theorem. The details of the proof are given in Problem 81 in the problem set that follows.

A formula for expanding binomials

The Binomial Theorem

For any positive integer n,

$$(a+b)^n = \dbinom{n}{0}a^n + \dbinom{n}{1}a^{n-1}b + \dbinom{n}{2}a^{n-2}b^2 + \dbinom{n}{3}a^{n-3}b^3 + \cdots + \dbinom{n}{n}b^n.$$

ENRICHMENT ESSAY

The Universality of Mathematics

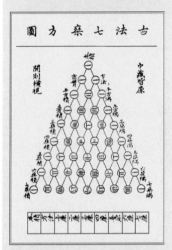

Reprinted with permission from *Science and Civilization in China*, Vol. 3, by Joseph Needham, 1959, Cambridge University Press.

"Pascal's" triangle, credited to French mathematician Blaise Pascal (1623–1662), appeared in a Chinese document printed in 1303. The Binomial Theorem was known in Eastern cultures prior to its discovery in Europe. The same mathematics is often discovered by independent researchers separated by time, place, and culture.

EXAMPLE 3 Using the Binomial Theorem

Expand: $(x + 2)^4$

Solution

$$(x + 2)^4 = \binom{4}{0}x^4 + \binom{4}{1}x^3 \cdot 2 + \binom{4}{2}x^2 \cdot 2^2 + \binom{4}{3}x \cdot 2^3 + \binom{4}{4}2^4$$

$$= \frac{4!}{0!4!}x^4 + \frac{4!}{1!3!}x^3 \cdot 2 + \frac{4!}{2!2!}x^2 \cdot 4 + \frac{4!}{3!1!}x \cdot 8 + \frac{4!}{4!0!} \cdot 16$$

$$= 1 \cdot x^4 + 4x^3 \cdot 2 + 6x^2 \cdot 4 + 4x \cdot 8 + 1 \cdot 16$$

$$= x^4 + 8x^3 + 24x^2 + 32x + 16$$ ∎

EXAMPLE 4 Using the Binomial Theorem

Expand: $(3x + 4y)^3$

Solution

$$(3x + 4y)^3 = \binom{3}{0}(3x)^3 + \binom{3}{1}(3x)^2(4y) + \binom{3}{2}(3x)(4y)^2 + \binom{3}{3}(4y)^3$$

$$= \frac{3!}{0!3!}(27x^3) + \frac{3!}{1!2!}(9x^2)(4y) + \frac{3!}{2!1!}(3x)(16y^2) + \frac{3!}{3!0!}(64y^3)$$

$$= 1(27x^3) + 3(36x^2y) + 3(48xy^2) + 1(64y^3)$$

$$= 27x^3 + 108x^2y + 144xy^2 + 64y^3$$ ∎

EXAMPLE 5 Using the Binomial Theorem

Expand: $(a - 2b)^6$

Solution

$$(a - 2b)^6 = [a + (-2b)]^6$$

$$= \binom{6}{0}a^6 + \binom{6}{1}a^5(-2b) + \binom{6}{2}a^4(-2b)^2 + \binom{6}{3}a^3(-2b)^3$$

$$+ \binom{6}{4}a^2(-2b)^4 + \binom{6}{5}a(-2b)^5 + \binom{6}{6}(-2b)^6$$

$$= \frac{6!}{0!6!}a^6 + \frac{6!}{1!5!}a^5(-2b) + \frac{6!}{2!4!}a^4(-2b)^2 + \frac{6!}{3!3!}a^3(-2b)^3$$

$$+ \frac{6!}{4!2!}a^2(-2b)^4 + \frac{6!}{5!1!}a(-2b)^5 + \frac{6!}{6!0!}(-2b)^6$$

$$= 1a^6 + 6a^5(-2b) + 15a^4(4b^2) + 20a^3(-8b^3)$$

$$+ 15a^2(16b^4) + 6a(-32b^5) + 1(64b^6)$$

$$= a^6 - 12a^5b + 60a^4b^2 - 160a^3b^3$$

$$+ 240a^2b^4 - 192ab^5 + 64b^6$$ ∎

3 Find a specified term in a binomial's expansion.

Finding a Particular Term in a Binomial Expansion

The Binomial Theorem can be used to write any single term of a binomial expansion.

Discover for yourself

Consider the terms in $(a + b)^4$:

$$(a + b)^4 = \binom{4}{0}a^4 + \binom{4}{1}a^3b + \binom{4}{2}a^2b^2 + \binom{4}{3}ab^3 + \binom{4}{4}b^4$$

Term 1   Term 2   Term 3   Term 4   Term 5

1. What is the relationship between the exponent on b and the term number? Using this pattern, what is the exponent on b for the rth term in the expansion of $(a + b)^n$?

2. What number is always obtained when you add the exponents in each term on a and b? What number would you obtain in the case of $(a + b)^n$? Use this observation and the exponent on b for the rth term in the expansion of $(a + b)^n$ (step 1) to obtain the exponent on a in this term.

3. Use the pattern of the binomial coefficients to write the binomial coefficient for the rth term in the expansion of $(a + b)^n$.

4. What is the rth term in the expansion of $(a + b)^n$?

In the Discover for yourself, were you able to write the following formula?

Finding a particular term in a binomial expansion

Formula for the rth term in a binomial expansion

The rth term of the expansion of $(a + b)^n$ is

$$\binom{n}{r-1}a^{n-r+1}b^{r-1}.$$

EXAMPLE 6 **Finding a Single Term of a Binomial Expansion**

Find the fourth term in the expansion of $(c^4 + d^3)^8$.

Solution

$$\binom{n}{r-1}a^{n-r+1}b^{r-1}$$ Use the formula for the rth term in a binomial expansion. Let $r = 4$, $n = 8$, $a = c^4$, and $b = d^3$.

$$= \binom{8}{4-1}(c^4)^{8-4+1}(d^3)^{4-1}$$

$$= \binom{8}{3}(c^4)^5(d^3)^3$$

$$= 56c^{20}d^9$$ This is the fourth term in the expansion of $(c^4 + d^3)^8$. ∎

PROBLEM SET 7.5

Practice Problems

In Problems 1–8, evaluate the given binomial coefficient. Use a graphing utility to verify your answer.

1. $\binom{6}{3}$ **2.** $\binom{7}{2}$ **3.** $\binom{12}{1}$ **4.** $\binom{11}{1}$

5. $\binom{6}{6}$ **6.** $\binom{15}{2}$ **7.** $\binom{100}{2}$ **8.** $\binom{100}{98}$

In Problems 9–38, use the Binomial Theorem to expand each binomial and express the result in simplified form.

9. $(x + 2)^3$ **10.** $(x + 5)^3$ **11.** $(3x + 2y)^3$ **12.** $(4x + 3y)^3$

13. $(2a - 1)^3$ **14.** $(3a - 2)^3$ **15.** $(x^2 + 2y)^4$ **16.** $(3x^2 + 2y^4)^4$

17. $(y - 3)^4$ **18.** $(y - 4)^4$ **19.** $(2x^3 - 1)^4$ **20.** $(2x^5 - 1)^4$

21. $(c + 2)^5$ **22.** $(c + 3)^5$ **23.** $(a - 2)^5$ **24.** $(3 - b)^5$

25. $(4x - 5y)^5$ **26.** $(3x - 2y)^5$ **27.** $\left(\dfrac{1}{a} + b\right)^5$ **28.** $\left(\dfrac{1}{a} + 2b\right)^5$

29. $(2a + b)^6$ **30.** $(x + 2y)^6$ **31.** $(x^2 - y^2)^6$ **32.** $(x - 2y^2)^6$

33. $(a^{1/2} + 2)^4$ **34.** $(a^{1/3} + 3)^3$ **35.** $(a^{-1} + b^{-1})^3$ **36.** $(a^{-1} - b^{-1})^3$

37. $4(x - 2)^4 + 5(x - 1)^3$ **38.** $3(x - 1)^4 + 4(x - 2)^3$

In Problems 39–48, write the first three terms in each binomial expansion, expressing the result in simplified form.

39. $(x^2 + x)^8$ **40.** $(a^2 + a)^{14}$ **41.** $(a^2 - b^2)^{16}$ **42.** $(a^2 - b^2)^{16}$

43. $(a + 3b)^9$ **44.** $(a - 2b)^8$ **45.** $(a + b)^{42}$ **46.** $(a - b)^{43}$

47. $\left(1 + \dfrac{1}{y}\right)^7$ **48.** $\left(y - \dfrac{1}{y}\right)^8$

Find the term indicated in each expansion in Problems 49–56.

49. $(2a + b)^6$; third term

50. $\left(x - \dfrac{1}{2}\right)^9$; fifth term

51. $\left(1 - \dfrac{a^2}{2}\right)^{14}$; eighth term

52. $(c^5 + d^7)^9$; third term

53. $(4a - b)^{10}$; term containing a^2 as a factor

54. $(3a - 2b)^9$; term containing a^4 as a factor

55. $(x^2 + 3)^{12}$; term containing x^8 as a factor

56. $(x^2 + 1)^{10}$; term containing x^8 as a factor

57. Show that $\binom{n}{n - 1} = n$.

58. Show that $\binom{n}{n} = \binom{n + 1}{n + 1}$.

59. Show that $\binom{n}{0} = \binom{n}{n}$.

60. Show that $\binom{n}{0} = \binom{n + 1}{0}$.

True–False Critical Thinking Problems

61. Which one of the following is true?

   a. The binomial expansion for $(a + b)^n$ contains n terms.

   b. The Binomial Theorem can be written in condensed form as $(a + b)^n = \sum_{r=0}^{n} \binom{n}{r} a^{n-r}b^r$.

   c. The sum of the binomial coefficients in $(a + b)^n$ cannot be 2^n.

   d. There are no values of a and b such that $(a + b)^4 = a^4 + b^4$.

62. Which one of the following is true?

   a. The relationship

$$\binom{n}{r} = \binom{n}{n - r}$$

shows that the coefficients of a binomial expansion increase and then decrease in a symmetric pattern.

   b. The second term in the expansion of $(3x - 7y)^5$ is $5(3x)^4(7y)$.

   c. No binomial expansion contains more than a quadrillion terms.

   d. If a and b represent the same algebraic expressions in both $(a + b)^n$ and $(a - b)^n$, then these two expansions do not have the same terms in any of their respective positions.

Technology Problems

In Problems 63–64, graph each of the functions in the same viewing rectangle. Describe how the graphs illustrate the Binomial Theorem.

63. $f_1(x) = (x + 2)^3$
$f_2(x) = x^3$
$f_3(x) = x^3 + 6x^2$
$f_4(x) = x^3 + 6x^2 + 12x$
$f_5(x) = x^3 + 6x^2 + 12x + 8$

64. $f_1(x) = (x + 1)^4$
$f_2(x) = x^4$
$f_3(x) = x^4 + 4x^3$
$f_4(x) = x^4 + 4x^3 + 6x^2$
$f_5(x) = x^4 + 4x^3 + 6x^2 + 4x$
$f_6(x) = x^4 + 4x^3 + 6x^2 + 4x + 1$

In Problems 65–67, use the Binomial Theorem to find a polynomial expansion for each function. Then use a graphing utility and an approach similar to the one in Problems 63 and 64 to verify the expansion.

65. $f_1(x) = (x - 1)^3$

66. $f_1(x) = (x - 2)^4$

67. $f_1(x) = (x + 2)^6$

68. Graphing utilities capable of symbolic manipulation, such as the TI-92, will expand binomials. On the TI-92,

to expand $(3a - 5b)^{12}$, input the following:

   $\boxed{\text{expand}}\,((3a\ \boxed{-}\ 5b)\ \boxed{\text{^}}\ 12)\ \boxed{\text{ENTER}}$

Use a graphing utility with this capability to verify some of the expansions you performed by hand in Problems 9–36.

Writing in Mathematics

69. Are there situations in which it is easier to use Pascal's triangle than binomial coefficients? Describe these situations.

70. Write 11^6 as $(10 + 1)^6$, and describe how the Binomial Theorem can be used to find 11^6 without the use of a calculator. Give some examples that are similar. Now give an example where this technique would not be particularly useful. Describe how this example differs from your previous examples.

71. Describe how you would use mathematical induction to prove that

$$(a + b)^n = \binom{n}{0} a^n + \binom{n}{1} a^{n-1}b + \binom{n}{2} a^{n-2}b^2$$

$$+ \cdots + \binom{n}{n - 1} ab^{n-1} + \binom{n}{n} b^n$$

What happens when $n = 1$? Write the statement that we assume true. Write the statement that we must prove. What must be done to the left side of the assumed equality to make it look like the left side of the statement that must be proved? (More detail on the actual proof is found in Problem 81.)

Critical Thinking Problems

72. Prove that if n is a positive integer,

$$\binom{n}{0} + \binom{n}{1} + \binom{n}{2} + \cdots + \binom{n}{n} = 2^n$$

[*Hint:* $(1 + 1)^n = 2^n$. How does this result compare to the statement in Problem 61(c)?]

73. Which is larger, $(101!)^{100}$ or $(100!)^{101}$?

74. Use patterns to write a conjectured formula, in the form of a single binomial coefficient, for

$$\binom{n}{0}^2 + \binom{n}{1}^2 + \binom{n}{2}^2 + \cdots + \binom{n}{n}^2.$$

75. Simplify:

$$\frac{\dfrac{(x+5)^{n+1}}{(n+1)!}}{\dfrac{(x+5)^n}{n!}}.$$

76. Prove that

$$\binom{n}{r} = \frac{n-r+1}{r}\binom{n}{r-1}.$$

77. Expand $(x^2 + x + 1)^4$ by writing $x^2 + x + 1$ as $x^2 + (x + 1)$.

78. Find a formula for $(a - b)^n$ using sigma notation.

79. Solve for y:

$$\sum_{i=0}^{5}\binom{5}{i}(-1)^i y^{5-i}3^i = 32.$$

80. Show that

$$\binom{n}{r} + \binom{n}{r+1} = \binom{n+1}{r+1}.$$

81. Follow the outline below to use mathematical induction to prove that:

$$(a + b)^n = \binom{n}{0}a^n + \binom{n}{1}a^{n-1}b + \binom{n}{2}a^{n-2}b^2$$

$$+ \cdots + \binom{n}{n-1}ab^{n-1} + \binom{n}{n}b^n.$$

a. Verify the formula for $n = 1$.

b. Replace n with k and $k + 1$, writing the statement that is assumed true and the statement that must be proved.

c. Multiply both sides of the statement assumed to be true by $a + b$. Add exponents on the left. On the right, distribute a and b, respectively.

d. Collect like terms on the right. At this point, you should have

$$(a + b)^{k+1} = \binom{k}{0}a^{k+1} + \left[\binom{k}{0} + \binom{k}{1}\right]a^k b$$

$$+ \left[\binom{k}{1} + \binom{k}{2}\right]a^{k-1}b^2 + \left[\binom{k}{2} + \binom{k}{3}\right]a^{k-2}b^3$$

$$+ \cdots + \left[\binom{k}{k-1} + \binom{k}{k}\right]ab^k + \binom{k}{k}b^{k+1}.$$

e. Use the result of Problem 80 to add the binomial sums in brackets. For example, since $\binom{n}{r} + \binom{n}{r+1} = \binom{n+1}{r+1}$ then $\binom{k}{0} + \binom{k}{1} = \binom{k+1}{1}$ and $\binom{k}{1} + \binom{k}{2} = \binom{k+1}{2}$.

f. Since $\binom{k}{0} = \binom{k+1}{0}$ (why?) and $\binom{k}{k} = \binom{k+1}{k+1}$ (why?), substitute these results and the results from part (e) into the equation in part (d). This should give the statement that we were required to prove in the second step of the mathematical induction process.

S E C T I O N   7 . 6 **Counting Principles, Permutations, and Combinations**

Solutions Manual   Tutorial   Video 13

Objectives

1 Solve counting problems using the Fundamental Counting Principle.

2 Solve counting problems involving permutations.

3 Solve counting problems involving combinations.

Solve counting problems using the Fundamental Counting Principle.

The Fundamental Counting Principle

Counting the number of ways an event can occur plays a major role in the study of probability. Example 1 describes a counting problem.

EXAMPLE 1 Using a Tree Diagram

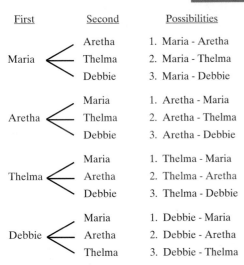

First	Second	Possibilities
Maria	Aretha	1. Maria - Aretha
	Thelma	2. Maria - Thelma
	Debbie	3. Maria - Debbie
Aretha	Maria	1. Aretha - Maria
	Thelma	2. Aretha - Thelma
	Debbie	3. Aretha - Debbie
Thelma	Maria	1. Thelma - Maria
	Aretha	2. Thelma - Aretha
	Debbie	3. Thelma - Debbie
Debbie	Maria	1. Debbie - Maria
	Aretha	2. Debbie - Aretha
	Thelma	3. Debbie - Thelma

Figure 7.8

There are 12 ways for four runners to finish first and second in a race.

Four runners are in a one-mile race: Maria, Aretha, Thelma, and Debbie. Points are awarded only to the women finishing first or second. In how many ways can these four finish first or second?

Solution

Observe that if Maria finishes first, then each of the other three runners could finish second. This observation might help us to construct a *tree diagram* in which each of the ways the runners can finish first or second is determined by reading along the branches of the diagram.

The tree diagram in Figure 7.8 shows that there are 12 possible ways for the runners to finish first and second in the race. ∎

Because constructing a tree diagram is a tedious task if the number being counted is large, there is another method of quickly and accurately counting the number of possibilities in more complex situations. This method is appropriately called the *Fundamental Counting Principle*.

The Fundamental Counting Principle

Let $E_1, E_2, E_3, \ldots, E_k$ be a sequence of events. The first event, E_1, can occur in m_1 different ways. After E_1 has occurred, E_2 can occur in m_2 different ways. After E_2 has occurred, E_3 can occur in m_3 different ways, and so on. The total number of ways that all k events can occur is

$$m_1 \cdot m_2 \cdot m_3 \cdot \cdots \cdot m_k.$$

We can apply the Fundamental Counting Principle, also called the Multiplication Principle, to the situation of the four runners in Example 1. Two events are involved, finishing first and second. The first event, finishing first, can occur in four different ways, meaning that any one of the four runners can finish in first place. The second event, finishing second, can occur in only three ways, meaning that any one of the remaining three runners can achieve second place.

Event	E_1 (coming in first)	E_2 (coming in second)
Number of Ways the Event Can Occur	4	3

By the Fundamental Counting Principle, there are $4 \cdot 3$ or 12 ways that first and second place can be taken in a race involving four runners. This result agrees with the number of possibilities obtained in the tree diagram in Example 1.

EXAMPLE 2 **Applying the Fundamental Counting Principle**

Telephone numbers in the United States begin with three-digit area codes followed by seven-digit local telephone numbers. Area codes and local telephone numbers cannot begin with 0 or 1. How many different telephone numbers are possible?

Colleen Browning "Telephones." Photo courtesy The Butler Institute of American Art, Youngstown, Ohio.

Solution

This solution involves ten different events, shown as follows:

Area Code			**Local Telephone Number**						
E_1	E_2	E_3	E_4	E_5	E_6	E_7	E_8	E_9	E_{10}
8 ways	10 ways	10 ways	8 ways	10 ways	10 ways	10 ways	10 ways	10 ways	10 ways

Notice that because E_1 and E_4 cannot use 0 or 1, there are only eight choices for these digits (2–9). The other events can be chosen in ten ways (0–9). By the Fundamental Counting Principle, the total number of telephone numbers possible is

$$8 \cdot 10 \cdot 10 \cdot 8 \cdot 10 \cdot 10 \cdot 10 \cdot 10 \cdot 10 \cdot 10 = 6,400,000,000.$$

Since the population of the United States is approximately 270,000,000, there seems to be no need to worry about running out of telephone numbers in the immediate future. ■

2 Solve counting problems involving permutations.

Permutations

Examples 1 and 2 involved arrangements of runners and arrangements of telephone numbers. The Fundamental Counting Principle can be used to determine the number of arrangements of distinct objects in a sequential manner. An ordering of n elements is called a *permutation* of the elements.

> **Definition of permutation**
>
> A permutation of n elements is an ordering of the n elements such that one element is first, one is second, one is third, and so on.

EXAMPLE 3 **Listing Permutations**

List all permutations of the letters in the word SUM. How many different permutations are there? Verify the result using the Fundamental Counting Principle.

Solution

The letters S, U, and M can be arranged in the following six different ways:

S, U, M   U, S, M   M, S, U

S, M, U   U, M, S   M, U, S

The list indicates that these three letters have six different permutations. Now let's verify this result using the Fundamental Counting Principle.

> 1st position: Any of the *three* letters (S, U, or M)
> 2nd position: Any of the remaining *two* letters
> 3rd position: The *one* remaining letter

There are three choices for the first position, two choices for the second position, and one choice for the third position, shown as follows:

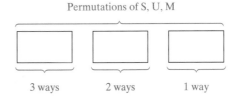

Permutations of S, U, M

3 ways   2 ways   1 way

The number of permutations of three letters is

$3 \cdot 2 \cdot 1 = 3!$ or   6. ∎

EXAMPLE 4 **Finding the Number of Permutations Without Making a List**

How many different permutations are possible for the letters in the word PRODUCT?

Solution

Two of the possible permutations are

P, R, O, D, U, C, T   P, R, O, D, U, T, C.

You can see that there are too many different permutations to attempt listing them all. We can, however, use the Fundamental Counting Principle.

> 1st position: Any of the *seven* letters
> 2nd position: Any of the remaining *six* letters
> 3rd position: Any of the remaining *five* letters
> 4th position: Any of the remaining *four* letters
> 5th position: Any of the remaining *three* letters
> 6th position: Any of the remaining *two* letters
> 7th position: The *one* remaining letter

The situation is illustrated as follows:

Permutations of P, R, O, D, U, C, T

| 7 ways | 6 ways | 5 ways | 4 ways | 3 ways | 2 ways | 1 way |

By the Fundamental Counting Principle, the total number of permutations of the seven letters is given by

$$7 \cdot 6 \cdot 5 \cdot 4 \cdot 3 \cdot 2 \cdot 1 = 7! \quad \text{or} \quad 5040.$$ ■

In Examples 3 and 4, we considered the number of permutations of n objects where all of the n objects were used in the arrangements; this is denoted $_nP_n$. Since there are n choices for first position, $n - 1$ choices for second position, $n - 2$ choices for third position, and so on down to only one choice for last position,

$$_nP_n = n \cdot (n - 1) \cdot (n - 2) \cdot \cdots \cdot 3 \cdot 2 \cdot 1 = n!$$

Equivalently, there are $n!$ different ways in which n elements can be arranged.

Permutations of n elements taken n at a time

The notation $_nP_n$ represents the number of permutations of n things taken n at a time.

$$_nP_n = n!$$

Let's now turn our attention to permutations in which not all of the elements are used. Our next example deals with $_nP_r$, the number of permutations of n elements taken r at a time, where $r < n$.

EXAMPLE 5 **Permutations of n Elements Taken r at a Time**

A club with 20 members is to choose three officers—president, vice-president, and secretary-treasurer. In how many ways can these offices be filled?

Solution

We have the following possibilities:

President: 20 choices (any one of the 20 members)
Vice-president: 19 choices (any one of the 19 remaining members)
Secretary-treasurer: 18 choices (any one of the 18 remaining members)

We now apply the Fundamental Counting Principle.

Different orders of officers

| President | Vice-President | Secretary-Treasurer |

| 20 ways | 19 ways | 18 ways |

Thus, there are $20 \cdot 19 \cdot 18 = 6840$ ways in which the three offices can be filled. Equivalently, we can write

$$_{20}P_3 = 6840 \qquad \text{The number of permutations of 20 elements taken 3 at a time is 6840.} \qquad \blacksquare$$

We can generalize the result of Example 5 and use a formula to find $_nP_r$.

S tudy tip

There are three different notations for the number of permutations of n elements taken r at a time. One is $_nP_r$, the one in the box. Also used are $P(n, r)$ and P_r^n.

Permutations of n elements taken r at a time

The number of permutations of n elements taken r at a time is

$$_nP_r = \frac{n!}{(n-r)!} = n(n-1)(n-2) \cdots (n-r+1).$$

U sing technology

Graphing utilities have a key for permutations. For example, to find $_{20}P_3$, many utilities require the sequence

20 $\boxed{_nP_r}$ 3 $\boxed{\text{ENTER}}$

and 6840 is displayed. Consult your manual and provide numerical support for all permutation results in this section.

With this formula, we can solve Example 5 to find that the number of permutations of 20 members taken 3 at a time is

$$_{20}P_3 = \frac{20!}{(20-3)!} = \frac{20!}{17!} = \frac{20 \cdot 19 \cdot 18 \cdot 17!}{17!} = 20 \cdot 19 \cdot 18 = 6840.$$

EXAMPLE 6   Counting Permutations

How many different five-letter code words can be arranged using the 26 letters of the alphabet if no letter is repeated?

Solution

We must find the number of permutations of 26 elements taken 5 at a time.

$$_{26}P_5 = \frac{26!}{(26-5)!} \qquad \text{Use } _nP_r = \frac{n!}{(n-r)!}$$

$$= \frac{26!}{21!}$$

$$= \frac{26 \cdot 25 \cdot 24 \cdot 23 \cdot 22 \cdot 21!}{21!} \qquad \text{You many find that this step is done mentally.}$$

$$= 26 \cdot 25 \cdot 24 \cdot 23 \cdot 22$$

$$= 7{,}893{,}600$$

D iscover for yourself

Solve Example 6 using the Fundamental Counting Principle. At what point in the solution process shown on the right is the work the same as yours?

Thus, 7,893,600 different five-letter code words can be arranged. $\blacksquare$

The formula for $_nP_r$ cannot be used if restrictions are placed on the arrangements. In this case, we can count permutations using the Fundamental Counting Principle.

EXAMPLE 7   Using the Fundamental Counting Principle with Restricted Arrangements

How many four-digit even numbers greater than or equal to 7000 can be formed using the digits 0, 1, 2, 3, 4, 5, 6, 7, 8, and 9 if digits may be repeated?

Solution

Examples of numbers in the arrangements include

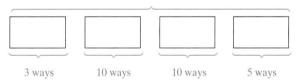

Once again, there are too many permutations to attempt listing them all, so we turn to the Fundamental Counting Principle.

> 1st position: *Three* choices (7, 8, or 9), since the number is to be greater than or equal to 7000
> 2nd position: *Ten* choices (any one of 0–9)
> 3rd position: *Ten* choices (any one of 0–9)
> 4th position: *Five* choices (0, 2, 4, 6, or 8), since the number is to be even

The situation is illustrated as follows:

Even arrangements greater than or equal to 7000

3 ways	10 ways	10 ways	5 ways

By the Fundamental Counting Principle, the number of four-digit even numbers greater than or equal to 7000 that can be formed is

$$3 \cdot 10 \cdot 10 \cdot 5 \quad \text{or} \quad 1500.$$ ■

3 Solve counting problems involving combinations.

Combinations

A permutation takes into account the order in which elements are selected. Change the order and a new permutation results. For example, 7234 is a different permutation than 7432 because of the different place-values of the digits.

We now consider cases in which the order of selection is not important. For example, let's return to our four runners: Maria, Aretha, Thelma, and Debbie. Suppose that only two of the runners are allowed to represent the college in a conference track meet. Listed below are all the possible pairs of two runners that could be selected.

Maria and Aretha	Aretha and Thelma
Maria and Thelma	Aretha and Debbie
Maria and Debbie	Thelma and Debbie

There are six possible groups of two that can be formed from the runners. In other words, there are six subjects of size two that can be formed. The pair consisting of Maria and Aretha is the same pair as the one consisting of Aretha and Maria. In considering these pairs, the order in which the names appear is not important.

The six choices of runners are the *combinations* of four elements (the four runners) taken two at a time. A *combination* of n elements taken r at a time is one of the ways in which r elements can be chosen from n elements.

In the example above, each combination of two conference runners forms 2! permutations (Maria and Aretha, and Aretha and Maria, for example). The number of combinations can therefore be found without making a list. We need only divide the number of permutations of four things taken two at a time by 2!. This gives us

$$\frac{{}_4P_2}{2!} = \frac{\dfrac{4!}{(4-2)!}}{2!} = \frac{\dfrac{4!}{2!}}{2!} = \frac{4!}{2!2!} = \frac{4 \cdot 3 \cdot 2 \cdot 1}{2 \cdot 1 \cdot 2 \cdot 1} = 6$$

which agrees with the answer that we obtained by listing all combinations.

In general, we can find the number of combinations of n elements taken r at a time by dividing the number of permutations of n elements taken r at a time by $r!$ to get

$$\frac{{}_nP_r}{r!} = \frac{\dfrac{n!}{(n-r)!}}{r!} = \frac{n!}{(n-r)!r!}.$$

We will use the symbol $\binom{n}{r}$ to represent the number of combinations of n elements taken r at a time. Notice that $\binom{n}{r}$ is also the symbol for a binomial coefficient, giving the coefficients in the Binomial Theorem. Using this symbol, we have the following formula.

S tudy tip

1. Since $\binom{n}{r}$ gives the number of combinations of n elements taken r at a time, it is sometimes read as "n pick r."

2. There are several notations for the number of combinations of n elements taken r at a time. One is $\binom{n}{r}$, the one in the box. Also used are $C(n, r)$, C_r^n, and ${}_nC_r$.

Combinations of n elements taken r at a time

The number of combinations of n elements taken r at a time is

$$\binom{n}{r} = \frac{n!}{(n-r)!r!}.$$

EXAMPLE 8 **Counting Combinations**

From a group of 25 people, 5 are to be selected to write a committee report.

a. In how many different ways can the people be selected?
b. If it is decided that one particular person must be on the committee, in how many ways can the group of 5 be chosen?
c. If the group writing the committee report is to contain at most 5 people, in how many ways can the group be selected?

Solution

a. We must find the number of combinations of 25 elements taken 5 at a time.

$$\binom{25}{5} = \frac{25!}{(25-5)!5!} \qquad \binom{n}{r} = \frac{n!}{(n-r)!r!}$$

$$= \frac{25!}{20!5!}$$

U sing technology

Graphing utilities have a key for combinations. For example, to find $\binom{25}{5}$, many utilities require the sequence

25 ${}_nC_r$ 5 ENTER

and 53130 is displayed. Consult your manual and provide numerical support for all combination results in this section.

$$= \frac{25 \cdot 24 \cdot 23 \cdot 22 \cdot 21}{5 \cdot 4 \cdot 3 \cdot 2 \cdot 1} \qquad \text{Cancel factors of 20! in the numerator and denominator.}$$

$$= 53{,}130$$

There are 53,130 ways to select the group that will write the committee report. Notice that this is a combination, not a permutation, situation because the order within the group of 5 does not matter.

b. One person must be on the committee, so the problem becomes the number of ways of selecting 4 additional people from the 24 remaining members of the group.

$$\binom{24}{4} = \frac{24!}{(24 - 4)!4!} \qquad \binom{n}{r} = \frac{n!}{(n - r)!r!}$$

$$= \frac{24!}{20!4!}$$

$$= \frac{24 \cdot 23 \cdot 22 \cdot 21}{4 \cdot 3 \cdot 2 \cdot 1} \qquad \text{Cancel factors of 20! in the numerator and denominator.}$$

$$= 10{,}626$$

Since one person must be on the committee, the remaining 4 people can be selected in 10,626 ways.

c. If at most 5 people are to be chosen, then the committee writing the report will contain 1, 2, 3, 4, or 5 people.

Case	Number of Combinations
1 person selected	$\binom{25}{1} = \frac{25!}{24!1!} = \frac{25 \cdot 24!}{24! \cdot 1} = 25$
2 people selected	$\binom{25}{2} = \frac{25!}{23!2!} = \frac{25 \cdot 24 \cdot 23!}{23!2!} = 300$
3 people selected	$\binom{25}{3} = \frac{25!}{22!3!} = \frac{25 \cdot 24 \cdot 23 \cdot 22!}{22!3!} = 2300$
4 people selected	$\binom{25}{4} = \frac{25!}{21!4!} = \frac{25 \cdot 24 \cdot 23 \cdot 22 \cdot 21!}{21!4!} = 12{,}650$
5 people selected	$\binom{25}{5} = 53{,}130$ (from part (a))

The total number of ways to select at most 5 people to write the committee report is

$$25 + 300 + 2300 + 12{,}650 + 53{,}130 \quad \text{or} \quad 68{,}405 \text{ ways}. \qquad \blacksquare$$

In our next example, we use both the Fundamental Counting Principle and the formula for counting combinations.

EXAMPLE 9 **Combinations and the Fundamental Counting Principle**

The U.S. Senate of the 104th Congress consisted of 54 Republicans and 46 Democrats. How many subcommittees can be formed if each must have 3 Republicans and 2 Democrats?

Jenny Holzer Selections from "Truisms" 1982. Spectacolor board, Times Square, New York. Sponsored by The Public Art Fund. Photo credit: Lisa Kahne. Courtesy Barbara Gladstone Gallery. © Jenny Holzer.

Solution

The order in which the members are selected does not matter. Thus, this is a problem in combinations.

The number of ways of selecting 3 Republicans out of 54 without regard to order is

$$\binom{54}{3} = \frac{54!}{51!3!} = \frac{54 \cdot 53 \cdot 52}{3 \cdot 2 \cdot 1} = 24{,}804.$$

The number of ways of selecting 2 Democrats out of 46 without regard to order is

$$\binom{46}{2} = \frac{46!}{44!2!} = \frac{46 \cdot 45}{2 \cdot 1} = 1035.$$

By the Fundamental Counting Principle, the total number of subcommittees possible is

$$\binom{54}{3} \cdot \binom{46}{2} = (24{,}804)(1035) = 25{,}672{,}140. \qquad \blacksquare$$

Discover for yourself

Give an example of a real world situation that is a permutation. Now give an example of one that is a combination. Describe the difference between your two examples.

Study tip

The major problem generally encountered in the study of permutations and combinations is the ability to distinguish between each in a given real world situation. Remember that *order is important* when considering a *permutation*. Different orderings or arrangements produce different permutations. On the other hand, *order is not important* when considering a *combination*. A change in order does not produce a new combination.

Permutations	**Combinations**
$_nP_r = \dfrac{n!}{(n-r)!}$	$\dbinom{n}{r} = \dfrac{n!}{(n-r)!r!}$
Different orderings produce different permutations.	Different orderings do not produce different combinations.
Helpful words: Arrangement, Order, Schedule	**Helpful words:** Selection, Group, Committee, Subcommittee, Subset

PROBLEM SET 7.6

Practice and Application Problems

Solve Problems 1–14 using the Fundamental Counting Principle.

1. In how many ways can 10 people finish a one-mile race in first, second, and third position, assuming there are no ties?

2. From a club of eight members, a chairperson, vice-chairperson, and secretary are to be elected. Each per-son can fill only one position. In how many ways can this be done?

3. The local seven-digit telephone numbers in Coral Gables, Florida, have 661 as the first three digits. How many different telephone numbers are possible?

4. How many seven-digit local telephone numbers are possible if the first two digits must be 45?

5. How many different four-letter radio station call letters can be arranged if the first letter must be W or K and letters may be repeated?

6. How many different license plates are there that consist of one letter followed by a three-digit number? Repetitions of digits are allowed.

7. How many four-digit odd numbers greater than 3000 can be formed using the digits 0, 1, 2, 3, 4, 5, 6, 7, 8, and 9 if digits may be repeated?

8. How many four-digit odd numbers greater than 6000 can be formed using the digits 0, 1, 2, 3, 4, 5, 6, 7, 8, and 9 if digits may be repeated?

9. How many four-digit even numbers greater than 5000 can be formed using the digits 0, 1, 2, 3, 4, and 5 if digits may not be repeated?

10. How many four-digit odd numbers greater than 8000 can be formed using the digits 0, 1, 2, 3, 4, and 8 if digits may not be repeated?

11. If no questions are omitted, in how many ways can a ten-question true-false exam be answered?

12. If no questions are omitted, in how many ways can a five-question multiple-choice quiz be answered, where each question contains four choices?

13. Four women and three men have reserved seats in a row for a concert. In how many ways can the seven people be seated if the men and women are to be alternated?

14. Three men and two women walk through a doorway single-file. In how many ways can this be done if the women and men are to be alternated?

In Problems 15–22, evaluate $_nP_r$.

15. $_6P_6$

16. $_9P_9$

17. $_9P_4$

18. $_{20}P_5$

19. $_{10}P_2$

20. $_{50}P_2$

21. $_7P_6$

22. $_8P_7$

In Problems 23–28, express the answer using permutation notation ($_nP_r$) and evaluate.

23. A club consisting of 20 members wishes to elect a president, vice-president, secretary, and treasurer. In how many ways can this be done?

24. A baseball team has 20 players. The team's manager intends to select the starting 9 players by assigning a batting order. In how many ways can this be done?

25. In how many ways can six paintings be displayed in each of six different places in a row on the walls of a museum?

26. In how many ways can seven children be seated in seven chairs in the front row of an auditorium?

27. How many arrangements can be made using three of the letters of the word DOLPHIN if no letter is to be used more than once?

28. How many arrangements can be made using four of the letters of the word COMBINE if no letter is to be used more than once?

In Problems 29–36, evaluate $\binom{n}{r}$.

29. $\binom{8}{1}$

30. $\binom{7}{1}$

31. $\binom{6}{6}$

32. $\binom{4}{4}$

33. $\binom{10}{6}$

34. $\binom{12}{6}$

35. $\binom{30}{3}$

36. $\binom{25}{4}$

In Problems 37–40, express the answer using combination notation, $\binom{n}{r}$, and evaluate.

37. An election ballot asks voters to select three city commissioners from a group of six candidates. In how many ways can this be done?

38. A four-person committee is to be elected from an organization's membership of 20 persons. How many different committees are possible?

39. To win at LOTTO in the state of Florida, one must correctly select six numbers from a collection of 49 numbers (1 through 49). The order in which the selection is made does not matter. How many different selections are possible?

40. How many distinct triangles can be formed using six points lying on a circle, where the vertices of each triangle are chosen from the six points?

In Problems 41–59, solve each problem using permutation notation, combination notation, or the Fundamental Counting Principle.

41. The lock on a briefcase consists of three dials, each marked with the digits 0–9, as shown in the figure. Only by selecting the correct digit on each of the three dials can the briefcase be opened. If repetitions of digits are allowed, how many different number sequences are possible?

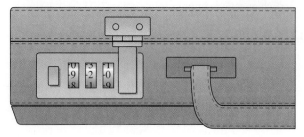

42. A student located eight books needed at the library, but only three books can be checked out at a time. How many sets of three books can be selected?

43. A state lottery is played by selecting five different numbers from 1 through 25. If the order of selection is not important, in how many ways can this be done?

44. How many subjects of 6 elements can be formed from a set of 20 elements?

45. How many subjects of 5 elements can be formed from a set of 25 elements?

46. A person can order a new car with a choice of six possible colors, with or without air-conditioning, with or without automatic transmission, with or without power windows, and with or without a radio. In how many different ways can a new car be ordered in terms of these options?

47. At the Greasy Spoon Restaurant a person can order a dinner with or without salad, with or without a beverage, with or without dessert, have the meat broiled rare, medium, or well done, and have the potato baked, mashed, or french fried. In how many different ways can a person order a dinner?

48. In a production of *Bye Bye Birdie*, seven actors are considered for the male roles of Albert, Mr. MacAfee, Hugo, and Conrad. In how many ways can the director cast the male roles?

49. In a production of *West Side Story*, eight actors are considered for the male roles of Tony, Riff, and Bernardo. In how many ways can the director cast the male roles?

50. Of the seven actors considered for the male roles in *Bye Bye Birdie* (Problem 48), four are to be randomly selected to attend a workshop on acting in musical theater. In how many ways can the selections be made?

51. Of the eight actors considered for the male leads in *West Side Story* (Problem 49), three are to be randomly selected to attend a workshop on Stephen Sondheim's contributions to musical theater. In how many ways can the selections be made?

52. A carton contains ten apples, three of which are rotten. How many selections consisting of five apples each can be formed if each is to contain three good apples and two rotten apples?

53. A basketball team consists of three centers, six forwards, and five guards. In how many ways can the coach select one center, two forwards, and two guards for the starting team?

54. In a group of 20 people, how long will it take each person to shake hands with every other person in the group, assuming that it takes three seconds for each shake and only two people can shake hands at a time? What if the group is increased to 40 people?

55. Six people are to be selected from a group of six couples. In how many ways can this be done if there is to be at least one couple in the group of six?

56. Three identical Chevrolets, four identical Fords, and five identical Toyotas are assigned to 12 salespeople. In how many ways can this be done?

57. A club has a membership of 9 men and 12 women. A five-member committee is to be selected to attend a conference. How many committees with no more than three women can be formed?

58. From a club of 20 people, in how many ways can a group of at most three members be selected to attend a conference?

59. From a theater group of 40 people, five are to be selected to attend the opening night of a new Andrew Lloyd-Weber musical. If it has already been decided that a certain group member will attend the performance, in how many ways can different groups be selected?

As illustrated in the figure, a standard 52-card bridge deck consists of two red suits, diamonds and hearts, and two black suits, clubs and spades. Each suit has 13 cards of different face values—A (ace), 2, 3, 4, 5, 6, 7, 8, 9, 10, J (Jack), Q (Queen), and K (King).

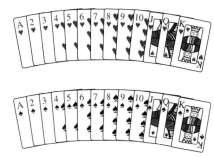

Use this information to answer Problems 60–64.

60. How many five-card poker hands are possible?

61. Five cards such as K–K–6–6–6, consisting of two cards of the same face value from different suits and three cards of the same face value from different suits, is called a *full house*. How many different full houses are possible?

62. In how many ways can a five-card poker hand be dealt so that all five cards are of the same suit (called a *flush*)?

63. In how many ways can a 5-card poker hand be dealt so that 4 of the 5 cards have the same face value (called *four of a kind*)?

64. In bridge, each player is dealt a 13-card hand. How many different bridge hands are there? If a hand can be dealt in two seconds, how many years would it take to deal every possible combination?

True–False Critical Thinking Problems

65. Which one of the following is true?

a. The diagram shows five towns that are connected by roads. Starting at any one town, there are 140 different routes so that each town is visited exactly once.

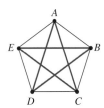

b. According to NASA, the space shuttle food menu for dinner provides 10 options for the main dish, eight vegetable options, and 13 dessert options. Assuming that each astronaut selects one option from the main dish, one from the vegetable dish, and

one from the dessert dish, there are $10 + 8 + 13$ menus that are possible.

c. $\dbinom{n}{r} = \dfrac{{}_nP_r}{r!}$

d. As a group increases in size, the number of two-person relationships increases in a linear manner.

66. Which one of the following is true?

a. The number of ways to choose four questions out of ten questions on an essay test is ${}_{10}P_4$.

b. If $r > 1$, ${}_nP_r$ is less than $\dbinom{n}{r}$.

c. ${}_7P_3 = 3! \dbinom{7}{3}$

d. The number of ways to pick a winner and first runner-up in a piano recital with 20 contestants is $\dbinom{20}{2}$.

Writing in Mathematics

In Problems 67–69, write a word problem that can be solved by evaluating the given expression.

67. $5!$

68. ${}_7P_3$

69. $\dbinom{10}{3} \cdot \dbinom{7}{2}$

70. Simplify the following binomial coefficients and describe what each result means in terms of combinations.

a. $\dbinom{n}{1}$

b. $\dbinom{n}{n-1}$

71. The lock on the briefcase in Problem 41 is often called a *combination* lock. In terms of the mathematical definition of combination, is this an accurate word choice? Why or why not? Give examples of the everyday use of the word "combination" in which the word should be replaced by "permutation."

72. State whether each of the following is a problem involving a permutation or a combination. Explain your answer.

a. How many student identification numbers can be assigned in your college?

b. How many faculty committees with five faculty members who make decisions about student peti-

tions can be chosen from the full-time faculty of your college?

c. How many ways can a student government president, vice-president, and treasurer be elected from all full-time students of your college?

d. How many ways can a five-card hand be dealt from an ordinary 52-card bridge deck?

73. What does the answer in Problem 39 tell you about winning the LOTTO game in the Florida lottery? How many LOTTO tickets would one have to purchase to have a 50/50 chance of winning? Describe how you arrived at this number.

74. List all distinguishable three-letter permutations using the letters in the word TOO. Explain why the number of permutations is not 3!.

75. For a long time New York City had an area code of 212 followed by seven-digit local telephone numbers. In terms of the population of New York City and the average number of phones per household, explain why the city now uses more than one area code.

Critical Thinking Problems

76. Two permutations that differ only in the arrangement of identical objects are said to be *indistinguishable*. For example, the total number of permutations of the four letters B, O, B, S is not 4! because there are two B's in the list. We can find the number of distinguishable permutations using the following idea:

If a collection of n objects has n_1 identical objects of the same type, n_2 identical objects of a second kind, n_3 of a third kind, and so on for a total of $n = n_1 + n_2 + \cdots + n_k$ objects, the number of distinguishable permutations of the n objects is given by

$$\frac{n!}{n_1!\,n_2!\,n_3!\cdots n_k!}.$$

a. Use this formula to find the number of distinguishable permutations of the letters B, O, B, S. List all the distinguishable permutations.

b. How many letter patterns can be obtained by arranging the letters in MISSISSIPPI?

c. How many different signals consisting of eight flags can be made using three white flags, four red flags and one blue flag?

d. In how many different ways can a^4bc^6 be written without using exponents?

77. A student has five English books, three history books, two mathematics books, and one chemistry book. In how many ways can these 11 books be arranged on a shelf if those in a given subject are to be placed together?

78. In how many ways can five people be seated around a circular table? (Two seating arrangements are identical if each person moves one place to the right (or left).) In general, how many circular permutations of n objects are there?

79. You are standing on a street corner and decide to take the following *random walk*: You toss a coin and if the result is heads, you walk one block north. If the result is tails, you walk one block south. Once you reach the new corner, you apply the same rules. If you toss the coin ten times, in how many ways will you return to the original corner?

SECTION 7.7

Solutions Manual   **Tutorial**   **Video 13**

An Introduction to Probability

Objectives

1 Find the probability of an event.
2 Compute probabilities with permutations and combinations.
3 Find probabilities involving unions, intersections, and complements.

Many things in life have uncertainty associated with them. Regrettably, only taxes and death appear certain! In spite of this, we are frequently required to make decisions based on what is likely rather than on what is certain. In this section, we consider some of the basic ideas in the theory of probability, which enables us to study uncertainties mathematically.

1 Find the probability of an event.

Squeak Carnwath "Equations" 1981, oil on cotton canvas 96 in. h × 72 in. w.

The Probability of an Event

Let's consider a situation in which a student does not study for a quiz that contains multiple-choice questions. Each question has five possible choices, with exactly one correct choice. By guessing at the answer to each question, the student is engaged in an *experiment*. In the study of probability, an experiment is any occurrence for which the outcome is uncertain. It is uncertain if the guess is right or wrong. The *sample space* of the experiment is the set of all possible outcomes, in this case the set consisting of the five choices. Using S for the sample space, we can write

$$S = \{a, b, c, d, e\}$$

(assuming that the choices are designated from a through e). The event of interest is getting the question correct. In general, an *event* is any subcollection, or subset, of the sample space. The probability of success of any one question is $\frac{1}{5}$ because there are five choices (five elements in the sample space) and only one is correct.

When all outcomes in an experiment are equally likely, we can find the probability of an event E by dividing the number of outcomes that make up the event, $n(E)$, by the number of outcomes in the sample space, $n(S)$. For the student who is guessing on the multiple-choice items, we have

$$\frac{n(E)}{n(S)} = \frac{1}{5}.$$

The probability of a correct guess is $\frac{1}{5}$.

We can generalize from this situation and define the probability of any event whose occurrence is uncertain.

The probability of an event

Suppose that E is an event in the sample space S, where all outcomes in E and S are equally likely. If event E has $n(E)$ outcomes and its sample

Figure 7.9

When two six-sided dice are rolled, there are 36 equally likely outcomes.

space S has $n(S)$ outcomes, then the *probability* of E, denoted by $P(E)$, is defined by

$$P(E) = \frac{n(E)}{n(S)}.$$

EXAMPLE 1   Finding the Probability of an Event

Two ordinary six-sided dice are rolled (Figure 7.9). What is the probability of getting a sum of 8?

Solution

Since there are six equally likely outcomes for each die, by the Fundamental Counting Principle, there are $6 \cdot 6$ or 36 equally likely outcomes in the sample space. The 36 outcomes are shown as ordered pairs and the five ways of rolling an 8 appear in the highlighted diagonal as follows.

	Second Die					
First Die	(1, 1)	(1, 2)	(1, 3)	(1, 4)	(1, 5)	(1, 6)
	(2, 1)	(2, 2)	(2, 3)	(2, 4)	(2, 5)	(2, 6)
	(3, 1)	(3, 2)	(3, 3)	(3, 4)	(3, 5)	(3, 6)
	(4, 1)	(4, 2)	(4, 3)	(4, 4)	(4, 5)	(4, 6)
	(5, 1)	(5, 2)	(5, 3)	(5, 4)	(5, 5)	(5, 6)
	(6, 1)	(6, 2)	(6, 3)	(6, 4)	(6, 5)	(6, 6)

$$S = \{(1, 1), (1, 2), (1, 3), (1, 4), (1, 5), (1, 6),$$
$$(2, 1), (2, 2), (2, 3), (2, 4), (2, 5), (2, 6),$$
$$(3, 1), (3, 2), (3, 3), (3, 4), (3, 5), (3, 6),$$
$$(4, 1), (4, 2), (4, 3), (4, 4), (4, 5), (4, 6),$$
$$(5, 1), (5, 2), (5, 3), (5, 4), (5, 5), (5, 6),$$
$$(6, 1), (6, 2), (6, 3), (6, 4), (6, 5), (6, 6)\}$$

The phrase "getting a sum of 8" describes the event

$$E = \{(6, 2), (5, 3), (4, 4), (3, 5), (2, 6)\}.$$

This event has 5 outcomes, so $n(E) = 5$. Thus, the probability of getting a sum of 8 is

$$P(E) = \frac{n(E)}{n(S)} = \frac{5}{36}.$$ ■

John Wilde "Oct., 1985 w/a Bierstadtian Landscape" 1991. Oil/panel, 12 × 14 in. Courtesy of Schmidt Bingham Gallery, NY.

EXAMPLE 2   Random Selection

The number of people in various regions of the contiguous 48 states of the United States is shown in the rather unusual map in Figure 7.10. The size of each state is drawn according to its population rather than its geographic area. Each small square grid in the figure represents 100,000 people. One person from the 48 states is selected at random, meaning that each person has an equal chance of being selected. What is the probability that the selected person lives in the Northeast?

Solution

From the figure, there are 51.6 million people living in the Northeast. There are a total of

$$51.6 + 89.6 + 56.1 + 61.5 = 258.8 \text{ million people.}$$

The probability of selecting a person who lives in the Northeast is

$$P(E) = \frac{n(E)}{n(S)} = \frac{51.6}{258.8} \approx 0.2 \text{ or } 20\%.$$ ■

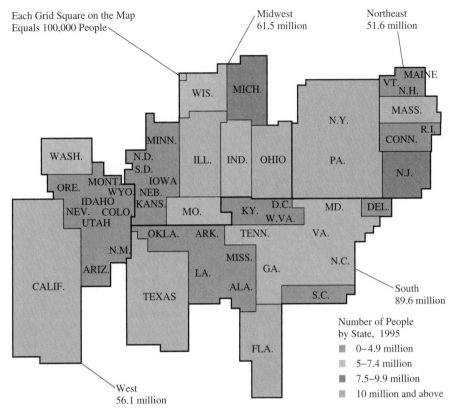

Each Grid Square on the Map Equals 100,000 People

Midwest 61.5 million

Northeast 51.6 million

South 89.6 million

West 56.1 million

Number of People by State, 1995
- 0–4.9 million
- 5–7.4 million
- 7.5–9.9 million
- 10 million and above

Figure 7.10
U.S. population for the 48 contiguous states
Source: U.S. Bureau of the Census

2 Compute probabilities with permutations and combinations.

Probability with Permutations and Combinations

In Section 7.6, we learned how to count the number of permutations and combinations without having to make long lists. These counting techniques also can be applied to obtain the numbers in the numerator and denominator of probability fractions, as illustrated in Examples 3–4.

EXAMPLE 3 **Probability and Permutations**

Find the probability that a seven-digit telephone number chosen at random has no repeated digits.

Solution

Let

E = the set of all seven-digit telephone numbers with no repeated digits

S = the set of all seven-digit telephone numbers

We must determine

$$P(E) = \frac{n(E)}{n(S)}$$

where both the numerator and denominator of the probability fraction can be found using the Fundamental Counting Principle. Begin with $n(S)$. Recall that a telephone number cannot begin with a 0 or 1, so the first digit can be chosen in eight ways (2, 3, 4, 5, 6, 7, 8, or 9). There are ten choices for each of the remaining digits (0, 1, 2, 3, 4, 5, 6, 7, 8, or 9). The situation is illustrated as follows:

Permutations of a seven-digit telephone number

There are $8 \cdot 10 \cdot 10 \cdot 10 \cdot 10 \cdot 10 \cdot 10$ or 8,000,000 possible seven-digit telephone numbers.

Now let's determine $n(E)$, the number of seven-digit telephone numbers with no repeating digits. Again, there are eight choices for the first digit (2 through 9). Since digits are not to be repeated, whatever digit we used in the first position cannot be used in the second position, although we can now use 0 or 1. Thus, digit 2 can be chosen in nine ways (0 through 9 excluding the digit in position 1). Similarly, digit 3 can be chosen in eight ways (0 through 9, excluding the two digits used in positions 1 and 2), digit 4 can be chosen in seven ways, and so on. The situation is illustrated as follows:

Permutations of a seven-digit telephone number with no repeating digits

There are $8 \cdot 9 \cdot 8 \cdot 7 \cdot 6 \cdot 5 \cdot 4$ or 483,840 possible seven-digit telephone numbers with no repeating digits. Therefore,

$$P(E) = \frac{n(E)}{n(S)} = \frac{483,840}{8,000,000} = 0.06048.$$

The probability that a seven-digit telephone number chosen at random has no repeating digits is 0.06048. ∎

Karen Furth

EXAMPLE 4 **Probability and Combinations: Winning the Lottery**

Florida's lottery game, LOTTO, is set up so that each player chooses six different numbers from 1 to 49. If the six numbers chosen match the six numbers drawn randomly each Saturday evening, the player wins (or shares) the top cash prize. (As of this writing, the top cash prize has ranged from $7 million to $106.5 million.) What is the probability of winning this prize:

a. With one LOTTO ticket?
b. With ten LOTTO tickets?

Solution

Since the order of the numbers does not matter, the sample space consists of all possible combinations of six numbers chosen from the numbers 1 to 49. Thus,

$$n(S) = \binom{49}{6}$$

$$= \frac{49!}{(49-6)!6!} \qquad \text{Recall that } \binom{n}{r} = \frac{n!}{(n-r)!\,r!}$$

$$= \frac{49!}{43!6!}$$

$$= \frac{49 \cdot 48 \cdot 47 \cdot 46 \cdot 45 \cdot 44}{6 \cdot 5 \cdot 4 \cdot 3 \cdot 2 \cdot 1}$$

$$= 13{,}983{,}816$$

a. If a person buys one LOTTO ticket, that person has selected only one combination of the six numbers.
   The probability of winning is

$$P(E) = \frac{n(E)}{n(S)} = \frac{1}{13{,}983{,}816} \approx 0.0000000715.$$

b. If a person buys ten LOTTO tickets, that person has selected ten different combinations of the six numbers, so the probability of winning is

$$P(E) = \frac{n(E)}{n(S)} = \frac{10}{13{,}983{,}816} \approx 0.000000715. \qquad \blacksquare$$

> **D**iscover for yourself
>
> If a person purchases 13,983,816 LOTTO tickets at $1 per ticket (all possible combinations), isn't this a guarantee of winning the lottery? Since the probability in this situation is 1, with a top cash prize of $106.5 million, what's wrong with doing this?

3 Find probabilities involving unions, intersections, and complements.

Intersections, Unions, and Complements of Events

We turn now to the problem of finding the probability of two or more events from the same sample space. These situations involve the intersection, union, and complement of events, whose definitions are summarized as follows.

> **S**tudy tip
>
> To find the union of events, *unite* (or join) the elements in their sets. To find their intersection, write the common elements in their sets.

Definitions of intersection, union, and complement

Let A and B be events from the same sample space S. Then

1. The *intersection* of A and B, denoted by $A \cap B$ (read: "A and B"), is the event that both A and B occur.
2. The *union* of A and B, denoted by $A \cup B$ (read: "A or B"), is the event that either A or B (or both) occur.
3. The *complement* of A, denoted by A' (read: "A complement") is the event that A does *not* occur.

EXAMPLE 5 **Finding Intersections, Unions, and Complements**

Suppose a fair die is tossed, with $S = \{1, 2, 3, 4, 5, 6\}$. Let A be the event that the outcome is even, and let B be the event that the outcome is greater than 2.

ENRICHMENT ESSAY

A Brief History in the Study of Uncertainty

The study of uncertain events has fascinated humankind since the beginning of history. Artifacts show that games of chance were played by ancient civilizations. Although probability theory can easily be applied to a card game, probabilities play a part in nearly all aspects of everyday life, ranging from calculating the health risks of smoking a pack of cigarettes a day to deciding on the best investments for your money.

The first important work in the area of probability was done by mathematicians who were more interested in finding methods for winning at gambling than in the creation of a theory. For example, one of the first persons to apply mathematics to probability was an Italian, Girolamo Cardano (1501–1576), whose textbook for gamblers, *The Book on Games of Chance*, contains a section with tips on how to succeed in cheating.

Mathematical probability was further developed by the gambler and amateur mathematician Chevalier de Méré, who was interested in the following problem: If a game of chance is interrupted before completion, how should the players divide up the money that is on the table? Méré contacted two French mathematicians, Blaise Pascal (1623–1662) and Pierre de Fermat (1601–1665), resulting in a correspondence that served as the starting point for the modern theory of probability.

Todd Gilens "Calender" 1996, comprising fifty-two cut playing cards from seven different decks. © Todd Gilens.

Find expressions for each of the following events, and then compute the probability of the event.

a. $A \cap B$ **b.** $A \cup B$ **c.** A'

Solution

We are given that

$$S = \{1, 2, 3, 4, 5, 6\}$$ There are six outcomes in the sample space.

$$A = \{2, 4, 6\}$$ This is the event that the outcome is even.

$$B = \{3, 4, 5, 6\}$$ This is the event that the outcome is greater than 2.

a. $A \cap B$ is the event that *both A and B* occur. Since sets A and B have 4 and 6 in common, $A \cap B = \{4, 6\}$. This is the event that the outcome is both even and greater than 2. We are now ready to compute the probability for this event, which has two outcomes.

$$P(A \cap B) = \frac{n(A \cap B)}{n(S)} = \frac{2}{6} = \frac{1}{3}$$

b. $A \cup B$ is the event that *A or B* or both occur. $A \cup B$ consists of all numbers from 1 to 6 that are either even or greater than 2. Thus, $A \cup B = \{2, 3, 4, 5, 6\}$. The probability for this event, which has 5 outcomes, is given by

$$P(A \cup B) = \frac{n(A \cup B)}{n(S)} = \frac{5}{6}.$$

> **S**tudy tip
>
> $A = \{2, 4, \quad 6\}$
> $B = \{3, 4, 5, 6\}$
> $A \cap B = \{4, 6\}$
>
> 4 and 6 are the common elements.

> **S**tudy tip
>
> $A = \{2, 4, \quad 6\}$
> $B = \{3, 4, 5, 6\}$
> $A \cup B = \{2, 3, 4, 5, 6\}$
>
> Join all the elements.

Study tip

$A = \{\ 2,\quad 4,\quad 6\}$
$S = \{1, 2, 3, 4, 5, 6\}$
$A' = \{1,\quad 3,\quad 5\ \}$

Take elements in S that are not in A.

c. A' is the event that A does not occur. Since A is the event that the outcome is even, then A' is the event that the outcome is not even, and $A' = \{1, 3, 5\}$. Using the probability definition for this event, which has three outcomes, we obtain

$$P(A') = \frac{n(A')}{n(S)} = \frac{3}{6} = \frac{1}{2}. \qquad \blacksquare$$

Study tip

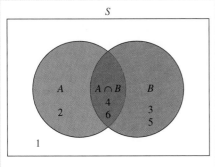

$$S = \{1, 2, 3, 4, 5, 6\}$$
$$A = \{2, 4, 6\}$$
$$B = \{3, 4, 5, 6\}$$
$$A \cap B = \{4, 6\}$$
$$A \cup B = \{2, 3, 4, 5, 6\}$$

The ideas of probability can be illustrated using *Venn diagrams*. Figure 7.11 is a Venn diagram for Example 5. The rectangle in Figure 7.11 represents the sample space for the experiment, $\{1, 2, 3, 4, 5, 6\}$.

The area inside circle A represents event A, $\{2, 4, 6\}$, and the area inside circle B represents event B, $\{3, 4, 5, 6\}$. The intersection of the two circles represents $A \cap B$, $\{4, 6\}$. Finally, the area inside the rectangle but outside circle A represents event A', $\{1, 3, 5\}$. Similarly, event B' is shown by the area outside circle B but inside the rectangle, $\{1, 2\}$.

Figure 7.11

Illustrating sets with a Venn diagram

Discover for yourself

Look again at Figure 7.11 in the Study tip. Is the number of outcomes in the event $A \cup B$ equal to the sum of the number of outcomes in events A and B? That is, is the following statement true?

$n(A \cup B) = n(A) + n(B)$

If the statement is not true, describe what's wrong with it.

In the Discover for yourself, did you observe that $n(A \cup B)$ is not equal to $n(A) + n(B)$?

$$A \cup B = \{2, 3, 4, 5, 6\}, \text{ so } n(A \cup B) = 5$$
$$A = \{2, 4, 6\}, \text{ so } n(A) = 3$$
$$B = \{3, 4, 5, 6\}, \text{ so } n(B) = 4 \qquad 5 \neq 3 + 4$$

Thus $n(A \cup B)$ is not equal to $n(A) + n(B)$ because $n(A) + n(B)$ results in counting the outcomes in $A \cap B$—4 and 6—twice. To correct for this double counting, we must subtract the number of outcomes in the set $A \cap B$.

$$n(A \cup B) = n(A) + n(B) - n(A \cap B)$$

A similar formula holds for the probability of the union of two events.

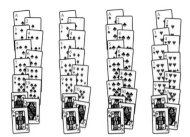

The set of 52 playing cards in the standard deck has four suits:

♠ spades

♦ diamonds

♣ clubs

♥ hearts

Ace is the unit card. Jacks, queens, and kings are "face cards." Each suit contains 13 denominations: ace, 2, 3, ..., 10, jack, queen, king.

A Venn diagram

A: Diamonds

$n(A \cap B) = 3$

B: Face Cards

Figure 7.12

A Venn diagram, with
A = selecting a diamond and
B = selecting a face card

Probability of the union of two events

Let A and B be events from the same sample space S.
The probability of A or B occurring is given by

$$P(A \cup B) = P(A) + P(B) - P(A \cap B).$$

EXAMPLE 6 **Finding the Probability of the Union of Two Events**

A single card is selected at random from a standard 52-card deck. Find the probability that the card selected is either a diamond or a face card (jack, queen, or king).

Solution

Let

A = the event that a diamond is selected

B = the event that a face card is selected

We must find $P(A \cup B)$, the probability that the card selected is either a diamond or a face card.

The deck has 13 diamonds, so the probability of selecting a diamond (event A) is

$$P(A) = \frac{13}{52} \qquad \text{See Figure 7.12.}$$

The deck has 12 face cards, so the probability of selecting a face card (event B) is

$$P(B) = \frac{12}{52} \qquad \text{See Figure 7.12.}$$

Figure 7.12 indicates that we have counted the cards that are diamonds and face cards twice. The probability of selecting both a diamond and a face card (event $A \cap B$) is

$$P(A \cap B) = \frac{3}{52} \qquad \text{Three of the cards are diamonds and face cards.}$$

We now compute $P(A \cup B)$ as follows:

$$P(A \cup B) = P(A) + P(B) - P(A \cap B)$$

$$= \frac{13}{52} + \frac{12}{52} - \frac{3}{52}$$

$$= \frac{22}{52} = \frac{11}{26} \approx 0.423$$

The probability of selecting either a diamond or a face card is $\frac{11}{26}$. ∎

In Example 6, events A and B had three outcomes in common. In some situations, however, it is impossible for two events to both occur. For example, when a pair of dice is rolled once, getting a sum of 7 excludes getting a sum of 8. These events have no outcomes in common and are said to be *mutually exclusive*. In this case, $A \cap B = \emptyset$ and $P(A \cap B) = 0$. This gives rise to a special case for probability involving union.

Probability of the union of mutually exclusive events

If A and B are mutually exclusive events, then $P(A \cap B) = 0$ and it follows that $P(A \cup B) = P(A) + P(B)$.

EXAMPLE 7   **Probability of the Union of Mutually Exclusive Events**

Two ordinary six-sided dice are rolled. What is the probability of getting a sum of 7 or 8?

Solution

Let

A = the event that the sum is 7

B = the event that the sum is 8

	Second Die					
	1	*2*	*3*	*4*	*5*	*6*
1	(1, 1)	(1, 2)	(1, 3)	(1, 4)	(1, 5)	(1, 6)A
2	(2, 1)	(2, 2)	(2, 3)	(2, 4)	(2, 5)	(2, 6)B
3	(3, 1)	(3, 2)	(3, 3)	(3, 4)	(3, 5)	(3, 6)
4	(4, 1)	(4, 2)	(4, 3)	(4, 4)	(4, 5)	(4, 6)
5	(5, 1)	(5, 2)	(5, 3)	(5, 4)	(5, 5)	(5, 6)
6	(6, 1)	(6, 2)	(6, 3)	(6, 4)	(6, 5)	(6, 6)

First Die

Figure 7.13

Rolling two dice, with A = their sum is 7 and B = their sum is 8

The 36 outcomes are shown as ordered pairs in Figure 7.13 and the highlighted diagonals indicate outcomes whose sums are 7 and 8, respectively. The probability of a sum of 7 (event A) is $\frac{6}{36}$ and the probability of a sum of 8 (event B) is $\frac{5}{36}$. Since both events cannot occur simultaneously, we compute $P(A \cup B)$ as follows:

$$P(A \cup B) = P(A) + P(B) = \frac{6}{36} + \frac{5}{36} = \frac{11}{36}.$$

The probability of getting a sum of 7 or 8 is $\frac{11}{36}$. ∎

Probability of Independent Events

Two events are *independent events* if the occurrence of one has no effect on the occurrence of the other. For example, when tossing a fair coin, the occurrence of tails on the first toss does not affect the outcome of the second toss (although it is difficult to convince some people of that fact).

ENRICHMENT ESSAY

Independent Events

The run of nine girls in a row in a family has a probability of $\frac{1}{512}$. (Verify this.) If another child is born into the family, this is an independent event and the probability of a girl is still $\frac{1}{2}$.

UPI /
Corbis – Bettmann

If two or more events are independent, we can find the probability of them all occurring by multiplying the probabilities.

Probability of independent events

If $A_1, A_2, A_3, \ldots, A_n$ are independent events, then the probability that they will all occur is

$$P(A_1 \text{ and } A_2 \text{ and } A_3 \text{ and } \ldots \text{ and } A_n) = P(A_1)P(A_2)P(A_3) \cdots P(A_n).$$

U.S. Racial and Ethnic Makeup, 1995

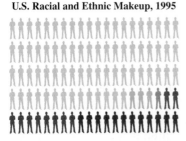

▪ White	74%
▪ Asian	3
▪ Native American	1
▪ Hispanic	10
▪ African-American	12

Figure 7.14

Source: U.S. Bureau of the Census

EXAMPLE 8 **Probability of a Sequence of Independent Events**

The racial and ethnic makeup of the United States in 1995 is shown in Figure 7.14. If four people are chosen at random from this population, what is the probability that all four are African-American?

Solution

Let A represent the event of selecting an African-American. Figure 7.14 indicates that $P(A) = 12\% = 0.12$. The probability that all four people selected are African-American is

$$P(A) \cdot P(A) \cdot P(A) \cdot P(A) = [P(A)]^4 = (0.12)^4 = 0.00020736. \quad \blacksquare$$

PROBLEM SET 7.7

Practice and Application Problems

In Problems 1–4, find each probability in an experiment of selecting a family with three children. Let M represent a male child, F a female child, and use the sample space

$$S = \{MMM, MMF, MFM, MFF, FMM, FMF, FFM, FFF\}.$$

1. The probability of selecting a family with exactly one female child.

2. The probability of selecting a family with exactly two male children.

3. The probability of selecting a family with at least one male child.

4. The probability of selecting a family with at least two female children.

In Problems 5–10, find each probability in an experiment in which a six-sided die is rolled twice. The 36 possible outcomes are shown as follows:

	Second Roll					
First Roll	⚀	⚁	⚂	⚃	⚄	⚅
⚀	(1, 1)	(1, 2)	(1, 3)	(1, 4)	(1, 5)	(1, 6)
⚁	(2, 1)	(2, 2)	(2, 3)	(2, 4)	(2, 5)	(2, 6)
⚂	(3, 1)	(3, 2)	(3, 3)	(3, 4)	(3, 5)	(3, 6)
⚃	(4, 1)	(4, 2)	(4, 3)	(4, 4)	(4, 5)	(4, 6)
⚄	(5, 1)	(5, 2)	(5, 3)	(5, 4)	(5, 5)	(5, 6)
⚅	(6, 1)	(6, 2)	(6, 3)	(6, 4)	(6, 5)	(6, 6)

5. The probability that the sum is 5.

6. The probability that the sum is less than 5.

7. The probability that the sum is 3 or 4.

8. The probability that the sum is at least 6.

9. The probability that the sum is odd or divisible by 5.

10. The probability that the sum is even or divisible by 3.

11. The numbers of violent crimes per 100,000 population in various regions of the United States in 1992 are shown in the figure. If a violent crime occurs, what is the probability that it occurred in

   a. the Northeast?

   b. the West or the Midwest?

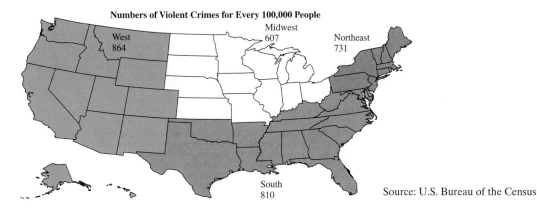

Numbers of Violent Crimes for Every 100,000 People

West 864
Midwest 607
Northeast 731
South 810

Source: U.S. Bureau of the Census

12. The numbers of property crimes per 100,000 population in various regions of the United States in 1992 are shown in the figure. If a property crime occurs, what is the probability that it occurred in

   a. the South?

   b. the South or the Northeast?

Numbers of Property Crimes for Every 100,000 People

West 5524
Midwest 4368
Northeast 4105
South 5345

Source: U.S. Bureau of the Census

13. Find the probability that a seven-digit telephone number chosen at random begins with 2 and ends with 9.

14. Find the probability that a seven-digit telephone number chosen at random begins with an even number, has alternating even and odd digits, and does not contain 0 in any of its digits.

In Problems 15–16, Martha, Lee, Marilyn, Paul, and Ann have all been invited to a dinner party. They arrive randomly and at different times.

15. What is the probability that Marilyn will arrive first and Lee last?

16. What is the probability they will arrive in the following order: Paul, Martha, Marilyn, Lee, and Ann?

17. To play the California lottery, a person has to correctly select 6 out of 51 numbers, paying $2 for each six-number selection. If the six numbers picked are the same as the ones drawn by the lottery, mountains of money are bestowed. What is the probability that a person with one combination of six numbers will win? What is the probability of winning if 100 lottery tickets are purchased? How much must a pigeon, er, person, spend so that the probability of winning is $\frac{1}{2}$?

18. A state lottery is designed so that a player chooses five numbers from 1 to 30 on one lottery ticket. What is the probability that a player with one lottery ticket will win? What is the probability of winning if 100 lottery tickets are purchased? How many lottery tickets must be purchased so that the probability of winning is $\frac{1}{4}$?

19. At an international meeting of 100 people, 30 speak English, 40 speak Spanish, and 20 speak both English and Spanish. If a delegate is randomly selected, what is the probability that the delegate speaks English or Spanish?

20. A student is selected at random from a group of 100 students in which 45 take math, 47 take English, and 27 take both. What is the probability that the selected student takes math or English?

Use these figures for 1995 resident U.S. population by age to answer Problems 21–28.

Age	0–19	20–39	40–59	60–74	75–84	85 and older
Population (in thousands)	75,791	81,004	62,383	28,805	11,145	3,628

Source: U.S. Bureau of the Census

If a resident of the United States is chosen at random, find the probability that:

21. The resident is under the age of 20.

22. The resident is at least 85.

23. The resident is at least 60.

24. The resident is at most 39.

25. The resident is in the 20–39 or 60–74 age group.

26. The resident is in the 40–59 or 75–84 age group.

27. The resident is in the 20–29 age group if it is known that 44,117 thousand residents are in the 30–39 age group.

28. The resident is in both the 20–39 and 60–74 age groups.

Projections for the racial and ethnic makeup of the United States in 2050 are shown in the figure. Use these projections to answer Problems 29–34.

U.S. Racial and Ethnic Makeup, 2050

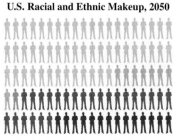

White	53%
Asian-American	8
Native American	1
Hispanic-American	24
African-American	14

Source: U.S. Bureau of the Census

29. If one person is chosen at random from this population, what is the probability of selecting an Asian-American or a Hispanic-American?

30. If one person is chosen at random from this population, what is the probability of selecting a Native American or an African-American?

31. If two people are randomly selected, what is the probability that both are Hispanic-American?

32. If two people are randomly selected, what is the probability that both are African-American?

33. If two people are randomly selected, what is the probability that neither is Hispanic-American?

34. If two people are randomly selected, what is the probability that neither is African-American?

The graph shows the number of Americans (in thousands) ages 12 or older who have used illicit drugs in their lifetime. In 1994, the population of the United States ages 12 or older was approximately 203,000 thousand. Use the graph and this figure to answer Problems 35–38.

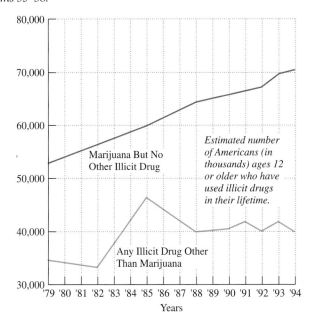

Source: U.S. Department of Health and Human Services

35. In 1994, estimate the probability of randomly selecting an American 12 or older who used marijuana, but no other illicit drug, in his or her lifetime.

36. In 1994, estimate the probability of randomly selecting an American 12 or older who used an illicit drug other than marijuana, but not marijuana, in his or her lifetime.

37. In 1994, estimate the probability of randomly selecting an American 12 or older who used marijuana but no other illicit drug or any illicit drug other than marijuana in his or her lifetime.

38. In 1994, two Americans 12 or older are selected at random. Estimate the probability that both used marijuana, but no other illicit drug, in their lifetime.

Problems 39–50 deal with a deck of 52 cards, pictured below.

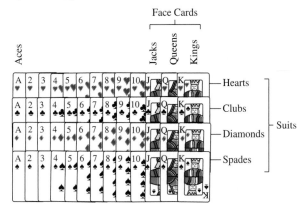

If a single card is drawn from the deck, find the probability:

39. Of drawing a heart.

40. Of not getting an ace.

41. Of getting a 3 or a 5.

42. That the card will be a number card 7 or more.

43. Of getting an ace or a heart.

44. Of getting a face card and a spade.

45. Of getting a heart or a face card.

46. Of getting a diamond or a number card greater than 3 but less than 8.

If you are dealt five cards from the deck, find the probability:

47. Of getting no hearts.

48. Of getting no picture cards.

49. Of getting all hearts.

50. Of getting all cards from the same suit.

True–False Critical Thinking Problems

51. Which one of the following is true?

 a. If S is a sample space in which all outcomes are equally likely and E is a subset of S, then $P(E) = n(E)$.

 b. If an experiment consists of rolling two ordinary six-sided dice, then the sample space consists of $2 \cdot 6$ or 12 equally likely outcomes.

 c. Suppose that two balls are randomly selected from a collection of basketballs that contains some defective basketballs. The complement of selecting at least one ball that is defective is selecting two good balls.

 d. If a coin is tossed four times in succession, the probability of getting four tails is $\frac{4}{16}$.

52. Which one of the following is true?

 a. The probability of an event is calculated by dividing the number of outcomes in the sample space by the number of equally likely outcomes for the event.

 b. Theoretically, it is not possible to purchase enough tickets to guarantee winning a state lottery in which a player selects six different numbers from 1 to 40 trying to match the six numbers drawn by the lottery commission.

 c. Probability theory points out that no events are certain.

 d. If it is known that 3 darts have hit the dart board shown in the figure, the probability that all 3 landed in the A ring is $\frac{1}{729}$. (Assume that no dart hits on the lines between the rings.)

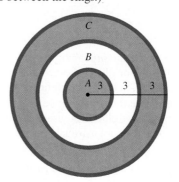

Technology Problem

53. Use the random generator feature on a graphing utility to generate 20 random numbers greater than 0 and less than 1. What is the theoretical probability that the number generated is less than or equal to 0.5? What is the probability obtained from the numbers generated by your graphing utility? If you increase the number of random numbers from 20 to 50, does your experimental result get closer to the theoretical result?

Writing in Mathematics

54. If the probability of rain on Saturday is $\frac{3}{5}$ and on Sunday is $\frac{2}{5}$, does this mean that there is a 100% probability of rain for the weekend? Explain.

55. The president of a large company with 10,000 employees is considering mandatory cocaine testing for every employee. The test that would be used is 90% accurate, meaning that it will detect 90% of the cocaine users who are tested, and that 90% of the nonusers will test negative. This also means that the test gives 10% false positive. Suppose that 1% of the employees actually use cocaine. Find the probability that someone who tests positive for cocaine use is, indeed, a user. *Hint:* Find $\dfrac{n(E)}{n(S)}$, where

$n(E)$ = the number of employees who test positive and are cocaine users
   = 90% of 1% of 10,000

and

$n(S)$ = the number of employees who test positive
   = the number who test positive who actually use cocaine plus the number who test positive who do not use cocaine

What does this probability indicate in terms of the percent of employees who test positive who are not actually users? Discuss these numbers in terms of the issue of mandatory drug testing. Write a paper either in favor of or against mandatory drug testing, incorporating the actual percentage accuracy for such tests.

Critical Thinking Problems

56. An experiment consists of randomly selecting one of the 100 points in the *xy*-plane shown in the figure.

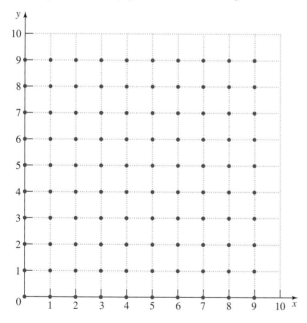

Determine the probability for each of the following events.

a. The point lies within but not on the circle $x^2 + y^2 = 36$.

b. The point lies on the circle $x^2 + y^2 = 36$ or on the line $y = x$.

c. The point lies on $y = |x - 2|$.

d. Describe an event such that the probability is $\frac{17}{100}$.

57. If a monkey were to randomly strike the 48 keys on a typewriter (which include the space bar), what is the probability the monkey will type the phrase "to be or not to be" as the first six words?

Robert Arneson "Typewriter" 1965, ceramic, 6 1/2 × 11 3/8 × 12 1/2 in. University of California, Berkeley Art Museum; gift of the artist. © Estate of Robert Arneson / Licensed by VAGA, New York, NY.

CHAPTER PROJECT
Named Numbers

Many families of numbers arise again and again in pure mathematics and applications. Often, they are named after the mathematicians who worked with them. The first section of this chapter introduced one example of a famous sequence of "named" numbers: the Fibonacci sequence. The Fibonacci numbers are closely related to another set of numbers called the *Lucas numbers,* denoted by l_n. Lucas numbers are generated the same way as Fibonacci numbers. However, Lucas numbers have a different starting point.

1. Lucas numbers begin: 2, 1, 3, 4, 7, 11, 18, ... Give the next five Lucas numbers.
    Lucas numbers may be defined recursively as:

    $$l_0 = 2 \qquad l_1 = 1 \qquad l_{n+1} = l_{n-1} + l_n$$

    Compare that definition to the recursive definition for Fibonacci numbers:

    $$f_0 = 1 \qquad f_1 = 1 \qquad f_{n+1} = f_{n-1} + f_n$$

2. Write down the first 20 Fibonacci numbers and the first 15 Lucas numbers in a chart. Verify the following relationships between Fibonacci numbers and Lucas numbers:

    $$f_{2n} = f_n l_n \qquad l_n = f_{n-1} + f_{n+1} \qquad 2f_{m+n} = f_m l_n + f_n l_m$$

    Edward Lucas, for whom the Lucas numbers are named, is also well known as the creator of the Tower of Hanoi puzzle given in Problem 31 on page 813. He also created many other mathematical games found in his four volume work on recreational mathematics.

    Studying Pascal's Triangle will also lead us to a new set of numbers called *Catalan numbers.* If we look at the middle numbers in Pascal's Triangle, we will see the following sequence:

    $$1, 2, 6, 20, 70, 252, 924, 3432, \dots$$

3. If we form a sequence by writing down one divisor of each number in turn, there is an interesting pattern. Can you discover the sequence of divisors? If we divide each number in turn by its divisor, the result is the Catalan numbers. Verify you have found the correct numbers by using the formula for the *n*th Catalan number:

    $$\frac{(2n)!}{n!(n+1)!}.$$

    Catalan numbers may be used to describe the number of ways a polygon with $n + 2$ sides can be cut into n triangles, or they may also be used to describe the number of ways in which parentheses can be placed in a sequence of numbers to be multiplied, two at a time.

4. Show that a 4-sided polygon may be cut into two triangles in two ways, a 5-sided polygon may be cut into three triangles in five ways, and a 6-sided polygon may be cut into four triangles in fourteen ways.

Worldwide Web Resources

Go to the Prentice Hall website (http://www.prenhall.com/blitzer) to access other locations on the Internet that will allow you to further explore the concepts presented in this project.

Chapter Review

SUMMARY

1. Sequences and Factorial Notation

a. An infinite sequence $\{a_n\}$ is a function whose domain is the set of positive integers. The function values, or terms, are represented by

$$a_1, a_2, a_3, a_4, \ldots, a_n, \ldots$$

b. *Factorial Notation*

$$n! = 1 \cdot 2 \cdot 3 \cdot 4 \cdot \cdots \cdot (n-1)n \quad \text{and} \quad 0! = 1$$

c. Sequences can be defined using recursion formulas that define the nth term as a function of the previous term.

2. Series

a. The nth partial sum of the infinite sequence

$$a_1, a_2, a_3, \ldots, a_n, \ldots$$

is the sum of the first n terms of the sequence.

$$S_n = a_1 + a_2 + a_3 + \cdots + a_n$$

b. An infinite series is the sum of the terms

$$a_1 + a_2 + a_3 + \cdots + a_n + \cdots.$$

c. Series are frequently expressed in summation, or sigma, notation using Σ.

$$\sum_{i=1}^{n} a_i = a_1 + a_2 + a_3 + \cdots + a_n$$

$$\sum_{i=1}^{\infty} a_i = a_1 + a_2 + a_3 + \cdots + a_n + \cdots$$

3. Arithmetic Sequences and Series

a. An arithmetic sequence, or arithmetic progression, is one in which each term after the first differs from the preceding term by a constant amount. The difference between consecutive terms is called the common difference of the sequence.

b. *Formula for the nth term of an arithmetic sequence:* The nth term (the general term) of an arithmetic sequence with first term a_1 and common difference d is given by $a_n = a_1 + (n-1)d$.

c. *Formula for the sum of the first n terms of an arithmetic sequence:* The sum, S_n, of the first n terms of a finite arithmetic sequence is given by

$$S_n = \frac{n}{2}(a_1 + a_n)$$

where a_1 is the first term and a_n is the nth term.

4. Geometric Sequences and Series

a. A geometric sequence, or geometric progression, is one in which each term after the first is obtained by multiplying the preceding term by a nonzero constant. This common multiple is called the common ratio of the sequence, obtained by dividing any term by its predecessor.

b. *Formula for the nth term of a geometric sequence:* The nth term (the general term) of a geometric sequence with first term a_1 and common ratio r is given by $a_n = a_1 r^{n-1}$.

c. *Formula for the sum of the first n terms of a finite geometric sequence:* The sum, S_n, of the first n terms of a geometric sequence with first term a_1 and the common ratio r is given by

$$S_n = \frac{a_1(1 - r^n)}{1 - r} \quad (r \neq 1)$$

d. *Formula for the sum of an infinite geometric series:* The sum of an infinite geometric series in which a_1 is the first term and r is the common ratio, where $|r| < 1$, is given by

$$S_\infty = \frac{a_1}{1 - r}.$$

Equivalently,

$$\sum_{n=1}^{\infty} a_1 r^{n-1} = \frac{a_1}{1 - r}, \quad |r| < 1.$$

5. Proof by Mathematical Induction

To prove that S_n is true for all positive integers n:

a. Show that S_1 is true.

b. Show that if S_k is assumed true, then S_{k+1} is also true, for every positive integer k.

6. The Binomial Theorem

a. *Definition of a Binomial Coefficient*

$$\binom{n}{r} = \frac{n!}{r!(n-r)!}$$

for nonnegative integers n and r, with $n \geq r$

b. *Binomial Theorem*

$$(a+b)^n = \binom{n}{0}a^n + \binom{n}{1}a^{n-1}b$$

$$+ \binom{n}{2}a^{n-2}b^2 + \cdots + \binom{n}{n}b^n$$

c. *Formula for the rth term in a binomial expansion*

$$\binom{n}{r-1} a^{n-r+1} b^{r-1}$$

7. Counting Principles, Permutations, and Combinations

a. *The Fundamental Counting Principle*
   If event E_1 can occur in m_1 ways, event E_2 in m_2 ways, ... and event E_k in m_k ways, then the total number of ways that all k events can occur is $m_1 \cdot m_2 \cdot m_3 \cdot \cdots \cdot m_k$.

b. *Permutations*

   1. A permutation of n elements is an ordering of the elements such that one element is first, one is second, one is third, and so on.

   2. $_nP_n$, the number of permutations of n things taken n at a time, is given by $_nP_n = n!$

   3. $_nP_r$, the number of permutations of n elements taken r at a time, is given by

   $$_nP_r = \frac{n!}{(n-r)!} = n(n-1)(n-2)\cdots(n-r+1).$$

c. *Combinations*

   1. A combination of n elements taken r at a time is one of the ways in which r elements can be chosen from n elements.

   2. $\binom{n}{r}$, the number of combinations of n elements taken r at a time, is given by

   $$\binom{n}{r} = \frac{n!}{(n-r)!r!}.$$

8. Probability

a. *Basic Definitions*

   1. *Experiment:* An occurrence with an uncertain outcome

   2. *Sample Space (S):* The set of all possible outcomes in an experiment

3. *Event (E):* A subset of the sample space

b. *Probability of an Event:* $P(E) = \dfrac{n(E)}{n(S)}$, where $n(E) =$ the number of outcomes in the event and $n(S) =$ the number of outcomes in the sample space.

c. *Probability, Permutations, and Combinations*

   1. The probability of a permutation is the number of ways the permutation can occur divided by the total number of possible permutations. Use the Fundamental Counting Principle or the formula for $_nP_r$ to compute the numbers in the numerator and denominator of the probability fraction.

   2. The probability of a combination is the number of ways the combination can occur divided by the total number of possible combinations. Use the formula for $\binom{n}{r}$ to compute the numbers in the numerator and denominator of the probability fraction.

d. *Intersections, Unions, and Complements of Events*

   1. $A \cap B$ (A intersection B) is the event that both A and B occur.

   2. $A \cup B$ (A union B) is the event that either A or B (or both) occur.

   3. A' (the complement of A) is the event that A does not occur.

e. *Probability of the Union of Two Events*

   1. $P(A \cup B)$, the probability of A or B occurring, is given by

   $$P(A \cup B) = P(A) + P(B) - P(A \cap B).$$

   2. If A and B are mutually exclusive events with no outcomes in common, then

   $$P(A \cup B) = P(A) + P(B).$$

f. *Independent Events*

   1. Two events are independent if the occurrence of one has no effect on the occurrence of the other.

   2. The probability that a sequence of independent events will occur is found by multiplying their individual probabilities.

REVIEW PROBLEMS

In Problems 1–3, write the first five terms of each sequence whose general term is given.

1. $a_n = \dfrac{(-1)^{n+1}}{2^n + 1}$

2. $a_n = \dfrac{(-2)^n}{(n+1)!}$

3. $a_1 = 4$ and $a_n = -2a_{n-1} + 5$ for $n \geqslant 2$

For Problems 4–5, write an expression for the general, or nth term, of each sequence. More than one expression may be possible.

4. $1, -4, 9, -16, 25, \ldots$

5. $-\dfrac{1}{3}, \dfrac{1}{4}, -\dfrac{1}{5}, \dfrac{1}{6}, -\dfrac{1}{7}, \ldots$

In Problems 6–7, find the indicated partial sum for the given sequence.

6. $a_n = (-1)^n n!; S_5$

7. $a_n = \dfrac{(n+3)!}{n!}; S_6$

In Problems 8–9, find each indicated sum.

8. $\displaystyle\sum_{i=1}^{5} (2i^2 - 3)$

9. $\displaystyle\sum_{i=0}^{4} \dfrac{(-1)^{i+1}}{i!}$

10. Find a formula for the *n*th partial sum of the following series. Then find the indicated sum.

$$\sum_{i=1}^{100} [i^2 - (i+1)^2]$$

For Problems 11–12, write each series in summation notation. Use the index i and let i begin at 1.

11. $1 + 8 + 27 + 64 + 125$

12. $-\dfrac{1}{2} + \dfrac{1}{4} - \dfrac{1}{8} + \dfrac{1}{16} - \dfrac{1}{32} + \dfrac{1}{64} + \cdots$

13. The finite sequence whose general term is $a_n = 0.0012n^2 - 0.027n + 1.09$ models the ratio of men to women in the United States, where $n = 1$ corresponds to 1910, $n = 2$ to 1920, $n = 3$ to 1930, and so on.

a. Find and interpret $a_3, a_4, a_9,$ and a_{10}.

b. The population projection for the United States in the year 2000 is 275 million. How many men and how many women will there be at that time?

In Problems 14–16, write a formula for the nth term of each arithmetic sequence. Do not use a recursion formula. Then use the formula for a_n to find a_{20}, the 20th term of the sequence.

14. $-7, -3, 1, 5, \ldots$

15. $a_1 = 200, d = -20$

16. $a_1 = 8\dfrac{1}{2}, a_{n+1} = a_n - \dfrac{1}{4}$

17. In 1911, the world record time for the men's mile run was 1043.04 seconds. The world record has decreased by approximately 0.4118 seconds each year since then.

a. Write the general term for the arithmetic sequence modeling record times for the men's mile run, where $n = 1$ corresponds to 1911.

b. Use the model to predict the record time for the men's mile run for the year 2000.

Use the formula for the sum of the first n terms of an arithmetic sequence to answer Problems 18–20.

18. Find the sum of the first 28 terms of the arithmetic sequence 5, 12, 19, 26, ….

19. Find the sum of the first 56 terms of an arithmetic sequence whose first term is 15 and whose common difference is 3.

20. Find the sum of the first 200 positive multiples of 5.

In Problems 21–22, write out the first three terms and the last term. Then use the formula for the sum of the first n terms of an arithmetic sequence to find the sum indicated.

21. $\sum_{i=1}^{40} (4i - 7)$

22. $\sum_{i=1}^{80} (5 - 3i)$

23. The model $a_n = 7.7n + 55$ describes the annual advertising expenditures in billions of dollars by U.S. companies beginning with the year 1980, where $n = 0$

corresponds to 1980. Find $\sum_{n=0}^{20} a_n$ and describe what this number represents in practical terms.

24. A theater has 40 seats in the first row, 45 seats in the second row, 50 seats in the third row, and so on. Find the total number of seats if there are 25 rows.

In Problems 25–26, find the general term (the nth term) and the eighth term for each geometric sequence.

25. 100, 105, 110.25, 115.7625, …

26. $2, \dfrac{2}{3}, \dfrac{2}{9}, \dfrac{2}{27}, \cdots$

27. Find the sum of the first 15 terms of the geometric sequence 5, −15, 45, −135, ….

Use the formula for S_n, the sum of the first n terms of a finite geometric sequence, to find each sum in Problems 28–29.

28. $\sum_{i=1}^{20} 12\left(-\dfrac{1}{3}\right)^i$

29. $\sum_{i=7}^{15} 12\left(\dfrac{3}{2}\right)^i$

$$A = 400\left(1 + \dfrac{0.05}{12}\right) + 400\left(1 + \dfrac{0.05}{12}\right)^2$$
$$+ \cdots + 400\left(1 + \dfrac{0.05}{12}\right)^{60}$$

30. A job pays a salary of $18,000 the first year, with a guaranteed raise of 5% each year. The salaries for the first five years are given by the sequence

18,000, 18,000(1.05), 18,000(1.05)^2,

18,000(1.05)^3, 18,000(1.05)^4.

What is the salary at the end of year 20? What is the total salary over this 20-year period?

31. A deposit of $400 is made at the beginning of each month for 5 years in an account that pays 5% annual interest compounded monthly. What is the balance in the account at the end of 5 years? Use the sum shown in the next column above Problem 32.

32. A ball is dropped from a height of 6 feet. Each time the ball strikes the ground, it bounces up one-third of its previous height. Find the total distance traveled by the ball by the time it strikes the ground for the fifth time.

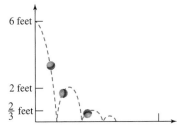

In Problems 33–36, find the sum of each infinite geometric series.

33. $\sum_{n=1}^{\infty} \left(\dfrac{1}{3}\right)^n$

34. $\sum_{i=0}^{\infty} 8\left(\dfrac{1}{2}\right)^i$

35. $\sum_{j=1}^{\infty} 5(-0.8)^j$

36. $\sum_{i=1}^{\infty} \dfrac{2^i}{3^{i+1}}$

37. Express $0.\overline{36}$ as a fraction in lowest terms.

38. A factory in an isolated town has an annual payroll of $4 million. It is estimated that 70% of this money is spent within the town, that people in the town receiving this money will again spend 70% of what they receive in

the town, and so on. What is the total annual economic impact of the factory on the town?

39. The first swing for the pendulum shown in the figure is 20 inches. Each successive swing has a length that is 90% that of the length of the previous swing. What is the total distance traveled by the pendulum by the time it comes to rest?

20 inches

18 inches

In Problems 40–45, use mathematical induction to prove that the given formula is true for every positive integer n.

40. $1 + 3 + 5 + \cdots + (2n - 1) = n^2$

41. $\dfrac{1}{1 \cdot 3} + \dfrac{1}{3 \cdot 5} + \dfrac{1}{5 \cdot 7} + \cdots + \dfrac{1}{(2n-1)(2n+1)} = \dfrac{n}{2n+1}$

42. $1^2 + 3^2 + 5^2 + \cdots + (2n-1)^2 = \dfrac{n(2n-1)(2n+1)}{3}$

43. $2^n < (n + 2)!$

44. Find a formula for the following product and verify the result using mathematical induction.

$$\frac{2}{1} \cdot \frac{3}{2} \cdot \frac{4}{3} \cdot \cdots \cdot \frac{n+1}{n}$$

45. Evaluate: $\dbinom{6}{4} + \dbinom{25}{2}$. Use a graphing utility to verify your answer.

In Problems 46–49, use the Binomial Theorem to expand each binomial and express the result in simplified form.

46. $(3x + y)^4$ **47.** $(x - 2y)^5$ **48.** $(x^2 + 2y^3)^6$ **49.** $(2x - y^4)^7$

In Problems 50–51, write the first three terms in each binomial expansion, expressing the result in simplified form.

50. $\left(\dfrac{x}{2} + 3y\right)^8$

51. $(2x - 3y)^9$

52. Find the seventh term in the expansion of $(3c + d)^9$.

53. What is the coefficient of x^6 in the expansion of $(x^2 + 5)^{10}$?

54. How many seven-digit local telephone numbers can be formed if the first two digits are 45?

55. In how many different ways can six airplanes line up for departure on a runway?

56. A club with 15 members is to choose four officers—president, vice-president, secretary, and treasurer. In how many ways can these offices be filled?

57. How many different five-letter code words can be arranged using the seven letters of the word GERMANY if no letter is repeated?

58. How many four-digit odd numbers less than or equal to 3000 can be formed using the digits 0, 1, 2, 3, 4, 5, 6, 7, 8, and 9 if digits may be repeated?

59. From a club with 15 members, four are to be selected to attend a convention.

    a. In how many different ways can the members be selected?

    b. If it is decided that one particular member must attend the convention, in how many ways can the group of four be chosen?

    c. If the group attending the convention is to contain at most four people (and at least one person), in how many ways can the group be selected?

60. A political discussion group consists of 12 Republicans and 8 Democrats. In how many ways can 5 Republicans and 4 Democrats be selected for a subcommittee?

61. How many different ways are there to mark the answers to a ten-question multiple-choice quiz in which each question has four possible answers?

62. An experiment consists of tossing a single six-sided die. Find the probability that:

    a. The number showing is less than 5.

    b. The number showing is prime.

63. What is the probability of getting a sum greater than 5 if two ordinary six-sided dice are rolled?

College enrollment figures by gender and age for 1994 are given in the table. Use this data to answer Problems 64–67.

College Enrollment, 1994

Characteristic	Enrollment (in thousands)
Total	15,022
Male	6764
18 to 24 years	4152
25 to 34 years	1589
35 years old and over	958
Female	8258
18 to 24 years	4576
25 to 34 years	1830
35 years old and over	1766

Source: U.S. Bureau of the Census

If a college student is selected at random, find the probability that the student is

64. female   **65.** age 35 or older

66. a male 18 to 24 years or a female 35 years and older

67. male or female

68. Find the probability that a seven-digit telephone number chosen at random has 4 in the first position, alternating even and odd numbers, and no zeros for any of its digits. When setting up the number of seven-digit telephone numbers in the denominator, remember that a telephone number cannot begin with a 0 or a 1.

69. A lottery game is set up so that each player chooses five different numbers from 1 to 20. If the five numbers match the five numbers drawn in the lottery, the player wins (or shares) the top cash prize. What is the probability of winning the prize:

   a. With one lottery ticket?

   b. With 100 lottery tickets?

70. A six-sided die is tossed. Let A be the event that the outcome is odd, and let B be the event that the outcome is greater than 4. Find expressions for each of the following events and then compute the probability of the event.

   a. $A \cap B$

   b. $A \cup B$

   c. A'

   d. B'

71. In a mathematics class consisting of 40 students, 15 students are older than 30, 11 students are majoring in business, and 5 students are business majors older than 30. If one student is selected at random, find the probability that:

   a. The student is older than 30 or a business major.

   b. The student is older than 30 and a business major.

   c. The student is not older than 30.

   d. The student is not a business major.

   e. The student is neither older than 30 nor a business major.

72. A single card is selected at random from a standard 52-card deck. Find the probability that the card selected is either a heart or a number card greater than 6.

73. The well-balanced spinner shown in the figure is spun three times. What is the probability it will stop at red on the first spin, green on the second spin, and yellow on the third spin?

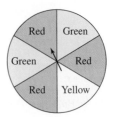

74. A quiz consisting of ten questions contains only true-false items. If a person guesses at every item, what is the probability of answering all items correctly?

75. Six books are carried to a bookshelf. Since the books have titles only written in Arabic, a language unfamiliar to the person shelving the books, what is the probability that the books will be arranged on one shelf in the correct order?

76. The circle graphs show the percent breakdown of the U.S. military by gender and race. If one person is randomly selected from the U.S. military, what is the probability that the person will be female and African-American?

The U.S. Military by Gender and Race, 1994

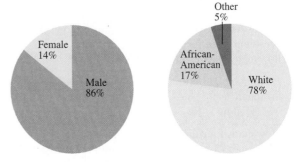

Source: U.S. Department of Defense

77. Describe what is wrong with the following statement:

There are 50 states, so the probability of being born in North Dakota is $\frac{1}{50}$.

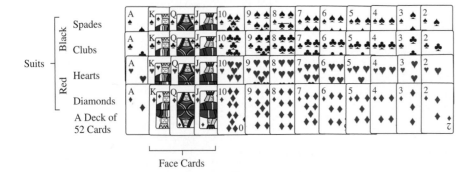

Suits — Black — Spades / Clubs — Red — Hearts / Diamonds — A Deck of 52 Cards

Face Cards

CHAPTER 7 TEST

1. Write the first five terms of the sequence whose general term is
$$a_n = \frac{(-1)^{n+1}}{n^3 - 2}.$$

2. Write an expression for the general, or nth term, of the sequence
$$\frac{1}{4}, -\frac{4}{9}, \frac{9}{16}, -\frac{16}{25}, \ldots .$$

3. Find S_6, the sixth partial sum of
$$a_n = (-1)^{n+1}(n-1)!$$

4. Find the indicated sum: $\sum_{i=1}^{5} (-2i^2 + 30)$.

5. Write in summation notation. Use the index i and let i begin at 1.
$$\frac{2}{3} + \frac{4}{5} + \frac{6}{7} + \frac{8}{9} + \frac{10}{11}$$

6. Write a formula for the nth term of the arithmetic sequence
$$-11, -9.5, -8, -6.5, \ldots .$$
Do not use a recursion formula. Then use the formula for a_n to find the 27th term.

Use the formula for the sum of the first n terms of an arithmetic sequence to solve Problems 7–9.

7. Find the sum of the first 26 terms: $5, 9, 13, 17, \ldots$.

8. Find the sum: $\sum_{i=1}^{60} (3i - 5)$.

9. Use the following salary schedule to determine which company will pay more over a ten-year period and by how much.

	First Year Salary	Yearly Raise
Company A	$24,000	$1500
Company B	$27,000	$1000

10. Find the nth term and the eighth term for the geometric sequence
$$2, 10, 50, 250, \ldots .$$

11. Find the sum of the first ten terms of the geometric sequence
$$5, 15, 45, 135, \ldots .$$

12. Find the sum: $\sum_{i=1}^{10} 104 \left(\frac{5}{6}\right)^{i-1}$.

13. A company pays a salary of $30,000 the first year with a guaranteed raise of 4% each year. What is the total salary over the first eight years?

14. Find the sum: $\sum_{i=1}^{\infty} 6(-0.8)^{i-1}$.

15. Express $0.\overline{823}$ as a fraction in lowest terms.

16. The first swing of a pendulum is 40 inches. Each successive swing has a length that is 75% that of the previous swing. What is the total distance traveled by the pendulum by the time it comes to rest?

17. Use mathematical induction to prove that for all natural numbers n,
$$1 + 4 + 7 + \cdots + (3n - 2) = \frac{n(3n - 1)}{2}.$$

18. Use the Binomial Theorem to expand and simplify $(x^2 + 2y)^6$.

19. How many seven-digit local telephone numbers can be formed if the first three digits are 279?

20. A human resource manager has 11 applicants to fill four different positions. Assuming that all applicants are equally qualified for any of the four positions, in how many ways can this be done?

21. The premise of a game show is that if the digits 3, 4, 5, and 6 are arranged in the correct order to form the price of an item, a player wins the item. Each digit is used exactly once. What is the probability of winning an item that the player knows is worth at least $5000?

22. A political discussion group is made up of six Republicans and four Democrats. Three people are selected at random to attend a convention. What is the probability that all three people selected are Democrats?

23. A box contains five white balls numbered 1 through 5 and five green balls numbered 1 through 5. A ball is chosen at random. Find the probability that it is white or odd-numbered.

24. The well-balanced spinner in the figure shown is spun twice. Assuming that it always stops in one of the colored sectors, what is the probability it will stop on red on the first spin and blue on the second spin?

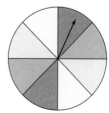

Appendix

Review Problems Covering the Entire Book

Use your graphing utility to verify as many answers as possible.

Solving Equations and Inequalities

Systematic procedures for solving certain equations and inequalities are an important component of algebra. Problems 1–21 give you the opportunity to review these procedures. Solve each problem, expressing irrational solutions in simplified form and imaginary solutions in the form a + bi.

1. $\dfrac{2}{x + 1} + \dfrac{3}{2x - 3} = \dfrac{6x + 1}{2x^2 - x - 3}$

2. $x(4x - 11) = 3$

3. $(x^2 - 5)^2 - 3(x^2 - 5) - 4 = 0$

4. $\sqrt{2x + 3} - \sqrt{4x - 1} = 1$

5. $x^{1/2} - 6x^{1/4} + 8 = 0$

6. $\dfrac{x - 3}{4} + \dfrac{x + 2}{3} \leqslant 2$

7. $-2 < 8 - 5x < 7$

8. $|2x + 1| \leqslant 1$

9. $6x^2 - 6 < 5x$

10. $\dfrac{1 - x}{3 + x} < 4$

11. $x^3 - x^2 - 4x + 4 = 0$

12. $3x^3 + 4x^2 - 7x + 2 = 0$

13. $e^{14 - 7x} - 53 = 24$

14. $e^{2x} - 10e^x + 9 = 0$

15. $\log_2(x + 1) + \log_2(x - 1) = 3$

16. $\ln(3x) + \ln(x + 2) = \ln 9$

17. (Solve using matrices.)
$$x - y + z = 17$$
$$-4x + y + 5z = -2$$
$$2x + 3y + z = 8$$

18. (Solve for x only using Cramer's rule.)
$$x + 2y - z = 1$$
$$x + 3y - 2z = -1$$
$$2x - y + z = 6$$

19. $\dfrac{x^2}{4} + \dfrac{y^2}{16} = 1$
$$x - y = 2$$

20. $4x^2 + 3y^2 = 48$
$$3x^2 + 2y^2 = 35$$

21. $3x^2 - 2xy + 3y^2 = 34$
$$x^2 + y^2 = 17$$

Graphs and Graphing

Throughout the book, we have used graphs to help visualize a problem's solution. We have also studied graphing equations and inequalities in the rectangular coordinate system. Problems 22–36 focus on problem solving with graphs.

22. Use the graphs shown in the figure to describe how the graph of $h(x) = (x + 2)^3 + 1$ is obtained from the graph of $f(x) = x^3$. Include the graph of g in your description.

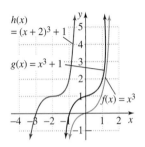

Gender Ratios in the U.S.

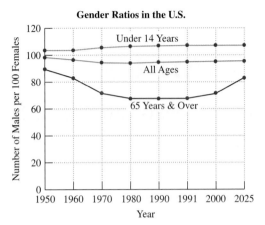

Source: U.S. Bureau of the Census, Statistical Abstract of the United States

23. The graph shown in the figure is based on the distance traveled by a car over a 30-minute period.

a. Describe what happened to the car over the six-minute interval from $t = 14$ to $t = 20$.

b. What is a reasonable estimate for how far the car traveled during the last five minutes?

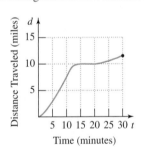

25. Sketch the graph of the function f defined by

$$f(x) = \begin{cases} x^2 & \text{if} & x \le 0 \\ x + 1 & \text{if} & 0 < x < 4 \\ \sqrt{x} & \text{if} & x \ge 4 \end{cases}$$

26. **a.** List all possible rational roots to
$$32x^3 - 52x^2 + 17x + 3 = 0.$$

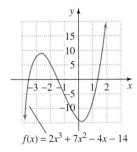

b. The graph of $f(x) = 32x^3 - 52x^2 + 17x + 3$ is shown in the figure. Use the graph of f and synthetic division to solve the equation in part (a).

27. Use the graph of $f(x) = 2x^3 + 7x^2 - 4x - 14$ shown in the figure to find the smallest positive integer that is an upper bound and the largest negative integer that is a lower bound for the roots of $2x^3 + 7x^2 - 4x - 14 = 0$. Then use synthetic division to verify your observation.

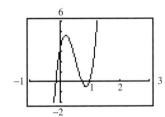

$f(x) = 2x^3 + 7x^2 - 4x - 14$

24. The graph on the top of the right column shows gender ratios in the United States, with future projections. For males age 65 and over:

a. In what time interval is the number of males per 100 females constant?

b. In what time interval is the number of males per 100 females increasing?

c. In what time interval is the number of males per 100 females decreasing?

For all ages:

d. Write a constant function $f(x)$ that approximately models the data shown for x in the interval [1950, 2025].

e. What is misleading about the scale on the horizontal axis?

Graph each rational function in Problems 28–29.

28. $f(x) = \dfrac{x^2 + 3x - 40}{x^2 - x - 20}$

29. $f(x) = \dfrac{x^2 - 1}{x - 2}$

30. Graph $y = 2^x$ and $y = 2^{x-2} + 1$ in the same rectangular coordinate system.

31. Graph $y = \log_2 x$ and $y = \log_2(x + 2)$ in the same rectangular coordinate system.

32. Graph the solution set for this system of inequalities:

$$5x + y \leqslant 10$$

$$y \geqslant \frac{1}{4}x + 2.$$

In Problems 33–35, identify each conic section and sketch its graph. For ellipses and hyperbolas, find the foci. For parabolas, identify the vertex, focus, and directrix.

33. $100x^2 + y^2 = 25$

34. $4x^2 - 9y^2 - 16x + 54y - 29 = 0$

35. $x + 2 = 4(y - 3)^2$

36. Graph the solution set for this system of inequalities:

$$x^2 + 9y^2 \leqslant 9$$

$$x^2 - y^2 \leqslant 1.$$

Mathematical Models

Describing the world compactly and symbolically using formulas is one of the most important aspects of algebra. Problems 37–44 concentrate on mathematical models.

37. The model $d = 0.01v^2 + 0.16v$ describes the average stopping distance, (d, in meters) for a car traveling v kilometers per hour. If a car stops in 45.6 meters, how fast was it traveling?

38. The function $f(x) = 3.14x + 64.98$ models a woman's height ($f(x)$, in centimeters) in terms of the length of the humerus, also in centimeters. If a woman is 159.18 centimeters tall, find the length of her humerus.

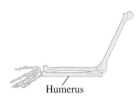

Humerus

39. The data points (3, 5.13) and (8, 8.03), shown and described below, fall on a straight line.

x (Number of Years after 1985)	3	8
y (Total Prize Money [in millions of dollars] Awarded at the Indianapolis 500 Car Race)	5.13	8.03

a. Write the equation of the line on which these measurements fall in point-slope form and slope-intercept form.

b. Use the slope-intercept model from part (a) to find the anticipated total prize money in the Indianapolis 500 race for the year 2010.

40. The model

$$C = \frac{350p}{100 - p}$$

describes the cost (C, in millions of dollars) for removing p percent of the industrial pollutants discharged into a river.

a. Find the cost of removing 60% of the pollutants.

b. According to this model, would it be possible to remove 100% of the pollutants? Explain.

41. Use the compound interest model, $A = Pe^{rt}$, to solve this problem. At what annual rate, compounded continuously, must $7000 be invested to grow to $14,000 in 10 years?

42. The perceived loudness of a sound (D, in decibels) at the eardrum is described by

$$D = 10 \log \frac{I}{I_0}$$

where

I = the intensity of the sound
    (in watts per meter squared)

I_0 = the intensity of a sound
    barely audible to the human ear

How much more intense is a sound of 80 decibels than a sound of 40 decibels?

43. Use the exponential growth model, $A = A_0 e^{kt}$, to solve this problem. In 1930, world population was approximately 2 billion, and by 1960 it had grown to approximately 3 billion.

a. Write the exponential growth function that models this data t years after 1930.

b. Use the model from part (a) to predict the year in which the world will reach its carrying capacity of 20 billion.

44. A ball is thrown directly upward. The ball's height above the ground at various times is given as follows.

Time t (in seconds)	1	2	3
Position $s(t)$ above the ground (in feet)	108	140	140

a. Use the three data values and the position function $s(t) = \frac{1}{2}at^2 + v_0 t + s_0$ to find values for a, v_0, and s_0, thereby modeling the height of the ball above the ground at any time t.

b. What is the ball's maximum height above the ground? After how many seconds does this occur?

c. How long does it take for the ball to strike the ground?

d. Use parts (b) and (c) to graph the position function.

A Potpourri of Skills

Problems 45–65 give you the opportunity to review a number of course objectives presented throughout the book.

45. Factor completely: $x^4 + 3x^3 - 8x - 24$.

46. Simplify: $\dfrac{x^2 - 9}{x^2 - x - 20} \div \dfrac{4x^2 - 12x}{4x^2 - 20x}$.

47. Simplify: $\dfrac{\dfrac{x+2}{x-2} - \dfrac{x}{x+2}}{3 - \dfrac{4}{x+2}}$.

48. Solve for a: $S = \dfrac{a - ar^n}{1 - r}$.

49. If $f(x) = 17x^2 - 11x + 23$, find $\dfrac{f(a+h) - f(a)}{h}$.

50. Write the point-slope, slope-intercept, and general forms of the equation of the line passing through $(-1, 7)$ and perpendicular to the line whose equation is $x - 6y - 4 = 0$.

51. Find $f^{-1}(x)$ if $f(x) = \sqrt[3]{x + 4}$.

52. If $f(x) = x^2 - 3x + 9$ and $g(x) = 5x - 7$, find

a. $f \circ g$

b. $g \circ f$.

53. Expand and evaluate logarithmic expressions where possible: $\log_8 \dfrac{\sqrt[3]{x^5}}{64y^2}$.

54. Write as a single logarithm: $3 \ln x + \frac{1}{2} \ln y - 6 \ln z$.

55. If $A = \begin{bmatrix} 4 & 2 \\ 1 & -1 \\ 0 & 5 \end{bmatrix}$ and $B = \begin{bmatrix} 2 & 4 \\ 3 & 1 \end{bmatrix}$, find $AB - 4A$.

56. Find the partial fraction decomposition for $\dfrac{2x^2 - 10x + 2}{(x - 2)(x^2 + 2x + 2)}$.

In Problems 57–59, find the standard form of the equation for each conic section with the given conditions.

57. Ellipse: foci: $(-12, 0)$, $(12, 0)$; vertices: $(-13, 0)$, $(13, 0)$

58. Hyperbola: vertices: $(-7, 0)$, $(7, 0)$; asymptotes: $y = \pm \frac{2}{7}x$

59. Parabola: focus: $(0, -6)$; directrix: $y = 6$

60. Write a formula for the nth term of the arithmetic sequence $2, -3, -8, -13, \ldots$. Do not use a recursion formula. Then use the formula for a_n to find the 19th term.

61. Use the formula for the sum of the first n terms of an arithmetic sequence to find $\displaystyle\sum_{i=1}^{50} (4i - 25)$.

62. Find the sum: $\displaystyle\sum_{i=1}^{10} 20(\frac{3}{2})^{i-1}$.

63. Find the sum: $\displaystyle\sum_{i=1}^{\infty} 10(-0.4)^{i-1}$.

64. Use mathematical induction to prove that, for all natural numbers n,

$$3 + 7 + 11 + 15 + \cdots + (4n - 1) = n(2n + 1).$$

65. Use the Binomial Theorem to expand and simplify: $(x^3 + 2y)^5$.

Problem Solving _____

Problem solving is the central theme of algebra. Some problems can be solved by modeling conditions with linear equations, systems of equations, functions, or equations of conics. Problems 66–79 give you a chance to review some of the problem-solving situations that have appeared throughout the book.

66. Yearly income for men increases by $1600 with each year of education. Men with no education can expect to earn about $6300 yearly. For women, a comparable model suggests a $1200 increase for each year of education. Women at the zero-education level earn about $2100 yearly. How many years of education, to the nearest tenth of a year, must a woman achieve to earn the same yearly income as a man with 14 years of education?

67. A rectangular room measures 15 by 20 feet. A rectangular rug with an area of 126 square feet is to be placed in the center of the room so that a strip of flooring of uniform width is left uncovered around the room's edges. How wide will this strip be?

68. A chemical is sold in bulk to manufacturers at $800 per ton for orders up to 12 tons. If an order exceeds 12 tons, the charge is discounted 2% per ton based on the number of tons sold. How many tons should be sold to produce a maximum income of $10,000?

69. The price of a VCR is reduced by 20%. When the VCR still does not sell, it is reduced by 20% of the reduced price. If the price of the VCR after both reductions is $224, what was the original price?

70. A box with no top is to be made from a rectangular piece of aluminum. Squares of the same size are cut from each corner and folded up the sides. The box is to be 2 inches deep and the length of its base is to be 5 inches more than the width. If the volume is to be 352 cubic inches, what dimensions should the piece of aluminum have?

71. The rectangular plot of land shown in the figure is to be fenced along three sides using 800 feet of fencing. No fencing is to be placed along the river's edge. Find the dimensions that maximize the enclosed area. What is the maximum area?

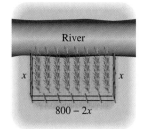

72. The following nutritional values are given for three types of foods.

Nutritional Component per Ounce	Food A	Food B	Food C
Calories	200	50	10
Protein (in grams)	0.2	3	1
Vitamin C (in milligrams)	0	10	30

Suppose a dietician wants to prepare a meal that consists of foods A, B, and C with 600 calories, 20 grams of protein, and 200 milligrams of vitamin C. Model the situation with a system of three linear equations and determine how many ounces of each type of food must be used.

73. You are about to take a test that contains computation problems worth 6 points each and word problems worth 10 points each. You can do a computation problem in 2 minutes and a word problem in 4 minutes. You have forty minutes to take the test and may answer no more than 12 problems. Assuming you answer all the problems attempted correctly, how many of each type of problem must you do to maximize your score? What is the maximum score?

74. The figure shows a semielliptic archway that has a height of 10 feet at the center and a width of 30 feet. How high is the arch 8 feet from the center?

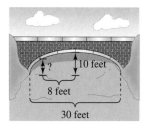

75. A small television has a picture with a diagonal measure of 10 inches and a viewing area of 48 square inches. Find the length and width of the screen.

76. Carpets are purchased for two square rooms with a combined area of 65 square yards. The price of the first carpet was $16 per square yard, and the carpet for the larger room was $20 per square yard. If the total cost was $1236, what are the dimensions of the smaller and larger square rooms?

77. A college graduate is considering a job that pays $28,000 the first year, with a 6% raise each year after that. The graduate plans to put aside half of all money earned toward the purchase of a house and wants at least $160,000 at the end of 10 years for this purpose. What is the total salary that will be paid by this job over a 10-

year period? Can the desired amount be saved for purchasing a house?

78. You have forgotten a friend's telephone number, but you do remember the first three digits and the last two digits, namely, 451-??71. If you randomly dial a number with these digits and select any two digits where the question marks now appear, what is the probability that you will dial your friend?

79. A committee contains seven women and three men. Three people are randomly selected to write the committee report. Find the probability of selecting three men for this task.

Using Technology

Use of a graphing utility enhances problem solving and exploration, often building greater appreciation of algebraic techniques. Problems 80 and 81 integrate technology into problem solving and developing mathematical models.

80. The rate of a stream is 4 miles per hour. A boat is rowed 9 miles downstream (with the current) and then returns 9 miles upstream (against the current) to the starting point of the excursion.

   a. If x represents the rowing rate in still water, write a rational function containing the sum of two rational expressions that models the total time for the round trip.

   b. Use a graphing utility to graph the function. Then trace along the curve and determine the rate in still water that one must row to make the round trip in 10 hours.

   c. Verify part (b) algebraically.

81. The data below indicate national expenditures for health care in billions of dollars for eight selected years.

Year	1960	1965	1970	1975	1980	1985	1990	1993
Health-Care Expenditures (in billions of dollars)	27.1	41.6	74.3	132.9	251.1	434.5	696.6	884.2

Source: U.S. Dept. of Health and Human Services

   a. Let $x = 0$ correspond to 1960 and let y correspond to each of the figures for health-care expenditure. Take the natural logarithm of each y-value.

   b. Graph the eight data points $(x, \ln y)$. Notice that the points lie nearly on a straight line, suggesting the use of a linear model.

   c. Use the linear regression option on a graphing utility to write the equation of the line that best fits the $(x, \ln y)$ transformed data.

   d. Write both sides of the equation in part (c) as exponents on base e. Simplify and obtain a model for health-care expenditures x years after 1960.

   e. Use the model from part (d) to predict health-care expenditures for the year 2000.

Answers to Selected Exercises

Chapter P

1. a. $\sqrt{4}, 7, \dfrac{18}{2}, 100$ **b.** $0, \sqrt{4}, 7, \dfrac{18}{2}, 100$ **c.** $-10, 0, \sqrt{4}, 7, \dfrac{18}{2}, 100$ **d.** $-10, -\dfrac{3}{4}, 0, \dfrac{4}{5}, \sqrt{4}, 7, \dfrac{18}{2}, 100$

e. $-\sqrt{2}, \pi$ **2. a.** $\dfrac{30}{2}, 90$ **b.** $0, \dfrac{30}{2}, 90$ **c.** $-5, -\sqrt{9}, 0, \dfrac{30}{2}, 90$ **d.** $-5, -\sqrt{9}, -\dfrac{4}{3}, 0, \dfrac{2}{3}, \dfrac{30}{2}, 90$ **e.** $\sqrt{7}, 2\pi$

3. a. None   **b.** $\dfrac{0}{3}$ **c.** $-\sqrt[3]{8}, \dfrac{0}{3}$ **d.** $-\sqrt[3]{8}, \dfrac{0}{3}, \sqrt{\dfrac{4}{9}}, 1.\overline{126}$ **e.** $\sqrt[3]{7}$

4. a. None   **b.** $\dfrac{0}{6}$ **c.** $-\sqrt[3]{125}, \dfrac{0}{6}$ **d.** $-\sqrt[3]{125}, \dfrac{0}{6}, \sqrt{\dfrac{9}{25}}, 3.\overline{47}$ **e.** $\sqrt[4]{7}$

5. $1 < x \le 6$

6. $-2 < x \le 4$

7. $-5 \le x < 2$

8. $-4 \le x < 3$

9. $-3 \le x \le 1$

10. $-2 \le x \le 5$

11. $2 < x < \infty$

12. $3 < x < \infty$

13. $-3 \le x < \infty$

14. $-5 \le x < \infty$

15. $-\infty < x < 3$

16. $-\infty < x < 2$

17. $-\infty < x < 5.5$

18. $-\infty < x \le 3.5$

19. $\{x \mid x < 6\}; (-\infty, 6)$

20. $\{x \mid x > 2\}; (2, \infty)$ **21.** $\{x \mid x \ge -1\}; [-1, \infty)$ **22.** $\{x \mid x \le -3\}; (-\infty, -3]$ **23.** $\{x \mid 5 < x < 12\}; (5, 12)$

24. $\{x|-4 < x < 7\}; (-4, 7)$ **25.** $\{x|2 < x \le 13\}; (2, 13]$ **26.** $\{x|-7 \le x < 2\}; [-7, 2)$ **27.** $\{x|x \le 6\}; (-\infty, 6]$

28. $\{x|x \ge 3\}; [3, \infty)$ **29.** $\{x|2 \le x \le 5\}; [2, 5]$ **30.** $\{x|-3 \le x \le 2\}; [-3, 2]$ **31.** $\{x|x \le 60\}; (-\infty, 60]$

32. $\{x|x \le 32\}; (-\infty, 32]$ **33.** $\{x|-2 \le x < 0\}; [-2, 0)$ **34.** $\{x|-5 \le x \le 0\}; [-5, 0]$ **35.** $[2, 5)$ **36.** $[2, 3)$

37. $(5, \infty)$ **38.** $(7, \infty)$ **39.** $(-\infty, 2)$ **40.** $(-\infty, 1)$ **41.** $(-\infty, 3]$ **42.** $(-\infty, 7]$ **43.** $\varnothing$ **44.** $\varnothing$

45. $(1, 7)$ **46.** $(1, 6]$ **47.** $(-1, \infty)$ **48.** $(-3, \infty)$ **49.** $(-\infty, 4)$ **50.** $(-\infty, 5)$ **51.** $(-\infty, 6)$

52. $(-\infty, 10)$ **53.** $(-3, -1) \cup [2, 4]$ **54.** $(-2, 0) \cup [3, 5]$ **55.** 300   **56.** 705   **57.** 203   **58.** 109

59. $12 - \pi$ **60.** $7 - \pi$ **61.** $12 - \pi$ **62.** $7 - \pi$ **63.** $5 - \sqrt{2}$ **64.** $13 - \sqrt{5}$ **65.** -1 **66.** -1

67. 4   **68.** 8   **69.** 15   **70.** 11   **71.** 7   **72.** 14   **73.** 15   **74.** 23   **75.** 2.2   **76.** 4.2   **77.** $\dfrac{7}{10}$

78. $\dfrac{3}{2}$ **79.** $\dfrac{11}{4}$ **80.** $9\dfrac{9}{20}$ **81.** $|x| = 7$ **82.** $|x| = 13$ **83.** $|x| \ge 6$ **84.** $|x| \le 6$

85. $|72 - 99| = 27$ miles   **86.** $|54 - 76| = 22$ miles   **87.** $|y - 7| \le 3$ **88.** $|y + 4| \ge 10$ **89.** 80; 100

90. 80; 20   **91.** 65; 35   **92.** 45; 55   **93.** 27; 25   **94.** 8; 77

95. General medical examination, cough, routine prenatal examination, progress visit.

96. General medical examination, cough.

97. Ear ache or infection, back symptoms, vision dysfunctions, skin rash.

98. Throat, postoperative visit, ear ache or infection.

99. General medical examination, cough, routine prenatal examination, progress visit, throat, postoperative visit.

100. Postoperative visit, ear ache or infection, back symptoms, vision dysfunctions, skin rash.

101. Routine prenatal examination, progress visit, throat, postoperative visit, ear ache or infection.

102. Throat and postoperative visit.   **103.** d   **104.** d   **105.** $\dfrac{112}{80}, \dfrac{873}{618}, \sqrt{2}, \dfrac{4243}{3000}, \dfrac{55}{37}$

106. $\sqrt{17} - \sqrt{13}, \dfrac{111,525}{50,000}, 2.23\overline{20}, \dfrac{37,961}{17,000}, \sqrt{5}$ **107.** $\dfrac{a}{b}$ gets close to zero.

108. Answers may vary, $\left|\dfrac{a}{b}\right|$ gets vary large.   **109.** Answers may vary.   **110.** No

PROBLEM SET P.2

1. 45   **2.** 196   **3.** 35   **4.** 23   **5.** -36 **6.** -12 **7.** 0   **8.** $-\dfrac{4}{5}$ **9.** $-\dfrac{2}{3}$ **10.** 0   **11.** $\dfrac{5}{2}$ **12.** $\dfrac{1}{3}$

13. Commutative Property of Multiplication   **14.** Associative Property of Addition

15. Commutative Property of Multiplication   **16.** Commutative Property of Addition

17. Distributive Property   **18.** Multiplicative Inverse Property   **19.** Distributive Property

20. Additive Inverse Property   **21.** Additive Identity Property   **22.** Multiplicative Identity Property

23. Multiplicative Identity Property   **24.** Multiplicative Identity Property   **25.** $7x + 21y$ **26.** $30a - 24b$

27. $-4x$ **28.** $-y$ **29.** $12x^5 - 8x^3$ **30.** $-30y^7 + 12y^4$ **31.** $13x$ **32.** $-26xy$ **33.** $20ab$ **34.** ab

35. $-2x^2 + 5x + 6$ **36.** $6x^4 + 9x - 17$ **37.** x **38.** y **39.** 50   **40.** 72   **41.** -32 **42.** 16   **43.** -16

44. -125 **45.** 32   **46.** 27   **47.** 64   **48.** 243   **49.** 64   **50.** 729   **51.** 32   **52.** -2 **53.** $\dfrac{1}{3}$

54. $\dfrac{1}{4}$ **55.** 1   **56.** 1   **57.** $64x^6$ **58.** $36x^6$ **59.** $6x^{11}$ **60.** $99x^{17}$ **61.** $5a^6$ **62.** $3a^{10}$ **63.** $\dfrac{2}{b^7}$

64. $\dfrac{2}{b^{10}}$ **65.** $2x^3$ **66.** $\dfrac{x^2}{4}$ **67.** $-\dfrac{x^6}{729}$ **68.** $-\dfrac{x^{12}}{216}$ **69.** $-32x^{15}y^{20}$ **70.** $-27x^6y^9$

71. $24x^3y^8$ **72.** $-x^3y^8$ **73.** $-5a^{11}b$ **74.** $-5a^7b^3$ **75.** $-32x^3$ **76.** $\dfrac{12}{x^2}$ **77.** $\dfrac{y^4}{25x^6}$ **78.** $\dfrac{y^{28}}{81x^{16}}$

79. $\dfrac{y^{10}}{9x^4}$ **80.** $-\dfrac{x^9}{64y^6}$ **81.** $-75x^{13}y^{19}$ **82.** $-\dfrac{40r^{10}}{s^2}$ **83.** $-\dfrac{192b^{17}}{a^2}$ **84.** $\dfrac{3x^8}{y^3}$ **85.** $-\dfrac{2s^{21}}{r^3}$ **86.** $-\dfrac{y^{28}}{x^{18}z^{21}}$

87. $-\dfrac{b^6c^{13}}{a^5}$ **88.** $\dfrac{x^{10}}{4y^6}$ **89.** $-\dfrac{729z^{17}}{x^{11}y^{22}}$ **90.** $-\dfrac{12b^8}{5a^8}$ **91.** $-\dfrac{9z^5}{5x^{12}}$ **92.** $-2x^7y^{14}$ **93.** $16x^{10}y^6$ **94.** $16ab^{13}$

95. 713,000 **96.** 5,024,000,000 **97.** 0.0000000307 **98.** 0.0000006573 **99.** 9.65×10^7

100. 1.673×10^8 **101.** 7.361×10^{12} **102.** 5.024×10^{15} **103.** 7.53×10^0 **104.** 9.04×10^0

105. 1.6×10^{-4} **106.** 3.7×10^{-6} **107.** 7.253×10^{-3} **108.** 9.621×10^{-3} **109.** 3.071×10^3

110. 2.35×10^{11} **111.** 3×10^8 **112.** 4×10^{-16}

113. 4.466; A person who earns \$10,000 annually contributes 4.466% to charity.

114. 5294; In 1990, the world population was 5,294,000,000.

115. 192π in.^{2} **116.** $1\dfrac{1}{3}$ hours; 4 km upstream from C. **117.** 5.88×10^{12}

118. 2.6×10^7 **119.** 1.016×10^{-7} **120.** 3.937×10^{-5} **121.** 3.6×10^{51} hydrogen atoms

122. 6.67×10^9 times as large **123.** 1.5×10^9 trees **124.** 2.473×10^4 seconds **125.** 25,350 years

126. 2.53×10^3 cm/sec **127.** c **128.** d **129.** c **130.** c **131–133.** Answers may vary.

134. $A = C + D$ **135.** The pair 10^7 and 10^{43} is closer together.

PROBLEM SET P.3

1. a.

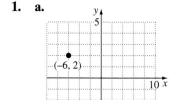

b. $\sqrt{125} \approx 11.18$

c. $\left(-1, -\dfrac{1}{2}\right)$

2. a.

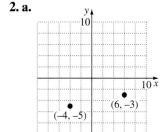

b. $\sqrt{104} \approx 10.20$

c. $(1, -4)$

3. a.

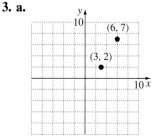

b. $\sqrt{34} \approx 5.83$

c. $\left(\dfrac{9}{2}, \dfrac{9}{2}\right)$

4. a.

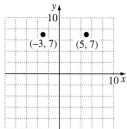

b. 8

c. $(1, 7)$

5. a.

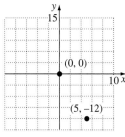

b. 13

c. $\left(\dfrac{5}{2}, -6\right)$

6. a.

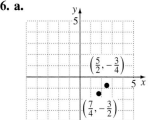

b. $\sqrt{\dfrac{9}{8}} \approx 1.06$

c. $\left(\dfrac{17}{8}, -\dfrac{9}{8}\right)$

7. a.

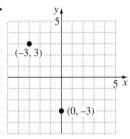

b. $\sqrt{45} \approx 6.71$

c. $\left(-\dfrac{3}{2}, 0\right)$

8. a.

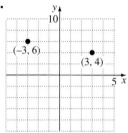

b. $\sqrt{40} \approx 6.32$

c. $(0, 5)$

9. a.

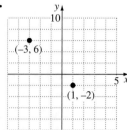

b. $\sqrt{80} \approx 8.94$

c. $(-1, 2)$

10. a.

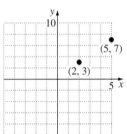

b. 5

c. $\left(\dfrac{7}{2}, 5\right)$

11.

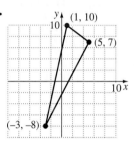

40.44 units

12.

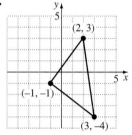

Two sides have length 5 units.

13. 8.49 units   **14.** 5.59 units

15.

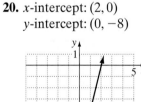

Area = 7.5 square units

16.

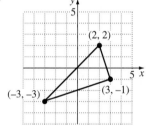

Area = 10 square units

17. $AB = \sqrt{61},\ BC = 2\sqrt{61},\ AC = 3\sqrt{61}$ so $AB + BC = AC$ **18.** $d = 17; d = 13$

19. x-intercept: $(2, 0)$
y-intercept: $(0, -6)$

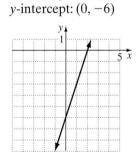

20. x-intercept: $(2, 0)$
y-intercept: $(0, -8)$

21. x-intercept: $(2, 0)$
y-intercept: $(0, 4)$

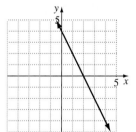

22. x-intercept: $(3, 0)$
y-intercept: $(0, 6)$

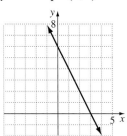

23. x-intercepts: $(1, 0), (-1, 0)$
y-intercept: $(0, -1)$

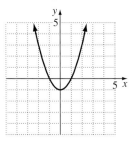

24. x-intercepts: $(-3, 0), (3, 0)$
y-intercept: $(0, -9)$

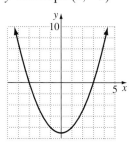

25. x-intercepts: $(-3, 0), (3, 0)$
y-intercept: $(0, 9)$

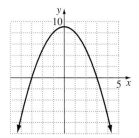

26. x-intercepts: $(-1, 0), (1, 0)$
y-intercept: $(0, 1)$

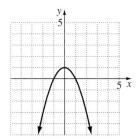

27. x-intercepts: $(-3, 0), (2, 0)$
y-intercept: $(0, -6)$

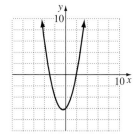

28. x-intercepts: $(1, 0), (3, 0)$
y-intercept: $(0, 3)$

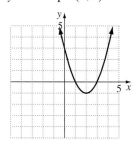

29. x-intercepts: $(0, 0), (2, 0)$
y-intercept: $(0, 0)$

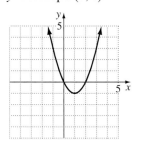

30. x-intercepts: $(0, 0), (-1, 0)$
y-intercept: $(0, 0)$

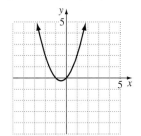

31. x-intercepts:
$$\left(\frac{1 - \sqrt{5}}{2}, 0\right), \left(\frac{1 + \sqrt{5}}{2}, 0\right)$$
y-intercept: $(0, 1)$

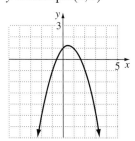

32. x-intercept: none
y-intercept: $(0, -2)$

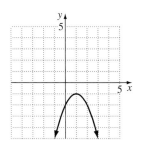

33. x-intercept: $(0, 0)$
y-intercept: $(0, 0)$

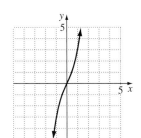

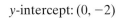

34. x-intercepts: $(\sqrt{3}, 0)$, $(0, 0)$, $(-\sqrt{3}, 0)$
y-intercept: $(0, 0)$

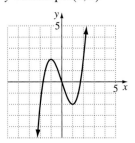

35. x-intercept: $(0, 0)$
y-intercept: $(0, 0)$

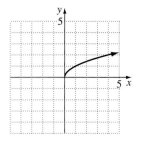

36. x-intercept: $(1, 0)$
y-intercept: none

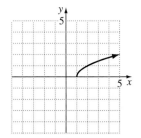

37. x-intercept: $(2, 0)$
y-intercept: none

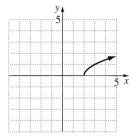

38. x-intercept: $(-4, 0)$
y-intercept: $(0, 2)$

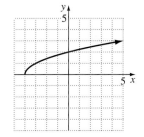

39. x-intercept: $(0, 0)$
y-intercept: $(0, 0)$

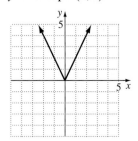

40. x-intercept: $(0, 0)$
y-intercept: $(0, 0)$

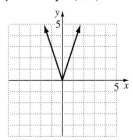

41. x-intercept: $(-1, 0)$
y-intercept: $(0, 1)$

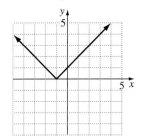

42. x-intercept: $(-2, 0)$
y-intercept: $(0, 2)$

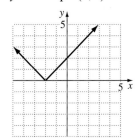

43. x-intercept: $(-1, 0)$
y-intercept: $(0, 1)$

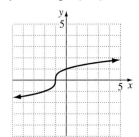

44. x-intercept: $(-2, 0)$
y-intercept: $(0, \sqrt[3]{2})$

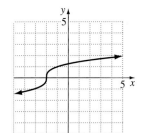

45. *x*-intercept: $(1, 0)$
y-intercept: $(0, 1)$

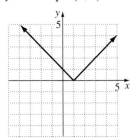

46. *x*-intercept: $(2, 0)$
y-intercept: $(0, 2)$

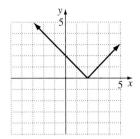

47. *x*-intercept: $(1, 0)$
y-intercept: $(0, -1)$

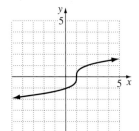

48. *x*-intercept: $(2, 0)$
y-intercept: $(0, -\sqrt[3]{2})$

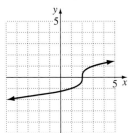

49. $(1977, 2200)$; In 1977 about 2,000,000 women participated in high-school athletics.

50. a. 1250 feet   **b.** 8.8 seconds   **51.** 33,000; 1945   **52.** 19,000; 1992   **53.** Decreasing
54. 70 feet; after 2 seconds   **55.** $x \approx 4.1$; the ball hits the ground after approximately 4.1 seconds.

56. $y = 6$; the *y*-intercept gives the height from which the ball is thrown.   **57.** $\dfrac{1}{4}$ **58.** d   **59.** c

60. Answers may vary.

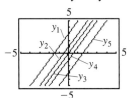

61. Answers may vary.

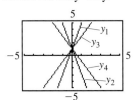

62. a.

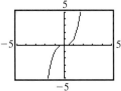

b.

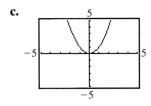

c.

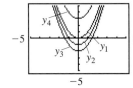

d. Each pair of equations has the same graph.
$$x^a \cdot x^6 = x^{a+b}; \frac{x^a}{x^6} = x^{a-b}$$

63. $[-10, 10]$ by $[-50, 50]$ **64.** $[-8, 16]$ by $[-16, 8]$ **65.** $[-50, 50]$ by $[-10, 10]$ **66.** $[-10, 10]$ by $[-50, 100]$
67.

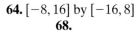

68.

69.

70. a. $y = 2$ **b.** $x = 2$ **71. a.** $y = 5.25$ **b.** $x = 0, x = 4$

72. a. $y = 0$ **b.** $x \approx -2.21, x \approx 0.54, x \approx 1.68$ **73. a.** $y \approx 7.81$ **b.** $x \approx -1.13, x \approx 2.35$

74. a.

x	y
0	4025
5	4204
10	4229
15	4098
20	3813
30	2777
40	1121

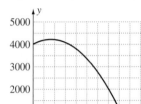

b.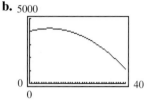

c. 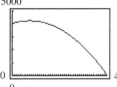   No; equations may be unreliable over long periods of time.

75. a. 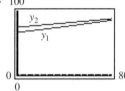   **b.** $y_1 \approx 75.79$; $y_2 \approx 81.24$; a 40-year-old white male can expect to live about 76 years; a 40-year-old white female can expect to live about 81 years.

c. The corresponding value for y increases in each graph as the value for x increases.

76. Answers may vary   **77.** Answers may vary.   **78.** Answers may vary.

79. $AB = 2\sqrt{2}$; $BC = 3\sqrt{2}$; $AC = 5\sqrt{2}$ **80.** Verification is left to the student.

PROBLEM SET P.4

1. 7   **2.** 10   **3.** 3   **4.** 5   **5.** -2 **6.** -4 **7.** Not a real number   **8.** Not a real number   **9.** $\dfrac{2}{5}$

10. $\dfrac{2}{3}$ **11.** -1 **12.** -1 **13.** $-\dfrac{2}{3}$ **14.** $-\dfrac{3}{5}$ **15.** $-\dfrac{1}{2}$ **16.** -3 **17.** -8 **18.** $\dfrac{1}{2}$ **19.** $2\sqrt[3]{2}$

20. $5\sqrt[3]{2}$ **21.** $2\sqrt[5]{2}$ **22.** $2\sqrt[4]{2}$ **23.** $10|y|\sqrt{2y}$ **24.** $2|x|\sqrt{10x}$ **25.** $2|y|\sqrt{5xy}$ **26.** $3x^2\sqrt{5y}$ **27.** $5|y|z^2\sqrt{3xz}$

28. $3|xy|\sqrt{2yz}$ **29.** $2x\sqrt[3]{4}$ **30.** $2y\sqrt[3]{2x}$ **31.** $-2yz^2\sqrt[3]{4xy^2}$ **32.** $-2xy\sqrt[3]{2xy^2z^2}$ **33.** $2|y|\sqrt[4]{3y^3}$

34. $2x^2\sqrt[4]{2x^3}$ **35.** $\dfrac{\sqrt{7}}{7}$ **36.** $\dfrac{\sqrt{10}}{5}$ **37.** $\dfrac{\sqrt{15}}{5}$ **38.** $\dfrac{\sqrt{21}}{7}$ **39.** $\dfrac{\sqrt[3]{28}}{2}$ **40.** $\dfrac{\sqrt[5]{14}}{2}$ **41.** $3\sqrt[3]{4}$ **42.** $-4\sqrt[3]{4}$

43. $-3\sqrt[5]{4}$ **44.** $-8\sqrt[5]{8}$ **45.** $\dfrac{50 + 10\sqrt{3}}{11}$ **46.** $\dfrac{6 + \sqrt{2}}{2}$ **47.** $\dfrac{13\sqrt{11} - 39}{2}$ **48.** $\dfrac{9 - 3\sqrt{7}}{2}$

49. $3\sqrt{5} - 3\sqrt{3}$ **50.** $3\sqrt{7} - 3\sqrt{3}$ **51.** $\dfrac{11\sqrt{7} + 11\sqrt{3}}{4}$ **52.** $\dfrac{13\sqrt{5} + 13\sqrt{3}}{2}$ **53.** $8\sqrt{2}$ **54.** $5\sqrt{7}$

55. $-16\sqrt{2}$ **56.** $-2\sqrt{3}$ **57.** $20\sqrt{2} - 5\sqrt{3}$ **58.** $\sqrt{6} + 12\sqrt{7}$ **59.** $3\sqrt{2}$ **60.** $\dfrac{14\sqrt{5}}{15}$ **61.** $7\sqrt{10}$ **62.** $\dfrac{8\sqrt{3}}{3}$

63. 6   **64.** 11   **65.** 2   **66.** -3 **67.** 25   **68.** 4   **69.** 81   **70.** $\dfrac{81}{625}$ **71.** $\dfrac{1}{16}$ **72.** $\dfrac{1}{1024}$ **73.** $14y^{7/12}$

74. $12x^{17/12}$ **75.** $-15x^{1/4}$ **76.** $48x^{5/12}$ **77.** $4x^{1/4}$ **78.** $8y^{5/12}$ **79.** $\dfrac{8}{y^{1/12}}$ **80.** $-\dfrac{3}{y^{5/12}}$ **81.** $32xy^{10}z^2$

82. $81xy^{12}z^3$ **83.** $5x^2y^3$ **84.** $5x^3y^2$ **85.** $2x^{1/4}y^{1/16}z^{1/6}$ **86.** $3x^{1/12}y^{2/9}z$ **87.** $\dfrac{8x^{3/4}}{125y}$ **88.** $\dfrac{49x^{4/3}}{9y}$ **89.** $\dfrac{y^{5/2}}{x^{3/2}}$

90. $\dfrac{1}{ab^{4/5}}$ **91.** $\dfrac{4a}{3b}$ **92.** $\dfrac{5x^6}{9y^8}$ **93.** $\sqrt{b}$ **94.** $\sqrt[3]{b^2}$ **95.** $\sqrt[3]{(x-1)^2}$ **96.** $\sqrt[4]{x-1}$ **97.** $\sqrt{xy}$ **98.** $\sqrt{xy}$

99. $\sqrt[3]{2xy^2}$ **100.** $\sqrt[3]{2xy^2}$ **101.** $\sqrt[3]{3xy^2}$ **102.** $3|y|\sqrt{xy}$ **103.** $11\sqrt[3]{2}$ **104.** $7\sqrt{5}$ **105.** $\sqrt{5}$ **106.** $2\sqrt[4]{2}$

107. $f = \dfrac{\sqrt{Pm}}{2Lm}$ **108. a.** 880 variations per second   **b.** 262 vibrations per second   **109.** 275.52 mg

110. 1.89 hours   **111.** 5.9%   **112.** d   **113.** c

114. 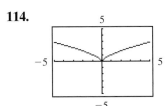   **115.** 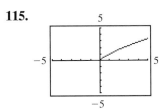   **116.** No; $y_1 = x^{2/3}$ is graphed only for values $x \geq 0$.

117. 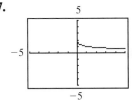   **118.** 4   **119.** Numbers are decreasing and are approaching 1.

120. a. $>$ **b.** $>$ **c.** $<$ **d.** $>$ **121.** $\dfrac{7 + 3\sqrt{3} - \sqrt{5} - 2\sqrt{15}}{11}$ **122.** 1

PROBLEM SET P.5

1. $11x^3 + 7x^2 - 12x - 4$; degree 3   **2.** $12x^3 - 5x^2 - 4x - 4$; degree 3
3. $12x^3 + 4x^2 + 12x - 14$; degree 3   **4.** $9x^4 + 4x^3 - 2x + 1$; degree 4   **5.** $6x^2 - 6x + 2$; degree 2
6. $6x^3 + 10x^2 + 11x - 8$; degree 3   **7.** $3x^2 - 2xy + 5y^2$; degree 2
8. $5x^3 - 7x^2y + 11xy^2 + 8y^3$; degree 3   **9.** $x^3 + 1$ **10.** $x^2 + 125$
11. $3y^4 + 20y^3 - 14y^2 + 17y - 6$ **12.** $2y^4 - y^3 - 15y^2 + 22y - 8$ **13.** $2z^5 + z^4 + 2z^2 - 8z$
14. $4z^5 - z^4 + 15z^3 - 3z^2 + 9z$ **15.** $x^3y^3 + 8$ **16.** $x^6y^{12} + 27$ **17.** $35x^2 + 26x + 3$ **18.** $27x^2 + 33x - 20$
19. $18y^6 + 53y^3 - 35$ **20.** $18y^6 + 3y^3 - 28$ **21.** $6x^4y^2 - 7x^3y^2 - 20x^2y^2$ **22.** $35x^4y^2 + 3x^3y^2 - 2x^2y^2$
23. $81x^5 + 36x^3 - 36x^2 - 16$ **24.** $81y^5 + 45y^3 - 45y^2 - 25$ **25.** $-2x^3 + 5x^2 - x - 2$
26. $-2x^3 - x^2 + 13x - 6$ **27.** $25y^2 - 9$ **28.** $9y^2 - 49$ **29.** $16x^4y^2 - 25x^2$ **30.** $25x^6y^2 - 9x^2$
31. $y^2 - 4x^2$ **32.** $-9a^8b^4 + 25c^2$ **33.** $x^2 + 12x + 36$ **34.** $x^2 - 10x + 25$
35. $x^3 - 18x^2 + 108x - 216$ **36.** $x^3 + 15x^2 + 75x + 125$ **37.** $9x^4y^2 + 12x^3y^2 + 4x^2y^2$
38. $16x^2y^4 + 8x^2y^3 + x^2y^2$ **39.** $27x^3 + 54x^2y + 36xy^2 + 8y^3$ **40.** $8x^3 - 60x^2y + 150xy^2 - 125y^3$
41. $8x^6y^3 - 12x^5y^3 + 6x^4y^3 - x^3y^3$ **42.** $64x^6y^3 + 48x^5y^3 + 12x^4y^3 + x^3y^3$ **43.** $8x^{15} + 12x^{14} + 6x^{13} + x^{12}$
44. $-x^{21} + 9x^{20} - 27x^{19} + 27x^{18}$ **45.** $9x^2 - 25y^2 + 42x + 49$ **46.** $25x^2 - 49y^2 - 28y - 4$
47. $-4x^2 + 25y^2 - 12x - 9$ **48.** $-9x^2 + 64y^2 + 42x - 49$ **49.** $4x^2 + 4xy + y^2 + 4x + 2y + 1$
50. $25x^2 + 60xy + 36y^2 + 10x + 12y + 1$ **51.** $9x^2 + 6xy + y^2 - 6x - 2y + 1$
52. $4x^2 - 4xy + y^2 + 32x - 16y + 64$ **53.** $9x^2 - y^2 - 6x + 1$ **54.** $4x^2 - 4xy + y^2 - 64$
55. $x^2 - y^2 + 6y - 9$ **56.** $x^4 - 9y^2 + 6y - 1$ **57.** $5x^2 + 9x$ **58.** $22x + 29$ **59.** $20x^2 + 16x$
60. $x^2 + 2x$ **61.** $x^2 + 7x + 6$ **62.** $3x^2 + x - 4$ **63.** $6x^2 + 26x$ **64. a.** $x^2 - 2x$ **b.** $4x - 4$

65. $4x^3 - 32x^2 + 80x - 80$ **66.** $\frac{2}{3}t^3 - 2t^2 + 4t$ **67.** $13.3t^2$ **68.** d   **69.** b   **70.** No

71. They are equivalent expressions.

72. Let $y_1 = (x - 1)(x + 3)$ and   **73.** Let $y_1 = (x - 1)^3$ and   **74.** c   **75.** $x^3 + 7x^2 - 3x$
   $y_2 = x^2 + 2x - 3$. They give $y_2 = x^3 - 1$
   the same graph.

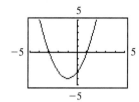   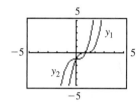

76. $2x^3 + 12x^2 + 12x + 10$ **77. a.** $24, 35, 48; x^2 - 1$ **b.** $130, 222, 350; x^3 + x$ **78.** $6y^n - 13$

PROBLEM SET P.6

1. $9(2x + 3)$ **2.** $8(2x - 3)$ **3.** $3x(x + 2)$ **4.** $4x(x - 2)$ **5.** $9x^2(x^2 - 2x + 3)$ **6.** $6x^2(x^2 - 3x + 2)$
7. $4xy(3x - 2y)$ **8.** $11x(3x - 2y^4)$ **9.** $7xy(x^2y + 2x - 6x^4y^2 + 3y^3)$ **10.** $12pq^2(3q - p^2 + 5pq^2 - 2p^3q)$
11. $(x - 3)(x^2 + 12)$ **12.** $(2x + 5)(x^2 + 17)$ **13.** $(x - 5)[2x(x - 5) - 3y]$ **14.** $(x - 3)[4x(x - 3) - 2y]$
15. $(x^2 + 5)(x - 2)$ **16.** $(x - 3)(x^2 + 4)$ **17.** $(x - 1)(x^2 + 2)$ **18.** $(x + 6)(x^2 - 2)$ **19.** $(3x - 2)(x^2 - 2)$
20. $(x - 1)(x^2 - 5)$ **21.** $(a + 5)(ac + 2)$ **22.** $(3y + 4z)(y + 8)$ **23.** $(x + 4)(4xy - 1)$
24. $(x^2 + 3)(5 - y)$ **25.** $(x - 10)(x + 10)$ **26.** $(x - 12)(x + 12)$ **27.** $(6x - 7)(6x + 7)$
28. $(8x - 9)(8x + 9)$ **29.** $(3x - 5y)(3x + 5y)$ **30.** $(6x - 7y)(6x + 7y)$ **31.** $(3y - 2x - 1)(3y + 2x + 1)$
32. $(4y - 3x + 1)(4y + 3x - 1)$ **33.** $(x^2 + 9)(x + 3)(x - 3)$ **34.** $(x^2 + 4)(x + 2)(x - 2)$
35. $(3x^2 - y^3)(3x^2 + y^3)$ **36.** $(9x^2 - 7y^3)(9x^2 + 7y^3)$ **37.** $(x - 2)(x^2 + 2x + 4)$ **38.** $(x - 4)(x^2 + 4x + 16)$
39. $(y + 3)(y^2 - 3y + 9)$ **40.** $(y + 5)(y^2 - 5y + 25)$ **41.** $(4d + 5)(16d^2 - 20d + 25)$
42. $(2d + 3)(4d^2 - 6d + 9)$ **43.** $(5 - 2pd)(25 + 10pd + 4p^2d^2)$ **44.** $(4 - 3yz)(16 + 12yz + 9y^2z^2)$
45. $(x + 5)(x + 3)$ **46.** $(x + 8)(x + 2)$ **47.** $(x - 5)(x - 3)$ **48.** $(x - 3)(x - 2)$ **49.** $(x + 6)(x - 5)$
50. $(x + 16)(x - 2)$ **51.** $(x - 7)(x + 4)$ **52.** $(x - 7)(x + 3)$ **53.** $(x + 2)^2$ **54.** $(x + 3)^2$ **55.** $(x - 4)^2$
56. $(x - 5)^2$ **57.** $(5x - 2)^2$ **58.** $(3x - 2)^2$ **59.** $(2x + 7)(x + 1)$ **60.** $(5x + 1)(x + 11)$
61. $(4y + 1)(y + 2)$ **62.** $(4y + 3)(2y + 1)$ **63.** $(5x + 2)(2x + 3)$ **64.** $(5x + 2)(x + 3)$
65. $(4y - 3)(2y - 3)$ **66.** $(3y - 5)^2$ **67.** $(4y - 3)(y - 6)$ **68.** $(6y - 5)(y - 3)$ **69.** $(4y + 3)(3y - 7)$
70. $(8y + 9)(2y - 3)$ **71.** Prime   **72.** Prime   **73.** $2y(y + 2)^2$ **74.** $3y(y - 1)^2$ **75.** $4(x - 5)(x + 1)$
76. $3(x - 6)(x + 2)$ **77.** $3x(x + 2)(x^2 - 2x + 4)$ **78.** $2x(x - 5)(x^2 + 5x + 25)$ **79.** $(x - 5)(x - 2)(x + 2)$
80. $(x + 3)(x - 9)(x + 9)$ **81.** $(x + 5 - 6a)(x + 5 + 6a)$ **82.** $(x - 4 - 7a)(x - 4 + 7a)$
83. $c(c - 4)(c + 4)$ **84.** $y(y - 1)(y + 1)$ **85.** $(3b - 4)(3b + 4)(x + y)$ **86.** $(cx + dy)(x + 1)$ **87.** Prime
88. Prime   **89.** $(y - 2)^2(y + 2)$ **90.** $(y - 2)(y - 1)(y + 1)$ **91.** $-2x(x - 2)(x^2 + 2x + 4)$
92. $7xy(x^2 + y^2)(x - y)(x + y)$ **93.** $2(2x + 5)^2$ **94.** $-9b^3(b - 1)(b^2 + b + 1)$
95. $(x + y)(x^2 - xy + y^2)(x - z)(x + z)$ **96.** $12ab(a - b)(a + b)$ **97.** $(3a + 4b)(x - 5)$
98. $3(7y - 6)(3y + 4)$ **99.** $(r - s)(r^2 + rs + s^2 + 1)$ **100.** $7b(y^2 + 1)(y + 1)(y - 1)$
101. $a^3b^3(ab - 1)(a^2b^2 + ab + 1)$ **102.** $x^3(y + 1)(y - 1)(y^2 - y + 1)(y^2 + y + 1)$ **103.** $x^{1/2}(x - 1)$
104. $x^{1/4}(x^{1/2} - 1)$ **105.** $\dfrac{4(2x + 1)}{x^{2/3}}$ **106.** $\dfrac{6(x + 2)}{x^{3/4}}$ **107.** $\dfrac{3(x - 2)(x - 1)}{x^{1/2}}$ **108.** $\dfrac{(x + 1)^2}{x^{3/2}}$
109. $-(x + 3)^{1/2}(x + 2)$ **110.** $(x^2 + 4)^{3/2}(x^4 + 8x^2 + 17)$ **111.** $\dfrac{x + 4}{(x + 5)^{3/2}}$ **112.** $\dfrac{x^2 + 4}{(x^2 + 3)^{5/3}}$
113. $-\dfrac{4}{3}(4x - 1)^{1/2}(x - 1)$ **114.** $\dfrac{2(100x^2 + 95x + 11)}{(4x + 3)^2}$ **115.** $\dfrac{x(x^2 + 6)}{(x^2 + 3)^{3/2}}$ **116.** $\dfrac{5}{(x - 5)^{3/2}(x + 5)^{1/2}}$

117. d   **118.** b   **119.** Not correct   **120.** Not correct   **121.** Correct   **122.** $(x + 4)(x - 1)$
123. $(x - 3)^2$ **124.** $(x^2 + 1)(x - 1)(x + 1)$ **125.** $(x + 1)(x^2 - x + 1)$ **126–130.** Answers may vary.
131. a. $(x^2 + y^2)^2$ **b.** $(x^2 - xy + y^2)(x^2 + xy + y^2)$ **132.** $(x + y)(x - y)(x^2 - xy + y^2)(x^2 + xy + y^2)$
133. $(x + y)(x - y)^3$ **134.** $3(y + z)(x + z)(x + y)$ **135.** $(x^2 + 1)(2x^2 + x + 2)$

PROBLEM SET P.7

1. $\dfrac{3}{x - 3}$ $(x \neq 3)$ **2.** $\dfrac{4}{x - 2}$ $(x \neq 2)$ **3.** $\dfrac{x - 6}{4}$ $(x \neq 6)$ **4.** $\dfrac{x - 4}{3}$ $(x \neq 4)$ **5.** $\dfrac{y + 9}{y - 1}$ $(x \neq 1, 2)$

6. $\dfrac{y - 5}{y + 4}$ $(y \neq -4, -1)$ **7.** $\dfrac{y - 6}{y^2 + 3y + 9}$ $(y \neq 3)$ **8.** $\dfrac{y^2 + 2y + 4}{y + 4}$ $(y \neq -4, 2)$ **9.** $\dfrac{x - 4}{x - 3}$ $(x \neq -5, 0, 3)$

10. $\dfrac{x - 3}{x + 2}$ $(x \neq -3, -2, 0)$ **11.** $\dfrac{x - 1}{x + 2}$ $(x \neq -2, -1, 2, 3)$ **12.** $\dfrac{x + 3}{x - 2}$ $(x \neq -3, -2, 2, 3)$

13. $\dfrac{x^2 + 2x + 4}{3x}$ $(x \neq -2, 0, 2)$ **14.** $\dfrac{1}{x^2 - 3x + 9}$ $(x \neq -3)$ **15.** $x^2 + x + 1$ $(x \neq -1, 1)$

16. $x - 1$ $(x \neq -1)$ **17.** 1 $\left(x \neq -7, -3, -\dfrac{1}{3}, \dfrac{5}{2}\right)$ **18.** $\dfrac{x^2 + 6x + 8}{x - 5}$ $(x \neq -6, -3, -1, 3, 5)$

19. $\dfrac{x - 5}{2}$ $(x \neq -5, 1)$ **20.** 1 $(x \neq -5, -3, -2, 2)$ **21.** $x^2(x + 5)(x - 5)$ $(x \neq -5, 0, 5)$

22. $(y - 1)(y + 3)$ $(y \neq -3, 0\ 3)$ **23.** $\dfrac{(x - 1)(x + 1)}{2}$ $(x \neq 0)$ **24.** $-\dfrac{1}{x + 1}$ $(x \neq 1)$ **25.** 3

26. $\dfrac{x - 2}{x + 2}$ $(x \neq 3)$ **27.** $\dfrac{9x + 39}{(x + 4)(x + 5)}$ **28.** $\dfrac{10x - 28}{(x - 2)(x - 3)}$ **29.** $-\dfrac{3}{x(x + 1)}$ **30.** $\dfrac{x + 12}{x(x + 3)}$

31. $\dfrac{3x^2 + 4}{(x + 2)(x - 2)}$ **32.** $\dfrac{2x^2 + 5x + 12}{(x - 3)(x + 2)}$ **33.** $\dfrac{2x^2 + 50}{(x - 5)(x + 5)}$ **34.** $\dfrac{2x^2 + 18}{(x - 3)(x + 3)}$

35. $\dfrac{4x + 16}{(x + 3)^2}$ **36.** $\dfrac{20x - 6}{(5x - 2)(5x + 2)}$ **37.** $\dfrac{x^2 - x}{(x - 2)(x + 3)(x + 5)}$ **38.** $-\dfrac{5x}{(x - 6)(x - 1)(x + 4)}$

39. $\dfrac{7y + 1}{(y - 2)(y + 1)^2}$ **40.** $-\dfrac{2(3y + 5)}{(y + 1)(y + 2)(y + 3)}$ **41.** $\dfrac{3y^2 - 25y + 26}{(y - 2)^2(y - 1)(y + 2)}$ **42.** $\dfrac{y - 12}{2y - 5}$ $(y \neq 2)$

43. $\dfrac{x - 1}{x + 2}$ $(x \neq -1)$ **44.** $\dfrac{x^2 + 40x - 25}{(x - 4)(x + 5)}$ **45.** $\dfrac{1}{3}$ $(x \neq 3)$ **46.** $\dfrac{1}{4}$ $(x \neq 4)$ **47.** $\dfrac{x + 1}{3x - 1}$ $(x \neq 0)$

48. $\dfrac{8x + 1}{4x - 1}$ $(x \neq 0)$ **49.** $\dfrac{1}{xy}$ $(x \neq y)$ **50.** $\dfrac{x - 1}{x^2 y}$ **51.** $\dfrac{c}{c + 3}$ $(c \neq -2)$ **52.** $\dfrac{b - 2}{b + 1}$ $(b \neq 2, 3)$

53. $-\dfrac{y - 14}{7}$ $(y \neq -2, 2)$ **54.** $\dfrac{2(y + 2)}{y + 1}$ $(y \neq 1)$ **55.** $\dfrac{x - 3}{x + 2}$ $(x \neq -1, 3)$ **56.** $\dfrac{x - 1}{(x - 3)(x + 6)}$ $(x \neq -5)$

57. $\dfrac{1}{x^2 + xy + y^2 - x + y}$ $(x \neq y)$ **58.** $\dfrac{x + 5}{5x + 50}$ $(x \neq -5, 5)$ **59.** $-\dfrac{2x + h}{x^2(x + h)^2}$ $(h \neq 0)$

60. $\dfrac{1}{(x + 1)(x + h + 1)}$ $(h \neq 0)$ **61.** $\dfrac{x + 2y}{x^2 y^2}$ **62.** $-\dfrac{x^2 - 2x}{x + 2}$ $(x \neq 0)$ **63.** $\dfrac{\sqrt{x} - \sqrt{x + h}}{h\sqrt{x}\sqrt{x + h}}$ **64.** $\dfrac{4y - 1}{4y}$

65. $-\dfrac{2}{y^2\sqrt{y^2 + 2}}$ **66.** $\dfrac{5}{\sqrt{(5 - y^2)^3}}$ **67.** $\dfrac{2(y^2 - 9)(y^2 + 9)}{y^{5/3}}$ **68.** $-\dfrac{y^2 - 2}{y^3(1 - y^2)^{1/2}}$

69. $\dfrac{540t^2 + 12{,}640t + 107{,}100}{-0.14t^2 + 0.51t + 31.6}$ (in dollars per person); $5997 per person, $6833 per person, $7843 per person,

$9086 per person   **70.** $\dfrac{2r_1 r_2}{r_1 + r_2}$; 24 mi/h   **71.** $\dfrac{R_1 R_2 R_3}{R_2 R_3 + R_1 R_3 + R_1 R_2}$; $2.\overline{18}$ ohms   **72.** $\dfrac{8y + 4}{x}$

73. $\dfrac{8y + 4}{\pi y}$ **74.** d   **75.** d

76. a. $\dfrac{x^2 - x - 6}{x + 2} = \dfrac{(x - 3)(x + 2)}{(x + 2)} = x - 3$ **b.** Yes

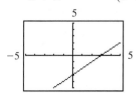

77. $\dfrac{2x + 1}{6}$ **78.** $\dfrac{x + 2}{2x}$ **79.** $-x - 2$ **80.** $\dfrac{x + 1}{2}$ **81.** $\dfrac{7x}{10}$ **82.** $\dfrac{6x + 13}{x + 3}$ **83.** $\dfrac{x + 5}{x - 5}$ **84.** $\dfrac{1}{x^{2n} - 1}$

85. -8 **86.** $\dfrac{1}{200}$ **87.** $(1 + b^{x - y})^{-1} + (1 + b^{y - x})^{-1} = 1$ **88.** Group activity

PROBLEM SET P.8

1. 1   **2.** 1   **3.** $-i$ **4.** i **5.** i **6.** i **7.** $8 - 2i$ **8.** $2 + 5i$ **9.** $-2 + 9i$ **10.** $2 + 16i$

11. $24 + 7i$ **12.** $33 + 4i$ **13.** $-10 + 18i\sqrt{2}$ **14.** $14 - 4i\sqrt{2}$ **15.** $21 + 15i$ **16.** $16 + 56i$

17. $-43 - 23i$ **18.** $60 - 60i$ **19.** $-29 - 11i$ **20.** $12 + 84i$ **21.** 34   **22.** 53   **23.** 34   **24.** 65

25. $-5 + 12i$ **26.** $21 - 20i$ **27.** $-\dfrac{1}{2} + \dfrac{\sqrt{3}}{2}i$ **28.** $\dfrac{1}{2} - \dfrac{\sqrt{3}}{2}i$ **29.** $\dfrac{3}{5} + \dfrac{1}{5}i$ **30.** $\dfrac{12}{17} - \dfrac{3}{17}i$ **31.** $1 + i$

32. $-1 + 2i$ **33.** $-\dfrac{24}{25} + \dfrac{32}{25}i$ **34.** $-\dfrac{12}{13} - \dfrac{18}{13}i$ **35.** $\dfrac{7}{5} + \dfrac{4}{5}i$ **36.** $-i$ **37.** $-\dfrac{27}{29} - \dfrac{34}{29}i$ **38.** $\dfrac{3}{5} - \dfrac{1}{5}i$

39. $-\dfrac{1}{2} - \dfrac{3}{2}i$ **40.** $\dfrac{1}{3} - \dfrac{4}{3}i$ **41.** $-\dfrac{7}{5} - \dfrac{4}{5}i$ **42.** $\dfrac{1}{3} - i$ **43.** $23 + 10i$ **44.** $-11 - 5i$ **45.** $\dfrac{14}{25} - \dfrac{2}{25}i$

46. $\dfrac{9}{5} + \dfrac{3}{5}i$ **47.** $-26 + 18i$ **48.** $-11 - 2i$ **49.** $1 - i$ **50.** $12 - 6i$ **51.** $-i$ **52.** $-i$ **53.** $3i$

54. $-3i$ **55.** $47i$ **56.** $19i\sqrt{2}$ **57.** $-8i$ **58.** $16 + 30i$ **59.** $2 + 6i\sqrt{7}$ **60.** $-7 - 4i\sqrt{11}$

61. $-\dfrac{1}{3} + \dfrac{\sqrt{2}}{6}i$ **62.** $-\dfrac{3}{8} + \dfrac{\sqrt{7}}{16}i$ **63.** $-\dfrac{1}{8} - \dfrac{\sqrt{3}}{24}i$ **64.** $-\dfrac{5}{11} - \dfrac{\sqrt{2}}{11}i$ **65.** $-2\sqrt{6} - 2i\sqrt{10}$

66. $-4\sqrt{3} - 2i\sqrt{6}$ **67.** $24\sqrt{15}$ **68.** $-12\sqrt{14}$

69.

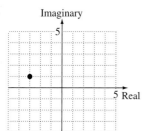

70.

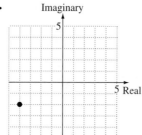

71.

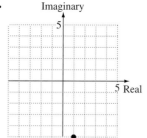

72.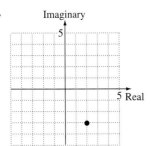

73.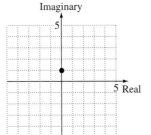

74.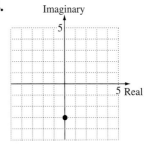

75.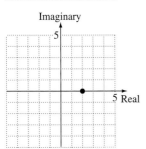

76.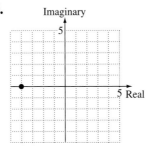

77. d

78. d   **79.** $\sqrt{-100} = 10i$ **80.** $(\sqrt{-9})^2 = -9$ **81.** $\sqrt{-9} + \sqrt{-16} = 7i$ **82.** $(-i)^2 = -1$

83. $\left(-\dfrac{1}{2} + \dfrac{\sqrt{3}}{2}i\right)^3 = \left(-\dfrac{1}{2} + \dfrac{\sqrt{3}}{2}i\right)\left(-\dfrac{1}{2} + \dfrac{\sqrt{3}}{2}i\right)\left(-\dfrac{1}{2} + \dfrac{\sqrt{3}}{2}i\right)$

$\qquad = \left(\dfrac{1}{4} - \dfrac{\sqrt{3}}{2}i + \dfrac{3}{4}i^2\right)\left(-\dfrac{1}{2} + \dfrac{\sqrt{3}}{2}i\right)$

$\qquad = \left(-\dfrac{1}{2} - \dfrac{\sqrt{3}}{2}i\right)\left(-\dfrac{1}{2} + \dfrac{\sqrt{3}}{2}i\right)$

$\qquad = \dfrac{1}{4} - \dfrac{3}{4}i^2 = \dfrac{1}{4} + \dfrac{3}{4} = 1$

84. $\dfrac{6}{5}$ **85.** $-i$ when n is odd; i when n is even   **86.** $-\dfrac{13}{10} + \dfrac{11}{10}i$ **87. a.** Correct   **b.** Incorrect

c. Incorrect   **d.** Correct   **e.** Incorrect   **f.** Correct   **g.** Correct

CHAPTER P REVIEW PROBLEMS

1. a. $\sqrt[4]{16}$ **b.** $0, \sqrt[4]{16}$ **c.** $-13, -\sqrt{16}, 0, \sqrt[4]{16}$ **d.** $-13, -\sqrt{16}, -\dfrac{2}{3}, 0, 1.\overline{27}, \sqrt[4]{16}$ **e.** $-\sqrt{5}, \dfrac{\pi}{3}$

2. $-2 < x \le 3$ **3.** $-1.5 \le x \le 2$ **4.** $x > -1$

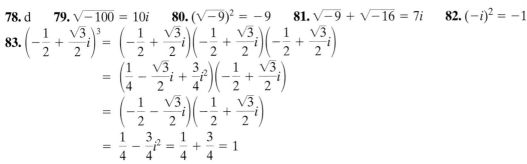

5. $[2, 3)$ **6.** $(5, \infty)$ **7.** $(1, 7]$ **8.** $(-1, \infty)$ **9. a.** $\sqrt{2} - 1$ **b.** $-3 + \sqrt{17}$ **10.** $|-17 - 4| = 21$

11. -45 **12.** Multiplicative inverse   **13.** Associative property of addition   **14.** 7   **15.** $-8x^{12}y^9$

16. $\dfrac{250y^3}{x^2}$ **17.** $\dfrac{81x^8y^{12}}{16}$ **18.** $-\dfrac{8}{x^3y^9}$ **19.** 9.8×10^{13} **20.** 3.62×10^{-4}

21. a. $(1985, 50{,}000)$; 50,000 juveniles were arrested for marajuana possession in 1985.   **b.** 1990; 18,000

22.

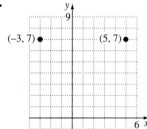

$8; (1, 7)$ **23.**

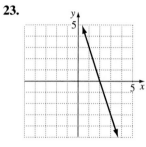

24.

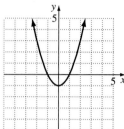

25.

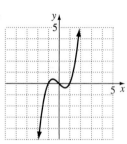

26.

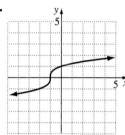

27.

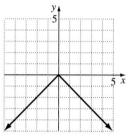

28. -1 **29.** $2xy^2\sqrt[3]{2y^2}$ **30.** $\dfrac{3\sqrt{5}}{2}$ **31.** $\dfrac{\sqrt[3]{9}}{3}$ **32.** $7\sqrt{7} + 7\sqrt{5}$ **33.** $20\sqrt{2}$ **34.** $7\sqrt{10}$ **35.** $\dfrac{63}{16}$

36. $28x^{7/12}$ **37.** $-4y^{11/20}$ **38.** $\sqrt[4]{4x^3y}$ **39.** $12\sqrt[4]{2}$ **40.** $21x^3 - 8x^2 - 10x + 5$, degree 3

41. $2x^4 + 5x^3 - 9x^2 + 17x - 20$ **42.** $36x^2 - x - 21$ **43.** $64x^3 - 144x^2y + 108xy^2 - 27y^3$

44. $-4x^2 + 25y^2 - 4x - 1$ **45.** $10xy^2(6x^5y - 2x^3 + y^3)$ **46.** $(x - 7)(x^2 + 9)$

47. $(x^2 + 9y^2)(x - 3y)(x + 3y)$ **48.** $(x - 10)(x^2 + 10x + 100)$ **49.** $(2x - 5)^2$

50. $(x + 3 - 2a)(x + 3 + 2a)$ **51.** $(4x - 5)(2x + 3)$ **52.** $4(a + 2)(a^2 - 2a + 4)$ **53.** $(x + 3)(x - 3)^2$

54. $-2xy(x - 1)(x^2 + x + 1)(x^6 + x^3 + 1)$ **55.** $(x - 1)^2(x^2 + x + 1)$ **56.** $\dfrac{16(1 + 2x)}{x^{3/4}}$

57. $\dfrac{7(x - 2)(x + 2)}{(x^2 + 3)^{3/2}}$ **58.** $\dfrac{3x + 2}{x - 5}\left(x \neq -\dfrac{1}{2}\right)$ **59.** $4(y + 3)$ $(y \neq 3, 5)$ **60.** $\dfrac{4}{x}$ $(x \neq -5, -4, 4)$

61. $\dfrac{4x^2 - 4x}{(x + 2)(x - 2)}$ **62.** $\dfrac{2x^2 - 3}{(x - 3)(x + 3)(x - 2)}$ **63.** $\dfrac{5a^2 - 7a + 6}{(a + 2)(a - 1)(a - 2)^2}$ **64.** $-\dfrac{y^2 + 3y - 18}{y - 1}$

65. $\dfrac{25}{\sqrt{(25 - x^2)^3}}$ **66.** $-i$ **67.** $-9 - 7i$ **68.** $29 + 11i$ **69.** $\dfrac{1}{5} - \dfrac{11}{10}i$ **70.** $\dfrac{1}{3} - \dfrac{5}{3}i$ **71.** $i\sqrt{2}$

72. $-7 - 4i\sqrt{11}$ **73.** $-\dfrac{3}{5} - \dfrac{\sqrt{43}}{30}i$ **74.**

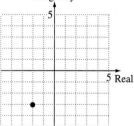

CHAPTER P TEST

1. $x \leq 3$
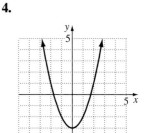

2. $[3, 5)$ **3.** $2\sqrt{13}$ **4.**

5. Commutative Property of Addition   **6.** $\dfrac{27y^6}{x^6}$ **7.** $\sqrt{2} - 1$ **8.** $2\sqrt[3]{4}$ **9.** $11\sqrt{2}$ **10.** $\dfrac{4\sqrt{3}}{3}$

11. $\sqrt[3]{4x^2y}$ **12.** $15xy + 39y^2$ **13.** $49y^2 - 16x^2 - 8x - 1$ **14.** $-\dfrac{x}{3x + 5}$ $(x \neq 4)$ **15.** $(x - 2)(x + 2)(x^2 + 4)$

16. $2(x + 4y)(x^2 - 4xy + 16y^2)$ **17.** $(x - 4)(x - 1)(x + 1)$ **18.** $\dfrac{(x - 3)(x + 3)x^2}{(x^2 + 1)^{3/2}}$ **19.** $\dfrac{(3y - 7)(y - 2)}{3}$

20. $-\dfrac{2}{x(x - 1)(x + 1)}$ **21.** $\dfrac{2x}{(x + 1)(x + 2)}$ **22.** $\dfrac{10x}{\sqrt{(x^2 + 5)^3}}$ **23.** $47 + 16i$ **24.** $2 + i$ **25.** -3

Chapter 1

PROBLEM SET 1.1

1. $\{16\}$ **2.** $\{11\}$ **3.** $\{7\}$ **4.** $\left\{\dfrac{25}{3}\right\}$ **5.** $\{13\}$ **6.** $\{8\}$ **7.** $\{2\}$ **8.** $\{-19\}$ **9.** $\{9\}$ **10.** $\{-1\}$

11. $\{-5\}$ **12.** $\{-4\}$ **13.** $\{6\}$ **14.** $\{3\}$ **15.** $\{-2\}$ **16.** $\left\{-\dfrac{81}{11}\right\}$ **17.** $\{-15\}$ **18.** $\{-20\}$

19. $\{-12\}$ **20.** $\{-19\}$ **21.** $\left\{\dfrac{46}{5}\right\}$ **22.** $\left\{\dfrac{25}{7}\right\}$ **23.** $\left\{\dfrac{1}{2}\right\}$ **24.** $\left\{\dfrac{5}{12}\right\}$ **25.** $\{20\}$ **26.** $\left\{\dfrac{36}{7}\right\}$

27. $\{4\}$ **28.** $\{8\}$ **29.** $\{-2\}$ **30.** $\left\{\dfrac{15}{2}\right\}$ **31.** $\{0\}$ **32.** $\{0\}$ **33.** $\{-9\}$ **34.** $\{7\}$ **35.** $\{4\}$ **36.** $\{2\}$

37. $\varnothing$ **38.** $\varnothing$ **39.** $\varnothing$ **40.** $\varnothing$ **41.** $\{-4, 9\}$ **42.** $\{-7, 8\}$ **43.** $\{-10, 7\}$ **44.** $\left\{-\dfrac{3}{2}, 4\right\}$

45. $\left\{-\dfrac{16}{5}, \dfrac{32}{5}\right\}$ **46.** $\left\{-\dfrac{8}{3}, \dfrac{16}{3}\right\}$ **47.** $\{-1, 9\}$ **48.** $\{-4, 4\}$ **49.** $\{-1, 5\}$ **50.** $\left\{-\dfrac{5}{2}, 0\right\}$

51. $\left\{-\dfrac{5}{4}, \dfrac{3}{2}\right\}$ **52.** $\left\{-\dfrac{1}{2}, 1\right\}$ **53.** $P = \dfrac{S}{1 + rt}$ **54.** $m = \dfrac{Ft}{v - v_0}$ **55.** $n = \dfrac{IR}{E - Ir}$ **56.** $R_1 = \dfrac{RR_2}{R_2 - R}$

57. $y = \dfrac{e(c - d)}{cd}$ **58.** $y = \dfrac{2cd}{c + d}$ **59.** $f = \dfrac{p - s}{s(p - 1)}$ **60.** $\dfrac{1}{a + b}$ **61.** 2036   **62.** 60 yards

63. Yes   **64.** 64.171   **65.** 205; 125,000   **66.** 1989   **67.** c   **68.** a   **69.** $\{-5\}$ **70.** $\{-2\}$ **71.** $\{4.5\}$
72. $\{0.3\}$ **73.** $\{-2, 5\}$ **74.** $\{5\}$ **75.** $\varnothing$ **76.** $\{0, 2\}$
77. a. The equation has infinitely many solutions.   **b.** The equation has infinitely many solutions.

c.

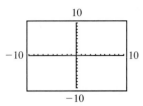

Only the axes appears on the screen.

d.

This is the graph of the *x*-axis.

The equation is true for all values of *x*.

78. Find the point where the graphs of the two equations intersect.   **79–81.** Answers may vary.   **82.** $\left\{\dfrac{3}{2}\right\}$

83. $\varnothing$ **84.** $\{C + D + 1\}$ **85.** 2007   **86.** 2068   **87.** The answer checks.

PROBLEM SET 1.2

1. $\{-2, 5\}$ **2.** $\{4, 9\}$ **3.** $\{3, 5\}$ **4.** $\{-10, -1\}$ **5.** $\left\{-\dfrac{5}{2}, \dfrac{2}{3}\right\}$ **6.** $\left\{-\dfrac{2}{3}, -\dfrac{1}{3}\right\}$ **7.** $\left\{-\dfrac{4}{3}, 2\right\}$

8. $\left\{\dfrac{1}{4}, 3\right\}$ **9.** $\left\{-\dfrac{7}{3}, 0\right\}$ **10.** $\left\{0, \dfrac{9}{5}\right\}$ **11.** $\left\{0, \dfrac{1}{3}\right\}$ **12.** $\left\{\dfrac{5}{4}\right\}$ **13.** $\{-3, 1\}$ **14.** $\left\{\dfrac{1}{2}, 1\right\}$ **15.** $\{-3, 3\}$

16. $\{-3, 3\}$ **17.** $\{-\sqrt{10}, \sqrt{10}\}$ **18.** $\{-4, 4\}$ **19.** $\{-7, 3\}$ **20.** $\{-3, 9\}$ **21.** $\left\{-\dfrac{5}{3}, \dfrac{1}{3}\right\}$ **22.** $\left\{-\dfrac{3}{4}, \dfrac{5}{4}\right\}$

23. $\left\{\dfrac{1 - \sqrt{7}}{5}, \dfrac{1 + \sqrt{7}}{5}\right\}$ **24.** $\left\{\dfrac{3 - \sqrt{5}}{8}, \dfrac{3 + \sqrt{5}}{8}\right\}$ **25.** $\left\{\dfrac{4 - 2\sqrt{2}}{3}, \dfrac{4 + 2\sqrt{2}}{3}\right\}$

26. $\left\{\dfrac{-8 - 3\sqrt{3}}{2}, \dfrac{-8 + 3\sqrt{3}}{2}\right\}$ **27.** $\{-3 - 4i, -3 + 4i\}$ **28.** $\{-5 - 3i, -5 + 3i\}$ **29.** $\{2 - \sqrt{15}, 2 + \sqrt{15}\}$

30. $\{1 - \sqrt{6}, 1 + \sqrt{6}\}$ **31.** $\left\{\dfrac{7 - \sqrt{17}}{4}, \dfrac{7 + \sqrt{17}}{4}\right\}$ **32.** $\left\{\dfrac{1 - \sqrt{5}}{3}, \dfrac{1 + \sqrt{5}}{3}\right\}$ **33.** $\left\{-\dfrac{1}{2}, \dfrac{3}{2}\right\}$ **34.** $\left\{-\dfrac{3}{2}, 2\right\}$

35. $\left\{\dfrac{3 - i}{2}, \dfrac{3 + i}{2}\right\}$ **36.** $\left\{\dfrac{-1 - 3i}{2}, \dfrac{-1 + 3i}{2}\right\}$ **37.** $\left\{\dfrac{3 - \sqrt{57}}{6}, \dfrac{3 + \sqrt{57}}{6}\right\}$

38. $\left\{\dfrac{-1 - \sqrt{41}}{10}, \dfrac{-1 + \sqrt{41}}{10}\right\}$ **39.** $\left\{\dfrac{1 - \sqrt{29}}{4}, \dfrac{1 + \sqrt{29}}{4}\right\}$ **40.** $\left\{\dfrac{3 - \sqrt{6}}{3}, \dfrac{3 + \sqrt{6}}{3}\right\}$ **41.** $\{3 - i, 3 + i\}$

42. $\{1 - 4i, 1 + 4i\}$ **43.** $\{2, 4\}$ **44.** $\{-6, 8\}$ **45.** $\{1 - 2\sqrt{2}, 1 + 2\sqrt{2}\}$ **46.** $\{1 - 2\sqrt{3}, 1 + 2\sqrt{3}\}$

47. $\left\{-\dfrac{\sqrt{105}}{3}, \dfrac{\sqrt{105}}{3}\right\}$ **48.** $\{-3.25, 0.875\}$ **49.** $\{-\sqrt{21}, \sqrt{21}\}$ **50.** $\left\{-\dfrac{4}{3}, \dfrac{5}{2}\right\}$ **51.** $\left\{-\dfrac{\sqrt{6}}{4}, \dfrac{\sqrt{6}}{4}\right\}$

52. $\left\{-\dfrac{\sqrt{15}}{6}, \dfrac{\sqrt{15}}{6}\right\}$ **53.** $\{-1, 0\}$ **54.** $\{1, 2\}$ **55.** $\{-\sqrt{6}, \sqrt{6}\}$ **56.** $\left\{-\dfrac{2\sqrt{5}}{5}, \dfrac{2\sqrt{5}}{5}\right\}$ **57.** $\left\{-\dfrac{5}{3}, \dfrac{13}{3}\right\}$

58. $\{-4, 7\}$ **59.** $\{0.3, 5\}$ **60.** $\{0.699, 5.018\}$ **61.** $\left\{\dfrac{2 - i\sqrt{2}}{3}, \dfrac{2 + i\sqrt{2}}{3}\right\}$ **62.** $\{1 - i\sqrt{3}, 1 + i\sqrt{3}\}$

63. $\{-\sqrt{7}, \sqrt{7}\}$ **64.** $\{-\sqrt{10}, \sqrt{10}\}$ **65.** $\left\{\dfrac{5 - \sqrt{73}}{6}, \dfrac{5 + \sqrt{73}}{6}\right\}$ **66.** $\left\{\dfrac{1 - \sqrt{97}}{4}, \dfrac{1 + \sqrt{97}}{4}\right\}$

67. $\left\{-\dfrac{13}{4}, -2\right\}$ **68.** $\left\{\dfrac{-1 - i\sqrt{15}}{2}, \dfrac{-1 + i\sqrt{15}}{2}\right\}$ **69.** $\{-10\}$ **70.** $\{-15 - 5\sqrt{5}, -15 + 5\sqrt{5}\}$

71. $\left\{\dfrac{-5\sqrt{2}}{2}, 2\sqrt{2}\right\}$ **72.** $\left\{-\dfrac{2\sqrt{5}}{5}, \dfrac{\sqrt{5}}{2}\right\}$ **73.** $\{1\}$ **74.** $\left\{-\dfrac{3}{2}\right\}$ **75.** $\left\{-\dfrac{1}{4}, \dfrac{1}{2}\right\}$ **76.** $\left\{-\dfrac{2}{3}, \dfrac{2}{5}\right\}$ **77.** $\{-3, 0\}$

78. $\left\{-\dfrac{10}{3}, 4\right\}$ **79.** $\{-1, 3\}$ **80.** $\left\{-\dfrac{3}{2}, \dfrac{1}{3}\right\}$ **81.** $\{-8, -6, 4, 6\}$ **82.** $\{-7, -3, 1\}$ **83.** 2000   **84.** 1996

85. a. $D = 20$ **b.** No

86. 1840; the model has the restriction that $0 \le t \le 9$ so it should not be used to predict the present population.

87. 2024   **88.** 54   **89. a.** No   **b.** Yes; twice; going up and going down.   **c.** Yes; once; at the maximum height
90. Discriminant is negative; may not remain employed.   **91.** No   **92.** c   **93.** b
94.   1958; 12.4 mi/gal   **95.**

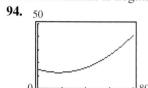

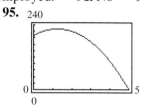

96.   Yes   **97–104.** Answers may vary.

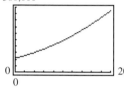

105. b. The product of the solutions is $\dfrac{c}{a}$. The sum of the solutions is $-\dfrac{b}{a}$.

c. Check that the product is $\dfrac{c}{a} = 4$, and the sum is $-\dfrac{b}{a} = 7$; yes   **106.** {8}   **107.** $\left\{\dfrac{72}{25}\right\}$

108. $x^2 - 2x - 15 = 0$ **110. c.** $(x + 4 - 8i\sqrt{2})(x + 4 + 8i\sqrt{2}); 3\left(x + \dfrac{5}{3} - \dfrac{\sqrt{7}}{3}\right)\left(x + \dfrac{5}{3} + \dfrac{\sqrt{7}}{3}\right)$

111. $\left\{\sqrt{\dfrac{d}{e}}, \sqrt{\dfrac{e}{d}}\right\}$ **112.** $b = 1$ and $c = -2$ or $b = 0$ and $c = 0$ **113.** Students should show equalities.

PROBLEM SET 1.3

1. 2003   **2.** 28   **3.** 115.4 months; $1620   **4.** 64.8 months; $3784   **5.** $489.72   **6.** $760.87

7. a. $-10x^2 + 40x + 3200$ **b.** 17 or 19 trees   **8. a.** $-100x^2 + 2000x + 30{,}000$ **b.** 6 or 14   **9.** $1350

10. $1200   **11.** Black male: $34,340; white male: $43,690   **12.** Black female: $28,130; white female: $30,520

13. *Coach*: $290,000; *Seinfeld*: $295,000; *Home Improvement*: $325,000

14. Cobb: 2245; Ruth: 2174; Aaron: 2174; Rose: 2165   **15. a.** $2\dfrac{1}{2}$ seconds   **b.** 5 seconds   **c.** $\left(\dfrac{5}{2}, 200\right); (5, 0)$

16. a. 5 seconds   **b.** Yes   **17.** 5 or 7 ovens   **18.** 500 cells   **19.** Width: 6 feet; length: 9 feet

20. Width: 7 feet; length: 30 feet   **21.** 3 yards   **22.** 20 feet   **23.** 2 yards   **24.** 5 yards   **25.** 3 yards

26. 2.00 yards   **27.** 3 inches   **28.** Width: 10 feet; length: 30 feet   **29.** 2 cm   **30.** 12 cm wide and 16 cm long

31. 8.18 feet   **32.** 5 feet   **33.** 33.51 cubic feet   **34.** 7.07 m³   **35.** 6 meters   **36.** 5 meters   **37.** 5 m

38. 3.42 m   **39.** 522 cm   **40.** 2 meters   **41.** 24 feet   **42.** About 0.247 meters

43. a. $\dfrac{1 + \sqrt{5}}{2}$ **b.** (d)   **44. a.** $12x - 2x^2$ **b.** 2 or 4 centimeters

45. a. 300 m by 400 m or 200 m by 600 m   **b.** 15,000

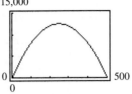

c. 125,000 m²; 250 meters by 500 meters   **d.** No   **46. a.** $75x - x^2$ **b.** $0 < x < 75$

c. 1600   **d.** Answers may vary.   **47. a.** 225

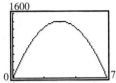

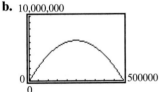

b. The moon graph shows a larger parabola due to lower gravity.   **c.** 119.7 feet

48. a. 50   As the price decreases, the demand increases.

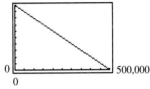

b. 10,000,000   **c.** 250,000 units; $25.00   **d.** $R \leq 6,250,000$

e. 480,000 units at $2 per unit, 20,000 units at $48 per unit   **49.** No   **50–53.** Answers may vary.

54. $373.33; 25%   **55. a.** $\dfrac{(4 + \pi)x^2 - 32x + 64}{16\pi}$ **b.** Side of the square is 0 m; Radius is $\dfrac{2}{\pi}$ m.

56. a. $-\left(\dfrac{1}{2} + \dfrac{\pi}{8}\right)x^2 + 8x$ **b.** $x \approx 3.01, y \approx 4.13$ or $x \approx 5.95, y \approx 0.35$ **57.** 144 ft/sec   **58.** Group activity

---

PROBLEM SET 1.4

1. $\{-4, 0, 4\}$ **2.** $\{-2, 0, 2\}$ **3.** $\left\{-2, -\dfrac{2}{3}, 2\right\}$ **4.** $\left\{-\dfrac{3}{2}, \dfrac{3}{2}, 3\right\}$ **5.** $\left\{-\dfrac{1}{2}, \dfrac{1}{2}, \dfrac{3}{2}\right\}$ **6.** $\left\{-1, -\dfrac{1}{3}, \dfrac{1}{3}\right\}$

7. $\left\{-2, -\dfrac{1}{2}, \dfrac{1}{2}\right\}$ **8.** $\left\{\dfrac{-2 - 2\sqrt{19}}{9}, \dfrac{-2 + 2\sqrt{19}}{9}\right\}$ **9.** $\{-3, -1, 1, 3\}$ **10.** $\{-\sqrt{5}, -\sqrt{3}, \sqrt{3}, \sqrt{5}\}$

11. $\left\{-\dfrac{\sqrt{6}}{3}, -\dfrac{\sqrt{2}}{2}, \dfrac{\sqrt{2}}{2}, \dfrac{\sqrt{6}}{3}\right\}$ **12.** $\left\{-\dfrac{\sqrt{6}}{2}, -\dfrac{1}{2}, \dfrac{1}{2}, \dfrac{\sqrt{6}}{2}\right\}$ **13.** $\{-1, 0, 5\}$ **14.** $\{-1, 0, 3\}$ **15.** $\{-3, -1, 2, 4\}$

16. $\{-2, -1, 3, 4\}$ **17.** $\{-8, -2, 1, 4\}$ **18.** $\{-10, -2, 1, 5\}$ **19.** $\left\{-9, \dfrac{3}{4}\right\}$ **20.** $\{-3, 5\}$ **21.** $\{6\}$ **22.** $\{2\}$

23. $\{10\}$ **24.** $\{5\}$ **25.** $\{2, 6\}$ **26.** $\varnothing$ **27.** $\left\{\dfrac{13 + \sqrt{105}}{6}\right\}$ **28.** $\{0, 4\}$ **29.** $\left\{\dfrac{1}{2}\right\}$ **30.** $\varnothing$ **31.** $\{7\}$

32. $\{7\}$ **33.** $\{8\}$ **34.** $\{-1\}$ **35.** $\{-60, 68\}$ **36.** $\{-13, 3\}$ **37.** $\{-4, 5\}$ **38.** $\{1, 2\}$

39. $\{-\sqrt{3494}, \sqrt{3494}\}$ **40.** $\{2, \sqrt{3}\}$ **41.** $\left\{\dfrac{9}{2}\right\}$ **42.** $\{-11\}$ **43.** $\left\{-8, \dfrac{1}{8}\right\}$ **44.** $\left\{-125, \dfrac{27}{8}\right\}$ **45.** $\left\{\dfrac{1}{4}, 1\right\}$

46. $\{1\}$ **47.** $\{18, 6563\}$ **48.** $\{626\}$ **49.** 208.33 feet   **50.** 18 feet   **51.** 161,081 miles/second

52. 180,094 miles/second   **53.** 161,081 miles/second   **54.** 22.94 seconds   **55. a.** \$29   **b.** 402 CD sets

56. 9%   **57.** 12%   **58.** 1.22 ft or 7.53 ft away from the pole of length 6 ft.   **59. a.** $\dfrac{12}{7}$ miles or 12 miles

b. Answers may vary.   **60.** 6 in. by 6 in. by 6 in.   **61.** $\sqrt{3}$ ft by $(\sqrt{3} + 3)$ ft by $(\sqrt{3} - 1)$ ft

62. $\sqrt[3]{91}$ in.   **63.** $6(\sqrt[3]{2} - 1)$ in.   **64.** 3 in. by 3 in. by 5 in.   **65.** 7.5 miles or about 7.13 miles   **66.** d   **67.** a

68. $\{-1, -0.33, 0.4\}$ **69.** $\{-2, -1, 2\}$ **70.** $\{-2, 0.5, 3\}$ **71.** $\{-4, -0.33, 2\}$ **72.** $\{-5, 2\}$

73. $\{-2.28, -0.32\}$ **74.** $\{0.80\}$ **75. a.**   **b.** Increased each year   **c.** 1990

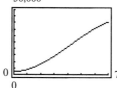

d.   **e.** Answers may vary.

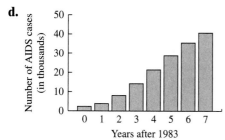

76. Approximately 4.29 ft away from the pole of length 6 ft   **77.** $x = 7.5$ miles (road length ≈ 14.4 miles)
78. The cost given in Problem 65 is very close to the minimum cost.
79. a. Verification is left to the student.   **b.** $x = 4.5$; The person should come ashore 4.5 miles from the house. This path gives the minimum amount of time.   **c.** Verification is left to the student.

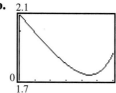

80. No   **81–84.** Answers may vary.   **85.** $\{3, 4\}$ **86.** $\{0, 1\}$ **87.** $\{8\}$ **88.** $\{-2\}$ **89.** $\{c + d\}$

90. a. Area $= 4xy$ **b.** $y = \sqrt{1 - x^2}$ **c.** $4x\sqrt{1 - x^2}$ **d.** $x = \dfrac{\sqrt{2}}{2}$; $\sqrt{2}$ ft by $\sqrt{2}$ ft
e. The maximum area is 2 square feet.

PROBLEM SET 1.5

1. $(-\infty, 3)$ **2.** $(-\infty, 6)$ **3.** $\left[\dfrac{20}{3}, \infty\right)$

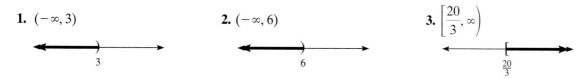

4. $[2, \infty)$

5. $(-\infty, -4]$

6. $[-6, \infty)$

7. $\left(-\infty, -\dfrac{2}{5}\right]$

8. $\left(-\infty, -\dfrac{53}{6}\right]$

9. $[0, \infty)$

10. $(5, \infty)$

11. $(-\infty, 1)$

12. $(-\infty, -4)$

13. $[6, \infty)$

14. $[2, \infty)$

15. $\left[-\dfrac{32}{5}, \infty\right)$

16. $[-2, \infty)$

17. $(-\infty, -6)$

18. $(8, \infty)$

19. $(-\infty, \infty)$

20. $\varnothing$

21. $(3, 5)$

22. $(2, 6)$

23. $[-1, 3)$

24. $(-2, 5]$

25. $(-3, -2]$

26. $[-6, -5)$

27. $[2, 3]$

28. $[5, 6]$

29. $(-5, -2]$

30. $\left[\dfrac{3}{2}, \dfrac{11}{2}\right)$

31. $[3, 6)$

32. $[-4, 2)$

33. $\left(-\dfrac{8}{3}, \dfrac{10}{3}\right)$

34. $[-2, 2]$

35. $\left(-\dfrac{15}{2}, 4\right]$

36. $[-10, -8)$

37. $\left[-\dfrac{1}{2}, 1\right)$

38. $\left(\dfrac{19}{9}, \dfrac{13}{6}\right]$

39. $(-3, 3)$

40. $(-5, 5)$

41. $[-1, 3]$

42. $[-7, 1]$

43. $(-1, 7)$

44. $\left(-\dfrac{22}{3}, 4\right)$

45. $[-5, 3]$

46. $\left[-\dfrac{19}{3}, 7\right]$

47. $(-6, 0)$

48. $(-7, 9)$

49. $(-\infty, -3) \cup (3, \infty)$

50. $(-\infty, -5) \cup (5, \infty)$

51. $(-\infty, -1] \cup [3, \infty)$

52. $(-\infty, -7] \cup [1, \infty)$

53. $\left(-\infty, \dfrac{1}{3}\right) \cup (5, \infty)$

54. $\left(-\infty, -\dfrac{11}{5}\right) \cup (3, \infty)$

55. $(-\infty, -5] \cup [3, \infty)$

56. $(-\infty, -2] \cup [4, \infty)$

57. $(-\infty, -3) \cup (12, \infty)$

58. $(-\infty, -8) \cup (16, \infty)$

59. $(-\infty, -1] \cup [3, \infty)$

60. $[2, 6]$

61. $(-\infty, -1) \cup (2, \infty)$

62. $[-1, 9]$

63. $\left(-\infty, -\dfrac{75}{14}\right) \cup \left(\dfrac{87}{14}, \infty\right)$

64. $(-\infty, \infty)$

65. $(-\infty, -6] \cup [24, \infty)$

66. $[0, 8]$

67. $[-11, 1]$

68. $\left[\dfrac{2}{5}, \dfrac{58}{25}\right]$

69. $25,000   **70.** $6650

71. $16,400   **72.** $740   **73.** 61 or more   **74.** 7 or more   **75.** 82 hours or more   **76.** 199 or less
77. 5001 or more   **78.** 1001 or more   **79.** $2 \le k \le 3$ **80.** $4 \le x \le 10$
81. 2,695,000 barrels; 2,425,000 barrels   **82.** 59 or more, or 41 or less   **83.** $2 \le x \le 5$ **84.** $8 \le x \le 12$
85. Severe cognitive impairment, substance abuse disorders, and depressive: manic, major depression
86. Wednesday, Thursday, Friday   **87.** $500 \le \text{cost} \le 2500$ **88.** $0.1559 < B < 0.2058$ **89.** c   **90.** c

91. $(-\infty, 2)$ **92.** $(-5, \infty)$ **93.** $\left(-\infty, -\dfrac{5}{2}\right] \cup [4, \infty)$ **94.** $(-8, 0)$ **95.** $(-\infty, -1]$ **96.** $\left(-4\dfrac{2}{3}, 6\dfrac{2}{3}\right)$

97. $(-\infty, 3)$ **98.** Possible answer: Determine where the graph of y_1 is below the graph of y_2.   **99.** $\varnothing$

100. $(-\infty, \infty)$ **101.** $C > 25$ **102.** 15,000 or less   **103–105.** Answers may vary.   **106.** $(-5, 6)$

107–108. Verification is left to the student.

PROBLEM SET 1.6

1. $(-\infty, 3) \cup (5, \infty)$

2. $(-\infty, -2) \cup (5, \infty)$

3. $\left[-3, \dfrac{1}{2}\right]$

4. $\left[1, \dfrac{5}{2}\right]$

5. $(-\infty, -4 - \sqrt{3}] \cup [-4 + \sqrt{3}, \infty)$ **6.** $(-\infty, 1 - 3\sqrt{2}] \cup [1 + 3\sqrt{2}, \infty)$

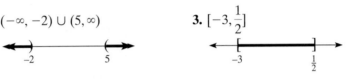

7. $\left(\dfrac{4 - \sqrt{6}}{2}, \dfrac{4 + \sqrt{6}}{2}\right)$

8. $\left(\dfrac{3 - \sqrt{3}}{2}, \dfrac{3 + \sqrt{3}}{2}\right)$

9. $[-\sqrt{6}, \sqrt{6}]$

10. $[1 - \sqrt{5}, 1 + \sqrt{5}]$

11. $(-\infty, \infty)$

12. $(-\infty, \infty)$

13. $\varnothing$

14. $\varnothing$

15. $(-\infty, -3) \cup (6, \infty)$

16. $(-\infty, 0) \cup (1, \infty)$

17. $(-\infty, \infty)$

18. $(-\infty, \infty)$

19. $[-2, -1] \cup [1, \infty)$

20. $\{-2\} \cup [2, \infty)$

21. $(-\infty, -3)$

22. $(-\infty, -7) \cup (-1, 1)$

23. $\left(-\dfrac{2}{3}, \dfrac{2}{3}\right) \cup (2, \infty)$

24. $\left(-2, -\dfrac{1}{2}\right) \cup \left(\dfrac{1}{2}, \infty\right)$

25. $(-1, \infty)$

26. $(1, \infty)$

27. $(2, \infty)$

28. $(3, \infty)$

29. $(-\infty, -3) \cup (7, \infty)$

30. $(-\infty, -3) \cup (4, \infty)$

31. $(-4, -3)$

32. $(-5, -2)$

33. $(-\infty, 2) \cup [7, \infty)$

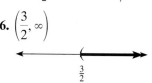

34. $(-\infty, -6] \cup (-2, \infty)$

35. $\left(\dfrac{1}{2}, \dfrac{7}{5}\right]$

36. $\left(\dfrac{3}{2}, \infty\right)$

37. $(-3, 2)$

38. $(-\infty, -3) \cup (-1, 1)$

39. $(-1, 1 - \sqrt{3}) \cup (1, 1 + \sqrt{3})$

40. $(-2, -1) \cup (0, 1)$

41. $(-1, 1) \cup (3, 4)$

42. $(-5, 0) \cup \left(\dfrac{15}{7}, 5\right)$

43. $[-\sqrt{10}, -1) \cup (2, \sqrt{10}]$

44. $[-1, 1) \cup [6, \infty)$

45. $(-\infty, -1) \cup (1, 2) \cup (3, \infty)$

46. $(-\infty, -1) \cup (1, 2) \cup (3, \infty)$

47. $(-\infty, -8) \cup (-6, 4) \cup (6, \infty)$

48. $(-\infty, -7) \cup (1, \infty)$

49. Between 2 and 3 seconds, exclusive.   **50.** Between $4 - 2\sqrt{2}$ and $4 + 2\sqrt{2}$ seconds, exclusive.

51. $2\sqrt{3}$ seconds   **52.** $\dfrac{\sqrt{17}}{4}$ seconds   **53.** $\{1, 2, 3, \dots, 9, 10\}$ **54.** No possible hourly output

55. Less than 60% **56.** At least $100,000$ **57.** No more than 3 feet.

58. $\dfrac{19 - \sqrt{209}}{4}$ cm $\le x < 4$ cm   **59.** c   **60.** c   **61.** $(-1.5, 1) \cup (1, \infty)$

62. $[-1, 2]$ **63.** $(-4, -3)$ **64.** $[-2, 0) \cup [2, \infty)$ **65.** $(-\infty, -1] \cup [1, \infty)$ **66.** $[-2, 0) \cup (1, 3]$

67. a. Less than 80% **b.** 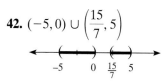   **c.** It is impossible to remove all of the pollutants.

68. 1.7 mm to 3.5 mm   **69. a.** 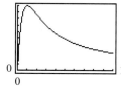   **b.** $(0.27, 3.73)$ **c.** Never reaches zero

70. $(-\infty, -1] \cup [2, \infty)$ **71.** $(-\infty, 2] \cup [5, \infty)$ **72.** $(-\infty, -1) \cup (2, \infty)$ **73.** $(-\infty, 2) \cup (5, \infty)$ **74.** $(-\infty, \infty)$

75. $(-\infty, \infty)$ **76.** $(-\infty, -1] \cup (3, \infty)$ **77.** $(-\infty, -2] \cup (5, \infty)$ **78.** $(-1, 0) \cup (0, \infty)$ **79.** $(-\infty, -3) \cup (0, \infty)$

80. $(-\infty, \infty)$ **81.** $\left(-\infty, -\dfrac{3}{4}\right) \cup \left(-\dfrac{3}{4}, \infty\right)$ **82.** $(-\infty, 2) \cup (2, \infty)$ **83.** $(-\infty, 2) \cup (2, \infty)$ **84.** $\varnothing$

85. $(-\infty, \infty)$ **86.** No values of x; Answers may vary.   **87.** $(-5, \infty)$ **88.** $(-\infty, 0) \cup (4, \infty)$

89–90. Answers may vary.

91. $(-\infty, 1) \cup (1, \infty)$ **92.** If $b < 0$: $(-\infty, b) \cup (b, 0) \cup (0, \infty)$; if $b > 0$: $(-\infty, 0) \cup (0, b) \cup (b, \infty)$

CHAPTER 1 REVIEW PROBLEMS

1. $\{1\}$ **2.** $\left\{\dfrac{3}{2}\right\}$ **3.** $\left\{\dfrac{22}{3}\right\}$ **4.** $\{2\}$ **5.** $\varnothing$ **6.** $(-\infty, -1) \cup (-1, 1) \cup (1, \infty)$ **7.** $\{3\}$ **8.** $\left\{-\dfrac{2}{3}, 2\right\}$

9. $\left\{-\dfrac{8}{3}, \dfrac{16}{3}\right\}$ **10.** $\left\{-4, \dfrac{1}{4}\right\}$ **11.** $\{-4, -3\}$ **12.** $\{-3, 4\}$ **13.** $\{0, 2\}$ **14.** $\left\{-\dfrac{\sqrt{30}}{6}, \dfrac{\sqrt{30}}{6}\right\}$

15. $\{2 - \sqrt{3}, 2 + \sqrt{3}\}$ **16.** $\left\{\dfrac{1 - \sqrt{11}}{2}, \dfrac{1 + \sqrt{11}}{2}\right\}$ **17.** $\left\{\dfrac{-5 - \sqrt{33}}{2}, \dfrac{-5 + \sqrt{33}}{2}\right\}$

18. $\left\{\dfrac{-1 - i\sqrt{5}}{2}, \dfrac{-1 + i\sqrt{5}}{2}\right\}$ **19.** $\{-5, 0, 5\}$ **20.** $\left\{-3, \dfrac{1}{2}, 3\right\}$ **21.** $\{-2, -1, 1, 2\}$

22. $\{-3, -1 - \sqrt{3}, -1 + \sqrt{3}, 1\}$ **23.** $\{13\}$ **24.** $\{5\}$ **25.** $\{0\}$ **26.** $\{-27, 27\}$ **27.** $\{16\}$ **28.** $\{-4, 5\}$

29. $\{16\}$ **30.** $\left\{-1, -\dfrac{2\sqrt{6}}{9}, \dfrac{2\sqrt{6}}{9}, 1\right\}$ **31.** $\{2, 3, 4, 6\}$

32. $(-\infty, 4]$

33. $\left(-\infty, \dfrac{15}{4}\right)$

34. $\left(-\infty, -\dfrac{21}{2}\right)$

35. $(2, 3]$

36. $\left(-\dfrac{8}{5}, \dfrac{2}{5}\right]$

37. $\left(-\dfrac{19}{3}, 7\right)$

38. $(-\infty, -3] \cup \left[\dfrac{9}{2}, \infty\right)$

39. $\left(-\dfrac{9}{2}, \dfrac{15}{2}\right)$

40. $\left[-4, \dfrac{1}{2}\right]$

41. $\left(-\infty, \dfrac{3 - \sqrt{3}}{2}\right) \cup \left(\dfrac{3 + \sqrt{3}}{2}, \infty\right)$

42. $\varnothing$

43. $(-\infty, \infty)$

44. $(-3, -2) \cup (3, \infty)$

45. $(-3, -1)$

46. $(-\infty, 4) \cup \left(\dfrac{23}{4}, \infty\right)$

47. $\left(-3, -\dfrac{5}{3}\right] \cup (-1, \infty)$

48. $\dfrac{A - 2LW}{2(W + L)}$

49. $\dfrac{R_2 R_T}{R_2 - R_T}$

50. a. $\dfrac{NDPM}{NDP - MC}$ **b.** Answers may vary. **51.** 2012 and 2030 **52.** 50 mph **53.** 2 seconds

54. a. $s = -16t^2 + 64t + 80$ **b.** No   **c.** 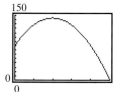   **55.** 17.5 years; \$24,250   **56.** 7.1 years

57. \$400 per week   **58.** 19 or 31   **59. a.** About 21.4 million   **b.** About 114.2 million   **60.** 6000 books
61. 12 ft by 27 ft   **62.** 30 ft by 75 ft   **63. a.** 600 ft by 1200 ft   **b.** Yes   **64.** 12 in. by 12 in.

65. 21.5 cm by 21.5 cm   **66.** 285 in^3 **67.** 56.5 ft^3 **68.** 182,200 miles/second   **69.** $33\frac{1}{3}$ seconds

70. 800 units   **71. a.** $\sqrt{25 + x^2}$; 6   **b.** $x \approx 4.8$; 1.2 miles   **72. a.** Males
b. $A \approx 75.1$ or $A \approx 92.4$; a person who has 10 years remaining to live in 1993 is about 75 years old.   **73.** \$6250
74. 1987   **75.** From 1 to 2 seconds   **76.** $(-\infty, -5] \cup [-2, \infty)$ **77.** Canada, Former Soviet Union
78. Australia, Canada, U.S.

CHAPTER I TEST

1. $\{-1\}$ **2.** $\left\{-\dfrac{4}{3}, 2\right\}$ **3.** $\left\{-\dfrac{1}{2}, 2\right\}$ **4.** $\left\{\dfrac{1 - 5\sqrt{3}}{3}, \dfrac{1 + 5\sqrt{3}}{3}\right\}$ **5.** $\{1 - \sqrt{5}, 1 + \sqrt{5}\}$

6. $\{-2 - \sqrt{5}, -2 + \sqrt{5}\}$ **7.** $\left\{\dfrac{2 - i}{2}, \dfrac{2 + i}{2}\right\}$ **8.** $\{-1, 1, 4\}$ **9.** $\{-1, 1, 2, 4\}$ **10.** $\{23\}$ **11.** $\{1, \ 262,144\}$

12. $\{\sqrt[3]{4}\}$

13. $\left(-\infty, -\dfrac{1}{3}\right]$ **14.** $\left[-7, \dfrac{13}{2}\right)$ **15.** $\left(-\infty, -\dfrac{5}{3}\right] \cup \left[\dfrac{1}{3}, \infty\right)$

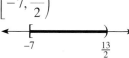

16. $(-3, 4)$ **17.** $(3, 10)$

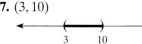

18. a. $125B - 50w$ **b.** 24,000 years!   **19.** 2050   **20.** 60 hours   **21.** \$32,000   **22.** 15 in. by 15 in.

23. 100 m and 240 m   **24.** $\dfrac{9\pi}{4}$ ft^3 **25.** $19 \le x \le 23$; the length for x must be between 19 ft and 23 ft, inclusive.

Chapter 2

PROBLEM SET 2.1

1. Function; domain: $\{1, 3, 5\}$; range: $\{2, 4, 5\}$ **2.** Function; domain: $\{4, 6, 8\}$; range: $\{5, 7, 8\}$
3. Not a function; domain: $\{3, 4\}$; range: $\{4, 5\}$ **4.** Not a function; domain: $\{5, 6\}$; range: $\{6, 7\}$
5. Function; domain: $\{-3, -2, -1, 0\}$; range: $\{-3, -2, -1, 0\}$
6. Function; domain: $\{-7, -5, -3, 0\}$; range: $\{-7, -5, -3, 0\}$ **7.** Not a function; domain: $\{1\}$; range: $\{4, 5, 6\}$
8. Function; domain: $\{4, 5, 6\}$; range: $\{1\}$ **9. a.** -6 **b.** -14 **c.** -4 **d.** $4a - 6$ **e.** $-4a - 6$ **f.** $8a - 6$
10. a. -5 **b.** -17 **c.** -7 **d.** $3a - 5$ **e.** $-3a - 5$ **f.** $6a - 5$ **11. a.** 7   **b.** 11   **c.** 35
d. $x^2 + 3x + 7$ **e.** $a^2 + 2ah + h^2 - 3a - 3h + 7$ **12. a.** 6   **b.** 48   **c.** 36   **d.** $4x^2 + 2x + 6$

e. $4a^2 + 8ah + 4h^2 - 2a - 2h + 6$ **13. a.** 2   **b.** $\dfrac{17}{4}$ **c.** $a^2 + \dfrac{1}{a^2}$ **d.** $a^2 + \dfrac{1}{a^2}$ **14. a.** 2   **b.** $-\dfrac{65}{8}$

c. $a^3 + \dfrac{1}{a^3}$ **d.** $-a^3 - \dfrac{1}{a^3}$ **15. a.** $4a + 2$ **b.** $4a + 4h + 2$ **c.** 4   **d.** $4a + 4h + 4$ **16. a.** $3a - 1$

b. $3a + 3h - 1$ **c.** 3   **d.** $3a + 3h - 2$ **17. a.** $2a^2 - 3a + 5$ **b.** $2a^2 + 4ah + 2h^2 - 3a - 3h + 5$

c. $4a + 2h - 3$ **d.** $2a^2 + 2h^2 - 3a - 3h + 10$ **18. a.** $3a^2 - 7a - 2$ **b.** $3a^2 + 6ah + 3h^2 - 7a - 7h - 2$

c. $6a + 3h - 7$ **d.** $3a^2 + 3h^2 - 7a - 7h - 4$ **19. a.** 6   **b.** 6   **c.** 0   **d.** 12   **20. a.** 7   **b.** 7   **c.** 0   **d.** 14

21. a. $\dfrac{1}{a}$ **b.** $\dfrac{1}{a + h}$ **c.** $-\dfrac{1}{a(a + h)}$ **d.** $\dfrac{a + h}{ah}$ **22. a.** $\dfrac{1}{a + 1}$ **b.** $\dfrac{1}{a + h + 1}$ **c.** $-\dfrac{1}{(a + 1)(a + h + 1)}$

d. $\dfrac{a + h + 2}{(a + 1)(h + 1)}$ **23. a.** $-6x - 5$ **b.** $6x + 6h - 5$ **c.** 6   **24. a.** $4x + 3$ **b.** $-4x - 4h + 3$ **c.** -4

25. a. $2x^2 + 4x - 1$ **b.** $2x^2 + 4xh + 2h^2 - 4x - 4h - 1$ **c.** $4x + 2h - 4$ **26. a.** $-3x^2 - 2x + 8$

b. $-3x^2 - 6xh - 3h^2 + 2x + 2h + 8$ **c.** $-6x - 3h + 2$ **27. a.** 3   **b.** 3   **c.** 0   **d.** $\pi - 3$

28. a. 5   **b.** 1   **c.** 0   **d.** $-\sqrt{10} + 5$ **29.** $(-\infty, \infty)$ **30.** $(-\infty, \infty)$ **31.** $(-\infty, 4) \cup (4, \infty)$

32. $(-\infty, -5) \cup (-5, \infty)$ **33.** $(-\infty, -8) \cup (-8, -3) \cup (-3, \infty)$ **34.** $\left(-\infty, -\dfrac{2}{3}\right) \cup \left(-\dfrac{2}{3}, \dfrac{1}{2}\right) \cup \left(\dfrac{1}{2}, \infty\right)$

35. $(-\infty, \infty)$ **36.** $(-\infty, \infty)$ **37.** $[3, \infty)$ **38.** $[-2, \infty)$ **39.** $(-\infty, \infty)$ **40.** $(-\infty, \infty)$ **41.** $(3, \infty)$

43. $(-\infty, -4] \cup [2, \infty)$ **44.** $\left(-\infty, -\dfrac{5}{2}\right] \cup \left[\dfrac{1}{3}, \infty\right)$ **45.** $[1, 3) \cup (3, \infty)$ **46.** $[2, 5) \cup (5, \infty)$

47. $\left(-\infty, \dfrac{1}{3}\right) \cup (2, \infty)$ **48.** $\left(-\infty, -\dfrac{1}{2}\right] \cup \left(\dfrac{2}{3}, \infty\right)$ **49.** $[-1, 1]$ **50.** $[-4, 4]$

51. 28,771; the number of new AIDS cases in the United States in 1988 was 28,771.

52. 353.4; the carbon dioxide concentration in 1990 was 353.4 ppm.

53. 170; the difference in income for a diving boat with 40 and 50 passengers is $170.

54. -48; the difference in the position of a free-falling body after 4 and 5 seconds is 48 feet.

55. 240; the length of a 400-meter tall starship moving at 148,800 miles/second from the perspective of an observer is 240 meters.   **56.** 693; the length of an 800-meter tall starship moving at 93,000 miles/second from the perspective of an observer is approximately 693 meters.   **57.** 33; the average velocity between 2 and 4 seconds is 33 units/seconds   **58.** 18; the average velocity between 2 and 5 seconds is 18 units/ second

59. 8507.84; the difference in the average price of a mobile home in the United States between 1986 and 1996 is $8507.84.   **60.** 9202.5; the difference in the taxes owed by a married person filing separately with an adjusted gross income of $40,000 and $70,000 is $9202.50.   **61.** d   **62.** d   **63.** c   **64.** c   **65.** d

66. $(-\infty, 3) \cup (3, 5) \cup (5, \infty)$ **67.** $(-\infty, 2) \cup (2, 6) \cup (6, \infty)$ **68.** $(-\infty, 3]$ **69.** $(-\infty, 5]$ **70.** $(-\infty, 2) \cup (5, \infty)$

71. $(-\infty, -4) \cup (-1, \infty)$ **72.** $(-\infty, \infty)$ **73.** $(-\infty, 2) \cup (2, \infty)$ **74. a.** $-2, -1, 0, \dfrac{1}{2}, 1, 2$

b.   **c.** $L(x) = \log x$ **75. a.** $P(x) = x(6 - x)$ **b.**

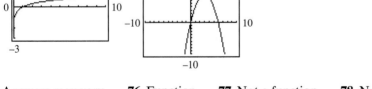

c. Answers may vary.   **76.** Function   **77.** Not a function   **78.** Not a function   **79.** Function

80. a–b. Answers may vary.   **81. a.** No   **b.** Answers may vary.   **82.** Answers may vary.

83. a–c. Answers may vary.   **84.** 0   **85.** x **86.** Answers may vary.   **87.** $g(t) = t + 1$

PROBLEM SET 2.2

1. -6 **2.** 1 **3.** 2 **4.** $-\dfrac{6}{5}$ **5.** 0 **6.** 0 **7.** $L_1: \dfrac{2}{3}; L_2: -\dfrac{3}{4}; L_3: 0$ **8.** $L_1: -1; L_2: \dfrac{5}{6}; L_3: 0$

9. a. m_1, m_2, m_4, m_3 **b.** b_3, b_2, b_4, b_1 **10. a.** m_1, m_3, m_2, m_4 **b.** b_2, b_1, b_4, b_3

11. $y - 6 = 5(x + 5); y = 5x + 31; 5x - y + 31 = 0$ **12.** $y + 4 = -2(x - 3); y = -2x + 2; 2x + y - 2 = 0$

13. $y + 1 = -(x - 6); y = -x + 5; x + y - 5 = 0$ **14.** $y + 8 = 4(x - 3); y = 4x - 20; 4x - y - 20 = 0$

15. $y + 3 = \dfrac{1}{8}(x + 1); y = \dfrac{1}{8}x - \dfrac{23}{8}; x - 8y - 23 = 0$ **16.** $y - 4 = \dfrac{1}{3}(x - 1); y = \dfrac{1}{3}x + \dfrac{11}{3}; x - 3y + 11 = 0$

17. $y + 10 = -4(x + 8); y = -4x - 42; 4x + y + 42 = 0$

18. $y + 7 = -5(x + 2); y = -5x - 17; 5x + y + 17 = 0$ **19.** $y + 7 = -2(x - 4); y = -2x + 1; 2x + y - 1 = 0$

20. $y + 9 = 7(x - 5); y = 7x - 44; 7x - y - 44 = 0$ **21.** $y - 0 = 3(x + 3); y = 3x + 9; 3x - y + 9 = 0$

22. $y + 2 = -\dfrac{2}{3}(x - 0); y = -\dfrac{2}{3}x - 2; 2x + 3y + 6 = 0$

23. $m = 2, b = 1$ **24.** $m = 3; b = 2$ **25.** $m = -2; b = 1$ **26.** $m = -3; b = 2$

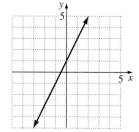   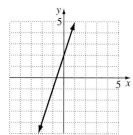   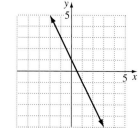   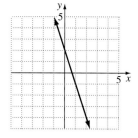

27. $m = \dfrac{3}{4}; b = -2$ **28.** $m = \dfrac{3}{4}; b = -3$ **29.** $m = -3; b = 2$ **30.** $m = -4; b = 3$

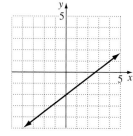   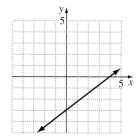   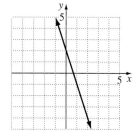   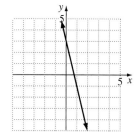

31. $m = -\dfrac{3}{2}; b = 0$ **32.** $m = -\dfrac{4}{3}; b = 0$ **33.** $m = 3; b = 2$ **34.** $m = 4; b = 3$

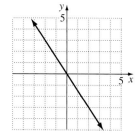   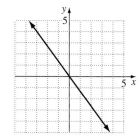   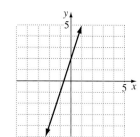   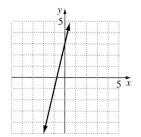

35. Slope is undefined; no *y*-intercept

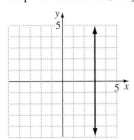

36. Slope is undefined; no *y*-intercept

37. $m = 0; b = -2$ **38.** $m = 0; b = -3$ **39.** $m = 0; y = 0$ **40.** $m = 0; y = 0$

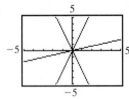 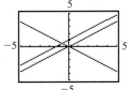 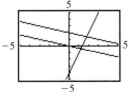 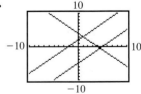

41. a. 23.4; U.S. health-care costs between 1975 and 1980 increased at a rate of $23.4 billion per year.
b. 50.2; U.S. health-care costs between 1985 and 1990 increased at a rate of $50.2 billion per year

42. a. 289; 28; the number of arrests per 100,000 juveniles between 1965 and 1970 increased at a rate of 28
per year   **b.** 340 and 300; -8; the number of arrests per 100,000 juveniles between 1980 and 1985
decreased at a rate of -8 per year

43. a. Answers may vary: 1975 and 1980 or 1985 and 1987; 1500; the number of motor-vehicle deaths between
1975 and 1980 and between 1985 and 1987 increased at a rate of approximately 1500 per year.
b. Answers may vary: 1988 and 1989; -2000; the number of motor-vehicle deaths between 1988 and 1989
decreased at a rate of 2000 per year.   **c.** The horizontal axis is not evenly scaled.

44. Hispanic: 1985–1987; Black: 1985–1989; White: 1985–1991; the dropout rate did not change between
these years.   **45.** $y = \dfrac{424{,}000}{3}x - \dfrac{843{,}752{,}000}{3}; 1{,}416{,}000$ **46.** $y = -\dfrac{10{,}000}{63}x + \dfrac{20{,}497{,}000}{63}; 2050$

47. $y = 0.109x - 215.579$; $2.421 billion; $1.113 billion   **48.** $y = 3991x - 7{,}734{,}753; 247{,}247; 215{,}319$

49. $y = 2500x + 26{,}500$ **50.** $V = -1200t + 9300, 0 \le t \le 6$ **51. a.** $C = \dfrac{4}{30}M$ **b.** 150 miles

52. a. $T = 0.021I$ **b.** $756   **53.** 7.5 miles   **54.** 16.25 inches   **55.** d   **56.** c

57. **58.** **59.** **60.**

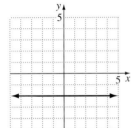 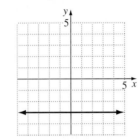 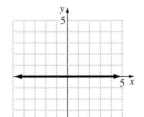 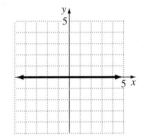

61. a. $y = -\dfrac{2}{3}x + 2$ **b.** $y = -\dfrac{2}{3}x - 2$ **c.** $y = -\dfrac{7}{4}x + 7$ **d.** $y = -\dfrac{7}{4}x - 7$

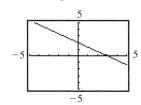
x-intercept: $(3, 0)$
y-intercept: $(0, 2)$

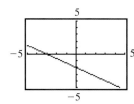
x-intercept: $(-3, 0)$
y-intercept: $(0, -2)$

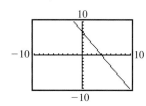
x-intercept: $(4, 0)$
y-intercept: $(0, 7)$

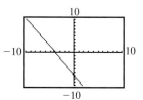
x-intercept: $(-4, 0)$
y-intercept: $(0, -7)$

e. Line with x-intercept $(a, 0)$ and y-intercept $(0, b)$

62. b. 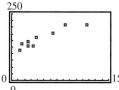   **c.** $y = 9.2739x + 114.6816$
$r = 0.9268$
d.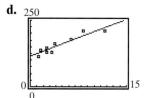

63. b. 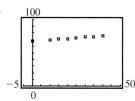   **c.** $y = 0.1907x + 67.7404$
$r = 0.9882$
d.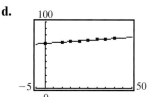

e. 79.2   **f.** 2067

64. b. $r = -0.9766$; the points can be described by a linear relationship and the regression line has a negative slope.   **c.** $r = -0.9716$; the points can be described by a linear relationship and the regression line has a negative slope.   **d.** Approximately $(2047.5, 38.5)$ **e.** Answers may vary.   **65–66.** Answers may vary.

67. Yes; no; it is not reasonable to expect that jumping ability is directly proportional to body length.

68. a. $y = -\dfrac{11}{7}x + \dfrac{81}{7}$ **b.** $y = -\dfrac{5}{3}x + \dfrac{37}{3}$ **c.** $y = -\dfrac{5}{3}y + 12$ **d.** $y = \dfrac{5}{2}x - \dfrac{23}{2}$; $y = -\dfrac{1}{11}x + \dfrac{35}{11}$

e. $y = \dfrac{3}{2}x - \dfrac{9}{2}$; $y = \dfrac{1}{7}x + \dfrac{19}{7}$ **f.** $y = \dfrac{3}{2}x - \dfrac{13}{2}$; $y = \dfrac{1}{7}x + \dfrac{10}{7}$ **69. a.** $y = \dfrac{b}{a - 2c}x - \dfrac{bc}{a - 2c}$

b. $y = \dfrac{a}{b}x - \dfrac{ac}{b}$ **c.** $y = \dfrac{a}{b}x - \dfrac{a^2 - b^2}{2b}$ **d.** $y = -\dfrac{a + c}{2b}x + b$; $y = \dfrac{b}{a - 2c}x - \dfrac{ab}{c - 2a}$ **e.** $x = 0$;

$y = \dfrac{c}{b}x - \dfrac{ac}{b}$ **f.** $x = \dfrac{a + c}{2}$; $y = \dfrac{a}{b}x - \dfrac{b^2 - c^2}{2b}$ **70.** $(1, 5), (3, 13)$ **71.** $y - 5 = -\dfrac{5}{6}(x - 0)$; $y = -\dfrac{5}{6}x + 5$

72. The x-intercept and y-intercept are given in the equation.   **73.** Group activity

PROBLEM SET 2.3

1.

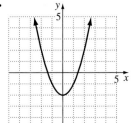

Domain: $(-\infty, \infty)$
Range: $[-2, \infty)$

2.

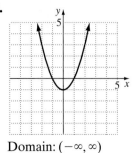

Domain: $(-\infty, \infty)$
Range: $[-1, \infty)$

3.

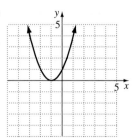

Domain: $(-\infty, \infty)$
Range: $[0, \infty)$

4.

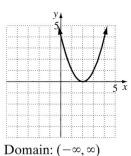

Domain: $(-\infty, \infty)$
Range: $[0, \infty)$

5.

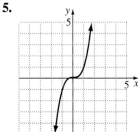

Domain: $(-\infty, \infty)$
Range: $(-\infty, \infty)$

6.

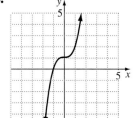

Domain: $(-\infty, \infty)$
Range: $(-\infty, \infty)$

7.

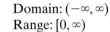

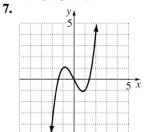

Domain: $(-\infty, \infty)$
Range: $(-\infty, \infty)$

8.

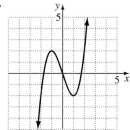

Domain: $(-\infty, \infty)$
Range: $(-\infty, \infty)$

9.

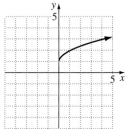

Domain: $[0, \infty)$
Range: $[1, \infty)$

10.

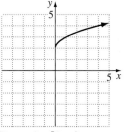

Domain: $[0, \infty)$
Range: $[2, \infty)$

11.

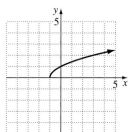

Domain: $[-1, \infty)$
Range: $[0, \infty)$

12.

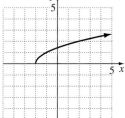

Domain: $[-2, \infty)$
Range: $[0, \infty)$

13.

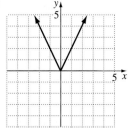

Domain: $(-\infty, \infty)$
Range: $[0, \infty)$

14.

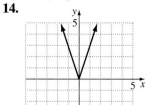

Domain: $(-\infty, \infty)$
Range: $[0, \infty)$

15.

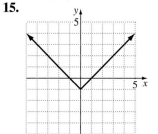

Domain: $(-\infty, \infty)$
Range: $[-1, \infty)$

16.

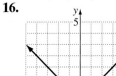

Domain: $(-\infty, \infty)$
Range: $[-2, \infty)$

17.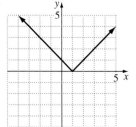

Domain: $(-\infty, \infty)$
Range: $[0, \infty)$

18.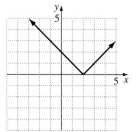

Domain: $(-\infty, \infty)$
Range: $[0, \infty)$

19.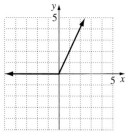

Domain: $(-\infty, \infty)$
Range: $[0, \infty)$

20.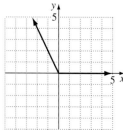

Domain: $(-\infty, \infty)$
Range: $[0, \infty)$

21.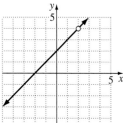

Domain: $(-\infty, 2) \cup (2, \infty)$
Range: $(-\infty, 4) \cup (4, \infty)$

22.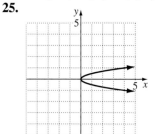

Domain: $(-\infty, 4) \cup (4, \infty)$
Range: $(-\infty, 8) \cup (8, \infty)$

23.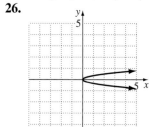

Domain: $(-\infty, -1) \cup (-1, \infty)$
Range: $(-\infty, -2) \cup (-2, \infty)$

24.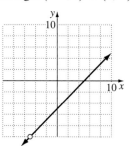

Domain: $(-\infty, -5) \cup (-5, \infty)$
Range: $(-\infty, -10) \cup (-10, \infty)$

25.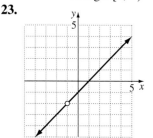

Domain: $[0, \infty)$
Range: $(-\infty, \infty)$

26.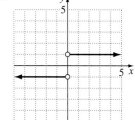

Domain: $[0, \infty)$
Range: $(-\infty, \infty)$

27.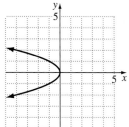

Domain: $(-\infty, 0]$
Range: $(-\infty, \infty)$

28.

Domain: $(-\infty, 0]$
Range: $(-\infty, \infty)$

29.

Domain: $(-\infty, 0) \cup (0, \infty)$
Range: $\{-1, 1\}$

30.

Domain: $(-\infty, 0) \cup (0, \infty)$
Range: $\{-1, 1\}$

31. Not a function
Domain: $[-5, 5]$
Range: $[-2, 2]$

32. Not a function
Domain: $[-3, \infty)$
Range: $(-\infty, \infty)$

33. Function
Domain: $(-\infty, \infty)$
Range: $(-\infty, \infty)$

34. Function
Domain: $(-\infty, \infty)$
Range: $(-\infty, \infty)$

35. Function
Domain: $(-\infty, \infty)$
Range: $(-\infty, 3]$

36. Function
Domain: $(-\infty, \infty)$
Range: $(0, \infty)$

37. Not a function
Domain: $\{2\}$
Range: $\{-2, 1, 3, 5\}$

38. Function
Domain: $\{-5, -2, 1, 3\}$
Range: $\{2\}$

39. Function; Domain: $\{-6, -5, -4, -3, -2, -1, 0, 1, 2, 3, 4, 5, 6\}$; Range: $\{0, 1, 2, 3, 4, 5, 6\}$

40. Not a function
  Domain: $\{0, 1, 2, 3, 4, 5, 6\}$
  Range: $\{0, 1, 2, 3, 4, 5, 6\}$

41. Function
  Domain: $(-\infty, \infty)$
  Range: $(-\infty, -2]$

42. Function
  Domain: $[-4, 4]$
  Range: $[0, 3]$

43. $x^2 + y^2 = 49$ **44.** $x^2 + y^2 = 64$ **45.** $(x - 3)^2 + (y - 2)^2 = 25$ **46.** $(x - 2)^2 + (y + 1)^2 = 16$

47. $(x + 1)^2 + (y - 4)^2 = 4$ **48.** $(x + 3)^2 + (y - 5)^2 = 9$ **49.** $(x + 3)^2 + (y + 1)^2 = 3$

50. $(x + 5)^2 + (y + 3)^2 = 5$ **51.** $(x + 4)^2 + (y - 0)^2 = 0$ **52.** $(x + 2)^2 + (y - 0)^2 = 36$

53. Center: $(0, 0)$; radius: 4

54. Center: $(0, 0)$; radius: 7

55. Center: $(3, 1)$; radius: 6

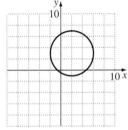

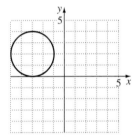

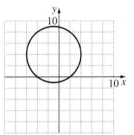

56. Center: $(2, 3)$; radius: 4

57. Center: $(-3, 2)$; radius: 2

58. Center: $(-1, 4)$; radius: 5

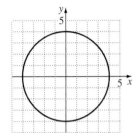

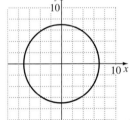

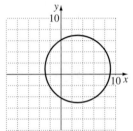

59. Center: $(-2, -2)$; radius: 2

60. Center: $(-4, -5)$; radius: 6

61. $(x + 3)^2 + (y + 1)^2 = 4$
  Center: $(-3, -1)$; radius: 2

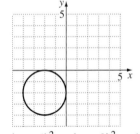

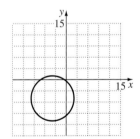

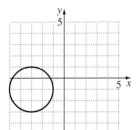

62. $(x + 4)^2 + (y + 2)^2 = 4$
  Center: $(-4, -2)$; radius: 2

63. $(x - 5)^2 + (y - 3)^2 = 64$
  Center: $(5, 3)$; radius: 8

64. $(x - 2)^2 + (y - 6)^2 = 49$
  Center: $(2, 6)$; radius: 7

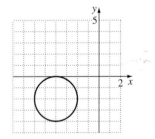

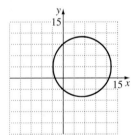

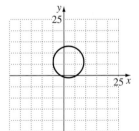

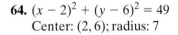

65. $(x + 4)^2 + (y - 1)^2 = 25$
Center: $(-4, 1)$; radius: 5

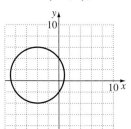

66. $(x + 6)^2 + (y - 3)^2 = 49$
Center: $(-6, 3)$; radius: 7

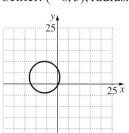

67. $(x - 1)^2 + (y - 0)^2 = 16$
Center: $(1, 0)$; radius: 4

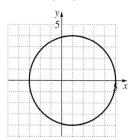

68. $(x - 0)^2 + (y - 3)^2 = 16$
Center: $(0, 3)$; radius: 4

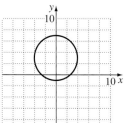

69.

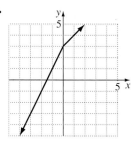

70.

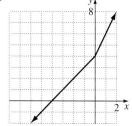

71.

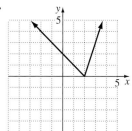

72.

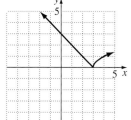

73.

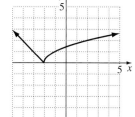

74.

75.

76.

77.

78.

79. 0, 85, 112, 117, 112, 85, 0

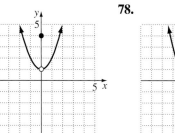
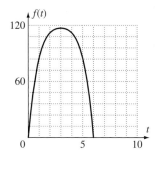
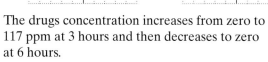

The drugs concentration increases from zero to 117 ppm at 3 hours and then decreases to zero at 6 hours.

80.

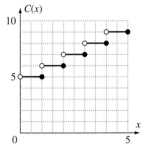

81.

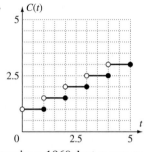

82. c   **83.** c   **84.** d   **85.** d

86. a.

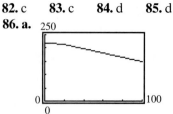

b. Cholesterol levels have been decreasing since 1960, but more rapidly after 1980.   **c.** Yes, $C(40) = 197$

87.

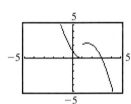

88.

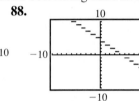

89.

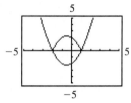

90.

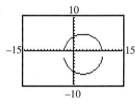

91.

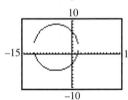

92.

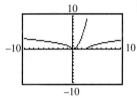

93.

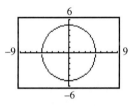

94.

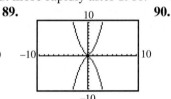

95.

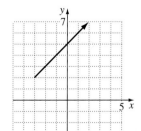

96. Answers may vary.   **97.** No; consider the left or right semicircle.
98. No; the point $(3, 5)$ **99.** No; there is no graph.   **100.** (h)   **101.** (a)
102. (d)   **103.** (g)   **104.** 11π

PROBLEM SET 2.4

1. a. $f(-6) = 6; f(0) = 1; f(3) = 4$ **b.** Increasing: $(0, \infty)$; decreasing: $(-\infty, 0)$
c. Minimum value is 1 at $x = 0$.   **d.** $[1, \infty)$ **2. a.** $g(0) = 0; g(1) = 3$
b. Increasing: $(-2, 2)$; decreasing: $(-4, -2) \cup (2, 4)$ **c.** Minimum value is -5 at $x = -2$.
d. Maximum value is 4 at $x = 2$.   **e.** $[-5, 4]$ **3. a.** $f(-2) = 0; f(-1) = 2; f(5) = 3$
b. Increasing: $(4, 6)$; decreasing: $(-5, -1)$; Constant: $(-1, 4)$ **c.** $(-2, 4]$ **4. a.** $g(-1) = 3; g(2) = 5$
b. Increasing: $(-6, 2)$; decreasing: $(2, 6)$ **c.** $[0, 5]$

5. Possible answer: **6.** Possible answer: **7.** Possible answer: **8.** Possible answer:

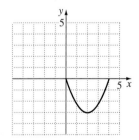

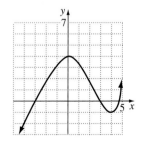

9. Possible answer:

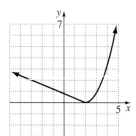

10. Possible answer:

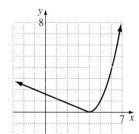

11. Possible answer:

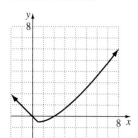

12. Possible answer:

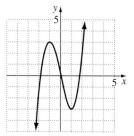

13. Possible answer:

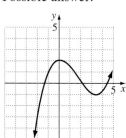

14. Possible answer:

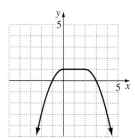

15. a. 1967–1975, 1980–1990   **b.** 1975–1980   **c.** 1975; approximately 13%   **d.** 1978; 15%
16. a. 1985–1986, 1989–1992   **b.** 1986–1987, 1988–1989, 1992–1994   **c.** 1987–1988
d. 1989; approximately 50,000 fires   **e.** 1992; approximately 86,000 fires
17. a. Increasing: $(45, 74)$; decreasing: $(16, 45)$ **b.** 45; 190   **c.** $[190, 526.4]$
18. a. Increasing: $(20, 30)$; decreasing: $(12, 20)$ **b.** 20; 100   **c.** The power expended is approximately
between 100 and 158 calories.
19. Possible answer:

20. Possible answer:

21. $A(x) = x(50 - x)$; (b)

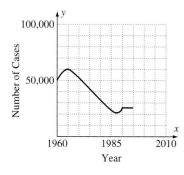

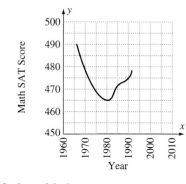

22. $R(x) = (60 - x)(35 + x)$; (a)   **23.** d   **24.** d
25.

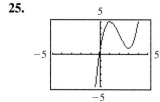

Increasing: $(-\infty, 1) \cup (3, \infty)$
Decreasing: $(1, 3)$
Relative minimum: 1
Relative maximum: 5

26.

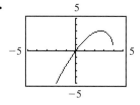

Increasing: $(-\infty, 2.67)$
Decreasing: $(2.67, 4)$
Relative maximum: 3.08
Relative minimum: 0

27.

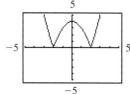

Increasing: $(-2, 0) \cup (2, \infty)$
Decreasing: $(-\infty, -2) \cup (0, 2)$
Relative maximum: 4
Relative minimum: 0

28.

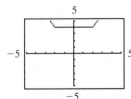

Increasing: $(2, \infty)$
Decreasing: $(-\infty, -2)$
Constant: $(-2, 2)$
Relative maximum: 4

29.

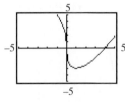

Increasing: $(1, \infty)$
Decreasing: $(-\infty, 1)$
Relative minimum: -3

30.

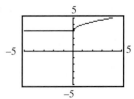

Increasing: $(0, \infty)$
Constant: $(-\infty, 0)$
Relative minimum: 3

31.

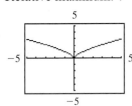

Increasing: $(0, \infty)$
Decreasing: $(-\infty, 0)$
Relative minimum: 0

32.

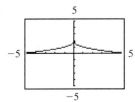

Increasing: $(-\infty, 0)$
Decreasing: $(0, \infty)$
Relative maximum: 2

33. a.

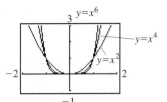

b.

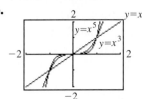

c. Increasing: $(0, \infty)$; decreasing: $(-\infty, 0)$

d. $f(x) = x^n$ is increasing from $(-\infty, \infty)$ when n is odd.
e. As n increases, the steepness increases.

34. a.

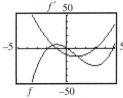

b. f' is positive when f is increasing.
c. f' is negative when f is decreasing.
d. f' equals zero when f has a relative maximum or minimum.

35. a. $P(x) = x(9 - x)$ **b.**

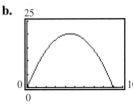

c. 20.25   **36. a.** $V(x) = 4x^3 - 26x^2 + 40x$

b.

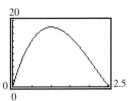

c. 18 cubic inches   **37. a.** $P(x) = -0.01x^2 + 150x - 20{,}000$

b.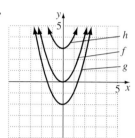

c. 7500 items; $5425   **38–42.** Answers may vary.   **43.** Group activity

P R O B L E M   S E T   2 . 5

1.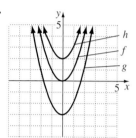

2.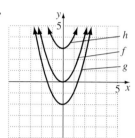

3.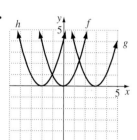

4.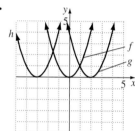

5.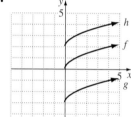

6.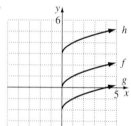

7.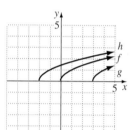

8.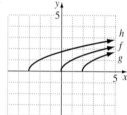

9.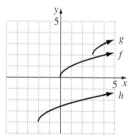

10.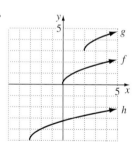

11.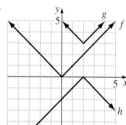

12.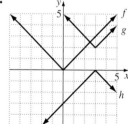

13.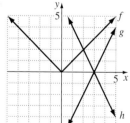

14.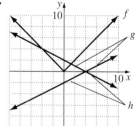

15.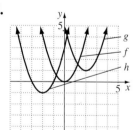

16.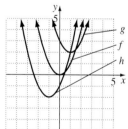

17.

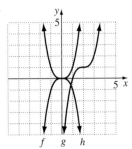

18.

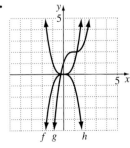

19.

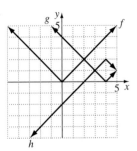

20.

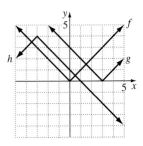

21.

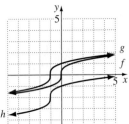

22.

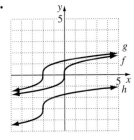

23.

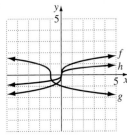

24.

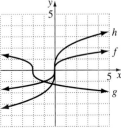

25.

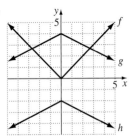

26.

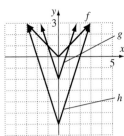

27. $g(x) = -(x - 3)^2$

28. $g(x) = -(x + 4)^2$

29. $g(x) = -\sqrt{x - 2} + 2$ **30.** $g(x) = -\sqrt{x + 1}$ **31.** $g(x) = -|x - 4| + 1$ **32.** $g(x) = -|x - 5| + 1$

33. $g(x) = -\sqrt{9 - (x - 1)^2} - 1$ **34.** $g(x) = -\dfrac{1}{4}\sqrt{16 - x^2} - 1$

35.

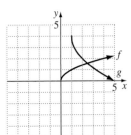

36.

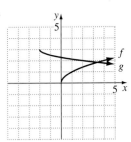

37.

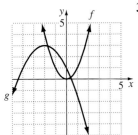

38.

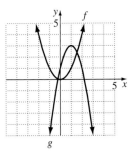

39.

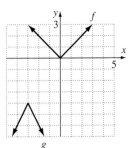

40.

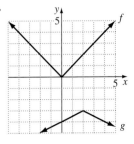

41.

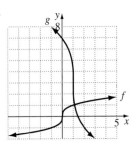

42.

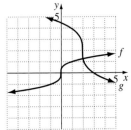

43.

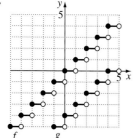

44.

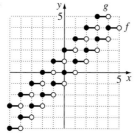

45. a.

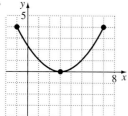

b.

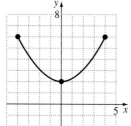

c.

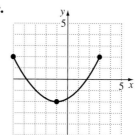

d.

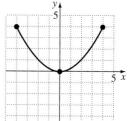

e.

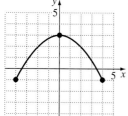

f.

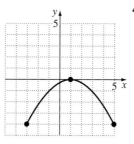

46. a.

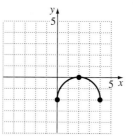

b.

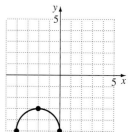

c.

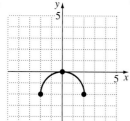

d.

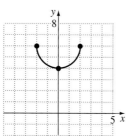

e.

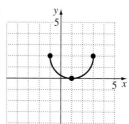

f.

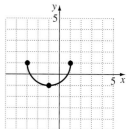

47. a.

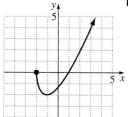

b.

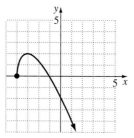

c.

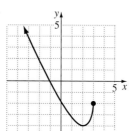

d.

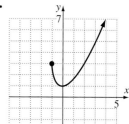

48. a. 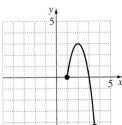   **b.** 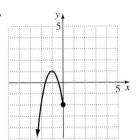   **c.** 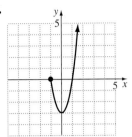   **d.**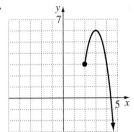

49. a. 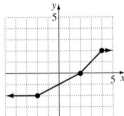   **b** 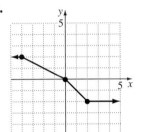   **c.** 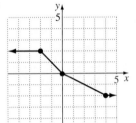   **d.**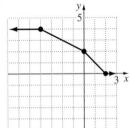

e. 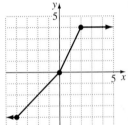   **f.** 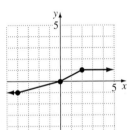   **50. a.** 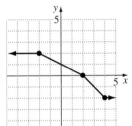   **b.**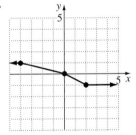

c. 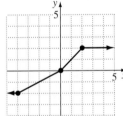   **d.** 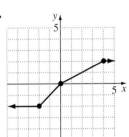   **e.** 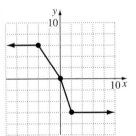   **f.**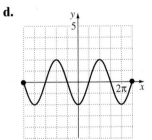

51. a. 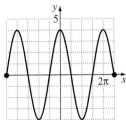   **b.** 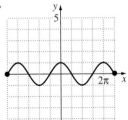   **c.** 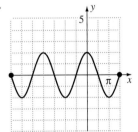   **d.**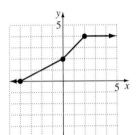

52. a.
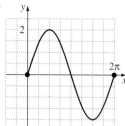

b.
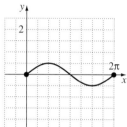

c.
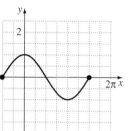

d.
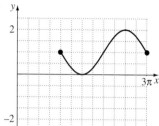

53. Even   **54.** Odd   **55.** Odd   **56.** Even   **57.** Neither   **58.** Neither   **59.** Even   **60.** Even
61. Even   **62.** Neither
63. Possible answer:

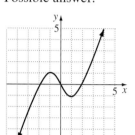

64. Possible answer:

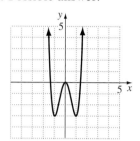

65. Possible answer:

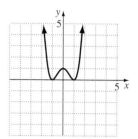

66. Possible answer:

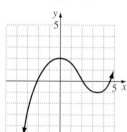

67. Possible answer:

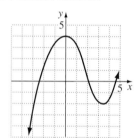

68. d
69. c

70.
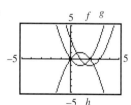

71.
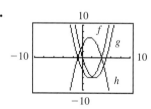

72.
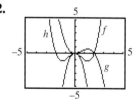

73.
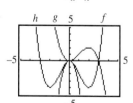

74.
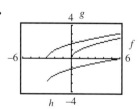

75.
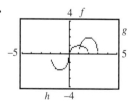

76. a, b.
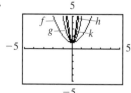

c. Answers may vary.

d. Shrink the graph of f by multiplying each of its x-coordinates by $\dfrac{1}{c}$.

e. Answers may vary.

77. a, b.

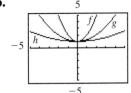

c. Answers may vary.

d. Stretch the graph of f by multiplying each of its x-coordinates by $\dfrac{1}{c}$.

e. Answers may vary.

78. a.

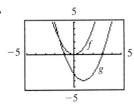

b. $(-1, 3)$ **c.** $x^2 - 2x - 3 < 0$ **79–81.** Answers may vary.   **82.** No

83. No   **84.** $(-a, b)$ **85.** $(a, 2b)$ **86.** $(a + 3, b)$ **87.** $(a, b - 3)$ **88.** $(a - 2, b - 4)$ **89.** $(a + 5, -b)$

90. $(a - 3, -b + 2)$ **91.** Increasing on $(-\infty, -6)$; decreasing on $(-6, -1)$; increasing on $(-1, \infty)$

92. Increasing on $(-\infty, 1)$; decreasing on $(1, 6)$; increasing on $(6, \infty)$

93. Decreasing on $(-\infty, -6)$; increasing on $(-6, -1)$; decreasing on $(-1, \infty)$

94. a. Both f and g even or both f and g odd.   **b.** f odd and g even or f even and g odd.

95. $mx - y - mx_1 + y_1 = 0$

PROBLEM SET 2.6

1. $(f + g)(x) = 2x^2 - 2$, domain: $(-\infty, \infty)$; $(f - g)(x) = 2x^2 - 2x - 4$, domain: $(-\infty, \infty)$;

$(fg)(x) = 2x^3 + x^2 - 4x - 3$, domain: $(-\infty, \infty)$; $\left(\dfrac{f}{g}\right)(x) = 2x - 3$, domain: $(-\infty, -1) \cup (-1, \infty)$

2. $(f + g)(x) = 6x^2 + x - 2$, domain: $(-\infty, \infty)$; $(f - g)(x) = 6x^2 - 3x$, domain: $(-\infty, \infty)$;

$(fg)(x) = 12x^3 - 8x^2 - x + 1$, domain: $(-\infty, \infty)$; $\left(\dfrac{f}{g}\right) = 3x + 1$, domain: $\left(-\infty, \dfrac{1}{2}\right) \cup \left(\dfrac{1}{2}, \infty\right)$

3. $(f + g)(x) = \sqrt{x + 4} + \sqrt{x - 1}$, domain: $[1, \infty)$; $(f - g)(x) = \sqrt{x + 4} - \sqrt{x - 1}$, domain: $[1, \infty)$;

$(fg)(x) = \sqrt{x^2 + 3x - 4}$, domain: $[1, \infty)$; $\left(\dfrac{f}{g}\right)(x) = \dfrac{\sqrt{x + 4}}{\sqrt{x - 1}}$, domain: $(1, \infty)$

4. $(f + g)(x) = \sqrt{x + 6} + \sqrt{x - 3}$, domain: $[3, \infty)$; $(f - g)(x) = \sqrt{x + 6} - \sqrt{x - 3}$, domain: $[3, \infty)$;

$(fg)(x) = \sqrt{x^2 + 3x - 18}$, domain: $[3, \infty)$; $\left(\dfrac{f}{g}\right) = \dfrac{\sqrt{x + 6}}{\sqrt{x - 3}}$, domain: $(3, \infty)$

5. $(f + g)(x) = \dfrac{x^2 + 1}{x^3}$, domain: $(-\infty, 0) \cup (0, \infty)$; $(f - g)(x) = \dfrac{x^2 - 1}{x^3}$, domain: $(-\infty, 0) \cup (0, \infty)$;

$(fg)(x) = \dfrac{1}{x^4}$, domain: $(-\infty, 0) \cup (0, \infty)$; $\left(\dfrac{f}{g}\right)(x) = x^2$, domain: $(-\infty, 0) \cup (0, \infty)$

6. $(f + g)(x) = \dfrac{x - 1}{(x - 2)^2}$, domain: $(-\infty, 2) \cup (2, \infty)$; $(f - g)(x) = \dfrac{x - 3}{(x - 2)^2}$, domain: $(-\infty, 2) \cup (2, \infty)$;

$(fg)(x) = \dfrac{1}{(x - 2)^3}$, domain: $(-\infty, 2) \cup (2, \infty)$; $\left(\dfrac{f}{g}\right)(x) = x - 2$, domain: $(-\infty, 2) \cup (2, \infty)$

7. $(f + g)(x) = \dfrac{x\sqrt{2x - 6} + 1}{x}$, domain: $[3, \infty)$; $(f - g)(x) = \dfrac{x\sqrt{2x - 6} - 1}{x}$, domain: $[3, \infty)$;

$(fg)(x) = \dfrac{\sqrt{2x - 6}}{x}$, domain: $[3, \infty)$; $\left(\dfrac{f}{g}\right)(x) = x\sqrt{x - 6}$, domain: $[3, \infty)$

8. $(f + g)(x) = \dfrac{x\sqrt{3x - 12} + 1}{x}$, domain: $[4, \infty)$; $(f - g)(x) = \dfrac{x\sqrt{3x - 12} - 1}{x}$, domain: $[4, \infty)$;

$(fg)(x) = \dfrac{\sqrt{3x - 12}}{x}$, domain: $[4, \infty)$; $\left(\dfrac{f}{g}\right)(x) = x\sqrt{3x - 12}$, domain: $[4, \infty)$

9.

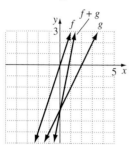

10.

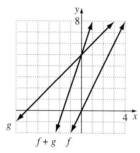

11.

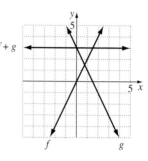

12.

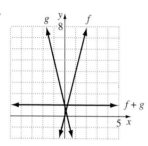

13. $\left(\dfrac{f}{g}\right)(x) = 2x - 5\ (x \neq -2)$

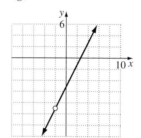

14. $\left(\dfrac{f}{g}\right)(x) = x + 5\ (x \neq -3)$

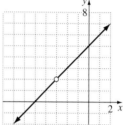

15. a. $(f \circ g)(x) = x^2 + 4x + 4$ **b.** $(g \circ f)(x) = x^2 + 2$ **c.** $(f \circ f)(x) = x^4$ **16. a.** $(f \circ g)(x) = x^2 - 6x + 9$
b. $(g \circ f)(x) = x^2 - 3$ **c.** $(f \circ f)(x) = x^4$ **17. a.** $(f \circ g)(x) = -2x + 11$ **b.** $(g \circ f)(x) = -2x + 1$
c. $(f \circ f)(x) = 4x + 9$ **18. a.** $(f \circ g)(x) = 11 - 4x$ **b.** $(g \circ f)(x) = 4 - 4x$ **c.** $(f \circ f)(x) = 16x - 5$
19. a. $(f \circ g)(x) = x$ **b.** $(g \circ f)(x) = x$ **c.** $(f \circ f)(x) = 256x^9 - 192x^6 + 48x^3 - 5$ **20. a.** $(f \circ g)(x) = x$

b. $(g \circ f)(x) = x$ **c.** $(f \circ f)(x) = x^9 - 6x^6 + 12x^3 - 10$ **21. a.** $(f \circ g)(x) = \sqrt{2x + 1};\ \left[-\dfrac{1}{2}, \infty\right)$

b. $(g \circ f)(x) = 2\sqrt{x} + 1;\ [0, \infty)$ **22. a.** $(f \circ g)(x) = \sqrt{3x - 1};\ \left[\dfrac{1}{3}, \infty\right)$ **b.** $(g \circ f)(x) = 3\sqrt{x} - 1;\ [0, \infty)$

23. a. $(f \circ g)(x) = 2x;\ (-\infty, 0) \cup (0, \infty)$ **b.** $(g \circ f)(x) = \dfrac{x}{2};\ (-\infty, 0) \cup (0, \infty)$

24. a. $(f \circ g)(x) = 3x;\ (-\infty, 0) \cup (0, \infty)$ **b.** $(g \circ f)(x) = \dfrac{x}{3};\ (-\infty, 0) \cup (0, \infty)$ **25. a.** $(f \circ g)(x) = -x^2;\ [-3, 3]$

b. $(g \circ f)(x) = \sqrt{-x^4 + 18x^2 - 72};\ [-\sqrt{12}, -\sqrt{6}] \cup [\sqrt{6}, \sqrt{12}]$ **26. a.** $(f \circ g)(x) = -x^2;\ [-1, 1]$

b. $(g \circ f)(x) = \sqrt{-x^4 + 2x^2};\ [-\sqrt{2}, \sqrt{2}]$ **27. a.** $(f \circ g)(x) = x^2;\ (-\infty, \infty)$ **b.** $(g \circ f)(x) = x^2;\ (-\infty, \infty)$
28. a. $(f \circ g)(x) = x^2;\ (-\infty, \infty)$ **b.** $(g \circ f)(x) = x^2;\ (-\infty, \infty)$ **29. a.** $(f \circ g)(x) = |3x - 1|;\ (-\infty, \infty)$
b. $(g \circ f)(x) = 3|x| - 1;\ (-\infty, \infty)$ **30. a.** $(f \circ g)(x) = |2x + 3|;\ (-\infty, \infty)$ **b.** $(g \circ f)(x) = 2|x| + 3;\ (-\infty, \infty)$
31. a. $(f \circ g)(x) = x;\ (-\infty, 3) \cup (3, \infty)$ **b.** $(g \circ f)(x) = x;\ (-\infty, 1) \cup (1, \infty)$
32. a. $(f \circ g)(x) = x;\ (-\infty, -1) \cup (-1, \infty)$ **b.** $(g \circ f)(x) = x;\ (-\infty, 1) \cup (1, \infty)$
33. a. $(f \circ g)(x) = x;\ (-\infty, \infty)$ **b.** $(g \circ f)(x) = x;\ (-\infty, \infty)$ **34. a.** $(f \circ g)(x) = x;\ (-\infty, \infty)$
b. $(g \circ f)(x) = x;\ (-\infty, \infty)$ **35.** undefined   **36.** -6 **37.** 4.5 **38.** Undefined   **39.** 4.5 **40.** -2 **41.** 2
42. -3 **43.** -3 **44.** -10 **45.** $f(x) = x^4; g(x) = 3x - 1$ **46.** $f(x) = x^3; g(x) = 2x - 5$
47. $f(x) = \sqrt[3]{x}; g(x) = x^2 - 9$ **48.** $f(x) = \sqrt{x}; g(x) = 5x^2 + 3$ **49.** $f(x) = |x|; g(x) = 2x - 5$

50. $f(x) = |x|; g(x) = 3x - 4$ **51.** $f(x) = \dfrac{1}{x}; g(x) = 2x - 3$ **52.** $f(x) = \dfrac{3}{x^2}; g(x) = 5x - 1$

53. $f(x) = x^{2/3}; g(x) = 2x^2 - 5x + 1$ **54.** $f(x) = x^{3/4}; g(x) = 3x^2 - 7x + 4$ **55.** $f(x) = 3x^2 + 5x; g(x) = x - 1$
56. $f(x) = 4x^2 + 6x; g(x) = x - 3$ **57.** $k(x) = (g \circ f)(x)$ **58.** $k(x) = (f \circ g)(x)$ **59.** $k(x) = (f \circ h)(x)$
60. $k(x) = (h \circ f)(x)$ **61.** $k(x) = (g \circ f \circ h)(x)$ **62.** $k(x) = (g \circ h \circ f)(x)$ **63.** $f^3(x) = 8x - 21$
64. $f^3(x) = 27x + 26$ **65.** $f^3(x) = x^8 + 4x^6 + 8x^4 + 8x^2 + 5$ **66.** $f^3(x) = 128x^8 - 256x^6 + 160x^4 - 32x^2 + 1$

67. $f^3(x) = x^{27}$ **68.** $f^3(x) = -8192x^{27}$ **69.** $f^4(x) = \dfrac{5x + 3}{3x + 2}$ **70.**

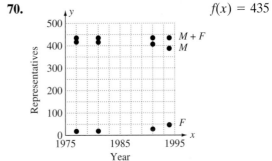

$f(x) = 435$

71. a. -0.44; profit is decreasing.   **b.** 0.51; profit is increasing.
  c. $(f + g)(x) = 0.07x + 24.76; 0.07$; profit is increasing.
72. a. $T(C(F)) = 0.15F + 22.6$; skin temperature (in degrees Celsius) expressed as a function of Fahrenheit temperature.
  b. 28.75; when the temperature of the environment is $41°$ Fahrenheit, skin temperature is $28.75°$ Celsius.
73. $(A \circ r)(t) = 0.16\pi t^2$; area of the circle as a function of time
74. $(V \circ r)(t) = \dfrac{32}{3}\pi t^3$; volume of the balloon as a function of time
75. a. The cost of the car to the consumer is the selling price with a $2000 rebate.
  b. The cost of the car is 85% of the selling price, a 15% discount to the consumer.
  c. $(f \circ g)(x) = 0.85x - 2000$; the cost of the car is 85% of the selling price with a $2000 rebate.
  d. $(g \circ f)(x) = 0.85x - 1700$; the cost of the car is 85% of the selling price less $1700.
  e. The cost in part (c) is less.
76. $f^3(x) = 0.830584x$; area of rain forest at the start of the year 2000 is approximately 83% of the area at the start of 1997.   **77.** $0, 0, 0, 0$; yes   **78.** $2, 6, 42, 1806$; no   **79.** $-i, -1 -i, -1 + i, -1 -i$; yes
80. $\dfrac{i}{2}, -\dfrac{1}{4} + \dfrac{1}{2}i, \dfrac{7}{16} + \dfrac{1}{4}i, -\dfrac{79}{256} + \dfrac{1}{32}i$; yes   **81.** $1 - i, 1 - 3i, -7 - 9i, -39 + 117i$; no
82. $2 - i, 5 - 5i, 5 - 55i, -2995 - 605i$; no   **83.** c   **84.** b
85. Domain of $f: (-\infty, \infty)$ **86.** Domain of $f: [1, \infty)$ **87.** Domain of $f: (-\infty, -1] \cup [1, \infty)$
  Domain of $g: [-2, 2]$ Domain of $g: (-\infty, \infty)$ Domain of $g: (-\infty, 1]$
  Domain of $f \circ g: [-2, 2]$ Domain of $f \circ g:$ Domain of $f \circ g: (-\infty, 0]$
    $(-\infty, -1) \cup (1, \infty)$

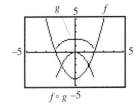

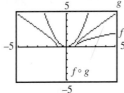

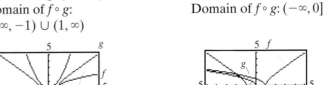

88. Domain of f: $(0, \infty)$
Domain of g: $(-\infty, \infty)$
Domain of $f \circ g$:
$(-\infty, -2) \cup (2, \infty)$

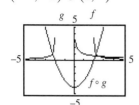

89. Domain of f: $[1, \infty)$
Domain of g: $(-\infty, 1]$
Domain of $f \circ g$: $(-\infty, 0]$

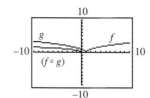

90. Domain of f: $[-1, 1]$
Domain of g: $(-\infty, -1] \cup [1, \infty)$
Domain of $f \circ g$:
$[-\sqrt{2}, -1] \cup [1, \sqrt{2}]$

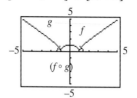

91. a.

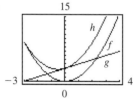

b. The y-coordinates of $f + g$ are equal to the sum of the corresponding y-coordinates of f and g.
c. Answers may vary.

d.

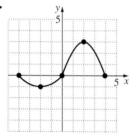

92. a. $A(r(t)) = 25\pi t^2 + 30\pi t + 9\pi$ **b.** 3 seconds

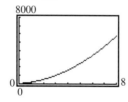

93. $f''(0.96)$ approaches zero. $f''(1.04)$ gets very large.   **94. a.** Answers may vary.   **b.** $f''(0.2)$ approaches 0.5.

95. $g(x) = 4x - 1$ or $g(x) = -4x - 4$ **96.** $(f \circ g \circ h) = \dfrac{1}{9}x^2 + \dfrac{4}{3}x + 4$ **97.** Answers may vary.

98. $4[(x + a) + 1]$ **99.** Verification is left to the student.   **100.** Even   **101.** Odd   **102.** Even

103. Odd   **104.** Odd   **105.** $m = -\dfrac{1}{2}; b = \dfrac{1}{2}$ **106–107.** Group activity

PROBLEM SET 2.7

1. Function; not one-to-one   **2.** Function; one-to-one   **3.** Not a function
4. Function; not one-to-one   **5.** Not a function   **6.** Function; not one-to-one
7. Function; not one-to-one   **8.** Function; one-to-one   **9.** Function; one-to-one
10. Function; one-to-one function   **11.** $[1, \infty)$ **12.** $[-2, \infty)$ **13.** $[2, \infty)$ **14.** $[-1, \infty)$ **15. a.** 13
b. 7   **16. a.** $\dfrac{\pi}{2}$ **b.** $\dfrac{\sqrt{3}}{2}$ **17.** 5   **18.** -2 **19.** $(-4, 0), (0, 2), (4, 4)$

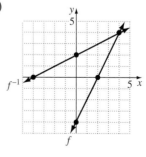

20. $(0, -3), (2, -1), (3, 5)$

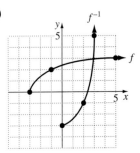

21. $(1, 0), (2, 1), (4, 2)$

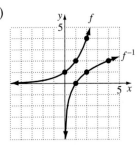

22. $(1, 0), (2, -1), (4, -2)$

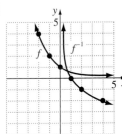

23–28. Verification is left to the student.

29. $f^{-1}(x) = \dfrac{x - 7}{3}$ **30.** $f^{-1}(x) = \dfrac{x + 5}{2}$ **31.** $f^{-1}(x) = 6 - 2x$ **32.** $f^{-1}(x) = -3x - 12$ **33.** $f^{-1}(x) = \dfrac{x^2 - 7}{2}$

34. $f^{-1}(x) = \dfrac{x^2 - 5}{3}$ **35.** $f^{-1}(x) = \dfrac{x^3 + 3}{2}$ **36.** $f^{-1}(x) = \dfrac{x^3 + 5}{4}$ **37.** $f^{-1}(x) = \sqrt[3]{\dfrac{x + 7}{5}}$

38. $f^{-1}(x) = \sqrt[3]{\dfrac{x + 4}{3}}$ **39.** $f^{-1}(x) = \dfrac{3x + 1}{x - 2}$ **40.** $f^{-1}(x) = \dfrac{-x - 3}{x - 2}$ **41.** $f^{-1}(x) = \sqrt{x} + 3$

42. $f^{-1}(x) = \sqrt{x} + 1$ **43.** $f^{-1}(x) = (x - 3)^3 + 4$ **44.** $f^{-1}(x) = (x + 4)^3 + 7$ **45.** $f^{-1}(x) = \dfrac{1}{x - 3}$

46. $f^{-1}(x) = \dfrac{1 - 3x}{x}$ **47.** $f^{-1}(x) = x^{5/3}$ **48.** $f^{-1}(x) = (x + 1)^{7/5}$

49.

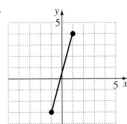

50.

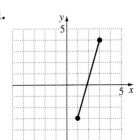

51.

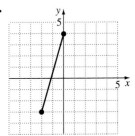

52.

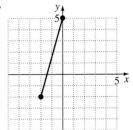

53.

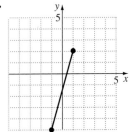

54.

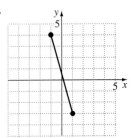

55.

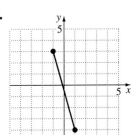

56.

57.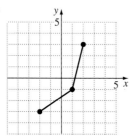

58.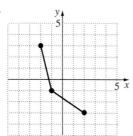

59.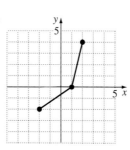

60.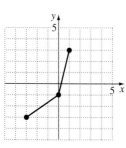

61.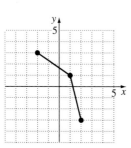

62.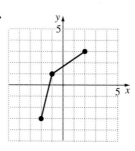

63. $(f^{-1} \circ g^{-1})(x) = \dfrac{x - 5}{2}$

64. $(g^{-1} \circ f^{-1})(x) = \dfrac{x - 2}{2}$ **65.** $(f \circ g)^{-1}(x) = \dfrac{x - 2}{2}$ **66.** $(g \circ f)^{-1}(x) = \dfrac{x - 5}{2}$

67. a. f is increasing on its domain.   **b.** Number of people in a room for a 25% probability of two people sharing a birthday; number of people in a room for a 50% probability of two people sharing a birthday; number of people in a room for a 70% probability of two people sharing a birthday

68. a. f is decreasing on its domain   **b.** The number of years after the study begins when the river to contain 14 parts of DDT per million   **c.** 5 years   **69. a.** Yes   **b.** No; time is not a function of arts endowment.

70. a. Answers may vary.   **b.** $(f \circ g)(x) = 0.6x - 5$ **c.** $(g \circ f)(x) = 0.6x - 3$ **d.** $f \circ g$ **e.** $f^{-1}(x) = x + 5$

71. $5D + 5$ **72. a.** $t = \dfrac{24 \pm \sqrt{576 - s}}{4}$ **b.** No   **c.** $t = 3$ or $t = 9$; the ball's height is 432 feet after 3 seconds and 9 seconds.   **74.** b   **75.** a

76. $f^{-1}(x) = \dfrac{x + 4}{3}$

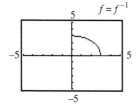

77. $f^{-1}(x) = \sqrt[3]{x - 2}$

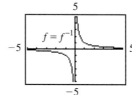

78. $f^{-1}(x) = -\sqrt{x}$

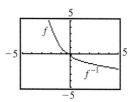

79. $f^{-1}(x) = \sqrt{9 - x^2}, 0 \le x \le 3$ **80.** $f^{-1}(x) = \dfrac{1}{x}$

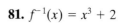

81. $f^{-1}(x) = x^3 + 2$

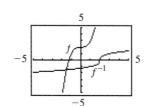

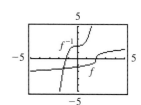

82. $f^{-1}(x) = x^{5/3}$

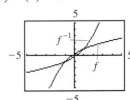

83. Not one-to-one

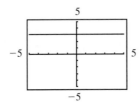

84. Not one-to-one

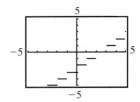

85. Not one-to-one

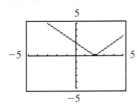

86. Not one-to-one

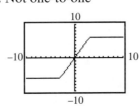

87. One-to-one

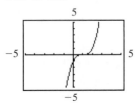

88. Not one-to-one

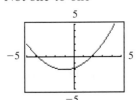

89. Not one-to-one

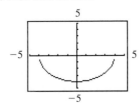

90. Not one-to-one

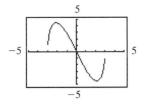

91. One-to-one

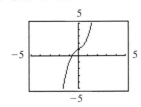

92. Inverses

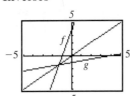

93. Inverses

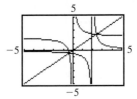

94. Inverses

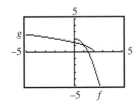

95. Not inverses

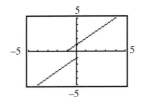

96. Inverses

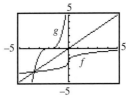

97. a.

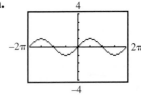

b. No; $\left[-\dfrac{\pi}{2}, \dfrac{\pi}{2}\right]$ **c.** Answers may vary.   **98–100.** Answers may vary.

101. $f^{-1}(x) = \dfrac{Ax + B}{Cx - A}$ **102.** 7   **103.** 8   **104–105.** Verification is left to the student.

CHAPTER 2 REVIEW PROBLEMS

1. 824.9; from the perspective of an observer at rest, the length is approximately 824.9 meters.
2. -0.095; the average distance each automobile was driven in 1950 was greater than in 1985 by 95 miles.
3. a. -3 **b.** -17 **c.** $7a - 3$ **d.** $7a + 7h - 3$ **e.** 7   **f.** $7a + 7h - 6$ **g.** $14a - 3$ **4. a.** 11   **b.** 37
c. $4a^2 - 5a + 11$ **d.** $4a^2 + 8ah + 4h^2 - 5a - 5h + 11$ **e.** $8a + 4h - 5$ **f.** $4a^2 - 5a + 4h^2 - 5h + 22$
g. $16a^2 - 10a + 11$ **5.** $(-\infty, 4]$ **6.** $(-\infty, \infty)$ **7.** $(-\infty, 1] \cup [4, \infty)$ **8.** $(-\infty, 1) \cup (4, \infty)$ **9.** $[-2, 2]$
10. $[2, 5) \cup (5, \infty)$ **11.** $y - 2 = -6(x + 3); y = -6x - 16; 6x + y + 16 = 0$
12. $y - 6 = 2(x - 1)$ or $y - 2 = 2(x + 1); y = 2x + 4; 2x - y + 4 = 0$
13. $y + 7 = -3(x - 4); y = -3x + 5; 3x + y - 5 = 0$ **14.** $y - 6 = -3(x + 3); y = -3x - 3; 3x + y + 3 = 0$
15. a. 6.6 million; $m = 0.12$; from 1975 to 1980, the number of victims increased at an average rate of 120,000
per year.   **b.** $m = -0.08$; from 1980 to 1985, the number of victims decreased at an average rate of 80,000 per
year.   **16.** Answers may vary: $y = 16x - 31,108; 892$ nurses
17. a. $y - 666.2 = 60.1(x - 5)$ or $y - 546 = 60.1(x - 3); y = 60.1x + 365.7$ **b.** 485.9 billion
c. 1267.2 billion   **18.**

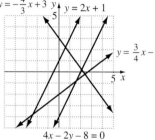

19. $\dfrac{1}{50}; \dfrac{1}{100}; \dfrac{1}{150}$ **20. a.** Drive On Rental Company: $y = 0.15x + 30$; Hit the Road Rental: $y = 0.10x + 40$

b.

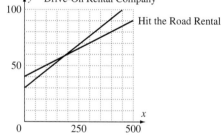

21. a. $y = \dfrac{2}{115}x$ **b.** 608.70

22.

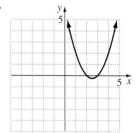

Domain: $(-\infty, \infty)$

Range: $\left[-\dfrac{1}{4}, \infty \right)$

23.

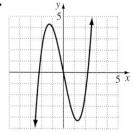

Domain: $(-\infty, \infty)$

Range: $(-\infty, \infty)$

24.

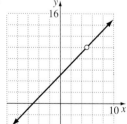

Domain: $(-\infty, 5) \cup (5, \infty)$

Range: $(-\infty, 10) \cup (10, \infty)$

25.

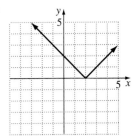

Domain: $(-\infty, \infty)$
Range: $[0, \infty)$

26.

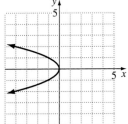

Domain: $[-3, \infty)$
Range: $[0, \infty)$

27.

Domain: $(-\infty, 0]$
Range: $(-\infty, \infty)$

28.

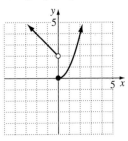

Domain: $(-\infty, \infty)$
Range: $[0, \infty)$

29.

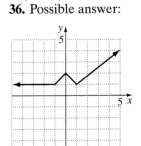

Domain: $(-\infty, \infty)$
Range: $[0, \infty)$

30. a. $[-3, 5)$
b. $[-5, 0]$
c. 3

31. Center: $(2, -1); r = 3$ **32.** Center: $(-1, 2); r = 3$ **33. a.** $-5; 0; -3$
b. Increasing: $(-5, 0)$; decreasing: $(-\infty, -5) \cup (0, \infty)$ **c.** Relative maximum: 3 at $x = 0$; relative minimum: -6
at $x = -5$ **d.** Not one-to-one; $[0, \infty)$ **34. a.** Answers may vary.   **b.** $(3, 12)$ **c.** $(0, 3), (12, 17)$
d. $(17, 30)$ **e.** No   **35. a.** Maximum $= 13.5$ at $x = 25$ **b.** Range: $[1, 13.5]$
36. Possible answer: **37.** Possible answer: **38. a.** Possible answer: **b.** 4
c. $(-\infty, -2)$ and $(2, 3.5)$
d. $(-2, 2)$ and $(3.5, \infty)$

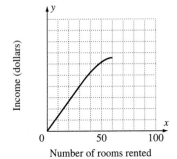

39. Possible answer: **40.** Possible answer: **41.**

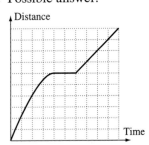

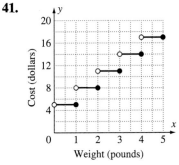

42–43. Answers may vary.

44.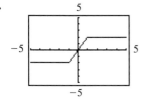

a. Increasing: $(-1, 1)$
Constant: $(-\infty, -1), (1, \infty)$
b. Relative maximum: 1
Relative minimum: -1
c. Odd function

45.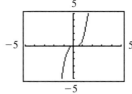

a. Increasing: $(-\infty, 0), (0.33, \infty)$
Decreasing: $(0, 0.33)$
b. Relative maximum: 0
Relative minimum: -0.04
c. Neither

46.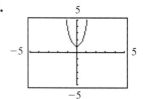

a. Increasing: $(0, \infty)$
Decreasing: $(-\infty, 0)$
b. Relative minimum: 1
c. Even function

47.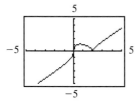

a. Increasing: $(-\infty, 0.67)$ and $(2, \infty)$
Decreasing: $(0.67, 2)$
b. Relative maximum: 1.06
Relative minimum: 0
c. Neither

48. a. $f(x) = x(18 - 2x)^2$
b.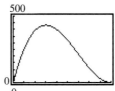

c. 3; 432 cubic inches

49. a. $f(x) = (200 + 25x)(400 - x)$
b.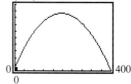

c. 204 people; $1,040,400

50. a. $A = 4x$ **b.** $A = \dfrac{x\sqrt{64 - x^2}}{4}$ **c.** $f(x) = 4x + \dfrac{x\sqrt{64 - x^2}}{4}$

d. 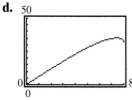 $x \approx 7.44$; 35.23 square feet

51. a. Shift 3 units up   **b.** Shift left 3 units   **c.** Stretch by a factor of 2
d. Shift 1 unit to the right and 3 units down   **e.** Reflect in the y-axis
f. Reflect in the x-axis   **g.** Reflect in the y-axis and then in the x-axis   **h.** Reflect in the line $y = x$

52.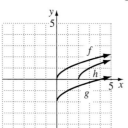

53.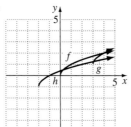

54.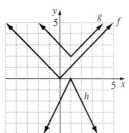

55.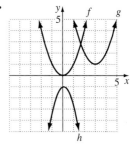

56.

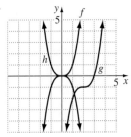

57.

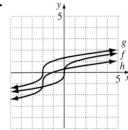

58.

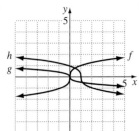

59. $g(x) = |x - 3| - 1$

$h(x) = -\dfrac{1}{2}|x| + 2$

60.

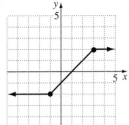

61.

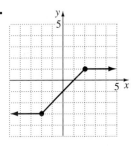

62.

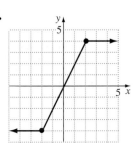

63.

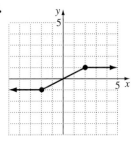

64.

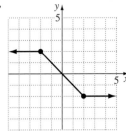

65.

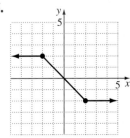

66.

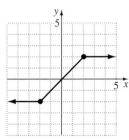

67.

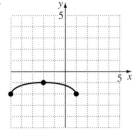

68. $g(x) = -\dfrac{1}{2}\sqrt{16 - x^2} - 1$ **69.** Odd   **70.** Even   **71.** Odd   **72. a.** $(f + g)(x) = x^2 - 9x + 13$

b. $(f - g)(x) = x^2 - 5x + 3$ **c.** $(fg)(x) = -2x^3 + 19x^2 - 51x + 40$ **d.** $\left(\dfrac{f}{g}\right)(x) = \dfrac{x^2 - 7x + 8}{5 - 2x}$

e. $(f \circ g)(x) = 4x^2 - 6x - 2$ **f.** $(g \circ f)(x) = -2x^2 + 14x - 11$ **73. a.** $(f + g)(x) = \sqrt{x + 4} + \sqrt{x - 1}; [1, \infty)$

b. $(f - g)(x) = \sqrt{x + 4} - \sqrt{x - 1}; [1, \infty)$ **c.** $(fg)(x) = \sqrt{x^2 + 3x - 4}; [1, \infty)$ **d.** $\left(\dfrac{f}{g}\right)(x) = \dfrac{\sqrt{x + 4}}{\sqrt{x - 1}}; (1, \infty)$

74. a. $(f \circ g)(x) = \sqrt{2x + 1}$ **b.** $(g \circ f) = 2\sqrt{x} + 1$ **c.** 3   **d.** 11

e. Domain of $f \circ g$: $\left[-\dfrac{1}{2}, \infty\right)$; domain of $g \circ f$: $[0, \infty)$ **75.** $(f \circ g)(x) = -x^2; [-5, 5]$

76. a. $(L \circ p)(t) = 0.7\sqrt{0.0001t^6 + 0.02t^3 + 4}$ **b.** 7.8; the level of carbon monoxide in 2000 is approximately
7.8 ppm.   **77.** $f(x) = x^4; g(x) = 2x + 7$ **78.** $f(x) = \sqrt{x}; g(x) = 3x^2 - 7$
79. $f(x) = 3x^2 + 5x; g(x) = x - 4$ **80.** $f^3(x) = 27x - 26$ **81.** $f^3(x) = x^8 - 4x^6 + 4x^4 - 1$
82. $f^3(x) = 0.857375x$; at the start of 2000 the area of the rain forest will be approximately 86% of the area at
the start of 1997.   **83.** Not a function   **84.** Function; not one-to-one   **85.** Function; one-to-one

86. Not a function   **87.** $f^{-1}(x) = 2x + 8$ **88.** $f^{-1}(x) = \dfrac{x + 3}{4}$ **89.** $f^{-1}(x) = x^2 - 2$ **90.** $f^{-1}(x) = \dfrac{\sqrt[3]{x - 1}}{2}$

91. $f^{-1}(x) = \sqrt{x + 4}$ **92.** $f^{-1}(x) = \dfrac{x^3 + 1}{2}$

93. a. Yes   **b.** The number of compounding periods needed to get a balance of $11,250
c. The statement is false.

94. a.

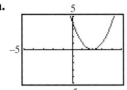

b. $[2, \infty)$ **c.** $f^{-1}(x) = \sqrt{x} + 2$ **d.**

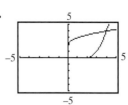

95. a.

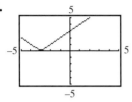

b. $[-3, \infty)$ **c.** $f^{-1}(x) = x - 3$ **d.**

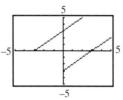

96. a.

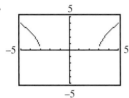

b. $[3, \infty)$ **c.** $f^{-1}(x) = \sqrt{x^2 + 9}$ **d.**

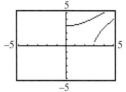

97.

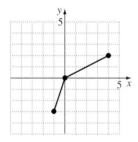

98.

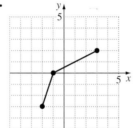

99.

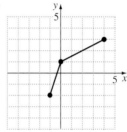

100.

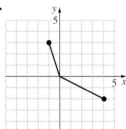

101. a. $f(T) = 4T - 160$

b. 200; at a temperature of 90°, the cricket chirps 200 times per minute.

c. $(40, 136)$

d. $(0, 384)$

e. $T = \dfrac{C + 160}{4}$

102.

x	$f(x)$	$g(x)$	$h(x)$
-3	0	0	0
-2	5	5	undefined
-1	5	5	undefined
0	0	0	0
1	-5	5	undefined
2	-5	5	undefined
3	0	0	0

CHAPTER 2 TEST

1. 11.64; 11.64% of the U.S. population graduated from college in 1970
2. $14a + 7h - 9$ **3.** $(-\infty, 4]$ **4.** $y - 1 = 3(x - 2)$ or $y + 8 = 3(x + 1)$; $y = 3x - 5$; $3x - y - 5 = 0$
5. $y - 6 = 4(x + 4)$; $y = 4x + 22$; $4x - y + 22 = 0$
6. a. $y - 401.1 = 14.9(x - 4)$ or $y - 475.6 = 14.9(x - 9)$; $y = 14.9x + 341.5$ **b.** $639.50
7.   **8.**   **9. a.** 5   **b.** $(-1, 2)$
        c. $(-\infty, -1), (2, \infty)$
        d. 5 at $x = 2$
        e. -3 at $x = -1$

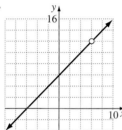 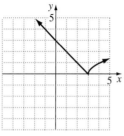

10. 128 cubic inches
11. Possible answer: **12.** $g(x) = (x - 2)^3$
        13. The graph of f is shifted 3 units to the right to obtain the graph of g. Then the graph g is stretched by a factor of two and reflected in the x-axis to obtain the graph of h.

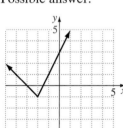

14. Even; the graph in the figure does not have y-axis symmetry.
15. $(f - g)(x) = x^2 - 2x - 2$ **16.** $(f \circ g)(x) = 25x^2 - 5x - 6$ **17.** $(g \circ f)(x) = 5x^2 + 15x - 22$ **18.** 84
19. 28   **20.** $f^{-1}(x) = x^2 + 2$ **21. a.** Yes; the graph of f passes the horizontal line test.
b. The height of a man who has the recommended weight of 170 pounds.   **22. a.** No   **b.** Neither
c. Relative maximum: 3.23 at $x \approx 1.22$; relative minimum: 0 at $x = 0$; relative minimum: -9.91 at $x \approx 3.28$
d. $[-9.91, \infty)$ **e.** $(0, 1.22), (3.28, \infty)$ **f.** $(-\infty, 0), (1.22, 3.28)$

CUMULATIVE REVIEW PROBLEMS (CHAPTERS P–2)

1. $\dfrac{x^9}{8y^{12}}$ **2.** $-14\sqrt{2y}$ **3.** $(x - 1 - y)(x - 1 + y)$ **4.** $\dfrac{1}{(x - 3)(x - 2)}$ **5.** $\sqrt{x + 1} + \sqrt{2}$
6. $\left\{-\dfrac{1}{4}, \dfrac{5}{3}\right\}$ **7.** $\{-1 \pm \sqrt{6}\}$ **8.** $\{4\}$ **9.** $\{-8, 27\}$ **10.** $\left[-\dfrac{11}{2}, \dfrac{7}{2}\right)$
11. $y - 5 = -2(x + 2)$; $y = -2x + 1$; $2x + y - 1 = 0$ **12.**

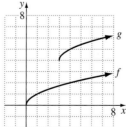

13. $f^{-1}(x) = \sqrt[5]{2x + 3}$ **14.** $[1, \infty)$ or $(-\infty, 1]$ **15.** m **16.** $r = \dfrac{G - a}{G}$

17. 15 sides   **18.** 1993   **19.** 3 ft by 8 ft   **20.** $3.00 or $9.50

Chapter 3

PROBLEM SET 3.1

1. Vertex: $(4, -1)$
x-intercepts: $(3, 0)$ and $(5, 0)$
y-intercept: $(0, 15)$
axis of symmetry: $x = 4$

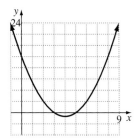

2. Vertex: $(1, -2)$
x-intercepts: $(1 + \sqrt{2}, 0)$ and
$(1 - \sqrt{2}, 0)$
y-intercept: $(0, -1)$
axis of symmetry: $x = 1$

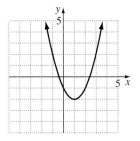

3. Vertex: $(1, 2)$
x-intercepts: none
y-intercept: $(0, 3)$
axis of symmetry: $x = 1$

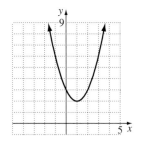

4. Vertex: $(3, 2)$
x-intercepts: none
y-intercept: $(0, 11)$
axis of symmetry: $x = 3$

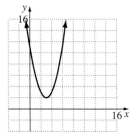

5. Vertex: $(3, 1)$
x-intercepts: none
y-intercept: $(0, 10)$
axis of symmetry: $x = 3$

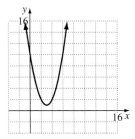

6. Vertex: $(1, 3)$
x-intercepts: none
y-intercept: $(0, 4)$
axis of symmetry: $x = 1$

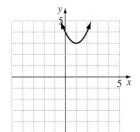

7. Vertex: $(-2, -1)$

x-intercepts: $\left(-2 + \dfrac{\sqrt{2}}{2}, 0\right)$

and $\left(-2 - \dfrac{\sqrt{2}}{2}, 0\right)$

y-intercept: $(0, 7)$

axis of symmetry: $x = -2$

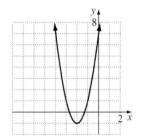

8. Vertex: $\left(\dfrac{1}{2}, \dfrac{5}{4}\right)$

x-intercepts: $\left(\dfrac{1 + \sqrt{5}}{2}, 0\right)$

and $\left(\dfrac{1 - \sqrt{5}}{2}, 0\right)$

y-intercept: $(0, 1)$

axis of symmetry: $x = \dfrac{1}{2}$

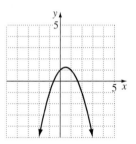

9. Vertex: $(1, 4)$

x-intercepts: $(-1, 0)$ and $(3, 0)$

y-intercept: $(0, 3)$

axis of symmetry: $x = 1$

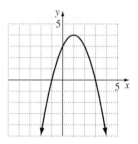

10. Vertex: $(3, 1)$
x-intercepts: $(2, 0)$ and $(4, 0)$
y-intercept: $(0, -8)$
axis of symmetry: $x = 3$

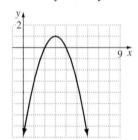

11. Vertex: $(1, -4)$
x-intercepts: $(-1, 0)$ and $(3, 0)$
y-intercept: $(0, -3)$
axis of symmetry: $x = 1$

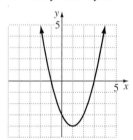

12. Vertex: $(1, -16)$
x-intercepts: $(-3, 0)$ and $(5, 0)$
y-intercept: $(0, -15)$
axis of symmetry: $x = 1$

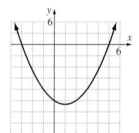

13. Vertex: $\left(-\dfrac{3}{2}, -\dfrac{49}{4}\right)$

x-intercepts: $(2, 0)$ and $(-5, 0)$

y-intercept: $(0, -10)$

axis of symmetry: $x = -\dfrac{3}{2}$

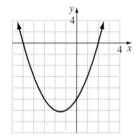

14. Vertex: $\left(\dfrac{7}{4}, -\dfrac{81}{8}\right)$

x-intercepts: $\left(-\dfrac{1}{2}, 0\right)$ and $(4, 0)$

y-intercept: $(0, -4)$

axis of symmetry: $x = \dfrac{7}{4}$

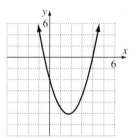

15. Vertex: $(1, 4)$

x-intercepts: $(-1, 0)$ and $(3, 0)$

y-intercept: $(0, 3)$

axis of symmetry: $x = 1$

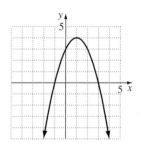

16. Vertex: $(-2, 9)$
x-intercepts: $(-5, 0)$ and $(1, 0)$
y-intercept: $(0, 5)$
axis of symmetry: $x = -2$

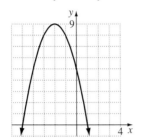

17. Vertex: $(1, -1)$
x-intercepts: none
y-intercept: $(0, -2)$
axis of symmetry: $x = 1$

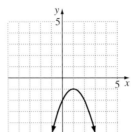

18. Vertex: $(2, 2)$
x-intercepts: none
y-intercept: $(0, 6)$
axis of symmetry: $x = 2$

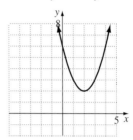

19. Year: 1968   **20.** Year: 1978   **21.** Pair: 8 and 8; Product is 64.   **22.** Pair: $\frac{9}{2}$ and $\frac{9}{2}$; Product is $\frac{81}{4}$.
23. Dimensions: 5 yds × 5 yds   **24.** Dimensions: 20 ft × 20 ft
25. Dimensions: 120 ft × 260 ft; Area: 31,200 sq ft   **26.** Dimensions: 100 yds × 150 yds; Area: 15,000 sq yds
27. $\left(\frac{1}{2}, \frac{\sqrt{2}}{2}\right)$ **28.** $\left(\frac{12}{5}, \frac{6}{5}\right)$ **29.** $\frac{1}{2}$ **30.** 25, -25 **31.** 20, 20   **32. a.** $s(t) = -16t^2 + 64t + 1053$
b. 2 seconds; 1117 feet   **33.** 200 people, \$800   **34.** 2 weeks, \$16 profit   **35.** 5 inches
36. $r \approx 1.68$ ft, $h \approx 1.68$ ft   **37.** $r \approx 35.01$ yards, $x = 110$ yards   **38.** a   **39.** c   **40.** b   **41.** d   **42.** c   **43.** a

44.

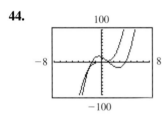

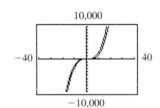

45. a.

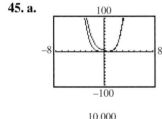

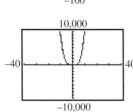

b.

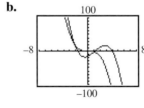

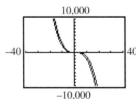

c.

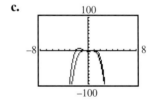

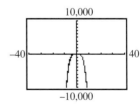

46. a. c   **b.** a   **c.** b   **d.** d   **47.** The graph is unrealistic because after peaking around $x = 8$ (1991), the graph rapidly drops below the x-axis indicating that a negative number of cases would be reported. Polynomial functions are best used to model behavior over short periods of time.   **48.** The graph suggest that consumption of cigarettes will decrease steadily over time. Polynomial functions do not model change precisely over extended periods.   **49.** The midpoint and the value of $-\frac{b}{2a}$ are equal when $b^2 - 4ac > 0$. When $b^2 - 4ac = 0$, only one x-intercept exists; when $b^2 - 4ac < 0$, no x-intercept exists.   **50.** $x = -2; (-3, -2)$
51. $x = 3; (0, 11)$ **52.** $y = -2(x + 1)^2 + 5$ **53.** $y = -(x - 3)^2 + 1$ **54.** $\frac{9}{2}$

55–56. Verification is left to the student.   **57.** 3 square units   **58.** $c = -\dfrac{55}{6}$

59. Vertex: $[-(d + e), (d^2 - 2de + e^2)]$; No, $b^2 - 4ac \le 0$ **60.** $(x - 3)^2 + (y - 2)^2 = 13$

61. Verification is left to the student.   **62.** $K = \pm\sqrt{2}$

P R O B L E M   S E T   3 . 2

1. d   **2.** c   **3.** f   **4.** b   **5.** e   **6.** a

7. **a.** Rises to right, falls to left
    b. $x = -2, x = -1, x = 2$
    c.

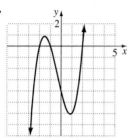

8. **a.** Rises to right, falls to left
    b. $x = -2, x = -1, x = 1$
    c.

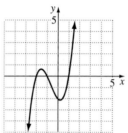

9. **a.** Rises to the left and the right
    b. $x = -1, x = 0, x = 1$
    c.

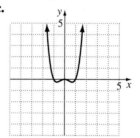

10. **a.** Rises to the left and the right
    b. $x = -3, x = 0, x = 3$
    c.

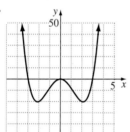

11. **a.** Falls to the left and the right
    b. $x = 0, x = -2, x = 2$
    c.

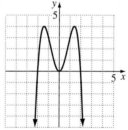

12. **a.** Falls to the left and the right
    b. $x = 0, x = -4, x = 4$
    c.

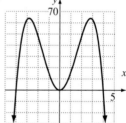

13. **a.** Rises to the left and the right
    b. $x = 0, x = 3$
    c.

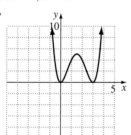

14. **a.** Rises to the left and the right
    b. $x = 0, x = 1$
    c.

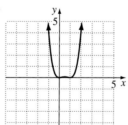

15. **a.** Falls to the left and the right
    b. $x = 0, x = 1$
    c.

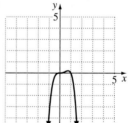

16. a. Falls to the left and the right
b. $x = 0, x = 2$
c.
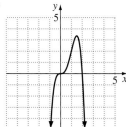

17. a. Rises to the left and falls to the right
b. $x = 0, x = -\sqrt{2}, x = \sqrt{2}$
c.
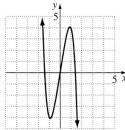

18. a. Rises to the left and falls to the right
b. $x = 0, x = -\sqrt{3}, x = \sqrt{3}$
c.
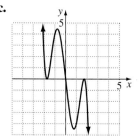

19. a. Falls to the left and the right
b. $x = -1, x = 1$
c.
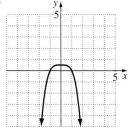

20. a. Rises to the left and falls to the right
b. $x = 0, x = 3$
c.
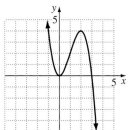

21. a. Falls to the left and the right
b. $x = 4, x = -5, x = 5$
c.
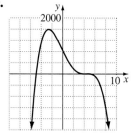

22. a. Falls to the left and the right
b. $x = 1, x = -2, x = 2$
c.
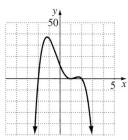

23. a. Most of the data points lie very close to the model.
b. For 1988, $f(5) = 104{,}706.$
c. For 1991, $f(8) = 256{,}568.1$
d. Leading coefficient is negative.

24. a. The population will decline and the elk will die off.
b.
c.   ; after 5 years
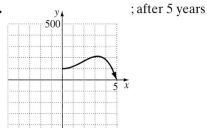

25. a. $V(x) = 216x - 60x^2 + 4x^3$ **b.** 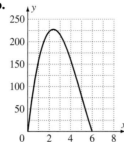 **26.** ≈ 555.6 feet **27.** 320 feet

28. $y = \dfrac{4\pi}{3}x^3$; This is the formula for the volume of a sphere, where x is the radius and y is the volume.

29. $y = 0.1x^5$ **30.** b **31.** b **32.** Possible answer: $f(x) = x^6$ **33.** Possible answer: $f(x) = -x^6$
34. Possible answer: $f(x) = -x^5$ **35.** Possible answer: $f(x) = x^5$ **36.** $f(x)$ rises to the left and falls to the right.
37. $g(x)$ falls to the left and to the right. **38.** $h(x)$ falls to the left and rises to the right.
39. $h(t)$ rises to the left and to the right.

40. a. 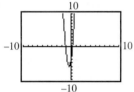 **b.** The graph does not fall to the left. **c.**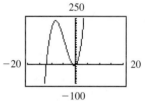

41. a. 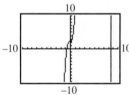 **b.** The graph does not fall to the right. **c.**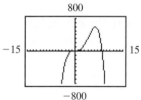

42. a. 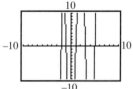 **b.** The graph does not show local minimums and maximums. **c.**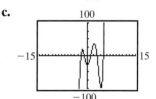

43. a. 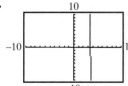 **b.** The graph does not rise to the left. **c.**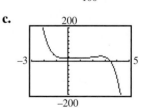

44. 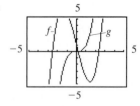 **45.** 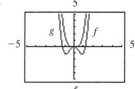 **46.**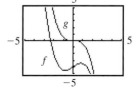

47.

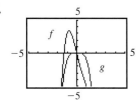

48.

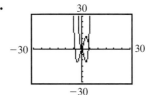

49.

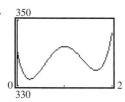

f and *g* are not the same.   Answers may vary.

50. The graph represents $f(x) = (x - 2)^4$ because it rises to the left and to the right, but the portion of the graph near $x = 2$ is too "flat" to represent a parabola.   **51.** A fifth-degree polynomial function will never rise on both ends. **52.** The maximum number of turns a polynomial can have is $n - 1$ where n equals the degree of the polynomial. Since the graph displays 4 turns, the degree of the polynomial is at least 5.

53. a. $N(t) = -0.25t^4 - 0.5t^3 + 7t^2 - 4t + 5$ **b.** The model's graph indicates that the patient will feel sickest on the 69th hour.   **54.** Group activity

---

PROBLEM SET 3.3

1. $x^2 - 3x + 2; x = -1, x = 2, x = 1$ **2.** $x^2 - 5x + 6; x = -1, x = 2, x = 3$ **3.** $\left\{-2, \frac{1}{2}, 3\right\}$ **4.** $\left\{-\frac{1}{2}, 1, 2\right\}$

5. $\left\{-4, -\frac{1}{3}, 2\right\}$ **6.** $\left\{-\frac{3}{2}, -\frac{1}{3}, \frac{1}{2}\right\}$ **7.** $\pm 1, \pm 2, \pm 3, \pm 6, \pm\frac{1}{3}, \pm\frac{2}{3}$ **8.** $\pm 1, \pm 3, \pm 5, \pm 15, \pm\frac{1}{2}, \pm\frac{3}{2}, \pm\frac{5}{2}, \pm\frac{15}{2}$

9. $\pm 1, \pm 2, \pm 3, \pm 6, \pm\frac{1}{2}, \pm\frac{1}{4}, \pm\frac{3}{2}, \pm\frac{3}{4}$ **10.** $\pm 1; \pm 2, \pm 4, \pm 8, \pm\frac{1}{3}, \pm\frac{2}{3}, \pm\frac{4}{3}, \pm\frac{8}{3}$

11. No positive real roots exist; 3 or 1 negative real roots exist.

12. 3 or 1 positive real roots exist; No negative real roots exist.

13. 3 or 1 positive real roots exist; No negative real roots exist.

14. 2 or 0 positive real roots exist; 2 or 0 negative real roots exist.

15. 2 or 0 positive real roots exist; No negative real roots exist.

16. 1 positive real root exists; No negative real roots exist.

17. a. $\pm 1, \pm 2, \pm 3, \pm 4, \pm 6, \pm 12, \pm\frac{1}{5}, \pm\frac{1}{2}, \pm\frac{2}{5}, \pm\frac{3}{2}, \pm\frac{3}{5}, \pm\frac{4}{5}, \pm\frac{6}{5}, \pm\frac{12}{5}$ **b.** 2 is a zero.

c. $2, \dfrac{-5 + \sqrt{265}}{20}, \dfrac{-5 - \sqrt{265}}{20}$ **18. a.** $\pm 1, \pm 3, \pm 5, \pm 15, \pm\frac{1}{3}, \pm\frac{5}{3}$ **b.** 3 is a zero.   **c.** $3, \dfrac{2 + \sqrt{19}}{3}, \dfrac{2 - \sqrt{19}}{3}$

19. a. $\pm 1, \pm 3, \pm\frac{1}{2}, \pm\frac{1}{4}, \pm\frac{3}{2}, \pm\frac{3}{4}$ **b.** Answers may vary.   **c.** $1, \frac{1}{2}, \frac{3}{2}$ **20. a.** $\pm 1, \pm 3, \pm 5, \pm 15, \pm\frac{1}{2}, \pm\frac{3}{2}, \pm\frac{5}{2}, \pm\frac{15}{2}$

b. Answers may vary.   **c.** $5, -3, -\dfrac{1}{2}$ **21–22.** Verification is left to the student. (Use the Upper and Lower Bound Theorem.)   **23. a.** $\pm 1, \pm 2, \pm 3, \pm 4, \pm 6, \pm 12$ **b–e.** 1 is not a root; -3 is not a root; Remaining anwers may vary.   **24. a.** $\pm 1, \pm 3, \pm 9, \pm\frac{1}{2}, \pm\frac{3}{2}, \pm\frac{9}{2}$ **b–e.** $\frac{3}{2}$ is a root; -3 is a root; Remaining answers may vary.

25. $\{1, 12\}$ **26.** $\{-3, 1\}$ **27.** $\left\{\frac{1}{6}, \frac{1}{2}, -3\right\}$ **28.** $\left\{-2, -\frac{2}{3}, \frac{3}{2}\right\}$ **29.** $\{-4, -2, -1, 1\}$ **30.** $\{-6, -1, 1, 3\}$

31. $\left\{-1, -\frac{2}{3}, 1\right\}$ **32.** $\left\{-4, \frac{2}{3}, -\sqrt{2}, \sqrt{2}\right\}$ **33.** $\left\{-2, -1, \frac{1}{5}, 1, 2\right\}$ **34.** $\left\{\frac{1}{2}, 2\right\}$

35. $\left\{-2, \frac{1}{3}, -\sqrt{3}, \sqrt{3}\right\}$ **36.** $\left\{-\frac{3}{2}, -1, 1, 4\right\}$ **37.** $\left\{-1, -\frac{1}{2}, \frac{3}{2}, 2, 4\right\}$ **38.** $\left\{-3, -\frac{\sqrt{3}}{3}, \frac{\sqrt{3}}{3}, 3\right\}$

39. $\left\{-4, \frac{1}{2}, 6\right\}$ **40.** $\left\{-1, -\frac{1}{3}, -\frac{2}{3}\right\}$ **41.** $x = 1$ inch or $x \approx 1.1$ inch   **42.** 4 ft $\times$ 4 ft $\times \dfrac{5}{8}$ ft or approximately

0.83 ft $\times$ 0.83 ft $\times$ 14.57 ft   **43.** $x = 3$ in.   **44.** $x = 65$ cars   **45.** 2 days   **46.** $n = 14$ **47.** d   **48.** b

49. $\pm 1, \pm 3, \pm 5, \pm 15, \pm\dfrac{1}{2}, \pm\dfrac{3}{2}, \pm\dfrac{5}{2}, \pm\dfrac{15}{2}$; solutions: $-\dfrac{1}{2}, 3, 5$

50. $\pm 1, \pm 2, \pm 4, \pm\dfrac{1}{2}, \pm\dfrac{1}{3}, \pm\dfrac{2}{3}, \pm\dfrac{4}{3}, \pm\dfrac{1}{6}$; solutions: $\dfrac{1}{2}, \dfrac{2}{3}, 2$

51. $\pm 1, \pm 2, \pm 3, \pm 6, \pm 9, \pm 18, \pm\dfrac{1}{2}, \pm\dfrac{3}{2}, \pm\dfrac{9}{2}$; solutions: $-3, -\dfrac{3}{2}, -1, 2$ **52.** $\pm 1, \pm 2, \pm\dfrac{1}{2}, \pm\dfrac{1}{4}$; solutions: $\pm\dfrac{1}{2}$

53. $f(x)$ has no sign variations, so $f(x)$ has no positive real roots. $f(-x)$ has no sign variations, so no negative roots exist. The polynomial graph doesn't intersect the x-axis.   **54.** $f(x)$ has 5 sign variations, so either 5, 3 or 1 positive real roots exist. $f(-x)$ has no sign variations, so no negative real roots exist.   **55.** Polynomials of odd degree must cross the x-axis (leading term test). Polynomials of even degree may sometimes have no real zeros.

56.

-1	1	-53	103	-51
		-1	54	-157
	1	-54	157	-208

60	1	-53	103	-51
		60	420	31,380
	1	7	523	31,329

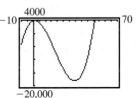

57. -3 is a lower bound; 3 is an upper bound.   **58.** -3 is a lower bound; 4 is an upper bound. Note, 5 is the lowest integer value which can be verified as an upper bound using synthetic division.
59. Answers may vary.   **60.** f and g share the same zeros; g is a "stretched" version of f.
61. $x^5 + x + 1 = 0$ has one real solution; Answers may vary.   **62–63.** Answers may vary.
64. a–b. Verification is left to the student.   **c.** p is a factor of a_0.
d. $a_{n-1}p^{n-1}q + a_{n-2}p^{n-2}q + \ldots + a_1pq^{n-1} + a_0q^n = -a_np^n$; Verification is left to the student.
65. By rational root theorem, the possible rational roots of the equation are ± 1 and ± 3. Since $\sqrt{3}$ is a root of the equation, it cannot be a rational number.   **66.** For $p = 3$: $\{-1 + i\sqrt{2}, -1 - i\sqrt{2}, 1\}$

PROBLEM SET 3.4

1. $f(1) = -1$, $f(2) = 5$; To the nearest tenth, the zero is 1.3.
2. $f(0) = 2$, $f(1) = -1$; To the nearest tenth, the zero is 0.8.
3. $f(-1) = -1$, $f(0) = 1$; To the nearest tenth, the zero is -0.5.
4. $f(2) = -8$, $f(3) = 81$; To the nearest tenth, the zero is 2.2.
5. $f(-3) = -11$, $f(-2) = 1$; To the nearest tenth, the zero is -2.1.
6. $f(1) = -1$, $f(2) = 23$; To the nearest tenth, the zero is 1.2.
7. $f(-3) = -42$, $f(-2) = 5$; To the nearest tenth, the zero is -2.2.
8. $f(2) = -4$, $f(3) = 14$; To the nearest tenth, the zero is 2.4.
9. $f(x) = x^3 - 3x^2 - 15x + 125$ **10.** $f(x) = 16x^3 - 20x^2 - 4x + 15$ **11.** $f(x) = x^4 + 10x^2 + 9$

12. $f(x) = 2x^4 + 5x^3 + 4x^2 + 5x + 2$ **13.** $\{-2i, 2i, 2\}$ **14.** $\{-3i, 3i, -2i, 2i\}$ **15.** $\left\{1 + i, 1 - i, \dfrac{1}{3}\right\}$

16. $\{3 - i, 3 + i, 1\}$ **17.** $\{2 - i, 2 + i, -2 + i, -2 - i\}$ **18.** $\{1 - 2i, 1 + 2i, -4, 3\}$

19. $\left\{1 - i, 1 + i, -3, \dfrac{1}{3}\right\}$ **20.** $\left\{3 - 4i, 3 + 4i, \dfrac{1}{2} + i, \dfrac{1}{2} - i\right\}$ **21.** $\{3 - i\sqrt{3}, 3 + i\sqrt{3}, 2, 4\}$

22. $\left\{-1 - i\sqrt{3}, -1 + i\sqrt{3}, -\dfrac{4}{3}, \dfrac{4}{3}\right\}$ **23.** $\{2 - 3i, 2 + 3i, -i, i, -2\}$ **24.** $\{-1 + 2i, -1 - 2i, 1, -2i, 2i\}$

25. $f(x) = (x - 1)(x - 5i)(x + 5i)$ **26.** $g(x) = (x - 2)(x - 4 - i)(x - 4 + i)$
27. $p(x) = (x - 6i)(x + 6i)(x - i)(x + i)$ **28.** $p(x) = (x + 5)(x + 5)(x - 1 - i\sqrt{3})(x - 1 + i\sqrt{3})$
29. $p(x) = (x + 2)(4x - 3)(2x + 1 - 2i)(2x + 1 + 2i)$ **30.** $p(x) = (x - 1)(2x + 1)(x - 2i)(x + 2i)$

31. $g(r) = (r + 2)(3r - 1)(r + 1 + i\sqrt{3})(r + 1 - i\sqrt{3})(r - 1 + i\sqrt{3})(r - 1 - i\sqrt{3})$

32. $g(x) = (x + 1)(x - \sqrt{7})(x - \sqrt{7})(x + \sqrt{7})(x + \sqrt{7})$ **33.** $x \approx 1.7$ **34.** $x \approx 0.8, x \approx -1.5$

35. a. 10.3%   **b.** 1980 and 1990   **36.** Side of square base = 23 dekameters; Pyramid's height = 15 dekameters

37. 9 cm × 9 cm × 9 cm   **38.** Approximately 3.9 in. × 3.9 in. × 3.9 in.   **39.** $f(x) = -\dfrac{1}{3}x^3 + \dfrac{20}{3}x^2 - \dfrac{121}{3}x + 70$

40. c   **41.** b   **42. a.** The graph suggests that people visit a physician more often as they age.   **b.** 60 years of age
c. Use Trace feature.   **43.** Relative minimum ≈ −6.9 feet; Relative maximum ≈ 4.2 feet
44. 1 real zero; 2 complex zeros   **45.** 3 real zeros (counting multiplicity); 2 complex zeros   **46.** 3 real zeros;
2 complex zeros   **47.** 2 real zeros; 2 complex zeros   **48.** 2 real zeros; 4 complex zeros
49. 1 real zero; 4 complex zeros (counting multiplicity)   **50.** 4 real zeros (counting multiplicity)
51. 3 real zeros; 2 complex zeros   **52.** 3   **53.** 3   **54.** 5   **55.** 6   **56.** Answers may vary.
57. Because of the leading term test which states the polynomial rises to the left and the right, if the polynomial crosses the x-axis once, it must cross it again. A polynomial of degree 20 may be tangent to the x-axis in one place however.   **58.** One such polynomial is $f(x) = x^5 - 3x^4 - 6x^3 - 18^2 + 25x - 75$; Verification is left to the student.   **59.** Yes, a third-degree polynomial has at most 2 turns. A relative maximum occurs at one of these turns.   **60.** Answers may vary.   **61. a–d.** Answers may vary.   **e.** No, the polynomial falls to the left and rises to the right. It must cross the x-axis at least once.   **62.** $[-1, 0] \cup [1, +\infty)$ **63.** $(-\infty, -1] \cup [1, 4]$
64. $(-\infty, -2.1) \cup (0.3, 1.9)$; Note that the zeros of $x^3 - 4x + 1$ are irrational. Answers are approximated to the nearest tenth.   **65. a–b.** Verification is left to the student.   **c.** The conjugate roots theorem only refers to polynomials having real coefficients.   **66.** $f(x) = x^4 + 2$ **67.** Group activity

PROBLEM SET 3.5

1.

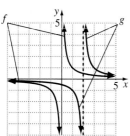

Vertical asymptote: $x = 2$
Horizontal asymptote: $y = 0$

2.

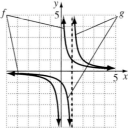

Vertical asymptote: $x = 1$
Horizontal asymptote: $y = 0$

3.

Vertical asymptote: $x = -2$
Horizontal asymptote: $y = 0$

4.

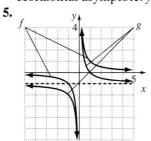

Vertical asymptote: $x = -1$
Horizontal asymptote: $y = 0$

5.

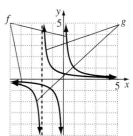

Vertical asymptote: $x = 0$
Horizontal asymptote: $y = -1$

6.

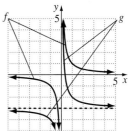

Vertical asymptotes: $x = 0$
Horizontal asymptote: $y = -3$

7.

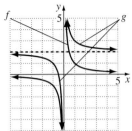

Vertical asymptote: $x = 0$
Horizontal asymptote: $y = 2$

8.

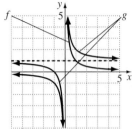

Vertical asymptote: $x = 0$
Horizontal asymptote: $y = 1$

9.

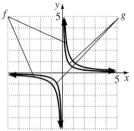

Vertical asymptote: $x = 0$
Horizontal asymptote: $y = 0$

10.

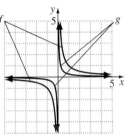

Vertical asymptote: $x = 0$
Horizontal asymptote: $y = 0$

11.

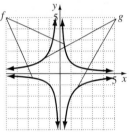

Vertical asymptote: $x = 0$
Horizontal asymptote: $y = 0$

12.

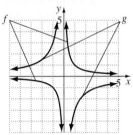

Vertical asymptote: $x = 0$
Horizontal asymptote: $y = 0$

13.

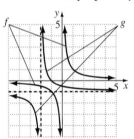

Vertical asymptote: $x = -2$
Horizontal asymptote: $y = -1$

14.

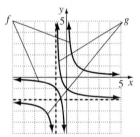

Vertical asymptote: $x = -1$
Horizontal asymptote: $y = -2$

15.

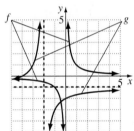

Vertical asymptote: $x = -2$
Horizontal asymptote: $y = -1$

16.

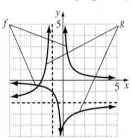

Vertical asymptote: $x = -1$
Horizontal asymptote: $y = -2$

17.

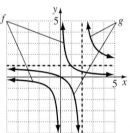

Vertical asymptote: $x = 2$
Horizontal asymptote: $y = 1$

18.

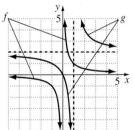

Vertical asymptote: $x = 1$
Horizontal asymptote: $y = 2$

19.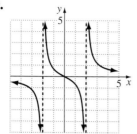

20.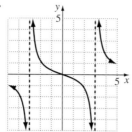

21.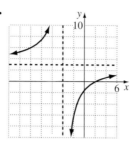

22.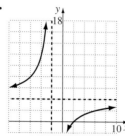

23.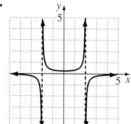

24.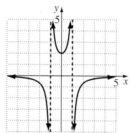

25.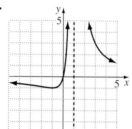

26.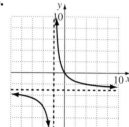

27.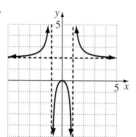

28.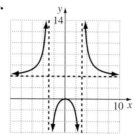

29.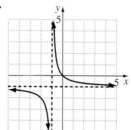

30.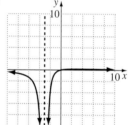

31.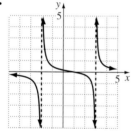

32.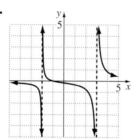

33.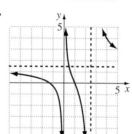

34.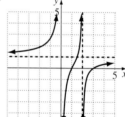

35.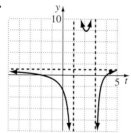

36.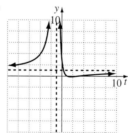

37.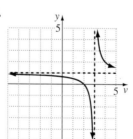

38.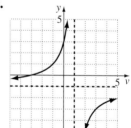

39.

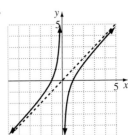

40.

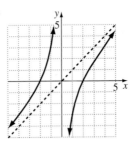

41.

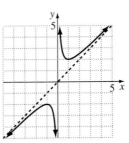

42.

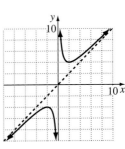

43.

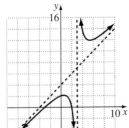

44.

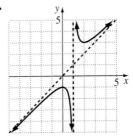

45.

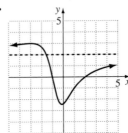

46.

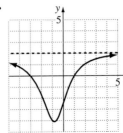

47.

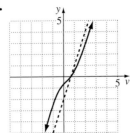

48.

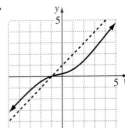

49.

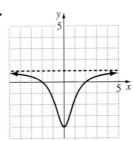

50.

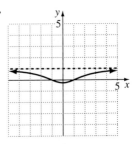

51. d   **52.** c   **53.** Anwers may vary.   **54.** Answers may vary.

55. a.

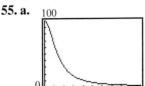

b. The higher the level of education, the lower the unemployment.

c. $f^{-1}(3.6) \approx 13.97$; The unemployment rate was 3.6% for people with about 14 years of education.

d. Because the x-axis is an asymptote, zero unemployment cannot be fully achieved in spite of higher education levels.

56. No   **57.** No; the graphing utility indicates the domain is $(-\infty, \infty)$.

58. a. Neither   **b.** The quotient is $x^2 + 2x + 3$ and indicates the graph seen in part (a).

59. The graph of f has a "hole" at $x = a$ if $p(a) = 0$, provided that the factor $(x - a)$ occurs at least as many times in the factored form of $p(x)$ as it occurs in the factored form of $q(x)$. The graph of f has a vertical asymptote whose equation is $x = a$ if $p(a) \neq 0$ or if the factor $(x - a)$ occurs more often in the factored form of $q(x)$ than in the factored form of $p(x)$.

60.

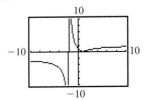

The graph has two horizontal asymptotes, $y = -2$ and $y = 2$.

61.

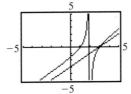

Answers may vary.

62. a. 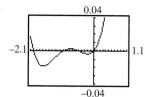   **b.** 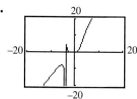   **c.** Answers may vary.

63. a. 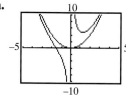   **b.** f will have a parabolic asymptote if the degree of p is two greater than the degree of q.

c. Answers may vary.

64. 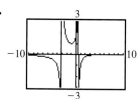   **65.** The graph of g will approach the x-axis more rapidly as values of x increase.

66. Answers may vary.   **67.** No; Yes; Verification is left to the student.

68. Domain: $\left(-\infty, -\dfrac{d}{c}\right) \cup \left(-\dfrac{d}{c}, \infty\right)$; range: $\left(-\infty, \dfrac{a}{c}\right) \cup \left(\dfrac{a}{c}, \infty\right)$, provided $ad \neq bc$

69. $f(x) = \dfrac{3}{x-3}$ **70.** $f(x) = \dfrac{2x^2 - 18}{x^2 - 4}$ **71.** $f(x) = \dfrac{x^2 - x - 2}{x - 1}$

72. Possible answer: $f(x) = \dfrac{x^2 + 1}{2}$

PROBLEM SET 3.6

1. a. \$220; \$40; \$22; \$20.2   **b.** $y = 20$; As more canoes are manufactured, the average cost approaches \$20.

2. a. $\overline{C}(x) = \dfrac{100x + 100,000}{x}$ **b.** \$300; \$200; \$150; \$125; The average cost decreases as the number of bicycles manufactured increases.   **c.** $y = 100$; As greater numbers of bicycles are manufactured, the average cost approaches \$100 per bicycle.   **3. a.** \$100,000   **b.** No; the model indicates that no amount of money can remove 100% of the pollutants since $C(p)$ increases without bound as p approaches 100.

4. a. $433\dfrac{1}{3}$; the difference in cost to inoculate 40% and 80% of the population is $\$433\dfrac{1}{3}$ million.

b. 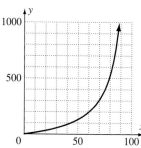   **c.** No amount of money can inoculate 100% of the population against the flu.

5. a. $F(0) = 80$; When the dessert was placed in the icebox, its temperature was 80°F.

b. Approximately 13.3°F; 6.2°F; 3.6°F; 2.4°F; 1.7°F   **c.** $y = 0$; the temperature will approach but not reach 0°F.

d.

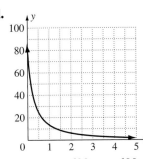

6. 0.5 ohms   **7.** Approximately 8885.4 newtons   **8.** Approximately 159.6 lbs

9. $y = \dfrac{400}{x^2}$; The intensity of the illumination on a surface varies inversely with the square of the distance of the light source from the surface.

10. $y = \dfrac{1600}{x^2}$ **11.** $y = \dfrac{216{,}000}{x^3}$ **12.** $y = \dfrac{200}{x^3}$

13. a. $T(x) = \dfrac{600}{x} + \dfrac{600}{x - 10}$ **b.**

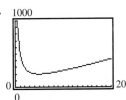

c. $T^{-1}(22) = 60$; If the total time was 22 hours, then the average rate on the outgoing trip was 60 mph.
d. Answers may vary.

14. a. $P(x) = \dfrac{2x^2 + 5000}{x}$ **b.**

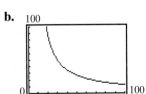

50 feet by 50 feet

15. a.

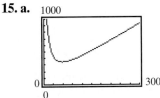

b. 58 cases; Approximately $346   **16.** Answers may vary.
17. Answers may vary.   **18.** Use one-third resistance.
19. Verification is left to the student.   **20.** Group activity

CHAPTER 3 REVIEW PROBLEMS

1.

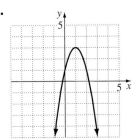

2.

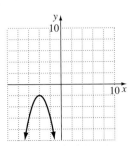

3.

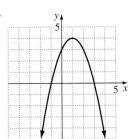

4.

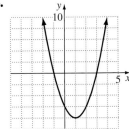

Axis of symmetry: $x = 1$ Axis of symmetry: $x = -4$ Axis of symmetry: $x = 1$ Axis of symmetry: $x = 1$

5. $54,607   **6.** (17.84, 12.413008); between 1957 and 1958 fuel efficiency of passenger cars was at an all time low of approximately 12.4 miles per gallon.

7. a. $s(t) = -16t^2 + 64t + 80$ **b.** 2 seconds; 144 feet   **c.**

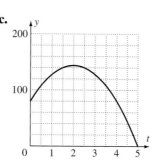

8. 250 ft by 500 ft; 125,000 ft^2 **9.** 40 divers; $360   **10.** $\left(\dfrac{4}{5}, \dfrac{2}{5}\right)$

11. a. Falls to the left and rises to the right   **c.**   **d.**
  b. $x = -1, 1, 5$

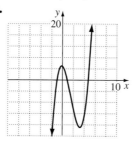

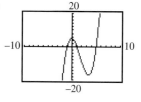

12. a. Falls to the left and to the right   **b.** $-5, 0, 5$ **c.**   **d.**

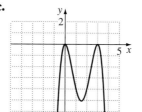

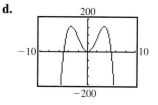

13. a. Falls to the left and to the right   **b.** $0, 3$ **c.**   **d.**

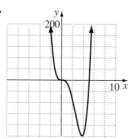

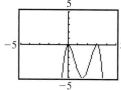

14. a. Rises to the left and to the right   **b.** $0, 5$ **c.**   **d.**

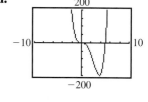

15. Because the degree is odd and the leading coefficient is negative, the graph falls to the right. Therefore, the model indicates that the percentage of families below the poverty level will eventually be negative, which is impossible.

16. c   **17.** b   **18.** a   **19.** d   **20.** $f(x) = 30{,}000x^4$ **21.** 2 or 0 positive solutions; no negative solutions

22. 3 or 1 positive solutions; 2 or 0 negative solutions   **23.** $f(x) = f(-x) = 2x^4 + 2x^2 + 8$ has no sign changes.

24. a. $\pm 1, \pm 3, \pm 5, \pm 15, \pm\dfrac{1}{2}, \pm\dfrac{1}{4}, \pm\dfrac{1}{8}, \pm\dfrac{3}{2}, \pm\dfrac{3}{4}, \pm\dfrac{3}{8}, \pm\dfrac{5}{2}, \pm\dfrac{5}{4}, \pm\dfrac{5}{8}, \pm\dfrac{15}{2}, \pm\dfrac{15}{4}, \pm\dfrac{15}{8}$

b. 3 or 1 positive real solutions; no negative real solutions   **d.** $\left\{\dfrac{1}{2}, \dfrac{3}{2}, \dfrac{5}{2}\right\}$ **25. a.** $\pm 1, \pm 2, \pm 3, \pm 6$

b. 2 or 0 positive real solutions; 2 or 0 negative real solutions   **d.** $\{-2, -1, 1, 3\}$

26. a. $\pm 1, \pm 2 \pm 5 \pm 10, \pm\dfrac{1}{2}, \pm\dfrac{5}{2}$ **b.** 3 or 1 positive real solutions; 1 negative real solution   **d.** $\left\{\dfrac{1}{2}, 2, \pm\sqrt{5}\right\}$

27. a. $\pm 1, \pm 2, \pm 3, \pm 4, \pm 5, \pm 6, \pm 10, \pm 12, \pm 15, \pm 20, \pm 30, \pm 60, \pm\dfrac{1}{2}, \pm\dfrac{1}{4}, \pm\dfrac{3}{2}, \pm\dfrac{3}{4}, \pm\dfrac{5}{2}, \pm\dfrac{5}{4}, \pm\dfrac{15}{2}, \pm\dfrac{15}{4}$

b. 2 or 0 positive real solutions; 2 or 0 negative real solutions   **d.** $\left\{-\dfrac{5}{2}, -1, \dfrac{3}{2}, 4\right\}$

28. a. $\pm 1, \pm 2, \pm 4, \pm 8, \pm\dfrac{1}{3}, \pm\dfrac{2}{3}, \pm\dfrac{4}{3}, \pm\dfrac{8}{3}$ **b.** 3 or 1 positive real solutions; 2 or 0 negative real solutions

d. $\left\{-2, -1, \dfrac{2}{3}, 1, 2\right\}$ **29. a.** $\pm 1, \pm\dfrac{1}{2}, \pm\dfrac{1}{3}, \pm\dfrac{1}{6}$ **b.** 2 or 0 positive real zeros; 1 negative real zero

d. $-1, \dfrac{1}{3}, \dfrac{1}{2}$ **30.** $\pm 1, \pm 3, \pm\dfrac{1}{2}, \pm\dfrac{3}{2}$ **b.** 1 positive real zero; 3 or 1 negative real zeros

d. $-1, \dfrac{3}{2}, \dfrac{-1 + i\sqrt{3}}{2}, \dfrac{-1 - i\sqrt{3}}{2}$ **31.** $\pm 1, \pm 2, \pm 3, \pm 4, \pm 6, \pm 8, \pm 12, \pm 24, \pm\dfrac{1}{3}, \pm\dfrac{2}{3}, \pm\dfrac{4}{3}, \pm\dfrac{8}{3}$

b. 3 or 1 positive real zeros; 1 negative real zero   **d.** $-\dfrac{2}{3}, 3, \pm 2i$ **32. a.** $\pm 1, \pm 2, \pm 3, \pm 4, \pm 6, \pm 12, \pm\dfrac{1}{2}, \pm\dfrac{3}{2}$

b. 2 is not a root but is an upper bound.   **c.** -2 is not a root but is a lower bound.   **d.** $\pm 1, \pm\dfrac{1}{2}, \pm\dfrac{3}{2}$

33. $\pm 1, 2, 3, 4, \pm\dfrac{1}{2}, \pm\dfrac{3}{2}$ **34.** 7   **35.** 3 mm   **36.** 4   **37.** $f(1) = -2, f(2) = 3$; 1.6

38. $f(-3) = -32, f(-2) = 7$; -2.3 **39.** $f(x) = x^3 - 6x^2 + 21x - 26$ **40.** $g(x) = 2x^4 + 12x^3 + 20x^2 + 12x + 18$

41. $f(x) = x^4 - 3x^3 + 6x^2 + 2x - 60$ **42.** $\left\{-\dfrac{1}{4}, 6 \pm 5i\right\}$ **43.** $\{1 \pm i, 1 \pm 3i\}$ **44.** $\left\{-\dfrac{1}{2}, 1, 4 \pm 7i\right\}$

45. $\{-3, -1, 1 \pm 2i\}$ **46.** $-2, \dfrac{1}{2}, \pm i$; $f(x) = (x - i)(x + i)(x + 2)\left(x - \dfrac{1}{2}\right)$ **47.** $4, -1$; $g(x) = (x + 1)^2(x - 4)^2$

48. 4 real zeros, one with multiplicity two   **49.** 3 real zeros; 2 nonreal complex zeros

50. 2 real zeros, one with multiplicity two; 2 nonreal complex zeros   **51.** 1 real zero; 4 nonreal complex zeros

52. Verification is left to the student; 0.47, -1.40 **53.** Original length of cube = 8 inches.

54.

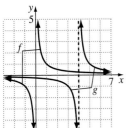

55.

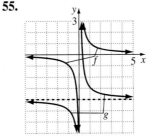

56.

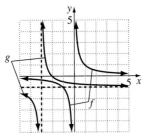

Horizontal asymptote: $y = 0$
Vertical asymptote: $x = 4$

Horizontal asymptote: $y = -4$
Vertical asymptote: $x = 0$

Horizontal asymptote: $y = -1$
Vertical asymptote: $x = -3$

57.

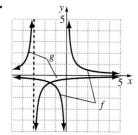

Horizontal asymptote: $y = 0$
Vertical asymptote: $x = -3$

58.

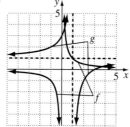

Horizontal asymptote: $y = 1$
Vertical asymptote: $x = 1$

59.

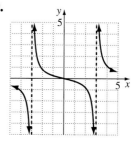

60.

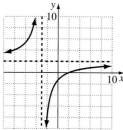

61.

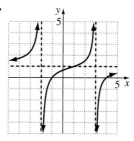

62.

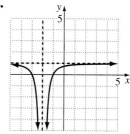

63.

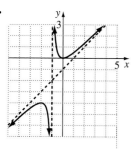

64.

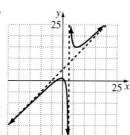

65.

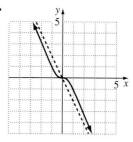

66.

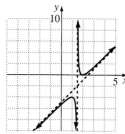

67. $y = 60$; The maximum number of words a person can type per minute approaches 60.

68. $y = 3000$; The number of fish available in the pond approaches 3,000,000.

69. $y = 0$; As the number of years of education increases the percentage rate of unemployment approaches zero.

70. a. $\overline{C}(50) = \$1620$; $\overline{C}(100) = \$820$; $\overline{C}(1000) = \$100$; $\overline{C}(100{,}000) = \$20.80$

b. $y = 20$; as more calculators are manufactured, the cost per calculator approaches $20.

c.

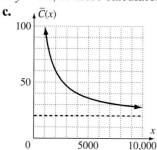

71. 1600; the difference in cost of removing 90% versus 50% of the contaminants is 16 million dollars.

b.

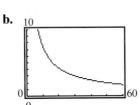

c. No amount of money can remove 100% of the contaminants since $C(x)$ increases without bound as x approaches 100.

72. Answers may vary.   **73.** Approximately 1.8×10^{20} newtons

74. $y = \dfrac{1200}{x^2}$; 1.3 lumens; Intensity decreases to zero.

75. a. $T(x) = \dfrac{60}{x} + \dfrac{60}{x + 30}$ **b.**

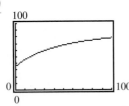

c. 30 mph

76. a. $C(x) = \dfrac{100x + 1750}{x + 50}$

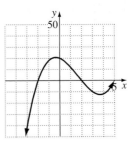

b. Answers may vary.   **c.** 80 ounces

CHAPTER 3 TEST

1. Axis of symmetry: $x = 1$

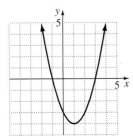

2. Axis of symmetry: $x = 1$

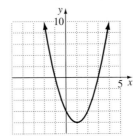

3. 23 VCRs; $16,900   **4.** 150 ft by 300; 45,000 sq ft   **5. a.** $5, 2, -2$ **b.**

6. The graph should fall to the left and rises to the right; x-intercepts should be -1 and 1.   **7. a.** 2   **b.** $\dfrac{2}{3}, \dfrac{1}{2}$

8. $\pm 1, \pm 2, \pm 3, \pm 6, \pm\dfrac{1}{2}, \pm\dfrac{3}{2}$ **9.** 3 or 1 positive real solutions; no negative real solutions   **10.** $\{-5, -3, 2\}$

11. a. $\pm 1, \pm 3, \pm 5, \pm 15, \pm\dfrac{1}{2}, \pm\dfrac{3}{2}, \pm\dfrac{5}{2}, \pm\dfrac{15}{2}$ **b.** $\left\{-1, \dfrac{3}{2}, \pm\sqrt{5}\right\}$ **12.** -3 is a lower bound; 2 is an upper bound; Verification is left to the student.   **13.** $\{2, 3, 1 \pm i\}$ **14.** $(x + 2)^2(x - 1)$

15. 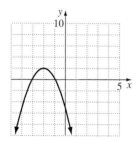   **16.** 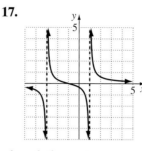   **17.** 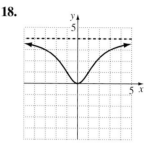   **18.**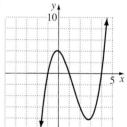

19. a. 750 manuals   **b.** $y = 6$; As more manuals are produced, the cost per manual approaches \$6.
20. $H = 81{,}000$ **21. a.** $-0.92, -0.38, 0.38, 0.92$ **b.** Minimum: $(0.71, -1)$ and $(-0.71, -1)$, maximum: $(0, 1)$

CUMULATIVE REVIEW PROBLEMS (CHAPTERS P–3)

1. $2 + \sqrt{3}$ **2.** $26\sqrt[3]{2}$ **3.** $(x - y)(x + y)^2$ **4.** $\{-2, 2, -i\sqrt{2}, i\sqrt{2}\}$ **5.** $\{-4, -2, -1, -3\}$

6. $(-\infty, -1)\cup\left(\dfrac{5}{3}, \infty\right)$ **7.** $\{-3, -1, 2\}$ **8.** $t = \dfrac{15C - 15V}{c}$ **9.** $(-\infty, 5]$ **10.** $g(x) = \sqrt{x - 1}$

11. $f^{-1}(x) = \sqrt[3]{x - 4}$ **12.** $(f \circ g)(x) = 49x^2 - 91x + 42; (g \circ f)(x) = 7x^2 - 35x + 38$

13. Possible answer: **14.**   **15. a.** $-1, 1, 4$ **b.**

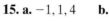

16.   **17.**   **18.** 3 ft   **19.** 5 trees; 12,250 apples
20. 11.1 foot-candles

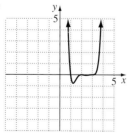

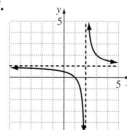

Chapter 4

PROBLEM SET 4.1

1.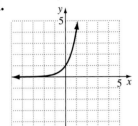

2.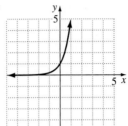

3.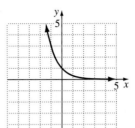

4.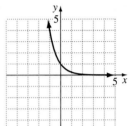

5.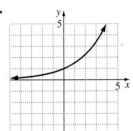

6.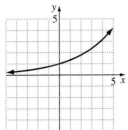

7.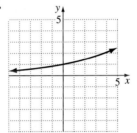

8.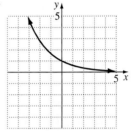

9. Domain: $(-\infty, \infty)$;
Range: $(0, \infty)$

10. Domain: $(-\infty, \infty)$;
Range: $(0, \infty)$

11. Domain: $(-\infty, \infty)$;
Range $(3, \infty)$

12. Domain: $(-\infty, \infty)$;
Range: $(-1, \infty)$

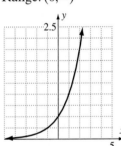

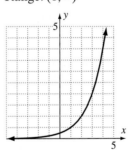

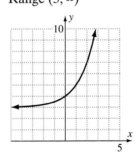

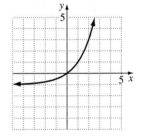

13. Domain: $(-\infty, \infty)$;
Range: $(-1, \infty)$

14. Domain: $(-\infty, \infty)$;
Range: $(-1, \infty)$

15. Domain: $(-\infty, \infty)$;
Range: $(-2, \infty)$

16. Domain: $(-\infty, \infty)$;
Range: $(-2, \infty)$

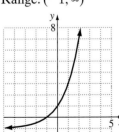

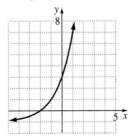

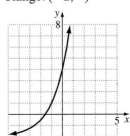

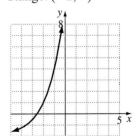

17. Domain: $(-\infty, \infty)$;
Range: $(-\infty, 0)$

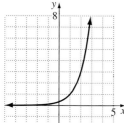

18. Domain: $(-\infty, \infty)$;
Range: $(-\infty, 0)$

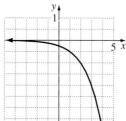

19. Domain: $(-\infty, \infty)$;
Range: $(3, \infty)$

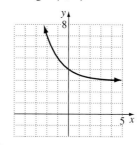

20. Domain: $(-\infty, \infty)$;
Range: $(4, \infty)$

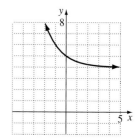

21.

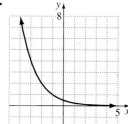

22.

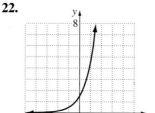

23.

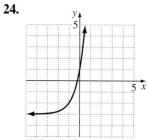

24.

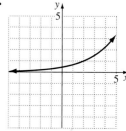

25.

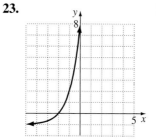

26.

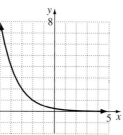

27.

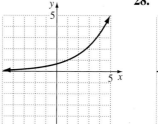

28.

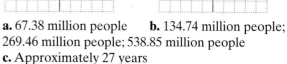

29.

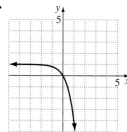

30.

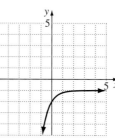

31. a. 67.38 million people   **b.** 134.74 million people;
269.46 million people; 538.85 million people
c. Approximately 27 years

32.

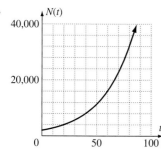

33. a. $3,052,428,614
   b. $3,169,289,064
   c. $3,173,350,575

34. 7% compounded monthly
35. 8.25% compounded quarterly

36. Compounded semiannually: $5279, $6558, $8602, $14,799;
Compounded monthly: $5282, $6579, $8655, $14,983;
Compounded 365 times per year: $5283, $6583, $8666, $15,020;
Compounded continuously: $5286, $6583, $8666, $15,021

37. Compounded semiannually: $31,982, $41,307, $56,875, $107,826;
Compounded monthly: $32,009, $41,485, $57,366, $109,693;
Compounded 365 times per year: $32,015, $41,520, $57,463, $110,066;
Compounded continuously: $32,015, $41,521, $57,466, $110,079

38. a. 241,786 cases   **b.** 16,123,834 cases   **39.** $A = 9.98350115$ billion

40. a. 50.00 mg. after 140 days; 35.36 mg after 210 days; 25.00 mg after 280 days; 12.50 mg after 420 days; 4.42 mg after 630 days

b.

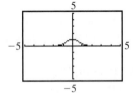

41. No, they cannot safely visit.   **42. a.** 0.4
b. about 0.70
c. 0.8 is approached.

43. a. about 0.05   **b.** Horizontal asymptote: $y = 100$. This means that the percent of pilots who suffer from bends will approach 100% as the altitude increases.

44. a. 68.5%   **b.** 30.8%   **c.** $y = 20$ is the horizontal asymptote, so 20% is ultimately retained.
d. 100% is retained immediately after learning.   **e.**

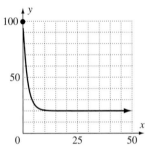

45. $S = \$143,605.71$ **46.** d   **47.** d   **48.**

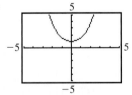

Increasing: $(-\infty, -2) \cup (0, \infty)$
Decreasing: $(-2, 0)$
Relative max: about 0.54
Relative min: 0

49.

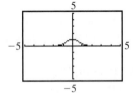

Increasing: $(-\infty, 0)$
Decreasing: $(0, \infty)$
Relative max: 1

50.

Decreasing: $(-\infty, 0)$
Increasing: $(0, \infty)$
Relative min: 1

51.

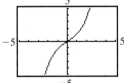

Increasing: $(-\infty, \infty)$
No relative max. or min.

52.

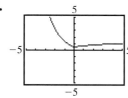

Decreasing: $(-\infty, 0)$
Increasing: $(0, \infty)$
Relative min: 0.5

53. a.

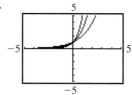

b. $x < 0$ **c.** $x > 0$ **d.** Yes, $x = 0$
e. When graphing several functions of the form $y = b^x$ where $b > 1$, the graphs will be ordered with the largest b on the bottom in the interval $(-\infty, 0)$ and the largest b on top in the interval $(0, \infty)$. At $x = 0$ the graphs will intersect.

f.

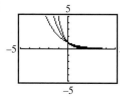

$x > 0, x < 0, x = 0$. When graphing several functions of the form $y = b^x$ where $0 < b < 1$, the graphs will be ordered with the largest b on the bottom in the interval $(-\infty, 0)$ and the largest b on the top in the interval $(0, \infty)$. At $x = 0$ graphs will intersect.

54. a. 4 O-rings   **b.**

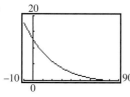

Answers may vary.

55. a. $f(-x) = f(x)$ **b.** $y = 0$; When $|x|$ gets large, the likely occurrence approaches zero.

c.

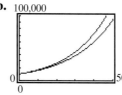

56. If $|a| > 1$, the $|a|$ shrinks the graph. If $a < 0$, the graph is reflected in the x-axis.

57. a. $A_1 = 10,000\left(1 + \dfrac{0.05}{4}\right)^{4t}$, $A_2 = 10,000\left(1 + \dfrac{0.045}{12}\right)^{12t}$ **b.**

The bank paying 5% interest compounded quarterly offers the better return.

58. a. 20 people   **b.** 1080 people   **c.** Horizontal asymptote: $y = 100,000$. The maximum number of people with the flu will approach 100,000.   **d.**

e. The epidemic growth increases over time until it begins to level off at 100,000 people.

59.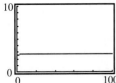

y approaches e as x
approaches infinity.

60. a.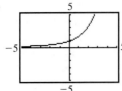

b. y approaches 1.

61.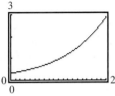

After ten days the tumor is nearly 1 cubic centimeter.
After twenty days the tumor is nearly 3 cubic centimeters.

62. a.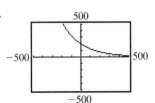

b. On the 280th day, the mass will be 50 milligrams.

63. a.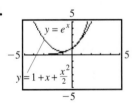

b.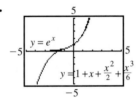

c.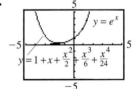

d. For values of $x < 0$, the second function alternates between being an increasing and decreasing function.
As the degree of the polynomial increases, the two functions appear to approach the same values for $x > 0$.

64. $y = e^x$ has been shifted upward 2 units and to the right 2 units.

65. A means for isolating the variable y must be used.

66. Use the graph to approximate the value of y when $x = \dfrac{1}{2}$.

67. $y = 3^x$ is graph d, $y = 5^x$ is graph c, $y = \left(\dfrac{1}{3}\right)^x$ is graph a, $y = \left(\dfrac{1}{5}\right)^x$ is graph b, Use the facts that when $b > 1$,

$f(x)$ is increasing and when $0 < b < 1$, $f(x)$ is decreasing.   **68.** $h(x)$ is graph a, $g(x)$ is graph b and $f(x)$ is graph c

69. A problem involving the decay of a substance would be appropriate.

70. A problem involving the growth of the incidence of a disease would be appropriate.

71. A problem involving heights of a population would be appropriate.

72. $\dfrac{1}{b^x + 1} + \dfrac{1}{b^{-x} + 1}$ simplifies to 1.   **73.** $\dfrac{3^{x+h} - 3^x}{h}$ simplifies to $3^x\left(\dfrac{3^h - 1}{h}\right)$.

74. a has a half life of 1 year; b has a half life of 1 year; c has a half life of 2 years.   **75.** Approximately $2\dfrac{1}{2}$ years

76. a. $(\cosh x)^2 - (\sinh x)^2 = \left(\dfrac{e^x + e^{-x}}{2}\right)^2 - \left(\dfrac{e^x - e^{-x}}{2}\right)^2 = \dfrac{e^{2x} + 2 + e^{-2x}}{4} - \dfrac{e^{2x} - 2 + e^{-2x}}{4}$

$= \dfrac{e^{2x} + 2 + e^{-2x} - e^{2x} + 2 - e^{-2x}}{4} = \dfrac{4}{4} = 1$ **b.** $\cosh(-x) = \dfrac{e^{(-x)} + e^{-(-x)}}{2} = \dfrac{e^{-x} + e^x}{2} = \dfrac{e^x + e^{-x}}{2} = \cosh(x)$

c. $\sinh(-x) = \dfrac{e^{-x} - e^{-(-x)}}{2} = \dfrac{e^{-x} - e^x}{2} = \dfrac{-(-e^{-x} + e^x)}{2} = \dfrac{-(e^x - e^{-x})}{2} = -\sinh x$

d. $(\cosh x)^2 = \left(\dfrac{e^x + e^{-x}}{2}\right)^2 = \dfrac{(e^x + e^{-x})(e^x + e^{-x})}{4} = \dfrac{e^{2x} + 1 + 1 + e^{-2x}}{4} = \dfrac{2 + e^{2x} + e^{-2x}}{4}$

$= \dfrac{1 + \frac{e^{2x} + e^{-2x}}{2}}{2} = \dfrac{1 + \cosh 2x}{2}$

e. $\sinh(x + y) = \dfrac{e^{x+y} - e^{-(x+y)}}{2} = \dfrac{2e^{x+y} - 2e^{-(x+y)}}{4}$

$= \dfrac{e^{x+y} + e^{x-y} - e^{-x+y} - e^{-(x+y)}}{4} + \dfrac{e^{x+y} + e^{-x+y} - e^{x-y} - e^{-(x+y)}}{4}$

$= \left(\dfrac{e^x - e^{-x}}{2}\right)\left(\dfrac{e^y + e^{-y}}{2}\right) + \left(\dfrac{e^x + e^{-x}}{2}\right)\left(\dfrac{e^y - e^{-y}}{2}\right) = \sinh x \cosh y + \cosh x \sinh y$

P R O B L E M   S E T   4 . 2

1. 2 **2.** 2 **3.** 6 **4.** 3 **5.** $\dfrac{1}{2}$ **6.** $\dfrac{1}{2}$ **7.** -5 **8.** -4 **9.** $\dfrac{1}{2}$ **10.** $\dfrac{1}{2}$ **11.** 1 **12.** 1 **13.** 0

14. 0 **15.** $\dfrac{3}{4}$ **16.** $\dfrac{1}{3}$ **17.** 4 **18.** 5 **19.** -2 **20.** -3 **21.** 7 **22.** 13 **23.** $\dfrac{1}{3}$ **24.** $\dfrac{1}{5}$ **25.** 7

26. 6 **27.** 19 **28.** 34 **29.** 7 **30.** -2 **31.** 0 **32.** 1

33.

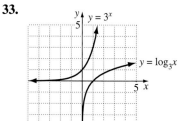

34.

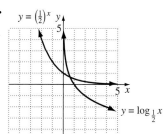

35. Domain: $(0, \infty)$; Range: $(-\infty, \infty)$

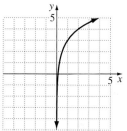

36. Domain: $(0, \infty)$; Range: $(-\infty, \infty)$

37. Domain: $(-3, \infty)$; Range: $(-\infty, \infty)$

38. Domain: $(1, \infty)$; Range: $(-\infty, \infty)$

39. Domain: $(1, \infty)$; Range: $(-\infty, \infty)$

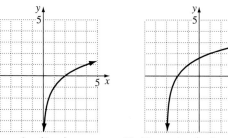

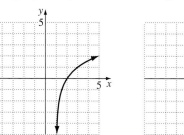

40. Domain: $(2, \infty)$; Range: $(-\infty, \infty)$

41. Domain: $(-1, \infty)$; Range: $(-\infty, \infty)$

42. Domain: $(-2, \infty)$; Range: $(-\infty, \infty)$

43. Domain: $(-\infty, 0)$; Range: $(-\infty, \infty)$

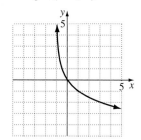

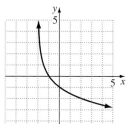

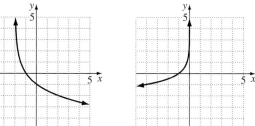

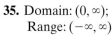

44. Domain: $(-\infty, 0)$;   **45.** d   **46.** b   **47.** a   **48.** c   **49.** b   **50.** d   **51.** c   **52.** a   **53.** -1 **54.** 1   **55.** $\dfrac{1}{2}$
Range: $(-\infty, \infty)$ **56.** 1   **57.** -3 **58.** 1

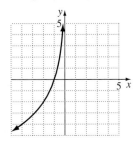

59. Domain: $(-2, \infty)$
Vertical asymptote: $x = -2$
x-intercept: $(-1, 0)$

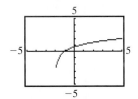

60. Domain: $(1, \infty)$
Vertical asymptote: $x = 1$
x-intercept: $(2, 0)$

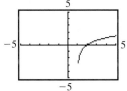

61. Domain: $(-\infty, 2)$
Vertical asymptote: $x = 2$
x-intercept: $(1, 0)$

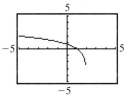

62. Domain: $(-\infty, 1)$
Vertical asymptote: $x = 1$
x-intercept: $(0, 0)$

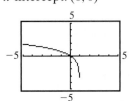

63. Domain: $(-\infty, 0) \cup (0, \infty)$
Vertical asymptote: $x = 0$
x-intercept: $(-1, 0)$ and $(1, 0)$

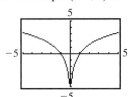

64. Domain: $(-\infty, -1) \cup (-1, \infty)$
Vertical asymptote: $x = -1$
x-intercept: $(-2, 0)$ and $(0, 0)$

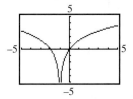

65. Domain: $(2, \infty)$
Vertical asymptote: $x = 2$
x-intercept: $(3, 0)$

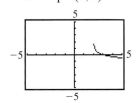

66. Domain: $(-3, \infty)$
Vertical asymptote: $x = -3$
x-intercept: $(-2, 0)$

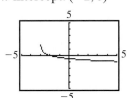

67. 0.94 feet per second
68. 0.55 feet per second
69. 28 minutes
70. around 11:38 p.m.
71. a. about 6.4; acid
    b. about 3.55; acid
72. a. about 4.5; acid
    b. about 3.03; acid

73. a. 5, moderate   **b.** 6, large   **c.** 7, major   **d.** 8, greatest   **74. a.** 7.7   **b.** 8.25   **c.** k
75. about 65.1 decibels   **76.** about 85 decibels
77. a. 88%   **b.** about 71.5%; about 63.9%; about 58.8%; about 55%; about 52%; about 49.5%
c.

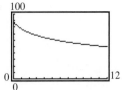

The amount of material retained decreases with time.

78. $\dfrac{x}{\ln x}$: 22; 1086; 72,382; 5,428,681; 48,254,942; 434,294,482; $\dfrac{x}{\ln(x) - 1.08366}$: 28; 1231; 78,543; 5,768,004;

50,917,519; 455,743,004; $\dfrac{x}{\ln(x) - 1.08366}$ gives a better approximation.

79. 479,001,600 actual value; 475,687,486.5 (by Stirling's formula)   **80.** d   **81.** d   **82.** c

83.

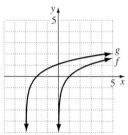

Shift the graph of f to
the left 3 units.

84.

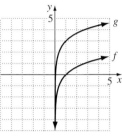

Shift the graph of f
upward 3 units.

85.

Shift the graph of f down two
units and reflect the resulting
graph across the y-axis.

86.

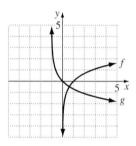

Shift the graph of f to the
left 1 unit and reflect the
resulting graph across the
x-axis.

87.

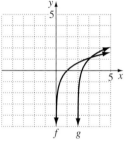

Shift the graph of f to
the right 2 units and
upward 1 unit.

88. Domain: $(0, \infty)$
Range: approximately $(-0.4, \infty)$
Decreasing: approximately
$(0, 0.4)$
Increasing: approximately
$(0.4, \infty)$

89. Domain: $(0, \infty)$
Range: approximately $(-\infty, 0.4)$
Increasing: $(0, e)$
Decreasing: (e, ∞)

90. Domain: $(-\infty, \infty)$
Range: $(0, \infty)$
Decreasing: $(-\infty, 0)$
Increasing: $(0, \infty)$

91. Domain: $(1, \infty)$
Range: $(-\infty, \infty)$
Increasing: $(1, \infty)$

92. Domain: $(1, \infty)$
Range: $(0, \infty)$
Increasing: $(1, \infty)$

93. Domain: $(-\infty, 0) \cup (0, \infty)$
Range: $(-\infty, \infty)$
Decreasing: $(-\infty, 0)$
Increasing: $(0, \infty)$

94. 10 months

95. a.

b.

c.

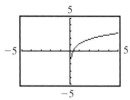

d. $\log_b(MN) = \log_b(M) + \log_b(N)$ **e.** "...the sum of the logs of each of the product's factors."

96. a. 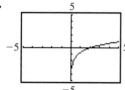   **b.** 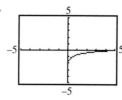   **c.**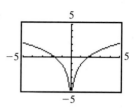

d. $\log_b\left(\dfrac{M}{N}\right) = \log_b(M) - \log_b(N)$

e. "...the difference between the log of the numerator and the log of the denominator."

97. a. 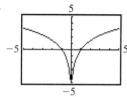   **b.** 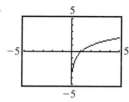   **c.**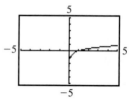

d. $\log_b M^p = p\log_b M$ **e.** "...the value of the exponent times the log of the number."

98. False   **99.** False   **100.** True   **101.** False   **102.** True   **103.** True   **104.** True   **105.** False   **106.** True

107. a. 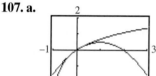   **b.** 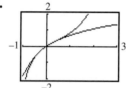   **c.**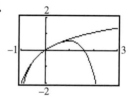

d. As terms are added, the end behavior of the graph of $y = x - \dfrac{x^2}{2} + \dots$ is flipping above and below the graph of $\ln(1 + x)$.   **108.** (1) $y = \ln x$, (2) $y = \sqrt{x}$, (3) $y = x$, (4) $y = x^2$, (5) $y = e^x$, (6) $y = x^x$

109. The city's population, in thousands, is less than one. We need to multiply our answer by 1000 to find the population (about 379).   **110.** Answers may vary.   **111.** Answers may vary.   **112.** 0.8   **113.** 3
114. $x = 8$ **115.** $y = 2x - 4$ **116.** $f^{-1}(x) = e^{e^x}$

PROBLEM SET 4.3

1. $1 + \log_7 x$ **2.** $2 + \log x$ **3.** $3\ln x + 2\ln y$ **4.** $5\ln x + 4\ln y$ **5.** $\dfrac{1}{2}\log_4 x - 3$ **6.** $\dfrac{1}{3}\log_5 x - 2$

7. $2 - \dfrac{1}{2}\log(3x - 2)$ **8.** $2 - \dfrac{1}{2}\log_8(x^2 + 1)$ **9.** $\ln 9 - 4$ **10.** $\ln 23 - 7$ **11.** $\dfrac{1}{3}\log x - \dfrac{1}{3}\log y$

12. $\dfrac{1}{5}\log x - \dfrac{1}{5}\log y$ **13.** $\dfrac{1}{2}\log_b x + 3\log_b y - 3\log_b z$ **14.** $\dfrac{1}{3}\log_b x + 4\log_b y - 5\log_b z$ **15.** $\dfrac{2}{3}\log_5 x + \dfrac{1}{3}\log_5 y - \dfrac{2}{3}$

16. $\dfrac{1}{5}\log_2 x + \dfrac{4}{5}\log_2 y - \dfrac{4}{5}$ **17.** $\dfrac{1}{4}\ln x + 3\ln y$ **18.** $\dfrac{1}{2} + \dfrac{1}{3}\log_4 x + \dfrac{1}{6}\log_4 y$ **19.** $\ln(7x)$ **20.** $\ln(3y)$

21. $\log_5\left(\dfrac{x}{y}\right)$ **22.** $\log_4\left(\dfrac{7}{y}\right)$ **23.** $\log_3(x + 1)^2$ **24.** $\log_3(x - 1)^2$ **25.** $\log_2\sqrt{(5x)}$ **26.** $\log_2\sqrt[3]{7x}$

27. $\log_3(x + 3)$ **28.** $\log_5(x^2 - 2x + 4)$ **29.** $\log(\sqrt{x}y^3)$ **30.** $\log\sqrt[3]{x}y^2$ **31.** $\log_4[\sqrt[3]{x}(3x + 2)^2]$

32. $\log_6[\sqrt[4]{x}(x - 7)^2]$ **33.** $\ln\left(\dfrac{x^3 y^5}{z^6}\right)$ **34.** $\ln\left(\dfrac{x^4 y^7}{z^3}\right)$ **35.** $\log\sqrt{xy}$ **36.** $\log_4\sqrt[3]{\dfrac{x}{y}}$ **37.** $\log_5\left[\dfrac{\sqrt{xy}}{(x + 1)^2}\right]$

38. $\log_4\left[\sqrt[3]{\dfrac{x}{y}}\,(x + 1)^2\right]$ **39.** $\log c + 9t$ **40.** $\log B - 14a$ **41.** $\ln A + 7x^2$ **42.** $3 \ln 10 - 7$

43. $\ln A + \dfrac{3}{4}$ **44.** $\dfrac{4}{7} - \ln B$ **45.** $18x^6$ **46.** $14x^8$ **47.** $9y^3$ **48.** $5x^3y^2$ **49.** $\dfrac{y^2}{e^2}$ **50.** e^5y^4

51. 2.579 **52.** 0.565 **53.** -2.585 **54.** -0.311 **55.** -0.125 **56.** 2.122 **57.** 1.947 **58.** 2.661

59. a. $\text{pH} = 6.1 + \log\left(\dfrac{B}{C}\right)$ **b.** $\text{pH} \approx 7.197$ **60.** c   **61.** b

62. a.

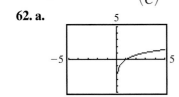

b. $y = \log_3 x$ has been shifted up 2 units to obtain $y = (\log_3 x) + 2$; $y = \log_3 x$ has been shifted left two units to obtain $y = \log_3(x + 2)$; $y = \log_3 x$ has been reflected across the x-axis to obtain $y = -\log_3 x$.

63. a.

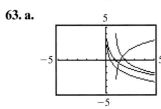

b. $y = \log_{1/2} x$ has been shifted down 1 unit to obtain $y = (\log_{1/2} x) - 1$; $y = \log_{1/2} x$ has been shifted 1 unit to the right to obtain $y = \log_{1/2}(x - 1)$; $y = \log_{1/2} x$ has been reflected across the x-axis, shifted 1 unit to the right and 1 unit upward to obtain $y = -\log_{1/2}(x - 1) + 1$.

64. The graphs are shifted vertically. The product rule property accounts for the change in the graph's appearance.

65. a.

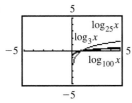

b. top graph is $\log_{100} x$; bottom graph is $y = \log_3 x$
c. top graph is $y = \log_3 x$; bottom graph is $y = \log_{100} x$
d. $(0, 1)$: top graph is $y = \log_{1/2} x$, bottom graph is $y = \log_{1/10} x$; $(1, \infty)$: top graph is $y = \log_{1/10} x$, bottom graph is $y = \log_{1/2} x$

e. Comparing graphs of $\log_b x$ for $b > 1$, the graph of the equation with the largest b will be on the top in the interval $(0, 1)$ and on the bottom in the interval $(1, \infty)$. Comparing graphs of $\log_b x$ for $b < 1$, the graphs of the equations with the largest b will be on the top in the interval $(0, 1)$ and on the bottom in the interval $(1, \infty)$.
66. a. $-3\log_4(x - 2)$ **b.** $y = \log_4 x$ has been reflected about the x-axis, stretched, and shifted to the right two units.   **c.**   **67. a.**   **b.** $y = \log_2 x$ has been shrunk by a factor of $\dfrac{1}{2}$ and shifted downward.

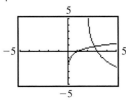

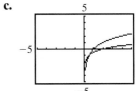

68. a.

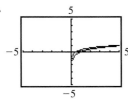

b.

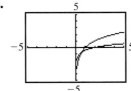

c.

69. $k = 2$

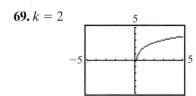

70. No, this does not constitute a proof. A proof must demonstrate that the statement is true for <u>all</u> possible values of y.

71. Answers may vary.

72. $\log\left(\dfrac{1}{3}\right)$ is negative. When you multiply an inequality by a negative number, you must flip the inequality.

73. $\log e = \dfrac{\ln e}{\ln 10} = \dfrac{1}{\ln 10}$ **74.** $13 - e^{-1}$

75. Since $R = \log_b M$ and $S = \log_b N$, $b^R = M$ and $b^S = N$. $\dfrac{M}{N} = \dfrac{b^R}{b^S}$;

$\log_b \dfrac{M}{N} = \log_b\left(\dfrac{b^R}{b^S}\right) = \log_b(b^{R-S}) = R - S = \log_b M - \log_b N$

76. Let $R = \log_b M$ so $b^R = M$; $M^P = (b^R)^P$; $M^P = b^{RP}$; $\log_b M^P = \log_b b^{RP} = RP = P\log_b M$

77. Nap $\log x = 10^7\log_{1/e}\left(\dfrac{x}{10^7}\right) = 10^7(\log_{1/e}x - \log_{1/e}10^7) = 10^7\left(\dfrac{\ln x}{\ln \frac{1}{e}} - \dfrac{\ln 10^7}{\ln \frac{1}{e}}\right) = 10^7\left(\dfrac{\ln x}{\ln e^{-1}} - \dfrac{\ln 10^7}{\ln e^{-1}}\right)$

$= 10^7\left(\dfrac{\ln x}{-\ln e} - \dfrac{7\ln 10}{-\ln e}\right) = 10^7(-\ln x + 7\ln 10) = 10^7(7\ln 10 - \ln x)$

78. 0   **79.** $\ln\left(\dfrac{\sqrt{3} + \sqrt{2}}{\sqrt{3} - \sqrt{2}}\right) = \ln\left[\dfrac{(\sqrt{3} + \sqrt{2})(\sqrt{3} + \sqrt{2})}{(\sqrt{3} - \sqrt{2})(\sqrt{3} + \sqrt{2})}\right] = \ln[(\sqrt{3} + \sqrt{2})^2] = 2\ln(\sqrt{3} + \sqrt{2})$

PROBLEM SET 4.4

1. $\left\{\dfrac{\ln 3.91}{\ln 10}\right\}$; 0.592   **2.** $\left\{\dfrac{\ln 8.07}{\ln 10}\right\}$; 0.907   **3.** $\{\ln 5.7\}$; 1.740   **4.** $\{\ln 0.83\}$; -0.186 **5.** $\left\{\dfrac{\ln 17}{\ln 5}\right\}$; 1.760

6. $\left\{\dfrac{\ln 143}{\ln 19}\right\}$; 1.685   **7.** $\left\{\ln\dfrac{23}{5}\right\}$; 1.526   **8.** $\left\{\ln\dfrac{107}{9}\right\}$; 2.476   **9.** $\left\{\dfrac{\ln 659}{5}\right\}$; 1.298   **10.** $\left\{\dfrac{1}{7}\ln\left(\dfrac{10{,}273}{4}\right)\right\}$; 1.122

11. $\left\{\dfrac{1 - \ln 793}{5}\right\}$; -1.135 **12.** $\left\{\dfrac{1 - \ln 7957}{8}\right\}$; -0.998 **13.** $\left\{\dfrac{\ln 10{,}476 + 3}{5}\right\}$; 2.451

14. $\left\{\dfrac{\ln 11{,}250 + 5}{4}\right\}$; 3.582   **15.** $\left\{\dfrac{\ln 410}{\ln 7} - 2\right\}$; 1.092   **16.** $\left\{3 + \dfrac{\ln 137}{\ln 5}\right\}$; 6.057   **17.** $\left\{\dfrac{\ln 813}{0.3\ln 7}\right\}$; 11.478

18. $\left\{\dfrac{7\ln 0.2}{\ln 3}\right\}$; -10.255 **19.** $\left\{\dfrac{\ln 3}{\ln 5 - \ln 3}\right\}$; 2.151   **20.** $\left\{\dfrac{\ln 11}{\ln 4 + \ln 11}\right\}$; 0.634   **21.** $\left\{\dfrac{-3\ln 5 - \ln 3}{2\ln 5 - \ln 3}\right\}$; -2.795

22. $\left\{\dfrac{2\ln 3 - \ln 7}{2\ln 7 - \ln 3}\right\}$; 0.090   **23.** $\{0, \ln 2\}$; 0, 0.693   **24.** $\{\ln 3\}$; 1.099   **25.** $\left\{\dfrac{\ln 3}{2}\right\}$; 0.549   **26.** $\left\{\dfrac{\ln 6}{2}\right\}$; 0.896

27. $\{81\}$ **28.** $\{125\}$ **29.** $\{59\}$ **30.** $\{32\}$ **31.** $\left\{\dfrac{109}{27}\right\}$ **32.** $\left\{-\dfrac{97}{49}\right\}$ **33.** $\left\{\dfrac{62}{3}\right\}$ **34.** $\left\{\dfrac{31}{4}\right\}$ **35.** $\left\{\dfrac{5}{4}\right\}$

36. $\{4\}$ **37.** $\{6\}$ **38.** $\{3\}$ **39.** $\{6\}$ **40.** $\{2\}$ **41.** $\{3\}$ **42.** $\{3\}$ **43.** $\{3\}$ **44.** $\{2\}$ **45.** $\{2\}$ **46.** $\left\{\dfrac{5}{3}\right\}$

47. $\{1\}$ **48.** $\{3\}$ **49.** $\{4\}$ **50.** $\{1\}$ **51.** $\{6\}$ **52.** $\left\{\dfrac{3}{2}\right\}$ **53.** $\{20\}$ **54.** $\{50\sqrt{2}\}$ **55.** $\{6\}$ **56.** $\left\{\dfrac{4}{3}\right\}$

57. 8.7 years   **58.** 20.3%   **59.** 15.7%   **60.** 8.5 years   **61.** 8.2 years   **62.** 11.2 years   **63.** 16.8%
64. 14.7%   **65.** 2050   **66.** 2108   **67.** 7.4 years after 1995; 81.9%   **68.** 5.1 years after 1995; 24.9 million people
69. a. 2035   **b.** 2019   **70.** 8.7 years   **71.** 0.004 moles per liter
72. 4.3×10^{-8} moles per liter; 3.6×10^{-8} moles per liter   **73. a.** $10^{7.1}$ times   **b.** $10^{8.9}$ times   **c.** 63 times
d. $10^{22.05}$ ergs; $10^{24.75}$ ergs   **74.** 3.7   **75. a.** 100 times   **b.** 6.3 times   **c.** 15.9 times   **76.** 25 times
77. 158,489 times   **78.** b   **79.** c   **80.** a   **81.** $\{-1.3, 1.3\}$ **82.** $\{2.1\}$ **83.** $\{-1.4, 1.7\}$ **84.** $\{-1.3, 1.3\}$

85.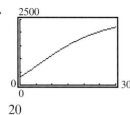
20

86.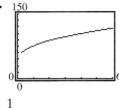
1

87.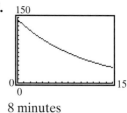
8 minutes

88. {4}   **89.** {1.0001, 5.67}   **90.** {0.08}   **91.** {0.14}

92. a. 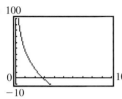   **b.** 9; after 9 days, zero percent of the students could recall the important features of a classroom lecture.
c. 2.8; after approximately 3 days, 50% of the students could recall the important features of a classroom lecture.

93–95. Answers may vary.   **96.** 36 years   **97.** 29 days
98. Isolate e^{kt} on one side of the equation. Then take the natural logarithms of both sides and solve for t.

99. 90.4   **100.** $\{1, e^2\}$ **101.** $\{e^{(3-\sqrt{5})/2}, e^{(3+\sqrt{5})/2}\}$ **102.** $\left\{\dfrac{1}{100}, 10\sqrt{10}\right\}$ **103.** $\{e\}$ **104.** Group activity

---

PROBLEM SET 4.5

1. a. $k = \dfrac{\ln 2.565}{4}$ **b.** $A = 200{,}000e^{[(\ln 2.565)/4]t}$ **c.** 1998   **d.** No   **e.** $y = 78{,}250x + 200{,}000$; 2001

f. Answers may vary.   **2. a.** $k = \dfrac{1}{10}\ln\dfrac{7{,}415{,}096}{6{,}739{,}571}$ **b.** $A = 13{,}479{,}142e^{[(1/10)\ln(7{,}415{,}096/6{,}739{,}571)]t}$

c. 22,577,647   **d.** Answers may vary.   **3. a.** $k = \dfrac{1}{28}\ln\dfrac{1}{2} \approx -0.024755$ **b.** 72.4 years

c. 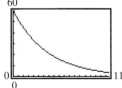   **4. a.** $k = \dfrac{1}{1.31}\ln\dfrac{1}{2} \approx -0.52912$ **b.** 107 million years
5. 34.4 years from today
6. a. 51.3°F   **b.** 17.6 minutes

c. 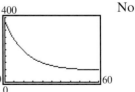 No   **7. a.** 102°F   **b.** 6.9 minutes   **c.** No

8. 86.4 feet   **9.** $y = 1.4597e^{1.0745x}$ **10.** $y = 0.2990e^{0.873x}$

11. a. 5.6348; 5.7366; 5.8021; 5.8230; 5.8608

b.

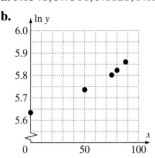

c. $\ln y = 0.00246x + 5.6273$

d. $y = 277.91e^{0.00246x}$

e. 2213

12. a. 6.810; 6.9056; 7.0405; 7.2278; 7.3602; 7.5083; 7.6748; 7.7623; 7.8642; 7.8467; 8.0813; 8.1610; 8.2946; 8.3783; 8.4433; 8.5095

b.

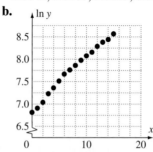

c. $\ln y = 0.1165x + 6.8752$

d. $y = 967.97e^{0.1165x}$

e. \$9949 billion

13. a.

y-axis: 0.26, 0.24, 0.22, 0.20, 0.18; x-axis: 0, 50, 100

b.

y-axis: 0.26, 0.24, 0.22, 0.20, 0.18; ln x-axis: 0, 1, 2, 3, 4, 5

Logarithmic function

c. $y = 0.06044 \ln x + 0.00131$

d. 0.26 seconds

14. $y = -4.2801 \ln x + 18.5776$; 4 rings

15. Linear model: $y = 3513.8x + 15{,}833.2$; exponential model: $y = 17{,}235e^{0.1326x}$; logarithmic model: $y = 9971.42 \ln x + 17{,}197.42$; logarithmic model.   **16.** Linear model: $y = 700.23x + 1222.2$; exponential model: $y = 1732.1e^{0.1987x}$; logarithmic model: $y = 1874.57 \ln x + 1617.45$; exponential model

17. Linear model: $y = 229.43x + 1978.33$; exponential model: $y = 2053.7e^{0.08375x}$; logarithmic model: $y = 630.49 \ln x + 2089.97$; linear model   **18.** $y = 88.1828 - 15.0371 \ln (x + 1)$; 55; 53.5; 52; 51; 50   **19.** b   **20.** d

21. a. 450 ... 250 ... 0 ... 100   **b.** $y = 277.9195(1.0025)^x$ **c.** 700 ... 0 ... 350   **d.** 313 years after 1900

22. a. 10,000 ... 0 ... 20   **b.** $y = 967.96(1.1235)^x$ **c.** 10,000 ... 0 ... 20   **d.** 9944 billion

23. a. 0.3 ... 0 ... 100   **b.** $y = 0.001313 + 0.06044 \ln x$ **c.** 0.3 ... 0 ... 100   **d.** 0.26 seconds

24. a.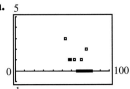
b. $y = 18.578 - 4.2801 \ln x$
c.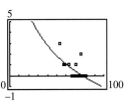
d. 4

25. a.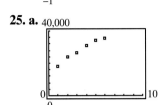
b. Linear model: $y = 3513.8x + 15{,}833.2$;
exponential model: $y = 17{,}235.6(1.1418)^x$;
logarithmic model: $y = 17{,}197.4 + 9971.4 \ln x$
c. Logarithmic model; 36,601 deaths

26. a. 6000
b. Linear model: $y = 700.23x + 1222.2$; exponential model: $y = 1732.1(1.2199)^x$;
logarithmic model: $1617.5 + 0.9451 \ln x$
c. Exponential model; 6962 deaths

27. a. 4000
b. Linear model: $y = 229.4x + 1978.3$; exponential model: $y = 2053.7(1.0874)^x$;
logarithmic model: $y = 2090 + 630.49 \ln x$
c. Linear model; \$3584

28. a.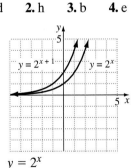
b. Linear model: $y = 279{,}887x + 274{,}373$;
exponential model: $y = 857{,}932(1.1239)^x$;
logarithmic model: $y = -251{,}211 + 1{,}515{,}198 \ln x$
c. Exponential model; \$6,250,596

29–33. Answers may vary.   **34.** Group activity

CHAPTER 4 REVIEW PROBLEMS

1. d   **2.** h   **3.** b   **4.** e   **5.** f   **6.** a   **7.** g   **8.** c

9.

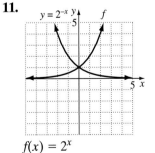

$y = 2^x$
  Domain: $(-\infty, \infty)$
  Range: $(0, \infty)$
$y = 2^{x + 1}$
  Domain: $(-\infty, \infty)$
  Range: $(0, \infty)$

10.

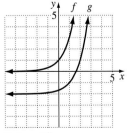

$f(x) = 3^x$
  Domain: $(-\infty, \infty)$
  Range: $(0, \infty)$
$g(x) = 3^{x - 1} - 2$
  Domain: $(-\infty, \infty)$
  Range: $(-2, \infty)$

11.

$f(x) = 2^x$
  Domain: $(-\infty, \infty)$
  Range: $(0, \infty)$
$y = 2^{-x}$
  Domain: $(-\infty, \infty)$
  Range: $(0, \infty)$

12.

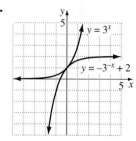

$y = 3^x$

   Domain: $(-\infty, \infty)$
   Range: $(0, \infty)$

$y = -3^{-x} + 2$

   Domain: $(-\infty, \infty)$
   Range: $(-\infty, 2)$

13.

$y = \left(\dfrac{1}{2}\right)^x$

   Domain: $(-\infty, \infty)$
   Range: $(0, \infty)$

$y = \left(\dfrac{1}{2}\right)^{-x} - 1$

   Domain: $(-\infty, \infty)$
   Range: $(-1, \infty)$

14.

$y = \log_2 x$

   Domain: $(0, \infty)$
   Range: $(-\infty, \infty)$

$y = \log_2(x - 2) + 3$

   Domain: $(2, \infty)$
   Range: $(-\infty, \infty)$

15.

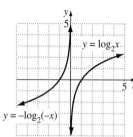

$y = \log_2 x$
   Domain: $(0, \infty)$
   Range: $(-\infty, \infty)$
$y = -\log_2(-x)$
   Domain: $(-\infty, 0)$
   Range: $(-\infty, \infty)$

16. 5.5% compounded semiannually   **17.** 7% compounded monthly
18. China: 2443 million; India: 3009 million; yes
19. a. 100 people   **b.** 127 people   **c.** $y = 200,000$; 200,000 is the limiting size of the population that becomes ill.   **d.**

20. 7   **21.** -8 **22.** 1
23. Domain: $(2, \infty)$ **24.** Domain: $(-\infty, 4)$ **25.** Domain: $(-\infty, 2) \cup (2, \infty)$
   Vertical asymptote: $x = 2$ Vertical asymptote: $x = 4$ Vertical asymptote: $x = 2$
   x-intercept: $(3, 0)$ x-intercept: $(3, 0)$ x-intercepts: $(1, 0), (3, 0)$

26. 9 weeks   **27.** $\dfrac{1}{2}\log_7 x - 2 - 2\log_7 y$ **28.** $4 - \dfrac{1}{3}\log(3x + 1)$ **29.** $\dfrac{3}{2}\ln x + \dfrac{1}{2}\ln y - 2$ **30.** $\ln 9 + 5t$

31. $-\dfrac{3}{10}\log_b x - \dfrac{1}{10}\log_b y$ **32.** $\log\dfrac{7(x - 4)^5}{x^3}$ **33.** $\ln\dfrac{\sqrt{x}}{\sqrt[7]{y}}$ **34.** $\ln(x - 3)^{5/3}$ **35.** $23x^3$ **36.** 1.760

37. 2.957   **38.** 6.7   **39.** 31.6 times as intense   **40. c.** 1.45×10^{17} joules   **41.** Answers may vary.
42. Yes   **43.** $\left\{\dfrac{\ln 12{,}143}{\ln 8}\right\}; x = 4.523$ **44.** $\left\{\dfrac{1}{5}\ln\dfrac{1268}{9}\right\}; 0.990$ **45.** $\left\{\dfrac{12 - \ln 130}{5}\right\}; 1.426$

46. $\left\{\dfrac{2\ln 5 + 3\ln 7}{\ln 7 - 4\ln 5}\right\}; -2.016$ **47.** $\{\ln 3\}; 1.099$ **48. a.** 24.4 years   **b.** 8.7%   **49.** 7.3 years
50. 30.7 years   **51.** 2038   **52.** $\{23\}$ **53.** $\{5\}$ **54.** $\varnothing$ **55.** $\{5\}$ **56.** $\{-1 + \sqrt{6}\}$ **57.** 8338 thousand
58. 91 times more intense   **59. a.** 0.041   **b.** $f(t) = 14{,}609e^{0.041t}$ **c.** 40,717; 49,981; 92,447   **60.** 15,642 years
61. a. 200°F   **b.** 119°F   **c.** 70°; the room temperature is 70°.   **62.** 7:30 A.M.

63. High projection: exponential function; medium projection: linear function; low projection: quadratic function   **64. a.** $y = 3.46e^{0.0233x}$ **b.** 114 million   **c.** 12% in 1990; 29% in 2050

65. a.

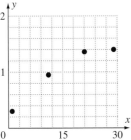

Logarithmic function
b. $y = 0.2738 + 0.3308 \ln x$
c. 2029

66. a.

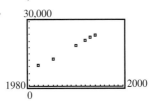

b. Linear model: $y = 1182.47x - 2,334,123.42$
Exponential model: $y = 1.07 \times 10^{-6}(1.08)^x$
Logarithmic model:
$y = -17,836,653 + 2,350,698 \ln x$
The linear model best describes book expenditures as a function of time.
c. Using the linear model, when $x = 2005$, $y \approx 36726$. Consumers will spend $36,725 million on books in the year 2005.

CHAPTER 4 TEST

1.

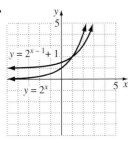

2. 6.5% compounded semiannually; $663
3. a. 140°F   **b.** $y = 72$; the temperature of the room is 72°F.   **4.** 6

5.

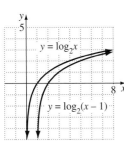

6. $(-\infty, 3)$ **7.** $\frac{1}{3}\log_5 x - 3 - 2\log_5 y$ **8.** $\ln 10 + 7t$ **9.** $\log(x^6 y^2)$ **10.** $\ln\left(\dfrac{7}{x^3}\right)$

11. $\left\{\dfrac{9 - \ln 42}{3}\right\}$; 1.754   **12.** $\{0, \ln 5\}$; 0, 1.609   **13.** 11.6 years   **14.** 8.7 years   **15.** $\left\{\dfrac{16}{7}\right\}$ **16.** $\left\{\dfrac{1}{2}\right\}$
17. 10^{12} times louder   **18. a.** $A = 5.5e^{0.0177t}$ **b.** 2022   **19. a.** $T = 28 + 40e^{(\ln 0.9)}$ **b.** 6.6 hours

20. 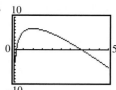   **a.** $x = 0$ **b.** 6.51 at $x = 10$ **c.** Range $(-\infty, 6.51]$ **d.** $\{1.12, 35.77\}$

CUMULATIVE REVIEW PROBLEMS (CHAPTERS P–4)

1. $\dfrac{-6x^2 + 2x - 2}{(2x + 1)^2(x - 1)}$ **2.** $\dfrac{x}{x - 1}$ **3.** $\left\{\dfrac{2}{3}, 2\right\}$ **4.** $\{3, 7\}$ **5.** $\{-1, -2, 1\}$ **6.** $\left\{\dfrac{11 - \ln 128}{5}\right\}$ **7.** $\{3\}$

8. $9y^2 - 4x^2 - 30y - 25$ **9.**

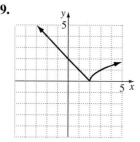

10.

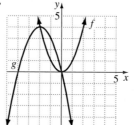

11. $\{2 - i, 2 + i, i, -i\}$ **12.**

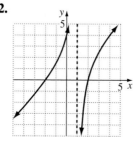

13. $\dfrac{0.5}{\ln 4}; 0.3$ **14.** $4 + \dfrac{1}{3}\log_3 x$ **15.** $(x - 5)^2(x + 5)$

16. $n = \dfrac{IR}{E - Ir}$ **17.** 2004   **18.** $\dfrac{9}{4}\pi$ ft^3 **19.** $\dfrac{1}{2}$

20. $p = 100$; it is impossible to remove 100% of the river's pollutants.

Chapter 5

PROBLEM SET 5.1

1. $\{(-4, -11)\}$ **2.** $\left\{\left(-\dfrac{1}{2}, -1\right)\right\}$ **3.** $\{(-1, 2)\}$ **4.** $\{(-3, -2)\}$ **5.** $\{(3, -4)\}$ **6.** $\{(-7, 3)\}$

7. $\left\{\left(-\dfrac{19}{5}, -3\right)\right\}$ **8.** $\{(6, -8)\}$ **9.** $\{(3, 2)\}$ **10.** $\{(2, -5)\}$ **11.** $\{(3, -4)\}$ **12.** $\{(1, 2)\}$ **13.** $\left\{\left(\dfrac{5}{3}, -4\right)\right\}$

14. $\{(2,4)\}$ **15.** $\{(8,-1)\}$ **16.** $\{(4,-1)\}$ **17.** $\left\{\left(\dfrac{41}{7},\dfrac{36}{7}\right)\right\}$ **18.** $\left\{\left(\dfrac{37}{7},\dfrac{19}{7}\right)\right\}$ **19.** $\left\{\left(2,\dfrac{9}{2}\right)\right\}$

20. $\{(-3,-4)\}$ **21.** $\{(2,-1)\}$ **22.** $\{(-14,11)\}$ **23.** Dependent system; $\left\{\left(x,\dfrac{x}{3}-2\right)\right\}$

24. Dependent system; $\{(x,2x-4)\}$ **25.** $\{(1,2)\}$ **26.** $\{(-4,2)\}$ **27.** $\left\{\left(-\dfrac{9}{4},1\right)\right\}$ **28.** $\{(-6,2)\}$

29. Inconsistent system   **30.** Inconsistent system   **31.** $\left\{\left(\dfrac{34}{31},-\dfrac{33}{31}\right)\right\}$ **32.** $\left\{\left(x,-\dfrac{15x+35}{7}\right)\right\}$ **33.** $\{(3,8)\}$

34. $\{(7,3)\}$ **35.** $\left\{\left(\dfrac{13}{22},\dfrac{2}{11}\right)\right\}$ **36.** $\left\{\left(-\dfrac{6}{37},\dfrac{98}{37}\right)\right\}$ **37.** $\{(2,6)\}$ **38.** $\{(2,8)\}$ **39.** $\{(2,3,3)\}$ **40.** $\{(-1,4,0)\}$

41. $\{(3,-1,4)\}$ **42.** $\{(2,1,4)\}$ **43.** $\{1,-1,-1\}$ **44.** $\{(2,-2,-1)\}$ **45.** $\{(1,1,2)\}$ **46.** $\{(1,0,-3)\}$

47. $(0,0,4)$ **48.** $\{(1,-5,-6)\}$ **49.** $\varnothing$ **50.** $\{(2,2,2)\}$ **51.** $\varnothing$ **52.** $\varnothing$ **53.** $\{(4,6,8)\}$ **54.** $(3,-1,2)$
55. 38 executions in 1993; 31 executions in 1994
56. 5.8 million pounds of potato chips; 4.6 million pounds of tortilla chips   **57.** 90 feet by 70 feet
58. 100 feet by 80 feet   **59.** Quarter Pounders: 77 mg; Whoppers: 122 mg
60. Sponge cake: 162 mg; pound cake: 68 mg   **61.** Glass: 36 Btu per square foot; plaster: 8 Btu per square foot
62. 450 bags; 300 jackets   **63.** $x=55; y=35$ **64.** $x=79; y=40$ **65.** Japan: 16%; Germany: 15%; France: 14%
66. 118°; 25°; 37°   **67.** 100°; 20°; 60°   **68.** 200 $8 tickets; 150 $10 tickets; 50 $12 tickets
69. Pulley with center A: 5 in.; pulley with center B: 3 in.; pulley with center C: 6 in.
70. $3000 at 6%; $6000 at 10%; $1000 at 8%   **71.** $A=-8; B=50; C=0; y=156$ **72.** d   **73.** d
74–76. Verifications left for the student.   **77–81.** Answers may vary.

82. $y-3=-3(x+2); y=-3x-3; 3x+y+3=0$ **83.** $\left\{\left(\dfrac{b_2c_1-b_1c_2}{a_1b_2-a_2b_1},\dfrac{a_1c_2-a_2c_1}{a_1b_2-a_2b_1}\right)\right\}$ **84.** $\{(3,4)\}$

85. 500 students   **86.** 5 adult, 7 adult with one child, 3 adult with two children

87. $\left\{\left(\dfrac{a-b+c}{2},\dfrac{a+b-c}{2},\dfrac{-a+b+c}{2}\right)\right\}$ **88.** $\left\{\left(\dfrac{A+B}{AB},-\dfrac{1}{AB}\right)\right\}; A\neq 0, B\neq 0, A\neq B$
89. $\{(-AB, A+B)\}; A\neq B$ **90.** Group activity

PROBLEM SET 5.2

1. $\left\{\left(3,7,-\dfrac{1}{2}\right)\right\}$ **2.** $\{(1,-2,3)\}$ **3.** $\{(1,1,4)\}$ **4.** $\{(5,-2,0)\}$ **5.** $\{(2,1,-1,3)\}$ **6.** $\{-1,2,1,3)\}$

7. $\begin{bmatrix} 1 & -2 & 3 & \bigm| & 9 \\ -1 & 3 & 0 & \bigm| & -4 \\ 2 & -5 & 5 & \bigm| & 17 \end{bmatrix}$ **8.** $\begin{bmatrix} 1 & 3 & -3 & \bigm| & 0 \\ 5 & -2 & 1 & \bigm| & 2 \\ 2 & -4 & 6 & \bigm| & -2 \end{bmatrix}$ **9.** $\begin{bmatrix} 5 & 8 & 1 & \bigm| & 50 \\ 0 & 6 & -5 & \bigm| & -50 \\ 9 & 10 & 17 & \bigm| & 23 \end{bmatrix}$ **10.** $\begin{bmatrix} 6 & 7 & 3 & \bigm| & 40 \\ 0 & 5 & -10 & \bigm| & -3 \\ 7 & 3 & 9 & \bigm| & 11 \end{bmatrix}$

11. $\begin{bmatrix} 1 & -3 & 2 & \bigm| & 0 \\ 0 & 10 & -7 & \bigm| & 7 \\ 0 & 4 & -3 & \bigm| & 3 \end{bmatrix}$ **12.** $\begin{bmatrix} 1 & -1 & 5 & \bigm| & -6 \\ 0 & 6 & -16 & \bigm| & 28 \\ 0 & 4 & -3 & \bigm| & 11 \end{bmatrix}$ **13.** $\begin{bmatrix} 1 & 3 & 4 & \bigm| & 10 \\ -5 & -20 & -35 & \bigm| & -88 \\ 0 & -4 & -12 & \bigm| & -31 \end{bmatrix}$

14. $\begin{bmatrix} 1 & 1 & -1 & | & 6 \\ 0 & -3 & 3 & | & -15 \\ 0 & -4 & 2 & | & -14 \end{bmatrix}$
  15. $\begin{bmatrix} 1 & -1 & 1 & 1 & | & 3 \\ 0 & 1 & -2 & -1 & | & 0 \\ 0 & 2 & 1 & 2 & | & 5 \\ 0 & 6 & -3 & -1 & | & -9 \end{bmatrix}$
  16. $\begin{bmatrix} 1 & -5 & 2 & -2 & | & 4 \\ 0 & 1 & -3 & -1 & | & 0 \\ 0 & 15 & -4 & 5 & | & -6 \\ 0 & -19 & 12 & -6 & | & 13 \end{bmatrix}$

17. $\{(1, -1, 2)\}$ **18.** $\{(1, -1, 1)\}$ **19.** $\{(3, -1, -1)\}$ **20.** $\{(-3, 0, 1)\}$ **21.** $\{(1, 2, -1)\}$ **22.** $\{(2, 0, -1)\}$

23. $\{(1, 2, 3, -2)\}$ **24.** $\{(2, 1, -1, 3)\}$ **25.** $\{(0, -3, 0, -3)\}$ **26.** $\{(1, 3, 0, -2)\}$ **27.** $\{(-1, 2, 3, 4)\}$ **28.** $\{(4, 3, 2, 1)\}$

29. $\{(1, -1, 2, -2, 0)\}$ **30.** $\{(1, 1, 2, 2, 3, 3,)\}$ **31. a.** $s(t) = -16t^2 + 128t + 6$

b. $s(0) = 6$; $s(4) = 262$; $s(7) = 118$ **c.** 128; 6 **32. a.** $s(t) = -16t^2 + 48t + 8$ **b.** 44 feet; 3.1 seconds

33. $y = \dfrac{1}{2}x^2 - \dfrac{1}{2}x$ **34.** $f(x) = \dfrac{209}{2}x^2 - \dfrac{3003}{2}x + 6016$; 1121; 769; 2144   **35.** $\overline{AB} = 34$, $\overline{AD} = 19$, $\overline{AC} = 12$

36. 4 ounces of Food A; $\dfrac{1}{2}$ ounce of Food B; 1 ounce of Food C **37.** 3 servings of Food A; 2 servings of

Food B; 4 servings of Food C **38.** 15 children's models; 10 office models; 20 deluxe models   **39.** d

40. c   **44. a.** $y = -7.2x^3 + 286x^2 - 3339x + 25{,}130$

b. The model indicates that the number of bachelor degrees conferred will be zero before the year 2000.

45–48. Answers may vary.   **49.** $(3, 5)$ **50.** $\left\{ \left(\ln A, \ln (2A), \ln \left(\dfrac{A}{2} \right) \right) \right\}$ **51.** Group activity

---

PROBLEM SET 5.3

1. No solution   **2.** No solution   **3.** $\left\{ \left(-2t + 2, 2t + \dfrac{1}{2}, t \right) \right\}$ **4.** $\{(t - 2, t - 2, t)\}$ **5.** $\{(-3, 4, -2)\}$

6. $\{(2, -3, 7)\}$ **7.** $\{(-2t + 5, t - 2, t)\}$ **8.** $\{(7t - 5, 3t + 1, t)\}$ **9.** $\{(-1, 2, 1, 1)\}$ **10.** No solution

11. $\{(1, 3, 2, 1)\}$ **12.** $\left\{ \left(\dfrac{1}{3}t, \dfrac{2}{3}t, -\dfrac{1}{3}t, t \right) \right\}$ **13.** $\{(1, -2, 1, 1)\}$ **14.** $\{(1, -2, 3, -4)\}$ **15.** $\left\{ \left(\dfrac{1}{3}t + 1, \dfrac{1}{3}t, t \right) \right\}$

16. $\left\{ \left(2, \dfrac{1}{2}t - \dfrac{1}{2}, t \right) \right\}$ **17.** $\{(-13t + 5, 5t, t)\}$ **18.** $\{(-2t + 3, -2t + 1, t)\}$ **19.** $\left\{ \left(2t - \dfrac{5}{4}, \dfrac{13}{4}, t \right) \right\}$

20. $\{(98, 2t - 43, t)\}$ **21.** $\{(1, -t - 1, 2, t)\}$ **22.** $\{(t + 3, 5t + 5, 3t + 4, t)\}$

23. $\left\{ \left(-\dfrac{2}{11}t + \dfrac{81}{11}, \dfrac{1}{22}t + \dfrac{10}{11}, \dfrac{4}{11}t - \dfrac{8}{11}, t \right) \right\}$ **24.** $\left\{ \left(\dfrac{1}{3}t - \dfrac{2}{3}s, \dfrac{1}{3}t + \dfrac{1}{3}s, t, s \right) \right\}$

25. 0 ounces of Food 1, 10 ounces of Food 2, and 8 ounces of Food 3 or 4 ounces of Food 1, 0 ounces of Food 2, and 10 ounces of Food 3. (Other answers are possible.)
26. There are many combinations of the foods that satisfy the new requirements.
27. One solution is $x = 330$, $y = 270$, $z = 100$, $w = 50$ (Other answers are possible.)
28. 7 of product A, 2 of product B, 2 of product C or 7 of product A, 1 of product B, 4 of product C
(Other answers are possible.)   **29.** a   **30.** d
31. a. $x_1 - x_6 = 100$; $x_2 - x_6 + x_7 = 600$; $x_3 + x_7 = 900$; $x_1 - x_4 = -200$; $x_2 - x_4 + x_5 = 100$; $x_3 + x_5 = 700$
b. $\{(t + 100, t - s + 600, -s + 900, t + 300, s - 200, t, s)\}$ **c.** $200 \le x_7 \le 600$ **d.** Answers may vary.
e. Group activity   **32–34.** Answers may vary.   **35.** $a = 1$ or $a = 3$ **36.** Group activity

---

PROBLEM SET 5.4

1. a. 2×3 **b.** a_{32} does not exist: $a_{23} = -1$ **2. a.** 2×3 **b.** a_{32} does not exist; $a_{23} = \dfrac{1}{2}$

3. a. 3×4 **b.** $a_{32} = \dfrac{1}{2}; a_{23} = -6$ **4. a.** 3×4 **b.** $a_{32} = 0; a_{23} = \pi$ **5.** $x = -5, y = -2$ **6.** $x = -2, y = 5$

7. $x = -3, y = -4, z = -7, w = 1$ **8.** $x = -6, y = 6, z = 7$

9. a. $\begin{bmatrix} 3 & 2 \\ 6 & 2 \\ 5 & 7 \end{bmatrix}$ **b.** $\begin{bmatrix} -1 & 4 \\ 0 & 6 \\ 5 & 5 \end{bmatrix}$ **c.** $\begin{bmatrix} 4 & 12 \\ 12 & 16 \\ 20 & 24 \end{bmatrix}$ **d.** $\begin{bmatrix} -1 & 18 \\ 6 & 26 \\ 25 & 27 \end{bmatrix}$

10. a. $\begin{bmatrix} 5 & -2 & 7 \\ -4 & 3 & 1 \end{bmatrix}$ **b.** $\begin{bmatrix} 1 & 4 & -5 \\ 2 & 1 & 9 \end{bmatrix}$ **c.** $\begin{bmatrix} 12 & 4 & 4 \\ -4 & 8 & 20 \end{bmatrix}$ **d.** $\begin{bmatrix} 9 & 14 & -13 \\ 4 & 7 & 37 \end{bmatrix}$

11. a. $\begin{bmatrix} -3 \\ -1 \\ 0 \end{bmatrix}$ **b.** $\begin{bmatrix} 7 \\ -7 \\ 2 \end{bmatrix}$ **c.** $\begin{bmatrix} 8 \\ -16 \\ 4 \end{bmatrix}$ **d.** $\begin{bmatrix} 25 \\ -29 \\ 8 \end{bmatrix}$

12. a. $[10 \quad 0 \quad 0]$ **b.** $[2 \quad 4 \quad -6]$ **c.** $[24 \quad 8 \quad -12]$ **d.** $[18 \quad 16 \quad -24]$

13. a. $\begin{bmatrix} 3\sqrt{2} & 0 \\ -1 & 6 \end{bmatrix}$ **b.** $\begin{bmatrix} \sqrt{2} & 2 \\ 5 & -6 \end{bmatrix}$ **c.** $\begin{bmatrix} 8\sqrt{2} & 4 \\ 8 & 0 \end{bmatrix}$ **d.** $\begin{bmatrix} 7\sqrt{2} & 8 \\ 19 & -18 \end{bmatrix}$

14. a. $\begin{bmatrix} 8 & 0 & -4 \\ 14 & 0 & 6 \\ -1 & 0 & 0 \end{bmatrix}$ **b.** $\begin{bmatrix} -4 & -20 & 0 \\ 14 & 24 & 14 \\ 9 & -4 & 4 \end{bmatrix}$ **c.** $\begin{bmatrix} 8 & -40 & -8 \\ 56 & 48 & 40 \\ 16 & -8 & 8 \end{bmatrix}$ **d.** $\begin{bmatrix} -8 & -80 & -4 \\ 70 & 96 & 62 \\ 35 & -16 & 16 \end{bmatrix}$

15. a. $[30]$ **b.** $\begin{bmatrix} 1 & 2 & 3 & 4 \\ 2 & 4 & 6 & 8 \\ 3 & 6 & 9 & 12 \\ 4 & 8 & 12 & 16 \end{bmatrix}$ **16. a.** $\begin{bmatrix} -1 & -2 & -3 \\ -2 & -4 & -6 \\ -3 & -6 & -9 \end{bmatrix}$ **b.** $[-14]$

17. a. $\begin{bmatrix} 0 & 16 \\ 12 & 13 \end{bmatrix}$ **b.** $\begin{bmatrix} -7 & 3 \\ 29 & 15 \end{bmatrix}$ **18. a.** $\begin{bmatrix} -10 & 12 \\ 25 & -30 \end{bmatrix}$ **b.** $\begin{bmatrix} 0 & 0 \\ 9 & -40 \end{bmatrix}$

19. a. $\begin{bmatrix} 4 & -5 & 8 \\ 6 & -1 & 5 \\ 0 & 4 & -6 \end{bmatrix}$ **b.** $\begin{bmatrix} 5 & -2 & 7 \\ 17 & -3 & 2 \\ 3 & 0 & -5 \end{bmatrix}$ **20. a.** $\begin{bmatrix} 3 & 4 & -3 \\ -1 & 7 & -4 \\ 7 & 9 & -6 \end{bmatrix}$ **b.** $\begin{bmatrix} 6 & -1 & -1 \\ -4 & -11 & 19 \\ 4 & -7 & 9 \end{bmatrix}$

21. a. $\begin{bmatrix} 6 & 8 & 16 \\ 11 & 16 & 24 \\ 1 & -1 & 12 \end{bmatrix}$ **b.** $\begin{bmatrix} 38 & 27 \\ -16 & -4 \end{bmatrix}$ **22. a.** $\begin{bmatrix} 2 & -8 & 20 \\ 8 & 3 & 5 \\ 10 & 2 & 10 \end{bmatrix}$ **b.** $\begin{bmatrix} 12 & 14 \\ 9 & 3 \end{bmatrix}$

23. a. $\begin{bmatrix} 0 & 0 \\ 0 & 0 \end{bmatrix}$ **b.** $\begin{bmatrix} 4 & -1 & -3 & 1 \\ -1 & 4 & -3 & 2 \\ 14 & -11 & -3 & -1 \\ 25 & -25 & 0 & -5 \end{bmatrix}$ **24. a.** $\begin{bmatrix} 18 & 1 \\ -1 & 15 \end{bmatrix}$ **b.** $\begin{bmatrix} 0 & 1 & -7 & 0 \\ 3 & -1 & 1 & 3 \\ 2 & -3 & 17 & 2 \\ 17 & -6 & 8 & 17 \end{bmatrix}$

25. $\begin{bmatrix} 17 & 7 \\ -5 & -11 \end{bmatrix}$ **26.** $\begin{bmatrix} -5 & -7 \\ -1 & 9 \end{bmatrix}$ **27.** $\begin{bmatrix} 11 & -1 \\ -7 & -3 \end{bmatrix}$ **28.** $\begin{bmatrix} 24 & 0 \\ -33 & -5 \\ -3 & -1 \end{bmatrix}$ **29.** Not defined **30.** Not defined

31. $\begin{bmatrix} 16 & -16 \\ -12 & 12 \\ 0 & 0 \end{bmatrix}$ **32.** $\begin{bmatrix} 28 & 12 \\ -56 & -24 \\ -7 & -3 \end{bmatrix}$ **33.** $\begin{bmatrix} 2 & -\frac{31}{2} \\ \frac{7}{2} & 0 \\ -\frac{29}{2} & -1 \end{bmatrix}$ **34.** $\begin{bmatrix} -\frac{4}{3} & -\frac{49}{3} \\ \frac{13}{3} & 0 \\ -\frac{31}{3} & -6 \end{bmatrix}$ **35.** $\begin{bmatrix} -3 & -\frac{7}{4} \\ 1 & 0 \\ 3 & -\frac{19}{4} \end{bmatrix}$

36. $\begin{bmatrix} -\frac{4}{3} & -1 \\ \frac{1}{2} & 0 \\ \frac{7}{6} & -\frac{13}{6} \end{bmatrix}$ **37.** $\begin{bmatrix} -4 & -5 \\ 2 & 0 \\ 2 & -7 \end{bmatrix}$ **38.** $\begin{bmatrix} -12 & 5 \\ 1 & 0 \\ 21 & -16 \end{bmatrix}$ **39.** $\begin{bmatrix} 8 \\ 5 \end{bmatrix}$ **40.** $\begin{bmatrix} 2 \\ -1 \end{bmatrix}$ **41.** $\begin{bmatrix} 2 \\ 3 \\ -2 \end{bmatrix}$ **42.** $\begin{bmatrix} 10 \\ -2 \\ 6 \end{bmatrix}$

43. a. $\begin{bmatrix} \$109,000 & \$127,000 \\ \$66,000 & \$10,000 \end{bmatrix}$; This gives the combined June and July sales with row 1 being person x's sales (by model), and row 2 being person y's sales. The columns give the sales for each model broken down by salesperson. **b.** $\begin{bmatrix} \$19,000 & \$3,000 \\ \$6,000 & \$10,000 \end{bmatrix}$; This gives the increase in sales from June to July. **c.** $\begin{bmatrix} \$2700 & \$3720 \\ \$1800 & \$0 \end{bmatrix}$; This represents the commissions paid to each salesperson for the sales of each model in June.

44. a. $\begin{bmatrix} 52,202 & 11,662 & 476 \\ 16,674 & 9168 & 90 \\ 25,670 & 12,240 & 79 \end{bmatrix}$; This represents the number of each type of account at each branch at the end of the first quarter. **b.** $0.96(A + B - C) = \begin{bmatrix} 50,114 & 11,196 & 457 \\ 16,007 & 8801 & 86 \\ 24,643 & 11,750 & 76 \end{bmatrix}$ **c.** $B - C$ represents the increase or decrease in the number of each type of account at each branch during the first quarter.

45. a. $\begin{bmatrix} 15,800 & 18,600 \\ 12,500 & 14,800 \\ 3700 & 4600 \end{bmatrix}$ **b.** AB represents the distribution of voters by gender and political party registration. In this county, there are 15,800 men registered as Republicans, 12,500 men registered as Democrats, etc. **c.** 14,800

46. a. $\begin{bmatrix} 589.5 & 586 \\ 1556 & 1521 \\ 1234.5 & 1183 \end{bmatrix}$ **b.** AB represents the distribution of students by gender and state of health. On this campus, there are $589.5 \approx 590$ males who are well, 1556 males who are sick, etc. **c.** 1521; 1235

47. a. $\begin{bmatrix} 45 & 18 & 9 \\ 65 & 26 & 13 \\ 13 & 6 & 1 \end{bmatrix}$ **b.** AB represents the number of grams of protein, carbohydrate, and fat (from milk) ingested by children, adolescents, and adults. A child gets 45 grams of protein, 65 grams of carbohydrate, and 13 grams of fat per day from milk, etc. **c.** 45 grams

48. a. $\begin{bmatrix} 16.8 & 12.8 \\ 27.2 & 20.8 \\ 39.6 & 30.4 \end{bmatrix}$ **b.** AB represents the amount it costs to manufacture the different types of boats in the two factories. It costs $16.80 to manufacture a one-person boat in the domestic factory and $12.80 in the overseas factory, etc. **c.** $30.40   **49.** a   **50.** d   **51.** d   **52.** b

54. a. A^n describes the transition probabilities from the first election in the community to the $(n + 1)$st

election.   **b.** As n increases, A^n approaches $\begin{bmatrix} \frac{1}{3} & \frac{1}{3} & \frac{1}{3} \\ \frac{1}{3} & \frac{1}{3} & \frac{1}{3} \\ \frac{1}{3} & \frac{1}{3} & \frac{1}{3} \end{bmatrix}$. After a great number of elections, $\frac{1}{3}$ of the voters

registered with each party will not change parties while $\frac{1}{3}$ will switch to each of the other two parties.

55. Answers may vary.   **56.** Answers may vary.   **57.** AB is not defined when A is $m \times n$ and B is $p \times q$ where $n \neq p$.   **58.** A is an $n \times m$ matrix and B is an $m \times n$ matrix.

59. a. System 1: The midterm and final both count for 50% of the course grade.; System 2: The midterm counts for 30% of the course grade and the final counts for 70%.

b. $\begin{bmatrix} 84 & 87.2 \\ 79 & 81 \\ 90 & 88.4 \\ 73 & 68.6 \\ 69 & 73.4 \end{bmatrix}$; Student 1: B; Student 2: C or B; Student 3: A or B; Student 4: C or D; Student 5: D or C (System 1 grades are listed first.)   **c.** Answers may vary.

60. $\begin{bmatrix} 1 & 1 & 1 & 1 & 1 \\ 2 & 2 & 2 & 1 & 1 \\ 0 & 0 & 1 & 1 & 1 \end{bmatrix}$; The rows correspond to the colleges, the columns to the math courses. The entries in a

row tell how many of the degree programs offered by that rows' college require the math course

corresponding to the column.   **61–62.** Verification is left to the student.   **63–65.** Answers may vary.

66. $A^n = \begin{bmatrix} 1 & n \\ 0 & 1 \end{bmatrix}$ **67.** $A^n = \begin{bmatrix} 2^{n-1} & 2^{n-1} \\ 2^{n-1} & 2^{n-1} \end{bmatrix}$ **68.** Answers may vary.   **69–70.** Group activity

---

PROBLEM SET 5.5

1. Inverses   **2.** Inverses   **3.** Not inverses   **4.** Not inverses   **5.** Inverses   **6.** Inverses   **7.** Inverses

8. Inverses   **9.** $\begin{bmatrix} \frac{2}{7} & -\frac{3}{7} \\ \frac{1}{7} & \frac{2}{7} \end{bmatrix}$ **10.** $\begin{bmatrix} \frac{1}{6} & \frac{1}{4} \\ \frac{1}{3} & 0 \end{bmatrix}$ **11.** $\begin{bmatrix} 1 & \frac{1}{2} \\ 2 & \frac{3}{2} \end{bmatrix}$ **12.** $\begin{bmatrix} -1 & 3 \\ -\frac{1}{2} & 1 \end{bmatrix}$ **13.** No inverse   **14.** No inverse

15–16. Verification is left to the student.   **17.** $\begin{bmatrix} 1 & 0 & 1 \\ 1 & 1 & 2 \\ 3 & 2 & 6 \end{bmatrix}$ **18.** $\begin{bmatrix} 3 & 3 & -1 \\ -2 & -2 & 1 \\ -4 & -5 & 2 \end{bmatrix}$ **19.** $\begin{bmatrix} -3 & 2 & -4 \\ -1 & 1 & -1 \\ 8 & -5 & 10 \end{bmatrix}$

20. $\begin{bmatrix} 1 & 2 & -2 \\ -1 & 3 & 0 \\ 0 & -2 & 1 \end{bmatrix}$ **21.** $\begin{bmatrix} 1 & 0 & 0 & 0 \\ 0 & -1 & 0 & 0 \\ 0 & 0 & \frac{1}{3} & 0 \\ -1 & 0 & 0 & 1 \end{bmatrix}$ **22.** $\begin{bmatrix} \frac{1}{2} & 0 & 0 & -\frac{1}{4} \\ 0 & 1 & 0 & 0 \\ 0 & 0 & -1 & 0 \\ 0 & 0 & 0 & \frac{1}{2} \end{bmatrix}$ **23.** $\{1, 2, -1)\}$

24. $\{(-2, -3, 2)\}$ **25.** $\{(2, -1, 5)\}$ **26.** $\{(1, -7, -3)\}$ **27.** $\{(2, 3, -1, 0)\}$ **28.** $\{(1, 3, -1, 2)\}$

29. a. Verification is left to the student.   **b.** $\{(2, -1, 1)\}$ **30.** $3, -2$ **31.** $-3, -2$
32–35. Verification is left to the student.   **36.** The encoded message is $27, -19, 32, -20$.
37. The encoded message is $33, -21, 83, -61$.
38. The encoded message is $71, -52, 52, -38, -3, 3, -15, 16, 32, -24$.
39. The encoded message is $56, -37, -21, 22, -23, 23, 8, -3, 48, -36$.
40. $-17, 39, -19, -13, 37, -14, 4, 50, -23, -5, 13, -6, -13, 36, -12$.

41. $-17, 43, -21, -5, 28, -14, 9, 60, -21, 3, 62, -27$.   **42. c**   **43. d**   **44.** $\begin{bmatrix} 0 & -1 & 0 & 1 \\ -1 & -5 & 0 & 3 \\ -2 & -4 & 1 & -2 \\ -1 & -4 & 0 & 1 \end{bmatrix}$

45. $\begin{bmatrix} \frac{3}{5} & 0 & -\frac{2}{5} & \frac{1}{5} \\ \frac{1}{5} & 0 & \frac{1}{5} & -\frac{1}{10} \\ 0 & 1 & 0 & 0 \\ -\frac{6}{5} & 0 & \frac{4}{5} & \frac{1}{10} \end{bmatrix}$ **46.** $\{(2, 3, -5)\}$ **47.** $\{(1, 2, -1)\}$ **48.** $\{(1, 2, -1)\}$ **49.** $\{(5, 4, -1)\}$

50. $\{(2, 1, 3, -2, 4)\}$ **51.** $\left\{ \left(-\frac{22}{5}, \frac{17}{5}, \frac{11}{5}, \frac{14}{5} \right) \right\}$ **52. a.** $\begin{bmatrix} 18.04 \\ 80 \\ 64.04 \end{bmatrix}$; In order to produce 100 million dollars

worth of agricultural goods, 120 million dollars worth of manufactured goods, and 120 million dollars worth of transportation, $18,040,000 worth of agricultural goods, $80,000,000 worth of manufactured goods, and

$64,040,000 worth of transportation are consumed.   **b.** $\begin{bmatrix} 81.96 \\ 40 \\ 55.96 \end{bmatrix}$ **53–55.** Answers may vary.   **56.** Yes   **57.** No

58. $(A \circ B) \circ C = \begin{bmatrix} 3 & 12 \\ -6 & -3 \end{bmatrix}$; $A \circ (B \circ C) = \begin{bmatrix} -12 & 12 \\ -12 & 12 \end{bmatrix}$ **59.** Answers may vary. For example, $\begin{bmatrix} 0 & 1 \\ 1 & 0 \end{bmatrix}$.

60. Answers may vary. For example, $\begin{bmatrix} 1 & 2 \\ 2 & 3 \end{bmatrix}$ and $\begin{bmatrix} 2 & 4 \\ 0 & 1 \end{bmatrix}$.   **61.** $\begin{bmatrix} 1 & a & b \\ 0 & 1 & c \\ 0 & 0 & 1 \end{bmatrix}^{-1} = \begin{bmatrix} 1 & -a & ac-b \\ 0 & 1 & -c \\ 0 & 0 & 1 \end{bmatrix}$

62. $\begin{bmatrix} \frac{1}{a} & 0 & 0 & 0 \\ 0 & \frac{1}{b} & 0 & 0 \\ 0 & 0 & \frac{1}{c} & 0 \\ 0 & 0 & 0 & \frac{1}{d} \end{bmatrix}$ **63.** $A^{-1} = \begin{bmatrix} \frac{1}{a} & -\frac{1}{ab} & -\frac{1}{ac} & -\frac{1}{ad} \\ 0 & \frac{1}{b} & 0 & 0 \\ 0 & 0 & \frac{1}{c} & 0 \\ 0 & 0 & 0 & \frac{1}{d} \end{bmatrix}$; $abcd \neq 0$ **64.** $\begin{bmatrix} a^{-n} & 0 \\ 0 & b^{-n} \end{bmatrix}$

65. There must be exactly one nonzero entry in each row and each column.
66. To find the inverse of the product of two matrices, find the inverses of the individual matrices and multiply them in the reverse order of the original product.

PROBLEM SET 5.6

1. -13 **2.** 28 **3.** -14 **4.** -31 **5.** 72 **6.** 28 **7.** -65 **8.** -76 **9.** 0 **10.** -20 **11.** -200
12. 78 **13.** 195 **14.** 48 **15.** 48 **16.** a^5 **17.** -24 **18.** -42 **19.** -60 **20.** 600 **21.** 2400

22. 1260   **23.** 25,740   **24.** 23,250   **25.** $\{(9, -5)\}$ **26.** $\{(4, -1)\}$ **27.** $\left\{\left(\dfrac{40}{29}, \dfrac{41}{29}\right)\right\}$ **28.** $\{(10, 5)\}$

29. $\{(-5, -2, 7)\}$ **30.** $\{(-2, 3, 4)\}$ **31.** $\{(2, -3, 4)\}$ **32.** $\{(2, 3, 5)\}$ **33.** $\{(3, -1, 2)\}$ **34.** $\{(-1, 3, 2)\}$
35. $\{(2, 3, 1)\}$ **36.** $\{(0, 4, 2)\}$ **37.** $\{(0, 2, -2, 1)\}$ **38.** $\{(1, 0, -10, 2)\}$ **39.** $\{(2, -1, 1, -3)\}$

40. $\{(1, -1, 1, -1, 1)\}$ **41.** $\dfrac{4}{3}$ **42.** $\dfrac{77}{254}$ **43. a.** $f(x) = 125x^2 - 500x + 12{,}000$ **b.** 2 years; \$11,500; no

44. $f(x) = \dfrac{1}{200}x^2 - \dfrac{7}{20}x + 25$; about 27 cents/mile   **45. a.** 28 square units   **b.**

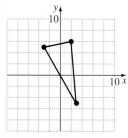

46. 26 square units   **47.** The points are on the line $y = -2x + \dfrac{17}{2}$ or $y = -2x - \dfrac{3}{2}$.   **48.** 31 square units

49. Yes   **50.** Yes   **51.** $\begin{vmatrix} x & y & 1 \\ 3 & -5 & 1 \\ -2 & 6 & 1 \end{vmatrix} = 0$; $-11x - 5y + 8 = 0$ **52.** $\begin{vmatrix} x & y & 1 \\ -1 & 3 & 1 \\ 2 & 4 & 1 \end{vmatrix} = 0$; $-x + 3y - 10 = 0$

55. b   **56.** b   **57.** -2100 **58.** 13,200   **59–61.** Answers may vary.
62. Verification is left to the student; the system is dependent   **63.** Answers may vary.

64. The points are collinear.   **65.** Answers may vary.   **66.** $\{-4, -1, 4\}$ **67.** $\left\{-2, -\dfrac{3}{2}, 0, 1\right\}$

70. $\{(0, 0, ..., 0)\}$ **71.** 24

PROBLEM SET 5.7

1. $-\dfrac{1}{x + 2} + \dfrac{1}{x - 2}$ **2.** $\dfrac{1}{x} - \dfrac{1}{x + 1}$ **3.** $\dfrac{4}{x - 2} + \dfrac{3}{x - 1}$ **4.** $\dfrac{2}{x + 2} + \dfrac{5}{x - 4}$ **5.** $\dfrac{2}{x + 3} + \dfrac{3}{x - 1}$

6. $\dfrac{3}{3x + 4} - \dfrac{1}{2x - 3}$ **7.** $\dfrac{3}{x} + \dfrac{2}{x - 1} - \dfrac{1}{x + 3}$ **8.** $\dfrac{3}{x} - \dfrac{1}{x + 1} + \dfrac{2}{x - 5}$ **9.** $\dfrac{6}{x - 1} - \dfrac{5}{(x - 1)^2}$

10. $\dfrac{1}{x + 1} - \dfrac{1}{(x + 1)^2}$ **11.** $\dfrac{1}{x - 2} - \dfrac{2}{(x - 2)^2} - \dfrac{5}{(x - 2)^3}$ **12.** $\dfrac{2}{x + 1} + \dfrac{4}{(x + 1)^2} - \dfrac{3}{(x + 1)^3}$

13. $\dfrac{7}{x} - \dfrac{6}{x - 1} + \dfrac{10}{(x - 1)^2}$ **14.** $\dfrac{1}{x} + \dfrac{2}{x + 7} - \dfrac{28}{(x + 7)^2}$ **15.** $\dfrac{-2}{x + 1} + \dfrac{3}{(x + 1)^2} + \dfrac{7}{x - 3}$

16. $-\dfrac{1}{x + 1} + \dfrac{1}{x + 2} + \dfrac{2}{(x + 2)^2}$ **17.** $\dfrac{3}{x} + \dfrac{4}{x + 1} - \dfrac{3}{x - 1}$ **18.** $\dfrac{3}{x} - \dfrac{5}{x - 2} + \dfrac{4}{x + 2}$ **19.** $\dfrac{3}{x - 1} + \dfrac{2x - 4}{x^2 + 1}$

20. $\dfrac{3}{x - 4} + \dfrac{2x - 1}{x^2 + 5}$ **21.** $\dfrac{4}{x + 1} + \dfrac{2x - 3}{x^2 + 1}$ **22.** $\dfrac{3}{x + 2} + \dfrac{-2}{x^2 + 4}$ **23.** $\dfrac{2}{x + 1} + \dfrac{3x - 1}{x^2 + 2x + 2}$

24. $\dfrac{2}{x - 2} + \dfrac{-2x + 1}{x^2 + 2x + 2}$ **25.** $\dfrac{-1}{x} + \dfrac{2x}{x^2 + 1} + \dfrac{2x - 1}{(x^2 + 1)^2}$ **26.** $\dfrac{1}{x} + \dfrac{2x}{x^2 + 3} - \dfrac{3}{(x^2 + 3)^2}$

27. $\dfrac{1}{x - 1} + \dfrac{2x + 1}{(x^2 + 1)^2}$ **28.** $\dfrac{-1}{x + 1} + \dfrac{x}{x^2 + 1} + \dfrac{x}{(x^2 + 1)^2}$ **29.** $\dfrac{x - 2}{x^2 - 2x + 3} + \dfrac{2x + 1}{(x^2 - 2x + 3)^2}$

30. $\dfrac{3x}{x^2 - 2x + 2} + \dfrac{x - 2}{(x^2 - 2x + 2)^2}$ **31.** $\dfrac{1}{x - 1} + \dfrac{2x + 1}{x^2 - 2x + 2} + \dfrac{x}{(x^2 - 2x + 2)^2}$ **32.** $\dfrac{1}{x} - \dfrac{2}{x^2} + \dfrac{4}{(x^2 + 1)^2}$

33. $\dfrac{1}{x^2 + x + 1} - \dfrac{1}{x^2 + x + 2}$ **34.** $\dfrac{-2}{x^2 + 4} + \dfrac{x}{x^2 + 1}$ **35.** $x^2 - \dfrac{x + 1}{x^2 + 1} + \dfrac{3}{x - 2}$

36. $x^2 + 3x + 1 + \dfrac{5}{x - 2} + \dfrac{3}{x + 1}$ **37.** $1 + \dfrac{2}{x} - \dfrac{2x + 4}{x^2 + 4}$ **38.** $x + 2 + \dfrac{3}{x - 1} + \dfrac{1}{(x - 1)^2}$ **39.** $\dfrac{99}{100}$

40. $\dfrac{100}{101}$ **41.** b   **42.** d   **43.** $-\dfrac{27}{16x} + \dfrac{13}{8x^2} - \dfrac{1}{8x^3} - \dfrac{4x + 9}{3(x^2 + x + 1)} + \dfrac{145x + 2}{48(x^2 + 2)} + \dfrac{38x + 17}{12(x^2 + 2)^2} + \dfrac{9x + 14}{4(x^2 + 2)^3}$

44. $\dfrac{1}{ax} + \dfrac{b}{a(a - bx)}$ **45.** $\dfrac{1}{a(ax + b)} - \dfrac{b}{a(ax + b)^2}$ **46.** Answers may vary.

47. By adding the rational expressions   **48.** Answers may vary.   **49.** $\dfrac{a}{x - c} + \dfrac{b + ac}{(x - c)^2}$

50. $\dfrac{2}{x - 3} + \dfrac{2x + 5}{x^2 + 3x + 3}$ **51.** $\dfrac{3}{\ln x + 1} + \dfrac{5}{\ln x^2 - 1}$ **52.** Group activity

PROBLEM SET 5.8

1.

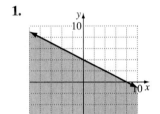

2.

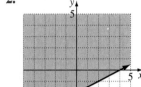

3.

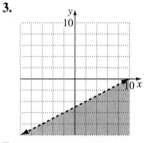

4.

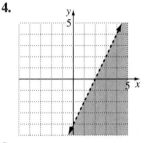

5.

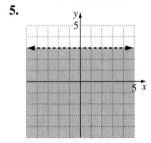

6.

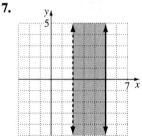

7.

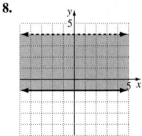

8.

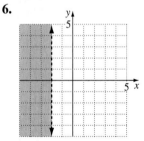

9.

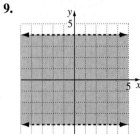

10.

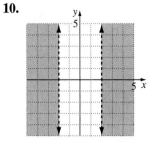

11.

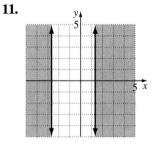

12.

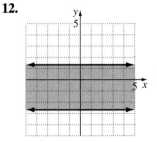

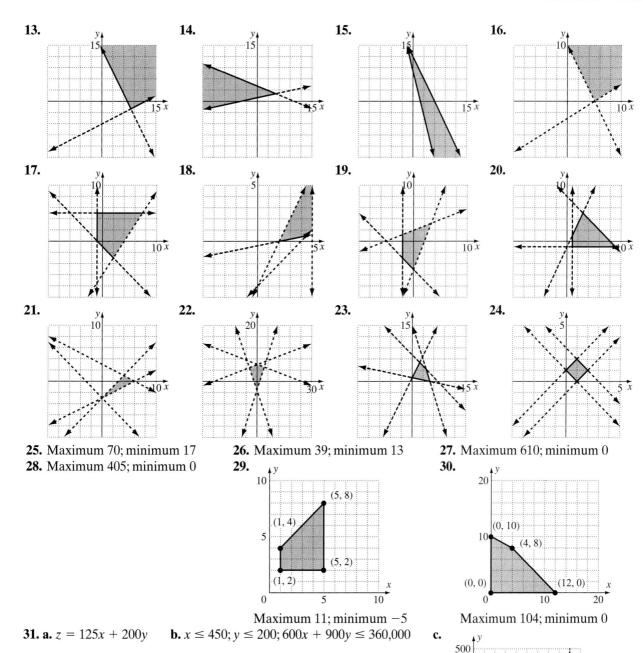

13.   **14.**   **15.**   **16.**

17.   **18.**   **19.**   **20.**

21.   **22.**   **23.**   **24.**

25. Maximum 70; minimum 17   **26.** Maximum 39; minimum 13   **27.** Maximum 610; minimum 0

28. Maximum 405; minimum 0   **29.**   **30.**

Maximum 11; minimum −5   Maximum 104; minimum 0

31. a. $z = 125x + 200y$ **b.** $x \leq 450; y \leq 200; 600x + 900y \leq 360,000$ **c.**

d. (0, 0): 0; (0, 200): 40,000; (300, 200): 77,500; (450, 100): 76,250; (450, 0): 56,250   **e.** 300; 200; $77,500

32. 40 model A bicycles and no model B bicycles   **33.** No cartons of food and 600 cartons of clothing

34. $5000 in stocks and $5000 in bonds   **35.** 20 American planes and 24 British planes

36. 50 students and 100 parents   **37.** b   **38.** b   **40–42.** Answers may vary.   **43.** The maximum value of 30 occurs at both $(2, 6)$ and $(6, 3)$, and at any point on the line segment connecting $(6, 3)$ and $(2, 6)$.

44.

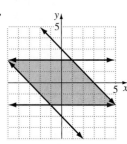

45.

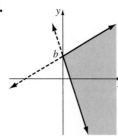

46. $x + y \le \dfrac{a}{2}, x + y \ge -\dfrac{a}{2}, x - y \le \dfrac{a}{2}, x - y \ge -\dfrac{a}{2}$

CHAPTER 5 REVIEW PROBLEMS

1. $\left\{ \left(4, \dfrac{5}{2} \right) \right\}$ **2.** $\{(2000, 5000)\}$ **3.** Inconsistent   **4.** $\{(8, 6)\}$ **5.** Dependent   **6.** $\{(0, 1, 2)\}$

7. $\{(2, 1, -1)\}$ **8.** Shrimp: 42 mg of cholesterol per ounce; scallops: 15 mg of cholesterol per ounce
9. $A = 40, B = 25$ **10.** New York City: 30,531; Chicago: 12,598; Los Angeles: 7631   **11.** $37°, 40°, 103°$
12. 3 ounces of Food A, 2 ounces of Food B, 4 ounces of Food C **13.** $\{(1, 3, -4)\}$ **14.** $\{(2, -2, 3, 4)\}$
15. a. $y = -2x^2 + 32x + 42$ **b.** 2 P.M.; 170 parts per million   **16.** No solution   **17.** $\{(2t + 4, t + 1, t)\}$
18. $\{(-37t + 2, 16t, -7t + 1, t)\}$ **19.** $\{(7t + 18, -3t - 7, t)\}$
20. a. $x + z = 750; y - z = -250; x + y = 500$ **b.** $\{(-t + 750, t - 250, t)\}$ **c.** $x = 350; y = 150$

21. $\begin{bmatrix} 2 & 3 & 8 \\ 21 & 5 & 5 \end{bmatrix}$ **22.** $\begin{bmatrix} 4 & -10 & -8 \\ -22 & 2 & -14 \end{bmatrix}$ **23.** $\begin{bmatrix} 0 & -10 & -15 \\ -40 & -5 & -15 \end{bmatrix}$ **24.** $\begin{bmatrix} -1 & -16 \\ 8 & 1 \end{bmatrix}$

25. $\begin{bmatrix} -10 & -6 & 2 \\ 16 & 3 & 4 \\ -23 & -16 & 7 \end{bmatrix}$ **26.** $\begin{bmatrix} -6 & 4 & -8 \\ 0 & 5 & 11 \\ -17 & 13 & -19 \end{bmatrix}$ **27.** $\begin{bmatrix} 10 & 5 \\ -2 & -30 \end{bmatrix}$ **28.** $\begin{bmatrix} -4 & 10 & 10 \\ -4 & 3 & 1 \\ -4 & 2 & 2 \end{bmatrix}$

29. $\begin{bmatrix} -6 & -22 & -40 \\ 9 & 43 & 58 \\ -14 & -48 & -94 \end{bmatrix}$ **30.** Not possible   **31.** $\begin{bmatrix} 7 & 6 & 5 \\ 2 & -1 & 11 \end{bmatrix}$ **32.** $\begin{bmatrix} -\frac{13}{4} & -\frac{1}{4} \\ -3 & \frac{19}{4} \end{bmatrix}$

33. a. $\begin{bmatrix} 30 & 58 & 78 \\ 50 & 175 & 308 \end{bmatrix}$ This gives the company's costs and retail sales broken down by division.

b. $\begin{bmatrix} 4 & 16 & 16 \\ 6 & 17 & 18 \end{bmatrix}$ This gives the amount by which the costs and retail sales of branch store 1 exceed those of branch store 2.   **c.** $[20 \quad 117 \quad 230]$

34. a. $\begin{bmatrix} 260{,}000 & 344{,}000 \\ 404{,}000 & 533{,}000 \\ 130{,}000 & 170{,}500 \end{bmatrix}$ **b.** AB gives the wholesale and retail prices of the inventory at each outlet.

Thus, the wholesale cost of the inventory of outlet 1 is $260,000 and the retail price is $344,000, etc.
c. $260,000   **d.** $533,000   **e.** $40,500   **35.** Verfication is left to the student.

36. $\begin{bmatrix} \frac{5}{13} & -\frac{3}{13} \\ \frac{1}{13} & \frac{2}{13} \end{bmatrix}$ **37.** $\begin{bmatrix} 3 & 0 & -2 \\ -6 & 1 & 4 \\ 1 & 0 & -1 \end{bmatrix}$ **38.** The system is $AX = B$ where $A = \begin{bmatrix} 1 & 1 & 2 \\ 0 & 1 & 3 \\ 3 & 0 & -2 \end{bmatrix}$, $X = \begin{bmatrix} x \\ y \\ z \end{bmatrix}$,

and $B = \begin{bmatrix} 7 \\ -2 \\ 0 \end{bmatrix}$ **b.** $\begin{bmatrix} -2 & 2 & 1 \\ 9 & -8 & -3 \\ -3 & 3 & 1 \end{bmatrix}$ **c.** $\{(-18, 79, -27)\}$ **39.** $\{(2, 1, -1)\}$

40. The encoded message is $3, 13, 24, 101$; the decoded message is $2, 1, 19, 5$ or BASE.   **41.** 18   **42.** -86

43. 4   **44.** $\left\{ \left(\frac{7}{4}, -\frac{25}{8} \right) \right\}$ **45.** $\{(23, -12, 3)\}$ **46.** $\{(0, 3, 1, -1)\}$

47. $f(x) = \frac{5}{8}x^2 - 50x + 1150$; 30- and 50-year-olds are involved in an average of 212.5 automobile accidents per

day.   **48.** $\frac{2}{x} + \frac{3}{x + 2} - \frac{1}{x - 1}$ **49.** $\frac{1}{x} + \frac{3}{(x - 1)^2} - \frac{4}{(x - 1)^3}$ **50.** $\frac{x}{x^2 + 3} + \frac{1}{x^2 + 2}$

51. $-\frac{2}{x^2 + 1} + \frac{3}{(x^2 + 1)^2} + \frac{2}{x^2 + 2}$ **52.** $1 + \frac{2}{x + 2} - \frac{3x - 1}{x^2 - 2x + 2}$

53.
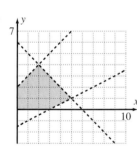

54.
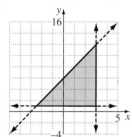

55.
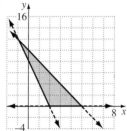

56.
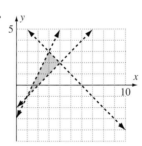

57.
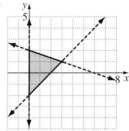

58. maximum 70; minimum 0
58. maximum 70; minimum 0
59. maximum 26; minimum 13
60. 8 hours tutoring and 12 hours as a teacher's aid.
61. 480 of model A and 240 of model B.

CHAPTER 5 TEST

1. $\{(4, -2)\}$ **2.** $\left\{ \left(\frac{14}{3}, -\frac{10}{9} \right) \right\}$ **3.** $\{(1, 3, 2)\}$ **4.** $25°, 155°,$ and $25°$ **5.** $\left\{ \left(-3, \frac{1}{2}, 1 \right) \right\}$ **6.** $\{(t, t - 1, t)\}$

7. $\begin{bmatrix} 5 & 4 \\ 1 & 11 \end{bmatrix}$ **8.** $\begin{bmatrix} 5 & -2 \\ 1 & -1 \\ 4 & -1 \end{bmatrix}$ **9.** $\begin{bmatrix} \frac{3}{5} & -\frac{2}{5} \\ \frac{1}{5} & \frac{1}{5} \end{bmatrix}$ **10.** $\begin{bmatrix} -1 & 2 \\ -5 & 4 \end{bmatrix}$ **11.** Verification is left to the student.

12. a. The system is $AX = B$ where $A = \begin{bmatrix} 3 & 5 \\ 2 & -3 \end{bmatrix}$, $X = \begin{bmatrix} x \\ y \end{bmatrix}$, and $B = \begin{bmatrix} 9 \\ -13 \end{bmatrix}$.   **b.** $\begin{bmatrix} \frac{3}{19} & \frac{5}{19} \\ \frac{2}{19} & -\frac{3}{19} \end{bmatrix}$ **c.** $\{(-2, 3)\}$

13. 18   **14.** $x = 2$ **15.** $\dfrac{3}{4(x + 2)} + \dfrac{5}{4(x - 2)}$ **16.** $\dfrac{5}{x + 3} - \dfrac{3x + 5}{x^2 + 2x + 4}$ **17.** $\dfrac{x}{x^2 + 1} - \dfrac{x}{(x^2 + 1)^2}$

18.

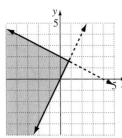

19. a. $y = 2x^2 + 22x + 320$
    b. 2780
    c. Answers may vary.
20. 8 ounces of food A, 0 ounces of food B, and 2 ounces of food C.
21. a. $-x + y + z = 10; x - y + z = 50; x + y + z = 180$
    b. The measure of angle A is 85°, angle B is 65°, and angle C is 30°.
22. 120 units of writing paper; 80 units of newsprint; $88,000

CUMULATIVE REVIEW PROBLEMS (CHAPTERS P–5)

1. $\dfrac{2x - 5}{2x + 3}$ **2.** $(-\infty, -19]$ **3.** $1 < x < 7$ or $(1, 7)$ **4.** $x = -3, x = \dfrac{1}{2}, x = 2$ **5.** $x = 3$ **6.** $22a + 11h - 19$

7.

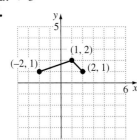

8. The graphs are reflections about the line $y = x$.

9.

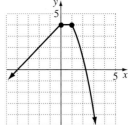

increasing on $(-\infty, 0)$
decreasing on $(1, \infty)$
constant on $(0, 1)$

10. a. $x = -1, x = 0, x = 1$

b.

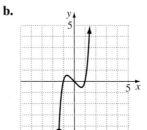

11. Verification is left to the reader. Hint: look for a pattern in results from synthetic division.

12.

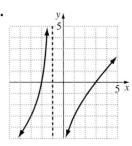

13. $\dfrac{11}{2}$ **14.** $\ln 27 - 5t$ **15.** 10.99%   **16.** $\{(-2, 1, 0, 4)\}$ **17.** $\begin{bmatrix} 2 & 2 \\ -1 & 4 \end{bmatrix}$

18.

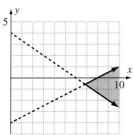

19. 2 feet   **20.** 25 feet by 50 feet; 1250 square feet

Chapter 6

PROBLEM SET 6.1

1. Foci at $(-\sqrt{21}, 0)$ and $(\sqrt{21}, 0)$

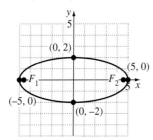

2. Foci at $(-\sqrt{5}, 0)$ and $(\sqrt{5}, 0)$

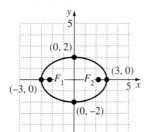

3. Foci at $(0, 2\sqrt{3})$ and $(0, -2\sqrt{3})$

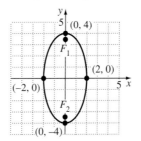

4. Foci at $(0, \sqrt{6})$ and $(0, -\sqrt{6})$

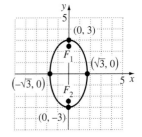

5. Foci at $(2 - \sqrt{5}, 1)$ and $(2 + \sqrt{5}, 1)$

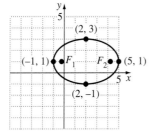

6. Foci at $(1 - \sqrt{7}, -2)$ and $(1 + \sqrt{7}, -2)$

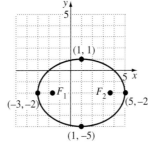

7. Foci at $(0, 2 + \sqrt{11})$ and $(0, 2 - \sqrt{11})$

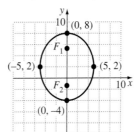

8. Foci at $(4, \sqrt{21})$ and $(4, -\sqrt{21})$

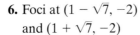

9. Foci at $(-2, -1)$ and $(6, -1)$

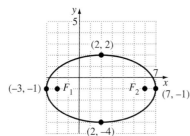

10. Foci at $(4 - \sqrt{5}, -2)$ and $(4 + \sqrt{5}, -2)$

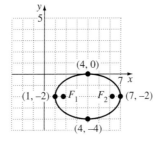

11. Foci at $(1 - \sqrt{7}, -2)$ and $(1 + \sqrt{7}, -2)$

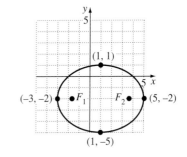

12. Foci at $(-5 - 2\sqrt{3}, 1)$
and $(-5 + 2\sqrt{3}, 1)$

13. Foci at $(-2, 3 + 4\sqrt{3})$
and $(-2, 3 - 4\sqrt{3})$

14. Foci at $(3 - \sqrt{21}, -2)$
and $(3 + \sqrt{21}, -2)$

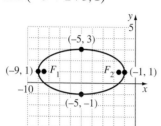

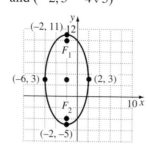

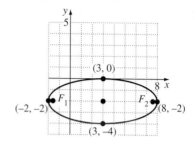

15. $\dfrac{x^2}{25} + \dfrac{y^2}{16} = 1$; foci at $(-3, 0)$ and $(3, 0)$ **16.** $\dfrac{x^2}{4} + \dfrac{y^2}{9} = 1$; foci at $(0, \sqrt{5})$ and $(0, -\sqrt{5})$

17. $\dfrac{(x - 3)^2}{1} + \dfrac{(y - 1)^2}{9} = 1$; foci at $(3, 1 + 2\sqrt{2})$ and $(3, 1 - 2\sqrt{2})$

18. $\dfrac{(x + 1)^2}{16} + \dfrac{(y - 2)^2}{4} = 1$; foci at $(-1 - 2\sqrt{3}, 2)$ and $(-1 + 2\sqrt{3}, 2)$ **19.** $\dfrac{x^2}{64} + \dfrac{y^2}{39} = 1$ **20.** $\dfrac{x^2}{32} + \dfrac{y^2}{36} = 1$

21. $\dfrac{x^2}{4} + \dfrac{y^2}{25} = 1$ **22.** $\dfrac{x^2}{10} + \dfrac{y^2}{1} = 1$ **23.** $\dfrac{x^2}{16} + \dfrac{4y^2}{1} = 1$ **24.** $\dfrac{9x^2}{1} + \dfrac{y^2}{4} = 1$ **25.** $\dfrac{x^2}{36} + \dfrac{5y^2}{36} = 1$

26. $\dfrac{21x^2}{400} + \dfrac{y^2}{25} = 1$ **27.** $\dfrac{x^2}{16} + \dfrac{y^2}{7} = 1$ **28.** $\dfrac{x^2}{100} + \dfrac{y^2}{75} = 1$ **29.** $\dfrac{x^2}{25} + \dfrac{y^2}{16} = 1$ **30.** $\dfrac{4x^2}{81} + \dfrac{4y^2}{17} = 1$

31. $\dfrac{x^2}{81} + \dfrac{y^2}{45} = 1$ **32.** $\dfrac{x^2}{16} + \dfrac{y^2}{12} = 1$ **33.** Yes, the truck will clear.   **34.** Yes, the truck will clear.

35. $\dfrac{x^2}{2500} + \dfrac{4y^2}{6561} = 1$; Distance between foci $= 2c = 58.6$ million miles   **36.** $e \approx 0.249$ **37.** $e \approx 0.0543$

38. At each focus, 40 feet from the center along the major axis   **39.** 35 AU   **40.** d   **41.** b

42.

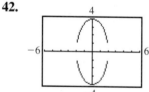

43.

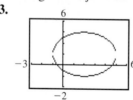
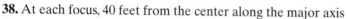
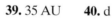

44.

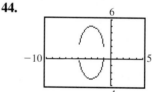

45.

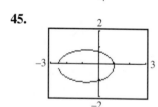

46. $(-1.54, 0.62), (1.54, 0.62)$

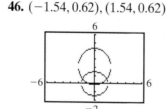

47.

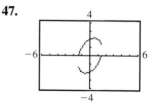

48.

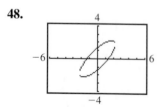

49.

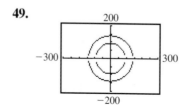

50.

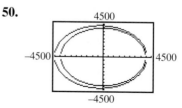

51. Answers may vary.   **52.** The graphs are not ellipses. Answers may vary.   **53–55.** Answers may vary.

56. $\dfrac{(x-2)^2}{9} + \dfrac{(y-2)^2}{5} = 1$ **57.** $\dfrac{(x-1)^2}{25} + \dfrac{(y+4)^2}{29} = 1$ **58.** $\dfrac{4x^2}{9} + \dfrac{y^2}{25} = 1$ **59.** $\dfrac{(x-5)^2}{9} + \dfrac{8(y-2)^2}{9} = 1$

60. $\dfrac{x^2}{9} + \dfrac{(y-2)^2}{8} = 1$ **61.** $\dfrac{(x+1)^2}{25} + \dfrac{(y-3)^2}{16} = 1$ **62.** $\dfrac{(x+1)^2}{7} + \dfrac{(y-4)^2}{16} = 1$

63. $\dfrac{x^2}{3{,}763{,}600} + \dfrac{y^2}{3{,}753{,}196} = 1$ **64. a.** Verification is left to the student.

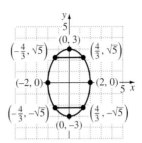

65. a. Verification is left to the student.   **b.** As e approaches zero the equation approaches that of a circle of radius a centered at (h, k).   **66.** Verification is left to the student.

67. $\dfrac{4x^2}{17} + \dfrac{y^2}{17} = 1$ **68.** Verification is left to the student.

PROBLEM SET 6.2

1. Foci: $(\pm\sqrt{13}, 0)$ **2.** Foci: $(\pm\sqrt{29}, 0)$ **3.** Foci: $(0, \pm\sqrt{34})$ **4.** Foci: $(0, \pm 5)$

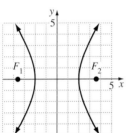

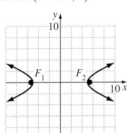

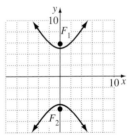

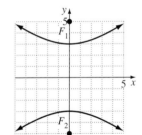

5. Foci: $(0, \pm\sqrt{2})$ **6.** Foci: $(\pm\sqrt{2}, 0)$ **7.** Foci: $(-9, -3), (1, -3)$ **8.** Foci: $(-2 \pm \sqrt{34}, 1)$,

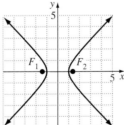

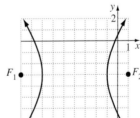

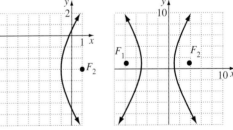

9. Foci: $(-3 \pm \sqrt{41}, 0)$ **10.** Foci: $(-2 \pm \sqrt{34}, 0)$ **11.** Foci: $(1, -2 \pm 2\sqrt{5})$ **12.** Foci: $(-1, -2 \pm \sqrt{85})$

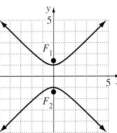

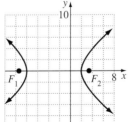

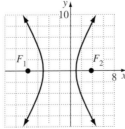

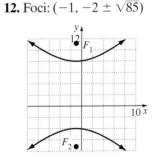

13. Foci: $(1 \pm \sqrt{2}, -2)$ **14.** Foci: $(-4 \pm 2\sqrt{5}, 3)$ **15.** Foci: $(-2, -1 \pm \sqrt{4.25})$ **16.** Foci: $(3, 1 \pm \sqrt{13})$

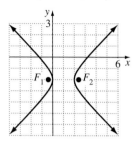   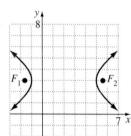   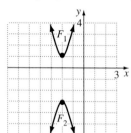   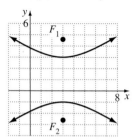

17. Foci: $(2 \pm \sqrt{13}, 3)$ **18.** Foci: $\left(-1 \pm \dfrac{\sqrt{13}}{6}, -1\right)$ **19.** Foci: $(4, \pm\sqrt{29})$ **20.** Foci: $(2, -7), (2, 3)$

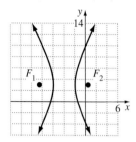   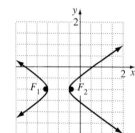   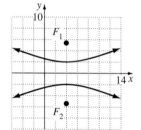   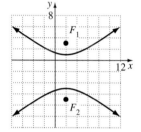

21. $\dfrac{x^2}{4} - \dfrac{y^2}{9} = 1$; foci: $(\pm\sqrt{13}, 0)$ **22.** $\dfrac{y^2}{4} - \dfrac{x^2}{16} = 1$; foci: $(0, \pm 2\sqrt{5})$

23. $(y - 3)^2 - (x - 1)^2 = 1$; foci: $(1, 3 \pm \sqrt{2})$ **24.** $\dfrac{(x - 2)^2}{25} - \dfrac{(y + 1)^2}{4} = 1$; foci: $(2 \pm \sqrt{29}, -1)$

25. $\dfrac{x^2}{9} - \dfrac{y^2}{7} = 1$ **26.** $y^2 - \dfrac{x^2}{8} = 1$ **27.** $x^2 - \dfrac{y^2}{9} = 1$ **28.** $y^2 - 4x^2 = 1$ **29.** $\dfrac{x^2}{36} - \dfrac{y^2}{9} = 1$

30. $\dfrac{y^2}{9} - \dfrac{x^2}{4} = 1$ **31.** $\dfrac{y^2}{9} - \dfrac{16x^2}{36} = 1$ **32.** $\dfrac{x^2}{4} - \dfrac{y^2}{\frac{12}{5}} = 1$ **33.** $\dfrac{x^2}{16} - \dfrac{y^2}{144} = 1$ **34.** $\dfrac{y^2}{4} - x^2 = 1$

35. $\dfrac{x^2}{1{,}210{,}000} - \dfrac{y^2}{5{,}759{,}600} = 1$; If M_1 is located 2640 ft to the right of the origin on the x-axis, the explosion is located on the right branch of the hyperbola given by the equation above.

36. a. $\dfrac{x^2}{5625} - \dfrac{y^2}{4375} = 1$ **b.** $y = 58.3$ km   **37.** $\dfrac{x^2}{9} - \dfrac{4y^2}{9} = 1$ **38.** 40 yds apart

39. $\dfrac{x^2}{576} - \dfrac{y^2}{144} = 1$; 43.9 mi   **40.** $y = 3 + \left(\dfrac{3}{5 + 4\sqrt{2}}\right)(x - 5)$ **41.** a   **42.** d

43.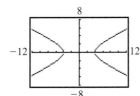

44.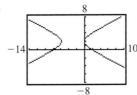

45.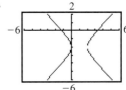

46.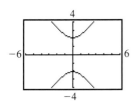

47.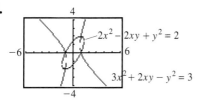

$(\pm 6.3, 4.9), (\pm 6.3, -4.9)$

48.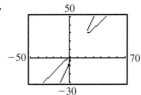

49.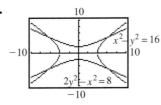

$(1, 2), (1, 0), (-1, 0), (-1, -2)$

50.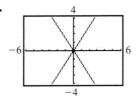

No. Two intersecting lines.

51. Answers may vary depending on the choice for a and b. For $a = 2, b = 3$, a graph is shown.

The two graphs open right/left and up/down, sharing a common set of asymptotes given by $y = \pm\dfrac{b}{a}x$.;

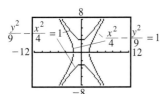

52.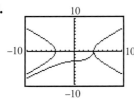

53–55. Answers may vary.

56. a. $\dfrac{x^2}{a^2} = 1 + \dfrac{y^2}{b^2} \geq 1$ since $\dfrac{y^2}{b^2} \geq 0$ **b.** $x^2 \geq a^2; x \geq |a|$ or $x \leq -|a|$ **c.** Domain: $(-\infty, -|a|] \cup [|a|, \infty)$

57. $1 < e$ **58.** As $e \to \infty$, the hyperbola gets wider. As $e \to 1$, the hyperbola gets narrower.

59. $\dfrac{x^2}{16} - \dfrac{y^2}{20} = 1$ **60.** It is impossible to determine which way the hyperbola opens.

61. $\dfrac{4(y + 1)^2}{121} - \dfrac{(x - 3)^2}{9} = 1$ **62.** $(x - 4)^2 - \dfrac{(y + 1)^2}{8} = 1$ **63.** $\dfrac{y^2}{36} - \dfrac{(x - 5)^2}{20} = 1$

64. $\dfrac{x^2}{9} - \dfrac{(y + 3)^2}{7} = 1$ **65.** $\dfrac{(x - 4)^2}{2} - \dfrac{(y - 2)^2}{2} = 1$ **66–68.** Verification is left to the student.

69. a. Vertices: $(\pm\sqrt{2}, 0)$; asymptotes: $y' = \pm x'$
  b. Vertices: $(-1, -1), (1, 1)$; asymptotes: the x-axis and the y-axis
  c. The graph will be a hyperbola with transverse axis $y = x$ and its two branches in the first and third quadrants. Vertices: $(-\sqrt{c}, -\sqrt{c}), (\sqrt{c}, \sqrt{c})$; asymptotes: x-axis, y-axis
  d. The graph will be a hyperbola with transverse axis $y = -x$ and its two branches in the second and fourth quadrants. Vertices: $(-\sqrt{c}, \sqrt{c})$ and $(\sqrt{c}, -\sqrt{c})$; asymptotes: x-axis, y-axis

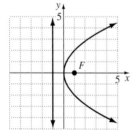

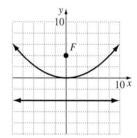

70. Possible answer: $\dfrac{x^2}{2} - \dfrac{y^2}{2} = 1; a = b$ **71–72.** Verification is left to the student.

PROBLEM SET 6.3

1. Vertex: $(0, 0)$
 focus: $(1, 0)$
 directrix: $x = -1$

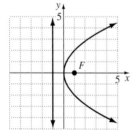

2. Vertex: $(0, 0)$
 focus $(-2, 0)$
 directrix: $x = 2$

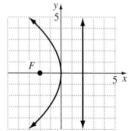

3. Vertex: $(0, 0)$
 focus: $(0, -3)$
 directrix: $y = 3$

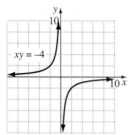

4. Vertex: $(0, 0)$
 focus: $(0, 4)$
 directrix: $y = -4$

5. Vertex: $(2, 1)$
 focus: $(2, 3)$
 directrix: $y = -1$

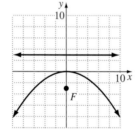

6. Vertex: $(-2, -1)$
 focus: $(-2, 0)$
 directrix: $y = -2$

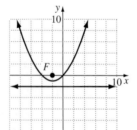

7. Vertex: $(-1, -3)$
 focus: $(-5, -3)$
 directrix: $x = 3$

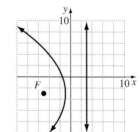

8. Vertex: $(1, 2)$
 focus: $(-2, 2)$
 directrix: $x = 4$

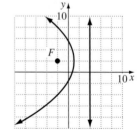

9. Vertex: $(-1, 0)$

focus: $(-1, -2)$

directrix: $y = 2$

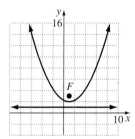

10. Vertex: $(2\sqrt{2}, -\sqrt{2})$

focus: $\left(2\sqrt{2}, \dfrac{1}{4} - \sqrt{2}\right)$

directrix: $y = -\left(\dfrac{1}{4} + \sqrt{2}\right)$

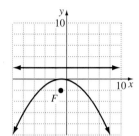

11. Vertex: $(22, 23)$

focus: $(-4, -3)$

directrix: $x = 0$

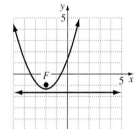

12. Vertex: $(3, 1)$

focus: $(0, 1)$

directrix: $x = 6$

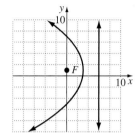

13. Vertex: $(1, 2)$

focus: $(1, 3)$

directrix: $y = 1$

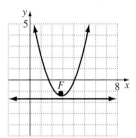

14. Vertex: $(-3, 1)$

focus: $(-3, -1)$

directrix: $y = 3$

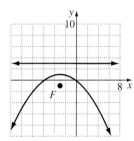

15. Vertex: $\left(-2, -\dfrac{4}{3}\right)$

focus: $\left(-2, -\dfrac{23}{24}\right)$

directrix: $y = -\dfrac{41}{24}$

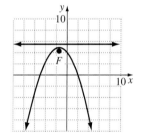

16. Vertex: $\left(-\dfrac{3}{2}, \dfrac{49}{10}\right)$

focus: $\left(-\dfrac{3}{2}, \dfrac{171}{40}\right)$

directrix: $y = \dfrac{221}{40}$

17. Vertex: $\left(1, \dfrac{3}{2}\right)$

focus: $\left(\dfrac{7}{16}, \dfrac{3}{2}\right)$

directrix: $x = \dfrac{25}{16}$

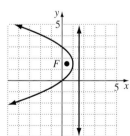

18. Vertex: $\left(-3, -\dfrac{1}{2}\right)$

focus: $\left(-\dfrac{13}{4}, -\dfrac{1}{2}\right)$

directrix: $x = -\dfrac{11}{4}$

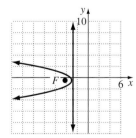

19. $(x - 1)^2 = -\dfrac{1}{2}(y - 6)$ **20.** $(x - 3)^2 = \dfrac{1}{2}(y + 4)$

21. $y^2 = -12x$; directrix: $x = 3$ **22.** $x^2 = -16y$; directrix: $y = 4$ **23.** $(x + 4)^2 = -4(y - 3)$; directrix: $y = 4$

24. $(y - 2)^2 = 8(x - 1)$; directrix: $x = -1$ **25.** $y^2 = 12x$ **26.** $x^2 = -16y$ **27.** $x^2 = \dfrac{1}{2}y$ **28.** $y^2 = \dfrac{9}{4}x$

29. $x^2 = -\dfrac{49}{10}y$ **30.** $y^2 = (16 - 4\sqrt{7})x$ **31.** $y^2 = 8x$ **32.** $x^2 = 6y$ **33.** $(y - 3)^2 = -12(x - 2)$

34. $(y-4)^2=-12(x-5)$ **35.** $(x+4)^2=-8(y-3)$ **36.** $(y-1)^2=12(x+1)$
37. $(x+3)^2=-\frac{1}{3}(y-2)$ **38.** $(y+5)^2=64(x-3)$ **39.** $(x+1)^2=-8(y-2)$
40. $(x-1)^2=-6\left(y-\frac{7}{2}\right)$ **41.** $(y+2)^2=6\left(x-\frac{5}{2}\right)$ **42.** $(y+3)^2=-8(x+1)$
43. Hyperbola **44.** Ellipse **45.** Parabola **46.** Circle **47.** Ellipse **48.** Hyperbola
49. Parabola **50.** Parabola **51.** $\left(\frac{15}{4},0\right)$ **52.** $(0,3)$ **53.** $y^2=230.4x$; 0.006250 m **54.** $\frac{25}{6}$ ft
55. 75.625 m **56.** Yes **57.** b **58.** d

59. Vertex: $(3,0)$

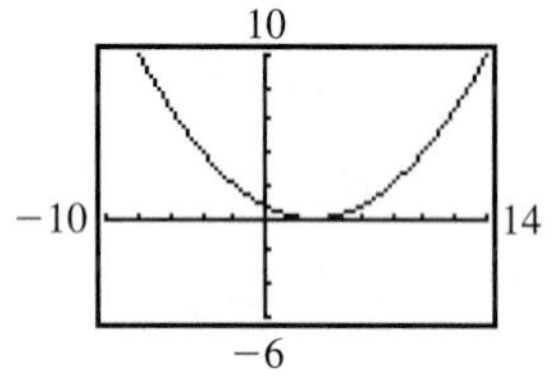

60. Vertex: $(2,-1)$

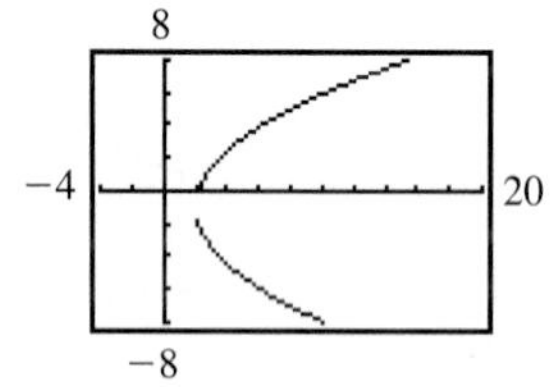

61. Vertex: $(0,-5)$

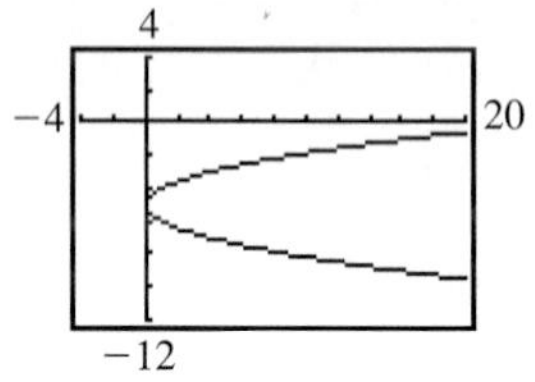

62. Vertex: $\left(\frac{1}{24},\frac{673}{144}\right)$

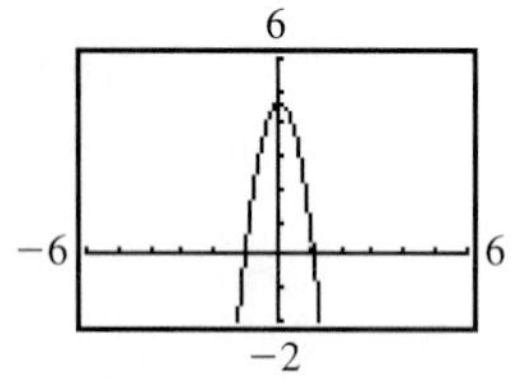

63. $y=\dfrac{12x+40\pm10\sqrt{15x-7}}{9}$

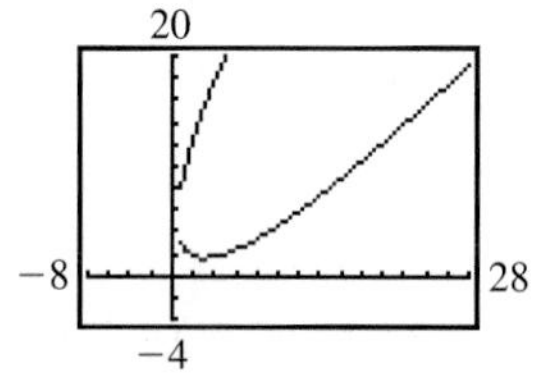

64. $y=-\dfrac{\sqrt{3}x+4\pm4\sqrt{-2\sqrt{3}x-5}}{3}$

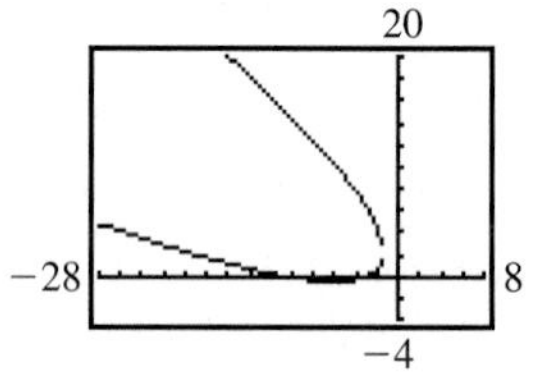

65. Expect a hyperbola. Graph is two lines.

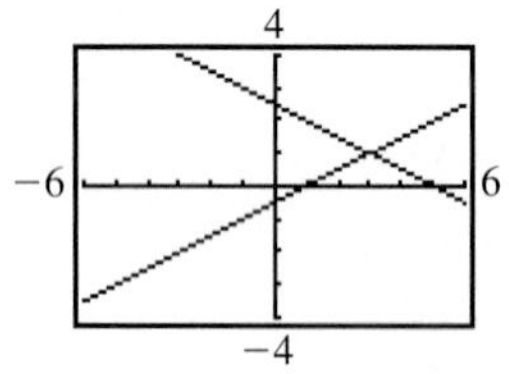

66. Expect a hyperbola. Graph is two lines.

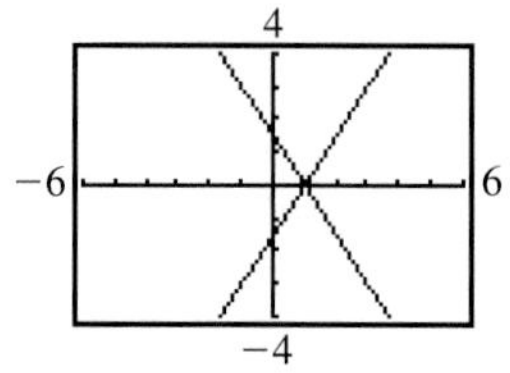

67. Expect an ellipse. No graph. **68.** Expect an ellipse. Graph is the point $(0,3)$.
69. Expect a circle. Graph is the point $(-2,3)$.

70.

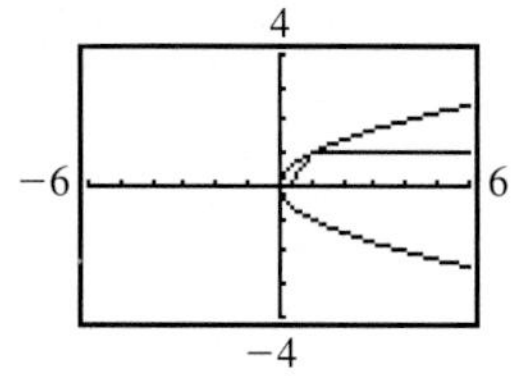

71. $y=x-1$

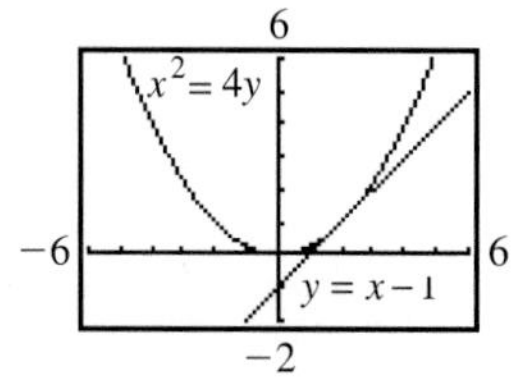

72. An infinite number of parabolas are possible when given only two speceifications.

73–74. Answers may vary. **75.** $\dfrac{(B-A)^3}{8}$ **76.** Verification is left to the student.

77. $A\left(x + \dfrac{D}{2A}\right)^2 + C\left(y + \dfrac{E}{2C}\right)^2 = -F + \dfrac{D^2}{4A} + \dfrac{E^2}{4C}$

$AC > 0, \left(-F + \dfrac{D^2}{4A} + \dfrac{E^2}{4C}\right) = 0$ yields a point.

$AC < 0, \left(-F + \dfrac{D^2}{4A} + \dfrac{E^2}{4C}\right) = 0$ yields two intersecting lines.

78. vertex: $\left(-\dfrac{b}{2a}, c - \dfrac{b^2}{4a}\right)$; focus: $\left(-\dfrac{b}{2a}, c - \dfrac{b^2}{4a} + \dfrac{1}{4a}\right)$; directrix: $y = c - \dfrac{b^2}{4a} - \dfrac{1}{4a}$

79. Verification is left to the student.   **80.** Group activity

PROBLEM SET 6.4

1. $(-3, 5), (2, 0)$ **2.** $(0, 3), (3, 0)$ **3.** $(-2, 3)$ **4.** $(0, 2), (3, 0)$ **5.** $(-6, 2)$ **6.** $(-3, 2)$

7. $\left(-\dfrac{3}{2}, -4\right), (2, 3)$ **8.** $(-2, 6), (-12, 1)$ **9.** $(4, 3), (-3, -4)$ **10.** $(3, 9), (-1, 1)$ **11.** $(3, 5), (-7, -15)$

12. $(1, 1), \left(-\dfrac{4}{7}, -\dfrac{15}{7}\right)$ **13.** $(3 + i, 1 + i), (3 - i, 1 - i)$ **14.** $(2 + i, 8 + 4i), (2 - i, 8 - 4i)$

15. $(-2\sqrt{2}, -\sqrt{2}), (-1, -4), (1, 4), (2\sqrt{2}, \sqrt{2})$ **16.** $(2\sqrt{2}, \sqrt{2}), (-2\sqrt{2}, -\sqrt{2}), (\sqrt{2}, 2\sqrt{2}), (-\sqrt{2}, -2\sqrt{2})$

17. $(1, 0), (-1, 0)$ **18.** $(3, 2), (3, -2), (-3, 2), (-3, -2)$ **19.** $(2, 1), (2, -1), (-2, 1), (-2, -1)$

20. $(\sqrt{5}, 2i), (\sqrt{5}, -2i), (-\sqrt{5}, 2i), (-\sqrt{5}, -2i)$ **21.** $(2, -3), (-2, -3), (3, 2), (-3, 2)$

22. $(-3, -1), (3, -1), \left(-\dfrac{7}{3}, \dfrac{5}{3}\right), \left(\dfrac{7}{3}, \dfrac{5}{3}\right)$ **23.** $(-\sqrt{6}, 2), (\sqrt{6}, 2), (-1, -3), (1, -3)$

24. $(0, 5), (-i\sqrt{11}, -6), (i\sqrt{11}, -6)$ **25.** $(-2, -2), (2, 2), (4, 1), (-4, -1)$

26. $(2, 1), (-2, -1), (i\sqrt{2}, -i\sqrt{2}), (-i\sqrt{2}, i\sqrt{2})$ **27.** $(-3i, -5), (3i, -5), (0, 4)$

28. $(-i, -6), (i, -6), (-\sqrt{2}, 3), (\sqrt{2}, 3)$ **29.** $(-\sqrt{2}, 5), (\sqrt{2}, 5), (0, -3)$ **30.** $(0, 0), (-2, 2), (2, 2)$

31. $(-4, 1), \left(-\dfrac{5}{2}, \dfrac{1}{4}\right)$ **32.** $(-2, 0), (2, 0)$ **33.** $(-3, -4), (-3, 4), (3, -4), (3, 4)$ **34.** $(2, 2), (2, 4)$

35. $(5, -4), (5, 4), (3, 0)$ **36.** $\left(\dfrac{12}{5}, -\dfrac{29}{5}\right), (-2, 3)$ **37.** $(5i, -3i), (-5i, 3i), (3, 5), (-3, -5)$

38. $(2, 1), (-2, -1), (1, 2), (-1, -2)$ **39.** $(2, -1), (-2, 1), (1, -2), (-1, 2)$

40. $(4, -1), (-4, 1), (1, -4), (-1, 4)$ **41.** $(-1.4927, -1.6691), (1.9927, -2.8309)$ **42.** $(0, 5), (4, 3)$

43.
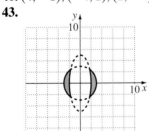

44.
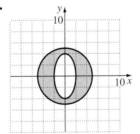

45.
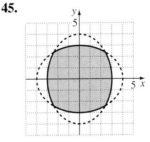

46.
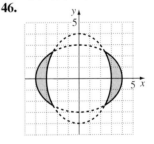

47.
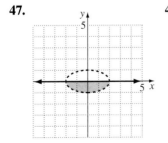

48.
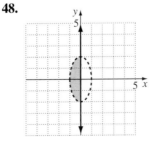

49.
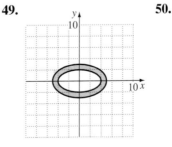

50.
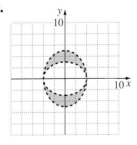

51.
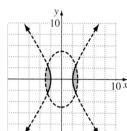

52.
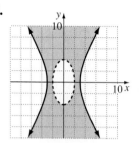

53.
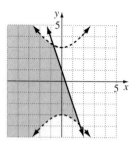

54.
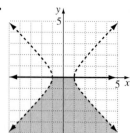

55.
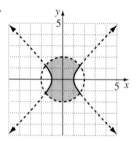

56.
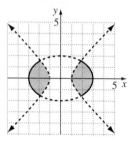

57.
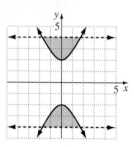

58.
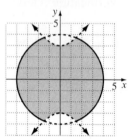

59.
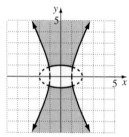

60.
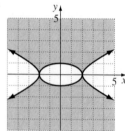

61.
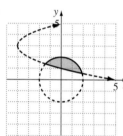

62.
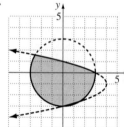

63.
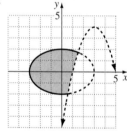

64.
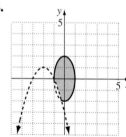

65.
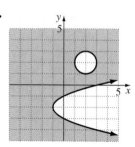

66.
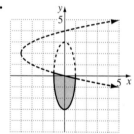

67. Location: $\left(\dfrac{4}{7}\sqrt{14}, \dfrac{3}{7}\sqrt{7}\right)$

68. $(0, -4), (-2, 0), (2, 0)$

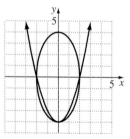

69. $x = 9\,\text{m}, y = 6\,\text{m}$ **70.** Either $x = \dfrac{9}{2}\,\text{m}, y = \dfrac{13}{2}\,\text{m}$ or $x = \dfrac{13}{2}\,\text{m}, y = \dfrac{9}{2}\,\text{m}$ **71.** $L = 12\,\text{ft}, W = 9\,\text{ft}$

72. $L = 20\,\text{ft}, W = 15\,\text{ft}$ **73.** $y = 5\,\text{in}, x = 3\,\text{in}$ **74.** $x = 24\,\text{ft}, y = 7\,\text{ft}$ **75.** $x = 5\,\text{m}, y = 2\,\text{m}$

76. $L = 25$ ft, $W = 12$ ft   **77.** $R = 3.5$ in, $r = 0.5$ in   **78.** $x = 3, y = 9$ **79.** b   **80.** b

81.

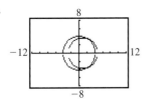

$(2, -3.61), (2, 3.61)$

82.

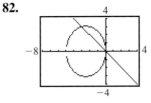

$(0, 0), (-3, 3)$

83.

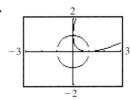

There are two crossing points: $(0, 0.9849)$ and $(1.0149, 0)$

84–88. Answers may vary.   **89.** 18 square units

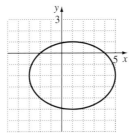

90–92. Verification left to the student.   **93.** Ship's location on grid: $(2.1052, 1.1382)$ **94.** Group activity

CHAPTER 6 REVIEW PROBLEMS

1. parabola
vertex: $(0, 0)$
focus: $(-2, 0)$
directrix: $x = 2$

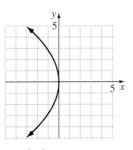

2. ellipse
foci: $(0, \pm 2\sqrt{3})$

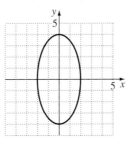

3. hyperbola
foci: $(\pm 5, 0)$

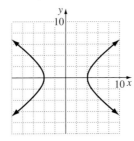

4. parabola
vertex: $(0, 0)$
focus: $(0, -4)$
directrix: $y = 4$

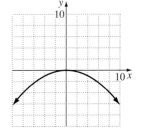

5. parabola
vertex: $(0, 2)$
focus: $(-4, 2)$
directrix: $x = 4$

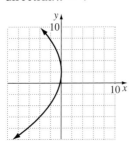

6. ellipse
foci: $(1 \pm \sqrt{7}, -2)$

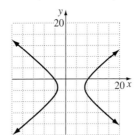

7. hyperbola
foci: $(2 \pm \sqrt{41}, -3)$

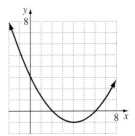

8. parabola
vertex: $(4, -1)$
focus: $(4, 0)$
directrix: $y = -1$

9. hyperbola
foci: $(1 \pm 2\sqrt{5}, -2)$

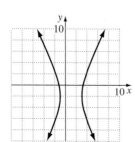

10. parabola
vertex: $(5, 2)$
focus: $\left(5, \dfrac{33}{16}\right)$
directrix: $y = \dfrac{31}{16}$

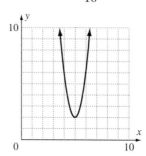

11. ellipse
foci: $(-3 \pm \sqrt{5}, 2)$

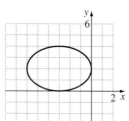

12. parabola
vertex: $(-1, 5)$
focus: $(0, 5)$
directrix: $x = -2$

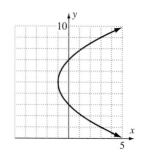

13. hyperbola
foci: $(-1, 2 \pm \sqrt{5})$

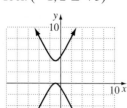

14. $\dfrac{(x+2)^2}{49} + \dfrac{(y-3)^2}{25} = 1$

15. $\dfrac{(x+5)^2}{9} - \dfrac{(y+2)^2}{25} = 1$

16. $(y+1)^2 = -\dfrac{1}{3}(x+2)$ **17.** $(x+4)^2 = 2(y+2)$ **18.** $\dfrac{x^2}{9} + \dfrac{y^2}{25} = 1$ **19.** $\dfrac{x^2}{25} + \dfrac{y^2}{25} = 1$

20. $\dfrac{x^2}{100} + \dfrac{y^2}{64} = 1$ **21.** $\dfrac{y^2}{9} - \dfrac{x^2}{16} = 1$ **22.** $\dfrac{y^2}{(5.4)^2} - \dfrac{x^2}{(7.2)^2} = 1$ **23.** $\dfrac{x^2}{8} - \dfrac{y^2}{1} = 1$ **24.** $y^2 = -8x$

25. $y^2 = \dfrac{1}{2}x$ **26.** $(x-4)^2 = -8(y-2)$ **27.** Yes   **28.** 94.60 million miles   **29. a.** $\dfrac{x^2}{2162.25} - \dfrac{y^2}{7837.75} = 1$

b. Approximately $(-56.2, 60)$ **30.** $x^2 = 12y$; at $(0, 3)$; answers may vary.   **31. a.** $x^2 = 40y$
b. 7.625 in   **32.** 103 feet

33. $(-2, 1), (1, 4)$

34. $\left(\dfrac{5}{2}, \dfrac{3}{2}\right)$

35. $(3, 1), (3, -1), (-3, 1), (-3, -1)$

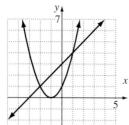

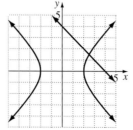

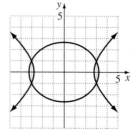

36. $(-3, 4), (4, -3)$

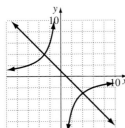

37. $(0, 1), (-3, 4)$

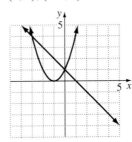

38. $(2, 2), (-2, -2)$

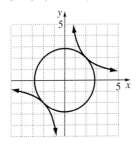

39. $(-\sqrt{14}, -1), (\sqrt{14}, -1),$
$(-\sqrt{13}, -2), (\sqrt{13}, 2)$

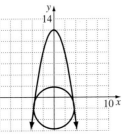

40. $(\sqrt{7}, 3 + \sqrt{7}), (-\sqrt{7}, 3 - \sqrt{7}),$
$(-1, -4), (1, -2)$

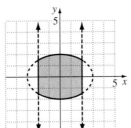

41. $(3, 5), (-3, -5), (-5i, 3i),$
$(5i, -3i)$

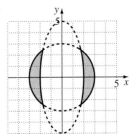

42. $x = 100, y = 50$; Advertisement is not accurate.   **43.** Screen: 24 in wide, 10 in high; Television will not fit.

44. $b = 5, a = 12$ **45.** $(1, 6), (3, 2)$ **46.** For Neptune: $\dfrac{x^2}{(4498)^2} + \dfrac{y^2}{0.999936(4498)^2} = 1;$

For Pluto: $\dfrac{x^2}{(5900)^2} + \dfrac{y^2}{0.937999(5900)^2} = 1;$ No collision is possible.

47.

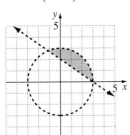

48.

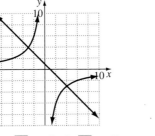

49.

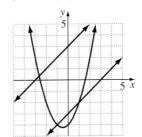

50.

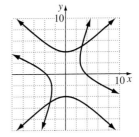

51.

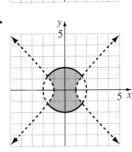

52.

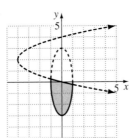

53. Foci must be farther away from the center than the vertices.

54. The hit ball will collide with the other ball.

CHAPTER 6 TEST

1. Hyperbola
foci: $(\pm\sqrt{13}, 0)$

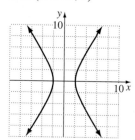

2. Parabola
vertex: $(0, 0)$
focus: $(0, -2)$
directrix: $y = 2$

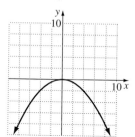

3. Ellipse
foci: $(\pm 4, 0)$

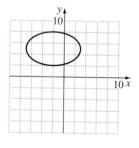

4. Hyperbola
foci: $(-1, 1 \pm \sqrt{5})$

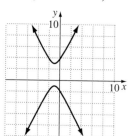

5. Parabola
vertex: $(-5, 1)$
focus: $(-5, 3)$
directrix: $y = -1$

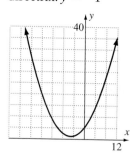

6. $\dfrac{x^2}{4} + \dfrac{y^2}{9} = 1$

7. $\dfrac{x^2}{4} - \dfrac{y^2}{25} = 1$ **8.** $y^2 = 6x$ **9.** Yes   **10. a.** $x^2 = 3y$ **b.** Light is placed $\dfrac{3}{4}$ inch above the vertex.

11. $(-2, 0), \left(\dfrac{5}{2}, \dfrac{9}{4}\right)$ **12.** $(2, -\sqrt{2}), (2, \sqrt{2}), (-2, -\sqrt{2}), (-2, \sqrt{2})$

13. $(3, 4), (-3, -4), (4, 3), (-4, -3)$ **14.**

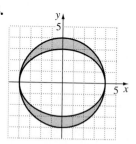

15. The rectangle is 6.5 feet by 10 feet.   **16.** The rectangle is 9 feet by 12 feet.

17. Solutions: $(-1, -2.83), (-1, 2.83), (1, -2.83), (1, 2.83)$

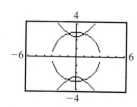

CUMULATIVE REVIEW PROBLEMS (CHAPTERS P–6)

1. $\dfrac{x}{x+3}$ $(x \neq 2)$ **2.** $\left\{ \dfrac{-2-\sqrt{10}}{2}, \dfrac{-2+\sqrt{10}}{2} \right\}$ **3.** $\left(\dfrac{4}{3}, \infty \right)$ **4.** $\{6\}$ **5.** $\left\{ -4, \dfrac{1}{3}, 1 \right\}$ **6.** $\{\ln 5, \ln 9\}$

7. $\{4\}$ **8.** $\{(7, -4, 6)\}$ **9.** -1 **10.** $3x - 2y = 0$ **11.** $g(x) = -\sqrt{x} + 1$ **12.** $f^{-1}(x) = \dfrac{x^2 + 7}{4}, x \geq 0$

13.

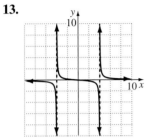

14. $(x+2)(x-3)(2x+1)(2x-1)$ **15.**

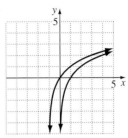

16. $10^{3.6}$ times as intense   **17. a.** $A = 900e^{\frac{\ln(1/2)}{40}t}$ **b.** 757 grams   **18.** $\begin{bmatrix} 2 & -1 \\ 13 & 1 \end{bmatrix}$

19. $\dfrac{4}{x} - \dfrac{4x-2}{x^2+1} - \dfrac{4x+5}{(x^2+1)^2}$ **20.** Hyperbola

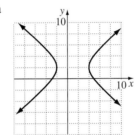

Chapter 7

PROBLEM SET 7.1

1. $5, 8, 11, 14, 17$ **2.** $3, 7, 11, 15, 19$ **3.** $3, 9, 27, 81, 243$ **4.** $\dfrac{1}{3}, \dfrac{1}{9}, \dfrac{1}{27}, \dfrac{1}{81}, \dfrac{1}{243}$ **5.** $-3, 9, -27, 81, -243$

6. $-\dfrac{1}{3}, \dfrac{1}{9}, -\dfrac{1}{27}, \dfrac{1}{81}, -\dfrac{1}{243}$ **7.** $-4, 5, -6, 7, -8$ **8.** $5, -6, 7, -8, 9$ **9.** $\dfrac{2}{5}, \dfrac{2}{3}, \dfrac{6}{7}, 1, \dfrac{10}{9}$ **10.** $\dfrac{1}{2}, \dfrac{6}{7}, \dfrac{9}{8}, \dfrac{4}{3}, \dfrac{3}{2}$

11. $1, -\dfrac{1}{3}, \dfrac{1}{7}, -\dfrac{1}{15}, \dfrac{1}{31}$ **12.** $\dfrac{1}{3}, -\dfrac{1}{5}, \dfrac{1}{9}, -\dfrac{1}{17}, \dfrac{1}{33}$ **13.** $1, 2, \dfrac{3}{2}, \dfrac{2}{3}, \dfrac{5}{24}$ **14.** $2, \dfrac{3}{2}, \dfrac{8}{3}, \dfrac{15}{2}, \dfrac{144}{5}$

15. $4, 12, 48, 240, 1440$ **16.** $0, -2, -4, -12, -48$ **17.** $2, 3, 4, 5, 6$ **18.** $\dfrac{1}{6}, \dfrac{1}{20}, \dfrac{1}{42}, \dfrac{1}{72}, \dfrac{1}{110}$

19. $4, 11, 25, 53, 109$ **20.** $-5, -16, -49, -148, -445$ **21.** $2, 0, -4, 12, 140$

22. $-11, 116, 13{,}451, 1.81 \times 10^8, 3.27 \times 10^{16}$ **23.** $1, 1, 2, 3, 5$ **24.** $2, -1, 3, -4, 7$ **25.** $a_n = 2n - 1$

26. $a_n = 3n - 2$ **27.** $a_n = \dfrac{n+1}{n}$ **28.** $a_n = \dfrac{1}{2^n}$ **29.** $a_n = (-1)^n$ **30.** $a_n = (-1)^{n+1}$

31. $a_n = \dfrac{(-1)^{n+1}}{n^2}$ **32.** $a_n = (-1)^{n+1}2n$ **33.** $a_n = n!$ **34.** $a_n = \dfrac{1}{n!}$ **35.** 105   **36.** 147   **37.** $\dfrac{31}{16}$

38. 12   **39.** $\dfrac{29}{6}$ **40.** $\dfrac{23}{12}$ **41.** 9   **42.** 10   **43.** 19   **44.** 100   **45.** 20   **46.** 166   **47.** 105

48. 147   **49.** 60   **50.** 225   **51.** 115   **52.** -4 **53.** $-\dfrac{5}{16}$ **54.** $\dfrac{7}{81}$ **55.** 55   **56.** 60   **57.** $\dfrac{3}{8}$

58. $-\dfrac{19}{30}$ **59.** $\dfrac{29}{20}$ **60.** 110   **61.** $\dfrac{250}{501}$ **62.** $\dfrac{125}{502}$ **63.** $\dfrac{100}{101}$ **64.** $-2{,}097{,}150$ **65.** $a_1 - a_{1001}$

66. $-\ln 10{,}000$ **67.** $\dfrac{1}{3}$ **68.** $\dfrac{1}{11}$ **69.** $\displaystyle\sum_{i=1}^{5} i$ **70.** $\displaystyle\sum_{i=1}^{5} 2i$ **71.** $\displaystyle\sum_{i=1}^{6} \dfrac{1}{2^i}$ **72.** $\displaystyle\sum_{i=1}^{7} \dfrac{1}{i^3}$ **73.** $\displaystyle\sum_{i=1}^{6} i^2$

74. $\displaystyle\sum_{i=}^{6} i^3$ **75.** $\displaystyle\sum_{i=1}^{8} \dfrac{(-1)^{n+1}}{2^{n-1}}$ **76.** $\displaystyle\sum_{i=1}^{11} \dfrac{1}{i(i+2)}$ **77.** $\displaystyle\sum_{i=1}^{4} \dfrac{x}{x+i}$ **78.** $\displaystyle\sum_{i=1}^{5} \dfrac{x^{2i-1}}{2i-1}$ **79.** $\displaystyle\sum_{i=1}^{\infty} 8i$ **80.** $\displaystyle\sum_{i=1}^{\infty} 6i$

81. 30.15; Americans spent \$30.15 billion on recreational boating from 1991 through 1995.

82. 149; Weekly earnings for U.S. workers increased by \$149 between 1985 and 1995.

83. -55.79; The percapita federal debt decreased by \$55.79 between 1981 and 1995.

84. 26,063; The total number of AIDS cases reported for 1984, 1985, and 1986 approximated by this model is 26,063.   **85.** b   **86.** d   **87.** c   **88–91.** Verification left to the student.

92. $a_n = \dfrac{1}{n}$

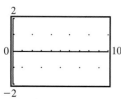

limit = 0

93. $a_n = \dfrac{2n}{n+1}$

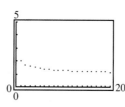

limit = 2

94. $a_n = \dfrac{3n^2 + 4n}{2n + 1}$

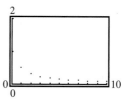

no limit, increases without bound

95. $a_n = (-1)^n$

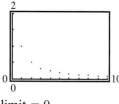

no limit, alternates between -1 and 1

96. $a_n = (1 + n)^{1/n}$

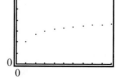

limit = 1

97. $a_n = \dfrac{2n + 1}{2n^2 + 4n}$

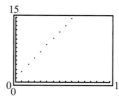

limit = 0

98–100. Answers may vary   **101.** $\dfrac{3}{2}, \dfrac{15}{8}, \dfrac{35}{16}, \dfrac{315}{128}, \dfrac{693}{256}$ **102.** $9, 23, 51, 107, 219$ **103.** $1, 0, -1, 0, 1, 0, -1, 0$

104. Possible answers: Sequence of primes, which continues..., 19, 23, 29, 31, 37, 41, 43, 47, 53, 59, ...

105. Answers may vary.

PROBLEM SET 7.2

1. $11, 6, 1, -4, -9, -14$ **2.** $17, 11, 5, -1, -7, -13$ **3.** $-9, -3, 3, 9, 15, 21$ **4.** $-7, -3, 1, 5, 9, 13$

5. $-1.6, -2, -2.4, -2.8, -3.2, -3.6$ **6.** $-1.7, -2, -2.3, -2.6, -2.9, -3.2$ **7.** $a_n = 4n - 3; a_{20} = 77$

8. $a_n = 5n - 3; a_{20} = 97$ **9.** $a_n = 11 - 4n; a_{20} = -69$ **10.** $a_n = 11 - 5n; a_{20} = -89$

11. $a_n = 3.20 - 0.05n; a_{20} = 2.20$ **12.** $a_n = 2.25 + 0.75n; a_{20} = 17.25$ **13.** $a_n = \frac{e}{6}n; a_{20} = \frac{10e}{3}$

14. $a_n = \frac{\pi}{12}n; a_{20} = \frac{5\pi}{3}$ **15.** $a_n = 7 + 2n; a_{20} = 47$ **16.** $a_n = -\frac{2}{3} + \frac{1}{3}n; a_{20} = 6$ **17.** $a_n = 1 + 3n; a_{20} = 61$

18. $a_n = 8 - 2n; a_{20} = -32$ **19.** $a_n = -\frac{2}{3} + \frac{n}{3}; a_{20} = 6$ **20.** $a_n = \frac{23}{4} - \frac{n}{4}; a_{20} = \frac{3}{4}$ **21.** 10th

22. 10th   **23.** 96th   **24.** 21st   **25.** 23rd   **26.** 44th   **27.** 5430   **28.** 2730   **29.** 8763   **30.** 7590
31. $2 + 8 + 14 + \cdots + 356 = 10{,}740$ **32.** $8 + 13 + 18 + \cdots + 303 = 9330$
33. $5 + 3 + 1 + \cdots + (-73) = -1360$ **34.** $11 + 8 + 5 + \cdots + (-106) = -1900$
35. $5 + 10 + 15 + \cdots + 500 = 25{,}250$ **36.** $3 + 6 + 9 + \cdots + 300 = 15{,}150$
37. $-2 - 4 - 6 - \cdots - 100 = -2550$ **38.** $-4 - 8 - 12 - \cdots - 200 = -5100$
39. $0.7 + 0.6 + 0.5 + \cdots + (-0.2) = 2.5$ **40.** $0.4 + 0 + (-0.4) + \cdots + (-6) = -47.6$ **41.** 3660   **42.** 6480
43. 2304   **44.** 1764   **45.** 57,250   **46.** 10,800   **47. a.** $a_n = 19{,}858 + 2350n$ **b.** \$59,808 million   **c.** 2016
48. a. $a_n = 125{,}159 + 1265n$ **b.** 139,074 thousand   **c.** 2011   **49. a.** $a_n = 10{,}850 + 1150n$ **b.** 14 years
50. a. $a_n = 5 + 14n$ **b.** 173 oranges   **51. a.** $a_n = 3.204 + 0.576n$ **b.** 574.236 million tons
52. a. $a_n = 83.96 + 3.14n$ **b.** 5723.64 million tons   **53.** \$172,500   **54.** Company B   **55.** 320 degree days
56. 26,733; The total expenditures of the federal government from 1980 to 2000 is \$26,733 billion.
57. a   **58.** d   **59.** d   **60.** d   **61–62.** Verify results.   **63.** Possible answer: 7, 8, 9, 10, ... and 4, 6, 8, 10, ...
Explanations may vary.   **64. a.** 1, 3, 5   **b.** 3, 6, 9   **c.** Either the underlined or the not–underlined values

always contain a three–term arithmetic sequence.   **65.** 650   **66.** \$120   **67.** 102   **68.** $\frac{1}{2}$

69. $\left(\sum\limits_{i=1}^{n} i\right) + n^2$ and $\sum\limits_{i=n+1}^{2n} i$ both equal $\frac{n}{2} + \frac{3n^2}{2}$.   **70.** Group activity

---

PROBLEM SET 7.3

1. $20, 10, 5, \frac{5}{2}, \frac{5}{4}$ **2.** $24, 8, \frac{8}{3}, \frac{8}{9}, \frac{8}{27}$ **3.** $-\frac{1}{3}, 1, -3, 9, -27$ **4.** $-\frac{1}{20}, \frac{1}{4}, -\frac{5}{4}, \frac{25}{4}, -\frac{125}{4}$

5. $\frac{x^2}{y}, 2x, 4y, \frac{8y^2}{x}, \frac{16y^3}{x^2}$ **6.** $c, b, \frac{b^2}{c}, \frac{b^3}{c^2}, \frac{b^4}{c^3}$ **7.** $a_n = -3(4)^{n-1}; a_7 = -12{,}288$ **8.** $a_n = 18\left(-\frac{1}{3}\right)^{n-1}; a_7 = \frac{2}{81}$

9. $a_n = 1.5(-2)^{n-1}; a_7 = 96$ **10.** $a_n = 5\left(-\frac{1}{5}\right)^{n-1}; a_7 = \frac{1}{3125}$ **11.** $a_n = -2(-\sqrt{3})^{n-1}; a_7 = -54$

12. $a_n = -4\left(-\frac{\sqrt{2}}{2}\right)^{n-1}; a_7 = -\frac{1}{2}$ **13.** $a_n = 0.0004(-10)^{n-1}; a_7 = 400$ **14.** $a_n = \frac{2000}{9}\left(\frac{1}{10}\right)^{n-1}; a_7 = \frac{2}{9000}$

15. $a_n = a^6\left(\frac{b}{a}\right)^{n-1}; a_7 = b^6$ **16.** $a_n = a^7 b^6\left(\frac{1}{ab^2}\right)^{n-1}; a_7 = \frac{a}{b^6}$ **17.** $a_n = 3(3^d)^{n-1}; a_7 = 3^{6d+1}$

18. $a_n = 1(a^{2/7})^{n-1}; a_7 = a^{12/7}$ **19.** $a_n = -3(-2)^{n-1}; a_7 = -192$ **20.** $a_n = -4(-2)^{n-1}; a_7 = -256$
21. $a_1 = 7$ **22.** $a_1 = 486$ **23.** $a_1 = 16$ **24.** $a_1 = \pm 81$ **25.** 531,440   **26.** 12,285   **27.** 2049

28. 177,148   **29.** $\frac{16{,}383}{2}$ **30.** $\frac{5461}{24}$ **31.** 3280   **32.** -85 **33.** 341.3125   **34.** -172 **35.** 0.400024

36. 0.500229   **37.** 5866.01   **38.** 1.81826   **39.** 1784.56   **40.** 2230.29   **41.** $\frac{3}{2}$ **42.** $\frac{4}{3}$ **43.** $\frac{2}{3}$ **44.** $\frac{3}{5}$

45. 4   **46.** 6   **47.** 6.15385   **48.** 7.05882   **49.** $2\frac{1}{4}$ **50.** $2\frac{2}{3}$ **51.** $\frac{2}{9}$ **52.** $\frac{7}{9}$ **53.** $\frac{47}{99}$ **54.** $\frac{83}{99}$

55. $\frac{347}{999}$ **56.** $\frac{659}{999}$ **57.** $\frac{41}{11}$ **58.** $\frac{4727}{999}$ **59.** $\frac{3221}{990}$ **60.** $\frac{227}{55}$ **61.** $\frac{6752}{4995}$ **62.** $\frac{32{,}683}{9990}$

63. a. $r \approx 1.15$ **b.** $a_n = 7.00(1.15)^{n-1}$ **c.** $a_{41} \approx 1875$; Model's prediction is too high.

64. $a_n = 40,000(0.9)^{n-1}$; $19,132   **65. a.** $a_n = 200(2)^{n-1}$ **b.** 1,677,721,600   **66.** Company A

67. $793,582.90   **68.** Company B   **69.** $19,143.92   **70.** $27,791.65   **71.** $14,284.02   **72.** $66,241.17

73. a. $S_n = \dfrac{R[1 - (1 + i)^n]}{1 - (1 + i)} = R\dfrac{(1 + i)^n - 1}{i}$ **b.** $S_n = (2000)\dfrac{(1 + 0.06)^n - 1}{0.06}$ **74.** $32,767

75. 134.07 inches   **76.** 150.6688 feet   **77.** $\dfrac{1}{3}$ **78.** 320 inches   **79.** $9 million   **80.** $15 million

81. $11\dfrac{1}{9}$ meters; $1\dfrac{1}{9}$ seconds   **82.** 300 feet   **83.** 24 feet   **84.** $a + \dfrac{2ar}{1 - r} = \dfrac{a(1 + r)}{1 - r}$

85. Release 2,000 flies each day.   **86.** d   **87.** c

88. a. $f(x) = \dfrac{1 - 0.5^x}{1 - 0.5}$

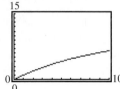

$f(x) = \dfrac{1 - 0.9^x}{1 - 0.9}$

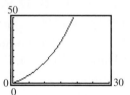

$f(x) = \dfrac{1 - 1.1^x}{1 - 1.1}$

$f(x) = \dfrac{1 - 3^x}{1 - 3}$

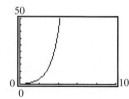

b. $r = 0.5$ and $r = 0.9$ **c.** When $0 < r < 1$

89. Horizontal asymptote at $y = 3$; $\displaystyle\sum_{n=0}^{\infty} 2\left(\dfrac{1}{3}\right)^n = 3$

90. Horizontal asymptote at $y = 10$; $\displaystyle\sum_{n=0}^{\infty} 4(0.6)^n = 10$

91. No; there is no sum.

92. The second option pays much more: $\approx$ $11 million.

93. Answers may vary.   **94.** The sum gets very large; $|r| > 1$ **95.** Possible answer: 1, 2, 4 and 1, –3, 9.

96. $60,000   **97.** $\dfrac{41}{24}$ **98.** $7\dfrac{1}{3}$ **99.** 1   **100.** Group activity

---

PROBLEM SET 7.4

1. S_k implies S_{k+1} because $k(k + 1) + 2(k + 1) = (k + 1)(k + 2)$.

2. S_k implies S_{k+1} because $\dfrac{3k(k + 1)}{2} + 3(k + 1) = \dfrac{3(k + 1)(k + 2)}{2}$.

3. S_k implies S_{k+1} because $2k(k + 2) + (4k + 6) = 2(k + 1)(k + 3)$.

4. S_k implies S_{k+1} because $k(2k + 1) + 4(k + 1) - 1 = (k + 1)(2k + 3)$.

5. S_k implies S_{k+1} because $2^{k+1} - 2 + 2^{k+1} = 2 \cdot 2^{k+1} - 2 = 2^{k+2} - 2$.

6. S_k implies S_{k+1} because $1 - \dfrac{1}{2^k} + \dfrac{1}{2^{k+1}} = 1 + \dfrac{-2 + 1}{2^{k+1}}$.

7. S_k implies S_{k+1} because $\dfrac{k^2(k + 1)^2}{4} + (k + 1)^3 = \dfrac{(k + 2)^2(k + 1)^2}{4}$.

8. S_k implies S_{k+1} because $\dfrac{2k(k + 1)(2k + 1)}{3} + [2(k + 1)]^2 = \dfrac{2(k + 2)(2k + 3)(k + 1)}{3}$.

9. S_k implies S_{k+1} because $\dfrac{k(k+1)(k+2)}{3} + (k+1)(k+2) = \dfrac{(k+1)(k+2)(k+3)}{3}$.

10. S_k implies S_{k+1} because $\dfrac{k(k+1)(2k+7)}{6} + (k+1)(k+3) = \dfrac{(k+1)(k+2)(2k+9)}{6}$.

11. S_k implies S_{k+1} because $\dfrac{k(k+3)}{4(k+1)(k+2)} + \dfrac{1}{(k+1)(k+2)(k+3)} = \dfrac{(k+1)(k+4)}{4(k+2)(k+3)}$.

12. S_k implies S_{k+1} because $6 - \dfrac{2k+3}{2^{k-1}} + \dfrac{2(k+1)-1}{2^k} = 6 - \dfrac{2(k+1)+3}{2^k}$.

13. S_k implies S_{k+1} because $\dfrac{k(k+1)(2k+1)(3k^2+3k-1)}{30} + (k+1)^4$

$$= \dfrac{(k+1)(k+2)(2k+3)[3(k+1)^2 + 3(k+1) - 1]}{30}.$$

14. S_k implies S_{k+1} because $\dfrac{k^2(k+1)^2(2k^2+2k-1)}{12} + (k+1)^5$

$$= \dfrac{(k+1)^2(k+2)^2[2(k+1)^2 + 2(k+1) - 1]}{12}.$$

15. $S_1 = \dfrac{1}{6}$, $S_2 = \dfrac{1}{4}$, $S_3 = \dfrac{3}{10}$, $S_4 = \dfrac{1}{3}$, $S_5 = \dfrac{15}{42}$, $S_n = \dfrac{n}{2n+4}$

16. $S_1 = \dfrac{1}{3}$, $S_2 = \dfrac{2}{5}$, $S_3 = \dfrac{3}{7}$, $S_4 = \dfrac{4}{9}$, $S_5 = \dfrac{5}{11}$, $S_n = \dfrac{n}{2n+1}$

17. $S_2 = \dfrac{1}{4}$, $S_3 = \dfrac{1}{3}$, $S_4 = \dfrac{3}{8}$, $S_5 = \dfrac{2}{5}$, $S_6 = \dfrac{5}{12}$, $S_n = \dfrac{n-1}{2n}$

18. $S_1 = 2$, $S_2 = 3$, $S_3 = 4$, $S_4 = 5$, $S_5 = 6$, $S_n = n + 1$ **19.** $S_1 = \dfrac{1}{2}$, $S_2 = \dfrac{1}{3}$, $S_3 = \dfrac{1}{4}$, $S_4 = \dfrac{1}{5}$, $S_5 = \dfrac{1}{6}$, $S_n = \dfrac{1}{n+1}$

20. a. $4; 9; 16; 25; 36$ **b.** n^2 **c.** S_k implies S_{k+1} because $k^2 + k + (k+1) = (k+1)^2$.

21. S_k implies S_{k+1} because $k + 1 + 1 = k + 2$.   **22.** S_k implies S_{k+1} because $9 \leq 3^{k+1}$ implies $3 \leq 3^{k+1}$.

23. S_k implies S_{k+1} because $2^{k+1} > 2k$ implies $2^{k+1} > k + 1$.

24. S_k implies S_{k+1} because $1 + 2(k+1) \leq 3^k + 2$ implies $1 + 2(k+1) \leq 3^{k+1}$.

25. S_k implies S_{k+1} because $2^k \leq 2(k!)$ implies $2^k \leq (k+1)!$

26. S_k implies S_{k+1} for $k \geq 6$ because $e^{n+1} < e(n!)$ implies $e^{n+1} < (n+1)!$.

27. S_k implies S_{k+1} because $(ab)^{k+1} = a^{k+1}b^{k+1}$.   **28.** S_k implies S_{k+1} because $\left(\dfrac{a}{b}\right)^{k+1} = \dfrac{a^{k+1}}{b^{k+1}}$.

29. S_k implies S_{k+1} because if $k^2 + k$ is divisible by 2, then so is $k^2 + k + 2k + 2$.

30. S_k implies S_{k+1} because if $k^3 - k$ is divisible by 3, then so is $k^3 - k + 3k^2 + 3k + 1 - 1$.

31. a. $3; 7; 15; 1$ **b.** Moving $k + 1$ disks requires first moving the first k disks (in $2^k - 1$ moves), then moving the $(k+1)$th disk, then moving the first k disks again ($2^k - 1$ more moves, for a total of $2^{k+1} - 1$).

32. a. Number of sides: $3, 12, 48, 192, \ldots$ $3 \cdot 4^{n-1}$; Perimeter: $3, 4, \dfrac{16}{3}, \dfrac{64}{9}, \ldots, 3\left(\dfrac{4}{3}\right)^{n-1}$;

Area: $\dfrac{\sqrt{3}}{4}, \dfrac{\sqrt{3}}{3}, \dfrac{10\sqrt{3}}{27}, \dfrac{94\sqrt{3}}{243}, \ldots, \dfrac{\sqrt{3}}{80}\left[32 - 27\left(\dfrac{4}{9}\right)^n\right]$ **33.** d   **34.** d   **35.** Answers may vary.

36. Possible answer: What's true for lots of n in a row need not be true for the next.   **37.** Answers may vary.

38. S_k implies S_{k+1} because $a_{k+2} - 1 + a_{k+1} = a_{k+3} - 1$.

39. S_k implies S_{k+1} because $a_k a_{k+1} + a_{k+1}^2 = a_{k+1}(a_k + a_{k+1})$.

40. S_k implies S_{k+1} because $a_{2k+1} - 1 + a_{2k+2} = a_{2k+3} - 1$.

41. S_k implies S_{k+1} because $\dfrac{k+1}{2k}\left(1 - \dfrac{1}{(k+1)^2}\right) = \dfrac{(k+1)^3 - (k+1)}{2k(k+1)^2} = \dfrac{k+2}{2(k+1)}$.

42. S_k implies S_{k+1} because $2^k + 2^2 < 2(2^k)$.   **43.** S_k implies S_{k+1} because $k^2 + 1 < (k+1)^2$.

44. S_k implies S_{k+1} because $\dfrac{\sqrt{k}\sqrt{k+1}+1}{k+1} > \dfrac{\sqrt{k}\sqrt{k}+1}{k+1} = 1$.

P R O B L E M   S E T   7 . 5

1. 20   **2.** 21   **3.** 12   **4.** 11   **5.** 1   **6.** 105   **7.** 4950   **8.** 4950   **9.** $x^3 + 6x^2 + 12x + 8$

10. $x^3 + 15x^2 + 75x + 125$ **11.** $27x^3 + 54x^2y + 36xy^2 + 8y^3$ **12.** $64x^3 + 144x^2y + 108xy^2 + 27y^3$

13. $8a^3 - 12a^2 + 6a - 1$ **14.** $27a^3 - 54a^2 + 36a - 8$ **15.** $x^8 + 8x^6y + 24x^4y^2 + 32x^2y^3 + 16y^4$

16. $81x^8 + 216x^6y^4 + 216x^4y^8 + 96x^2y^{12} + 16y^{16}$ **17.** $y^4 - 12y^3 + 54y^2 - 108y + 81$

18. $y^4 - 16y^3 + 96y^2 - 256y + 256$ **19.** $16x^{12} - 32x^9 + 24x^6 - 8x^3 + 1$ **20.** $16x^{20} - 32x^{15} + 24x^{10} - 8x^5 + 1$

21. $c^5 + 10c^4 + 40c^3 + 80c^2 + 80c + 32$ **22.** $c^5 + 15c^4 + 90c^3 + 270c^2 + 405c + 243$

23. $a^5 - 10a^4 + 40a^3 - 80a^2 + 80a - 32$ **24.** $243 - 405b + 270b^2 - 90b^3 + 15b^4 - b^5$

25. $1024x^5 - 6400x^4y + 16{,}000x^3y^2 - 20{,}000x^2y^3 + 12{,}500xy^4 - 3125y^5$

26. $243x^5 - 810x^4y + 1080x^3y^2 - 710x^2y^3 + 240xy^4 - 32y^5$ **27.** $\dfrac{1}{a^5} + \dfrac{5b}{a^4} + \dfrac{10b^2}{a^3} + \dfrac{10b^3}{a^2} + \dfrac{5b^4}{a} + b^5$

28. $\dfrac{1}{a^5} + \dfrac{10b}{a^4} + \dfrac{40b^2}{a^3} + \dfrac{80b^3}{a^2} + \dfrac{80b^4}{a} + 32b^5$ **29.** $64a^6 + 192a^5b + 240a^4b^2 + 160a^3b^3 + 60a^2b^4 + 12ab^5 + b^6$

30. $x^6 + 12x^5y + 60x^4y^2 + 160x^3y^3 + 240x^2y^4 + 192xy^5 + 64y^6$

31. $x^{12} - 6x^{10}y^2 + 15x^8y^4 - 20x^6y^6 + 15x^4y^8 - 6x^2y^{10} + y^{12}$

32. $x^6 - 12x^5y^2 + 60x^4y^4 - 160x^3y^6 + 240x^2y^8 - 192xy^{10} + 64y^{12}$ **33.** $a^2 + 8a^{3/2} + 24a + 32a^{1/2} + 16$

34. $a + 9a^{2/3} + 27a^{1/3} + 27$ **35.** $\dfrac{1}{a^3} + \dfrac{3}{a^2b} + \dfrac{3}{ab^2} + \dfrac{1}{b^3}$ **36.** $\dfrac{1}{a^3} - \dfrac{3}{a^2b} + \dfrac{3}{ab^2} - \dfrac{1}{b^3}$

37. $4x^4 - 27x^3 + 81x^2 - 113x + 59$ **38.** $3x^4 - 8x^3 - 6x^2 + 36x - 29$ **39.** $x^{16} + 8x^{15} + 28x^{14} + \dots$

40. $a^{28} + 14a^{27} + 91a^{26} + \dots$ **41.** $a^{32} - 16a^{30}b^2 + 120a^{28}b^4 + \dots$ **42.** $a^{32} - 16a^{30}b^2 + 120a^{28}b^4 + \dots$

43. $a^9 + 27a^8b + 324a^7b^2 + \dots$ **44.** $a^8 - 16a^7b + 112a^6b^2 + \dots$ **45.** $a^{42} + 42a^{41}b + 861a^{40}b^2 + \dots$

46. $a^{93} - 93a^{92}b + 4{,}278a^{91}b^2 + \dots$ **47.** $y^7 + 7y^5 + 21y^3 + \dots$ **48.** $y^8 + 8y^6 + 28y^4 + \dots$ **49.** $240a^4b^2$

50. $\dfrac{63x^5}{8}$ **51.** $-\dfrac{429a^{14}}{16}$ **52.** $36c^{35}d^{14}$ **53.** $720a^2b^8$ **54.** $-326{,}592a^4b^5$ **55.** $3{,}247{,}695x^8$

56. $210x^8$ **57.** $\dfrac{n!}{(n-1)!(n-n+n)!} = \dfrac{n!}{(n-1)!}$

58. $\dfrac{n!}{n!(n-n)!} = \dfrac{n!}{n!} = \dfrac{(n+1)!}{(n+1)!} = \dfrac{(n+1)!}{(n+1)![(n+1)-(n+1)]!}$ **59.** $\dfrac{n!}{0!(n-0)!} = \dfrac{n!}{n!} = \dfrac{n!}{n!(n-n)!}$

60. $\dfrac{n!}{0!(n-0)!} = \dfrac{n!}{n!} = \dfrac{(n+1)!}{(n+1)!} = \dfrac{(n+1)!}{0!(n+1)!}$ **61.** b   **62.** a

63.

$f_2, f_3,$ and f_4 are approaching $f_1 = f_5$.

64.

$f_2, f_3, f_4,$ and f_5 are approaching $f_1 = f_6$.

65. $x^3 - 3x^2 + 3x - 1$

66. $x^4 - 8x^3 + 24x^2 - 32x + 16$

67. $x^6 + 12x^5 + 60x^4 + 160x^3 + 240x^2 + 192x + 64$ **68.** Verify results.   **69–71.** Answers may vary.

72. Consider the expansion of $(1 + 1)^n$ **73.** $(101!)^{100}$ **74.** $\binom{2n}{n}$ **75.** $\dfrac{x + 5}{n + 1}$

76. $\binom{n}{r} = \dfrac{n!}{r!(n - r)!} = \dfrac{(n - r + 1)n!}{r(r - 1)!(n - r + 1)!} = \dfrac{n - r + 1}{r}\binom{n}{r - 1}$

77. $x^8 + 4x^7 + 10x^6 + 16x^5 + 25x^4 + 16x^3 + 4x^2 + 4x + 1$ **78.** $\sum_{r=0}^{n}\binom{n}{r}a^{n-r}(-b)^r$ **79.** $y = 5$

80. Add $\dfrac{n!}{r!(n - r)!}$ and $\dfrac{n!}{(r + 1)!(n - r - 1)!}$ by finding the LCD.   **81.** Verification left to the student.

PROBLEM SET 7.6

1. 720   **2.** 336   **3.** 10,000   **4.** 100,000   **5.** 35,152   **6.** 26,000   **7.** 3500   **8.** 2000   **9.** 36   **10.** 24
11. 1024   **12.** 1024   **13.** 144   **14.** 12   **15.** 720   **16.** 362,880   **17.** 3024   **18.** 1,860,480   **19.** 90
20. 2450   **21.** 5040   **22.** 40,320   **23.** 116,280   **24.** 6.0949×10^{10} **25.** 720   **26.** 5040   **27.** 210
28. 840   **29.** 8   **30.** 7   **31.** 1   **32.** 1   **33.** 210   **34.** 924   **35.** 4060   **36.** 12,650   **37.** 20   **38.** 4845
39. 13,983,816   **40.** 20   **41.** 1000   **42.** 56   **43.** 53,130   **44.** 38,760   **45.** 53,130   **46.** 96   **47.** 72
48. 840   **49.** 336   **50.** 35   **51.** 56   **52.** 105   **53.** 450   **54.** 9.5 minutes; 39 minutes   **55.** 1260   **56.** 27,720
57. 15,102   **58.** 1351   **59.** 82,251   **60.** 2,598,960   **61.** 3744   **62.** 5148   **63.** 624
64. 6.3501×10^{11}; About 40,272 years   **65.** c   **66.** c   **67–69.** Answers may vary.   **70. a.** n **b.** n
71. No; answers may vary.   **72. a.** Permutation   **b.** Combination   **c.** Permutation   **d.** Combination
73. $\dfrac{1}{13,983,816}$; 6,991,908   **74.** TOO, OTO, OOT   **75.** Not enough phone numbers
76. a. 12; BOBS, BOSB, BBSO, BBOS, BSOB, BSBO, OBBS, OBSB, OSBB, SBBO, SBOB, SOBB
b. 34,650   **c.** 280   **d.** 2310   **77.** 34,560   **78.** 24   **79.** 252

PROBLEM SET 7.7

1. $\dfrac{3}{8}$ **2.** $\dfrac{3}{8}$ **3.** $\dfrac{7}{8}$ **4.** $\dfrac{1}{2}$ **5.** $\dfrac{1}{9}$ **6.** $\dfrac{1}{6}$ **7.** $\dfrac{5}{36}$ **8.** $\dfrac{13}{18}$ **9.** $\dfrac{7}{12}$ **10.** $\dfrac{2}{3}$ **11. a.** $\dfrac{731}{3012}$ **b.** $\dfrac{1471}{3012}$

12. a. $\dfrac{5345}{19,342}$ **b.** $\dfrac{4725}{9671}$ **13.** $\dfrac{1}{80}$ **14.** $\dfrac{1}{250}$ **15.** $\dfrac{1}{20}$ **16.** $\dfrac{1}{120}$ **17.** $\dfrac{1}{18,009,460}$; $\dfrac{5}{900,473}$; $18,009,460$

18. $\dfrac{1}{142,506}$; $\dfrac{50}{71,253}$; 35,627   **19.** $\dfrac{1}{2}$ **20.** $\dfrac{13}{20}$ **21.** $\dfrac{75,791}{262,756}$ **22.** $\dfrac{907}{65,689}$ **23.** $\dfrac{21,789}{131,378}$ **24.** $\dfrac{156,795}{262,756}$

25. $\dfrac{109,809}{262,756}$ **26.** $\dfrac{1414}{5053}$ **27.** $\dfrac{36,887}{262,756}$ **28.** 0   **29.** 32%   **30.** 15%   **31.** 5.76%   **32.** 1.96%   **33.** 57.76%

34. 73.96%   **35.** 0.35   **36.** 0.20   **37.** 0.54   **38.** 0.12   **39.** $\dfrac{1}{4}$ **40.** $\dfrac{12}{13}$ **41.** $\dfrac{2}{13}$ **42.** $\dfrac{4}{13}$ **43.** $\dfrac{4}{13}$

44. $\dfrac{3}{52}$ **45.** $\dfrac{11}{26}$ **46.** $\dfrac{25}{52}$ **47.** $\dfrac{2109}{9520}$ **48.** $\dfrac{2109}{8330}$ **49.** $\dfrac{33}{66,640}$ **50.** $\dfrac{33}{16,660}$ **51.** c   **52.** d

53. $P(\leq 0.5) = 0.5$. Experiment should approach theory for larger numbers of trials.   **54.** No   **55.** $\dfrac{1}{12}$

56. a. $\dfrac{33}{100}$ **b.** $\dfrac{3}{25}$ **c.** $\dfrac{1}{10}$ **d.** Answers may vary.   **57.** $\left(\dfrac{1}{48}\right)^{18}$

CHAPTER 7 REVIEW PROBLEMS

1. $\dfrac{1}{3}, -\dfrac{1}{5}, \dfrac{1}{9}, -\dfrac{1}{17}, \dfrac{1}{33}$ **2.** $-1, \dfrac{2}{3}, -\dfrac{1}{3}, \dfrac{2}{15}, -\dfrac{2}{45}$ **3.** $4, -3, 11, -17, 39$ **4.** $a_n = (-1)^{n+1} n^2$ **5.** $a_n = \dfrac{(-1)^n}{n+2}$

6. -101 **7.** 1254 **8.** 95 **9.** $-\dfrac{3}{8}$ **10.** $-10{,}200$ **11.** $\displaystyle\sum_{i=1}^{5} i^3$ **12.** $\displaystyle\sum_{i=1}^{\infty}\left(-\dfrac{1}{2}\right)^i$

13. a. $a_3 = 1.0198; a_4 = 1.0012; a_9 = 0.9442; a_{10} = 0.94;$ The ratio of men to women in the U.S. is 1.0198, 1.0012, 0.9442, and 0.94 in years 1930, 1940, 1990, and 2000, respectively.   **b.** Approximately 141.75 million women and 133.25 million men   **14.** $a_n = 4n - 11; a_{20} = 69$ **15.** $a_n = 220 - 20n; a_{20} = -180$

16. $a_n = 8\dfrac{3}{4} - \dfrac{1}{4}n; a_{20} = 3\dfrac{3}{4}$ **17. a.** $a_n = 1043.4518 - 0.4118n$ **b.** 1006.3898 seconds   **18.** 2786 **19.** 5460

20. $100{,}500$ **21.** $-3 + 1 + 5 + \cdots + 153 = 3000$ **22.** $2 - 1 - 4 - \cdots - 235 = -9320$

23. $2772;$ The total annual advertising expenditures in billions of dollars by U.S. companies from 1980 to 2000 is \$2772 billion.   **24.** 2500 seats   **25.** $a_n = 100(1.05)^{n-1}; a^8 \approx 140.71$ **26.** $a_n = 2\left(\dfrac{1}{3}\right)^{n-1}; a_8 = \dfrac{2}{2187}$

27. $17{,}936{,}135$ **28.** -3 **29.** $15{,}354.12$ **30.** \$45,485.10; \$595,187.17   **31.** \$27,315.78   **32.** 11.9 feet

33. $\dfrac{1}{2}$ **34.** 16 **35.** $-2.\overline{2}$ **36.** $\dfrac{2}{3}$ **37.** $\dfrac{4}{11}$ **38.** $\$9\dfrac{1}{3}$ million   **39.** 200 inches

40. S_k implies S_{k+1} because $k^2 + [2(k+1) - 1] = (k+1)^2.$

41. S_k implies S_{k+1} because $\dfrac{k}{2k+1} + \dfrac{1}{(2k+1)(2k+3)} = \dfrac{(2k+1)(k+1)}{(2k+1)(2k+3)}.$

42. S_k implies S_{k+1} because $\dfrac{k(2k-1)(2k+1)}{3} + (2k+1)^2 = \dfrac{(k+1)(2k+1)(2k+3)}{3}.$

43. S_k implies S_{k+1} because $2^{k+1} < 2(k+2)!$ implies $2^{k+1} < (k+3)!$ **44.** $S_n = \dfrac{2}{1} \cdot \dfrac{3}{2} \cdot \dfrac{4}{3} \cdot \ldots \cdot \dfrac{n+1}{n} = n+1$

45. 315 **46.** $81x^4 + 108x^3y + 54x^2y^2 + 12xy^3 + y^4$ **47.** $x^5 - 10x^4y + 40x^3y^2 - 80x^2y^3 + 80xy^4 - 32y^5$

48. $x^{12} + 12x^{10}y^3 + 60x^8y^6 + 160x^6y^9 + 240x^4y^{12} + 192x^2y^{15} + 64y^{18}$

49. $128x^7 - 448x^6y^4 + 672x^5y^8 - 560x^4y^{12} + 280x^3y^{16} - 84x^2y^{20} + 14xy^{24} - y^{28}$

50. $\dfrac{x^8}{256} + \dfrac{3x^7y}{16} + \dfrac{63x^6y^2}{16} + \cdots$ **51.** $512x^9 - 6912x^8y + 41{,}472x^7y^2 + \cdots$ **52.** $2268c^3d^6$ **53.** $9{,}375{,}000$

54. $100{,}000$ **55.** 720 **56.** $32{,}760$ **57.** 2520 **58.** 1000 **59. a.** 1365 **b.** 364 **c.** 1940 **60.** $55{,}440$

61. $1{,}048{,}576$ **62. a.** $\dfrac{2}{3}$ **b.** $\dfrac{1}{2}$ **63.** $\dfrac{13}{18}$ **64.** $\dfrac{4129}{7511}$ **65.** $\dfrac{1362}{7511}$ **66.** $\dfrac{2959}{7511}$ **67.** 1 **68.** $\dfrac{1}{1000}$

69. a. $\dfrac{1}{15{,}504}$ **b.** $\dfrac{25}{3876}$ **70. a.** Outcome is 5; $\dfrac{1}{6}$ **b.** Outcome is 1, 3, 5, or 6; $\dfrac{2}{3}$ **c.** Outcome is even; $\dfrac{1}{2}$

d. Outcome is 4 or less; $\dfrac{2}{3}$ **71. a.** $\dfrac{21}{40}$ **b.** $\dfrac{1}{8}$ **c.** $\dfrac{5}{8}$ **d.** $\dfrac{29}{40}$ **e.** $\dfrac{19}{40}$ **72.** $\dfrac{25}{52}$ **73.** $\dfrac{1}{36}$ **74.** $\dfrac{1}{1024}$

75. $\dfrac{1}{720}$ **76.** 2.38% **77.** The probability of a person being born in a state is not equal for each sate.

CHAPTER 7 TEST

1. $-1, -\dfrac{1}{6}, \dfrac{1}{25}, -\dfrac{1}{62}, \dfrac{1}{123}$ **2.** $a_n = \dfrac{(-1)^{n+1}n^2}{(n+1)^2}$ **3.** -100 **4.** 40 **5.** $\displaystyle\sum_{i=1}^{5} \dfrac{2i}{2i+1}$

6. $a_n = -12.5 + 1.5n; a_{27} = 28$ **7.** 1430 **8.** 5190 **9.** Company B; \$7500   **10.** $a_n = 2(5)^{n-1}; a_8 = 156{,}250$

11. $147{,}620$ **12.** 523.22 **13.** \$276,426.79   **14.** $3\dfrac{1}{3}$ **15.** $\dfrac{823}{999}$ **16.** 160 inches

17. S_k implies S_{k+1} because $\dfrac{k(3k-1)}{2} + (3k+1) = \dfrac{(k+1)(3k+2)}{2}$.

18. $x^{12} + 12x^{10}y + 60x^8y^2 + 160x^6y^3 + 240x^4y^4 + 192x^2y^5 + 64y^6$ **19.** $10{,}000$ **20.** 7920 **21.** $\dfrac{1}{12}$ **22.** $\dfrac{1}{30}$

23. $\dfrac{4}{5}$ **24.** $\dfrac{1}{16}$

Appendix

APPENDIX

1. $\{4\}$ **2.** $\left\{-\dfrac{1}{4}, 3\right\}$ **3.** $\{-3, -2, 2, 3\}$ **4.** $\left\{\dfrac{1}{2}\right\}$ **5.** $\{16, 256\}$ **6.** $\left(-\infty, \dfrac{25}{7}\right]$ **7.** $\left(\dfrac{1}{5}, 2\right)$ **8.** $[-1, 0]$

9. $\left(-\dfrac{2}{3}, \dfrac{3}{2}\right)$ **10.** $(-\infty, -3) \cup \left(-\dfrac{11}{5}, \infty\right)$ **11.** $\{-2, 1, 2\}$ **12.** $\left(\dfrac{2}{3}, -1 \pm \sqrt{2}\right)$ **13.** $\left\{\dfrac{14 - \ln 77}{7}\right\}$

14. $\{0, \ln 9\}$ **15.** $\{3\}$ **16.** $\{1\}$ **17.** $\{(7, -4, 6)\}$ **18.** 2 **19.** $\left\{\left(-\dfrac{6}{5}, -\dfrac{16}{5}\right), (2, 0)\right\}$

20. $\{(3, 2), (3, -2), (-3, 2), (-3, -2)\}$ **21.** $\left\{\left(\dfrac{\sqrt{34}}{2}, \dfrac{\sqrt{34}}{2}\right), \left(-\dfrac{\sqrt{34}}{2}, -\dfrac{\sqrt{34}}{2}\right)\right\}$

22. The graph of f is shifted 1 unit up to get the graph of g. Then the graph of g is shifted 2 units to the right to get the graph of h.   **23. a.** The car is not moving.   **b.** 1 mile   **24. a.** $1980 - 1991$ **b.** $1991 - 2025$ **c.** $1950 - 1980$ **d.** $f(x) = 98$ **e.** The scale is not uniformly spaced.

25.

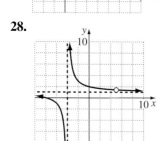

26. a. $\pm 1, \pm 3, \pm\dfrac{1}{2}, \pm\dfrac{3}{2}, \pm\dfrac{1}{4}, \pm\dfrac{3}{4}, \pm\dfrac{1}{8}, \pm\dfrac{3}{8}, \pm\dfrac{1}{16}, \pm\dfrac{3}{16}, \pm\dfrac{1}{32}, \pm\dfrac{3}{32}$

b. $\left\{-\dfrac{1}{8}, \dfrac{3}{4}, 1\right\}$ **27.** Upper bound: 2; lower bound: -4

28.

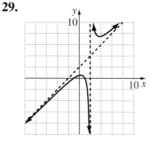

29.

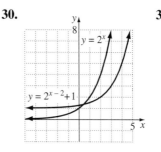

30.

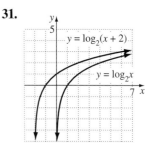

31.

32.

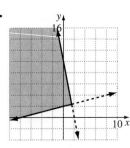

33. Ellipse; Foci: $\left(0, -\dfrac{3\sqrt{11}}{2}\right), \left(0, \dfrac{3\sqrt{11}}{2}\right)$

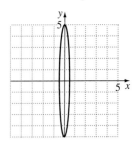

34. Hyperbola;

Foci: $(2, 3 - \sqrt{13}), (2, 3 + \sqrt{13})$

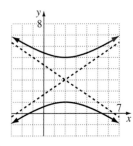

35. Parabola; Vertex: $(-2, 3)$,

Focus: $\left(-\dfrac{31}{16}, 3\right)$, Directrix: $x = -\dfrac{33}{16}$

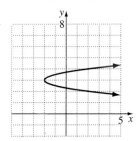

36.

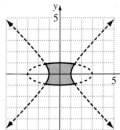

37. 60 km/h   **38.** 30 cm

39. a. $y - 5.13 = 0.58(x - 3)$ or $y - 8.03 = 0.58(x - 8)$; $y = 0.58x + 3.39$
   b. \$17.89 million

40. a. \$525 million   **b.** No; the cost increases without bound as p gets close to 100.

41. 6.9%   **42.** 10,000 times as intense

43. a. $A = 2e^{[(1/30)\ln(3/2)]t}$ **b.** 2100

44. a. $s(t) = -16t^2 + 80t + 44$ **b.** 144 feet; 2.5 seconds   **c.** 5.5 seconds

45. $(x + 3)(x - 2)(x^2 + 2x + 4)$ **46.** $\dfrac{x + 3}{x + 4}, (x \neq 0, x \neq 3, x \neq 5)$

47. $\dfrac{2}{x - 2}, \left(x \neq -2, x \neq -\dfrac{2}{3}\right)$ **48.** $\dfrac{S(1 - r)}{(1 - r^n)}$ **49.** $34a + 17h - 11, (h \neq 0)$

50. $y - 7 = -6(x + 1); y = -6x + 1; 6x + y - 1 = 0$ **51.** $f^{-1}(x) = x^3 - 4$

52. a. $(f \circ g)(x) = 25x^2 - 85x + 79$ **b.** $(g \circ f)(x) = 5x^2 - 15x + 38$ **53.** $\dfrac{5}{3}\log_8 x - 2 - 2\log_8 y$

54. $\ln \dfrac{x^3\sqrt{y}}{z^6}$ **55.** $\begin{bmatrix} -2 & 10 \\ -5 & 7 \\ 15 & -15 \end{bmatrix}$ **56.** $\dfrac{-1}{x - 2} + \dfrac{3x - 2}{x^2 + 2x + 2}$ **57.** $\dfrac{x^2}{169} + \dfrac{y^2}{25} = 1$ **58.** $\dfrac{x^2}{49} - \dfrac{y^2}{4} = 1$

59. $x^2 = -24y$ **60.** $a_n = -5n + 7; -88$ **61.** 3850   **62.** $\dfrac{290,125}{128}$ **63.** $\dfrac{50}{7}$

64. S_k implies S_{k+1} because $3 + 7 + 11 + \ldots + (4k - 1) + (4k + 3) = (k + 1)(2(k + 1) + 1)$.

65. $x^{15} + 10x^{12}y + 40x^9y^2 + 80x^6y^3 + 80x^3y^4 + 32y^5$ **66.** 22.2 years   **67.** 3 feet   **68.** 25 tons   **69.** \$350

70. 15 inches by 20 inches   **71.** 200 ft by 400 ft; 80,000 ft^2 **72.** Food A: 1.53 oz, Food B: 4.89 oz, Food C: 5.04 oz

73. 4 computation problems, 8 word problems; 104 points   **74.** 8.46 feet   **75.** Length: 8 inches; width: 6 inches

76. 4 yards by 4 yards; 7 yards by 7 yards   **77.** 369,062.26; yes   **78.** $\dfrac{1}{100}$ **79.** $\dfrac{1}{120}$ **80. a.** $\dfrac{9}{x + 4} + \dfrac{9}{x - 4}$

b–c. 5 mph   **81. a.** 3.2995; 3.7281; 4.3081; 4.8896; 5.5259; 6.0742; 6.5462; 6.7847

b.

c. $\ln y = 0.1096x + 3.2543$ **d.** $y = 25.9e^{0.1096x}$ **e.** \$2076 billion

Index